D1697945

Temperature and Life

By
H. Precht · J. Christophersen
H. Hensel · W. Larcher

With Contributions by
K. Brück · D. M. Gates · B. Havsteen · U. Heber · J. L. Ingraham
H.-D. Jankowsky · H. Laudien · K. Napp-Zinn · A. Pisek
P. Raths · K. A. Santarius · A. Vegis

With 263 Figures

Springer-Verlag Berlin · Heidelberg · New York 1973

Professor Dr. HERBERT PRECHT, Zoologisches Institut der Universität,
   D-2300 Kiel, Federal Republic of Germany

Professor Dr. JES CHRISTOPHERSEN, Universität Hohenheim, Fachgruppe Lebensmitteltechnologie, D-7000 Stuttgart 70, Garbenstraße 25, Federal Republic of
   Germany

Professor Dr. HERBERT HENSEL, Institut für Physiologie der Universität,
   D-3550 Marburg, Deutschhausstraße 2, Federal Republic of Germany

Professor Dr. WALTER LARCHER, Institut für Allgemeine Botanik der Universität,
   A-6020 Innsbruck, Sternwartestraße 15, Austria

ISBN 3-540-06441-9 Springer-Verlag Berlin Heidelberg New York
ISBN 0-387-06441-9 Springer-Verlag New York Heidelberg Berlin

Preface

The book by PRECHT, CHRISTOPHERSEN and HENSEL referred to in the text as the first edition was published in German in 1955 with the title *Temperatur und Leben*.

The present volume is a revised version of this book, constructed along the same lines, but it cannot properly be called the second edition because it is in English. Yet another difference is in the number of contributors, who now include two microbiologists, seven botanists, three zoophysiologists, one biochemist, and three human physiologists.

We have again endeavored to treat as many problems as possible but the main theme is still the adaptation of organisms to changing temperatures. What was conceived as a chapter on physical and chemical aspects by Professor L. LUMPER of Gießen will be published later as a supplementary volume. A special effort has been made to cover the copious literature published since 1955 though not, of course, exhaustively. The various chapters were completed at different times and those written earlier have footnotes referring to subsequent literature.

The botanical contributions by W. LARCHER, K. NAPP-ZINN and A. PISEK were translated by Mrs. JOY WIESER; Dr. J. M. AUGENFELD was the translator of those on poikilotherms by H. D. JANKOWSKY, H. LAUDIEN and H. PRECHT as well as of those on homeotherms by H. HENSEL, K. BRÜCK and P. RATHS. The section on limiting temperatures by H. PRECHT was translated by HAZEL PROSSER. We are grateful to them for undertaking this work.

It is with deep regret that we have to record the sudden death of Dr. A. VEGIS in June 1973. We are most grateful to his department head, Professor N. FRIES, for taking on the task of seeing Dr. VEGIS' work through the press.

The editors also wish to thank Dr. KONRAD F. SPRINGER for his sustained interest and for the care and attention which his staff devoted to the presentation of this work.

Autumn 1973

H. PRECHT
J. CHRISTOPHERSEN
H. HENSEL
W. LARCHER

Contents

Homeothermic Organisms

List of Authors

BRÜCK, K. Physiologisches Institut der Justus-Liebig-Universität, D-6300 Gießen (Federal Republic of Germany)

CHRISTOPHERSEN, J. Fachgruppe Lebensmitteltechnologie der Universität, D-7000 Stuttgart-Hohenheim (Federal Republic of Germany)

GATES, D. M. The University of Michigan Biological Station, Ann Arbor, MI 48104 (USA)

HAVSTEEN, B. Institut für Physiologische Chemie und Physikochemie, D-2300 Kiel (Federal Republic of Germany)

HEBER, U. Botanisches Institut der Universität, Abteilung Biochemische Pflanzenphysiologie, D-4000 Düsseldorf (Federal Republic of Germany)

HENSEL, H. Institut für Physiologie der Universität, D-3550 Marburg (Lahn) (Federal Republic of Germany)

INGRAHAM, J. L. College of Letters and Science, Department of Bacteriology, University of California, Davis CA 95616 (USA)

JANKOWSKY, H.-D. Zoologisches Institut der Universität, D-2300 Kiel (Federal Republic of Germany)

LARCHER, W. Institut für Allgemeine Botanik der Universität Innsbruck A-6020 Innsbruck (Austria)

LAUDIEN, H. Zoologisches Institut der Universität, D-2300 Kiel (Federal Republic of Germany)

NAPP-ZINN, K. Botanisches Institut der Universität, D-5000 Köln-Lindenthal (Federal Republic of Germany)

PISEK, A. Institut für Allgemeine Botanik der Universität Innsbruck A-6020 Innsbruck (Austria)

PRECHT, H. Zoologisches Institut der Universität, D-2300 Kiel (Federal Republic of Germany)

RATHS, P. Physiologisches Institut der Universität, DDR-402 Halle (Saale) (German Democratic Republic)

SANTARIUS, K. A. Botanisches Institut der Universität, Abteilung Biochemische Pflanzenphysiologie, D-4000 Düsseldorf (Federal Republic of Germany)

VEGIS, A. † formerly of Uppsala Universitets Institution för fysiologisk Botanik, S-75121 Uppsala (Sweden)

Poikilothermic Organisms

Microorganisms

I. Basic Aspects of Temperature Action on Microorganisms

J. CHRISTOPHERSEN

A. Influence of Temperature on the Growth and Multiplication of Microorganisms

When discussing the influence of temperature on the "growth" of microorganisms, one has to visualize that growth may refer to the cell as well as to a population of cells. In the latter case multiplication or reproduction would be the correct expression.

1. Growth of Cells

It is well established that cell growth and reproduction are governed by different cellular activities in spite of the fact that both processes are related by mechanisms whose nature is supposed to be rather complex. But it can be stated that a cell has to reach a certain size and physiological condition to develop the power to divide into two daughter cells. It has been demonstrated by HOFFMANN (1967) that cell division of *E. coli* comes to a standstill at 7.3° C whereas cell growth continues and large filamentous cells are produced.

A generalized and sophisticated theoretical treatise about the growth of unicellular organisms has been advanced by BERTALANFFY (1951). According to this theory one has to differentiate between growth processes which are surface dependent on the one hand and growth processes which are mass dependent on the other hand. In simpler cases the increase of cell substance is proportional to mass of cell material initially present. Such conditions hold true for rod-shaped organisms where the proportion between cell surface and cell volume remains constant during the growth of the cell. Therefore the length l at the time t can be expressed by the function

$$l = l_0 \cdot e^{(\mu - \varkappa)t} , \tag{1}$$

where l_0 is the length of the rod-shaped cell at the beginning of growth, $\varkappa$ is a constant denoting the dissimilative processes, and μ is a constant denoting the synthetic processes. The growth rate is expressed by the difference between both constants. The increase of mass of a rod-shaped cell therefore proceeds with a constant growth rate

$$\frac{\mathrm{d}l}{\mathrm{d}t} = l \cdot \text{constant} . \tag{2}$$

If cells have the shape of a sphere, as cocci and yeast cells, the growth relations are more complicated. In this case the increase of the volume is limited by size of the cell surface, which governs the influx of metabolites. Then the growth rate is given by the equation of SCHMALHAUSEN and BORDZILOWSKAJA (1948):

$$\frac{dg}{dt} = \mu g^{2/3} - \varkappa g \tag{3}$$

expressing that the increase of cell weight (g) results from synthetic processes (μ) which are proportional to the size of the cell surface ($g^{2/3}$), and that the increase of cell weight is opposed by dissimilative processes which are proportional to the cell weight ($\varkappa g$). Consequently the surface/weight ratio changes with growth to the disadvantage of the cell surface (proportional growth) until an equilibrium between input of metabolites and dissimilation is obtained. In this case the maximum size of the cell is reached.

Assuming that the temperature coefficient of the influx of metabolites is significantly smaller than the temperature coefficients of the metabolic processes inside the cell, the equilibrium between cell surface and cell volume must depend therefore on the growth temperature, i.e. the cell size must decrease with increasing growth temperatures.

This theory is supported by the frequent observation that organisms growing at high temperatures are significantly smaller than organisms from environments with low temperatures. This is valid especially for thermophilic bacteria and algae, the cells of which measure about $0.3-0.5\,\mu$ in diameter (COPELAND, 1936). But most thermophilic bacteria have a threadlike shape, and their morphology is a genetic peculiarity which is not to be compared with that of mesophilic species of bacteria. In a few instances, however, the influence of the temperature on the size of one and the same species has been investigated. MARGALEFF (1953) has shown that the size of *Chlorophyceae* decreases with increasing incubation temperature. On the other hand larger yeast cells were obtained at higher incubation temperatures (CHRISTOPHERSEN and PRECHT, 1954), as is demonstrated in Fig. 1. Similar observations have been made by DOWBEN and WEIDEMÜLLER (1968) with *Bacillus*

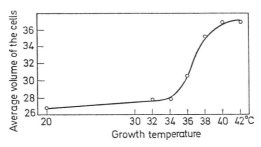

Fig. 1. The influence of the growth temperature on the size of yeast cells (CHRISTOPHERSEN and PRECHT, 1954)

subtilis. The cell volume of organisms grown at 63° C was approximately 2- to 3-fold greater than that of the same strain grown at 37° C using an enriched medium, and about 3.5- to 4.5-fold greater at 63° C than at 37° C when a minimal medium was used.

These few examples may indicate that a general growth theory as given by
BERTALANFFY cannot be applied to microorganisms. One point of criticism may be
that the temperature coefficient of the transport processes through the cellular
membrane is in a range of magnitude which is normally encountered in biological
reactions (active transport). The growth rate may therefore not be limited through
the size of the cell surface. Another point which should be considered is that even
when the temperature coefficients of the input and of the breakdown of metabolites
are equal, the energy yield of one and the same metabolic reaction increases with
increasing temperature because of the temperature dependence of the entropy in
$\Delta G = \Delta H - T\Delta S$.

2. Effects of Temperature on Reproduction

The reproduction rate of microorganisms or the growth rate of a culture of microorganisms in a given volume of culture medium depends on various factors. Some
of these factors are inherent in the kind of organism whereas other factors are the
result of environmental conditions. Temperature, of course, is only one of the
environmental conditions.

When we plot the logarithm of the number of cells against time, we obtain a
growth curve like that in Fig. 2, in which eight different growth phases can be
distinguished: a) lag phase, b) accelerated growth phase, c) exponential growth
phase (log phase), d) decelerated growth phase, e) stationary phase, f) accelerated
death phase, g) death phase and h) stationary phase.

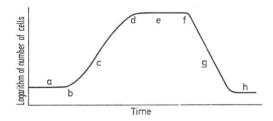

Fig. 2. The phases of the log growth curve

The three major phases are the lag phase, the exponential (log) phase and the
stationary phase, but one has to envision that any of the eight phases is affected
by temperature in a specific manner. Furthermore, temperature not only influences
the rate of cellular reactions involved in the afore-mentioned growth phases, but
also exerts an influence on other environmental conditions. Amongst these are
pH-value, water activity, ion activity, viscosity, hydration and aggregation of
macromolecules, toxic action of metabolic products and other unfavorable components of the medium, solubility of gases (especially oxygen and carbon dioxide),
etc. Therefore the temperature effect on microbial populations is very complex. It
cannot be denied that in a number of experiments reported in the literature some
of the indirect temperature effects have been neglected.

As to the yield in cell material it has been stated that usually in cultures of meso-philic and psychrophilic bacteria, maximal cell numbers are obtained at temper-atures considerably below those at which the rate of growth is most rapid (GRA-HAM-SMITH, 1920; HESS, 1934; FOTER and RAHN, 1936; DORN and RAHN, 1939; SPICER, 1940; UPADHYAY and STOKES, 1962). It has been suggested that utilization of food by bacteria may be more efficient, or that toxic metabolic products may be inactive at low temperatures. Experimental data of SINCLAIR and STOKES (1963) indicate, however, that the lower temperatures act only indirectly by increasing the solubility and therefore the supply of oxygen. Equal cell crops could be pro-duced at high and low temperatures when cultures were vigorously aerated so that oxygen was not limiting. Also the cell yields of facultative bacteria were equal when they were grown anaerobically at different temperatures.

3. Growth Curves

The reproduction rate of an exponentially growing culture can be expressed by the specific rate constant (k) in

$$\frac{dN}{dt} = kN \qquad or \qquad k = \frac{2.3\,(\log N_2 - \log N_1)}{t_2 - t_1} \tag{4}$$

in which N_2 and N_1 are the number of cells in the culture at times t_2 and t_1. In practice N is usually given by turbidity readings. The constant k depends con-siderably on the growth temperature, but also on any other environmental factor. Another parameter conveniently used to express the exponential growth rate is the generation time (G). If n is the number of generations produced in the total time elapsed (t), the generation time is $G = t/n$. Since bacteria multiply by binary fission the number of cells at n generations (N_n) is

$$N_n = N_0 \cdot 2^n \qquad or \quad n = \frac{\log N_n - \log N_0}{\log 2} \tag{5}$$

where N_0 is the initial number of cells. Hence

$$G = \frac{t \cdot \log 2}{\log N_n - \log N_0}. \tag{6}$$

Also the number of cell divisions per unit of time (r) can be estimated. Since

$$N_t = N_0 \cdot 2^{r(t_2 - t_1)} \tag{7}$$

r can be expressed by using the $\log_2$ as

$$r = \frac{\log_2 N_t - \log_2 N_0}{t_2 - t_1} \tag{8}$$

or by replacing $\log_2$ by $3.32 \cdot \log_{10}$

$$r = \frac{3.32\,(\log N_t - \log N_0)}{t_2 - t_1} = 1/G. \tag{9}$$

Temperature effects on growth of bacterial populations can be demonstrated by plotting the growth rate constant or the generation time against the temperature of incubation. In the latter case a curve is obtained like that in Fig. 3. In such curves three cardinal temperatures are conveniently distinguished, namely the temperature minimum at which reproduction comes to a standstill, the temper-ature optimum at which the generation time has the lowest value, and the temperature maximum at which the multiplication is inhibited by the height of the temperature.

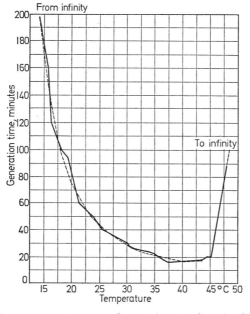

Fig. 3. Influence of temperature on growth rate (generation time) of *E. coli* (BARBER, 1908)

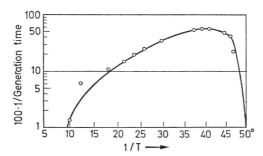

Fig. 4. Influence of temperature on growth rate. Plot of log of the generation time against temperature (BUCHANAN and FULMER, 1930)

Early observations have revealed that microorganisms can be found in environments in which the temperatures range from the freezing point of water to some degrees below the boiling point of water. This phenomenon has led bacteriologists to divide the microbial world into three groups according to their temperature preferendum, namely psychrophiles, mesophiles and thermophiles. It should be noted that any species apparently has a range of about 30° C which may overlap to a certain extent. Such classifications are found particularly in older handbooks, e.g., the curve presented in Fig. 4 (BUCHANAN and FULMER, 1930).

Despite the fact that the above-mentioned classification may be useful from a practical point of view, namely to characterize microorganisms with regard to their ecological dominance in nature, a strict separation of microbes into psychrophiles, mesophiles, and thermophiles has in the meantime turned out to be generally not correct. It was shown that a significant number of organisms which according to their ability to grow at 0° C are classified as psychrophiles, have a temperature optimum near 37° C, as should be characteristic for mesophiles (INGRAHAM and STOKES, 1959). This problem has been thoroughly discussed by INGRAM (1965). Similar observations have been made with thermophiles.

Most of the data on the influence of temperature on biological activities are presented in the well-known form of Arrhenius plots, not to stress the thermodynamic implications of the data but rather as a matter of convenience (see p. 19). Most enzymatically catalyzed reactions give a linear Arrhenius plot over a wide temperature range, as one would expect for a simple chemical reaction. On the other hand, for phenomena as complex as bacterial growth, respiration etc., this is true only in a limited interval, usually between 15° C and 35° C. When the temperature maximum and the temperature minimum are approached, the slope of the curve increases progressively until it becomes infinite.

The simplest, though not entirely precise, interpretation of the fact that Arrhenius plots of complex reactions give a linear curve, is that under such conditions the rate of the total process, even though involving a series of consecutive reactions, is largely limited by the slowest one of the series. The change in slope of the line, however, could not be accounted for so readily. Various hypotheses were expressed, prominent among which was CROZIER's (1925), to the effect that the change in slope is actually a sudden, though sometimes very slight, break, resulting from the replacement of a limiting reaction at the one temperature by a different member of the series, having a different temperature coefficient, at the other temperature. Thus, it was once thought possible that a "master reaction" occurring in diverse biological phenomena could be identified by its temperature characteristics, and a great deal of literature now exists on the discussion of this question.

The values of the activation energies in the normal range of growth are in the order of 3—20 kcal, corresponding to enzyme-catalyzed reactions. On the other hand, the activation energies calculated from the data in the range of the temperature minimum are in the order of 30—100 kcal, indicating that denaturation processes for certain enzymes may take place at low temperatures (KAVENAU, 1950; HULTIN, 1955). The temperature characteristics of the denaturation at the temperature maximum are even higher, in the order of 100—300 kcal, which is typical for the heat inactivation of proteins.

A most striking fact in the influence of temperature on growth rates of organisms, as well as on the rates of other activities, is that these activities are most rapid at a distinct temperature which is well below temperatures at which a measurable inactivation of proteins takes place. The majority of enzymes have maximum activities more or less far above the growth optimum. Despite this observation it is generally assumed that inactivation processes are responsible for the diminution in rate above the temperature optimum. The question is still open, however, whether enzymes themselves are inactivated or whether an inhibition of one or several control mechanisms of the enzyme synthesis limits the efficiency of biochemical reactions contributing to optimal growth rate.

A mathematical description of growth curves of microorganisms has been tried by different authors. The simplest one is that of HINSHELWOOD (1947), which assumes that growth (G) results from synthesis of cellular matter (A) and irreversible denaturation of unknown protein components (B):

$$G = A - B. \tag{10}$$

A theoretical growth curve can be obtained when A and B are replaced by expressions which describe the specific rate constants of A and B. According to the Eyring theory of absolute rate processes (EYRING, 1935; EYRING and MAGEE, 1942), phenomena occurring in living cells as well as denaturation processes of enzymes can be described by the equation:

$$k = k' \frac{KT}{h} e^{\frac{\Delta S}{R}} e^{-\frac{\Delta H}{RT}} \tag{11}$$

where k' is the transmission coefficient, representing the probability that the formation of an activated complex will lead to its decomposition into the products of the reaction, K is the Bolzmann constant and h is Plank's constant. ΔH is the change of the enthalpy of the reaction, ΔS the entropy of activation, R is the gas constant and T the absolute temperature.

Substituting A and B in Eq. (10) by Eq. (11) the following expression for a theoretical growth curve is obtained:

$$G = \left(k_1' \frac{KT}{h} e^{\frac{\Delta S_1}{R}} e^{-\frac{\Delta H_1}{RT}} \right) - \left(k_2' \frac{KT}{R} e^{\frac{\Delta S_2}{R}} e^{-\frac{\Delta H_2}{RT}} \right). \tag{12a}$$

Since the expressions KT/h appear in both members of the equation, they can be neglected. The unknown constants k_1' and k_2' can be safely assumed to be unit. Thus for practical calculations the expression

$$G = \frac{e^{\Delta S_1/R}}{e^{\Delta H_1/RT}} - \frac{e^{\Delta S_2/R}}{e^{\Delta H_2/RT}} \tag{12b}$$

can be applied. Substituting the thermodynamic constants of the enzyme-catalyzed synthetic reactions with average numerical values, namely $\Delta S_1 = 24$ cal and $\Delta H_1 = 5000$ cal, and substituting the thermodynamic constants of the denaturation reaction with $\Delta S_2 = 200$ cal and $\Delta H_2 = 62000$ cal, a growth curve can be drawn.

Another procedure for developing a growth curve for microorganisms is that of JOHNSON and LEWIN (1946), proceeding on the assumption that above the temperature optimum of growth an equilibrium exists between normal enzyme (E_n) and denatured enzyme (E_d). The total amount of enzyme is E_0 and the corresponding equilibrium constant is denoted by k_1. The substrate concentration is $[X]$. Applying the equation of EYRING the reaction rate can be written as

$$k = k_x \frac{KT}{h} [X] [E_n] e^{-\Delta E/RT} e^{-pV/RT} e^{\Delta S/R}. \tag{13}$$

The concentration of normal enzyme is given by

$$E_n = \frac{E_0}{1 + k_1}.$$

When k_1 is replaced by the equation of the reaction rate at which the turnover of the enzyme is supposed to occur, one obtains

$$E_n = \frac{E_0}{1 + e^{-\Delta H_1/RT}\, e^{\Delta S_1/R}} \,. \tag{14}$$

After substitution of E_n in Eq. (13) by the latter expression, the growth rate can be described by

$$G = \frac{s\,\dfrac{KT}{h}\,[X]\,[E_0]\,e^{-\Delta H_2/RT}\,e^{\Delta S_2/R}}{1 + e^{-\Delta H_1/RT}\, e^{\Delta S_1/R}} \tag{15a}$$

where s is an unknown turnover coefficient. Also ΔS_2 is not known because it can only be calculated from the concentration of the reacting molecules inside the cell. Assuming that $[X]$ and $[E_0]$ remain unchanged and summarizing all constants including k/h to a single constant c, the growth rate is now given by

$$G = \frac{c\,T\,e^{-\Delta H_2/RT}}{1 + e^{-\Delta H_1/RT}\, e^{\Delta S_1/R}} \,. \tag{15b}$$

If the following numerical values are chosen for the thermodynamic constants: $\Delta H_1 = 150\,000$ cal, $\Delta H_2 = 15\,000$ cal, and $\Delta S_1 = 476.46$ cal, a hypothetical growth curve can be drawn (see Fig. 5). The constant c can be calculated when the growth rate at a given temperature is expressed in per cent of the maximal growth rate at the optimum temperature. With *E. coli* the growth optimum has been found by JOHNSON and LEWIN (1946) to be at 39° C. In this case a value of $0.3612 \cdot e^{24.04}$ has been obtained. Assuming that the growth optimum is at 37° C, c becomes $0.331\, e^{24.36}$.

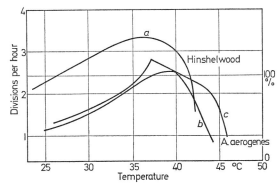

Fig. 5. Theoretical growth curve after HINSHELWOOD (1947) (*a*), JOHNSON and LEWIN (*b*), and experimental growth curve (*A. aerogenes*) (*c*)

Recent investigations by VAN UDEN and MADEIRA-LOPES (1970) have led to the conclusion that the decline of the specific growth rate of yeast between the temperature optimum and the temperature maximum cannot merely be explained by irreversible or reversible enzyme denaturation. In the suboptimal temperature range (20—37.7° C) growth rate did not vary with time when the exponential growth rate was reached. However, at superoptimal temperatures above 37.7° C the growth rates decreased after a period of time, the duration of which decreased with increasing temperature, and were no longer measurable above the maximum of 43.8° C. Therefore in Arrhenius plots of growth rates two branches in the superoptimal temperature range are obtained (Fig. 6). In this temperature range exponential death concurs with exponential growth, which decreases the apparent specific rate. Above the temperature maximum the specific death rate is greater than the specific growth rate, and the population suffers net exponential death and will eventually die out.

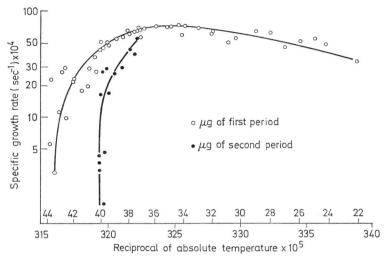

Fig. 6. Arrhenius plot of experimental specific growth rates of a strain of *S. cerevisiae* (VAN UDEN and MADEIRA-LOPES, 1970)

Reservations regarding the cardinal temperatures of bacterial multiplication are that some anomalies have been observed. OPPENHEIMER and DROST-HANSEN (1960) showed that in plots of cell yield against growth temperature, multiple optima near 11°, 25° and 39° C and multiple minima at 16°, 31° and 43° C occur. They attributed this atypical behavior to higher order phase transitions of water at specific temperatures (DROST-HANSEN, 1965). These changes occur within a

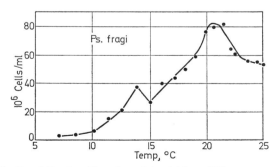

Fig. 7. Number of cells of *Ps. fragi* incubated for 24 h at different temperatures (DAVEY et al., 1966)

rather narrow range near 15°, 30°, 45° and 60° C and are believed to affect biological systems. Corresponding anomalies are also reported from other bacteria (DAVEY et al., 1966), in which growth was suppressed at the predicted temperatures (Figs. 7—10). There are several possible explanations for the anomalies in biological

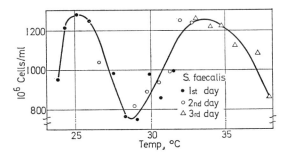

Fig. 8. Number of cells of *S. faecalis* incubated for 16 h at different temperatures (DAVEY et al., 1966)

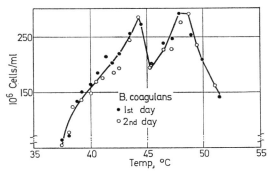

Fig. 9. Number of cells of *B. coagulans* incubated for 16 h at different temperatures (DAVEY et al., 1966)

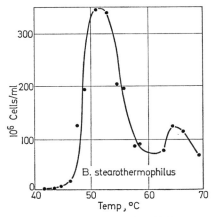

Fig. 10. Number of cells of *B. stearothermophilus* 1518 smooth incubated for 8 h at different temperatures (DAVEY et al., 1966)

systems. The effect could be external to the biological body or cell itself. The movement of fluids across membranes could be affected by the change in water structure. Or, it could be that water will allow efficient formation of enzymes or coupling of enzymes and substrates when one phase of water predominates between the anomalous temperatures, but at or near the specific temperatures, the formation or coupling is inefficient, and the biological activity is reduced.

B. Heat Killing of Microorganisms

1. Time Course of Heat Killing of Microorganisms

The effect of heat in the destruction of microbial populations is important in practical sterilization, especially in the preservation of foods. Although microorganisms can be destroyed by heating temperatures which are high enough and heating times which are long enough, the heat treatment of many menstrua is limited because of their sensitivity to high temperatures. Therefore a minimal heat treatment in many cases is necessary, e.g. in food preservation, to avoid destruction of valuable constituents. Consequently the individual heat resistance of microorganisms as well as the influence of environmental conditions on the heat resistance of microbial populations to be killed is necessarily to be considered in sterilization procedures.

The first quantitative studies of the killing of suspensions of bacteria at elevated temperatures were carried out in the first decade of this century (MADSEN and NYMAN, 1907; CHICK, 1908, 1910). These reports emphasize the time factor in inactivation processes and draw attention to the importance of survival curves, which are obtained by plotting the logarithm of the number of survivors in the suspension against the exposure in time units. Such survival curves are often exponential, since a straight line is obtained. This has led to the concept that death of microorganisms by lethal agents generally obeys first-order kinetics, because destruction by chemical disinfectants and radiation (RAHN, 1945) shows corresponding survival curves. Furthermore the common occurrence of exponential curves has led to the hypothesis that killing of microbial cells by all disinfectants is due to a single lethal event occurring at random, the so-called target theory. The hypothesis that the death of microorganisms by heat and other lethal agents is caused by a monomolecular reaction may, however, apply only to radiation, where the lethal event could be a single change produced in an essential molecule, such as the deoxyribonucleic acid of the bacterial genome. But in general it is assumed nowadays that heat and chemicals kill the cell following the interaction of larger numbers of molecules. When an inactivation process is described by a straight survival curve, it simply indicates that a constant fraction of the population is destroyed per unit of time.

Nevertheless the concept of first-order kinetics of thermal death of microorganisms could be usefully employed to describe their heat resistance and to predict extinction of a population of microorganisms under specified treatment conditions. The survival curve, i.e. the loss of numbers of cells per unit of time, is directly proportional to their number:

$$\frac{-\,\mathrm{d}N}{\mathrm{d}t} = kN \quad \text{or} \quad \frac{N_0}{N_t} = \mathrm{e}^{-kt} \tag{16}$$

where t = time, N_0 = number of cells at the beginning of the heating process, N_t = number of surviving cells after the heating time t, and k = constant of integration which designates the individual heat resistance of the organisms under the conditions of heat treatment. In other words, k depends on the temperature, the kind of organism under investigation and on the environmental conditions which influence the heat resistance of organisms [pH, electrolyte concentration, presence or absence of proteins, carbohydrates and other factors which influence the heat stability of living systems (see also p. 20—24)].

Consequently k may be applied to characterize the heat resistance of microorganisms under specified conditions. k is calculated (using Brigg's logs) from

$$k = \frac{2.3}{t_1 - t_0} \log N_0/N_t \cdot \qquad (17)$$

Since k is rather abstract for practical purposes, the time required to reduce the number of viable microorganisms by 90% (= D, or decimal reduction time) is usually applied to characterize the heat resistance of a microbial population at a given temperature. Then the equation is written

$$k = \frac{2.3}{D} \cdot \log 10 \qquad (18)$$

or

$$D = \frac{2.3}{k} \cdot \qquad (18a)$$

To calculate D from convenient heating times and the corresponding number of survivals the equation

$$D = \frac{t_1 - t_0}{\log N_0/N_t} \qquad (19)$$

is developed from Eqs. (18a) and (17).

As mentioned above, in many cases linear curves have been observed in plots of logs of survivors versus units of time when bacterial populations were submitted to heat treatment. This observation led to the hypothesis that the lethal effect of heat is the result of a monomolecular reaction. In addition to the fact that this concept has since been questioned, it has frequently been observed that non-exponential survivor curves are obtained, the different types of which are shown in Fig. 11. Curve A represents the exponential order of death, indicating that a

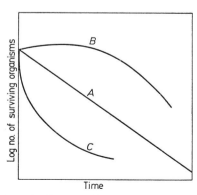

Fig. 11. Different types of survival curves for the death of microorganisms in a population. See text for explanation (FARRELL and ROSE, 1967)

constant fraction of cells dies per unit of time. A convex deviation (curve *B*) was originally explained as a multitarget event of killing occurring in populations of multinuclear organisms. A similar curve may, however, result when the organisms in the population are clumped, because each surviving cell in the clump creates only one colony on the counting plate regardless of the fraction of survivors in a cell aggregate. On the other hand, concave curves (Type *C*) are obtained when the heat resistance of the population is unequal. In the latter case susceptible cells die at a faster rate at the beginning of the heating procedure, while the resistant cells die at a slower rate when a majority of susceptible organisms have been destroyed.

As to the multitarget theory of inactivation, RAHN (1930) in cooperation with HUR-WITZ developed an equation expressing the dependence of the number of destroyed cells upon the number of susceptible molecules (nuclei) per cell:

$$Dn^{(r)} = a \left[1 - q^n - nq^{n-1}p - \frac{n(n-1)}{2} - \cdots \right.$$
$$\left. \cdots - \frac{n(n-1) \cdots (n-r+2)}{(r-1)!} \cdot q^{n-r+1} p^{r-1} \right] \tag{20}$$

where Dn = number of inactivated cells after the time n, r = number of molecules per cell to be destroyed, $q = 1 - m/a$ und $p = 1 - q = m/a$, a = number of cells per volume under investigation and m = total number of molecules inactivated per unit of time.

The application of the law of first-order kinetics of inactivation implies that the culture is uniformly susceptible to an injurious agent. This presupposition may be valid for populations of resting cells but does not hold true for multiplying cells. It has been shown that in log phase cultures of synchronously dividing cells *(E. coli)* fluctuations in heat resistance occur (CHRISTOPHERSEN, 1964). Cells are more sensitive immediately after division than are mature cells before division (Fig. 12).

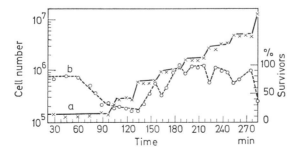

Fig. 12. Alterations of heat resistance in a synchronously growing population of *E. coli*, *a* log growth curve, *b* surviving cells after heat treatment (CHRISTOPHERSEN, 1964)

The occurrence of nonlinearity of thermal death rate curves in log-linear co-ordinates may be taken as a further indication that the amount of heat sufficient to inactivate a part of the population of microorganisms causes some injury to the remainder of the population, thus rendering it more susceptible to continued heating. In this case convex curves (Type *B*) should be obtained, indicating that the killing rate decreases. The function (19) may then be written as

$$D = \beta \left[\frac{t_1 - t_0}{\log N_0/N_t} \right] \tag{21}$$

accounting for linear or logarithmic variation of D with time, depending on the interpretation of the factor β (TISCHER and HURWICZ, 1954).

Another complication in applying D-values to describe the heat resistance of microorganisms has been encountered in bacterial spores. Nonlogarithmic thermal death curves for certain strains of bacterial spores have been reported by several investigators (HALVORSON, 1958; STERN and PROCTER, 1954; SHULL and ERNST, 1962). These spores may behave as a bimodal population with respect to heat resistance for which the death curve is described by the sum of two exponential functions, or they may exhibit a lag followed by logarithmic decline in the population of viable spores. HALVORSON (1958) suggested that the initial lag could be attributed partly to heat activation of dormant spores (see p. 58) and partly to a "multiple hit" inactivation process (see above). SHULL and ERNST (1962) demonstrated that heat activation can be responsible for a significant portion of the initial lag in the thermal death curves of *Bacillus stearothermophilus* spores. A mathematical treatment of this phenomenon has been developed by SHULL, CARGO, and ERNST (1963). They assume that both the inactivation and the activation of spores are processes of first-order kinetics with individual velocity constants. Then the number of activated spores after the time t is

$$N_t = N_0 \, \mathrm{e}^{-\alpha t} \tag{22}$$

and the number of surviving activated spores at the same time is

$$A_t = A_0 \, \mathrm{e}^{-kt} \tag{23}$$

where $N_0 =$ initial number of non-activated spores, $N_t =$ number of non-activated spores at time t, $\alpha =$ activation rate constant, $A_0 =$ initial number of activated spores, $A_t =$ surviving number of activated spores at time t, and $k =$ inactivation rate constant. The number of living spores at time t (L_t) is the sum of activated spores at time t and survivors of activated spores at the same time, namely

$$L_t = N_t + A_t. \tag{24}$$

An examination of the behavior of A_t and L_t over a short interval of time yields the following differential equations:

$$A'_t = - k A_t + \alpha N_0 \, \mathrm{e}^{-\alpha t} \tag{25}$$

$$L'_t = - k A_t. \tag{26}$$

When $\alpha \neq k$, the solution of Eq. (25) subject to the initial condition is

$$A_t = A_0 \, \mathrm{e}^{-kt} + \frac{\alpha N_0}{k - \alpha} \, (\mathrm{e}^{-\alpha t} - \mathrm{e}^{-kt}) \,. \tag{27}$$

Upon integration, Eq. (26) becomes

$$L_{t_2} - L_{t_1} = - k \int_{t_1}^{t_2} A_1 \, \mathrm{d}t \,. \tag{28}$$

Equation (27) describes the behavior of the activated spore population in the hypothetical system as the sum of two terms; the first term accounts for the first-order death of the initial activated spore population; the second describes the activation and subsequent death of nonactivated spores. Equation (28) indicates that the total number of deaths occurring between time t_1 and time t_2 is equal to the product of the death rate constant and the area under the death curve for the time period in question.

After determination of A_0, N_0, k, and α from experimental data, a hypothetical curve has been drawn according to Eq. (27) (broken line in Fig. 13) together with an experimental curve (solid line in Fig. 13). Although the curves are similar in shape, there are still significant differences in the beginnings of the curves. Assuming that the experimental curve is valid, it appears that corrections are to be made in the ratio of α to k.

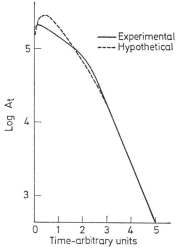

Fig. 13. Heat activation and thermal death of B. stearothermophilus (SHULL and ERNST, 1962)

2. The Time-Temperature Relation of Heat Killing

The alteration of biological reaction rates including killing rates of microorganisms with the temperature can be described by the temperature coefficient Q_{10}. But it should be kept in mind that such a mere description has no relation to the actual biochemical mechanisms which are involved in the heat inactivation of biological systems. The Q_{10} value is just the factor by which the killing rate increases per 10 centigrade degrees rise in temperature

$$Q_{10} = \frac{k_T}{k_{T+10}} \tag{29}$$

where k is the reaction rate constant according to Eq. (17). But any other parameter of the killing rate can be used, provided it correspondes to k, e.g., the decimal reduction time D [see Eq. (19)]. Since D is the inverse value of k, Q_{10} must necessarily be calculated from

$$Q_{10} = \frac{D_{T+10}}{D_T} . \tag{30}$$

Another convenient parameter in practical sterilization is the total time necessary to kill a given number of microorganisms per volume unit, the so-called thermal death time.

The temperature coefficient Q_{10} is calculated from arbitrary experimental data by the equation

$$\log Q_{10} = \frac{10}{T_2 - T_1} \log \frac{k_2}{k_1},$$ (31a)

or

$$\log Q_{10} = \frac{10}{T_2 - T_1} \log \frac{D_1}{D_2}.$$ (31b)

In applied bacteriology and in the canning industry the temperature dependence of killing rates of bacteria and bacterial spores is commonly expressed in terms of the temperature difference by which the killing rate is altered by the factor of 10, the z value. Of course, this value has a more practical meaning than the Q_{10} value. By definition Eq. (31b) becomes

$$\log Q_{10} = \frac{10}{z} \log 10,$$ (32a)

or

$$z = \frac{10}{\log Q_{10}}.$$ (32b)

To calculate z from arbitrary experimental data, Eqs. (32b) and (31b) (since D values are more common in practice) are combined:

$$z = \frac{T_2 - T_1}{\log D_1/D_2}.$$ (33)

The observed z values for killing vegetative bacteria or spores by moist heat, which correspond to temperature coefficients as high as 10—20, are known to be rather small, whereas temperature coefficients of biochemical reactions are normally in the region of 2—3.

The heat denaturation of proteins as well as the heat killing of microorganisms have large temperature coefficients, consequently the destruction of microorganisms by moist heat should be due to similar mechanisms. Furthermore Q_{10} values of killing rates of microorganisms are far from constant. As shown in Table 1, considerable variations are normally observed and are marked by an average decrease in value with increasing temperatures. Therefore calculations of Q_{10} values and z values are only applicable within narrow temperature ranges when the heat inactivation of a bacterial population is concerned.

Table 1. Change in Q_{10} values for the heat killing of *Salmonella paratyphi* with the heating temperature (ØRSKOV, 1926)

Heating temperature (°C)	Killing time (min)	Q_{10}
55	140	1370
57	33	37
59	16	44
61	7.5	99
63	3	32
65	1.5	

The dependence of Q_{10} values upon the temperature at which they are obtained is valid for all chemical and biological reactions, and it has already been indicated that Q_{10} values have no basis in terms of thermodynamics. But it is well established that the Arrhenius equation accounts for the activation of reaction rates by rising temperatures. Considering the theory of absolute reaction rates, a reaction is assumed to be possible only when the reacting molecules enter a transient activated state to form an activated complex. One has to imagine that this activating energy is derived from the kinetic movement of the molecules participating in the reaction, which in turn depends on the temperature expressed in $°K$. The minimum energy difference between the reactants and the activated fraction of the reacting molecules is denoted by E. In biological systems, E is replaced by the term μ, which is referred to as the temperature characteristic of the process. Furthermore the reaction rate depends on the frequency of collisions of the reacting molecules and a probability factor. Both these constants are combined in a constant denoted as A. The Arrhenius equation, in which R is the gas constant and k the velocity constant, is written

$$\ln k = \ln A - \frac{E}{RT}, \qquad \text{or} \quad k = A\,e^{-E/RT}. \tag{34a, b}$$

E (or μ) can be calculated from the velocity constants k_1 and k_2 measured at the corresponding temperatures T_1 and T_2. By subtraction of

$$\left. \begin{aligned} \ln k_1 &= \ln A - \frac{E}{RT_1} \\ \ln k_2 &= \ln A - \frac{E}{RT_2} \end{aligned} \right\}$$

one obtains

$$\ln k_1 - \ln k_2 = -\frac{E}{RT_1} + \frac{E}{RT_2}, \tag{35a}$$

or

$$E = \frac{2.3 \cdot R\,(\log k_1 - \log k_2)}{\dfrac{1}{T_2} - \dfrac{1}{T_1}}. \tag{35b}$$

In plots of $\log k$ against $1/T$, linear curves are obtained as long as E is constant. The slope of such straight lines is $-E/2.3 \cdot R$. In bacteriological practice one usually plots thermal death times or D values on a log ordinate versus the reciprocal of the absolute temperature. With homogeneous populations linear Arrhenius plots should be obtained, but nonlinear plots are encountered more often in practice, indicating that the population is heterogeneous.

Another similarity which is encountered in denaturation of proteins as well as in heat inactivation of microorganisms is the large values of the change in entropy. According to classical equations of thermodynamics, the change in the free energy when one mole of native protein is transformed into one mole of denatured protein is given by

$$\Delta F = \Delta H - T\Delta S$$

where ΔH is the change in the enthalpy of the reaction per mole and ΔS is the change in the entropy of the reaction per mole.

The values of ΔH measured during inactivation of proteins by moist heat are extremely large. The order of magnitude shows variations between approximately

50 and 150 kcal/mole. If the rate of denaturation were solely governed by ΔH, the rate constant would be extremely small and would be almost negligible under ordinary conditions. However, the large ΔH found in denaturation by moist heat is offset by the accompanying large values of ΔS, giving mean values for ΔF of about 22 ± 5 kcal/mole (JOLY, 1965). The common concept is that large positive values of ΔS indicate a great increase in randomness, as would result from the breakup of many labile bonds, e.g. hydrogen bonds, accompanying protein denaturation. Similar conditions are suggested to hold true in the killing reactions of microorganisms by moist heat.

Inactivation by dry heat, however, produces little or no change in ΔS, indicating no radical changes in the protein structure.

Activation entropy can be calculated using the equation for absolute rate processes of EYRING (1935):

$$k = \frac{KT}{h}\,\mathrm{e}^{-\Delta F/RT} \tag{36}$$

where h ist the Plank constant ($6.62 \cdot 10^{-27}$ erg $\cdot$ sec), and K is the Bolzmann constant ($1.38 \cdot 10^{-16}$ erg/degree). This equation leads to

$$\Delta F = RT \ln \frac{KT}{hk} \tag{37}$$

for the free energy of activation. The enthalpy of activation is given by

$$\Delta H = RT^2 \frac{\mathrm{d}\ln k}{\mathrm{d}T} - RT\,. \tag{38}$$

By substituting ΔF and ΔH in $\Delta F = \Delta H - T\Delta S$ with the right-hand terms of the Eqs. (37) and (38), one obtains

$$RT \ln \frac{KT}{hk} = RT^2 \frac{\mathrm{d}\ln k}{\mathrm{d}T} - RT - T\Delta S \tag{39a}$$

and

$$\Delta S = -R\left(\ln \frac{KT}{hk} - T\frac{\mathrm{d}\ln k}{\mathrm{d}T} + 1\right). \tag{39b}$$

In spite of the fact that a treatise on the heat killing of microorganisms on the basis of unique thermodynamic principles must be regarded as a simplification, such procedures have been valuable for the understanding and for the practical application of thermal processes. Actually the reaction of living organisms and microbial populations to thermal stress is thought to be extremely complex and is probably not explainable by the assumption of the single-hit hypothesis. Experimental results in this direction have been obtained by DIMMICK (1965).

3. Factors Affecting the Heat Resistance of Vegetative Bacteria

a) The Effect of Water

It is well known that the heat resistance of micoorganisms depends greatly on the humidity. The heat stability of living systems is generally greater when water is absent or when the water content is low. An increase in heat resistance can usually be observed as the available water (water activity) in the heating menstruum is reduced (GOEPPERT et al., 1970; BAIRD-PARKER et al., 1970; GARIBALDI, 1968; RIEMANN, 1968; NG et al., 1969; MURELL and SCOTT, 1966; CALHOUN and FRA-

ZIER, 1966). The effect of water can to a certain extent be explained by the dependence of the heat stability of proteins upon hydration. The water of hydration is attached to groups within, or at the surface of, protein molecules having free charges and also, though less strongly, to dipoles such as $-CO-$ and $-NH-$ of peptide chains. Some water is bound osmotically by the Donnan effect, and some is present simply as inclusion water in the three-dimensional network of peptide chains.

The stability of protein molecules, on the other hand, is maintained by intramolecular bonds such as hydrogen bonds, salt bridges, $-S-S-$ bonds, and hydrophobic bonds (HAUROWITZ, 1950; NEMETHY and SCHERAGA, 1962). It must be assumed that such intramolecular bonds are weakened when a protein molecule is "blown up" by water. This holds especially for hydrogen bonds, as has been indicated by CHRISTOPHERSEN and KAUFMANN (1952). Therefore less energy is required to unfold the peptide chains; this gives rise to decreased heat resistance.

b) The Effect of Salts and Ions

Cations may exert different effects on the heat resistance of microorganisms. In most cases univalent cations decrease the heat resistance whereas bivalent cations lead to an increase in heat resistance. Furthermore the stabilizing or destabilizing effect of ions depends on the concentration, as has been shown by NAKAMURA (1898), as well as on other environmental conditions such as suspending media, pH etc. At least to a certain extent the effect of ions can be explained by their influence on the hydration of proteins. NANNINGA (1957) showed that proteins such as casein and pepsin had high affinity for calcium and magnesium. BIER and NORD (1951) and GORINI (1951) demonstrated that calcium salts induced the thermostability of trypsin, serum albumin and certain bacterial proteases. It has been shown by VOR DEM ESCHE (1953) that electrolytes which cause proteins to swell decrease the heat resistance of microorganisms, whereas electrolytes which cause proteins to de-swell increase the thermostability. Optimal stability can be obtained by a defined proportion of monovalent and divalent cations. A stabilizing effect of divalent cations such as Ca^{++} and Mg^{++} may also be due to a "linkage" of proteins together to give more stable complexes, as has been shown by BOGEN (1948) in plant cells.

Finally high concentrations of salts can increase the heat resistance by osmotic water depletion and thereby exert a mechanism which is similar to the effect of drying.

c) The Effect of Proteins

An example of the protecting effect of proteins in the heating menstruum is given in Fig. 13a which gives a comparison of the killing rate of *Streptococcus lactis* in milk and in Ringer's solution (WHITE, 1952). The mechanism of protein protection is difficult to explain. The common opinion that coagulated proteins are precipitated on the cell surface and thus furnish an isolating barrier (TOWNSEND et al., 1938) is not in accordance with calculations about the heat flow through ultrathin layers. AMAHA and SAKAGUCHI (1954) suppose that some of the activated water molecules are consumed by native proteins. In terms of thermodynamics, it may thus be assumed that a considerable part of the heat energy is consumed by the denaturation of the protecting proteins. This explanation is in accordance with the extraordinary high values for the activation entropy of the denaturation reaction of proteins (see p. 20).

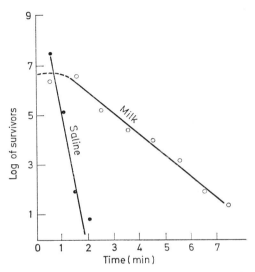

Fig. 13a. Effect of the heating menstruum on the heat resistance of *Str. lactis* (WHITE, 1952)

d) The Effect of Oils and Fats

An increase in heat resistance of vegetative microorganisms as well as bacterial spores in the presence of fatty materials has been observed particularly in food microbiology. The mechanism of fat protection is supposed to depend on a localized absence of water. But an isolating effect against heat flow has also been considered. Since it was observed that long-chain fatty acids have a protective effect on proteins (SUGIYAMA, 1951), it also seems that these acids have a stabilizing effect on intramolecular bonds, e.g., hydrophobic bonds.

e) The Effect of pH

The well-known fact that proteins are most stable at their isoelectric point led to the opinion that the heat resistance of proteins in living cells also reaches an optimum at pH values in the region of the isoelectric point of the protein or the plasma, respectively. It has turned out, however, that the action of pH on the rate of heat denaturation is rather complex (JOLY, 1965). Nevertheless microorganisms show a maximal heat resistance near pH 7.0. Sometimes on both sides of the value of maximal heat resistance a sharp increase of heat sensivity is observed, as has been demonstrated with *Streptococcus faecalis* in buffer solutions (WHITE, 1963; see

Fig. 14). It must be emphasized, however, that the pH value inside the cells may differ from that of surrounding buffer solutions. This may be the cause of conflicting results with regard to the effect of pH on heat resistance of microorganisms.

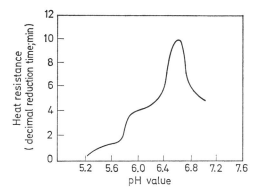

Fig. 14. Effect of pH value on the heat resistance of *S. faecalis*. The bacteria were grown at 37° C and suspended in citrate-phosphate buffers at 60° C (WHITE, 1963)

f) The Effect of Age

The relation between heat resistance and the growth phase has been known since SCHULTZ and RITZ (1910) observed that growing microorganisms are generally less resistant to heat than organisms in the stationary phase of growth. SHERMAN and ALBUS (1923) coined the term physiological youth to characterize the properties of young cells. Such cells are generally more labile in the presence of injuring agents. Detailed studies of the effect of the age of the culture on the heat resistance of *Streptococcus faecalis* have been made by WHITE (1953). She found two pronounced maxima in the heat resistance during the growth cycle of a culture (see Fig. 15).

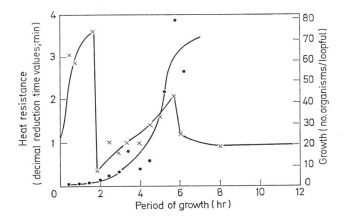

Fig. 15. Effect of age of the culture on the resistance of *S. faecalis* to heating at 60° C (WHITE, 1953)

During the lag phase a steady increase of heat resistance was observed until the beginning of the exponential growth of the culture. At this point a maximal heat sensivity was observed. During the exponential growth of the culture the heat resistance gradually increased until a second maximum was reached at the end of the exponential growth phase. The degree of heat resistance during the latter growth phase, however, was dependent upon the temperature of incubation, as is discussed below.

g) The Effect of Growth Temperature

It has been found that the heat resistance of many bacteria increases with increasing incubation temperature. But the same condition may also lead to a decrease of resistance (ELLIKER and FRAZIER, 1938; CHRISTOPHERSEN and THIELE, 1952; CHRISTOPHERSEN and PRECHT, 1950a; ANDERSON and MEANWELL, 1936). In the above-mentioned experiments of WHITE (1953) with *Streptococcus faecalis*, the maximal heat resistance (expressed as D values) at the end of the exponential growth phase was at an incubation temperature of $27°$ C $= 1.5$, at $37°$ C $= 2$, and at $45°$ C $= 6.45$. Very young cultures grown at $27°$ C, however, were more resistant than those at $37°$ C, which in turn were slightly more resistant than $45°$ C-cultures. Obviously the age of the culture should be controlled when studying the effect of incubation temperature on heat resistance.

4. Resistance Adaptation

a) General Remarks

It has generally been observed with cells from the exponential growth phase that a rapid increase in resistance of from $3°-5°$ C occurs when the growth temperature is elevated from about $20°$ C to the temperature optimum of about $37°$ C. In the case of yeasts, *E. coli*, *P. fluorescens* and *Serratia marcescens*, however, not only does the entire cell became more resistant but individual enzymatic activities are destroyed only at higher temperatures. Surprisingly, though, the opposite observation has also been made. In lactic acid streptococci, the heat resistance of which decreases with increasing incubation temperatures (ANDERSON and MEANWELL, 1936), the dehydrogenase activity showed a corresponding behavior (see Fig. 16).

We studied the phenomenon of resistance increase mainly in yeasts (CHRISTOPHER-SEN and PRECHT, 1950b, 1951; CHRISTOPHERSEN and THIELE, 1952; RING and CHRISTOPHERSEN, 1964; CHRISTOPHERSEN, 1963). Stabilization obtained by adaptation to higher temperatures was observed, not only in intact cells and their activities (respiration, dehydrogenase, permeability), but also in cell extracts. These studies included the resistance of dehydrogenase, transaminase and hexokinase.

The individual heat resistance of various enzymatic activities turned out to be different; the inactivation in yeasts took place at between approximately $50°$ and $60°$ C. Transaminase and peroxydase were in the lower range for inactivation, whereas catalase and permeability of the cell membrane to dyes were in the upper range. Hexokinase and dehydrogenases exhibited medium resistance. The question, however, as to which enzyme limits the viability of the entire cell was not studied and is obviously not easily answered. The viability of the cell may possibly depend on structures other than enzyme proteins and may be sought in control fractions such as nucleic acids.

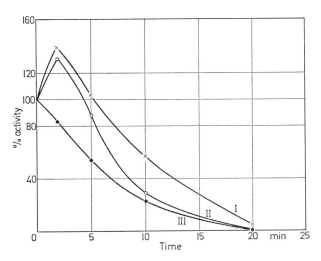

Fig. 16. Heat inactivation and heat activation of succinic dehydrogenase of *S. faecalis* after adaptation to 29°, 37° and 45° C. The cells were heated for the stated times (abscissa) to 55° C

b) Mechanisms of Resistance Adaptation

The mechanisms of resistance alteration of enzyme proteins are poorly under-stood. In early investigations, we observed that aqueous solutions of commercial enzyme preparations exhibited a significant increase in thermostability when they were kept at higher temperatures for about 24—48 h. At first, this was shown by proteolytic activity of pancreatin and diastase preparations (CHRISTOPHERSEN and THIELE, 1952). Pure trypsin, however, did not exhibit this property, whereas aqueous solutions of crystallized alcohol dehydrogenase showed increased resist-ance. In the latter enzyme, we also observed a different infrared absorption spec-trum after it had been adapted to different temperatures. Apparently, this is caused by changes in the molecular structure. It would be valuable to investigate the temperature action on protein structure in connection with alterations of protein stability.

The molecular mechanisms of resistance alterations are, doubtless, various. Hydra-tion is a decisive factor for protein stability. There was a correlation between this and increase of resistance achieved by adaptation or aging, on the one hand, and the free water content, on the other hand (CHRISTOPHERSEN and PRECHT, 1952a). But it is not possible to explain every alteration of resistance of enzyme proteins by means of hydration. It must be emphasized that specific differences of structure must also be regarded as being responsible for the stability of proteins (CHRISTO-PHERSEN and PRECHT, 1952b), especially in genetically induced resistance altera-tions. Much attention has been drawn to the concept developed by HEILBRUNN (1924), BĚLEHRÁDEK (1931), GAUGHRAN (1947), and others, according to which lipids may be responsible for the stability of cell structures. The first two authors assert that the structures in the protoplasm or in the plasma membrane collapse on melting of the cellular fat, whereas GAUGHRAN assumes that active cell processes are maintained as long as the cellular lipids are liquid. The fact that degree of satu-ration, length of chains and melting point of lipids increase with increasing growth temperature can be in accord with both interpretations.

Lipid alterations as mentioned above have been found in yeast and in a thermo-philic spore-former after adaptation to various temperatures (CHRISTOPHERSEN and KAUFMANN, 1955). Further experiments revealed that fatty acids of different cell constituents of yeasts, such as free and bound lipids and phosphatides, show a corresponding dependence on adaptation temperatures. But it is not certain that this is evidence of a relationship to heat and cold resistance; investigations of MARR and INGRAHAM (1962) on *E. coli* have raised further doubts on this point. Furthermore, it should be considered that fatty acids bound in lipoproteins and phosphatides possess properties that differ completely from free acids, the melting point of which is often regarded as a measure of protoplasmic stability.

Nevertheless, the composition of the lipid fraction of the cell membrane may determine one character of the cell, namely, its permeability. It is reasonably certain that phosphatides are important components of the permeability barrier. RING (1965) proved a relationship between cold stability of the permeability barrier of Streptomyces cells and adaptation temperature. After these cells were enriched with radioactive α-aminoisobutyric acid and were subsequently cooled, a rapid efflux took place in the 0° C range.

It is suggested that the permeability barrier, which, according to DAWSON and DANIELLI (1943), consists of a double layer of lipids, collapses at low temperatures, owing to a phase transition within the lipid layer. When the intracellular radio-activity is plotted against cooling temperature after 60 min of cooling, sigmoid curves are obtained, the flexion points of which are dependent on the adaptation temperature of the cells (Fig. 17). The points of flexion, or abrupt rise in retention, are between −2° C in the 18° C-adapted cells and +6° C in the 35° C-adapted cells (Fig. 18). According to the lipid theory, it could be concluded that the lower melting point of the membrane lipids in the cold-adapted cells is responsible for the higher stability of the membrane at lower temperatures. Thus, a common molecular basis would exist for the dependence of membrane permeability on temperature.

Morphologic alterations have also been observed in bacterial cells adapted to low temperatures. A temperature shift from 37−15° C caused, in logarithmically growing cells of *B. subtilis*, structural modifications involving mesosome deteriora-tion and double cell wall formation (NEALE and CHAPMAN, 1970). Such cells re-covered when they were kept at 15° C and regained their normal morphologic appearance. When *B. subtilis* cells were transferred from 37 −12° C no recovery was observed. In contrast, cultures transferred from 15−12° C continued to grow and retained their morphologic appearance. It is suggested that the 15° thermal history of the population has enabled it to respond differently to 12° C than a culture adapted to 37° C.

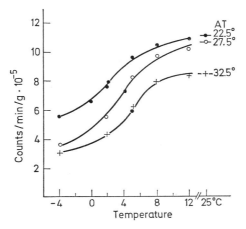

Fig. 17. Efflux of α-aminoisobutyric acid as a function of the cooling temperature. Cells of *Streptomyces hydrogenomonas* that had been grown at three adaptation temperatures (AT) were preloaded with ¹⁴C-α-aminoisobutyric acid at 30° C. After cells had been cooled to various temperatures for 60 min, the cellular radioactivity was determined (RING, 1965)

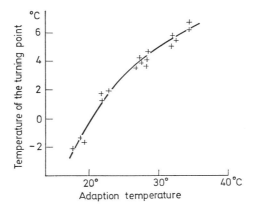

Fig. 18. Flexion points, or temperatures for beginning of abrupt rise in retention, in the experiments described in Fig. 17, as a function of the adaptation temperature (RING, 1965)

c) Recovery of Heat-Injured Microorganisms

The usual criterion of death of microorganisms is the failure to multiply on plate count media. It was soon recognized, however, that this criterion was insufficient (RAHN and BARNES, 1933). The ability of heat-treated microorganisms to reproduce depends to a greater or lesser extent upon the environmental conditions into which they are transferred after heating (reactivation). In general, heated microorganisms (as well as radiation-injured cells) require supplementary nutrients for recovery (NELSON, 1943, 1944; HEATHER and VANDERZANT, 1957; HANSEN and RIEMANN, 1963; HARRIS, 1963; CLARK and ORDAL, 1969; DABBAH et al., 1969).

Delayed growth of bacteria following heat treatment has been explained on the basis of attenuation (ALLEN, 1923) or dormancy (DAHLBERG, 1946). Other workers have concluded that the bacterial cells are injured by heat and suggested that the nature of damage is enzyme inactivation (CURRAN and EVANS, 1937), damage to the osmotic barrier and disruption of the balance of reaction rates (MITCHELL, 1951; POSTGATE and HUNTER, 1963), or structural changes in the protoplasm (HEDEN and WYCKOFF, 1949).

A precondition of reactivation of heat-treated cells seems to be that the heating be sublethal and lead to a delay only of multiplication. The recovery is thought to involve the repair of injuries directed by enzymatic activity (HILLS and SPURR, 1952; HEINMETS et al., 1954; IANDOLO and ORDAL, 1966). Reactivation of partially inactivated cellular enzymes has also been considered (LEMBKE et al., 1951, 1952; CHRISTOPHERSEN and KAUFMANN, 1952).

C. Adaptive Temperature Responses

The basic principles of temperature adaptation have been thoroughly discussed by PRECHT (see p. 302ff). As far as microorganisms are concerned, adaptive responses to temperature alterations occur exclusively on cellular and molecular levels.

It is necessary to distinguish (a) adaptations of the viability of organisms exposed to temperature extremes from (b) adaptations of metabolic activities measured over a wide range of temperatures. These two types of adaptation have been called resistance adaptation and capacity adaptation, respectively (CHRISTOPHERSEN and PRECHT, 1953). Both types of adaptation include heat and cold adaptation. In the case of heat-resistance adaptation, the limits of the tolerated temperature shift towards a higher level under the influence of elevated temperatures. If the sensitivity to cold increases simultaneously, we call this a reasonable resistance adaptation. For adaptations to low temperatures, the resistance to cold increases and the resistance to high temperatures decreases. In many cases, however, adaptation to one extreme temperature range coincides with an increase in resistance to the opposite direction as well. This is called paradoxical adaptation.

As to the biological mechanisms of temperature adaptation, it is well known that adaptations may be based on genetic variations. Such adaptations require several weeks or months of subculture and stepwise alterations of the temperature beyond the maximum or minimum. Efforts to alter the temperature range of microorganisms, with few exceptions, however, have not been successful. Since genetic mechanisms of temperature adaptation will not be discussed in this chapter, see the contribution of INGRAHAM (p. 60).

Genetic strains that exhibit high stability maintain their properties during a series of subcultures. On the other hand, in environmentally induced short-term adaptations, the alterations are more labile. This latter type of adaptation is reversible, as can be demonstrated by restoration to the original temperature conditions.

This chapter is confined exclusively to short-term environmentally induced temperature adaptations. The usual procedure in such experiments was to subculture organisms that had been maintained at an intermediate temperature. The subcultures were grown with or without the substrate whose utilization was to be tested, at various experimental temperatures. Cells were harvested in the second third of the exponential growth phase.

1. Capacity Adaptation

The term capacity adaptation originates from observations on animals (PRECHT, 1949) in which metabolic and locomotor activities increase with increasing temperature according to the rate-temperature rule; they may, however, reduce to a tolerable or compensated rate after prolonged exposure in the new temperature range. After the shift in temperature has taken place, many activities show an initial overshoot or undershoot before they reach a more or less extended induction period. In some cases even oscillations in the flow of intermediates can be observed immediately after the temperature shift (see also "Time course of capacity adaptation", p. 38). The subsequent curves of performance alteration are generally sigmoid (Fig. 19).

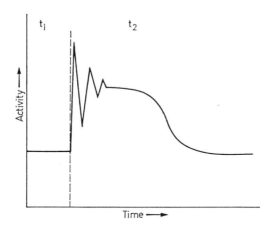

Fig. 19. The effect of changes of temperature on a metabolic reaction. Hypothetical curve

There are five different patterns of the metabolism-time curves (presented greatly simplified in Fig. 20). We speak of an ideal compensation when the original activity is restored (curve *a* in Fig. 20). The compensation is partial when the original activity is not fully restored (curve *b*). The original activity may also be overshot; in this case, we speak of a supraoptimal compensation (curve *c*). A further increase of activity may occur; it is termed an inverse compensation (curve *d*). Finally, there may be no compensation at all (curve e). In the case of temperature decreases, the corresponding phenomena may take place in the reverse direction.

For practical purposes, the type of adaptation can be deduced from rate-temperature curves of differentially adapted organisms by comparison of the activities at

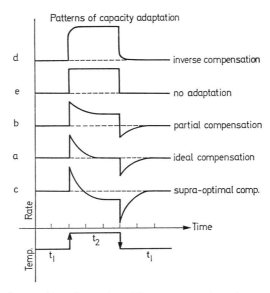

Fig. 20. Patterns of capacity adaptation. Time-course of performance activities after temperature increase from t_1 to t_2 and after temperature decrease from t_2 to t_1

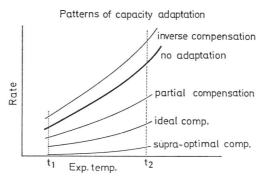

Fig. 21. Patterns of capacity adaptation. Rate function of temperature after adaptation to t_1 (heavy line) and after adaptation to t_2 (thin lines) (PRECHT, 1949)

two or more adaptation temperatures. In Fig. 21 the heavy line represents the rate-temperature curve of an organism's activity after adaptation to t_1. The curves with thin lines may be obtained after acclimation to t_2. A line connecting the activity after acclimation to t_2 with the activity after acclimation to t_1 indicates the type of compensation. This is shown by broken lines in Figs. 22–28. In the case of no adaptation both activity curves, after adaptation to t_1 and to t_2, are identical.

In capacity adaptation of microorganisms, metabolic reactions such as respiration, fermentation, enzymatic activities and transport of metabolites are involved. Before turning to molecular mechanisms of such temperature-induced capacity alterations, some characteristic adaptation phenomena of microbial performances are described.

2. Examples of Capacity Adaptation of Microbial Activities

A frequently observed partial compensation of a metabolic activity is shown in Fig. 22, i.e., the respiration of *Pseudomonas fluorescens*. Similar curves were also obtained for the succinic dehydrogenase activity of yeast (CHRISTOPHERSEN and

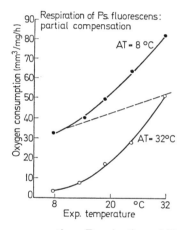

Fig. 22. Pattern of partial compensation: Respiration of *P. fluorescens* at different experimental temperatures and at adaptation temperatures (AT) of 8° and 32° C

PRECHT, 1950a, 1950b). Ideal compensations are illustrated in Fig. 23, which shows titratable acid formation by *P. fluorescens*, in Fig. 24, which shows lactic acid production by *Streptococcus cremoris*, and in Fig. 25, which shows the respiration of *Leuconostoc citrovorum*. Fig. 26 demonstrates the pattern of a supraoptimal compensation as it occurs in the proteolytic activity of *Lactobacillus helveticus* (CHRISTOPHERSEN and THIELE, 1952). The pronounced decrease in the ability to split peptide bonds in the case of adaptation to higher temperatures apparently occurs frequently in microorganisms. ALFORD (1960) observed in addition to the loss of proteolytic activity, an inhibition of lipolysis, citrate utilization, fermentation of xylose and arabinose, and action on litmus milk when strains of *Pseudomonas* and *Achromobacter* were cultivated at elevated temperatures.

An inverse compensation was observed in the case of peroxydase of yeast (CHRISTOPHERSEN and PRECHT, 1950a). An additional typical example is the glucose uptake of *P. fluorescens* (Fig. 27). This difference in glucose consumption indicates that the metabolism of bacteria adapted to 8° C and 32° C apparently differs in enzymatic pathways. This point is discussed in detail later. Also, the rate of phosphate

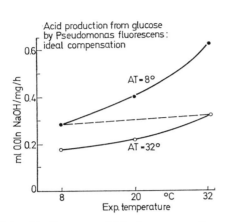

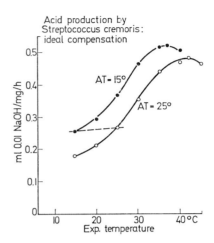

Fig. 23. Pattern of ideal compensation: Acid production from glucose by *P. fluorescens* at three experimental temperatures and two adaptation temperatures (*A T*)

Fig. 24. Pattern of ideal compensation: Acid production from glucose by *S. cremoris* at several experimental temperatures and at adaptation temperatures (*A T*) of 15° and 25° C

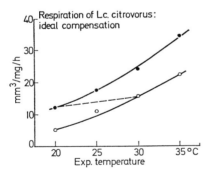

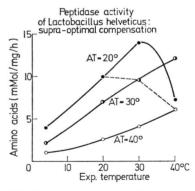

Fig. 25. Pattern of ideal compensation: Respiration of *Leuconostoc citrovorum* in mm³ of oxygen per mg each hour at adaptation temperatures of 20° (upper curve) and 37° C (lower curve)

Fig. 26. Pattern of supraoptimal compensation: Peptidase activity of *Lbc. helveticus* in mM of amino acids per mg of bacteria each hour at several experimental temperatures and at the adaptation temperatures (*A T*) of 20°, 30° and 40° C

uptake by yeast cells shows the pattern of inverse compensation. The initial rate of uptake of phosphate ions is significantly higher in the case of organisms adapted to 37° C compared with those adapted to 20° C (Fig. 28). An example of little or no compensation of metabolic activities is the respiration of yeast cells (CHRISTO-PHERSEN and PRECHT, 1950b).

From the point of view of suitability, it appears to be reasonable that an ideal compensation (curve *a* in Fig. 20) is the most meaningful biologically. Also, an

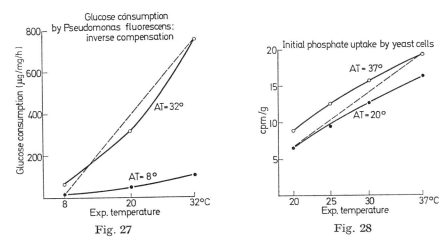

Fig. 27

Fig. 28

Fig. 27. Pattern of inverse compensation: Glucose consumption by *P. fluorescens*

Fig. 28. Pattern of inverse compensation: Initial phosphate uptake by yeast cells in counts per minute for each gram of cells

inverse compensation may be interpreted as meaningful — for instance, the increase of peroxydase activity in yeasts and in other organisms during adaptation to higher temperatures. The biological benefit in this case is the elimination of toxic hydrogen peroxide, which may, perhaps, be of particular importance at higher temperatures.

3. Mechanisms of Capacity Adaptation

Various factors are involved in the regulation of cellular processes. The most important are quantity and concentration of enzymes, coenzymes and other enzyme activators, electrolytes, and enzyme inhibitors. Transport reactions must also be considered. Another factor is the water content, which may be important in the transport of metabolites (CHRISTOPHERSEN and PRECHT, 1953). Finally, the possibility that the cell may shift to an alternate metabolic pathway should be emphasized as a regulatory factor.

The decrease in hexokinase activity at higher temperatures is typical for the alteration of the enzyme concentration as deduced from the activity of the cell-free extracts of yeast (CHRISTOPHERSEN, 1963). In one case, it was also found that an enzyme — leucine-oxaloacetic acid transaminase — apparently disappeared at high temperatures (RING and CHRISTOPHERSEN, 1964). But, judging from the reducing activity, as measured by the rate of electron transport to methylene blue or triphenyl tetrazoliumchloride, one cannot necessarily conclude that, in the case of intact yeast cells, differences in dehydrogenase activity result from differences in enzyme concentration. In intact yeast cells, for example, a decrease of reducing activity was found with increasing temperature. These differences, however, were not observed in cell-free extracts (CHRISTOPHERSEN and PRECHT, 1952a).

For the participation of coenzymes in regulation, an example is given by transaminases (RING and CHRISTOPHERSEN, 1964). The asparaginic acid-α-ketoglutaric acid transaminase of yeast was more active in cells adapted at 40° C than in cells adapted at 20° C. This is another case of inverse adaptation. The lower activity of this enzyme obviously results from the decreased saturation with pyridoxal phosphate. Addition of this coenzyme to the reaction mixture activated the 20° C-adapted yeast by 82% and the 40° C-adapted yeast by only 20%. At present, the biological significance of the adaptation of asparaginic acid-α-ketoglutaric acid transaminase cannot be explained. It is tempting to suppose that the cell is able to regulate enzymes that depend on the same coenzyme (pyridoxal phosphate) through the apoenzyme, on the one hand (as in leucin-oxaloacetic acid transaminase), and through the coenzyme, on the other hand (as in asparaginic acid-α-ketoglutaric acid transaminase).

Presently, there are no indications about the participation of electrolytes in capacity adaptation of microorganisms. The water content, however, must be considered a significant factor. The example of the dehydrogenase activity of yeast has just been mentioned. The fact that the reduction times of cell-free extracts of differently adapted cells were equal was regarded as support of the notion that the water content differs in warm- and cold-adapted yeast cells.

There are few indications of the different rates of the transport of metabolites through the cell membrane that are associated with temperature adaptations. Apparently, transport increases with adaptation to higher temperatures. This was the case in the phosphate uptake by yeast cells, as is shown in Fig. 28. To inhibit respiration, uranylic ions, which are said to inhibit sites of active transport, were required in a higher concentration in warm-adapted than in cold-adapted *Pseudomonas* cells. This indicates a more active transport mechanism in the warm-adapted cells. It has also been suggested that with an increase in lipid, especially in the cell wall, an organism could compensate for an impaired incorporation of building materials (WELLS et al., 1963). Reduction of growth temperature did not cause lipid changes in psychrophilic bacteria but did cause profound increase in lipid content in a mesophilic organism. Other information about the influence of the adaptation temperature on the quality of the permeability barrier was obtained by RING (1965) from a strain of *Streptomyces*, which has been discussed in the context of resistance adaptation (see p. 24).

Different metabolic pathways appear to play an important role in capacity adaptation. Suggestions in this direction have already been made by PROSSER (1958). Alterations in metabolic pathways have frequently been observed in microorganisms. Earlier investigations in this respect are those by DORN and RAHN (1939) and FOTER and RAHN (1936) who found that the amount of lactose consumed by lactic streptococci during the doubling of one cell was constant at low and medium temperatures, but that it increased towards the optimal temperature. They concluded that bacteria utilize their food more economically at low temperatures. A decrease of dry weight yield of *E. coli* cells formed per dry weight of substrate has been found by NG et al. (1962). We found a striking difference in the metabolism of *P. fluorescens*: the cells adapted to 8° C exhibit a very low glucose consumption and a high respiratory activity (Figs. 22, 23). After adaptation to 32° C, the cells behave in the reverse manner, and glucose consumption is very high. Another peculiarity is seen from the "Arrhenius plot" for the respiration rates. In most cases the Arrhenius plots for the respiration rates of different microorganisms adapted to different temperatures showed the same slope. In *P. fluorescens*, however, significantly diverging lines were obtained (Fig. 29); the temperature coefficient was much larger in cells from 32° C than in those from 8° C. The different sensitivity to inhibition by iodoacetate, as was demonstrated by EKBERG (1958) with the respiration of gill tissues of fish that were adapted to 10° and to 30° C, was also evident in the oxygen consumption of *P. fluorescens*.

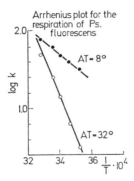

Fig. 29. Arrhenius plot for the respiration rate of *P. fluorescens*. Log of respiratory rate plotted as a function of the reciprocal of the absolute temperature $(T) \cdot 10^{-4}$

These findings, however, may only be applied to cultures which grow in non-limiting concentrations of glucose. When *P. fluorescens* was grown in continuous culture with limiting concentrations of glucose, 86% of the glucose was metabolized via the Entner-Doudoroff pathway at 30°, 20° and 8° C. The remaining glucose, 14%, was metabolized via the hexose monophosphate pathway, as has been shown by PALUMBO and WITTER (1969) using the radiorespirometry method.

Investigations by ALFORD (1960) dealt with the alteration of enzymatic activities of *P. fluorescens* and *Achromobacter* species. Lipase and peptidases virtually disappeared at higher growth temperatures, and metabolism of citrate and pentoses was impaired when the cells had been cultivated at elevated temperatures. Investigations of *Pseudomonas* and yeast cells have shown that the enzymes of the Krebs cycle and the hexose monophosphate (HMP) shunt are less active after adaptation to elevated temperatures. Among the key enzymes of the HMP, however, only glucose-6-phosphate dehydrogenase was affected; the activity of the 6-phosphogluconate dehydrogenase was unchanged. The adaptation temperatures were 8° and 32° C with Pseudomonas, 20° and 37° C with yeast. Results are given in Table 2.

Table 2. Activity of enzymes of the hexose monophosphate shunt

Enzyme	Ps. fluorescens		Candida pseudotropicalis	
	8° C	32° C	20° C	37° C
Glucose-6-phosphate dehydrogenase	0.1829	0.0805	0.097	0.058
6-Phosphogluconate dehydrogenase	0.0108	0.0110	0.025	0.021

Enzyme activity was measured as E_{340} per mg each minute with nicotinamide adenine dinucleotide phosphate.

Furthermore, aconitase was more active in *Pseudomonas* cells adapted to 8° C, because, for the inhibition of this enzyme, a higher concentration of transaconitic acid was required than in cells adapted to 32° C. It was concluded from investigations with enzyme inhibitors that *Pseudomonas* cells adapted to 8° C are inhibited more easily in anaerobic metabolism and not so easily in aerobic metabolism. The cells that were adapted to 32°C, however, were more resistant to inhibitions of the anaerobic pathways.

Evidence has also been obtained that, in the case of adaptation to low temperatures, the HMP and the Krebs cycle are more active and, in the case of adaptation to elevated temperatures, anaerobic pathways play a greater role. Supporting evidence came from an investigation of alcohol dehydrogenase in yeast. Cell-free extracts from yeast adapted to 37° C showed 2—3 times more alcohol dehydrogenase activity than yeast adapted to 15° C. An even more pronounced decrease of the activity of this enzyme was observed with adaptation temperatures lower than 15° C. The alcohol dehydrogenase activity of 37° C yeast was 3—5 times greater than that of 7° C yeast.

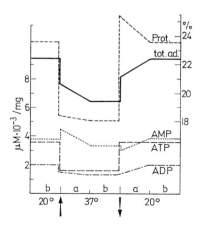

Fig. 30. Change of the adenylic acid and protein content of yeast cells with temperature shifts. Phase *a* begins 2 hours after the temperature alteration and indicates the concentration change during adaptation. Phase *b* indicates the level in adapted cells. Left ordinate: adenylic acids ($\mu M \cdot 10^{-3}$/mg). Right ordinate: protein content (percentage of dry matter); *Prot.* = protein; *tot. ad.* = total adenylic acid

The role of the adenosine triphosphate (ATP) concentration as a regulatory factor of metabolism was also examined (NAGUIB and CHRISTOPHERSEN, 1965). A clear decrease in ATP was observed after a sudden rise in temperature (Fig. 30). After incubation of the cells for 2 h at high temperature, however, a small decrease in adenosine diphosphate (ADP) was also observed. Adenosine monophosphate (AMP) increased initially. Cells adapted to 37° C had the same ATP content as 20° C cells 2 h after they had been transferred to 37° C, but ADP had further decreased and AMP had also decreased. Thus, the 37° C cells had an overall decrease of adenylic acids. With a transition from 37°–20° C, the reverse changes were observed.

The higher rate of phosphate uptake by 37° C cells, as shown in Fig. 28, does not fit into this lower phosphate demand. If the adenylic acids, which are important as regulatory factors, are reduced at higher temperatures, then a lower phosphate uptake should be expected. But when the uptake was plotted against time, an earlier drastic decrease of phosphate uptake by 37° C cells was observed (Fig. 31). Thus it can be stated that, with adaptation of yeast to elevated temperatures, the initial transport rate of phosphate is higher, but the over-all uptake is lower.

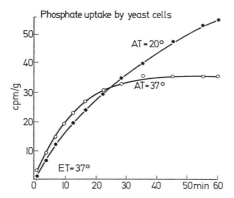

Fig. 31. Phosphate uptake by yeast cells in counts per minute per gram. The slopes of the curves indicate the different phosphate demand after adaptation to 20° and 37° C. Experimental temperature (ET) was 37° C

Decreased protein synthesis may be a consequence of the decreased availability of ATP; hence, the energy supply may be considered to be a link between temperature effect, on the one hand, and performance alteration, on the other. Similar observations have been made on other organisms (RAO, 1963). A decreased enzymatic activity may be the consequence of decreased protein synthesis. It should again be emphasized that low levels of activity or even lack of some individual enzymes at high growth temperatures may occur, as in the case of peptidases in *Pseudomonas*, of amino acid decarboxylase (GALE, 1940), and of lipase (NASHIF and NELSON, 1953; ALFORD and ELLIOT, 1960).

The regulation of energy-rich phosphates certainly is only one of the various possibilities for adjustment of cellular activities. Regulations involving repression and induction of enzymes obviously play an equally important role in temperature adaptation (see INGRAHAM, p. 60).

4. Time Course of Capacity Adaptation

In microorganisms, as well as in other organisms, we have to distinguish several stages of temperature action. Opinions as to which of the responses of an organism exposed to temperature changes are to be termed adaptations are varied. Adaptive responses may not be easily defined in every case. We have to consider immediate responses on the one hand, and slowly developing changes of the over-all biological systems on the other hand; the distinction between them is not necessarily sharp.

The immediate responses of cell systems to temperature changes occur in a matter of seconds. They are expressions of changes of steady state. Such shifts from one metabolic state to another have been investigated by observation of the pyridine nucleotide kinetics of cell-free glycolytic systems after the concentration of meta-bolites has been altered. These shifts are frequently accompanied by overshoots and undershoots. Furthermore, there may be oscillations in the flow of inter-mediates, as was described by GOSH and CHANCE (1964) for glucose-6-phosphate and fructose-1,6-diphosphate in connection with the control function of phospho-fructokinase. Oscilliations in the flow of amino acids and nucleic acid compounds have also recently been observed in yeast cells immediately after temperature changes (Fig. 32, BÖCKELMANN and CHRISTOPHERSEN, 1973). Overshoots appear

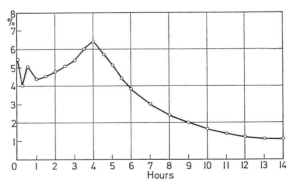

Fig. 32. Change of RNA content of yeast cells with temperature shift from 17° to 37° C. Ordinate: RNA content in percentage of dry matter. Abscissa: time after temperature shift in hours

to occur frequently when there are steady state changes. Very brief overshoots, measured in fractions of seconds, also occur in sensory receptors. More extended overshoots are observed in more complex processes, such as rate of heart beat or respiratory movements (for example, frequency of gill cover movements in fishes) and oxygen demand of animals. GRAINGER (1956, 1958), who has contributed many experimental data to this problem, emphasized the fact that the mechanisms of overshoot differ from one case to another.

ZERBST (1963) classified the adaptation patterns of PRECHT (1949) as shifts of energetic steady states in a given enzymatic system and reproduced the patterns in an electric model. He succeeded in deducing an analogous model for each pattern. According to observations on cellular responses in temperature adaptation, one can hardly imagine that a capacity adaptation should exist only in readjustment of the steady state of flow rate of metabolites, including feedback controls. In our opinion, an altered steady state resulting from a stepwise change in temperature is only the stimulus to adaptation and not the adaptation itself. The adaptation consits in long-term alterations of the cellular systems, including quantitative and even qualitative changes in the enzymatic equipment of the cell. In the interaction between temperature and cell mechanisms, two phases must be separated: a rapid alteration of the steady state of energy flow, and a moderately slow successive readjustment of the cellular system by increased or decreased synthesis of enzymes, coenzymes, etc. The boundary of transition from the first phase to the second phase is not fixed. The second phase, however, should be understood as capacity adaptation in the proper sense.

The short-term overshoots after temperature alterations, which may be tentatively defined as "unsteady states", probably represent a metabolic stress that will be extraordinarily strong in the case of sharp temperature increase. It is possible that such stress at higher temperatures could lead to irreversible cell injuries and thus could be responsible for inactivation of the cell. There are reasonable doubts concerning the opinion that the killing of cells by heat is generally and necessarily due to protein inactivation, since inactivation of cells has frequently been observed far below the usual inactivation temperatures of proteins, as was shown by ALEXANDROV and his colleagues (ALEXANDROV, 1964). Moreover, it should be tested whether metabolic stress can lead to changes of steady state that provide the cell with a high resistance to such temperature shocks. In this way, we may perhaps explain the "hardening" that results from short-term temperature shocks.

D. Response of Microorganisms to Subzero Temperatures

There are convincing reasons for discussing the influence of freezing temperature, freezing time and freezing rate together, because these factors are more or less correlated with each other and cannot in every case be regarded separately. When studying, for instance, the influence of very low temperatures on the survival of microorganisms, one has to pass through the whole temperature scale down to the experimental temperature. Consequently one must also consider the influence of temperature on the specimen in the upper temperature range, especially when large samples are to be frozen. This difficulty can be overcome only with small particles and fluids, which allow a very rapid heat exchange; for instance by dispersion of droplets in liquid nitrogen or by freezing ultrathin layers between cover glasses, etc. Such techniques have been of considerable value in laboratory research in elucidating the mechanism of frost killing.

1. Influence of Freezing Temperature, Freezing Time, and Freezing Rate

Regarding the lethal effect of freezing temperatures, it has been stated that relatively high freezing temperatures are more lethal than low temperatures. In the range from $-4°$ C to $-10°$ C many more microorganisms are inactivated than at $-15°$ C, and at $-30°$ C frost killing is even less pronounced. During prolonged storage under frozen conditions a further gradual destruction of microorganisms takes place. This rate of killing is also high at relatively high freezing temperatures, whereas at low freezing temperatures the destruction rate is slowed down. At very low temperatures, i.e., in the range of liquid nitrogen and especially near absolute zero, the destruction rate may become nearly zero. Under such conditions, time may cease to exist for living systems, and consequently living systems can be described as potentially immortal. These statements, however, are not to be generalized. As to the immediate frost killing, the percentage survival of yeast cells gradually decreases as the freezing temperature is lowered. In experiments by ARAKI and NEI (1962), for example, 90% of the cells survived at $-5°$ C, whereas only about 20% remained at $-40°$ C. On the other hand, slow freezing is more harmful to bacteria. When yeast suspensions were frozen at various rates, however, cells frozen at relatively slow rates of cooling showed a higher survival rate than a quick-frozen suspension. The storage death of yeast, on the other hand, showed the same behavior as bacteria, namely a slower decrease in the cell survival rate at lower temperatures.

2. Influence of Defrosting

The rate of defrosting (i.e., the rate at which frozen samples are warmed up to the melting temperature as well as the rate of thawing itself) also effects the viability of microorganisms in a way similar to freezing. The mechanisms of cell injury, however, are not necessarily the same in freezing and defrosting.

Like the responses to freezing rate, different organisms respond to defrosting in different ways. STILLE (1950) has found that rapid warming was more harmful than slow warming to the viability of frozen cells of *Pseudomonas fluorescens, P. aeruginosa, Serratia marcescens* and *Saccharomyces cerevisiae*. There are also instances reported in the literature in which different warming rates had no effect (ULRICH and HALVORSON, 1946/47). Finally there are also reports according to which slow warming may be more harmful than rapid warming. The latter situation applies to spores of the fungi *Alternaria, Fusarium, Penicillium* (HASKINS, 1957), *Aspergillus flavus* (MAZUR, 1956), and *Pasteurella tularensis* (MAZUR et al., 1957).

3. Influence of Repeated Freezing and Thawing

When plotting the log number of survivors versus time one would obtain a straight line when a constant fraction of cells dies per unit of time (see p. 14). Also after repeated freezing and thawing similar curves may be obtained when the log number of survivors after each freeze-thaw procedure is plotted against the number of freeze-thaw procedures. Experimental results on the effect of repeated freezing and thawing on the death curves of microorganisms are not unequivocal. In early studies of GOETZ and GOETZ (1938) on yeast cells which were frozen in liquid air, a steady decrease of the killing rate was observed after each successive freezing. Hence, the sensitivity of the cells increased. STILLE (1942/43), however, found a constant rate of decrease of surviving cells when repeatedly freezing *Pseudomonas aeruginosa* and *Bact. rubidaeum*. Similar results were obtained by HOLLANDER and NELL (1954) with *E. coli* and *Diplococcus pneumoniae*. Apparently there are additional factors involved which are responsible for susceptibility to killing by repeated freezing and thawing. The first studies to elucidate this question were carried out by HARRISON (1955, 1956), who reported that aerobically grown cultures of *E. coli* were more resistant than anaerobically grown cultures. Also the storage intervals between the successive freezing cycles turned out to have an influence on the shape of the death curves. With *E. coli* and *Serratia marcescens*, sigmoidal curves were obtained when the storage intervals were long (24 h). When the storage intervals were only 0.5 h then the curves became straight lines. This latter finding has also been confirmed by investigations of PACKER et al. (1965) on factors which are responsible for the frost stability of *E. coli*, which will be dealt with later.

4. Sensitivity of Microorganisms to Freezing

It is a well-known fact that the inherent resistance of microorganisms to injurious effects varies considerably. This holds also for the resistance of microorganisms to freezing and thawing. As a general rule one can say that organisms which exhibit higher resistance to other damaging activities such as heat, radiation or poisons, also exhibit higher resistance to cold injury and vice versa. There may, of course, be exceptions. It is difficult, however, to establish a strict order of cold resistance of microorganisms because their stability differs greatly, depending on the medium.

In view of this reservation, it may be possible to classify microorganisms in three groups based on their resistance to frost: (a) susceptible, (b) moderately resistant and (c) insensitive organisms.

Especially sensitive are vegetative cells of yeast and moulds. Also gram-negative bacteria succumb easily at subfreezing temperatures. To this group belong Coliforms, *Pseudomonas*, *Achromobacter*, and *Salmonella* species. A distinctly more pronounced resistance against cold injury is exhibited by gram-positive microorganisms. Food poisoning cocci, such as *Staphylococcus aureus*, have received special interest in this respect. Enterococci are more resistant than *E. coli*. Soil bacteria belong to the more resistant organisms, because they are able to endure the freezing temperatures occurring in their natural environment. The third group consists chiefly of spore-formers, the spores of which are relatively unaffected by freezing conditions. Spores of *Bacilli* and *Clostridia* are highly insensitive, whereas spores of fungi are grouped amongst the moderately resistant germs when frozen in water.

5. Recovery of Frost-Injured Microorganisms

Destruction of microorganisms by freezing is characterized particularly by intermediate stages of injury. This injury is considered to be metabolic in nature, since cells become more dependent on certain nutritional factors after freezing. Generally, synthetic media which contain inorganic nitrogen compounds are poor recovery media; however, when complex organic substances are added, recovery is greatly improved. It has been reported that enzymatic protein digests, such as trypticase, tryptone and yeast extract, contain factors responsible for the recovery of meta-bolically injured cells (Moss and SPECK, 1963; STRAKA and STOKES, 1959; ARPAI, 1964). The active factors which cause the reactivation of partially damaged cells have not been positively identified. Probably peptides are required by injured cells for resynthesis of essential proteins, such as enzymes denatured by freezing.

6. Influence of Freezing on the Metabolism of Microorganisms

In spite of the fact that freezing apparently interferes with certain activities of the cell which are responsible for catabolic reactions, it should not be overlooked that many enzymes are not destroyed at subzero temperatures. There may be a decrease of metabolic activity due to freezing. For instance a drop in respiration and glyco-lytic activity has been observed in yeast cells (LUND and LUNDBERG, 1949). But on the other hand activation of dehydrogenase activity under the same freezing conditions has been observed. The peptidase activity of various bacteria was stimulated by freezing (ARPAI, 1961). The mechanisms of such activations may be different. They may be caused by a destruction of the permeability barrier of the cell, leading to enhanced diffusion of metabolites on the one hand, or to an acti-vation of the enzyme on the other hand. The latter assumption has received sup-port from the observation that cell-free enzymes can also be activated (ARPAI, 1961).

7. Interaction between Microorganisms and Menstrua during Freezing and Defrosting

As indicated above, the degree of killing of microorganisms subjected to subzero temperatures depends first on the freezing temperature and the freezing time. Since there is a preferential killing of the more freeze-sensitive varieties, the type of dominating bacteria in mixed populations is of considerable significance for the bacteriological condition after thawing, when, for instance, practical freezing of foodstuffs is concerned. Furthermore, the physiological state of the microorganisms has a decisive influence on susceptibility. Finally the degree of killing depends on environmental conditions such as pH, presence or absence of protecting substances as well as of injurious agents. Conflicting claims regarding the relative sensitivities of microorganisms are the result of these factors.

a) The Phases of Frost Killing

When a population of microorganisms is subjected to freezing, the killing rate cannot be described by first-order reaction kinetics as in the case of heat killing. In frost killing three distinct phases can be distinguished (Fig. 33). During the first phase a rapid decline in numbers takes place as a result of the immediate effect of freezing. Then the so-called storage death begins, which is characterized by a gradual destruction of microorganisms. Finally, a residual number of germs is reached, which consists of resistant survivors, the number of which depends upon the freezing conditions and upon the individual resistance of the kinds of micro-organisms. This behavior is especially pronounced in mixed populations, which are encounterd in foods. Here the second phase leads to a selection of the most frost resistant varieties of germs.

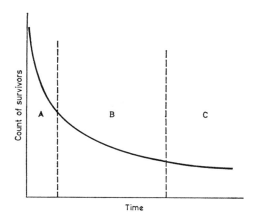

Fig. 33. Pattern of frost killing of mixed bacterial populations. *A* immediate freezing effect. *B* phase of storage death, selection phase. *C* phase of maintenance of a character-istic surviving flora

b) Protective Action against Frost Killing

The most important factor governing the survival of microorganisms in natural environments, e.g., in foods, is the protective action of specific constituents such as proteins, peptides, sugar and fat as well as of substances which are still unidentified. On the other hand one must also consider constituents which appear to function as enhancing factors in the frost injury of cells.

The influence of electrolytes on the resistance to freezing of yeast cells has been investigated by STILLE (1950), who compared killing in distilled water with that in diluted (0.2 M) salt solutions. An increase in resistance was found in solutions of KCl, KNO_3, NaCl (very low) and $NaNO_3$ (also very low), whereas solutions of lithium, calcium and magnesium salts exhibited a decrease in resistance. It can thus be deduced that the cations Na^+ and K^+ provide a certain degree of stabilization for the cells. The same holds true to a lesser extent for the anions Cl^- and NO_3^-. NaCl, an additive commonly used in the preservation of foods, has received special attention with regard to low temperature treatment. In spite of the afore-mentioned observations as well as other findings, e.g., by DEVIK and ULRICH (1948/49), there are many contradictory results suggesting that NaCl has a sensitiz-ing rather than a protecting effect on cold injury. The salt concentration and man-ner of freezing, of course, are decisive factors. With increasing concentration the sensitivity of the cells increases as well. *E. coli* and other gram-negative bacteria, however, exhibit less resistance in a physiological solution of NaCl than in distilled water.

The use of glycerol as a protective agent has been of interest since KEITH (1913) first showed that the killing of *E. coli* during frozen storage of a suspension is greatly reduced by addition of this substance. Normally glycerol is applied as an aqueous solution (5—15%) or as an additive to broth in the same concentration (POLGE et al., 1949; LOVELOCK, 1953; FOX and HOTCHKISS, 1957; HOWARD, 1956; SQUIRES and HARTSELL, 1955; POSTGATE and HUNTER, 1961). A combination of phosphate buffer and glycerol also offers good protection. Glycerol is reported to have a similar or even superior protecting effect as compared to related substances such as glycols, sugar alcohols and sugars. In any event high survival is also obtained when glucose, sucrose, erythrol, diglycol or polyethylene glycol are present in the menstrua in which microorganisms are frozen. The mode of the protective action has been considered extensively (e.g. POSTGATE and HUNTER, 1961) and can only be understood in terms of the effect of the mechanisms of freeze injury on living systems. The protective activity of the mentioned substances evidently is not necessarily due to penetration into the cells, because some of them do not penetrate the cell membrane, e.g. polyethylene glycol. Also similarities in the molecular structure such as the high content of OH-groups cannot be regarded as a decisive factor, since other soluble substances, such as urea and acetamide, behave in a similar way (KEANE, 1953). Therefore, only the interference of these substances with intracellular or extracellular freezing can thus far be considered responsible for the protective action.

Other substances found to offer protection to microorganisms against injury from freezing and thawing are natural substrates, especially broth and other infusions, milk and serum. In these menstrua different protecting substances are involved, particularly proteins and protein-related compounds. The protective action of such media is variable. Broth has a significant but relatively small effect, whereas milk, serum and solutions of proteins such as albumin display a pronounced protection. Since colloids of different origins are also protective agents, the colloidal character of high molecular substances may be responsible for their function. It is, however, not clear how colloids stabilize living systems against cold injury. Doubtless very complex phenomena are involved.

It is a common experience that cold-resistant organisms are also resistant to other injurious agents such as heat and radiation. Furthermore protective agents against frost injury also exhibit a certain protection against heat inactivation and sometimes against radiation and poison injury as well. A credible theory about the inactivation of living systems is based on the assumption that protein denaturation takes place when cells are injured. Agents which stabilize proteins are therefore supposed to protect organisms from any injurious activity. With relatively weak destructive influences, such as temperatures in the lower range of protein denaturation, limited doses of radiation, and probably of freezing, a rupture of intramolecular bonds in the protein molecules is regarded as the causative factor. This leads to the loss of specific activities of the protein, for instance, of an enzyme. Hydrogen bonds are of considerable importance for the stability of the structure of protein molecules. Therefore compounds which are able to build up such bonds exhibit a pronounced stabilizing effect against different injurious influences on protein inactivation. Amino acids and related substances of low molecular weight are among these agents. The question is still pending whether these substances display their protective action at the metabolic level or whether they are of a more physico-chemical nature.

As indicated previously, the experimental data on the protection of microorganisms against low temperature treatment were mainly obtained by people working in the freeze-drying field. MILLER and GOODNER (1952/53) were among the first to show that sodium glutamate and aspartate protected tubercle bacilli during freeze-drying. Meanwhile many researchers have drawn attention to glutamate as an additive to freezing media.

In spite of the extensive studies on the application of glutamate in the process of freeze-drying, the mechanism of its protective effect remains obscure. Since we obtained reasonable evidence that substances which are capable of forming H-bonds are involved in the recovery of microorganisms from heat injury (CHRISTO-PHERSEN and KAUFMANN, 1952), MORICHI et al. (1963, 1965a, b) reported similar findings, thus supporting our concept. The Japanese team found that in addition to glutamate, chemical compounds which resemble glutamic acid in chemical structure prevent the death of bacteria from frost killing. Considerable protection was given by aspartic acid, malic acid, cysteic acid, pyrrolidone carboxylic acid, α-aminopimilic acid, acetyl glycine, DL-threonine and DL-allothreonin. The reasoning that the protective effects of these compounds are considered to be physico-chemical in nature is based on various facts. Firstly, among the effective compounds there are some, such as α-aminopimelic, cysteic and pyrrolidone carboxylic acids, which can hardly be thought to be metabolized by bacteria. Secondly, there was no correlation with the optical activity of glutamic acid. Thirdly, there were related compounds, namely glutamine and asparagine, that were metabolized but had no protective effect. Modification of the $-COOH$ group of glutamic acid to a hydrophobic structure (e.g. ester and amide) appeared to destroy the protective activity. But the $-NH_2$ group could be replaced by some other hydrogen-bond generating groups, such as $=NH$, $-OH$ or $=O$. It is suggested that the effectiveness has a close relationship to the presence of two $-COOH$ groups in the α- and γ-position and one electronegative group on the α-carbon, e.g. $-NH_2$. There are additional findings supporting the view that hydrogen bonds are responsible for the stabilizing effect, but they need further conformation. The above-mentioned workers suggested that the protective amino acids and related compounds display their stabilizing effect especially during the drying process. But there can be no doubt that frost injury is also prevented to a considerable extent by such stabilizing agents.

Of course there are various other substances which protect microorganisms against frost injury, whose natures are still completely obscure. AMBROSINI and BRITT (1963) reported that the mortality of *E. coli* is diminished if the cells are frozen in lysates of *E. coli*, indicating that constituents of the cells exhibit a protective activity. Investigations of PACKER et al. (1965) have shown, however, that both the optical density and the number of survivals decreased linearly with the number of freeze-thaw cycles, thus clearly establishing that the level of the products of lysis released does not affect the killing rate. On the other hand, presence of spent growth medium (a filtrate of a stationary culture) in the freezing medium protected the cells even in high dilution. Substances with similar effectiveness could be obtained by heating an acidified salt medium without a carbon source. Therefore an alteration of the ionic composition of the medium is drawn into consideration in addition to the formation of specific substances.

8. Metabolic Activities of Microorganisms and Microbial Enzymes at Subzero Temperatures

A decisive factor for enzymatic reactions is the presence of water. It is well established that even at temperatures as low as $-40°$ C there may be water in tissues which can be described as liquid. From a theoretical point of view the interactions between enzymes and substrates during freezing of living systems are governed by two main events. Firstly, they are governed by the decrease of the reaction rate of enzymatic processes. The dependence of the rate of biological reactions on temperature can be described roughly by the RT-rule, according to which a temperature decrease of 10 centigrade degrees leads to a decrease of the reaction rate by a factor of 2—3. The second event governing the interaction of enzymes with substrates during freezing is the formation of ice. The freezing out of water leads to a concentration of enzymes as well as to a concentration of substrates. Under certain conditions this should favor the reaction rate. It has indeed been observed that an increase of enzymatic reactions at freezing temperatures may occur (KIERMEIER, 1948). In addition to these two main factors, there are many additional factors which may either enhance or inhibit the reaction rate. Inhibiting factors include, for example, the partial inactivation of enzymes, the increase of viscosity leading to a decrease of diffusion rates, the concentration of enzyme inhibitors and inhibiting ions, etc. Enhancing factors include especially the precipitation of inhibiting substances and the activation of enzymes, as mentioned above.

9. Mechanisms of Freezing Injury

In spite of the fact that a considerable amount of work has been done to elucidate the mechanisms of injury to living systems from freezing, the actual mechanisms of killing of cells and inactivation of cellular constituents are still obscure. Recent data on the biological effects of freezing, supercooling and thawing are summarized by MAZUR (1966), SMITH (1961), and MERYMAN (1960). Information about the specific behavior of microorganisms is scattered and has been discussed primarily in connection with observations on plant and animal tissues.

The main event in injury from freezing is the formation of ice, but killing without ice formation, i.e., by lowering the temperature itself, may also occur. The formation of ice may take place only outside the cell (extracellular freezing) or both outside and inside the cell (intracellular freezing). Only in the latter case is the deposition of ice crystals likely to participate in the killing of the cell in a more mechanical way, that is, by disrupting structures, especially the plasma membrane and/or the cell wall. On the other hand, ice formation may cause a concentration of intra- and extracellular solutes which in turn leads to chemical reactions which denature enzymes and/or essential structures such as the permeability barrier. Extracellular formation of ice may be injurious in that the cell becomes dehydrated. On the basis of these concepts, any one of which has been supported by experimental evidence, the following theories of injury from freezing have been advanced.

Injury from freezing as a result of the mechanical effects of ice crystals on the living cell is characterized by deformation and rupture of intracellular structures.

This concept was postulated in the early days of biological research on plants (HAMEL, SENNELIER, cit. from BĚLEHRÁDEK, 1935) and has also been discussed in connection with the frost killing of bacteria (KEITH, 1913). The theory agrees with the observation that intracellular ice formation usually results in death. The frequently observed death of cells from extracellular freezing cannot be explained by this theory. The inhibition of freeze killing by protective compounds such as glycerol could be due to the preventive measure of seeding the cell with ice crystals.

Another important theory argues that when water is removed from a cell to form extracellular ice, the intracellular solutes are progressively concentrated. Intracellular freezing, however, may also lead to a concentration of solutes in the unfrozen parts of the cellular lumen. When nucleation starts, for instance, in a vacuole of a larger cell, the progressive growth of an ice crystal will lead to progressive dehydration of the plasma and of macromolecular cell structures such as the plasma membrane. The concomitant concentration of solutes, especially H ions and salts, is regarded as the causative factor of cell injury. When 80% of the water is removed, the concentration of acids and salts can become sufficient to denature proteins. In this case the protective action of compounds such as glycerol and sugars is attributed to a reduction of the freezing point of the cell contents, thereby reducing the amount of ice formed and stopping the concentration of electrolytes short of the lethal level.

A special theory has been advanced by LEVITT (1962, 1967), according to which the precipitation of proteins due to cell dehydration during freezing results from the mutual approach of the protein molecules until they are close enough to form intracellular SS bonds either by oxidation of adjacent SH groups or by SH—SS interchange. When the cells thaw, rehydration occurs and the water pushes the newly attached protein molecules apart. But because the originally separate proteins are now joined by strong SS bonds, thawing and rehydration will rupture or distort protein chains. The protective action of glycerol and sugars can be explained by the hydrogen bonding of these compounds to the SH groups of proteins, thereby keeping the proteins too far apart for interaction. According to LEVITT and DEAR (1970), the locus of freezing injury is the semipermeable plasma membrane, in which "holes" are formed. The "holes" are produced either by ice crystals penetrating the membrane during intracellular freezing or by tension on the membrane due to cell collapse during extracellular freezing. The "holes" are irreversibly fixed by intramolecular SS bonding of the membrane proteins, which leads to loss of semipermeability, efflux of the cell solution and subsequent death.

It should not be overlooked, however, that intracellular dehydration leading to physical contact of cell structures which are normally separated from each other, may also permit the formation of cross-linkages other than specific SS bonds.

Many such possibilities have been discussed in the literature (LOVELOCK, 1957; MERYMAN, 1967; ASAHINA, 1962).

All of these theories of freezing injury provide explanations for many experimental observations, yet no single theory can explain all cellular freezing phenomena. The theories are not mutually exclusive, so perhaps each proposed mechanism is operating to some degree during freezing injury. In fact, it is highly probable that no single mechanism for freezing injury exists that will explain all cases. One mechanism may be operative in one cellular system while a completely different mechanism may predominate in other types of cells. Also the freezing conditions, viz. cooling rate, freezing temperature, and freezing time, may have a considerable influence on the predominating mechanism for freezing injury. In any case, the mechanisms of freezing injury and of freezing protection are still highly controversial.

10. Cold Shock

Injury from cold without ice formation (cold shock) has been observed in organisms rapidly cooled at temperatures near 0° C. The first observations about the sensitivity of "physiologically young" cells of *E. coli* to rapid cooling were made by SHERMAN and ALBUS (1923) and SHERMAN and CAMERON (1934). HEGARTY and WEEKS (1940) showed that both low osmotic pressure and cold were necessary to kill organisms exposed to these conditions. The organisms showed maximum sensitivity during the exponential growth phase. MEYNELL (1958) observed that the number of organisms which appeared to be killed by cold shock depended upon the medium in which they were subsequently plated. More detailed observations in this respect have been reported by GORRILL and McNELL (1960). LOVELOCK (1954) suggested that cold shock results from alterations in the lipids in the cell membrane. It seemes probable, therefore, that the injury involves damage to the permeability barrier (see also the experiments of RING, p. 26).

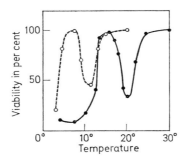

Fig. 34. Critical temperature zones of cold shock treatment (SATO and TAKAHASHI, 1967)

SMEATON and ELLIOTT (1967) indicated that, when subjected to cold shock treatment, *Bac. subtilis* releases a ribonuclease inhibitor. When the cells are rapidly chilled through a narrowly defined critical temperature zone, at about 14—16° C, the inhibitor is released. SATO and TAKAHASHI (1967) reported that in *E. coli*, two critical temperature zones exist, the positions of which depend upon the initial temperature (Fig. 34). When the initial temperature of the cell suspension submitted to cold shock was lowered by 10° C, both critical zones moved to temperatures which were approximately 10° C lower. Magnesium, calcium and manganese ions as well as sucrose showed a protective effect against cold shock. Similar results were obtained with *Bac. subtilis* and *Ps. fluorescens* (SATO and TAKAHAHI, 1968). Further investigations on *E. coli* provided evidence that upon cold shock, a nuclease is induced. Possibly, the damage of DNA by nuclease, which can no longer be repaired enzymatically, is reposible for the death of cold-shocked cells (SATO and TAKAHASHI, 1970).

E. Thermophilic Organisms

1. General Remarks

The ability of organisms to grow at temperatures far above the maximum usually found for mesophilic organisms was initially recognized by MIQUEL (1888), who isolated an aquatic bacterium which grew at 73° C. Strikingly this ability is confined to simple unicellular organisms which do not contain a typical nucleus, such as bacteria, algae and fungi. The majority of thermophilic microorganisms grow best between 50 and 70° C. It has been claimed that blue-green algae are able to grow even at 93.3° C, which is the boiling temperature of water at the elevation of Yellowstone Park, where these organisms have been observed (BREWER, 1866). But subsequent investigations have shown that the maximum temperature is barely above 75° C (KEMPNER, 1963). For reviews of the early literature, the texts of ROBERTSON (1927), GAUGHRAN (1947), ALLEN (1953), CHRISTOPHERSEN (1955), and KOFFLER (1957) should be perused. In the past ten years an appreciable amount of work has been done on this subject (FARREL and ROSE, 1967). The following chapter will therefore be limited to some aspects of cellular mechanisms of thermophilism.

2. Cellular Mechanisms of Growth at High Temperatures

Different theories have been advanced about how organisms obtain the ability to grow at high temperatures. On the one hand the concept has been presented that thermophilic organisms represent an archetype of biological entities which originated from warmer periods of terrestrial development (WEED, 1889; BARDOU, 1907; SCHNETZLER, 1889; AMBROZ, 1910/11) or even from warmer planets like Venus (ARRHENIUS, 1927). On the other hand explanations have been offered which are based to a certain extent on experimental results. Two general theories have been presented. The first and most obvious is that the essential cell components of thermophilic organisms are more heat stable than those of mesophilic organisms. The second is that in thermophilic cells the resynthesis of cellular components takes place at a higher rate than thermal destruction or inactivation (BĚLEHRÁDEK, 1935).

The first approach to understanding the heat stability of cells has been through observations that the thermal stability of organisms is correlated with the melting point of their lipids. HEILBRUNN (1924) and BĚLEHRÁDEK (1931, 1935) developed the concept that the structures in the protoplasm or in the plasma membrane collapse upon the melting of the cellular fat, whereas GAUGHRAN (1949) assumed that active cell processes are maintained as long as the cellular lipids are liquid. Below the solidification point of their lipids, cells are not able to grow. The fact that degree of saturation, length of chains and melting point of lipids increase with increasing growth temperatures is in accord with both interpretations.

Observations that lipids formed by living organisms at high temperatures are more solid than lipids formed at lower temperatures have been reported from studies on plants, insect larvae (FRAENKEL and HOPF, 1940) and microorganisms (TERROINE et al., 1930; GAUGHRAN, 1947; CHRISTOPHERSEN and KAUFMANN, 1955). MARR and INGRAHAM (1962), however, found also that growth of E. coli at a particular temperature did not lead to a unique fatty acid composition. Alteration of nutrition independently of temperature also resulted in significant changes in fatty acid composition.

With regard to thermophilism it seems obvious that characteristic differences exist in the fatty acid composition of thermophiles and mesophiles. In a thermophilic Bacillus (stearothermophilus?) the most abundant fatty acids were those having 16 or 17 carbon atoms (DARON, 1970). They accounted for 80—90% of the total fatty acids in all extracts examined regardless of the growth conditions of the cells. In contrast, fatty acids with 15 carbon atoms predominated in all of the mesophilic Bacillus species examined by KANEDA (1967, 1969), and fatty acids with 17, 18, and 19 carbon atoms were the major constituents in some extremely thermophilic bacteria (BAUMAN and SIMMONDS, 1969).

Another characteristic feature of thermophilic organisms appears to be the predomination of branched fatty acids. The lipid extracts of two strains of extremely thermophilic bacteria, both isolated from super-heated water pools of Yellowstone National Park, contained a large amount of branched-chain fatty acids when they were grown at maximum temperatures of 80 and 91° C respectively. In one strain (Thermus aquaticus) grown at 50° C, two additional unsaturated fatty acids were formed, while at 80° C only saturated and branched fatty acids were present. HEINEN (1970) concludes therefore that branched compounds could well be of major importance for the thermostability of thermophilic organisms. Under most conditions, branched-chain fatty acids were more abundant in Bacillus stearothermophilus than normal fatty acids (DARON, 1970). Only when the cells were grown in glucose medium at 60° C, did palmitic acid alone account for over 60%. This seems to support KANEDA'S contention that the preponderance of branched-chain fatty acids is a characteristic feature of the genus Bacillus. KANEDA (1963a, b, 1966) has provided evidence that the biosynthesis of the branched-chain fatty acids is related to the biosynthesis of the branched-chain amino acids. The common intermediates for the synthesis of valine, leucine and isoleucin on the one hand and for branched-chain fatty acids on the other hand are α-ketoisovalerate, α-keto-isocapronate and α-keto-β-methylvalerate. The reversibility of the transamination reaction is in accordance with the observation that the added branched-chain amino acids act as precursors for the afore-mentioned amino acids (ALBRO and DITTMER, 1969; KANEDA, 1963a, b; TORNABENE and ORÓ, 1967). When the demands of protein synthesis are not adequately met by the supply of branched-chain amino acids, it seems possible that branched-chain fatty acids or their immediate precursors are channeled into protein synthesis. It has been observed that the percentage of branched-chain fatty acids decreased from 79—57% in the case of acetate-grown cells and from 56—30% in the case of glucose-grown cells as the growth temperature was increased from 40—60° C (DARON, 1970). This is in contradiction to the concept that branched-chain fatty acids are essential for the maintenance of high cellular thermostability.

Another point of interest is that the fatty acid composition of thermophilic blue-green algae and thermophilic bacteria is significantly different. Both bacteria and blue-green algae are found in hot springs in which temperatures of about 90° C are reached. But the algae are limited to growth temperatures of 73—75° C or less, evidently because of evolutionary limits to the photosynthetic mechanisms (BROCK, 1967). Filamentous flexibacteria-like organisms can grow even in water at temperatures of 85—88° C. Blue-green algae may be distinguished from bacteria in terms of their fatty acids because algal fatty acids are usually straight-chain saturated or unsaturated (HOLTON et al., 1968; PARKER et al., 1967), whereas the bacterial fatty acids contain significant amounts of branched and normal chains but rarely polyunsaturated acids (KANEDA, 1963a; MOSS and CHERRY, 1968; O'LEARY, 1962; TORNABENE et al., 1967a, b). Furthermore the polar lipids of bacteria include phosphatidylethanolamine and occasionally lecithin (IKAWA, 1967), whereas these substances are absent from blue-green algae (NICHOLS et al., 1965).

The reports in the literature on the effect of environmental temperature on the composition of the fatty acids of fungal lipids are not unequivocal. In some fungi, lipids become more unsaturated at higher temperatures (COONEY and EMERSON, 1964; PEARSON and RAPER, 1927; SHAW, 1966), while in others the lipids become more unsaturated at lower growth temperatures (GREGORY and WOODBINE, 1953; KATES and BAXTER, 1962; SALMONOWICZ and NIEWIADOMSKI, 1965; SHAW, 1966; SINGH and WALKER, 1956). In still other fungi there appears to be little relationship between temperature and fatty acid composition (BOWMAN and MUMMA, 1967; PRILL et al., 1935). Studies on the influence of temperature on the fatty acid composition of mesophilic, psychrophilic, thermotolerant and thermophilic fungi in the order *Mucorales* revealed that the fatty acid composition depends on the temperature as well as on the growth cycle (SUMNER et al., 1968). The lipids of the psychrophilic, thermotolerant and thermophilic fungi were generally more unsaturated when grown at a lower temperature. But in the psychrophilic *Mucor oblongisporus* more unsaturated fatty acids were found at 25° C growth temperature than at 10° C. As the culture aged, however, the lipids became more unsaturated at the lower incubation temperature. Similar behavior was observed in young cultures of the mesophilic strain of *Phytium ullinum* grown at 20 and 30° C (BOWMAN and MUMMA, 1967).

The mechanisms by which a change in temperature regulates fatty acid synthesis are not yet clear. KATES and BAXTER (1962) have proposed that the rates of synthesis and of degradation of unsaturated acids are both temperature dependent, and that these rates have different temperature coefficients. They assume that at higher temperatures the synthesis of unsaturated acids is less retarded than their degradation, so that under these conditions linoleic acid accumulates in the lipids, with a consequent increase in the lipid unsaturation. On the other hand, it was demonstrated by HARRIS and JAMES (1969) that the degree of lipid unsaturation is influenced by the oxygen concentration of the environment. In non-photosynthetic plant tissue the rate of synthesis of unsaturated fatty acids increased with decreasing temperatures under normal oxygenation conditions, but when the supply of oxygen was increased, the rate of synthesis of unsaturated fatty acids actually increased with temperature.

The conversion of saturated into unsaturated fatty acids is known to be regulated by desaturative enzymes which require oxygen as a cofactor together with acetyl coenzyme A, acyl carrier protein, $NADH_2$ and $NADPH_2$ (BLOOMFIELD and BLOCH, 1960; NAGAI and BLOCH, 1966). Therefore the synthesis of unsaturated fatty acids will be retarded when the oxygen concentration becomes rate-limiting for the desaturation reaction. Because of the temperature dependence of oxygen solubility in water the lower degree of unsaturation of lipids found in fungi grown at high temperatures may be due to the low concentration of available oxygen at that temperature.

The O_2-dependent long-chain fatty acid desaturation systems in bacilli have especially been studied by BLOCH, FULCO and others. Two distinct systems have been detected (FULCO, 1969). One system introduces a cis-double bond into palmitic acid in positions 8, 9 or 10 in a reaction that is relatively insensitive to the growth temperature of the culture. A second system results in the desaturation of palmitic acid exclusively to cis-5-hexadecenoic acid and is under strict temperature control. The Δ^5-desaturating enzyme is not present at 30—35° C but is rapidly induced at 20° C. A transfer of an active 20°-culture to 30°—35° C results in rapid and complete loss of desaturating activity (FULCO, 1970).

The thermostability of a variety of enzymes and other macromolecular structures in thermophilic bacteria is now well established (AMELUNXEN and LINS, 1968; BROCK, 1967; FRIEDMAN, 1968). In a recent study, AMELUNXEN and LINS (1968) compared the thermostability of eleven enzymes from a thermophilic and a mesophilic species of *Bacillus*; with two exceptions, the enzymes from the mesophile showed considerably greater heat lability than the same enzymes from the thermophiles. These authors also confirmed earlier work by KOFFLER (1957) that the proteins of thermophiles showed greater resistance to coagulation by heat than those of mesophiles. There is now good evidence from studies on purified enzymes that the enzymes of thermophiles are inherently thermostable and not merely stabilized by external factors (SINGLETON et al., 1969).

According to CAMPBELL and PACE (1968) cellular enzymes and proteins of thermophilic bacteria can be placed into three general groups:

(a) the enzymes that are stable at the temperature of production (usually 55 to 60° C), but are inactivated at slightly higher temperatures. Some examples of this group are malic dehydrogenase (MILITZER et al., 1949; MARSH and MILITZER, 1952), adenosin triphosphatase (MILITZER and TUTTLE, 1952; MARSH and MILITZER, 1956a), inorganic pyrophosphatase (MARSH and MILITZER, 1956b; BROWN et al., 1957), aspartokinase (SAIKI and ARIMA, 1970), aldolase (THOMPSON et al., 1958; THOMPSON and THOMPSON, 1962) and certain peptidases (O'BRIEN and CAMPBELL, 1957).

(b) the enzymes that are inactivated at the temperature of production in the absence of substrate. Examples of this group are asparagin deaminase (MANNING and CAMPBELL, 1957), catalase (NAKAMURA, 1966), pyruvic acid oxydase (MILITZER and BURNS, 1954), isocitrate lyase (DARON, 1967) and certain membrane-bound enzymes (MILITZER et al., 1950; DOWNEY et al., 1962).

(c) the high heat-resistant enzymes and proteins such as α-amylase (CAMPBELL and MANNING, 1961), protease (ENDO, 1962), glyceraldehyde-3-phosphate dehydrogenase (AMELUNXEN, 1966, 1967), amino acid-activating enzymes (ACRA et al., 1964), proteolytic enzymes from *Actinomyces* (DESAI and DHALA, 1969), and flagellar proteins (KOFFLER, 1957; KOFFLER et al., 1957).

The question as to whether differences in protein stability depend upon differences in the primary, or in the secondary and tertiary structures, is under investigation in only a few laboratories. At the present state of knowledge one must assume that the basic composition and arrangement of amino acids in the molecule governs the secondary and tertiary structures.

Enzymes from mesophilic organisms may also be extraordinarily heat stable, as has been shown in intracellular proteinase of lactic acid streptococci (CIBLIS and CHRISTOPHERSEN, 1970) and in extracellular proteinase from *E. coli* (WILLIAMSON et al., 1964). The pyrophosphatase of the mesophile *Azotobacter agilis* exhibited a remarkable resistance when heated to 75° C for 2 hrs in the presence of Mg^{++} but not in its absence (JOHNSON and JOHNSON, 1959). Pyrophosphatase from *E. coli* showed a resistance to heat denaturation nearly the same as the resistance of enzyme isolated from thermophiles (SCHITO and PESCE, 1965).

The main problem of bacterial thermophily should be reflected by intracellular enzymes. AMELUXEN (1966, 1967) attempted a correlation of heat stability and structure of intracellular glyceraldehyde-3-phosphate dehydrogenase from *B. stearothermophilus*. He found that the heat stable enzyme has a high molecular weight and is not inactivated by 8 M urea (SINGLETON et al., 1969). Aldolase from *B. stearothermophilus* could be changed by cysteine and other sulfhydryl reagents to a new form which was less heat resistant but exhibited elevated activity (THOMPSON et al., 1958). Similar behavior from the influence of cysteine was demonstrated with an aldolase from *Thermus aquaticus*, an extremely thermophilic organism from hot springs (FREEZE and BROCK, 1970). The optimal activity of this enzyme occurred at about 95° C. Differences in structure are indicated by the observation that the *T. aquaticus* aldolase is activated by both cysteine and Fe^{++}, substances which also activate the *C. perfringens* and *Anacystis nidulans* enzymes. On the other hand, the yeast aldolase, a zinc metaloenzyme, is activated by K^+ but not by exogenous divalent metals, and is not activated, although it is stabilized, by cysteine. The *B. stearothermophilus* enzyme is thought not to be metal ion-dependent.

Among thermostable enzymes isolated from thermophilic bacteria, α-amylase in particular has been studied (OGASAHARA et al., 1970a—c). α-amylases from *Bac. subtilis* and from *B. stearothermophilus*, both denatured with 8 M urea, could be renatured by elevated temperatures. The temperature of reactivation was closely related to the temperature optimum of the amylolytic activities of the different enzymes. The rate of reactivation of denatured thermophilic α-amylase was most rapid at 40—50° C, whereas *B. subtilis* amylase exhibited a temperature optimum of reactivation at 25–30° C. These properties are presumably inherent in the primary structure of the enzyme proteins (AFFINSON, 1962). The tertiary structure may be constituted by secondary and tertiary bonds (KAUTZMANN,1956) at elevated temperatures.

Structural differences may also be reflected by differences in the kinetic data of enzymes. ISONO (1970) showed that α-amylase purified from the culture filtrate of *B. stearothermophilus* grown at 55° C exhibited remarkable stability at higher temperatures and had smaller K_m values for starch than that from the same bacterium grown at 37° C.

It may be of interest, however, that kinetic data of glucose-6-phosphate dehydrogenase purified from a thermophilic fungus are different from those from mesophilic organisms. But the heat resistance of the thermophilic enzyme does not appear to be elevated when compared with the same enzyme from mesophilic organisms (FREEZE and BROCK, 1970).

HOCHACHKA (1965, 1967) demonstrated with fish tissues, that metabolic compensation may be achieved by the use of a given isozymal form of an enzyme in a certain temperature range (see p. 344). Recent findings indicate that bacteria also contain isoenzymes with different temperature optima. In *B. stearothermophilus* three aminopeptidases were recognized by RONCARI and ZUBER (1969). The quantitative relationship between the three isoenzymes seems to depend on temperature.

Further evidence for the uniquely heat stable macromolecules has been reported in studies of ribosomes and ribosomal RNA (ACRA et al., 1964; FRIEDMAN and WEINSTEIN, 1965; MANGIANTINI et al., 1965; SAUNDERS and CAMPBELL, 1966;

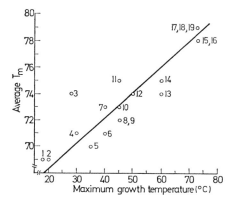

Fig. 35. Melting temperatures (T_m) and maximum growth temperatures of 19 selected organisms. The points are numbered corresponding to the numbers cf the organisms in Table 3 (PACE and CAMPBELL, 1967)

STENESH and HOLAZO, 1967; CAMPBELL and PACE, 1968). These cellular components from thermophilic organisms were more heat stable when their thermal denaturation profile was compared to that from similar components of *E. coli* (see Table 3 and Fig. 35). STENESH and HOLAZO (1967) suggest that the more heat

Table 3. Ribosome melting and maximal growth temperatures of 19 selected organisms (from PACE and CAMPBELL, 1967)

Organism and strain no.		Max. growth temp. (°C)	Ribosome T_m, (°C)
(1)	*V. marinus* (15381)	18	69
(2)	7 E-3	20	69
(3)	1 − 1	28	74
(4)	*V. marinus* (15382)	30	71
(5)	2 − 1	35	70
(6)	*D. desulfuricans (cholinicus)*	40	73
(7)	*D. vulgaris* (8303)	40	73
(8)	*E. coli* (B)	45	72
(9)	*E. coli* (Q 13)	45	72
(10)	*S. itersonii* (SI − 1)	45	73
(11)	*B. megaterium* (Paris)	45	75
(12)	*B. subtilis* (SB − 19)	50	74
(13)	*B. coagulans* (43 P)	60	74
(14)	*D. nigrificans* (8351)	60	75
(15)	Thermophile 194	73	78
(16)	*B. stearothermophilus* (T − 107)	73	78
(17)	*B. stearothermophilus* (1503 R)	73	79
(18)	Thermophile (Tecce)	73	79
(19)	*B. stearothermophilus* (10)	73	79

resistant ribosomal RNA from thermophiles is stabilized by more extensive hydrogen bonding due to a larger proportion of more heat stable helical segments, rich in guanine and cytosine.

F. Effect of Temperature on Bacterial Spores

1. Properties and Heat Resistance of Spores

The spores of the bacterial genera *Bacillus* and *Clostridium* are usually refractive and are quite noticeable in the light microscope. They are capable of remaining alive under conditions where vegetative bacterial cells would be killed. Most of the spores are characterized by considerable heat resistance.

The sporulation and the increase in heat resistance of bacterial spores are related to peculiar biochemical and anatomical changes. HASHIMOTO et al. (1960) distinguished a sequence of recognizable stages that occurs during sporulation of a culture of *Bac. cereus*. The main features in the development of a typical spore are qualitatively summarized in Table 4.

The appearance of thermostability is obviously related to the formation of the cortex, which is an electron-transparent region surrounding the core. This structure first appears in the early transitional stage, before measurable heat resistance, but concomitant with the start of the synthesis of dipicolinic acid (DPA). It is, however, suggested that the maximum level of DPA is reached before that of heat resistance. Furthermore, heat resistance is related to refractivity and this, to nonstainability. Refractivity of spores under phase contrast illumination is acquired

Table 4. Qualitative summary of distinguishing characteristics during sporogenesis (from HASHIMOTO et al., 1960)

Stage of development	Fine structure	Refractivity	Stainability	% of maximum thermo-stability	% of maximum DPA content
Filamentous vegetative cell	Homogeneous	None	Full	0	0
Granular sporangium	Granular	None	Full	0	0
Forespore	Single outer coat	None	Full	0	0
Early transi-tional spore	Exosporium and primordial cortex	Rapidly increasing	Rapidly decreasing	0	0→40
Late transi-tional spore	Thickened cortex and inner coat	High	Low	0→100	40→100
Mature spore	Complete differentiation	Full	None	100	100

well before the time that thermostability increases. In germinating spores, on the other hand, thermostability is lost earlier than the fall in refractivity (POWELL, 1957). There is also a correlation between structural development of the cortex and the inner coat on the one hand and the acquisition of refractivity on the other (YOUNG, 1958).

In the form of its calcium salt, DPA is invariably a major constituent of bacterial spores. Undetectable in vegetative cells, it is rapidly synthesized during the course of sporulation (POWELL, 1953). The physiological function of DPA is not yet clear, but several observations show that the appearance of mature spores coincides with the synthesis of DPA, the incorporation of a higher calcium level, and the attainment of higher thermoresistance (PERRY and FOSTER, 1955; COLLIER and MURTY, 1957; CHURCH and HALVORSON, 1959; BLACK et al., 1960; HASHIMOTO et al., 1960; ARONSON et al., 1967). The synthesis of DPA proceeds on a pathway connected to the pathway of the synthesis of diaminopimelic acid and L-lysin (CHASIN and SZULMAJSTER, 1967).

Calcium and probably other divalent cations play a role in the heat stability of spores. CURRAN et al. (1943) found that bacterial spores contained higher levels of divalent cations, particularly calcium, than did their homologous vegetative forms. They also reported that higher concentrations of calcium were associated with higher thermoresistance. VAS and PROSZT (1957) found that intact spores of Bac. cereus contained 4.7% of their dry weight calcium whereas the germinated spores contained only 0.86%. The higher calcium level in the intact spores was associated with higher thermoresistance. SLEPECKY and FOSTER (1959) also showed that the content of individual metals in spores was flexible within a wide range and was dependent on the relative concentration of the particular metal in the growth medium. They inferred that the various cations accumulated by the spores were interchangeable. Spores with maximal or minimal metal content were indifferent in their morphology, staining, refractivity and resistance to killing by desiccation, phenol and UV radiation. Higher thermoresistances, however, were associated with higher levels of calcium. Spores with maximal manganese or zinc content possessed minimum calcium, and therefore were thermosensitive. EL BISI and ORDAL (1956) obtained higher thermal death rates of spores of Bac. coagulans produced in the presence of higher levels of phosphate. It was postulated that higher levels of phosphate interfered with the availability to the sporulating cell of the divalent cations. AMAHA and ORDAL (1957) demonstrated that enriching the sporulating medium with calcium and manganese produced spores with higher contents of calcium and manganese and consequently of higher thermoresistance. They showed also that the presence of chelating agents with high affinities for calcium and manganese, such as ethylendiaminetetraacetic acid, tris-hydroxy-methylaminomethane and glycylglycine in the heating menstruum accelerated the death reaction and thus caused a considerable reduction in the apparent thermal resistance of the spores. EL BISI and ORDAL (1958) further demonstrated that the higher levels of phosphate or glycylglycine in the heating menstruum increased the rate of thermal destruction and that the original destruction rates were partially restored by addition of calcium and manganese salts. Magnesium and monovalent cations were ineffective in this respect.

The heat resistance of spores depends greatly on the humidity of the surrounding medium (ANGELOTTI et al., 1970). In the absence of liquid water, i.e., in the "dry state", spores are remarkably more heat resistant than in wet conditions. The thermostability of a system which does not contain liquid water is also affected considerably by the relative humidity as has been demonstrated by MURRELL and SCOTT (1966). They showed that the heat resistance was maximal at environmental water activity (a_w) values between 0.2 and 0.4. ALDERTON and SNELL (1970)

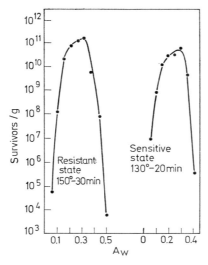

Fig. 36. Heat survival versus environmental water activity for the chemical forms of B. stearothermophilus spores (ALDERTON and SNELL, 1970)

furthermore were able to manipulate mature spores chemically into different states for the property of resistance to heat at intermediate water activity. The pre-treatment was performed by equilibrating a sample of spores with a defined amount of water in sealed thermal death time tubes. An example of the heat survival of resistant and sensitive spores is given in Fig. 36. Significant differences in the thermostability were obtained as is shown by the difference in temperature-time product of $150°-30$ min for resistant spores and $130°-20$ min for sensitive ones.

Whether the fatty acid composition (see p. 51) of spores plays a role in heat stability has not yet been determined. PHEIL and ORDAL (1967) estimated the fatty acid composition of spores. The rough variant of *Bac. stearothermophilus* was more heat resistant than the smooth variant. A relationship to the concentrations of minerals and DPA in the different variants was, however, not found (ROTMAN and FIELDS, 1969).

2. Heat Activation of Spores

Most spores germinate poorly or not at all when they are transferred to a favorable medium. They require an activation to induce germination. That sublethal heating is able to do so has been reported by CURRAN and EVANS (1945). The presence of water is required for this activation (HYATT and LEVINSON, 1968; POWELL and HUNTER, 1955). A number of workers have evaluated the time-temperature relationship of heat activation (BUSTA and ORDAL, 1964; DESROSIER and HEILIG-MAN, 1956; MURRELL, 1961). The heating time required to achieve a given germination rate usually increases with decreasing temperature. LEVINSON and HYATT (1969), GRECZ and OLEJNIKOW-KURITZA (1966), KEYNAN et al. (1964) and BUSTA and ORDAL (1964) have performed calculations concerning the temperature co-efficient Q_{10} and other apparent average thermodynamic constants. Their values indicate high activation energy, large increase in entropy and high temperature coefficients. Hence it is suggested that denaturation of one or more proteins is involved in the activation process of dormant spores.

3. Heat Killing of Spores

The destruction of spores by heat appears to be closely associated with the release of DPA in the surrounding menstruum (POWELL, 1957; STEDMAN, 1956; RODE and FOSTER, 1960). EL BISI et al. (1962) carried out investigations to assess quantitatively the behavior of cellular DPA and certain divalent cations during exposure of spores to moist heat and to correlate such behavior with the thermal death rates. A typical sequence of events is shown in Fig. 37. In all of their experiments, it could be shown that the thermal death reaction is associated with the release of DPA and certain cations such as calcium and manganese. Factors that accelerated the death reaction also accelerated the release of the preceding cellular components. Whether such an exudate is the cause or the effect of thermal death of the spore cell remained difficult to determine. From the kinetic data of both phenomena, it appeared that death preceded the release of DPA and divalent cations. The possibility exists, however, that in the latter case one may be dealing with a two-step reaction: (1) the disruption within the spore cell of a certain critical stereostructure that relies on components such as DPA and the divalent cations to act as the cementing material; and (2) the release of such widely varied DPA-divalent cation molecular combinations from the disrupted, particular type of structure into the surrounding menstruum. This latter step could in itself be composed of two successive steps instead of one: (a) the release or freeing of such small molecular fragments (DPA-cation) from the disrupted structure, and (b) the exudation of such free fragments into the surrounding medium.

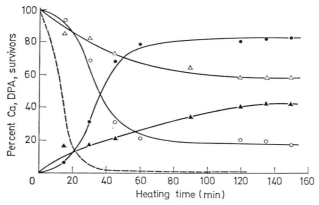

Fig. 37. Chemical changes during thermal destruction of spores of *B. coagulans (thermoacidurans)* at 99° C. Spores were heated in 5 mM phosphate buffer at pH 7.0. ○ DPA in heated spores; ● DPA in supernatant; △ calcium in heated spores; ▲ calcium in supernatant; - - -, per cent survivors (EL BISI et al., 1962)

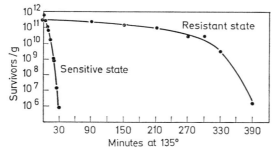

Fig. 38. Heat resistance at 135° C for the chemical state of *B. stearothermophilus* spores at intermediate water activity ($a_w = 0.28$) (ALDERTON and SNELL, 1970)

Another phenomenon is the deviation of the death curve in plots of the log of survivors against heating time at a given temperature (Fig. 38), indicating a non-exponential order of heat killing. From the kinetic point of view one may suggest that at the beginning of the time course of inactivation, most of the heat is consumed for reactions other than inactivation of the monomolecular genetic structure of the spore cell (ALDERTON and SNELL, 1970).

References to this section, see p. 78.

II. Genetic Regulation of Temperature Responses

J. L. INGRAHAM

A. Introduction

Growth of microorganisms is the consequence of a complex and highly inter-dependent set of chemical reactions through which cells are able to replicate themselves efficiently and quickly. Hence, if we are to inquire as to the effect of temperature on microbial growth, we should first consider how chemical reactions are affected. Temperature affects both the equilibrium constant and the rate of chemical reactions in known and predictable ways; the velocity of a reaction is a logarithmic function of the reciprocal of absolute temperature according to the equation:

$$v = se^{-\Delta E^*/RT}$$

where $v =$ the velocity of the reaction, $s =$ a constant, ΔE^* is the activation ener-gy, $R =$ the gas constant, and $T =$ temperature in $^\circ$K. Upon integration we have the form:

$$\ln v = \frac{-\Delta E^*}{R} \cdot \frac{1}{T} + \text{constant} .$$

Thus, the logarithm of the velocity of a chemical reaction is a linear function of the reciprocal of absolute temperature ($^\circ$K) (Fig. 1a).

If a similar plot (frequently called an Arrhenius plot after the Swedish chemist who first discovered the relationship between temperature and the rate of a chemical reaction) is made of the growth rate (k) of a microorganism as a function of the reciprocal of absolute temperature, a different response is seen (Fig. 1b). In the

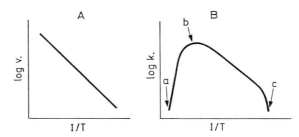

Fig. 1. Effect of temperature (T) on: A the velocity (v) of a chemical reaction, and B on the growth rate (k) of a microorganism. Arrows indicate: a the maximum temper-ature of growth; b the optimum temperature of growth, and c the minimum temper-ature of growth

midrange of temperatures, normal chemical kinetics seems to apply; below that range, the slope of the curve increases and eventually becomes vertical at the minimum temperature of growth. As the temperature is increased above the midrange, growth rate again decreases, and very rapidly falls to zero at the maximum temperature of growth.

The parameters, maximum, optimum, and minimum temperatures of growth, are clearly genetically determined, and their actual values vary over wide ranges among various microorganisms. It is now rather convincingly established that certain bacteria are capable of growing at temperatures as high as 95° C (BOTT and BROCK, 1969; BROCK, 1967), and as low as − 10° C (LARKIN and STOKES, 1967).

On the basis of the temperature range of growth, bacteria are frequently divided into three broad classifications: Those that grow at high temperatures are called thermophiles; those that grow at moderate temperatures, mesophiles; and those that grow at low temperatures, psychrophiles or rhigophiles. Such a simple tripartate classification has proved to be, in many respects, inadequate, because the maximum and minimum temperatures of growth appear to be independent biological variables, i.e. certain bacteria which grow well in the psychrophilic range of growth (e.g. 0° C) also grow at temperatures in the upper regions of the mesophilic range (e.g. 40°C) (INGRAHAM, 1962) while other psychrophiles which grow well at 0°C have maximum temperatures of growth of less than 20°C (EIMHJELLEN,c.f., STOKES, 1963; MORITA, 1966; HARDER and VELDKAMP, 1966; STANLEY and ROSE, 1967; and SIEBURTH, 1967). Similarly, some bacteria which grow well in the thermophilic range can grow at temperatures as low as 30°C (c.f. INGRAHAM, 1962). In attempts to resolve this classification problem, additional terms have come into use: "stenothermophiles" for those thermophiles which cannot grow in the mesophilic range, and "eurithermophiles" for those which can; "obligate" is used to describe those psychrophiles which cannot grow above 20° C, and "facultative" for those that can.

Thus the absolute values of the maximum and minimum temperatures of growth and the temperature range which separates them seem to be set by independent genetic determinants. This hypothesis is strengthened by the existence of temperature-sensitive (ts) mutants; ts mutants of the heat-sensitive (hs) class have lowered maximum temperatures of growth, often with no corresponding change in the minumum temperature of growth; the class of cold-sensitive (cs) mutants has increased minimum temperatures of growth, but usually no change in the maximum.

B. Genetic Determination of Heat Stability of Proteins

Genes code for the primary structure of proteins, which, in turn, determines their heat stability as well as their functional activity, be it catalytic or structural. What portion of the protein is required for heat stability as contrasted with functional activity remains an unanswered question. Certainly heat stability is a particularly sensitive property of a protein in that it seems to be more readily changed by mutations affecting primary structure than is functional activity. Several lines of experimental evidence support this contention.

The most detailed data in this respect are those of LANGRIDGE (1968), who examined 52 mutationally altered forms of the enzyme β-glactosidase from *Escherichia coli*. The set of altered proteins was generated by isolating a number of non-allelic amber mutations in the gene coding for β-galactosidase, and then introducing a suppressor mutation which causes a serine residue to be introduced at the site of the amber codon. Thus, he was able to obtain a series of mutationally altered enzymes without having to depend on loss of catalytic activity for their selection. He found that over 70% of these mutant proteins showed a distinct loss of heat stability (Table 1) while (as judged by the ability of lactose to inhibit competitively the binding of the analogue o-nitrophenyl-β-D-galactopyranoside) only one showed loss of catalytic function.

Table 1. Frequency of half-lives in minutes at 57° C of 52 serine-substituted β-galactosidases[a]

Half-life classes[b] (min at 57° C)	Frequency
10	16
10— 20	5
20— 30	4
30— 40	3
40— 50	4
50— 60	1
60— 70	1
70— 80	1
80— 90	1
90—100	3
100—110	15

[a] After LANGRIDGE, 1968.
[b] Wild-type enzyme has a half-life of 104 minutes.

These results suggest that most naturally occurring mutations are counterselected on the basis of loss of heat stability rather than on the basis of loss of function.

It follows from the above, that a microorganism which produces an essential but heat labile protein precluding its growth at high temperatures, would, in the absence of the challenge of the elevated temperature, rapidly accumulate mutations in genes coding for other proteins and cause them to lose heat stability. Certainly bacteria capable of growing at elevated temperatures must contain heat-stable proteins, a fact that has been established by many studies (see for example, MARCH and MILITZER, 1952, 1956; MATHEMEIER and MORITA, 1964; AKAGI and CAMPBELL, 1961). KOFFLER and GALE (1957), on the other hand, made the very interesting corollary observation that bacteria which are incapable of growing at high temperatures contain very few thermostabile proteins (Table 2). Whereas most of the total cytoplasmic proteins from representative mesophiles are precipitated by an 8 min heat treatment of 60° C, only a very small percentage of the proteins from representative thermophiles is precipitated by the same treatment.

Table 2. Stability of cytoplasmic proteins from mesophilic and thermophilic bacteria at 60° C (KOFFLER and GALE, 1956)

Organism	Temperature class	% of Proteins denatured[a]
Proteus vulgaris	Mesophile	55
Escherichia coli	Mesophile	55
Bacillus megaterium	Mesophile	58
Bacillus subtilis	Mesophile	57
Bacillus stearothermophilus	Thermophile	3
Bacillus sp. (Purdue CD)	Thermophile	0
Bacillus sp. (Texas 11330)	Thermophile	4
Bacillus sp. (Nebraska 1492)	Thermophile	0

[a] % of total trichloracetic acid — precipitable material from a sonic extract of cells which is coagulated by an 8 min heat treatment at 60° C.

The thesis that mutations conferring thermolability tend to accumulate in organisms not subjected to the challenge of exposure to high temperature is also supported by the accumulating evidence regarding marine psychrophiles. Whereas at one time it was generally agreed that psychrophilic bacteria which are unable to grow at temperatures as high as 20° C (obligate psychrophiles) were rare or possibly nonexistent (STOKES, 1963), more recently it has been shown that these organisms are abundant and, indeed, where the temperature always remains low, constitute the major microbial population (MORITA, 1966; HARDER and VELDKAMP, 1968), suggesting again that in the absence of selective pressures, protein stability at temperatures even as low as 20° C is lost by random mutational events. Recently, HARDER and VELDKAMP (1971) presented evidence that obligate psychrophiles grow more rapidly at 4° C than do facultative psychrophiles which were isolated from the same environment. At first glance, one might conclude that this observation is inconsistent with the hypothesis of loss of heat stability through random mutations, i.e., the observations seem to indicate that selection for maximal growth rate at 4° C itself precludes the ability of the cells to grow above 20° C. However, these facultative psychrophiles were probably (during long periods of natural selection) occasionally exposed to temperatures above 20° C, and, hence, the primary selective pressure might have been to withstand these temperatures even at the expense of growing more slowly at 4° C.

Virologists studying genetics of bacteriophages have come to accept, as a matter of course, that any missense mutation has a high probability of decreasing the heat-stability of the product protein. If the mature mutant virion is more sensitive to killing by heat than its parent, it is a virtual certainty that the mutation in question must lie in a gene coding for an essential protein present in the virion; but further it is expected that any missense mutations in genes coding for these proteins must necessarily affect the heat stability of the virion (DOWELL, C. E., personal communication).

C. Loss of Function at Low Temperatures

The chemical basis for loss of function of a protein at high temperatures is self-evident; those chemical bonds which maintain the proper secondary and tertiary structure of proteins become weakened at elevated temperatures, resulting in denaturation and loss of function of the protein. At low temperatures, loss of function is more difficult to explain; but it is clear that weakening of hydrophobic bonds (as a consequence of physical changes in the structure of the solvating water) accounts for the denaturation and changes in protein structure that occur at low temperatures (see, for example BRANDTS, 1963).

The selective pressures for ability to grow at low temperatures seem dramatically different from those for growth at high temperatures. We have discussed evidence indicating that exposure to the challenge of high temperatures is essential if a microorganism is to retain its ability to grow at high temperatures, but such does not seem to be the case for growth at low temperatures. Gram negative enteric bacteria must have grown at temperatures quite close to 37° C for millions of years, yet both *Salmonella typhimurium* and *Escherichia coli* have retained their ability to grow at temperatures as low as 8° C (SHAW, MARR, and INGRAHAM, 1971; HOFFMANN, 1967), although it is not readily apparent why this ability is of selective advantage.

D. Mutations which Decrease the Temperature Range for Growth

Mutants which differ from the parent only in the ability to grow over a narrower range of temperatures are called "temperature-sensitive" (*ts*) mutants. It is convenient to divide these mutants into two subclasses: "heat-sensitive" (*hs*) mutants or those with decreased maximum temperatures of growth, and "cold-sensitive" (*cs*) mutants, or those with increased minimum temperatures of growth. The use of *ts* mutants has played a major role in our understanding of biological processes because they allow one to analyze by the mutant technique, cellular reactions, the product of which cannot be supplied from the medium. Thus, any cellular reaction, can, in theory, be studied using temperature-sensitive mutants, and most studies employing *ts* mutants have been done with the aim of determining the functional role of a given protein. In this respect, the *ts* mutant technique has been spectacularly successful. The role of acyl-t-RNA synthetase in metabolism and regulation, the biochemistry of DNA and RNA polymerization, and the mechanism of translation have all been elucidated to a considerable extent using *ts* mutants as a tool.

The biochemical basis of *hs* mutants was first established by MAAS and DAVIS (1952), and their results serve as a general explanation of biochemical bases of other classes of *hs* mutants. They isolated a mutant of *Escherichia coli* which, below 30° C, is able to grow in an unsupplemented medium, but which, above 30° C, requires pantophenate for growth. They established that the metabolic block imposed by the elevated temperature was the condensation of alanine and pantoic acid to form pantothenic acid, and further that the mutant produced a form of the enzyme catalyzing this reaction which was much more sensitive to heat than that which the wild type produced; whereas undetectable levels of the wild-type enzyme (under the conditions employed) were inactivated in two hours at 35° C, over 90% of the mutant enzyme was inactivated in one hour at 30° C.

Thus, MAAS and DAVIS were able to establish unequivocally that the temperature-sensitivity of the mutant was a consequence of a mutation (presumably a missense mutation) which decreased the thermostability of the protein product of the gene.

Heat-sensitivity of certain other mutant strains of bacteria seems to have a slightly different biochemical basis, namely, in certain cases the enzyme catalyzing the thermostable reaction is found to be stable at the restrictive temperature once it has been synthesized at the permissive temperature; but the enzyme cannot be synthesized at the restrictive temperature. SADLER and NOVICK (1965) have called this class of mutants, *tss* (for temperature-sensitive-synthesis).

Several classes of such mutants have been described. SADLER and NOVICK (1965) reported a mutationally generated temperature-sensitive-synthesis mutant of the repressor of the *Escherichia coli* lactose operon; KORNBERG and SMITH (1966) have isolated a temperature-sensitive-synthesis mutant of the *E. coli* isocitrate lyase; and CONDON and INGRAHAM (1967) have discovered temperature-sensitive-synthesis of the muconate lactonizing enzyme.

Cold-sensitive mutants have been studied much less extensively than have heat-sensitive mutants, and hence such studies have so far provided only a limited amount of information about biochemical processes, but they have provided important information concerning the genetic basis for loss of function at low temperature. Typical growth response of a cold-sensitive mutant of *Escherichia coli* (O'DONOVAN, KEARNEY, and INGRAHAM, 1965) as affected by temperature is shown in Fig. 2. At 37° C the mutant strain grows almost as fast as its parent, but as temperature is lowered, the growth rate of the mutant decreases more rapidly than the parent, and stops growing completely at about 20° C, while the parent continues to grow down to about 8° C. The cold-sensitive mutant depicted in Fig. 2 then differs from its parent by having a minimum temperature of growth at 20° C rather than 8° C. That this particular mutant stops growing at 20° C probably merely reflects the fact that the penicillin counterselection procedure used to isolate it was done at 20° C — the temperature selected as the restrictive temperature for the isolation of most cold-sensitive mutants. One would imagine that the minimum temperature for the isolation of most cold-sensitive mutants of enteric bacteria could be higher than 20° C (and very probably also lower) but no systematic studies have been done to determine those possibilities, also most studies on cold-sensitive mutants have been limited to enteric bacteria.

It is not difficult to isolate cold-sensitive mutants even if no prior enrichment procedure is used, and from the somewhat limited data available, it appears that generation of a mutation which can lead to cold-sensitivity is about equally as probable as the generation of a mutation which can lead to heat-sensitivity, although, as we will see later, the total number of genes which can mutate to yield a cold-sensitive phenotype is probably much smaller than the number of genes which can mutate to yield heat-sensitivity. KNUTE RASMUSSEN (personal communication), in a limited but convincing experiment, mutagenized a culture, and without employing any enrichment procedure, simply scored the number of cold-sensitive and heat-sensitive mutants among the survivors. Of the several hundred mutants scored, he found the cold-sensitive class to be slightly more frequent than the heat-sensitive class.

The same general types of mutants exist among the cold-sensitive class, as exist among the temperature indepedent class: i) cold-sensitive auxotrophic mutants

(O'DONOVAN and INGRAHAM, 1965; ABD-EL-AL and INGRAHAM, 1969; HOFFMANN and INGRAHAM, 1970), ii) cold-sensitive carbon source mutants (CONDON and INGRAHAM, 1967; SQUIRES and INGRAHAM, 1969), and iii) cold-sensitive conditional lethal mutants (i.e., those which are unable to grow even on complex medium at low temperature) (TAI, KESSLER, and INGRAHAM, 1969; GUTHRIE, NASHIMOTO, and MOMURA, 1969). In spite of the frequent occurrence of cold-sensitive mutants and their general distribution among the various classes of mutants, it is quite clear that, in distinct contrast to heat-sensitivity, mutations conferring cold-sensitivity are restricted to a small portion of the total number of genes of an organism. O'DONOVAN and INGRAHAM (1965) reported the isolation of 7 independent mutant strains of *E. coli* which were cold-sensitive for the biosynthesis of histidine (i.e., were prototrophs at 37° C but were histidine auxotrophs at 20° C). Although there are 9 genes coding for the enzymes required for the biosynthesis of histidine, one might expect the lesions conferring cold-sensitivity to be randomly distributed among these 9 genes. In fact, all the mutations were found to lie in the same gene (*hisG*). Similarly, there seems to be a lack of randomness of the mutations conferring the conditional lethal type of cold-sensitivity; TAI, KESSLER, and INGRAHAM (1969) estimated that about 40% of such mutations lay in genes which are cotransducible with the streptomycin-resistance locus.

Studies on mutant phages support the conclusion that mutations conferring cold-sensitivity are restricted to a limited number of genes (SCOTTI, 1968). On analyzing 75 cold-sensitive mutants of phage T_4D it was found that only 9 complimentation groups contained the remaining 65 mutants. These results are in sharp contrast to those of EDGAR and his colleagues (EPSTEIN et al., 1963; EDGAR et al., 1964; EDGAR and LIELAUSES, 1964) who in studies with the same phage, T_4D, showed that heat-sensitive and amber type nonsense mutations were randomly scattered over the genome; 382 heat-sensitive mutants were caused by mutations distributed among 37 genes.

E. Biochemical Basis of Cold-Sensitivity

Clearly, cold-sensitive mutants provide a system for analyzing the biochemical basis for the minimum temperature of growth. Since a single genetic change increases the minimum temperature of growth, the protein alteration resulting from the genetic change must be the biochemical basis for the minimum temperature of growth of the particular mutant being studied. It can be argued that these artificially induced defects might not be representative of the types of loss of function that account for the minimum temperature for growth of wild-type organisms in nature; and, therefore, knowledge concerning the genetic and biochemical basis for the lower temperature limits of growth in nature can only be gained by the direct study of wild-type organisms; but several practical considerations indicate that such a direct approach holds very little hope of success. As will be discussed later, convincing evidence exists that in most microorganisms, many independent biochemical functions cease simultaneously at the minimum temperature of growth. Thus, in naturally-occurring microorganisms, there is not a single cause of the minimum temperature of growth, and this fact makes biochemical analysis and even the distinction between cause and effect, quite difficult. Moreover, there is no reason to believe that artificially-induced mutations would preferentially affect functions different from those affected by the naturally accumulated mutations that set the minimum temperature of growth of wild-type microorganisms. At the very least, a study of cold-sensitive mutants gives information as to the types of lesions which can selectively prevent function at low temperature.

One such type of lesion is that which alters the sensitivity of regulated proteins to their small molecule effectors, and hence results in a level of distortion of the regulated system such that function at low temperature is prevented. Cold-sensitive histidine mutants are an example of this type. A number of such mutants from *Escherichia coli* (O'DONOVAN and INGRAHAM, 1965) and *Salmonella typhimurium* (unpublished data) have been isolated and studied. The growth response of one of these mutants growing in minimal medium is shown in Fig. 2, a growth condition

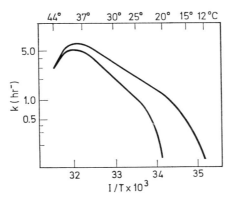

Fig. 2. Arrhenius plot of the specific growth rate (k) of a cold-sensitive mutant (lower curve) of *Escherichia coli*, and its parent, growing in a minimal medium. The ordinate is the specific growth rate (log scale) and the abscissa is the reciprocal of the absolute temperature times 1000. (O'DONOVAN, KEARNEY, and INGRAHAM, 1965)

where the mutant has a minimum temperature of growth some 12° C higher than its parent; but in minimal medium, enriched with histidine, the parent and mutant grow at identical rates at all temperatures. The functional block at low temperatures, as identified by the accumulation of biochemical intermediates of the pathway, was found to be the first reaction of the pathway — that catalyzed by phosphoribosyl adenosine triphosphate pyrophosphorylase (PR-ATPppase). In spite of the fact that compelling evidence showed that this enzyme did not function *in vivo* at 20° C, cell-free preparations of the enzyme were completely active at the low temperature. Since PR-ATPppase is the enzyme of the histidine pathway which is sensitive to feedback inhibition (MARTIN, 1963) by the end product of the pathway, histidine, studies were made to determine if altered feedback properties of the enzyme could account for loss of function at low temperature, and, indeed, it was found that the enzyme produced by the mutant was remarkably (almost 1000-fold) more sensitive to feedback inhibition than was the enzyme produced by the parent (wild-type). It was also found that the mutant and wild-type enzymes were both about 10 times more sensitive to feedback inhibition at 20° C than at

37° C. Thus, altered feedback inhibition seemed to be a feasible biochemical explanation for the cold-sensitivity. The increased sensitivity of the mutant enzyme
to feedback inhibition, plus the added sensitivity imposed by the low temperature,
could reasonably be expected to create conditions whereby the intracellular concentration of histidine sufficient to prevent its own further biosynthesis by feedback inhibition of PR-ATPppase would be insufficient to allow the biosynthesis of
proteins. Thus the mutant would be unable to grow at 20° C in the absence of an
exogenous source of histidine. Although this explanation for the cold-sensitivity
of the mutants at first appears to be based only on circumstantial evidence,
strong support for the explanation can be derived from the data presented in
Table 3.

Table 3. Growth of wild-type and mutant strains of *E. coli* and properties of phosphori-
bosyl-adenosine triphosphate pyrophosphorylase produced by them (from O'DONOVAN
and INGRAHAM, 1965)

Strain No.	Feedback inhibitor[a]	Growth (°C)		Phenotype [f]
		37 (k, h^{-1})	20	
1) C-600-1	$6 \cdot 10^{-5}$	0.53	+	Parental
2) K-II-27[b]	$8 \cdot 10^{-8}$	0.49	—	Cold sensitive
3) K-II-E27	$1 \cdot 10^{-7}$	0.51	—	Cold sensitive
4) K-II-A28	$3 \cdot 10^{-6}$	0.50	—	Cold sensitive
5) K-II-A29	$1 \cdot 10^{-7}$	0.47	—	Cold sensitive
6) K-II-M30	$1 \cdot 10^{-7}$	0.46	—	Cold sensitive
7) K-II-E31	$3 \cdot 10^{-7}$	0.48	—	Cold sensitive
8) K-II-E32	$2 \cdot 10^{-7}$	0.49	—	Cold sensitive
9) K-II-27-R	$2 \cdot 10^{-5}$	0.53	+	Parental[c]
10) K-II-27/TA1	$5 \cdot 10^{-3}$	0.43	+	TA-resistant[d]
11) K-II-27/TA2	$1 \cdot 10^{-2}$	0.32	+	TA-resistant[d]
12) K-II-27/TA3	$1 \cdot 10^{-2}$	0.41	+	TA-resistant[d]
13) K-II-27/TA4	$5 \cdot 10^{-2}$	0.41	+	TA-resistant[d]
14) K-II-27/TA5	$1 \cdot 10^{-2}$	0.38	+	TA-resistant[d]
15) K-II-27/TA6	$1 \cdot 10^{-2}$	0.39	+	TA-resistant[d]

[a] Molarity of L-histidine giving 50% inhibition of PR-ATPppase at 37° C.
[b] Ultraviolet light used as a mutagen.
[c] E, A, and N indicate ethyl methanesulfonate, 2-aminopurine, and N-methyl, N-
nitro, N-nitrosoguanidine, respectively, used as mutagens.
[d] Resistant to the histidine analogue, 2-thiazole-DL-alanine.

All the cold-sensitive mutants (lines 2 through 8), regardless of the mutagen used
to induce them, produce species of PR-ATPppase that are dramatically more
sensitive to histidine than that produced by the parent. Thus, the properties of
cold-sensitivity and increased sensitivity to feedback inhibition seem to be vitally
linked. If so, revertants selected as being able to grow at 20° C should produce a
PR-ATPppase that is normally sensitive to feedback inhibition, and, conversely,
revertants for producing a PR-ATPppase with normal (or less) sensitivity to feed-

back inhibition (this can easily be accomplished by selecting for resistance to the histidine analogue, 2-thiazole-DL-alanine, a compound which is active as a feedback inhibitor but which does not serve as a substrate for amino acyl-t-RNA synthesis, and hence cannot be incorporated into protein) should grow at 20° C. Both predictions are fulfilled (Table 3); revertants which can grow at 20° C produce an enzyme with normal sensitivity to feedback inhibition and revertants selected for being resistant to 2-thiazole-DL-alanine, and hence to feedback inhibition, are capable of growing normally at 20° C (lines 10 through 14, Table 3). Thus, it seems clear that this class of mutants owes its cold-sensitivity to a slight alteration in the structure of PR-ATPppase, rendering it more sensitive to feedback inhibition at all temperatures; the specific loss of function at low temperatures is dependent on the natural augmentation of this inhibition at low temperatures. One must assume that without this natural gradient effect of temperature on the inhibition of PR-ATPppase by histidine, cold-sensitive mutants affecting histidine biosynthesis would not exist.

If the biochemical explanation found for cold-sensitivity of histidine biosynthesis is of any general significance for explaining other examples of loss of function at low temperatures, then the degree of inhibition of many other allosteric proteins might be expected to be changed significantly by temperature. Accumulating evidence supports the contention that changes with temperature of sensitivity of allosteric proteins to their effectors is the rule rather than the exception, but predictions are not possible as to the sense of change with temperature, i.e. does it increase or decrease at low temperature. Some proteins become more sensitive to inhibition as temperature is decreased and others become less sensitive.

As an example we may consider the detailed data concerning the effect of temperature on the inhibition of fructose-1,6-diphosphatase from rat liver by adenosine monophosphate (AMP), which was provided by TAKITA and POGELL (1965). They showed that the enzyme exhibits dramatically increased sensitivity to AMP as temperature is decreased. In the absence of AMP the effect of temperature follows normal chemical kinetics, i.e., the Arrhenius plot is completely linear over the temperature range of 46° C to 2° C, but in the presence of the inhibitor, the reaction is selectively inhibited at low temperatures (Fig. 3). APM at 10^{-4} M has no inhibiting effect on the enzyme at 46° C, but this concentration inhibits more than 90% of its activity at 20° C, and 10^{-5} M AMP is sufficient to give 90% inhibition. From these data it is clear that since enzymes from homeothermic organisms also exhibit strong temperature effects on the sensitivity to inhibition, such effects most probably reflect fundamental properties of allosteric proteins. Low temperature sensitivity to AMP cannot be a product of natural selection, because natural selection for optimal regulation of an allosteric protein must occur at the temperature of growth; and we see in the case of fructose-1,6-diphosphatase that, if sensitivity to inhibition is optimal at a given temperature, it will be less than optimal at all other temperatures. If a mutation were to occur making fructose-1,6-diphosphatase more sensitive to feedback inhibition by AMP, the result would necessarily be a selective loss of function at low temperatures.

The regulation of carbamyl phosphate synthetase from *Salmonella typhimurium* has also been shown to be highly temperature dependent (ABD-EL-AL and INGRA-

HAM, 1969). The product of this enzyme, carbamyl phosphate, is required both for the biosynthesis of pyrimidines and of arginine (Fig. 4). Enzyme activity is under dual regulation, being inhibited by the pyrimidine intermediate, uridine monophosphate (UMP), and being activated by the intermediate of arginine biosynthesis,

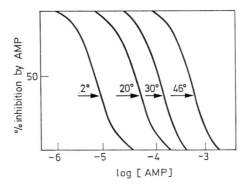

Fig. 3. Percentage inhibition of fructose-1,6-diphosphatase by different concentrations of AMP at different temperatures (from TAKITA and POGELL, 1965)

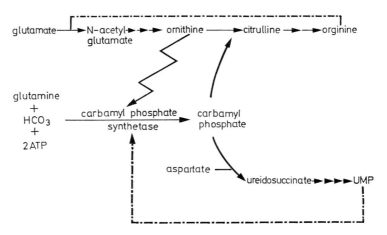

Fig. 4. Pathways of biosynthesis of arginine and uridine monophosphate by enteric bacteria

ornithine, which would tend to accumulate if the cell were starved for carbamyl phosphate. Thus, through the dual control mechanisms and the inhibition of the first step of arginine biosynthesis by arginine itself, intracellular levels of carbamyl phosphate can be maintained in the face of changing concentrations of end products of the two reaction sequences. For example, if intracellular levels of UMP were to rise and inhibition of carbamyl phosphate synthetase would occur, resulting in a drop of the levels of carbamyl phosphate, thus inhibiting the conversion of ornithine to citrulline, ornithine would accumulate and would in turn activate the enzyme to insure an adequate supply of carbamyl phosphate for the synthesis of arginine. Conversely, if the intracellular levels of arginine were to rise and inhibit the first enzyme of the pathway, ornithine would not be produced, activation would not occur, and the production of carbamyl phosphate would diminish.

The potency of UMP as an inhibitor of carbamyl phosphate synthetase is virtually unchanged by temperature, but ornithine becomes a more powerful activator of the reaction as temperature is decreased (Table 4). The physiological implications of these temperature effects can be emphasized by considering the effect of the simultaneous addition of $1 \cdot 10^{-4}$M UMP and $1 \cdot 10^{-3}$M ornithine on the velocity of the reaction. At 37° C this combination of effectors results in distinct inhibition, while at 20° C it results in distinct activation.

Table 4. Effect of ornithine and UMP on carbamyl phosphate synthetase activity at 37° C and 20° C (ABD-EL-AL and INGRAHAM, 1969)

Effector		Specific activity		Relative activity	
UMP	Ornithine	37° C	20° C	37° C	20° C
None	None	45.0	7.0	100	100
None	$5 \cdot 10^{-6}$	50.0	—	111	—
None	$1 \cdot 10^{-5}$	55.8	—	124	—
None	$1 \cdot 10^{-3}$	88.2	18.0	196	257
None	$6 \cdot 10^{-3}$	89.1	18.3	198	261
$1 \cdot 10^{-5}$	None	28.1	3.9	62.5	55.7
$1 \cdot 10^{-4}$	None	6.3	0.9	8.5	12.8
$1 \cdot 10^{-3}$	None	3.8	0.4	8.5	5.7
$1 \cdot 10^{-5}$	$1 \cdot 10^{-3}$	87.8	19.3	195	276
$1 \cdot 10^{-4}$	$1 \cdot 10^{-3}$	38.4	8.8	85.5	126
$1 \cdot 10^{-3}$	$1 \cdot 10^{-3}$	13.7	—	30.4	—

Aspartic transcarbamylase from *Escherichia coli* and *Salmonella typhimurium* is an example of an enzyme that is less sensitive to its effector at low temperatures. The enzyme (which catalyzes the first reaction in the pathway of biosynthesis of pyrimidines) is feedback inhibited by cytidine triphosphate (CTP) and activated by adenosine triphosphate (ATP). CTP is a less effective inhibitor of the enzyme below 20° C and at very low temperatures (4° C) the velocity-substrate plots exhibit a hyperbolic response typical of an unregulated enzyme rather than the sigmoid response that one observes at higher temperatures (O'DONOVAN and NEUHARD, 1970). In contrast, aspartic transcarbamylase from *Saccharomyces cerviseae* (which is regulated by uridine triphosphate) is more sensitive to feedback inhibition at low temperatures (3° C) than at higher temperatures (30° C) (KAPLAN, JUPHILL, and LACRONTE, 1967). Thus, these studies on aspartic transcarbamylase bring clear focus on the generalizations about the effect of temperature on allosteric proteins: there is a high probability that sensitivity to inhibition or activation changes with temperature, but the sense of this change is unpredictable.

We can tentatively conclude from the above that mutations which alter the sensitivity of allosteric proteins to their effectors would have a high probability of being expressed as a cold-sensitive phenotype. ABD-EL-AL and INGRAHAM (1969) described another example of cold-sensitivity resulting from altered sensitivity of an allosteric protein to regulation. They isolated a mutant of *Salmonella typhimurium* which required arginine for growth at low temperatures, and they showed that the mutation lay in the gene *pyrA*, which codes for carbamyl phosphate synthetase. The mutant enzyme is 5-fold more sensitive to the inhibition by UMP at 20° C than at 37° C and also the stimulatory effect of ornithine is less at 20° C than at 37° C; the combined effect of the presence of both allosteric ligands is less activation and more inhibition at lower temperatures, resulting in the observed cold-sensitivity.

Cold-sensitivity need not result exclusively from loss of function at low temperatures; alternatively, the ability, only at low temperatures, to produce a toxic compound would exhibit a similar phenotype. An example of such a basis for cold-sensitivity was reported by SQUIRES and INGRAHAM (1969). They isolated and studied a mutant of *E. coli* which was able at 37°C, but unable at 20°C, to utilize lactose as a source of carbon and energy. They established that, in fact, the mutant strain had suffered two mutations in the galactose operon; one in *galK*, which rendered galactokinase heat-sensitive (nonfunctional at 37° C) and one in *galT*, rendering galactose-1-phosphate uridyl transferase nonfunctional at all temperatures. Thus, at 37° C, the mutant strain was able to grow on the glucose moiety of lactose (the galactose moiety was not metabolized because of the lack of a functional galactokinase) but at 20° C growth was inhibited by the toxic compound, galactose-1-phosphate (KURAHASHI and WAHBA, 1958) which accumulated because it could be synthesized by the cell (owing to the active galactokinase) but could not be further metabolized (owing to the lack of an active transferase enzyme).

Mutants cold-sensitive for the biosynthesis of tryptophan (HOFFMANN and INGRAHAM, 1970) produce a species of tryptophan synthetase which is reversibly inactivated at low temperature. The biochemical basis of the reversible inactivation is not yet understood.

A major class of cold-sensitive mutants includes those which are unable to synthesize ribosomes at low temperatures. TAI et al. (1969), studying *Salmonella typhimurium* and later GUTHRIE et al. (1969) studying *Escherichia coli*, found that many mutations in genes coding for ribosomal proteins are expressed as a cold-sensitive phenotype. The rationale used by these two groups in searching for ribosomal mutants among the class of cold-sensitive mutants was quite distinct: the former reasoned, on the basis of their studies establishing that allosteric proteins are particularly frequent targets for mutations expressing cold-sensitivity, that other proteins whose proper functions are critically dependent on a precise conformation might also be subject to mutational alterations which would preclude their function at low temperatures. Proteins which are clearly dependent on being in a definite conformation in order to function must include those proteins which comprise building blocks of cell organelles, and particularly critical organelles in this respect are ribosomes. Accordingly, they sought and found mutants unable to synthesize ribosomes at low temperatures.

The second group based their search on a different rationale. They had previously shown that the *in vitro* assembly of 30s ribosomal subunits is strongly temperature dependent (TRAUB and NOMURA, 1969) with an Arrhenius activation energy of about 40 kcal/mole. The rate-determining reaction step is a structural rearrangement of an intermediate particle, the "RI" particle, which is deficient in some ribosomal proteins. The group reasoned that even *in vivo*, such a rate-limiting step would exist which has a high activation energy and, therefore, that many mutational defects which affect the assembly process should be intensified at lower temperatures.

The distinction between these rationales is probably more apparent than real. Weakening of hydrophobic bonds at the low temperature most probably accounts for the loss of function.

Studies on cold-sensitive ribosomal mutants have contributed materially to our understanding of ribosomal assembly and function, and they have also provided important genetic markers which have aided the mapping of genes coding for structural components of ribosomes (NASHIMOTO and NOMURA, 1970; TAI and INGRAHAM, 1971; NASHIMOTO et al., 1971), but for our purposes here, they have provided important information about the genetic determination of temperature limits of growth, because certain of the mutational defects appear to mimic properties of wild-type organisms growing near their minimal temperature of growth. Certain cold-sensitive mutants accumulate a preribosomal particle which is strikingly similar in composition to the 21s "RI" particles which accumulate during the *in vitro* reconstitution reaction of the 30s ribosomal subunits at low temperatures (10° or less). These particles (as judged by sedimentation properties only) can also be detected in wild-type strains after a temperature shift down to 20° C. Thus it appears that this class of cold-sensitive mutants (unable to synthesize ribosomes at low temperature) represents an exaggeration of one of the functional defects which might determine the minimum temperature of growth of the wild type.

F. Adaptation to New Temperature Ranges for Growth

Heat- and cold-sensitive mutants allow the precise determination of the genetic factors which set their particular temperature growth limits, and, by analogy, intelligent guesses can be made concerning the factors which determine these limits for wild-type strains. An alternative, and possibly more direct, approach to the problem would be to isolate and study mutants which have an extended rather than a more restricted temperature range for growth. Clearly, the selective pressure for isolating such mutants is very powerful; one need only incubate a mutagenized culture above the maximum or below the minimum growth temperature. The very limited number of reports of such mutants is in itself convincing evidence that they are, indeed, quite rare. No such mutants of gram negative enteric bacteria have been reported, although many investigators (including the author of this chapter) have tried, and it is reasonable to conclude that it is, in fact, not possible. We can only conclude that a number of mutations are required to extend significantly the temperature-growth range of these organisms. But such mutants have been reported among other groups of bacteria.

Aerobic spore formers seem readily to undergo changes in their minimum temperature for growth during cultivation in the laboratory. ALLEN (1953) compared the minimum growth temperatures of 21 strains of bacilli, when they were freshly isolated from nature, with the same strains after only 4 transfers in laboratory media at 55° C. Initially, only 4 strains grew at 30° C and 13 at 35° C, but after the 4 transfers, 8 could grow at 30° C and 13 at 35° C. Also, she was able to isolate thermophilic variants from certain mesophilic organisms. The *Bacillus* species, *subtilis, cereus, megaterium* and *circulans*, consistently yielded thermophilic variants, but she found it impossible to isolate thermophilic variants of *Pseudomonas, Escherichia, Aerobacter, Lactobacillus, Streptococcus, Mycobacterium, Actinomyces,* and *Bacillus mascerans*. The frequency of appearance of thermophilic variants from *Bacilli* is remarkably high; 1 thermophile was isolated/10^6cells of *B. circulans*.

In addition to these results, there is independent support for the thesis that *Bacillus* species readily give rise to variants with altered temperature limits of growth. MEDFERD and CAMPBELL (1952) were able to isolate thermophilic variants of *B. globigii*, and KLUYVER and BAARS (1932) reported the isolation of variants that could grow at both higher and lower temperatures.

From these studies the generalization appears to develop that species of *Bacilli* are unique in their ability to adapt to a new temperature range, but whether this adaptation is mutational or physiological remains quite unclear. The second possibility is supported by two general types of studies: (1) those showing that properties of proteins produced by these organisms are dependent on the temperature at which they were synthesized and (2) those showing that by proper cultural manipulation the majority of cells in the population can be induced to grow in the new temperature range.

CAMPBELL's (1955) data on the heat stability of α-amylasis produced by the facultative thermophiles *Bacillus stearothermophilus* and *B. coagulans* illustrate the first point. He grew both organisms at 35° and 55° C, purified the α-amylases produced by them 450—600 fold, and determined various properties of the four purified enzymes. He found that the only significant difference between the enzymes produced at 35° and those produced at 55° C was the thermal stability at 90° C. Preparations from cultures grown at 55° C decayed only 6—10% after 1 h exposure to the elevated temperature, whereas preparations from cultures grown at 35° C decayed 90—92% during the same treatment. Similar results have been obtained by others (STARK and TATRAULT, 1951; MILITZER et al., 1949, 1950, 1951) including BROWN et al. (1957) who showed that in crude preparations the heat stability of a pyrophosphatase from *Bacillus stearothermophilus* increases progressively as the growth temperature is increased. Half-lives at 80° C of enzyme preparations from cultures grown at 40°, 50°, 60°, and 70° C are 14, 34, 41, and 62 minutes respectively. LANGRIDGE (1963) pointed out that this sort of adaptive synthesis may be more apparent than real; he suggested that the known phenomenon of enzymes existing in heat-stabile and heat-sensitive forms (DILLON and O'COLLA, 1950; HEPPEL and HILMORE, 1953; LAWRENCE and HALVORSON, 1954) could account for the apparent adaptive synthesis, i.e. with an increase in temperature the proportion of heat stable molecules would also increase, leading to an apparent adaptation of the enzyme to higher temperatures.

On the other hand DOWBEN and WEIDENMULLER (1968) described how they were able to adapt *Bacillus subtilis* (which normally cannot grow above 55° C) to be able to grow at 72° C by slow stepwise increments of temperature during exponential growth. They found that heat-adapted organisms are larger and contain more protein than those grown at ordinary temperature, but contain approximately the same amount of RNA and DNA. As proof that the heat-adapted cultures were, in fact, *B. subtilis*, they showed that DNA isolated from the heat-adapted strains was able to transform unadapted strains at normal frequency. Adapted strains were found to deadapt rapidly during growth at 37° C, exponential growth commenced at the lower temperature after a variable period of lag which lasted from 1.5–15 h. After growth began, samples were removed and tested for their ability to grow at 63° C, with the result that after two doublings in cell density at 37° C, the ability to grow at elevated temperatures was largely lost.

FRIEDMAN and WEINSTEIN (1964, 1966) suggested a possible biochemical basis for temperature adaptation by a mechanism that might enable a cell to synthesize two types of proteins. They found a temperature-dependent ambiguity in translation using synthetic polyribonucleotides and a subcellular protein-synthesizing system, from *B. stearothermophilus*. Thus, one might imagine that these cells could synthesize one class of proteins at ordinary temperatures, but at higher temperatures one or more amino acid substitutions might occur which would result in greater heat stability of proteins.

Alternatively, one might imagine that organisms susceptable to temperature adaptation might have duplicate genes for certain essential functions and that one gene is expressed over the lower range of temperatures, and that the other is expressed over the higher temperature range.

It is surprising that we are unable to answer even some of the most basic questions relating to temperature adaptation of bacteria. The questions are clearly defined, the techniques to answer them are now available, and the answers appear to have a high probability of producing results of fundamental significance. As a start, it would be very important to know the primary structure of the proteins produced at the various temperatures.

G. Mutations which Extend the Temperature Range of Growth

There are also reports in which mutations appear to be the most probable cause of extending the temperature range for growth. AZUMA et al. (1962) reported the isolation of a psychrophilic mutant of *Pseudomonas aeruginosa* following mutagenesis by ultraviolet light, and their results have been confirmed by OLSEN and METCALF (1968) who also isolated psychrophilic variants of *P. aeruginosa*. The latter group noted that both the upper and lower limits of growth were changed by the mutation, i.e., the parent strain grew over the range of 44°–11° C, while the mutant grew from 32°–0° C. They also noted that not all strains of *P. aeruginosa* served equally as a source of psychrophilic mutants; with some strains they were unable to recover mutants at any time, whereas with other strains they regularly produced psychrophiles.

P. aeruginosa seems to be a particularly wise choice of organism in which to look for psychrophilic variants because it is almost unique among the various species of *Pseudomonas* in not itself being a psychrophile. Certainly, it is quite closely related to psychrophilic species, and presumably has evolved from them. Logically, one might expect the number of mutations required to regain the ability to grow at low temperatures to be small, and the results of AZUMA et al. (1962) and OLSEN and METCALF (1968) seem to establish this point. In fact, OLSEN and METCALF (1968) found that those strains which are capable of undergoing mutation to psychrophily, do so at a relatively high rate. Following ultraviolet irradiation sufficient to kill more than 99% of the cells, they found approximately 1 psychrophilic mutant per 10^8 bacteria in the original population. The results are certainly compatible with only a small number of mutations being required to allow the cells to grow in the psychrophilic range. To support this contention, OLSEN and METCALF (1968) attempted and were successful in transducing psychrophily from *P. flourescens* into *P. aeruginosa.*

H. Direct Temperature Effects on Nucleic Acids

Temperature also has well-known direct effects on DNA and RNA. As temperature is increased, the hydrogen bonds joining the complementary bases are broken, resulting in an increased absorbancy (or hyperchromicity) in the ultraviolet range at about 260 nm. Hydrogen bonds joining guanine (G) and cytosine (C) are stronger than those joining adenine (A) and thymine (T) (or uracil) so DNA molecules which are rich in G + C separate into complementary single strands (or "melt" is the term frequently used to describe this process) at a higher temperature than DNA molecules which are relatively poor in G + C (MARMUR and DOTY, 1959). If melting of nucleic acids were to occur within the cell, they would certainly be rendered nonfunctional. In this respect, it is interesting to inquire whether there is any relationship between the G + C content of a particular bacterial DNA and the temperature range of growth of the organism which produces it, that might suggest a role in natural selection. From first principles, such a relationship would sound unlikely because measured values of T_m (the temperature at which the absorbance due to hyperchromicity has increased to half its maximum value) tend to be higher than the upper temperature limit for growth of most bacteria. MARMUR (1960) compared the T_m of the DNA from the thermophilic *B. stearothermophilus* with the mesophilic *Escherichia coli* and found, in fact, that the mesophilic DNA melted at a higher temperature (90° C) than the thermophilic DNA (88° C). The melting point of DNA seems to be unrelated to the growth temperature range of the organism which produces it.

In the case of ribosomes, however, melting point (temperature at which they undergo denaturation) does seem to correlate with growth range. MANGIANTINI et al. (1965) and also SAUNDERS and CAMPBELL (1965) showed that ribosomes isolated from *Bacillus stearothermophilus* undergo denaturation at a higher temperature than that necessary to denature the ribosomes of *Escherichia coli*. Later PACE and CAMPBELL (1967) compared the thermal stability of ribosomes from 19 different species of bacteria, and found the melting points to correlate well with the maximum growth temperatures of the organisms, suggesting very strongly that for many bacteria, thermal stability of ribosomes may well be the limiting factor which establishes the maximum temperature of growth. On the other hand, there is no clear relationship between thermal denaturation of ribosomal RNA (gross base composition of ribosomal RNA) and heat-induced disintegration of intact ribosomes as revealed by optical properties; also denaturation of free ribosomal RNA begins at temperatures substantially below those at which hyperchromicity of whole ribosomes occurs. It appears, therefore, that the thermal stability of ribosomes lies in the protein-RNA interaction, and that the stability of ribosomes to disruption by heat might well be primarily a function of the structure and assembly of ribosomal proteins.

Of fundamental importance in an inquiry of the relation between genetic regulation and temperature, is the question of the relationship between temperature and mutation rate. In their now classic study of the natural mutation rate of *Escherichia coli*, RYAN and KERITANI (1951) showed that mutation rate has a very similar temperature characteristic to growth rate; thus the rate of mutation per mutable unit per generation is the same at all temperatures of growth. When one considers temperatures above those at which growth can occur, a clear mutagenic effect is apparent (ZAMENHOF, 1960; and CHAISSON and ZAMENHOF, 1966). The proportion of mutants formed in spores of *Bacillus subtilus* increased as the exposure temperature was increased from 105°–115° C, but decreased with a further rise in the temperature to 155° C.

I. Summary

Our knowledge of the genetic regulation of temperature responses of microorganisms has reached a state of development where the broad outlines of the subject are generally understood, but we remain almost completely ignorant of the biochemical details of the response of any particular microorganism. It seems clear that the upper temperature limit for the growth of microorganisms is set by the thermal stability of some limiting component of the cell, and also that ribosomes are a prime candidate (in many wild-type organisms) to be this limiting component. The continual demand for counterselection of mutations which would otherwise decrease the thermal stability of essential cellular components also seems well established.

The genetic determination of the lower temperature limits of growth is less well established, but from studies on cold-sensitive mutants, genes coding for allosteric proteins or proteins which must be assembled into some essential cellular organelle appear to be the most frequent targets for mutations which increase the minimal temperature of growth, and we can reason, by analogy, that the class of proteins coded for also plays a significant role in determining both temperature limits of growth of microorganisms.

Our knowledge of those factors, both genetic and physiological, which are involved in the adaptation of microorganisms to grow in a new range of temperatures, remains very rudimentary, and it seems very likely that a study of them will contribute significantly to basic questions of gene expression as well as provide information about the mechanisms of this fascinating phenomenon.

References to Section "Microorganisms"

ABD-EL-AL, A., INGRAHAM, J. L.: J. Biol. Chem. **244**, 4033—4038 (1969).
ABD-EL-AL, A., INGRAHAM, J. L.: J. Biol. Chem. **244**, 4039—4045 (1969).
ACRA, M., CALVORI, C., FRONTALI, L., TECCE, G.: Biochim. Biophys. Acta **87**, 440—448 (1964).
AFFINSON, C. B.: Brookhaven Symp. Biol. **15**, 184—198 (1962).
AKAGI, J. M., CAMPBELL, L. L.: J. Bacteriol. **82**, 927—932 (1961).
AKAGI, J. M., CAMPBELL, L. L.: J. Bacteriol. **84**, 1194—1201 (1962).
ALBRO, P. W., DITTMER, J. L.: Biochemistry **8**, 953—959 (1969).
ALDERTON, G., SNELL, N.: Appl. Microbiol. **19**, 565—572 (1970).
ALEXANDROW, V. Y.: Quart. Rev. Biol. **39**, 35—77 (1964).
ALFORD, J. A.: J. Bacteriol. **79**, 591—593 (1960).
ALFORD, J. A., ELLIOT, L. E.: Food Res. **26**, 326—337 (1960).
ALLEN, M. B.: Bacteriol. Rev. **17**, 125—175 (1953).
ALLEN, P. W.: J. Bacteriol. **8**, 555—566 (1923).
AMAHA, M., ORDAL, Z. J.: J. Bacteriol. **74**, 596—604 (1957).
AMAHA, M., SAGAGUCHI, K. T.: J. Bacteriol. **68**, 338—345 (1954).
AMBROSINI, R. A., BRITT, H. W.: Bacteriol. Proc. **63**, 6 (1963).
AMBROZ, A.: Zbl. Bakt. I. Abt. Ref. **48**, 257—270, 289—321 (1910/11).
AMELUNXEN, R. E.: Biochim. Biophys. Acta **122**, 175—181 (1966).
AMELUNXEN, R. E.: Biochim. Biophys. Acta **139**, 24—32 (1967).
AMELUNXEN, R., LINS, M.: Arch. Biochem. Biophys. **125**, 765—769 (1968).
ANDERSON, E. B., MEANWELL, L. J.: J. Dairy Res. **7**, 182—191 (1936).
ANGELOTTI, R., MARYANSKI, J. H., BUTLER, T. F., PEELER, J. T., CAMPBELL, J. E.: Appl. Microbiol. **16**, 735—745 (1968).
ARAKI, T., NEI, T.: Low Temp. Sci. B. **20**, 57—68 (1962).
ARONSON, A. I., HENDERSON, J. R., TINCHER, A.: Biochem. Biophys. Res. Commun. **26**, 454—460 (1967).
ARPAI, J.: Experientia **17**, 170—171 (1961).
ARPAI, J.: Z. Allg. Mikrobiol. **4**, 105—113 (1964).
ARRHENIUS, S.: Z. Physik. Chem. **130**, 516—519 (1927).
ASAHINA, E.: Nature **196**, 445—446 (1962).
AZUMA, Y., NEWTON, S. B., WITTER, L. D.: J. Dairy Sci. **45**, 1529—1530 (1962).
BAIRD-PARKER, A. C., BOOTHROYD, M., JONES, E.: J. Appl. Bacteriol. **33**, 512—522 (1970).
BARBER: J. Inf. Dis. **5**, 379—385 (1908).
BARDOU, P.: Zbl. Bakt. I. Abt. Ref. **39**, 744—745 (1907).
BAUMANN, A. J., SIMMONDS, P. G.: J. Bacteriol. **98**, 528—531 (1969).
BÊLEHRÁDEK, J.: Protoplasma **12**, 406—434 (1931).
BÊLEHRÁDEK, J.: Temperature and living matter. Berlin 1935.
BERTALANFFY, L. v.: Theoretische Biologie II. Bern 1951.
BIER, M., NORD, J.: Arch. Biochem. Biophys. **33**, 320—332 (1951).
BLACK, S. H., HASHIMOTO, T., GERHARDT, P.: Can. J. Microbiol. **6**, 213—224 (1960).
BLOOMFIELD, D. K., BLOCH, K.: J. Biol. Chem. **235**, 337—345 (1960).
BOCKELMANN, G., CHRISTOPHERSEN, J.: 1973 in press.
BOGEN, H. J.: Planta (Berlin) **36**, 298—340 (1948).
BOTT, T. L., BROCK, T. D.: Science **164**, 1411—1412 (1969).
BOWMAN, R. D., MUMMA, R. O.: Biochim. Biophysica Acta **144**, 501—510 (1967).

BRANDTS, J. F.: J. Am. Chem. Soc. **86**, 4291—4314 (1964).
BREWER: Am. J. Sci. **42**, 429 (1866).
BROCK, T. D.: Science **158**, 1012—1019 (1957).
BROWN, D. K., MILITZER, W., GEORGI, C. E.: Arch. Biochem. Biophys. **70**, 248—256 (1957).
BUCHANAN, R. E., FULMER, E. J.: Physiology and Biochemistry of Bacteria. London 1930.
BUSTA, F. F., ORDAL, Z. J.: J. Food Sci. **29**, 345—353 (1964).
CALHOUN, C. L., FRAZIER, W. C.: Appl. Microbiol. **14**, 416—420 (1966).
CAMPBELL, L. L.: Arch. Biochem. Biophys. **54**, 154—161 (1955).
CAMPBELL, L. L., MANNING, G. B.: J. Biol. Chem. **236**, 2762—2765 (1961).
CAMPBELL, L. L., PACE: J. Appl. Bacteriol. **31**, 24—35 (1968).
CHAISSON, L. P., ZAMENHOF, S.: Can. J. Microbiol. **12**, 43—46 (1966).
CHASIN, L. A., SZULMAJSTER, J.: J. Appl. Bacteriol. **29**, 648—654 (1967).
CHICK, H.: Z. Hyg. 8, 92—158 (1908); **10**, 237—286 (1910).
CHRISTENSEN, J. R., SAUL, S. H.: Virology 29, 497—499 (1966).
CHRISTOPHERSEN, J.: Mikroorganismen. In: PRECHT, H., CHRISTOPHERSEN, J., HENSEL, H.: Temperatur und Leben, S. 268—282. Berlin-Heidelberg-New York: Springer 1955.
CHRISTOPHERSEN, J.: Arch. Mikrobiol. **45**, 58—64 (1963).
CHRISTOPHERSEN, J.: Naturwissenschaften 51, 370—371 (1964).
CHRISTOPHERSEN, J., KAUFMANN, W.: Naturwissenschaften **39**, 67—68 (1952).
CHRISTOPHERSEN, J., KAUFMANN, W.: Kieler Milchw. Forsch. Ber. **7**, 323—335 (1955).
CHRISTOPHERSEN, J., PRECHT, H.: Biol. Zbl. **69**, 240—256 (1950a).
CHRISTOPHERSEN, J., PRECHT, H.: Biol. Zbl. **69**, 300—323 (1950b).
CHRISTOPHERSEN, J., PRECHT, H.: Biol. Zbl. **70**, 261—274 (1951).
CHRISTOPHERSEN, J., PRECHT, H.: Arch. Mikrobiol. **18**, 32—48 (1952).
CHRISTOPHERSEN, J., PRECHT, H.: Biol. Zbl. **72**, 104—119 (1953).
CHRISTOPHERSEN, J., PRECHT, H.: Zbl. Bakt. II. Abt. **108**, 1—6 (1954).
CHRISTOPHERSEN, J., THIELE, H.: Kieler Milchw. Forsch. **4**, 683—700 (1952).
CHURCH, B. D., HALVORSON, H. O.: Nature 183, 124—125 (1959).
CIBLIS, E., CHRISTOPHERSEN, J.: Z. Symp. Techn. Mikrobiol pp. 299—309. Berlin 1970.
CLARK, C. W., ORDAL, Z. J.: Appl. Microbiol. **18**, 332—336 (1969).
COLLIER, R. E., MURTY, G. G. K., Bacter. Proc. Soc. Am. Bacteriol. **32**, (1957).
CONDON, S., INGRAHAM, J. L.: J. Bacteriol. **94**, 1970—1981 (1967).
COONEY. D. G., EMERSON, R.: Thermophilic fungi. London: W. H. Freeman and Comp. 1964 p. G.
COPELAND, J. J.: Ann. N. Y. Acad. Sci. **36**, 1—232 (1936).
CROZIER, W. J.: J. Gen. Physiol. **7**, 189—250 (1925).
CURRAN, H. R., BRUNNSTETTER, B. C., MYERS, A. T.: J. Bacteriol. **45**, 484—494 (1943).
CURRAN, H. R., EVANS, F. R.: J. Bacteriol. **34**, 178—189 (1937).
CURRAN, H. R., EVANS, F. R.: J. Bacteriol. **49**, 335—346 (1945).
DABBAH, R., MOATS, W. A., MATTIK, J. F.: J. Dairy Sci. **52**, 608—614 (1969).
DAHLBERG, A. C.: J. Dairy Sci. **29**, 651—655 (1946).
DARON, H. H.: J. Bacteriol. **93**, 703—710 (1967).
DARON, H. H.: J. Bacteriol. **101**, 145—151 (1970).
DAVEY, C. B., MILLER, R. J., NELSON, L. A.: J. Bacteriol. **91**, 1827—1830 (1966).
DAWSON, H., DANIELLI: The permeability of natural menbrandes. Cambridge: Univ. Press 1943.
DESAI, A. J., DHALA, S. A.: J. Bacteriol. **100**, 149—155 (1969).
DESROSIER, N. W., HEILIGMAN, F.: Food Res. **21**, 54—62 (1956).
DEVIK, O., ULRICH, J. A.: Hormel Inst. Univ. Minnesota Ann. Rep. **40** (1948/49).
DILLON, T., O'COLLA, P.: Nature 166, 67 (1950).
DIMMICK, R. C.: J. Bacteriol. **89**, 791—798 (1965.)
DORN, F. L., RAHN, O.: Arch. Mikrobiol. **10**, 6—12 (1939).
DOWBEN, R. M., WEIDENMÜLLER, R.: Biochim. Biophys. Acta **158**, 255—261 (1968).
DOWELL, C. E.: Proc. Nat. Acad. Sci. U. S. **58**, 958—961 (1967).

DOWNEY, R. J., GEORGI, C. E., MILITZER, W. E.: J. Bacteriol. **83**, 1140—1146 (1962).

DROST-HANSEN, W.: Int. Symp. Water Desanilation. Washington D. C. 1965.

EDGAR, R. S., DENHARDT, G. H., EPSTEIN, R. H.: Genetics **59**, 635—648 (1964).

EDGAR, R. S., LIELAUSIS, I.: Genetics **49**, 649—662 (1964).

EKBERG, D. R.: Biol. Bull. **114**, 308—316 (1958).

EL BISI, H. M., LECHOWICH, R. V., AMAHA, M., ORDAL, Z. J.: J. Food Sci. **27**, 219—231 (1962).

EL BISI, H. M., ORDAL, Z. J.: J. Bacteriol. **71**, 1—9 (1956).

EL BISI, H. M., ORDAL, Z. J.: Bact. Proc. Soc. Am. Bacteriol. **1958**, 11.

ELLIKER, P. R., FRAZIER, W. C.: J. Dairy Sci. **12**, 801—813 (1938).

ENDO, S.: J. Ferment. Technol. **40**, 346—353 (1962).

EPSTEIN, R. H., BOLLE, A., STEINBERG, C. M., KELLENBERGER, E., BOY DE LA TOUR, E., CHEVALLEY, R., EDGAR, R. S., SUSMAN, S., DENHARDT, G. H., LIELAUSIS, A.: Cold Spring Harbor Symp. Quant. Biol. **28**, 375—394 (1963).

ESCHE, P. VOR DEM: Arch. Hyg. **137**, 397—414 (1953).

EYRING, H.: J. Chem. Phys. **3**, 107—115 (1935).

EYRING, H., MAGEE, J. L.: J. Cell. Comp. Physiol. **20**, 169—177 (1942).

FARREL, J., ROSE, A.: Ann. Rev. Microbiol. **21**, 102—120 (1967).

FOTER, M. J., RAHN, O.: J. Bacteriol. **32**, 485—499 (1936).

FOX, M. S., HOTCHKISS, R. D.: Nature **179**, 1322—1325 (1957).

FRAENKEL, G., HOPF, H. S.: Biochem. J. **34**, 1085—1092 (1940).

FRECZE, H., BROCK, T. D.: J. Bacteriol. **101**, 541—550 (1970).

FRIEDMANN, S. M.: Bacteriol. Rev. **32**, 27—38 (1968).

FRIEDMAN, S. M., WEINSTEIN, I. B.: Proc. Natl Acad. Sci. U. S. **52**, 988—996 (1964).

FRIEDMANN, S. M., WEINSTEIN, I. B.: Biochim. Biophys. Acta **114**, 593—605 (1966).

FULCO, A. J.: J. Biol. Chem. **244**, 889—895 (1969).

FULCO, A. J.: Biochim. Biophys. Acta **218**, 558—560 (1970).

GALA, E. F.: Biochem. J. **34**, 392—413 (1940).

GARIBALDI, I. A.: Food Technol. **22**, 1031—1033 (1968).

GAUGHRAN, E. R. L.: Bacteriol. Rev. **11**, 189—225 (1947a).

GAUGHRAN, E. R. L.: J. Bacteriol. **53**, 506 (1947b).

GOEPPERT, J. M., ISKANDER, I. K., AMUNDSON, C. H.: Appl. Microbiol. **19**, 429—433 (1970).

GOETZ, A., GOETZ, S. S.: Proc. Am. Phil. Soc. **79**, 361—388 (1938).

GORINI, L.: Biochim. Biophys. Acta **7**, 318 (1951).

GORRILL, R. H., McNEIL, E. M.: J. Gen. Microbiol. **22**, 437—442 (1960).

GOSH, A., CHANCE, B.: Biochem. Biophys. Res. Commun. **16**, 174—181 (1964).

GRAHAM-SMITH, G. S.: J. Hyg. **9**, 239—248 (1920).

GRAINGER: Nature **178**, 930 (1956); In Physiol. Adaptation. C. L. PROSSER, Ed., p. 79. Washington, D. C. 1958.

GRECZ, M., OLEJNIKOW-KURITZA, H.: Bacteriol. Proc. **1966**, 14.

GREGORY, M. E., WOODBINE, M.: J. Exptl. Botany **4**, 314—318 (1953).

GUTHRIE, C., NASHIMOTO, H., NOMURO, M.: Proc. Nat. Acad. Sci. U. S. **63**, 384—392 (1969).

HALVORSON, H. O.: The Physiology of the bacterial Spore. Techn. Univ., Trodheim, Norway (1958).

HANSEN, N. H., RIEMANN, H.: J. Appl. Bacteriol, **27**, 314—333 (1963).

HARDER, W., VELDKAMP, H.: J. Appl. Bacteriol. **31**, 12—23 (1968).

HARDER, W., VELDKAMP, H.: Antonie van Leeuwenhoek **37**, 51—63 (1971).

HARRIS, N. D.: J. Appl. Bacteriol. **26**, 387—397 (1963).

HARRIS, P., JAMES, A. T.: Biochem. J. **112**, 325—333 (1969).

HARRISON, A. P.: J. Bacteriol. **70**, 711—715 (1955).

HARRISON, A. P.: J. Microbiol. Serol. **22**, 407—418 (1956).

HASHIMIOTO, T., BLACK, S. H., GERHARDT, P.: Can. J. Microbiol. **6**, 203—212 (1960).

HASKINS, R. H.: Can. J. Microbiol. **3**, 477—485 (1957).

HAUROWITZ, F.: Chemistry and Biology of Proteins. New York 1950.

HEATHER, C. D., VANDERZANT, W. C.: Food Res. **22**, 164—169 (1957).

HEDEN, G. G., WYCKOFF, R. W. G.: J. Bacteriol. **58**, 153—160 (1949).
HEGARTY, C. P., WEEKS, O. B.: J. Bacteriol. **39**, 475—484 (1940).
HEILBRUNN, L. V.: Am. J. Physiol. **69**, 190—199 (1924).
HEINEN, W.: Antonie Leuwenhock, **36**, 582—584 (1970).
HEINMETS, F. W., TAYLOR, W., LEHMANN, J. J.: J. Bacteriol. **67**, 5 (1954).
HEPPEL, L. A., HILMOE, R. J.: J. Biol. Chem. **202**, 217—226 (1953).
HESS, E.: Contrib. Canad. Biol. Fish. Ser. C8, 491—505 (1934).
HILLS, G. M., SPURR, E. D.: J. Gen. Microbiol. **6**, 64—73 (1952).
HINSHELWOOD, C. N.: The Chemical Kinetics of the Bacterial Cell. Oxford 1947.
HOCHACHKA, P. W.: Arch. Biochem. Biophys. **111**, 96—103 (1965).
HOCHACHKA, P. W.: Molecular Mechanisms of Temperature Adaptation. p. 177—203,
    C. L. Prosser (Ed.) Washington, D. C.: 1969.
HOFFMANN, B.: Arch. Mikrobiol. **58**, 302—304 (1967).
HOFFMANN, B., INGRAHAM, J. L.: Biochim. Biophys. Acta **201**, 300—308 (1970).
HOLLANDER, D. H., NELL, E. E.: Appl. Microbiol. **2**, 164—170 (1954).
HOLTON, R. W., BLECKER, H. H., STEVENS, T. S.: Science **160**, 545—547 (1968).
HOWARD, D. H.: J. Bacteriol. **71**, 625 (1956).
HULTIN, E.: Acta Chem. Scand. **9**, 1700—1710 (1955).
HYATT, M. T., LEVINSON, H. S.: J. Bacteriol. **95**, 2090—2101 (1968).
IANDOLO, J. J., ORDAL, Z. J.: J. Bacteriol. **91**, 134—142 (1966).
IKAWA, M.: Bacteriol. Rev. **31**, 54—64 (1967).
INGRAM, M.: Ann. Inst. Pasteur Lille **16**, 111—118 (1965).
INGRAM, J. L.: In: GUNSALUS, I. C., STANIER, R.. : The Bacteria, Vol. **4**, pp. 265—296.
    New York: Academic Press.
INGRAHAM, J. L., STOKES, J. L.: Bact. Rev. **23**, 97—108 (1959).
ISONO, K.: Biochem. Biophys. Res. Commun. **41**, 852—857 (1970).
JOHNSON, E. J., JOHNSON, M. K.: J. Bacteriol. **78**, 792—795 (1959).
JOHNSON, F. H., LEWIN, L.: J. Cell. Comp. Physiol. **28**, 47—97 (1946).
JOLY, M.: A. Physico-chemical approach to the denaturation of proteins. London-
    New York: Academic Press 1965.
KANEDA, T.: J. Biol. Chem. **238**, 1222—1228 (1963a).
KANEDA, T.: J. Biol. Chem. **238**, 1229—1235 (1963b).
KANEDA, T.: Biochim. Biophysica Acta **125**, 43—45 (1966).
KANEDA, T.: J. Bacteriol. **93**, 844—903 (1967).
KANEDA, T.: J. Bacteriol. **98**, 143—146 (1969).
KAPLAN, J. G., DUPHIL, M., LACROUTE, F.: Arch. Biochem. Biophys. **119**, 541—551
    (1967).
KATES, M., BAXTER, R. M.: Can. J. Biochem. Physiol. **40**, 1213—1227 (1962).
KAUTZMAN, W.: J. Cell Comp. Physiol. **47**, 113—131 (1956).
KAVENAU, J. L.: J. Gen. Physiol. **34**, 193—209 (1950).
KEANE, F. K.: Biodynamica **7**, 157—169 (1953).
KEITH, S. C.: Science **37**, 877—879 (1913).
KEMPNER, E. S.: Science **142**, 1318—1319 (1963).
KEYNAN, A., EVENCHIK, S., HALVORSON, H. O., HASTINGS, J. W.: J. Bacteriol. **88**,
    313—318 (1964).
KIERMEIER, F.: Biochem. Z. **318**, 275—296 (1948).
KOFFLER, H.: Bacteriol. Rev. **21**, 227—240 (1957).
KOFFLER, H., BALLETT, G. E., AYDE, J.: Proc. Natl. Acad. Sci. U. S. **43**, 446—477
    (1957).
KOFFLER, H., GALE, G. O.: Arch. Biochim. Biophys. **67**, 249—251 (1957).
KORNBERG, H. L., SMITH, J.: Biochem. Biophys. Acta **123**, 654—657 (1966).
KURAHASHI, K., WAHBA, A.: Biochim. Biophys. Acta **30**, 298—302 (1958).
LANGRIDGE, J.: Ann. Rev. Plant Physiol. **14**, 441—462 (1963).
LANGRIDGE, J.: Genetics **108**, 116—126 (1968).
LARKIN, J. M., STOKES, J. L.: Can. J. Microbiol. **14**, 97—101 (1968).
LAWRENCE, N. L., HALVORSON, H.: J. Bacteriol. **48**, 334—337 (1954).
LECHOWICH, R. V., ORDAL, Z. J.: Bacter. Proc. Am. Bacteriol. 44 (1960).

LEMBKE, A., KAUFMANN, W., LAGONI, H., GANTZ, H.: Kieler Milchw. Forsch. Ber. **3**, 679—689 (1951).

LEMBKE, A., KAUFMANN, W., LAGONI, H., GANTZ, H.: Kieler Milchw. Forsch. Ber. **4**, 233—241 (1952).

LEVINSON, H. S., HYATT, M. T.: Biochem. Biophys. Res. Commun. 37, 909—916 (1969).

LEVITT, J.: Theoret. Biol. 3, 355—299 (1962).

LEVITT, J.: In: Molecular Mechanisms of Temperature Adaptation AAAS. pp. 41—51 Washington D. C. 1967.

LEVITT, J., DEAR, J.: In: The Frozen Cell. Ciba Foundation Symp. London, pp. 149—173, 1970.

LINDENBERG, G., LODGE, A.: Can. J. Microbiol. 9, 523—530 (1963).

LOVELOCK, J. E.: Biochim. Biophys. Acta 11, 28—36 (1953).

LOVELOCK, J. E.: Nature 173, 659—661 (1954).

LOVELOCK, J. E.: Proc. Roy. Soc. B. 147, 427—433 (1957).

LUND, A. J., LUNDBERG, W. O.: Hormel. Inst. Univ. Minnesota Ann. Rep. 50, 47—51 (1949).

MAAS, W. K., DAVIS, B. D.: Proc. Natl. Acad. Sci. U. S. 38, 785—797 (1952).

MADSEN, M., NYMAN, E.: Z. Hyg. 57, 380—398 (1907).

MANGIANTINI, M. T., TECCE, G., TOSCHI, G., TRENTALANCE, A.: Biochim. Biophys. Acta 103, 252—274 (1965).

MANNING, G. B., CAMPBELL, L. L.: Can. J. Microbiol. 3, 1001—1009 (1957).

MARGALEFF, R.: Pupl. Inst. Biol. Aplicada 12, 5—78 (1953).

MARMUR, J.: Biochim. Biophys. Acta. 38, 342—343 (1960).

MARMUR, J., DOTY, P.: Nature 183, 1427—1429 (1959).

MARR, A. G., INGRAHAM, J. L.: J. Bacteriol. 84, 1260—1267 (1962).

MARSH, C., MILITZER, W.: Arch. Biochem. Biophys. 36, 269—275 (1952).

MARSH, C., MILITZER, W.: Arch. Biochem. Biophys. 60, 439—451 (1956).

MARSH, C., MILITZER, W.: Arch. Biochem. Biophys. 60, 433—438 (1956a).

MARSH, C., MILITZER, W.: Arch. Biochem. Biophys. 60, 9—451 (1956b).

MARTIN, R. G.: J. Biol. Chem. 238, 257—268 (1963).

MATHEMEIER, P. F., MORITA, R. Y.: J. Bacteriol. 88, 1661—1666 (1956).

MAZUR, P.: J. Gen. Physiol. 39, 869—888) (1956).

MAZUR, P.: Biophys. J. 1, 247—264 (1961).

MAZUR, P.: In: Cryobiology; pp. 213—315. London-New York: 1966.

MAZUR, P., RHIAN, M. A., MAHLANDT, B. G.: Arch. Biochem. Biophys. 71, 31—51 (1957).

MEDFERD, R. B., CAMPBELL, L. L.: Texas Repts. Biol. Med. 10, 419—420 (1952).

MERYMAN, H. T. (Ed.): Am. N. Y. Acad. Sci. 85, 501—734 (1960).

MERYMAN, H. T.: In: The cell and enviromental temperature, pp. 113—117. New York: 1967.

MEYNELL, G. G.: J. Gen. Microbiol. 19, 380—389 (1958).

MILITZER, W., BURNS, L.: Arch. Biochem. Biophys. 52, 66—73 (1954).

MILITZER, W., SONDEREGGER, T. B., TUTTLE, L. C., GEORGI, C. E.: Arch. Biochem. Biophys. 24, 75—82 (1949).

MILITZER, W., SONDEREGGER, T. B., TUTTLE, L. C., GEORGI, C. E.: Arch. Biochem. Biophys. 26, 299—306 (1950).

MILITZER, W., TUTTLE, L. C.: Arch. Biochem. Biophys. 39, 379—451 (1952).

MILITZER, W., TUTTLE, L. C., GEORGI, C. E., Arch. Biochem. Biophys. 31, 416—423 (1951).

MILLER, R., GOODNER, K.: Yale J. Biol. Med. 25, 262—283 (1952/53).

MIQUEL, P.: Ann. Micrograph. 1, 4—10 (1888).

MITCHELL, P.: In Bacterial Physiology. WERKMAN, C. H., and WILSON, P. W., (eds). New York: Academic Press 1951.

MORITA, R. Y.: Oceanogr. Mar. Biol. Ann. Rev. 4, 105—121 (1966).

MOSS, C. W., CHERRY, W. B.: J. Bacteriol. 95, 241—242 (1968).

MOSS, C. W., SPECK, M. L.: Appl. Microbiol. 11, 326—329 (1963).

MURELL, W. G.: Symp. Soc. Gen. Microbiol. 11, 100—150 (1961).

MURELL, W. G., SCOTT, W. J.: J. Gen. Microbiol. **43**, 411—425 (1966).
NAGA, J., BLUCH, K.: J. Biol. Chem. **241**, 1925—**1927** (1966).
NAGUIB, M., CHRISTOPHERSEN: J.: Naturwissenschaften **52**, 193—194 (1965).
NAKAMURA, T.: Bakt. II. Abt. Orig. **4**, 777—778 (1898).
NAKAMURA, Y.: J. Biochem. **48**, 295—307 (1960).
NANNINGA, C. B.: Arch. Biochem. Biophys. **70**, 346—366 (1947).
NASHIF, S. A., NELSON, F. E.: J. Dairy Sci. **36**, 471—480 (1953).
NASHIMOTO, H., NOMURA, M.: Proc. Natl. Acad. Sci. U. S. **67**, 1440—1447 (1970).
NASHIMOTO, W., KALTSCHMIDT, E., NOMURA, M.: J. Molec. Biol. **62**, 121—138 (1971).
NEALE, E. K., CHAPMAN, G. B.: J. Bacteriol. **104**, 518—528 (1970).
NELSON, F. E.: J. Bacteriol. **45**, 395—403 (1943).
NELSON, F. E.: J. Bacteriol. **48**, 473—477 (1944).
NEMETHY, G., SCHERAGA, H. A.: J. Chem. Phys. **36**, 3382—3400; 3401—3412 (1962).
NG, H., BAYNE, H. G., GARIBALDI, J. A.: Appl. Microbiol. **17**, 78—82 (1969).
NG, H., INGRAHAM, J. L., MARR, A. G.: J. Bacteriol. **84**, 331—339 (1962).
NICHOLS, B. W., HARRIS, R. V., JAMES, A. T.: Biochem. Biophys. Res. Commun. **20**, 256—262 (1965).
O'BRIEN, R. T., CAMPBELL, L. L.: Arch. Biochem. Biophys. **70**, 432—441 (1957).
O'DONOVAN, G. A., INGRAHAM, J. L.: Proc. Natl. Acad. Sci. U. S. **54**, 451—457 (1965).
O'DONOVAN, G. A., KEARNEY, C. L., INGRAHAM, J. L.: J. Bacteriol. **90**, 611—616 (1965).
O'DONOVAN, G. A., NEUHARD, J.: Bacteriol. Rev. **34**, 278—343 (1970).
OGASAHARA, K., IMANISHI, A., ISEMURA, T.: J. Biochem. **67**, 65—75 (1970a).
OGASAHARA, K., IMANISHI, A., ISEMURA, T.: J. Biochem. **67**, 77—82 (1970b).
OGASAHARA, K., YUTANI, K., IMANISHI, A., ISEMURA, T.: J. Biochem. **67**, 83—89 (1970c)
O'LEARY, W. M.: Bacteriol. Rev. **26**, 256—262 (1962).
OLSEN, R. H., METCALF, E. S.: Science **162**, 288—289 (1968).
OPPENHEIMER, C. W., DROST-HANSEN, W.: J. Bacteriol. **80**, 21—24 (1960).
PACE, B., CAMPBELL, L. L.: Proc. Natl. Acad. Sci. U. S. **57**, 1110—1116 (1967).
PACKER, E. L., INGRAHAM, J. L., SCHER, S.: J. Bacteriol. **89**, 718—724 (1965).
PALUMBO, S. A., WITTER, L. D.: Canad. J. Microbiol. **15**, 995—1000 (1969).
PARKER, P. L., VAN BAALEN, C., MAURER, K., Science **155**, 707—708 (1967).
PEARSON, L. K., RAPER, H. S.: Biochem. J. **21**, 875—891 (1927).
PERRY, J. J., FOSTER, J. W.: J. Bacteriol. **69**, 337—346 (1955).
PHEIL, C. G., ORDAL, Z. J.: J. Bacteriol. **93**, 1727—1728 (1967).
POLGE, C., SMITH, A. U., PARKES, A. S.: Nature **164**, 666 (1949).
POSTGATE, J. R., HUNTER, J. R.: J. Gen. Microbiol. **24**, 367—378 (1961).
POSTGATE, J. R., HUNTER, J. R.: J. Appl. Bacteriol. **26**, 295—306 (1963).
POWELL, E. O.: J. Appl. Bacteriol. **20**, 342—348 (1957).
POWELL, J. F.: Biochem. J. **54**, 210—211 (1953).
POWELL, J. F.: J. Appl. Bacter. **20**, 349—358 (1957).
POWELL, J. F., HUNTER, J. R.: J. Gen. Microbiol. **13**, 59—67 (1955).
PRECHT, H.: Z. Naturforsch. **46**, 26—35 (1949).
PRILL, E. A., WENK, R. R., PETERSON, W. H.: Biochem. J. **29**, 21—33 (1935).
PROSSER, C. L.: The nature of physiological adaptation. In: C. C. Prosser (Ed.): Physiological adaptation, p. 167. Wash. D. C.: Am. Physiol. Soc., 1958.
RAHN, O.: J. Gen. Physiol. **13**, 395—407 (1930).
RAHN, O.: Biodynamica **4**, 81—130 (1943).
RAHN, O.: Bacteriol. Rev. **9**, 1—47 (1945).
RAHN, O., BARNES, M. N.: J. Gen. Physiol. **16**, 579—592 (1933).
RAO, K. P.: Intern. Symp. Cytoecology, Leningrad 1963.
RIEMANN, H.: Appl. Microbiol. **16**, 1621—1622 (1968).
RING, K.: Biochem. Biophys. Res. Commun. **95**, 598—600 (1965).
RING, K., CHRISTOPHERSEN, J.: Arch. Mikrobiol. **48**, 50—65 (1964).
ROBERTSON, A. H.: N. Y. State Agr. Bull. **130** (1927).
RODE, L. J., FOSTER, J. W.: J. Bacteriol. **79**, 650—656 (1960).
RONCARI, G., ZUBER, H.: Int. J. Protein Res. **1**, 45—61 (1969).
ROTMAN, Y., FIELDS, L.: J. Food Sci. **34**, 345—346 (1969).

RYAN, F. J., KIRITANI, K.: J. Gen. Microbiol. **20**, 644—653 (1959).
SADLER, J. R., NOVICK, A.: J. Molec. Biol. **12**, 305—327 (1965).
SAIKI, T., ARIMA, K.: Agr. Biol. Chem. **34**, 1762—1764 (1970).
SALMONOVICZ, J., NIEWIADOMSKI, H.: Rev. Franc. Corps Gras **12**, 309—316 (1965).
SATO, M., TAKAHASHI, H.: Agr. Biol. Chem. **31**, 1100—1101 (1967).
SATO, M., TAKAHASHI, H.: Agr. Biol. Chem. **32**, 259—260 (1968).
SATO, M., TAKAHASHI, H.: J. Gen. Microbiol. **16**, 279—290 (1970).
SAUNDERS, G. F., CAMPBELL, L. L.: J. Bacteriol. **91**, 332—339 (1966).
SCHITO, G. C., PESCE, A.: Giorn. Microbiol. **13**, 145—156 (1965).
SCHMALHAUSEN, J., BORDZILOWSKAJA, N.: Roux'Arch. **121**, 726—754 (1948).
SCHMETZLER, B.: Arch. Sci. Phys. Nat. **21**, 240—246 (1889).
SCHULTZ, J. H., RITZ, H.: Zbl. Bakt. I. Abt. **54**, 283 (1910).
SCOTTI, P. D.: Mutation Res. **6**, 1—14 (1968).
SHAW, M. K., MARR, A. G., INGRAHAM, J. L.: J. Bacteriol. **105**, 683—684 (1971).
SHAW, R.: Comp. Biochem. Physiol. **18**, 325—331 (1966).
SHERMAN, J. M., ALBUS, W. R.: J. Bacteriol. **8**, 127—139 (1923).
SHERMAN, J. M., CAMERON, G. M.: J. Bacteriol. **27**, 341—348 (1934).
SHULL, J. J., CARGO, G. T., ERNST, R. R.: Appl. Microbiol. **11**, 485—487 (1963).
SHULL, J. J., ERNST, R. R.: Appl. Microbiol. **10**, 452—457 (1962).
SIEBURTH, J. M.: J. Exp. Mar. Biol. Ecol. **1**, 98—121 (1967).
SINCLAIR, N. A., STOKES, J. L.: J. Bacteriol. **85**, 164—167 (1963).
SINGH, J., WALKER, T. K.: Res. Bull. Panjab. Univ. Sci. **92**, 135—138 (1956).
SINGLETON, R., KIMMEL, I. R., AMELUNXEN, R. E.: J. Biol. Chem. **244**, 1623—1630 (1969).
SLEPECKY, R., FOSTER, J. W.: J. Bacteriol. **78**, 117—123 (1959).
SMEATON, J. R., ELLIOTT, W. H.: Biochem. Biophys. Res. Commun. **26**, 75—84 (1967).
SMITH, A. U.: Biological Effect of Freezing and Supercooling. Baltimore 1961
SPICER, S.: J. Bacteriol. **39**, 517—526 (1940).
SQUIRES, C. K., INGRAHAM, J. L.: J. Bacteriol. **97**, 488—494 (1969).
SQUIRES, R. W., HARTSELL, S. E.: Appl. Microbiol. **3**, 40—45 (1955).
STANLEY, S. O., ROSE, A. H.: Phil. Trans. Roy. Soc. London, Ser. B. **252**, 199 (1967).
STARK, E., TETRAULT, P. A.: J. Bacteriol. **62**, 247—251 (1951).
STEDMAN, R. L.: Am. J. Pharm. **128**, 84—98 (1956).
STENESH, J., HOLAZO, A. A.: Biochim. Biophys. Acta **138**, 286—295 (1967).
STERN, J. A., PROCTOR, B. E.: Food Technol. **8**, 139—143 (1954).
STILLE, B.: Arch. Mikrobiol. **14**, 554—587 (1950).
STOKES, J. L.: General biology and nomenclature of psychrophilic microorganisms. 187—192, In: GIBBONS, N. E. (Ed.) Recent Progress in Microbiology, VIII. Toronto: University of Toronto Press.
STRAKA, R. P., STOKES, I. L.: J. Bacter. **78**, 181—185 (1959).
SUGIYAMA, H.: J. Bacteriol. **62**, 81—96 (1951).
SUMNER, J. E., MORGAN, E. D., EVANS, H. C.: Canad. J. Microbiol. **15**, 515—520 (1968).
TAI, P.-C., INGRAHAM, J. L.: Biochim. Biophys. Acta **232**, 151—166 (1971).
TAI, P.-C., KESSLER, D. P., INGRAHAM, J. L.: J. Bacteriol. **97**, 466—468 (1968).
TAKETA, K., POGELL, M.: J. Biol. Chem. **240**, 651—662 (1965).
TERROINE, E. F., HATTERER, C., ROEHRIG, P.: Bull. Soc. Chim. Biol. **12**, 682—702 (1930).
THOMPSON, P. J., THOMPSON, T. L.: J. Bacteriol. **84**, 694—900 (1962).
THOMPSON, T. L., MILITZER, W., GEORGI, C. E.: J. Bacteriol. **76**, 337—341 (1958).
TISCHER, R. G., HURWICZ, H.: Food. Res. **19**, 80—91 (1954).
TORNABENE, T. G., GELP, E., ORO, J.: J. Bacteriol. **94**, 333—343 (1967 a).
TORNABENE, T. G., GELP, E., ORO, J.: J. Bacteriol. **94**, 344—348 (1967 b).
TORNABENE, T. G., ORO, J.: J. Bacteriol. **94**, 349—358 (1967).
TOWNSEND, G. T., ESTY, J. R., BASELT, F. C.: Food Res. **3**, 323—335 (1938).
TRAUB, P., NOMURA, M.: J. Mol. Biol. **40**, 391—413 (1969).
ULRICH, J. A., HALVORSON, H. O.: Hormel. Inst. Univ. Minnesota Ann. Rep. 44—46 (1946/47).
UPADAHYAY, J., STOKES, J. L.: J. Bacteriol. **83**, 270—275 (1962).

Van Uden, N., Madeira-Lopes, A.: Z. Allg. Mikrobiol. **10**, 515—526 (1970).

Vas, K., Proszt, C.: J. Appl. Bacteriol. **21**, 431—441 (1957).

Weed, W. H.: Am. Naturalist **23**, 344—357 (1889).

Wells, F. E., Hartsell, S. E., Stadelman, W. J.: J. Food Sci. **28**, 140—144 (1963).

White, H. R.: Proc. Soc. Appl. Bacteriol. **15**, 8—14 (1952).

White, H. R.: J. Gen. Microbiol. **8**, 27—37 (1953).

White, H. R.: J. Appl. Bacteriol. **26**, 91—99 (1963).

Williamson, W. T., Tove, S. B., Speck, M. L.: J. Bacteriol. **87**, 49—53 (1964).

Young, I. E.: Chemical and morphological changes during sporulation in variants of *Bacillus cereus*. Ph. D. Thesis, London (Canada): Univ. West. Ont., 1958

Zamenhof, S.: Proc. Natl. Acad. Sci. U. S. **46**, 101—105 (1960).

Zerbst, E.: Pflügers Arch. Ges. Physiol. **277**, 434—457 (1963).

Plants

I. Plant Temperatures and Energy Budget

DAVID M. GATES

A. Introduction

The temperature of an organism is an extremely important factor in its life history, growth, physiological response, and viability. The average temperature of the earth is about 15° C and although organisms have evolved and adapted to this temperature, they possess a considerable tolerance for variations from the mean (cf. p. 222). Bacteria have been reported living at the boiling point of water in alkaline hot springs and BOTT and BROCK (1969) have measured rapid growth rates of bacteria in hot springs at temperatures exceeding 90° C.

Vast numbers of observations of leaf temperature have been made and reported in the scientific literature (see reviews by HUBER, 1956 and BIEBL, 1962). Aquatic

Table 1. Highest temperatures in plants at the natural habitat (from BIEBL, 1962)

Plants	Maximal temperature (summer, noon) ° C	Air temperature ° C	Difference plant temp. − air temp. ° C	Author
Island in the Baltic Sea (Hiddensee):				
Tussilago farfara	34.5	26.0	8.5	G. FRITZSCHE
Eryngium maritimum	33.9	27.0	6.9	
Salsola kali	31.2	26.9	4.3	
Xerothermic habitat (near Vienna):				
Globularia cordifolia	48.7	33.0	15.7	M. DÖRR
Sempervivum hirtum	50.2	34.8	15.4	
Helianthemum canum	46.2	33.0	13.2	
Teucrium chamaedrys	49.9	37.3	12.6	
Hieracium pilosella	49.7	39.8	9.9	
Onosma visianii	42.7	34.2	8.5	
Bupleurum falcatum	40.1	33.3	6.8	
Xerothermic habitat (Southern Tyrol):				
Sempervivum schottii	52.0	30.5	21.5	B. HUBER
Sempervivum arachnoides	56.2	34.7	21.5	
Hieracium pilosella	47.9	34.7	13.2	
Thymus spec.	42.9	34.7	8.2	

Table 1. continuation

Plants	Maximal temperature (summer, noon) °C	Air temperature °C	Difference plant temp. — air temp. °C	Author
Mediterranean Macchia (Rovigno):				
Quercus ilex	41.9	26.2	15.7	E. Rouschal
Viburnum tinus	43.5	28.2	15.3	
Laurus nobilis	37.6	27.8	9.8	
Phillyrea media	39.8	30.1	9.7	
Mediterranean Maccia (Jerusalem):				
Rhamnus alaternus	52.5			E. Konis
Arbutus andrachne	49.6			
Laurus nobilis	49.5			
Ceratonia siliqua	47.2			
Quercus calliprinos	47.0			
Phillyrea media	46.5			
Algerian desert, Oued:				
Zollikoferia arborescens	31.5	34.0	7.5 windy	R. Harder
Zollikoferia arborescens	40.6	36.5	4.1 still air	
Zilla macroptera, shoot	40.1	37.0	3.1	
SW-Sahara desert (Mauretania):				
Salvadora persica	50.7	41.8	8.9	O. L. Lange
Zygophyllum cf. fontanesii	40.8	33.3	7.5	
Nitraria schoberi	42.4	37.7	4.7	
Phoenix dactylifera	53.3	49.3	4.0	
Boscia senegalensis	51.4	47.6	3.8	
Aristida pungens	45.8	43.1	2.7	
Capparis decidua	45.4	44.6	0.8	
Ziziphus mauretania	33.5	33.3	0.2	
Leptadenia pyrotechnica	46.1	47.6	−1.5	
Chrozophera senegalensis	43.0	44.8	−1.8	
Calotropis procera	45.7	47.6	−1.9	
Cucumis melo var. agrestis	40.1	42.6	−2.5	
Abutilon muticum	43.9	46.6	−2.7	
Pergularia tomentosa	40.9	44.4	−3.5	
Citrullus colocynthis	42.5	51.4	−8.9	
Desert of Arizona:				
Parkinsonia microphylla	46.9	37.5	9.4	F. Shreeve
Opuntia blakeana	55.0			J. M. McGee
Opuntia discata	65.0			C. Schratz

plants are generally at the same temperature as the water in which they are immersed, although the upper surfaces of water-lily pads may have a different temperature from that of the water. Terrestrial plants, especially their massive organs, often have temperatures which differ considerably from the temperature of the surrounding air, sometimes being higher and sometimes lower (Table 1). The roots of all plants have essentially the same temperature as the soil.

Temperature is a thermometric property which describes the energy content of matter. Energy is essential to all life. Energy is the ability to do work and all processes of life require that work to be done. According to the first law of thermodynamics energy can neither be created nor destroyed but only transformed. Energy is transferred by radiation, convection, conduction, evaporation or condensation, and by electrical, chemical, and mechanical means. Generally the temperature of a plant is affected by the first four methods of energy transfer and little or not at all by the last three. A few instances are known where plant temperatures are affected by the energy released during the chemical events of respiration, e.g. with the Arums, see PRIME (1960), but usually the temperature effects of respiration within a leaf are negligible. It is in the context of energy exchange that the two variables, leaf temperature and transpiration rate, are expressed as functions of the variables, radiation, air temperature, wind speed, and humidity, which describe the environment nearby the plant.

The two most important processes which result from the interaction of a plant leaf with its environment are photosynthesis and transpiration. Two requirements must be fulfilled for a leaf to transpire water vapor, i.e. energy and a vapor-pressure difference between the water vapor in the leaf and in the air. Photosynthesis within a leaf is dependent upon many factors, among which are leaf temperature, intensity of illumination, and carbon dioxide concentration. A characteristic relationship between net photosynthesis and leaf temperature is shown in Fig. 1 for various values of total resistance to the diffusion of carbon dioxide into the leaf. If the diffusion resistance is high there is relatively little temperature effect on net photosynthesis since the process is diffusion limited. On the other hand, if the diffusion resistance is low then net photosynthesis is strongly temperature dependent since the process is not diffusion limited, but is highly dependent upon the chemical kinetics of the assimilation and respiration events taking place in the protoplasm.

Recently, LOMMEN et al. (1971) formulated a model of leaf photosynthesis and transpiration based on the physical laws of energy exchange, gas diffusion, and chemical kinetics. The result of their calculations for transpiration rate and net photosynthesis as a function of the air temperature and the carbon dioxide concentration is shown in Fig. 2. A hypothetical leaf of $5 \cdot 5$ cm dimensions is considered with minimum diffusion resistance of 2.5 sec cm^{-1}, for full sunlight, still air conditions ($v = 10$ cm sec^{-1}), a relative humidity of 50%, and a normal concentration of oxygen in the air. At a given air temperature an increase of photosynthetic efficiency (smaller values of N) occurs with increasing carbon dioxide concentration. N is the ratio of number of water molecules transpired to the number of carbon dioxide molecules assimilated. Normal carbon dioxide concentration is 13 nanomoles per cm^3. At this concentration minimum N occurs at

Plant Temperatures and Energy Budget

air temperatures lower than 20° C, while the maximum photosynthetic rate occurs at air temperatures near 27° C.

The energy exchange between a plant leaf and its environment involving radiation, convection, and transpiration provides a basis for estimating leaf temperatures

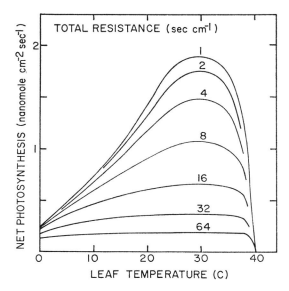

Fig. 1. Net photosynthesis as a function of the leaf temperature and of the total diffusion resistance to carbon dioxide

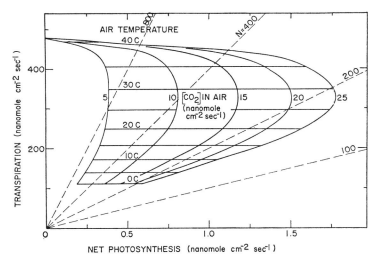

Fig. 2. Calculated transpiration rate and net photosynthesis for a leaf in full sunlight in still air of relative humidity 50% as a function of the carbon dioxide concentration and the temperature of the air. From the model of LOMMEN et al. (1971). The leaf size is $5 \cdot 5$ cm and its internal diffusion resistance to water vapor is 2.5 sec cm^{-1}

for any given set of environmental conditions. The major input of energy to a leaf is in the form of radiation. A leaf absorbs a certain fraction of incident radiation, the total quantity of radiation absorbed being designated as Q_{abs}. The leaf reradiates some of this energy, R, according to the fourth power of the absolute surface temperature of the leaf. Energy is exchanged between the leaf and the air

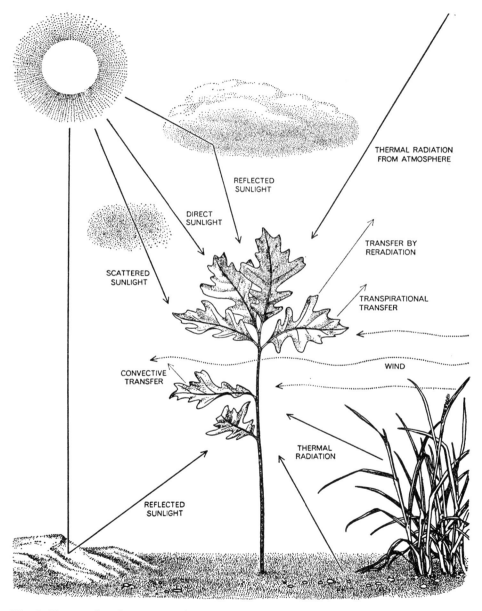

Fig. 3. Energy flow between a plant and the environment

by convection in a manner proportional to the temperature differences between leaf and air. If the leaf is warmer than the air it will be cooled by convection as heat is lost to the latter, but if the air is warmer than the leaf convective heat exchange will deliver energy to the leaf and warm it. The convective heat exchange is designated C. If transpiration occurs then energy is consumed by this process and the leaf is cooled. Approximately 580 calories per gram are required to evaporate water at a temperature of 30° C. The amount of water transpired per unit surface area of leaf per unit time is designated by E. When multiplied by the latent heat of vaporization L, which is temperature dependent, the product LE is the energy lost by the leaf to the transpirational stream. The amount of energy consumed by assimilation or contributed by respiration is very small and can be neglected in the consideration of the total energy budget of a plant leaf. The energy budget for a leaf is written as follows:

$$Q_{abs} = R + C + LE. \tag{1}$$

Each term is expressed in erg cm^{-2} sec^{-1} or in cal cm^{-2} min^{-1}, as is the convention. The conversion factor is 1 erg cm^{-2} sec^{-1} = $1.433 \cdot 10^{-6}$ cal cm^{-2} min^{-1} or 1 cal cm^{-2} min^{-1} = $6.98 \cdot 10^5$ erg cm^{-2} sec^{-1}. For all calculations the total surface area of a leaf is used; for a broad leaf this is $2A$, which is essentially twice the surface area A of one side of the leaf. Even if transpiration occurs from only one surface of a leaf, as in a hypostomatus leaf, transpirational cooling is computed, for energy budget purposes, as the average energy lost per total surface area (upper and lower surfaces) per unit time. The various streams of energy exchange between a plant leaf and its environment are illustrated in Fig. 3.

B. Radiation

Often the most dominant factor affecting leaf temperature is the radiation flux incident upon the leaf. Streams of radiation are incident upon aerial parts of plants throughout the day and night. Radiation is by far the most complex of the environmental factors and by the same token the most difficult to evaluate. Direct sunlight has a wavelength distribution from the ultraviolet to the infrared, with approximately half of the energy distributed on either side of 700 nm. The total intensity S of direct sunlight varies with the elevation of the sun in the sky and with the transparency of the atmosphere. Some of the direct solar beam is scattered by the molecules, dust, and aerosols of the atmosphere to reach the ground surface as skylight. Because of the size of molecules in the air, skylight is relatively richer in ultraviolet and blue wavelengths of light than in red and infrared. The intensity of skylight is designated s and reaches the ground surface from a hemispherical source rather than from an almost point source as does the stream of direct sunlight. A leaf receives reflected sunlight and skylight from the ground and other nearby surfaces. The reflected light is in part spectral and in part diffuse, but is treated as if diffusely reflected from a hemispherical source. The intensity of reflected sunlight and skylight is $r(S + s)$ where r is the mean reflectance of the ground and other surfaces.

Water vapor and carbon dioxide gases in the atmosphere emit radiation in specific broad bands at infrared wavelengths. This thermal emission comes from the total hemispherical source of atmosphere and is designated R_a. All surfaces emit thermal radiation with an intensity proportional to the fourth power of the absolute surface temperature. The ground surface emits upward with an intensity R_g which is comprised entirely of infrared wavelengths greater than 4.0 micron. The ground surface is considered as a hemispherical source.

A leaf surface absorbs some fraction, a, of the incident flux of radiation and presents to each source of radiation a part of its total surface area. As an example, if the upper surface area A of a leaf is normal to the direction of the solar rays then the total amount of direct sunlight absorbed by a leaf is aAS. If the solar rays strike the leaf surface at an angle of incidence i, then the amount of direct sunlight absorbed is $aAS\cos i$. The absorptivity of the leaf to skylight differs from that to direct sunlight, or to reflected light or to the streams of thermal radiation. Consider that the upper and lower surfaces of a leaf both have a surface area A, and the total leaf area is $2A$, then the total quantity of radiation absorbed is given by:

$$2A\,Q_{abs} = a_1 AS\cos i + a_2 As + a_3 A\bar{r}\,(S+s) + a_4 AR_a + a_5 AR_g. \tag{2}$$

The absorbances $a_1, a_2, \ldots, a_5$ are determined as follows:

$$a_1 = \frac{\Sigma\, a_\lambda\, s_\lambda}{S}; \qquad a_2 = \frac{\Sigma\, a_\lambda\, s_\lambda}{s}; \qquad a_3 = \frac{\Sigma\, a_\lambda\, r_\lambda\,(S_\lambda + s_\lambda)}{r\,(S+s)}$$

$$a_4 = \frac{\Sigma\, a_\lambda\, R_{a\lambda}}{R_a}; \qquad a_5 = \frac{\Sigma\, a_\lambda\, R_{g\lambda}}{R_g} \tag{3}$$

where S_λ, s_λ, etc. are the monochromatic intensities of each radiation stream. Actually $a_4 = a_5$ and $a_1 \cong a_3$. The actual surface area A of one side of the leaf cancels out of Eq. 2 and with the other simplifications there results:

$$2Q_{abs} = a_1 S \cos i + a_2 s + a_1 \bar{r}\,(S+s) + a_4\,(R_a + R_g). \tag{4}$$

Sometimes it is possible to simplify this calculation even further, particulary if one considers $a_1 \cong a_2$. Then

$$2Q_{abs} = a_1\,[S \cos i + s + r\,(S+s)] + a_4\,(R_a + R_g). \tag{5}$$

The calculation of values of Q_{abs} for any known set of radiation fluxes is straightforward. The simplest situation prevails for a plant leaf in a dark room or under a dense canopy at night in which only black-body radiation exists. The amount of radiation received by the leaf is exactly given by the black-body radiation according to the absolute surface temperature of the walls of the room or of the ground and canopy. The air within such a cavity is essentially at the same temperature as that at the surfaces. Black-body radiation can be looked up in tables, such as LIST (1963), and an exact graphical representation can be made as shown in Fig. 4. The absorbance of leaf surfaces to all thermal radiation is about 0.96 according to GATES and TANTRAPORN (1952). A leaf in the open air at night receives blackbody radiation R_g from the ground and gray-body radiation R_a from the night sky. The amount of flux from a clear sky is calculated according to SWINBANK's (1963) formula as follows:

$$R_a = 5.31 \cdot 10^{-14}\, T_a{}^6\,(\text{mW cm}^{-2})$$
$$= 7.60 \cdot 10^{-16}\, T_a{}^6\,(\text{cal cm}^{-2}\,\text{min}^{-1}) \tag{6}$$

where T_a is in °K.

If the sky emitted no radiation to the leaf, then $Q_{abs} = R_g/2$ and the ground surface temperature would be considered as sufficiently close to the air temperature at screen height to compute R_g as the equivalent black-body radiation for this temperature. Since, in fact, the sky emits some flux of thermal radiation then $Q_{abs} > R_g/2$ (left-hand line in Fig. 4). If the night sky is cloudy rather than clear, the amount of radiation received and absorbed by a leaf is somewhere between the left-hand line and the black-body line, but generally closer to the black-body line. The cloud base will emit radiation according to its absolute temperature, which is colder than the temperature of the ground surface.

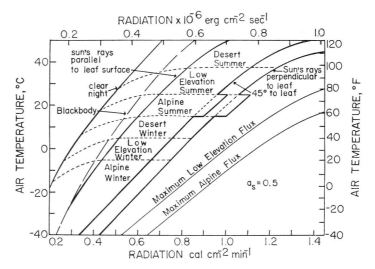

Fig. 4. Relationship between the amount of radiation absorbed by a plant leaf and the air temperature. The leaf has an absorptivity to sunlight, skylight, and reflected light of 0.5 and to infrared thermal radiation of 1.0. The approximate conditions for typical habitats, summer and winter, are indicated

During the daytime a leaf receives not only the same streams of radiation as it does at night but, in addition, direct sunlight, scattered skylight, and reflected sunlight. Assuming a leaf absorptivity to sunlight, skylight, and reflected light of 0.5, we calculate Q_{abs} for a leaf at various angles to the direct rays of the sun. These values are given as the right-hand lines in Fig. 4 which are labelled according to the orientation of the leaf. Figure 4 is now complete and is a reasonable guide to the total amount of radiation absorbed by a leaf in various habitats, expressed as a function of the air temperature. The black-body line is very precise, the night-sky left-hand limit is rather exact, and the daytime radiative fluxes in sunshine are less accurate. Nevertheless the diagram does offer a fairly good guide to the radiative fluxes of the world in which we live.

C. Convective Heat Exchange

A detailed description of convective heat exchange between a leaf and air is given by PARKHURST et al. (1968) and by GATES (1962). For given conditions of wind speed, air pressure, density, etc., convective heat transfer is proportional to the leaf area and to the temperature difference between the leaf and the air. The proportionality factor is known as the convection ˙coefficient, denoted by h_c. The rate of convective heat transfer C per unit leaf area (including both surfaces of a leaf) per unit time is given by:

$$C = h_c \, (T_1 - T_a) \tag{7}$$

where T_1 and T_a are the leaf- and air-temperatures respectively.

The convection coefficient depends upon the wind speed, the size, shape, orientation and roughness of the leaf and upon the air density, viscosity, etc. The effects are described in the articles cited and in well-known texts on the subject of heat transfer. GATES, ALDERFER, and TAYLOR (1968) derived experimentally empirical values for h_c in order that reasonable approximations could be made of the value of convective heat transfer to or from a leaf without too much involvement in the complexities of laminar or turbulent flow, free or forced convection, etc. The results of their work gave:

$$h_c = k_1 \, (V/D)^{1/2} \tag{8}$$

where

$$k_1 = 0.0100 \text{ if } W >> D$$
$$k_1 = 0.0162 \text{ if } W << D$$

where D is the characteristic dimension of the leaf in the direction of wind flow across the leaf and W is the dimension transverse to the direction of flow. A detailed analysis of characteristic dimensions is given in PARKHURST et al. (1968). If D is in cm and V in cm sec^{-1} then k_1 has units such that h_c is in cal cm^{-2} min^{-1} $°C^{-1}$. Still air is represented by a wind speed of 10 cm sec^{-1}. Very seldom in nature is the air actually still and wind speeds of approximately 100 cm sec^{-1} (2.2 mph) are very frequent. It is shown later that most of the effects of wind speed on leaf temperature and transpiration rate occur at wind speeds under 200 cm sec^{-1}.

The temperature of very small leaves is tightly coupled to the air temperature by virtue of large convective heat-transfer coefficients. The temperature of large leaves can depart considerably from air temperature when in strong radiation fields because of the relatively small convective heat-transfer coefficients which exist for leaves of large characteristic dimensions. Many non-succulent desert plants have small leaves and their temperature does not depart from that of the air by more than about 2° C as shown by GATES, ALDERFER, and TAYLOR (1968). This is of considerable advantage to plants growing where water is of limited supply and the intensity of radiation is enormous. Figure 5 demonstrates how leaf temperature and transpiration rate depend upon the leaf size and internal leaf resistance to diffusion of water vapor when the air temperature is 30° C, the amount of radiation is typical of a clear sunny summer day at noon (1.2 cal cm^{-2} min^{-1}), the wind speed moderate (100 cm sec^{-1}), and the air very dry (r.h. = 20%). TAYLOR and SEXTON (1972) were able to show a dramatic drop in temperature of frayed banana leaves as compared with the temperatures of intact banana leaves. The significance of the above effects on leaf temperature lies in the subsequent effect on photosynthesis, transpiration, and the viability of the leaf for various conditions of heat stress.

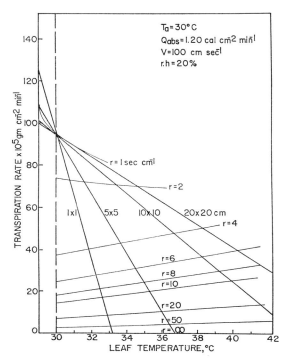

Fig. 5. Transpiration rate and leaf temperature as functions of the internal diffusion resistance to water vapor, r, and the leaf size for a leaf in air at a temperature of 30° C, relative humidity of 20%, and with air speed of 100 cm sec⁻¹. The leaf absorbs radiation of 1.20 cal cm⁻² min⁻¹

D. Transpiration

The evaporation of liquid water within a leaf and its removal by gaseous diffusion affect leaf temperature since every gram of water evaporated requires an amount of energy equivalent to the heat of vaporization of water. The amount of cooling produced by transpiration is readily calculated and measured for any set of environmental conditions.

The rate of water-vapor loss from a leaf, E, is proportional to the difference in water-vapor density between the leaf mesophyll and the free air beyond the boundary layer, and is inversely proportional to the resistance offered by the diffusion pathway through the substomatal cavity, the stomatal channel, and the boundary adhering to the leaf surface. Hence,

$$E = \frac{{}_s d_1(T_1) - \text{r.h.} \cdot {}_s d_a(T_a)}{r_1 + r_a} \qquad (9)$$

where $_sd_1(T_1)$ and $_sd_a(T_a)$ are the saturation water-vapor densities in the leaf meso-phyll and in the free air, r.h. is the relative humidity of the air, r_1 is the diffusion resistance of the substomatal and stomatal channels and r_a is the diffusion resistance of the boundary layer adhering to the leaf surface. The resistances are in sec cm^{-1} and the water-vapor densities in gm cm^{-3}. Hence E is in gm cm^{-2} sec^{-1}.

The latent heat of vaporization $L(T_1)$ of water is 540 cal g^{-1} at 100°C, 600 cal g^{-1} at 0° C and with intermediate values at temperatures in between. The total energy absorbed by a leaf, usually in the form of radiation, is partitioned between re-radiation, convection, and transpiration. If convective heat transfer puts energy into a leaf then the total energy is partitioned between re-radiation and transpiration.

The boundary layer resistance to the transfer of water vapor is closely related to the convective heat transfer coefficient for a leaf and, in fact, is inversely proportional to it. GATES, ALDERFER, and TAYLOR (1968) showed the following for the boundary layer resistance to water-vapor diffusion:

$$r_a = k_2 \frac{W^{0.20} D^{0.35}}{V^{0.55}} \tag{10}$$

where

$$k_2 = 35 \cdot 10^{-3} \text{ if } W = D > 5 \text{ cm}$$
$$k_2 = 26 \cdot 10^{-3} \text{ if } W = D \leq 5 \text{ cm.}$$

E. Energy Budget Equation

The total energy budget for a broad (approximately flat) leaf is given by the following:

$$Q_{abs} = \varepsilon \sigma T_1^4 + k_1 (V/D)^{1/2} (T_1 - T_a) + L(T_1) \frac{_sr_1(T_1) - \text{r.h.} \cdot {}_sr_a(T_a)}{r_1 + k_2 \frac{W^{0.20} D^{0.35}}{V^{0.55}}} \tag{11}$$

where ε is the emissivity of the leaf surface to thermal infrared radiation, usually about 0.96.

In order to understand the relative role of the various heat transfer mechanisms as they affect the temperature of a leaf, the calculations presented in Table 2 should be considered. First, if a leaf is in a vacuum and does not transpire, so that

Table 2

Q cal cm^{-2} min^{-1}	Radiation only	Radiation and convection V (cm sec^{-1})			Radiation, convection and transpiration V (cm sec^{-1})		
		10	100	500	10	100	500
0.6	20	28.4	29.4	29.7	25.2	27.2	28.5
1.0	60	40.9	34.3	32.0	33.4	31.1	30.5
1.4	89	53.0	39.2	34.4	40.5	34.8	32.5

Air temperature 30° C; Diffusion resistance 2 sec cm^{-1}; Relative humidity 50%; Leaf dimension 5 · 5 cm.

energy is only ejected by re-radiation, then the temperatures it will assume are shown in the column entitled "radiation only". The leaf will become very hot indeed if the quantity of absorbed radiation is high and if heat is only dissipated by re-radiation.

If the air surrounding the leaf is cooler than the leaf itself then convective heat transfer will remove energy from it. The leaf temperatures at air speeds of 10, 100, and 500 cm sec^{-1} are given in columns three to five of Table 2. Note that at moderate (1.0 cal cm^{-2} min^{-1}) and at high (1.4 cal cm^{-2} min^{-1}) amounts of absorbed radiation the leaf temperature is greatly reduced by convective cooling, but if the amount of absorbed radiation is low (0.6 cal cm^{-2} min^{-1}) the leaf temperature is increased. This is true because the quantity of radiation absorbed is less than black-body radiation (0.8 cal cm^{-2} min^{-1}) corresponding to a temperature of 30° C. Now allow the leaf to transpire, let it have an internal diffusive resistance of 2 sec cm^{-1}, and let it be in air with a relative humidity of 50% at 30° C. Substantial cooling of the leaf results, depending upon the wind speed and the quantity of radiation

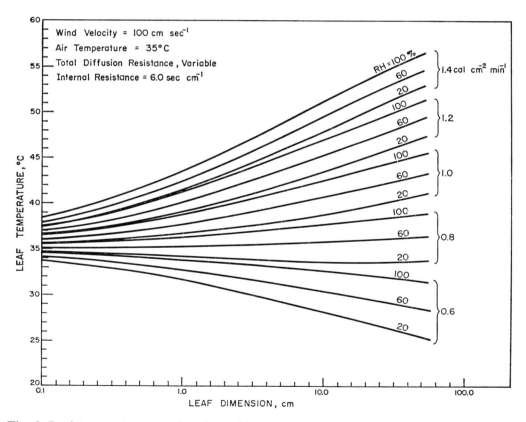

Fig. 6. Leaf temperature as a function of leaf characteristic dimension to air flow and as a function of the amount of radiation absorbed by the leaf and of the relative humidity. The air temperature is 35° C, the air speed 100 cm sec^{-1}, and the internal diffusion resistance to water vapor is 6.0 sec cm^{-1}

absorbed by the leaf. The leaf temperatures with transpirational cooling in addition to re-radiation and convection are given in the last three columns of Table 2. It can have a substantial effect on leaf temperature if water is available to the plant, if the diffusion resistance is minimal, and if the air is not too humid.

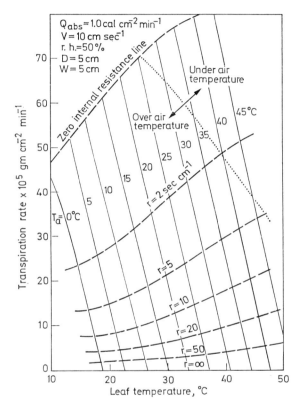

Fig. 7. The transpiration rate and the leaf temperature as functions of the internal diffusion resistance to water vapor and as functions of the air temperature. Leaf size is $5 \cdot 5$ cm, the air speed is 10 cm sec^{-1}, the relative humidity is 50%, and the amount of radiation absorbed by the leaf is 1.0 cal cm^{-2} min^{-1} (typical of a clear summer day at noon)

Transpiration rate and leaf temperature are dependent variables. Their values result from the action of the independent variables of radiation, air temperature, wind speed, and humidity as coupled to the leaf by energy exchange by means of leaf properties of absorptivity, surface area, orientation, dimensions, and internal diffusive resistance. With two dependent variables, four independent variables, and five or more leaf parameters to consider it is evident that the dimensions of the space in which one must navigate in order to understand these phenomena is very great indeed. Unfortunately only limited cross sections through these many variables and parameters can be presented and several variables must be held constant simultaneously. The following figures are examples of certain cross sections in this multidimensional space, showing how transpiration rate and leaf temperature respond to various conditions.

Referring once again to Fig. 5, it is interesting to note how the transpiration rate is influenced by the leaf dimension or by the internal resistance to diffusion. The effect of leaf dimension on transpiration rate is generally small for a given internal resistance whereas the effect of resistance is very large for any given leaf size. For a fixed internal resistance of 6 sec cm⁻¹, the influence of leaf dimension on leaf temperature is given in Fig. 6 as a function of the relative humidity and the

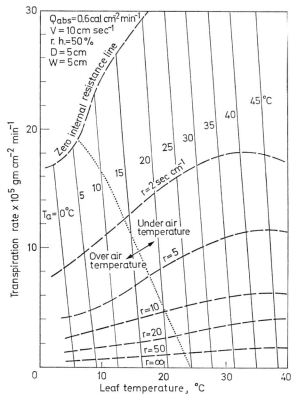

Fig. 8. The transpiration rate and the leaf temperature as functions of the internal diffusion resistance to water vapor and as functions of the air temperature. Leaf size is 5 · 5 cm, the air speed is 10 cm sec⁻¹, the relative humidity is 50%, and the amount of radiation absorbed by the leaf is 0.6 cal cm⁻² min⁻¹

amount of radiation absorbed. Once again it is seen that the temperature of small leaves is tightly coupled to the air temperature, but that the temperature of large leaves can differ widely from the air temperature.

Leaf temperature and transpiration rate are shown in Fig. 7 as functions of the air temperature and the internal resistance of the leaf to diffusion of water vapor.

Figure 8 is the same graph except that the amount of radiation absorbed is low (0.6 cal cm^{-2} min^{-1}) and typical of a cool, cloudy day. It is interesting to note the way in which the line separating the domains in which leaf temperature is above or below air temperature shifts dramatically with the amount of radiation absorbed. "Still" air has an air movement of 10 cm sec^{-1}. Note in Fig. 7 that for $Q_{abs} = 1.0$ cal cm^{-2} min^{-1} at low air temperatures (0—15° C) a non-transpiring leaf will be 15—18° C above air temperature, but at high air temperatures (30—45° C) a non-transpiring leaf will be only 7—11° C above the air temperature.

References to this section, see p. 264.

II. The Normal Temperature Range

A. Pisek, W. Larcher, A. Vegis, and K. Napp-Zinn

A. Effect of Temperature on Metabolic Processes

A1. Carbohydrate Metabolism

Carbohydrates play a central role in the metabolism of the plant as energy carriers, as reserves and, above all, as building materials. Plants can in fact be considered as carbohydrate organisms: more than 60% of the dry weight of higher plants consists of carbohydrates.

1. Photosynthesis

A. Pisek

a) Basic Aspects of Photosynthesis

Two fundamentally different kinds of reaction are involved in the process of photosynthesis, one drawing its energy from either natural or artificial light, the other independent of light as a source of energy. In the former, so-called light reactions, light is exploited by means of absorption by chlorophyll and accessory pigments: approximately that portion of the spectrum is utilized which is appreciated by the human eye. Up to the point where excess light is present (light saturation) photosynthesis is dependent upon the intensity of the light (BOYSEN-JENSEN, 1932). The second type of reaction, or dark reaction, although independent of light, is dependent upon the temperature with which, in contrast to the light reactions, the intensity of photosynthesis rises and falls. Since plants are capable of utilizing much larger quantities of CO_2 than are normally present in the air, so that CO_2 normally represents a minimum factor in nature, photosynthesis is, thirdly, also dependent upon the CO_2 content of the air (PONOMAREVA, 1960; KOCH, 1969, LUDLOW and WILSON, 1971). To complete the list of the most important factors influencing the fixation of CO_2 it should be mentioned that this process can only function maximally if the state of hydration of the assimilatory tissue is nearly optimum. A large water saturation deficit presents an indirect obstacle to photosynthesis since the hydrolabile stomates shut down (STALFELT, 1935, 1936; PISEK and WINKLER, 1956; ZELITCH, 1963; MEIDNER and MANSFIELD, 1968). If this happens scarcely any CO_2 enters the leaf and photosynthesis is then confined to the reassimilation of CO_2 released internally by respiration, bringing with it no material gain. Nevertheless, even small deficits can lead to a direct reduction in photosynthesis, as was shown by SLAVIK (1963) on the liverwort *Conocephalum conicum* which has rigid pores instead of stomates.

It must not be overlooked that apart from utilizing the CO_2 of the air, plants also take up CO_2 from the soil via their roots, from where it is shifted by transpiration (HÄRTEL, 1938). The amount of CO_2 taken up in this manner, however, never amounts to more than 5% of the total (VOZNESENSKY, 1958).

At the same time as the CO_2 taken up in light from the environment is being incorporated into the anabolic processes in the green tissues, the plant is respiring and releasing CO_2, part of which is in turn reassimilated. In light, it is only possible to measure directly the difference between total photosynthesis or *gross photosynthesis* and the concurrent *respiration*. In other words it is the actual gain achieved by gross photosynthesis, termed *apparent* or *net photosynthesis*, which is measured. Fortunately it is this latter value which is of immediate interest in the production of material. Figure 1 shows the relationships between the three values and their variations with temperature. Gross photosynthesis has so far usually been calculated as the sum of the net gain and respiration in the dark, since respiration can be measured directly only in the dark. The results of such simple calculations should nevertheless be treated with a certain amount of caution until it is ascertained whether respiration is identical in light and darkness. For literature on this topic JACKSON and VOLK, 1970 and ZELITCH, 1971, among others, should be consulted. If, as is usually the case nowadays, photosynthesis of terrestrial plants is measured by recording the variations in the CO_2 content of the air passing over the plant in a container, then a drop in the CO_2

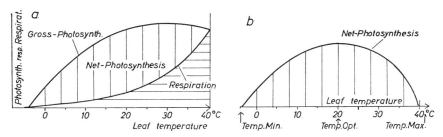

Fig. 1a. Simple schematic representation of the connections between net photosynthesis and gross photosynthesis with increasing and decreasing temperature. For more detail see LARCHER (1969 and 1973b), Fig. 5

Fig. 1b. Scheme of the temperature curve of net photosynthesis. Cardinal temperatures

content of this air indicates that the gross assimilation of the plant exceeds its respiration at the given instant, or that net assimilation is positive. An increase in CO_2 content, on the other hand, indicates that respiration predominates and net assimilation is negative. If there is neither a decrease nor an increase, indicating that gaseous exchange with the exterior is zero, then the two processes are in a state of equilibrium (compensation point).

b) Cardinal Temperatures of Photosynthesis in General

If the temperature is raised over that range of the temperature scale capable of supporting life, the above-mentioned state of equilibrium is encountered twice.

Firstly, a certain minimum temperature must be attained before the intensity of photosynthesis reaches that of the less cold-sensitive respiration (low temperature limit, i.e. *temperature minimum* of net photosynthesis). On the other hand, at high temperatures the intensity of respiration increases rapidly and finally catches up with the concurrent gross carbon fixation. With rising temperature, therefore, respiratory CO_2 is reassimilated at steadily increasing partial pressures and the CO_2 gradient in the leaf consequently becomes smaller and smaller (EGLE and SCHENK, 1952). Thus in the higher temperature region there must be a point at which no further absolute gain is possible and net photosynthesis is zero (high temperature limit, i.e. *temperature maximum* of net photosynthesis). The productive temperature range, bounded by the two limits, spans a relatively wide interval, permitting in the middle range the highest possible production *(temperature optimum)*.

c) Temperature Curves for Net Photosynthesis

LUNDEGARDH and his school (1924a, b, 1927) were the first to systematically measure CO_2 fixation in flowering plants under simultaneously controlled conditions of temperature and light. Their aim was to plot curves of the dependence of photosynthesis upon temperature, to seek an ecological interpretation for their results and to apply them in production analyses. They used both strong and weak light, and normal as well as high CO_2 concentrations in the surrounding air. Herbaceous cultivated plants such as potatoes, tomatoes, sugar beet, some varieties of barley and beans were chosen as test objects. As a result of the enormous technical difficulties prevailing at that time the curves (Fig. 2), based upon few

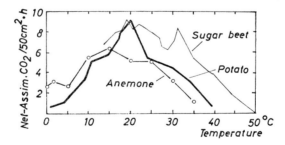

Fig. 2. Examples of temperature curves of net photosynthesis in full light and with normal CO_2 content of the air (0.03%). (From LUNDEGARDH, 1927)

and not always reliable measurements, often fluctuated strikingly and misled LUNDEGARDH into reading several maxima into them. His attempts long remained the only ones of their kind for flowering plants. Only in the past fifteen years has his work been taken up again and pursued more extensively. The use of modern, efficient and precise instruments such as the infrared gas analyzer (IRGA) makes it possible to take many more and closer readings. Leaf temperature itself, and this is important, can now be measured with much greater accuracy.

Figures 3, 4, and 5 provide typical examples of temperature curves plotted more recently using measurements made under standardized laboratory conditions at a normal CO_2 concentration. It can be seen (Fig. 3, *Ranunculus glacialis*) that for one and the same object the curves rise and fall more steeply and their peak is higher the greater the intensity of illumination used, until the specific saturation value is reached (light dependence). Under identical conditions of illumination the

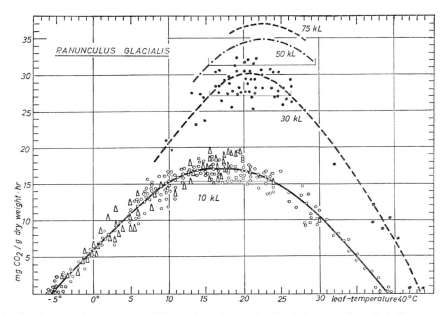

Fig. 3. Temperature curves of the net photosynthesis of *Ranunculus glacialis* at various intensities of illumination (10—75 kLx) and normal CO_2 content of the air (approx. 300 ppm), IRGA. With increasing intensity of illumination, photosynthesis and its temperature optimum and maximum move upwards. MOSER in PISEK et al. (1969)

intensity of photosynthesis varies somewhat from species to species. Plants with a C_4-pathway of carboxylation, cultivated herbaceous plants and some wild herbs are the most efficient with respect to photosynthesis, whereas the reverse is true of sclerophyllous, broad-leaved evergreens and evergreen conifers (collected data in LARCHER, 1969a and 1973b). If the specific differences in photosynthetic capacity are eliminated by equating the highest value for each species to 100 and recalculating the remaining values accordingly (Fig. 5) it becomes clear that the temperature optima and maxima (not so much the minima) vary according to species and habitat. The optimal range is sometimes approximately in the center of the productive temperature limits, sometimes displaced to one side or the other. On the whole, however, the curves of all species investigated are very similar: when the temperature minimum has been exceeded the intensity of photosynthesis increases almost linearly and after a flat central portion the curve finally falls off steeply (optimum curve). This appears to hold also for curves plotted elsewhere and for other species

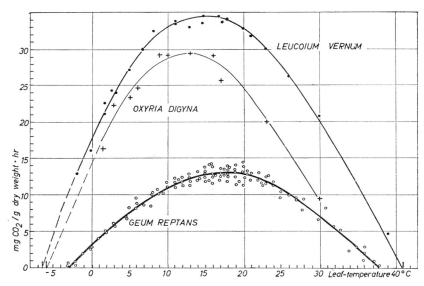

Fig. 4. Temperature curves of net photosynthesis in various plants. Unlike Fig. 3, however, uniform intensity of illumination (10 kLx). Low and high temperature limits (temperature minimum and maximum) in the curves for *Oxyria* and *Leucojum* obtained using the Ålvik method. MOSER and PACK, from PISEK et al. (1969)

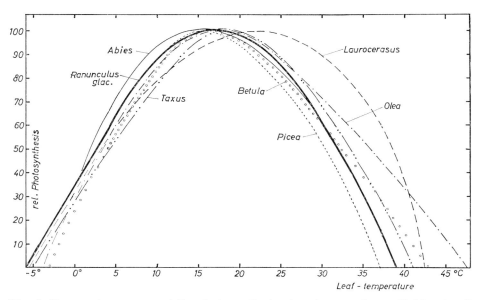

Fig. 5. Temperature curves of the photosynthesis of various species at 10 kLx, in all cases the maximum being equated to 100

[e.g. KUSUMOTO (1957a, b, 1961), warm-temperate evergreens in southern Japan; NEGISI (1966), one-year seedlings of three economically important conifers; HELLMUTH (1968, 1969), two plants in a west Australian habitat, Fig. 19; NEILSON, LUDLOW and JARVIS (1972), *Picea sitchensis*].

An exception is presented by *Citrus limon* (Fig. 6) whose photosynthesis takes remarkably long to start up after the temperature minimum has been exceeded. The first part of the temperature dependence curve is definitely concave: this is understandable when it is considered that *Citrus* occurs only in regions with relatively warm winters and can therefore afford to dispense to a large extent with the exploitation of temperatures below zero. Not all plants of the warm temperate regions, however, can afford to behave like *Citrus* in dispensing with a property which can only be to their advantage.

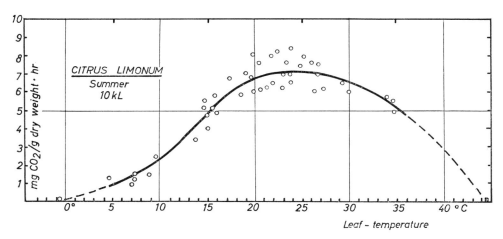

Fig. 6. *Citrus limon*. Temperature curve of photosynthesis from a specially wind-sheltered plantation in Limone on Lake Garda, current year shoots, CO_2-content of the air 300 ppm; 10 kLx. UNTERHOLZNER, in PISEK et al. (1969)

d) Cardinal Temperatures of Photosynthesis in Various Plants

α) *Plants from Different Altitudes in the Central European Alps and from their Sub-Mediterranean Southern Foot*

The Temperature Optimum

The Range of the Temperature Optimum. The highest absolute gain from photosynthesis is achieved at a certain medium temperature of the assimilatory organs: this appears in the temperature curve as a relatively flat summit. The flatter the curve (weak light) the more difficult it becomes to detect accurately the highest point, especially if the individual measurements for CO_2 uptake are very scattered. As limit of the optimum temperature range was defined those values within which the variations in photosynthesis, as given by the compensation curve, are less than 10% to either side of the maximum. In the linear representations in Figs. 7 and 9 the mean value is also indicated. This does not always lie midway between the upper and lower limits since the curves are often slightly asymmetrical. At 10 kLx, the most propitious temperature region extends over at least 8° C in the plants investigated by the authors *(Picea abies* and *Pinus cembra* at the upper timberline), commonly over 10—12° C and in *Abies alba, Leucojum vernum* and *Laurocerasus officinalis* 13—14° C. Thus the old observation is confirmed that material gain is largely independent of temperature over a broad range of moderate temperatures. In other words, plants of temperate latitudes best exploit a wide range of low to moderate temperatures for their CO_2 assimilation.

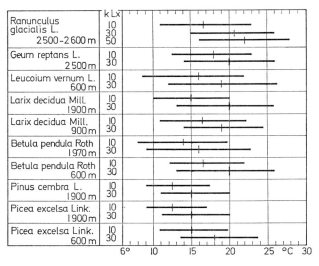

Fig. 7. Optimum temperature ranges of photosynthesis, taken from curves like Figs. 3—6, 10 kLx (~0.10 cal cm^{-2} min^{-1}, measured with starpyranometer) and 30 kLx (~0.36 cal cm^{-2} min^{-1}). If the samples measured are taken from the same altitude the temperature optimum in all species moves upwards with increasing light. At the same light intensity the optimum of trees from the timberline is somewhat lower than that of those in the valley. (From Pisek et al., 1969). — *Picea excelsa* Link = *P. abies* Karst.

Before considering the extent to which plants characteristic of different climatic regions have adapted in respect to their optimum productive range, the factors capable of displacing the temperature optimum must be discussed briefly.

Variability of the Temperature Optimum. On account of their active respiration, young leaves at first achieve only a small material gain from photosynthesis (Pisek and Winkler, 1958; Neuwirth, 1959; Schulze, 1970). Furthermore, photosynthetic capacity increases and finally declines with age, especially when the leaves begin to change color. In evergreen plants photosynthetic capacity decreases year by year (Clark, 1961). The temperature optimum, however, appears to remain fairly constant throughout the course of the development of the organs responsible for assimilation (Pisek et al., 1969; Schulze, 1970).

With increasing *light*, however, the temperature range most suitable for production is displaced upwards. In principle, this was already known to LUNDEGARDH (1924) and has since been confirmed repeatedly (MÜLLER, 1928; PISEK and WINKLER, 1959; PISEK and LARCHER et al., 1969). MOSER has plotted comprehensive temperature curves for photosynthesis at 10 kLx and 30 kLx in *Ranunculus glacialis* and has made measurements in the peak region of the curves at 50 and 75 kLx. Fig. 3 shows the displacement quite clearly: it is seen to be at its greatest at lower values of illumination (10–30 kLx) but is negligible at the highest values (50 to 75 kLx). Apparently the temperature optimum becomes stationary when the strength of illumination approaches the saturation point, the latter varying according to species and individual state of the test material (sun or shade forms). The saturation value of *Ranunculus* is particularly high. Figure 7 shows that the optimum temperature rises by about 2–4° C when the illumination increases from 10 to 30 kLx.

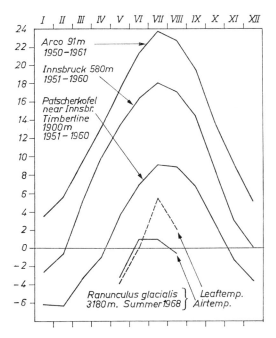

Fig. 8. Monthly mean air temperatures at the three meteorological stations connected with the work carried out from Innsbruck research group, PISEK et al. 1967—1969. To illustrate the situation in the nival region the leaf temperatures of *Ranunculus glacialis* is given in addition to the mean monthly air temperatures on the Nebelkogel ridge

This appears to be a widespread phenomenon which has to be taken into consideration when comparing results of authors who have used different intensities of illumination. Since strong light such as direct sunlight is usually accompanied by considerable heating up, the significance of the above phenomenon is obvious. According to SCHULZE (1970) the beech does not quite conform in this respect, although it may simply be that in his investigations in the field the displacement was not so striking as in the laboratory.

Figure 9 also shows that the temperature optimum of trees at the subalpine timberline (1900 m in the case investigated) is 2—3° C lower than that of trees

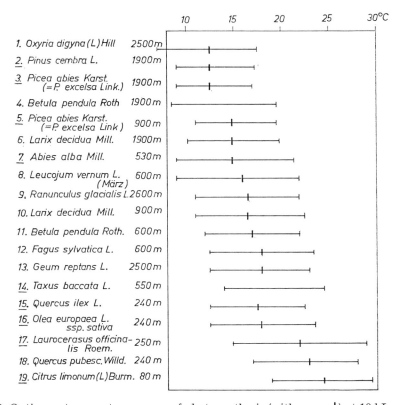

Fig. 9. Optimum temperature ranges of photosynthesis (with mean, |) at 10 kLx (~ 0.1 cal cm⁻² min⁻¹). The samples of the last five species originate from the region around Arco-Riva on Lake Garda (80 m—240 m); the other species were taken from the valleys (530 m—900 m), the timberline (1900 m), and the high mountains (2500—2600) near Innsbruck. Summer VII—VIII. The underlined numbers denote evergreens. Of these only the most recent growth was investigated. (From PISEK et al., 1969, rearranged; for values on *Fagus* in the field see SCHULZE, 1970). — The mean of optimum range of Nr. 4 is at 14° C, of Nr. 14 at 19° C

lower down (600—900 m). This no doubt reflects a certain degree of *adaptation* to the *cooler climate at the timberline*[1]. The mean monthly air temperatures in both localities, recorded under meteorological conditions in weather stations 2 m above

---

[1] In seedlings of *Abies balsamea* which have been cultivated under the same conditions from seeds of trees from various elevations (730 m to 1468 m above sea level) the temperature optimum is shifted to lower temperatures the higher the elevation was at which the parent tree had grown (FRYER and LEDIG, 1972).

ground level, are 7.5—10° C lower from April to September at the timberline than in the valley (Fig. 8).

In areas with a mild *winter* the capacity of evergreens to assimilate at this time of year, depending upon the degree of cold, is lower than in summer under identical conditions of light and temperature. According to LARCHER (1961, Figs. 2 and 4) the temperature optimum and maximum of *Olea europaea and Quercus ilex*, for example, in the frost-free part of the winter sink by up to 2° C and in the periods of slight frosts by a further 2° C. Broad-leaved evergreens in southernmost warm temperate Japan (KUSUMOTO, 1957, 1961) and the conifer seedlings described by NEGISI (1966) behave in a similar manner.

Only sufficiently frost resistant plants, of which many examples are found among the northern and alpine conifers, are capable of survival in regions with severe winters. It is known of *Pinus cembra, Picea abies* and *Abies alba* that their photosynthesis can come to a complete although temporary standstill for weeks or even months in the face of continuous if only slight frost. This phenomenon will be discussed later. Photosynthesis can at any time be reactivated by the action of continuous warmth of 10° C or more over a period of several days, depending upon the duration and degree of the preceding cold (TRANQUILLINI, 1957; TRANQUILLINI and MACHL, 1971; PISEK and WINKLER, 1958; PISEK and KEMNITZER, 1968; SCHWARZ, 1970). In such cases the activity is not only weaker than in summer but is, in relative terms, at its best at lower temperatures.

Optimum Temperature Ranges of Different Plant Species. In Fig. 8 the species investigated are arranged according to their optimum ranges. The spectrum of the mean values extends from 12—24° C, which is also the interval between the lower and upper limits of the optimum temperature ranges. The list is headed by *Oxyria digyna*, a perennial widespread in circumpolar arctic regions and in the high alps where it prefers a wet habitat, and by two evergreen conifers of the central alpine timberline, *Picea abies* and *Pinus cembra*. In its natural distribution the latter does not extend far below the timberline. It is significant that the third tree characteristic of this altitude, the deciduous *Larix*, needs higher temperatures for photosynthesis. *Oxyria* is the least demanding as regards temperature: even at around 0° C it achieves half of the maximum photosynthesis possible at 10 kLx and 7° C suffices for full photosynthesis (Fig. 4). The fact that *Oxyria* remains highly active up to about 18° C and that its activity only falls to half of the maximum at 27° C enables it to exploit moderate warmth to the full. Nevertheless, like all plants that have to cope principally with leaf temperatures only a few degrees above zero during their season of growth, the lower limit of the optimum temperature span is much more important than the upper limit. The latter, conversely, is more important to species living in warm or hot environments.

Betula pendula of the timberline represents a transition to a heterogeneous, centrally situated group in which the mean of the optimum range lies between 15° C and 18° C. This group also includes *Larix* from both timberline and valley, evergreen conifers and deciduous angiosperms of the valleys, besides the early geophyte *Leucojum*. According to WINKLER and PREGENZER (1970) the most common pasture grasses and clover species of the permanent meadows are to be included in this group. Oddly enough, the other two of the three high alpine perennials investigated, *Ranunculus glacialis* and *Geum reptans*, although they would be expected to behave similarly to *Oxyria*, also fit in here. Much more warmth is in fact provided by sheltered, sunny corners and hollows in the high alps than the meteorological data suggest. It is to these climatically favorable miniature habitats that the pioneer vegetation becomes more and more confined with increasing altitude, at the same time becoming progressively more dwarfed and finally disappearing altogether. The mean monthly leaf

temperatures of *Ranunculus* in Fig. 8 are several degrees higher than the air temperature 2 m above ground level in July and August (there was still snow on the ground
in May). On isolated occasions the leaves may be as much as 10° C warmer than the
air (MOSER, 1970).

The optimum range for *Taxus baccata* (14—25° C, mean 19° C) is decidedly higher
than that of any of the species previously mentioned, although unexpectedly this is
not true of *Quercus ilex* and *Olea*, two evergreens typical of the Mediterranean belt.
It should be added, however, that the climate on the northern edge of Lake Garda
whence these samples were taken is not quite typically Mediterranean (see Fig. 8,
"Arco"), the winters being somewhat more severe and the summers less hot and dry
than in the true Mediterranean region. Both species have reached their northern
European limit in this locality. The measurements on *Quercus pubescens* were made
in the same place, and it became obvious that, presumably because deciduous, it is
adapted to higher temperatures than is its evergreen relative *Quercus ilex* for which
the temperature conditions over the entire year are decisive (LARCHER, 1961). *Laurocerasus*, *Citrus* and *Quercus pubescens* constitute the group with the highest optimum
temperature range, its mean being 22—25° C and its upper limit from 28° C to almost
30° C. It is this upper limit that is of primary importance, and in strong light (see
p. 109) it can be pushed up even further. At still higher temperatures the efficiency
declines rapidly and at 36—38° C is reduced to half of its maximum. On the other hand,
the southern evergreens profit from the fact that in the cold half of the year the
entire productive range, including its lower limit, sinks by a few degrees, as mentioned
previously on p. 119.

High Temperature Limit (Temperature Maximum of Net Photosynthesis)[2]

Increases in intensity of illumination bring about an upward displacement of both
temperature optimum and temperature maximum until light saturation is achieved. The temperature maximum of both sun and shade leaves of beech trees
(Fig. 10) was found, for example, to be approximately the same at 30 kLx. At
1 kLx the shade leaves were already light saturated so that the further rise in net
photosynthesis was negligible for additional increases in intensity of illumination.
The sun leaves, on the other hand, were not even saturated at 3 kLx and the
maximum temperature did not remain constant until 10 kLx and more.

The temperatures at which net photosynthesis ceases when the test plants are
exposed to a uniform illumination of 10 kLx for one hour are marked with a cross
in Fig. 11. All of the conifers investigated, headed by *Pinus cembra* from the
timberline (the isolated examples of this species found lower down have been
planted), compensate at 36—38° C. The only exception is provided by *Taxus* which,
like *Leucojum*, compensates at 41° C. Following the conifers come the three high-
alpine herbaceous species and *Viscum* (38/39° C); the temperature maximum of
eight of the 23 species listed, as well as of *Acer pseudoplatanus* (BAUER, 1972)
and of the typically Mediterranean *Quercus ilex* and *Arbutus unedo* is 42/43° C.
The four remaining Mediterranean and southern trees achieve no further absolute
gain after the leaf temperature has reached 44/45° C whereas the corresponding
value in *Olea*, which is also the most heat resistant species listed, is 48° C.

---

[2] Determinations of temperature maximum (CO_2 lost to the exterior $= 0$) sometimes
using IRGA (high alpine!), usually, however, performed using a simple closed vessel
with bicarbonate solution which comes into equilibrium with the CO_2 content of the
enclosed air. If the plant samples give off CO_2 the pH of the solution moves in the
acid direction whereas if they take up CO_2 the solution becomes more alkaline, as
revealed by the indicator in the solution (Ålvik technique; details in PISEK, LARCHER
et al., 1968 a).

Concurrently with measuring the temperature maximum for photosynthesis an attempt was made to find the temperature at which approximately half of the leaf surface is killed after exposure for one hour (the "heat resistance value" of

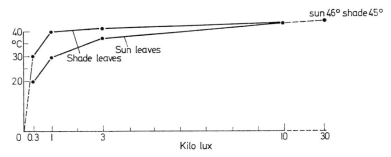

Fig. 10. Dependence of the temperature maximum of photosynthesis on light, at normal CO_2 content of the air (approx. 300 ppm). *Fagus sylvatica*, June. 3 kLx in the cases presented indicates ~ 0.02 cal cm^{-2} min^{-1}; 10 kLx ~ 0.1 cal and 30 kLx ~ 0.36 cal (From PISEK et al., 1969)

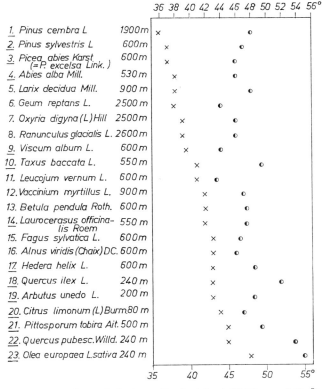

Fig. 11. Temperature maxima of photosynthesis (x) at 10 kLx and the heat resistance values (50% damage, ◑). List of species as in Fig. 9, plus four more. From PISEK et al., 1969, changed. See BAUER, 1972 for *Acer pseudoplatanus* (temperature maximum 42° C, heat resistance 47° C). — *Picea excelsa* Link = *P. abies* Karst

LEVITT (1958) and LANGE (1961), both of whom, however, only exposed the leaves for half an hour; Fig. 11, small circles). If these values are compared with the temperature maxima it is found that heat compensation temperature and heat resistance[3] values in conifers are about 8—12° C apart, 9° C in *Quercus* species, 6—7° C in the three high-alpine species, and in other cases even closer. Depending upon light saturation the values would be closer still in some cases if the upper compensation temperature had been measured under stronger illumination (see p. 109).

An explanation for the large discrepancy between temperature maximum and lethal temperature in conifers might lie in the possibility of plugging of the stomata by resin exudations. This explanation, however, is weakened by the fact that *Taxus* behaves similarly even though its leaves possess no resin channels. In all conifers investigated there was wax in the cavities at the base of which the stomata are situated, thus blocking the view into the pores. At a temperature of about 35° C this wax becomes sticky and might present an irreversible obstacle to the entry of CO_2 into the leaf, so that the upper temperature limit is lowered. Nevertheless, this cannot be the entire explanation since, according to BAUER (1972, *Abies alba*), the original photosynthetic capacity is restored if the needles are cooled down again. It has to be mentioned that there are also some conifers which have a higher temperature maximum than those listed in Fig. 11 (see NEGISI, 1966).

Low Temperature Limit (Temperature Minimum of Net Photosynthesis)[4]

Summer. Significantly, it is the high-alpine plants *Oxyria* and *Ranunculus glacialis*, both of which also occur in the Arctic (*Oxyria* is circumpolar), that have the lowest compensation temperatures (Fig. 12, black dots). They are closely followed by the central European bulbous geophyte *Leucojum vernum* which flowers in the valleys in the very early spring at a time when ground frosts with temperatures several degrees below zero are still frequent. At high altitudes such low temperatures are often encountered in the middle of summer. The ecological significance of the low temperature minimum in all three cases is thus obvious. The leaves of the high-alpine (but not arctic) *Geum reptans* are more sensitive to frost. Only *Fagus sylvatica* leaves have a temperature minimum as low as the three above-mentioned plants although this low minimum is not exploited. At the other end of the list are *Citrus* and *Laurus*, both of which cease to function profitably at the occurrence of even the slightest frost (about −1° C). This is logical if it is recalled that certain tropical greenhouse plants suffer damage if kept for some hours or days at only a few degrees above zero (SEIBLE, 1939; SPRANGER, 1941).[5]

The temperature minimum of the majority of species investigated lies between −3° C and −5° C; the margin is so small that no clear specific differences can be detected. In some cases it was possible to investigate leaves of approximately the same age from one and the same species from the valley and from the timberline: the temperature minimum of leaves taken from the timberline, where the buds

---

[3] See p. 113.
[4] Method analogous to that used for determination of temperature maximum; see footnote p. 112).
[5] See p. 221.

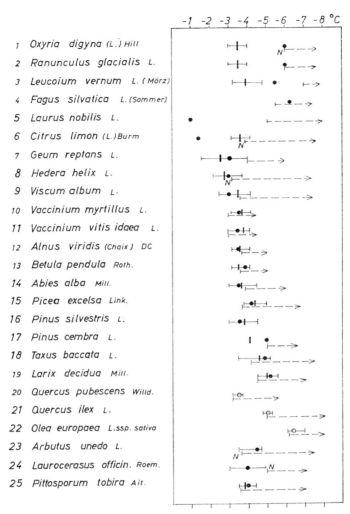

Fig. 12. Temperature minima of photosynthesis (●) at 3 kLx, range of freezing temperatures of the leaves (with mean value —+—), and frost temperature causing first damage and at least 50% damage (——→). If the veins freeze before surface damage occurs this is indicated by a large N. In evergreens (large numbers) only current year shoots are used. Summer 1962. (From PISEK et al., 1967, rearranged)

burst approximately a month later than in the valley, was definitely if only slightly lower (1—2° C) (Fig. 13).

Winter. When, in late autumn and in winter, the temperature sinks to −5° C or less the assimilatory capacity of evergreens declines. This impairment of activity is the more lasting depending upon severity and frequency of frost. A valuable series of investigations by TRANQUILLINI (1957) shows that in the course of the late autumn the amount of CO_2 taken up by young autochthonous *Pinus cembra*

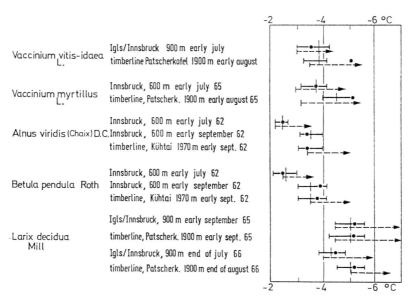

Fig. 13. Comparison of temperature minima of photosynthesis of one and the same species from valley and mountains. Explanation as in Fig. 12. (From PISEK et al., 1967)

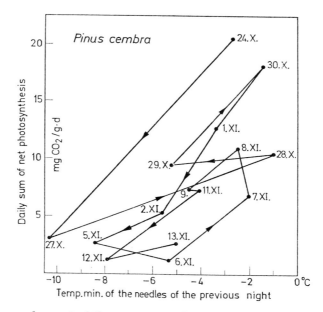

Fig. 14. *Pinus cembra*; autochthonous young plants from the central alpine timberline (2070 m) in late autumn. Decline in the daily sum of net photosynthesis with increasing frequency and severity of frost in the preceding night. The consecutive readings are linked by arrowed lines and dated. Measurements in the open air. Days are not taken into consideration when the light intensity was the limiting factor for the photosynthesis. (From TRANQUILLINI, 1957)

at the timberline rises and falls with the degree of frost in the preceding night. With increasing frequency and severity of frost the CO_2 uptake values gradually approach zero (Fig. 14). Production sinks even if the temperature of the needles rises above zero during the daytime. Finally at best, photosynthesis can only compensate for the concurrent losses due to respiration. Winter dormancy usually lasts for several months at the timberline: brief periods of warmer weather only result in a loss of material since respiration starts up more rapidly than photosynthesis.

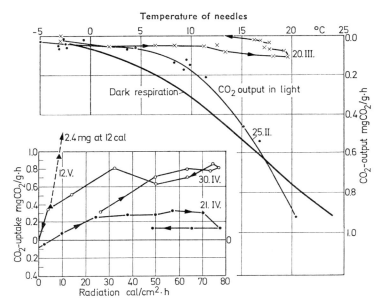

Fig. 15. Winter dormancy and commencement of photosynthesis at beginning of spring in *Pinus cembra*. Daily course in the open air. On February 25 and March 20 CO_2 is given off continuously in light. On February 25 only at low needle temperatures is the quantity lost less than for respiration in the dark, at 15 and 20° C it is approximately equal to this; on March 20 the CO_2 losses are only small, indicating that although photosynthesis is going on it is still unable to compensate the simultaneous loss of CO_2 due to respiration. Only from April 21 onwards does photosynthesis bring a rapidly increasing absolute gain. Simplified from Figs. 13 and 14a, TRANQUILLINI (1957)

Not until May are the temperature conditions at the timberline such that, apart from a few setbacks, photosynthesis is again in full action, as shown in Fig. 15 (note the sequence of days upon which measurements were made)[6]. In the

---

[6] The sun side of the crown of the tree comes into action later than the other parts of the crown (twigs): chlorophyll is more sensitive to irradiation in winter (more photolabile) than in the warmer seasons which leads to its partial destruction in winter at timberline altitudes and to a noticeable discoloration of the needles. Not until the chlorophyll has regenerated and the needles are quite green again can the photosynthetic apparatus function as it did prior to the onset of winter (TRANQUILLINI, 1957; for review article on photosynthesis in winter see PISEK, 1960).

valleys, however, the winter usually begins later and ends earlier and may be interrupted by longer periods of relatively warm weather, so that the non-productive cold period is much shorter than at the timberline. The assimilatory capacity of *Picea* trees from both levels (600 m and 1840 m) has been compared repeatedly over the course of a year. The tests on the twig samples were performed under standardized conditions and immediately after collection. Fig. 16 shows

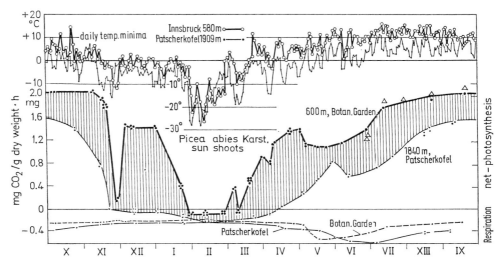

Fig. 16. Difference between CO_2 assimilation capacity (mg CO_2/g dry weight/hour, 12°C, 10 kLx) of sun shoots of *Picea abies* in the botanical garden in Innsbruck (600 m, thick curve) and the assimilation capacity of sun shoots from a tree at the alpine timberline on the Patscherkofel near Innsbruck (1840 m, thin curve) over the course of the year. Above: daily minima from the weather stations at both sites. (From PISEK and WINKLER, 1958). The large clefts in the annual curves of assimilation capacity of the trees from the botanical garden correspond to the cold periods, during which the daily minima of the air temperature in Innsbruck were usually or continually below $-5/-6°$ C (freezing temperature of the needles). On the Patscherkofel these periods link up to form a continuous, longer winter; the assimilation capacity is thus zero or negative from the end of November to the middle of April. The broader, later depression in both curves is due to the fact that when the buds began to sprout the older of the two previous years growth, in use until then, was discarded and replaced by the new shoots, on which the young needles respire very strongly at first

the results, including the temperature minima throughout the year at both altitudes. It stands out at once that productivity at the timberline is greatly hampered by the long, uninterrupted winters: the nights remain cold even when in late winter the days are sunny. The slow growth of trees at this altitude is understandable for that reason alone. The situation is particularly bad in the continental arctic where the severe, uninterrupted winter is even longer. The degree to which the evergreens of temperate latitudes with cold winters are capable of exploiting the advantage offered by the fact that their photosynthetic apparatus is available all the year round, depends entirely upon the time of onset, continuity, duration and severity of frost. If the assimilatory organs are cooled only a few degrees below their freezing point they become incapable of CO_2 uptake.

Plant samples can be released artificially from such inactivity even in the middle of a severe winter if kept at a constant temperature of about 10° C or slightly more for several days (see PISEK and WINKLER, 1958; PARKER, 1963; PISEK and KEMNITZER, 1968). Obviously, this can only result in success if, in the foregoing

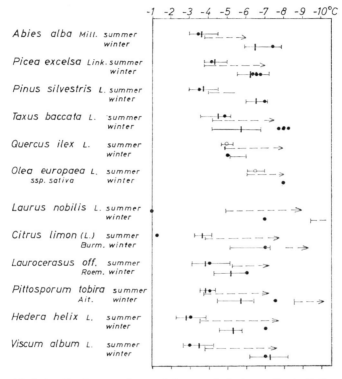

Fig. 17. In midwinter the temperature minimum of photosynthesis (following warmth treatment), the freezing range of the assimilatory organs and commencement of frost damage undergo a downward displacement as compared with summer. Symbols as in Fig. 12. (From PISEK et al., 1969). — *Picea excelsa* Link = *Picea abies* Karst.

period, the frost has not exceeded the limits of resistance of the plants: this holds true in midwinter for the highly resistant northern alpine conifers tested. Material treated in this manner revealed, according to species, compensation temperatures of 2—4° C lower than in summer (Fig. 17). Only in *Citrus* and *Laurus* was the difference between summer and winter values greater because both species, as already mentioned, stop assimilating carbon with profit even at the slightest summer frost. In cold seasons, however, their behavior is similar to that of the other species. The freezing ranges of the assimilatory organs are also lower in winter (Fig. 17), the extent to which they and the temperature minima for photosynthesis sink depending, among other factors, upon the severity of the winter.

The Low Temperature Limit of Photosynthesis in Relation to Freezing and Frost Damage

A comparison of the temperature minimum of photosynthesis of *Pinus cembra* needles (TRANQUILLINI, 1957) with the temperature at which they freeze (TRANQUILLINI and HOLZER, 1958) shows that the two lie very close together. This has given rise to the suggestion that photosynthesis stops when the assimilatory organs begin to freeze. Results obtained using various evergreens (PISEK and REHNER, 1959; LARCHER, 1961) also pointed in the same direction. In order to pursue this idea more thoroughly not only the temperature minimum of photosynthesis but also the *freezing temperature* of the assimilatory organs was measured[7]. In addition, the leaves were exposed to various degrees of cold for 2—3 h and the depth of frost required to kill 15% and 50% of the leaf surface, or to cause corresponding damage to the veins and thus death of the area served by them, was recorded.

The results (Fig. 12) indicate that in most species the temperature minimum of photosynthesis lies within the range of variation (1—2° C) and close to the mean value of the freezing temperatures of the assimilatory organs. They also show that frost of a magnitude equal to the freezing temperature or slightly lower, invariably leads to considerable damage to the leaf surface and to the veins. Exceptions to this rule are provided by the few plants with the lowest (ca. −6° C) and the highest (ca. −1° C) minima. As already mentioned, *Ranunculus* and *Oxyria* belong to the former group, their leaves only freezing if cooled to 3° C below their freezing point. *Fagus* is another example: the blades of its leaves are divided up into minute intercostal areas, so well separated from one another that if ice nuclei form they are unable to spread rapidly. The areas freeze individually and in small groups one after the other, which is reflected in the appearance of a beech leaf after frost damage has set in. Sharply delineated, tiny brown dots and patches are scattered at random over the leaf surface (Fig. 4 in PISEK et al., 1967, for *Laurus* see LARCHER, 1955, Fig. 3). The freezing "point" of the individual leaves of *Fagus* and *Laurus* is therefore almost impossible to detect since the heat of crystallization connected with freezing is successively released in minute quantities. This explains why the exotherm in the otherwise perfectly smooth leaf temperature curve, so obvious in other species, is missing in *Fagus* and *Laurus*.[8]

Apart from the exceptions cited, the temperature minimum of photosynthesis and the temperature at which the assimilatory organs freeze and suffer damage lie, as a rule, close together. The suspicion that photosynthesis ceases when frost results in the formation of considerable quantities of ice in the tissues is thus justified. This was confirmed in an especially clear case: 1. PACK (see PISEK et al., 1967, p. 253) inserted a bunch of *Larix decidua* needles (short shoots) containing a small resistance thermometer into each of several glass tubes and the temperature was recorded continuously. Beneath the needles bicarbonate solution containing an indicator reflected by its color changes the decrease or increase in CO_2 content of the air in the closed test tubes or, in other words, the CO_2 uptake (assimilation) or output (respiration) of the samples. At some point in the course of the gradual cooling to about 1° C below the freezing point of the needles, the temperature curve for the majority of the tubes showed the simple peak already mentioned, indicating that the needles had reached their freezing point. The CO_2 balance, usually positive at first, was zero or negative subsequent to freezing. At the end of the experiment when the test tubes were opened the penetrating smell characteristic of frost-bitten *Larix* needles streamed out of the tubes containing the frozen samples. In addition, in contrast to the healthy needles, the frozen samples were discolored and later changed color even further. This ex-

---

[7] If the assimilatory organs are gradually cooled to 1—2° C below the approximately recorded freezing temperature the point at which freezing actually occurs reveals itself as a sudden peak in an otherwise perfectly smooth temperature curve (exotherm).
[8] The freezing temperature of the beech leaf (Fig. 12) is deduced from the frost damage picture.

periment demonstrates clearly that even in summer ice formation in the leaf causes irreversible damage to the assimilatory tissue and terminates photosynthesis. As long as a leaf remains unfrozen it can assimilate profitably.

The question now to be considered is why the assimilatory tissue is unable to cope with massive ice formation in summer.

Without going into the question of frost damage it will only be mentioned that ice formation in tissues, whether intra- or intercellular, means a sudden and large withdrawal of water for the cells involved. This leads to a rapid and passive rise in concentration of the cell solutions and of osmotic values both in vacuoles and plasma. Finally, the plasma is dehydrated (MAXIMOV, 1914; ILJIN, 1935), which must also have a severe effect upon photosynthesis.[9] 2. It must be borne in mind that the withdrawal of water connected with ice formation leads to a direct or indirect hydroactive closing of the stomates, with the result that photosynthesis is limited to the reassimilation of respiratory CO_2. The little that is known about the behavior of stomates in winter supports this view (e.g. KEMNITZER in PISEK and co-workers, 1967, p. 256; TRANQUILLINI and MACHL-EBNER, 1971). 3. The freezing of large quantities of cell water, even if it is mainly "readily mobile" or vacuolar water, can drastically cut down the CO_2 supply to the chloroplasts as well as the transport of all materials for which it is a vehicle.

The two last considerations gain in significance when it is realized that even in *winter* the assimilatory organs can at best do no more than compensate as soon as they freeze, whereas the leaves of various evergreens, especially the conifers of the northern alps, by no means suffer frost damage upon freezing as is common in summer. At the height of a severe winter, for example, the needles of freshly gathered, frost-hardened samples of *Pinus cembra* and *Picea abies* easily tolerate several hours at $-35°$ C or $-40°$ C if cautiously tested (ULMER, 1937; PISEK and SCHIESSL, 1947; SCHWARZ, 1970).

Having acquired a state of true frost resistance in the course of the gradual transition to winter they are now able to freeze stiff without suffering damage, i.e. their tissues can withstand massive ice formation. As already mentioned, as long as severe frost prevails these plants are completely inactive.

The high resistance to winter frost apparently depends upon the fact that the plasma is in a condition to tolerate a greater degree of dehydration than in the warmer seasons (PISEK and LARCHER, 1954). A possible connection between frost temperature and the quantity of ice formed (indicating the degree of dehydration) in both summer and winter is shown in Fig. 18 from PISEK and KEMNITZER, 1968. In summer, cooling to only a few degrees below zero suffices to transform about 40% of the total water in the leaf into ice. Large areas of assimilatory tissue are thus damaged and photosynthesis ceases. In winter, photosynthesis is arrested at only a few degrees lower than in summer and pre-

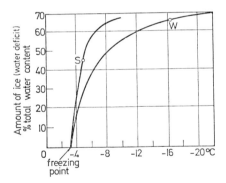

Fig. 18. Connection between degree of frost, quantity of ice formed in the leaf and frost damage in summer (*S*) and winter (*W*). (From PISEK and KEMNITZER, 1968)

sumably after only slightly greater water losses due to ice formation. Permanent damage in winter, however, only occurs when the gradual and continuous ice formation resulting from long, severe frost, reaches much higher levels than in summer, varying from species to species.

[9] Concerning the mechanism of dehydration on cell structures see SANTARIUS (1971) and LEVITT (1962, 1972).

*β) Additional Facts Concerning Temperature Dependence and Cardinal Temperatures
   of Photosynthesis*

The earliest experiments were aimed at discovering the degree of frost at which
the photosynthesis of northern conifers ceases to function profitably in winter.
IVANOW and ORLOWA (1931), for example, investigated *Pinus silvestris* and *Picea
abies* from the neighborhood of Leningrad; ZELLER (1951) worked with central
European *Picea abies, Laurocerasus*, winter wheat and spinach; NEGISI (1966)
obtained values from seedlings of three conifers important in forestry in Japan;
SALAGEANU and ATANASIU (1962a, b) and ATANASIU (1964, 1965) studied in
particular winter wheat, *Vinca minor, Ilex quifolium, Hedera helix* and conifers
in Rumania.[10] The values obtained, some of which are very approximate, vary between $-3°$ C and $-7°$ C and agree well with the author's own results. It is all the
more remarkable, therefore, that GODNEW and ROTFARB (1960) reported *Picea* as
taking up CO_2 labelled with radioactive carbon at a temperature of $-14°$ C and
incorporating it into soluble carbohydrate in the cell. Only four of thirty lichens
from different regions examined by LANGE (1965) behaved similarly.

Two of these, *Cladonia convoluta* and *Cladonia alcicornis*, achieved a measurable
net gain at 10 kLx even with thallus temperatures of $-22°$ C and $-24°$ C. Light-
dependent incorporation of ^{14}C was demonstrable at about $-10°$ C (LANGE and
METZNER 1965). On the other hand, comprehensive temperature curves for the
photosynthesis of *Cladonia coreyi* and *Cladonia alcicornis* show that below 0° C,
net assimilation falls steeply with decreasing temperature and stops altogether
at about $-10°$ C. Where a minimum net gain could be detected at even lower
temperatures this must be attributed to the very slow decline in CO_2 uptake at
$-10°$ C and below. In lichens, too, the temperature optima seem to be of greater
ecological significance than the minima: two species from the north Brazilian
catinga and the tropical *Cora pavonia* require at least 20° C at 10 kLx for maximum
efficiency, whereas *Ramalina maciformis* of the Negev desert in Israel needs merely
10° C. The latter confines assimilation during the long dry season to the cool early
hours of the morning when the water vapor pressure of the air is high. A number
of central European species function best at 0–10° C: *Hypogymna physodes*
(SCHULZE and LANGE, 1968), in the field, during a period of cold weather in March
at about 0° C, and antarctic species of *Neuropogon* at 0 to $-5°$ C. The temperature
maxima are usually relatively low, at most slightly above 30° C, and about 10° C
in the last-mentioned species. All of these data refer to investigations at 10 kLx,
which is worthy of mention since the temperature optima and maxima of lichens
rise with increasing intensity of illumination (LANGE, 1969, Fig. 9).

The unpretentiousness of many of these lichens is unrivalled by any flowering
plant, although admittedly little is known, for example, about the flowering
plants of the Arctic.

The values given by UNGERSON and SCHERDIN (1962, 1964) for the *temperature
optima* of *Betula nana* and *Betula tortuosa* in the Arctic (13/14° C, and 12/17° C
resp.) agree well with the results on trees of the alpine timberline. MOONEY
and BILLINGS (1961) cultivated *Oxyria digyna* from Alaska and from southern
alpine populations in the climate chamber under arctic conditions. Their meas-

---

[10] NEILSON, LUDLOW, and JARVIS (1972), *Picea sitchensis*.

urements were made using a light intensity of 23000 Lx over temperature ranges of 10−34° C and 10−44° C respectively. From the curves obtained it seems that the former is most efficient at a temperature of 15−24° C and the latter at 25−34° C. Measurements made in the open in the Medicine Bow Mountains put the optimum at about 22−28° C. These values are remarkably high as compared with optimum temperatures measured in the European central alps. In two-year-old needles of *Pinus halepensis*, the largest gain is at 20° C (2700 Lx, WHITEMAN and KOLLER, 1964), which is also the optimum temperature for one-year-old seedlings of *Pinus densiflora*, *Cryptomeris japonica* and *Chamaecyparis obtusa* (light saturated) in Japan (NEGISI, 1966), as well as for two varieties of *Pseudotsuga menziesii* (54000 Lx, KRUEGER and FERELL, 1965). One-year-old needles of *Picea glauca* and *Abies balsamea*, on the other hand, achieve their largest absolute gain at 26 kLx and 22−24° C (CLARK, 1961). A dozen broad-leaved evergreen dicotyledonous trees of subtropical southernmost Japan, according to KUSUMOTO's curves (1957a, b, 1961), function best at 25° C (20° C in winter), and only *Ficus retusa* assimilated at a maximum from 25−30° C. Three dicotyledonous trees investigated by STOCKER (1935) in their native habitat in Java gave light-saturated maximum net rates at 30° C, which coincides with values recently obtained in the field by HELLMUTH (1968) on *Rhagodia baccata*, a shrub of the arid to semi-arid Mulga bush zone of western Australia. *Acacia craspedocarpa* (HELL-MUTH, 1969), from the same habitat as *Rhagodia*, only reaches its temperature optimum in the dry, late summer when the temperature of the phyllodes attains 38° C (34−40° C), and its high temperature compensation point when their temperature is 53° C (Fig. 19). A further rise of only 2° C causes 50% damage to the phyllodes, *in situ* (LANGE's "heat resistance value"). Some typical plants of the Negev desert in Israel, such as *Salsola inermis*, *Haloxylon articulatum*, as well as

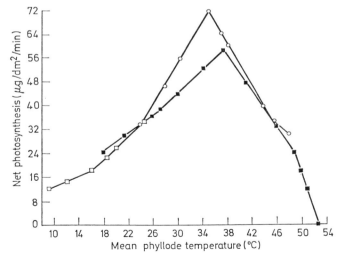

Fig. 19. Relationship between net photosynthesis and mean temperature of mature phyllodes of *Acacia craspedocarpa* in early summer dry season (○), in late summer dry season (■), and early winter wet season (□). (From HELLMUTH, 1969)

the cultivated *Citrullus colocyntis* exhibit comparable behavior at the end of the dry season. The two last-mentioned plants compensate at a temperature of 55° C (LANGE, KOCH, SCHULZE, 1969, Fig. 18). Approximately the same figures for temperature optimum and maximum were obtained by CHMORA and OYA (1967) in experiments carried out in Tadshikistan on maize with an irradiation of 0.4 cal/cm² min (= 33 kLx) (Fig. 20).[11] Due to its environment, the C_4 plant *Tidestromia oblongifolia* (Amaranthaceae) surpasses all plants. This low growing perennial of sandy habitats on the floor of the Death Valley, California, reaches the daily maximum of its photosynthesis at noon at a leaf temperature of 43 − 50° C (HARRISON and MOONEY, 1972).

A comparison of all data on the *temperature maxima* of other authors shows that only a few species like the arctic *Oxyria digyna* (BILLINGS et al., 1971) and grasses from Alaska (TIESZEN, 1970) cease their net-photosynthesis at 35−40° C as do the northern-alpine conifers (see Fig. 11). The majority of plants, however, stops binding CO_2 with a gain between 40 and 50° C. The net gain of a number of desert plants and tropical C_4 grasses becomes zero only at temperatures between 50 and 60° C (EL SHARKAWY and HESKETH, 1964; HELLMUTH, 1971; LUDLOW and WILSON, 1971; SCHULZE et al., 1972).

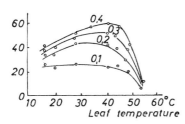

Fig. 20. Temperature curves of photosynthesis of maize in Tadshikistan at various intensities of irradiation (cal/cm² min) and normal CO_2 content of the air. The optimum heat range at 0.4 cal/cm². min extends from about 35−45° C, the upper compensation temperature limit is 53−55° C. (From CHMORA and OYA, 1967)

The differences in behavior of individuals of *Picea*, *Larix* and *Betula* (Fig. 7) from the valley compared with those from the timberline are not so large as would be expected from the meteorological temperature data (Fig. 8). This is due to the following: only the leaf temperature itself is important in photosynthesis and this is by no means always as low as meteorological records suggest. Second, photosynthesis, as emphasized at the outset, is largely buffered against variations in temperature by the breadth of its optimum temperature range. The coolness of higher locations in overcast weather and especially at night, brings with it the advantage of curbing respiration. In the far north and in high mountainous localities assimilation is limited less by the coolness of the summers than by the briefness of the snow-free period during which the temperature of the assimilatory organs is above the point at which they freeze.

e) Experimental Adaptation: Extent and Speed of the Reactions Involved

Very little is known at present concerning the extent and speed of the adaptation of photosynthesis to thermal changes in the environment. Here, as in other instances, certain limits are probably genetically determined. It is difficult to imagine that a typically Mediterranean or tropical plant, even if allowed plenty of time to become accustomed (apart from other obstacles) to an extreme northern environment, would be capable of achieving the required degree of carbon assimilation, or *vice versa*. Besides this, the margin of adaptation varies from case to

[11] Photosynthetic capacity is said to vary with the time of day, being at its highest at midday.

case, being sometimes wider, sometimes narrower (eurytopic and stenotopic species). Races of the same species growing in climatically different localities do not necessarily have to be adapted with regard to photosynthesis. Six races of *Mimulus cardinalis*, for example, investigated by MILNER and HIESEY (1964) in diverse habitats (45–2220 m) in California differed with regard to light saturation values which increase with altitude, but not with regard to the temperature dependence of photosynthesis or its temperature optimum. On the other hand, individuals of several species, all of which, with the exception of *Encelia*, were

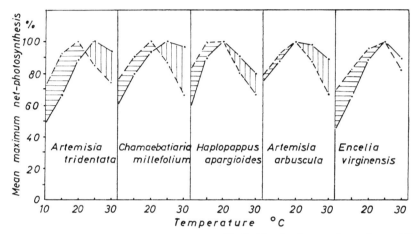

Fig. 21. Photosynthesis-temperature curves of individuals of several species, cultivated under identical conditions, after 3–7 weeks in the open at stations at various altitudes. Net photosynthesis as a percentage of the mean maximum value for any given acclimation treatment. Desert acclimated 1400 m (——), subalpine acclimated, 3100 m (.------.). (From MOONEY and WEST, 1964, simplified)

found up to 3000 m in the White Mountains of California, and which had been reared as seedlings at uniform temperature and photoperiod and then set out at stations of various altitudes and temperatures at their site of origin, revealed certain differences in behavior after 3–7 weeks of exposure (MOONEY and WEST, 1964). In all species, samples that were subsequently exposed to a cooler subalpine climate (3094 m, subalpine forest) assimilated more at lower temperatures than those that were transplanted into a warmer environment at lower altitudes (1440 m, desert scrub). The latter, on the other hand, proved to be superior at higher temperatures (Fig. 21).

The same trend was revealed in further experiments performed by MOONEY and SHROPSHIRE (1966) using *Encelia* and *Polygonum bistortoides*, and by ROOK (1969) on *Pinus radiata*: exposure to cold displaces the most productive temperature range downwards (with concurrent reduction of capacity to assimilate) and, conversely, the temperature optimum moves upwards, accompanied by a relative increase in intensity of photosynthesis, following exposure to higher temperatures. In both sets of experiments a remarkably brief period of time sufficed to bring about the alterations in photosynthesis. The photosynthesis temperature curve on the left in Fig. 22 was obtained from *Encelia* immediately following 12 days in a cold chamber (16 hours' light, 9000 Lx at plant level, 15° C; 8 hours' dark, 2° C). After only 23 hours in light of the above intensity at 30° C the curve on the right in Fig. 22 was obtained. ROOK found that the adaptation of *Pinus* seedlings from day/night temperatures of 33/28° C to 15/10° C and the reverse

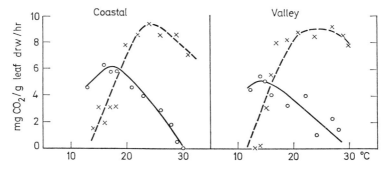

Fig. 22. Temperature curves of photosynthesis of coastal and valley clones of *Encelia californica*. ○ indicate values measured immediately upon removal from 12 days in the cold acclimation chamber (15/2° C) and **x** values measured after the same plants were maintained 23 hours in the light at 30° C. (From MOONEY and SHROPSHIRE, 1968)

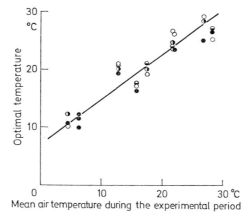

Fig. 23. *Triticum aestivum*. Relationship between optimal temperature of photosynthesis (light saturation: 60 kLx) and the mean air temperature during the experimental periods. (From SAWADA, 1970)

were effected within 7 and 4 days respectively. Finally, it should be mentioned that SAWADA (1970)[12] has plotted temperature curves of the photosynthesis of *Triticum aestivum* seedlings germinated at different times of the year. An almost linear relationship exists between the temperature optima for CO_2 uptake and the mean air temperature during the corresponding period of germination (Fig. 23).

---

[12] Based, like ROOK's results, on measurements of CO_2 uptake at only four different temperatures.

Seedlings of alfalfa, dallis grass, teosinte, Sudan grass and new sorgo, on the other hand, following 10 days at 15° C, 20° C and 25° C, exhibited no detectable adaptation to change of temperature (MURATA, ref. by SAWADA, 1970).

2. Respiration

A. PISEK

The net gain due to photosynthesis is continually being tapped for the respiration of the non-green tissues and organs, as well as by the C-autotrophic green tissues when insufficient light is available (below the light compensation point), or in the dark. In the following, only this so-called 'dark respiration' will be discussed: it constitutes the largest negative quantity in the C-relations of the plant. In a Danish forest of 25−46-year-old *Fagus sylvatica* of average quality, for example, it accounted for the loss of 40−43% of the gross material production (MOELLER, MÜLLER, and NIELSEN, 1954).

It is an old saying that one is alive as long as it breathes[13] and in fact the temperature limits of respiration for the individual cell coincide with the temperature limits governing its viability. It is therefore not surprising that the respiratory apparatus in the leaf is, in certain cases, more resistant to extremes of temperature than its photosynthetic equipment. This certainly holds true for higher temperatures. According to BAUER (1972), the respiration of needles and leaves of *Abies alba* and *Acer pseudoplatanus* is only slightly or not affected at all even after heat stress which leads to the subsequent *blocking* of photosynthesis; it does not even stop at higher temperatures. TRANQUILLINI (1957) has also pointed out that the respiratory system of *Pinus cembra* is less sensitive to cold than is its photosynthetic system. SEMIKHATOVA (1953, 1956, 1959; summarized and reviewed by PISEK, 1960) demonstrated that the susceptibility of the respiration of plants of the Pamir to heat and cold varies from species to species, as well as with the stage of development of the individual.

The intensity of respiration, like that of photosynthesis, also varies with the stage of development. Young beech leaves, for example, respire several times more strongly than mature leaves (MÜLLER, 1954; SCHULZE, 1970) and a second, temporary maximum is observable prior to leaf fall (SCHULZE, 1970). It has also long been known that shade leaves respire less than sun leaves of the same individual, particularly if release of CO_2 is calculated on leaf surface area (MÜLLER, 1954; LÖHR, 1969). Needless to say the intensity can also vary from species to species, some being typified by stronger, others by weaker respiration. On the whole, the well-hydrated leaves of the majority of rapidly growing wild and cultivated herbaceous plants are characterized by relatively active respiration per unit of dry weight, as compared with evergreen hard leaves and needles of conifers. In part, this is attributable to the small amount of dry substance in the former plants. Table 1 shows a small cross section of the respiratory capacity of leaves of various plants at 20° C in illustration of the foregoing. Other organs will be considered later.

---

[13] Only in a state of complete rest, such as in seeds with a minimum water content, can a latent viability be retained without accompanying release of CO_2 (HUBER and ZIEGLER, 1960).

Table 1. Respiration (mg CO_2) in midsummer per g dry weight and hour at 20° C

Deciduous		Evergreen	
Veronica persica	8.3	*Vaccinium vitis idaea*	1.1
Convolvulus arvensis (sun)	8.1	*Pinus cembra*	0.8−1.1
Urtica dioica	5.5	*Picea abies* (sun)	0.52−0.75
Solanum tuberosum	4.0	*Albies alba*	0.4
Leucojum vernum	4.0	*Taxus baccata*	0.4−0.6
Vaccinium myrtillus	2.7	*Laurocerasus officinalis*	0.4
Ranunculus glacialis	2.5	*Citrus limon*	0.4
Larix decidua	2.2	*Olea europaea sativa*	0.2−0.4
Quercus robur	1.7	*Quercus ilex*	0.3
Betula verrucosa	1.72		
Fagus sylvatica (sun)	3.0−1.5		
(shade)	0.95		
Quercus pubescens	0.8		

Data mainly from PISEK and KNAPP (1959), LARCHER (1961). More extensive list in STOCKER (1935) and especially in LARCHER (1969).

The shapes of the temperature dependence curves are essentially similar. The curves in Fig. 24 serve as examples: they were obtained from densely needled sun shoots of *Picea abies*. They are very similar to those obtained by KUIJPER (1910) 60 years ago using pea seedlings over approximately the same experimental temperature range. Respiration rises exponentially with increasing temperature and the curve runs more or less concavely to the abscissa. The slope in one and the same curve may vary, and can be described by the temperature coefficient (Q_{10}) or some other unit. This is dealt with comprehensively by FORWARD (1960). The longer the plant is exposed to heat the weaker the respiration gradually becomes, until it finally breaks down altogether after a certain time (Fig. 25).

In Fig. 24, apart from the basic shape of the respiration/temperature curve, it can be seen that 1. the shoots of *Picea* taken from trees in the valleys respire more weakly over the entire temperature range tested, both summer and winter, than do those from trees at the timberline, and 2. that both respire more weakly in winter than in summer. The differences between valley and mountain are particularly large in winter between 0° C and 5° C. At such low temperatures the respiration of mountain spruce is only half that of the same species in the valleys. Older axial parts that have long since lost their needles behave in an identical manner.

The intensity of respiration of bare twigs[14] of approximately the same diameter (3.5–3.8 cm) from the few subalpine conifers at the timberline and any deciduous trees found at this altitude varies from species to species, the order remaining the same irrespective of whether the CO_2 released is calculated from dry weight or per unit surface area (TRANQUILLINI and SCHÜTZ 1970, Fig. 26). The main cause of the variations lies in species differences in diffusion resistance to CO_2: the

---

[14] For approximate annual mean bark respiration of 8 woody species in their habitat see GEURTEN (1950).

quotient, CO_2 lost from open cross-sectioned surfaces/CO_2 lost from sealed surfaces ("degree of depression", ZIEGLER, 1956/57), offers a useful standard for
measurement. It is usually larger for deciduous than for coniferous trees because
the axial structures of the latter are more permeable. According to ZIEGLER the

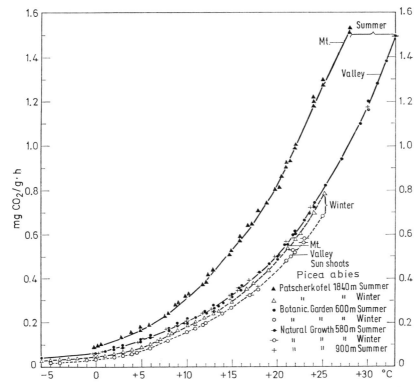

Fig. 24. Respiration (mg CO_2/g dry weight and hour) of sun shoots (the previous two
years growth) of *Picea abies* Karst. from valley (580 – 900 m) and timberline (1840 m)
related to temperature. 1. In summer in the valley in mid August, at the timberline
at beginning of September. 2. In winter (February). In both localities *Picea* respires
more in summer than in winter; it respires more at the tree line than in the valley.
(From PISEK and WINKLER, 1958)

diffusion resistance to gaseous exchange in the intact trunk resides not in the
continuous cambium cylinder but principally in the periderm. The same author considers that access of O_2 to the internal structures is impeded by this diffusion
resistance so that normal respiration must sometimes be replaced by fermentation
processes. These are not so much the aerobic type of process which is characteristic
of meristem tissues, according to RUHLAND and RAMSHORN (1938), as anaerobic
processes resulting from O_2 deficiency[15]. As ZIEGLER points out, such fermentation

---

[15] LÖHR (1957) reported respiratory quotients (CO_2/O_2) of 0.9 for sealed pieces of
Fagus trunk of 30–35 cm diameter, so that at least in this case diffusion and convection sufficed for normal respiration.

processes should not be regarded from the energetic point of view alone: they also provide the building units for specific syntheses (lignin).

Fermentation processes also take place in the root (see JENSEN, 1960). Comprehensive investigations by EIDMANN (1943) showed that the roots of 1—4-year-old conifer and deciduous seedlings give off, according to the species, about 20

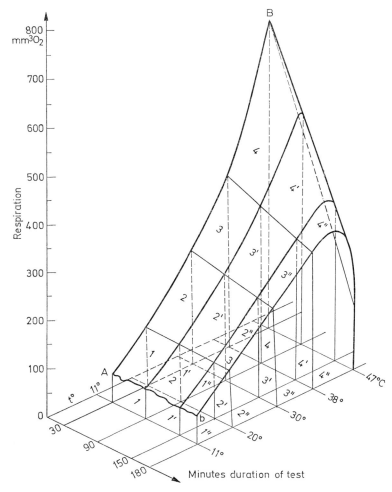

Fig. 25. *Podophyllum peltatum*, detached leaves. Temperature curves of respiration (measured by O_2 consumption), steep or flat according to length of exposure. (From SEMIKHATOVA and DENKO, 1960)

(Abies alba) to more than 100 *(Betula verrucosa)* mg CO_2 daily per g dry root over the growth period. In well-aerated soil roots respire particularly actively and make an important contribution to soil respiration as a whole. EIDMANN also gives curves demonstrating the connection between root respiration and mean

daily temperature in the culture vessels. The curves show that in general a mean daily temperature of about 22° C is the most propitious for root respiration. CO_2 production sinks at higher temperatures.

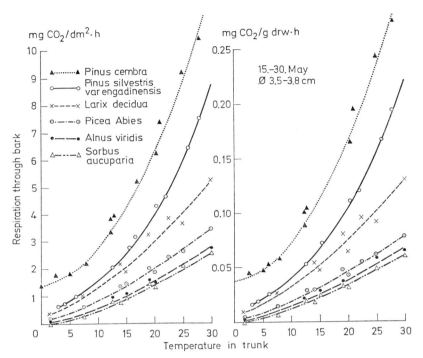

Fig. 26. Temperature curves of the respiration of the axial structures (mainly bark) of trees occurring at the timberline on the Patscherkofel near Innsbruck. Left: respiration related to surface, right: related to dry weight of stem samples. All samples taken in May, with a diameter of 3.5–3.8 cm. Cross-sectional surface sealed. (From TRAN-QUILLINI and SCHÜTZ, 1970)

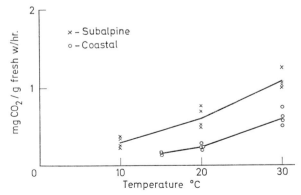

Fig. 27. Leaf respiration rates of plants from the Cape Mendocino (150 m) and Winne-mucca Lake populations (2730 m) of *Polygonum bistortoides*. (From MOONEY, 1963)

It was mentioned above that the shoots of sun twigs of *Picea* at the timberline (1900 m) respire more intensely at a given temperature than those from trees in the valleys (600 m, Fig. 24). A comparable difference was noticed decades ago by STOCKER (1935) in comparing the respiration of the leaves of his three Javan test trees with earlier data of other authors who had used plants from the arctic and from temperate latitudes. He found that the leaves of tropical trees did not give off more CO_2 per unit of leaf area in the dark at 30° C than was given off, on an average, by the leaves of northern trees at 10° C. Using Mediterranean sclerophylls, LARCHER (1961) found higher rates for 20° C respiration in winter than in the warmer season. It has been reported repeatedly that a cool environment promotes respiration. Not until recently have results become available from investigations involving uniform material, that is to say, using members of one and the same species of the same degree of ploidy but taken from areas varying with respect to temperature. MOONEY and BILLINGS (1961) recorded the intensity of respiration

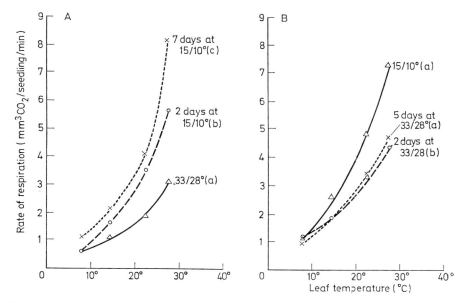

Fig. 28. The effect of temperature on respiration of *Pinus radiata* seedlings. A: Seedlings raised at 33/28° C (*a*) and transferred to 15/10° C for 2 (*b*) or 7 days (*c*). B: Seedlings raised at 15/10° C (*a*) and transferred to 33/28° C for 2 (*b*) or 5 days (*c*). (From ROOK, 1969)

over several temperature intervals in *Oxyria digyna* from Arctic populations and from southern alpine locations in the USA, but reared under identical (arctic) conditions. They found that the output of CO_2 in the dark was definitely larger in the former (especially at 10° −30° C) than in the individuals of southern origin. The respiration of *Polygonum bistortoides* (MOONEY, 1963) from Sierran subalpine habitats (Winnemucca Lake, California, 2730 m), at the same experimental temperatures, was quite definitely higher than that of individuals from the warmer coast (Cape Mendocino, 150 m; both diploid, Fig. 27). Rook (1969) cultivated seedlings of *Pinus radiata* for two months under uniform illumination but varying thermoperiod: the respiration of the plants which had been kept in the cold (15/10° C) was about twice that of those cultivated in warmth (33/28° C) over the temperature range 0–28° C (Fig. 28, curve *a*). If the thermal conditions were then reversed the effect of the change from cold to warmth and *vice versa* was seen almost in full after about two days (Fig. 28, curves *b* and *c*). Adaptation to thermal (in this case very drastic) changes in environment is achieved within a remarkably short period of time, just as in the corresponding photosynthesis experiments cited at the end of the previous section. Whether or not older plants react equally quickly has still to be investigated.

Finally, the favorable effect of a cool environment on respiration has recently been confirmed using *Atriplex polycarpa*. Plants grown from seeds taken from three geographically isolated localities of different altitude and temperature were used. In addition to confirming earlier findings it was also shown that the climate of the site of origin plays a part in so far as plants originating in warmer climes (within the distributional area of the species) adapt less well to cold (16/5° C, respiration ceased at an experimental temperature of 5° C) than plants from cooler regions (Chatterton et al., 1970).

3. Transformations and Translocation of Carbohydrates

W. Larcher

a) Temperature-Dependent Conversions of Carbohydrates

Autotrophic plants build up starch and other storage forms of carbohydrate from monosaccharides at the site of photosynthesis itself. These products are only temporarily deposited, soon to be mobilized, redistributed, consumed in metabolic reactions or polymerized to storage forms at other sites in the plant (for the chemistry of storage carbohydrates see relevant literature such as Encyclopedia of Plant Physiology. Vol. VI).

Apart from short-term carbohydrate transformations and displacements which are closely linked with the primary carbon metabolism, other rearrangements connected with the stage of development and season occur, often involving the storage processes. The time and place of occurrence of these phenomena vary from species to species.

Carbohydrate conversions have been investigated with especial accuracy and thoroughness in trees. Classical publications on this subject are those of A. Fischer (1890) and Lidforss (1896); the older literature was reviewed in detail by Jeremias in 1964. Special attention is merited by the comprehensive investigations of Gäumann (1935), Jeremias (1964), and Kimura (1969) on a variety of woody plants. Critical reviews of the prolific literature are provided by H. Fischer (1958), Wanner (1958), Kramer and Kozlowski (1960), Biebl (1962), Jeremias (1954), Ziegler (1964), Kozlowski and Keller (1966), Lyr, Polster, and Fiedler (1967), Steinhübel (1967); cf. also p. 184. In summer and early autumn starch is stored in the wood and bark of the stems and roots of most perennial plants in the temperate zone, and sometimes even in evergreen leaves. In many species, at least a part of this starch is dissolved in late autumn and early winter. In late winter and spring soluble carbohydrate is usually reconverted into starch, which when growth commences is broken down and consumed (see Fig. 29).

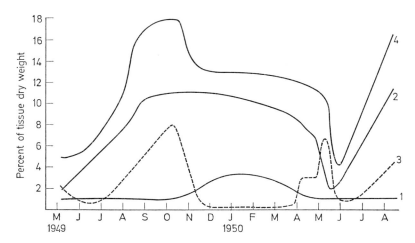

Fig. 29. Seasonal variation in reducing sugars (*1*), sucrose (*2*), starch (*3*), and total reserve carbohydrates (*4*) in the living bark of *Robinia pseudacacia* trees in Ontario. (From SIMINOVITCH, WILSON, and BRIGGS, 1953)

In addition to other environmental factors and endogenous control mechanisms, temperature appears to be an essential factor in triggering carbohydrate conversions. The lower the temperature (above freezing) to which the plant is exposed the more pronounced is the rise in sugar concentration at the expense of the stored carbohydrates, particularly oligosaccharides. The temperature threshold for the conversion of starch into soluble carbohydrate varies from species to species: the critical temperature range for trees and shrubs of the temperate zone is from 0−10° C (below +5° C in *Pinus mugo* and *Populus*: JEREMIAS, 1964; SAUTER, 1957; similar values for herbaceous plants: KACPERSKA-PALACZ et al., 1969; about 8°C in *Betula* and *Tilia*: JEREMIAS, 1964). Glasshouse plants (COVILLE, 1920), and roots which are protected from severe cold (WEBER, 1909) break down little or no starch. In angiosperm trees of the temperate climatic zone a rise of more than 5° C above the critical temperature leads to reaccumulation of starch (SAUTER, 1967). Starch-sugar transformations also occur in tropical trees, but are endogenously controlled and less pronounced (see KOZLOWSKI and KELLER, 1966, p. 111).

Low temperatures lead not only to starch-sugar conversions, but also to proportional changes in the nature and quantity of carbohydrates dissolved in the cell sap (detailed literature in JEREMIAS, 1964; cf. Fig. 30). Significant changes in concentration are mainly found among the various mono-, di- and oligosaccharides (STEINER, 1933; occurrence of sugars of the raffinose group see chiefly PARKER, 1957, 1962; JEREMIAS, 1962), but apart from this the content of sugar alcohols (SAKAI, 1961) and pentosans is subject to variation (HENZE, 1959).

Temperature-dependent carbohydrate conversions can also be observed during the ripening of fruit and vegetables, and especially in cold storage. In ripening fruit, high temperatures accelerate both the conversion of starch into sugar and

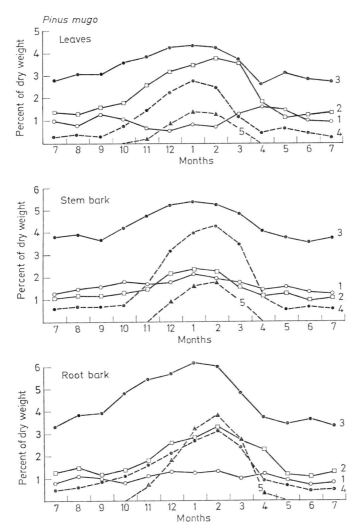

Fig. 30. Seasonal variation in glucose (*1*), fructose (*2*), sucrose (*3*), raffinose (*4*), and stachyose (*5*) in leaves, stems, and roots of *Pinus mugo* in southern Germany. (From JEREMIAS, 1964)

the degradation of hemicelluloses and pectins (for details see WOLF, 1958). The cold-induced shift in the equilibrium between starch and sugar in potatoes, Swedish turnips and iris rhizomes was reported in detail as early as 1882 by MÜLLER-THURGAU. At temperatures between $+1°$ C and $+5°$ C sugars, especially saccharose, accumulate in potatoes at the expense of starch. Here, too, the process can be reversed by raising the temperature (Fig. 31). The transformation of starch is very sensitive to temperature: both cold and warm stimuli elicit a physiological response within a short period of time.

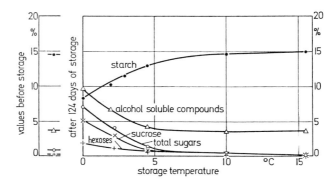

Fig. 31. Changes in the carbohydrate composition in potato tubers during storage as influenced by temperature. (After WRIGHT, PEACOCK, WHITEMAN, and WHITEMAN, 1936, from WOLF, 1958)

Little is known concerning the way in which temperature influences the bio-chemical processes involved in the conversions. The accumulation of sugar would be understandable if the speed of starch breakdown had a smaller temperature coefficient than the respiration of the sugar formed. The same effect could be achieved by a temperature-dependent enzyme inhibition. Changes in acidity, too, resulting from temperature-dependent alterations in the intensity of respiration might cause specific changes in enzyme activity (see ARREGUIN-LOZANO and BONNER, 1949; EWART, SIMINOVITCH, and BRIGGS, 1954; JEREMIAS, 1964). The biochemical literature (e.g. WHELAN, 1958) should be consulted regarding the temperature dependence of the enzymes involved in carbohydrate conversions.

b) The Influence of Temperature on the Translocation of Carbohydrates

The transport of carbohydrates in cells, both within the protoplasts and between adjacent cells, obeys the usual rules governing the temperature dependence of diffusion, osmosis and facilitated transport across protoplasmic membranes.
The translocation of carbohydrates between individual organs in the plant usually follows an optimum curve, with increasing rates from 5–20° C, optimum efficiency between 20° C and 30° C, followed by a sharp drop above 40° C. Results obtained from various plants are to be found in CURTIS and CLARK (1950), H. FISCHER (1958), KRAMER and KOZLOWSKI (1960), LANGRIDGE and MCWILLIAM (1967), and WARDLAW (1968). The various processes involved in translocation differ in the degree of their temperature dependence, which accounts for the striking differences observed. The exudation of carbohydrates into the sieve tubes, for example, and their analogous uptake from the phloem by parenchyma cells at the other end are both processes which are closely coupled with cellular metabolism and therefore

strongly dependent upon temperature. The mass flow of assimilates in the vessels, on the other hand, is scarcely dependent upon temperature at all (see FORD and PEEL, 1966; further literature in LANGRIDGE and McWILLIAM, 1967). The formation of the assimilates and their consumption are very much dependent upon temperature. In addition, growth and correlations between the organs of the plant play an important role (survey in FISCHER, 1958; WARDLAW, 1968). The entire process is geared to the speed of the slowest reaction involved (the "speed-limiting step").

In order to investigate the actual process of translocation independently of other concurrent reactions the rate of translocation of ^{14}C compounds through the petiole and shoot nodes was measured at strictly controlled and usually low temperatures (THROWER, 1965; WEBB and GORHAM, 1965; WILLENBRINK, 1966; SWANSON and GEIGER, 1967; WEATHERLEY and WATSON, 1969; for further literature and review see GEIGER 1969). Even using this procedure translocation in many plants followed a noticeably optimum curve with the highest transport rates between 20° C and 30° C (see Fig. 32). Sugar transport in cold-sensitive plants such as soybean, squash and geranium is almost completely but reversibly blocked below

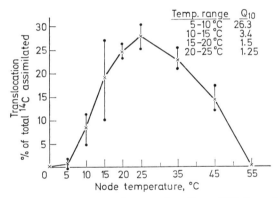

Fig. 32. Effect of node temperature on the translocation of ^{14}C from the photosynthetic active leaves through the temperature-treated node of squash (*Cucurbita melopepo torticollis*). (From WEBB and GORHAM, 1965)

5° C. On return to a more favorable temperature the translocation mechanism soon recovers and is back to normal, for example, within one hour in the squash (WEBB and GORHAM, 1965). The long-distance transport of cold-adapted plants such as sugar beet and northern ecotypes of *Cirsium* is less strongly affected by low temperatures; in *Salix viminalis* it is not blocked until a temperature of −4° C is reached (then, however, irreversibly; WEATHERLEY and WATSON 1969).

A2. Temperature Dependence of other Metabolic Processes

W. LARCHER

As a result of the influence exerted by temperature on membrane transport and on enzyme-controlled reactions, the entire cellular metabolism is strictly temper-

ature dependent. The manner in which temperature influences protein and fat
metabolism and the manifold secondary metabolic reactions in higher plants is
basically the same as in other poikilothermic organisms (see p. 298 ff, and the relevant
reference books).

In the following, only the water relationships of terrestrial plants and their uptake
of nutrients will be considered, since these two aspects of their metabolism present
some special features.

1. The Influence of Temperature on the Water Balance of Higher Plants

The water relations of terrestrial cormophytes depend on the balance existing
between water uptake, water transport and water losses (guttation and transpiration, the latter being by far the more important). The critical step in the water
supply of plants is usually its uptake from the soil.

The three components of the water balance of the plant vary in the degree of their
temperature dependence.

a) Water Uptake

Soil temperature is a critical factor in water uptake: it influences the mobility of
the soil water (GARDNER, 1968; TAYLOR, 1968) and affects the intensity with
which the roots absorb water. Each species is characterized by its own temperature optimum for water absorption via the roots (Fig. 33). The influence of
temperature on water uptake, however, is obscured by soil water stress (BABA-
LOLA et al., 1968). At suboptimum temperatures the process of absorption is
greatly enhanced by raising the temperature (Q_{10} of about 3—4): it effectively
improves the permeability to water, reduces the viscosity and accelerates the

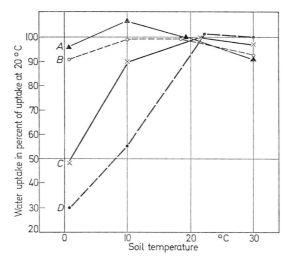

Fig. 33. Water uptake by roots of intact seedlings of *Populus nigra* (*A*), *Salix fragilis*
(*B*), *Fagus silvatica* (*C*), and *Fraxinus excelsior* (*D*) at different soil temperatures.
(After DÖRING, 1935)

metabolic and growth activities in root tissue (KRAMER, 1940; KUIPER, 1963, 1964; for mathematical expression of the temperature dependence of water uptake by root cells see KUIPER, 1964). Preconditioning of plants in soils of various temperatures leads to an adaptation in efficiency at the temperature employed (see KRAMER, 1965, and Fig. 34). A close connection exists between the temperature prevailing in the natural habitat of a plant and the temperature optimum of water uptake (DÖRING, 1935).

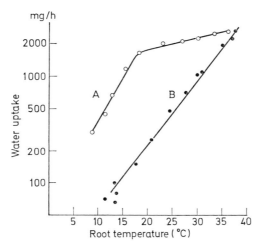

Fig. 34. The effect of root temperature on water uptake by intact bean plants grown at root temperatures of 17° C (*A*) and 24° C (*B*). (From KUIPER, 1964)

Special attention has been devoted to water uptake from cold soils. The observation that a number of herbaceous plants began to wilt at soil temperatures slightly above the freezing point led to the suspicion that insufficient water was being taken up (SACHS, 1860; GOEBEL, 1908 e.g.). The problem of impeded water uptake from cold soils has been investigated by FIRBAS (1931), DÖRING (1935), ROUSCHAL (1935), KRAMER (1940, 1942), DADYKIN (1950), GREB (1957), COX and BOERSEMA (1957); a survey of the relevant literature is to be found in KRAMER, 1956; WALTER, 1960; LYR, POLSTER, and FIEDLER, 1967. It has been shown that water uptake in many woody and herbaceous plants is considerably depressed at a few degrees above zero, but despite this, after an initial shock phase, most plants are able to recover, even if not completely. In addition, marked specific differences in the sensitivity of water uptake to cold, not only between different families and genera, but also between species of one and the same genus, have been detected and accurately measured. DÖRING found *Salix glauca* to be well adapted to cold, *Salix purpurea* somewhat sensitive and *Salix triandra* highly sensitive to cold. The absorption capacity of species that begin to develop early is less impeded by cold soil than that of species developing later. Besides this, many plants appear to be well adapted to the soil temperature prevailing in their natural biotype: species from the bogs of northern latitudes still absorb adequate amounts of water at temperatures slightly above zero (DÖRING, 1935) and plants of the Siberian tundra can even absorb water from partially frozen soil (DADYKIN, 1950).

b) Water Transport

The temperature dependence of water transport results from the physical laws connected with the influence of temperature on suction potentials, root pressure and viscosity. Tension gradients and root pressure are the driving forces for the long-distance transport of the transpiration stream. The root pressure, is, via the cellular metabolism, temperature dependent (root pressure of terrestrial plants is reviewed by KRAMER, 1956, and of aquatic plants by GESSNER, 1956). The effect of temperature on the flow of water brought about by suction potentials can be defined by thermodynamic equations (SLATYER, 1967; TAYLOR, 1968; WALTER and KREEB, 1970). The viscosity of the water in the xylem vessels, in accordance with the Hagen-Poiseuille law, is approximately doubled when the temperature sinks from $10-0°C$. However, at normal and even low temperatures, if above the freezing point, such effects play scarcely any part in the water relations of the plant (reviewed by KRAMER, 1956, see also JOHNSTON, 1959 and ZIMMERMANN, 1964).

At very low temperatures the water balance even of large trees may be severely affected by freezing of the water in the vessels. It can be expected that the xylem water of the branches and trunk of the tree begins to freeze at from $-1.5°$ C to $-3°$ C. Supercooling evidently plays a role here because the freezing-point depression of the pure water in the vessels ought at most to be $-0.02°$ C (measured by ZIMMERMANN, 1964, in *Ulmus*). Values of -5 or $-6°$ C or lower, reported in the

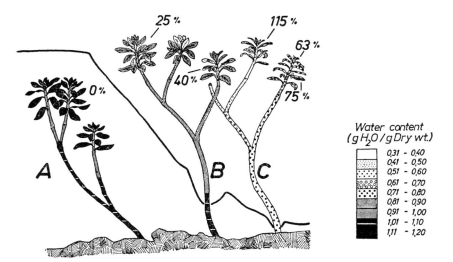

Fig. 35. Decrease in the water content of leaves and stems of *Rhododendron ferrugineum* at the alpine tree line depending on soil freezing and snow cover. *A* Permanent snow cover; *B* Twigs without snow cover during winter, in February; *C* Twigs in March. The percentages indicate the degree of loss of available water (initial minus sublethal water percentage) during winter drought. (From LARCHER, 1963b)

literature (DIXON, 1914, *inter alia*) as the freezing point for vessel water, are apparently based upon insufficiently accurate measurements (for detailed discussion see ZIMMERMANN, 1964). When the water in the conducting vessels freezes the water flow in the stem comes to a standstill, but under favorable conditions it can begin to move again as soon as the ice thaws (HAMMEL, 1967; SUCOFF, 1969). This is sometimes associated with the formation of small bubbles of gas in the conducting vessels which as a result are blocked for a considerable time or even permanently (SCHOLANDER, LOVE, and KANWISHER, 1955; LYBECK, 1959; KOZLOWSKI, 1961; HYGEN, 1965). Subsequently, the water supply to the shoots gradually deteriorates and they dry out (Fig. 35: "winter drought"; see EBERMEYER, 1873; F. C. GATES, 1914; WALTER, 1929; MICHAELIS, 1934; THREN, 1934; LARCHER, 1957, 1973; MICHAEL, 1966, 1967 with detailed review of the literature).

c) Transpiration

Transpiration from plant surfaces is strictly dependent on the physical laws of evaporation, but the plant is nevertheless able to control water losses to a considerable extent by regulating its stomatal aperture (detailed reports in SEYBOLD, 1929 a, b, 1930; HUBER, 1930; STÅLFELT, 1956, and various contributions in Encyclopedia of Plant Physiology, Vol. III). The physical components of transpiration can be calculated from the parameters of radiant energy budget and convection (RASCHKE, 1958; KUIPER, 1961; MELLOR, SALISBURY, and RASCHKE, 1964;

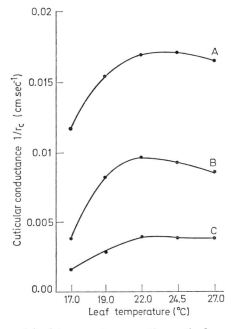

Fig. 36. The influence of leaf temperature on the cuticular conductance for water vapor on the upper surface of leaves of *Lamium galeobdolon* (*A*), *Betula verrucosa* (*B*), and *Acer platanoides* (*C*). (From HOLMGREN, JARVIS, and JARVIS, 1965)

HOLMGREN, JARVIS and JARVIS, 1965; TANNER, 1968; D. M. GATES, 1968, see
p. 96). The effect of regulation of the stomatal aperture can also be calculated if
the stomatal diffusion resistance is known (BANGE, 1953; GAASTRA, 1959; JARVIS,
ROSE, and BEGG, 1967; SLATYER, 1967).
Temperature influences transpiration *directly* via the temperature-dependent exchange processes at the plant surfaces (temperature effect on the cuticular conductance of leaves see Fig. 36), and *indirectly* due to the temperature dependence
of the stomatal aperture (STÅLFELT, 1928; GÄUMANN and JAAG, 1936, 1939; WIL-
SON, 1948; RUFELT, JARVIS, and JARVIS, 1963; DRAKE, RASCHKE, and SALISBURY,

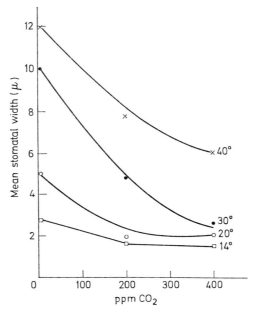

Fig. 37. Temperature effect on mean stomatal width of *Zea mays* at different carbon
dioxide concentrations in the air. (From Moss, 1963)

1970; WUENSCHER and KOZLOWSKI, 1971; SCHULZE et al., 1973; further literature
in STOCKER, 1956; MEIDNER and MANSFIELD, 1968). The degree to which the
stomatal apparatus opens in light depends primarily upon the intensity of illumination and the degree of water saturation of the plant. Elevated temperature accelerates the process of *opening*, especially at low partial pressures of CO_2 (STÅL-
FELT, 1962; MOSS, 1963; ZELITCH, 1963; RASCHKE, 1970; see also Fig. 37). Stomatal
closure seems to be largely independent of temperature, occurring equally rapidly
at 10° C and 30° C (ZELITCH, 1963). Possible mechanisms of stomatal movement
and its dependence upon temperature are discussed in detail by STÅLFELT (1956),
ZELITCH (1963), LEVITT (1967) and MEIDNER and MANSFIELD (1968).
Stomata of most plant species do not open to their full width at temperatures near
0° C and opening occurs slowly; in frozen leaves the stomata are closed (PISEK

et al., 1967; NELSON, LUDLOW, and JARVIS, 1972). Also for these reasons transpiration of evergreen plants is remarkably reduced in winter (KILLIAN, 1932; CARTELLIERI, 1935; review by LARCHER, 1973).

2. Influence of Temperature on the Mineral Nutrition of Higher Plants

Under natural conditions higher terrestrial plants take up their inorganic nutrients from the soil via the roots (for aquatic plants see GESSNER, 1959). Mineralization processes and the mobility of nutrient ions in the soil are temperature dependent (for the dynamics of nutrients in the soil see relevant soil science literature).

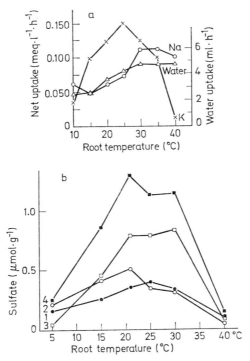

Fig. 38. a Uptake of water and Na and K ions by roots of intact plants of *Ricinus communis* as a function of root temperature; test solution: 2 meq/l KNO₃ and NaNO₃. (From NWACHUKU, 1968). b Effect of different root temperatures on the components of sulfate absorption in wheat plants. (*1*) Amount of labile-bound sulfate in the roots, (*2*) retention of sulfate in the roots, (*3*) transport of sulfate to the shoots, (*4*) total active uptake of sulfate by the roots. Test solution: 0.5 mM MgSO₄. (From PERSSON, 1969)

Nutrients move passively, dissolved in soil water, into the outer tissue layers of the root ("apparent free space", ROBERTSON, 1958). Their uptake at this point is achieved by surface absorption following diffusion along a concentration gradient, by electrodiffusion along a potential gradient, and by ion exchange. These steps in salt uptake take place spontaneously and are almost temperature independent but closely linked with the intensity of the water uptake (details in KISSER, 1956; EPSTEIN, 1956, 1960; KRAMER and KOZLOWSKI, 1960). The uptake of nutrient ions by the protoplasm requires energy from respiration and is thus subject to the temperature dependence characteristic of metabolic processes. A voluminous literature deals with the mechanism of active salt uptake; surveys are given by e.g. LEGGETT (1968), LATIES (1969), and in particular by BAUMEISTER and BURG-HARDT (1969). Classical publications include those of LUNDEGÅRDH (1932), LUNDE-GÅRDH and BURSTROM (1933) and HOAGLAND (1944). The Q_{10} between 15° C and 25° C for the entire process is low and lies at about 1.2 on the average; the value for anions is slightly larger and for cations somewhat smaller. At low temperatures, however, the energy demands are larger so that the Q_{10} may rise to 2 or 3 or even more (see ROBERTSON, 1958). The various nutrient ions each have their own temperature range for optimum absorption (see Fig. 38).

Uptake of nutrients from cold soils can be strongly inhibited, just as is water uptake (see p. 139), so that the salt metabolism of plants in habitats where the soil remains cold for long periods is often disturbed. Nitrogen supplies are particularly impaired (MOTHES, 1932; SCHMIDT, 1936; DADYKIN, 1950; GREEB, 1957).

B. Effect of Temperature on Growth and Development

B1. Dependence of the Growth Processes on Temperature

A. VEGIS

1. Dormancy of Buds and Seeds

For some decades the majority of investigators believed that the dormant condition originated autonomously in plants during a particular season. They conceded, however, that external conditions might accelerate or delay the onset of dormancy to a certain extent, but never prevent it. The results of recent research (see VEGIS, 1961, 1963, 1964, 1965 and WAREING, 1969) using many plant species have verified the opinion of KLEBS (1903, 1914, 1917) and GASSNER (1918, 1921, 1925, 1933) that the dormant condition in plants is induced by external factors. Among plants which normally have an annual dormant period, the number which have been induced to grow and develop more or less continuously, given certain conditions, has increased steadily during recent years. In almost all plants which periodically become dormant, dormancy can be induced even in a season during which there is normally active growth in the open, by changing the external conditions. Without overstatement, we may say that we are on the way to controlling arbitrarily the growth activity of plants. Among the factors which greatly influence growth activity are temperature, photoperiod, quality of light, temperature during the daily light and dark periods, nutritional conditions and water supply. The growth responses of plants are usually a result of the combined effect of these factors. For references see VEGIS (1961, 1965b), WAREING (1968, 1969) and WAREING and PHILLIPS (1970).

The environmental temperature often has a very important influence not only on cessation of growth and onset of dormancy but also on the termination of dormancy. Before discussing in detail the influence of temperature on dormancy we must consider the different stages or phases of the rest period. Dormant organs of plants often differ in their response to different temperatures at various stages of this period.

a) Phases of the Rest Period and the Various Kinds of Dormancy

Until recently, comparatively little attention had been paid to the various phases of the rest period and the different degrees of dormancy in plants. In general, the terms *true dormancy*, resting condition, deep dormancy or organic dormancy refer to a condition in buds, seeds and other plant organs where no growth occurs, irrespective of the external conditions. However, the organs do not remain in this state during the entire rest period but only in its middle phase, called the "main rest" or "middle rest". For some time it has been known that the state of true dormancy does not start or terminate abruptly (JOHANNSEN, 1900, 1933; HOWARD, 1906, 1911). The transition from the state of full growth activity to true dormancy occurs gradually. This transition occurs during the initial phase of the rest period, called "early rest", and the state of plant organs at that time is known as "predormancy" or "preliminary dormancy". Characteristic of organs in this state is that they have not completely lost the ability to grow, although growth is only possible within a narrower range of external conditions than at the time of full growth activity. As predormancy progresses this range becomes narrower and narrower and growth becomes progressively slower. External factors unnecessary for initiating growth at full growth activity, as for example a daily light or dark period of a certain duration and light of a definite wavelength, are often essential for the continuation of growth during this phase (see BASS, 1954; KOLLER, 1955; EVENARI, 1956, 1965; E. H. TOOLE, HENDRICKS, BORTHWICK et al., 1956; DE LOUCHE, 1958; ROLLIN, 1959; VEGIS, 1961; KOLLER, MAYER, POLJAKOFF-MAYBER et al., 1962; MAYER and POLJAKOFF-MAYBER, 1963, and EVENARI, 1965). Many observations have shown that the ability of seeds to germinate and of buds to continue growth and development disappears under the external conditions that prevail in the original habitat of the species or variety in question. Character-istically, at the beginning of the rest period these organs show no growth under the prevailing environmental conditions, but retain for a time the ability to grow under other conditions. This is exemplified by the natural distribution of many species and varieties of *Caryophyllaceae* and other plant species around the Mediterranean. THOMPSON (1968, 1970a, b, c, 1973) has found that exposure of im-bibed seeds of the aforementioned species to high temperatures (for example 31° C) corresponding to the summer temperatures in their region of growth, reduced the subsequent temperature maximum for germination of these seeds by about 4° C. By this mechanism seeds were prevented from germinating during the summer before the hot dry season unfavorable for the survival of young seedlings. However, these seeds still maintain the ability to germinate over a temperature range lower than that prevailing during the above-mentioned season. Such a dormant state, observable only within a certain range of external conditions, is often called "relative dormancy" (BORRISS, 1940; VEGIS, 1949a, b, 1961, 1964, 1965a). This is essentially an imposed dormancy, growth being rendered impossible by a lack of the necessary environmental factors. Relative dormancy should not be mistaken for the state of quiescence or arrest of active growth of *non-dormant* organs due to unfavorable environmental conditions. Unfortunately this restriction of growth is also often called "imposed dormancy". Unlike relative dormancy this type of restrained growth can be observed during maximum growth activity in organs which are potentially capable of growth over a wide range of external conditions.

The external conditions required by plants for undisturbed development must also be considered. Conditions for the formation and growth of the organ primordia may change during ontogenesis: external conditions which are optimal for one phase of ontogenesis may be unfavorable for another. These changing require-ments often correspond with the seasonal changes of external conditions in the natural environment. If the external conditions do not change to meet the require-ments of the plant in question, growth and development cease. Under certain external conditions, cessation of growth is followed by the onset of dormancy.

The transition of predormancy to true dormancy marks the entrance to the middle phase of the rest period, the main or the middle rest. At this stage the depression of growth activity is strongest and the dormancy deepest. Truly dormant resting buds and seed embryos cannot be induced to immediate normal growth by any means. Upon termination of the true dormancy state, the phase of main rest passes on to the final phase of the rest period, which is called "after-rest" for buds and "after-ripening" for seeds. The internal condition of dormant organs from the time of completion of true dormancy until maximum growth activity is attained is called "post-dormancy". During this phase of the rest period the organs, as in predormancy, are in the state of relative or conditional dormancy and, within the limits of certain external conditions, such dormant organs are able to resume growth. At the beginning of after-rest these limits are very narrow, but widen progressively as the phase progresses. The after-rest and the rest period as a whole are terminated, and the state of post-dormancy disappears, when growth activity reaches its maximum and the limits of external conditions within which growth can take place are at their widest. These limits are characteristic for each species and variety. However, growth can start before the completion of the after-rest if the dormant organs have already attained the ability to grow within that range of external conditions which prevails in their environment. In this way, the duration of after-rest is highly dependent on the external environment. Plants which have not completed their rest period can, under certain unfavorable conditions, even after growth has begun, be obliged to re-enter the true dormant state or secondary dormancy. For further details concerning the rest period and dormancy in seeds, which are of great diversity, see BALDWIN (1942); CROCKER (1948); CROCKER and BARTON (1953); BARTON and CROCKER (1948); BARTON (1961, 1965a, b); EVENARI (1949, 1956, 1957); E. H. TOOLE, HENDRICKS, BORTHWICK et al. (1956); VEGIS (1961); KOLLER, MAYER, POLJAKOFF-MAYBER et al. (1962); MAYER and POLJAKOFF-MAYBER (1963); AMEN (1963, 1968); LANG (1965); STOKES (1965); WAREING (1965, 1969); WAREING and PHILLIPS (1970) and NIKOLAEVA (1969).

Until now, most investigations of the dormant state have primarily been concerned with the middle phase of the rest period, or the condition of true dormancy. The methods used to determine whether the experimental material is in a state of true dormancy before the experiments begin are often inadequate. Inability to grow under conditions which at full growth activity are favorable for growth is not always a valid indication of true dormancy. Significant results have been obtained in recent investigations of the changes in plant responses to temperature during early rest and, in particular, after-rest. Several ecological types have been revealed among plants which become dormant during certain seasons. This kind of variation again shows how closely growth cessation at the beginning of the rest period is dependent on the external conditions regularly prevalent before the onset of the unfavorable season. Moreover, it appears that decrease in growth activity is not always followed by development of true dormancy. The early phase of the rest period often passes directly into the after-rest phase. In this case, even at the time of lowest growth activity, or the strongest depression of growth, plants are only in a state of relative or conditional dormancy. This means that during the whole of the rest period they have retained the ability to grow, although only within a narrow range of external conditions. However, these conditions do not normally prevail in the natural environment during the rest period, and thus growth cessation is usually observed. For many species and varieties this seems to be a normal phenomenon. For others, the presence or absence of true dormancy appears to be dependent on the climatic conditions during the year in question. In several cases the reduction of growth activity during the unfavorable season is manifested not as cessation of growth but only as a decrease in growth rate within certain temperature limits.

b) Temperature Range for Seed Germination and Bud Break in Relation to the Depth of Dormancy during the Rest Period

The work of LEFEBURE (1801), SACHS (1860) and HABERLANDT (1874, 1875a, b) laid the foundation for the concept that germination and growth usually take place over a defined temperature range which is characteristic for the different organs of each species and variety. They introduced the idea of three cardinal points of temperature: minimum, optimum and maximum. However, this concept is not entirely correct. Changes in growth activity can bring about a narrowing or widening of this temperature range, besides a shift in the minimum as well as the optimum and maximum temperatures for germination or other growth processes. SACHS and HABERLANDT studied essentially the dependence of growth processes on temperature during the period of most active growth, when seed germination, bud break and growth occur comparatively quickly over the widest range of temperatures.

The idea of a narrowing and a widening of the limits of external conditions under which growth is possible during early rest and after-rest, is based mainly on observations of the behavior of plants at various temperatures and at different phases of the rest period. The pioneer observations on seeds were made by ATTERBERG (1899, 1907) and have been confirmed and extended more recently by other investigations (for references see VEGIS, 1961, 1964 and 1965b). With regard to the behavior of buds, and especially of bulbs, ASKENASY (1877) and BLAAUW et al. respectively (see PURVIS, 1937, 1938; WENT, 1948; HARTSEMA, 1954, 1961) show that temperature requirements of buds change somewhat during the rest period. Widening of the temperature range for bud break during the after-rest was clearly demonstrated first in 1949 (VEGIS, 1949a, b). For further references see VEGIS (1961, 1963, 1964, 1965b, and 1967).

In a number of cases a narrowing of the temperature range occurs which makes a species or variety incapable of growth and thus enables it to survive an unfavorable season.

Fig. 1a and b. Diagrammatic representation of the various types of narrowing and widening of the temperature range for germination and bud break at times of change in growth activity. *A* Narrowing of the temperature range through decrease of the maximum temperature at which germination and bud break can occur; and widening through increase. *B* Narrowing of the temperature range through increase of the minimum temperature at which germination and bud break can occur; and widening through decrease. *C* Narrowing of the temperature range through simultaneous decrease of the maximum and increase of the minimum temperature at which germination and bud break can occur and widening through increase of the maximum and decrease of the minimum. A_1, B_1 Predormancy is followed directly by postdormancy, with no true dormancy. A_2, B_2, *C* A period of true dormancy separates pre- and postdormancy. *Min.*: Minimum temperature for germination and bud break. *Max.*: Maximum temperature for germination and bud break. *1* ▦ The temperature range within which germination and bud break occur. *2* ▨ The temperature range within which seeds and buds in a state of predormancy do not germinate or break. *3* ▩ The range of true dormancy. *4* ▧ The temperature range within which seeds and buds in a state of postdormancy do not germinate or break

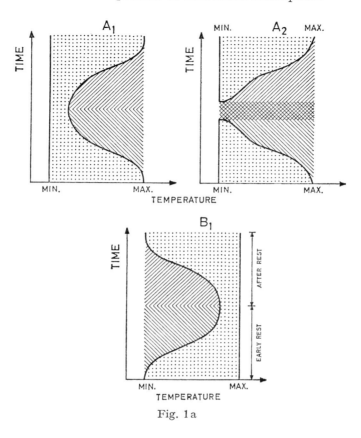

Fig. 1a

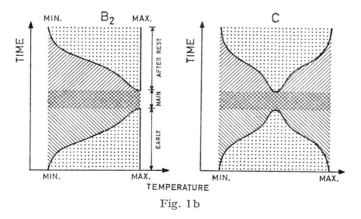

Fig. 1b

Legend, see opposite page.

1. Retention of ability of the seeds and buds to develop further at low temperatures only. In this case the ability to grow at high temperatures is temporarily lost (Fig. 1, A_1 and A_2).

2. Loss of ability to develop at low temperatures (Fig. 1, B_1 and B_2).

3. Loss of ability to develop at both high and low temperatures. This loss does not occur over both temperature ranges simultaneously. Growth continues only over a narrow range of moderate temperatures (Fig. 1, C).

The seeds of some species which are adapted for survival under very restricted environmental conditions, for example desert plants, germinate within a narrow temperature range even while exhibiting maximum growth activity (WENT, 1948b, 1949; WENT and WESTERGAARD, 1949).

Most observations concerning the narrowed temperature range for seed germination and bud break and its widening have been made in the phase of after-rest during the course of after-ripening. At the beginning of this phase both seeds and buds retain their ability to develop within only a narrow temperature range. In such a condition, small temperature differences can be decisive for germination and break of resting buds. At a temperature outside this narrow range neither germination nor growth occurs, even though the temperature might be optimum for development at a stage of higher growth activity. Seeds and buds remain in a condition of relative or conditional dormancy until the temperature range has widened to include the prevailing temperature.

α) Narrowed Range for Germination and Bud Break at Low Temperatures

The earliest observations at low temperatures were made by ATTERBERG (1899, 1907). He was the first to show that cereal caryopses that were not after-ripened, often regarded as incapable of germination, could in fact germinate at temperatures as low as 10° C. He further showed that during after-ripening, which he succeeded in accelerating by cautious drying at high temperatures, the caryopses gradually became capable of germinating at increasingly high temperatures. He pointed out that the maximum germination temperature of these cereals can be used as a measure of the degree of after-ripening. ATTERBERG was also the first to emphasize the gradual widening of the germination temperature range as characteristic of after-ripening processes in caryopses (see also WALDEN, 1910; MUNERATI, 1920, 1925; 1926; HARRINGTON, 1923a; H. E. TOOLE, 1923; DORPH-PETERSEN, 1925; GASSNER, 1926; KROSBY, 1926; CRESCINI, 1928; BYTSCHIKHINA, 1930; KEARNS and H. E. TOOLE, 1939; KOBLET, 1943).

It has long been known that cereal embryos can germinate at a very early developmental stage, 7—14 days after fertilization (HARLAN and POPE, 1922; POPE and BROWN, 1943; POPE, 1949; GRABE, 1956). More recently FUCHS (1941) has shown that the germination of immature seeds can occur over a comparatively wide temperature range. During the ripening process the caryopses lose their ability to germinate at high temperatures. WELLINGTON (1956a, b) and WELLINGTON and DURHAM (1961) also showed that during the ripening of grains, the germination response drops to a minimum at the time of cell-wall thickening in the epidermis. In contrast, during after-ripening, the caryopses gradually recover their ability to germinate at high temperatures. After a long period of cold and damp weather, as in autumn, germination of caryopses that have not after-ripened can be seen on the ear. In rainy but warm weather, however, as long as the caryopses are not after-ripened they are incapable of growth on the ear.

This type of alteration in temperature range, a widening due to an increase of the maximum temperature compatible with germination at a relatively constant minimum temperature (Fig. 1 A), seems to be very common. Seeds of many species and varieties of fruit trees that normally retain a state of true dormancy for a fairly long time, behave in the same way. During stratification in moisture at lower temperatures and during the process of after-ripening, these seeds begin to germinate at lower temperatures and continue gradually at higher temperatures (HARRINGTON and HITE, 1923; KOBLET, 1937; POTAPENKO and ZAKHAROVA, 1939; DE HAAS and SCHANDER, 1952; SCHANDER, 1955a, b, c; ABBOTT, 1956; VISSER, 1956a, b). Similar behavior is also shown by seeds of many other perennials and winter annuals (R. C. ROSE, 1919; KOBLET, 1932; R. C. THOMPSON, 1938; KEARNS and E. H. TOOLE, 1939; E. H. TOOLE and V. K. TOOLE, 1939; E. H. TOOLE and HOLLOWELL, 1939; BORRISS, 1940; BORRISS and ARNDT, 1956; EVENARI, 1952 and recently P. A. THOMPSON, 1968, 1970a, b, c). Examples of a number of weed species are to be found in the work of LAUER (1953), see also Fig. 2. For more references see VEGIS (1961) and STOKES (1965).

The seeds of perennials and winter annuals of this group normally ripen during summer when the temperature is relatively high and steady. Immediate germination of seeds of these species during the summer would be disadvantageous, because either the unhardened seedlings might not survive the dry, hot season (see P. A. THOMPSON, 1968, 1970a, b, c) or, after a damp summer, the unhardened seedlings would winter badly under the snow. The ability retained by the seeds to germinate at low temperatures allows germination during late autumn. Seedlings which have developed only few leaves by the beginning of winter are more winter resistant than those with many leaves. The best known examples are the caryopses of many *Gramineae*, including most cereal crops.

Resting buds of various trees probably also belong to this group, particularly those of species and varieties which form their resting buds early in summer and begin growth again very early in spring when temperatures are still low. Such is the case in *Prunus persica* and *P. armeniaca*, where bud break begins at as low a temperature as 5° C (for references see VEGIS, 1961). However, even a moderate rise in temperature at an early stage of postdormancy can induce secondary dormancy in the resting buds of these and many other *Rosaceae* trees (BENNETT, 1949, 1950a, b; OVERCASH and CAMPBELL, 1955). In this connection it may be mentioned that ASKENASY (1857), from his experiments on the dormant flower buds of cherry, has directed attention to the fact that during the rest period the response of buds to high temperatures undergoes a change. During the rest period such temperatures inhibit the expansion of the buds, but finally promote growth.

At the end of the 18th, and the beginning of the 19th century, British gardners had already observed that in greenhouses where grapes, cherries and peaches were cultivated for early forcing, the temperature at the end of winter had to be raised gradually over a longer period and not abruptly. If this procedure is not followed, flowering and leaf expansion are greatly delayed and some of the buds do not open at all (for references see VEGIS, 1961). This phenomenon, called "delayed foliation", has been frequently observed outdoors in woody plants and especially

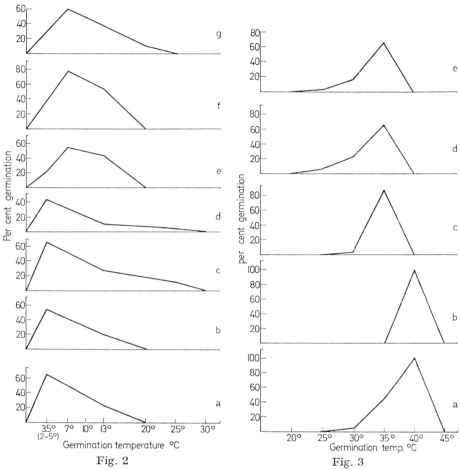

Fig. 2. Maximum germination of some arable weed seeds at constant temperatures. Species in which the non-after-ripened seeds have maximum germination at low temperatures: a *Juncus bufonius*, b *Veronica hederifolia*, c *Polygonum convolvulus*, d *Campanula rapunculoides*, e *Delphinium consolida*, f *Fumaria officinalis*. g *Arenaria serpyllifolia*. (Redrawn after LAUER, 1953)

Fig. 3. Maximum germination of some arable weed seeds at constant temperatures. Species in which the non-after-ripened seeds have maximum germination at high temperatures: a *Chenopodium rubrum*, b *Chenopodium filicifolium*, c *Datura stramonium*, d *Polygonum persicaria*, e *Gnaphalium uliginosum*. (Redrawn after LAUER, 1953)

in fruit trees originating from countries with a rather cold and long winter, but cultivated in countries with a mild, short winter.

β) Narrowed Range for Germination and Bud Break at High Temperatures

Less well known are the narrowing of the temperature range due to an increase in the minimum temperature and its widening due to a decrease in the minimum temperature for germination and bud break. The maximum temperature in this case remains more or less unchanged and thus germination and bud break at the high temperatures are possible (Fig. 1, B_1 and B_2). This type of widening of the temperature range for seed germination was first discovered in seeds of *Amaranthus* species that do not require after-ripening (BAAR, 1912; CROCKER, 1916; EVANS, 1922) and somewhat later in *Thlaspi arvense* (WEHSARG, 1918). Freshly harvested *A. retroflexus* seeds do not germinate at temperatures lower than 40° C but will germinate at this temperature. This relative dormancy gradually disappears in dry storage, as is shown by a continual lowering of the minimum temperature for germination. The seeds of *Thlaspi arvense* germinate only within a very narrow temperature range of about 28–30° C. During stratification at low temperature they gradually acquire the ability to germinate at progressively lower temperatures.

WEHSARG also observed a narrow high-temperature range for germination of seeds of some other species, e.g., *Plantago major*, *Polygonum nodosum*, *Erysium cheiranthoides*, and *Rumex crispus*. Detailed investigations in this group have been made, for example, with seeds of *Betula* sp. (WEISS, 1926; JOSEPH, 1929). Non-after-ripened seeds of the *Betula* species germinate only at temperatures of about 28–30° C. However, after several months of stratification at low temperatures they gradually achieve the ability to germinate at temperatures slightly above 0° C. Other examples are cited by BARTON (1935) and POPCOV (1935, 1954). POPCOV was one of the first to emphasize that the decrease in the germination activity of seeds at ripening is connected with a narrowing, and the increase of this activity at after-ripening, with a widening of the temperature range for germination (see also VEGIS, 1961; LANG, 1965; STOKES, 1965). Other examples may be found in LAUER (1953). Some examples from her work are given in Fig. 3.

At present, only a few species are known in which resting buds retain, for any length of time, the ability to break only at higher temperatures during the narrowing of the temperature range within which bud break may occur. One such species is potato, *Solanum tuberosum*, where eyes in the tuber have this ability. Whether true dormancy occurs in potato tubers has not been definitely ascertained. There is much evidence that freshly harvested potato tubers can sprout within a narrow temperature range at high temperatures (for references see VEGIS, 1961).

Particularly in experiments designed to develop a practical method for sprouting early potatoes immediately after harvest, in order to obtain a second crop during the same year, it has become apparent that a number of varieties have no true dormancy. At least under some conditions predormancy is followed directly by postdormancy (Fig. 1, B_1). Even during the period of least growth activity the tubers of these varieties retain their ability to sprout within a narrow temperature range above 30° C, if well aerated.

In warm climates potato tubers may frequently start to sprout before harvesting. Further, in cases where they do not sprout before harvesting, they begin to sprout very early when brought into an insufficiently cooled store room. Such tubers are unsuitable for planting, because plants developed from them are readily prone to diseases and give a poor crop. In such cases it is often better to import seed potatoes from cooler regions. During after-rest potato tubers become capable of sprouting at progressively lower temperatures.

Similar phenomena have been demonstrated in winter buds (hibernacles) of some *Utricularia* species (*U. vulgaris*, *intermedia* and *minor*). Years ago it was observed that at least during some years, these buds were able to sprout shortly after their formation (GÖBEL, 1893; KLEBS, 1903; GLÜCK, 1906). The author of this paper has studied the temperature dependence of sprouting of these winter buds collected at different times during the autumn and winter. It seems that, at least in Sweden, true dormancy does not occur. At sufficiently high temperatures, such as 30° to 32.5° C, these buds could sprout at all times. In early autumn they sprout only at high temperatures. The later they were collected in autumn, i.e. the longer they had been exposed to the effect of naturally occurring low temperatures, the lower was the minimum temperature at which they could break (VEGIS, 1961, 1965a). The action of high temperatures here should not be confused with the role of warm water baths in the early forcing of truly dormant plants.

This group appears to consist of species which are especially adapted to a climate with a periodically recurring cold season. The formation of their seeds and buds is normally completed in the autumn when high temperatures no longer prevail or occur only for a short time daily. The potato, for example, originates from the highlands in South America where the temperature has already dropped by the time tuber formation is complete. Similarly, *Utricularia* winter buds are normally formed during autumn when the water temperature begins to decrease, especially at night, and night frosts begin. Under these conditions the loss of the ability to sprout at low and moderate temperatures prevents premature sprouting and death of young plants: there is no need for a mechanism to stop sprouting at high temperatures since this does not occur in autumn.

Buds of some trees, such as *Fagus sylvatica*, the *Betula* species, and possibly the *Quercus* species, may break at high temperatures even during the period of lowest growth activity. At this stage, however, they cannot break in the dark or under short-day conditions. During long days or in continuous light at a sufficiently high temperature, bud break can occur and, if other external conditions are favorable, growth may be continuous (KLEBS, 1914; KRAMER, 1936, 1937; POTAPENKO and ZAKHAROVA, 1940b; GULISASHVILI, 1948; LEMAN, 1948a, b; WAREING, 1953, 1954, 1956; VEGIS, 1953, 1955; WAXMAN, 1955, 1957; WAXMAN and NITSCH, 1956; NITSCH, 1957a, b; DOWNS and BORTHWICK, 1956a, b; DOWNS, 1958; WENT, 1957, p. 169; see also VEGIS, 1965b). Such a premature bud break can often be observed outdoors during midsummer. The shoots which develop thereafter are called "lammas shoots" (SPÄTH, 1912).

It is probable that buds of *Fraxinus excelsior*, which for a certain time are obviously in a state of true dormancy, can break only within a narrow high-temperature range just after true dormancy has ended. It is striking that bud break in this species under natural conditions begins very late in spring when the air temperature is rising rapidly. In this connection it should be mentioned that in another species of *Oleaceae*, *Syringa vulgaris*, early forcing by warm-bath treatment before Christmas is only successful if the plants are subsequently kept in a greenhouse at high temperatures (MOLISCH, 1908, 1909a, b). At low temperatures the buds start to break, but later cease to grow, remaining "seated". Nevertheless, some weeks later, when post-dormancy is over, i.e. in the early spring, the buds of this species break, flower and form shoots at considerably lower temperatures. Apparently the temperature range for growth has widened considerably in the direction of low temperatures.

γ) Narrowed Range for Germination of Seeds and Bud Break at Intermediate Temperatures

Both of the previously discussed types of narrowing and widening of temperature range for seed germination and bud break originate from a close adaptation either to a hot, arid or to a cold, unfavorable season. In addition, a third type is known. In seeds and buds of plants in this group the narrowing of the temperature range takes place by increase of the minimum temperature and decrease of the maximum temperature for germination and bud break (Fig. 1, *C*). At the time of reduction in growth activity during early rest, the ability to germinate and to grow at certain moderate temperatures is retained longest. After the termination of true dormancy, this ability to germinate and grow first reappears at the same moderate temperatures. It seems possible that this type of narrowing and widening of the temperature range has originated from adaptation to two unfavorable seasons: a hot, arid summer and a cold winter. Apparently, at the beginning of predormancy in the summer the temperature range narrows mainly by a decrease in the maximum temperature. The ability to grow at low temperatures seems to disappear later in autumn, because low environmental temperatures do not occur in summer. In this type of plant, or at least for indigenous plants under such natural external conditions, predormancy invariably seems to be followed by true dormancy. However, under certain external conditions, if the plants are cultivated at a different latitude or altitude, the development of true dormancy can also be prevented.

WEHSARG (1918) showed that the freshly harvested seeds of *Rumex crispus* germinated only at moderate temperatures, but during after-ripening they could germinate both at higher and lower temperatures. This has also been observed for the seeds of *Viola tricolor* v. *arvensis, Chenopodium album* (KRUG, 1929), and *Taraxacum megalorhison* (POPCOV, 1935). Many other examples among the weed plants are to be found in LAUER (1953). More recently, similar observations have been reported for the seeds of some other plants (for further references see VEGIS, 1961).

The winter buds of *Stratiotes aloides* also first attain their ability to sprout at a moderate temperature (15° C) during after-rest. The widening of the temperature range in the direction of higher temperatures occurs more rapidly than in the direction of low temperatures which prevail in the environment at the time (VEGIS, 1949 a, b; see also Table 1).

Table 1. Sprouting of Postdormant Buds of *Stratiotes aloides*[a]

Time of collection	% of buds which sprout at cultivation temperatures of						
	5° C	10° C	15° C	20° C	25° C	27.5° C	30° C
Oct. 22, 1948	0	0	0	0	0	0	—
Nov. 1, 1948	0	0	21.9	9.5	0	0	—
Nov. 8, 1948	0	35.0	69.2	49.4	2.5	0	—
Nov. 15, 1948	0	65.0	85.6	78.7	33.1	0	—
Dec. 8, 1948	0	100.0	100.0	100.0	73.3	3.6	0
Jan. 15, 1949	0	100.0	100.0	100.0	92.9	62.9	12.1
Feb. 15, 1949	0	100.0	100.0	100.0	100.0	100.0	64.2
March 30, 1949	100.0	100.0	100.0	100.0	100.0	100.0	80.0

[a] Collected out-of-doors at various times during autumn and winter and cultivated at various constant temperatures in the dark (after VEGIS, 1949 a, b).

The resting buds of *Stratiotes aloides* are usually formed in leaf axils in autumn. In this case they are typical organs of hibernation. However, in the shallow shore zone of the overgrown lakes the water level often sinks drastically in hot, dry summers and the plants finally lie in dark mud where they dry out and die. But prior to this they rapidly form "winter" buds which ensure survival of the species through the dry, hot season.

δ) Two Narrow Germination Temperature Ranges

One final type of widening of the temperature range for germination, the ecological significance of which is difficult to understand, must be mentioned. The only known example is provided by the seeds of *Taraxacum cocsaghiz* (POPCOV, 1952, 1960): if not after-ripe, they at first show two separate temperature ranges for germination, and later one temperature range with two maxima, one at 25−30° C and the other at 3−6° C. Between the two maxima is a minimum, which is most pronounced between 10−12° C, over which temperature range the seeds remain in the state of relative dormancy for the longest time.

c) Effect of Alternating Temperatures

We have thus far discussed only the effect of constant temperatures. The objection may be raised that under natural conditions constant temperatures do not prevail. However, it is established that in order to exert its maximum effect on the growth activity, a favorable temperature need not act continuously. Several hours a day may be satisfactory, provided that during the rest of the day the changes brought about by the favorable temperature are not completely reversed by other temperatures. To obtain a successful result characteristic for the temperature in question not only the daily duration of this temperature is important but also the difference between the favorable and unfavorable temperatures. This pertains to the promotion of germination and bud break as well as to their inhibition and the induction of secondary dormancy (VEGIS, 1948a, b; 1949a, b; BENNETT, 1949, 1950a, b; OVERCASH and CAMPBELL, 1955).

In woody plants a diurnal alternation of high day temperatures and low night temperatures usually proves to be much more favorable than a constant temperature for the maintenance of continued growth of terminal apices. However, alternating high temperature with low day temperature can lead to an earlier cessation of growth and onset of dormancy (see Table 2). For references see KRAMER (1957, 1958); HELLMERS and SUNDAHL (1959) and PERRY (1962). See also the results of investigations of POTAPENKO and ZAKHAROVA (1938, 1940a, b) concerning the influence of daily variation of temperature on the growth of many fruit trees.

The favorable influence of daily alternating temperatures on seed germination is known from the work of v. LIEBENBERG (1884) and more recently GASSNER (1911, 1915), PICKHOLZ (1911), HARRINGTON (1923b), E. H. TOOLE and GOSS (1923), MORINAGA (1926), WARINGTON (1936), SCHROEDER and BARTON (1939), E. H. TOOLE and V. K. TOOLE (1941), E. H. TOOLE, V. K. TOOLE, BORTHWICK et al. (1955b) and ZAIN UL ABIDIN (1956). Alternating temperatures may increase the

Table 2. Increase in height of tree seedlings grown with various combinations of day and night temperatures, expressed as percentage of height at the beginning of the experiment (from KRAMER, 1958)

(The results are averages of 8 pines and 15 oaks in each treatment)

Loblolly pine				Northern red oak		
Day temp.				Day temp.		
17° C	50	35	—	17° C	—	33	—
23° C	97	69	51	23° C	86	98	78
30° C	—	106	62	30° C	—	103	—
	11° C	17° C	23° C		11° C	17° C	23° C
	Night temp.				Night temp.		

germination capacity and speed, and reduce variations. Some seeds, e. g. *Cynodon dactylon* and *Typha latifolia*, show no germination at constant temperatures, but do so under alternating temperatures (MORINGA, 1926). If the seed coat is ruptured, however, germination can occur at constant temperatures. In some other cases alternating temperatures seem to exert their effect on the embryo itself. The favorable effect of temperature fluctuation on germination is found in many seeds which show some degree of dormancy, for instance, *Rumex species* (GILL, 1938) and *Bidens tripartitus* (ROLLIN, 1956). Temperature changes may also have a favorable effect on the germination of seeds with light requirements (v. LIEBENBERG, 1884; PICKHOLZ, 1911; GASSNER, 1911, 1915; TOOLE, TOOLE, BORTHWICK et al., 1955 a). For more data and references concerning the interaction between alternating temperatures and irradiation on germination see the reviews of EVENARI (1956, 1965). In some plants it has been found that a single temperature change greatly promotes germination (GASSNER, 1910; TOOLE, TOOLE, BORTHWICK et al., 1955 a; KNAPP, 1956). More facts and a discussion of the action of alternating temperatures are to be found in the review of LANG (1965).

d) The Role of Enclosing Structures in the Induction of Dormancy and in Narrowing the Temperature Range for Germination of Seeds and Bud Break

The inability of relatively dormant, non-after-ripened seeds to germinate within a wide range of temperatures, as compared with totally after-ripened seeds, is often due to the character of the enclosing structures, such as fruit and seed coats and endosperm. These can limit the supply of oxygen to the embryo and also generally restrict the exchange of gases between the embryo and surrounding atmosphere. Within certain temperature ranges (usually at high temperatures) there arises a disproportionality between the requirement for and the accessibility of oxygen, resulting in inhibition of seed germination (CROCKER, 1906, 1916; SHULL, 1911; W. E. DAVIS and ROSE, 1912; W. E. DAVIS, 1930a; ZADE, 1912; ATWOOD, 1914; HARRINGTON, 1916, 1917, 1919, 1923a, b; HARRINGTON and CROCKER, 1923; THORNTON, 1935, 1945; STIER, 1937; DELOUCHE, 1954; VISSER, 1954; ROBERTS, 1961; EDWARDS, 1969; NIKOLAEVA, 1969, and others). The narrowing of the temperature range for seed germination can often be eliminated by removal or injury of these structures, thus facilitating oxygen supply to the embryo, and germination may be obtained over a wide range of temperatures (Fig. 4), see HILTNER (1901), BEHRENS (1906), ATTERBERG (1907), KIESSLING (1911) and others. Such germination occurs at temperatures at which the intact seeds are normally only able to germinate after the completion of the after-ripening period.

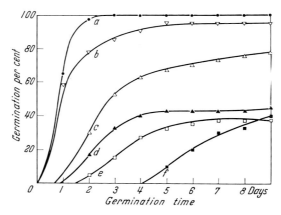

Fig. 4. Influence of complete or partial removal of the surrounding coats on the germination of incompletely after-ripened apple seeds. Germination temperature = 25° C. *a* excised embryos; *b* seed coat removed, endosperm perforated at basal end; *c* as *b* but endosperm perforated at apical end; *d* coat removed, endosperm intact; *e* only a narrow strip of the seed coat removed; *f* intact seeds. (After VISSER, 1956a)

It seems that the scales and the secretions of glandular hairs on the surfaces of resting buds play a similar role in the narrowing of the temperature range for bud break (POLLOCK, 1953). During pre- and postdormancy resting, buds can be brought to foliation by removal of scales (SPÄTH, 1912; PORTHEIM and KÜHN, 1914).

Truly dormant embryos cannot normally be induced to germinate merely by removal of seed coats (CROCKER, 1906; DAVIS and ROSE, 1912; ECKERSON, 1913). It seems clear that the inability of seeds to germinate within a certain temperature range depends at first only on the covering structures; but upon transition from predormancy to true dormancy it becomes progressively dependent on the condition of the embryo itself.

e) The Role of Low Temperatures in the Abolition of Dormancy

Under natural conditions low temperatures play the most important role in the termination of dormancy and widening of the temperature range for seed germination and bud break. On the other hand, high temperatures usually act in the opposite manner and induce dormancy. This is especially true for the origin of secondary dormancy.

α) *Stratification of Seeds*

Seeds and buds of many different plants of the temperate zone must be exposed
for a certain period of time to low temperatures in order to overcome dormancy.
Besides temperature, other factors important for the success of low-temperature
treatment are sufficient humidity in the seeds and good aeration. This moist, low-
temperature treatment is often called stratification. In the old nursery practice
seeds and layers of moist soil or sand were arranged in alternating horizontal
layers and exposed outdoors to low temperatures during the winter. The term
"stratification" originates from this treatment. In more recent years the seeds are
usually mixed with moist sand, granulated peat, vermiculite or some other moist
medium and exposed for the desired time to low temperature. Low-temperature
stratification thus imitates the natural after-ripening of seeds in the temperate
zone. Artifical stratification under controlled laboratory conditions has the advantage that it is possible to hold the temperature, moisture and oxygen supply at
optimum levels during the whole procedure whereas outdoors these factors,
especially temperature, are optimum only for a limited time.

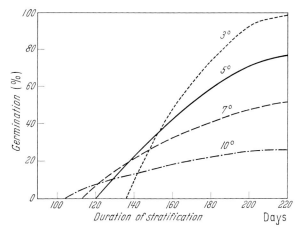

Fig. 5. Progress of germination of apple seeds stratified at various temperatures. (From
SCHANDER, 1955c)

The changes which are involved in seed after-ripening during moist, low-temperature treatment occur at temperatures above the freezing point, usually ranging
from 1—8° C, but sometimes also at slightly higher temperatures (Fig. 5). During
stratification the temperature range of germination gradually widens (Fig. 6). A
small rise in temperature above the maximum during stratification will suspend
the process of abolition of seed dormancy, but does not induce reversion. Thus at
temperatures within certain narrow limits, dormancy is neither removed nor
induced. ABBOTT (l.c.) called this neutral temperature the "compensation point".
But at higher temperatures, 14° C and above, this stratification is interrupted and
secondary dormancy is induced. After this, stratification must start anew. Ex-

perience has shown that the action of low temperatures can be reversed by high
temperatures and vice versa. This reversal can be repeated many times. The
reversal of the stratification effects can be observed at all steps of the process, and
only when stratification is completed does the induction of secondary dormancy

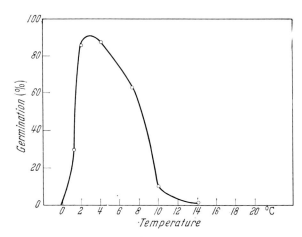

Fig. 6. Percentage germination of apple seeds after 85 days stratification at various
constant temperatures. (From SCHANDER, 1955a)

become impossible (NIKOLAEVA, KOZLOVA, and YUDIN, 1959). There is reason to
believe that the action of low and high temperatures during seed stratification is
analogous to that observed during vernalization and devernalization (VEGIS, 1961)
and that it is very similar to the action of red and infrared light (NIKOLAEVA,
1969). The rate of reversal of dormancy is closely dependent upon temperature:
the higher the temperature during the interruption of low-temperature stratification, the faster does dormancy develop.
The minimum duration of stratification required for removal of dormancy can
vary from a few days to a few months depending upon plant species, conditions of
stratification, seed ripeness and prior storage. Relatively dormant seeds require
the shortest stratification. After-ripening of seeds at low temperatures can also
proceed within mature, fleshy fruits (HARRINGTON and HITE, 1923; LESLEY and
BONNER, 1952; SCHANDER, l.c.).
In certain species morphological dormancy (NIKOLAEVA, 1969) is abolished under
moist conditions at a temperature of 15—25° C or slightly higher. This is called
warm stratification. Some species require warm stratification followed by cold
stratification, and some prefer stratification under alternating conditions.
More detailed information on stratification requirements of seeds of different plant
species with different kinds of dormancy, and references concerning stratification
are to be found in BALDWIN (1942), CROCKER (1948), DeHAAS and SCHANDER
(1952), TURKEVICH (1952), CROCKER and BARTON (1953), OKNINA and BARSKAYA
(1956), NEKRASOV (1960), VEGIS (1961, p. 175), STOKES (1965), and NIKOLAEVA
(1969).

β) Chilling Requirements of Buds

Buds of most deciduous trees, shrubs, perennial herbs and aquatic plants of the temperate zone pass through a rest period. During autumn, somewhat prior to or at the time of leaf-fall, the early rest phase of the resting buds passes on to main rest. It then becomes impossible to bring about the resumption of bud growth by different treatments, thus indicating that the buds have entered into the state of true dormancy. The duration of main rest is dependent not only on the species and variety but also on the climatic conditions during late autumn and winter. During the winter main rest is succeeded by after-rest, true dormancy thereby gradually passing on to a state of post-dormancy. This occurs latest in the buds of deciduous trees, in the later part of winter. At this time the dormancy of buds may be broken by different physical and chemical treatments (for a review on different methods of early forcing see VEGIS, 1965 b). During after-rest the post-dormant buds are evidently able to break within a very narrow temperature range which, however, progressively widens. As long as the outdoor temperatures lie below the minimum of this range relatively dormant buds exhibit no growth. Even if the rest period is terminated, as long as the external temperature has not reached the minimum level necessary for starting growth, the buds will remain in a state of imposed dormancy or quiescence.

When maintained in a greenhouse where the temperature is kept constant at the level typical for late spring or early summer, deciduous trees with truly dormant buds show no bud growth. The dormant condition often persists as long as the temperature is high. However, this dormancy may be gradually broken by subjecting the plants to temperatures near the freezing point for a period of time. On the other hand, in buds of trees kept outdoors at low winter temperatures, dormancy gradually diminishes and finally disappears (ASKENASY, 1877; MÜLLER-THURGAU, 1885; HOWARD, 1906, 1910; MOLISCH, 1909; WEBER, 1916, 1924; COVILLE, 1920; for further references see VEGIS, 1961).

The first observations on this phenomenon were made by KNIGHT (1801, quoted by Treviranus 1811, p. 112) with *Vitis vinifera*. If a branch of the plant was exposed to natural low winter temperatures by allowing it to project through a small hole in a greenhouse while the rest of the plant remained inside the greenhouse for the winter, only the buds on the exposed branch broke and formed new shoots during the following spring. Similar observations were also made by COVILLE (1920) on *Vaccinium corymbosum* (Fig. 7). There is no doubt that the dormancy of buds, like that of seeds, is terminated by low temperatures.

The necessity of a period of low temperature for termination of dormancy and normal bud break, expanding of leaves, and normal development of the new shoots is most clearly demonstrated by the behavior of many fruit trees of the temperate zones with short and mild winters. The opening of the blossom and leaf buds during the spring is both delayed and irregular. Often some of the buds and unopened blossoms die and are shed. These trees, the yield of which is greatly reduced, often bear unexpanded flowers and leaf buds, open flowers, fruits, leaves and new shoots simultaneously. The difficulties in bud break and growth can be avoided by providing an additional chilling period. The most effective temperatures for eliminating dormancy are 1−8° C, as for the stratification of seeds.

Fig. 7. Breaking of bud dormancy in blueberry by low temperatures. *A* Plant was kept through the winter in a warm greenhouse. *B* Branch on the right was exposed to low temperatures by allowing it to project through a small hole in the greenhouse during the winter. The branch on the left remained within the greenhouse. Figure shows appearance of the plant in May. (From COVILLE, 1920)

CHANDLER (1907) was the first who considered the phenomenon of delayed foliation of trees to be a consequence of extension of the rest period (see also HODGSON, 1924, and CHANDLER, 1945). WELDON (1934), on the basis of his 15 years of observations in Southern California, came to the conclusion that delayed foliation is connected with high daytime temperatures during the months of December and January. The correctness of this conclusion has been confirmed by numerous subsequent observations. CHANDLER and TUFTS (1934) and CHANDLER, KIMBAL, PHILP et al. (1937) suggested that prolongation of the rest period in cases of delayed foliation is a result of too short an exposure to low temperatures (see also CHANDLER and BROWN, 1951; BROWN, 1952, 1957, 1960; BROWN and ABI-FADEL, 1953; BROWN and KOTOB, 1957). In this connection the term "chilling requirement" is used. Through outdoor observations an attempt has been made to determine approximately in hours the minimum duration of chilling required, at temperatures not higher than 7° C for example, for normal foliation and flowering. The chilling requirement varied greatly according to species and variety. In different varieties of peach it fluctuates between 750 and 1150 h (WEINBERGER, 1950a); see OVERCASH and LOOMIS (1959). Furthermore, for the results with different varieties of pear, the chilling requirement of a particular variety often changes from year to year (WEINBERGER, 1950b). It is conceivable that the required hours of chilling need not be consecutive, but may have a cumulative effect. However, this assumption is not entirely correct. In instances where the number of chilling hours required by a particular variety at the same site differed from year to year, the temperatures, especially the day temperatures during the months of December and January, were also different. In the years with a high temperature during these months the plants required a greater number of chilling hours. BENNETT (1950) and WEINBERGER (1954, 1956) found that high temperatures during the winter have the opposite effect from low temperatures in the elimination of dormancy. Just as in cold stratification of seeds they induce secondary dormancy. During December and January clear and sunny weather often prevails in California. In the middle of the day the buds may be warmed considerably. In an experiment of OVERCASH and CAMPBELL (1955), when plants of *Prunus persica* were exposed

Fig. 8. Upper: Redhaven peach trees on February 24, 1953, one month after the chilling periods at 39° F (4° C) ended. *A* 750 h continuous chilling; *B* 750 h intermittent; *C* 950 h continuous; *D* 950 h intermittent. Lower: Sunhigh peach trees, on February 24, 1953, one month after the chilling periods at 39° F (4° C) ended. *A* 750 h continuous chilling; *B* 750 h intermittent; *C* 950 h continuous; *D* 950 h intermittent. Intermittent daily treatment: 8 h daily during the days in the greenhouse at 70° F (21.6° C) or more, and during the night 16 h at 39° F (4° C) in a refrigerated room. (From OVERCASH and CAMPBELL, 1955)

8 h daily to a temperature of 21° C and 16 h daily to 4° C, the high temperature partially nullified the favorable action of low temperature. Obviously under such temperature conditions many more chilling hours are required for the removal of dormancy than if the diurnal temperature cycle included no such high temperature periods or if their daily duration was shorter (see Fig. 8). Similar observations have also been made with *Malus mandshurica* (MOROZ, 1940).

The phenomenon of delayed foliation of fruit trees has also been observed in South Africa (REINECKE, 1931, 1936; M. W. BLACK, 1936, 1947, 1953; M. W. BLACK and MICKLEM, 1939; DE VILLIERS, 1943a, b, 1946; JACKSON, 1947, 1957; HILL and CAMPBELL, 1949), in Israel (LACEY, 1944, and SAMISH, 1945), in India (JAVARAYA, 1943), in the USSR (RJADNOVA, 1950), in Australia (WICKENS, 1931, and HARRIS, 1950) and in South America (BURGOS, 1943, and LEDESMA, 1950). For additional references concerning the chilling requirement and delayed foliation of fruit trees see VEGIS (1961, 1965b).

Dormancy in underground organs of perennial herbaceous plants, i.e. rhizomes, tubers, corms, bulbs and bulbils, may also be removed by exposure to low temperatures. Rhizomes of *Convallaria majalis* which become dormant in summer after the formation of buds, can begin to grow during the autumn after a one-week period at 0.5–2° C or three weeks at 5° C (HARTSEMA and LUITEN, 1933). The favorable effect of low temperatures on the removal of dormancy has been observed also in corms of *Gladiolus* (DENNY, 1936, 1937, 1938, 1942), tubers of *Helianthus tuberosus* (STEINBAUER, 1932; STELZNER, 1942), rhizomes of *Veratrum* species (WENT, 1957, p. 178), in bulbs and bulbils of *Ranunculus ficaria* (WINKLER, 1925; MUDRAK, 1935; AUGSTEN, 1957), and in bulbs of *Lilium longiflorum* (BRIERLEY, 1941; STUART, 1954a, b), *Poa bulbosa* (TOOLE, 1941), and *Allium sativum* (MANN and LEWIS, 1956).

In addition, the dormancy of the winter buds of the aquatic plants *Hydrocharis morsus ranae* (LORENZ, 1903; WISNIEWSKI, 1913; SIMON, 1928; MATSUBARA, 1931), *Stratiotes aloides* (WISNIEWSKI, 1931; VEGIS and VEGIS, 1935, 1948a, b, 1949a, b), *Spirodela polyrrhiza* (JACOBS, 1947; HENSSEN, 1954) and *Utricularia* species (VEGIS, 1965a, p. 550) is overcome by low temperatures. For other references see VEGIS (1961, 1965b).

γ) Induction of Flower-Bud Opening by Decrease in Temperature

Certain plants in the tropics, particularly orchids such as *Dendrobium crumenatum*, *Zephyranthes rosea* and a number of other species, *Chusquea abietifolia* and *Corypha umbraculifera*, exhibit the phenomenon of gregarious flowering. All the plants of such species growing over a wide area open their flowers simultaneously on the same day. It was found that the flowering usually occurs 8–11 days after a heavy rainfall. Investigations (COSTER, 1926; KUJPER, 1933) with *Dendrobium crumenatum* showed that it was not the rain as such, but the subsequent rapid drop in temperature that caused the opening of flower buds. Only rainfall lasting the whole day and resulting in a considerable drop in temperature was effective. An artificial, rapid drop in temperature also provoked flowering while mere wetting of buds without a drop in temperature did not. It was found that the flower buds developed gradually up to a certain stage and subsequently became dormant, probably under the influence of high temperature, while low temperature removed the dormancy as in other species. For further references see NITSCH (1965, p. 1572).

2. Temperature Requirements for the Growth and Development of Plants

SACHS (1860) was the first to study the temperature conditions needed for the growth and development of plants. Most of his work was with seeds and seedlings but he also considered the temperature relationships of growing plants up to maturity. SACHS was of the opinion that each growth stage in plant ontogenesis has its own temperature characteristics which he defined as minimum, optimum and maximum temperatures. The stages of ontogenesis he considered were germination, seedling growth, vegetative growth, flowering and fruiting. SACHS pointed out that if all of the temperature characteristics for the different developmental stages were known, it would be possible to determine whether a given climate could provide the necessary growth conditions for a particular plant. He also considered it important to clarify the specific time requirements for each stage. SACHS was convinced that with such data it would be possible to ascertain the shortest time necessary for complete development of the plant in question.

In order to clarify the effect of several growth factors on organ formation and enlargement, BLAAUW et al. (1936, 1940) investigated the annual periodicity of the development of bulbs and buds from initiation to the completion of development, i.e. up to flowering. The annual developmental cycle of numerous cultivated plants, such as *Hyacinthus, Tulipa, Narcissus, Iris, Hippeastrum convallaria, Syringa, Rhododendron, Azalea, Prunus avium, P. domesticus, Malus silvestris, Pyrus communis* and numerous others has been investigated under the usual cultivation conditions. The most important investigations concern the bulbs of *Hyacinthus, Tulipa* and *Narcissus*. The work of BLAAUW and collaborators has been reviewed by PURVIS (1937, 1938), WENT (1948) and HARTSEMA (1954, 1961).

In the annual developmental cycle of bulbs BLAAUW and his collaborators distinguished three periods: 1. initiation and formation of leaves, 2. formation of flowers and 3. elongation growth. The first two developmental processes occur within the bulbs at the time when they externally appear to be dormant and, in fact, these two periods are the most important. At this time, through active cell division within the bulbs, all the organs are formed. However, these become visible externally only after several months subsequent to active cell elongation. This period of stretching is striking externally and has been observed and studied extensively. At this stage, however, only the organs that have been formed within the bulbs during the preceding two periods appear above the ground.

a) Development of Tulipa Bulbs

The experiments of HARTSEMA, LUITEN, and BLAAUW (1930) are concerned with bulbs after lifting at the beginning of July. At that time the growing apex as well as some leaf primordia were developed and initiation of flower primordia was about to start. During subsequent storage (approximately 3 months) the dried bulbs appeared to have a rest period. However, it turned out that during the first 3 weeks of storage after formation of 3—5 leaf primordia, the growing point increased in size and initiation of flower primordia commenced. The optimum temperature for organ initiation was found to be 17—20° C. After all the flower primordia were initiated and had acquired a definite size the optimum temperature for further development fell abruptly to 8° C.

After another 3 weeks it increased slightly to 9° C and persisted at that level for a considerable time. Development was further accelerated if the 3 weeks at 8° C were replaced by 3 weeks at 5° C. However, this temperature was too low for the production of flowers of good quality. This developmental stage at low temperature, strictly speaking, had no morphological significance except that the flower primordia increased in size and formed complete flowers. The stem elongated only slightly during this stage, called by WENT (1948) preparation for elongation. When the main bud had elongated, the optimum temperature for stem elongation increased slowly to 13° C. When the main bud was 3 cm long the optimum temperature shifted gradually to 17° C, and at a bud length of 6 cm it increased to 23° C. For good flowering this temperature must be replaced by 20° C at the time when the flower becomes visible between the leaves.

It has been demonstrated that each stage of development has its own optimum temperature reference. A high temperature is needed for organ initiation, but for stem elongation somewhat lower temperatures are necessary, at least at first. Low temperature for a comparatively long time is required for the preparation for stem elongation. These changes in temperature requirement during the development of tulip bulbs var. W. COPLAND are represented schematically in Fig. 9.

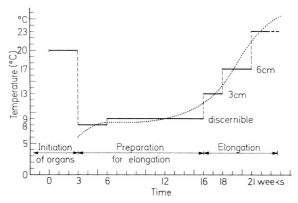

Fig. 9. Optimum temperatures (ordinate, in degrees centigrade) of the development of tulip bulbs var. W. Copland from the time of lifting from ground to flowering (abscissa: time in weeks). Step-curve: experimentally determined optimal temperatures. Stippled curve: most likely actual optimum temperatures. (From HARTSEMA, LUYTEN, and BLAAUW, 1930, p. 33)

Thus the development of *Tulipa* as well as of *Hyacinthus* and *Narcissus* is controlled by variations in temperature. The annual developmental cycle of these plants is synchronized with seasonal temperature changes in their native steppe countries which have a short spring, a hot, arid summer and a cold winter. In the seasons which are unfavorable for the growth of aerial parts, the plants utilize the underground organs, bulbs, for developmental processes: during the summer for initiation and formation of organ primordia and during the cold season for the preparation for elongation and for a gradual beginning of extension growth. Under the climatic conditions prevailing outdoors the annual developmental cycle of these plants lasts 12 months since the natural temperature conditions exclude any other possibility. On the other hand, under optimum temperature conditions corresponding to the requirements of each developmental stage of the plant in question, the developmental cycle can be completed in 8–9 months. Furthermore, it can start and finish during any season.

b) Development of Iris Bulbs

Dutch bulbous *Iris* form a hybrid group which originated from crosses of South-Spanish *Iris xiphium precox* with many other Southern European and North African species. This *Iris* exhibits a mode of development remarkably different from that of *Tulipa, Hyacinthus* and *Narcissus*. Whereas *Tulipa* and *Hyacinthus* form their flowers after being lifted in July and August, and *Narcissus* earlier in May and June, irises lifted in August do not show flower formation even in late October. The growing point forms only leaf primordia at a slow rate, but this occurs most rapidly at 13° C. Leaves, however, are formed slowly at 20° C or higher. The upper limits lie at 25–28° C. Flower formation outdoors does not usually start until March and it occurs rather rapidly at a temperature as low as 6° C.

Iris xiphium praecox "Imperator". BLAAUW (1933, 1941) and BLAAUW, LUITEN and HARTSEMA (1936, 1940) performed numerous experiments concerning the influence of temperature on the actual flower formation and subsequent flowering. The earliest flowering was obtained by a pretreatment at 31° C for one week followed by treatment at 9° C for 2.5–3 months. Flower initiation began when the foliage was about 6 cm above the bulbs. For further flower formation a temperature of 15° C was the most favorable. For *Imperator*, in contrast to *Tulipa, Hyacinthus* and *Narcissus*, the same temperature of 15° C was also optimum for extension growth (see Fig. 10).

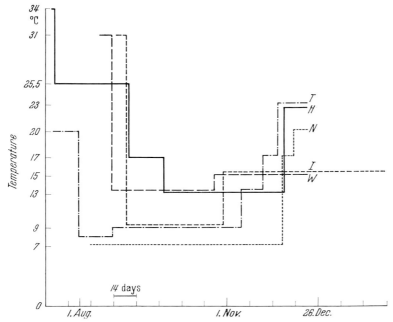

Fig. 10. Comparison of temperature treatments for early flowering of 1 tulip "W. Copland" (*T*), 2 hyacinth "l'Innocence" (*H*), 3 daffodil "King Alfred" (*N*), 4 iris "Imperator" (*I*) and 5 iris "Wedgwood" (*W*). (From HARTSEMA, 1961)

Iris tingitana. This *Iris* grows wild in Morocco, where it flowers in the open at the end of February. Flower formation occurs during the rainy period from November to January. For early flowering, according to LUITEN and BLAAUW (1934), the bulbs were first exposed to 28° C for 24 days, followed by 9° C until the foliage attained a length of approximately 6 cm, when they were transferred to a greenhouse at 17° C where flowering occurred.

Iris wedgwood. This *Iris* originated from crosses of the Moroccan *Iris tingitana*. In the Netherlands bulbs of this variety flower around the middle of November. Changes in the optimum temperature during the development of bulbs have been studied by HARTSEMA and LUITEN (1940, 1955). As in the bulbs of *Iris "Imperator"*, flower formation in *Iris wedgwood* was greatly favored by pretreatment for 1—3 weeks at 31° C followed by incubation for 6 weeks at a lower temperature. The quickest extension growth after planting occurred at 13° C, a temperature which also proved to be optimum for flower formation. The formation of leaves before flowering occurred at this temperature. Although 17° C, 20° C and particularly 23° C, were more favorable, flower formation at these temperatures proceeded slowly. At 25.5° C and higher no flower primordia were formed. However, these temperatures may be used for storage of bulbs for several months. By the time the foliage of the plants at 13° C had reached a length of 6 cm, flower formation was almost over. For acceleration of this, flowering plants must be transferred to a greenhouse at 15° C.

c) Development of *Hippeastrum*

Hippeastrum species are inhabitants of tropical and subtropical America, from Mexico and the West Indies to Chile and South Brazil. Some of the species from which the hybrids have been obtained grow in moist, shaded localities in subtropical forests, while others grow in regions with very dry seasons. In both cultivated as well as wild varieties, flowering occurs once annually. The appearance of another period with an occasional flower stalk is an exception. However, BLAAUW (1931) observed several flowering periods with 2, 3, or 4 stalks in his cultures in one year. BLAAUW (1931), who studied the periodicity of development in bulbs of *H. hybridum*, found that organ formation was not restricted to a definite season: the formation of leaves alternated with flower formation. He found that in *Hippeastrum* bulbs a new growing point was formed two or three times in one year, and after the development of 4 leaves and a terminal inflorescence it was replaced by the next. It seems that under optimum growth conditions *Hippeastrum hybridum* plants may be capable of growth without interruption and exhibit 3—4 developmental cycles per year. *Hippeastrum* bulbs always show all developmental stages. In contrast, in other plants examined, the developmental stages recur only after 12 months.

In the greenhouse from February to the middle of September 8—12 leaves and 2—3 inflorescences may be formed. Flowering periods in *Hippeastrum* seem to be independent of the period of actual flower formation and are often induced by a period of dry storage or, in its native geographic locality, by dryness. Early flowering, according to LUITEN (1946), can be obtained by storage at 15° C or 17° C for 4.5 weeks followed by 4 weeks at 23° C, after which the bulbs should be kept in a greenhouse at 20—24° C. For normal flowering in February-March LUITEN advised storage of bulbs from mid-October to mid-January at 15° C, 17° C or 20° C. This treatment was based on the principle that the inflorescence has first to develop inside the bulb before its emergence can be accelerated by a higher temperature. A temperature of 23° C for 4 weeks has an accelerating and regulating influence on the appearance of flower stalks. Low temperatures of 9° C or 13° C produce no accelerating effect in *Hippeastrum*, unlike *Tulipa*,

Hyacinthus, Narcissus and bulbous *Iris. Hippeastrum* bulbs can be stored at 3.3° to 7.2° C from February until July. Such bulbs flower abundantly four weeks after planting.

d) Thermoperiodicity

WENT (1944a) directed attention to the fact that in many plants the growth rate remains rather constant both day and night, but in others the greater part of the growth occurs during the night. In the latter case the night temperature can be expected to have a significant influence on the growth of the plant as a whole. In the majority of plants investigated maximum growth occurred when the night temperature was considerably lower than the day temperature. In such cases the alternation of a higher day temperature and a lower night temperature is neces- sary for maximum growth. WENT introduced the term thermoperiodicity for this diurnal alternation of temperature requirements. In his investigations of the optimum temperature requirements for the growth of seedlings of *Solanum lycopersicum*, WENT (1944a) found that the optimum daytime temperature was approximately 26° C, whereas the optimum night temperature was 17—18° C (see Fig. 11). The optimum night temperature decreased during development. In the

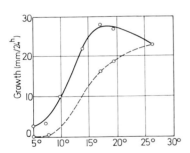

Fig. 11. Relationship between stem growth rate (ordinate, in mm per day) and tem- perature (abscissa, in degrees centigrade) of tomato plants. The dotted line repre- sents plants kept both day and night at 26.5° C. The dark line shows growth rates of plants kept during eight day hours at 26.5° C and during night at the temper- atures indicated on abscissa. (Redrawn after WENT, 1944a)

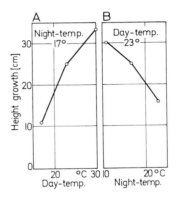

Fig. 12. Increase in height growth of *Pinus taeda* seedlings at different day and night temperatures. *A* Increasing growth with increasing day temperature and constant night temperature (17° C). *B* Decreasing growth with increasing night temperature and constant day temper- ature (23° C). (From KRAMER, 1957)

seedling stage the night temperature was 30° C, but during subsequent develop- ment it gradually fell to 8—18° C. This optimum also depended on the day temper- ature and on light intensity during the daily light period. High night temperatures, especially following relatively low day temperatures may be injurious, but this effect may be lessened by shortening the daily light period. In commercial green- house culture it is well known that the temperatures should be kept at different levels during the day and the night (see Table 3). For tree seedlings see Fig. 12.

Following his investigation of the temperature dependence of different metabolic and growth processes, WENT (1944b, 1945b) suggested that during the daily light period, photosynthesis seems to be a limiting factor for growth only when the night temperature lies within the optimum range for the efficient utilization of the photo-synthetic products. Photosynthesis does not limit growth when there is a drop in the night temperature (WENT, 1945a). WENT suggested that at night the limiting process may be the movement of assimilatory products. With an increase in temperature this becomes progressively slower, although there is an increase in the growth rate of defoliated shoots and isolated roots.

Table 3. Optimal temperatures for growth (from LAURIE and KIPLINGER, 1944)

	Day temp. (in °C)	Night temp. (in °C)
Violet	8.5—14	4.5—10
Snapdragon	14 —16	7 —9
Lathyrus	13 —15.5	9 —10
Roses	21 —23	14.5—16.5
Orchids:		
Seedlings in general		21 —29
Seedlings of *Odontoglossum*		13.5—15.5
Mature plants of *Cattleya*		15.5—18.5
Mature plants of *Odontoglossum*		10

Changes in the optimum night temperature may be associated with an increase of the distance between the sites of assimilation and growth. WENT (1948) pointed out that in tomato plants a low optimum night temperature resulted from the competition between two different processes: growth and translocation of assimi-lates. He indicates that the growth processes have a temperature coefficient of $Q_{10} > 2$, and the optimum is around 30° C. On the other hand, the rate of trans-location has a temperature coefficient of $Q_{10} > 1$. In consequence of this ". . . at higher night temperatures less sugar reaches the growing region and the food supply becomes limiting" (WENT, 1948, p. 156). In this connection WENT showed that in young, small tomato plants, when translocation takes place only over a short distance, the optimum night temperature for the process corresponds to the optimum temperature for the growth processes (30° C). However, as the plants grow the distance between the photosynthesizing and the growing parts increases and the movement of assimilates becomes progressively limiting for growth (at higher temperatures). Depending on the variety, the optimum night temperature decreases to as low as 13–17° C. LEWIS and WENT (1945) established a similarly important role for the night temperature (with several annuals from California). Some species, for instance *Baeria chrisostoma* and *Maedia elegans*, die at a night temperature of 26° C (see also LOO, 1946). For further details and references see Went (1948, 1953, 1957).

B2. Low Temperature Effect on Flower Formation: Vernalization

K. NAPP-ZINN

1. Introduction

The majority of stages in the course of the development of plants, as well as the underlying biochemical processes, are accelerated by a higher temperature within the range $0-40°$ C. In some plants, however, phases such as germination and/or flower formation appear to provide exceptions to this generalization: their "normal" development is made possible or enhanced by exposure to cold (i.e. to temperatures below the growth optimum). In the case of germination, cold treatment is known as stratification, whereas the term vernalization is applied with respect to flower formation.

Cold is, of course, only one of many *external factors* exerting an influence upon plants. It was KLEBS (1903, 1917) who pointed out that the genetically determined *specific structure* (nowadays termed "genetic constitution") and the external factors are in a constant state of interplay. This leads to a gradual alteration in *internal conditions*, in turn resulting in a change in appearance of the plant. In accordance with KLEBS' distinction between these three groups of factors, Chapter 2 deals with the effect of cold as an external factor, Chapter 3 reviews the results so far obtained concerning the genetic analysis of the cold requirements of various plants and finally, in Chapter 4, alterations in internal conditions under the influence of cold (and the genes responsible) are considered. Owing to limitations of space only a few examples have been given in most cases and it has been found necessary to make use of summarizing statements[1]. More detailed information is to be found in the Encycl. of Pl. Physiology (NAPP-ZINN, 1961a, for Chapters 2 and 3, PURVIS, 1961b, and LANG, 1965, for Chapter 4), as well as in AUGSTEN (1964) and PICARD (1968). For information concerning individual plant species such as *Chrysanthemum morifolium*, *Arabidopsis thaliana*, and *Sinapis alba*, the collective volume edited by EVANS (1969) should be consulted. The sources of data used in the following pages, unless otherwise stated, are to be sought in the literature listed in the above-mentioned publications.

2. The Kinetics of Vernalization

Strictly speaking, vernalization refers only to the (accelerated) induction of the *ripeness to flower*[2] by exposure to cold (Section a). Chilling also has an advantageous effect on the two other main stages in flower formation (flower initiation and flower unfolding; Section b) distinguished by KLEBS (1918). Comparable phenomena are also to be seen in lower plants (Section c).

a) Vernalization in the Strict Sense

α) *Criteria of Vernalization*

From what has already been said it is clear that ripeness to flower and thus a vernalization effect, if present, is not *immediately* manifest. In order to study vernalization, therefore, favorable conditions must be provided for the *ensuing* stages of flower formation (and kept uniform for all experimental groups): any

---

[1] At the time of writing (1970), publications on the subject of vernalization number about 2000.

[2] Ripeness to flower (aptitude à fleurir) is said to have been attained when all of the prerequisites for flower initiation, with the exception of one last condition, are fulfilled. The latter is usually a certain day length and therefore ripeness to flower has also been termed "competence to photoperiod".

influence of exposure to cold on the *age at flowering* and the *number of leaves* on the main axis, below the first flower, can then be recorded. Only when both of these criteria are altered in the same direction can it be assumed that the process of vernalization has been specifically involved.

Changes in *leaf form*, such as a reduction in the length of the primary leaves, for some time held to be a criterion for vernalization, have now proved to be non-specific effects of cold (PURVIS and HATCHER, 1959: winter rye; KREKULE, 1961: winter wheat). The *yield* of cold-requiring plants, although of great importance from the agricultural point of view, is not a reliable criterion since the variety of factors involved in the yield may respond to cold in opposite directions.

β) Cold-Requiring Phanerogams

Vernalization requirements have been investigated thoroughly in crop plants, which are both of practical importance and scientific interest. Winter cereals, such as varieties of wheat, barley and rye, have pronounced cold requirements. Even in summer cereals, however, which do not bloom particularly early, the time of shooting, ear emergence, full bloom and fruit formation can be advanced by vernalization, with an accompanying decrease in the number of leaves. Other crop, fodder and ornamental plants which can be successfully influenced in these respects by cold treatment are: certain varieties of *Bromus-, Festuca-, Lolium-,* and *Poa*-species (see FYODOROV, 1958; BELLINI, 1959; MATHON and NEHOU, 1960; BOMMER, 1961; BARENDSE, 1965; BEAN, 1970), *Cyperus rotundus, Allium cepa, Dianthus barbatus,* and *D. caryophyllus* (G. P. and J. E. HARRIS, 1962), numerous cabbage- and beet-races of the genera *Beta* and *Brassica, Raphanus sativus,* certain species of *Lupinus, Vicia* and *Trifolium* (BEATTY and GARDNER, 1961; HAGGAR, 1961; WELLENSIEK, 1966), *Daucus carota* (MATHON and LUCIANI-GRESTA, 1968), *Apium graveolens* and other Umbelliferae, various Campanulaceae (MATHON, 1959, 1960b), varieties of *Lactuca sativa* and *Chrysanthemum morifolium.*

From the scientific point of view the most comprehensive and instructive vernalization experiments have been carried out on *Secale cereale* (*Petkus* winter rye), *Triticum, Hyoscyamus niger, Chrysanthemum morifolium,* and *Arabidopsis thaliana.* Furthermore, interesting results with respect to more specialized problems in vernalization have been obtained with species of *Lunaria, Streptocarpus, Oenothera, Geum,* and many others (for detailed references see, for example, NAPP-ZINN, 1961a).

γ) Cold Requirement and Cold Resistance

Most plants requiring vernalization are resistant to cold since they would otherwise soon become extinct. Nevertheless, there are other forms which, although equally insensitive to cold, do *not* require any appreciable degree of cold. An example is provided by *Lin Calel* (RUDORF, 1935), a summer wheat important in the breeding of resistant forms. Conflicting results have been obtained with regard to the influence of vernalization on resistance to cold, and particularly to frost, in the course of the development of individual plants. This is to be accounted for, at least in part, by the varying conditions under which resistance was tested. In the majority of cases a diminished resistance to cold was observable following vernalization,

probably resulting from the transition to photophase and an associated enhancement of the propensity to develop (for further information see, among others,
MÜLLER, 1959: barley and wheat; GRAHL, 1960a, b: wheat; ANDREWS, 1960: rye;
see also p. 208—212).

δ) *Locus of Perception*

Transplantation of non-vernalized embryos to cold-treated endosperm and of
vernalized embryos to untreated endosperm (in winter wheat for example,
GERHARD, 1940) revealed that the essential factor in seed vernalization was the
exposure of the *embryo* to low temperatures. PURVIS (1940) went a step further and
exposed fragments of winter-rye embryos to cold. She found that regeneration was
followed by flowering only if the fragments included the shoot apex. This confirmed
what was already known to hold true for various rosette plants such as *Apium
graveolens* (O. F. CURTIS and CHANG, 1930), *Beta vulgaris* var. *esculenta* (CHROBO-
CZEK, 1934) and biennial *Hyoscyamus niger* (MELCHERS, 1936, 1937), in which
single organs can be chilled for extensive periods of time. Namely, exposure of the
shoot apex was essential for vernalization.

This view prevailed for decades before it could be demonstrated that it was, in
fact, possible to vernalize the macrocotyledons of *Streptocarpus wendlandii* and
S. wendlandii × *S. grandis* (OEHLKERS, 1956), leaf cuttings of *Lunaria biennis*
(WELLENSIEK, 1961, 1962, 1964a), and even root segments of *Cichorium intybus*
(WELLENSIEK, 1964b). The vernalization effect manifested itself in the formation
of flowers on the regenerated shoots arising from the cold-treated organs or fragments thereof. Furthermore, WELLENSIEK demonstrated that in leaf cuttings of
Lunaria the vernalization effect is initiated primarily at the base of the petiole,
which is also the site of shoot regeneration, and came thus to the conclusion that
actively dividing cells are a prerequisite for the success of cold treatment. Nevertheless, vernalization can also take place under conditions which practically exclude cell division, such as, for example, negative temperatures (see p. 175). In
other plants, such as *Geum urbanum* (CHOUARD and WEBER, 1956), shoot regenerates on cold-treated leaf cuttings exhibited no vernalization effect.

ε) *Plants with and without a Juvenile Phase*

The connection between the *age* of cold-requiring plants at the beginning of exposure to cold and the *effect* of (otherwise similar) cold treatment varies greatly
from case to case. Some plants which are by nature perennial or biennial and have
very pronounced cold requirements go through a juvenile phase in which they are
not yet responsive to cold treatment. The duration of this juvenile phase amounts
to 10 days in *Hyoscyamus niger* (F_1-hybrid $\odot$ × $\odot$) according to SARKAR (1958),
8 weeks in *Lunaria biennis* (WELLENSIEK, 1958), and 15—18 weeks in *Dianthus
barbatus* (WATERSCHOOT, 1957). In some cases, the juvenile phase is terminated
only when the plant has attained a certain size (number of leaves, weight etc.):
2 g or 15 mm diameter or 12 leaves in *Allium cepa* bulbs, a pencil-thick shoot
(BOSWELL, 1929) in cabbage *(Brassica oleracea* var. *capitata)*, 2—3 juvenile leaves
in *Geum urbanum* (CHOUARD and WEBER, 1956), 20 leaves in short-day specimens
or 10 leaves in long-day specimens of *Oenothera biennis* (PICARD, 1965). The juvenile
phase is responsible for the fact that in the temperate zone biennials growing in

the open merely germinate and develop a leaf rosette after the first winter. The second winter induces ripeness to flower and this is succeeded by the second period of growth in which flower initiation and unfolding, fructification and ripening of seeds occur, thus completing the two-year cycle.

Cold-requiring plants lacking a juvenile phase include a whole series of variously reacting types. Although it was long believed that perennials and biennials were, without exception, characterized by a juvenile phase, it is now recognized that there are examples of both types, especially among the Gramineae (BOMMER, 1961b) and species of *Beta* and *Brassica,* in which a vernalization effect can be demonstrated (albeit usually relatively slight) following exposure of the seeds to cold. The majority of plants lacking a juvenile phase are annuals. On the one hand there are the winter annuals, some of which, such as *Carstens Dickkopf V* variety of winter wheat (NAPP-ZINN, 1957a), are characterized by an almost absolute cold requirement. On the other hand, there are intermediary forms such as some strains of *Arabidopsis* and numerous "alternative" cereals, which, according to the prevailing conditions, can be sown either in autumn or in early spring. Finally, there

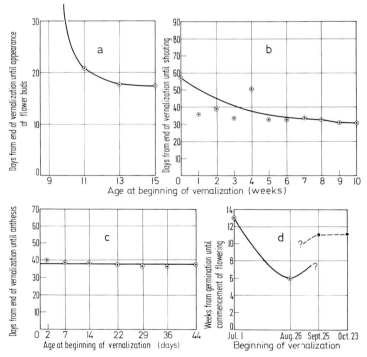

Fig. 1. Dependence of the vernalization effect on the age of the plants at commencement of cold treatment. a) Cauliflower *(Brassica oleracea* var. *gemmifera), Kolom* variety, redrawn from DE ZEEUW and LEOPOLD, 1955; b) beetroot *(Beta vulgaris* ssp. *vulgaris* convar. *crassa* provar. *conditiva),* variety "*Egyptian flat round*", taken from a table in WELLENSIEK and HAKKAART, 1955; c) winter wheat *(Triticum aestivum),* variety *Winter Minflor,* from a table in GOTT, 1957; d) *Ranunculus ficaria,* redrawn from AUGSTEN, 1957 [●- - -● 30-day vernalization (1952); ⊙——⊙ 60-day vernalization (1954)]

are summer annuals with very low cold requirements and which, even *without* vernalization, flower only slightly later. The vernalizability or, in other words, the effectiveness of a certain cold exposure may increase in the days or weeks succeeding germination (as in the *Egyptian flat round* variety of beetroot, according to WELLENSIEK and HAKKAART, 1955), or first decrease and later increase again (for example the *St* race and mutant *F* of *Arabidopsis thaliana*, NAPP-ZINN, 1957d, 1965, or, under certain conditions, *Petkus* winter rye, PURVIS, 1961a) or remain constant (*Winter Minflor* variety of wheat), until finally flowering occurs even without cold (GOTT, 1957), see Fig. 1. It should also be mentioned that the ripening embryos of many plants can be vernalized while still on the mother plant. This is true of cereals (see recent reports by RIDDELL and GRIES, 1958; GRZESIUK, 1962, 1967) and of *Vicia faba* (EVANS, 1959) but apparently not of *Arabidopsis thaliana* (NAPP-ZINN, unpublished).

ζ) The Role of Soaking in the Vernalization of Seed

Vernalization at the seed stage, mentioned in the preceding section, is only possible if the seed is adequately moistened. In the case of cereals, 30—40 l of water per 100 kg of air-dried caryopses usually suffice. With such a low water content germination and growth are scarcely observed even after several weeks of vernalization. This is important in mechanized sowing of cold-treated winter cereals in areas where vernalization is used as an emergency measure in the event of the unavailability of either sufficient cold-resistant winter cereals or of rapidly-growing, productive summer cereals. In seeds in which the main reserve materials are fat and/or protein, as for example in Cruciferae and Papilionaceae, the minimum amount of water necessary for vernalization is greater and amounts to between 60 and 100 g water per 100 g of seed.

The homogeneity of the vernalization effect can, in many cases, be enhanced by soaking the seed in the appropriate quantity of water at approximately room temperature before exposure to cold. The optimum duration for such a procedure is usually 1—2 days. A longer period of soaking sometimes has an anti-vernalizing effect (see p. 181).

η) Effective Temperatures

The greatest vernalization effect is usually achieved by employing temperatures immediately above the freezing point of water as shown when cold treatments of equal duration are compared. In isolated cases temperatures as low as $-8°$ C (for cereals), and as high as $+17°$ C (for *Allium cepa* and *Hyoscyamus niger*), have proved effective if of sufficient duration. Occasionally it has been observed that the most propitious temperature is a few degrees higher if the duration of cold treatment is suboptimal than if it is optimal (see Table 1). Some plants from warmer regions, such as *Rhipsalidopsis graeseri* (RÜNGER, 1960) and *Olea europaea* (HACKETT and H. T. HARTMANN, 1967), exhibit the greatest vernalization effect in response to temperatures of about $+10°$ C extending over a period of about 2—4 months. Cold treatment lasting only 4—10 days is successful almost solely in summer annuals with low cold requirements and in older individuals of winter annuals (KOHL, 1958: *Matthiola*). An isolated observation by PARLEVLIET (1967) records the fact that flowering of *Anthriscus cerefolius* was advanced by 13 days after exposure of the seeds to a temperature of 9° C for only 1 day. For many winter races of cereal the minimum duration of vernalization is about 7–10 weeks, and the largest possible effect is achieved only if cold treatment extends over 100 days or more. *Oenothera biennis* var. *sulfurea* produces flowers only after one year at a

a

c

b

Fig. 2. *Arabidopsis thaliana.* a), b) Race *St.* State of development of the plants 36 days after end of various vernalization treatments: a) 70 days at $+2°$ C; b) 70 days at $-3.5°$ C. c) Summer annual control (race Li_5) sown at end of vernalization of seeds or seedlings of winter annual race (scale in cm)

constant temperature of $+3$ or $+7°$ C with subsequent return to warmth. This is the sole case where alternating temperatures (for example $+3/+11°$ C alternating semi-diurnally to weekly) have proved to be advantageous. The minimum duration of vernalization could be reduced to about 10 weeks by such means (PICARD, 1960, 1965).

Table 1. Optimum temperatures for vernalization (in °C) according to duration of exposure to cold

Objects	Duration of vernalization in weeks					Authors
	2	3	4	5	6	
Secale cereale (*Petkus* winter rye)	+7	+5	+1	+1	+1	Purvis (1948)
Hyoscyamus niger ☉	+10		+6		+3—6	Lang (1951)
Arabidopsis thaliana race *St*	+4				+2	Napp-Zinn (1957 d)
Lactuca sativa cv. *Imperial 456*			+2		+0.5	Adamson (1961)

A so-called "Yarovisation[3] by warmth" is said to reduce the period of growth in some cultivated plants of the warmer zones. So far, however, the results obtained by keeping soaked seeds of rice, maize, some species of millet, cotton and soy beans for 6–30 days in darkness at temperatures between 20 and 35° C have not been convincing.

ϑ) *Vernalization and Light Conditions*

The degree of effectiveness of any vernalization treatment is to a large extent dependent upon the light conditions prevailing before (see, for example, Napp-Zinn, 1960b, 1962b), during (Pierik, 1967: *Cardamine pratensis*) and, above all, after exposure to cold. Many plants for which vernalization is favorable react most rapidly to cold treatment if, in imitation of the seasonal variations in temperature and day length in temperate zones, short day prevails prior to, long day subsequent to vernalization (see e.g. Blondon and Chouard, 1965: *Dactylis glomerata*). Some varieties of chrysanthemum, however, prefer a few weeks of long day immediately after vernalization, followed later by short day.

Two types of plants requiring long day subsequent to vernalization can be distinguished. One, including many varieties of winter cereal, *Hyoscyamus niger* and *Daucus carota*, only flowers in long day. The other, including beets and some races of *Arabidopsis* with low cold requirements, although initiating flowers sooner in long day, exhibits the largest *relative* vernalization effect in short day. In some cases this is a true photoperiodic effect but is in others due purely to the quantity of light. This view finds support in the fact that members of some races of *Arabidopsis*, following incomplete vernalization, flower much later in short day or under continuous illumination of low intensity than in long day or when exposed to a continuous strong illumination. Prolonged exposure to cold causes these plants to flower equally early (Napp-Zinn, 1957b) under either type of light conditions. It is noteworthy that some cold-requiring plants such as *Geum urbanum* and certain varieties of winter wheat and winter barley (Trân Thanh Vân-

---

[3] The term Yarovisation, as originally coined by Russian authors, denoted what is nowadays termed "vernalization". It later came to be used on many occasions where developmental phenomena were influenced by certain temperatures.

LÊ KIÊM NGOC, 1965; INOUYE, TASHIMA, and KATAYAMA, 1964; INOUYE and ITO, 1968; ADACHI, INOUYE, and ITO, 1970; further examples in NAPP-ZINN, 1961 a) flower even in total darkness following adequate vernalization. These phenomena will be considered later in the course of the physiological analysis of vernalization including the substitution of certain degrees of exposure to light for cold treatment.

b) The Promotion of Later Stages of Flower Formation by Cold

α) *Flower Initiation*

Since ripeness to flower is not immediately recognizable the later stages of flower formation have to be used as criteria for successful vernalization. Thus it is often difficult to decide whether a continued exposure to cold after onset of ripeness to flower has not also exerted a favorable influence on flower initiation. Such is unequivocally the case where flowers are only initiated under the direct influence of cold, as in different varieties of *Iris* (optimum $+7°$ to $+9°$ C; for further details see HARTSEMA, 1961) and in *Brassica oleracea* var. *gemmifera*. Vernalization experiments with the late-flowering pea *Greenfeast* have led PATON (1969) to the conclusion that long day is responsible for "induction", whereas exposure to cold favors the subsequent "evocation" leading to flower initiation.

β) *Sex Determination*

Not only does temperature determine in certain cases when, if at all, flowers are to be formed, it can also sometimes influence their quality, especially with respect to sex, as has been shown in maize among others. Maize flowers, like those of other grasses, are monoclinic (bi-sexual) at initiation, and only later is the further development of the stamens in the lateral cobs and of the carpels in the terminal panicles suppressed. The timely influence of cold can bring about a transition of what are normally purely male inflorescences into female, fruit-bearing inflorescences (for references and further examples see NAPP-ZINN, 1967 a).

γ) *Flower Unfolding*

Many Liliiflorae require cold to bring about the shooting which is essential if the flowers already initiated in the preceding warm season are to unfold. This holds true for cultivated members of the genera *Tulipa, Hyacinthus, Convallaria*, and *Narcissus* (see HARTSEMA, 1961). In some varieties of cereal the interval between commencement of sprouting and full bloom can be reduced by chilling.
The optimum temperature for shooting in Brussels sprouts is about $+4°$ C, the maximum above $+14°$ C. Other dicotyledons in which unfolding of the flowers or inflorescences is promoted or even rendered possible at all by cold are *Hydrangea macrophylla, Sempervivum funckii, Viola palustris*, and *Streptocarpus grandis*. Two groups of plants which should be mentioned here are alpine plants and fruit trees: under natural conditions their flowers are initiated in the warmth of summer or autumn but do not open until the following spring. The former group includes species of *Saxifraga, Gentiana, Androsace, Primula*, and *Soldanella*. LONA (1968) was able to demonstrate that at $+15°$ C or more only pollen mother cells are

formed in the anthers of *Soldanella minima* while pollen formation and complete flowering occur only after exposure to cold (20 days at $0°$ C or $+5°$ C).

Although the cold requirements of fruit trees with respect to germination have been thoroughly investigated, the part played by cold in flower formation has not been adequately considered. In some varieties of apple, plum and peach the flower buds initiated in the previous year either fail to open or open only after considerable delay if the plants have spent the winter in a greenhouse. In other varieties – at least in *Pirus malus* – flowering can occur without the influence of cold (see, for example, NESTEROV, 1961; POPLAVSKY and GOLUBKOVA, 1961).

c) Similar Phenomena in Lower Plants

In numerous algae, fungi and liverworts the formation of spores or gametes is enhanced by cold: it promotes the formation of zoospores in *Bumilleria sicula* and of apothecia in *Rhytisma acerinum*, the development of sporangia in *Rhizopus sexualis* and of archegonia and antheridia in *Lunularia* (BENSON-EVANS, 1961) and *Marchantia*.

3. The Genetic Basis of Vernalization Requirements

The fact that in hundreds of varieties or local races of wheat, barley and *Arabidopsis* a more or less marked graduation in vernalization requirements is observable suggests the involvement of several genes. The number and nature of genes manifesting themselves following a cross depend, of course, upon the partners originally chosen for the cross. The difference between *Petkus* winter and summer rye has been shown to be unifactorially conditioned: the cold requirement of the winter rye is evoked by the recessive allele and the summer character by the dominant allele. The annual and biennial forms of *Hyoscyamus niger* also differ in one locus only, but the F_2 generation exhibits a ratio of 1 annual : 3 biennial, from which it may be assumed that two different physiological mechanisms are responsible for the cold requirements of the two species.

In most cases, including the three genera mentioned above, the situation is greatly complicated by the fact that in certain cold-requiring genotypes several genes, some dominant, some recessive, are present and localized on different chromosomes. Each gene alone is capable of eliciting a stronger or weaker need for vernalization (for *Arabidopsis* see NAPP-ZINN, 1957c, 1961b, 1962a). A further complication arises from the heterogeneous epistatic and cryptomeric connections existing between the genes involved, in cereals on the one hand and in *Arabidopsis* on the other. It is therefore not surprising if in wheat (ANDERSON and McGINNIS, 1960) and barley, as a result of crossing two summer varieties, winter forms appear in the F_2 or if, after crossing annual forms of rye, perennial forms appear (NÜRNBERG-KRÜGER, 1961), or, conversely, if following a cross between two cold-requiring strains of *Arabidopsis* a form appears which is not cold-requiring (VAN DER VEEN, 1965). The significance of these results for the physiological analysis of the process of vernalization has already been discussed at length elsewhere (NAPP-ZINN, 1961b). In view of the tremendous variability in genetic constitution it is impossible, on the basis of data drawn from a single object, to arrive at conclusions as to the situation in other genotypes of the same species, still less of different species.

Genetic analyses, including the localization of vernalization genes on definite chromosomes, have been performed in recent years mainly on wheat (MORRISON, 1960, who also lists the older literature; TSUNEWAKI and JENKINS, 1961) and barley (YASUDA, 1963), sometimes utilizing monosomic strains. Such analyses are complicated by the fact that, apart from the true vernalization genes, there are also separate, independent genes that are responsible for late flowering. Furthermore, the possibility of inducing a certain degree of vernalizability by means of short day or weak light in summer annuals (NAPP-ZINN, 1962b: *Arabidopsis*) or by amputation of the cotyledons (HAUPT and NAKAMURA, 1970: *Pisum sativum*) must have a genetic foundation.

Transformation of summer to winter forms and the reverse can only be brought about by selection of the required genotypes from an initially heterogenic population (the normal situation for *Arabidopsis thaliana* under natural conditions; NAPP-ZINN, 1964a; CETL, DOBROVOLNÁ, and EFFMERTOVA, 1968), or by spontaneous or induced mutation from an initially pure strain. This has been demonstrated convincingly by STUBBE (1955) in 20 varieties of wheat, which makes it all the more difficult to comprehend the continued belief of disciples of LYSSENKO, such as STROUN, MATHON, and PUGNAT (1960: rye) or RAJKI (1962, 1967: wheat) that they had brought about designed "transformations" or "conversions" with the help of specially selected external factors. Apart from a few "dauer modifications", external factors only influence the individuals exposed to them (see SCHWABE, 1963: winter rye), and occasionally the embryos still ripening on the mother plant (see above).

4. Physiological Analysis

In the foregoing pages the vernalizing effect of cold as an external factor has been considered in various connections and the genetic basis of the cold requirements of different plants has been discussed briefly. We will now consider the way in which "internal conditions" can be so modified by cold that flower formation finally occurs.

Unfortunately, there is not a single plant in which the causal sequence of events leading from cold to flower formation has been completely elucidated. It must therefore suffice to mention the various ways in which attempts have been made to solve this problem.

a) Kinetic Analysis

One of the reasons for the intense interest in vernalization lies in the paradox that cold — in deviation from the general rule — exerts a favorable effect upon one of the stages in plant development. The experiments which have been carried out under the heading of "kinetic analysis" have at least rendered this paradoxical situation understandable. The key to the situation lies in the fact that the cold-induced ripeness to flower can, in the majority of cold-requiring plants, be annulled under certain conditions by the action of heat (about $3-10$ days at $25-40°$ C). The usual prerequisites for this so-called "devernalization" are that the heat treatment follows immediately upon the cold treatment and that the vernalization should not have been of too long a duration. The vernalization effect is irreversible, however, if the plants are held at "neutral" temperatures of $13-20°$ C (a so-called stabilization) for about 5 days between cold and heat treatment or, in some cases at least, if the plants have achieved a certain age before vernalization and if vernalization is carried out in light (NAPP-ZINN, 1960a). Following successful devernalization the plants can be revernalized. Plants of the *St* race of *Arabidopsis* if devernalized for 5 days at 30° C all bloom at the same time irrespective of the duration of the foregoing vernalization; only after uniform revernalization (40 days at +2° C) does the length of the initial cold treatment manifest itself again (NAPP-ZINN, 1953, 1957 e).

In many plants it is possible to delay flower formation by heat treatment alone (without preceding cold treatment), independently of whether it is followed by vernalization or not. This so-called antivernalization has also been discovered recently in *Petkus* winter rye by FRIEND and PURVIS (1963) and termed pre-devernalization. This process can be exploited in order to prevent onions from sprouting.

The interpretation of all these phenomena revolves around the argument, that in the course of successful de- or antivernalization as well as at "normal" non-vernalizing temperatures, a thermolabile intermediary product is inactivated. This intermediary product is capable of reacting in two ways: at low temperatures it is mainly involved in the synthesis of, for example, a hypothetical thermostable flowering hormone such as vernalin (so-named by MELCHERS, 1939), whereas at higher temperatures another process, characterized by a considerably higher Q_{10}, dominates, in which the intermediary product is rendered useless for synthesis of the flowering hormone. This explanation circumvents the above-mentioned paradox and eliminates the necessity for assuming a $Q_{10} < 1$ for inducing ripeness to flower. Instead, two competing processes with a "normal" Q_{10} between 2 and 4 are postulated. In plants not requiring cold the allele responsible would stabilize the intermediary product or prevent its inactivation (MELCHERS, 1950).

Explanations of this nature, developed for different plants such as *Hyoscyamus niger* (LANG and MELCHERS, 1947), *Secale cereale* (PURVIS and GREGORY, 1952; PURVIS, 1954; FRIEND and PURVIS, 1963) and *Arabidopsis thaliana* (NAPP-ZINN, 1953, 1957 e), differ mainly in two respects. First they differ in the number of thermolabile (and intercalated thermostable) intermediate stages involved in the process of vernalization, and second, as to whether the labile intermediary product can be reconverted to its thermostable precursor under the influence of warmth or not.

A number of phenomena, however, are not rendered plausible by such explanations: 1. A short period of exposure to 25° C subsequent to vernalization has a devernalizing effect on winter rye; a longer period of exposure (14 days) to 25° C, on the other hand, promotes flower formation (FRIEND and GREGORY, 1953; FRIEND and PURVIS, 1963)[4]. This latter observation is interpreted as meaning that longer exposure to warmth favors the autocatalytic multiplication of the thermostable end product of vernalization. An autocatalytic multiplication of this nature is also postulated in the propagation of the flowering stimulus in the plant. 2. Devernalization can be brought about in *Cheiranthus allionii* (BARENDSE, 1963), *Teucrium scorodonia* (JOUGLARD, 1965) and other perennials, not only by warmth, but also by short-day conditions which are, however, ineffective in the case of *Hyoscyamus niger* (SARKAR, 1958).

b) Grafting and Decapitation

The nature of the thermostable condition brought about by vernalization has in part been elucidated by means of grafting experiments, the first of which were

---

[4] *Scrophularia alata* also flowers after exposure to 3° C, but especially rapidly following temperatures of 27—32° C (CHOUARD and LARRIEU, 1964); it remains vegetative at a constant temperature of 17° C. Certain clones of *Festuca arundinacea* behave in a similar manner (BLONDON, 1970).

carried out in the 1890's. DANIEL (1898) was the first to succeed in advancing appreciably the time of flowering of a reputedly biennial, and thus definitely cold-requiring, variety of lettuce by grafting the plants onto *Tragopogon*. Unfortunately, the reports of this and other experiments leave much to be desired and the results were already long forgotten by the time MELCHERS (1936, 1937) successfully brought about flower formation without cold treatment in a biennial race of *Hyoscyamus niger* possessing obligatory vernalization requirements, by grafting onto them annual, or cold-treated biennial, shoots of *Hyoscyamus* which were ripe to flower. MELCHERS (1938) also succeeded in getting cold-requiring *Hyoscyamus* plants to flower under long-day conditions by grafting shoots of the short-day variety (not cold requiring), *Maryland Mammoth*, of *Nicotiana tabacum* onto them. From this it could be concluded that the decisive step is in fact the transmission of vernalin and not of florigen, the hypothetical end-product of photoperiodic induction. The transmissible substance does not pass through a fine filter, which suggests that it is either very labile, or high-molecular, or both. All attempts to isolate such a substance have so far been in vain.

Similar grafting experiments with *Beta vulgaris altissima* (STOUT, 1945; G. J. CURTIS, 1964), *Cheiranthus cheiri*, *Brassica oleracea gemmifera* (ZEEVAART, 1956), *B. napus*, *Daucus*, *Anethum*, *Petroselinum*, and *Pastinaca* (KRUZHILIN and SHVEDSKAYA, 1961) were in some cases unsuccessful; in other cases success was achieved only under particular conditions such as the removal of leaves or preliminary treatment with certain photoperiods. This led CURTIS to postulate a substance localized in the leaves and inhibitory to flowering, while ZEEVAART suspects that, at least in some cases, an intermediary product is only transmissible following the end stage of vernalization.

In still other grafting experiments a definite inhibition of flowering was transmitted from the cold-requiring to the non-cold-requiring partner in the graft, as, for example, in experiments carried out by VOECHTING (1892: *Beta*) and DANIEL (1898: *Scorzonera* spp.), which led v. DENFFER (1950) to discuss at some length the question as to whether a flowering hormone or an inhibition of flowering is involved. PATON (1969) assumes that in the late *Greenfeast* variety of *Pisum sativum* there is a flowering inhibitor which is transmissible by grafting and the synthesis of which is arrested by vernalization. Prior to this, TAKIMOTO, TASHIMA, and IMAMURA (1960) had arrived at the conclusion that lower temperatures ($+ 10°$ C) rendered *Pharbitis nil* plants insensitive to the inhibitory effect of light upon flowering.

The idea that vernalization involves the removal of an inhibition of flowering rather than a positive flowering stimulus is also suggested by the fact that amputation of the shoot tip renders cold treatment superfluous in some plants. Good examples of this are provided by *Chrysanthemum morifolium* (*Sunbeam* and *President* varieties, VINCE and MASON, 1954, 1957) and *Geum urbanum* (CHOUARD and TRÂN THANH VÂN, 1965). According to MARGARA (1963b) the lowest buds on the stem of winter rape are devernalized if the sprouting plants are not decapitated soon enough. In this context the question of a relationship between apical dominance and flower formation has been considered by a variety of authors.

c) Nutrition

In a paper published in 1954, CHAILAKHYAN suggested that flower formation should be considered as a function of metabolic processes, particularly of assimilation and breakdown. The number of investigations with this aim is already legion. Unfortunately, some of the most basic prerequisites for such experiments have repeatedly been neglected. First, only plants in which the genetic basis of the cold requirements is known, and which differ from the corresponding non-cold-requiring form in only a single gene locus (a vernalization gene), should be used. Second, comparison with a non-cold-requiring form is essential and, third, the effect of designed variations in internal conditions upon metabolic processes as well as upon flower formation should be noted.

Apart from indicating a possible connection between respiration and vernalization in *Petkus* winter rye (p. 184), experiments performed on the few plants for which exact genetic information is available, and in which the cold requirements are known to be unifactorially controlled (*Petkus* summer and winter rye, annual and biennial *Hyoscyamus niger* and various forms of *Arabidopsis*), have so far produced few conclusive results. There is scarcely one observation of any interest that cannot be countered by conflicting results obtained using a different plant. It is generally agreed that ample nutrition via the roots accelerates flower formation and augments the effect of incomplete vernalization (ABOLINA, 1938, 1951: winter wheat, P and K especially effective; MANN, 1951: *Beta*, use of stable manure and ammonium sulphate fertilizer increased per cent of sprouting plants; EGUCHI, MATSUMURA, and ASHIZAWA, 1958: *Raphanus sativus* and *Brassica chinensis*, acceleration of flowering by NPK-fertilizers only after incomplete vernalization; CHOUARD and TRÂN THANH VÂN, 1964: *Geum urbanum;* BLONDON and CHOUARD, 1965: *Dactylis glomerata;* BÜHRING, 1967: *Brassica oleracea* convar. *acephala* var. *medullosa*).

High intensities of light, either alone or combined with a generous supply of inorganic nutrients, can render treatment with cold unnecessary. This has been observed in some varieties of winter wheat and winter rye (FYODOROV, 1960: 50000–70000 lux) and in *Geum urbanum* (CHOUARD and TRÂN THANH VÂN, 1964; TRÂN THANH VÂN-LÊ KIÊM NGOC, 1965: 10000–70000 lux). FYODOROV (1963) interprets this as an indirect effect via photosynthesis and general energy relationships, rather than due to a direct influence upon vernalization. This would also offer an explanation for the reduced effect of subsequent cold treatment in *Geum urbanum* following amputation of the larger leaves (TRÂN THANH VÂN-LÊ KIÊM NGOC, 1965).

It is therefore surprising that a cold-requiring clone of *Dactylis glomerata* reacts more favorably when exposed to strong light (14000 lux) for 16 h daily than if subjected to uninterrupted light of the same intensity (BLONDON and CHOUARD, 1965; BLONDON, 1966). To this can be added the remarkable observation that in *Petkus* winter rye (PURVIS and GREGORY, 1937), in some but not all varieties of winter wheat (KREKULE, 1961 c, 1964; W. HARTMANN, 1968), and in various strains of some species of *Lolium* (COOPER, 1960) cold can be replaced by short day up to a certain limit, despite the fact that, as previously mentioned, the latter can have a devernalizing effect in other plants.

d) Breakdown of Reserve Materials

Changes in the levels of reserve materials are obviously of great interest in studying
the metabolic processes taking place in the cold in plants and seeds undergoing
vernalization. The degradation of high-molecular substances such as polysaccharides, fats and proteins into oligosaccharides, fatty acids and amino acids is less
inhibited in the cold than is the further breakdown of the low-molecular compounds
to CO_2, water and ammonia. A resulting accumulation of low-molecular substances
has been observed by DAVID (1945a: wheat: aleuron and starch, 1947: wheat:
lipids, 1949: wheat: carbohydrates, 1951: wheat: lipids), DAVID and SÉCHET (1947:
barley: protein/amino acids, 1948: wheat: carbohydrates), SÉCHET (1949: oats,
barley, maize: carbohydrates), DUPÉRON (1946: wheat: lipids, 1949: *Sinapis alba:*
carbohydrates, 1950: *Raphanus sativus:* carbohydrates, 1952/53), BLAGOVE-
SHCHENSKIJ and KIRILLOVA (1955: wheat: protein), KRUZHILIN and SHVEDSKAYA
(1957: *Daucus carota* and species of *Brassica:* carbohydrates) and SPARMANN (1961:
barley), among others. As a rule, the ratio of starch to sugar undergoes the largest
alteration and the ratio of protein to amino acids the smallest. The rise in the C/N
quotient during vernalization, as measured by the ratio of reducing sugars to free
amino acids (and already considered by KLEBS, in a different context, to be an important factor in the transition to the reproductive phase) is thus mainly due to the
breakdown of starch to di- and monosaccharides. However, in the non-obligatorily
cold-requiring winter barley *Mammoth II*, according to SPARMANN, the content of
soluble protein and of reducing sugars declines steeply, the former especially at the
commencement of sprouting. The level of soluble amino acids, on the other hand,
rises and the C/N quotient sinks until, when shooting begins, it is lower than in
the non-vernalized control plants. If cold treatment is followed by UV irradiation,
development is inhibited and the concentrations of the groups of substances under
consideration approximates to some extent those found in unvernalized individuals
of this barley variety in which UV irradiation has no appreciable effect. Nevertheless, the question as to whether such quantitative changes[5] are, as DUPÉRON
(1952/53) would like to assume, a necessary prerequisite for, or merely a secondary
phenomenon of, or even the result of the attainment of ripeness to flower, flower
initiation and shooting, remains unanswered.

A series of observations of different kinds supports the view that carbohydrates,
even if only as a non-specific source of energy, play a part in vernalization. It has
already been mentioned that strong light can, in certain cases, replace cold. GRE-
GORY and PURVIS (1938a and b) demonstrated that vernalization of caryopses of
Petkus winter rye is possible only in the presence of oxygen (1/500 of the normal O_2
pressure suffices)[6] and that the embryos of winter rye have to be provided with
sufficient sugar, sucrose being the most effective. If, prior to vernalization,
isolated embryos were exhausted of their own carbohydrates by means of a 5-day
"residual growth" treatment in the imbibed state at 20° C, subsequent exposure
to cold only had a vernalizing effect if sugar was added to the substrate (0.5%

---

[5] Similar changes also occur in non-vernalizable objects: for example, the sweetening
of potatoes in winter.

[6] Devernalization, on the other hand, can also take place under anaerobic conditions.

saccharose sufficed). The effect was proportional to the sugar content of the sub-strate (PURVIS, 1944, 1947). The findings of KRUZHILIN and SHVEDSKAYA (1957) using *Daucus* and *Brassica* pointed in the same direction: amputation of the root only diminished the vernalization effect if performed prior to exposure to cold. The root could, just as the endosperm of cereal caryopses, be replaced by a solution of glucose or saccharose (KRUZHILIN and SHVEDSKAYA, 1958).

By exposing seedlings of radish and winter wheat to increased air pressure (3.6 kg/cm²) for 10 days TOMITA (1964) was able to produce a vernalization effect of about the same magnitude as that resulting from exposure to $+5°$ C for an equal length of time. This is the only report of its kind in the literature, and whether or not the effect on the gaseous exchange was specific is unknown.

Comparative measurements of respiration in pairs of summer and winter annuals have revealed, in many cases, a greater respiratory activity in the partner for which cold is not essential (SISAKYAN and FILIPPOVICH, 1953: 4 winter- and 2 summer-cereals, NAPP-ZINN, 1954: *Petkus* summer and winter rye, and a summer and winter annual race of *Arabidopsis;* COÏC and DURANTON, 1959: summer and winter wheat). Investiga-tions on 13 varieties of wheat and 7 races of *Arabidopsis*, however, revealed *no* corre-lation between intensity of respiration of the seeds or caryopses and their flowering behavior (NAPP-ZINN, 1957a).

In 1933 RICHTER, RANCAN and PEKKER had already arrived at the conclusion, after investigating changes in the intensity of respiration and of catalase and peroxidase activity during the vernalization of caryopses of two varieties of winter wheat, that changes in enzyme activities are not strictly correlated with qualitative steps in plant development. In spite of this, innumerable authors have since studied the variations in enzyme activity (principally of respiratory enzymes) in the course of vernalization and subsequent development. The investigations of SISAKYAN and FILIPPOVICH (1953) on the six varieties of cereal cited above deserve particular mention. They measured peroxidase, cytochrome oxidase, ascorbic acid oxidase, polyphenol oxidase, succinic-, citric- and malic-acid dehydrogenase activities over the course of 50 days of vernaliza-tion at $0—2°$ C, as well as during germination and later development. During vernaliza-tion the terminal oxidation was switched from cytochrome- to ascorbic acid-oxidase, and still later the two enzymes appeared to complement one another: the enzyme activity of winter cereals came to resemble more closely that of the summer cereals as a result of vernalization. Nevertheless, GÜNTHER (1959), using one summer and two winter varieties of wheat, was able to detect a rise in only ascorbic acid oxidase activity at the cost of cytochrome oxidase activity if germination had already set in during vernalization (water content of the caryopses was higher than 55% at the commence-ment of cold treatment). Calculated on the basis of protein-N, the cytochrome oxidase content in fact remained constant. GÜNTHER's experiments with enzyme inhibitors also seem to indicate that ascorbic acid plays no vital role in vernalization. Earlier, SÉCHET (1953a) had shown that there is a progressive rise in the rate of formation of ascorbic acid in the course of vernalization at $2°$ C in *Lupinus albus*. CHINOY, NANDA, and GARG (1957), besides advancing the flowering of *Trigonella foenum-graecum* and *Brassica chinensis* by about 10 days by treating the seedlings with ascorbic acid, also found that in wheat, oats and barley, vernalization influenced the rise in ascorbic acid content and the transition to the reproductive stage in a parallel manner. Finally, DÉVAY (1965a), using a Hungarian variety of winter wheat, showed that ascorbic acid oxidase activity usually rises with temperature. Further communications concerning a connection between vernalization and the activity of various enzymes can be found, among others, in SISAKYAN (1937: wheat and cotton: synthetic/hydrolytic activity of invertases), SISAKYAN and FILIPPOVICH (1951: winter wheat and winter barley: poly-phenol- and cytochrome-oxidase), AUGSTEN (1956: non-obligatorily cold-requiring *Haisa* variety of summer barley: amylase, catalase, phosphatase), INOUE and YAHIRO (1956: *Raphanus sativus* var. *raphanistroides:* catalase and peroxidase) and FEIER-

ABEND (1966: similar kind of increase in glucose-6-phosphate dehydrogenase in both winter and summer barley in the cold).

Another means of investigating the connection between vernalization and respiration suggested by the results of GREGORY and PURVIS (see above) lies in the use of enzyme poisons. A number of authors (CHOUARD and POIGNANT, 1951: *Vilmorin 27* winter wheat; GÜNTHER, 1961: *Derenburger Silber* winter wheat; KREKULE, 1961b: *Hodoninská holice* winter wheat; SHVEDSKAYA and KRUZHILIN, 1961: turnips and early cabbage) have been able to prevent or delay flower formation and shooting (as compared with only vernalized controls) by applying respiratory poisons such as diethyl thiocarbamate, malonic acid, sodium azide, K_3AsO_3, KCN and so on, before or during vernalization. Unfortunately, since they all failed to count the leaves on the experimental plants, the inhibition of flowering may have been entirely non-specific. Indeed, further experiments of this nature with *Vilmorin 27* wheat and with *Petkus* summer and winter rye led as a rule to no perceptible increase in the number of leaves despite an often considerable delay in anthesis (NAPP-ZINN, 1963a). This would suggest that inhibition of flowering was non-specific. Even using respiratory inhibitors, PURVIS (1961a) was unable to influence appreciably the effect of vernalization on the time of flowering in *Petkus* winter rye. W. HARTMANN and BUSCHBECK (1965a) were the first to achieve a parallel rise in age at flowering and number of leaves in isolated cases by soaking the caryopses of winter rye in ethyl urethane or malonic acid solution. In other cases, however (sown on May 23, 1964 following 30-day vernalization), malonic acid led to a rise in the percentage of shooting plants and a reduction in the number of leaves as compared with controls soaked in water. A similar observation had previously been made by TAN KE-WEI (1959) using the *Ukrainka* variety of winter wheat: soaking the grains in 0.005% malonic acid solution (or aconitic-, fumaric- or succinic-acid) advanced flower formation, as compared with water controls, only if vernalization had been suboptimal (25—30 days), but not in cases where the material had either not been vernalized at all or vernalized for 50 days. But, since a similar effect could be achieved in plants which had only undergone suboptimal vernalization by spraying with the above acids following cold treatment the positive effect on growth was probably non-specific and not connected with vernalization.

e) Nucleic Acids, Proteins and Amino Acids

The question of a possible connection between respiratory enzymes and vernalization suggests a possible influence of cold on the synthesis of specific amino acids. Experiments on embryos of *Petkus* summer and winter rye and on seeds of both a summer and winter annual race of *Arabidopsis* revealed a number of changes in the amino-acid spectrum. None of these variations, however, was such as to suggest that substances appeared in the course of the vernalization of cold-requiring forms which, in the corresponding summer annual form were already present before exposure to cold (NAPP-ZINN, 1956). Neither did cold bring about the disappearance from the winter forms of ninhydrin-positive substances which were absent from the summer forms. Changes in the amino-acid picture under the influence of cold were, in many cases, investigated only in winter annuals or in biennials and not in the corresponding non-cold-requiring forms. Demonstrable quantities of tryptophan appeared in *Hydrangea macrophylla* only during the course of the exposure to cold which is essential for the opening of the flowers (ASEN and STUART, 1958). In vernalized individuals of *Mammoth II* winter barley the proportion of aspartic acid in the total amino acids before sprouting was relatively high and that of aminobutyric acid relatively low, whereas the reverse was true in non-vernalized controls (SPARMANN, 1961). MARKOWSKI, MYCZKOWSKI, and LEBEK (1962) were unable to demonstrate a connection between the process of vernalization and the total nitrogen in winter wheat[7]. Such a connection could be demonstrated only for proline in agreement with the subsequent findings of SHVEDSKAYA and KRUZHILIN (1966) on *Beta* and *Daucus*. In *Rideau* winter wheat vernalization lowered the proportion of arginine, which makes up a large part of the soluble amino-acid fraction of unvernalized caryopses (WEINBERGER and GODIN, 1966). TRIONE, YOUNG, and YAMAMOTO (1967) recorded changes in the content of 17 identified amino acids and 7 unidentified ninhydrin-positive substances in both a summer and a winter variety of wheat during 0—5-week cultures at $+2°$ C and at $+25°$ C. Despite the fact that in some cases marked differences in dependence upon variety, temperature and duration of the culture were detectable, the observations do not permit the drawing of any conclusions as to the chemistry of vernalization. The same holds true for the analyses performed by ELLIS and TRIONE (1967) on changes in content of substances containing sulfhydryl groups or disulfide bridges in the two varieties of wheat under cold and warm conditions.

Earlier investigations concerning a connection between vernalization and nucleic acid metabolism were of a quantitative nature. KONAREV (1954) was apparently the first to detect a rise in the ribonucleic acid content in the embryos of soaked rye and wheat grains following vernalization, the rise being larger in winter forms than in summer forms. In the course of vernalization of winter wheat, UZAMI (1956) observed an increased content not only of ribo- but also of deoxyribonucleic acid. This finding was confirmed by SÉCHET (1962) in *Vilmorin 27* winter wheat: if embryos of equal coleoptile length were compared, then the DNA- and RNA-content was always higher in vernalized individuals than in those which had germinated in warmth. In the meantime other results had been obtained by FINCH and CARR (1956) with *Petkus* winter rye: the content of acid-soluble phosphorus, RNA- and DNA-P varied only slightly with vernalization and devernalization and there was no indication of a connection with the induction or annulment of ripeness to flower. These results of FINCH and CARR do not, however, exclude the possibility of the production of a *specific* nucleic acid as a result of vernalization. In a variety of *Raphanus sativus* with a low, but nevertheless "obligate" requirement for cold, there appears to be very little necessity for such a specific nucleic acid, since TASHIMA and IMAMURA (1954) were able to induce flower initiation without cold and even in darkness by adding a commercial preparation of RNA to the substrate.

The investigations of HESS (1961a, b, c) and of DÉVAY (1965b, 1966, 1967a), however, are dominated by the idea of a specific RNA. *Streptocarpus wendlandii*, used by HESS, has the disadvantage of requiring simultaneous short day and cold for the induction of flower formation, so that observations related to the latter may be connected with photoperiod rather than vernalization. Under inductive conditions, RNA synthesis, taken as a whole and measured by the incorporation of ^{32}P, is lowered. HESS explains this as indicating an inhibition of the synthesis of "vegetative" RNA, the constructional elements of which are thus made available for the production of, albeit smaller quantities of, "reproductive" RNA. DÉVAY, carrying these ideas further, has detected variations in ribonuclease activity in

---

[7] Neither did the experiments of MARKOWSKI and MADEJ (1962a, b) reveal a correlation between vernalization and changes in total phosphorus and the various phosphorus fractions.

the course of vernalization of winter-wheat caryopses. This she discovered by following spectrophotometrically the liberation of bases from ribonucleic acid under the influence of embryo homogenates. The RNAase activity, with an optimum at about $+2°$ C, passed through several maxima in the course of a 7-week cold treatment. According to Dévay, these maxima are attributable to different RNAases. "RNAase I" is thought to be responsible for the first maximum occurring approximately 14 days after the commencement of vernalization, and is apparently connected with the inhibition of synthesis of "vegetative" RNA as well as with its destruction. Various inhibitors such as 2.4-dinitrophenol, trypa-flavine, acridine orange and chloramphenicol suppress this peak and the vernalization effect. The latter, however, was only measured by the number of leaves. "RNAase I" arises in the shoot tips and also in homogenates of whole embryos, apparently in two stages, the second of which can be inhibited by dinitrophenol, silver nitrate and mercury acetate, and enhanced by dehydroascorbic acid. In other experiments Dévay sprayed winter-wheat plants with inhibitors such as 2-thiouracil, thioacetamide, digitonin and chloramphenicol as follows: during the preculture period (up to the two-leaf stage) at $15°$ C and 8-h short day; during the ensuing 4-week vernalization at $+2°$ C and 8-h short day; during the subsequent 6-day stabilization treatment at $15°$ C and, finally, during the continued culture at $17°$ C and 16-h long day. All four poisons brought about a 20% inhibition of growth at the most, but drastically prevented the vernalization effect if applied during stabilization, although the criteria used are not mentioned. 2-Thiouracil was particularly effective if applied at the beginning of vernalization, digitonin at the end of vernalization and chloramphenicol in the middle of cold treatment as well as at the commencement of exposure to long day. From this it is concluded that the various phases are associated with different synthetic processes (nucleic acid and protein).

Experiments carried out by other authors on different objects, using the same compounds as Dévay as well as other substances inhibitory to syntheses, have sometimes led to identical effects, sometimes to the reverse, and in some cases have had no significant effect at all. Soaking caryopses of *Petkus* winter rye, for example, in a 0.05% solution of chloramphenicol instead of in water markedly enhanced the effect of subsequent incomplete vernalization (4 weeks at $+2°$ C) while other inhibitors of protein- or RNA synthesis proved to be less effective or completely ineffective (Feierabend and Bünsow, 1967). Ethionine, an antagonist of methionine, blocks protein synthesis. In quantities of $0.125-2$ mg per plant and week of induction it inhibits the induction of flowering in *Streptocarpus wendlandii*: "the most plausible explanation of the selective inhibition of the induction of flowering by ethionine would, according to these results, be that it blocks proteins which are only necessary for flower formation". The variable composition of RNA in induced and in vegetative plants is probably "as far as the processes involved in the induction of flowering are concerned . . . . . . of secondary importance" (Hess, 1961b). Earlier (1959), Hess had succeeded in inhibiting induction of flowering in *Streptocarpus* by the use of 2-thiouracil, which is incorporated into RNA in the place of uracil. Suge and Yamada (1965b) obtained the same effect in winter wheat whereas the application of 2-thiouracil during the vernalization of rosettes of the winter annual *St* race of *Arabidopsis* did not appreciably influence flower formation, but had at most a necrotic effect (NAPP-ZINN, 1963b)[8]. In *Oenothera biennis* var. *sulfurea* in which large quantities of RNA normally accumulate in the leaves in the first weeks of vernalization, PICARD (1967a) was able to inhibit the vernalization effect by the addition of 5-fluorouracil in the course of exposure to cold. Simultaneous application of orotic acid (a constituent of RNA) in small quantities terminated this inhibition. PICARD, however, does not believe that fluorouracil specifically influenced the process of vernalization but rather that it inhibited shooting, here a prerequisite for flower formation.

f) Additional Observations of a Cell-Physiological Nature

Many authors believed that they had succeeded in demonstrating a connection between vernalization and certain cell-physiological phenomena. A shift in the isoelectric point, alterations in the cell of the redox potential, in plasma permeability and viscosity, in its ability to coagulate and in stainability were among the observations made. In most cases, however, such alterations may have been merely secondary effects of vernalization (i.e. non-specific effects of cold). Most results of this nature were obtained some years ago (RICHTER, 1934; BASSARS-KAYA, 1934a, b, 1936; FILIPPENKO, 1936; SISAKYAN, 1937; DAVID, 1945b; STOUT, 1949; MEKHANIK, 1961, 1962; WEINBERGER and KU, 1966). Using a new method SPECTOROV (1957) was unable to detect any influence of vernalization upon the isoelectric point of winter wheat, and CHEL'TSOVA and LEBEDEVA (1966) came to the conclusion that any such alterations were connected with germination rather than with vernalization. Further references to cytological, and especially to karyological, changes accompanying vernalization can be found in SÉCHET (1953b) and BES-NARD-WIBAUT (1970).

g) Growth Substances

Following the recognition, in the 1930's, of the diverse effects of growth substances upon growth and development of higher plants it obviously became necessary to consider a connection between such substances and flower formation and especially vernalization. In the ensuing years many authors have devoted themselves to this problem and after what has been said already concerning the varied genetic basis of cold requirements in different plants (for many of the test objects used so far, little reliable genetic information is available), it is hardly surprising that their results are partly conflicting, partly irrelevant.

α) Auxins and Antiauxins

Reviews of a number of papers concerned with the connections between auxins, auxin antagonists and inhibitors on the one hand and flower formation on the other, are to be found in AUDUS (1953: p. 295ff.) and LANG (1961: p. 909ff.). Therefore only a limited number of examples will be given here.

---

[8] The later observation by BESNARD-WIBAUT (1970) of an increased DNA synthesis in the shoot apical region (measured by the increase in the per cent of nuclei into which ^{3}H-labelled thymidine was incorporated) during the vernalization of older individuals of this race of *Arabidopsis* was probably a secondary phenomenon due to the mitoses immediately connected with flower initiation.

Several authors have reported that treatment with growth substances has a beneficial effect on cold-requiring plants which have either not been vernalized at all or only incompletely vernalized, for example, MUNERATI (1938: *Beta* roots, dipped in growth-substance solution following amputation of the root, β-indole acetic acid most effective), LEOPOLD (1952) and LEOPOLD and GUERNSEY (1953a, b, c, 1954: *Alaska* pea and cold-requiring *Wintex* summer barley, after the seed had been soaked in solutions of α-naphthalene acetic acid, β-naphthalene oxyacetic acid or 2,3,5-triiodobenzoic acid (!) followed by brief exposure to cold, a so-called chemical vernalization), CHAKRAVARTI and PILLAI (1955: *Brassica campestris*, after soaking the seeds in solutions of β-indole acetic acid, β-indole butyric acid or α-naphthalene acetic acid with subsequent cold treatment) as well as BALLANTYNE and LINK (1961: varieties of *Rhododendron*, lifting of dormancy of flower buds by injection of α-naphthalene acetic acid or 2,3,5-triiodobenzoic acid).

KHOLODNY (1935, 1936a) believed that in the course of vernalization a hormone ("blastanin") was transferred from the grain endosperm to the embryo. However, in the experiments in which he soaked caryopses of wheat, rye, barley and millet in indole acetic acid there was practically no effect (see also CHAILAKHYAN and ZHDANOVA, 1938a, b). Ultraviolet irradiation, which destroys growth substances, delayed flower formation in vernalized sugar beet, but spraying such plants with β-indole acetic acid or with 2,3,5-triiodobenzoic acid was ineffective (SCHNEIDER, 1960). Spraying unvernalized plants of a biennial race of *Hyoscyamus* and *Petkus* winter rye had no effect either (MELCHERS, 1936; BRUINSMA and PATIL, 1963).

Inhibition of flower formation by means of the application of either growth substances or their antagonists, or of inhibitory substances, before or instead of cold treatment has been reported by CHAKRAVARTI (1954: *Linum usitatissimum*), CHAKRAVARTI and PILLAI (1955: *Brassica campestris*, 2,4-dichlorophenoxyacetic acid and 2,3,5-triiodobenzoic acid), KOJIMA, YAHIRO, and ÊTÔ (1957: *Raphanus sativus* var. *raphanistroides*), and NAPP-ZINN (1963a: winter wheat and rye) among others.

It is likely that all of these effects are non-specific. This view is supported by the observation that often only the time of flowering *or* the number of leaves preceding the first flowers was significantly altered. Furthermore, as a rule, no convincing connection could be detected between the progress of vernalization and changes in autochthonous auxins in the course of exposure to cold (see, for example, HESS, 1958: *Streptocarpus wendlandii*; SPARMANN, 1961: *Hordeum*; KREKULE and TELTSCHEROVÁ, 1963: *Triticum*; TRÂN THANH VÂN-LÊ KIÊM NGOC, 1965: *Geum urbanum*). On the other hand, FILIPPENKO (1937) found a threefold rise in growth-substance content in two varieties of winter wheat following vernalization. Conversely, non-vernalized seedlings of *Lactuca sativa*, *Great Lakes* variety, contained twice as much β-indole acetic acid as vernalized plants (FUKUI, WELLER, WITTWER, and SELL, 1958).

β) Gibberellins and Antigibberellins

With LANG'S (1956a, b, c) success in bringing about shooting and flowering in unvernalized individuals of a biennial race of *Hyoscyamus* in long day, but without exposure to cold, by dropping onto them a mixture of gibberellins A_1 and A_3

(10−50 mg/l), it seemed that the long-existing doubts of a connection between vernalization and metabolism of growth substances were about to be resolved in a biochemical explanation. LANG wrote (1956c) that gibberellin "replaces the 'ver-nalin' but not the 'florigen' in biennial *Hyoscyamus*". Similar results were obtained, mostly with gibberellic acid alone (GA₃), in other normally cold-requiring plants such as a *Daucus carota* variety (LANG, 1956c), *Campanula medium* (CHOUARD, 1956/57), *Cichorium endivia* (HARRINGTON, RAPPAPORT, and HOOD, 1957; RAPPAPORT and BONNER, 1960), *Arabidopsis thaliana* races *St* (SARKAR, 1958), *Kru* and *Öst*, as well as *A. suecica* and *Cardaminopsis arenosa* (LAIBACH, 1958), *Chrysanthemum morifolium* (HARADA and NITSCH, 1959), *Rudbeckia bicolor* and *R. speciosa* (CHAILAKHYAN, KHLOPENKOVA, and LOZHNIKOVA, 1964), *Althaea rugosa* (GULKANYAN and KHAKHATRYAN, 1964), and *Scrophularia alata* (LARRIEU, 1966). For further examples see LANG (1965: p. 1499: Table 38). Using *Hyoscyamus niger*, LANG, SANDOVAL, and BEDRI (1957) obtained the same effects with extracts from the endosperm of *Echinocystis macrocarpa*, which is rich in gibberellins, as with purified preparations of gibberellin from *Fusarium moniliforme*. Further experiments carried out by MICHNIEWICZ and LANG (1962) showed that gibberellins A_1, A_4, A_5, A_6, A_7 and A_9, as well as A_3, were capable of producing more or less copious flower formation in *Centaurium umbellatum*, whereas A_2 induced merely a slight elongation of the stem and A_8 had no effect whatsoever[9]. However, in *Myosotis alpestris*, which also requires vernalization, only those individuals treated with A_7 and some of those treated with A_1 flowered, while the remaining 7 gibberellins elicited only sprouting.

The many cases in which gibberellic acid elicits shooting (often correlated with flower formation), but not flower formation itself, must now be considered. Another variety of carrot and one variety each of beet and parsley (LANG, 1956d), *Oenothera biennis* (PICARD, 1958), *Beta vulgaris* (MARGARA, 1959, 1960a), numerous Campanulaceae (MATHON, 1960a, b, c) and *Anthriscus cerefolius* (PARLEVLIET, 1967), to mention only a few, began to shoot but did not flower following treatment with gibberellins. This was usually most clearly noticeable under long-day conditions. Gibberellic acid had no effect whatever in several perennial species of *Phacelia* (GILLET, 1963).

In many of the above-mentioned plants gibberellic acid is readily capable of enhancing the flowering effect of incomplete vernalization. This was seen by CHOUARD and TRÂN THANH VÂN (1962) in *Geum urbanum* and led to their distinguishing between "ripeness to flower", which can be induced only by cold, and "ripeness to shoot", which may be induced by cold or gibberellic acid. Some grasses behave in a similar manner: for example, certain varieties of winter wheat (PAULI, WILSON, and STICKLER, 1962; POP, BĂRBAT, and OCHEŞANU, 1963; REJOWSKI, 1965) and *Lolium temulentum* (PETERSON and BENDIXEN, 1963). Of the other authors who have been concerned with the influence of gibberellic acid on flower formation in cold-requiring cereals and fodder grasses only the following will be mentioned here: FEJER (1960: *Lolium perenne*), BOMMER (1961a: *L. perenne, L. multiflorum* and *Dactylis glomerata*), JAMES and LUND (1960, 1965: winter barley),

---

[9] Various races of *Arabidopsis* with pronounced cold requirements also respond to most if not all of the gibberellins mentioned (NAPP-ZINN, 1963c).

KOLLER, HIGHKIN, and CASO (1960: 2 varieties of barley and *Petkus* winter rye), PURVIS (1960: *Petkus* winter rye), BRUINSMA and PATIL (1963: ditto), HURD and PURVIS (1964: ditto), FYODOROV and SHINKAREVA (1966: winter rye and winter wheat), RAZUMOV, LIMARJ, and KE-WEI TAN (1960: winter wheat) and SUGA and YAMADA (1964: winter wheat and barley). The results diverge partly according to species and variety, partly on account of the criteria chosen ("double ridge" stage, initiation of anthers, shooting, ear emergence, anthesis etc.). Most authors, however, conclude that the effect of gibberellin treatment does not exactly imitate that produced by chilling. They find that the time elapsing between flower initiation and anthesis can be shortened, but not that the induction of ripeness to flower or flower initiation can be accelerated[10].

The questions now arising concern the specificity of the action of gibberellin and whether "vernalin" is identical with a gibberellin. In attempting to provide answers only those plants will be considered in which gibberellin treatment induces flower formation without either previous or subsequent exposure to cold. A fundamental difference between vernalization and treatment with gibberellin lies in the fact that only the former is often already effective at the seed stage (p. 173—175). References to such experiments can be found in SARKAR, 1958: *Arabidopsis thaliana;* RAPPAPORT and BONNER, 1960: *Cichorium endivia;* BACHMANN, CURTH, and RÖSTEL, 1960: beets. Nevertheless, gibberellins do readily penetrate into seeds and in *Arabidopsis* they can render after-ripening and light unnecessary for germination. A further difference is to be seen in certain secondary phenomena: in cold-requiring races of *Arabidopsis*, for example, the number of stem leaves is lowered by vernalization but raised by gibberellins (SARKAR, 1958). Further, the per cent of empty ("sterile") axils on the stem is decreased by vernalization but is increased by gibberellins in the *Zü* race of *Arabidopsis*. Gibberellins can even evoke the appearance of sterile axils in races not usually exhibiting this phenomenon (NAPP-ZINN, 1957f, 1963d).

Conflicting reports exist concerning the variations in level of endogenous gibberel. lins during vernalization. According to MARGARA (1960a, b, 1963a) various biological tests revealed no difference between vernalized and unvernalized individuals- Only in long day following vernalization did the content of gibberellins rise noticeably when the plants began to shoot; the R_f values suggested A_1- and A_3-like substances. Similar observations were made on *Hyoscyamus niger* (REINHARD and LANG, 1961). Comparative investigations carried out by KREKULE and TELT-SCHEROVÁ (1963) on vernalized and unvernalized caryopses of summer and winter wheat failed equally to show a definite connection between gibberellin content and vernalization. In a winter rye variety studied by CHAILAKHYAN (1967) the level of gibberellin also rose in seedlings and leaves only during long day *subsequent* to vernalization and was greater the longer the exposure to cold. CHAILAKHYAN concluded from this that only a *precursor* of gibberellin is formed in the course of vernalization.

In the foregoing cases the gibberellin content rose only when the days were long enough for flower initiation. It is not surprising, therefore, that following vernalization a gibberellin-like substance could readily be isolated from the cold-requiring but day-neutral *Chrysanthemum* variety *Shuokan*, and that it was capable of eliciting flower formation even in non-vernalized individuals (HARADA, 1960,

---

[10] In alpine plants such as *Soldanella minima* (see p. 178/79) gibberellin treatment replaces only the cold exposure essential for pollen formation and flower unfolding (LONA, 1968).

1962). The gibberellin content of caryopses of the *Dańkowska Selekcyjna* variety of winter wheat studied by KENTZER (1967) was normally raised by cold treatment. Treatment with a 10^{-3} molar solution of NaN_3 either before or during the first phase of exposure to cold completely prevented a vernalization effect, and the inhibition induced by NaN_3 could be lifted by treatment with gibberellins. The content of endogenous gibberellins in the caryopses treated with NaN_3 was radically lowered during the first 40 days of vernalization, both absolutely as well as in comparison with caryopses vernalized without NaN_3. If cold treatment was extended to 50–70 days the gibberellin level rose steeply again without any accompanying vernalization effect.

The influence on vernalization of the so-called antigibberellins such as CCC, Amo-1618, Phosphon-D and B-995 that block the synthesis of gibberellins will now be considered. In winter annual races of *Arabidopsis* the effect of a subsequent cold treatment can be diminished by soaking the seeds in CCC solution. Since, however, soaking the seeds also of summer annual, non-cold-requiring races of *Arabidopsis* increases the age at flowering, as well as the number of leaves, it is probably not vernalization but rather a later stage of flower induction that is affected (NAPP-ZINN, 1964b). Similar results were obtained by GÜNTHER (1966) with summer and winter wheat, rye and barley: treatment with high concentrations of CCC at various intervals before and after exposure to cold delayed ear emergence even in summer cereals. Since, however, in this case the number of leaves was never appreciably influenced GÜNTHER deduces that a "specific inhibitory effect upon vernalization can be excluded". In similar experiments carried out by SUGE and OSADA (1966) a summer annual control was omitted and the number of leaves was not recorded. The application of B 995 to *Oenothera biennis* in low concentrations following adequate cold treatment prevented only shooting without influencing either rate of leaf formation or time of flowering, whereas in higher concentrations it also prevented flowering. PICARD (1967b) considered this to be merely an indirect effect upon flower formation. Conversely, CCC even stimulates stem growth in *Oenothera biennis* if dropped onto the rosette (PICARD, 1967c). Similarly, if applied to the young rosettes of the *Zü* race of *Arabidopsis thaliana*, CCC can also lower the age at flowering as well as the number of leaves (NAPP-ZINN, 1971). In so far as it also lowers the number of sterile leaf axils in the *Zü* race of *Arabidopsis*, CCC elicits an effect comparable to vernalization (see p. 192 and NAPP-ZINN, 1967b).

γ) Other Growth Substances

BRUINSMA (1963) and BRUINSMA and PATIL (1963) successfully produced ear initiation in long day by spraying non-vernalized individuals of *Petkus* winter rye with either a 70 ppm solution of a mixture of 8 synthetic stereoisomers of α-tocopherol (vitamin E) or with a 270 ppm potassium gibberellate solution (see p. 192). A similar flower-inducing effect was achieved in *Cichorium intybus* by MICHNIE-WICZ and KAMIEŃSKA (1967) with tocopherol as well as with kinetin, although the latter had otherwise nearly always been proved ineffective. The endogenous gibberellin level was, in this case, lowered, at least at the end of the treatment with kinetin or tocopherol. In *Geum urbanum*, too, flower formation was triggered in young axillary buds by kinetin treatment without vernalization, or by decapitation

of the main axis (Trân Thanh Vân and Chouard, 1968). According to Feierabend (1970) the cytokinin level in *Petkus* winter rye is optimum at temperatures
below 17° C; temperatures of 7–12° C lead to a selective repression of plastid
development and of the formation of photosynthetic enzymes. In their place non-
photosynthetic glucose-6-phosphate dehydrogenase, the synthesis of which is
arrested at higher temperature, continues to be formed.

h) Phytochrome

Since many plants fail to respond to the inductive photoperiod until after vernalization it is very reasonable to assume that the phytochrome essential for perception
of day length is formed in the course of the cold treatment. Friend (1965), on the
basis of comprehensive experiments with summer and winter rye, arrived at the
conclusion that the phytochrome system has nothing to do with the flower-
inducing effect of low temperatures, but that irradiation during cold treatment
may contribute to the stabilization of the vernalization effect. On the other hand,
Dévay (1967b), using winter wheat, found that phytochrome is not formed in
caryopses until vernalization is in progress. In the leaves of seedlings from caryopses
vernalized for 5–56 days at 0° C phytochrome could be demonstrated only if
vernalization had taken place for at least 21 days in darkness, 35 days in long day,
or 49 days in short day. Treatment for 5 h at 30° C following 21–42 days of vernalization in the dark reduced the phytochrome content to zero.

i) Concluding Remarks

As far as the physiological analysis of vernalization is concerned it appears from
the foregoing review that scarcely a single observation has been made that has
not later been contradicted by the results of other authors using different test
objects. This is hardly surprising considering the diversity of the genetic structure
upon which the need for vernalization is based. The present author feels unable to
share the optimistic view of Augsten (1964) that "a basically uniform mechanism
can be assumed". The search for a single flowering hormone, representing the
universal end product of vernalization, does not at present appear to be very
promising. It is more usual nowadays to assume that ripeness to flower is connected
to a definite quantitative relationship between several or many hormonelike substances, in the syntheses of which (or of the enzymes responsible for them) intermediary thermolabile stages as mentioned on p. 181 might be intercalated. This
would provide an explanation for the need for cold exhibited by many plants on
their way to ripeness to flower.

References to this section see p. 264.

III. Limiting Temperatures for Life Functions

W. Larcher, U. Heber, and K. A. Santarius

A. Gradual Progress of Damage Due to Temperature Stress

W. Larcher

Temperature stress strains a plant excessively; it may, however, not necessarily cause lethal injuries. The reactions of the plant depend on the degree and the duration of the stress, and on the resistance and adaptability of the stressed plant. Temperature stress, therefore, may present a limiting factor for plant life.
Cell structure and function may be drastically injured when their critical temperature threshold is exceeded, with the result that a necrotic picture immediately

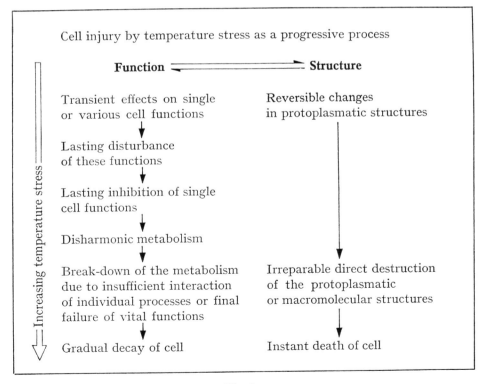

Fig. 1

appears ("direct injury", LEVITT, 1956). Apart from this abrupt type of destruction another kind of damage may occur. It proceeds imperceptibly at first, involving only certain cell functions, and increases so gradually towards a final phase that the changeover from normal to pathological behavior is difficult to recognize (cf. Fig. 1). As extreme temperature ranges are approached, only transient interruption of various cell functions occurs at first. Later, however, the disturbance becomes permanent: finally cellular activity suffers lasting impairment, vital functions are gradually extinguished and the cell dies. ALEXANDOV et al. (1970) consider the degree of damage at a given moment to result from the interplay of destructive and reparative processes in the protoplasm. Immediate structural and functional damages are regarded as the destructive processes. As long as it is able, the plant responds rapidly to such injury by resynthesizing and, somewhat more slowly, by adapting to a new equilibrium with the resulting increase in resistance.

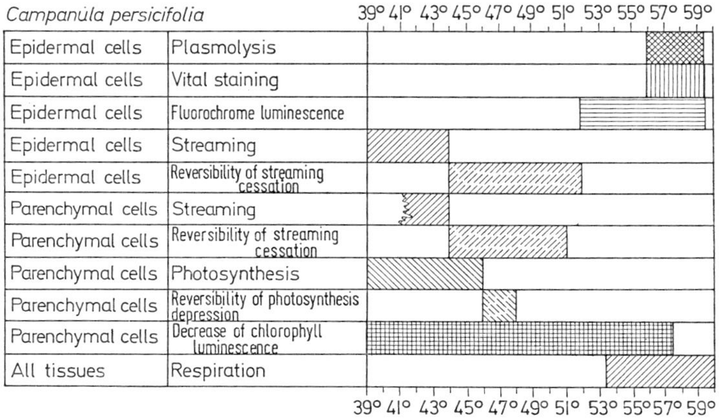

Campanula persicifolia   39° 41° 43° 45° 47° 49° 51° 53° 55° 57° 59°

Epidermal cells	Plasmolysis	
Epidermal cells	Vital staining	
Epidermal cells	Fluorochrome luminescence	
Epidermal cells	Streaming	
Epidermal cells	Reversibility of streaming cessation	
Parenchymal cells	Streaming	
Parenchymal cells	Reversibility of streaming cessation	
Parenchymal cells	Photosynthesis	
Parenchymal cells	Reversibility of photosynthesis depression	
Parenchymal cells	Decrease of chlorophyll luminescence	
All tissues	Respiration	

39° 41° 43° 45° 47°49° 51° 53° 55°57° 59°

Fig. 2. Injury in leaf cells of *Campanula persicifolia*. The left of the shaded areas indicates the first deviation from the normal behavior, the right corresponds to definite, irreversible damage after 5 min heating. (From ALEXANDROV, 1964)

Vital functions differ in their sensitivity to temperature (see examples in ALEXANDROV, 1964; BARABALCHUK, 1969; ALEXANDROV et al., 1970; cf. Fig. 2). Functional disturbances have mainly been observed following brief heating to temperatures close to the thermal death point, in cold storage of cold-sensitive plants (chilling between $+10°$ C and $0°$ C) and in connection with freezing processes. The temperature level of and the temperature span between partial and total cell damage depend upon the species and the condition of the individual plant as well as on the cell type.

1. Effect of Extreme Temperatures on Protoplasmic Streaming

Protoplasmic streaming is the most sensitive indicator of the beginning of temperature-conditioned disturbances in cell function. Its speed depends directly upon the energy supplied by oxidative processes and upon the availability of energy-rich phosphate (ZURZYCKI, 1951; LEWIS, 1956; THIMANN and KAUFMANN, 1958; KAMIYA, 1959).

Both SACHS (1864) and ALEXANDROV (1964) have given exact descriptions of the events taking place upon heating: at first protoplasmic streaming slows down and then stops but, if the samples are cooled it can recover almost immediately. Further stress, however, leads to the irreversible arrest of streaming. The temperature span between the onset of reversible and irreversible cessation varies from 4—9° C, depending upon species, tissue and experimental procedure (ALEXANDROV, 1964). In most plants protoplasmic movement ceases finally at a few degrees below or

above 50° C, e.g. between 48° C and 56° C in leaf-epidermis cells of various phanerogams following 5 minutes' heating (ALEXANDROV, 1964), at 42° C in cambium cells of *Pinus* after brief heating (THIMANN and KAUFMANN, 1958).

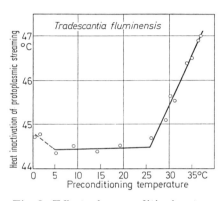

Fig. 3. Effect of preconditioning temperature on the heat sensitivity of protoplasmic streaming in leaf cells of *Tradescantia fluminensis*. Exposure time for preconditioning temperatures: 16—18 h, exposure time for test temperatures: 5 min. (After ALEXANDROV and FELDMAN, 1958)

Cooling can also lead to the suspension of protoplasmic streaming. In chilling-sensitive plants such as tomato, melon, marrow, tobacco, sweet potato *(Ipomea)* and *Nitella (Characeae)*, this occurs at +10° C (SACHS, 1864; LEWIS, 1956; KAMIYA, 1959). In cold-resistant plants such as *Pinus, Raphanus, Daucus, Erodium* etc. (LEWIS, 1956; THIMANN and KAUFMANN, 1958) cold immobilization sets in only below 0° C and is always reversible.

The temperature limits for cold and heat inactivation of protoplasmic movements can be shifted by hardening. Hardening is the adaptive process by which sensitive cells are transformed into tolerant or hardy cells. The reaction level of algae changes within a few hours of exposure to a different environmental temperature. Cells of higher plants can also be hardened but exposure to the stressing temperatures must extend over hours or even days (ALEXANDROV and FELDMANN, 1958; ALEXANDROV, 1967, 1969; ALEXANDROV et al., 1970, Fig. 3).

2. Respiratory Behavior

Temperature stress and temperature shock can lead to deviations from normal respiratory behavior. These involve the intensity of respiration and, in cases of greater stress, the respiratory quotient. LANGE (1953), SEMIKHATOVA (1953), TERUMOTO (1965b), and SCHRAMM (1968) used the extent of respiratory disturbance during and following exposure to extreme temperatures as a measure of the degree of damage caused. Plants injured by temperature stress were recognizable by strongly fluctuating and imbalanced respiration after their return to their original conditions.

During exposure to great *heat* the intensity of respiration is strongly increased, this being a heat-promoted process (algae: MONTFORT, RIED, and RIED, 1955; lichens: LANGE, 1965b), but at still higher temperatures respiration finally ceases altogether due to plasma injury (e.g. SEMIKHATOVA and DENKO, 1960; KINBACHER and SUL-LIVAN, 1967, ref. FORWARD, 1960). *Aftereffects* on respiration have been observed following temporary *heat exposure* (LANGE, 1965b; WAGENBRETH, 1965): if the stress was not too severe, respiration, being a relatively heat-resistant process, very rapidly returned to normal following restoration of favorable temperature conditions (MONTFORT, RIED, and RIED, 1955; LYUTOVA, 1962; BAUER, HUTER, and LARCHER, 1969; BAUER, 1972).

Many respiratory changes have been recorded *during cold stress:* cold-sensitive plants such as cucumber, tomato, bean, banana and papaya undergo an extraordinary diminution in respiratory activity in the temperature range $0-10°$ C (PLATENIUS, 1942; JONES, 1942; EAKS and MORRIS, 1956). In cold storage below $10°$ C the Q_{10} for the metabolic activity of isolated mitochondria of chilling-sensitive species rises rapidly from the normal level $(1.3-1.6)$ to values between 2.2 and 6.3.

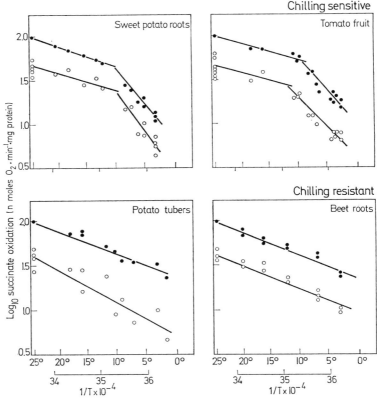

Fig. 4. Arrhenius plots of succinate oxidation by mitochondria of chilling-sensitive *Ipomoea* roots and *Lycopersicum* fruits and of chilling-resistant *Solanum* tubers and *Beta* roots. The upper curves represent state 3 respiration (high ADP-level), the lower curves show state 4 (high ATP-level). (From LYONS and RAISON, 1970a)

Mitochondria of chilling-resistant species such as *Beta* and *Brassica* maintain Q_{10} values of 1.7–1.8 over the entire temperature range $+1.5$ to 25° C (LYONS and RAISON, 1970a; cf. Fig. 4). Below 10–11° C the activation energy for succinate oxidation increases in chilling-sensitive species up to seven times its value at 20° C, but remains unchanged in chilling-resistant species. The behavior of poikilothermic animals such as trout and catfish is analogous to that of chilling-resistant plants, whereas that of homeothermic animals such as squirrels and rats is analogous to the behavior of chilling-sensitive plants, except that the critical temperature threshold for such animals is 23–24° C (LYONS and RAISON, 1970b). After several weeks at low temperatures the oxidative and phosphorylative capacities of the mitochondria of chilling-sensitive plants come to a standstill (in *Ipomea* from 5th week onwards; LIEBERMANN et al., 1958).

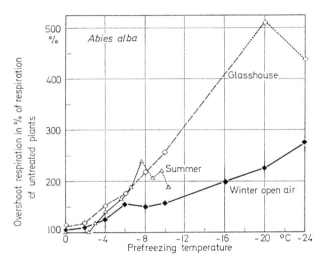

Fig. 5. Enhancement of the respiration (CO_2-output) of fir needles measured at $+20°$ C following previous freezing at various temperatures. (From BAUER, HUTER, and LARCHER, 1969)

The tissues of frost-hardy plants respire normally until freezing processes set in. If the plants have been frozen, however, respiration exhibits a striking overshoot: after the end of the frost, the respiratory gaseous exchange rises to 3 to 5 times its normal level during 4–12 h. The less well adapted to cold and the further the plants have been cooled, the greater is this respiratory overshoot (SEMIKHATOVA, 1953; SEMIKHATOVA et al., 1962; TRANQUILLINI, 1957; PISEK and WINKLER, 1958; RAKITINA, 1960; LARCHER, 1961; WEISE and POLSTER, 1962; TERUMOTO, 1965a; PISEK and KEMNITZER, 1968; BAUER, HUTER, and LARCHER, 1969; KALLIO and HEINONEN, 1971; cf. Fig. 5). Similar respiratory reactions have also been observed in precooled chilling-sensitive plants stored at temperatures below $+10$ to 15° C after their return to room temperature (citrus fruits: EAKS, 1960, Fig. 6; cotton plants: AMIN, 1969).

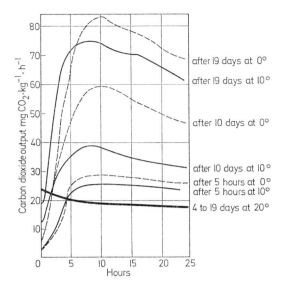

Fig. 6. Respiration rates of mature orange fruits at 20° C after exposures of various durations to chilling temperatures. The heavy line indicates the CO₂ output of control samples. (From EAKS, 1960)

3. Depression of Photosynthetic Capacity

Photosynthesis is a cell function especially sensitive to *heat*. The linkage between the plastid pigments and the carrier proteins apparently becomes unstable at higher temperatures (ENGELBRECHT and MOTHES, 1960, 1964; LYUTOVA, 1963). An exception is provided by the thermophilic blue algae, in which carbon assimilation proceeds optimally at 40° C, and which can maintain photosynthesis up to their thermal death point (BÜNNING and HERDTLE, 1946; for literature see BIEBL, 1962, and BROCK, 1967). The photosynthetic apparatus of desiccated lichens is also extremely thermostable: xerophytic lichens tolerate strong heating in a dry state with no resultant depression of photosynthesis (LANGE, 1953, 1969). In their imbibed state on the other hand, lichens suffer an early heat-inhibition of photosynthesis, which can last many days and even be irreversible if the stress is too great (LANGE, 1966, 1969).

The temperature span between the initial transitory depression of photosynthesis and its permanent damage by heat usually amounts to several degrees centigrade (about 4° C in *Usnea dasypoga* and *Ramalina maciformis*, about 8° C in *Cladonia rangiferina*). An exception is provided by the Atlantic coastal lichen *Rocella fucoides*, in which the photosynthetic apparatus is damaged to such an extent at only 34° C that it is already incapable of recovery even at the first sign of a depression of photosynthesis.

The response of marine algae to brief rises in temperature varies according to species and individual conditions. Eurythermic intertidal algae such as *Chaetomorpha linum* and *Fucus vesiculosus* suffer no appreciable reduction in CO₂ uptake if heated rapidly from 14 to 32° C, whereas the photosynthetic capacity of cold stenothermic sublittoral algae such as *Furcellaria fastigiata, Delesseria sanguinea*

·and *Rhodomela subfusca* breaks down within a couple of hours at 32° C and recovers only partially, if at all, long after restoration to the initial temperature (MONTFORT, RIED, and RIED, 1955; BIEBL, 1962). If heat treatment is prolonged the photosynthetic apparatus is irreversibly damaged even at only moderately elevated temperatures: after 6 h at 34° C in *Fucus vesiculosus* as compared to 1/2 h at 38° C (SCHRAMM, 1968).

In higher plants photosynthesis is inhibited both during the heat treatment (PISEK et al., 1968; see p. 112) and subsequently (demonstrated by LYUTOVA, 1958, 1962; SEMIKHATOVA et al., 1962; KAMENTSEVA, 1969 in various herbaceous plants, and by FELDMAN, 1968; WAGENBRETH, 1968; BAUER, HUTER, and LARCHER, 1969a; BAUER, 1972 in woody plants). Inactivation occurs gradually: initially

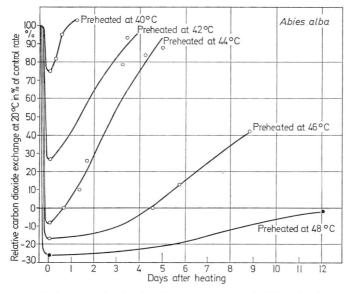

Fig. 7. Recovery of photosynthesis after heat treatment of *Abies alba* in summer. Heat exposure 30 min, illuminance 10 kLx. ○ Needles without necrotic damage. ● Needles with necrotic damage. (According to BAUER, 1970)

photosynthesis can still recover within a few hours after return to a more propitious temperature range. Stronger heating leads to a more prolonged depression of photosynthesis (see Fig. 7), and the degree of injury can be appreciably increased by the action of simultaneous strong illumination (LOMAGIN and ANTROPOVA, 1966). Finally, photosynthesis ceases completely, before the appearance of direct tissue damage in some species like *Abies* and concurrently with the first signs of necrosis in others like *Acer* (BAUER, HUTER, and LARCHER, 1969). The photosynthetic apparatus of plants can be hardened to heat, so that the inhibition of photosynthesis sets in at higher temperatures, the resultant decrease in efficiency is less, and the plants are capable of a more rapid recovery (LYUTOVA, 1958, 1962; FELDMAN, 1968; BAUER, 1972; Fig. 8).

Photosynthesis of many mosses and lichens is rather insensitive to *frost:* Mosses of the temperate zone exhibit a positive CO_2 balance to about $-8°$ and $-10°$ C (ATANASIU, 1971a; KALLIO and KÄRENLAMPI, 1973), and some lichens of the temperate and polar regions to -8 and $-13°$ C (LANGE, 1965a; ATANASIU, 1969b, 1971b; KALLIO and HEINONEN, 1971). *Cladonia alcicornis, Cladonia convoluta,*

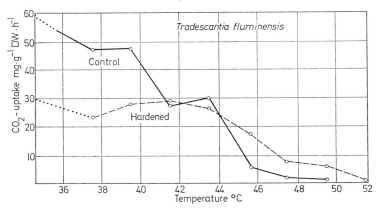

Fig. 8. Effect of heat hardening on the heat stability of photosynthesis of *Tradescantia fluminensis.* Heat exposure in the test 5 min. *Control* Unhardened control plants (solid line). *Hardened* Heat-hardened plants; hardening at $35.5°$ C for 18 h (dashed line). (From LYUTOVA, 1958)

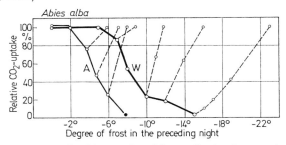

Fig. 9. Depression of carbon dioxide uptake of fir needles in the morning after 12 h frost exposure, and recovery of photosynthetic capacity in the following days (dotted lines) in autumn (*A*) and winter (*W*). Illuminance: 10 kLx. (From PISEK and KEMNITZER, 1968)

Stereocaulon alpinum, and even the Mediterranean lichen *Cladonia endiviaefolia* still show CO_2 uptake at -22 to $24°$ C (LANGE, 1963b, 1965a). The photosynthetic incorporation of CO_2 in the frozen state was proved using ^{14}C (LANGE and METZNER, 1965). Furthermore, even longer lasting frost causes only little or no aftereffect on the photosynthetic activity of mosses (ATANASIU, 1971a; KALLIO and KÄRENLAMPI, 1973) and of lichens (LANGE, 1963a, 1966, 1969a; KALLIO and HEINONEN, 1971; review by KALLIO and KÄRENLAMPI, 1973).
In higher plants frost brings about a rapid cessation of CO_2 uptake. As soon as ice begins to form in assimilatory tissue the net photosynthesis sinks to zero (PISEK et al., 1967; PISEK and KEMNITZER, 1968; review by BAUER et al., 1973, cf. also p. 114). This immediate obstruction of CO_2 uptake probably results from impeded

gaseous diffusion due to freezing. In many plants CO_2 uptake remains inhibited even after thawing, and the longer and stronger the frost has been the more deeply and longer does the inhibition continue. Repeated freezing has the same effect as more severe cold. Depression of photosynthesis is still reversible even after the most severe frost as long as the assimilatory organs have suffered no direct frost damage (ZELLER, 1951; TRANQUILLINI, 1957; WEISE, 1961; WEISE and POLSTER, 1962; LARCHER, 1961; PISEK and KEMNITZER, 1968; BAUER, HUTER, and LARCHER, 1969; PHARIS et al., 1970; MÜNTZ, 1971). The degree and duration of the after-effect on CO_2 uptake depends largely upon the species as well as on the degree of hardening and the state of activity of the plant (Fig. 9). Strong light immediately following or during cold stress can, just as with heat stress, enhance the degree and duration of the disturbance of photosynthesis (MÜNTZ, 1971; TAYLOR and ROWLEY, 1971).

B. Temperature Resistance and Survival

W. LARCHER

Temperature resistance is the ability of an organism to survive the direct action of extreme temperatures without suffering permanent damage.

The temperature resistance of a plant results from the ability of the protoplasm to tolerate cold or heat ("tolerance"; LEVITT, 1958, 1972) and the effectiveness of the mechanisms aimed at delaying or preventing damage ("avoidance").
Thus

$$Resistance = Avoidance + Tolerance$$

The components involved in temperature resistance are shown in Fig. 10. Table 1 gives an example of the complex phenomenon of the cold resistance of winter-hardened leaves of various woody evergreen plants.

Table 1. Frost resistance (temperature T_i at the first occurrence of injuries), frost avoidance (as essentially limited by the tissue freezing temperature T_f), and frost tolerance (expressed as difference between T_i and T_f) of evergreen leaves in winter. Data from PISEK and SCHIESSL, 1947; TRANQUILLINI and HOLZER, 1958; PISEK, LARCHER, and UNTERHOLZNER, 1967; LARCHER, 1970, 1971 and unpublished data

Object	Resistance (T_i)	Avoidance (T_f)	Tolerance $(T_i - T_f)$	Ratio avoid. : tol.
Eucalyptus globulus	— 3° C	— 3° C	nil	1 : 0
Citrus limon	— 5	— 5	nil	1 : 0
Ceratonia siliqua	— 5	— 5	nil	1 : 0
Nerium oleander	— 7	— 7	nil	1 : 0
Olea europaea	—10	—10	nil	1 : 0
Arbutus unedo	— 9	— 6	3°	1 : 0.5
Pinus pinea	—11	— 7	4°	1 : 0.6
Quercus ilex	—13	— 8	5°	1 : 0.6
Cupressus sempervivum	—14	— 5	9°	1 : 1.8
Cedrus deodara	—15	— 6	9°	1 : 1.5
Taxus baccata	—20	— 6	14°	1 : 2.3
Abies alba	—30	— 7	23°	1 : 3.3
Picea abies	—38	— 7	31°	1 : 4.4
Pinus cembra	—42	— 7	35°	1 : 5

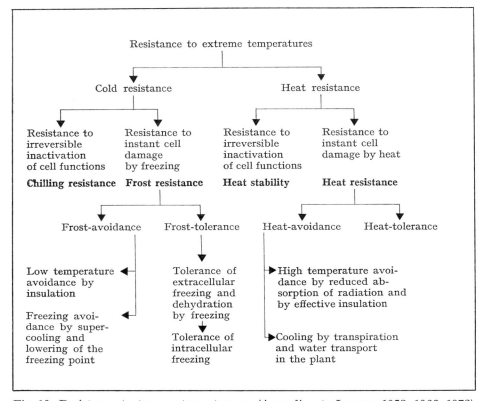

Fig. 10. Resistance to temperature stresses. (According to LEVITT, 1958, 1966, 1972)

1. Avoidance

The various ways in which protoplasm can be protected from temperature damage have been discussed extensively by LEVITT (1958, 1972). In the following discussion two especially important avoidance mechanisms will be considered in some detail:

a) Frost Avoidance by Delay or Prevention of Ice Formation in Tissues

Unlike insects, which are able to survive the winter and its continuous low temperatures in a supercooled state (ref. ASAHINA, 1966, 1969; see also p. 411ff.), plants are protected to only a limited extent by mechanisms delaying or preventing ice formation. Depending upon histological structure, water content, cell-sap concentration, degree of maturity and state of hardening, the leaves, buds and storage organs can be supercooled to $-5°$ C or $-7°$ C, and in exceptional cases to $-12°$ C, before ice begins to form spontaneously in the tissues (MOLISCH, 1897; ASAHINA, 1956; LEVITT, 1956, 1966; LARCHER, 1963a; KAKU, 1964, 1966; KITAURA, 1967; PISEK, LARCHER, and UNTERHOLZNER, 1967; MCLEESTER et al., 1969; YELENOSKY and HORANIC, 1969 inter alia). The supercooled state is extremely labile and can seldom be maintained for more than a few hours. It is therefore probably of little use to plants in the open (CHANDLER, 1954).

The effective extent of frost avoidance is limited by the freezing temperature of the tissues. The tissue freezing point is the highest temperature at which ice formation is observable in the tissues. Freezing is recognized by the release of latent heat (MAXIMOV, 1914; JACCARD and FREY-WYSSLING, 1934; ULLRICH and MÄDE, 1940; LEVITT, 1957; TRANQUILLINI and HOLZER, 1958; HATAKEYAMA, 1961; HUDSON and IDLE, 1962; LARCHER, 1963a; PISEK, LARCHER, and UNTERHOLZNER, 1967; SALT and KAKU, 1967; see Fig. 11).

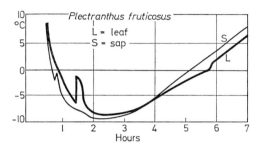

Fig. 11. Freezing curves of leaves (L) and cell sap (S) of *Plectranthus fruticosus*. (From ULLRICH and MÄDE, 1940)

Young leaves, fruits, and succulent storage organs freeze at − 1° to − 2° C (MO-LISCH, 1897; SHIFTAN, 1945; LEVITT, 1956, 1966; HATAKEYAMA, 1957, 1961) whereas differentiated leaves usually remain frozen down to temperatures of − 3° to − 5° C (list of species in LEVITT, 1956; PISEK, LARCHER, and UNTERHOLZNER, 1967). Drying lowers the freezing point by about 1 to 2° C (MAXIMOV, 1914; LAR-CHER, 1954; PISEK, LARCHER, and UNTERHOLZNER, 1967; SALT and KAKU, 1967). As winter approaches the freezing temperature is lowered by a further 2−4° C as a result of the accumulation of osmotically active cell-sap constituents (KAWANO et al., 1960; HATAKEYAMA and KATO, 1965; LARCHER, 1963a; PISEK, LARCHER, and UNTERHOLZNER, 1967). The natural limit of freezing protection of the cell is then about −5° to −7° C, in exceptional cases −10° to −12° C (TYURINA, 1953; LARCHER, 1963a, 1970; KAKU, 1971; MITTELSTÄDT, 1971b).

Depression of the freezing point provides a moderate but reliable degree of protection against frost, sufficient for plants in regions with a mild winter. Frost avoidance thus constitutes the mechanism of cold resistance of the evergreen leaves of many Mediterranean woody plants in winter (LARCHER, 1963a, 1970, cf. Table 1: species list from *Ceratonia* to *Olea*). It is very likely that the protection of leaves and probably also of the shoots of subtropical and tropical woody plants from frost injury is exclusively based upon avoidance mechanisms (LARCHER, 1971). Cold resistance in summer is entirely attributable to frost avoidance.

b) Heat Avoidance by Means of Reduction of Radiation Absorption and by Transpirational Cooling

Under natural conditions dangerously high temperatures occur in plant organs only if the irradiation is particularly strong. A reduction of the radiation absorption by profile positioning of the leaves (SEYBOLD, 1929b) or by the development of leafless cylindrical assimilatory shoots tends to prevent the occurrence of extremely high temperatures. The leafless twigs of *Leptadenia pyrotechnica* (cf. Fig. 13) and *Capparis decidua* in Mauretanian habitats exhibited, according to LANGE (1959), temperatures only about 4° C above air temperature, whereas the leaf temperature of foliage plants exposed to the same conditions was 8° to 12° C above air temperature. The connections between the radiation balance and thermal conditions of plant leaves and shoots are dealt with in detail by SEYBOLD, 1929a,b,c; SEYBOLD and BRAMBRING, 1933; HUBER, 1935, 1956; RASCHKE, 1956, 1958, 1960; GATES, 1962, 1963 and 1968 cf. also p. 87.

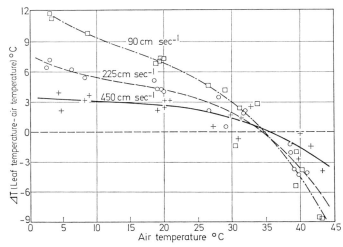

Fig. 12. Dependence of the temperature difference between a leaf of *Xanthium pennsylvanicum* and air upon air temperatures at wind speeds of 90 (□), 225 (○) and 450 (+) cm sec⁻¹. Irradiance:1 ly · min⁻¹. (From DRAKE, RASCHKE, and SALISBURY, 1970)

Another important protective mechanism against overheating in habitats with intense irradiation is transpirational cooling. The cooling effect of transpiration, which can be calculated (BROWN and ESCOMBE, 1905; SEYBOLD, 1929a; HUBER, 1935; RASCHKE, 1958; GATES, 1968; see also p. 96) and measured both in the open and under laboratory conditions (survey in CURTIS and CLARK, 1950; HUBER, 1956; LANGE, 1959; BIEBL, 1962), ensures that the overheating of the leaves remains within the limits of tolerance or even that the effects of irradiation are counterbalanced. Transpirational cooling is more effective at higher temperatures because the diffusion resistance within the leaf decreases with rising temperature (Fig. 12; DRAKE, RASCHKE, and SALISBURY, 1970). Wherever evaporation is extremely high, as it is in the desert, steppe or savanna, the leaf temperature of plants that can transpire strongly under such conditions drops about 4−6° C, and in some cases even 10−15° C below that of the air („Untertemperaturarten" like *Citrullus colocynthis, Solanum melongena, Convolvulus althaeoides;* LANGE, 1959, 1963). The extent of the cooling effect depends to a large degree upon the water supply of the plant (KARSCHON and PINCHAS, 1971) as can readily be demonstrated by obstructing water transport to the leaf (see Fig. 13) or by coating the leaves with antitranspirants (vaseline: LANGE, 1959; antitranspirants: THAMES, 1961; WILLIAMSON, 1963; GALE and POLJAKOFF-MAYBER, 1965).

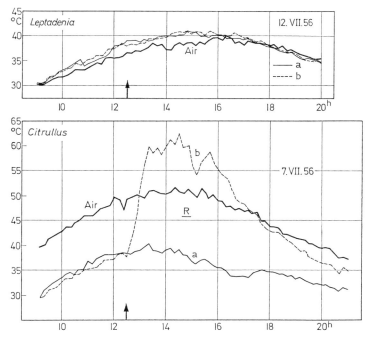

Fig. 13. Surface temperature of two shoots (a and b) of *Leptadenia pyrotechnica* and of two leaves of *Citrullus colocynthis* under strong radiation conditions in the natural desert habitat of the plants. Air: air temperature, *R* heat resistance of the leaves; the arrows indicate the moment when shoot and leaf b were cut in order to interrupt the water supply for transpiration. (From LANGE, 1959)

2. Tolerance

a) Variability of Heat and Frost Tolerance

Tolerance is not a quality possessed permanently and at a constant level by a plant species but is a temporary state. The level of tolerance attained by a plant at any given time results from its specific hardening capacity and the environmental conditions. *Hardening capacity* may vary from day to day, from season to season as well as over longer time intervals according to the state of activity, age and phase of development of the plant, its state of nutrition and other conditions. The magnitude of the hardening capacity at any particular time determines the effectiveness of the adaptive processes within the protoplasm. The resultant adaptation of resistance[1] at a given hardening capacity is governed by the *intensity of environmental stimuli*.

---

[1] Different authors distinguish between acclimatization, acclimation, adaption, adjustment etc. Details of terminology and definition will not be discussed (see p. 319, 419 ff.).

α) *Influence of Stage of Development and Age*

Stages in the life of the plant that are characterized by higher growth and developmental activity are, as a rule, particularly susceptible to extreme temperatures. This is apparent in higher plants during germination, seedling growth and spring sprouting, as well as in budding algae (SCHÖLM, 1968) and in actively multiplying cultures of microorganisms (see also FARRELL and ROSE, 1967, and CHRISTOPHERSEN, p. 23f).

At the first sign of germination most seeds lose their originally high tolerance proportionally to the elongation of the primary roots (RUDORF, 1938; KEMMER and THIELE, 1955; LARCHER, 1969b; cf. scheme in LEVITT, 1956). Seedlings are at their most sensitive following stem elongation when the leaves are beginning to develop. The temperature-hardening capacity increases gradually with increasing tissue maturity and young plants are capable of efficient hardening under thermal stress (THIELE, 1957; SAMYGIN et al., 1960; SALZER, 1969). A resistance gradient which is dependent on the age of the leaves has been observed (ANDREWS, 1960; ROBERTS and GRANT, 1968; COX and LEVITT, 1969).

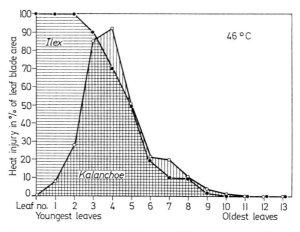

Fig. 14. Heat resistance of the leaves of *Ilex aquifolium* (●) and *Kalanchoe blossfeldiana* (○) depending upon development and age. Heat exposure to 46° C for 30 min. (Adapted from SCHWEMMLE and LANGE, 1959; and LANGE, 1961)

Similar gradients can be recognized in the sprouting shoot tips of perennial plants, where the most actively developing leaves are the most sensitive (Fig. 14 and Tab. 2). The same holds true for buds: MAIR (1968) observed a distinct relationship between the phenological stage and hardening capacity in a series of buds on ash shoots. The further a bud is developed the less the degree of frost hardening it achieves (Fig. 15). The various stem tissues develop a larger degree of tolerance only when mature. Twig tips are usually about 2–3° C, sometimes as much as 10° C, more sensitive to cold than mature shoot segments (examples and literature in LARCHER, 1970). Heat tolerance is also lower in shoot tips than in woody stems (KREEB, 1970; BAUER, 1970).

Table 2. Frost injury in leaves depending on the degree of their differentiation and age (TILL, 1956)

Age of leaves	Frost injury °C		
	Hedera helix	*Ilex aquifolium*	*Corylus avellana*
Expanding	—1.5	—1.5	—2.5
Young	—3.5	—1.5 to —3.5	
Mature current year	—4.5	—3.5	—4.5
Mature second year	—5.0	—4.5	

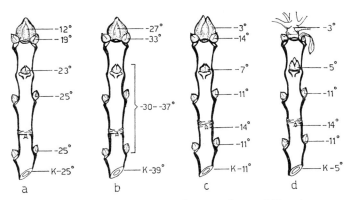

Fig. 15. Different frost resistance of the buds on a shoot of *Fraxinus ornus*, a at the end of November, b during winter, c at the beginning of bud swelling, and d at the time of shooting. K = shoot cambium. (From MAIR, 1968)

Young plants of some species are unable to develop the same degree of temperature tolerance as adult individuals. Examples are certain grasses (*Bromus:* LAUDE and CHAUGULE, 1953) and numerous woody plants (*Pinus:* SHIRLEY, 1936; *Larix:* HAMAYA et al., 1968; *Quercus:* LARCHER, 1969b; *Eucalyptus:* ASHTON, 1958; KREEB, 1970; see Fig. 16). In other species, however, full hardening capacity is acquired at the seedling stage (*Abies alba* and *Acer pseudoplatanus:* HARRASSER, 1969; BAUER, 1970).

Furthermore, changes in hardening capacity may also be involved in the transition from the juvenile to the reproductive stage. This was demonstrated by LANGE and SCHWEMMLE in 1960 on *Kalanchoe blossfeldiana:* its heat tolerance rose by almost 2° C at the onset of flowering.

β) *Seasonal Course of Resistance and Short-Term Adaptive Phenomena (Adjustments)*

Cold Resistance. In regions with cold seasons, terrestrial vascular plants acquire their high winter frost tolerance *in phases* (TUMANOV, 1960, 1967; SIMINOVITCH et al., 1967, 1968, 1969; WEISER, 1970; STEPONKUS, 1971; see Tab. 3). These phases can only be induced after IAA- and gibberellin-promoted processes are completed.

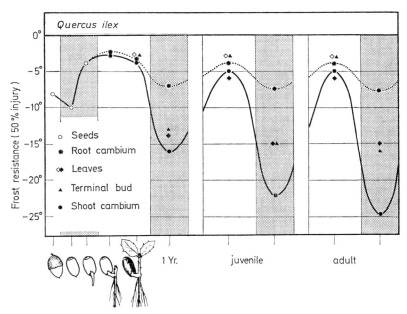

Fig. 16. Frost resistance of *Quercus ilex* during ontogenetic development and aging. Shaded areas indicate the frost-hardened state in winter. (From LARCHER, 1969 b)

The addition of metabolic and growth inhibitors (e.g. maleic hydrazide, 2-chloro-ethyltrimethylammonium chloride, N.N.-dimethylaminosuccinamic acid; survey in LEVITT, 1966) also preconditions the tissues for hardening. Conversely, readiness to deharden is present as soon as growth commences at any point in the plant. The seasonal course of cold tolerance thus mainly follows the plant's rhythm of development and activity (PISEK and SCHIESSL, 1947; COOPER et al., 1955a; OLDÉN, 1957; MURAWSKI, 1962; HAMAYA et al., 1968; LARCHER and MAIR, 1968; VAN DEN DRIESSCHE, 1969 inter alia; Fig. 17). The phases of development are, in turn, controlled both by endogenous stimuli (BÜNNING, 1943, 1956a, b; MESSERI, 1951; LEIKE, 1965, ref. LYR, POLSTER and FIEDLER, 1967) and by environmental factors (via temperature and photoperiod; ref. KRAMER and KOZLOWSKI, 1960; VEGIS, 1961 and Chapter B 1). Exposure to temperatures between +5° C and 0° C for several days or weeks preconditions the protoplasm of cereals and herbaceous plants for hardening. In some species short day is obligatory for hardening, whereas in others it can be replaced by low temperatures. Details concerning the mechanism of the photoperiodic induction of tolerance are to be found in TUMANOV, KUZINA, and KARNIKOVA (1965); STEPONKUS and LANPHEAR (1967); IRVING and LANPHEAR (1968); VAN HUYSTEE, WEISER, and LI (1967); HOWELL and WEISER (1970a); survey in WEISER (1970).

Readiness to acquire hardiness is essential to achieve tolerance, but does not determine the degree of frost resistance. A high degree of tolerance is the result of the *process of hardening*, which proceeds and is completed under the influence of increasingly severe frost. Plant parts that spend the winter protected by a snow cover are, on the average, 10° C less tolerant than those completely exposed to cold (TRANQUILLINI, 1958;

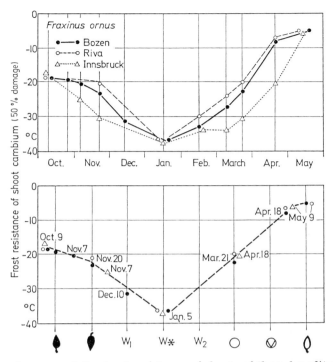

Fig. 17. Annual course of the frost resistance of shoots of the submediterranean tree *Fraxinus ornus* from various habitats as related to the date of sampling and to the phenological states. *Dark leaf symbol:* Coloring of leaves. *Inverse leaf symbol:* Abscission. *W 1* Winter before occurrence of continuous frost. *W** Winter during frost periods. ○ Bud swelling, ⊙ Bud opening. *Open leaf symbol:* shooting. Habitats: Bozen is situated at the northern distribution limit of submediterranean woody plants, Riva represents a warm habitat of the submediterranean region, and Innsbruck is situated beyond the natural distribution area of *Fraxinus ornus*. (From LARCHER and MAIR, 1968)

SCHNETTER, 1965). Woody plants kept in a greenhouse at constant positive temperatures may, under certain circumstances, enter a state of dormancy, but their degree of cold tolerance increases only minimally (LARCHER, 1954; SCHWARZ, 1970).

Hardening progresses stepwise, each phase preparing the way for the next. It is preceded by weeks of temperatures slightly below zero, and this phase of "prehardening" is chiefly associated with changes in the protein structure and the accumulation of sugars in the protoplasm. Fine structure changes in the protoplasm occur, rendering it resistant to extracellular ice formation. The plants can now safely enter the end stage of the hardening process (deep hardening), which takes place under continuous and increasing frost of $-10°$ C to $-15°$ C and below. In the course of this the protoplasm becomes extremely frost-tolerant as a result of progressive frost drying. Birch seedlings which had been frozen at $-15°$ C and $-20°$ C before commencement of the hardening process tolerated $-35°$ C at the end of the first phase of hardening, and survived temperatures of $-195°$ C at its completion (TUMANOV, 1967).

Table 3. Frost hardening in terrestrial plants as a phase process

---

1. Predisposition for hardening

Precondition:
Enzymes and membrane structures have to become resistant to chilling [31, 39, 41, 84].

Induced by:
Aging and/or influence of external factors such as temperature and day length [4, 13, 15, 71, 74, 83];
Application of growth regulators and inhibitors [3, 20, 42, 49, 82].

Response processes:
Changes in the levels of growth- promoting and -inhibiting phytohormones, quantitative and qualitative alterations in the metabolism [2, 12, 16, 59, 70, 72, 79].

Result:
Inhibition or cessation of growth.Beginning of a transitory rest period or of dormancy.

2. Prehardening

Precondition:
Predisposition for hardening has to be achieved.

Induced by:
Temperatures between —5 to +5° C lasting several days to weeks [5, 7, 23, 45, 46, 55, 58, 62, 64, 67, 71, 73, 74].

Response processes:
Changes in the ultrastructure of protoplasm and in metabolism [10—12, 32, 38, 39, 47, 50, 69, 76, 84, 86];
Increase in RNA, transformation of enzymes and membrane-proteins [19, 36, 37, 38, 44, 68, 69];
Accumulation of soluble carbohydrates and polyalcohols in the protoplasm and the vacuole [1, 6, 9, 17, 18, 50, 52, 61, 65, 66, 77, 78];
Increased permeability [14, 25, 28, 30, 35, 47, 48, 56, 81].

Result:
Lowering of the freezing point and improved supercooling capacity of the tissues [8, 21, 33, 34, 48, 54, 85];
Transition to the frost tolerant state.

3. Deep-hardening

Precondition:
Frost tolerance (i.e. ability to survive extracellular ice formation) has to be developed.

Induced by:
Continuous subfreezing temperatures [27, 57, 63, 73, 74, 75].

Response processes:
Stabilization of structured water in the protoplasm;
Translocation of movable water out of the protoplasm by extracellular freezing causing severe dehydration; increase in desiccation resistance [11, 22, 24, 30, 40, 43, 46—48, 53, 68, 69, 80].

Result:
Deep-hardiness (i.e. ability to withstand severe dehydration by ice formation in the tissues). Deep-hardened plants are able to survive at the lowest temperatures occurring on the earth [26, 29, 51, 60, 73, 74].

1. ÅKERMAN (1927)
2. BOLDUC et al. (1970)
3. COOPER et al. (1955 b)
4. COOPER and PEYNADO (1958)
5. COOPER and PEYNADO (1959)
6. GASSNER and GOEZE (1932)
7. HARVEY (1930)
8. HATAKEYAMA and KATO (1965)
9. HEBER (1959)
10. HEBER and SANTARIUS (1967)
11. HEBER (1968)
12. HENKE (1962)
13. HOWELL and WEISER (1970 a)
14. ILJIN (1934)
15. IRVING (1969)
16. IRVING and LANPHEAR (1968)
17. JEREMIAS (1956)
18. JEREMIAS (1964)
19. JUNG et al. (1967)
20. KACZPERSKA-PALACZ et al. (1969)
21. KAKU (1966)
22. KAPPEN (1966)
23. KAPPEN (1967)
24. KAROW and WEBB (1965)
25. KESSLER and RUHLAND (1938)
26. KRASAVTSEV (1960)
27. KRASAVTSEV (1967)
28. KRASAVTSEV (1968)
29. KRASAVTSEV (1969)
30. KRASAVTSEV (1970)
31. KURAISHI et al. (1968)
32. KWIATKOWSKA (1970 a,b)
33. LARCHER (1963 a)
34. LARCHER (1970)
35. LEVITT and SCARTH (1936)
36. LEVITT (1962)
37. LEVITT (1966)
38. LEVITT (1969)
39. LEVITT and DEAR (1970)
40. LING (1967)
41. LYONS and RAISON (1970 a, b)
42. MARTH (1965)
43. MAZUR (1966)
44. MC COWN et al. (1969)
45. MEYER (1928)
46. MITTELSTÄDT (1962)
47. MITTELSTÄDT (1968)
48. MITTELSTÄDT (1969, 1971 a)
49. MODLIBOWSKA (1965, 1968)
50. PARKER (1959)
51. PARKER (1960 a, b)
52. PISEK (1950)
53. PISEK and LARCHER (1954)
54. PISEK et al. (1967)
55. POGOSYAN (1967)
56. SAKAI (1955)
57. SAKAI (1956)
58. SAKAI (1958 a)
59. SAKAI (1959)
60. SAKAI (1960)
61. SAKAI (1962)
62. SAKAI (1964)
63. SAKAI (1965 a, b)
64. SAKAI (1966)
65. SAKAI and YOSHIDA (1968)
66. SAUTER (1967)
67. SCHEUMANN and HOFFMANN (1967)
68. SIMINOVITCH et al. (1967)
69. SIMINOVITCH et al. (1968)
70. STEPONKUS and LANPHEAR (1967)
71. TUMANOV (1955)
72. TUMANOV and TRUNOVA (1958)
73. TUMANOV (1960)
74. TUMANOV (1967 a)
75. TUMANOV and KRASAVTSEV (1959)
76. ULLRICH (1943)
77. ULLRICH and HEBER (1958)
78. ULMER (1937)
79. VAN HUYSTEE et al. (1967)
80. WEBB (1965)
81. WILLIAMS and MERYMAN (1970)
82. WÜNSCHE (1966)
83. YOUNG (1961)
84. Cf. Chapter A 2
85. Cf. Chapter B 1a
86. Cf. Chapter C

The references are only a choice of examples without claiming to be complete (literature considered till 1970). See also the publications of LEVITT (1956, 1958, 1972), BIEBL (1962), PARKER (1963), and MERYMAN (1966).

Under natural conditions frost hardiness adapts promptly to the external temperature conditions, and variations in tolerance of 5−10° C may occur (for examples see OLDÉN, 1957; PISEK, 1958; KOHN, 1959; PROEBSTING, 1959; SCHEUMANN, 1968). The degree to which tolerance can be influenced by temperature varies with the season and the species (PISEK and SCHIESSL, 1947; MITTELSTÄDT, 1965, 1966; SCHEUMANN and HOFFMANN, 1967; SCHEUMANN and SCHÖNBACH, 1968; HOWELL and WEISER, 1970 b). Cold is especially effective in promoting hardening at the beginning of the winter, and warm weather promotes dehardening especially following winter dormancy (Figs. 18 and 19).

The speed of the response and the degree of adaptation in tolerance that can be rapidly elicited by changes in external temperature have been determined ex-

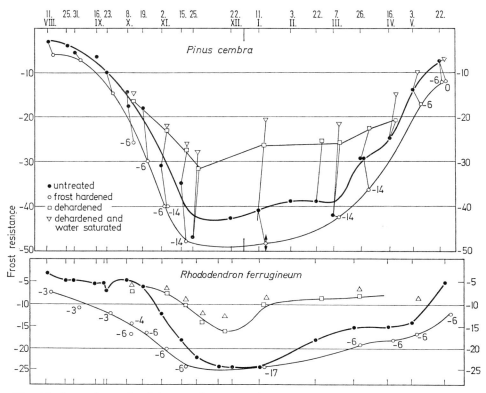

Fig. 18. Annual trends of frost resistance and of the ability to become hardened and dehardened (at +15°/17° C) in leaves of woody plants from the tree line in the Alps. (From PISEK and SCHIESSL, 1947)

perimentally on many occasions (WINKLER, 1913; PISEK and SCHIESSL, 1947; KRASAVTSEV, 1960; SCHEUMANN, 1965, 1968, inter alia). As a rule, an effective degree of cold hardening is achieved within one day, the final value gradually being reached after 4—7 days (Fig. 20). In various deciduous trees, SAKAI (1964) found that the primary cortex hardened most rapidly, whereas the xylem rays and the pith of the shoots were the slowest. Warmth, provided in nature by thaw temperatures, evokes a loss of tolerance equally or even more rapidly. At a dehardening temperature of about 10° C frost hardiness is lost within 2—3 days, and at temperatures above 20° C the loss may occur within a few hours (MICHAELIS, 1934a, b; PISEK and SCHIESSL, 1947; SCHEUMANN and HOFFMANN, 1967; SAKAI, 1968; SALZER, 1969; cf. Fig. 21). Not all terrestrial plants can be hardened and it appears that not even all plants that can be hardened are capable of working through all phases of the hardening process. This is probably the main reason why the maximum tolerance values vary so greatly from species to species (LARCHER, 1970, 1971; LEVITT and DEAR, 1970).

The cold tolerance of aquatic plants also follows an annual pattern if they are exposed to large temperature variations. This applies particularly to algae in

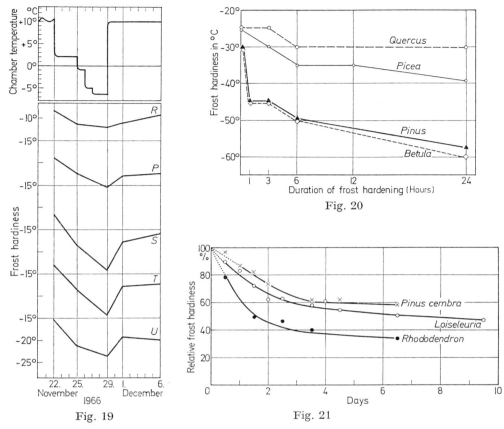

Fig. 19

Fig. 20

Fig. 21

Fig. 19. Frost hardening and dehardening characteristics of *Picea abies* of different regions in early winter. Samples from Roumania (R), Northern Poland (P), Sweden (S), the Tatra Mountains (T), and the Ural region (U). Preconditioning temperatures as indicated by "chamber temperature" (Adapted from SCHEUMANN and HOFFMANN, 1967)

Fig. 20. Rate of deep-hardening of oak, spruce, pine, and birch plants during continuous frost treatment at —10° C in November. (From KRASAVTSEV, 1960)

Fig. 21. Decrease in frost resistance of evergreen leaves of woody plants from the tree line in the Alps during dehardening at +15° to +17° C in winter. (From PISEK and SCHIESSL, 1947)

shallow inland water (TERUMOTO, 1959; SCHÖLM, 1968) and to marine algae of the tidal zone (PARKER, 1960a; BIEBL, 1962, 1970; FELDMAN et al., 1963; TERUMOTO, 1965b; LYUTOVA and DROBYSHEV, 1968). Little is known concerning the mode of hardening of algae, but it is probably a direct response to cold and not a phase process (ALEXANDROV et al., 1970).

Heat Resistance. Seasonal changes in heat resistance can be observed in many plant species. The fluctuations involved, however, seldom exceed 5° C and are thus considerably smaller than those seen in the annual pattern of cold resistance. Further-more, in contrast to the latter, heat resistance does not appear to follow the climatic rhythm. Many types of annual patterns of heat resistance are known.

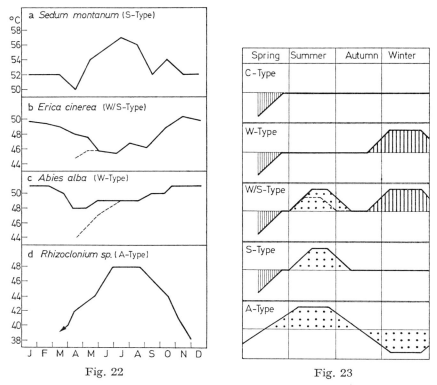

Fig. 22   Fig. 23

Fig. 22. Seasonal course of heat resistance of various plants. Heat exposure 30 min. Dashed lines: young leaves. From BIEBL and MAIER, 1969 (a); BANNISTER, 1970 (b); BAUER, 1970 (c); SCHÖLM, 1968 (d)

Fig. 23. Various types of the seasonal course of heat resistance among plants. ▥ Decrease in heat resistance during growth. ⬚ Specific adjustment of heat resist-ance to the environmental temperature. ▨ Unspecific increase in heat resistance connected with general enhancement of protoplasmic resistance and onset of dor-mancy. (For details see text)

These vary according to species and habitat: a selection is shown in Fig. 22. Figure 23 is an attempt at a schematic representation of the various types of heat resistance.

In all the terrestrial plants investigated heat resistance and growth activity have been found to depend strictly upon one another: during the main period of growth the heat resistance of all organs sinks to its annual minimum.

In the leaves of various species such as *Asplenium ruta-muraria* (KAPPEN, 1964), *Ilex aquifolium* (LANGE, 1961) and *Saxifraga aizoon* (BIEBL and MAIER, 1969) the only detectable alteration in resistance level over the entire year, is a drop in the spring. This type of constant behavior is denoted *"Type* C" in Fig. 23.

In species in which a deep winter dormancy occurs the leaves usually show a winter rise in heat resistance (*Type* "W"). This is linked with frost hardening and is an ecological paradox, controlled via the pattern of activity. In such plants a rise in heat resistance can also be elicited in summer after a period of short-day treatment (BIEBL, 1968). Examples of Type W are provided by various grasses such as *Dactylis glomerata, Elymus sp.* (ALEXANDROV et al., 1959) and *Bouteloua sp.* (JAMESON, 1961), dwarf shrubs such as *Calluna vulgaris* (ALEXANDROV et al., 1964; BANNISTER, 1970), *Vaccinium myrtillus* (BANNISTER, 1970) and *Rhododendron ferrugineum* (SCHWARZ, 1970), *Viscum album* (PISEK et al., 1968), *Picea abies* (PISEK et al., 1968), *Pinus cembra* (SCHWARZ, 1970) and *Abies alba* (PISEK et al., 1968; BAUER et al., 1972).

In contrast, the heat resistance of ferns such as *Phyllitis scolopendrium* and *Polypodium serratrum* (KAPPEN, 1964) and of flowering plants such as *Sedum montanum* (BIEBL and MAIER, 1969) and *Sedum spurium* (ALEXANDROV et al., 1964) exhibit an appreciable rise in heat resistance only in summer (*Type* S), despite the fact that they also develop resistance to cold in winter. This summer hardening is ecologically sound and can be regarded as a specific adaptation in resistance (cf. LANGE, 1967; LEVITT, 1969; ALEXANDROV et al., 1970).

The observation that the degree of heat hardening in summer depends upon the type of weather (ILLERT, 1924; LANGE, 1961) and on the local climate (LANGE, 1959; LANGE and LANGE, 1962) points to the conclusion that it is an adaptation to environmental temperature. Preconditioning to high temperature could be demonstrated experimentally by SAPPER (1935) in *Elodea callitrichoides,* but not in *Vallisneria spiralis,* by VAARTAJA (1954) in *Pinus* seedlings and by LANGE (1962) in *Commelina africana, Phoenix dactylifera* and *Veronica persica. Commelina* leaves that had developed in the course of 5—6 weeks of continuous cultivation at 28° C suffered 50% damage at 51.5° C whereas the leaves of plants cultivated at 20° C were already 50% damaged at 46.5° C.

In addition, there are intermediary types (*Type* W/S) in which the annual course of heat resistance exhibits two peaks, one coinciding with the winter rest period and one which represents a specific summer adaptation. This group includes certain mosses (LANGE, 1967), ferns (KAPPEN, 1964), dicotyledonous herbs (ALEXANDROV et al., 1964; BIEBL and MAIER, 1969), dwarf shrubs (*Erica tetralix;* LANGE, 1961 and BANNISTER, 1970), *Taxus baccata* (LANGE, 1961), *Pinus edulis* and various species of *Juniperus* from Arizona (JAMESON, 1961). The summer and winter peaks are often unequal and the summer peak, like that of Type S, is dependent upon the degree of climatic stress.

Algae constitute a type on their own. Both fresh-water and marine algae adjust their heat resistance to follow the seasonal changes in water temperature (*Type* A), so that it is at its highest in late summer and at its lowest in winter. Thus heat resistance and cold resistance behave in a diametrically opposite manner. The larger the difference between the temperature of the water in the habitat in summer and in winter, the greater is the annual amplitude of the heat resistance curve. This again illustrates the excellent thermal adaptability of algae (BIEBL, 1962, 1970; TERUMOTO, 1964; SCHÖLM, 1968).

Heat can initiate extraordinarily rapid adaptive processes in protoplasm. In the natural habitat high midday temperatures bring about a short-lived increase in heat resistance which, as a result, is measurably higher in the afternoon than in the morning. Using cessation of protoplasmic streaming as a criterion of viability ALEXANDROV and YASKULYEV (1961) found, in the hot summer season, a rise in resistance of about 1° C from morning to afternoon in the desert grass *Aristida karelini* and in *Catalpa speciosa* in Turkmenia. The resistance fell again overnight by the same amount. Such a hardening effect was observed only if the midday temperature rose above 38° C, and was therefore absent in spring and autumn and during periods of cold weather. Diurnal variations in heat resistance have repeatedly been observed in algae (LYUTOVA et al., 1967; SCHÖLM, 1968). Even when exposed to constant temperature conditions, the north Pacific intertidal alga *Chaetomorpha cannabina* still continues to follow the tidal rhythm for a few days, the higher heat resistance coinciding with the onset of the ebb tide (BIEBL, 1969b, Fig. 24). The remarkably rapid ability of plants to adapt to heat was demonstrated by KREEB (1970) in a heat-accumulation experiment on *Eucalyptus blakelyi* seedlings. Plants heated stepwise up to the test temperature suffered only slight damage at 48° C and 50% damage at 52° C, whereas those heated up immediately to the test temperature showed the first signs of necrosis at 42° C and 50% damage at 49° C.

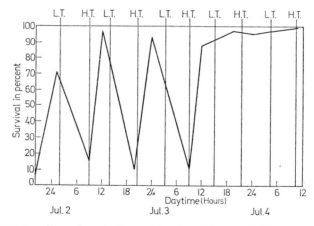

Fig. 24. Endogenous diurnal periodism of heat resistance of *Chaetomorpha cannabina* depending on tidal rhythm. Heat exposure 5 min at 38° C. *L.T.* low tide. *H.T.* high tide. (From BIEBL, 1969b)

The process of heat adaptation sets in sooner and is the more rapidly successful the higher the temperature at which hardening is brought about (for examples and comprehensive discussion see ALEXANDROV, 1964; ALEXANDROV et al., 1970; YARWOOD, 1967). Damage due to heat also follows the dosage law, i.e. less heat over a longer period of time is as destructive as greater heat for a short time (BĚLĚHRADEK, 1935, 1957). Hardening resulting from the action of very high but sub-injurious temperatures occurs after only a few minutes. Cessation of protoplasmic streaming is used as the criterion. The temperature limit for the necrotic effects of heat begins to rise about one hour after hardening treatment, depending upon species; the full hardening effect is achieved within hours or days (YARWOOD, 1961, 1967; SCHROEDER, 1963; SALZER, 1969). When the plant is returned to normal temperatures (+27° C is taken as the threshold value for germinating wheat grains: SALZER, 1969) the resistance begins to drop at once. The lower the dehardening temperature the more rapidly does this occur.

γ) The Influence of Various Environmental Stimuli on Temperature Resistance

Temperature resistance not only responds to thermal stimuli but can also be induced by other environmental factors. The most effective of these are the state of hydration, light and chemical influences.

The *state of hydration* of the plant (and also the ionic regulation in the cell, cf. BOGEN, 1948) influences temperature resistance via the degree of imbibition of the protoplasm. Water saturation elicits a drop in heat- and cold-resistance whereas a decrease in turgidity favors the hardening of the protoplasm to extreme temperatures. Turgid leaves of *Olea europea* freeze 3° C sooner than severely wilted samples (LARCHER, 1954), and various herbaceous plants are about 2° C more heat resistant following dry cultivation, or after wilting, than water-saturated specimens (SAPPER, 1935). Further examples of the influence of the state of hydration on temperature resistance can be found in ILJIN (1934), ULMER (1937), PISEK and SCHIESSL (1947), HAMMOUDA and LANGE (1962), KAPPEN (1966), KAPPEN and LANGE (1968), SALZER (1969) and KREEB (1970); see also Fig. 25.

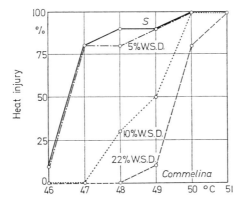

Fig. 25. Effect of drying on the heat resistance of leaves of *Commelina africana*. *S* Water-saturated leaves, *WSD* Water-saturation deficit in percent of water content at saturation point. Heat exposure 30 min. (From HAMMOUDA and LANGE, 1962)

A special case is provided by dry anabiotic stages and by poikilohydric plants that are well able to withstand desiccation. Thallophytes in a state of drought anabiosis are appreciably more resistant to cold and to heat than well-hydrated individuals (BIEBL, 1939; LANGE, 1953, 1955; MIGITA, 1966; Tab. 4). In poikilohydric cormophytes (some ferns, *Ramonda myconi* and *Myrothamnus flabellifolia* are examples of flowering plants) heat resistance and especially frost hardening rise steeply on drying (SAPPER, 1935; KAPPEN, 1966; VIEWEG and ZIEGLER, 1969). Dry seeds can be very resistant to extreme temperatures (DE CANDOLLE and PICTET, 1879; LIPMAN and LEWIS, 1934; VAARTAJA,

1954; Ben Zeev and Zamenhof, 1962; for further literature see Levitt, 1956 and Biebl, 1962). In swollen seeds resistance depends directly upon the water content (Bennett and Loomis, 1949; Rossmann, 1949; Agena, 1961; Schomer-Ilan, 1964). Most resistant of all, however, are anabiotic survival stages: dry spores of many species of bacteria and fungi survive immersion in liquid air as well as in boiling water (details in Christophersen, p. 39).

Table 4. Maximum temperature resistance of poikilohydric autotrophic plants in the hydrated and dry state

Plant	Low temperature injury at °C		Heat injury at °C		Author
	Hydrated	Dry	Hydrated	Dry	
Algae					
Bangia fuscopurpurea			+35	+ 42	Biebl (1939)
Various unicellular *Chlorophyceae*	—10 to —30	—196			Bequerel (1954)
Lichens					
Lichens of humid and shady habitats	—80 to —196	—196	about +35	+70 to +90	Lange (1953) Kappen and Lange (1970)
Soil lichens and epipetric lichens of xerothermic habitats	—80 to —196	—196	+43 to +46	about +100	Lange (1953) Bequerel (1954) Kappen and Lange(1970)
Mosses					
Forest mosses	—15 to —25	—196	+40 to +50	+ 80 to + 95	Irmscher (1912) Romose (1940)
Epipetric mosses	—30			+100 to +110	Bequerel (1954) Lange (1955)
Ferns (Leaves)					
Ceterach officinarum			+47	+100 to +120	Sapper (1935)
Polypodium vulgare	—18	—196	+47.5	+ 55	Kappen (1966)
Angiosperms (Leaves)					
Ramonda myconi	— 9	—196	+48	+ 56	Kappen (1966)
Myrothamnus flabellifolia		—196		+ 80	Vieweg and Ziegler (1969)

The *effect of light* on the level of resistance of autotrophic plants seems to be linked with photosynthesis. Light also exerts a controlling influence. Yarwood (1961, 1963) observed an increase in heat resistance of bean leaves under illumination and a decrease in the dark, whereas the reverse was found by Illert (1924) in *Oxalis acetosella*. In *Kalanchoe blossfeldiana*, a short-day plant with a diurnal rhythm in its organic acid metabolism, Schwemmle and Lange (1959) demonstrated an endogenous diurnal periodism of heat resistance with an amplitude of about 1° C controlled by photoperiod. Resistance is at a maximum in the middle of the dark phase and at a minimum in the middle of the light phase: in constant darkness the periodicity, although weakened, is still retained (Fig. 26). The periods of maximum heat sensitivity coincide with those of higher metabolic activity and high auxin content.

The rather varied effects on temperature resistance of mineral salts, osmotically active materials, metabolic and growth regulators, inhibitors and poisons are reviewed in detail by BIEBL (1962) and LEVITT (1966).

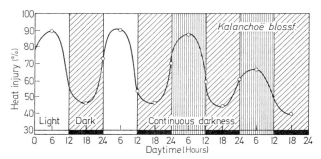

Fig. 26. Endogenous daily fluctuation of heat resistance in the leaves of *Kalanchoe blossfeldiana*. Heat exposure 30 min at 46° C. (After SCHWEMMLE and LANGE, 1959; from LANGE, 1967)

b) Specificity of Temperature Resistance

α) *Constitutional Differences*

Plants are found in all climatic regions capable of supporting life, and it is possible to compile exhaustive series ranging from the most temperature-sensitive species to the most resistant. If the various species are arranged according to their temperature resistances they fall into groups corresponding to their natural distribution (Tabs. 5 and 6). The eminent importance of temperature resistance as a selective factor immediately becomes apparent. The fact that plants form groups which are of ecological significance is due to the constitutionally specific limits governing the acquisition of resistance.

The following constitutional types may be distinguished with respect to **cold resistance:**

1. *Not Frost-Tolerant Plants.*

x) Chilling-Sensitive Plants.

These suffer irreversible damage at temperatures between +10° and 0° C. The group includes tropical algae, many vascular plants of the tropical rain forests, as well as tropical fruits and vegetables. The cold injury begins with functional disturbances and finally leads to cell necrosis (MOLISCH, 1897; SEIBLE, 1939; SPRANGER, 1941; JONES, 1942; WRIGHT et al., 1954; BIEBL, 1964, 1970; MURATA, 1969; TAYLOR and ROWLEY, 1971).

y) Freezing-Sensitive Plants.

Such plants tolerate low temperatures per se, but are damaged at once when ice begins to form in the tissues. Frost-avoidance mechanisms are the only protection

Table 5. Maximum temperature resistance in aquatic plants. Collected data and lists of species cited in BIEBL (1939, 1958, 1962, 1970), TERUMOTO (1964), ALTMAN and DITTMER (1966), SCHÖLM (1968)

Plant and distribution	Low temperature injury[a] in winter °C	Heat injury[b] in summer °C
Marine		
Tropical oceans		
Sublittoral algae	+16 to + 5	+32 to +40
Intertidal algae	+ 3 to − 2	+38 to +44
Cold oceans		
Sublittoral algae	about − 2	+25 to +32
Intertidal algae	−15 to −40	+36 to +42
Fresh-water		
Lakes, ponds, rivers		
Algae	− 5 to −10 (−20)	+40 to +45 (50)
Cormophytes	about −10	+38 to +42
Hot springs		
Bacteria		+70 to +90
Cyanophyceae		+70 to +75
Diatoms		+45 to +50

[a] Cooling at least 2 hours.
[b] Heating 30 minutes.

Table 6. Maximum temperature resistance of leaves and buds of autotrophic terrestrial plants. Lists of species with respect to cold resistance in: MOLISCH (1897), IRMSCHER (1912), ULMER (1937), LARCHER (1954, 1963b, 1970, 1971), TILL (1956), BIEBL (1964, 1968), KAPPEN (1964, 1969), PISEK et al. (1967), SAKAI and OTSUKA (1970). Lists of species for heat resistance in: SAPPER (1935), LANGE (1953, 1955, 1959), BIEBL (1964, 1968), ALEXANDROV et al. (1964), KAPPEN (1964), PISEK et al. (1968), BIEBL and MAIER (1969), BANNISTER (1970), and unpublished data. Collected data cited in: LEVITT (1956, 1966), BIEBL (1962), PARKER (1964), ALTMAN and DITTMER (1966), and ALEXANDROV et al. (1970)

Plant and distribution	Low temperature injury[a] in winter °C		Heat injury[b] in summer °C
	Leaves	Buds	Leaves
Tropical regions			
Evergreen rain forest			
Mosses	− 5 to −15		+44 to +50
Ferns, herbaceous angiosperms	+ 5 to − 2		+45 to +48
Lianes	+ 5 to 0		
Trees	+ 5 to − 2		+45 to +50
Mangroves	+ 5 to − 4		about +50
Drought deciduous woodlands	− 3 to − 7	− 5 to − 8	+45 to +55

Table 6 continuation

Plant and distribution	Low temperature injury[a] in winter °C		Heat injury[b] in summer °C
	Leaves	Buds	Leaves
Subtropical dry regions			
Succulents and C_4-plants			+50 to +60
Species with insufficient transpiration cooling („Übertemperaturpflanzen")			+50 to +57
Species with efficient transpiration cooling („Untertemperaturpflanzen")	— 8 to —12		+45 to +48
Warm temperate zone			
Mediterranean hard-leaved forests	— 6 to —13	— 8 to —16	+50 to +55
Warm temperate forests of coasts and islands	— 7 to —12	—12 to —15	
Temperate zone			
Lichens (wet)	below —80		+35 to +45
Mosses (wet)	—15 to —30		+40 to +50
Ferns, herbaceous angiosperms	—15 to —25	—10 to —20	
Sunny habitats			+48 to +52
Shady habitats			+40 to +45
Terrestrial halophytes	—14 to —20		
Dwarf shrubs of Atlantic heaths	about —20	about —20	+45 to +50
Submediterranean woody plants	about —20	—25 to —30	about 50
Trees and shrubs of wide distribution in the temperate zone	—25 to —35	—25 to —40	about +50
Winter-cold regions			
Herbaceous plants in high latitudes			+44 to +52
Herbaceous plants and cushion plants in high altitudes	—20 to —50	—40 to —196	+52 to +58
Alpine dwarf shrubs	—20 to —70	—20 to —50	+47 to +54
Evergreen conifers	—40 and below	—35 and below	+44 to +50
Deciduous boreal trees and shrubs	—40 and below	—40 and below	+45 to +48

[a] Cooling 2 hours and more.
[b] Heating 30 minutes.

these plants have against low temperature damage (see Fig. 10). In the cooler or drier seasons some of these species can avoid ice formation down to temperatures of — 10° C to — 12° C by concentrating the cell sap and by improved supercooling properties. The sublittoral algae of the cold seas, some fresh-water algae, tropical and subtropical cormophytes, various woody plants of warm temperature regions and many herbaceous plants remain sensitive to ice formation in the tissues (LARCHER, 1954, 1963a, 1970, 1971; PARKER, 1960b; BIEBL, 1964, 1970). All higher plants are frost-sensitive during the period of active metabolism and growth.

2. Frost-Tolerant Plants.

Frost tolerance is developed by fresh-water and intertidal algae, and especially by air algae and snow algae (KYLIN, 1917; KANWISHER, 1957; TERUMOTO, 1959, 1964; PARKER, 1960c; RIETH and SAGROMSKY, 1964; SCHÖLM, 1968), by mosses of all climatic zones (tropical also, BIEBL, 1967b), and perennial terrestrial cormophytes of the temperate zone. The ability to survive dehydration caused by extracellular freezing is developed to variable degrees by different species. A particular group of frost-tolerant plants is formed by species that develop a special degree of deep-hardening which enables them to survive prolonged, severe frost at temperatures down to $-60°$ C and below. Deep-hardening is developed by some algae (e.g. *Porphyra yezoensis:* TERUMOTO, 1965b), lichens (KAPPEN and LANGE, 1970) and various woody plants of winter-cold regions (SAKAI, 1960; PARKER, 1960b; TUMANOV, 1960, 1967; KRASAVTSEV, 1960, 1969; BAUER et al., 1971).

Concerning **heat resistance** a distinction can be made between constitutional heat sensitivity, pronounced heat tolerance, and heat stability.

1. Heat-Sensitive Species.

These have a habitually low resistance and suffer injury after only a half hour of heating to $30°$ C, $40°$ C, or at most to $45°$ C. This group comprises eukaryotic algae and submersed cormophytes, terrestrial lichens in their hydrated condition (which, however, become heat resistant when dried out by the sun), shade plants which are not exposed to great heat in their natural habitats, and various soft-leaved species with efficient transpirational cooling (SAPPER, 1935; BIEBL, 1939, 1958, 1970; LANGE and LANGE, 1963; ALEXANDROV et al., 1964, 1970; „Untertemperatur-arten" LANGE, 1959).

2. Heat-Tolerant Species („Übertemperaturarten", LANGE, 1959).

Representatives of this group, which is characterized by a pronounced ability to develop heat hardening, are to be found particularly among the sclerophyllous and the succulent plants of sunny and dry habitats, and among plants with C_4-dicarboxylic acid pathway of photosynthesis (BJÖRKMAN et al., 1972). They survive exposure to temperatures of $50°$ C to $60°$ C for half an hour (SAPPER, 1935; LANGE, 1959; LANGE and LANGE, 1963; LARCHER, 1961; BIEBL and MAIER, 1969).

3. Heat-Stable Species.

Some thermophilic prokaryotic organisms tolerate temperatures of up to $90°$ C even in a state of metabolic activity (bacteria), and all survive temperatures up to $60°$ and $75°$ C *(Cyanophyceae)*. Such organisms retain their heat tolerance even if cultivated at low temperatures (BÜNNING and HERDTLE, 1946; KEMPNER, 1963; LANGRIDGE, 1963; BROCK, 1967; BOTT and BROCK, 1969).

β) Differences in Resistance between Species and Varieties

The limits of a plant's resistance to extreme temperatures, the sensitivity of its reaction to environmental stimuli, as well as the course followed by the processes of hardening and dehardening are characteristics of the species and variety. This involves the specific reaction pattern by which the plant adapts, to a greater or lesser degree, to the climatic events in its environment (KARNATZ, 1956b; MURAWSKI, 1962, 1968; WILNER, 1965; EGUCHI et al., 1966; SCHÖNBACH and BELLMANN, 1967; MITTELSTÄDT, 1968).

The complex pattern of behavior involved in the acquisition of resistance seems to be polyfactorially inherited (in *Mirabilis jalapa*, however, chilling sensitivity is caused by one recessive gene; CORRENS, 1913). This was demonstrated in fruit trees, for example, by RUDOLF and NIENSTAEDT (1962), MURAWSKI (1968) and MITTELSTÄDT (1968), and in cereals by GOUJON et al. (1968), JENKINS (1969, 1970) and GROGAN (1970). GOUJON et al. and LAW and JENKINS (1970) succeeded in identifying 5 chromosomes determining cold resistance. This explains why crossing experiments have so often given discrepant results concerning the inheritance of resistance. On some occasions a negative heterosis effect occurred (e.g. *Pseudotsuga* hybrids: SCHÖNBACH and BELLMANN, 1967; *Eucalyptus* hybrids: MENDONZA, 1968; *Larix* hybrids: HAMAYA et al., 1968; *Avena* varieties: JENKINS, 1969).

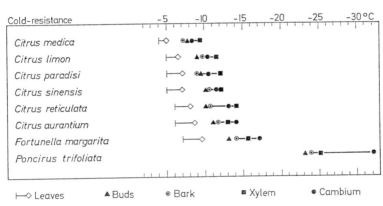

Fig. 27. Cold resistance of leaves, lateral buds, and mature shoots of *Citrus* varieties in winter. The line at the left border indicates the temperature at which leaves may become frost damaged. The other symbols represent 50% damage. (From LARCHER, 1971)

Species, and even varieties, of one and the same genus may vary markedly with respect to their temperature resistance. In some cases a characterization based on the individual maximum resistance values is possible; e.g. in plants from regions with a mild winter, which can be arranged according to their maximum cold resistance (*Cryptomeria* races: EGUCHI et al., 1966; *Eucalyptus*: MENDONZA, 1968; LARCHER, 1971; *Camellia*: SIMURA, 1957; *Coffea*: SODERHOLM and GASKINS, 1961; *Citrus*: KOCHERZENKO et al., 1951; PURCELL and YOUNG, 1963; LARCHER, 1971; see also Fig. 27). Characteristic differences also occur in the heat resistance of related species (*Cladonia* and *Ramalina*: LANGE, 1953; varieties of *Hypnum cupressiformae*: LANGE, 1955; *Acacia* species: LANGE, 1959; *Eucalyptus* species: KREEB, 1970).

In addition to giving the highest values of temperature resistance, a precise and exhaustive resistance characterization of species and varieties should also record specific hardening behavior (cf. Figs. 18 and 19). This involves a thorough analysis of resistance and is a time-consuming process (LARCHER, 1968), so that adequate data are at present available for only a few plants. Examples are provided by *Cryptomeria japonica* (EGUCHI et al., 1966), *Pinus cembra* (ULMER, 1937; PISEK and SCHIESSL, 1947; TRANQUILLINI, 1958; SCHWARZ, 1970), *Abies alba* (PISEK and KEMNITZER, 1968; BAUER, 1970; BAUER et al., 1971), *Larix* species (SCHÖNBACH et al., 1966; HAMAYA et al., 1968), *Morus bombycis* (SAKAI, 1955–1968; KITAURA, 1967 inter alia), Mediterranean *Quercus* species (LARCHER and MAIR, 1969), *Rosaceae* fruit trees (LOEWEL and KARNATZ, 1956; OLDÉN, 1957; PISEK, 1958; KOHN, 1959; LARCHER and EGGARTER, 1960; LAPINS, 1960, 1965; MITTELSTÄDT, 1965, 1966; HOWELL and WEISER, 1970a, b), *Camellia sinensis* (SIMURA and SUGIYAMA, 1965; SUGIYAMA and SIMURA, 1966, 1968), *Acer* species (IRVING and LANPHEAR, 1967, 1968; HARRASSER, 1969; BAUER, 1970), and *Cornus stolonifera* (HURST et al., 1967; VAN HUYSTEE et al., 1967).

Species-specific peculiarities in resistance behavior are not obscured by acclimatization to the habitat. Selection eliminates only those species in which the reaction pattern cannot cope with the demands of the habitat. Individual species of a plant community or a vegetational formation exhibit idiosyncracies in resistance by means of which they can be distinguished from one another and which permit representation by resistance spectra. Such resistance spectra are available for plants of tropical virgin forests (BIEBL, 1964), for Mediterranean maquis (LARCHER, 1954, 1970; LANGE and LANGE, 1963), for the vegetation of the summer-green deciduous forests (TILL, 1956), for high mountain plant communities (ULMER, 1937; TYURINA, 1953; SAKAI and OTSUKA, 1970) and for the vegetation of arctic rocks (BIEBL, 1968).

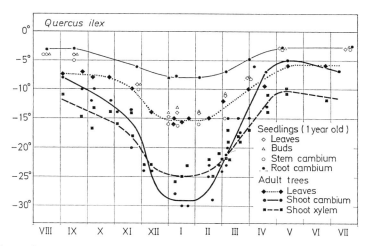

Fig. 28. Annual course of frost resistance of various parts of adult trees and seedlings of *Quercus ilex*. Measurements from different years, samples from different habitats, thus including the full individual variability and adapability of resistance. (From LARCHER and MAIR, 1969)

γ) Differences in Resistance between Organs and Tissues

Individual organs and even tissues behave differently with regard to resistance (Fig. 28) and reveal unequal maximum resistance values (Fig. 29, Table 7). Observations on various perennial woodland plants have been reported by TILL (1956), SAKAI (1956), LARCHER and MAIR (1969), LARCHER (1970); on fruit trees

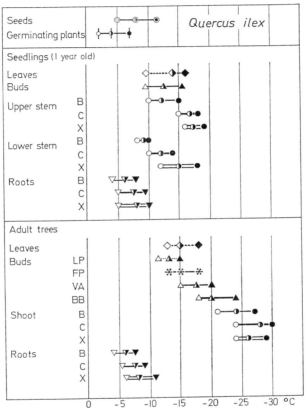

Fig. 29. Maximal frost resistance of various organs and tissues of *Quercus ilex* in winter. The symbols on the left indicate no injury, the middle symbols indicate 50% damage, and the symbols on the right indicate total damage. Abbreviations: *B* bark, *C* cambium, *X* xylem, *LP* leaf primordia, *FP* flower primordia, *VA* vegetative apex, *BB* bud base. (From LARCHER and MAIR, 1969)

by CHANDLER (1913), KEMMER and SCHULZ (1955), OLDÈN (1955, 1957), PISEK (1958), LARCHER and EGGARTER (1960), TYURINA (1967, 1968), LARCHER (1971); on grapevines by POGOSYAN (1967), POGOSYAN and SAKAI (1969) and on halophytes by KAPPEN (1969).

Cold Resistance. Roots are the organs most sensitive to cold although they are, in fact, capable of developing hardening (CHANDLER, 1954; TUMANOV and KHAVLIN,

1967; MITYGA and LANPHEAR, 1969). The leaves and buds are less sensitive, while the stems are the most resistant. Of the various types of tissues, resting or temporarily inactive meristem is able to develop the highest degree of hardening.

Table 7. Maximum frost resistance in winter of subterranean organs, leaves, and buds of various life forms of temperate forest plants. (After TILL, 1956)

Object	Maximum frost resistance in winter
Subterranean organs	
Roots, rhizomes, bud	— 6.0 to —13.5° C
Aerial organs	
Evergreen leaves of herbaceous plants	
3 to 5 cm above litter	—11.5 to —14.5° C
5 to 10 cm above litter	—11.5 to —18.0° C
10 to 20 cm above litter	—13.0 to —20.0° C
Buds of herbaceous plants	
below litter	— 7.0 to —11.5° C
close to litter	—12.5 to —18.0° C
3 to 20 cm above litter	—15.5 to —19.5° C
Shrubs and dwarf shrubs	
Sarothamnus scoparius, assimilating shoots	—18.5° C
Erica tetralix, leaves	—20.0° C
Erica tetralix, buds	—19.5° C
Trees	
Fagus silvatica, leaf buds	—29.0° C
Betula pendula, male flower	—38.0° C
Betula pendula, leaf buds	—40.0° C

In leaves the metabolically active parenchyma along the transporting vessels usually suffers damage first, giving rise to a typical necrosis of the transport system. However, there are also other patterns of distribution of leaf damage: intervascular damage in heterobaric leaves; zonal damage in conifer needles; more usual is a diffuse spread of damage. (Examples are to be found in GICKLHORN, 1936; SPRANGER, 1941; LARCHER, 1954, 1963c; PISEK et al., 1967, 1968.) In dormant buds the medullary meristem and procambium layers are usually the most resistant tissues (Fig. 30). Reproductive parts remain several degrees more sensitive to cold in winter than the apical meristem and the leaf primordia of the vegetative buds. In fully developed flowers the style and the ovule freeze first (KEMMER and SCHULZ, 1955; MODLIBOWSKA, 1956; PISEK, 1958; LARCHER, 1970), whereas ripe pollen and young fruits tolerate slightly more cold (PANKRATOVA, 1956; PISEK, 1958). In the stems the cambium is usually more resistant than the other tissues, and the wood lags far behind in its acquisition of resistance. However, plants that never achieve a large degree of frost hardening are exceptions to this rule (LARCHER, 1970, 1971). In apple seeds (KEMMER and THIELE, 1955) and oak seeds (LARCHER and MAIR, 1969) the radicle is usually the most cold-sensitive component; in wheat, rye and oat grains, radicle and scutellum are particularly frost sensitive (AGENA, 1961), and in cocoa seeds the cotyledons are the most sensitive (IBANEZ, 1964).

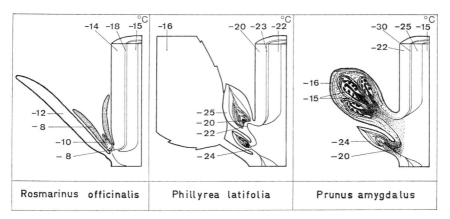

Fig. 30. Frost resistance of individual tissues in buds and shoots of Mediterranean woody plants in winter. (From LARCHER, 1970)

Heat Resistance. In winter, when they are in a state of maximum resistance, the shoots and roots of *Abies alba* and of *Acer pseudoplatanus* are the most heat resistant organs. The buds are less able to withstand heat, and all year round the leaves are more sensitive than the other organs (BAUER et al., 1971). The buds of young plants of various species of *Eucalyptus* were found by KREEB (1970) to be more resistant to heat than the leaves. Flower parts such as petals, stamens and pistil (but not the sepals: SAPPER, 1935), and fruits, are several degrees more resistant than the leaves in some herbaceous plants (SAPPER, 1935; ALEXANDROV, 1956). The pericarp tissue of avocado fruits is remarkable for its very high heat resistance (SCHROEDER and KAY, 1961).

In judging the temperature resistance of a particular plant species it is important to consider the specific vulnerability of the individual organs responsible for the functioning of the whole organism. The resistance of the roots and the leaves is decisive for the water relations, nutrition and dry matter production of the plant; the resistance of the vegetative buds and the twigs bearing them, for the growth of the shoot and the development of the foliage. The meristematic tissues of stem, branches and roots are responsible for the recovery of the plant as a whole from any injury suffered (KARNATZ, 1956a; LAPINS, 1965; LARCHER, 1970). The persistence of a certain species in any particular region is to a large extent determined by the vulnerability and temperature resistance of the flower primordia in the reproductive buds of the flowers, of the seeds and of the unprotected seedlings. These are the most sensitive stages in the life cycle and are determining factors (THIENEMANN's law) in the survival and spread of a species. Figure 31 offers a schematic representation of the zonation of cold resistance with respect to age and organ in a *Quercus ilex* population.

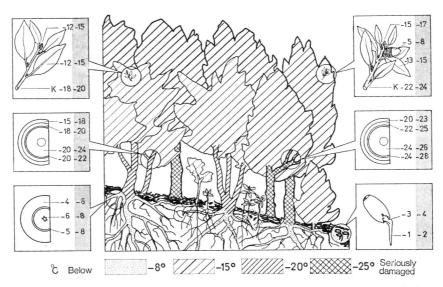

Fig. 31. Zonation of frost-resistance levels in winter in a community of *Quercus ilex*. Temperatures indicated in the unshaded part of the marginal pictures represent the extremes below which first frost damage is to be expected; the data in the shaded parts refer to 50% damage. (From LARCHER and MAIR, 1969)

3. Survival Limits of Plants

The survival capacity of a plant is its ability to withstand, without damage, the varied demands made by adverse weather conditions and by extreme habitats. Survival capacity is a much more complex property than temperature resistance (cf. "ecological resistance", BIEBL, 1952; "environmental resistance" LEVITT, 1958; FUCHS and ROSENSTIEL, 1958; OLIEN, 1967, 1969; SCHRAMM, 1968). During a period of heat, plants are not only endangered by high temperatures, but also by the accompanying dehydration. Winter offers, apart from the immediate effect of cold, further sources of danger in the form of winter drought and snow pressure.

Plants counter these manifold dangers with a variety of measures:

Evasion: Certain plants avoid the unfavorable effects of a season by processes of evasion such as a short life cycle, premature leaf fall, and reduction of vegetative parts to the minimum organs necessary for survival and storage. Short-term evasions are also known: according to ERNST (1971) certain African *Caesalpiniaceae* turn their leaf fronds upwards at temperatures above 34° C, despite adequate water supplies, and thus reduce by half the absorption of radiation.

Resistance: Plants that are exposed to great heat or cold in their natural habitats are obliged to develop avoidance mechanisms, or tolerance, or both, not only to the "direct injury" (LEVITT, 1956) due to high or low temperature but also to the "indirect injury" accompanying heat and frost.

Recovery: The important point in survival is not necessarily that the plant remains completely undamaged by the stress, but that the individual or species survives at all. In cormophytes isolated tissue damage does not necessarily lead to the death of the individual. According to the extent of the damage and the importance of the damaged tissue to the plant as a whole, partial or even complete repair is possible since plants have a regenerative capacity. In assessing the possibility of survival of an entire plant, therefore, regeneration from injury must also be taken into consideration (MÜNCH, 1928; HOLDHEIDE, 1941; KEMMER and SCHULZ, 1955; KARNATZ, 1956a; MODLIBOWSKA, 1956, 1962; PISEK and EGGARTER, 1959; MORET-TINI, 1961; JAHNEL and WATZLAWIK, 1961; LARCHER, 1963c, 1970; KREEB, 1970; for limits of regenerative capacity see KRAMER and KOZLOWSKI, 1960). Therefore the chance for survival of a plant species in a given environment can be represented as follows:

$$Survival = Evasion, Resistance, Recovery.$$

Survival is, in fact, the information sought in the investigation of resistance in an ecological or applied context. The complexity of the factors limiting survival renders a causal analysis of the connections between temperature resistance and plant distribution extremely difficult. For this purpose comprehensive physiological investigations of resistance as well as exact data concerning the microclimate in the habitat over longer periods of time are essential. At present only empirical information concerning survival limits is available (e.g. DE PHILIPPIS, 1937; SKINNER, 1962; USNA, 1965; OUELLET and SHERK, 1967). The most useful data so far refer to crop and ornamental plants, which as a rule are not exposed to natural competition. Such plants are therefore suitable indicators of the occurrence and local distribution of extreme temperatures (e.g. SCARAMUZZI and ANDREUCCI, 1957; LARCHER, 1963c; WELLER and SCHREIBER, 1965; KARSCHON, 1966), and they can undoubtedly be employed with success in future studies of the ecological aspects of resistance.

C. Cell Death by Cold and Heat and Resistance to Extreme Temperatures. Mechanisms of Hardening and Dehardening*

U. HEBER and K. A. SANTARIUS

1. Low Temperatures

a) Introduction

The purpose of this chapter is to analyze briefly the modes of injury and protection prevailing under the conditions to which plant cells are normally exposed. Since the earlier literature has been covered in often excellent reviews (LEVITT, 1956, 1969; PARKER, 1963; MERYMAN, 1966a; OLIEN, 1967; MAZUR, 1969, 1970; WEISER, 1970; ALDEN and HERMANN, 1971) this article will mainly deal with more recent advances in our knowledge.

b) Phenomenology of Cell Freezing

The first effect of lowering the temperature of cells is a slowing down of metabolism. A drastic decrease in temperature can result in metabolic disorder since steady state equilibria within the cell are disturbed owing to shifts in the balance between different enzyme-catalyzed reaction sequences. In fact, plants not adapted to large temperature changes, such as some tropical plants, exhibit the phenomenon of "chilling injury" even at temperatures above the freezing point of water (BIEBL, 1964 and others).

Plants of temperate climates tolerate a decrease of temperature towards and often below the freezing point. When the temperature decreases below the freezing point, cells and their surroundings supercool before freezing (cf. p. 204). Ice formation is usually, although not always, extracellular because in plant cell systems the concentration of solutes is lower outside than inside the cell, the freezing point depression accordingly smaller and conditions for seed formation more favorable. Once ice is formed, its formation will continue where it has begun until vapor pressure equilibrium between ice and unfrozen cell content has been established. If the original ice formation was extracellular and if the cell membranes are sufficiently permeable, intracellular water follows the gradient in vapor pressure during ice formation and, after passage through the cell membranes, crystallizes outside the cells. In consequence, the cells are dehydrated during extracellular freezing. The extent of dehydration is a function of temperature. Cell membranes prevent the growth of extracellular ice crystals into the intracellular space. The cell itself consists of several membrane-surrounded compartments. If the rate of freezing is sufficiently slow, all of the compartments are close to vapor pressure equilibrium. However, even though the extent of dehydration is the same in all cell compartments under equilibrium conditions, the effects of dehydration may be quite different according to the composition and structure of the individual cell compartments. This will be elaborated upon later.

---

* The literature pertaining to this chapter was completed by March 1971.

Extracellular ice formation is of normal occurrence and can usually be observed under natural conditions when cooling rates are slow. Intracellular ice formation takes place only when the rate of cooling is too high or the water permeability of the plasmalemma too low to permit a sufficiently rapid outflow of intracellular water. At a given temperature, the extent of dehydration of the system is the same during intracellular and during extracellular freezing. However, there is a decisive difference. While extracellular ice formation may or may not lead to injury, the formation of large intracellular ice crystals is always lethal (cf. p. 237).

From the known relation between the vapor pressure of solutions and solute concentration it follows that most of the cellular water will be converted into ice at relatively high temperatures (for the estimation cf. WILLIAMS and MERYMAN, 1965). Since part of the remaining water is bound to macromolecules and unavailable as a solvent the unfrozen solution remaining behind is highly concentrated: its composition differs in different cell compartments owing to the compartmentation not only of macromolecules but also of low-molecular-weight constituents of the cell.

If simple systems consisting of water and a solute are progressively frozen, water is converted to ice until the eutectic temperature is reached, when the whole system solidifies. Somewhat more complex systems behave similarly, in principle. However, the highly complex cell system contains components such as protein and sugars which do not readily crystallize when concentrated and which are capable of preventing the crystallization of other components of the system. It is therefore not surprising that eutectic behavior has not been observed in cells (LEVITT, 1966). Rather, a small part of the total water which is attached to solute molecules by sufficiently firm forces remains "fluid" even at very low temperatures, i.e. is not converted to ice. Thus even at very low temperatures a highly concentrated multiphase cell "solution" can exist along with ice.

If, after freezing, the temperature is increased, the vapor pressure over ice rises and the course of events observed in freezing is reversed. However, quasi-equilibrium conditions between vapor pressure over ice and vapor pressure of the cell system can be maintained only when the temperature rises slowly. If the temperature rises rapidly the water permeability of cell membranes and not the rate of melting determines the rate of water uptake by the dehydrated cell. Too rapid a rise in temperature, therefore, results in flooding of cells or tissues. As will be seen, this also affects survival.

c) Changes in a Cell System Brought about by Freezing

How do the events described above affect cells, tissues or organisms? The possibilities will be listed first, followed later by a critical evaluation.

1. Adverse effects of temperature changes per se may cause injury.

2. Since water occupies a larger volume in the form of ice than as a fluid, cells and tissues may be exposed to mechanical stress caused by the pressure or shearing forces produced by growing ice crystals. This may apply whether ice formation is extracellular or intracellular.

3. Since small ice crystals have, owing to their higher surface energy, a higher vapor pressure than larger crystals there is a tendency for the latter to grow at the

expense of the former. This so-called recrystallization is temperature-dependent and proceeds at significant rates only if the temperature is not too low. It also tends to produce mechanical stress.

4. Large water losses during extracellular freezing reduce the cell volume and result in considerable shrinkage which can easily be seen by microscopic examination of frozen tissues. This reduction in cell size may also be considered as a stress situation.

5. Another result of dehydration is the increase in the concentration of solutes. A number of low-molecular constituents of cells exhibit toxic effects at increased concentrations: dehydration during freezing might raise their concentration to and above the toxic level.

6. During concentration of solutes precipitation of salts can occur. This may cause changes in the pH and composition in the unfrozen phase of frozen biological materials.

7. The structure of some high-molecular cell constituents is known to be influenced by hydration. If dehydration irreversibly changes the structure of essential cell components it could constitute the injurious event which results in cell death.

8. In the course of dehydration during freezing, the removal of the intermolecular water to ice loci forces protein and other molecules to approach and come into contact with one another. This could lead to the formation of new bonds between chemical groups that were previously too far apart.

What are the tools available for distinguishing between different causes of freezing injury and for identifying a particular factor as being responsible for the damage?

Unfortunately a separation of factors is difficult in view of the complexity of living cells and only a few attempts have been made to study the effects of freezing on systems which are simpler than intact cells and still biologically relevant. Only a few examples will be cited.

Effects of temperature can be studied with intact cells and enzyme systems in the absence of freezing, i.e. under conditions that favor supercooling (ROBERTS, 1967). Mechanical injury caused by masses of extracellular ice may become manifest when the effects of freezing on water-saturated and wilted tissues are compared. It was found that particularly dehydrated cells, e.g. air-dried plant material such as mosses, lichens and others, are much more resistant to freezing than turgescent tissues (KAPPEN, 1966; BIEBL, 1968 cf. p. 220). Variations in the rates of cooling and warming can give very useful information on the cause of damage (MAZUR, 1969). When rates of thawing are held constant, increased damage with increased freezing rates may indicate injury due to intracellular freezing. Conversely, beneficial effects of faster freezing might be the result of the shorter exposure of the cells to increased levels of solutes at intermediate temperatures where solute effects are particularly pronounced. If the rate of freezing was uniform, variations of injury with the rate of thawing may indicate damage due to recrystallization (more damage caused by slow than by rapid thawing after ultrarapid freezing), due to solute concentration (more damage, after slow extracellular freezing, by very slow than by faster thawing because of prolonged exposure at intermediate temperatures to increased solute levels) or to osmotic effects (more damage, after slow extracellular freezing, by fast than by slow thawing because of cell "flooding").

In considering points 4 to 8 it is noteworthy that not only freezing results in dehydration and cell shrinkage and causes concentration of intracellular solutes. The same effects are brought about by drought. In fact, comparable injury can be caused by freezing and by drying, if the extent of dehydration is comparable, and the same changes that lead to the development of freezing tolerance may also result in increased drought resistance.

Injury caused by increased levels of toxic compounds can be investigated at temperatures above the freezing point by using highly concentrated solutions of these toxic solutes. The same holds true for alterations caused by changes in the pH.

d) Susceptibility of Cell Components

If freezing kills cells in different ways, the question arises whether it is justifiable to look for a common target of attack and whether there are specially susceptible cell components which when damaged by freezing cause subsequent death. Few attempts have been made to answer these important questions. There is no reason to assume that cell constituents of low molecular weight are altered by freezing and attention can be focused on a possible freezing sensitivity of constituents of high molecular weight involved in building cell structures. Of these, nucleic acids are apparently not damaged by freezing (MAZUR, 1966), and polymer carbohydrates are as little affected by freezing as is the cell wall structure of plant cells. However, recent evidence has shown that proteins are adversely affected not only by elevated, but also by lowered temperatures, with maximum stability occurring around or somewhat below room temperature (BRANDTS, 1967). Nevertheless, destabilizing effects of low temperatures do not necessarily become apparent, as is shown by the well-known fact that many enzymes can be preserved in the frozen state for long periods of time. Even in the frozen state some enzymes effectively catalyze substrate transformations (GRANT and ALBURN, 1967). On the other hand, low-temperature inactivation of some soluble enzymes has been observed *in vitro* (CHILSON, COSTELLO, and KAPLAN, 1965; BRANDTS, 1967): inactivation was temperature dependent and not related to ice formation. Whereas chilling injury is a temperature dependent process, it is well established that freezing injury occurs as a consequence of the dehydration accompanying ice formation. This makes it appear doubtful whether it is causally related to the cold inactivation of individual enzymes which is a temperature effect. In addition, in many cases, freezing injury becomes apparent immediately after thawing, while inactivation of soluble enzymes would be expected to result in metabolic disorder and not in the immediate breakdown of the cell.

During freezing the photosynthesis of moderately hardy leaves is influenced long before irreversible injury occurs. Even non-lethal freezing of hardy leaves results in a transient suppression of photosynthesis (PISEK, LARCHER, and UNTERHOLZNER, 1967; BÖRTITZ, FUCHS, and WEISE, 1967; PISEK and KEMNITZER, 1968; LARCHER, 1969a; BAUER, HUTER, and LARCHER, 1969 etc.). The extent of inhibition increases with the length of time the leaves were kept in the frozen state, and with the freezing temperature. Only some lichens do not show an inhibition in photosynthesis immediately after thawing (LANGE, 1966). Efficient photosynthesis requires close cooperation between stroma proteins and chloroplast membranes. In membranes, proteins are associated with lipids: membrane proteins may account for more than 50% of the total protein of the cell and the basic structural features seem to be identical in membranes of different origin or function (ROBERT-SON, 1960; STOECKENIUS, 1970). They appear to consist of two monomolecular layers of protein which enclose between them a bimolecular lipid film. The individual component molecules are linked by interactions involving hydrogen-bonding, ionic forces and van der Waals forces. A complex balance between these forces stabilizes the membranes in which, as the permeability characteristics indicate, hydrophilic phases and a hydrophobic phase exist side by side. Such structures would be expected to be sensitive to dehydration. In fact, LOVELOCK (1954a) has suggested that changes in membrane permeability are responsible for the sensitivity of red blood cells to freezing. SOUZU (1967) concluded that damage to protoplasmic membranes by freeze-thawing abolished compartmentation within living cells and thereby the control of metabolic reactions. Freezing of lysosomes released hydrolytic enzymes (DE DUVE, 1959).

Isolated membranes of mitochondria and chloroplasts are also damaged by freezing (cf. p. 239). The common observation that cells of microorganisms and plants which have been killed by freezing are readily permeable to normally non-penetrating solutes can also be explained on the grounds that freezing alters the permeability characteristics of cell membranes. Even if freezing kills cells in a variety of ways, membranes appear to be the common target. This is obvious if damage is caused mechanically or osmotically, i.e. by direct rupture of membranes. Evidence will be presented that it is also true in the much more important instances where damage is produced by dehydration or increased solute concentration. In agreement with this, DANIELL, CHAPPELL, and COUCH (1969) observed by light microscopy that the primary effect of lethal temperatures on the leaf cells is disintegration of the cellular membranes, e.g. a disorganization of the tonoplast membrane, plasmalemma and chloroplast membranes.

The vital importance of membranes for the cell is self-evident. Without a semipermeable system to maintain the internal composition and to regulate the import of substrates and export of end products cell life cannot continue.

e) Causes of Injury

From the preceding considerations it is apparent that low temperatures may affect a cell in different ways: metabolically, mechanically, by dehydration, by solute concentration or even osmotically. Which of these possibilities are realized under the conditions to which plants are exposed in nature?

α) Temperature Effects

It has already been mentioned that some tropical plants may be killed by temperatures even above the freezing point. Usually it takes hours or perhaps days for the injury to develop. The mechanism of chilling injury is not yet known although various suggestions have been made. Recent evidence points to phase changes in membrane lipids as a possible cause of damage. KUIPER (1970) determined changes in lipid composition of leaves of hardy and cold-sensitive Medicago varieties grown at different temperatures. LYONS and RAISON (1970) found that mitochondrial respiration is depressed, probably due to a physical effect of temperature on membrane components such as membrane lipids. An effect on phosphate metabolism has also been suggested by STEWART and GUINN (1969) and BUSCHBECK (1970a, b). It is also well known that some isolated and purified enzymes, although surprisingly stable at room temperature, rapidly lose their activity if exposed to 0° C (literature cf. FARRELL and ROSE, 1967; ROBERTS, 1967). Although no data from *in vivo* experiments are available, this cold inactivation of sensitive enzymes could also be involved in chilling injury.

Since hardy and non-hardy plants of temperate climates show no ill effects from exposure to temperatures around the freezing point or even from supercooling to sub-zero temperatures (LEVITT, 1966), whereas the non-hardy plants are killed after ice formation has set in (PISEK, LARCHER, and UNTERHOLZNER, 1967; BIEBL, 1968 etc.), it must be concluded that they are insensitive to changes of temperature per se. All available data indicate that freezing injury is not a temperature effect but is linked to the dehydration caused by ice formation or the ice formation itself.

There is yet another temperature effect, probably different from that causing chilling injury in sensitive plants and even less well understood. After a rapid drop in temperature red blood cells and cells of some microorganisms exhibit the phenomenon of thermal shock resulting in the immediate death of the cells (MAZUR, 1966). Thermal shock has recently also been observed with isolated chloroplast membranes (WILLIAMS and MERYMAN, 1970) and must therefore be a membrane effect. It does not seem to occur under the conditions to which plant cells are normally exposed.

β) Mechanical Damage by Ice Crystals

It has been mentioned that cells freeze intracellularly if the water permeability of cell membranes is low and the rate of lowering of the temperature is too fast to permit extracellular ice formation. The extent of dehydration is the same as if ice formation had taken place outside the cells. Still, with rare exceptions, intracellular freezing results in cell death, although the same cells frozen extracellularly would survive if hardy enough. From this it has been concluded that the formation of larger intracellular ice crystals causes death mechanically by disrupting structural elements within the cells (MERYMAN, 1966b; MAZUR, 1966, 1967; SAKAI, OTSUKA, and YOSHIDA, 1968; ASAHINA, HISADA, and EMURA, 1968 etc.). Only under special conditions, for instance during rapid cooling and very fast thawing (e.g. between − 50° C and 0° C), has high survival been observed for intact cells (SAKAI and YOSHIDA, 1967; SAKAI and OTSUKA, 1967; SAKAI, OTSUKA, and YOSHIDA, 1968; MAZUR and SCHMIDT, 1968; ROTTENBURG, 1968; MAZUR et al., 1969; ASAHINA, SHIMADA, and HISADA, 1970 and others), for nuclei from mouse liver (ARNOLD, YOUNG, and STOWELL, 1968) and for isolated mitochondria from tomato fruits (DICKINSON, MISCH, and DRURY, 1970). The data were interpreted to indicate that intracellular freezing is not inevitably harmful if it is fast enough to lead to the formation of many minute, invisible, intracellular ice crystals which are too small to cause mechanical damage. This "vitrification" takes place only under conditions of ultrarapid freezing. The harmful effects of slow thawing were explained on the grounds that during slow, but not during ultrarapid warming, recrystallization occurred above − 50° C, resulting in the formation of larger crystals at the expense of smaller ones. This was assumed to rupture cell membranes and cause death (LUYET and RASMUSSEN, 1967; MACKENZIE and LUYET, 1967; cf. also MERYMAN, 1966b). In contrast to fast freezing, slow freezing at the rate of 1° C/min resulted in extracellulai ice formation. In this instance percentages of survival were similar to those seen with very fast freezing combined with ultra-fast thawing even if thawing was slow (MAZUR, 1967).

Long after freezing, plant organs such as leaves may be killed by large masses of extracellular ice which can accumulate in certain regions of the stalks. This may lead to the disruption of vascular tissues and thereby to death even though most of the cellular material may have survived freezing, as judged by the reappearance of turgor after thawing.

Of much more practical interest than the effects of fast freezing and the rare cases where injury from extracellular freezing can be traced back to direct effects of ice, are the many observations on freezing injury which cannot be explained by the direct action of ice.

γ) Reduction of Cell Volume

General agreement prevails among workers in the field that cells are killed by the dehydration accompanying freezing rather than by temperature effects or ice crystals. Dehydration influences cells in several ways. If most of the water is removed cells collapse and this may, in itself, constitute a stress situation. ILJIN (1933) and LEVITT (1956) assumed that the dehydration accompanying freezing led to cell contraction, and deformation of the shrunken cell caused tension in the protoplasm which resulted in mechanical damage of the organized structures of the protoplasm, perhaps the membranes. But damage may also occur during thawing, when the cell expands due to absorption of water (cf. p. 241).

HUGGINS (1965) found that red blood cells were damaged with or without freezing whenever the external solute concentration rose above 1.5 osmolal. The nature of the external solute was of little importance, which suggests that either the internal solute concentration or the cell size are critical factors. MERYMAN (1968, 1970, 1971) and WILLIAMS and MERYMAN (1970) proposed that cells and even membrane vesicles isolated from leaves cannot survive shrinkage beyond a minimum tolerable volume because of the development of an osmotic pressure gradient across the membranes. The nature of this gradient and the mode of injury remain somewhat obscuie since the activity of water inside and outside a water-permeable membrane must be the same at equilibrium. Implications of the minimum volume theory will be discussed later (cf. p. 257).

δ) Toxicity of Cell Components

LOVELOCK (1953a) found that red blood cells suspended in isotonic saline were killed during freezing when the salt concentration in the unfrozen part of the system exceeded a mole fraction of 0.014. Freezing was not even needed for death to occur as the cells hemolyzed at 0° C if the salt concentration of the suspending medium was raised to the same level. LOVELOCK concluded that the freezing injury of erythrocytes could be traced back to the injurious effects of electrolytes. In the light of present knowledge this statement has to be somewhat modified and can now be extended to other cells and to other injurious agents. It has been mentioned that the available evidence incriminates biomembranes as the frost-sensitive components of cells. In higher plants a considerable part of the total protein resides in the inner membranes of the chloroplasts in the leaf cells. During photosynthesis the conversion of light energy into chemical energy takes place in the chlorophyll-containing thylakoids. A prominent part of this process is the energy-requiring and -conserving formation of ATP from ADP and phosphate. ATP plays a central role in the energy metabolism of the cell and is also produced in the final stages of respiration during electron transport in the inner mitochondrial membranes.

When leaves are damaged by freezing, photosynthesis appears to be affected more than other cellular activities. Evidence from [14]C- and ^{32}P-studies using intact leaves, and from enzyme measurements, indicates that only a part of this complex process, the part in which thylakoid membranes are engaged, is permanently injured by freezing (HEBER and SANTARIUS, 1964). Phosphorylation reactions appeared to be especially affected. Chloroplasts isolated from frost-damaged leaves which were still in the frozen state could not, or only much more slowly than unfrozen controls, phosphorylate ADP by photophosphorylation. In fact, when thylakoid membranes isolated from intact leaves were frozen under conditions similar to those which would kill the parent leaves, photophosphorylation was inactivated. The same effect was observed in ATP formation by mitochondria. The freezing sensitivity of chloroplast and mitochondrial membranes appears to be a general phenomenon. Thylakoids from different plant species and mitochondrial membranes of both plant and animal origin are affected by freezing. In the presence of suitable amounts of some compounds such as sugars, however, inactivation of the membranes by freezing was prevented (PORTER et al., 1953; FEWSON, BLACK, and GIBBS, 1963; HEBER and SANTARIUS, 1964).

Thus, these membranes can be considered to be simple models of intact cells as far as response to freezing is concerned. From their freezing sensitivity, which is expressed in the absence of cryoprotective compounds, they appear to be well suited as test systems for measuring the toxicity of cell constituents, such as salts, which are supposed to be involved in freezing damage of intact cells. The cells themselves are unsuited for such experiments since the entrance of test substances is hampered or prevented by membrane barriers and the response to additions during freezing is accordingly misleading.

When thylakoids protected against freezing damage by sucrose are frozen to −25° C in the presence of varying amounts of NaCl, the results depicted in Fig. 1 are obtained. Obviously NaCl can overcome the protective action of the sugar and inactivate membranes during freezing if the salt/sugar ratio exceeds a certain threshold value which is a function of the freezing temperature.

Freezing is not even required for the effect of NaCl to be deleterious. If its concentration is high enough it inactivates thylakoids at and above 0° C (SANTARIUS,

1969). This shows that ice formation is not directly involved in freezing injury. The sensitivity of thylakoids to NaCl is not very different from that of red blood cells and may constitute a property of cell membranes in general. Thus NaCl appears to be a membrane-toxic compound with the reservation that toxicity is manifest only at elevated concentrations.

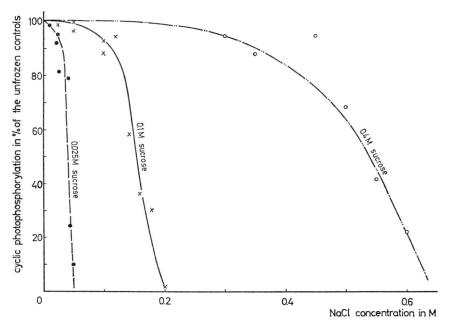

Fig. 1. Cyclic photophosphorylation of unwashed chloroplast membranes after freezing in the presence of 0.025, 0.1 or 0.4 M sucrose and NaCl as indicated on the abscissa. Freezing time was 3 h at —25° C. After thawing, the activity of cyclic photophosphory-lation was measured as an indicator of membrane integrity. The more sucrose present, the more NaCl was needed to overcome the protective effect of the sugar and to in-activate the membranes during freezing. (From HEBER and SANTARIUS, 1964)

A large number of different compounds, many of them cell constituents, possess similar characteristics. The membrane toxicities of chlorides of different alkali metals do not differ greatly whereas chlorides of bivalent cations such as Mg^{++} or Ca^{++} are much more toxic than the corresponding alkali chlorides. There is a steep toxicity gradient within different halogenides, with iodide the most and fluoride the least membrane-toxic anion. Nitrates are much more toxic than chlorides, sulfates less so (SANTARIUS, 1969).

Thus there are both highly toxic inorganic salts and others which can be tolerated up to fairly high concentrations. A similar situation exists for organic electrolytes and even for a number of neutral compounds (TYANKOVA, 1970; SANTARIUS, 1971; HEBER, TYANKOVA, and SANTARIUS, 1971). The hydrochloride of arginine is highly toxic, but that of lysine surprisingly nontoxic. Amino acids can be divided into groups whose members possess, at elevated concentrations, either little, con-

siderable or high toxicity (cf. p. 249). Partially apolar compounds such as phenylalanine or phenylpyruvate are highly toxic at elevated concentrations. Organic
acids form as diverse a group as amino acids: sodium succinate is toxic, salts of
other organic acids much less so, while sodium acetate is tolerated up to high concentrations without causing membrane injury (cf. p. 247). Polar neutral compounds such as sugars or sugar alcohols are largely non-toxic.

The analysis of factors involved in injury to membranes during freezing in the
presence of toxic electrolytes has shown that inactivation is a complicated function of temperature, the extent of dehydration and the time during which temperature and increased electrolyte concentration act on the membrane structures
(SANTARIUS and HEBER, 1970). It is occasionally difficult to decide whether, during
freezing of isolated membranes, inactivation is caused by the accumulation of a
potentially toxic compound in the unfrozen part of the system or, rather, by the
eutectic solidification of the whole system. In these cases investigation of the
kinetics of membrane inactivation and of the response of the membranes to high
concentrations of the inactivating compound at $0°$ C can provide evidence as to
the cause of membrane damage. Eutectic solidification can be prevented by
freezing in the presence of a substance such as sucrose which does not readily
crystallize, but the protective effects of the sucrose then have to be taken into
account.

Whether changes in pH during freezing are important for frost damage to cells, as
repeatedly claimed (VAN DEN BERG and SOLIMAN, 1969a), is not clear. Isolated
membranes are relatively insensitive to minor changes in pH so that smaller pH
shifts occurring in the highly-buffered cell system during freezing probably are
less injurious than the direct action of membrane-toxic compounds (SANTARIUS
and HEBER, 1970). Freezing concentrates the potentially toxic compounds which
are normal cell constituents. If only one of them reaches the toxic level, membrane
damage occurs, its extent determining whether or not the cell can survive freezing.

ε) Osmotic Injury

It is a commonly observed phenomenon that, in contrast to the cells of microorganisms or of animals, frozen tissues even of hardy plants may suffer damage
from fast thawing, while slowly thawed material recovers. Clearly, injury cannot
have resulted from freezing in these cases. Differentiated plant cells contain large
vacuoles which may occupy more than 90% of the cell volume. During slow extracellular freezing these vacuoles are dehydrated along with the cytoplasm and cell
organelles. If thawing is unphysiologically fast, a water permeability of the plasmalemma and of the membranes of cell organelles within the cytoplasm higher than
that of the tonoplast will result in a large water uptake by the cytoplasm and its
organelles, with the vacuole expanding only slowly. Rupture of the cytoplasmic
membranes due to osmotic shock in the swollen cytoplasm may then lead to cell
death. The sensitivity of vacuolated plant cells to rapid thawing appears to be a
fairly unique feature. As a rule, microorganisms and animal cells which have no
large central vacuoles show a higher rate of survival after fast than after slow
thawing (cf. p. 235). The injurious effect of slow thawing in these cells may be caus-

ed by a prolonged exposure to toxic levels of solutes (MAZUR, 1969) as discussed above.

However, there are exceptions. An isolated example of the injurious effect of fast thawing has been reported by LEIBO and MAZUR (1966) for the osmotically-sensitive bacteriophage T4B which survives only slow thawing. The mechanism of injury has been shown to be osmotic rupture.

f) Mechanisms of Injury

How does freezing kill cells or affect cellular membranes so as to alter vital membrane properties? How is phosphorylation uncoupled from electron transport during freezing of thylakoids?

If the piercing of a cellular membrane by an ice crystal were sufficient to cause death a discussion of the mechanism of injury would be pointless. Likewise the mechanism of osmotic rupture of a cell membrane which can occur during rapid thawing does not deserve much further consideration. However, it can easily be shown that during freezing of isolated thylakoids damage is neither caused mechanically by ice crystals rupturing the membranes nor by osmotic effects. Small amounts of sucrose added to the thylakoid suspension prior to freezing protect the membranes against damage but do not prevent the formation of ice. Damage to the membranes during freezing is more extensive if they are suspended in salt solutions rather than in water, although there is more ice formation in the latter case. Fast thawing is beneficial as compared to slow thawing (HEBER and SANTARIUS, 1964). Membrane inactivation is brought about by membrane-toxic compounds in the absence of any ice formation if the solute concentration is sufficiently high. For intact cells the situation may perhaps not be quite as clear, but still the available evidence points in the same direction. Probably the most important factor involved in freezing injury is the solute concentration which, during freezing, reaches abnormally high levels inside and outside the cells. Its composition differs inside and outside, and even within the cell it varies from one compartment to the other owing to the compartmentation of both high-molecular-weight and low-molecular-weight cell constituents. How are membranes inactivated by solute concentration effects?

The main components of cellular membranes are proteins and lipids. It is well known that proteins can be denatured by high concentrations of salts: ammonium sulfate is widely used for the "salting out" of proteins, which may or may not be denatured in the process. However, it has to be remembered that red blood cells and mitochondrial or chloroplast membranes are sensitive to concentrations of NaCl (0.8—1 M) which are far below those known to precipitate proteins. Furthermore, inactivation of individual membrane proteins has not been observed under the rather mild freezing conditions which suffice to arrest photophosphorylation in thylakoid membranes (HEBER, 1968). Finally, the membranes are sensitive not only to salts, but also to other compounds.

The formation of ATP from ADP and phosphate by thylakoids requires a suitable enzyme system in the membranes and, in addition, a chemical potential to drive the endergonic reaction. An enzyme involved in ATP synthesis has been isolated from the membranes and reported to be stable at room-, but labile at lower temperatures (McCarty and Racker, 1967). However, the suspicion that this enzyme is responsible for the freezing sensitivity of the membranes could not be confirmed since detailed analysis showed unequivocally that it withstands freezing which inactivates the membranes (Heber, 1967).

During light-induced electron transport, protons are taken up from the outside medium and are transferred across the membranes into the interior of thylakoid vesicles which are semipermeable and have a low permeability towards protons. A proton gradient as well as an electrical potential difference is thus established across the membranes, and constitutes, according to a modern theory (Mitchell, 1961, 1966), the driving force for phosphorylation.

After freezing, which inactivates photophosphorylation, thylakoids no longer display a net proton uptake from the medium on illumination (Heber, 1967). Since proton pumping appears to be closely associated with electron transport, which can even be stimulated by freezing, this observation suggests that the permeability properties of the membranes are altered so that a concentration gradient of protons across the membranes can no longer be maintained although pumping may be unimpaired. In fact, although unfrozen thylakoid vesicles behave as perfect osmometers over a large range of sucrose concentrations, they no longer respond after freezing to changes in the osmotic environment. Thus they are no longer semipermeable, and solutes which, before freezing, could not penetrate can pass in and out. Electron micrographs of frozen thylakoid vesicles suspended in a hypotonic solution show them to be collapsed, in contrast to the inflated unfrozen vesicles (Heber, 1967).

Cells whose ability to synthesize ATP is seriously impaired by freezing cannot survive. The same is true for cells whose membranes have become irreversibly leaky. In view of similarities in the basic structure of different cell membranes it is likely that not only chloroplastic and mitochondrial membranes, but other cell membranes are sensitive to freezing. The results suggest that the action of freezing on cells consists primarily in an irreversible change of the permeability characteristics of their membranes.

At present no direct information on the molecular nature of freeze-induced membrane alterations is available and an interpretation of the observations remains speculative. Deleterious effects of salts during freezing can be understood qualitatively as a suppression of ionic interactions among membrane components (cf. Lovelock, 1955). In the course of dehydration, the ionic strength in the unfrozen portion of the system increases and may finally reach a point where electrostatic interactions between solute and membrane components become stronger than the interactions among the latter. This must lead to a change in and finally the loss of the mutual orientation between membrane components. The resulting structural alterations are accompanied by permeability changes.

Similarly, an accumulation of apolar or partially apolar compounds, for instance phenylalanine (cf. p. 249), in the vicinity of the membrane may facilitate van der Waals interactions between these compounds and apolar membrane components. Although the mode of attack is entirely different from that of salts, the result will be the same: the membrane structure will be changed. As membranes are built of complex molecules and stabilized by different forces it can be expected that different compounds can destabilize them in different ways during freezing.

Another view has been expressed by LEVITT (1962) who proposed that the formation of disulfide bonds between adjacent proteins is responsible for freezing injury. During the dehydration accompanying freezing, cells shrink and reactive sulfhydryl groups of different proteins are envisaged as approaching one another and finally interacting, linking the proteins by stable covalent bonds. This would constitute the first stage of injury. In the second stage the proteins would be pulled apart during the cell expansion caused by water uptake during thawing. Since the covalent bonds holding the proteins together cannot yield, the secondary structure of the proteins would break down. The ensuing protein denaturation would result in cell death. Recently, LEVITT and DEAR (1970) postulated in their "membrane-hole hypothesis" the occurrence of breaks in the lipid layer of biological membranes during freezing. These breaks are supposed to lead to contacts between the protein layers of a membrane, which could result in the formation of covalent bonds between them due to the oxidation of SH groups to SS bridges. Such covalent bonds are presumed to render irreversible damage and cause membrane inactivation.

The interesting hypothesis of the irreversible oxidation of protein SH groups to SS bridges has stimulated a great deal of research. It has been supported by model studies using thiogel, an already denatured protein (LEVITT, 1965). At first sight it is also supported by the observation that the SH content of frost-killed cells is decreased. However, in a living cell the redox state of reactive SH groups is influenced, via suitable enzyme systems, by the redox state of pyridine nucleotides. The potential of the latter is strongly negative. If a cell is killed in an aerobic environment, a general shift in the redox status occurs towards positive values and it is to be expected that SH group oxidation will take place. This is a secondary event which is a consequence, not a cause, of cell death.

Other observations are inconsistent with the sulfhydryl theory of frost injury (HEBER and SANTARIUS, 1964; SANTARIUS and HEBER, 1967; MAZUR, 1969) which demands that cells be insensitive to freezing under conditions which prevent the formation of disulfide bonds. Added compounds of low molecular weight containing SH groups ought to protect cells and isolated membrane systems by competing with protein-SH for disulfide formation thus keeping the proteins in the reduced state. However, no such protection is observed. Likewise, anaerobic conditions, by shifting the redox potential to lower values and preventing direct oxidation by oxygen, should also be protective, which is not the case. Another drawback of the SH-theory of injury consists in its inability to explain solute effects which appear to play a dominant role in freezing injury.

g) Modes of Protection

As a matter of convenience, factors involved in protection during freezing may be classified as morphological, physiological and biochemical. It has to be kept in mind, however, that this distinction is rather artificial and that all factors have a molecular, that is to say, a biochemical basis.

A morphological factor, cell size, seems in many cases to be correlated to hardiness, although the number of exceptions is large (LEVITT, 1956). Permeability may be considered a physiological factor. Since extracellular freezing is often tolerated by hardy tissue whereas intracellular freezing is not, an efficient transfer of water across protoplasmic membranes is of obvious importance. Accordingly, an increase in water permeability has occasionally been assumed to be responsible for an increase in hardiness (LEVITT, 1956; MAZUR, 1969). KUIPER (1964, 1967) reported that decenylsuccinic acid and its amides increase water permeability and frost hardiness at the same time. However, refutation of these results has been claimed by other workers (NEWMAN and KRAMER, 1966). Another factor, protoplasmic viscosity, has been observed to increase with hardiness, but again contrary reports have appeared (LEVITT, 1966). An increased capability of the protoplasm to bind water, i.e. increased protoplasmic hydration, has also been considered as a cause of increased hardiness (LEVITT, 1956). There can be little doubt that changes in physical parameters such as viscosity and hydration reflect alterations in the composition of the cell during hardening.

In fact, changes in the frost hardiness of plant cells have been found to be positively and/or negatively correlated with changes in the intracellular concentrations of cell constituents such as sugars, organic acids, amino acids, soluble proteins, SH-containing compounds, lipids, RNA, DNA, adenine and pyridine nucleotides and others (MAZUR, 1969). The significance of these changes must now be considered.

α) Sugars and Sugar Derivatives

A positive correlation between frost hardiness and sugar content (LEVITT, 1956; JEREMIAS, 1969) or sugar-alcohol content (SEYBOLD, 1969) has often been observed. In many plants the levels of sugars and their derivates increase in the fall and decrease in the spring. Feeding sugars to plants may considerably increase their frost resistance (TUMANOV and TRUNOVA, 1963; SAKAI and YOSHIDA, 1968). But there are also data showing that sugar concentration and frost resistance are not correlated. Plants such as the sugar beet or sugar cane which accumulate sugars to levels higher than those in other plants do not become frost hardy. Therefore it has been doubted that sugars play a major role in frost hardiness (LEVITT, 1956). However, the question as to whether or not sugars serve as cryoprotectants cannot satisfactorily be resolved using intact cell material. A lack of correlation between hardiness and a hardiness factor can simply be caused by interference of other factors such as internal compartmentation. Positive correlations, on the other hand, can be accidental, since the conditions which bring about hardening of cells must also be expected to induce simultaneous biochemical changes unrelated to it. Further clarification can therefore come only from relevant work with systems simpler than intact cells.

Figure 2 shows that different sugars do, in fact, protect sensitive membranes such as thylakoids against freezing damage. On a molar basis, trisaccharides are more effective than disaccharides and these in turn are better than monosaccharides. Not only thylakoids, but mitochondrial membranes also are protected by sugars against freezing injury (PORTER et al., 1953; HEBER and SANTARIUS, 1964).

In addition to sugars, sugar alcohols such as sorbitol, mannitol, inositol and glycerol are protective substances in the thylakoid system. POLGE, SMITH, and PARKES (1949) found that glycerol protects red blood cells against freezing. During hardening, many insects capable of surviving the winter accumulate glycerol (SALT, 1961; ASAHINA, 1966). It is well known that spermatozoa can survive freezing when suspended in glycerol solutions. Tissue cultures (SMITH, 1961) and leaf tissue (RICHTER, 1968a, b) can also be protected against freezing damage by glycerol. While many cells are practically impermeable to hexoses and other sugars of comparable or higher molecular weight, and hence cannot be protected by these compounds if they are added to the medium surrounding the cells, glycerol usually

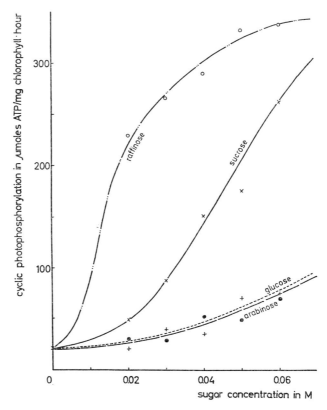

Fig. 2. Protection of unwashed chloroplast membranes by different sugars during freezing. Thylakoids were frozen for 3 h at —25° C in the presence of 0.1 M NaCl and various concentrations of sugars. After thawing, the activity of cyclic photophosphorylation was measured as an indicator of membrane integrity. Protection was complete at a photophosphorylation rate of about 350 μmoles ATP synthesized per mg chlorophyll per hour. Raffinose was more effective as a cryoprotectant than sucrose or glucose and arabinose. The latter sugars protected membranes completely during freezing only at concentrations much higher than those shown in the figure. (From HEBER and SANTARIUS, 1971)

manages to penetrate. Even though the results from work on the protection of intact cells by sugars are conflicting, the *in vitro* data establish that sugars are effective cryoprotective agents. A prerequisite of protection is, of course, that the sugar distribution inside the cell is such as to permit close contact between protective agents and sensitive membranes. Sugars present in the vacuole cannot, for obvious reasons, protect membranes of cytoplasmic organelles. In a number of cases frost-hardy cells contained sufficient sugar to explain their frost resistance as judged by the sugar concentration needed to prevent thylakoid inactivation. HEBER (1957, 1959), using a non-aqueous technique for isolation of chloroplasts, found that chloroplasts from frost-hardy leaves of spinach, wheat and other plants contained sufficient sugar for complete protection of the frost-sensitive structures. But sugars are not always available in sufficient amounts in frost-hardy cells. For instance, KAPPEN and ULLRICH (1970) found sodium chloride up to a concentration of 0.6 M in the chloroplasts of a frost-hardy halophyte. Even though their membranes were sensitive to chloride, the chloroplasts did not contain sufficient sugar to counteract the deleterious action of the salt, which becomes highly concentrated during freezing. Since the plants survived freezing the conclusion seems inevitable that compounds other than sugars must also contribute to protection.

β) Organic Acids

From the few reports available on changes in organic acids during cold acclimatization, e.g. an increase in the concentrations of malic and citric acids in needles of conifers and leaves of *Sempervivum* in winter (OECHSSLER, 1968; KULL, 1968), it is impossible to say whether these compounds are involved in protection or not. The response of thylakoid membranes to organic acids during freezing is complicated (SANTARIUS, 1971). Sodium succinate does not protect washed thylakoids against freeze inactivation; in fact, it has already been listed as a membrane-toxic compound. Other salts of organic acids prevent inactivation of washed thylakoids during freezing only at high concentrations and then often only partially (pyruvate, malate, tartrate, citrate), rarely almost completely (acetate). However, the situation changes dramatically in the presence of NaCl. In the presence of 0.1 M NaCl, very low concentrations of succinate, malate, tartrate or citrate (0.02 M) are sufficient for complete protection of the thylakoids (Fig. 3). It is especially interesting that, within a certain range, combinations of succinate and NaCl, both membrane-toxic compounds, effectively prevent freeze-inactivation of the membranes. Deviation from the optimal ratio on either side results in injury to the membranes (Fig. 4).

These effects are not specific. Sodium chloride can be replaced by other inorganic salts such as KCl, $MgCl_2$ or $NaNO_3$. The optimal ratio of inorganic to organic salt varies with the inorganic salt: it is lower for salts of bivalent cations such as $MgCl_2$ than for salts of monovalent cations. Increasing toxicity of the anionic part also shifts the optimal ratio to lower values.

Not only the inorganic but also the organic salts are replaceable. Substitution may also alter the optimal ratio for protection. The lower the toxicity of the organic salt, the broader is the range of protection. A combination of two organic acids may be even more effective in preserving the integrity of the membrane during freezing than either of the salts alone.

It is remarkable that a combination of inorganic salts usually does not result in protection during freezing. Only in rare cases has a small and not very significant preservation in activity been seen (HEBER, TYANKOVA, and SANTARIUS, 1971).

Salts of organic acids may occur in plant cells in considerable quantities. The results of the *in vitro* experiments demonstrate that their contribution to hardiness should not be neglected and may be quite significant.

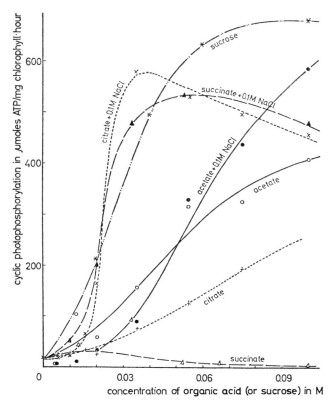

Fig. 3. The effect of salts of organic acids on washed chloroplast membranes during freezing in the presence and absence of NaCl. Thylakoids suspended in solutions containing various concentrations of sodium acetate, sodium succinate or sodium citrate, with or without 0.1 M NaCl, were frozen for 4 h at —25° C. After thawing, the activity of cyclic photophosphorylation was measured as an indicator of membrane integrity. Protection by sucrose is shown for comparison. Complete protection of the membranes was observed at about 680 μmoles ATP synthesized per mg chlorophyll per hour. NaCl, while decreasing the protection afforded by sugars (Fig. 1), increases protection by organic acids. (From HEBER and SANTARIUS, 1971)

γ) Amino acids

An increase in some amino acids, e.g. proline, during hardening has been observed by some authors (LE SAINT, 1966, 1969; LE SAINT-QUERVEL, 1969; BENKO, 1969). TYANKOVA (1969) found that proline and/or sugar added to pepper plants increased frost resistance. A causal relationship to hardiness has been postulated. Again, experiments with isolated membrane systems give evidence on the role of amino acids in hardiness (TYANKOVA, 1970; HEBER, TYANKOVA, and SANTARIUS, 1971).

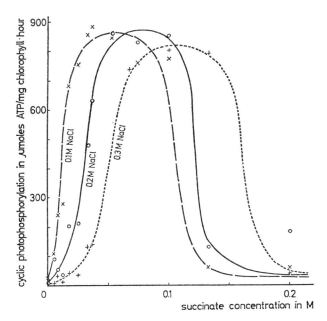

Fig. 4. The effect of freezing on unwashed chloroplast membranes, which were suspended in solutions containing 0.1, 0.2 or 0.3 M NaCl and succinate, as a function of the concentration of Na$_2$-succinate. Freezing time was 3 h at —25° C. After thawing, the
activity of cyclic photophosphorylation was measured as an indicator of membrane
integrity. Protection was complete at about 875 µmoles ATP synthesized per mg
chlorophyll per hour. For complete protection succinate had to be combined with
NaCl in a ratio of about 1/3. Departure from this ratio on either side resulted in progressive inactivation. (From SANTARIUS, 1971)

In the absence of inorganic salts only lysine, proline, threonine and γ-aminobutyric acid are able to prevent the inactivation of thylakoids during freezing. In
the presence of NaCl, effectiveness of the cryoprotective amino acids is decreased
as in the case of sucrose (Fig. 1).

Glycine, serine, α-alanine, hydroxyproline, sodium glutamate and sodium aspartate
are, in keeping with their membrane toxicity (p. 240), unable to prevent the inactivation of washed thylakoids during a 4-h period of freezing to — 25° C. However, like succinate, these compounds become highly protective during freezing in
the presence of an inorganic salt such as sodium chloride. As with succinate,
cryoprotection is expressed only within a certain range of ratios between the concentration of the inorganic salt and that of the amino acid. Departure from the
optimal ratio on either side results in progressive inactivation of the membranes
during freezing.

A third group of amino acids is unable to provide protection either alone or in
combination with other members of the group or with inorganic salts. It is represented, among others, by phenylalanine, leucine, isoleucine, valine, methionine,
cysteine, cystine, tyrosine and the hydrochloride of arginine. All amino acids with
apolar side chains belong to this group.

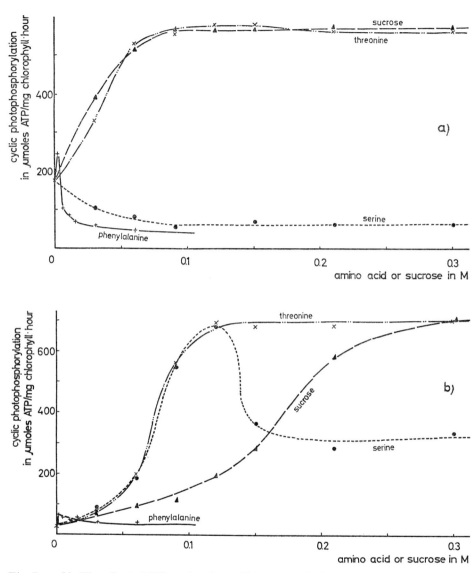

Fig. 5a and b. The effect of different amino acids compared with that of sucrose on washed chloroplast membranes during freezing in the absence (a) or presence (b) of 0.1 M NaCl. Freezing time was 4 h at —25° C. After thawing, the activity of cyclic photophosphorylation was measured as an indicator of membrane integrity. In the absence of NaCl only sucrose and threonine were protective, while serine and phenylalanine contributed to injury. In the presence of NaCl, the effectiveness of sucrose and threonine was decreased, whereas serine became protective over a limited range of concentrations. Only phenylalanine remained ineffective and was even injurious. (From TYANKOVA, 1970 and HEBER, TYANKOVA and SANTARIUS, 1971)

These relations are shown in Fig. 5. Threonine, as a member of the first group of amino acids, is protective in the absence of and, less so, in the presence of NaCl. Serine, from the second group, protects membranes during freezing only if NaCl is present. Phenylalanine, belonging to the third group, fails to provide significant protection under any conditions.

The results show that no general answer can be given to the question: do amino acids contribute, as cryoprotectants, to the frost resistance of cells? Some amino acids, such as proline, add to the protection irrespective of the composition of the cell contents. The response of the cell to other amino acids during freezing is somewhat more difficult to predict. Theoretically, amino acids such as glutamate could even contribute to damage during freezing. However, in view of the ionic composition of the cell contents it is much more likely that protective effects will predominate. Still other amino acids, especially those with apolar side chains, always tend to shift the balance between protective and injurious factors towards damage. Their effects on the cell must be counteracted by cryoprotective substances.

δ) Non-Physiological Solutes

A number of non-physiological compounds are also able to protect membranes, cell organelles, cells and even tissues against freezing injury. A well-known example is dimethyl-sulfoxide (LOVELOCK and BISHOP, 1959; GREIFF and MYERS, 1961). On a molar basis it is less protective than glycerol or sucrose in the thylakoid membrane test (HEBER and KEMPFLE, 1970). However, in contrast to many other cryoprotectants it penetrates rather easily into cells such as erythrocytes and can thus provide protection even where more active, but less permeable, compounds fail to produce significant effects.

Dimethylformamide, ethyleneglycol, methylglycol, propyleneglycol, low-molecular-weight polyethyleneoxides (Lutrol, Carbowax), polyvinylpyrrolidones, dextrans and ficoll and even methanol or ethanol are, to a certain extent, also protective (VOS and KAALEN, 1965; HUGGINS, 1965; NASH, 1966; DOEBBLER, 1966; FARRANT, 1969 and others). This holds true for isolated thylakoids and, in selected cases, for intact cells. Although protection by these compounds obviously is of little interest with regard to frost hardiness acquired under natural conditions, they may be used for the cryopreservation of cells. However, applicability is limited by their toxicity (RICHTER, 1968a).

A group of very different compounds has been reported to influence the resistance of plants to freezing. These include boron, trichlorophenoxypropionic acid, maleic hydrazide, decenylsuccinic acid and urea (SCHNELLE, 1965). There appears, however, to be little agreement as to their effectiveness and their action on plants seems to be indirect rather than direct.

ε) Proteins

An increase in the content of soluble proteins during hardening has been reported on several occasions (SIMINOVITCH and BRIGGS, 1949, 1953; SAKAI, 1958b; JUNG and SMITH, 1961; JUNG, SHIH, and SHELTON, 1967a, b; GERLOFF, STAHMANN, and SMITH, 1967), although contrary reports have also appeared (PARKER, 1962).

Direct evidence of the important role of proteins in frost resistance has become available only recently, when it was shown that protective proteins could be extracted from frost-hardy leaves of spinach and barley and from the bark of the frost-hardy poplar (HEBER and ERNST, 1967; HEBER, 1968, 1970; HEBER and KEMPFLE, 1970). These proteins (molecular weights 10000–20000) protected isolated thylakoids against freezing injury at surprisingly low concentrations. On a unit weight basis, effectiveness exceeded that of sucrose by a factor of 10–100, and on a molar basis by more than 1000. Salts decreased the protection afforded by the proteins. Interestingly, the proteins were not inactivated by heating, but by treatment with trypsin. In the leaves, for instance, they were located in the chloroplasts as shown by the fact that chloroplasts from hardy leaves were resistant to freezing although they had lost, during isolation, protective low-molecular-weight constituents such as sugars. Removal of the proteins by washing rendered chloroplasts from hardy leaves as sensitive to freezing as those from non-resistant tissue. As the protective proteins are highly active in the membrane test and could be found in significant quantities only in hardy material, there is little doubt that they play an important role in hardiness.

Other proteins from leaves were found to be ineffective. However, polypeptide fractions isolated from commercial peptone could also prevent the freeze inactivation of thylakoid membranes (HEBER and KEMPFLE, 1970). Even crude peptone and serum albumin protected thylakoids to some extent during freezing. DAVIES (1970) observed a considerable protection of red blood cells and microorganisms during freezing in the presence of various peptides of comparatively low molecular weight which had been isolated from peptone.

ζ) Other Compounds

There appears to be no group of cell constituents which has not in one way or another been related to hardiness. LI and WEISER (1966) and JUNG, SHI, and SHELTON (1967a) observed changes in DNA, and the same authors, SIMINOVITCH, GFELLER, and RHÉAUME (1967) and CRAKER, GUSTA, and WEISER (1969), changes in RNA. An application of certain purines and pyrimidines to *Medicago* varieties enhanced the development or maintenance of cold hardiness and the content of water-soluble protein and nucleic acids (JUNG, SHIH, and SHELTON, 1967b). KURAISHI et al. (1968) found alterations in NADP which seemed to be correlated to cold resistance. Purine and pyrimidine derivatives such as ATP, ADP etc. have also been correlated with winter hardiness (RAMMELT, 1967; RAMMELT and MÜLLER-STOLL, 1969). It has already been pointed out that parallelism between the concentration of cell constituents and hardiness does not necessarily indicate or even suggest a cause/effect relationship.

GERLOFF, RICHARDSON, and STAHMANN (1966) found that the cell content of fatty acids increased during hardening and the composition changed due to the preferential accumulation of polyunsaturated fatty acids. Lipid metabolism was apparently altered during hardening. SIMINOVITCH et al. (1968) found that lipids increased during hardening by 30–40%, phospholipids by even more than 100%. In some, but not all, cases the lipid content of protoplasmic membranes increased. Phospholipids are membrane constituents. As membranes appear to be the main target of freezing injury, these important observations may be an indication of membrane alterations during hardening.

A negative correlation between hardiness and soluble polysaccharides has often been observed (OLIEN, 1967). The starch content usually decreases during hardening and increases when hardiness is lost. The protective role of soluble sugars formed from starch has already been discussed.

h) Mechanisms of Protection

The available evidence clearly indicates that increased frost hardiness can, in most cases, be traced back to the formation of protective compounds in the cells or their addition to the cells. At present natural frost hardening can be explained by the synthesis and proper distribution of cryoprotective substances inside the cells.

As to the way in which these compounds prevent freezing injury two main effects are of importance: 1. The unspecific, colligative dilution of toxic compounds to non-toxic levels by cryoprotectants, which by definition must themselves be non-toxic at least under the particular freezing conditions which permit protection. Of the low-molecular-weight cell constituents, sugars and some amino acids meet the criteria of ideal cryoprotectants since they are non-toxic over a wide range of concentrations. Some limitations are connected with compounds such as succinate and glutamate or α-alanine which may exhibit toxic effects at higher concentrations (cf. p. 240). 2. Protective compounds can also directly stabilize sensitive cell compounds against the effects of freezing and thereby afford protection (SANTARIUS, 1971).

A clear separation of the two types of protection will rarely be possible since even compounds which directly influence membrane properties will, in addition, act colligatively. Furthermore, the freezing and thawing rates play an important role in the protective action of different compounds (RAPATZ and LUYET, 1968; RAPATZ, SULLIVAN, and LUYET, 1968; MAZUR et al., 1969; LEIBO et al., 1970).

The colligative theory of protection against freezing injury has been developed by LOVELOCK (1953b, 1954b; cf. also LOVELOCK and BISHOP, 1959) to account for the hemolysis of red blood cells. It has been mentioned that LOVELOCK (1953a) considered damage to erythrocytes during freezing to be a consequence of dehydration which permits the salt concentration to rise to toxic levels. The total solute concentration in the unfrozen part of a system which is in equilibrium with ice is solely a function of temperature. Addition of a non-toxic solute, such as glycerol, to the system before freezing therefore reduces the concentration of toxic solutes, such as salts, at any freezing temperature according to the osmolar ratio of neutral to toxic solute. In other words, the added non-toxic solute becomes concentrated during freezing at the expense of the salt. If the concentration of the latter falls below the toxic level, protection can occur. The theory obviously does not require protective compounds to act in a specific manner. They must be sufficiently soluble and permeable if added to cells which contain toxic solutes. For natural cryoprotectants synthesized inside the cells permeability is not a prerequisite, but a proper distribution inside the cell is necessary.

Some of the observations made on the protection of thylakoids during freezing can best be explained by the colligative action of the added cryoprotectants. Protection by a variety of structurally unrelated compounds such as sucrose, acetate, proline, lysine hydrochloride and γ-aminobutyric acid has been observed. They have in common only high solubility in aqueous solutions, low membrane toxicity and colligative properties. During freezing they can be concentrated along with the toxic components of the system and their concentration reduced in the unfrozen part of the system by "dilution", since only the concentration of the sum of all components, not that of any individual compound, is determined by the freezing temperature. Accordingly, protection is a function not of the initial concentration of cryoprotectant, but of the ratio of cryoprotectant to toxic solute. This is shown very clearly in Fig. 1 for protection by sucrose. Similar observations have been made with proline and γ-aminobutyric acid.

Even the surprising observation (Fig. 4) that combinations of two membrane-toxic compounds such as succinate and NaCl provide excellent protection can be explained by colligative action. In both the absence and presence of very low succinate concentrations NaCl reaches and exceeds the threshold of toxicity during freezing and inactivates the membranes. If more succinate relative to NaCl is present, freezing to the same temperature no longer results in the same accumulation of NaCl. Instead the concentration of succinate increases in the unfrozen portion of the system which is in equilibrium with ice, while that of NaCl decreases. At a certain ratio of succinate to NaCl toxic levels of NaCl are no longer reached and protection can be observed if the accumulation of succinate has not yet led to toxic effects. When the ratio of succinate to NaCl is raised further the threshold of succinate toxicity is exceeded during freezing and the membrane again suffers damage. Thus the inactivation of the membrane at low ratios of succinate/NaCl is caused by the toxic effects of the NaCl, at high ratios by the toxic effects of the succinate. At intermediate ratios balancing of the two compounds results in protection.

Similar relations exist for combinations of succinate and other inorganic salts, for combinations of inorganic salts and other membrane-toxic organic compounds such as α-alanine, serine, glycine, hydroxyproline, aspartate or glutamate, and for combinations of different organic compounds of limited toxicity, for instance serine and α-alanine.

It is important to note that combinations of different compounds of comparable membrane toxicity cannot result in significant protection during freezing if these compounds act on the membrane in the same manner. Obviously, partial replacement of a compound present at toxic levels by another compound of comparable toxicity cannot relieve stress on the membrane if the two compounds interact within the same sites in the same manner: their individual concentrations will merely add to one another to produce stress. The situation is very different if interaction takes place at non-identical sites or by different mechanisms. In these cases the combination of one toxic compound with another will be effective since the individual toxicities are not additive. Combinations of different inorganic salts and also of a number of toxic organic compounds which presumably inactivate membranes during freezing by the same mechanism have been found to be ineffective. In contrast to the above-mentioned examples they cannot prevent the freeze-inactivation of thylakoids but rather contribute to it.

Excellent protection of membranes by a combination of two compounds of comparable membrane toxicity, each of which alone causes inactivation during freezing, therefore indicates that different mechanisms of interaction with the membranes underlie inactivation during freezing. Alternatively, when the eutectic temperatures of the added solutes are higher than the freezing temperature, membrane inactivation during freezing may result from eutectic solidification of the whole system rather than from toxic action, or from both. In these cases, the protective effects or a combination of different compounds can also be caused by a shift in the eutectic temperature towards lower values, leaving part of the system unfrozen.

The colligative action of cryoprotective substances added to cells or produced during hardening not only decreases the concentration of toxic cell components during freezing, but also reduces other stress, for instance excessive volume reduction or pH changes (VAN DEN BERG and SOLIMAN, 1969b).

A detailed investigation of the kinetics of membrane inactivation by salts has revealed several different effects of temperature (SANTARIUS and HEBER, 1970). Lowering the temperature increases the concentration of toxic solutes in the surroundings of the membranes and thereby injury, since the crystallization of water is inversely related to temperature. On the other hand, membrane inactivation itself is temperature dependent with a definite activation energy: its rate is decreased by lowering the temperature. The two effects of temperature interplay (SANTARIUS and HEBER, 1970). At very low temperatures eutectic solidification of the system may cause mechanical damage.

Not all observations can be explained by the concepts of unspecific and colligative protection. Different protective compounds prevent inactivation of membranes under identical conditions at very different concentrations. This holds true for various sugars (Fig. 2), for sugar alcohols, dimethylsulfoxide, and dimethylformamide, for salts of organic acids (Fig. 3) and for amino acids (Fig. 5). A very dramatic example is the unusually high effectiveness of the protective proteins. These on a molar basis protect membranes better by a factor far higher than 1000 than low-molecular-weight cryoprotectants. It is difficult to explain protection in this case by colligative action although it is known that many compounds depart from ideal thermodynamic behavior at high concentrations (cf. SANTARIUS, 1969). FARRANT (1969) and FARRANT and WOOLGAR (1970b) have warned that polyvinylpyrrolidone could act colligatively, and FARRANT and WOOLGAR (1970a) have shown that a 30% solution of polyvinylpyrrolidone exerts a much greater freezing-point depression than would be expected from its molarity. MAZUR (1970) assumed that the anomalously high freezing-point depression exerted by polyvinylpyrrolidone results from the binding of water to the macromolecules.

A number of observations, however, permit the conclusion that specific interactions between cryoprotectants and membranes also contribute to protection and may in some cases even be primarily responsible for it. Interaction between membranes and certain solutes has been shown to decrease membrane stability and lead to inactivation. This property of solutes has been defined as toxicity (p. 238). It is not difficult to imagine that the opposite process can also take place and result in membrane stabilization.

Figure 6 shows that membrane inactivation by salt can be retarded by sucrose or succinate even at 0° C. Under these conditions a colligative reduction in the concentration of the toxic NaCl, which takes place during freezing, is not possible. Nevertheless protective effects of sucrose and succinate are observed and the conclusion is inevitable that these compounds stabilize membranes directly, in addition to their exerting a colligative action which becomes possible only during freezing (SANTARIUS, 1971).

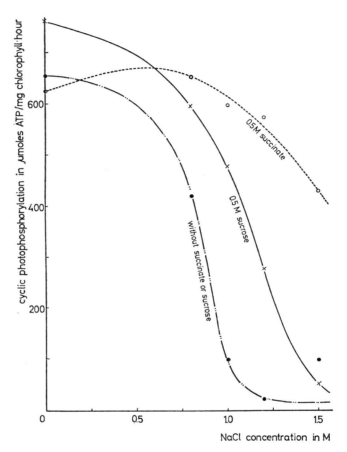

Fig. 6. The protective effect of Na_2-succinate or sucrose on unwashed chloroplast membranes which were suspended for 3 h at 0° C in NaCl solutions of different concentrations. The activity of cyclic photophosphorylation was measured after incubation under identical conditions in a reaction medium containing low levels of salt. Proper precautions revealed that the effect of sucrose and succinate was due to improved membrane stability, not to a stimulation of photophosphorylation during the assay. (From SANTARIUS, 1971)

In the absence of direct evidence one can only speculate on the nature of the direct interactions which lead to membrane stabilization. Hydrogen bonding has been assumed to be involved in protection (ULLRICH, 1962; HEBER and SANTARIUS, 1964; WEBB, 1965; HUGGINS, 1965; SANTARIUS, 1968 and others). More specifically, it has been suggested that hydroxyl groups of sugars or related compounds replace hydration water which is necessary for membrane stabilization and is removed during dehydration. Hydrogen bonding between sugars and proteins has, in fact, been observed (GILES and McKAY, 1962; cf. also STEPONKUS, 1969). Changes in membrane permeability have also been assumed to be responsible for increased stability (WILLIAMS and MERYMAN, 1970; MERYMAN, 1970, 1971). At any rate, the available information not only permits, but actually makes necessary an extension of LOVELOCK's (1953a, b, 1957) original concepts on freezing injury and freezing protection. Electrolytes are not indiscriminately toxic, but their toxicity varies over a wide range. Some inorganic cations and anions are highly toxic, others much less so. Non-toxic organic cations and anions may even function as effective cryoprotectants in cells. On the other hand, a number of organic compounds possessing apolar side chains or apolar structure have a high potential toxicity and can be assumed to play a significant role in freezing injury. Protection is not based solely on the colligative action of non-toxic solutes as proposed by LOVELOCK. Rather, direct interaction of cryoprotectants and membranes plays, in addition to colligative action, an important part in protection as evidenced by the existence in hardy plant cells of protective proteins which, in view of their high molecular weight, cannot exert much colligative influence.

According to WILLIAMS and MERYMAN (1970) and MERYMAN (1970, 1971) freezing injury develops when, during external freezing of cells or isolated membrane vesicles, dehydration causes shrinkage beyond a minimum tolerable volume. In their opinion impermeable compounds which are located outside membrane-surrounded spaces such as cells or cell organelles cannot be protective and may even contribute to damage since they dehydrate the cell or its organelles by osmosis. In contrast, all permeable solutes, whether salts or neutral compounds, are protective since they enter the membrane-surrounded spaces and resist osmotic dehydration, thereby reducing shrinkage. The protective action of sucrose (which is considered to be non-penetrating) on thylakoids has been explained on the grounds that the high concentrations of sucrose reached during freezing cause the membranes to become reversibly leaky. The leaky membranes can then be protected by the entering solutes. Evidence for sucrose uptake by the membrane vesicles during freezing has in fact been produced. It is postulated that other protective compounds which are normally non-penetrating cause similar changes in membrane permeability at the high concentrations reached during freezing thus permitting the membranes to leak "physiologically". This leakage is supposed to prevent damage.

This attractive theory is another modification of LOVELOCK's original concept. It does not try to explain why membranes are damaged if enclosing too small a space. The term "membrane-toxicity" as defined above (p. 240) is obviously related in this theory to the lack of ability to penetrate membranes or render them permeable under freezing conditions.

According to LEVITT (1962, 1969) and LEVITT and DEAR (1970) the main aspect of protection is the prevention of the oxidation of protein-SH to disulfide bridges during freezing. Low-molecular-weight cryoprotectants are supposed to function as "spacers" by preventing the contact of neighboring SH-groups of different proteins. In addition, increased resistance is also explained by temperature-dependent conformational changes of sensitive proteins, by degradation of sensitive proteins and by resynthesis of insensitive protein material. Unfortunately experimental support for these interesting speculations is largely missing. Washed thylakoids from hardy and from non-resistant spinach are not different as far as sensitivity to freezing is concerned. Their response to cryoprotectants and to membrane-toxic compounds is very similar. At the moment there appears to be little reason to assume that changes in protein conformation form a basis of hardiness.

WEISER (1970) suggested that freezing affects tender cells such as cereal coleoptiles, spinach and cabbage leaves, yeast, algae or red blood cells and woody cells in different ways. Experiments with woody plants have revealed the existence of a translocatable principle which has been called the "hardiness promoting factor". Its nature is still obscure, and although it has speculatively been termed a hormone, it is more likely to be a sugar or a similar "classical" cryoprotectant.

i) Development of Resistance

Hardening is defined as the cellular process or the sequence of cellular processes which transforms sensitive cells into hardy cells. Dehardening is the reversal of hardening. In the case of frost resistance hardening is induced by lowered temperatures and under conditions of short photoperiod (cf. p. 210). Light is necessary for synthesizing transportable compounds which act as protective agents (STEPONKUS and LANPHEAR, 1967).

In the natural environment hardening takes place in the fall when growth ceases and plants enter the dormant or semi-dormant stage (cf. p. 145). Growth-promoting compounds such as gibberellic acid or kinetins decrease, growth inhibitors increase or induce frost hardiness (IRVING and LANPHEAR, 1968). Starch is usually hydrolyzed and sugars accumulate during hardening (PARKER, 1963; OLIEN, 1967; HURST, HALL, and WEISER, 1967; MAZUR, 1969). Other compounds are also formed during hardening (cf. p. 245).

Hardening often proceeds in distinguishable stages (TUMANOV, 1967; cf. also p. 212). In the first stage, short days and low temperatures may function as the signals for those metabolic changes which result in a first large increase in resistance. The requirement for light and a supply of metabolites indicate that in this stage cryoprotectants such as sugars are formed.

The second stage of cold acclimatization seems to proceed at subzero temperatures and leads to a high freezing tolerance. The nature of the processes taking place in this phase is even less clear than that of the earlier events. It has been mentioned that cells are complicated systems subdivided into individual compartments bordered by semipermeable membranes. A number of natural cryoprotectants appear to be non-penetrating. Under these circumstances the mere production of cryoprotectants somewhere inside the cells is obviously insufficient for good protection. A proper redistribution within the cells is also an important aspect of hardening. Unfortunately there is still a large gap between the physiological observations and the investigations orientated towards an understanding of the molecular background of protection. Attempts to bridge this gap must, in the absence of new information, necessarily end in speculation.

During dehardening in the spring the processes that led to hardiness in the fall are reversed. Dehardening is induced by increased temperatures (cf. p. 214). The first signs of dehardening are often observed in February, when there is still danger of severe frosts. Loss of hardiness may occur while the total content of some cryoprotectants is still high. However, measurements of total content do not give information on the concentrations in the surroundings of sensitive membranes which alone are significant.

The detailed analysis of the response of unicellular organisms (MAZUR, 1969) and of isolated membrane systems to freezing has considerably advanced our knowledge of the mechanisms of freezing injury and freezing protection. However, we are still largely ignorant of the way a complex interplay between temperature, day length and other factors leads to the formation and disappearance of the protective materials which determine the resistance in higher plants.

2. High Temperatures

a) Introduction

Freezing may affect cells in different ways as has been discussed. When the temperature is reduced dehydration takes place, solutes are concentrated and mechanical stress may be exerted. Death from freezing may accordingly stem from different causes, the most important probably being the interaction of membrane-toxic compounds with sensitive membranes. The situation is much simpler if cells are killed by heat. During brief heating mechanical, dehydration and concentration effects can be ruled out leaving only the temperature increase. Thus the question may be limited to a discussion of which cell constituents are sensitive and which are insensitive to elevated temperatures.

b) Heat Sensitivity of Cell Components

Within a temperature range of practical interest, low-molecular-weight cell constituents are not sufficiently heat labile to warrant extensive consideration. This statement should contain a trace of reservation since cells are complicated multiphase systems and temperature changes may induce phase transitions which also involve low-molecular-weight material. It has been assumed that heat injury is related to the melting point of cellular lipids (CHAPMAN, 1967). In fact, lipids of organisms which are adapted to high environmental temperatures have higher melting points than those of organisms occurring in temperate climates. Lipids are membrane components and the question of their involvement in heat injury will be discussed further.

Among the high-molecular-weight cell constituents polysaccharides are rather thermostable, as are nucleic acids. Although the latter "melt" at higher temperatures owing to the breaking of the hydrogen bonds which hold base pairs together it does not appear that this process is responsible for heat injury (SZYBALSKI, 1967). Melting does not affect the base sequence and is a reversible process. There is a direct correlation between base composition and the melting point of DNA, but no similar correlation is known between base composition and thermophily (MARMUR, 1960; CAMPBELL and PACE, 1968).

Furthermore, thermolability of a compound or a group of compounds alone is insufficient reason for regarding it as a target of injury. This statement can be generalized and also applies to sensitivity to freezing. A cell will be damaged, and finally killed, if its least resistant component with a vital function succumbs to the stress. Only the first life-limiting injury is of interest as far as the question of survival or death is concerned. Any further damage to individual cell components occurring during increased stress caused by further raising the temperature constitutes "overkill" and is meaningless for a cell or an organism which is already mortally injured. Unfortunately, no final answer to the question of which is the most heat-labile cell component of vital function is available. The answer may even differ from species to species and under different physiological conditions.

It has long been known that proteins are denatured by heat. Different proteins exhibit different sensitivities to high temperatures. The heat denaturation of a highly sensitive protein may account for cell death at high temperatures (ALEX-ANDROV, LYUTOVA, and FELDMAN, 1959; HENCKEL, 1964). In fact, the Q_{10} for protein denaturation is unusually high, indicating a very high activation energy of denaturation. The Q_{10} of heat injury as measured by the cell survival rate is also very high. It has often been found that the resistance to heat denaturation of the enzymes which perform the same function in different organisms varies according to the thermophilic adaptation of these organisms (ALEXANDROV, 1969). Thus, a crystalline α-amylase isolated from the thermophilic *Bacillus stearothermophilus* was shown to retain 90% of its activity after 1 hour's exposure to 90° C (MANNING and CAMPBELL, 1961), while amylase from mesophilic organisms is rapidly inactivated at 65° C. The significance of these findings is not completely understood because the thermostability of individual enzymes is usually considerably higher than that of the intact cells containing these enzymes. This indicates that they cannot be directly involved in thermal death. However, it has been argued that even in these cases a somewhat indirect relation to heat injury does exist. Proteins in the cells are in a delicate state of balance between synthesis and enzymatic degradation. It has been assumed that thermal agitation, even where it does not lead to protein denaturation, results in a "weakening" of the flexible protein structure or, in other words, in transient unfoldings of the polypeptide chains which make the proteins more susceptible to stress situations such as enzymatic attack. Under these conditions the pressure of natural selection is supposed to have led to the evolution of "rigid" proteins with a thermostability which, even though greater than that of the intact organism, is still in a proper balance with it (ALEX-ANDROV, 1969).

The thermal stability of different proteins varies considerably. Many enzymes are inactivated by brief heating to 50 or 60° C, while some others, such as adenylate kinase, are resistant to prolonged heating at 100° C. There is no reason to assume that the enzymes whose temperature sensitivities are known are, in fact, the most heat sensitive proteins of the cell. It should be remembered that stress causes injury at the point of least resistance. In fact, many cells are killed by temperatures a little below 50° C, where sensitive enzymes become denatured.

c) Causes of Heat Injury

From these considerations it appears that heat injury to cells is caused by (directly or sometimes in a more circumstantial manner: ALTERGOTT, 1967; ALEXANDROV,

LOMAGIN, and FELDMAN, 1970) heat inactivation of proteins. There is no doubt that proteins perform vital functions in cell life. Enzyme destruction much in excess of synthesis must necessarily disturb cell metabolism. If repair mechanisms cannot cope with injury the cells are doomed. Experimental support for the view that a disturbance of the balance between the destruction and the synthesis of protein is responsible for heat injury has also been produced by ALLEN (1950).

However, there are instances where it is difficult to explain cell death as a consequence of the heat denaturation of proteins. It has been observed that psychrophilic bacteria grown at their temperature optimum of 5° C lysed when exposed to 19° C. According to present knowledge a thermal denaturation of proteins is unlikely at this temperature. In fact, it was observed that cytoplasmic membranes broke down after the onset of lysis. Death was assumed to be caused by the temperature-induced stimulation of lytic enzymes acting on vital cell structures (HAGEN, KUSHNER, and GIBBONS, 1964). A number of other psychrophilic or mesophylic organisms are also killed by temperatures too low for the denaturation of known proteins. As early as in the 19th century it was inferred from such evidence that protein coagulation, a secondary phenomenon often, but not necessarily, accompanying protein denaturation, cannot be involved in heat injury (SACHS, 1864). This certainly overestimates the weight of the evidence. But an explanation of injury still has to be sought in those instances where death is not likely to arise from protein denaturation.

In the section devoted to freezing resistance the question of membrane structure and sensitivity has been discussed briefly (cf. p. 257). Membrane stabilization occurs via secondary forces, as evidenced by the possible solubilization of membranes in certain solvents. Hydrogen bonding, ionic and dipole interaction and van der Waals forces hold membrane components in place. In comparison, the structure of individual proteins is stabilized by similar forces as well as by covalent links such as peptide bonds and disulfide bridges. It is to be expected that the forces stabilizing a special configuration of the polypeptide backbone of individual protein molecules are at least as strong as, if not stronger than, those responsible for the mutual attachment of membrane components. From these considerations it appears possible that destruction of membrane integrity, not denaturation of individual proteins, may often be the first step in heat injury.

The lipid-liberation theory envisages membrane alterations by heat-induced phase transitions of lipids as the cause of heat injury. It is a common observation that, with adaptation to increasing temperatures, unsaturated fatty acids within the lipid part of cell membranes are progressively replaced by saturated fatty acids which have higher melting points (BĚHLERÁDEK, 1931). This phenomenon has been thought to be causally linked to temperature acclimatization. MOLOTKOVSKY and ZHESTKOVA (1966, 1967) and MOLOTKOVSKY (1968) observed that heat treatment induced a swelling of isolated chloroplast membranes which could be inhibited by ATP, Mg^{++}, albumin or sucrose. The cause of swelling appeared to be the enzymatic liberation of fatty acids which resulted in a decrease of photochemical chloroplast activities. Fatty acids are known to have destructive effects on chloroplast membranes (McCARTY and JAGENDORF, 1965). Albumin was protective because of its ability to bind fatty acids.

Other experimental observations indicate that isolated membrane systems are, in fact, damaged by heating to temperatures which should not be high enough to result in marked protein denaturation. Isolated chloroplasts from *Euglena*, spinach and peas were, after brief exposure to 45 or 50° C, unable to evolve oxygen in the light in the presence of electron acceptors (KATOH and SAN PIETRO, 1967; CHENIAE and MARTIN, 1967; YAMASHITA and BUTLER, 1968). Only that part of the electron-transport chain close to the oxygen-evolving step was affected. This was shown by the failure of partial reactions such as ascorbate-, p-phenylendiamine-, hydroquinone- and dichlorophenolindophenol-dependent photoreduction of NADP or hydrogen production (STUART, 1971) to respond to heat treatment. The heat damaged oxygen evolution by the chloroplast membranes as well as photo-phosphorylation, which requires membranes with "normal" permeability properties (DE KIEWIET, HALL, and JENNER, 1965; EMMETT and WALKER, 1969; STUART, 1971). Thermal uncoupling of chloroplast membranes was assumed to proceed below 45° C also and to contribute to the marked decline of photosynthesis *in vivo* which occurs at temperatures somewhat lower than those causing cell death (LANGRIDGE and McWILLIAM, 1967).

Isolated mitochondria are also remarkably heat labile (SEMICHATOVA, BUSHUYEVA and NIKULINA, 1967). While thermolability is particularly pronounced in reconstituted electron-transfer systems, which lose activity rapidly and irreversibly at 38° C, it is also seen in non-fragmented respiratory particles (LUZIKOV, SAKS, and BEREZIN, 1970). In the latter case, however, reactivation of electron transport by NADH has been observed and it has been concluded that heat inactivation of NADH-oxidase activity is caused by heat-induced conformational changes which can be reversed by substrate.

d) Heat Resistance

When plants or plant organs such as leaves are exposed to superoptimal, but sublethal, temperatures their heat resistance increases (ALEXANDROV, 1956; FELDMAN et al., 1967; ALEXANDROV, LOMAGIN, and FELDMAN, 1970). Heat hardening has also been found to take place during low-temperature acclimatization (LANGE, 1961; KISLYUK, 1962). During heat hardening not only the resistance of the cell as a whole, but also that of photosynthesis and respiration are increased (LYUTOVA, 1962). The higher the temperature of exposure, the shorter the time required to produce a distinct increase in heat resistance. Interestingly, exposure times of 1–10 sec have been reported to be effective (ZAVADSKAYA, 1963). Heat resistance acquired during heat hardening is lost on reducing the temperature.

There is reason to assume that different factors are responsible for heat hardening. It is likely that the rise in heat resistance during low-temperature acclimatization occurs passively and is caused by the same biochemical changes which increase frost hardiness. The membrane-stabilizing action of a number of cryoprotectants has already been discussed. These compounds may also stabilize against heat stress. In fact, it has long been known that different low-molecular-weight compounds, among them sugars (BERGER et al., 1946), protect proteins against heat denaturation (BOYER, BALLOU, and LUCK, 1947; BURTON, 1951; PUTNAM, 1953). Little is known about the molecular mechanism of this effect. FELDMAN (1962) and

MOLOTKOVSKY and ZHESTKOVA (1964) have shown that sugars increase the heat resistance of leaves. The same sugars are good cryoprotectants and are known to accumulate in a number of plant species during frost hardening. These relations may explain, at least in part, the parallelism between frost, drought and heat resistance in the winter (LANGE, 1961; HENCKEL, 1964). In other cases complex formation between proteins and low-molecular-weight cell constituents plays a role in heat resistance. Dipicolinic acid has been found to combine with enzymes in some bacterial spores thus stabilizing them against heat stress (CHURCH and HALVORSON, 1959).

Synthesis and/or proper distribution within the cell of sufficient amounts of protective substances would be expected to take time. Frost hardiness, a process largely based on the formation of protective material, is acquired only slowly, over the course of weeks. On the other hand, the response of cells to heat-hardening conditions may be very rapid. Significant hardening may proceed within seconds. In this type of hardening, the time factor makes an important contribution of newly-formed protective substances rather unlikely. It also seems to rule out a de novo synthesis of proteins possessing a higher thermal stability. KINBACHER and KNULL (1967) and FELDMAN (1968) have compared the heat-inactivation of different enzymes extracted from hardened and non-hardened leaves. The temperature at which 50% of the enzyme activity was lost was a few degrees centigrade higher for enzymes from hardened as compared with non-hardened material. It is interesting that these differences persisted during purification of the enzymes. This rules out the participation of dialyzable small molecules in this type of protection. It is tempting to speculate on temperature-induced transitions in the conformation of protein molecules to a more stable state as a cause of resistance, but the basis for such speculation is narrow because the template for the primary structure of proteins, which in turn determines the secondary and tertiary structure, is fixed by the genotypes.

LEVITT (1969) has developed a detailed molecular theory of hardiness embracing cold, drought and heat resistance. In his view, injury is initiated by a transition from the native to a denatured state of the proteins. This step is thought to be reversible. A second, irreversible step leads from the denatured to the aggregated state completing injury. During heat hardening a breakdown of sensitive proteins followed by resynthesis or an opening of polypeptide chains is envisaged. It is postulated that the resynthesized or unfolded proteins whose primary structure is unaltered refold under the "hardening" influence of high temperature into a more stable secondary structure conferring resistance on the protein and thereby on the cell. Also the transition from the reversibly denatured to the irreversibly damaged, aggregated form of the protein is supposed to be prevented by appropriate mechanisms. Despite possessing the merit of provoking critical thought and experimentation, a major disadvantage of this theory is its lack of convincing experimental support.

Whereas it is almost generally agreed that heat injury is caused by thermal inactivation of proteins or protein- and lipid-containing membranes, a sound molecular explanation of protection against heat stress is at present unavailable. This is in contrast to the situation for frost hardiness, where at least some important aspects of protection are now well understood.

References to Section "Plants"

ABBOTT, D.L.: In: Rep. 14th Internat. Hort. Congr., The Hague-Scheveningen, 743—753 (1956).

ABOLINA, G.: Dokl. Akad. Nauk SSSR, N. S., **18**, 199—202 (1938).

ABOLINA, G.: Dokl. Akad. Nauk SSSR, N. S., **79**, 161—164 (1951).

ADACHI, K., INOUYE, J., ITO, K.: J. Fac. Agric. Kyushu Univ. **16**, 77—84 (1970).

ADAMSON, R.M.: Can. J. Plant Sci. **41**, 394—400 (1961).

AGENA, M.U.: Untersuchungen über Kältewirkungen auf lagernde Getreidefrüchte mit verschiedenem Wassergehalt. Diss. Bonn 1961.

ÅKERMAN, Å.: Studien über den Kältetod und die Kälteresistenz der Pflanzen. Belringska Bokr., Lund 1927.

ALDEN, J., HERMANN, R.K.: Bot. Rev. **37**, 37—142 (1971).

ALEXANDROV, V.YA.: Tr. Botan. Inst. Akad. Nauk SSSR, Ser. IV, 309—355 (1955).

ALEXANDROV, V.YA.: Botan. Zh. **41**, 939—961 (1956).

ALEXANDROV, V.YA.: Quart. Rev. Biol. **39**, 35—77 (1964).

ALEXANDROV, V.YA.: In: TROSHIN, A.S. (Ed.): The cell and environmental temperature, pp. 142—151, Oxford, NY., Toronto: Pergamon Press, 1967.

ALEXANDROV, V.YA.: Currents Mod. Biol. **3**, 9—19 (1969).

ALEXANDROV, V.YA., DENKO, E.I., KISLYUK, I.M., FELDMAN, N.L., SHUKHTINA, G.G.: In: Cytological aspects of adaptations of plants to environmental factors, pp. 103—127. Moscow, Leningrad: Nauka, 1964.

ALEXANDROV, V.YA., FELDMAN, N.L.: Botan. Zh. **43**, 194—213 (1958).

ALEXANDROV, V.YA., LOMAGIN, A.G., FELDMAN, N.L.: Protoplasma **69**, 417—458 (1970).

ALEXANDROV, V.YA., LYUTOVA, M.I., FELDMAN, N.L.: Tsitologiya **1**, 672—691 (1959).

ALEXANDROV, V.YA., YASKULIEV, A.: Tsitologiya **3**, 702—707 (1961).

ALLEN, M.B.: J. Gen. Physiol. **33**, 205—214 (1950).

ALTERGOTT, V.F.: The cell and environmental temperature, TROSHIN, A.S. (Ed.), pp. 275—282. Proc. Intern. Symp. Cytoecology, Leningrad 1963. Oxford: Pergamon Press 1967.

ALTMAN, PH.L., DITTMER, D.S. (Ed.): Environmental biology. Fed. Am. Soc. Exp. Biol. Bethesda 1966.

ÅLVIK, G.: Medd. Vestlandets Forst. Forsoksstat. **6**, 1 (1939).

AMEN, R.D.: Am. Sci. **51**, 408—424 (1963).

AMEN, R.D.: Botan. Rev. **34**, 1—31 (1968).

AMIN, J.V.: Physiol. Plant. **22**, 1184—1191 (1969).

ANDERSON, R.G., McGINNIS, R.C.: Can. J. Genet. Cytol. **2**, 331—335 (1960).

ANDREWS, J.E.: Can. J. Botany **38**, 353—363 (1960).

ANDREWS, J.E.: Can. J. Plant Sci. **40**, 94—103 (1960).

ARNOLD, E.A., YOUNG, D.E., STOWELL, R.E.: Exp. Cell Res. **52**, 1—12 (1968).

ARREGUIN-LOZANO, B., BONNER, J.: Plant Physiol. **24**, 720—738 (1949).

ASAHINA, E.: Contrib. Inst. Low Temp. Sci. Hokkaido Univ. **10**, 83—126 (1956).

ASAHINA, E.: In: MERYMAN, H.T. (Ed.). Cryobiology, pp. 451—486. London-New York: Academic Press, 1966.

ASAHINA, E.: Advan. Insect. Physiol. **6**, 1—49 (1969).

ASAHINA, E., HISADA, Y., EMURA, M.: Contrib. Inst. Low. Temp. Sci., Ser. B, Hokkaido Univ. **15**, 36—50 (1968).

ASAHINA, E., SHIMADA, K., HISADA, Y.: Exp. Cell Res. **59**, 349—358 (1970).

ASEN, S., STUART, N. W.: Proc. Am. Soc. Horticult. Sci. **71**, 563—567 (1958).
ASHTON, D. H.: Australian J. Botany **6**, 154—176 (1958).
ASKENASY, E.: Bot. Ztg. **35**, 793—815, 817—848 (1877).
ATANASIU, L.: Rev. Roumaine Biol., Ser. Botan. **9**, 341—359 (1964).
ATANASIU, L.: Ann. Univ. Bucureşti, Ser. Biol. **14**, 93—108 (1965).
ATANASIU, L.: Rev. Roumaine Biol., Ser. Botan. **14**, 165—169 (1969a).
ATANASIU, L.: Rev. Bryol. et Lichenol., Paris, **36**, 747—753 (1969b).
ATANASIU, L.: Bryologist **74**, 23—27 (1971a).
ATANASIU, L.: Rev. Roumaine Biol., Ser. Bot. **16**, 105—110 (1971b).
ATTERBERG, A.: K. Landbruksakad. Tidskr. **38**, 227—250 (1899a).
ATTERBERG, A.: Tidskr. f. Landtmän **1899**, 477—482 (1899b).
ATTERBERG, A.: Landw. Versuchsst. **67**, 127—143 (1907).
ATWOOD, W. M.: Botan. Gaz. **57**, 386—414 (1914).
AUDUS, L. J.: Plant growth substances. London: L. Hill 1953.
AUGSTEN, H.: Fermentaktivität und Atmung bei Sommergerste nach Kältebehandlung des Saatgutes. Dtsch. Akad. Landwirtschaftswiss. Berlin, Wiss. Abh. **20**, 1—87 (1956).
AUGSTEN, H.: Ber. Deut. Botan. Ges. **70**, 233—244 (1957).
AUGSTEN, H.: Biol. Rdsch. **1**, 241—258 (1964).
BAAR, H.: Sitzber. Akad. Wiss. Wien, Math-Naturw. Kl. Abt. I **121**, 667—705 (1912).
BABALOLA, O., BOERSMA, L., YOUNGBERG, C. T.: Plant Physiol. **43**, 515—521 (1968).
BACHMANN, L., CURTH, P., RÖSTEL, H.-J.: Züchter **33**, 50—57 (1963).
BALDWIN, H. I.: Forest tree seed. Waltham, Mass.: Chronica Botanica Co. 1942.
BALLANTYNE, D. J., LINK, C. B.: Proc. Am. Soc. Horticult. Sci. **78**, 521—531 (1961).
BANGE, G. G. J.: Acta Botan. Neerl. **2**, 255—297 (1953).
BANNISTER, P.: J. Ecol. **52**, 481—497 (1964).
BARABALCHUK, K. A.: Tsitologiya **11**, 1021—1032 (1969).
BARENDSE, G. W. M.: Proc. Koninkl. Ned. Akad. Wetenschap., Ser. C, **66**, 183—188 (1963).
BARENDSE, G. W. M.: Proc. Koninkl. Ned. Akad. Wetenschap. Ser. C, **68**, 81—85 (1965).
BARTON, L. V.: Contrib. Boyce Thompson Inst. **7**, 405—409 (1935).
BARTON, L. V.: Proc. Int. Seed Test Ass. **26**, 561—596 (1961).
BARTON, L. V.: Seed dormancy: In: RUHLAND, W. (Hrsg.): Handbuch der Pflanzenphysiologie, Bd. 15/II, S. 699—720, S. 727—745. Berlin-Heidelberg-New York: Springer 1965.
BARTON, L. V., CROCKER, W.: Twenty years of seed research at Boyce Thompson Institute for Plant Research. London: Faber & Faber Ltd. 1948.
BASS, L. N.: Iowa State Coll. J. Sci. **28**, 503—519 (1954).
BASSARSKAYA, M. A.: Tr. Prikl. Bot., Genet. Selekts., Ser. A, No. 11, 55—56 (1934a).
BASSARSKAYA, M. A.: Semenovodstvo **3**, 15—20 (1934b).
BASSARSKAYA, M. A.: Yarovizatsiya **6**, 101—108 (1936).
BAUER, H.: Hitzeresistenz und CO_2-Gaswechsel nach Hitzestress von *Abies alba* und *Acer pseudoplatanus*. Diss. Innsbruck 1970.
BAUER, H.: Photosynthetica **6**, 424—434 (1972).
BAUER, H., HARRASSER, J., BENDETTA, G., LARCHER, W.: Ber. Deut. Botan. Ges. **84**, 561—570 (1972).
BAUER, H., HUTER, M., LARCHER, W.: Ber. Deut. Botan. Ges. **82**, 65—70 (1969).
BAUER, H., LARCHER, W., WALKER, R. B.: In: COOPER, J. P. (Ed.): Photosynthesis and Productivity in Different Environments. Cambridge Univ. Press 1973.
BAUMEISTER, W., BURGHARDT: In: LINSER, H. (Hrsg.): Handbuch der Pflanzenernährung und Düngung, Bd. 1, S. 141—203. Berlin-Heidelberg-New York: Springer 1969.
BEAN, E. W.: Ann. Botany, N. S., **34**, 57—66 (1970).
BEATTY, D. W., GARDNER, F. P.: Crop Sci. **1**, 323—326 (1961).
BECQUEREL, P.: 8ème Cgr. Int. Botan. Paris, **11**, 269—270 (1954).
BEHRENS, J.: Ber. Grossh.-Bad. Versst. Augustenburg **1906**, 60—64 (1906).
BĚLEHRÁDEK, J.: Protoplasma **12**, 406—434 (1931).

BĚLEHRÁDEK, J.: Ann. Rev. Physiol. **19**, 59—82 (1957).
BELLINI, P.: Ann. Sper. Agr. **13**, 143—156 (1959).
BENKO, B.: Biol. Plant. (Praha) **11**, 334—337 (1969).
BENNETT, J. P.: Calif. Agr. **3** (11) **9**, 12 (1949).
BENNETT, J. P.: Calif. Agr. **4**, 1—11, **13**, 15—16 (1950a).
BENNETT, J. P.: Blue Anchor **27**, 17—31 (1950b).
BENNETT, N., LOOMIS, W. E.: Plant Physiol. **24**, 162 — 174 (1949).
BENSON-EVANS, K.: Nature **191**, 255—260 (1961).
BEN ZEEV, N., ZAMENHOF, S.: Plant Physiol. **37**, 1—5 (1962).
BERGER, L., SLEIH, M. W., COLOWICK, S. P., CORI, C. F.: J. Gen. Physiol. **29**, 379—391 (1946).
BESNARD-WIBAUT, C.: Compt. Rend. Acad. Sci. Paris, D **270**, 2932—2935 (1970).
BIEBL, R.: Jb. Wiss. Botan. **88**, 389—420 (1939).
BIEBL, R.: J. Marine Biol. Ass. **31**, 307—315 (1952).
BIEBL, R.: Protoplasma **50**, 217—242 (1958).
BIEBL, R.: Protoplasmatische Ökologie der Pflanzen. Wasser und Temperatur. Protoplasmatologia XII 1. Wien: Springer 1962.
BIEBL, R.: Protoplasma **59**, 133—156 (1964).
BIEBL, R.: Planta **75**, 77—84 (1967a).
BIEBL, R.: Flora B **157**, 25—30 (1967b).
BIEBL, R.: Flora B **157**, 327—354 (1968).
BIEBL, R.: Mikroskopie **25**, 3—6 (1969a).
BIEBL, R.: Protoplasma **67**, 451—472 (1969b).
BIEBL, R.: Protoplasma **69**, 61—83 (1970).
BIEBL, R., MAIER, R.: Österr. Botan. Z. **117**, 176—194 (1969).
BILLINGS, W. D., GODFREY, J., CHABOT, B. F., BOURQUE, D. P.: Arctic and Alpine Research **3**, 277—289 (1971).
BJÖRKMAN, O., PEARCY, R. W., HARRISON, A. T., MOONEY, H.: Science **175**, 786—789 (1972).
BLAAUW, A. H.: Verh. Koninkl. Ned. Akad. Wetenschap. Afd. Natuurk. Sect. II, **29**, (1), 1—90 (1931).
BLAAUW, A. H.: Proc. Koninkl. Ned. Akad. Wetenschap. **36**, 644—653 (1933).
BLAAUW, A. H.: Proc. Koninkl. Ned. Akad. Wetenschap. **44**, 513—520, 684—689 (1941).
BLAAUW, A. H., LUYTEN, I., HARTSEMA, A. M.: Proc. Koninkl. Ned. Akad. Wetenschap. **39**, 928—936 (1936a).
BLAAUW, A. H., LUYTEN, I., HARTSEMA, A. M.: Proc. Koninkl. Ned. Akad. Wetenschap. **39**, 1074—1078 (1936b).
BLAAUW, A. H., LUYTEN, I., HARTSEMA, A. M.: Proc. Koninkl. Ned. Akad. Wetenschap. **43**, 964—974 (1940).
BLACK, M. W.: J. Pomol. Hort. Sci. **14**, 175—202 (1936).
BLACK, M. W.: Farming S. Africa **22**, 645—656 (1947).
BLACK, M. W.: Proc. 13th Internat. Hort. Congr. (London 1952) **2**, 1122—1131 (1953).
BLACK, M. W., MICKLEM, T.: Union S. Africa Dep. Agr. Bull. No **194** (1939).
BLAGOVESHCHENSKIJ, A. V., KIRILLOVA, G. A.: Dokl. Akad. Nauk SSSR, N. S., **100**, 171—173 (1955).
BLONDON, F.: Compt. Rend. Acad. Sci. Paris, D **263**, 48—51 (1966).
BLONDON, F.: Compt. Rend. Acad. Sci. Paris, D **270**, 3063—3066 (1970).
BLONDON, F., CHOUARD, P.: Compt. Rend. Acad. Sci. Paris, D **260**, 6966—6969 (1965).
BÖRTITZ, S., FUCHS, S., WEISE, G.: Biol. Zbl. **86**, 67—77 (1967).
BOGEN, H. J.: Planta **36**, 298—340 (1948).
BOLDUC, R. J., CHERRY, J. H., BLAIR, B. O.: Plant Physiol. **45**, 461—464 (1970).
BOMMER, D.: Z. Pflanzenzüchtg. **45**, 105—120 (1961a).
BOMMER, D.: Z. Pflanzenzüchtg. **46**, 105—111 (1961b).
BONNER, J.: Proc. Am. Soc. Hort. Sci. **50**, 401—408 (1947).
BORRISS, H.: Jb. Wiss. Botan. **89**, 255—339 (1940).
BORRISS, H., ARNDT, M.: Flora **143**, 492—498 (1956).
BOSWELL, V. R.: Univ. Maryland Agric. Exp. Sta. Bull. **313** (1929).

BOTT, T. L., BROCK, T. D.: Science **164**, 1411—1412 (1969).
BOYER, P. D., BALLOU, G. A., LUCK, J. M.: J. Biol. Chem. **167**, 407—424 (1947).
BOYSEN-JENSEN, P.: Die Stoffproduktion der Pflanzen. Jena: Fischer 1932.
BRANDTS, J. F.: In: ROSE, A. H. (Ed.), Thermobiology, pp. 25—72. London-New York:
    Academic Press 1967.
BROCK, T. D.: Science **158**, 1012—1019 (1967).
BROWN, D. S.: Proc. Am. Soc. Hort. Sci. **59**, 111—118 (1952).
BROWN, D. S.: Plant Physiol. **32**, 75—85 (1957).
BROWN, D. S.: Proc. Am. Soc. Hort. Sci. **75**, 138—147 (1960).
BROWN, D. S., ABI-FADEL, J. F.: Proc. Am. Soc. Hort. Sci. **61**, 110—118 (1953).
BROWN, D. S., KOTOB, F. A.: Proc. Am. Soc. Hort. Sci. **69**, 158—164 (1957).
BROWN, H. T., ESCOMBE, F.: Proc. Roy. Soc. B **76**, 29—111 (1905).
BRUINSMA, J.: Chem. Weekbl. **59**, 599 (1963).
BRUINSMA, J., PATIL, S. S.: Naturwissenschaften **50**, 505 (1963).
BÜHRING, J.: Biol. Rdsch. **5**, 30—31 (1967).
BÜNNING, E.: Naturwissenschaften **31**, 493—499 (1943).
BÜNNING, E.: In: RUHLAND, W. (Hrsg.): Handbuch der Pflanzenphysiologie, Bd. 15/II,
    S. 878—907, Berlin-Göttingen-Heidelberg: Springer 1956a.
BÜNNING, E.: Ann. Rev. Plant Physiol. **7**, 71—90 (1956b).
BÜNNING, E., HERDTLE, H.: Z. Naturforsch. **1**, 93—99 (1946).
BURGOS, J. J.: Alm. Min. Agr. Buenos Aires **18**, 345—348 (1943).
BURTON, K.: Biochem. J. **48**, 458—467 (1951).
BUSCHBECK, R.: Biochem. Physiol. Pflanzen **161**, 14—25 (1970a).
BUSCHBECK, R.: Biochem. Physiol. Pflanzen **161**, 388—402 (1970b).
BYTCHIKHINA, E. A.: Tr. Prikl. Botan. Genet. Selekt. **23** (2), 299—347 (1960).
CARTELLIERI, E.: Jb. wiss. Bot. **82**, 460—506 (1935).
CASO, O. H., HIGHKIN, H. R., KOLLER, D.: Nature **185**, 477—479 (1960).
CETL, I., DOBROVOLNÁ, J., EFFMERTOVÁ, E.: Folia přir. Fak. Univ. Purkině Brno **9**,
    No. 5 (Biol. 18), 37—49 (1968).
CHAILAKHYAN, M. KH.: Zh. Obshch. Biol. **15**, 269—287 (1954).
CHAILAKHYAN, M. KH.: Wiss. Z. Univ. Rostock, Math.-Nat. R. **16**, 569—575 (1967).
CHAILAKHYAN, M. KH., KHLOPENKOVA, L. P., LOZHNIKOVA, V. N.: Akad. Nauk Armen.
    SSR Dokl. **38**, 45—51 (1964).
CHAILAKHYAN, M. KH., ZHDANOVA, L. P.: Dokl. Akad. Nauk SSSR, N. S., **19**, 303—306
    (1938a).
CHAILAKHYAN, M. KH., ZHDANOVA, L. P.: Izv. Akad. Nauk SSSR, Otd. mat. estestv.
    Nauk, Ser. biol., **1938**, 1281—1296 (1938b).
CHAKRAVARTI, S. C.: Nature **174**, 461—462 (1954).
CHAKRAVARTI, S. C., PILLAI, V. N. K.: Phyton (Vicente López) **5**, 1—17 (1955).
CHAMPBELL, L. L., PACE, B.: J. Appl. Bacteriol. **31**, 24—35 (1968).
CHANDLER, W. H.: Missouri Agr. Exp. Sta. Bull. **74**, 1—47 (1907).
CHANDLER, W. H.: Missouri Agr. Exp. Sta. Res. Bull. **8**, 143—309 (1913).
CHANDLER, W. H.: Trees in two climates. Berkeley Univ. Calif. Press 1945.
CHANDLER, W. H.: Proc. Am. Soc. Hort. Sci. **64**, 552—572 (1954).
CHANDLER, W. H., BROWN, D. S.: Calif. Agr. Ext. Serv. No. **179**, 1—38 (1951).
CHANDLER, W. H., KIMBALL, M. H., PHILP, G. L., TUFTS, W. P., WELDON, G. P.: Calif.
    Univ. Exp. Sta. Bull. **611**, 1—63 (1937).
CHANDLER, W. H., TUFTS, W. P.: Proc. Am. Soc. Hort. Sci. **30**, 180—186 (1934).
CHAPMAN, D.: In: ROSE, A. H. (Ed.): Thermobiology, pp. 123—146. London-New
    York: Academic Press 1967.
CHATTERTON, N. J., MACKELL, C. M., STRAIN, B. R.: Ecology **51**, 545—549 (1970).
CHEL'TSOVA, L. P., LEBEDEVA, N. I.: Fiziol. Rast. **13**, 525—527 (1966).
CHENIAE, G. M., MARTIN, I. F.: Brookhaven Symp. Biol. **19**, 406—417 (1967).
CHILSON, O. P., COSTELLO, L. A., KAPLAN, N. O.: Fed. Proc. **24**, Suppl. **15**, 55—65
    (1965).
CHINOY, J. J., NANDA, K. K., GARG, O. P.: Physiol. Plant. **10**, 869—876 (1957).
CHMORA, S. N., OYA, V. M.: Fiziol. Rast. **14**, 603—611 (1967).

CHOUARD, P.: Mém. Soc. Bot. France **1956/57**, 51—64 (1956/57).
CHOUARD, P., LARRIEU, C.: Compt. Rend. Acad. Sci. Paris **259**, 2121—2124 (1964).
CHOUARD, P., POIGNANT, P.: Compt. Rend. Acad. Sci. Paris **232**, 103—105 (1951).
CHOUARD, P., TRÂN THANH VÂN, M.: Bull. Soc. Botan. France **109**, 145—147 (1962).
CHOUARD, P., TRÂN THANH VÂN, M.: Compt. Rend. Acad. Sci. Paris **259**, 4783—4786 (1964).
CHOUARD, P., TRÂN THANH VÂN, M.: Compt. Rend. Acad. Sci. Paris **260**, 274—277 (1964).
CHOUARD, P., WEBER, M.-R.: Compt. Rend. Acad. Sci. Paris **243**, 1659—1661 (1956).
CHROBOCZEK, E.: Cornell Univ. Agric. Exp. Sta. Mem., Ithaca, N. Y. **154**, 1—84 (1934).
CHURCH, B. D., HALVORSON, H.: Nature **183**, 124—125 (1959).
CLARK, J.: Photosynthesis and respiration in white spruce and balsam fir. State Univ. Coll. For. Syracuse N. Y. 1961.
COÏC, Y., DURANTON, J.: Ann. Inst. Nat. Rech. Agron., A^{bis}, Ann. Physiol. végét. **1**, 113—120 (1959).
COOPER, J. P.: Ann. Botany, N. S., **24**, 232—246 (1960).
COOPER, W. C., PEYNADO, A.: Proc. Am. Soc. Hort. Sci. **72**, 284—289 (1958).
COOPER, W. C., PEYNADO, A.: Proc. Am. Soc. Hort. Sci. **74**, 333—347 (1959).
COOPER, W. C., PEYNADO, A., OTEY, G.: Proc. Am. Soc. Hort. Sci. **66**, 100—110 (1955).
COOPER, W. C., TAYLOE, S., MAXWELL, N.: Rio Grande Valley Hort. Inst. Proc. 7—1/25/55 (1955).
CORRENS, C.: Ztschr. indukt. Abstammungs- u. Vererbungslehre **10**, 130—135 (1913).
COSTER, CH.: Ann. Jard. Botan. Buitenzorg **35**, 125—162 (1926).
COVILLE, F. V.: J. Agric. Res. **20**, 151—160 (1920).
COX, L. M., BOERSMA, L.: Plant Physiol. **42**, 550—553 (1967).
COX, W., LEVITT, J.: Plant Physiol. **44**, 923—928 (1969).
CRAKER, L. E., GUSTA, L. V., WEISER, C. J.: Can. J. Plant Sci. **49**, 279—286 (1969).
CRESCINI, F.: Nuovi Ann. Agr. No. 8 (1928).
CROCKER, W.: Botan. Gaz. **42**, 265—291 (1906).
CROCKER, W.: Am. J. Botan. **3**, 99—120 (1916).
CROCKER, W.: Growth of plants. New York: Reinhold Publ. Corp. 1948.
CROCKER, W., BARTON, L. V.: Physiology of seeds. Waltham, Mass.: Chronica Botanica Co. 1953.
CURTIS, G. J.: Nature **203**, 201—202 (1964).
CURTIS, O. F., CHANG, H. T.: Am. J. Botan. **17**, 1047—1048 (1930).
CURTIS, O. F., CLARK, D. G.: An introduction to plant physiology. New York: McGraw-Hill 1950.
DADYKIN, V. P.: Dokl. Adad. Nauk SSSR **70**, 1073—1076 (1950).
DANIEL, L.: Ann. Sci. Nat., 8e Sér., Bot., **8**, 1—226 (1898).
DANIELL, J. W., CHAPPELL, W. E., CHOUCH, H. B.: Plant Physiol. **44**, 1684—1689 (1969).
DAVID, R.: Compt. Rend. Soc. Biol. **139**, 560—562 (1945a).
DAVID, R.: Compt. Rend. Soc. Biol. **139**, 643—645 (1945b).
DAVID, R.: Compt. Rend. Acad. Sci. Paris **224**, 146—147 (1947).
DAVID, R.: Compt. Rend. Acad. Sci. Paris **228**, 1242—1243 (1949).
DAVID, R.: Compt. Rend. Acad. Sci. Paris **233**, 428—430 (1951).
DAVID, R., SÉCHET, J.: Compt. Rend. Soc. Biol. **141**, 459—460 (1947).
DAVID, R., SÉCHET, J.: Compt Rend. Acad. Sci. Paris **227**, 537—539 (1948).
DAVIES, J. D.: Ciba Found. Symp. on the Frozen Cell, G. E. W. Wolstenholme and M. O'Connor (Eds.), pp. 213—233. London: J. & A. Churchill 1970.
DAVIS, W. E.: Am. J. Botan. **17**, 58—76 (1930a).
DAVIS, W. E.: Am. J. Botan. **17**, 77—87 (1930b)).
DAVIS, W. E., ROSE, R. C.: Bot. Gaz. **54**, 49—62 (1912).
DE CANDOLLE, C., PICTET, R.: Arch. Sci. Phys. Nat. Genève **3**, 629—632 (1879).
DE DUVE, C.: Subcellular particles, HAYASHI, T. (Ed.), pp. 128—156. New York: Ronald Press 1959.
DE HAAS, P. G., SCHANDER, H.: Z. Pflanzenzüchtg. **31**, 457—512 (1952).

DE KIEWIET, D. Y., HALL, D. O., JENNER, E. L.: Biochim. Biophys. Acta **109**, 284—292 (1965).

DELOUCHE, J. C.: Iowa State Coll. J. Sci. **30**, 348—349 (1956).

DELOUCHE, J. C.: Proc. Ass. Offic. Seed Analysts **48**A, 81—84 (1958).

DENFFER, D. v.: Naturwissenschaften **37**, 296—301, 317—321 (1950).

DENNY, F. E.: Contrib. Boyce Thompson Inst. **8**, 137—140 (1936).

DENNY, F. E.: Contrib. Boyce Thompson Inst. **8**, 351—353 (1937).

DENNY, F. E.: Contrib. Boyce Thompson Inst. **9**, 403—408 (1938).

DENNY, F. E.: Contrib. Boyce Thompson Inst. **12**, 375—386 (1942).

DE PHILIPPIS, A.: N. Giorn. Botan. Ital. **44**, 1—142 (1937).

DÉVAY, M.: Acta Agron. Acad. Sci. Hung. **14**, 93—97 (1965a).

DÉVAY, M.: Acta Agron. Acad. Sci. Hung. **14**, 275—287 (1965b).

DÉVAY, M.: Acta Agron. Acad. Sci. Hung. **15**, 85—94 (1966).

DÉVAY, M.: Acta Agron. Acad. Sci. Hung. **16**, 251—252 (1967a).

DÉVAY, M.: Acta Agron. Acad. Sci. Hung. **16**, 289—295 (1967b).

DICKINSON, D. B., MISCH, DRURY, R. E.: Plant Physiology **46**, 200—203 (1970).

DIXON, H. H.: Transpiration and the ascent of sap in plants. London, New York: MacMillian 1914.

DOEBBLER, G. F.: Cryobiology **3**, 2—11 (1966).

DÖRING, B.: Z. Botan. **28**, 305—383 (1935).

DORPH-PETERSEN, K.: Tidsskr. Planteavl. **31**, 122—127 (1925).

DOWNS, R. J.: In: THIMANN, K. W. (Ed.): The physiology of forest trees, pp. 527—537. New York: Ronald Press Comp. 1958.

DOWNS, R. J., BORTHWICK, H. A.: Botan. Gaz. **117**, 310—326 (1956a).

DOWNS, R. J., BORTHWICK, H. A.: Proc. Am. Soc. Hort. Sci. **68**, 518—521 (1956b).

DRAKE, B. G., RASCHKE, K., SALISBURY, F. B.: Plant Physiol. **46**, 324—330 (1970).

DUPÉRON, R.: Rev. Gén. Botan. **53**, 525—557 (1946).

DUPÉRON, R.: Compt. Rend. Acad. Sci. Paris **228**, 192—194 (1949).

DUPÉRON, R.: Compt. Rend. Acad. Sci. Paris **230**, 225—227 (1950).

DUPÉRON, R.: Rev. Gén. Botan. **59**, 580—631 (1952).

DUPÉRON, R.: Rev. Gén. Botan. **60**, 33—78, 90—122 (1953).

EAKS, I. L.: Plant Physiol. **35**, 632—636 (1960).

EAKS, I. L., MORRIS, L. L.: Plant Physiol. **31**, 308—314 (1956).

EBERMAYER, E.: Die physikalischen Einwirkungen des Waldes auf Luft und Boden. Resultate forstl. Vers. Sta. Kgr. Bayern, Vol. I. Aschaffenburg: Krebs 1873.

ECKERSON, S.: Botan. Gaz. **55**, 286—299 (1913).

EDWARDS, M. M.: J. Exp. Botan. **20**, 876—894 (1969).

EGLE, K., SCHENK, W.: Beitr. Biol. Pflanz. **29**, 75—105 (1952).

EGUCHI, T., MATSUMURA, T., ASHIZAWA, M.: Proc. Am. Soc. Horticult. Sci. **72**, 343—352 (1958).

EGUCHI, T., SAKAI, A., USUI, G., UEHARA, G.: Silvae Genetica **15**, 84—89 (1966).

EIDMANN, F.: Schr. H.-Göring-Akad. Deut. Forstwiss. **5**, 1—143 (1943).

ELLIS, R. W., TRIONE, E. J.: Physiol. Plant. **20**, 106—112 (1967).

EL-SHARKAWY, M. A., HESKETH, J. D.: Crop Sci. **4**, 515—518 (1964).

EMMET, J. M., WALKER, D. A.: Biochim. Biophys. Acta **180**, 424—425 (1969).

ENGELBRECHT, L., MOTHES, K.: Ber. Deut. Botan. Ges. **73**, 246—257 (1960).

ENGELBRECHT, L., MOTHES, K.: Flora **154**, 279—298 (1964).

EPSTEIN, E.: Ann. Rev. Plant Physiol. **7**, 1—24 (1956).

EPSTEIN, E.: Amer. J. Botan. **47**, 393—399 (1960).

ERNST, W.: Flora **160**, 317—331 (1971).

EVANS, C. R.: Botan. Gaz. **73**, 213—226 (1922).

EVANS, L. T.: Ann. Botany, N. S., **23**, 521—546 (1959).

EVANS, L. T. (Ed.): The induction of flowering. Some case histories. South Melbourne: Macmillan 1969.

EVENARI, M.: Botan. Rev. **15**, 153—194 (1949).

EVENARI, M.: Palestine J. Botan. Ser. Jerus. **5**, 138—160 (1952).

EVENARI, M.: In: HOLLAENDER, A. (Ed.): Radiation biology, Vol. 3, pp. 519—549. New York, Toronto, London: McCraw-Hill Book Co., Inc. 1956.

EVENARI, M.: In: The biological action of growth substances. 11th Symposium Soc. Biol. 1956, Alberystwyth Vol. 11, p. 21—43 (1957).

EVENARI, M.: In: RUHLAND, W. (Hrsg.): Handbuch der Pflanzenphysiologie Bd. 15/II, S. 804—847. Berlin-Heidelberg-New York: Springer 1965.

EWART, M. H., SIMINOVITCH, D., BRIGGS, D. R.: Plant Physiol. 29, 407—413 (1954).

FARRANT, J.: Nature 222, 1175—1176 (1969).

FARRANT, J., WOOLGAR, A. E.: Ciba Found. Symp. on the Frozen Cell, G. E. W. Wolstenholme and M. O'Connor (Eds.), pp. 97—119. London: J. & A. Churchill 1970.

FARRANT, J., WOOLGAR, A. E.: Cryobiology 7, 56—60 (1970b).

FARRELL, J., ROSE, A. H.: In: ROSE, A. H. (Ed.): Thermobiology, pp. 147—218. London, New York: Academic Press 1967.

FEIERABEND, J.: Planta 71, 326—355 (1966).

FEIERABEND, J.: Z. Pflanzenphysiol. 62, 70—82 (1970).

FEIERABEND, J., BÜNSOW, R.: Flora A 158, 153—156 (1967).

FEJER, S. O.: N. Z. J. agric. Res. 3, 734—743 (1960).

FELDMAN, N. L.: Tsytologiya 4, 633—643 (1962).

FELDMAN, N. L.: Planta 78, 213—225 (1968).

FELDMAN, N. L., ZAVADSKAYA, I. G., LYUTOVA, M. I.: Tsitologiya 5, 125—134 (1963).

FELDMAN, N. L., ALEXANDROV, V. Y., ZAVADSKAYA, I. G., KISLYUK, I. M., LOMAGIN, A. G., LYUTOVA, M. I., JASKULIEV, A.: The cell and environmental temperature, TROSHIN, A. S. (Ed.), pp. 152—160, Proc. Intern. Symp. Cytoecology, Leningrad 1963. Oxford: Pergamon Press 1967.

FEWSON, C. A., BLACK, C. C., GIBBS, M.: Plant Physiol. 38, 680—685 (1963).

FILIPPENKO, I. A.: Dokl. Akad. Nauk SSSR, N. S., 16, 185—189 (1936).

FILIPPENKO, I. A.: Dokl. Akad. Nauk SSSR, N. S., 17, 329—332 (1937).

FINCH, L. R., CARR, D. J.: Austral. J. Biol. Sci. 9, 355—363 (1956).

FIRBAS, F.: Jb. Wiss. Botan. 74, 459—696 (1931).

FISCHER, A.: Jb. Wiss. Botan. 22, 73—160 (1890).

FISCHER, H.: In: RUHLAND, W. (Hrsg.): Handbuch der Pflanzenphysiologie, Bd. 6, S. 952—977. Berlin-Göttingen-Heidelberg: Springer 1958.

FISCHER, H.: Flora 150, 416—426 (1961).

FORD, J., PEEL, A. J.: J. Exp. Botan. 17, 522—533 (1966).

FORWARD, D. F.: In: RUHLAND, W. (Hrsg.): Handbuch der Pflanzenphysiologie. Bd. 12/II S. 234—258. Berlin-Göttingen-Heidelberg: Springer 1960.

FRIEND, D. J. C.: Can. J. Botan. 43, 161—170 (1965).

FRIEND, D. J. C., GREGORY, F. G.: Nature 172, 667—668 (1953).

FRIEND, D. J. C., PURVIS, O. N.: Ann. Botany, N. S., 27, 553—579 (1963).

FRYER, J. H., LEDIG, F. TH.: Canad. J. Bot. 50, 1231—1235 (1972).

FUCHS, W. H.: Z. Pflanzenzüchtg. 24, 165—185 (1941).

FUCHS, W. H., ROSENSTIEL, K.: In: KAPPERT, H., RUDORF, W. (Hrsg.): Handbuch der Pflanzenzüchtg, Bd. 1, S. 365—442, 2nd Ed.

FUKUI, H. N., WELLER, L. E., WITTWER, S. H., SELL, H. M.: Am. J. Botan. 45, 73—74 (1958).

FYODOROV, A. K.: Agrobiologija 1958, No. 5 (113), 57—59 (1958).

FYODOROV, A. K.: Fiziol. Rast. 7, 686—694 (1960).

FYODOROV, A. K.: Fiziol. Rast. 10, 575—580 (1963).

FYODOROV, A. K., SHINKAREVA, V. G.: Fiziol. Rast. 13, 269—273 (1966).

GAASTRA, P.: Mededel. Landbouwhogeschool. Wageningen 59 (13), 1—68 (1959).

GALE, J., POLJAKOFF-MAYBER, A.: Plant Cell Physiol. 6, 111—115 (1965).

GANNUTZ, T. P.: Soc. Botan. France, Coll. Lichenes 1968, 169—179 (1968).

GARDNER, W. R.: In: KOZLOWSKI, T. T.: Water deficits and plant growth, Vol. I, p. 107—135. London-New York: Academic Press 1968.

GASSNER, G.: Jb. Hamb. Wiss. Anst. 29, 1—121 (1911).

GASSNER, G.: Ber. Deut. Botan. Ges. 33, 203—216 (1915).

GASSNER, G.: Z. Botan. 10, 417—480 (1918).

GASSNER, G.: Naturwiss. Umschau Chemiker-Ztg. **10**, 161—169 (1921).

GASSNER, G.: Mitt. deut. Landwirtsch. Ges. **40**, 950—955 (1925).

GASSNER, G.: Wien. Landw. Ztg. **76**, 157—158, 166—167 (1926).

GASSNER, G.: Handwörterbuch d. Naturwiss. 8, S. 646—654. Jena: G. Fischer 1933.

GASSNER, G., GOEZE, G.: Phytopath. Z. **4**, 387—413 (1932).

GATES, D. M.: Energy exchange in the biosphere, pp. 94—112, New York: Harper and Row 1962.

GATES, D. M.: Arch. Met. Geophys. Bioklim. B **12**, 322—336 (1963).

GATES, D. M.: Ann. Rev. Plant Physiol. **19**, 211—238 (1968).

GATES, D. M., ALDERFER, R., TAYLOR, E.: Science **159**, 994—995 (1968).

GATES, D. M., TANTRAPORN, W.: Science **115**, 613—616 (1952).

GATES, F. C.: Botan. Gaz. **57**, 445—489 (1914).

GÄUMANN, E.: Ber. Schweiz. Botan. Ges. **44**, 157—334 (1935).

GÄUMANN, E., JAAG, O.: Ber. Schweiz. Botan. Ges. **45**, 411—518 (1936).

GÄUMANN, E., JAAG, O.: Ber. Schweiz. Botan. Ges. **49**, 178—238 (1939).

GEIGER, D. R.: Ohio J. Sci. **69**, 356—366 (1969).

GERHARD, E.: J. Landw. **87**, 161—203 (1940).

GERLOFF, E. D., RICHARDSON, T., STAHMANN, M. A.: Plant Physiol. **41**, 1280—1284 (1966).

GERLOFF, E. D., STAHMANN, M. A., SMITH, D.: Plant Physiol. **42**, 895—899 (1967).

GESSNER, F.: In: RUHLAND, W. (Hrsg): Handbuch der Pflanzenphysiologie, Bd. 3, S. 854—901. Berlin-Göttingen-Heidelberg: Springer 1956.

GESSNER, F.: Hydrobotanik, Bd. II: Stoffhaushalt. Berlin: V. E. B. Deutscher Vlg. Wiss. 1959.

GEURTEN, I.: Forstwiss. Cbl. **69**, 704—743 (1950).

GICKLHORN, J.: Protoplasma **26**, 90—96 (1936).

GILES, C. H., McKAY, R. B.: J. Biol. Chem. **237**, 3388—3392 (1962).

GILL, N. T.: Ann. Appl. Botan. **25**, 447—456 (1938).

GILLETT, G. W.: Am. J. Botan. **50**, 798—801 (1963).

GLÜCK, H.: Biologische und morphologische Untersuchungen über Wasser- und Sumpfgewächse. II. Teil. Jena: G. Fischer 1906.

GODNEV, T. N., ROTFARB, R. M.: Ber. Akad. Wiss. USSR **134**, 963—964 (1960).

GOEBEL, K.: Pflanzenbiologische Schilderungen. II. Teil. 2. Aufl. Marburg: Elwertsche Verlagsbuchhandlung 1893.

GOEBEL, K.: Einleitung in die experimentelle Morphologie der Pflanzen. Leipzig-Berlin: Teubner 1908.

GORBAN, J. S.: Tsitologiya **4**, 182—192 (1962).

GOTT, M. B.: Nature **180**, 714—715 (1957).

GOUJON, C., MAIA, N., DOUSSINAULT, G.: Ann. Amélior. Plant. **18**, 49—57 (1968).

GRABE, D. F.: Agron. J. **48**, 253—256 (1956).

GRAHL, A.: Beitr. Biol. Pflanzen **35**, 413—446 (1960 a).

GRAHL, A.: Beitr. Biol. Pflanzen. **35**, 447—473 (1960 b).

GRANT, N. H., ALBURN, H. E.: Arch. Biochem. Biophys. **118**, 292—296 (1967).

GREB, H.: Planta **48**, 523—563 (1957).

GREGORY, F. G., PURVIS, O. N.: Ann. Botany, N. S., **2**, 237—251 (1938 a).

GREGORY, F. G., PURVIS, O. N.: Ann. Botany, N. S., **2**, 753—764 (1938 b).

GREIFF, D., MYERS, M.: Nature **190**, 1202—1204 (1961).

GROGAN, C. O.: Cryobiology **6**, 584—585 (1970).

GRZESIUK, S.: Bull. Acad. Pol. Sci., Sér. Sci. Biol., **10** 73—78 (1962).

GRZESIUK, S.: In: BORRISS, H. (Ed.): Proc. Symp. Physiology, Ecology, and Biochemistry of Germination, pp. 379—387. Greifswald: Ernst-Moritz-Arndt-Univ. 1967.

GÜNTHER, G.: Ber. Deut. Botan. Ges. **68**, (20)—(21) (1955).

GÜNTHER, G.: Ber. Deut. Botan. Ges. **72**, (25) (1959).

GÜNTHER, G.: Naturwissenschaften **48**, 385—386 (1961).

GÜNTHER, G.: Ber. Deut. Botan. Ges. **79**, 198—205 (1966).

GULISASHVILI, W. S.: Priroda **37**, 63—66 (1948).

GULKANYAN, V.O., KHAKHATRYAN, G.G.: Akad. Nauk Armen. SSR, Izv. biol. Nauki, 17, No. 12, 3—11 (1964).
HABERLANDT, F.: Landw. Versuchsstationen 17, 104—116 (1874).
HABERLANDT, F.: Haberlandts Wiss.-Prakt. Unters. Geb. d. Pflanzenb. Wien. 1, 109—116, 117—122 (1875).
HACKETT, W.P., HARTMANN, H.T.: Physiol. Plant. 20, 430—436 (1967).
HÄRTEL, O.: Jahrb. Wiss. Botan. 87, 173—210 (1938).
HAGEN, P.O., KUSHNER, D.J., GIBBONS, N.E.: Can. J. Microbiol. 10, 813—822 (1964).
HAGGAR, R.J.: Nature 191, 1120—1121 (1961).
HAMAYA, T., KURAHASHI, A., TAKANASHI, N., SAKAI, A.: Bull. Tokyo Univ. For. 64, 197—238 (1968).
HAMMEL, H.T.: Plant Physiol. 42, 55—66 (1967).
HAMMOUDA, M., LANGE, O.L.: Naturwissenschaften 49, 500 (1962).
HARADA, H.: Ann. Inst. Nat. Rech. Agron., A[bis], Ann. Physiol. Végét. 2, 249—254 (1960).
HARADA, H.: Rev. Gén. Botan. 69, 201—297 (1962).
HARADA, H., NITSCH, J.P.: Science 129, 777—778 (1959).
HARLAN, H.V., POPE, N.N.: J. Heredity 13, 72—75 (1922).
HARRASSER, J.: Die Kälteresistenz des Bergahorn (Acer pseudoplatanus L). Diss. Innsbruck 1969.
HARRINGTON, G.T.: J. Agric. Res. 6, 761—796 (1916).
HARRINGTON, G.T.: Proc. Ass. Seed Analysts N. Am. 9/10, 24—28 (1917a).
HARRINGTON, G.T.: Proc. Ass. Seed Analysts N. Am. 9/10, 71—76 (1917b).
HARRINGTON, G.T.: Proc. Ass. Seed Analytsts N. Am. 11, 58—63 (1919).
HARRINGTON, G.T.: J. Agric. Res. 23, 79—100 (1923a).
HARRINGTON, G.T.: J. Agric. Res. 23, 259—332 (1923b).
HARRINGTON, G.T., CROCKER, W.: J. Agric. Res. 23, 193—222 (1923).
HARRINGTON, G.T., HITE, B.C.: J. Agric. Res. 23, 153—161 (1923).
HARRINGTON, J.F., RAPPOPORT, L., HOOD, K.J.: Science 125, 601—602 (1957).
HARRIS, G.P., HARRIS, J.E.: J. Horticult. Sci. 37, 219—234 (1962).
HARRIS, W.B.: J. Dep. Agric. S. Australia 53, 356—358 (1950).
HARRISON, A.T., MOONEY, H.A.: Science 175, 786—789 (1972).
HARTMANN, W.: Flora A 159, 35—39 (1968).
HARTMANN, W., BUSCHBECK, R.: Wiss. Z. Päd. Inst. Güstrow, R. Biol. Chem. 3, 27—32 (1965a).
HARTMANN, W., BUSCHBECK, R.: Wiss. Z. Päd. Inst. Güstrow, R. Biol. Chem. 3, 33—35 (1965b).
HARTSEMA, A.M.: Bull. Inst. Intern. du Froid. Annexe 1 (1954).
HARTSEMA, A.M.: In: RUHLAND, W. (Hrsg.): Handbuch der Pflanzenphysiologie, Bd. 16, S. 123—167. Berlin-Göttingen-Heidelberg: Springer 1961.
HARTSEMA, A.M., LUYTEN, I.: Proc. Koninkl. Ned. Akad. Wetenschap. 36, 120—127 (1933a).
HARTSEMA, A.M., LUYTEN, I.: Proc. Koninkl. Ned. Akad. Wetenschap. 36, 210—216 (1933b).
HARTSEMA, A.M., LUYTEN, I.: Proc. Koninkl. Ned. Akad. Wetenschap. 41, 651—660, 800—809 (1938).
HARTSEMA, A.M., LUYTEN, I.: Proc. Koninkl. Ned. Akad. Wetenschap. 43, 879—890 (1940).
HARTSEMA, A.M., LUYTEN, I.: Proc. Koninkl. Ned. Akad. Wetenschap., Ser. C 58 (4), 1—27 (1955).
HARTSEMA, A.M., LUYTEN, I., BLAAUW, A.H.: Verh. Koninkl. Ned. Akad. Wetenschap Afd. Natuurk. Sect. II 27 (1), 1—46 (1930).
HARVEY, R.B.: J. For. 28, 50—53 (1930).
HATAKEYAMA, I.: Physiol. Ecol. (Kyoto) 7, 89—97 (1957).
HATAKEYAMA, I.: Mem. Coll. Sci. Univ. Kyoto B 28, 401—429 (1961).
HATAKEYAMA, I., KATO, J.: Planta 65, 259—268 (1965).
HAUPT, W., NAKAMURA, E.: Z. Pflanzenphysiol. 62, 270—275 (1970).

HEBER, U.: Ber. Deut. Botan. Ges. **70**, 371—382 (1957).
HEBER, U.: Protoplasma **51**, 284—298 (1959).
HEBER, U.: Plant Physiol. **42**, 1343—1350 (1967).
HEBER, U.: Cryobiology **5**, 188—201 (1968).
HEBER, U.: Ciba Found. Symp. on the Frozen Cell, pp. 175—188. G. E. W. Wolstenholme, M. O'Connor (Eds.). London: J. & A. Churchill 1970.
HEBER, U., ERNST, R.: In: ASAHINA, E. (Ed.): Cellular Injury and Resistance in Freezing Organisms. Proc. Internat. Conf. Low Temp. Sci. 1966, Vol. II, pp. 63—77. Sapporo: Inst. Low Temp. Sci., Hokkaido Univ., Sapporo 1967.
HEBER, U., KEMPFLE, M.: Z. Naturforsch. **25**b, 834—842 (1970).
HEBER, U., SANTARIUS, K. A.: Plant Physiol. **39**, 712—719 (1964).
HEBER, U., SANTARIUS, K. A.: In: TROSHIN, A. S. (Ed.) : The cell and environmental temperature, pp. 27—34. Oxford-New York-Toronto: Pergamon Press 1967.
HEBER, U., SANTARIUS, J. A.: Umschau **71** 930—936 (1971).
HEBER, U., TYANKOVA, L., SANTARIUS, K. A.: Biochim. Biophys. Acta **241**, 578—592 (1971).
HELLMERS, H., SUNDAHL, W. P.: Nature **184**, 1247—1248 (1959).
HELLMUTH, E. O.: J. Ecol. **56**, 319—344 (1968).
HELLMUTH, E. O.: J. Ecol. **57**, 613—634 (1969).
HELLMUTH, E. O.: J. Ecol. **59**, 225—259 (1971).
HENCKEL, P. A.: Ann. Rev. Plant. Physiol. **15**, 363—386 (1964).
HENKE, O.: Deut. Akad. Landw. Wiss. Sitzber. **11** (6), 5—18 (1962).
HENSSEN, A.: Flora **141**, 523—566 (1954).
HENZE, J.: Z. Botan. **47**, 42—87 (1959).
HESS, D.: Planta **50**, 504—525 (1958).
HESS, D.: Planta **54**, 74—94 (1959).
HESS, D.: Planta **56**, 229—232 (1961a).
HESS, D.: Planta **57**, 13—28 (1961b).
HESS, D.: Planta **57**, 29—43 (1961c).
HILL, A. G. G., CAMPBELL, G. K. G.: Empire J. Exp. Agr. **17**, 259—264 (1949).
HILTNER, L.: Mitt. Deut. Landwirtsch. Ges. **16**, 192—194 (1901).
HOAGLAND, D. R.: Lectures on the inorganic nutrition of plants. Waltham, Mass.; Chronica Botanica Co. 1944.
HODGSON, F. R.: Proc. Am. Soc. Hort. Sci. **20**, 151—155 (1924).
HOLDHEIDE, W.: Tharandter Forstl. Jb. **91**, 582—590 (1941).
HOLMGREN, P., JARVIS, P. G., JARVIS, M. S.: Physiol. Plant. **18**, 557—573 (1965).
HOWARD, W. L.: Untersuchungen über die Winterperiode der Pflanzen. Inauguraldissertation. Halle a. d. S. 1906.
HOWARD, W. L.: Missouri Univ. Agr. Exp. Sta. Res. Bull. **1**, 1—105 (1910).
HOWELL, G. S., WEISER, C. J.: Plant Physiol. **45**, 390—394 (1970a).
HOWELL, G. S., WEISER, C. J.: J. Am. Soc. Hort. Sci. **95**, 190—192 (1970b).
HUBER, B.: Z. Botan. **23**, 839—890 (1930).
HUBER, B.: Der Wärmehaushalt der Pflanzen, Freising: Datterer 1935.
HUBER, B.: In: RUHLAND, W. (Hrsg.): Handbuch der Pflanzenphysiologie, Bd. 3, S. 283—311. Berlin-Göttingen-Heidelberg: Springer 1956.
HUBER, B., ZIEGLER, H.: In: RUHLAND, W. (Hrsg).: Handbuch der Pflanzenphysiologie, Bd. XII/2, S. 150—184. Berlin-Göttingen-Heidelberg: Springer 1960.
HUDSON, M. A., IDLE, D. B.: Planta **57**, 718—730 (1962).
HUGGINS, C. E.: Fed. Proc. **24**, 190—195 (1965).
HURD, R. G., PURVIS, O. N.: Ann. Botany N. S., **28**, 137—151 (1964).
HURST, C., HALL, T. C., WEISER, C. J.: Hort. Sci. **2**, 164—166 (1967).
HYGEN, G.: In: SLAVIK, B. (Ed.): Water stress in plants, pp. 89—95. Praha: Academia 1965.
IBÁNEZ, M. L.: Nature **201**, 414—415 (1964).
ILJIN, W. S.: Protoplasma **20**, 105—124 (1933).
ILJIN, W. S.: Bull. Ass. Russe, Rech. Sci. Prague, Sect. Sci-Nat.-Math. **1** (6), 135—160 (1934).

ILLERT, H.: Botan. Arch. **7**, 133—141 (1924).

INOUE, S., YAHIRO, M.: Botan. Mag. (Tokyo) **69**, 215—218 (1956).

INOUYE, J., ITO, K.: Plant Cell Physiol. **9**, 137—142 (1968).

INOUYE, J., TASHIMA, Y., KATAYAMA, T.: Plant Cell Physiol. **5**, 355—358 (1964).

IRMSCHER, E.: Jb. Wiss. Botan. **50**, 387—449 (1912).

IRVING, R. M.: Plant Physiol. **44**, 801—805 (1969).

IRVING, R. M., LANPHEAR, F. O.: Plant Physiol. **43**, 9—13 (1968).

IVANOV, L. A.: Fiziol. Rast. **4**, 405—416 (1957).

IWANOFF, L. A., ORLOVA, J. M.: Russ. Bot. Ges. **16**, 139—160 (1931).

JACCARD, P., FREY-WYSSLING, A.: Jb. Wiss. Botan. **79**, 655—680 (1934).

JACKSON, T. H.: Rep. 14th Internat. Hort. Congr., The Hague-Scheveningen. 1955, **II**, 1463—1471 (1956).

JACKSON, T. H.: E. African Agr. J. **12**, 153—166 (1957).

JACKSON, W. A., VOLK, R. J.: Ann. Rev. Plant. Physiol. **21**, 385—432 (1970).

JACOBS, D. L.: Ecol. Monographs **17** (4), 437—469 (1947).

JAHNEL, H., WATZLAWIK, G.: Wiss. Z. TU. Dresden **10**, 1—3 (1961).

JAMES, N. I., LUND, S.: Agron. J. **52**, 508—510 (1960).

JAMES, N. I., LUND, S.: Am. J. Botan. **52**, 877—882 (1965).

JAMESON, D. A.: Botan. Gaz. **122**, 174—179 (1961).

JAVARAYA, H. C.: Indian J. Hort. **1**, 31—34 (1943).

JARVIS, P. G., ROSE, C. W., BEGG, J. E.: Agr. Meteorol. **4**, 113—117 (1967).

JENKINS, G.: J. Agric. Sci. Cambridge **73**, 477—482 (1969).

JENKINS, G.: Eucarpia Meeting Dijon, 1970, 1—3 (1970).

JENSEN, G.: Physiol. Plant. **13**, 822—830 (1960).

JEREMIAS, K.: Planta **47**, 81—104 (1956).

JEREMIAS, K.: Ber. Deut. Botan. Ges. **75**, 313—322 (1962).

JEREMIAS, K.: Über die jahresperiodisch bedingten Veränderungen der Ablagerungs-form der Kohlenhydrate in vegetativen Pflanzenteilen. Bot. Studien **15**. Jena: Fischer 1964.

JEREMIAS, K.: Ber. Deut. Botan. Ges. **82**, 87—97 (1969).

JOHANNSEN, W.: Das Ätherverfahren bei Frühtreiberei. Jena: Fischer 1900.

JOHANNSEN, W.: In: Handwörterbuch der Naturwissenschaften 2. Aufl., Bd. 8, S. 514—519. Jena: Fischer 1933.

JOHNSTEN, R. D.: Australian J. Botan. **7**, 97—108 (1959).

JONES, W. W.: Plant Physiol. **17**, 481—486 (1942).

JOSEPH, H. C.: Botan. Gaz. **87**, 127—151 (1929).

JOUGLARD, C.: Biol. Plant. **7**, 74—78 (1965).

JUNG, G. A., SMITH, D.: Agron. J. **53**, 359—366 (1961).

JUNG, G. A., SHIH, S. C., SHELTON, D. C.: Plant Physiol. **42**, 1653—1657 (1967).

JUNG, G. A., SHIH, S. C., SHELTON, D. C.: Cryobiology **4**, 11—16 (1967).

KACPERSKA-PALACZ, A., BLAZIAK, M., WCISLINSKA, B.: Botan. Gaz. **130**, 213—221 (1969).

KAKU, S.: Botan. Mag. Tokyo **77**, 283—289 (1964).

KAKU, S.: Botan. Mag. Tokyo **79**, 98—104 (1966).

KAKU, S.: Plant Cell Physiol. **12**, 147—155 (1971).

KALLIO, P., HEINONEN, S.: Rep. Kevo Subarctic Res. Sta., Turku, **8**, 63—72 (1971).

KALLIO, P., KÄRENLAMPI, L.: In: COOPER, J. P. (Ed.): Photosynthesis and Productivity in Different Environments. Cambridge Univ. Press 1973.

KAMENTSEVA, I. E.: Dok. Akad. Nauk SSSR **186**, 968—970 (1969).

KAMIYA, N.: Protoplasmic streaming. Protoplasmatologia, Bd. 8. Wien: Springer 1959.

KANWISHER, J.: Biol. Bull. **113**, 275—285 (1957).

KAPPEN, L.: Flora **155**, 124—166 (1964).

KAPPEN, L.: Flora B **156**, 427—445 (1966).

KAPPEN, L.: Allg. Forst- u. Jagd-Ztg. **138**, 181—185 (1967).

KAPPEN, L.: Flora B **158**, 232—260 (1969).

KAPPEN, L., LANGE, O. L.: Protoplasma **65**, 119—132 (1968).

KAPPEN, L., LANGE, O. L.: Beitr. Deut. Botan. Ges. N. F. **4**, 61—65 (1970).

KAPPEN, L., ULLRICH, W. R.: Ber. Deut. Botan. Ges. **83**, 265—275 (1970).

KARNATZ, H.: Züchter **26**, 178—187, 307—315 (1956).

KAROW, A. M., WEBB, W. R.: Cryobiology **2**, 99—108.

KARSCHON, R.: Volcani Inst. Agr. Res. Lfl. **28** (1966).

KARSCHON, R., PINCHAS, L.: Oecol. Plant **6**, 43—50 (1971).

KATOH, S., SAN PIETRO, A.: Arch. Biochem. Biophys. **122**, 144—152 (1967).

KAWANO, K., TUZII, R., HATAKEYAMA, I.: Bull. Fac. Text. Fib. Kyoto Univ. **3**, 32—36 (1960).

KEARNS, V., TOOLE, E. H.: U. S. Dept. Agr. Techn. Bull. **638**, 1—35 (1939).

KEMMER, E., SCHULZ, I. F.: Das Frostproblem im Obstbau. München: Bayerischer Landw. Verlag 1955.

KEMMER, E., THIELE, I.: Züchter **25**, 57—60 (1955).

KEMPNER, E.: Science **142**, 1318—1319 (1963).

KENTZER, T.: Acta Soc. Botan. Pol. **36**, 7—22 (1967).

KESSLER, W., RUHLAND, W.: Planta **28**, 159—204 (1938).

KHOLODNY, N. G.: Soviet Botan. **2** (1935).

KHOLODNY, N. G.: Dokl. Akad. Nauk SSSR, N. S., **3**, 391—394 (1936a).

KHOLODNY, N. G.: Dokl. Akad. Nauk SSSR, N. S., **3**, 439—442 (1936b).

KIESSLING, L.: Landw. Jb. Bayern **2**, 449—514 (1911).

KILLIAN, CH.: Bull. Soc. Bot. France **79**, 185—220 (1932).

KIMURA, M.: Bot. Mag. Tokyo **82**, 6—19 (1969).

KINBACHER, E. J., KNULL, H. R.: Crop Sci. **7**, 148—151 (1967).

KINBACHER, E. J., SULLIVAN, CH. Y.: Am. Soc. Hort. Sci. **90**, 163—168 (1967).

KISLYUK, I. M.: Botan. Zh. **47**, 713—715 (1962).

KISSER, J.: In: RUHLAND, W. (Hrsg.): Handbuch der Pflanzenphysiologie, Bd. 3, S. 668—684. Berlin-Göttingen-Heidelberg: Springer 1956.

KITAURA, K.: In: ASAHINA, E. (Ed.): Cellular injury and resistance in freezing organisms, Vol. II, pp. 143—156. ILTS Hokkaido Univ., Sapporo 1967.

KLEBS, G.: Willkürliche Entwicklungsänderungen bei Pflanzen. Ein Beitrag zur Physiologie der Entwicklung. Jena: Fischer 1903.

KLEBS, G.: Abh. Akad. Wiss. Heidelberg **3**, 1—113 (1914).

KLEBS, G.: Biol. Zbl. **37**, 373—415 (1917).

KLEBS, G.: Flora **111/112**, 128—151 (1918).

KNAPP, R., Naturwissenschaften **43**, 41—42 (1956).

KOBLET, R.: Ber. Schweiz. Botan. Ges. **41**, 199—283 (1932).

KOBLET, R.: Proc. Int. Seed Test Ass. **9**, 82—122 (1937).

KOBLET, R.: Ber. Schweiz. Botan. Ges. **53** A, 369—394 (1943).

KOCH, W.: Flora B **158**, 402—428 (1969).

KOCHERZHENKO, I. E., KHOLODNYI, N. G., SHUMAKOVA, R. R.: Tr. Glavn. Botan. Sada A. N. SSSR **2**, 26—58 (1951).

KOHL, H. C., jr.: Proc. Am. Soc. Hort. Sci. **72**, 481—484 (1958).

KOHN, H.: Gartenbauwissenschaften **24**, 314—329 (1959).

KOJIMA, H., YAHIRO, M., ÊTÔ, T.: J. Fac. Agric. Kyushu Univ. **11**, 25—35 (1957).

KOLLER, D.: Bull. Res. Council Israel Sect. D **5**, 85—108 (1955).

KOLLER, D., HIGHKIN, H. R., CASO, O. H.: Am. J. Botan. **47**, 518—524 (1960).

KOLLER, D., MAYER, A. M., POLJAKOFF-MAYBER, A., KLEIN, S.: Ann. Rev. Plant Physiol. **13**, 437—464 (1962).

KONAREV, V. G.: Biochimiya **19**, 131—136 (1954).

KOZLOWSKI, T. T.: Forest Sci. **7**, 177—192 (1961).

KOZLOWSKI, T. T., KELLER, T.: Botan. Rev. **32**, 294—382 (1966).

KRAMER, P. J.: Plant Physiol. **11**, 127—137 (1936).

KRAMER, P. J.: Plant Physiol. **12**, 881—883 (1937).

KRAMER, P. J.: Plant Physiol. **15**, 63—79 (1940).

KRAMER, P. J.: Am. J. Botan. **29**, 828—832 (1942).

KRAMER, P. J.: Forest Sci. **3**, 45—55 (1957).

KRAMER, P. J.: In: RUHLAND, W. (Hrsg.): Handbuch der Pflanzenphysiologie, Bd. 3, S. 124—159, S. 188—214. Berlin-Göttingen-Heidelberg: Springer 1965.

KRAMER, P. J.: In: THIMANN, K. V. (Ed.): The physiology of forest trees. New York: The Ronald Press Company 1958.

KRAMER, P. J., KOZLOWSKI, T. T.: Physiology of forest trees. New York: McGraw-Hill 1960.

KRASAVTSEV, O. A.: In: Physiology of hardiness of plants, pp. 229—234, Moscow: Nauka 1960.

KRASAVTSEV, O. A.: In: ASAHINA, E. (Ed.): Cellular injury and resistance in freezing organisms, Vol. II, pp. 131—142, ILTS Hokkaido Univ. Sapporo 1967.

KRASAVTSEV, O. A.: Fiziol. Rast. 15, 225—234 (1968).

KRASAVTSEV, O. A.: Fiziol. Rast. 16, 228—236 (1969).

KRASAVTSEV, O. A.: Fiziol. Rast. 17, 508—513 (1970).

KREEB, K.: Angew. Botan. 44, 167—177 (1970).

KREKULE, J.: Biol. Plant. 3, 85—88 (1961 a).

KREKULE, J.: Biol. Plant. 3, 107—114 (1961 b).

KREKULE, J.: Biol. Plant. 3, 180—191 (1961 c).

KREKULE, J.: Biol. Plant. 6, 299—305 (1964).

KREKULE, J., TELTSCHEROVÁ, L.: Biol. Plant. 5, 252—257 (1963).

KROSBY, P.: Meldinger Norg. Landsbruckshoegskole 6, 241 (1926).

KRUEGER, K., FERRELL, W.: Ecology 46, 794—801 (1965).

KRUG, H.: Botan. Arch. 27, 420—518 (1929).

KRUZHILIN, A. S., SHVEDSKAYA, Z. M.: Dokl. Akad. Nauk SSSR, N. S., 116, 870—873 (1957).

KRUZHILIN, A. S., SHVEDSKAYA, Z. M.: Dokl. Akad. Nauk SSSR, N. S., 121, 561—564 (1958).

KRUZHILIN, A. S., SHVEDSKAYA, Z. M.: Botan. Zh. 46, 936—948 (1961).

KUIJPER, J.: Rec. Trav. Botan. Néerl. 30, 1—22 (1953).

KUIPER, P. J. C.: Mededel. Landbouwhogeschool Wageningen 61 (7), 1—49 (1961).

KUIPER, P. J. C.: Conn. Agr. Exp. Sta. N. Haven Bull. 664, 59—67 (1963).

KUIPER, P. J. C.: Mededel. Landbouwhogeschool Wageningen 64 (4), 1—11 (1964).

KUIPER, P. J. C.: Science 146, 544—546 (1964).

KUIPER, P. J. C.: Mededel. Landbouwhogeschool Wageningen 67 (3), 1—23 (1967).

KUIPER, P. J. C.: Plant Physiol. 45, 684—686 (1970).

KULL, U.: Planta 79, 299—311 (1968).

KURAISHI, S., ARAI, N., USHIJIMA, T., TAZAKI, T.: Plant Physiol. 43, 238—242 (1968).

KUSUMOTO, T.: Japan J. Ecology 7, 126—139 (1957).

KUSUMOTO, T.: Botan. Mag. 70, 229—304 (1957).

KUSUMOTO, T.: Japan. J. Botan. 17, 307—331 (1961).

KUYPER, J.: Rec. Botan. Neerl. 7, 131—240 (1910).

KWIATKOWSKA, M.: Acta Soc. Botan. Polon. 39, 347—360 (1970a).

KWIATKOWSKA, M.: Acta Soc. Botan. Polon. 39, 361—371 (1970b).

KYLIN, H.: Ber. Deut. Botan. Ges. 35, 370—384 (1917).

LACEY, J. W.: Gard. Chron. 116, 6—7 (1944).

LAIBACH, F.: Planta 51, 148—166 (1958).

LANG, A.: Züchter 21, 241—243 (1951).

LANG, A.: Naturwissenschaften 43, 257—258 (1956 a).

LANG, A.: Plant Physiol. 31, XXXV (1956 b).

LANG, A.: Naturwissenschaften 43, 284—285 (1956 c).

LANG, A.: Naturwissenschaften 43, 544 (1956 d).

LANG, A.: In: RUHLAND, W. (Hrsg.): Handbuch der Pflanzenphysiologie, Bd. 14, S. 909—950. Berlin-Göttingen-Heidelberg: Springer 1961.

LANG, A.: In: RUHLAND, W. (Hrsg.): Handbuch der Pflanzenphysiologie, Bd. 15/1, S. 1380—1536. Berlin-Heidelberg-New York: Springer 1965.

LANG, A.: In: RUHLAND, W. (Hrsg.): Handbuch der Pflanzenphysiologie, Bd. 15/2, S. 848—893. Berlin-Heidelberg-New York: Springer 1965.

LANG, A., MELCHERS, G.: Z. Naturforsch. 2b, 444—449 (1947).

LANG, A., SANDOVAL, J. A., BEDRI, A.: Proc. Nation. Acad. Sci. 43, 960—964 (1957).

LANGE, O. L.: Flora 140, 39—97 (1953).

LANGE, O. L.: Flora **145**, 381—389 (1955).
LANGE, O. L.: Flora **147**, 595—651 (1959).
LANGE, O. L.: Planta **56**, 666—693 (1961).
LANGE, O. L.: Naturwissenschaften **49**, 20—21 (1962).
LANGE, O. L.: Ber. Deut. Botan. Ges. **75**, 351—352 (1963a).
LANGE, O. L.: Flora **153**, 387—425 (1963b).
LANGE, O. L.: Planta **64**, 1—19 (1965a).
LANGE, O. L.: In: Meth. Plant Eco-Physiology-Proc. UNESCO Symp. Montpellier, 1965, 399—405 (1965b).
LANGE, O. L.: Ber. Deut. Botan. Ges. **78**, 441—454 (1966).
LANGE, O. L.: Flora B **156**, 500—502 (1966).
LANGE, O. L.: In: TROSHIN, A. S. (Ed.): The cell and environmental temperature, pp. 131—141. Oxford, New York, Toronto: Pergamon Press 1967.
LANGE, O. L.: Flora B **158**, 324—359 (1969a).
LANGE, O. L.: Planta **89**, 90—94 (1969b).
LANGE, O. L.: Ber. Deut. Botan. Ges. **82**, 3—22 (1969c).
LANGE, O. L., KOCH, W., SCHULZE, E. D.: Ber. Deut. Botan. Ges. **82**, 39—61 (1969).
LANGE, O. L., LANGE, R.: Flora **152**, 707—710 (1962).
LANGE, O. L., LANGE, R.: Flora **153**, 387—425 (1963).
LANGE, O. L., METZNER, H.: Naturwissenschaften **52**, 1—2, 191 (1965).
LANGE, O. L. SCHULZE, E. D., KOCH, W.: Flora **159**, 38—62 (1969).
LANGE, O. L., SCHWEMMLE, B.: Planta **55**, 208—225 (1960).
LANGRIDGE, J.: Ann. Rev. Plant Physiol. **14**, 441—462 (1963).
LANGRIDGE, J., McWILLIAM, J. R.: In: ROSE, A. H. (Ed.): Thermobiology, pp. 231—292. London-New York: Academic Press 1967.
LAPINS, K.: Can. J. Plant Sci. **41**, 381—393 (1960).
LAPINS, K.: Can. J. Plant Sci. **45**, 429—435 (1965).
LARCHER, W.: Planta **44**, 607—635 (1954).
LARCHER, W.: Veröff. Ferdinandeum Innsbruck **37**, 49—81 (1957).
LARCHER, W.: Planta **56**, 575—606 (1961).
LARCHER, W.: Protoplasma **57**, 569—587 (1963a).
LARCHER, W.: Ber. Naturw.-Med. Ver. Innsbruck **53**, 125—137 (1963b).
LARCHER, W.: Veröff. Ferdinandeum Innsbruck **43**, 153—199 (1963c).
LARCHER, W.: Deut. Akad. Landw. Wiss., Tagungsber. **100**, 7—20 (1968).
LARCHER, W.: Photosynthetica **3**, 167—198 (1969a).
LARCHER, W.: Planta **88**, 130—135 (1969b).
LARCHER, W.: Oecol. Plant. **5**, 267—286 (1970).
LARCHER, W.: Oecol. Plant. **6**, 1—14 (1971).
LARCHER, W.: Ber. Dtsch. Bot. Ges. **85**, 315—327 (1973a).
LARCHER, W.: Ökologie der Pflanzen. Stuttgart: E. Ulmer 1973b.
LARCHER, W., EGGARTER, H.: Protoplasma **51**, 595—619 (1960).
LARCHER, W., MAIR, B.: Oecol. Plant. **3**, 255—270 (1968).
LARCHER, W., MAIR, B.: Oecol. Plant. **4**, 347—376 (1969).
LARRIEU, C.: Compt. Rend. Acad. Sci. Paris, D **262**, 1448—1451 (1966).
LATIES, G. G.: Ann. Rev. Plant Physiol. **20**, 89—116 (1969).
LAUDE, H. M., CHAUGULE, B. A.: J. Range Managemt. **6**, 320—324 (1953).
LAUER, E.: Flora **40**, 551—595 (1953).
LAURIE, A., KIPLINGER, D. C.: Commercial flower forcing, 4. ed., Philadelphia 1944.
LAW, C. N., JENKINS, G.: Genet. Res. Cambridge **15**, 197—308 (1970).
LEDESMA, N.: Rev. Fac. Agron. Univ. Nac. La Plata **27**, 181—196 (1950).
LEFÉBURE, E. A.: Expérience sur la germination des plantes. Strasbourg 1801.
LEGGETT, J. E.: Ann. Rev. Plant Physiol. **19**, 333—346 (1968).
LEIBO, S. P., FARRANT, J., MAZUR, P. HANNA, M. G., Jr., SMITH, L. H.: Cryobiology **6**, 315—332 (1970).
LEIBO, S. P., MAZUR, P.: Biophys. J. **6**, 747—772 (1966).
LEIKE, H.: Wiss. Z. Univ. Rostock **14**, 475—492 (1965).
LEMAN, V. M.: Dokl. Akad. Nauk. SSSR **59**, 777—780 (1948a).

LEMAN, V. M.: Dokl. Akad. Nauk SSSR **60**, 1261—1264 (1948 b).
LEOPOLD, A. C.: VI. intern. Grassland Congr. Doc. **5**, 9 (1952).
LEOPOLD, A. C., GUERNSEY, F. S.: Am. J. Botan. **40**, 46—50 (1953 a).
LEOPOLD, A. C., GUERNSEY, F. S.: Am. J. Botan. **40**, 603—607 (1953 b).
LEOPOLD, A. C., GUERNSEY, F. S.: Science **118**, 215—217 (1953 c).
LEOPOLD, A. C., GUERNSEY, F. S.: Am. J. Botan. **41**, 181—185 (1954).
LE SAINT, A. M.: Thèses Univ. Paris 1966.
LE SAINT, A. M.: Compt. Rend. Acad. Sci. Paris **268**, 310—313 (1969).
LE SAINT-QUERVEL, A. M.: Compt. Rend. Acad. Sci. Paris **269**, 1423—1426 (1969).
LEVIS, H., WENT, F. W.: Am. J. Botan. **32**, 1—12 (1945).
LEVITT, J.: The hardiness of plants. New York: Academic Press 1956.
LEVITT, J.: Protoplasma **48**, 289—302 (1957).
LEVITT, J.: Frost- dought-, and heat-resistance, Protoplasmatologia, Bd. VIII/6, Wien:
    Springer 1958.
LEVITT, J.: J. Theoret. Biol. **3**, 355—391 (1962).
LEVITT, J.: Cyrobiology **1**, 312—316 (1965).
LEVITT, J.: In: MERYMAN, H. T. (Ed.): Cryobiology, pp. 495—563. London, New York:
    Academic Press 1966.
LEVITT, J.: Planta **74**, 101—118 (1967).
LEVITT, J.: Symp. Soc. Exp. Biol. **23**, 395—448 (1969).
LEVITT, J.: Responses of Plants to Environmental Stresses. New York-London:
    Academic Press 1972.
LEVITT, J., DEAR, J.: In: WOLSTENHOLME, G. E. W., O'CONNOR, M. (Eds.): Ciba Foun-
    dation Symposium on the Frozen Cell, pp. 149—174. London: J & A. Churchill
    1970.
LEVITT, J., SCARTH, G. W.: Can. J. Res. C **14**, 267—305 (1936).
LEWIS, D. A.: Science **124**, 75—76 (1956).
LI, P. H., WEISER, C. J.: Proc. Am. Soc. Hort. Sci. **91**, 716—727 (1966).
LIDFORSS, B.: Botan. Cbl. **68**, 33—44 (1896).
LIEBENBERG, A. VON: Botan. Cbl. **18**, 21—26 (1884).
LIEBERMANN, M., CRAFT, C. C., AUDIA, M. V., WILCOX, M. S.: Plant Physiol. **33**, 307—311
    (1958).
LING, G. N.: In: ROSE, A. H. (Ed.): Thermobiology, pp. 5—24 London-New York:
    Academic Press 1967.
LIPMAN, C. B., LEWIS, G. N.: Plant Physiol. **9**, 392—394 (1934).
LIST, R. J.: Smithsonian Metereorological Tables: Smithsonian Institution, 411—413
    (1963).
LÖHR, E.: Dansk Skovfor. Tidskr. **52**, 321—327 (1967).
LÖHR, E.: Physiol. Plant. **22**, 86—93 (1969).
LOEWEL, E. L., KARNATZ, H.: Züchter **26**, 117—120 (1956).
LOMAGIN, A. G., ANTROPOVA, T. A.: Planta **68**, 297—309 (1966).
LOMAGIN, A. G., ANTROPOVA, T. A., SEMENIKHINA, L. V.: Planta **71**, 119—124 (1966).
LOMMEN, P. W., SCHWINTZER, C. R., YOCUM, C. S., GATES, D. M.: Planta **97**, 159—220
    (1971).
LONA, F.: Planta **82**, 145—152 (1968).
LOO, S. V.: Am. J. Botan. **33**, 382—389 (1946).
LORENZ, H.: Beiträge zur Kenntnis der Keimung der Winterknospen von Hydrocharis
    morsus ranae, Utricularia vulgaris und Myriophyllum verticallatum. Diss. Kiel.
    1903.
LOVELOCK, J. E.: Biochim. Biophys. Acta **10**, 414—426 (1953 a).
LOVELOCK, J. E.: Biochim. Biophys. Acta **11**, 28—36 (1953 b).
LOVELOCK, J. E: Proc. Roy. Soc. Med. **47**, 60—65 (1954 a).
LOVELOCK, J. E.: Biochem. J. **56**, 265—270 (1954 b).
LOVELOCK, J. E.: Brit. J. Haemat. **1**, 117—124 (1955).
LOVELOCK, J. E.: Proc. Roy. Soc. London, Ser. B, Biol. Sci. **147**, 427—433 (1957).
LOVELOCK, J. E., BISHOP, M. W. H.: Nature **183**, 1394—1395 (1959).
LUDLOW, M. M., WILSON, G. L.: Austral. J. Biolog. Sci. **24**, 449—470 (1971).

LUKNITSKAYA, A. F.: Tsitologiya **5**, 136—141 (1963).
LUNDEGÅRDH, H.: Biochem. Z. **154**, 194 (1924 a).
LUNDEGÅRDH, H.: Der Kreislauf der Kohlensäure in der Natur. Jena: G. Fischer 1924 b.
LUNDEGÅRDH, H.: Flora **121**, 273—300 (1927).
LUNDEGÅRDH, H.: Die Nährstoffaufnahme der Pflanzen. Jena: Fischer 1932.
LUNDEGÅRDH, H., BURSTRÖM, H.: Biochem. Z. **261**, 235—251 (1933).
LUYET, B. J., RASMUSSEN, D.: Biodynamica **10**, 137—147 (1967).
LUYTEN, I.: Mededel. Landbouwhogeschool Wageningen **48**, 3—31 (1946).
LUYTEN, I., BLAAUW, A. H.: Proc. Koninkl. Ned. Akad. Wetenschap. **37**, 132—138 (1934).
LUZIKOV, V. N., SAKS, V. A., BEREZIN, I. V.: Biochim. Biophys. Acta **223**, 16—30 (1970).
LYBECK, B. R.: Plant Physiol. **34**, 482—486 (1959).
LYONS, J. M., RAISON, J. K.: Plant Physiol. **45**, 386—389 (1970 a).
LYONS, J. M., RAISON, J. K.: Cryobiology **6**, 585—586 (1970 b).
LYONS, J. M., WHEATON, T. A., PRATT, H. K.: Plant Physiol. **39**, 262—268 (1964).
LYR, H., POLSTER, H., FIEDLER, H. J.: Gehölzphysiologie. Jena: Fischer 1967.
LYUTOVA, M. I.: Botan. Zh. **43**, 283—287 (1958).
LYUTOVA, M. I.: Botan. Zh. **47**, 1761—1774 (1962).
LYUTOVA, M. I.: Dokl. Akad. Nauk SSSR **149**, 1206—1208 (1963).
LYUTOVA, M. I., DROBYSHEV, V. P.: Tsitologiya **8**, 484—493 (1968).
LYUTOVA, M. I., LUKNITSKAYA, A. F., FELDMAN, N. L.: In: TROSHIN, A. S. (Ed.): The cell and environmental temperature, pp. 166—172. Oxford, New York, Toronto: Pergamon Press 1967.
McKENZIE, A. P., LUYET, B. J.: Biodynamica **10**, 95—122 (1967).
MAIR, B.: Planta **82**, 164—169 (1968).
MANN, H. H.: Ann. Appl. Biol. **38**, 435—443 (1951).
MANN, L. K., LEWIS, D. A.: Hilgardia **26** (3), 161—189 (1956).
MANNING, G. B., CAMPBELL, L. L.: J. Biol. Chem. **236**, 2952—2957 (1961).
MARGARA, J.: Compt. Rend. Acad. Sci. Paris **249**, 751—753 (1959).
MARGARA, J.: Ann. Inst. Nation. Rech. Agron. B, **10**, 361—371 (1960 a).
MARGARA, J.: Ann. Inst. Nation. Rech. Agron., A^{bis}, Ann. Physiol. Végét. **2**, 281—293 (1960 b).
MARGARA, J.: Physiol. Végét. **1**, 315—324 (1963 a).
MARGARA, J.: Compt. Rend. Acad. Sci. Paris **257**, 743—746 (1963 b).
MARKOWSKI, A., MADEJ, M.: Bull. Acad. Pol. Sci., Sér. Sci. Biol. **10**, 139—144 (1962 a).
MARKOWSKI, A., MADEJ, M.: Roczn. Nauk Pol., Ser. 4, **85**, 421—445 (1962 b).
MARKOWSKI, A., MYCZKOWSKI, J., LEBEK, J.: Bull. Acad. Pol. Sci. Sér. Sci. Biol. **10**, 145—150 (1962).
MAR-MÖLLER, C., MÜLLER, D., NIELSEN, J.: Forsgsvaesen **21**, 327—335 (1954).
MARMUR, J.: Biochim. Biophys. Acta **38**, 342—343 (1960).
MARTH, P. C.: Agr. Food Chem. **13**, 331—333 (1965).
MATHON, C.-C.: Phyton (Vicente López) **12**, 13—23 (1959).
MATHON, C.-C.: Compt. Rend. Soc. Biol. **153**, 1569—1571 (1960 a).
MATHON, C.-C.: Bull. Soc. Botan. France **106**, 454—456 (1960 b).
MATHON, C.-C.: Phyton (Vicente López) **14**, 167—174 (1960 c).
MATHON, C.-C., LUCIANI-GRESTA, F.: Compt. Rend. Soc. Biol. **162**, 962—964 (1968).
MATHON, C.-C., NEHOU, J.: Compt. Rend. Soc. Biol. **154**, 1056—1058 (1960).
MAXIMOV, N. A.: Jb. Wiss. Botan. **53**, 327—420 (1914).
MAXIMOV, N. A.: Protoplasma **7**, 259—288 (1929).
MAYER, A. M., POLJAKOFF-MAYBER, A.: The germination of seeds. Oxford-London-New York-Paris: Pergamon Press 1963.
MAZUR, P.: Cryobiology **2**, 181—192 (1966).
MAZUR, P.: In: MERYMAN, H. T. (Ed.): Cryobiology, pp. 213—315. London, New York: Academic Press 1966.
MAZUR, P.: Cellular injury and resistance in freezing organisms. Proc. Intern. Low Temp. Sci., Vol. II, 171—189. Sapporo: Inst. Low Temp. Sci., Hokkaido Univ., Sapporo 1967.

MAZUR, P.: Ann. Rev. Plant Physiol. **20**, 419—448 (1969).
MAZUR, P.: Science **168**, 939—949 (1970).
MAZUR, P., FARRANT, J., LEIBO, S. P., CHU, E. H. Y.: Cryobiology **6**, 1—9 (1969).
MAZUR, P., SCHMIDT, J. J.: Cryobiology **5**, 1—17 (1968).
MCCARTY, R. E., JAGENDORF, A. T.: Plant Physiol. **40**, 725—735 (1965).
MCCARTY, R. E., RACKER, E.: Brookhaven Symp. Biol. **19**, 202—214 (1967).
MCCOWN, B. H., HALL, T. C., BECK, G. E.: Plant Physiol. **44**, 210—216 (1969).
MCLEESTER, R. C., WEISER, C. J., HALL, T. C.: Plant Physiol. **44**, 37—44 (1969).
MEIDNER, H., MANSFIELD, T. A.: Physiology of stomata. London: McGraw Hill 1968.
MEKHANIK, F. YA.: Fiziol. Rast. **8**, 330—337 (1961).
MEKHANIK, F. YA.: Zh. Obshch. Biol. **23**, 265—275 (1962).
MELCHERS, G.: Biol. Zbl. **56**, 567—570 (1936).
MELCHERS, G.: Biol. Zbl. **57**, 568—614 (1937).
MELCHERS, G.: Naturwissenschaften **26**, 496 (1938).
MELCHERS, G.: Ber. Deut. Botan. Ges. **57**, 29—48 (1939).
MELCHERS, G.: Rc. Ist. Lombardo Sci. Lett., Cl. Sci. **83**, 1—29 (1950).
MELCHERS, G., LANG, A.: Z. Naturforsch. **36**, 105—107 (1948).
MELLOR, R. S., SALISBURY, F. B., RASCHKE, K.: Planta **61**, 56—72 (1964).
MENDONZA, L. A.: Idia, Suppl. Forestal 1968/69 (**5**,) 51—88 (1968).
MERYMAN, H. T.: Cryobiology, London-New York: Academic Press 1966.
MERYMAN, H. T.: Nature **218**, 333—336 (1968).
MERYMAN, H. T.: Ciba Found. Symp. on the Frozen Cell, G. E. W. Wolstenholme and
    M. O'Connor (Eds.), pp. 51—67. London: J. & A. Churchill 1970.
MERYMAN, H. T.: Cryobiology **8**, 173—183 (1971).
MESSERI, A.: N. Giorn. Botan. Ital. **58**, 535—549 (1951).
MEYER, B. S.: Am. J. Botan. **15**, 449—472 (1928).
MICHAEL, G.: Flora B **156**, 350—372 (1966).
MICHAEL, G.: Arch. Forstwesen **16**, 1015—1032 (1967).
MICHAELIS, P.: Beih. Botan. Cbl. **52**, B 333—377 (1934 a).
MICHAELIS, P.: Jb. Wiss. Botan. **80**, 337—362 (1934 b).
MICHNIEWICZ, M., KAMIEŃSKA, A.: Acta Soc. Botan. Pol. **36**, 67—72 (1967).
MICHNIEWICZ, M., LANG, A.: Planta **58**, 549—563 (1962).
MIGITA, S.: Bull. Fac. Fisheries Nagasaki Univ. **21**, 131—138 (1966).
MILNER, H. W., HIESEY, W. M.: Plant Physiol. **39**, 208—213 (1964).
MITCHELL, P.: Nature **191**, 144—148 (1961).
MITCHELL, P.: Biol. Rev. **41**, 445—502 (1966).
MITTELSTÄDT, H.: Sitzber. Deut. Akad. Landw. Wiss. **11**, 19—37 (1962).
MITTELSTÄDT, H.: Züchter **35**, 311—327 (1965).
MITTELSTÄDT, H.: Züchter **36**, 282—290 (1966).
MITTELSTÄDT, H.: Arch. Gartenbau **16**, 37—49 (1968).
MITTELSTÄDT, H.: Tagungsber. Deut. Akad. Landw. Wiss. **96**, 149—173 (1969).
MITTELSTÄDT, H.: Flora **160**, 195—216 (1971 a).
MITTELSTÄDT, H.: Arch. Züchtungsforsch. **1**, 59—63 (1971 b).
MITYGA, H. G., LANPHEAR, F. O.: Cryobiology **6**, 276—277 (1969).
MODLIBOWSKA, I.: Rapp. Cgr. Pomol. Int. Namur, **1956**, 83—112 (1956).
MODLIBOWSKA, I.: XVIth. Int. Hortic. Cgr. Brussels, **1962**, 180—189 (1962).
MODLIBOWSKA, I.: Nature **208**, 503—504 (1965).
MODLIBOWSKA, I: Cryobiology **5**, 175—187 (1968).
MOLISCH, H.: Untersuchungen über das Erfrieren der Pflanzen. Jena: Fischer 1897.
MOLISCH, H.: Sitzber. Akad. Wiss. Wien Math.-Naturw. Kl. Abt. I. **117**, 87—117 (1908).
MOLISCH, H.: Sitzber. Akad. Wiss. Wien Math.-Naturw. Kl. Abt. I. **118**, 637—691
    (1909 a).
MOLISCH, H.: Das Warmbad als Mittel zum Treiben der Pflanzen Jena: Fischer 1909 b.
MOLOTKOVSKY, Y. G.: Biochimiya **33**, 961—968 (1968).
MOLOTKOVSKY, Y. G., ZHESTKOVA, I. M.: Fiziol. Rast. **11**, 301—307 (1964).
MOLOTKOVSKY, Y. G., ZHESTKOVA, I. M.: Dokl. Akad. Nauk SSSR **166**, 488—491 (1966).
MOLOTKOVSKY, Y. G., ZHESTKOVA, I. M.: Fiziol. Rast. **14**, 367—371 (1967).

MONTFORT,C., RIED,A., RIED,I.: Beitr. Biol. Pflanz. **31**, 349—375 (1955).
MOONEY,H.A.: Ecology **44**, 812—816 (1963).
MOONEY,H.A., BILLINGS,W.D.: Ecol. Monog. **31**, 1—29 (1961).
MOONEY,H.A., JOHNSON,A.W.: Ecology **46**, 721—727 (1965).
MOONEY,H.A., SHROPSHIRE,F.: Oecol. Plant. **2**, 1—13 (1968).
MOONEY,H.A., STRAIN,B.R., WEST,M.: Ecology **47**, 490—491 (1966).
MOONEY,H.A., WEST,M.: Am. J. Botan **51**, 825—827 (1964).
MORETTINI,A.: Acc. Econ.-agr. Georgof. **8**, 1—40 (1961).
MORRISON,J.W.: Z. Vererbungsl. **91**, 141—151 (1960).
MOROZ,E.S.: Soviet. Botan. **1940**, 5/6, 233—241 (1940).
MOSS,D.N.: Conn. Agr. Exp. Sta. N. Haven Bull. **664**, 86—101 (1963).
MOTHES,K.: Biol. Zbl. **52**, 193—223.
MUDRACK,F.: Planta **23**, 71—104 (1935).
MÜLLER,D.: Forsøgsvaesen **21**, 303—318 (1954a).
MÜLLER,D.: Forsøgsvaesen **21**, 327—335 (1954b).
MÜLLER,F.: Angew. Botan. **33**, 159—162 (1959).
MÜLLER-THURGAU,H.: Landw. Jb. **11**, 751—828 (1882).
MÜLLER-THURGAU,H.: Landw. Jb. **14**, 851—907 (1885).
MÜLLER-THURGAU,H.: Landw. Jb. **15**, 453—610 (1886).
MÜNCH,E.: Mitt. Deut. Dendrol. Ges. **1928**, 175—184 (1928).
MÜNTZ,K.: Photosynthetica **5**, 32—37 (1971).
MUNERATI,O.: Atti R. Accad. Naz. Lincei. Ser. 5, **29**, 273—275 (1920).
MUNERATI,O.: Compt. Rend. Acad. Sci. Paris **181**, 1081—1083 (1925).
MUNERATI,O.: Compt. Rend. Acad. Sci. Paris **182**, 535—537 (1926).
MUNERATI,O.: Compt. Rend. Acad. Sci. Paris **206**, 762—763 (1938).
MURAWSKI,H.: Sitzber. Deut. Akad. Landw. Wiss. **11** (6), 39—59 (1962).
MURAWSKI,H.: Arch. Gartenbau **16**, 400—430 (1968).
NAPP-ZINN,K.: Ber. Deut. Botan. Ges. **66**, 362—367 (1953).
NAPP-ZINN,K.: Z. Naturforsch. **9b**, 218—229 (1954).
NAPP-ZINN,K.: Ber. Deut. Botan. Ges. **69**, 193—198 (1956).
NAPP-ZINN,K.: Planta **48**, 683—695 (1957a).
NAPP-ZINN,K.: Flora **144**, 403—419 (1957b).
NAPP-ZINN,K.: Z. Vererbungsl. **88**, 253—285 (1957c).
NAPP-ZINN,K.: Z. Botan. **45**, 379—394 (1957d).
NAPP-ZINN,K.: Planta **50**, 177—210 (1957e).
NAPP-ZINN,K.: Beitr. Biol. Pflanz. **34**, 113—128 (1957f).
NAPP-ZINN,K.: Planta **54**, 409—444 (1960a).
NAPP-ZINN,K.: Planta **54**, 445—452 (1960b).
NAPP-ZINN,K.: In: RUHLAND,W. (Hrsg.): Handbuch der Pflanzenphysiologie, Bd. **16**, S. 24—75. Berlin-Göttingen-Heidelberg: Springer 1961a.
NAPP-ZINN,K.: Züchter **31**, 128—135 (1961b).
NAPP-ZINN,K.: Z. Vererbungsl. **93**, 154—163 (1962a).
NAPP-ZINN,K.: Naturwissenschaften **49**, 473—474 (1962b).
NAPP-ZINN,K.: Züchter **33**, 201—212 (1963a).
NAPP-ZINN,K.: Z. Botan. **51**, 317—339 (1963b).
NAPP-ZINN,K.: Ber. Deut. Botan. Ges. **76**, 77—89 (1963c).
NAPP-ZINN,K.: Beitr. Biol. Pflanz. **38**, 161—177 (1963d).
NAPP-ZINN.K.: Über genetische und entwicklungsphysiologische Grundlagen jahreszeitlicher Aspekte von Pflanzengesellschaften. In: KREEB,K. (Hrsg.): Beiträge zur Phytologie, S. 33—49. Stuttgart: Ulmer 1964a.
NAPP-ZINN,K.: Antimetabolites in vernalization. In: Symp. "Differentiation of apical meristems, etc.", pp. 54—56. Praha-Nitra 1964b.
NAPP-ZINN,K.: In: RÖBBELEN,G. (Ed.): Arabidopsis Research (Symp. Göttingen), pp. 56—61. Göttingen: Arabidopsis Information Service 1965.
NAPP-ZINN,K.: In: RUHLAND,W. (Hrsg.) Handbuch der Pflanzenphysiologie, Bd. **18**, S. 153—213. Berlin-Heidelberg-New York: Springer 1967a.
NAPP-ZINN,K.: Ber. Deut. Botan. Ges. **80**, 218—226 (1967b).

NAPP-ZINN, K.: Z. Pflanzenphysiol. **65**, 351—358 (1971).
NASH, T.: In: MERYMAN, H.T. (Ed.): Cryobiology, pp. 179—211. London-New York: Academic Press 1966.
NEGISI, K.: Bull. Tokyo Univ. Forests **62**, 1—11 (1966).
NEILSON, R.E., LUDLOW, M.M., JARVIS, P.G.: J. appl. Ecol. **9**, 721—745 (1972).
NEKRASOV, V.I.: Presowing seed treatment of forest species at low temperatures. Acad. Sci. USSR, Moscow 1960.
NESTEROV, J.S.: Dokl. Akad. Nauk SSSR, N. S., **141**, 1243—1245 (1961).
NEUWIRTH, G.: Biol. Cbl. **78**, 559—584 (1959).
NEWMAN, E., KRAMER, P.: Plant Physiol. **41**, 606—609 (1966).
NIKOLAEVA, M.G.: (Physiology of deep dormancy). Acad. Sci. USSR. Sci. Transl. Jerusalem 1969.
NIKOLAEVA, M.G., KOZLOVA, L.M., YUDIN, V.G.: In: GREBINSKY, S.O. (Ed.): Growth of plants, pp. 178—182. L'vov Universitet L'vov 1959.
NITSCH, J.P.: Proc. Am. Soc. Hort. Sci. **70**, 512—525 (1957a).
NITSCH, J.P.: Proc. Am. Soc. Hort. Sci. **70**, 526—544 (1957b).
NITSCH, J.P.: In: RUHLAND, W. (Hrsg.): Handbuch der Pflanzenphysiologie, Bd. 15/1, S. 1537—1647. Berlin-Heidelberg-New York: Springer 1965.
NÜRNBERG-KRÜGER, U.: Züchter **31**, 197—202 (1961).
NWACHUKU, N.I.C.: Planta **83**, 150—160 (1968).
OECHSSLER, G.: Z. Pflanzenphysiol. **59**, 213—225 (1968).
OEHLKERS, F.: Z. Naturforsch. **11b**, 471—480 (1956).
OKNINA, E.Z., BARKSAYA, E.I.: Practical manual for the determination of the seed germinability of the most important fruit crops at stratification. Akad. Nauk SSSR, 1956.
OLDÉN, E.J.: Sver. Pomol. Fören. årsskr. **1954**, 36—50 (1955).
OLDÉN, E.J.: Medd. Fören. Växtförädl. Fruktträd Balsgård, **43/45**, 17—39 (1957).
OLIEN, CH.R.: Ann. Rev. Plant Physiol. **18**, 387—408 (1967).
OLIEN, CH.R.: Proc. Symp. on Barley Genetics, Pullman, II 356—363 (1969).
OUELLET, C.E., SHERK, C.L.: Can. J. Plant Sci. **47**, 231—238, 339—349, 351—358 (1967).
OVERCASH, J.P., CAMPBELL, J.A.: Proc. Am. Soc. Hort. Sci. **66**, 87—92 (1955).
OVERCASH, J.P., LOOMIS, N.H.: Proc. Am. Soc. Hort. Sci. **73**, 91—98 (1959).
PANKRATOVA, N.M.: Botan. Zh. **41**, 263—266 (1956).
PARKER, J.: Protoplasma **48**, 148—163 (1957).
PARKER, J.: Forest Sci. **5**, 56—63 (1959).
PARKER, J.: Protoplasma **52**, 223—229 (1960a).
PARKER, J.: Nature **187**, 1133 (1960b).
PARKER, J.: Biol. Bull. **119**, 474—478 (1960c).
PARKER, J.: For. Sci. **8**, 255—262 (1962).
PARKER, J.: Plant Physiol. **37**, 809—813 (1962).
PARKER, J.: Botan. Rev. **29**, 123—201 (1963).
PARKHURST, D.F., DUNCAN, P.R., GATES, D.M., KREITH, F.: Agr. Meteor. **5**, 33—47 (1968).
PARLEVLIET, J.E.: Z. Pflanzenphysiol. **58**, 76—83 (1967).
PATON, D.M.: Austral. J. Biol. Sci. **22**, 303—310 (1969).
PAULI, A.W., WILSON, J.A., STICKLER, F.C.: Crop Sci. **2**, 271—274 (1962).
PERRY, TH.O.: Forest Sci. **8**, 336—344 (1962).
PERSSON, L.: Physiol. Plant. **22**, 959—976 (1969).
PETERSON, M.L., BENDIXEN, L.E.: Crop Sci. **3**, 79—82 (1963).
PHARIS, R.P.. HELLMERS, H., SCHUURMANS, E.: Photosynthetica **4**, 273—279 (1970).
PICARD, C.: Compt. Rend. Acad. Sci. Paris **247**, 2184—2187 (1958).
PICARD, C.: Compt. Rend. Acad. Sci. Paris **250**, 573—575 (1960).
PICARD, C.: Ann. Sci. Nat., 12e Sér., Botan., **6**, 197—314 (1965).
PICARD, C.: Compt. Rend. Acad. Sci. Paris, D **264**, 303—306 (1967a).
PICARD, C.: Compt. Rend. Acad. Sci. Paris, D **264**, 603—606 (1967b).
PICARD, C.: Planta **74**, 302—312 (1967c).

PICARD, C.: Aspects et mécanismes de la vernalisation. Paris: Masson 1968.

PICKHOLZ, L.: Landw. Versuchsw. Österr. **14**, 124—151 (1911).

PIERIK, R. L. M.: Z. Pflanzenphysiol. **56**, 141—152 (1967).

PISEK, A.: Protoplasma **39**, 129—146 (1950).

PISEK, A.: Gartenbauwissenschaften **23**, 54—74 (1958).

PISEK, A.: Bull. Res. Counc. Israel **8** D, 285—289 (1960 a).

PISEK, A.: In: RUHLAND, W. (Hrsg.): Handbuch der Pflanzenphysiologie, Bd. 5/II, S. 376—459. Berlin-Göttingen-Heidelberg: Springer 1960 b.

PISEK, A.: Mitt. floristisch-sozial. Arbeitsgem. N. F. **10**, 34—41 (1963).

PISEK, A., EGGARTER, H.: Gartenbauwissenschaften **24**, 446—456 (1959).

PISEK, A., KEMNITZER, R.: Flora B **157**, 314—326 (1968).

PISEK, A., KNAPP, H.: Ber. Deut. Botan. Ges. **72**, 277—294 (1959).

PISEK, A., LARCHER, W.: Protoplasma **44**, 30—46 (1954).

PISEK, A., LARCHER, W., PACK, I., UNTERHOLZNER, R.: Flora B **158**, 110—128 (1968).

PISEK, A., LARCHER, W., UNTERHOLZNER, R.: Flora B **157**, 239—264 (1967).

PISEK, A., REHNER, G.: Ber. Deut. Botan. Ges. **71**, 188—192 (1958).

PISEK, A., SCHIESSL, R.: Ber. Naturw.-Med. Ver. Innsbruck **47**, 33—52 (1947).

PISEK, A., TRANQUILLINI, W.: Flora **141**, 237—270 (1954).

PISEK, A., WINKLER, E.: Protoplasma **46**, 597—611 (1956).

PISEK, A., WINKLER, E.: Planta **51**, 518—543 (1958).

PISEK, A., WINKLER, E.: Planta **53**, 532—550 (1959).

PLATENIUS, H.: Plant Physiol. **17**, 179—197 (1942).

POGOSYAN, K. S.: Fiziol. Rast. **14**, 109—116 (1967).

POGOSYAN, K. S., SAKAI, A.: Low Temp. Sci. B **44**, 125—142 (1969).

POLGE, C., SMITH, A. U., PARKES, A. S.: Nature **164**, 666 (1949).

POLLOCK, B. M.: Physiol. Plant. **6**, 47—77 (1953).

PONOMAREVA, M. M.: Botan. Exp. **14** (1960).

POP, E., BĂRBAT, I., OCHEŞANU, C.: Stud. Cercet. Biol. **14**, 11—17 (1963).

POPCOV, A. V.: Compt. Rend. Acad. Sci. URSS **2**, 593—597 (1935).

POPCOV, A. V.: Byul. Gl. Botan. Sada. Akad. Nauk SSSR **11**, 55—59 (1952).

POPCOV, A. V.: Byul. Gl. Botan. Sada. Akad. Nauk SSSR **19**, 67—72 (1954).

POPCOV, A. V.: Trudy Gl. Botan. Sada Akad. Nauk SSSR **7**, 174—218 (1960).

POPE, M. N.: J. Agr. Res. **78**, 295—309 (1949).

POPE, M. N., BROWN, E.: J. Am. Soc. Agron. **35**, 161—163 (1943).

POPLAVSKY, K. M., GOLUBKOVA, A. S.: Fiziol. Rast. **8**, 434—440 (1961).

PORTER, V. S., DENNING, N. P., WRIGHT, R. C., SCOTT, E. M.: J. Biol. Chem. **205**, 883—891 (1953).

PORTHEIM, L., KÜHN, O.: Österr. Botan. Z. **64**, 410—420 (1914).

POTAPENKO, J. I., ZAKHAROVA, E. I.: Za michurinskoe plodovodstvo 1938, 4, 7—23 (1938).

POTAPENKO, J. I., ZAKHAROVA, E. I.: Dokl. Vses. Akad. Sel'skokhoz. Nauk **21/22**, 3—8 (1939).

POTAPENKO, J. I., ZAKHAROVA, E. I.: Tr. Lab. Evoljuc. Ekol. Rast. Moskovskij Bot. Sad. Akad. Nauk SSSR **1**, 127—158 (1940 a).

POTAPENKO, J. I., ZAKHAROVA, E. I.: Compt. Rend. Acad. Sci. URSS **27**, 278—282 (1940 b).

PRIME, C. T.: Lords and ladies, pp. 168—171. London: Collins 1960.

PROEBSTING, E. L.: Proc. Am. Soc. Hort. Sci. **74**, 144—153 (1959).

PURCELL, A. E., YOUNG, R. H.: Proc. Am. Soc. Hort. Sci. **83**, 352—358 (1963).

PURVIS, O. N.: Sci. Hort. **5**, 127—140 (1937).

PURVIS, O. N.: Sci. Hort. **6**, 160—177 (1938).

PURVIS, O. N.: Nature **145**, 462 (1940).

PURVIS, O. N.: Ann. Botany, N. S., **8**, 285—314 (1944).

PURVIS, O. N.: Ann. Botany, N. S., **11**, 269—283 (1947).

PURVIS, O. N.: Ann. Botany, N. S., **12**, 183—206 (1948).

PURVIS, O. N.: In: 8e Congr. intern. Bot., Paris, Rapp. et Comm., Sect. **11/12**, 286—288 (1954).

PURVIS, O. N.: Nature **185**, 479 (1960).
PURVIS, O. N.: In: Recent Advances in Botany, pp. 1202—1205. Toronto: Univ. of Toronto Press 1961 a.
PURVIS, O. N.: In: RUHLAND, W. (Hrsg.): Handbuch der Pflanzenphysiologie, Bd. **16**, S. 76—122. Berlin-Göttingen-Heidelberg: Springer 1961 b.
PURVIS, O. N., GREGORY, F. G.: Ann. Botany, N. S., **1**, 569—592 (1937).
PURVIS, O. N., GREGORY, F. G.: Ann. Botany, N. S., **16**, 1—21 (1952).
PURVIS, O. N., HATCHER, E. S. J.: J. exper. Bot. **10**, 277—289 (1959).
PUTNAM, F. W.: In: NEURATH, H., BAILEY, K. (Eds.): The proteins, Vol. I B, pp. 807—892. New York: Academic Press 1953.
RAJKI, S.: Növénytermelés **11**, 125—146 (1962).
RAJKI, S.: Autumnization and its genetic interpretation. Budapest: Akadémiai Kiadó 1967.
RAKITINA, Z. G.: In: Physiology of hardiness of plants, pp. 278—284. Moscow: Nauka 1960.
RAMMELT, R.: Z. Pflanzenphysiol. **56**, 397—400 (1967).
RAMMELT, R., MÜLLER-STOLL, W. R.: Flora A **160**, 234—252 (1969).
RAPATZ, G., LUYET, B. J.: Cryobiology **4**, 215—222 (1968).
RAPATZ, G., SULLIVAN, J. J., LUYET, B. J.: Cryobiology **5**, 18—25 (1968).
RAPPAPORT, L., BONNER, J.: Plant Physiol. **35**, 98—102 (1960).
RASCHKE, K.: Planta **48**, 200—238 (1956).
RASCHKE, K.: Flora **146**, 546—578 (1958).
RASCHKE, K.: Ann. Rev. Plant Physiol. **11**, 111—126 (1960).
RASCHKE, K.: Planta **91**, 336—363 (1970).
RAZUMOV, V. I., LIMARJ, R. S., KE-WEI TAN: Botan. Zh. **45**, 1732—1738 (1960).
REINECKE, O. S. H.: Union S. Afr. Dep. Agr. Bull. **97**, 1—16 (1931).
REINECKE, O. S. H.: J. Pom. Hort. Sci. **14**, 164—174 (1936).
REINHARD, E., LANG, A.: Plant Physiol. **36**, XII—XIII (1961).
REJOWSKI, A.: Bull. Acad. Pol. Sci., Sér. Sci. Biol. **13**, 369—372 (1965).
RICHTER, A. A.: Priroda (Leningrad) **2**, 43—46 (1934).
RICHTER, A. A., RANCAN, V., PEKKER, M.: Dokl. Akad. Nauk SSSR, N. S., **2**, 72—75 (1933).
RICHTER, H.: Protoplasma **65**, 155—166 (1968 a.)
RICHTER, H.: Protoplasma **66**, 63—78 (1968 b).
RIDDELL, J. A., GRIES, G. A.: Agron. J. **50**, 743—746 (1958).
RIETH, A., SAGROMSKY, H.: Biol. Zbl. **83**, 489—500 (1964).
ROBERTS, E. H.: J. Exp. Bot. **12**, 430—445 (1961).
ROBERTS, D. W. A.: Can. J. Botan. **45**, 1347—1357 (1967).
ROBERTS, D. W. A., GRANT, M. N.: Can. J. Plant Sci. **48**, 369—376 (1968).
ROBERTSON, J. D.: Progr. Biophys. Biophys. Chem. **10**, 343—355 (1960).
ROBERTSON, R. N.: In: RUHLAND, W. (Hrsg.): Handbuch der Pflanzenphysiologie, Bd. 4, S. 243—279. Berlin-Göttingen-Heidelberg: Springer 1958.
ROLLIN, P.: Bull. Soc. Franc. Physiol. Végétale **5**, 24—26 (1959).
ROMOSE, V.: Dansk Botan. Arkiv **10**, 1—138 (1940).
ROOK, D. A.: N. Zealand J. Botan. **7**, 43—55 (1969).
ROSE, R. C.: Botan. Gaz. **67**, 281—308 (1919).
ROSSMANN, E. C.: Plant Physiol. **24**, 629—636 (1949).
ROTTENBURG, W.: Protoplasma **65**, 37—48 (1968).
ROUSCHAL, E.: Sitzber. Akad. Wiss. Wien, Math-Naturw. Kl. I, **144**, 313—348 (1935).
RUDOLPH, T. D., NIENSTAEDT, H.: J. For. **60**, 138—139 (1962).
RUDORF, W.: Züchter **7**, 193—199 (1935).
RUDORF, W.: Züchter **10**, 238—246 (1938).
RÜNGER, W.: Z. Botan. **48**, 381—397 (1960).
RUFELT, H., JARVIS, P. G., JARVIS, M. S.: Physiol. Plant. **16**, 177—185 (1963).
RUHLAND, W., RAMSHORN, K.: Planta **28**, 471—514 (1938).
RYADNOVA, I. M.: Agrobiologiya **1**, 130—134 (1950).
SACHS, J.: Jb. Wiss. Botan. **2**, 338—377 (1860).

SACHS, J.: Ber. Vhdl. Kgl. Sächs. Ges. Wiss. Leipzig, Math.-Physikal. Kl. **12**, 1—50 (1860).
SACHS, J.: Flora **47**, 5—12, 24—29, 33—39, 65—75 (1864).
SAKAI, A.: Low Temp. Sci. B **13**, 33—41 (1955).
SAKAI, A.: Low Temp. Sci. B **14**, 18—26 (1956).
SAKAI, A.: Low Temp. Sci. B **16**, 23—34 (1958a).
SAKAI, A.: Low Temp. Sci. B. **16**, 35—39 (1958b).
SAKAI, A.: Low Temp. Sci. B **17**, 43—49 (1959).
SAKAI, A.: Nature **185**, 393—394 (1960).
SAKAI, A.: Nature **189**, 416—417 (1961).
SAKAI, A.: Contr. Low. Temp. Sci. B **11**, 1—40 (1962).
SAKAI, A.: Low. Temp. Sci. B **22**, 29—50 (1964).
SAKAI, A.: Plant Physiol. **40**, 882—887 (1965).
SAKAI, A.: Low. Temp. Sci. B **23**, 27—36 (1965).
SAKAI, A.: Plant Physiol. **41**, 353—359 (1966).
SAKAI, A.: Contr. Low Temp. Sci. B **15**, 1—14 (1968).
SAKAI, A., OTSUKA, K.: Plant Physiol. **42**, 1680—1694 (1967).
SAKAI, A., OTSUKA, K.: Ecology **51**, 665—671 (1970).
SAKAI, A., YOSHIDA, S.: Plant Physiol. **42**, 1695—1701 (1967).
SAKAI, A., YOSHIDA, S.: Cryobiology **5**, 160—174 (1968).
SAKAI, A., OTSUKA, K., YOSHIDA, S.: Cryobiology **4**, 165—173 (1968).
SALAGEANU, N., ATANASIU, L.: Acad. Republ. Roumaine, Ser. Biol. Veget. 2, Tom. **14**, 153—160 (1962).
SALAGEANU, N., ATANASIU, L.: Acad. Republ. Roumaine, Rev. Biol. **7**, 507—512 (1962).
SALT, R. W.: Ann. Rev. Entomol. **6**, 55—74 (1961).
SALT, R. W., KAKU, S.: Can. J. Botan. **45**, 1335—1346 (1967).
SALZER, J.: D. Ak. Landw. Wiss. Tagungsber. **96**, 191—204 (1969).
SAMISCH, R.: J. Pomol. Hort. Sci. **21**, 164—179 (1945).
SAMYGIN, G. A., VARLAMOV, V. N., MATVEEVA, N. M.: Fiziol. Rast. **7**, 97—100 (1960).
SANTARIUS, K. A.: In: HAWTHORN, J., ROLFE, E. J. (Eds.): Low Temp. Biol. of Foodstuffs. Rec. Adv. Food Sci., pp. 135—151. Oxford: Pergamon Press 1968.
SANTARIUS, K. A.: Planta **89**, 23—46 (1969).
SANTARIUS, K. A.: Plant Physiol. **48**, 156—162 (1971).
SANTARIUS, K. A.: Ber. Deut. Bot. Ges. **84**, 425—436 (1971).
SANTARIUS, K. A., HEBER, U.: Planta **73**, 109—137 (1967).
SANTARIUS, K. A., HEBER, U.: Cryobiology **7**, 71—78 (1970).
SAPPER, I.: Planta **23**, 518—556 (1935).
SARKAR, S.: Biol. Zbl. **77**, 1—49 (1958).
SAUTER, J. J.: Z. Pflanzenphysiol. **56**, 340—352 (1967).
SAWADA, S. I.: J. Fac. Sci. Tokyo III, **10**, 233—263 (1970).
SCARAMUZZI, F., ANDREUCCI, E.: N. Giorm. Botan. Ital. **64**, 19—124 (1957).
SCHANDER, H.: Z. Pflanzenzücht. **34**, 421—440 (1955a).
SCHANDER, H.: Z. Pflanzenzücht. **35**, 89—97 (1955b).
SCHANDER, H.: Z. Pflanzenzücht. **35**, 179—198 (1955c).
SCHEUMANN, W.: Deut. Akad. Landw. Wiss. Tagungsber. **69**, 189—199 (1965).
SCHEUMANN, W.: Wiss. Z. Univ. Leipzig. Math.-Naturw. Reihe, **1968** (2), 417—420 (1968).
SCHEUMANN, W.: Deut. Akad. Landw. Wiss. Tagungsber. **100**, 45—54 (1968).
SCHEUMANN, W., HOFFMANN, K.: Arch. Forstwiss. **16**, 701—705 (1967).
SCHEUMANN, W., SCHÖNBACH, H.: Arch. Forstwiss. **17**, 597—611 (1968).
SCHMIDT, O.: Z. Botan. **30**, 289—334 (1936).
SCHNEIDER, G.: Planta **55**, 669—686 (1960).
SCHNEIDER, G.: Planta **56**, 322—347 (1961).
SCHNELLE, F.: Frostschutz im Pflanzenbau, Bd. II. München-Basel-Wien: BLV Verlagsgesellschaft 1965.
SCHNETTER, M. L.: Biol. Zbl. **84**, 469—487 (1965).
SCHÖLM, H. E.: Protoplasma **65**, 97—118 (1968).

SCHÖNBACH, H., BELLMANN, E.: Archiv Forstwiss. **16**, 707—711 (1967).
SCHÖNBACH, H., BELLMANN, E., SCHEUMANN, W.: Silvae Genet. **15**, 141—192 (1966).
SCHOLANDER, P. F., LOVE, W. E., KANWISHER, J. W.: Plant Physiol. **30**, 93—104 (1955).
SCHRAMM, W.: Int. Rev. Ges. Hydrobiol. **53**, 469—510 (1968).
SCHROEDER, E. M., BARTON, L. V.: Contrib. Boyce Thompson Inst. **10**, 235—255 (1939).
SCHROEDER, C. A.: Nature **200**, 1301—1302 (1963).
SCHROEDER, C. A.: In: PROSSER, C. L. (Ed.): Molecular mechanisms of temperature adaptation, pp. 61—72. Am. Soc. Advan. Sci. Publ. 84, Wash. DC. 1967.
SCHROEDER, C. A., KAY, E.: Calif. Avocado Soc. Yearb. **45**, 87—92 (1961).
SCHULZE, E. D.: Flora **159**, 177—232 (1970).
SCHULZE, E. D., LANGE, O. L.: Flora B **158**, 180—184 (1968).
SCHULZE, E. D., LANGE, O. L., KAPPEN, L., BUSCHBOM, U., EVENARI, M.: Planta **110**, 29—42 (1973).
SCHULZE, E. D., LANGE, O. L., KOCH, W.: Oecologia **9**, 317—340 (1972).
SCHWABE, W. W.: Ann. Botany, N. S., **27**, 671—683 (1963).
SCHWARZ, W.: Flora **159**, 258—285 (1970).
SCHWEMMLE, B., LANGE, O. L.: Planta **53**, 134—144 (1959).
SCHWENKE, H.: Kieler Meeresforsch. **15**, 34—50 (1959).
SÉCHET, J.: Compt. Rend. Acad. Sci. Paris **228**, 334—336 (1949).
SÉCHET, J.: Compt. Rend. Acad. Sci. Paris **237**, 434—436 (1953 a).
SÉCHET, J.: Botaniste **37**, 1—289 (1953 b).
SÉCHET, J.: Compt. Rend. Acad. Sci. Paris **254**, 3238—3240 (1962).
SEIBLE, D.: Beitr. Biol. Pflanz. **26**, 289—330 (1939).
SEMIKHATOVA, O. A.: Exp. Botan. **8**, 132—155 (1953).
SEMIKHATOVA, O. A.: Botan. Exp. **13**, 91—112 (1959).
SEMIKHATOVA, O. A., BUSHUYEVA, T. M., NIKULINA, G. N.: In: TROSHIN, A. S. (Ed.): The cell and environmental temperature, pp. 283—287. Proc. Intern. Symp. Cytoecology, Leningrad 1963. Oxford: Pergamon Press 1967.
SEMIKHATOVA, O. A., DENKO, E. I.: Exp. Botan. **14**, 112—137 (1960).
SEMIKHATOVA, O. A., SAAKOV, V. S., GORBATCHEVA, G. I.: Exp. Botan. **15**, 25—42 (1962).
SEYBOLD, A.: Planta **9**, 270—314 (1929 a).
SEYBOLD, A.: Die physikalische Komponente der pflanzlichen Transpiration, Monogr. Gesamtgeb. wiss. Bot., Bd. 2. Berlin: Springer 1929 b.
SEYBOLD, A.: Die pflanzliche Transpiration I, Erg. Biol., Bd. **5**. Berlin: Springer 1929 c.
SEYBOLD, A.: Die pflanzliche Transpiration II, Erg. Biol. Bd. **6**. Berlin: Springer 1930.
SEYBOLD, A., BRAMBRING, F.: Planta **20**, 201—230 (1933).
SEYBOLD, S.: Flora A **160**, 561—575 (1969).
SHIFTAN, S. L.: Pal. J. Botan. Rehovot Ser. **5**, 96—105 (1945).
SHIRLEY, H. L.: J. Agr. Res. **53**, 239—258 (1936).
SHOMER-ILAN, A.: Israel J. Botan. **13**, 93—100 (1964).
SHULL, C. A.: Botan. Gaz. **52**, 453—477 (1911).
SHVEDSKAYA, Z. M., KRUZHILIN, A. S.: Fiziol. Rast. **8**, 613—618 (1961).
SHVEDSKAYA, Z. M., KRUZHILIN, A. S.: Fiziol. Rast. **13**, 850—858 (1966).
SIMINOVITCH, D., BRIGGS, D. R.: Arch. Biochem. **23**, 8—17 (1949).
SIMINOVITCH, D., BRIGGS, D. R.: Plant Physiol. **28**, 177—200 (1953).
SIMONOVITCH, D., GFELLER, F., RHEAUME, B.: In: ASAHINA, E. (Ed.): Cellular injury and resistance in freezing organisms, Vol. II, pp. 93—118. ILTS Hokkaido Univ., Sapporo 1967.
SIMINOVITCH, D., RHEAUME, B., POMEROY, K., LEPAGE, M.: Cryobiology **5**, 202—225 (1968).
SIMINOVITCH, D., WILSON, C. M., BRIGGS, D. R.: Plant Physiol. **28**, 383—400 (1953).
SIMURA, T.: Cytologia, Suppl. Proc. Int. Genet. Symp. **1956**, 321—324 (1957).
SIMURA, T., SUGIYAMA, N.: Jap. J. Breeding **15**, 230—239 (1965).
SISAKYAN, N. M.: Biochimiya **2**, 263—273 (1937).
SISAKYAN, N. M., FILIPPOVICH, I. I.: Dokl. Akad. Nauk SSSR, N. S., **76**, 443—446 (1951).
SISAKYAN, N. M., FILIPPOVICH, I. I.: Zh. Obshch. Biol. **14**, 215—228 (1953).

SKINNER, H. T.: 15th Int. Cgr. Hort., Nice 1958, Proc. **3**, 485—491 (1962).
SLATYER, R. O.: Plant-water relationships. London-New York: Academic Press 1967.
SLAVIK, B.: Water stress in plants. Proc. Sympos. Prague. CS. Acad. Sci., pp. 195—202, 1963.
SMITH, A. U.: Biological effects of freezing and supercooling. London: E. Arnold 1961.
SODERHOLM, P. K., GASKINS, M. H.: Coffee **3**, 40— 46 (1961).
SOUZU, H.: Arch. Biochem. Biophys. **120**, 344—351 (1967).
SPÄTH, H. L.: Der Johannistrieb. Diss. Berlin (1912).
SPARMANN, G.: Planta **56**, 447—474, **57**, 176—201 (1961).
SPECTOROV, K. S.: Fiziol. Rast. **4**, 209—214 (1957).
SPRANGER, E.: Gartenbauwissenschaften **16**, 90—128 (1941).
STÅLFELT, M. G.: Planta **6**, 183—191 (1928).
STÅLFELT, M. G.: Planta **8**, 287—340 (1929).
STÅLFELT, M. G.: Planta **23**, 715—759 (1935).
STÅLFELT, M. G.: In: RUHLAND, W. (Hrsg.): Handbuch der Pflanzenphysiologie, Bd. 3, S. 351—426. Berlin-Göttingen-Heidelberg: Springer 1956.
STÅLFELT, M. G.: In: RUHLAND, W. (Hrsg.) Handbuch der Pflanzenphysiologie, Bd. 5/II, S. 100—118. Berlin-Göttingen-Heidelberg: Springer 1960.
STÅLFELT, M. G.: Physiol. Plant. **15**, 772—779 (1962).
STEINBAUER, C. E.: Proc. Am. Soc. Hort. Sci. **29**, 403—408 (1932).
STEINER, M.: Jb. Wiss. Botan. **78**, 564—622 (1933).
STEINHÜBEL, G.: Einführung in die ökologische Physiologie der Sempervirenz. Vydavatelstvo Slov. Akad. Vied, Bratislava 1967.
STELZNER, G.: Pflanzenbau **18**, 150—157 (1942).
STEPONKUS, P. L.: Cryobiology **6**, 285—286 (1969).
STEPONKUS, P. L.: Plant Physiol. **47**, 175—180 (1971).
STEPONKUS, P. L., LANPHEAR, F. O.: Plant Physiol. **42**, 1673—1679 (1967).
STEWART, J. McD., GUINN, G.: Plant Physiol. **44**, 605—608 (1969).
STIER, H. L.: Proc. Am. Soc. Hort. Sci. **35**, 601—605 (1937).
STOCKER, O.: Planta **24**, 402—445 (1935).
STOCKER, O.: In: RUHLAND, W. (Hrsg.): Handbuch der Pflanzenphysiologie, Bd. 3, S. 436—488, Berlin-Göttingen-Heidelberg: Springer 1956.
STOECKENIUS, W.: In: RACKER, E. (Ed.): Membranes of mitochondria and chloroplasts, pp. 53—90. New York: Van Nostrand-Reinhold Co. 1970.
STOKES, P.: In: RUHLAND, W. (Hrsg.): Handbuch der Pflanzenphysiologie, Bd. 15/II, S. 746—803. Berlin-Heidelberg-New York: Springer 1965.
STOUT, M.: Botan. Gaz. **107**, 86—95 (1945).
STOUT, M.: Botan. Gaz. **110**, 438—439 (1949).
STROUN, M., MATHON, C.-C., PUGNAT, C.: Ber. Schweiz. Botan. Ges. **70**, 440—447 (1960).
STUART, N. W.: Florists Exchange **123**, (12) 18—19, 24 (1954a).
STUART, N. W.: Proc. Am. Soc. Hort. Sci. **63**, 488—494 (1954b).
STUART, T. S.: Planta **96**, 81—92 (1971).
STUBBE, H.: Züchter **25**, 321—330 (1955).
SUCOFF, E.: Plant Physiol. **22**, 424—431 (1969).
SUGE, H., OSADA, A.: Plant Cell Physiol. **7**, 617—630 (1966).
SUGE, H., YAMADA, N.: Plant Cell Physiol. **6**, 147—160 (1965a).
SUGE, H., YAMADA, N.: Proc. Crop Sci. Soc. Japan **33**, 324—329 (1965b).
SUGIYAMA, N., SIMURA, T.: Jap. J. Breeding **16**, 165—173 (1966).
SUGIYAMA, N., SIMURA, T.: Jap. J. Breeding **18**, 229—233 (1968).
SWANSON, C. A., GEIGER, D. R.: Plant Physiol **42**, 751—756 (1967).
SWINBANK, W. C.: Quart. J. Roy. Meteor. Soc. **89**, 339—348 (1963).
SZYBALSKI, W.: In: ROSE, A. H. (Ed.): Thermobiology, pp. 73—122. London-New York: Academic Press 1967.
TAKIMOTO, A., TASHIMA, Y., IMAMURA, S.: Botan. Mag. Tokyo **73**, 377 (1960).
TAN KE-WEI: Botan. Zh. **44**, 1437—1444 (1959).

TANNER, C. B.: In: KOZLOWSKI, T. T.: Water deficits and plant growth, Vol. I, pp. 73—106. London-New York: Academic Press 1968.

TASHIMA, Y., IMAMURA, S.: Botan. Mag. Tokyo **67**, 281—282 (1954).

TAYLOR, A. O., CRAIG, A. S.: Plant Physiol. **47**, 719—725 (1971).

TAYLOR, A. O., ROWLEY, J. A.: Plant Physiol. **47**, 713—718 (1971).

TAYLOR, S. A.: In: KOZLOWSKI, T. T.: Water deficits and plant growth, Vol. I, pp. 49—72, London-New York: Academic Press 1968.

TAYLOR, S. E., SEXTON, O. J.: Ecology **53**, 143—149 (1972).

TERUMOTO, I.: Low Temp. Sci. B **17**, 1—7 (1959).

TERUMOTO, I.: Low Temp. Sci. B **22**, 19—28 (1964).

TERUMOTO, I.: Low Temp. Sci. B **23**, 1—9 (1965a).

TERUMOTO, I.: Low Temp. Sci. B **23**, 11—20 (1965b).

THAMES, J. L.: Plant Physiol. **36**, 180—182 (1961).

THIELE, I.: Züchter **27**, 161—172 (1957).

THIMANN, K. V., KAUFMANN, D.: In: THIMANN, K. V. (Ed.): The physiology of forest trees, pp. 479—492. New York: Ronald Press 1958.

THOMPSON, P. A.: Nature **217**, 1156—1157 (1968).

THOMPSON, P. A.: Nature **225**, 827—831 (1970a).

THOMPSON, P. A.: Ann. Botan. **34**, 427—449 (1970b).

THOMPSON, P. A.: Physiol. Plant. **23**, 739—746 (1970c).

THOMPSON, R. C.: U. S. Dep. Agric. Techn. Bull. **655**, 1—20 (1938).

THORNTON, N. C.: Contrib. Boyce Thompson Inst. **7**, 477—496 (1935).

THORNTON, N. C.: Contrib. Boyce Thompson Inst. **13**, 487—500 (1945).

THREN, R.: Z. Botan. **26**, 449—526 (1934).

THROWER, S. L.: Austral. J. Biol. Sci. **18**, 449—461 (1965).

TIESZEN, L. L.: Tundra-Biome-Meeting, Kevo 1—4 (1970).

TILL, O.: Flora **143**, 499—542 (1956).

TOOLE, E. H.: Proc. Ass. Off. Seed Analysts N. Am **14/15**, 80—83 (1923).

TOOLE, E. H., BORTHWICK, H. A., HENDRICKS, S. B., TOOLE, V. K.: Proc. Intern. Seed Testing Ass. **18**, 267—276 (1953).

TOOLE, E. H., GOSS, W. L.: Proc. Ass. Offic. Seed Analysts N. Am. **14/15**, 90 (1923).

TOOLE, E. H., HENDRICKS, S. B., BORTHWICK, H. A., TOOLE, V. K.: Ann. Rev. Plant Physiol. **7**, 299—324 (1956).

TOOLE, E. H., HOLLOWELL, E. A.: Am. Soc. Agr. **31**, 604—619 (1939).

TOOLE, E. H., TOOLE, V. K.: Proc. Intern. Seed Testing Ass. **11**, 51—56 (1939).

TOOLE, E. H., TOOLE, V. K.: J. Agric. Res. **63**, 65—90 (1941).

TOOLE, E. H., TOOLE, V. K., BORTHWICK, H. A., HENDRICKS, S. B.: Plant Physiol. **30**, 15—21 (1955a).

TOOLE, E. H., TOOLE, V. K., BORTHWICK, H. A., HENDRICKS, S. B.: Plant Physiol. **30**, 473—478 (1955b).

TOOLE, V. K.: J. Am. Sod. Agron. **33**, 1037—1045 (1941).

TRANQUILLINI, W.: Planta **46**, 154—178 (1955).

TRANQUILLINI, W.: Planta **49**, 612—661 (1957).

TRANQUILLINI, W.: Forstw. Cbl. **77**, 65—128 (1958).

TRANQUILLINI, W., HOLZER, K.: Ber. Deut. Botan. Ges. **71**, 143—156 (1958).

TRANQUILLINI, W., MACHL-EBNER, I.: Rep. Kevo Subarct. Res. Stat. **8**, 158—166 (1971).

TRANQUILLINI, W., SCHÜTZ, W.: Zbl. Ges. Forstwes. **87**, 42—60 (1970).

TRÂN THANH VÂN, M., CHOUARD, P.: Compt. Rend. Acad. Sci. Paris D, **267**, 181—184 (1968).

TRÂN THANH VÂN-LÊ KIÊM NGOG, M.: Ann. Sci. Nat., 12ᵉ Sér., Botan., **6**, 373—594 (1965).

TREVIRANUS, L. CH.: Beiträge zur Pflanzenphysiologie (Abhandlungen, die Pflanzenphysiologie betreffend von TH. A. KNIGHT.) Göttingen: Dietrich 1811.

TRIONE, E. J., YOUNG, J. L., YAMAMOTO, M.: Phytochemistry **6**, 85—91 (1967).

TSUNEWAKI, K., JENKINS, B. C.: Jap. J. Genet. **36**, 428—443 (1961).

TUMANOV, I. I.: Fiziol. Rast. **3**, 283—292 (1955).

TUMANOV, I. I.: In: Physiology of hardiness of plants, pp. 5—17. Moscow: Nauka 1960.
TUMANOV, I. I.: In: TROSHIN, A. S. (Ed.): The cell and environmental temperature, pp. 6—14. Proc. Intern. Symp. Cytoecology, Leningrad 1963. Oxford: Pergamon Press 1967.
TUMANOV, I. I., KHVALIN, N. N.: Fiziol. Rast. **14**, 908—918 (1967).
TUMANOV, I. I., KRASAVTSEV, O. A.: Fiziol. Rast. **6**, 654—667 (1959).
TUMANOV, I. I., KUZINA, G. V., KARNIKOVA, L. D.: Fiziol. Rast. **12**, 665—682 (1965).
TUMANOV, I. I., TRUNOVA, T. I.: Fiziol. Rast. **5**, 112—122 (1958).
TUMANOV, I. I., TRUNOVA, I.: Fiziol. Rast. **10**, 176—188 (1963).
TURKEVICH, N. C.: Nauk Sap. Kiivsjk. Derzh. Univ. **22**, 49—59 (1952).
TYANKOVA, L.: Compt. Rend. Acad. Sci. Agric. Bulg. **2**, 317—321 (1969).
TYANKOVA, L.: Ber. Deut. Botan. Ges. **83**, 491—497 (1970).
TYURINA, M. M.: Issledovanie morozostoikosti rastenii v uslovyakh vysokogorii Pamira. Akad. Nauk SSSR, Leningrad 1953.
TYURINA, M. M.: In: TUMANOV, I. I. (Ed.): Methods of frost resistance determination in plants, pp. 29—50. Moscow: Nauka 1967.
TYURINA, M. M.: Deut. Landw. Wiss. Tagungsber. **100**, 77—86 (1968).
ULLRICH, H.: Protoplasma **38**, 165—183 (1943).
ULLRICH, H.: Angew. Botan. **36**, 258—272 (1962).
ULLRICH, H., HEBER, U.: Planta **51**, 399—413 (1958).
ULLRICH, H., MÄDE, A.: Planta **31**, 251—263 (1940).
ULMER, W.: Jb. Wiss. Botan. **84**, 553—592 (1937).
UNGERSON, I., SCHERDIN, G.: Ann. Botan. Soc. Vanamo **32**, 1—32 (1962).
UNGERSON, J., SCHERDIN, G.: Ann. Botan. Soc. Vanamo **35**, 1—36 (1964).
US National Arboretum (USNA) Plant Hardiness Zone Map. US Gov. Printing Off., Washington Misc. Pbl. **814** (1965).
UZAMI, S.: Izv. Akad. Nauk SSSR, Ser. Biol., **1956**, No. 5, 97—98 (1956).
VAARTAJA, O.: Acta Forest. Fenn. **62**, 3—31 (1954).
VAN DEN BERG, L., SOLIMAN, F. S.: Cryobiology **6**, 10—14 (1969a).
VAN DEN BERG, L., SOLIMAN, F. S.: Cryobiology **6**, 93—97 (1969b).
VAN DEN DRIESSCHE, R.: Can. J. Plant. Sci. **49**, 159—172 (1969).
VAN DER VEEN, J. H.: In: RÖBBELEN, G. (Ed.): Arabidopsis Research (Symp. Göttingen), pp. 62—71. Göttingen: Arabidopsis Information Service 1965.
VAN HUYSTEE, R. B., WEISER, C. J., LI, P. H.: Bot. Gaz. **128**, 200—205 (1967).
VEGIS, A.: Symb. Botan. Upsal. **10**, 1—77 (1948a).
VEGIS, A.: Physiol. Plant. **1**, 216—235 (1948b).
VEGIS, A.: Physiol. Plant. **2**, 117—130 (1949a).
VEGIS, A.: Svensk Bot. Tidskr. **43**, 671—714 (1949b).
VEGIS, A.: Experientia **9**, 462 (1953).
VEGIS, A.: Symb. Botan. Upsal. **14** (1), 1—175 (1955).
VEGIS, A.: In: RUHLAND, W. (Hrsg,): Handbuch der Pflanzenphysiologie, Bd. **16**, S. 168—298. Berlin-Göttingen-Heidelberg: Springer 1961.
VEGIS, A.: In: EVANS, L. T. (Ed.): Environmental control of plant growth, pp. 265—287. New York-London: Academic Press 1963.
VEGIS, A.: Ann. Rev. Plant. Physiol. **15**, 185—224 (1964).
VEGIS, A.: In: RUHLAND, W. (Hrsg.): Handbuch der Pflanzenphysiologie, Bd. 15/II, S. 499—668. Berlin-Göttingen-Heidelberg: Springer 1965a.
VEGIS, A.: Biol. Rundschau **3**, 78—88 (1965b).
VEGIS, A.: Intern. Symp. Physiol. Ecol. Biochem. Germin. Greifswald 1963, pp. 367—378 (1967).
VEGIS, A., VEGIS, B.: Acta Hort. Botan. Latviensis **8**, 57—102 (1935).
VIEWEG, G. H., ZIEGLER, H.: Ber. Deut. Botan. Ges. **82**, 28—36 (1969).
VILLIERS, G. D. B. DE: Farming S. Africa **18**, 378—381 (1943a).
VILLIERS, G. D. B. DE: Chronica Botanica **7**, 388—390 (1943b).
VILLIERS, G. D. B. DE: Union S. Africa Dep. Agr. and Forest Sci. Bull. 250 (1946).
VINCE, D., MASON, D. T.: Nature **174**, 842—843 (1954).

VINCE, D., MASON, D. T.: J. Horticult. Sci. **32**, 184—194 (1957).

VISSER, T.: Proc. Koninkl. Ned. Akad. Wetenschap. Ser. C **57**, 175—185 (1954).

VISSER, T.: Proc. Koninkl. Ned. Akad. Wetenschap. Ser. C **59**, 211—222 (1956a).

VISSER, T.: Proc. Koninkl. Ned. Wetenschap. Ser. C **59**, 314—324 (1956b).

VÖCHTING, H.: Über Transplantation am Pflanzenkörper. Tübingen 1892.

VOS, O., KAALEN, M. C. A. C.: Cryobiology **1**, 249—260 (1965).

VOZNESENSKY, V. L.: Fiziol. Rast. **5**, 229—236 (1958).

WAGENBRETH, D.: Flora A **156**, 63—75 (1965a).

WAGENBRETH, D.: Flora A **156**, 116—126 (1965b).

WAGENBRETH, D.: Deut. Akad. Landw. Wiss. Tagungsber. **100**, 141—145 (1968).

WALDÉN, J. N.: Sveriges Utsädesfören. Tidskr. **20**, 88—110, 168—188, 354—379 (1910).

WALTER, H.: Ber. Deut. Botan. Ges. **47**, 338—348 (1929).

WALTER, H.: Einführung in die Phytologie, Bd. III/1: Standortslehre, 2. Aufl. Stuttgart: Ulmer 1960.

WALTER, H., KREEB, K.: Die Hydratation und Hydratur des Protoplasmas der Pflanzen und ihre öko-physiologische Bedeutung. Protoplasmatologia II C 6. Wien-New York: Springer 1970.

WANNER, H.: In: RUHLAND, W. (Hrsg.): Handbuch der Pflanzenphysiologie, Bd. 6, S. 841—870. Berlin-Göttingen-Heidelberg: Springer 1958.

WARDLAW, I. F.: Botan. Rev. **34**, 78—105 (1968).

WAREING, P. J.: Physiol. Plant. **6**, 692—706 (1953).

WAREING, P. J.: Physiol. Plant. **7**, 261—277 (1954).

WAREING, P. J.: Ann. Rev. Plant. Physiol. **5**, 183—204 (1956).

WAREING, P. J.: In: RUHLAND, W. (Hrsg.): Handbuch der Pflanzenphysiologie, Bd. 15/II, S. 909—924. Berlin-Heidelberg-New York: Springer 1965.

WAREING, P. J.: In: WILKINS, M. B. (Ed.): Physiology of plant growth and development, pp. 605—644. London: McGraw-Hill 1969.

WAREING, P. F., PHILLIPS, I. D. J.: The control of growth and differentiation in plants. Oxford: Pergamon Press 1970.

WARINGTON, K.: J. Ecol. **24**, 184—204 (1936).

WATERSCHOOT, H. F.: Proc. Koninkl. Ned. Akad. Wetenschap., Ser. C, **60**, 318—323 (1957).

WAXMAN, S.: Proc. Plant Prop. Soc. **5**, 47—49 (1955).

WAXMAN, S.: Diss. Abstr. Ann. Arbor, Mich. **17** (11), No. 2372 (1957).

WAXMAN, S., NITSCH, J. P.: Am. Nurseryman **105**, 11—12 (1956).

WEATHERLEY, P. E., WATSON, B. T.: Ann. Botany **33**, 845—853 (1969).

WEBB, S. J.: Bound water in biological integrity. Springfield: Thomas Publ. 1965.

WEBB, J. A., GORHAM, P. R.: Can. J. Botan. **43**, 1009—1020 (1965).

WEBER, F.: Sitzungsber. Akad. Wiss. Wien, Math.-Naturwiss. Kl. I, **118**, 967—1031 (1909).

WEBER, F.: Sitz. Ber. Akad. Wiss. Wien, Math-Naturwiss. Kl. Abt. I., **125**, 311—351 (1916).

WEBER, F.: In: Abderhaldens Handb. biol. Arbeitsmethoden **11** (2), 591—625 (1924).

WEHSARG, O.: Arb. dt. Landwirtsch. Ges. Nr. 297, 11—515 (1918).

WEINBERGER, J. H.: Proc. Am. Soc. Hort. Sci. **56**, 122—128 (1950a).

WEINBERGER, J. H.: Proc. Am. Soc. Hort. Sci. **56**, 129—133 (1950b).

WEINBERGER, J. H.: Proc. Am. Soc. Hort. Sci. **63**, 157—162 (1954).

WEINBERGER, J. H.: Proc. Am. Soc. Hort. Sci. **67**, 107—112 (1956).

WEINBERGER, P., GODIN, C.: Canad. J. Biochem. **44**, 1035—1049 (1966).

WEINBERGER, P., KU, T. E.: Canad. J. Bot. **44**, 633—644 (1966).

WEISE, G.: Biol. Zbl. **80**, 137—166 (1961).

WEISE, G., POLSTER, H.: Biol. Zbl. **81**, 129—143 (1962).

WEISER, C. J.: Science **169**, 1269—1278 (1970).

WEISS, F.: Am. J. Botan. **13**, 737—742 (1926).

WELDON, G. P.: Monthly Bull. Calif. Dep. Agr. **23** (7—9), 160—181 (1934).

WELLENSIEK, S. J.: Proc. Koninkl. Ned. Akad. Wetenschap., Ser. C, **61**, 561—571 (1958).

WELLENSIEK,S.J.: Nature **192**, 1097—1098 (1961).

WELLENSIEK,S.J.: Nature **195**, 307—308 (1962).

WELLENSIEK,S.J.: Plant Physiol. **39**, 832—835 (1964a).

WELLENSIEK,S.J.: Proc. Koninkl. Ned. Akad. Wetenschap., Ser. C, **67**, 307—312 (1964b).

WELLENSIEK,S.J.: Z. Pflanzenphysiol. **54**, 377—385 (1966).

WELLENSIEK,S.J., HAKKAART,F.A.: Proc. Konikl. Ned. Akad. Wetenschap., Ser. C, **58**, 16—21 (1955).

WELLER,F., SCHREIBER,K.F.: Phytoma **17**, 25—30 (1965).

WELLINGTON,P.S.: Ann. Botany **20**, 105—120 (1956a).

WELLINGTON,P.S.: Ann. Botany **20**, 481—500 (1956b).

WELLINGTON,P.S., DURHAM,V.M.: Ann. Botany **25**, 185—196 (1961).

WENT,F.W.: Am. J. Botan. **31**, 135—150 (1944a).

WENT,F.W.: Am. J. Botan. **31**, 597—618 (1944b).

WENT,F.W.: Am. J. Botan. Sci. **101**, 97—98 (1945a).

WENT,F.W.: Am. J. Botan. Sci. **32**, 469—479 (1945b).

WENT,F.W.: In: MURNEEK,A.E., WHYTE,R.O. (Eds.): Vernalization and photoperiodism, pp. 145—147. Waltham, Mass. USA. 1948a.

WENT,F.W.: Ecology **29**, 242—253 (1948b).

WENT,F.W.: Ecology **30**, 1—13 (1949).

WENT,F.W.: Ann. Rev. Plant Physiol. **4**, 347—362 (1953).

WENT,F.W.: The experimental control of plant growth. New York: Ronald Press Comp. 1957.

WENT,F.W., WESTERGAARD,M.: Ecology **30**, 26—38 (1949).

WHELAN,W.J.: In: RUHLAND,W. (Hrsg.): Handbuch der Pflanzenphysiologie, Bd. 6, S. 154—240. Berlin-Göttingen-Heidelberg: Springer 1958.

WHITEMAN,P.C., KOLLER,D.: Israel. J. Bot. **13**, 166—176 (1964).

WICKENS,G.W.: J. Dep. Agr. W. Australia **8**, 390—392 (1931).

WILHELM,A.F.: Phytopathol. Z. **8**, 337—361 (1935).

WILLENBRINK,J.: Z. Pflanzenphysiol. **55**, 119—130 (1966).

WILLIAMS,R.J., MERYMAN,H.T.: Cryobiology **1**, 317—325 (1965).

WILLIAMS,R.J., MERYMAN,H.T.: Plant Physiol. **45**, 752—755 (1970).

WILLIAMSON,R.E.: Soil Sci. Soc. Am. Proc. **27**, 106 (1963).

WILNER,J.: Can. J. Plant. Sci. **45**, 67—71 (1965).

WILSON,C.C.: Plant Physiol. **23**, 5—37 (1948).

WINKLER,A.: Jb. Wiss. Botan. **52**, 467—506 (1913).

WINKLER,E.: Flora **151**, 621—662 (1961).

WINKLER,E., PREGENZER,E.: Bodenkultur **21**, 22—44 (1970).

WINKLER,H.: Beitr. Biol. Pflanz. **14**, 219—230 (1925).

WISNIEWSKI,P.: Kosmos (Lemberg) **88**, 1376—1384 (1913).

WOLF,J.: In: RUHLAND,W. (Hrsg.): Handbuch der Pflanzenphysiologie, Bd. 6, S. 881—908. Berlin-Göttingen-Heidelberg: Springer 1958.

WRIGHT,R.C., ROSE,D.H., WHITEMAN,T.M.: The commercial storage of fruits, vegetables, and florist and nursery stocks. USDA Handbook 66, 1954.

WÜNSCHE,U.: Naturwissenschaften **53**, 386—387 (1966).

WUENSCHER,J.E., KOZLOWSKI,T.T.: Physiol. Plant. **24**, 254—259 (1971).

YAMASHITA,T., BUTLER,W.L.: Plant Physiol. **43**, 2037—2040 (1968).

YARWOOD,C.E.: Science **134**, 941—942 (1961).

YARWOOD,C.E.: Advan. Front. Plant Sci. **7**, 195—204 (1963).

YARWOOD,C.E.: In: PROSSER,C.L. (Ed.): Molecular mechanisms of temperature adaptation, pp. 75—89. Am. Assoc. Advan. Sci. Publ. 84, Wash. DC. 1967.

YASUDA,S.: Ber. Ohara Inst. Landw. Biol. **12**, 42—64 (1963).

YASUDA,S.: Ber. Ohara Inst. Landw. Biol. **12**, 197—215 (1964).

YELENOSKY,G., HORANIC,G.: Cryobiology **5**, 281—283 (1969).

YOUNG,R.H.: Proc. Am. Soc. Hort. Sci. **78**, 174—180 (1961).

ZADE,A.: Arb. Deut. Landwirtsch. Ges. Nr. **229**, 1—91 (1912).

ZAIN, U. A.: Z. Botan. **44**, 207—220 (1956).

ZAVADSKAYA, I. G.: Botan. Zh. **48**, 755—758 (1963).

ZEEVAART, J. A. D.: Bull. Soc. Franç. Physiol. Végét. **2**, 162—163 (1956).

ZELITCH, I.: Stomata and water relations in plants. Conn. Agr. Exp. Sta. New Haven Bull. **664**, 18—36 (1963).

ZELITCH, I.: Photosynthesis, Photorespiration and Plant Productivity. New York-London: Academic Press 1971.

ZELLER, O.: Planta **39**, 500—526 (1951).

ZIEGLER, H.: Flora **144**, 2 (229), 29—250 (1957).

ZIEGLER, H.: In: ZIMMERMANN, M. H. (Ed.): The formation of wood in forest trees, pp. 303—320. London-New York: Academic Press, 1964.

ZIMMERMANN, M. H.: Plant Physiol. **39**, 568—572 (1964).

ZURZYCKI, J.: Acta Soc. Botan. Pol. **21**, 241—264 (1951).

Animals

I. Body Temperature and External Temperature

H.-D. Jankowsky

Although the body temperature of poikilothermic animals changes in the same direction as the external temperature, the level of the body temperature is usually not the same as that of the external temperature. The temperature gradient between the body and its surroundings is not only a function of heat production and heat loss, but also of a number of environmental factors which determine the climatic conditions surrounding the organism, especially solar radiation, humidity, and air currents (cf. GATES, 1962, 1970; SCHWERDTFEGER, 1963; ed. 1, p. 133–139). The effects of external physical factors, endogenous heat production, and heat outflow via radiation, conduction, and convection will be discussed here only to the extent that they serve active temperature regulation.

True homeothermy is found only in the mammalia and aves. Nevertheless various species of poikilotherms are able to maintain their body temperatures at a more or less constant level while active. These animals have been designated as heterotherms. Heterothermy is endogenous in origin if it is based on increased heat production. Ectothermal organisms regulate their internal temperature by behavioral modification of external heat input (cf. COWLES, 1962).

A. Regulation of Body Temperature in Invertebrates

1. Body Temperatures above Environmental Temperatures

a) Individual Thermoregulation

Due to the low body weight of most invertebrates, resulting in low heat capacity and thermal inertia, their heat exchange with the environment is very rapid. In the case of diurnally active arthropods, which are protected from desiccation by a water-impermeable cuticle, the large surface-to-volume ratio favors the utilization of solar radiation for raising the body temperature above that of the environment. When external temperatures are low, many insects require solar heating of their body before they can be active (DIGBY, 1955, 1958; CLENCH, 1966; cf. p. 446). The coloration of lepidopteran wings, which function as heat exchangers, may play a role in the absorption of heat radiation. For example, the wings of butterflies of the genera *Erebia* and *Oeneis* living in arctic or high alpine regions are almost black. Among orthoptera, hymenoptera, and diptera coloration is less significant for raising the temperature; rather the difference between body temperature and external temperature increases with increasing body size (DIGBY, 1955;

STOWER and GRIFFITHS, 1966). Heat input can be regulated by changing the insolated body surface, by spreading or folding the wings, by turning the body axis in relation to the angle of incidence of the sun's rays, or by moving to locations with different intensities of solar radiation (FLITTERS, 1968). The body temperature of *Schistocerca gregaria* varies by 6° C, depending on whether the animal is orientated parallel or perpendicular to the sun's rays (STOWER and GRIFFITHS, 1966). Due to its size this locust can raise its resting body temperature by about 3° C by its metabolism without the aid of solar radiation. However in most insects basal metabolism produces only minimal increases over the external temperature (cf. CLARKE, 1960; ROTH, 1965). The relatively large surface areas of small animals are not favorable for increasing body temperature endothermally. Nevertheless nocturnal animals rely mainly on endogenous sources of warmth, since the flight muscles of many insects can function at their full capacity only above a certain minimum temperature. The needed heat energy is produced by muscular activity and is stored in localized body areas for which temperature increases have functional significance. For this reason heterothermy in nocturnal flying insects is generally restricted to the thorax, while the temperature of the rest of the body varies with the environmental temperature.

Many nocturnal lepidoptera stay in shaded places during their daytime rest period, to avoid overheating their bodies. At such times there is no temperature gradient between the interior of the body and the environment. At low light intensities and the consequently low external temperatures, spontaneous dawn or dusk activity begins. At air temperatures below 30° C the American mulberry silk moth, *Hyalophora cecropia*, heats its thorax to the minimum temperature of 34.8° C by flapping its wings or shivering before it flies off (HANEGAN and HEATH, 1970a; SOTAVALTA, 1954; cf. DORSETT, 1962; HEATH and ADAMS, 1967). While the moth shivers the temperature of the thorax increases almost linearly at a rate of 2−10° C/min even though at higher temperatures the gradient, and therefore the heat outflow, from the animal to its surroundings grows larger. Consequently the intensity of heat production through shivering does not go down with increasing body temperature, but increases. This has been confirmed by measurements of oxygen consumption (HEATH and ADAMS, 1967).

During flight the rate of oxygen consumption also changes in the same direction as the temperature gradient between the thorax and the air, while the temperature in the thorax is held between 33.4° C and 37.8° C. This temperature drops only slightly at lower external temperatures. Over the entire period of activity, which may amount to two hours per day for male *H. cecropia*, the thorax remains between 34.1° C and 38.2° C at air temperatures of 20−30° C. The temperature limits maintained by regulation during flight are not determined accidentally. At 34° C the moth can begin a new flight without a preliminary warming period; at ca. 39° C the body begins to be overheated, and cooler locations are actively sought. The upper temperature limit is surprisingly independent of the environmental thermal conditions. At air temperatures of 17° and 25° C, the maximal temperatures in the thorax are 38.5° and 38.3° C, respectively (HANEGAN and HEATH, 1970a, 1970b). The American mulberry silk moth takes no nourishment as an adult. Consequently the energy reserves available for metabolic heating are limited, and the frequency and duration with which it actively regulates its temperature are reduced at low external temperature (HANEGAN and HEATH, 1970c)[1].

Heterothermal regulation in lepidoptera is governed by the CNS. During the warming period the animals flap their wings with a smaller amplitude but higher frequency than during flight. The transition from one rhythm to the other is abrupt (MORAN and EWER, 1966; KAMMER, 1968). The change in temperature in the thorax seems to directly influence the relative co-ordination of central automatic mechanisms (WILSON, 1968). Warming only the thoracic ganglia of *Hyalophora cecropia* to 33°−40° C allows flight movements to be initiated even if the rest of the body is at 25°−27° C. Under natural conditions flight is impossible at thoracic temperatures below 34° C. If thoracic ganglia are cooled, the flight movements stop (HANEGAN and HEATH, 1970c). Therefore the temperature of the thoracic ganglia, and not that of the external air, determines whether shivering takes place in butterflies. In bees, on the other hand, thermogenesis is controlled by the brain (ROTH, 1965). Adaptation of warming behavior to temperature has been detected in the monarch butterfly, *Danaus plexippus;* a period of exposure to low temperatures increases the intensity and duration of shivering (KAMMER, 1971). The maintenance of high thoracic temperatures is made easier for many lepidoptera by a thick covering of hair or scales. The insulating layer absorbs the heat radiation of the body surface and hinders heat convection by maintaining a cushion of still air. Odonata have air sacs to insulate the thorax (cf. CHURCH, 1960b). In flight the intense heat production of a well-insulated thorax can have a detrimental effect by increasing the temperature to unfavorable levels. When air temperatures are high orthopterans, lepidopterans, dipterans and cicadae often fly intermittently, landing periodically to cool themselves (DIGBY, 1958; CHURCH, 1960a; HEATH and ADAMS, 1965 HEATH and WILKIN, 1970;). According to CHURCH (1960b), the heat loss of flying insects can be partitioned as follows; 5−10% evaporation, 10−15% radiation, 5−15% heat conduction to the interior of the body, 60−80% convection (cf. ESCH, 1960). The proportion lost by evaporation can be increased by hyperventilation (EDNEY and BARRAS, 1962; CLENCH, 1966).

The transport of heat into the interior of the body by convection through the hemolymph is probably frequently underestimated. Ligation of the dorsal blood vessels causes temperature regulation in the moth *Manduca sexta* to break down at air temperatures above 23° C. If the thoracic scales are removed, i.e. the ability to lose heat by convection and radiation is improved, interruption of the blood flow does not prevent flight (HEINRICH, 1970; HEINRICH and BARTHOLOMEW, 1971)[2].

---

[1] Recently a study of flight temperatures in representatives of Sphingidae, Noctuidae, Geometridae, Arctiidae, Saturniidae, Ctenuchidae, Pericopidae, Lasiocampidae, Notodontidae and Pyralidae has been published. In all cases body temperatures were higher than ambient, and thoracic flight temperatures were relatively uniform and independent of ambient temperatures from 7 to 17° C; except in Ctenuchidae and Arctiidae, flight temperature was independent of body weight within the various families [BARTHOLOMEW, G.A., HEINRICH, B.: J. Exp. Biol. 58, 123—135 (1973)].

[2] CREA, M. J. Mc., HEATH, J. E.: J. Exp. Biol. 54, 415—435 (1971).

b) Social Thermoregulation in Insects

Colonial hymenoptera have frequently developed a social heat economy, which specifically guarantees the constancy of the "nest temperature". The construction of well-insulated nests which keep heat exchange with the surroundings small is a significant adaptation for this purpose. But, well-insulated nests are more difficult to heat by direct sunlight at low temperatures. Nevertheless, ant colonies of *Formica rufa* can warm the brood chambers in the interior of their ant hills entirely by solar energy. A specialized group of ants takes over the task of heat transferral. During the day they alternately bask in front of the hill and move into the nest interior, thereby bringing the sun's warmth into the brood chamber via their warmed bodies. In the ant colony there is a thermotactic division of labor; in contrast to the condition of solitary insects, the heat requirements of the individual are of secondary importance (ZAHN, 1958). While the temperature in the ant colony is regulated predominantly by ectothermy, in wasp and hornet nests the inhabitants themselves produce adequate amounts of heat. A heat insulating nest cover is necessary for their temperature regulation also. If the nest cover of *Paravespula germanica* is destroyed, optimal temperatures of the brood can no longer be maintained (ROLAND, 1969). Even the larvae of the hornet, *Vespa crabro*, can generate temperatures within the nest that are 6° C higher than the external temperature by motor activity, but only the female adults can carry out true thermal regulation. Warming the pupae and the adult members of the colony is always the responsibility of the same female imagos which carry out telescoping pumping motions with their abdomens for this purpose. At a frequency of 200 pumping movements per minute the temperature of their bodies exceeds that of the air by 8° C, while the temperature gradient between the nest and the environment reaches a maximum of 15° C (ISHAY and RUTTNER, 1971). The highest stage of social thermoregulation is realized in the bee colony. Individual bees also regulate their heat production according to the environmental temperature outside the colony. If ample food is available, temperatures exceeding the external ones by 5—10° C are maintained for hours (ESCH, 1960). This thermogenesis is only effective during the bees' diurnal work periods (HEUSNER and STUSSI, 1964). It begins only at temperatures above 13° C and is reduced as the body temperature approaches the optimal value of 35° C. The ability of newly emerged bees to give off heat is slight, and the normal value is not reached till the 15th day (ROTH, 1965). The heat is produced by vibrations of the thoracic muscles, and the motor impulse frequency during warming behavior is the same as during flight (ESCH, 1964; cf. BASTIAN and ESCH, 1970). A group effect on shivering can be demonstrated in bees. As soon as two or more bees are brought together the heat production of the individual bee is reduced, so that less thermal work is required of each bee in the hive (ROTH, 1965; cf. SITBON, 1968). Due to the large population of a hive the temperature in the areas of the brood comb is maintained at exactly 36 ± 1° C. Heat outflow is retarded through honey-filled combs. The temperature of the brood nest is governed according to the principle of an oscillating two-point regulator. Each animal warms its thorax by muscular vibrations to somewhat over 36° C, and then cools down to the lower temperature limit, whereupon shivering begins again. The large number of individuals around the brood cluster prevents heat from flowing out to the surroundings. In winter the temperature of the cluster is regulated by the body temperature of the individual bee, making active movement possible. However heat is not produced rhythmically by shivering. Rather, the primary source of warmth is the uptake of food by the bees in the interior of the cluster, followed by augmented muscular vibration (cf. p. 468, FREE and SPENCER-BOOTH, 1958; ESCH, 1960; ROTH, 1965).

2. Heat Output at High Body Temperatures

Aquatic animals can avoid extreme, high temperatures only by moving away from them. The majority of terrestrial animals also move to shaded areas under these conditions, or live under the ground, or are active only in the cool night (cf. p. 448f, CLOUDSLEY-THOMPSON, 1961). On the other hand, sessile littoral animals are exposed to heat loads daily, almost without protection. Thus the body temperature of the cup shell *Patella vulgaris* changes by as much as 23° C in the course of a summer day (GRAINGER, 1969). Intertidal animals cannot lower their body temperature under intense solar radiation except by evaporation (SOUTH-WARD, 1958). When the mussel *Modiolus demissus* is exposed at low tides it opens its valves. Perhaps the increased water loss enables it to survive short-term exposure to extreme temperatures (LENT, 1968).

Evaporation is the most effective method of cooling. An organism can lower its body temperature at once by 5—6°C through the loss of only 1% of its body weight in water vapor since the heat of vaporization of water amounts to 580 cal/g. The intensity of evaporation is proportional to the difference in vapor pressure between the water-emitting surface and the water-absorbing air. In damp climates cooling by transpiration is more difficult. Since the water vapor must diffuse through the layer of air surrounding the body and the speed of diffusion is temperature dependent, evaporation increases with rising temperature. High external temperatures favor cooling by evaporation. Furthermore, the water permeability of the cuticle of many insects increases when they are warmed (BEAMENT, 1958; BEAMENT et al., 1964, p. 67—129). Perhaps certain specialized areas of the integument of Orthopterans — the Slifersian regions — serve to increase transpiration with temperature (MAKINS, 1968, cf. p. 463). All air movements accelerate body cooling by evaporation as well as by convection because they diminish the layer of still air on the surface. Isopods, whose cuticles are relatively highly permeable, often search out wind-exposed sites to increase the evaporation rate (EDNEY, 1957).

Cooling the body over longer periods by giving off water is only possible if the lost fluid can be replaced rapidly by mouth or parenterally. Therefore small arthropods tolerate high external temperatures better in moist environments since their body surface is highly permeable to water (WARBURG, 1965). Blood-sucking insects, e.g., the tsetse fly, also survive extreme, high temperatures because they can obtain adequate amounts of fluid for evaporation by feeding (EDNEY and BARRAS, 1962, cf. p. 463).

Since desert-dwelling invertebrates are extremely economical with water, evaporation can contribute little towards cooling their bodies (HADLEY, 1970). The inhabitants of arid regions have often developed behavioral protective measures against excessive temperatures. For example, at high temperatures scorpions walk as if on stilts. The elevated body posture improves heat loss by convection and reduces the transferral of heat from the substrate (ALEXANDER and EWER, 1958).

B. Regulation of Body Temperature in Poikilothermic Vertebrates

1. Fishes

Heat exchange between the bodies of fish and the surrounding medium takes place almost exclusively through conduction and convection. Since water has a high specific heat and thermal diffusion ensues rapidly, there is almost no difference in temperature between the surface of the fish and the water. However, the temperature of the interior of the body can differ from that of the surroundings. According to STEVENS and FRY (1970) the temperature of the brain of a 200 gram *Tilapia mossambica* was 0.44° C higher than that of the water, and the temperature of the red side trunk muscle was 0.8° C higher. Even steeper temperature gradients towards the center of the body are to be expected in heavier animals. The common tuna, *Thunnus thynnus*, which grows to a size of 820 kg, can maintain the temperature of its red side muscles between the limits of 26° and 32° C at environmental temperatures between 6° and 30° C. The ability to regulate the temperature of a single tissue — homeotherms regulate that of their body core — is based on the possession of a "rete mirabilis", formed from the arteries and veins which serve the red musculature. The retes work as countercurrent heat exchangers in which the outflow of heat can be controlled by changes in the blood supply (BARRETT and HESTER, 1964; CAREY and TEAL, 1966, 1969b; STEVENS and FRY, 1971). The high temperature in the red side muscle is due primarily to the intense metabolic heat formation of the tissue, which is amply supplied with sarcosomes (cf. GORDON, 1968). Sharks of the family *Isuridae* can also control the temperature of individual muscles (CAREY and TEAL, 1969a). The biological significance of high temperatures in the muscular tissue of constantly swimming predatory fish can be seen in the lack of dependence of swimming ability on the temperature conditions of waters at varying depths or in different geographical regions[3].

2. Amphibia

Amphibia can influence their body temperature only by their behavior, since their metabolic heat production is very slight (FROMM, 1956). Tadpoles move towards the optimal temperatures in their habitat (BRATTSTROM, 1963; LICHT and BROWN, 1967; LUCAS and REYNOLDS, 1967, cf. p. 463). *Rana pipiens* larvae utilize the warmth of solar radiation even in water. To absorb more of the warmth they gather in groups with their tails pointed towards the sun, thereby exposing the pigmented dorsal surface (BRATTSTROM, 1962). Adult anurans living near the shore can maintain a constant body temperature quite well during the day. In the bullfrog it fluctuates between 26° and 33° C. A number of behavioral modes may

[3] CAREY, F.G., TEAL, J.M., KANWISHER, J.W., LAWSON, K.D., BECKETT, J.S.: Am. Zoologist 11, 137—145 (1971).

be observed, e.g., basking on the water surface or on land; changing locations to spots with sunshine of different intensities; pressing the body to the ground; sitting with the body held upright, whereby the irradiated surface, the input of heat to the body, and the outflow via convection, radiation, and evaporation are changed (LILLYWHITE, 1970, cf. p. 466). The moist skin of amphibians enables them to lose heat by evaporation extremely effectively, and body temperatures as much as 13° C below those of the environment have been recorded (THORSON, 1955). *Rana pipiens* can replace the water lost through evaporation by absorbing ground moisture during the night, and therefore can exist at high air temperatures without access to bodies of water (BENTLEY, 1966; DOLE, 1967). The ethological regulation of body temperature by amphibia is aided by their coloration (EDGREN, 1954) and by a temperature dependent morphological adaptation of body shape (CALHOON and JAMESON, 1970; VOGT and JAMESON, 1970).

3. Reptiles

The body temperatures of reptiles are as varied as their requirements for environmental temperature conditions. The temperature of the rhyncocephalian *Sphenodon punctatus* lies between 6.2° C and 18° C (BOGERT, 1953), that of the desert iguana *Dipsosaurus dorsalis* at 42° C (NORRIS, 1953; KEMP, 1969). Varanids, iguanids, agamids, lizards, and snakes in particular maintain their body temperature at a relatively constant level in the region of their preferred temperature (PT), independent of the level of the external temperature (cf. p. 466).

Under intense solar radiation the body temperature may exceed the air temperature by 30°C, as in the iguanid *Liolaemus multiformis* (cf. SWAN, 1952; PEARSON, 1954) which inhabits the high Andes. Endothermal regulation of body temperature, utilizing metabolic heat, is possible only for a few large lizards, such as *Varanus varius* (BARTHOLOMEW and TUCKER, 1964; STEBBINS and BARWICK, 1968), or giant snakes (COGGER and HOLMES, 1961; GALVAO et al., 1965). However, metabolism increases the body temperature of the Australian painted monitor by only about 2° C, while a coiled boa may be 11° C warmer than the air. Brooding female *Python molurus* increase their endogenous heat formation still more by rhythmic, spasmodic muscular contractions (POPE, 1961). Most reptiles regulate their body temperature by behavioral mechanisms that control the exchange of heat between the animal's body and its surroundings.

Basking is widely distributed among aquatic reptiles such as crocodiles and many turtles. It has the effect of rapidly bringing the body temperature to an optimum, which lies between 32° C and 35° C for the American alligator and at various temperatures between 30° C and 32.7° C for different species of North American aquatic turtles (COLBERT et al., 1946; BOYER, 1965, cf. p. 467). Size, but not coloration, has a significant effect on the rate of warming. *Chrysemys picta* approaches the shore even in winter to capture the sun's rays under the ice cover (GIBBONS, 1967).

Diurnal terrestrial reptiles also maintain their body temperature within narrow limits inside their PT range by solar heating. For example, the PT of the Australian skink, *Tiliquia occipitalis*, as determined experimentally in a thermal gradient is 32.9° C (range = 29.1°−35.2° C) and the average body temperature of free-living skinks at an external temperature of 22° C is 33.7° C (range = 30.5°—35.5° C) (LICHT et al., 1966). Pigmentation plays no role in the thermoregulation of these reptiles, as skinks cannot change their coloration rapidly. The significance of coloration for heat exchange in reptiles has generally been overestimated. Even in the hot climate of a black lava landscape dark body colors predominate among many inhabitants due to their advantage for camouflage (cf. ed. I, p. 138; COWLES, 1967). The marine iguana, *Amblyrhynchus cristatus*, which is obliged to pass the entire day on the rocky shores of the Galapagos islands, exposed to the sun, controls its internal temperature at 35—37° C entirely by changing the position of its body. By pressing down on the substrate the lizard increases heat input, while an upright posture leads to cooling by the sea breeze (BARTHOLOMEW, 1966; cf. KRAMER, 1937; USCHAKOW, 1960; SCHMIDT-NIELSEN and DAWSON in DILL et al., BRADSHAW, 1968).

The body temperature of nocturnal reptiles seldom differs from that of their surroundings, though snakes gather at night on asphalt roads in American deserts if the pavement is somewhat warmer than the adjacent soil (BRATTSTROM, 1965). Gekkonidae also utilize temperature differences in the substrate to warm themselves (BUSTARD, 1967); some gekkos have lower PTs than other reptiles from the same habitat (LICHT et al., 1966).

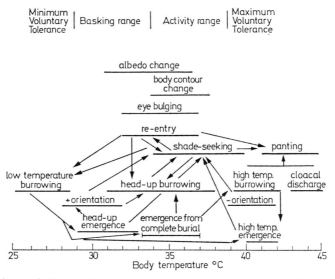

Fig. 1. The interrelations of temperature-regulating behavior in *P. coronatum*. The range of body temperatures for each pattern is given. (HEATH, 1965)

Many reptiles live in arid habitats with strong solar radiation where the danger of overheating is especially high. Therefore, under the extreme thermal load of the hottest times of day, desert-dwelling species seek shelter or withdraw below the surface of the earth[4]. But many reptiles can also counteract high external temper-

[4] McGINNIS, S. M., VOIGT, W. G.: Comp. Biochem. Physiol. **40** A, 119—126 (1971).

atures by increasing their rate of evaporation. Since they do not possess sweat glands, the water must be lost via the respiratory pathways, and the animals begin to pant. When the air temperature rises from 40 to 44° C, the desert iguana, *Dipsosaurus dorsalis*, increases its frequency of breathing from 19.4 to 58.9/min (TEMPLETON, 1960; cf. DAWSON and TEMPLETON, 1963; JONES, 1967; RICHARDS, 1970). The turtle *Terrapene ornata* uses salivation and urination for evaporative thermoregulation[5].

HEATH (1965) recorded the ethogram of the lizard *Phrynosoma coronatum* and analyzed it with respect to thermoregulation. Figure 1 shows the complicated interplay of the various modes of behavior that participate in the maintenance of a constant body temperature.

The ethological control of heat flow between the animal's body and the environment is aided by physiological adaptations. Within a temperature gradient, Galapagos island iguanas warm up twice as fast as they cool down (BARTHOLOMEW, 1966). The skink, *Tiliquia scincoides* also demonstrates rapid rising and slow sinking of body temperature (BARTHOLOMEW et al., 1965). MORGAREIDGE and WHITE (1969) showed, with the aid of radioisotopes, that blood flow through the skin of the Galapagos iguana is influenced directly by the temperature. The vessels dilate when warmed by radiation and constrict on cooling (BARTHOLOMEW and LASIEWSKI, 1965). Temperature dependent changes in blood circulation have been reported in various reptilian species (cf. LICHT, 1965; TUCKER, 1966; HEATH et al., 1968)[6], but it is difficult to judge their significance for body temperature.

---

[5] RIEDESEL, M. L., CLOUDSLEY-THOMPSON, J. L., CLOUDSLEY-THOMPSON, J. A.: Physiol. Zool. **44**, 28—32 (1971).
[6] WEATHERS, W. W.: Comp. Biochem. Physiol. **40** A, 503—515 (1971).

References to this section, see p. 470.

II. The Normal Temperature Range

H. Precht, H. Laudien, and B. Havsteen

The expression "normal" must of course be taken in a relative sense since this range may vary for different species. It is useful to distinguish biological processes (e.g. basal metabolism) that do not change appreciably during experiments (II.1, unchanging or constant systems), from changing systems; in the latter the time of a change serves directly as a measurement (for example, as the time needed for an imago to develop from the egg may be measured). Processes of development, growth, and reproduction fall into this category (II 2).

II 1. Constant Systems[1]

H. Precht *

Since the appearance of the first edition, which contains references to the older literature, the influence of "normal" temperatures on biological processes of poikilothermic animals has been investigated intensively (see Fry, 1958; Prosser,

---

* I am indebted to Prof. Dr. C. L. Prosser (Urbana, Ill.) and Dr. H. Laudien for critical reading of the Chapters II.1 and III and for their valuable suggestions. I wish to thank Mrs. H. Prosser and Dr. J. Augenfeld for the translation and for revising the English manuscript and Mrs. M. E. Vogt for her assistance in completing it. I am indebted to Dr. D. F. Alderdice (Nanaimo, B. C.), for his important suggestions, and to Dr. H. Künnemann for several new references, proofreading and completion of the index. Prof. Dr. B. Havsteen wrote three chapters (p. 310 ff.). — In addition to the rules mentioned in this chapter, exceptions are referred to for the purpose of further investigations.

[1] The following abbreviations will be used:

AT = adaptation (acclimation) temperature, before the tests; this may be constant (e.g. AT = 20° C) or during the day higher than during the night (e.g. AT = 20° ↔ 10° C),

ET = experimental (test) temperature,

CNS = central nervous system,

$K_m (= K_M)$ = Michaelis constant (V_{max} = maximal velocity) μ value ($= E_a$) = activation energy (cal/mol);

$$\frac{v_2}{v_1} = e^{\frac{(T_2 - T_1)\mu}{T_1 T_2 R}} , \quad \mu = \frac{4{,}574 \lg v_2/v_1}{1/T_1 - 1/T_2} .$$

In biological experiments, v_1 and v_2 are the measured velocities (e.g. the oxygen consumption) at the ETs T_1 and T_2 (R = gas constant). Although the ET dependence of biological processes is complex, because usually the total effect arises from several elementary steps, each of which is thermally sensitive, the Arrhenius equation may serve the purpose of empirical characterization of the system. However, since the measurements refer to a mesh of interrelated reactions and not to a single step, the

1958, 1967; PRECHT, 1961 a, 1964 b, 1968[2]; PROSSER and BROWN, 1961; HANNON and VIERECK, 1962; DILL et al., 1964; KINNE, 1964 b, 1965; ALTMANN and DITT-MER, 1966; PRECHT et al., 1966; ROSE, 1967; TROSHIN, 1967)[3].

Temperature is only one of the factors that influence the life of an organism, though a very important factor. The ecologist is interested in just those effects that are produced by the simultaneous operation of several factors, but polyfactorial investigations often present the physiologist with almost insuperable complications. Since considerations of multiple factors would go beyond the limits of this discussion, only the influence of temperature will be considered here. Essentially of interest here is the influence of temperature mentioned in point 1 by LAUDIEN (p. 454).

Most of the life processes of poikilothermic organisms respond to changes in experimental temperature (ET) as chemical reactions do, others as physical processes. The equilibrium adjustments (overshoots, etc.) following the changes can also be meaningfully considered as direct responses to temperature. In general the rates of biological processes (of constant systems) increase as the experimental temperature (ET) increases, up to a maximum, and then fall more or less steeply. This latter section of the curve will be treated separately.

II 1 a. The First, Usually Ascending, Sections of the Curves

A. Direct Responses

1. Relation of Biological Processes to Experimental Temperatures

In the first part of the discussion, direct responses will be emphasized, and the effects of the temperatures at which the animals were kept before the experiments (acclimation temperature = AT) will be considered; discussion of the adaptive processes themselves will come later (p. 334).

---

velocity constants in the Arrhenius equation must be substituted by the directly measured velocities. Therefore, the apparent activation energy, which may be calculated from the slope of the curve in the diagram, represents the process as a whole and generally not as a single, well-defined chemical reaction (HAVSTEEN); see 1st ed., p. 4, and PRECHT, 1968, p. 492.

Q_{10} value = temperature coefficient (the quotient of the measured values at ETs differing by 10° C); if this is not the case:

$$\lg Q_{10} \approx \frac{10}{T_2 - T_1} \cdot \lg \frac{v_2}{v_1} \quad \text{(see 1st ed., p. 4).}$$

(It is possible — but not usual — to calculate Q_{10} values for curves, showing the dependence on AT or for those in which ET = AT, as in Fig. 4.)

[2] This work formed the foundation of this chapter.

[3] Cf. KINNE in O. KINNE (Ed.): Marine ecology. Vol. I, part 2, pp. 407—514. London 1970; VERNBERG, W. B., VERNBERG, F. J.: Environmental physiology of marine animals. Springer-Verlag, Berlin, Heidelberg, New York 1972. Unfortunately the new book by HOCHACHKA, P. W., and SOMERO, G. N.: Strategies of Biochemical Adaptation. London: Saunders 1973, could not be taken into consideration.

a) Experimental Procedure

Several precautions should be observed, especially with animal experiments. For example, if one wants to measure the dependence of oxygen consumption of an animal on the ET, the conditions under which the animal was kept before the experiment can determine the amount measured as well as the shape of the curve. The AT, the photoperiod, the nutritional state of the animal, etc. can all have such an effect. During the measurements all factors except the one under investigation (here, the ET) must of course be held constant; according to Kasbohm (1967), even the maintenance of a definite day length can be important in longer experiments.

It is best to start with an ET that matches the AT. One must allow disturbances caused by the introduction of the animal into the experimental chamber to die down (1st ed., p. 17; Saunders, 1963). After the measurements have become constant, the ET should be changed in a single step. The speed and magnitude of the change should be selected so that overshoots, etc. falsify the picture as little as possible. Measurements should be retaken until constant values are reached. As a control, one can return to the original temperature and make more measurements, but it should be noted that increases and decreases in temperature, even over the same range, can have different effects (p. 306). If there is a danger that a shift in acclimation took place during the experiment, and/or stress effects have not diminished, the animal must be reintroduced to the AT for a sufficient period before the next measurement at a different ET. Of course, the procedure described here, with two ETs, may be changed with circumstances, and is probably not necessary in every case. If the temperature is changed very slowly and only small changes in temperature are used, measurements at several ETs can be made in one experiment, starting with an ET corresponding to the AT. Diurnal fluctuations of the values measured must be taken into account.

To make inter- or intraspecific comparisons between inhabitants of regions of different temperatures, one should maintain the experimental subjects, if possible, at the same AT (or sets of ATs) and also should make the measurements under the same experimental conditions. This is the only way to decide whether differences in the measurements, or in Q_{10} values, are determined genetically or by the different ATs of the habitats. One can measure the effect of ET on biological processes (using a constant AT) or the effect of AT (using a constant ET), or one can always set ET = AT, or investigate differently acclimated animals at various ETs, as in the experiment illustrated in Fig. 1. It has already been pointed out (1st ed., p. 21) that the presentation of the curves should also be considered in planning experiments.

From the standpoint of bio-ecology, the measurement of fully acclimated animals at an ET that corresponds to the AT may be especially important. However, such a restriction to ET = AT values narrows the possibilities for finding physiological meaning. Even in taking measurements at several ETs, it is best to begin, as has been said, with an ET that matches the AT. In this way one can include the ET = AT curve (sketched in as curve e in Fig. 1). Nevertheless, the subsequent changes in ET, especially when they are rapid, can cause transient though prolonged disturbances, which will be described later (p. 320 ff.).

The stated limitation is significant in the following case: Regions with low Q_{10} values that are often observed, can be of genetic origin (direct responses), or they may be due to regulation (p. 323 ff.), or to capacity adaptation in the form of compensation. A distinction between the third- and the first-named possibility cannot be made with the aid of ET = AT curves alone. (The first two possibilities can only be separated on the basis of further investigations.)

All these aspects are emphasized so strongly because, due to their neglect, the findings of many authors cannot be compared at all. This leads to the unfortunate result that really valid rules can be established in very few cases, quite apart from the fact that the number of species investigated is generally too small for the establishment of such rules.

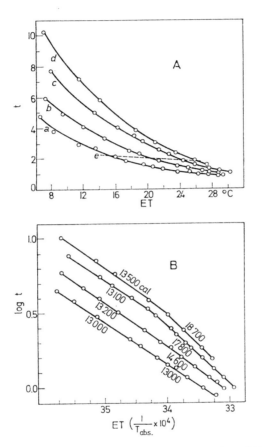

Fig. 1 A and B. A Pleopod beats of *Gammarus pulex*. t = duration of 10 beats (in seconds). Acclimation temperature for a = 15.5° C, for b = 20° C, for c = 25° C, for d = 26.3° C, e = measurements made at acclimation temperature (ET = AT). In B, the μ-values are written semilogarithmically. (From PRECHT, 1949)

b) Presentation of Q_{10} and μ Values

The use of temperature coefficients (Q_{10} values) has become established as an indication of the relation of biological processes to temperature, even though these values are not independent of the ET over the temperature range in question, while the activation energies (μ values) are. The Q_{10} values generally decline with increasing ET; less often they remain constant (see ROBERTS, 1957; DAWSON in

MILSTEAD, 1967). The Q_{10} values of respiration in plerocercoids of *Schistocephalus solidus* go down as temperatures go up to 30° C; at higher temperatures they rise (DAVIES and WALKEY, 1966). The straight-line slope of dependence on ET that has been found a number of times (1st ed., p. 23, RICHARDS, 1963; KANEHISA, 1965) also shows a declining Q_{10} value with increasing temperature. In comparing several reactions to the same temperature changes, the Q_{10} s are higher where the μ values are greater. The use of μ values is profitable when they are constant at least over fairly wide ET ranges; this is not always the case, even for defined enzyme reactions (see BEHRISCH, 1969, p. 695). On the other hand, constant μ values can often be given for more complex chemical or biological processes (see PRECHT, 1968, p. 492); defined enzyme reactions may show more than one constant μ value (see W. DROST-HANSEN).[4]

The magnitude of the μ value alone cannot be used to identify a "master reaction", as earlier authors assumed (1st ed., p. 21), as the apparent, often continuous change in the values with ΔT indicates (Fig. 1, see PRECHT, 1949; BENTHE, 1954). Furthermore, if the ET in an experiment is continually increased, the Q_{10} and μ values may differ — for the same life processes and ET ranges — from those obtained with continually decreasing ET; overshoots etc. may play a role in such cases.

For example NOPP and FARAHAT (1967) obtained higher Q_{10} values for oxygen consumption in cooling helicid tissues than in warming them over the same ET range (see LAGET et al., 1955; MORAWSKA and ZALESKA, 1958; BARTHOLOMEW and LASIEWSKI, 1965; ZERBST, 1966; WILSON and LEE, 1970; HUGHES and ROBERTS, 1970)[5]. This is not true for the heart frequency of *Periplaneta* (RICHARDS, 1963; see BURKHARDT, 1959; BOUDINOT, 1970).

The Q_{10} values of chemical reactions in general lie between 2 and 4, while those of physical processes (diffusion, osmosis, conductivity, etc.) are lower (1.1 to 1.4). Photochemical reactions are independent of temperature. But it is not possible to identify the nature of a biological process simply from the magnitude of its Q_{10} value. Higher temperature coefficients are not unequivocal signs of chemical reactions (see DANIELLI, 1954; VOGEL et al., 1958), and lower Q_{10} values do not rule them out.

The so-called Normal Curve of KROGH has long been regarded as an indicator of the relation of biological processes to the ET (1st ed., Fig. 6); it applies to the resting metabolism of various animals. As usual, the Q_{10} values of this curve decline with increasing ET; two fairly large temperature zones with constant μ values can be found. The curve also obeys the formula given by KRÜGER (1963, 1966, 1970). It was learned later that the metabolism of many animals did not correspond to KROGH's curve even under the specified normal conditions. This can result from genetic (BERG et al., 1962) or from external factors, e.g., preliminary treatment of the animals. KROGH himself referred to the deviation in oxygen consumption of *Tenebrio* pupae, and the different forms of relation to ET have been the subjects of numerous investigations.

---

[4] In: BROWN, H. D. (Ed.): Chemistry of the cell interface. New York: Acad. Press 1971; cf. ZEYLEMAKER, W. P., JANSEN, H., VEEGER, C., SLATER, E. C.: Biochem. biophys. Acta (Amst.) 242, 14—22 (1971).

[5] Cf. WEATHERS, W. W.: Comp. Biochem. Physiol. **40** A, 503—515 (1971); BETHEA, N. J.: Comp. Biochem. Physiol. **41** A, 301—305 (1972). It may be necessary to pay regard to the temperature of the body during such experiments (e.g. in turtles, cf. SPRAY, D. C., MAY, L.: Comp. Biochem. Physiol. **41** A, 507—522, 1972).

Even the different partial reactions comprising a single organ function can show different Q_{10} values (see STIEVE, 1963; WALTHER, 1965; HUGHES and ROBERTS, 1970 i.a.). Under the conditions used by SCHNEIDER (1964), dark adaptation of *Carausius morosus* is independent of ET, while light adaptation is not (see AUTRUM and HAMDORF, 1964; SEIBT, 1967). Measurements of the resting potential of frog and *Carcinus* muscle fibers belonging to the same animal gave different Q_{10} values (KERKUT and RIDGE, 1961; see KERKUT and TAYLOR, 1958). Q_{10} values for the heart beat of embryonal *Salmo irideus* depend on the initial basal frequency (HUGGEL, 1951). According to BURKHARDT (1959) the temperature coefficient of the stationary discharge of the stretch receptors of *Astacus* is a function of the state of tension on the muscle. The discharge frequency of strongly stretched receptors has a Q_{10} value of approximately 2; under weaker tension it drops to 1, and with complete relaxation even to less than 1. The Q_{10} value for the speed of adaptation of the receptors after stimulation is between 3 and 5 in all states of tension. The same organ functions in different species can show different Q_{10} values (for example, the frequency of discharge of muscle receptors; see WINTER, 1967; DALTON and HENDRIX, 1962; DUDEL and RÜDEL, 1968).

VERNBERG and MERINEY (1957) found genetically determined differences in metabolism between races of *Drosophila melanogaster* originating from a wild population. Over the ET range of $13-18°$ C the Q_{10} values amounted to 4.71 for the dark race, 4.32 for the lighter race, and 8.98 for the heterozygotes (for the range from 22° to 27° C, the values were: 3.39, 5.14 and 1.67). The temperature-function curves show plateaus which can of course exert an influence on Q_{10} values. According to SCHMIDT (1968) the Q_{10} values for oxygen consumption of ♂♂, ♀♀, and ⚥⚥ (pupae) of *Formica polyctena* and of the different pupal stages can be differentiated.

Hormones can influence the temperature coefficients, e.g. of sodium transport across frog skin (DALTON and SNART, 1968). Injection of vasopressin lowers the μ value for water uptake in *Bufo melatonicus* by 4000 cal (these animals had already been exposed to the ETs at night; DICKER and ELLIOT, 1967).

If the Q_{10} values for oxygen consumption of heart muscle and the supra-esophageal ganglion of *Uca* species are plotted against the ET, different curves are obtained for different tissues from the same species and for the same tissues from different species; the curves may show definite maxima and minima. The Q_{10} values for the nerve tissues come closest to approximating those for the respiration of the whole animal (W. B. VERNBERG and F. J. VERNBERG, 1966). KENNEDY and RICHARDS (1955) found that flight muscles of *Lethocerus americanus* displayed lower apyrase activity and smaller temperature coefficients than the swimming muscles.

HOCHACHKA and his co-workers investigated the effect of ET on the enzyme-substrate affinity (expressed as the Michaelis constant-K_m). The affinity between the substrate and LDH (as for other enzymes) approaches a maximum value (minimum K_m) near the lower extremes of the species habitat temperatures. At lower substrate concentrations, such as are usually found *in vivo*, the K_m values become important. Since they increase at higher ETs, the Q_{10} values must fall, as generally happens (HOCHACHKA and SOMERO, 1968; SOMERO et al., 1968).

c) Low Temperature Coefficients, and Other Special Cases

The temperature range usually found in the natural environment can show particularly low Q_{10} values, by which a more or less marked independence from external temperature is assured for life processes over this significant range. This is especially important when the external temperatures vary greatly. Though only direct responses are of interest here, further investigations are required to exclude the possibility that regulations by control systems play a part. This is also true for the examples given here. Plateaus resulting from capacity adaptation are recognized more easily (p. 341).

Komatsu and Feeney (1970, p. 314) give reasons for the advantage of high and low μ values (and concomitantly high and low Q_{10} values): "One is that the activation energy should be high, for any increase in temperature will give reactions catalyzed by any given enzyme a larger increase in activity than reactions with lower μ values". The other line of reasoning is that genetic or non-genetic adaptation to low temperatures should decrease the activation energy and reduce the sensitivity of the organism to temperature change. Furthermore, in reactions with lower activation energies there should be a larger number of molecules with sufficient energy to react. This should result in a faster rate of reaction at any given temperature (see Zuber, 1969; Behrisch, 1969, p. 695).

The reaction speed is increased by the addition of a suitable catalyst or by enzymes in biological processes; the activation energy is thereby reduced. Somero (1969b, p. 525) calls attention to the fact that in the determination of Q_{10} values of enzyme reactions often unnaturally high substrate concentrations are used; this leads to misleading, high values[6].

In the lower temperature range, the only one that is biologically significant, the oxygen consumption of overwintering *Melasoma populi*, for example, exhibits strikingly low μ values. These beetles, which must subsist on their food reserves, can prevent an overly abrupt increase in metabolism in this way, as well as by a capacity adaptation. Overwintering potato beetles that live in deeper burrows and so are less exposed to changes in temperature, do not show this phenomenon (Marzusch, 1952; see Clarke in Rose, 1967, p. 316). At temperatures below 15° C the resting metabolism of thermophilous lizards exhibits higher Q_{10} values than those of less heat resistant species; the Q_{10} values for heart beat frequency in the former group increases markedly below ET = 20° C, but in the latter group it does so only below 10° C (see Dawson in Milstead, 1967; Dawson and Templeton, 1966[7] i.a.). According to Woynárovich (1963) the oxygen consumption of embryonic and juvenile stages of fish such as *Silurus glanis* and *Cyprinus carpio*, which prefer warm water, decreases much more steeply at low ETs than is the case with those species *(Esox lucius, Lucioperca lucioperca)* that develop well in cold water. In parasites, Vernberg and Hunter (1961) have also reported that Q_{10} values at normally prevailing temperatures are small; the trematodes they investigated lived in a fish, in a turtle and in a bird, respectively (see Vernberg, 1961). Newell and Northcroft (1967; see Newell and Pye, 1970) making measurements in closed containers, observed fairly broad plateaus for "maintenance metabolism", while "active metabolism" showed Q_{10} values of ca. 2 *(Actina equina:* 7.5−20° C,

---

[6] If the activity of the LDH of *Idus idus* is measured at a small concentration of pyruvate, no ET-dependence of the velocity reaction is noted; whether this is also valid *in vivo* is, however, problematical [Wernick, A., Künnemann, H.: Mar. Biol. 18, 32—36 (1973)].

[7] It should be noted that the body temperature of reptiles is often raised during the day by behavioral regulation. Therefore it is not exactly the same as the AT of previous maintenance.

Nephthys hombergi: 3–20° C, *Littorina littorea*: 10–23° C). The fact that isolated mitochondria also show such plateaus may point to an innate reaction norm (adapted to the tidal rhythm). Such plateaus are said to be related to the temperature range in which the animals normally live (see Loosanoff, 1958; Sandison, 1967; Buffington, 1969; Aleksiuk, 1970)[8]. The Q_{10} values of the oxygen consumption of isolated mitochondria (of *Littorina littorea*) depend on the concentration of pyruvate used (Newell and Pye[9]).

The mentioned results give the impression that *extended* ranges of independence from ET are widespread, but the findings of many other authors (see Valen, 1958; Tribe and Bowler, 1968) contradict them.

According to Wieser and Nopp-Pammer (1968) independence from temperature (e.g. of melanin synthesis in *Triturus cristatus*) can arise because a participating enzyme has a reaction optimum at a low ET, and its activity therefore declines at higher ETs. However the activity of another enzyme may go up as usual with increasing ET, so that the combination of the two activities leads to a Q_{10} of ca. 1 (for Q_{10} values of less than 1, see p. 307, 436f.). Wieser et al. (1970) called attention to the seasonal influence on respiration in *Arianta arbustorum*. A highland population is distinguished from a lowland form by a higher Q_{10} value in spring (2.7, 1.9, respectively, range: 20–30° C) and a lower Q_{10} value during summer (1.3, 1.8, respectively). The low Q_{10} values have been attributed to complex mechanisms regulated by the central nervous system.

It is generally accepted that the major means of control of fructose 1,6-diphosphatase is through negative modulation by AMP. Fish (see Behrisch and Hochachka, 1969) may show a rather high affinity of FDPases for AMP and this may be sensitive to ET. In migrating salmon that take no nourishment, however, this affinity and the temperature sensitivity of this enzyme-modulator interaction are decreased. "Together with the thermal independence of enzyme-AMP affinity, the constant saturation of the enzyme by cofactor, and the increasing enzyme-FDP affinity with decreasing temperature, this temperature-independence of the enzyme saturated by substrate would seem to allow constant operation of liver FDPase in the migrating salmon, even in fluctuating temperatures" (Behrisch, 1969, p. 695).

The K_m values of an enzyme (such as the LDH of *Rhodeus amarus*) may increase with ET or (may increase) above a minimum value [such as the LDH of *Idus idus* (Wernick and Künnemann, p. 308) or of *Lepidosiren paradoxa*]. If the K_m values rise sharply and the maximum rate (V_{max}) increases with ET at a proportionately lower rate ($Q_{10} = 2.2$), then at lower substrate concentrations (when K_m is important in determining reaction velocity) catalytic rates are higher at low temperatures than at elevated ones. When substrate concentrations are raised above K_m values, catalytic rates are higher at higher temperatures. These properties can be important to metabolic control during changes in the environment, since they supply one mechanism by which a given reaction can be held relatively independent of temperature. During maintenance metabolism, when substrate levels would be

---

[8] Cf. Aleksiuk, M.: Comp. Biochem. Physiol. **39**, 495—503 (1971); Percy, A., Aldrich, F. A.: Nature (Lond.) **231**, 393—394 (1971).
[9] Newell, R. C., Pye, V. I.: Comp. Biochem. Physiol. **40**, 249—261 (1971).

low Q_{10} values may be low, in contrast to active metabolism, when substrate levels are high (HOCHACHKA and SOMERO, 1968; see HOCHACHKA and LEWIS, 1970)[10]. HOCHACHKA (1968a) reports another way in which low Q_{10} values may be obtained. Two major pathways for the further metabolism of acetyl Co A are the Krebs cycle and the pathway to lipid. As ET is elevated the Krebs cycle becomes less effective in competing for common substrates, so that its activity is not increased. Citrate and ATP inhibit acetate carbon flow through the Krebs cycle but favor conversion to lipid. These effects are accentuated at elevated ETs; hence, in the presence of high levels of citrate or ATP, the temperature coefficients for acetate oxidation are low (sometimes less than 1; see SOMERO and HOCHACHKA, 1968, p. 399).

SOMERO and HOCHACHKA (1969) found that two isozymes of LDH in the leg muscle of the crab *Paralithodes camtschatica* are especially important. At physiological concentrations only the "low K_m" LDH is active at $10-15°$ C. As the ET decreases, however, the K_m of the "high K_m" LDH drops sharply, activating the enzyme, and by $5°$ C both LDHs are active. As a result of the activation of this second LDH, the rate of catalytic activity at physiological substrate concentrations is higher at $5°$ C, the species' normal habitat temperature, than at $10°$ C, the upper extreme of its habitat temperature. SOMERO (1969a) later found a "cold" pyruvate kinase (minimal K_m at $5°$ C, active only below $10°$ C), and a "warm" variant (minimal K_m at about $12°$ C, inactive below $9°$ C) in the same animal. The combined activities of these two pyruvate kinases yield highly temperature-independent rates of catalysis over the range of habitat temperature ($4-12°$ C). SOMERO believes that both variants are formed by a temperature-dependent interconversion of one protein species[11].

O'CONNOR (1960, 1964) found irregularities in the respiration curves of mammalian and frog tissues. Different maxima were noted . . . "these maxima result from a surface action of fatty acids in molecular order". Peculiarities in insect respiration may occur because at lower ETs the periodic respiration may go over to a pure diffusion respiration, without opening of the spiracles (KANWISHER, 1966).

d) The Thermal Dependence of Michaelis' Constant

B. HAVSTEEN

Since the Michaelis constant, K_M, is a readily available[12] kinetic parameter for enzyme catalyzed reactions, the dependence of K_M upon external variables such as

---

[10] Cf. BEHRISCH, H. W.: Mar. Biol. **13**, 267—275 (1972), Biochem. J. **121**, 399—409 (1971); HOCHACHKA, P. W., and SOMERO, G. N. give a summary of their papers in HOAR, W. S., RANDALL, D. J.; Fish physiology. New York: Academic Press 1971, Vol. VI, pp. 100—156; cf. HOCHACHKA, P. W.: Am. Zool. **11**, 81—82, 425—435 (1971).

[11] The results concerning the relation between K_m and Q_{10} values are only reported so far; a detailed treatment is given by B. HAVSTEEN (p. 310—317).

[12] Abbreviations:

E	enzyme (inactive form)
E'	enzyme (active form)
ES	enzyme-substratum-complex (inactive form)
E'S	enzyme-substratum-complex (active form)
E_0	initial enzyme concentration (total enzyme concentration)
EP_2	enzyme-product-complex
ΔG^0	standard affinity
h	Boltzmann's constant

the temperature, pH, ionic strength, etc. is often measured. The prerequisite for a simple interpretation of the data is that the reaction follows the scheme:

$$E + S \underset{k_{-1}}{\overset{k_1}{\rightleftharpoons}} ES \overset{k_2}{\longrightarrow} E + P. \tag{1}$$

Unfortunately, the observations often indicate the existence of very complex relationships which may arise from the failure of one or more of the assumptions in the above scheme, e.g.:

1. The reaction consists of more than two steps. The additional steps are often rapid isomerizations of E, S or ES.
2. The formation of ES from E and P is significant ($k_{-2} \gg 0$).
3. ES is neither in thermodynamic equilibrium nor in the steady state.

If the conventional kinetic scheme (Eq. 1) is valid, then the temperature dependence of the Michaelis constant may be derived
a) by differentiation of the equation of definition:

$$K_M = \frac{k_{-1} + k_2}{k_1} \tag{2}$$

b) or by application of the Transition-State Theory of EYRING.

ad a) A convenient procedure is the logarithmic differentiation:

$$\ln K_M = \ln (k_{-1} + k_2) - \ln k_1 \tag{3}$$

$$\frac{d \ln K_M}{d T} = \frac{d \ln (k_{-1} + k_2)}{d T} - \frac{d \ln k_1}{d T}. \tag{4}$$

By rearrangement one obtains:

$$\frac{1}{K_M} \frac{d K_M}{d T} = \frac{1}{(k_{-1} + k_2)} \frac{d (k_{-1} + k_2)}{d T} - \frac{1}{k_1} \frac{d k_1}{d T} \text{ or}$$

$$\frac{d K_M}{d T} = \frac{1}{k_1} \frac{d (k_{-1} + k_2)}{d T} - \frac{K_M}{k_1} \frac{d k_1}{d T}. \tag{5}$$

In the right side of Eq. (5), $d T$ is substituted by $- T^2 d \frac{1}{T}$:

$$\frac{d K_M}{d T} = - \frac{1}{T^2 k_1 d \frac{1}{T}} \frac{d (k_{-1} + k_2)}{k_1 T^2} + \frac{K_M}{d \frac{1}{T}} \frac{d k_1}{d \frac{1}{T}}. \tag{6}$$

---

k_1, k_{-1} k_2, k_{-2} k_3, k_{-3}	velocity constants
K	Planck's constant
k'	barrier coefficient
K_M	Michaelis' constant
K_M^{app}	measured K_M-value (directly)
K_S	dissociation constant of the ES-complex
P	reaction product
P_1, P_2	various reaction products
R	gas constant
S	substratum
S_0	initial substrate concentration
T	absolute temperature
μ_1, μ_{-1} μ_2, μ_{-3}	various activation energies

The activation energies for the three steps are then introduced in Eq. (6):

$$\frac{\mathrm{d}\,K_\mathrm{M}}{\mathrm{d}\,T} = \frac{-k_{-1}\,\mathrm{d}\ln k_{-1}}{k_1\,T^2 \mathrm{d}\,\frac{1}{T}} - \frac{k_2\,\mathrm{d}\ln k_2}{k_1\,T^2 \mathrm{d}\,\frac{1}{T}} + \frac{K_\mathrm{M}\mathrm{d}\ln k_1}{T^2\,\mathrm{d}\,\frac{1}{T}}$$

$$= \frac{k_{-1}\,\mu_{-1}}{k_1\,R\,T^2} + \frac{k_2\mu_2}{k_1\,R\,T^2} - \frac{K_\mathrm{M}\,\mu_1}{R\,T^2} \tag{7}$$

ad b) The thermal dependence of the individual rate constants is, according to the theory of EYRING and EYRING (1963):

$$k = A_0\,\mathrm{e}^{-\frac{\mu}{RT}}$$

where A_0 is the pre-exponential factor:

$$A_0 = k'\,\frac{KT}{h}.$$

Introduction of this expression into Eq. (2) immediately gives the dependence of K_M upon the temperature:

$$K_\mathrm{M} = \frac{k_1'\,\mathrm{e}^{-\frac{\mu_{-1}}{RT}} + k_2'\mathrm{e}^{-\frac{\mu_2}{RT}}}{k_1'\,\mathrm{e}^{-\frac{\mu_1}{RT}}}. \tag{8}$$

Since the barrier coefficients are $> o$, K_M cannot in any temperature range become zero or negative if Eq. (1) is valid.

Differentiation of Eq. (8) with respect to the temperature shows that the temperature profile of K_M has no extrema but that the relationship:

$$k_{-1}'\,\mu_1\mathrm{e}^{-\frac{\mu_{-1}}{RT}} + k_2'\,\mu_2\mathrm{e}^{-\frac{\mu_2}{RT}} = \mu_1\left(k_1'\,\mathrm{e}^{-\frac{\mu_{-1}}{RT}} + k_2'\,\mathrm{e}^{-\frac{\mu_2}{RT}}\right)$$

$$k_1\mu_1 + k_2\mu_2 = \mu_1\,(k_1 + k_2) \quad \text{exists.}$$

A kinetic scheme which applies to many reactions catalyzed by enzymes is the following:

$$\mathrm{E} + \mathrm{S} \underset{k_{-1}}{\overset{k_1}{\rightleftharpoons}} \mathrm{ES} \overset{k_2}{\longrightarrow} \mathrm{EP}_2 \overset{k_3}{\longrightarrow} \mathrm{E} + \mathrm{P}_2 + \mathrm{P}_1.$$

We here assume that both ES and EP_2 are in the steady state, that the formation of the ES-complex is the fastest step of all in the process, that k_{-2} and $k_{-3}\cdot[\mathrm{E}]\cdot[\mathrm{P}_2]$ are negligible compared with k_2 and $k_3\cdot[\mathrm{EP}_2]$, respectively, and that $[\mathrm{S}_0] \gg [\mathrm{E}_0]$. Accordingly, we have:

$$[\mathrm{ES}]\cdot k_2 = [\mathrm{EP}_2]\cdot k_3 \tag{9}$$

and:

$$v = [\mathrm{EP}_2]\cdot k_3 = [\mathrm{ES}]\cdot k_2 = [\mathrm{E}_0]\cdot k_0 = [\mathrm{EP}_2]\left(1 + \frac{k_3}{k_2}\right)k_0 \tag{10}$$

where k_0 is the observed rate of formation of the product P_2.

Since the formation of ES is very fast compared with the other steps, the total enzyme concentration becomes:

$$[\mathrm{E}_0] = [\mathrm{ES}] + [\mathrm{EP}_2]. \tag{11}$$

Introducing Eq. (9) in Eq. (11) gives:

$$[\mathrm{E}_0] = \frac{k_3}{k_2}\,[\mathrm{EP}_2] + [\mathrm{EP}_2] = [\mathrm{EP}_2]\left(\frac{k_3}{k_2} + 1\right) \text{ which is entered into Eq. (10):}$$

$$v_0 = [\mathrm{E}_0]k_0 = [\mathrm{EP}_2]\left(1 + \frac{k_3}{k_2}\right)k_0. \tag{12}$$

By comparing Eqs. (10) and (12) we see that:

$$k_3 = k_0 \left(1 + \frac{k_3}{k_2}\right) \text{ or by rearrangement:}$$

$$k_0 = \frac{k_2 k_3}{k_2 + k_3} . \tag{13}$$

The steady state condition for ES requires that:

$$[S_0] ([E_0] - [ES] - [EP_2]) k_1 = [ES] (k_{-1} + k_2) .$$

Introducing the expression for ES, we get:

$$[S_0]([E_0] - \frac{k_3}{k_2}[EP_2] - [EP_2]) k_1 = \frac{k_3}{k_2}[EP_2] (k_{-1} + k_2)$$

or by rearrangement:

$$[S_0] \cdot [E_0] \cdot k_1 = [S_0] \cdot k_1 \cdot [EP_2] \left(1 + \frac{k_3}{k_2}\right) + \frac{k_3}{k_2}[EP_2] (k_{-1} + k_2)$$

$$= [EP_2] \left([S_0] \cdot k_1 \left(1 + \frac{k_3}{k_2}\right) + \frac{k_3}{k_2}(k_{-1} + k_2)\right) .$$

Hence, introducing the definition of $K_M \left(= \frac{k_{-1} + k_2}{k_1}\right)$:

$$[EP_2] = \frac{[E_0]}{1 + \frac{k_3}{k_2} + \frac{k_3}{k_2} \frac{K_M}{[S_0]}} .$$

This expression is introduced in Eq. (10):

$$v_0 = \frac{k_0 \left(1 + \frac{k_3}{k_2}\right) [E_0]}{\left(1 + \frac{k_3}{k_2}\right) \frac{k_3}{k_2} \frac{K_M}{[S_0]}} = \frac{k_0 [E_0]}{1 + \frac{k_3}{(k_2 + k_3)} \frac{K_M}{[S_0]}} = \frac{k_0 [E_0]}{1 + \frac{K_M^{app}}{[S_0]}} \tag{14}$$

where:

$$K_M^{app} = K_M \frac{k_3}{(k_2 + k_3)} = \frac{\left(k'_{-1} e^{-\frac{\mu_{-1}}{RT}} + k'_2 e^{-\frac{\mu_2}{RT}}\right)}{\left(k'_2 e^{-\frac{\mu_2}{RT}} + k'_3 e^{-\frac{\mu_1}{RT}}\right)} \frac{k'_3 e^{\frac{(\mu_1 - \mu_3)}{RT}}}{k'_1} .$$

Since Eq. (14) has the usual form of a Michaelis equation, the Lineweaver-Burk plot and similar diagrams will be linear. If Eq. (14) is differentiated with respect to temperature, we observe a more complicated dependence than the one found for the simple Michaelis equation ($\leq$ 4 maxima and/or minima can occur).
When the $E\,T$-dependence of K_M shows maxima or minima, which often is observed and has already been mentioned (p. 310), then the Michaelis equation is insufficient and must be extended by several steps[13]. Therefore, the observation of such a complicated K_M-$E\,T$-profile is useful and serves as a guide to the construction of the real reaction scheme. A sufficient condition for the appearance of a large number of maxima and minima in the K_M-$E\,T$-curve is the presence of a cyclic process, such as the one below (e.g. HAVSTEEN, 1967):

$$E + S \rightleftharpoons ES \rightleftharpoons E + P$$
$$E' + S \rightleftharpoons E'S \rightleftharpoons E' + P.$$

In this case as well as in the case of the simple Michaelis equation (Eq. 2) the measured K_M-value (K_M^{app}) is not identical with the dissociation constant

[13] Compare: LOW, P.S., BADA, J.L., SOMERO, G.N.: Proc. Nat. Acad. Sci. USA 70, 430—432 (1973).

$\left(K_S = \dfrac{k_{-1}}{k_1}\right)$ of the ES-complex. This requires at least that $k_2 \ll k_{-1}$. Hence, the affinity of the enzyme for the substrate $(\varDelta G^0 = -RT \ln K_S)$ is not always directly measured by K_M. This is often assumed in *in vivo* systems since so many significant variables are present that it rarely is possible to arrive at the parameters of the individual steps (LUMPER, 1964).

e) The Relationship between Q_{10} and the Kinetic Parameters
B. HAVSTEEN

Q_{10} is defined by the equation:

$$Q_{10} = \frac{k T_2}{k T_1}, \tag{15}$$

where $T_2 - T_1 = 10° C.$

This equation may be applied when the enzyme concentration is known, i.e. in isolated systems. Often, however, the catalytic activity is measured in the tissues where the concentration of membrane-bound enzymes is unknown. In the latter case, the rate constants of product formation commonly are substituted by the directly observed velocities. This leads to the following consequences:

$$Q'_{10} = \frac{v_{T_2}}{v_{T_1}} = \frac{k_2^{T_2} \cdot [ES_{T_2}]}{k_2^{T_1} \cdot [ES_{T_1}]} = \frac{k_2^{T_2} (K_M^{T_1} + [S_0])}{k_2^{T_1} (K_M^{T_2} + [S_0])} \tag{16}$$

where $k_2^{T_2}$ and $k_2^{T_1}$ are the rate constants for the product-forming steps at the two temperatures. When the enzyme is saturated with substrate, Q'_{10} is identical with Q_{10} as previously defined, but at low substrate concentration the following equation is valid:

$$Q'_{10} = \frac{k_2^{T_2}}{k_2^{T_1}} \cdot \frac{K_M^{T_1}}{K_M^{T_2}} = Q_{10} \frac{K_M^{T_1}}{K_M^{T_2}}.$$

Apart from the difficulties in precise experimental determination of K_M and in formulation of a reasonably simple but still fairly correct reaction scheme (see above), the following conclusions may with suitable caution be drawn:

When the K_M-values increase with increasing temperature, it follows from the definition equation for Q_{10} that this parameter is diminished [Eq. (16)]. This was observed very early (already by KROGH). Q_{10} even assumes the value of unity when the thermal dependence of K_M is compensated by that of the reaction velocity. The necessary condition is that $k_2 \gg k_1$, i.e. that the ES-complex only slowly dissociates in E + S, and that the association step (E + S → ES) is fast and therefore insensitive to thermal changes.

When K_M loses its temperature dependence, then Q_{10} also is rendered constant and normal. A special case is the observation of persistently oscillating K_M-values about a mean value in the K_M-ET-plot. This can arise if the enzymatic process includes one or more cyclic pathways, which is often the case with allosteric and hydrolytic enzymes.

f) Enthalpy-Entropy Compensation Phenomena in Aqueous Solutions. A Ubiquitous Property of Water?
B. HAVSTEEN

Most biochemical reactions occur in an aqueous environment. Therefore the properties of water are important to the understanding of physiological mechanisms.

Recently, a specific linear relationship between the entropy change and the enthalpy change in a variety of processes involving molecules of all sizes has been found. Such processes include solvation, hydrolysis, electron transfer, ionization of weak electrolytes, gas binding and denaturation of macromolecules. The value of the proportionality constant, the so-called compensation temperature, is $250-315°$ K in all these cases. The ubiquity of this phenomenon indicates that it could be a general property of water. The nature of the relationship between the enthalpy and entropy of apparently all reactions in water has not yet been definitely identified but probably rests upon phase changes (changes in heat capacity at constant temperature) in the solute, in the solvent (state of aggregation of the water molecules) or in both. The compensation phenomenon is found both for thermodynamic parameters and for activation parameters and may presumably be used as an indicator of the involvement of water in a reaction, e.g. in life processes. Proteins usually respond to changes in the state of aggregation of water by volume changes. Such changes have previously been measured in a different context, but these observations were then uninterpretable. However, in the light of the theory of structural compensation they become comprehensible[14].

α) *Observations*

The proportionality between the enthalpy and entropy has been found in a great variety of systems which only had the solvent (water) in common. The experiments consisted of determination of equilibrium constants of chemical reactions or of solubilities at various temperatures. Similarly, kinetic parameters for enzyme catalysis, such as K_M, V_{max} and k_{cat} $\left(= \dfrac{V_{max}}{[E_0]}\right)$, were determined as functions of the temperature. Besides, it was noted that at a certain temperature which coincided with the compensation temperature, the substrate specificity of the enzyme vanished. This phenomenon was only found under a unique set of conditions characterized by the pH (pH_{comp}), the ionic strength, etc. Such observations may be rationalized as the occurrence of a phase change under these specific compensation conditions. This hypothesis is supported by the measurement of a change in the total volume of the solution accompanying the passage of the compensation conditions.

β) *Characterization*

Since the compensation phenomena are widespread and appear to be a unique property of water they deserve careful investigation. Although the presence of water is obligatory for the sustenance of life as we know it, the structure and properties of this substance are only imperfectly known. Because of its great liability to form hydrogen bonds, water is one of the least known of all solvents. It consists, according to the current theories of dynamic equilibria, of hydrogen-bonded molecular clusters (NÉMETHY and SCHERAGA, 1962a, b, c). The number of water molecules in the clusters varies from a few to about a dozen. The majority of these aggregates are hexa- to nonamers, depending on the external parameters such as the temperature. Between the clusters are a few single water molecules; otherwise there is vacuum. The hole structure and the dynamic equilibria of hydro-

_____
[14] LUMRY, R., RAJENDER, S.: Biopol. 9, 1125—1227 (1970).

genbonded clusters are probably responsible for the compensation phenomenon since all other solvents are closely packed without holes and none of them display the phenomenon. Other characteristics of compensation are that the heat capacity, $\frac{d \triangle H}{dT}$, is large, which is a common property of phase changes, that the compensation temperature always falls in a rather narrow temperature range and that T_{comp} is independent of the nature of the perturbation.

γ) Thermodynamics

Since the laws of thermodynamics are completely general they must also be applicable to the compensation phenomenon. The thermodynamic function which is the most useful to the description of biological systems is the H-function because it emphasizes the entropy, the pressure and the molarity:

$$H = H\,(S,\,p,\,m_i).$$

Its total differential has proven to be:

$$dH = T\,dS + v\,dp + \sum \mu_i\,dm_i.$$

Since dH is compensated by $T\,dS$, the following condition must prevail:

$$v\,dp + \sum \mu_i\,dm_i = o$$

where μ_i is the chemical potential and m_i the molarity of the i'th component. If we introduce the equation for the chemical potential:

$$\mu_i = R\,T \ln a_i + \mu_{0i}$$

where a_i is the activity and μ_{0i} the standard potential of the i'th molecular species the relationship at $T = T_{comp}$ becomes:

$$\sum f_i \ln a_i \cdot dm_i = Q = -\frac{(v\,dp + \sum fi\,\mu_{0i}\,dm_i)}{R\,T_{comp}}.$$

The quantity on the right side of this equation is experimentally accessible and has been measured for a number of systems. Therefore, the equation which describes the molecular distribution at the compensation temperature[15] is:

$$\frac{1}{v}\left[\sum f_i \ln\left(\frac{f_i n_{w_i}}{N}\right) dn_w + \sum f_i \ln\,(f_j \cdot n_{L_j})\,dn_L\right] = Q$$

where N is Avogadro's number, f the activity-coefficient $(o - 1)$, n_{w_i} the number of clusters of water molecules of the size i per volume unit and n_{L_j} is the number of locks in the solvent of the size j.

Dissolution of a substance, e.g. a protein, in water therefore means that room is made for the foreign molecule by rearrangement of the molecular structure in the solvent until there is space enough for the solute. Similarly, unfolding of a macro-molecule, e.g. during denaturation, or conformational changes in an enzyme induced by the binding of a specific substrate requires a rearrangement of the water structure and is accompanied by a net volume change. The latter has often been observed but rarely adequately interpreted.

---

[15] T_{comp} = compensation temperature (not to be confused with AT).

δ) *Generalizations*

At this stage, the compensation phenomenon has not yet been sufficiently explored and precise measurements are rather scarce. However, the evidence which is available is now so diversified and plentiful that models which account for observations and put them on an acceptable thermodynamic and mechanistic basis must be tested. Such a working hypothesis which couples structural and entropic changes and incorporates the current view of the composition of water has been advanced. Time will show whether this model is crude or adequate and the necessary refinements will be elaborated. Meanwhile, since the compensation phenomenon is easy to detect, it may serve as a plausible diagnostic aid in the detection of water as an essential participant in life processes.

g) Body Size and Relation to Experimental Temperature

Comparisons can be made among related species or races of different sizes, or among different stages of growth of the same species. In the latter case differences will of course most probably be encountered if greatly differing stages (e.g., of holometabolic insects) are compared.

After a literature search RAO and BULLOCK (1954) came to the conclusion (in spite of the contrary opinion of some authors) that in general the Q_{10} value of the slope with respect to ET increases with increasing body size. Nevertheless, it is hardly possible to speak of a rule, though several later findings confirmed this view. Moreover the findings are not always in agreement for different weight categories (see MORRIS, 1962). On the contrary, several authors found either a decrease with increasing size, e.g. for the heart beat of *Asellus aquaticus* in the ET range of $9-22°$ C (SCHWARTZKOPFF, 1955), for the oxygen consumption of the scorpion *Heterometrus fulvipes* (see RAO, 1966), for the heart beat of the fish *Ophicephalus punctatus* (HASAN and QASIM, 1961, see JOB, 1957; ARMITAGE, 1962), or no dependence on size (e.g. of the oxygen consumption of *Tilapia mossambica*; JOB, 1969).

Related species may behave differently. The heart rates of *Salmo irideus, S. fario, Scyliorhinus canicula* and *Pristiurus melanostomus* increase as development progresses; at the same time the Q_{10} values decline *(S. irideus)* or remain almost constant (HUGGEL, 1959, 1961), as PICKENS (1965) also found for the heart rate of *Mytilus californicus* in relation to the inner shell volume (see DAVIES, 1966/67; DAVIES and WALKEY, 1966).

Differences have even been found over single ET ranges. According to ROBERTS (1957) the Q_{10} values for oxygen consumption of *Pachygrapsus crassipes* remain constant between $8.5°$ and $16°$ C with increasing body size. However between $16°$ and $23.5°$ C they change in direct proportion to body weight, an effect which may be connected with stress induced by the higher temperature (AT $= 16°$ C). Q_{10} values for oxygen consumption of laboratory adapted *Palaeomonetes varians* change only slightly with body weight. In freshly caught animals they vary more, but in summer animals not at all (ET $= 10°$ and $30°$ C; MCFARLAND and PICKENS, 1965). Twelve-day-old larvae of a holometabolic insect, *Calandra oryzae*, use far more oxygen and exhibit lower Q_{10} values than seven-day-old adult animals (AT $= 29°$ C; BIRCH, 1947).

Differences between geographical races are also found, e.g. in *Uca pugnax* according to TASHIAN (1956). Specimens from New York, North Carolina and Florida show a decrease in Q_{10} value for oxygen consumption with increasing body size; the opposite is true of animals from Trinidad (see TASHIAN and VERNBERG, 1958).

The cases mentioned so far have dealt with growth stages of different size. MINA-MORI (1957) measured the relation between ET and the oxygen consumption of certain embryonic stages of races of *Cobitis taenia striata* of different size. The Q_{10} values shown by three races increase with decreasing body size, but the magnitude of oxygen use declines (see Table 16 of the author). An interspecific comparison of fiddler crabs showed that smaller species had higher Q_{10} values for oxygen consumption in the ET range of $12-28°$ C or $30°$ C than larger species had (VERN-BERG, 1959). According to LOCKER (1958a, b, 1959) the μ values for respiration of liver and skin from *Rana esculenta* decrease with increasing body size. The degree of dependence varies with the season. However the μ values of tissue respiration are easily modified (e.g. by substrates).

h) Relation of the Temperature Coefficient of the ET Curve to the Acclimation Temperature

The question of the relation of Q_{10} values of the ET curve to AT can only be raised if capacity adaptation exists (this problem will be discussed later). In a few cases of adaptation in the form of compensation (types 1—3), the AT has no significant, or at least no consistent influence on the Q_{10} value. The value may rise as AT decreases; often it rises with increasing AT. In some cases of inverse compensation (type 5) Q_{10} values go down with rising AT; in other cases they go up (examples of all of these types are found in PRECHT, 1968, p. 499f.; see JUNGREIS and HOOPER, 1968; DAUSCHER and FLINDT, 1969; BUIKEMA and ARMITAGE, 1969; DANIELS and ARMITAGE, 1969).

Further complications can develop. According to KIRBERGER (1953) the Q_{10} values for oxygen consumption of *Lumbricus variegatus* are the same for constant ATs of $1°$ and $16°$ C, and for ATs that alternate every twelve hours between $15°$ and $23°$ C; only if the AT remains constant at $23°$ C do the Q_{10} values fall. The temperature coefficients for the rate of incorporation of marked leucine into protein in goldfish gill and liver are higher in animals acclimated to $25°$ C than in those acclimated to $5°$ C, but this holds true only for low ETs; the opposite is true in the upper range (DAS and PROSSER, 1967). Two constant μ values are found for each plot of the movement of *Gammarus pulex* pleopods (Fig. 1). In this case there is no dependence on AT at low ETs, but such a relation is found at high ETs. The Q_{10} maximum for CO_2 production in *Tinca tinca* shifts, with changing AT (PUNT, 1945).

Complications in cell metabolism can also appear. For example in the case of Q_{10} values of tissue respiration of species of *Uca* already mentioned above, the height and position of the minima and maxima depend on the AT. This is particularly noticeable in regard to the height for the heart tissue of *U. pugilator*. LOCKER and WEISH (1965) found that the μ value for oxygen consumption of some, but not all, isolated tissues of *Rana esculenta* increased with decreasing AT. This relation can be reversed by the addition of substances that increase metabolic rates (uncoupling agents). Seasonal influences may also have played a role (see RAO and BULLOCK, 1954).

PROSSER (1958) speaks of a "rotation" to designate a change in temperature co-efficient with AT; he uses the behavior of this coefficient to characterize capacity adaptation: I. No adaptation. II. Translation, (a) up or to left with cold, (b) down or to right with cold. III. Rotation, (a) clockwise, Q_{10} reduced in cold, (b) counter-clockwise, Q_{10} increased in cold. IV. Translation combined with rotation. PROSSER (in HANNON and VIERECK, 1962, p. 10) gives examples of these types (see PROSSER and BROWN, 1962).

If the different curves relating function to ET show different but constant Q_{10} values, they must intersect. Since Q_{10} values usually decline as ETs increase, only specific ET ranges of the curves for differently adapted animals can be compared. There is usually no intersection between the curves, as the change in Q_{10} values within the curves varies with the ET (PRECHT in PROSSER, 1958, p. 56).

It is also possible to compare the Q_{10} s of organisms held at constant ATs with Q_{10}s of those that had been exposed to varying preliminary treatment. BERKHOLZ (1966) made such measurements of oxygen consumption of muscle tissue from *Idus idus*. He found no consistent changes in Q_{10} values.

If ET-dependence is influenced by a stimulating temperature effect — which could have been the case in experiments of VON DALWIGH (1973)[16] — AT can have a definite influence on the Q_{10} values (discussed on p. 450, 454).

2. Effects of Temperature Changes

a) Overshoots and Other Responses

In the directions on experimental procedures it was suggested that measurements should be made at an ET equal to the AT, and that the ET should then be changed in a single step. As was said then, such a change, especially if it occurs relatively rapidly, can have undesirable effects, which can of course be avoided if one is satisfied with establishing a curve for ET = AT alone; this procedure however has disadvantages of its own (p. 304).

Increases or decreases in temperature have disturbing effects on many animals; a rise often causes an increase in activity which must be allowed to die down before proceed-ing with measurements. We are not concerned here with phenomena of this type, but with others that are not connected with changes in movement of the animal and that also appear in immobile organisms when relatively rapid temperature changes take place.

The initial direct responses can exceed the final end point, i.e., go too high when temperatures increase or go too low when they decrease (overshoots), or the final values may be reached only after a certain delay and not immediately once the animal encounters the new temperature. In some cases increasing (decreasing) temperatures cause at first a decrease (increase) of the measured values (under-shoots, which are also known as false starts; on this point see GRAINGER, 1960; some authors use "undershoot" to indicate an overshoot in the case of decreasing temperature). Sometimes the new value is established only after oscillating fluctua-tions occur (see GRAINGER in PROSSER, 1958; KARGER, 1962). In *Artemia* overshoots of oxygen consumption occurred even after anesthesia.

These phenomena have recently been thoroughly examined by ZERBST (1966; see HÜLSEN and ZERBST, 1964; ZERBST et al., 1966), who drew distinctions among

---

[16] Zool. Anz. **190**, 361—380 (1973).

five types, as for capacity adaptation, which he patterned on an electrical model. According to ZERBST oscillatory processes *(Einschwingprozesse)* are involved, that is to say, reactions that arise spontaneously from the characteristics of living systems, and therefore need no regulatory control. He worked with *Rana esculenta* hearts, isolated and *in situ*. When the temperature rose rapidly ($10 \rightarrow 25°$ C), the heart beat frequency *in situ* went up at first in an overshoot, until after six min a steady-state rate was reached; after a later decrease in temperature by $10°$ C, the frequency went down in another overshoot, but gradually established itself after 90 min at the initial value. With a slow temperature rise, the frequency did not overshoot but gradually established itself at a new value. Hearts from warm-acclimated frogs reached the new steady-state value sooner than hearts from cold-acclimated animals. SEGAL (1962) found overshoots to be the first response of *Acmaea limatula* heart rates to sudden temperature changes. Equal temperature elevations led to overshoots which were more pronounced at the higher range of temperature (see BLAŽKA, 1955).

Overshoots are also found in the speed of movement of insects (cf. KENNEDY, 1939). GÄRDEFORS (1964) observed undershoots in the activity of *Chortippus albomarginatus* after sudden lowering or raising of the temperature. In a central range the final activity was independent of ET. Typical undershoots also appear many times in electrophysiological studies. If the temperature is raised abruptly the frequency of discharge of *Astacus* stretch receptors under moderate to strong stimulation goes down suddenly; then, in the course of several minutes it begins to rise until the initial value goes up steadily, and finally (after ca. one hour) reaches a final value higher than the original one. If the temperature is lowered abruptly, the frequency of discharge rises rapidly at first, then goes down sharply, and finally continues to fall more slowly. Slow temperature changes, which have not been investigated as thoroughly, have a different effect. Intracellular conduction in *Astacus* also exhibits distinctly different responses to slow and fast temperature changes (BURKHARDT, 1959; see KERKUT and TAYLOR, 1958; WINTER, 1967; LARIMER, 1967; ROTH and SZABO, 1970).

Undershoots can even lead to changes in volume in Protozoa (Suctoria) "A sudden increase of temperature from below $15°$ C by $5°$ C or more causes a temporary fall in the rate of output, followed by a rise to a new level higher than the original. During the depression in activity the body swells slightly. The vacuolar frequency increases immediately but briefly when the temperature is raised, falls steeply when the depression sets in and when secretion is reestablished rises again to a level above the original" (KITCHING, 1954, p. 75).

SCHLIEPER and his students found that oxygen consumption (e.g., by *Astacus* and planarians) did not change at once after a temperature change, but only after a delay (1st ed., p. 38; however see VALEN, 1958). In contrast to the response to a large temperature change ($15° \rightarrow 25°$ C), the oxygen consumption of *Planaria gonocephala* and *P. alpina* does not rise at all within three hours after a change from $5° \rightarrow 10°$ C, and only later rises slowly to a final value (in ca. twelve hours). It is not certain that only direct responses are involved in this type of situation.

b) Transitory After-Effects

Many of the phenomena to be discussed in this section are not merely direct responses (since this is true for over- and undershoots too, which last only a short time), but most of them may not involve regulations or acclimations of the type discussed in later chapters; it is not yet possible to classify them precisely.

As noted earlier, different biological processes of an animal may have quite different Q_{10} values. Therefore temperature changes necessarily lead to changes in the interplay between processes that must in turn be treated as disturbances and partly overcome. E.g., the water intake and outflow of fresh water animals with permeable skins must be adjusted to each other. RAFFY (1954) reported transitory changes in weight taking place in fish after a temperature change (KITCHING'S (1954) investigation was cited). This recalls experiments on transfer of fresh water and sea water animals into brackish water, in which regulation often begins only after a substantial initial volume change (see REMANE and SCHLIEPER, 1958, p. 245 ff.). Certainly water influx is not limited only by an osmotic process with a low Q_{10} value. According to WIKGREN (1953) the temperature coefficients of urine formation and water permeability of e.g., *Petromyzon fluviatilis* were nearly equal when the animals were moved from room temperature to low temperatures (see REMANE and SCHLIEPER, 1958, p. 297 ff.; MACKAY and BEATTY, 1968; PARSONS and ALVARADO, 1968). The relation of Cl⁻ excretion to absorption is however disturbed at first; within ten to twelve hours a certain equilibrium is reached and water permeability is changed during this time.

KINNE (discussion in SCHLIEPER, 1966) has pointed out, correctly, that not only can the transition of organisms into sublethal temperatures cause stress, but so can temperature changes within the "normal" range, especially if the latter take place rapidly (see p. 420). However, it is often difficult to distinguish the transitory stress effects produced by temperature changes from lasting capacity adaptation. In that case one must make comparisons with other types of stress. If the animals remain at relatively extreme temperatures, one must reckon with longer-lasting stress effects.

HICKMAN et al. (1964), HEINICKE and HOUSTON (1965), HEITMANN in PRECHT (1968), REAVES et al. (1968), and MEINCKE (1970) and other authors investigated changes in the ion and water content of fish, following changes in AT; different results were obtained for the same species (e.g., *Salmo gairdneri*, used by HICKMAN et al., and REAVES et al.), depending on the degree and speed of alteration of temperature and on the AT range itself. When, after a shift in AT, a change occurs at first in the opposite direction from the change to be expected after full acclimation to the new AT, the operation of a stress effect is suggested. However it is not a sufficient indication since e.g., different Q_{10} values of the participating mechanisms could be responsible. Further indications of stress are found where rapid changes in AT produce more intense effects than slow changes, as e.g., MEINCKE found for several ions after transferring *Tinca tinca* from $16 \rightarrow 6°$ C. In addition the observations of REAVES et al., who found that temperature increases and decreases had the same effect, in many respects, on ion content, suggest the existence of a general stress effect, but this is not valid for every case. For *Tinca tinca* an AT fall from $16 \rightarrow 6°$ C, had an obviously stronger effect than an increase from $6 \rightarrow 16°$ C or even from $13.5 \rightarrow 26°$ C and also a different type of effect. However the $16 \rightarrow 6°$ C experiments did not give exactly opposite results from the $6 \rightarrow 16°$ C ones, which would suggest that stress was not involved (MEINCKE, 1970; for mussels see LÁBOS and LUKACSOVICS, 1968).

In nature, temperature usually follows the daily day-night cycle fairly slowly; perhaps this is related to the observation of TOEWS and HICKMAN (1969), who simulated such a slow alternation ($8 \leftrightarrow 18°$ C), and found no summation of shock effects. The Na^+ and K^+ content of muscle changed in opposite directions during the cycle. The measured electrolyte concentrations in plasma and muscle of the cycling fish more closely resembled the levels in the cold-acclimated than in the warm-acclimated fish (see p. 348).

HUGGEL et al. (1963) showed that even moving fish into their apparatus could act as a "stressor", and cause an increase in Na^+ and Cl^- content of the blood (see WEDEMEYER, 1969). For example, the concentrations of glutamic acid and γ-amino-butyric acid in the goldfish brain reach a maximum $24-28$ h after a warm or cold stress, but return to the initial value within two weeks (BASLOW in PROSSER, 1967).

LEHMANN (1970b) investigated the effects of an abrupt AT-change ($15 \rightarrow 5°$ C) on many enzyme activites in goldfish muscle over an extended period; he obtained many jagged curves that later became smoother. Similar experiments with *Idus idus* (KÜNNEMANN et al., 1970)[17] and *Rhodeus amarus* (BRAUN et al., 1950)[17] in which the AT was raised or lowered at various rates, showed a corresponding, very labile, phase for enzyme activity following temperature change. The fluctuations, appearing particularly after rapid changes, were again found only for certain enzymes, but eventually occurred for others after repeated trials. They could be significant in many cases, but were not always reproducible. After a rapid transfer of *Rhodeus* from $20 \rightarrow 30°$ C, in which both the change and the high temperature must have been strong stresses (p. 419ff.), the fluctuations hardly appeared (see RASMUSSEN and RASMUSSEN, 1967). The succinate respiration of *Idus* muscle tissue showed similar initial fluctuations in re-adaptation experiments *(Umadapta-tionen)* (KÜNNEMANN et al., 1970; see W. B. VERNBERG and F. J. VERNBERG, 1968a). A brief lowering of AT ($15 \rightarrow 0.5 \rightarrow 15°$ C) led to the same finding in LEHMANN's work.

In such experiments one must take into account the body's countermeasures that commonly follow stress. In mammals they include an increased corticotropin secretion leading to increased secretion of the adrenal cortex (especially of gluco-corticoids). In teleosts ACTH can produce increases in blood calcium and decreases in sodium content (see PARRY, 1966, p. 427). Interrenalectomy in *Raja radiata* led only to slight changes of the blood calcium level. Sudden temperature shocks ($1 \rightarrow 12°$ C) had only a slight effect on the ion level of operated and sham-operated animals (IDLER and SZEPLASKI, 1968). After a short cold shock the serum cortisol content of *Oncorhynchus kisutch* was elevated (WEDEMEYER, 1969; on hormones and stress, see MAHON et al., 1962; WEATHERLEY, 1963; HANE et al., 1966; FAGERLUND, 1967; and HILL and FROMM, 1968; for amphibians, see TOUTAIN, 1961).

According to LEHMANN (1970b) the RNA content of goldfish muscle is said to rise tem-porarily after sudden increases or decreases of AT. However, the method employed was not strictly RNA-specific, since disturbances by hexose and tissue proteins could have participated. After an increase of AT, GRONOW (unpubl.) found transitorily a fivefold increase in glucose content in muscle tissue of *Idus idus* relative to control animals

---

[17] The ET in these investigations (measurements of enzyme activities and of the oxygen consumption of tissues) was always $25°$ C.

under the same experimental conditions, while the RNA content decreased about 12%. When animals of this species, anesthetized by MS-222 Sandoz to exclude stress effects from catching and handling, were transferred into 5° C for 30 min and then back into their primary milieu of AT 20° C, glucose content increased after 10 h about one third, while RNA content did not change at all. — According to SMITH and MORRIS (1966), a transitory increase in the rate of incorporation of amino acids into goldfish gut mucosa that follows a change in AT (8 → 25° C) can be attributed to elevated protein synthesis, which can be inhibited.

A few more observations may be presented here, which can hardly be classified as direct responses, and probably do not belong in this chapter. After a fall in AT (20 → 8.6° C), oxygen use in *Porcellio scaber* (winter animals) rose for a few days, and then fell back to its old value (ET was always 20° C). Summer animals did not show this phenomenon, which can be useful in winter (WIESER, 1965 b). The oxygen consumption in *Blaberus craniifer* (motionless), also showing no capacity adaptation, rose very distinctly in the days following an increase or decrease of AT of about 10° C. It did not rise when AT was changed about 5° C (ET always 27° C) (RANDZIO, 1973)[18]. This may be due to shock effects[19]. The oxygen consumption of muscle tissue of *Blaberus* did not show similar shock effects. LANGE (in PRECHT et al., 1966) gradually transferred *Idus idus*, which had been kept for long periods at 8° or 25° C, respectively, to the midpoint temperature of 16.5° C, and left them there. Immediately after the temperature fell, the Q_{10} value for opercular movement was 2.9, but the reduction in frequency continued on the next day. The final, much higher, value was reached only slowly. An extended "overshoot" is present in this case (it might be possible to reserve this word for direct responses, but it is often very difficult to decide whether only direct responses are involved). When the temperature is raised, a corresponding effect is much less obvious among the widely fluctuating values that result. In this and other cases disturbances that appear initially diminish in the course of a few days (e.g., in the case of oxygen consumption of *Lacerta sicula*, according to GELINEO, 1967, see PRECHT, 1962, 1968, p. 507; PARVATHESWARARAO, 1968b).

Cooling of warm-adapted pupae of *Galleria mellonella* down to 4° C causes a decrease in number of hemocytes of the hemolymph until the 3rd day; then an increase takes place beyond the initial value (MAREK, 1970).

B. Regulation by Control Systems

Regulations may, in spite of the existence of some transitional forms, be distinguished from non-genetic adaptations (acclimations) by the fact that the latter produce true functional and structural modifications in the organism, require time for their development, and often are not restricted to specific organs, but concern more or less all levels of organization. Regulations, on the other hand, generally do not result in any true modification of the status quo ante. They usually pass off

---

[18] RANDZIO,G.: Zool. Anz. **189**, 1—26 (1972).
[19] Ann. by H. LAUDIEN: In cases where two (arbitrarily chosen) ATs give the same values (type 4; e.g. in Fig. 3 the ATs 10° C and 27.5° C), but not the ATs between these two (as in Fig. 3), other explanations are possible. However this is not true for *Blaberus*, since this animal does not show a capacity adaptation of the oxygen consumption over a wide temperature range.

rather quickly and are the result of routine activity of specific, pre-existing reaction systems (KINNE, 1965, p. 147; see WIESER, 1964; PRECHT and LINDNER, 1966).

The problem of the influence of AT is not significant for regulations, unless they occur together with capacity adaptations. However, it is possible to confuse regulations with cases of exceptionally rapid adaptations of control systems.

It should also be remembered that characteristics that are taken as innate reaction norms at first may be shown by more precise analysis to be regulations due to control systems. This is true for example, for regions of the ET-dependence curve with low Q_{10} values. Resting reptiles show a minimum respiratory rate at the temperature preferred by each species; it is abolished by urethane narcosis and by removal of the forebrain (HERTER, 1941, 1st ed., p. 24). KEISTER and BUCK (1961) found a plateau in the respiration curve of *Phormia regina* larvae (but not of pupae or imagos) in the ET range of $10-15°$ C that disappeared after decapitation. A plateau, or even a minimum, occurs in the temperature-respiration curve of some Carabidae because respiration is relatively intensive (presumably due to intensified ventilation) at ETs below this "zone of regulation"; beyond this zone, rise in oxygen consumption is particularly steep (SCHMIDT, 1956). BENTHE[20] (1954) measured the irritability of isolated feet of *Lymnaea stagnalis* (k-values according to BLAIR). He found a capacity adaptation of type 3, but the Q_{10} values of the ET-dependence curve for preparations from differently acclimated snails differed, depending on whether or not the pedal ganglion had been removed (PRECHT in PROSSER, 1958, p. 57). Of course, a change in activity of the animal can also produce a plateau in the metabolism curve (see GROMYSZ-KALKOWSKA and STOJA-LOWSKA, 1966).

It is difficult to classify the well-known, broad temperature-independence of biological rhythms. The endogenous components of biological rhythms in poikilothermic organisms must have this property in order to function meaningfully. Apparently there are two different "clocks", one which is synchronized with an external periodicity and a second, aperiodic one, which provides only an arbitrary (but species-specific) duration. The first type includes rhythms connected with daily, yearly and lunar cycles; the second probably includes clocks that determine the length of diapause[21] (see PRECHT, 1964a) or the duration of development of e.g., cicadas. This span (ca. 17 years) however is not completely independent of temperature (see HEATH, 1968).

The most thoroughly investigated, so-called endogenous circadian rhythm is widely independent of temperature, as is known; it is regulated by an external timer *(Zeitgeber)* (see REMMERT, 1965, i.a.). The activity of poikilothermic animals has often been investigated in this connection. The duration of activity depends on temperature, but the periodicity does not (see LOHMANN, 1967, i.a.). Rapid temperature changes can have transitory effects on endogenous rhythms of poikilotherms in the form of accelera-

---

[20] This work has been used to illustrate many conditions; not all results are significant and should be regarded only as models; further investigation is desirable.

[21] But in these investigations of diapause in the bug *Ischnodemus sabuleti* it is not yet known whether the resumption of development at the beginning of the year is related to the date (i.e. to a definite point in the endogenous annual rhythm), which cannot be displaced by temperature or day length (at least not during the time of the experiment), or whether the length of diapause is firmly fixed as an aperiodic process (see FISCHER, 1970). Further investigations are planned.

tion or retardation (literature may be found in PRECHT et al., 1966; ZIMMERMANN et al., 1968; PAVLIDIS et al., 1968, see also p. 450f.).

For example if a rapid capacity adaptation were to take place, it would have to be of type 2 (to be discussed later) in order to reset the clock correctly. The *Zeitgeber* could compensate for small deviations of types 1 and 3. Such an adaptation would involve only the clock, not the other life processes. It must be noted again that different organ functions and cell metabolic processes may behave very differently in terms of capacity adaptation. The clock's adjustment often fails at very low temperatures, which suggests that the regulation or adaptation mechanisms may "freeze up". However, the phase (light- or dark-phase) in which the temperature change or related stopping of the clock takes place may be important (see SAUNDERS, 1968, 1969; STEPHENS, 1957, i.a.).

It should be mentioned that temperature can act as a *Zeitgeber* (p. 450).

C. Acclimations, Acclimatizations, Adaptations

KINNE (in DILL et al., 1964) distinguishes three aspects of adaptation: (a) immediate responses, (b) the stabilization phase and (c) the new stable condition (see PROSSER in DILL et al., 1964, p. 19). PROSSER (1958, p. 169) separates acclimation from acclimatization; the former term involves adaptation to a single factor as in controlled experiments, the latter to a complex of environmental factors, as in seasonal or climatic changes. Complexes of factors usually are present in genetic adaptations in phylogeny, and non-genetic adaptations in nature (acclimatizations under uncontrolled conditions). Within this section only non-genetic acclimations will be dealt with. FRY (in ROSE, 1967) wanted to restrict the use of the term "adaptation" to genetically caused changes arising through phylogeny. Selection can play a role as well in experiments of long duration using rapidly multiplying organisms. In genetic adaptation, functional feedback systems are not involved (see ROZHAJA, 1963; LEAKE in DILL et al., 1964).

Sensory physiologists use the word "adaptation", which as used here includes both "acclimation" and "acclimatization", to signify mechanisms that are more like direct responses or regulations, according to KINNE's (p. 323f.) definition, and for which the expressions acclimation or acclimatization are inappropriate, inasmuch as these refer to climatic factors. To the extent that misunderstandings might arise, the two words acclimation or acclimatization should be given preference for the phenomena discussed here. Among non-genetic and genetic adaptations, capacity adaptation in the "normal" or "central" temperature range is differentiated from resistance adaptation (heat and cold adaptation, p. 428ff.) of cells, organs and intact animals.

The words adaptation, acclimation or acclimatization really imply a usefulness for the species, in the sense of a selective advantage, which also applies to those genetic adaptations that deserve the name (which is often difficult to determine). The identification of such usefulness in non-genetic acclimations would however require much additional investigation. Hardly one author has carried out experiments of this type.

In general such usefulness might well be present in non-genetic adaptations of intact organisms, but "accompanying circumstances" can complicate the picture (e.g. the increased heat resistance of frost-hardened plants). It is therefore advisable to exclude the problem of usefulness completely from consideration of non-genetic adaptations and to regard them as occurring when a change in AT has any consequences that cannot be considered as direct responses, but also not as regulations (PRECHT and CHRISTOPHERSEN, 1965; PRECHT and LINDNER, 1966).

1. Changes in Body Composition

In measuring biological processes (e.g., enzyme activity) one can start with differently acclimated organisms, draw curves showing the relations of the processes to ET, and thereby determine whether capacity adaptation has occurred and to which type it belongs. Not only enzyme activities but also enzyme concentrations may change with AT (CARLSEN, 1953; STANGENBERG, 1955; FREED, 1965; CALDWELL, 1969; VERNBERG and VERNBERG, 1968b) as well as the isoenzyme patterns (p. 344). Even the number of mitochondria may depend on AT (BOYCOTT et al., 1961; JANKOWSKY and KORN, 1965; THIESSEN and MUTCHMOR, 1967) and their size as well (FREWIN, 1969; compare SEMENOVA, 1967). CALDWELL (1969) suggests that in goldfish, mitochondrial electron transport systems compensate to the degree necessary to meet changing ATP demands. Particles from thorax tissue of *Schistocerca gregaria* (AT $= 35°$ C) were unable to oxidize higher fatty acids in contrast to particles from 45°-adapted animals (MEYER et al., 1960).

In describing amounts of cell substances found at different ATs one can only speak of relative quantities, i.e., whether the amounts go up or down with increasing AT. Measurements after changes of the AT are usually not made.

A change in the amount of a substance correlated with AT may be significant not only for normal temperature conditions, but also for resistance adaptation to extreme temperatures. Examples include glycerol and sorbitol levels in insects or changes in osmotic pressure of body fluids (p. 416). This is particularly likely to be true if the dependence on AT is limited to the higher part of the range, or in the case of cold resistance, to the lower.

A further complication can arise from the fact that a temperature change (perhaps working through differences in Q_{10} values of participating processes) may produce a new temperature equilibrium fairly rapidly that remains almost unchanged later, and therefore can scarcely be recognized as an adaptation. An example may be found in the water uptake of the frog *(R. temporaria)* after a fall in AT. Water is taken up into the body fluids and to a lesser extent into the tissues (GRAINGER, 1961, 1964; compare PARSONS and ALVARADO, 1968). Moderate pressure on the abdomen causes the urinary bladder of cold-acclimated, but not of warm-acclimated, frogs to empty (KASBOHM, 1967). According to JANKOWSKY (1960) these changes in water content have no significance for the capacity adaptation of oxygen consumption. On the other hand, identical values obtained from animals held at different ATs may be the result of an adaptation, e.g. in those cases, where (for example by different Q_{10} values of participating processes) after a change of AT the measured values are at first unequal and then identical after some days (GRIGO, unpubl.).

Genetic and non-genetic adaptations must be distinguished in the case of changes in composition as well. To separate the two possibilities one would have to maintain animals of related species or of the same species but from habitats differing in temperature (if possible) for longer periods under the same conditions (compare PRECHT, 1968, p. 510).

In general, only non-genetic adaptations have been investigated in experiments where the same experimental material was maintained for long periods at different ATs. It is not certain that these periods always were long enough because, as was stated, the temperature changes accompanying the transfer of the animal to the AT can have relatively long-term consequences. For this reason as well these problems have been looked at relatively briefly and incompletely. Seasonal changes, which often do not concern AT, or do not concern AT alone, will not be dealt with.

a) Water Content[22]

One possible way of modifying the intensity of metabolism in a cell consists of a change in the free water content, and thereby of the size of the reaction volume (*Lösungsraum*). The compensation of succinic dehydrogenase activity and its change with culture age in the yeast *Torulopsis kefyr* can be explained by changes in water content (1st ed., p. 32 ff.). But this principle by itself fails to explain all enzyme compensation even in this single-celled form, since peroxidase activity, for example, shows an inverse compensation. Changes in the partly non-solvent water bound to colloids must have an effect opposite to that of free water (for protists, see FREWIN, 1969).

Changes in water content following changes in AT have also been found many times in metazoan tissues. Related species or even races can react differently.

KANUNGO and PROSSER (1959), for instance, found that the water content of goldfish muscle and liver increased with increasing AT. MURPHY (1961, cited in DAS, 1967) and DAS were unable to confirm this for the liver, gills, and muscle of the same species (a slight decrease in the water content of liver and muscle in the cold-acclimated animals could not be completely verified; see HOAR and COTTLE, 1952; HEINICKE and HOUSTON, 1965; HOUSTON et al., 1968; TOEWS and HICKMAN, 1969; UMMINGER, 1969b; MEINCKE, 1970; on *Mya arenaria*, cf. DUPAUL and WEBB, 1970).

SUHRMANN (1955), who was probably working with *C. gibelio*, the ancestral form of the goldfish, could find no correlation between AT and the amount of free and bound water in muscle tissue, while the liver tissue of the same species showed a decrease in total water content with falling AT (CHRISTOPHERSEN and PRECHT, 1952b). The oxygen consumption of the muscle tissue and the activity of succinic dehydrogenase showed a type 5 capacity adaptation (inverse compensation) (SUHRMANN). The activity of the same enzyme in the intestinal gland of *Helix pomatia* showed compensation of type 3. Here too the content of free and bound water did not depend on AT (KIRBERGER, 1953), which shows that water content alone cannot be the critical factor (compare STANGENBERG, 1955; FRY, 1958).

b) Salts, Ions

In aquatic animals, as mentioned, temperature changes can lead to temporary or permanent changes in water and salt content in the course of capacity adaptation. The internal concentrations can thereby increase or decrease with temperature (cf. REMANE and SCHLIEPER, 1958; KINNE, PARREY i.a. in DILL et al., 1964; on the osmosis of urine in *Hemigrapsus*, cf. DEHNEL and STONE, 1964). LOCKWOOD (1960) maintained *Asellus aquaticus* for fairly long periods at different temperatures, but also investigated the effect of changes in AT. He found a minimum in hemolymph

---

[22] Cf. DROST-HANSEN, W., see p. 306.

concentration at ca. 5° C. The speed of Na^+ loss did not depend on temperature, but the rate of active Na^+ uptake did (compare p. 321). The active uptake of chloride can also be changed by the AT in some animals (RAO in PROSSER, 1967; cf. CORDIER and WORBE, 1954). Most changes in permeability accompanying changed AT affect the electrolyte content of the body. According to SMITH (1966a, b) qualitatively different carrier molecules for sodium transport are said to be formed in goldfish gut after such a change (cf. SMITH, 1967a). The difference in potential between the serosa and mucosa is related to AT. Puromycin injections prevented these changes from appearing when AT changed (SMITH, 1966c).

The blood plasma composition of *Rana esculenta* changes with AT (alkali reserves, Na^+ and Cl^- content, and pH decrease with increasing AT; STRAUB, 1957; cf. GAHLENBECK and BARTELS, 1968; MILLER et al., 1968; REEVES, 1969). According to RAO (1966, in PROSSER, 1967) and his co-workers, acclimations to low temperatures are generally accompanied by increases in calcium, potassium, and sodium in the body fluids, while the magnesium, sulfate, and chloride contents decrease ("these changes may result in increased muscle metabolism"), but they found exceptions to this rule (e.g., for the Na^+ content of blood in *Etroplus*). In *Lampitio mauritii* (ATs: 20°, 28° and 35° C) the Ca^{++} level is at a minimum at an AT of 28° C, but the Na^+ and Cl^- levels are at a maximum (SAROJA and RAO, 1965; PARVATHESWARARAO, 1967). As mentioned, the relation of salt content to AT can often play a role in resistance adaptation (especially to cold) (for fish, compare p. 321, and PLATNER, 1950; GORDON, 1959; SELVARAJAN, 1962; HOUSTON and MADDEN, 1968; RAO, 1969). HOUSTON et al. (1968, 1970) found seasonal variations in the relation between AT and blood plasma ions in *Salmo gairdneri* and carps (cf. TOEWS and HICKMAN, 1969).

It should be noted that ionic composition can influence enzyme activity, which could affect a change of the activity with AT; however, ion excesses are used in many of the methods for measuring them.

Special problems may also exist for terrestrial animals, e.g., after a blood meal hormonally induced diuresis takes place in *Rhodnius prolixus*, in which the speed of excretion and composition of the urine depend on the ET: "a fed insect kept at a high temperature will be left at the end of diuresis with a relative excess of water and an insect kept at a low temperature with a relative excess of salts" (MADDWELL, 1964, p. 172). Concerning the temperature dependence of the acid-base balance, see REEVES (1969) and HOUSTON[23].

c) Proteins, Fats, RNA, DNA, etc.

The rate of transport of amino acids (through the goldfish gut; SMITH, 1967a, 1970) can be altered by AT in the sense of a compensation. The fact, reported by RAO (1966) and others, that the level of free amino acids in the body fluids often decreases with decreasing AT, but that the amino acids bound to proteins, the total protein content in the cells and tissues, and the RNA content increase, would be consistent with a compensation (type 1, 2, 3) (cf. DEAN and BERLIN, 1969). Of course, changes in enzyme activity with AT need not be brought about exclusively by changes in apoenzyme concentration, but could also be caused by changes in

---

[23] HOUSTON, A. H.: Comp. Biochem. Physiol. **40 A**, 535—542 (1971); cf. SMITH, M. W., KEMP, P.: Comp. Biochem. Physiol. **39 B**, 357—365 (1971).

amounts of coenzymes (RING and CHRISTOPHERSEN, 1964). Furthermore the cellular environment, which depends on AT, may be significant for enzyme activity.
Reduction of the rearing temperature ($30° \to 24° \to 15°$ C) of *Drosophila melanogaster* leads to an increase in free amino acid concentration, but individual amino
acids such as glutamic acid can behave differently (ANDERS et al., 1964). Among
those amino acids in muscle tissue of *Eriocheir sinensis* not bound to proteins, the
concentration of proline in particular falls as AT decreases (DUCHÂTEAU and
FLORKIN, 1955). According to BASLOW (in PROSSER, 1967) some amino acids in
fish brain may increase with AT; others may decrease, and still others remain
unchanged[24]. In skeletal muscle of *Bufo boreas* the total protein content is hardly
influenced at all by AT (BISHOP and GORDON in PROSSER, 1967), which is not true
in other cases (cf. SHIELDS et al., 1960; MEISNER and HICKMAN, 1962; OHSAWA and
TSUKUDA, 1964; SOLOMON and ALLANSON, 1968; on the free amino acid pool of
Mya, cf. DuPAUL and WEBB, 1970). A ten-day stay in the cold leads to an enlargement of liver cell nuclei in *Rana pipiens*; this phenomenon is caused by an increase
in protein, while the RNA and DNA contents remain unchanged (ALVAREZ and
COWDEN, 1966; cf. HASCHEMEYER, 1968). The DNA content in goldfish is also
unchanged by AT. The protein increase of gill, liver and muscle associated with
falling AT does not lead to a corresponding decrease in free amino acids in these
three tissues. The total RNA content of the gills and the RNA concentrations in
the "nuclear" and "microsomal" fractions go up with adaptation to low temperatures; the RNA content of a "supernatant" fraction goes down with falling AT
in gill and liver, but increases in muscle (DAS, 1967, cf. FREWIN, 1969)[25].
HEITLINDEMANN (unpubl.) found recently a change with AT in concentration of
base composition of nuclear RNA of *Tenebrio molitor* larvae, an AT range where
the oxygen consumption did not show a capacity adaptation, and no heat adaptation but merely a reasonable cold adaptation of the larvae; however, this may be
a problem of development, although the larvae were measured at the same stage.
In species of *Drosophila*, the RNA content may go up or down with the rearing
temperature or be unaffected by it (BURR and HUNTER, 1969).
A few other substances (glycogen, etc.) whose concentration can be changed by
AT will not be discussed here (cf. e.g. DEAN and VERNBERG, 1965; McWHINNIE in
ROSE, 1967, p. 368; McWHINNIE and CORTELYOU, 1967; POLYANSKY and SUKHA-
NOVA in TROSHIN, 1967, p. 205; JUNGREIS, 1968). Even the titer of H-agglutinin
can be dependent on AT in reptiles (EVANS and COWLES, 1959).
The amount of hemoglobin of *Daphnia* goes up with increasing AT (FOX, 1955).
STRAUB (1957) found an increase in the erythrocyte count in the blood of *Rana
esculenta*, a decrease in their size and hemoglobin content, and an increase in total
blood cell volume with decreasing AT. At low ATs *Salmo gairdneri* showed relatively low erythrocyte counts, hematocrits and hemoglobin levels (DEWILDE and
HOUSTON, 1967; cf. ANTHONY, 1961; HUGGINS and PERCOCO, 1965; FRANKEL et al.,
1966; DEAN and VERNBERG, 1966; HOUSTON and DEWILDE, 1968, 1969; CAMERON,
1970)[26]. In the carp, the lymphocyte count goes down after AT goes down, as it
generally does after a stress (SERFATY and LAFFONT, 1955).

---

[24] Cf. HABIBULLA, M.: Comp. Biochem. Physiol. **39**, 499—502 (1971).
[25] But cf. FREED, J.M.: Comp. Biochem. Physiol. **39**, 747—764 (1971).
[26] Cf. KRISTOFERSSON, R., BROBERG, S.: Ann. zool. fenn. **8**, 427—433 (1971).

According to RAO (1966), the ratio ATP/ADP in *Lampitio* increases somewhat with AT. According to CHRISTOPHERSEN (p. 37), these changes are very significant for capacity adaptations (cf. HOCHACHKA in PROSSER, 1967; BEHRISCH and HOCH-ACHKA, 1969; HOCHACHKA in HOAR and RANDALL, I, 1969, p. 363; for the relation of inorganic and total phosphate goldfish tissues to AT, see ANDERSON, 1970).

It should be mentioned that changes in the substrate level of muscle which occur as a result of stress can also be observed after long adaptation to high ATs and less extensive after adaptation to low ATs (e.g. in muscle of *Idus idus*, GRONOW, unpubl.).

The differences (in %) to AT-20° C fishes are shown (L/P = lactate/pyruvate ratio, G-6-P = glucose-6-phosphate, Glu = glucose):

AT (° C)	L/P	G-6-P	Glu
30	+382	+108	+317
10	+ 30	+ 53	+ 28
6	+ 59	+192	+118

In many poikilotherms the total content of neutral fats and lipids goes up with decreasing rearing temperatures (cf. JOHNSTON and ROOTS, 1964 i.a.); however in nematodes it goes down (LOWER et al., 1970). Goldfish liver grows smaller with increasing AT, but its fat content goes up (PROSSER in HANNON and VIERECK, 1962, p. 40). On the other hand, the fat content of the beetle *Dendroctonus pseudo-tsuga* is lower when it is reared at 13° C than at 21° C (ATKINS, 1967). The Golgi apparatus of *Salmo gairdneri* liver cells and the material stored in them (probably lipoprotein) grow larger with increasing AT, according to BERLIN and DEAN (1967).

The fact that cold-acclimated organisms in general contain relatively more unsaturated fatty acids than warm-acclimated ones is probably related more to resistance than to capacity adaptation. It might, however, be significant for enzyme activities (that depend on bound lipids), for the permeability of cell membranes, etc. (cf. JOHNSTON and ROOTS, 1964).

The composition of the lipids and the number of double bonds can change with AT. The mitochondria are also affected by the changes (in goldfish and yellow bullhead, CALDWELL and VERNBERG, 1970; cf. HOCHACHKA in HOAR and RANDALL, I, 1969, p. 378). ROOTS (1968) investigated the phospholipids of the brains of differently acclimated goldfish. Greater relative unsaturation of C-16 fatty acids at the lower temperatures was mediated by an increase in the amount of palmitoleic acid from 5–10% at 30° C to 12–14% at 5° C. In contrast the greater degree of unsaturation of the C-18 fatty acids at 5° C is due to a decrease in stearic acid from 17–22% at 30° C to an average of 12% at 5° C. The oleic acid remained relatively constant at 24–27% (compare ANDERSON, 1970). Over a range of AT (5–30° C) the total amount of phospholipid and the amount of major phospholipids remain relatively constant. But the brain plasmologen content increases with AT (ROOTS and JOHNSTON, 1968). "Plasmologen content and species may well influence the rates of enzymatic reactions and provide a link between meta-

bolism and membrane structural changes occurring during temperature acclimation" (p. 558; cf. IRVINE et al., 1957; FARKAS and HERODEK, 1959; PROSSER,
1967, p. 367; CHAPMAN in ROSE, 1967; BĚLEHRÁDEK in TROSHIN, 1967, p. 437).
In the intestinal mucosa of the goldfish the major changes in composition with AT
were found (in contrast to the brain) in the long-chain unsaturated acids (KEMP
and SMITH, 1970).

KNIPPRATH and MEAD (1966) observed the changes over time of the fat saturation
level in *Gambusa affinis* and *Lebistes reticulatus* after a fall in AT. Different fatty
acids were affected in the two species (KNIPPRATH and MEAD, 1968).

Differences in differently adapted *Rana esculenta* were found mostly in the proportions of polyenoic and not in those of the monoenoic fatty acids (BARÁNSKA
and WLODAWER, 1969). The fats of hibernating mosquitoes were not affected by
the photoperiod or the temperature at which they were synthesized (v. HANDEL,
1967, but cf. HARWOOD and TAKATA, 1965; MADARIAGA et al., 1970).

2. Capacity Adaptations [27]

a) Genetic Adaptations

If related species or races, under the same conditions and with the same previous
history, exhibit differences in body composition or in the relation between the
speed of the response of life processes to changes in ET, these differences could
represent genetic adaptations to their habitat. It is sometimes difficult to say
whether such cases involve usefulness, in the sense of a selective advantage.

α) Organisms from Habitats of Different Temperature Regimes

It has long been known that both active and sluggish animals are found in very
cold regions as well as in the tropics. Also, the degree of movement of animals need
not vary with seasons so much as might at first be assumed, in spite of the great
differences in temperature. The oxygen consumption of different species from very
different regions may be approximately similar at the temperature of their own
habitat (for example, this is true of Mytilids, according to THORSON, 1952), or at
least not so different as one would expect if normal Q_{10} values were present (e.g.,
in the case of fish and crabs, cf. SCHOLANDER et al., 1953). When the same degree
of activity is observed even though there is incomplete adaptation of metabolism,
it remains to be learned whether arctic forms must use a greater percentage of
their energy for locomotion than more southerly forms do.

WOHLSCHLAG (1960, 1963, in LEE, 1964/65; compare POTTS and MORRIS, 1968)
examined antarctic fish that live permanently under a thick blanket of ice, except
for a few weeks in some years. The oxygen consumption of *Trematomus bernacchii*,
a stenothermal fish acclimatized to extreme cold, goes down above 0° C (ET = AT).
At 2° C the species used in the experiment suspended their swimming movements
(the greatest degree of activity took place at the lowest used temperature of
− 1.8° C). At least two of the four species of *Trematomus* have a relatively high
metabolic rate, while the rate of *Rhigophila dearborni* is lower. This species belongs

---

[27] This translation of *"Leistungsadaptation"* (by Prof. PROSSER) means capacity for
activity and performance.

to a group that is also found outside the antarctic. In summer *T. bernacchii* from deeper water have a lower metabolic rate than those from surface waters; similarly *T. lönnbergi* and *R. dearborni* from a depth of 680 m have a lower metabolic rate than do other *Trematomus* species from shallower depths. Most of the fish examined by WOHLSCHLAG were sluggish forms. There are other instructive cases of animals that remain active and metabolize relatively rapidly at low temperatures (e.g. 0° C), for example *Isotoma saltans* (compare 1st ed., p. 53; SCHALLER and ZINKLER, 1963). Some insects (Cynipidae) can be active at $-20°$ C (compare CLARKE in ROSE, 1967, p. 295). Many species are suited to extreme habitats. SCHOLANDER et al. (1953) could not find any genetic adaptations in arctic terrestrial insects which were able to avoid unfavorable temperatures and seek out better microclimatic conditions. Adaptations may also be absent in marine forms (THORSON, 1952; BULLOCK, 1955).

In examining adaptations, and especially in making inter- or intraspecific comparisons, one must always test the extent to which genetic adaptations under discussion here are present ab initio, and the extent to which they are elicited by external factors (especially, in this case, the AT) (cf. LEWIS, 1960). Species and races of genus *Uca* from habitats of different temperatures have been investigated very intensively. VERNBERG and COSTLOW (1966) for example found that young *U. pugilator* from Massachusetts had the highest metabolic rates at all ETs, those from North Carolina had lower rates and those from Florida had the lowest (AT = 25° C). W. B. VERNBERG and F. J. VERNBERG (1966) examined the oxygen consumption of various tissues of a tropical form *(U. rapax)* and of two species from the temperate zone (*U. pugilator* and *U. pugnax* from Beaufort), using two ATs and several ETs. *U. pugnax* had the greatest rate of consumption: this species does not range as far to the south as *U. pugilator*.

According to RAO and BULLOCK (1954) animals from cold regions exhibit not only relatively high metabolic rates, but also lower Q_{10} and/or μ values of their ET-dependence curves than their relatives from warmer zones, though there are exceptions (compare p. 306)[28]. As mentioned earlier (p. 308), low Q_{10} values can have various causes. For example the relation of the measured values and Q_{10}s of *Pachygrapsus crassipes* to the geographical latitude of its habitat is due entirely to a non-genetic capacity adaptation (ROBERTS, 1957). In experiments relating to this point, the animals must be held at the AT long enough. E.g., SEGAL et al. (1953) found higher heart rates in *Acmaea limatula* from populations taken in deeper waters than in animals from shallower tidal waters. The difference persisted when the animals were kept in the laboratory for several weeks under identical conditions. Nevertheless in long-term field experiments the phenomenon was seen to be reversible. However factors other than temperature might have played a role (compare SEGAL, 1961). According to TASHIAN (1956) and DÉMEUSY (1957) the Q_{10}s of oxygen consumption of geographical races of *Uca pugnax* (with the exception of animals from Florida) or *U. pugilator* decrease at higher latitudes (cf. VERNBERG, 1959, 1962). VERNBERG and COSTLOW (1966), working with young crabs, found the highest Q_{10} values (between 15° and 20° C) in the more northerly form of *U. pugilator*. The Q_{10} values of both of the warm water populations (*U. rapax*

---

[28] Cf. BEHRISCH, H. W. (1972); p. 310.

and *U. pugilator* from Florida) were highest between 20° and 25° C. At high ETs, values below 2 are typical for *U. pugilator*, but not for *U. rapax* (AT = 25° C, cf. VERNBERG and VERNBERG, 1964; MORRIS, 1961). Most of the direct responses with low temperature coefficients presented on p. 308, are probably also due to genetic adaptation.

The great difficulty of comparing the data of many authors and of postulating rules from them is shown, e.g. by Fig. 7 in WOHLSCHLAG in LEE 1964/65, in which the relationship of oxygen consumption to temperatures in tropical, temperate zone, arctic, and antarctic fish are presented in one setting. In those cases where stenothermal species could not tolerate maintenance at the same ATs as the others, an extrapolation (e.g., with the aid of Krogh's Normal Curve) was attempted in order to complete the comparison.

These investigations can be extended to enzyme activities; however, many quite unrelated species have been compared (cf. PRECHT, 1968, p. 516). LICHT (in PROSSER, 1967) found that the ET-dependence curves of ATPase activity for four species of lizards were displaced towards one another, as might have been expected, considering their original habitats (compare READ in PROSSER, 1967).

In the nematode genus *Heterodera*, the cyst wall is tanned by the activity of a polyphenol oxidase. At low temperatures *H. rostochienses*, a species which probably originated in the high Andean regions of South America, had the highest activity (ELLENBY and SMITH, 1967).

According to MUTCHMOR (in PROSSER, 1967) the Q_{10} values of ATPase activity in relatively cold-resistant insect species are higher than those from cold-sensitive species (taking the AT partly into consideration; however, cf. READ, 1964). The activation energy of the pyruvate kinase reaction of *Trematomus bernacchii* is about one third of that of the trout reaction (SOMERO and HOCHACHKA, 1968; cf. SOMERO et al., 1968; SOMERO, 1969b).

On the other hand, HOCHACHKA and SOMERO (1968) indicate that some enzymes from quite different species from different regions have similar μ values. According to them, enzymes are selected in the course of evolution whose K_m values (in the temperature range of the habitat of the species) are low. Such changes would be more meaningful in terms of selective advantage than of changes in Q_{10} or μ values or of quantities of enzymes that are no longer suitable (on non-genetic adaptations, cf. p. 334). The K_m values of phosphoenolpyruvate for the trout pyruvate kinase isozymes are e.g. minimal at about 15° C, while the K_m for *Trematomus* enzyme is minimal between 0° and −5° C (SOMERO and HOCHACHKA, 1968). However, evolutionary adaptation of enzymic rates does not appear to be effected by low absolute values of K_m (SOMERO, 1969b). According to BEHRISCH (1969), selection acts on the following: 1) enzyme affinity for unique substrates, 2) enzyme affinity for cofactors and/or modulators (p. 695; cf. BEHRISCH and HOCHACHKA, 1969). Most positive and negative modulators (activators) decrease or increase K_m. Modulator-induced changes in V_{max} (maximal reaction velocity) are far less important in controlling enzymatic activities (cited from SOMERO, 1969b, p. 518). For some enzymes, however, these rules are not valid (cf. KOMATSU and FEENEY, 1970).

β) Steno- and Eurythermal Forms

The expressions steno- and eurythermal have not been precisely defined. A species may be called eurythermal if it occurs in regions with greatly differing temperatures; but it still must be shown that it does not consist of genetically distinct stenothermal races. Furthermore a species may be given the name if it tolerates large seasonal temperature fluctuations, a phenomenon that may be influenced by temperature-independent factors or by special resting phases. Finally, the ability to endure temperature changes during one season must be considered. Genetic adaptations will be found oftener in stenothermal species or races, non-genetic ones in eurytherms.

From experiments (ET = AT) on planaria and fish, SCHLIEPER concluded that in general stenothermal species had higher Q_{10} values than eurythermal species. According to his co-workers' findings (1st ed., p. 37), this can be due to genetic factors, but it must also be considered that lower Q_{10} values may result from the eurythermal species' greater likelihood of showing non-genetic capacity adaptation, or of showing it to a greater extent than stenotherms do. The oxygen consumption of the eurythermal form *Palaeomonetes vulgaris* has low Q_{10} values (usually <2); non-genetic capacity adaptation is absent (MCFARLAND and PICKENS, 1965). Measurements of respiratory enzyme activities of *Drosophila* spec. do not add support to the theory that eurythermal species have a greater capacity for non-genetic temperature adaptation than stenothermal species, but for glutamate-aspartate transaminase the theory seems to be valid (HUNTER and CEDIEL, 1970; BURR and HUNTER, 1970).

b) Non-Genetic Adaptations

A non-genetic adaptation may be said to exist when the ET-dependence curve of an experimental organism is influenced by the AT. This type of adaptation has been investigated most intensively and is the type referred to when, for simplicity's sake, the term capacity adaptation is used without the qualification "non-genetic".

α) Experimental Design and Classification of Types

Fig. 2 presents the type classification of the author (1949; in PROSSER, 1958, 1964b, 1st ed., p. 26f.). PROSSER'S classification has already been presented on p. 319. Cold-adapted animals (AT = t_1) yield the ET-dependence curve A_1A_2. If, after re-adaptation to the higher AT, t_2, the value A_2 is still obtained when ET = t_2 (or if the curve A_2A_1 is maintained when several ETs are used) then there is no capacity adaptation (type 4). If a lower value is obtained, compensation[29] has occurred [type 3 = partial compensation; type 2 = ideal compensation (function value $A_1 = B_2$); type 1 = supraoptimal compensation]. The interpolated curve B_2B_1, whose corresponding Q_{10} values are equal to those of curve A_1A_2, indicates a type 3 compensation. In other cases the temperature coefficients are not the same (p. 499). Finally, there are cases in which B_2 lies above A_2 (type 5, inverse compensation). (For possible sources of error in the design of experiments cf. PRECHT, 1968, p. 518f., in PROSSER, 1958, p. 61 ff.; on the effects of starvation on frogs at different ATs, see JUNGREIS and HOOPER, 1970).

This classification is based in each case on the (two) ATs, that is, from the temperatures at which the groups of organisms to be compared actually lived and which therefore have special significance for them. Difficulties for the type classification

---

[29] When one speaks of a compensation only, the types 1—3 (without further differentiation) are concerned.

arise when the curves intersect between ETs t_1 and t_2, which seldom happens (examples are found in PRECHT, 1968, p. 499)[30].

This is also true for those cases in which the ET-dependence curves begin to go down slowly at relatively low temperatures, so that the declining parts of the curves still lie in the "normal" temperature range. Animal movements are often described by such curves (p. 449). Compensation of types 1 to 3 would be measured in the ET range to the left of the intersection of two such curves, and type 5 at each ET to the right of the intersection.

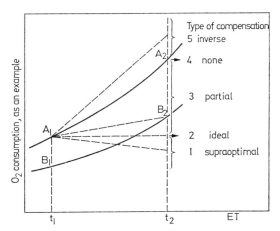

Fig. 2. Types of capacity adaptation (from PRECHT, 1949). AT for curve $A_1 A_2$: t_1, for curve $B_1 B_2$: t_2

It is a common practice (especially in investigating enzyme activity) to acclimate organisms at a high and at a low AT and then, for the sake of simplicity to measure the function rate at *one* ET, (often between the ATs). If the rates are higher in cold-acclimated than in warm-acclimated organisms then compensation of type 1, 2, or 3 has taken place. One cannot distinguish between these three possibilities. If the warm-acclimated rates are higher, type 5 compensation is present (it is implied that the classification is meaningful in every case). An example which again demonstrates the potential importance of ET is given by OHNESORGE and SCHMITZ (1968). They found that as AT increased, the degree of contraction of transverse striated gut muscle of *Tinca tinca* (after the addition of acetylcholine) decreased at an ET of 10° C but increased at an ET of 30° C.

LEHMANN (1970a) measured a large number of enzymes in *Idus idus* muscle after acclimation to many ATs between 0.5° C and 37° C (ET was always 25° C). Disregarding the high temperature extremes, several TCA cycle enzyme activities go down, up to an AT of ca. 21° C and then rise again to an AT of 29° C (and then

---

[30] A comparison of oxygen consumption of different sizes of *Spirostreptus athenes* showed an intersection of the ET curves of differently acclimated animals only in millipedes of high weight (DWARAKANATH, 1971).

begin to fall). Activities of the enzymes of glycolysis tend to go up with increasing AT. Viewed as a whole, however, the assignment of some enzymes to any category is questionable, since one would have to characterize each little section of their curves separately (Fig. 3; on the difficulty of obtaining reproducible measurements of enzyme activity, cf. KÜNNEMANN et al., 1970; BRAUN et al., 1970). Judging from present experience, such complications seem to be less marked in work with whole

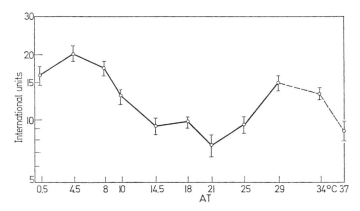

Fig. 3. Fumarase activity of lateral trunk muscle of *Idus idus* in relation to acclimation temperature; ET : 25° C. (From LEHMANN, 1968a)

animal respiration and organ functions of intact animals. Measurements of oxygen consumption of intact animals may show that capacity adaptation is restricted to biologically important AT ranges (p. 333), but this limitation leads to a meaningful subdivision of the curve into a small number of temperature ranges. Few measurements have been made of the oxygen consumption of tissues from animals that had been maintained at several ATs. PROSSER (in TROSHIN, 1967) kept goldfish at four ATs and found oxygen consumption and glucose uptake by the liver to be at a minimum in the middle of the range. This was not true for whole animal oxygen consumption. However, the oxygen use of intact *Limax flavus* was at a maximum when AT = 10° C (SEGAL, 1961). ET = AT curves may be different for related functions, e.g. the heart rate, blood pressure, cardiac output and cardiac work (flow x pressure) of sockeye salmons (BRETT[31]).

ALDERDICE recently drew attention to the fact that curves of the influence of ET and AT represent cuts or transects of multifactor response surfaces (BOX and WILSON, lit. in ALDERDICE)[32]. According to this author the properties of a reaction, deduced from a pair of curves of cold- and warm-adapted animals, can depend on the range of ETs selected and on the levels of ATs compared. The significance of these possibilities will be judged better when more data obtained as rate surfaces are available.

[31] BRETT, J. R.: Am. Zoologist 11, 99—113 (1971).
[32] ALDERDICE, D. F.: In: KINNE, O. (Ed.): Marine ecology. Vol. I, part 3, pp. 1659 to 1722 (1972).

β) Significance and Distribution of Non-Genetic Capacity Adaptations

The biological significance of compensation types 1 to 3 undoubtedly lies in the fact that they enable organisms to maintain their functions at more or less ideal levels in spite of changes in body temperature. Warm-blooded forms achieve the same effects by maintaining a constant body temperature. What is at stake is Claude Bernard's «fixité du milieu intérieur» or a "metabolic homeostasis" (ROZHAJA, 1963). These conclusions concern the intact animal; it does not matter by what means (e.g. enzyme levels, p. 344) the goal is attained. Compensations reduce potential differences in metabolism over long-term temperature differences within one season. Alone or together with mechanisms that change with AT or with other factors, such as day-length or an endogenous annual cycle, they may do the same for seasonal temperature changes. Non-genetic adaptation as well as genetic mechanisms can make life easier or even possible for animals in habitats of various temperatures.

The significance of type 5 compensation is harder to grasp[33]. Its occurrence in peroxidase activity in yeast may be connected with the enzyme's protective function, which may be especially important at high ATs (1st ed., p. 38). Adaptations of enzyme activity to higher levels of muscle activity are also known (1st ed., p. 32, HEATH and PRITCHARD, 1962; GLICK and BRONK, 1964). This suggests that type 5 compensation in these cases may be related to the higher activity level of warm-acclimated animals, as e.g. CARLSEN (1953) believed to be true for the increase of cocarboxylase level in eel muscle accompanying increased AT. It must be remembered in all of these interpretations, however, that the rates of biological processes usually go up directly with every increase in AT[34]. One can only speak of a type 5 compensation if a still greater effect (e.g., with respect to the protective function of an enzyme or to movement) is achieved. According to HAZEL and PROSSER (1970) enzymes for degradation of metabolic intermediates and products (such as peroxisomal and lysosomal enzymes, Mg-ATPase, and acetylcholine esterase) often show no compensation or inverse compensation (type 5), while in general enzymes concerned with energy liberation show compensatory acclimation. Of course a type 5 acclimation can result, especially in studies of enzyme activity, simply from the arbitrary selection of two ATs on an irregular AT dependence curve [assuming that the findings of LEHMANN (p. 335) are generally valid]. However, type 5 has also been found several times in measurements of oxygen consumption of intact animals (compare e.g. POCRNJIĆ, 1965; SØMME, 1968a; DUNLAP, 1969a, 1971), and WIESER's (1963) objection that cold-acclimated animals often stop feeding could not be taken as an explanation in all cases. Even in his earliest studies of this problem, the author (1939) found in pulmonary respiration of constantly fed *Lymnaea stagnalis* in November for one race type 3 and in June for another, type 5.

All the types of compensation have been illustrated by examples in earlier works, and others have been mentioned in this discussion. Recently so many examples, especially of type 3 compensation, have been described for intact organisms

---

[33] cf. FITZPATRICK, L. C., BRISTOL, J. R., STOKES, R. M.: Comp. Biochem. Physiol. 41 A, 89—96 (1972), also 40 A, 681—688 (1971).

[34] The respiration of mitochondria from red muscles of eels shows a compensation, while liver mitochondria show type 5; WODTKE (unpubl) believes that the cause is an increased gluconeogenesis at high ATs. In goldfish the intestinal transport *(in vitro)* of methionine and valine has been studied by KITCHIN, S. E., and MORRIS, D. (Comp. Biochem. Physiol. 40 A, 431—443, 1971). The rate of uptake of valine increased at low ATs; this was associated with an increased affinity for the transport mechanism. For methionine, the rate of uptake showed type 5; the affinity for the transport mechanism was temperature independent.

and for organ and cell functions, that it is no longer as necessary to collect more illustrative material as it is to gain an understanding of the general conditions under which capacity adaptation occurs, and of its mechanism. For this purpose, animals should be acclimated insofar as possible, to many more ATs than they usually have been in the past. Literature summaries are found in PRECHT, 1961a, 1964b, 1968; SEGAL, 1961 (mollusks); KINNE, 1963, 1964a, b, 1965, in DILL et al., 1964 (crayfish); PROSSER in HANNON and VIERECK, 1962, in TROSHIN, 1967; HAZEL and PROSSER, 1970 (enzyme activities); FRY, 1958; in ROSE, 1967; CHRISTO-PHERSEN in PROSSER, 1967 (microorganisms); I. PRECHT, 1967 (insects); GELINEO in DILL et al., 1964, 1969 i.a.). Only a few special cases of capacity adaptation will be referred to here.

As is known, the dissociation curve of blood respiratory pigments is displaced to the right during an increase in ET (to an extreme extent in antarctic fish; GRIGG, 1967)[35]. After frogs *(R. esculenta)* that had just been held at low temperatures were re-acclimated to higher ATs, this shift was partially reversed (STRAUB, 1957; compare GAHLENBECK and BARTELS, 1968; on fish ANTHONY, 1961; GRIGG, 1969). This adaptation effect is not exerted by the blood plasma, since it is still present after the plasma is replaced by a physiological salt solution, but it does depend on the integrity of the erythrocytes. The effect is absent in pure hemoglobin solutions prepared from the blood of differently acclimated frogs (KRÜGER, 1961). An adaptation effect need not appear in all species (cf. BLACK et al., 1966). The AT can also influence the speed of blood coagulation (e.g. in *Ligia*, NUMANOI, 1938). The blood coagulation time of cold-acclimated *Uca pugilator* is longer than that of warm-acclimated ones (DEAN and VERNBERG, 1966). Some of our own observations indicate that this is also true for the carp (cf. ZAIN-AL-UBEDIN and KA-BORSKI, 1966; for a species comparison see KOMATSU et al., 1970). Changes in AT may shift the critical point at which oxygen consumption becomes dependent on oxygen partial pressure (1st ed., p. 29); the AT may also change the response of ventilation and heart rate of fish to gradual hypoxia (SPITZER et al., 1969).

The relation of metabolism to the body size of animals is usually expressed as the b value ($b = 1$ for weight-dependence; $b = 0.67$ for surface area-dependence). These values may decline with increasing ET, as in the case of oxygen use by the scorpion *Heterometrus fulvipes* (which does not always happen, cf. DAVIES, 1967), but they usually rise with increasing AT (VIJAYALAKSHMI in RAO, 1966; cf. TSUKUDA and OHSAWA, 1958; ISTENIC, 1963; MORRIS, 1967). In the case of *Hemigrapsus*, body size can influence the extent of capacity adaptation as well as, at times, the type (DEHNEL, 1960; cf. MORRIS, 1962, 1965). The ability of the crayfish *Orconectes virilis* to undergo capacity adaptation depends on its stage of development (MCWHINNIE in ROSE, 1967, p. 358). In larvae of *Culex pipiens* the pattern of compensation changes with age too (BUF-FINGTON, 1969). Gastropods may also show a seasonal effect (RISING and ARMITAGE, 1969; cf. HALCROW, 1963). Apparently the extent of compensation of goldfish respiration is greater and the time needed for re-acclimation is longer when AT rises ($15 \rightarrow 30°C$) than when it falls ($30 \rightarrow 15°C$), but in these experiments of KLICKA (1965) in which the ET was also changed, other phenomena might have played a role (compare p. 321).

It is to be expected, as was said earlier, that eurytherms will probably compensate to a greater extent than stenothermal animals (compare PRECHT in PROSSER, 1958; ROBERTS in PROSSER, 1967). SEGAL (1961) asks the opposite question: "Are

---

[35] In some mollusks at higher temperature this trend may be reversed (COLLET, L. C., O'GOWER, A. K.: Comp. Biochem. Physiol. **41 A**, 843—850, 1972).

the animals eurytherm because they can compensate?" It is possible to use per cent change in an original value following re-acclimation as a measure of compensation (in Fig. 2 for example the value B_2 is circa 43% smaller than A_2). If the family of curves has the same Q_{10} values, as in Fig. 2, then the extent of compensation for all points on the curves is the same. If the Q_{10}s are not equal, the extent of compensation is different at each ET. Changes in Q_{10}s through compensation can also be considered as yardsticks, e.g. the differences between the temperature coefficients of curves A_1A_2 and A_1B_2 in Fig. 2 (EDNEY, 1964).

So far, Q_{10} values have been considered only in connection with ET-dependence curves (using constant ATs), but such values can also be calculated for ET = AT curves, or for AT-dependence curves (using constant ETs). If no specific designation is given, then only curves of the first type are meant by "Q_{10}".

According to the results of investigations on mussel gills, only types with a broad phenotypic resistance range may show individual, cellular, thermally induced capacity adaptations (SCHLIEPER in TROSHIN, 1967). But compensations are also shown by the extremely cold-stenothermal fish, *Trematomus bernacchii* (LITTLEPAGE in LEE, 1964, 1965; cf. SOMERO et al., 1968) as well as by the copepod, *Calanus hyperboreus* (CONOVER, cf. HALCROW, 1963). Only one type out of two dimorphic males of a copepod *(Euterpina acutifrons)* showed compensation for oxygen consumption (MOREIRA and VERNBERG, 1968).

The conjecture was raised in the first edition, p. 40, that compensation may develop phylogenetically, when an important, needed substance can easily become a limiting factor, as when high temperatures lead to rapid metabolism. A capacity adaptation is found in many aquatic or amphibious animals since oxygen, due to its slow rate of diffusion in water, can easily become a limiting factor. Furthermore, motile water animals are less able to evade unfavorable temperatures than terrestrial ones, though on the other hand temperature fluctuations in large bodies of water are less extensive than on land. For the latter reason, compensation seems less necessary (cf. WILBER and FRY in DILL et al., 1964, p. 661 ff., 715 ff.).

Compensations also occur in terrestrial and may be absent in aquatic animals (e.g. in *Eriocheir sinensis*, 1st ed. 1955, p. 36). Among the species of *Uca* examined by TEAL (1959), *Uca pugnax* for example showed a compensation, but *U. minax* did not, though both species are equally active in and out of the water. According to F. J. VERNBERG and W. B. VERNBERG (1966b) geographical races can be distinguished among fiddler crabs on the basis of capacity adaptation alone, even though the temperature conditions of their habitats are similar or even identical[36]. Some insects exhibit compensation, others do not. There seem to be no regularities (I. PRECHT, 1967; cf. SØMME, 1968a). It should be remembered that motile insects can simply avoid unfavorable temperatures, a behavior that could substitute for compensations. According to HUNTER (1964—1968) eurythermal species of *Drosophila* exhibit a more marked compensation than stenothermal species, but it is difficult to explain his findings that female *D. melanogaster* exhibit compensation with respect to oxygen use, while males do not. Lab-reared *D. immigrans*, collected

---

[36] Fiddler crabs acclimated to a low temperature and exposed to a high temperature may exhibit a different acclimation pattern than when acclimated to a high temperature and then subjected to low temperatures (VERNBERG, W. B., VERNBERG, F. J., 1972, p. 107 ff., see p 303); further investigations about this problem are necessary.

from warm and cold environments, showed a compensation of type 5 with respect to oxygen consumption (HUNTER, 1968). EDNEY (1964) found a compensation in both *Porcellio laevis* and *Armadillidium vulgare*. *Porcellio* showed a greater compensation with regard to heart rate than to standard metabolism, while *Armadillidium* showed a more distinct compensation of standard metabolism. According to WIESER (1965a, b), *P. scaber* showed no capacity adaptation.

The reserve material stored by resting stages should also be considered as a limiting factor, since it could easily be used up at high temperatures without being replaced. Two Chrysomelids exhibit compensation with respect to oxygen use only during resting stages, and the species with the longer resting period does so to the greatest extent, but this rule does not hold for *Chrysomela haemoptera*. Diapausing larvae of *Cephaleia abietis* also show no capacity adaptation of oxygen consumption or succinic dehydrogenase activity (KIRBERGER, 1953). *Helix pomatia* exhibit metabolic compensation not only during winter sleep but also during the feeding period (KIRBERGER, 1953; MEWS, 1957; cf. GELINEO and KOLENDIC, 1953; BLAŽKA, 1955; RISING and ARMITAGE, 1969). HARRI and HEDENSTAM[37] found that the oxygen consumption of *Rana temporaria* shows a compensation only in winter, not in summer. In winter adrenaline and noradrenaline cause an increase of oxygen consumption only in cold-adapted animals, in summer, however, in both cold- and warm-adapted frogs (cf. KASBOHM, 1967).

One could hypothesize that slow-moving animals, being less able to avoid unfavorable temperatures, would be more likely to develop compensations than faster ones. Nevertheless several lizards show compensation in oxygen consumption (SCHMIDT-NIELSEN and DAWSON and GELINEO in DILL et al., 1964; GELINEO, 1967, 1969; MURRISH and VANCE, 1968). SEGAL (1961) could not establish a correlation between occurrence of compensation and the habitats of the species of mollusks he examined. According to ROBERTS (in PROSSER, 1967) ability to compensate in fish is related to systematic affinity. It is present in Centrarchids, Ameiurids and Cyprinids but less so in Salmonids. But as stated previously, related species can behave differently (p. 339). SUHRMANN (1955) found a partial compensation in oxygen use of the crucian carps, such as is known for goldfish. She was probably working with the wild ancestral form of the goldfish *(C. gibelio)*. According to v. BUDDENBROCK (1960) and ROBERTS(1966) an inverse compensation is present in the closely related species *C. carassius* (cf. PRECHT, 1961a).

γ) Restriction of Capacity Adaptation to Specific Ranges of Acclimation Temperature

The extent of capacity adaptation and, as noted, even its type may differ in different parts of the AT range. Here only those cases will be considered in which the limitation of compensation to specific ranges seems to be explicable, and not the AT-dependence curves of enzyme activity that are difficult to explain (p. 337, compare FREED, 1965). For example the extent of compensation of the oxygen consumption of *Lumbriculus variegatus* (KIRBERGER, 1953) and the irritability of the foot of *Lymnaea stagnalis* (BENTHE, 1954) increase with AT (cf. EDNEY, 1964; SMIT, 1967; BUIKEMA and ARMITAGE, 1969), but that of oxygen consumption of

---

[37] HARRI, M., HEDENSTAM, R.: Comp. Biochem. Physiol. **41** A, 409—419 (1972); about Chrysomelids, see LÜHMANN and DREES (1 th ed; p. 40f.).

Pachygrapsus crassipes goes down, according to ROBERTS (1957). If the animals are maintained at several ATs, regions of strong compensation (partly of type 2) can be found in the form of plateaus. These can also be formed, as mentioned, through innate responses or through regulations. Such plateaus seem to appear in those ranges that are biologically important for the animal. For example McLEESE and WILDER (1958) found such a plateau, produced by compensation, in the lobster (Fig. 4, see PRECHT, 1968).

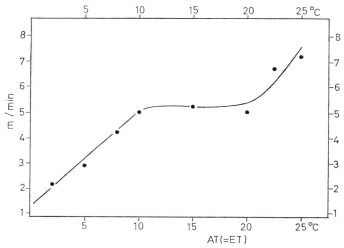

Fig. 4. A plateau, produced by capacity adaptation (compensation), in the speed of motion of *Homarus americanus*. Ordinate: speed of motion in m/min. (From McLEESE and WILDER, 1958)

In order to demonstrate that a plateau is the result of capacity adaptation it is not sufficient, as has been said, to present only ET = AT curves. But such regions of the curve probably are produced in this way, for example in fish, which often show meta-bolic compensation (cf. SERVATY and WAITZENEGGER, 1964; GRIGG, 1965 i.a.). Accord-ing to MEUWIS and HEUTS (1957, ET = AT curves), plateaus were progressively less noticeable in the metabolism of carp as the fish matured. ROBERTS (in PROSSER, 1967, ET = AT) presented evidence that the region of homeostasis in *Lepomis gibbosus* from Beaufort is shifted towards higher temperatures, compared with the more northerly animals from Amherst; its appearance depends on photoperiod and therefore on the season "and has been discussed as adaptive means for meeting the greater physiological and competitive demand of reproductive, compared with postbreeding animals". Ac-cording to SCHMEING-ENGBERDING (1953) the plateau in oxygen consumption of some fresh-water fish (ET = AT curves) disappears in narcotized animals, which may in-dicate that it is produced by regulation (p. 323). Of course it is still possible that com-pensation takes place at the level of control centers, whose operation is stopped by narcosis; this possibility remains to be investigated.

The AT range in which compensation is most marked can apparently differ for different functions of the same animal. KONISHI and HICKMAN (1964) transferred rainbow trout from an AT of 10° C to ATs of 4° C and 16° C, respectively. The midbrain response to direct electrical stimulation of the retina showed a compen-sation only after transfer to 4° C, but the tissue respiration of the brain showed a much more complete compensation between 10° C and 16° C than between 10° C and 4° C.

δ) The Time Course of Capacity Adaptation

Capacity adaptation of metabolism could generally be connected with a change in protein synthesis; such a change can take place relatively rapidly. But re-acclimation in metazoa generally requires more time (usually on the order of days) (PRECHT et al., 1966). Therefore it is a response to longer lasting temperature changes and not to short-term daily fluctuations. Among the more rapidly multiplying protists adaptation might well be faster. Capacity adaptation, like resistance adaptation, is itself often temperature-dependent.

The question has been raised whether resistance adaptation to very low temperatures is possible or whether the mechanism of adaptation "freezes up" in such conditions (p. 430). According to unpublished experiments of HELFENSTELLER the cirral rhythm of *Balanus crenatus* shows compensation (almost of type 2) and a reasonable heat adaptation. In an attempt at re-adaptation *(Umadaptation)* from 15 → 10.4°, 5.2°, and —1.3° C, respectively, the heat resistance followed a normal course of adaptation, in which the final levels were reached after 4, 6, and 8 days, respectively. Resetting of capacity adaptation took longer; at first there was a reduction in beat frequency extending over 6 days, which was especially evident in the —1.3° C animals, until the final, highest value was reached after about 3 weeks (ET was always 10° C; all animals were measured only once; PRECHT, 1968, p. 524). Antarctic animals (p. 331) may also show capacity adaptation at the very low temperatures of their habitat (cf. McWHINNIE in ROSE, 1967, p. 357; SOMERO et al., 1968). FÖH (unpubl.) carried out re-adaptation experiments of different rapidity (*a:* 20 → 5° C in one day, *b:* 20 → 5° C in 15 days, *c:* 10 → 25° C in one day, *d:* 10 → 25° C in 15 days) on *Idus idus* by measuring the succinate respiration of white muscle tissue (ET = 25° C). A new final value was reached: in *a* after 44, *b* after 52, *c* after 30 and in *d* after 40 days. The extent of alternation in the sense of compensation was: in *c* and *d* 60%, in *a* 15% and in *b* 45%. This extent can be different for different temperature ranges. The difference between *a* and *b* seems to indicate that a lower limit of adaptation capacity exists at 5° C, which is especially of consequence in quick re-adaptation. The fact that there may be no compensation at lower temperatures may simply mean that it is restricted to a specific part of the range (p. 340) and does not necessarily imply a "freeze-up".

Another example of a relatively long adaptation time may be mentioned here. Female *Xiphophorus helleri* underwent long-term acclimation to 19.5° C and 31° C, respectively (PRECHT et al., 1968). The ET was slowly (0.3° C/min) changed, starting from 25° C, so that the entire ET dependence curve of opercular movement was observed. When the AT was raised (19 → 31° C), re-adaptation in the upper part of the ET range was completed in thirteen days, but another twelve days were needed to complete re-adaptation in the lower range (below 19° C). Till the third day after the transfer, the rates at ETs between 19° C and 30° C were much the same as those for 19° C animals. Here too the process of compensation itself is temperature-dependent. Furthermore, different results are found in different sections of the ET range (also when AT is lowered). When the extent of compensation differs for different AT ranges, this must have an effect on the time course of adaptation (cf. e.g., KLICKA, 1965).

KRÜGER (1962) measured the whole-body respiration, the oxygen consumption of muscle tissue (with added succinate), and the succinic dehydrogenase activity (using Thunberg's methods), of *Rhodeus amarus*. Compensation was found in all cases, most markedly in whole-body respiration. When AT was changed (8 → 24° C) re-adaptation in all three cases was complete on the tenth day, but the time course of adaptation was not the same. According to F. J. VERNBERG and W. B. VERN- BERG (1966a) there is no correlation between the extent or time course of capacity adaptation of metabolism of whole animals (*Uca* sp.) and that of their tissues (gills, intestinal gland; type 3, partly type 5) (compare 1966b, W. B. VERNBERG and F. J. VERNBERG, 1966, 1968a, b). According to PICKENS (1965) the time needed for re-adaptation of the heart rate of *Mytilus californicus*, is shorter in small animals than in large ones. LAGERSPETZ and DUBITSCHER (1966) performed re-adaptation experiments (AT 5 → 21 °C) on compensation of the speed of particle transport by the frontal cilia of isolated gill segments of *Anodonta*. Seven to eight weeks were needed for complete compensation, but a reasonable heat resistance for the same function developed in only four days. As in *Balanus crenatus* (p. 342) resistance adaptation is finished sooner than capacity adaptation (cf. SCHLIEPER in TROSHIN, 1967). Insects that live in a relatively stable thermal environment, like *Tribolium confusum*, seem to acclimate slowly. *Musca domestica*, which lives in a highly variable thermal environment, acclimates more rapidly (the ATPase-activity only at low ATs; ANDERSON and MUTCHMOR[38]).

When the AT of *Rhodeus amarus* is changed abruptly (15 → 25° C), the K_m values are reduced in the beginning by about half (minimum after 24 hrs.), then they increase over the norm; finally after 10 days the former value (of the AT 5° C-animals) is gradually reached (enzyme: LDH, ET: 25° C; WERNICK and KÜNNEMANN, cf. p. 308).

ε) Capacity Adaptation on Various Levels

The examples mentioned so far show that capacity adaptation can be demonstrated on several levels — that of the whole animal, of single organ functions, or of cell metabolism, in which individual enzyme activities may react quite differently. BOYCOTT et al. (1961) found histological features of the central nervous system of lizards that varied with AT. JANKOWSKY (1968) mentions histological changes in fish musculature.

We have examined whole-body oxygen consumption, especially of fish, many times. That part of the respiratory chain that takes part in the oxidation of suc- cinate has been measured by the addition of succinate. MALESSA (1969) also ex- amined the activity of cytochrome oxidase. In the eel the relation to each other of both these parameters of oxidative capacity in muscle tissue can change with AT. In working with fish muscle one must distinguish between two layers — an outer red layer and an inner white one — that differ in function and, e.g., in the eel, differ in response to AT as well (MALESSA, 1969; cf. BOKDAWALA and GEORGE, 1967; STEVENS, 1968; GORDON, 1968; DEAN, 1969)[39]. The oxygen consumption of the muscle tissue and that of the whole animal often, but not always, show a similar relation to AT (cf. PRECHT, 1964b, 1968, p. 526).

In *Bufo boreas*, oxygen consumption and CO_2 production depend markedly on AT, while the respiratory quotient and heart rate do not (BISHOP and GORDON in PROS- SER, 1967). KANUNGO and PROSSER (1959) measured the oxygen consumption (per

---

[38] ANDERSON, R. L., MUTCHMOR, J. A.: Insect Physiol. **17**, 2205—2219 (1971).
[39] cf. BOSTRÖM, S.-L., JOHANSSON, R. G.: Comp. Biochem. Physiol. **42 B**, 533—542 (1972).

unit nitrogen content) and the P/O ratio of isolated goldfish liver mitochondria. The mitochondrial respiration rate was somewhat higher in either warm- or cold-acclimated subjects, depending on the substrate used; the P/O ratio went up with AT. The oxygen consumption of liver homogenates showed a distinct compensation. "It is concluded that the activities of dehydrogenases associated with the Krebs cycle in mitochondria are not significantly altered on acclimation. The phosphorylating system in the mitochondria of cold-acclimated fish is somehow decreased in efficiency so that less ATP is synthesized." Mitochondria of eels (without food) increased with rising AT, too; however, the weight of the liver and (probably) the number of mitochondria increased with decreasing AT (WODTKE, unpubl., cf. DAS, 1967; PRECHT, 1968, p. 526). The protein content of the whole liver did not depend on AT. Experimental data suggest that the P/O ratio in mitochondria of cold-acclimated eels is diminished at phosphorylation sites II and III (cytochrome chain) but not altered at site I (flavoprotein chain).

It should always be remembered that separate tissues of the same animal can behave differently. For example, according to VON BUDDENBROCK (1960), oxygen consumption of brain tissue of *Carassius carassius* shows compensation of type 3, but that of gill tissue (like the whole animal) shows a type 5. EVANS et al. (1962) measured the oxygen consumption of intact *Salmo gairdneri* and of its various tissues. The whole animal and the gills gave more or less the same response (type 4 or weak type 3 compensation); brain showed a type 2 and liver a type 5 (however, only ET = AT curves were used). In goldfish, the oxygen consumption of the gills shows a distinct compensation, that of the muscles a lesser one, that of the liver a still smaller one, and that of brain and heart none (EKBERG, 1958; MURPHY, cited by PROSSER in TROSHIN, 1967; compare FREEMAN, 1950; ROBERTS, 1957, 1966; CALDWELL, 1969; HAZEL and PROSSER, 1970).

It has been stated previously that the enzymes of the same tissues can behave differently. Since, according to LEHMANN (1970a), the activities of individual mitochondrial enzymes of *Idus idus* muscle can change in different ways with AT (cf. BRAUN et al., 1970), more than a change in the number of mitochondria must be involved (p. 335). Even the behavior of individual isozymes may depend on the AT. HOCHACHKA (in PROSSER, 1967) found fourteen isozymes of LDH in the white trunk musculature of the trout. "Warm adaption causes a striking loss in the activity of LDHs 1–9 (A–B–C-series) in relation to the LDHs 10–14 (D–E-series). The first are located in either the nucleus or the mitochondria or both; the D–E-series is definitely in the soluble fraction . . . and can be expected to display properties suited for anaerobic metabolism. . . . In cold adaptation of trout extramitochondrial is favored over mitochondrial metabolism in both liver and muscle (glycolysis, gluconeogenesis, glycogen synthesis, lipogenesis, and shunt participation; the complexity of the changes was greater in liver than in muscle" (p. 184f., 179). HOCHACHKA found five distinct and active isozymes of LDH in the liver of cold-acclimated goldfish and four in the warm-acclimated fish. "In cold adaptation, the isozymes that are being induced are regulatory enzymes and are markedly more sensitive to metabolic control than those that are unchanged during the adaptive process" (HOCHACHKA in PROSSER, 1967, p. 190).

In general, enzyme activities are measured under more or less optimal conditions, a procedure which essentially measures enzyme concentrations (compare PROSSER in HANNON and VIERECK, 1962; in PROSSER, 1967; CHRISTOPHERSEN in PROSSER, 1967; MALESSA, 1969; LEHMANN, 1970a i.a.). Changes in enzyme concentration with AT can also be determined directly (p. 326). HOCHACHKA and his co-workers[40] concluded, on the basis of other authors' findings, that changed levels of rate-limiting enzymes, altered concentrations and also modulators may be involved in capacity adaptation. But they believed that it would be more meaningful for the organism, after adaptation to a new AT, not merely to change the quantity or μ value of an enzyme that was no longer fully suited for the new conditions, but rather to change the isozyme pattern; by this the effective, low K_m range is displaced in the same direction as the AT. For example the K_m of pyruvate kinase of cold-adapted rainbow trout is at a minimum at $5°$ C, but that of warm-acclimated trout has a minimum at $15°$ C (SOMERO, 1969a).

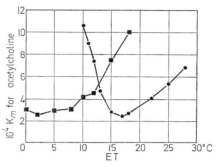

Fig. 5. Effect of ET on K_m of AChE for acetylcholine from trout acclimatized to $17°$ C ($\bullet$) and to $2°$ C ($\blacksquare$). (From BALDWIN and HOCHACHKA, 1970)

In Fig. 5 acclimation to intermediate ATs resulted in the occurrence of both types of acetylcholinesterase of the brain of trouts. The often observed compensation of the metabolism of whole animals cannot be explained only by a change of the quality of enzymes (K_m values) by AT. When measured at the same ETs, the decrease of metabolism with rising AT must be caused by other mechanisms (e.g. a change in the quantity of enzymes); however the reduced metabolism of warm-adapted animals may work with very efficient enzymes, due to a qualitative change with an increase of AT as in Fig. 5. According to HEBB et al. (1969) the K_m values of brain acetyl transferase of cold-acclimated goldfish are higher at low ETs than those of warm-acclimated fish, but smaller at high ETs.

In addition to changes of enzyme quantity and quality (of isozyme pattern) with AT, PASSIA (in print)[41] found for an enzyme (the $NADP^+$-dependent isocitrate dehydrogenase of the white dorsal muscular system of *Idus idus*) another possibility to act at its best in several ETs. The activity of this enzyme showed a distinct capacity adaptation in the sense of a compensation. The curve of the K_m value dependence on ET of almost purified enzyme extracts of animals from different ATs showed two minima (W-figure), which shifted with increasing AT only slightly towards higher ETs. Arrhenius plots of $\log V_{max}$ resulted always in $3\ \mu$ values. Thus the enzyme is always present with an optimum of action at 2 ETs. — It was mentioned that SOMERO found cold and warm "variants" of enzymes, which do not appear to be isozymes in the conventional sense; electrophoretic and electrofocus analysis revealed only single peaks of activity (p. 310).

SMITH (1967b) reported on components of goldfish gut ATPase. The ouabain-sensitive enzyme activity was high and the ouabain-insensitive activity low in

---

[40] cf. also SOMERO, G. N., HOCHACHKA, P.: Amer. Zool. 11, 157—165 (1971).
[41] Mar. Biol.; cf. MOON, T. W., HOCHACHKA, P. W.: Biochem. J. 123, 695—705 (1971).

membrane fractions prepared from fish acclimated to 8° C; the opposite was true for AT-30° C-fish. Incubation of intestinal membranes with phospholipase C enhances the Mg^{++} stimulated ATPase activity from cold-acclimated goldfish, but not the activity of warm-acclimated fish. The membranes from AT-30° C-fish contained more Mg^{++} and less $Na^+ + K^+$-activated ATPase/mg of protein than did membranes from 8° C-acclimated fish (SMITH and KEMP, 1969; compare SMITH et al., 1968; ELLORY and SMITH, 1969). The activity of the soluble succinic dehydrogenase of goldfish muscle was increased by incubation with soluble lipid preparations; a total mitochondrial lipid extract from 5° C-acclimated fish was a more effective reactivator of the enzyme than a comparable extract from 25° C-acclimated animals. This difference was referable to a higher degree of unsaturation of the 5° C lipid extract (HAZEL[42], see p. 330). Changes in membrane composition may be important for capacity adaptation by changing the activity rather than the quantity or the quality of an enzyme; thus mediate and immediate influences of AT can be distinguished.

Earlier experiments on goldfish have led to the supposition that the pentose shunt pathway is enhanced by decreases in AT (EKBERG, 1958; KANUNGO and PROSSER, 1959; HOCHACHKA and HAYES, 1962; but cf. PROSSER in HANNON and VIERECK, 1962; BISHOP and GORDON in PROSSER, 1967). This phenomenon may be related to the formation of NADPH and to lipid synthesis (p. 330, cf. KRÜGER, 1962; SOMERO et al., 1968; but cf. BARÁNSKA and WLODAWER, 1969). From a purely energetic standpoint this path is less efficient than the Embden-Meyerhof path. In general the pentose shunt does not play a great role in muscle tissue, although in fish liver it probably does (FRIED et al., 1969). BROWN (1960) however found only slight activity of two pentose shunt enzymes (glucose-6-phosphate- and 6-phosphogluconic acid dehydrogenase) in carp liver homogenate. LEHMANN (1970a) found similar results for both enzymes in the lateral trunk musculature of *Idus idus*. The tail muscles of fish are said to have greater activity (ZEBE, 1961). In *Carassius carassius* gill tissue, only the latter of the two enzymes shows compensation under the influence of AT (two ATs, EKBERG, 1962). LEHMANN (1970a) measured the relation of both enzyme activities in lateral trunk muscles of *Idus idus* to AT after acclimation to many temperatures (ET = 25° C). The activity of G6PDH goes down steadily as AT rises from $0.5 \rightarrow 14.5$° C. Then it goes up to relatively high levels (maximum at AT = 29° C). Only at low ATs, 6-PGDH has relatively high activities which then decline as AT rises. So, the activity curves of the two pentose shunt enzymes of *Idus* agree only in the lower part of the AT range. However the observed enzyme activities are not absolute criteria for the significance of a metabolic path, especially because this depends on the activities of the rate-limiting enzymes. Substrate supply at different ATs can be more decisive for the metabolic pathways than the activity of enzymes (e.g., PALUMBO and WITTER, 1969; for reports on the participation of the shunt, see HOCHACHKA in HOAR and RANDALL I, 1969, p. 353 ff.).

According to POLYANSKY and SUKHANOVA (in TROSHIN, 1967, p. 204) glycolysis becomes more important in Paramecia as AT falls (compare IRLINA in TROSHIN, 1967). KRÜGER (1962) found an inverse compensation (type 5), especially in fruc-

---

[42] HAZEL, J. R.: Comp. Biochem. Physiol. 43 B, 863—882 (1972).

tose-1,6-diphosphataldolase of *Rhodeus amarus* muscle. However, the succinate respiration of this tissue and the respiration of the total animal showed compensation (two ATs were used). He refers to the fact that aldolase is also involved in fat synthesis. The irregularities in enzyme activity-AT dependence curves found by LEHMANN (1970a) make it difficult to form conclusions on metabolic paths. EKBERG (1962) measured the anaerobic tissue metabolism of *Carassius carassius* by determining the amount of CO_2 given off by the tissues and the amount that could be released from tissue carbonates by acid treatment before and after the experiment; only the latter showed a compensation, which was more marked in animals exposed to long day length than in those exposed to short photoperiods (cf. BLAŽKA, 1958). The tissue metabolism of *Lepomis gibbosus* brain showed a relation to AT (of type 5) only after the addition of uncoupling substances (2,4-dinitrophenol, ROBERTS, 1964).

PROSSER (in TROSHIN, 1967) believes that compensation may lead to an accumulation of metabolites that could induce enzyme formation by action on the nucleus. Some possible ways in which such accumulations could come about are given: "if there are alternate and cross-linked pathways having different temperature characteristics, one may slow more than another in the cold . . . It can be suggested that the action of an intermediate or product of a biochemical reaction may show feedback control by direct inhibition, by repression of the synthesis of particular messenger RNA or by induction as by a substrate, the concentration of the controlling products or substrates would be dependent on the temperature characteristics of the particular steps" (p. 382).

ζ) Pre-Treatment under Varying Conditions

In the experiments presented so far, the animals were acclimated to constant temperatures. This is unnatural, inasmuch as the temperature in nature varies with the time of day, but experiments can simulate the natural conditions. When AT was varied (15↔23° C) in twelve hour cycles, *Lumbriculus variegatus* used at identical ETs the same amount of oxygen as when it was maintained at 19° C (KIRBERGER, 1953; cf. BIRK et al., 1962). According to BERKHOLZ (1966) the oxygen consumption of *Idus idus*, which shows a distinct compensation in animals maintained at constant temperatures, exhibits average values after exposure to varying ATs (15↔22° C, twelve hour cycles), approximating the values obtained after maintenance at a constant temperature of 22° C. The approximation was especially evident in small animals, but less evident in the oxygen consumption of muscle tissues. An attempt can be made to understand this type of shift, that is particularly marked in heat-resistance adaptation, on the basis of the temperature dependence of adaptational processes.[43] In addition, the time course of re-adaptation *(Umadaptation)* must be taken into consideration. If the speed of re-adaptation of animals maintained at constant temperature continually declines from an initial high rate, and if the rate depends on the extent to which the final conditions have been reached, then the final value reached under constant conditions will not be reached if the maintenance temperature varies, in spite of the often prevailing

---

[43] If $Q_{10} > 1$, the gain of heat resistance during the warm day is greater than the loss during the colder night.

temperature dependence of adaptational processes. This is so because the second effect will work more and more against the first effect, so that a final value will be reached that is displaced in the direction of that of the animals maintained under constant warm conditions. This interpretation assumes that a temperature-dependent compensation is established at once and takes effect within the twelve hours, and is not delayed for several days, as in the case of the cirral beat of several balanids (CRISP and RITZ, 1967). Stress phenomena, overshoots, etc., may enter into and complicate the picture. The whole-body respiration of *Idus*, under a twelve hour cyclic change in AT, was the same, regardless of whether the higher temperature prevailed during the day or the night. As was mentioned the oxygen consumption of *Blaberus craniifer* continues to increase in re-adaptation experiments during the following two days, specifically after an increase in AT (p. 323). If the constant pre-treated animals were exposed to widely fluctuating ATs (14↔27° C, less pronounced at 18↔27° C) or to fluctuations at high temperatures (27↔37° C), oxygen consumption increased only during the following day and then swung back to its normal value. The animal becomes accustomed to the stress (ET = 27° C, RANDZIO, see p. 323). Larvae of *Tenebrio molitor* show no capacity adaptation of the oxygen consumption (constant ATs); however after a pretreatment with varying ATs (22↔14° C), oxygen consumption, measured under the same conditions, is lower (HEITLINDEMANN, unpubl.). BRAUNE and GRONOW (unpubl.) also found that *Idus idus* become accustomed to changing temperatures. Animals which had been acclimated to constant 15° C for a long period of time showed, in relation to constant 25° C control animals, a higher lactate-pyruvate quotient of the white muscle in the first days following transfer to an environment of changing temperatures (15↔25°C; 12^h : 12^h). On the 4th and 5th days, the quotient was below that of the constant 15° C control; however, after one week, the value approached that of the constant 25° C control.

η) Further Investigations on the Mechanism of Capacity Adaptation

The findings discussed thus far make it clear how difficult it is to analyze the mechanism of capacity adaptation, especially because a unified explanation can hardly be given, inasmuch as protists as well as metazoa can show such an adaptation. In the case of the metazoa one must examine the extent of the role played by control systems (central nervous system, hormones, etc.) and whether tissues can become adapted directly.

If control systems play a role in acclimation, then secondary effects must be distinguished from direct effects (see the scheme).

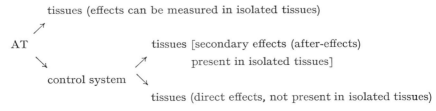

The influence of AT on tissues (and organ functions)

The former can be detected even in isolated organs or tissues, the latter cannot be. Hormones may, as mentioned, e.g., influence synthesis of proteins by genes and therefore the concentration of enzymes (a secondary or after-effect), or they may have direct effects by stimulating or inhibiting enzyme activity or altering membrane permeability (KARLSON, 1951). We have tried to clarify these problems, especially by work on the eel (PRECHT, 1951, 1961b, 1968, p. 530ff.; CARLSEN, 1953; SCHULTZE, 1965; JANKOWSKY, 1966, 1968; MALESSA, 1969; LEICHT, 1969; SCHEIL, 1970).

If one exposes the anterior and posterior sections of an eel to different temperatures for long periods, the effect on succinate respiration of the muscle tissue of each section is the same as it would be if the entire animal were exposed to that AT (PRECHT, 1961b; SCHULTZE, 1965; PROSSER et al., 1965): In addition, SCHULTZE found that the oxygen consumption of the whole animal depends only on the treatment of the head region. The control of total respiration by the head probably means that the flow of blood to the muscles is governed from there by either hormonal or nervous means (JANKOWSKY). For the oxygen consumption of the whole animal it is not important whether the posterior muscles *(in vitro)* use relatively little oxygen due to acclimation to high AT or use more oxygen due to acclimation to low ATs. It may be important for the mentioned results of SCHULTZE that the wet weight of the eel liver increases with decreasing AT (WODTKE, see p. 344). This author also found that the oxygen consumption of liver mitochondria of the eel (with several different substrates) showed type 5; however, the mitochondria of the red muscle showed a compensation [44].

According to PROSSER it is still not certain whether the acclimation of the tissues is caused by the direct action of AT on the tissue or by indirect action through the spinal cord, which is, after all, also affected by the different temperatures of the anterior and posterior parts. The latter possibility must be considered since compensations have been discovered in the spike trains going from the cord to the muscles in the front and hind ends of eels treated in this way (PROSSER et al., 1965). According to SCHEIL the AT seems to work directly on the tissues.

Central Nervous System (CNS) Influences[45]: The head (i.e. brain) of the eel seems to play a significant role in capacity adaptation. Opercular movements, whose compensation was studied in *Xiphophorus helleri* by THIEDE (1965), are controlled by the respiratory center (cf. PRECHT, 1959). A function may show a compensation in the absence of central nervous system action, but such action may still have an enhancing effect. BENTHE (1954) found such a case while working on irritability of the isolated foot of *Lymnaea stagnalis*. He observed greater compensation when the pedal ganglion remained with the preparation than when he removed it. Central nervous system changes take on a special significance inasmuch as the CNS is often the first system to respond, e.g. with respect to hormone production also; thus, compensation in the CNS would suffice. LAGERSPETZ and TIRRI (1968; cf. LAGERSPETZ et al., 1970) refer to the effect of AT on transmitter substances (e.g. in *Anodonta*). Acetylcholine and eserine inhibit at high and stimulate at low con-

---

[44] cf. MAYERLE, J. A., BUTLER, D. C.: Comp. Biochem. Physiol. 40 A, 1087—1095 (1971).
[45] Concerning the adaptation of the fish brain, see BASS, E. L.: Comp. Biochem. Physiol. 40 A, 833—849 (1971).

centrations the activity of the cilia; the effects of serotonine were not definite[46]. Acetylcholine greatly increased the heat resistance of the cilia. LAGERSPETZ and his co-workers believe that transmitter substances are important for capacity and resistance adaptation; they also have investigated the significance of adrenaline, noradrenaline, and serotonine for frogs.

In eels vagotomy causes tachycardia (more pronounced in 15° C-acclimated than in 25° C-acclimated fishes). Elimination of vagal influence upon the heart cannot prevent partial compensation of the heart frequency (WEGENER, unpubl.).

Attempts have been made to draw conclusions about the functions of whole animals (especially of fish) from measurements of the relation of respiration and enzyme activity of brain tissue to AT, but their significance is probably limited (EVANS et al., 1962; JOHNSTON and ROOTS, 1964; HICKMAN et al., 1964; KONISHI and HICKMAN, 1964; BASLOW in PROSSER, 1967).

When *Carassius gibelio* were kept during adaptation in water with low oxygen concentration (3.3. ml/l), an increase of AT (15 → 25° C) did not change the oxygen consumption of white lateral muscles and gills (with succinate); however, the same tissues from control fishes from well-aerated water showed a marked compensation (KREBS, unpubl.). The CO_2 output of *Tubifex tubifex* (measured under anaerobic conditions) showed no compensation when the animals were kept in aerated water (ATs: 15°, 20° and 25° C); however a compensation could be found in worms kept in water with low oxygen concentration in the range between 15° and 20° C (type 5 between 20 and 25° C; BRANDT, unpubl.).

Effects of Body Fluids: Body and circulatory fluids can also exert direct effects. Extracts or blood of cold-adapted animals *(Lampitio mauritii, Heterometrus swammerdami, H. fulvipes, Etroplus maculatus)* stimulate the oxygen consumption of tissues from normal or warm-acclimated animals, even when they are added in small quantities in *in vitro* experiments. Extracts or blood of warm-acclimated animals have the opposite effect (RAO in TROSHIN, 1967). PRECHT (1964c, 1965) and JANKOWSKY (1964a) were able to confirm the results of the Indian authors for the muscle tissues of *Idus idus* and *Tinca tinca* (and carp blood). The stimulating effects of serum from cold-acclimated carp appear only if the oxygen consumption of the muscle tissue of *Idus* has not already been raised to very high levels by its own compensation effects (as for the tissue of cold-acclimated fish). The blood serum of warm-acclimated carp, in contrast, acts precisely to lower the high oxygen use of muscles of cold-acclimated *Idus*, and has hardly any effect on animals from high ATs. As mentioned, the various tissues of one animal can behave quite differently; the direct effects exerted by the body fluids may restore a certain conformity, but this remains to be investigated.

Hormonal Influences: RAO and his students have carried on further investigations on the body fluid effects discussed in the last section. The supra- and subesophageal ganglion of *Lampitio* contain neurosecretory cells whose activity changes with AT (minimum at 28° C, according to SAROJA and RAO, 1965). Even the type of active cell changes with AT. A student of RAO's examined the effect of injected nerve extract from cold- and warm-acclimated worms on 28° C animals. Three hours after the injection of extract from cold-acclimated *Lampitio*, increases in respiration,

[46] LAGERSPETZ, K.Y.H., LÄNSIMIES, H., IMPIVAARA, H., SENIUS, K.: Comp. gen. Pharmacol. 1, 152—154 (1970), about frogs: ibid. 3, 11—18, 101—112, 226—234 (1972).

RNA, lipase activity, degree of unsaturation of fatty acids, and Mg^{++} content were noted, in other words, those symptoms that appear when AT falls. Extracts from warm-acclimated worms had the opposite effect. Similar result are said to be found in scorpions. Sterol fractions extracted from nervous tissues of differently acclimated worms are also said to have had the same effect.

PARVATHESWARARAO (1968a) believed thyroid hormone to be responsible for the described effects in *Etroplus*. RAO had already found that thyroid activity in this fish rose as AT fell. A problem is touched on here that has been investigated many times, but which has not been greatly clarified. The role of metabolically active vertebrate endocrine glands in capacity adaptation, which could naturally be different in various species, has been an obvious target for inquiry. The differences in activity of enzymes within one tissue, and of the same enzyme in different tissues show that one must not oversimplify the problem, even though it is known that the same hormone may stimulate the activity of one hormone and reduce the activity of another (compare PRECHT, 1961a). According to TISHLER (1963), L-thyroxin stimulates the oxidation of succinate in frog myocardium, but inhibits the oxidation of malate. Logical difficulties also arise in interpreting the participation of hormones in the results of Russian authors (compare PRECHT in TROSHIN, 1967, p. 261).

Although special interest has been focused on the role of the vertebrate pituitary-thyroid system in capacity adaptation, only a brief report will be given here since many gaps remain in our knowledge, in spite of many investigations (PRECHT in PROSSER, 1958, 1961a, 1964d; PROSSER in HANNON and VIERECK, 1962; JANKOWSKY, 1964a, b).

It should be noted at the outset, that the effects of thyroxin on the metabolism of poikilotherms is not always as clearly discernible as it is in mammals (cf. PARKER, 1967; SAGE, 1968; MASSEY and SMITH, 1968 i.a.). Measurements can be made of the oxygen consumption of whole animals or of individual tissues, or of individual enzyme activities after hormone injections or after removal and/or re-implantation of the gland. The question of whether thyroxin also stimulates metabolism in *in vitro* experiments has been particularly disputed. PRECHT (1965) could not confirm the often cited findings of HAARMANN (1936) on carp, but newer, positive findings have been made, using other animals (e.g., isolated toad urinary bladder, according to THORNBURN and MATTY, 1966; cf. MARUSIC et al., 1966; TAYLOR and BARKER, 1967; for steroid hormones see PORA et al., 1967).

This problem is less important for our considerations because it touches only on the type of effect (direct or secondary) exerted by thyroxin. It must be remembered that the various tissues can respond differently to the hormone (cf. MAHER, 1964) and that the response can change with the season. The effect of hormone injection depends on the environmental temperature in lizards (MAHER, 1965; WILHOFT, 1966), in goldfish (THORNBURN and MATTY, 1963), and in *Rana pipiens* in contrast to *Bufo woodhousii* (MAHER, 1967). According to HOCHACHKA (1962) high concentrations of thyroxin are said to influence the pathways for carbohydrate metabolism by increasing the activity of the HMP-shunt.

The question of whether AT can affect thyroid activity in ways that can be investigated histologically or through measurements of function (e.g., of ^{131}I uptake) has stood in the forefront of research. The conclusions from such attempts have not always been drawn correctly (see PRECHT, 1964a). If the thyroid were responsible for metabolic compensation of type 3, its epithelial cells would have to increase (in height) with falling AT: the measured function at equal ETs decreases with rising AT, but in AT = ET experiments the function would have to grow as the AT rises (cf. DRURY and EALES, 1968).

A very inconsistent picture has emerged from studies on fish. Thyroid activity can rise with AT, fall with it, or be fairly independent (see GORBMAN in HOAR and RANDALL, 1969 II, p. 258; McNAPP and PICKFORD, 1970; McERLEAN and BRINK-LEY, 1971)[47]. Cold-acclimated amphibians very often have higher thyroid gland activity than warm-acclimated ones, but the opposite is true for some reptiles (Literature in PRECHT, 1964d; cf. LYNN et al., 1965; DRURY and EALES, 1968; also BANERJEE and DE, 1969). Species differences between goldfish and trout are also found in the effect of thyroxin on NH_3 production (THORNBURN and MATTY, 1963). Seasonal changes in thyroid activity that arise in connection with the spawning season or migration in fish have been observed often, but factors other than temperature are also involved (such as photoperiod, see SINGH and SATHYANESAN, 1968; YARON, 1969; CHIU et al., 1969)[48].

The remarkably inactive thyroid of the goldfish is not affected by AT; nevertheless it responds to treatment with TSH, which would indicate that this pituitary hormone does not depend on AT either (FORTUNE, 1956, 1958; DELSOL and FLATIN, 1956)[49]. KLICKA (1965) could find no indication of participation by the thyroid in goldfish capacity adaptation. In *Umbra limi* hypophysectomy had no effect on the time-course or degree of compensation of the metabolism following changes of AT (HANSON and STANLEY, 1970).

Positive findings were reported by SUHRMANN (1955) for the oxygen consumption of crucian carp and by JANKOWSKY (1960) for the participation of the pituitary-thyroid system in compensation of respiration in *Rana temporaria*. Suppression of hormone production by thiourea or thyroidectomy can lead to suppression or a substantial reduction in the extent of compensation (see JANKOWSKY, 1964b; but also GALTON and INGBAR, 1962). Anti-thyroid agents can of course have undesirable side effects. More species would have to be investigated (compare AUERBACH, 1957; KASBOHM, 1967).

It may be mentioned once more that hypophysectomized *Fundulus heteroclitus* become very anemic at an AT of 10° C. Transfer to 20° C reverses this condition. Treatment with cortisol does not prevent the pancytopenia at 10° C (SLICHER and PICKFORD, 1968; on the effect of temperature on adrenal function, cf. SCHMIDT and SANTA, 1969; on the effect of parathormone on amphibia at different ATs, cf. McWHINNIE and CORTELYOU, 1967).

Histological changes in the corpora allata related to changes in AT have been noted. In *Locusta migratoria* subjected to fluctuating temperatures, the corpora allata show increasing signs of hyperactivity and exhaustion as the amplitude of the temperature fluctuation is increased (CLARKE, 1966, in ROSE, 1967, p. 328).

---

[47] SMITH, C. D., EALES, J. G.: Can. J. Zool. **49**, 783—786 (1971).
[48] cf. BELSARE, D. K.: Zool. pol. **21**, 17—24 (1971).
[49] cf. PETER, R. E.: Gen. comp. Endocr. **15**, 88—94 (1970).

II 1b. Inflection Points and Descending Parts of the Curves

So far only the ascending part of the ET-dependence curve, the only one to be
widely investigated, has been discussed. When the inflection point lies at very high
temperatures and the descent (as in the case of denaturation processes) is steep,
it can be used at most as a measure of heat resistance. But the descent above the
inflection point is often not much steeper than the rate of increase below it, e.g.,
in Fig. 16 of the first edition (relation of light intensity of luminescent bacteria to
ET, taken from ROOT), or in the EEG frequency of the frog (LOCKER, 1962; cf.
MURRAY, 1966). A very gradual decline is often found in curves relating motion or
oxygen consumption at specific speeds of motion to ET. As in Fig. 4 the position
of the inflection point of such curves often depends on the AT. High levels are seen
here for the AT = ET curves. In the curves relating ET to the endogenous activity
of isolated *Periplaneta* nerve cord the maxima were precisely at each AT (KERKUT
and TAYLOR, 1958; cf. D'AJELLO et al., 1967). In the curve representing ATPase
activity of *Periplaneta* and *Musca*, the level of enzyme activity at the inflection
point, but not its location on the ET scale, depends on AT (Fig. 3 of MUTCHMOR in
PROSSER, 1967; cf. USHAKOV and KUSAKINA, 1961).

Attempts have been made to discover the extent to which protein denaturations
are responsible for the decline in organ and cell functions above the inflection
point. There are unequivocal cases in which the damage caused by high temperatures is recognized through the presence of two μ values, of which only the second
indicates denaturation by its magnitude.

If one measures the oxygen consumption of a tissue (e.g. frog muscle), first at a
normal ET, then at a high ET, and then returning to the original temperature, one
sees that the inflection point of the respiratory curve is caused by reversible
denaturation processes, while the decrease in respiration during the third part of
the experiment, compared to the first, results from irreversible denaturations that
attack a different enzyme (PRECHT, 1960b; OHLENBUSCH and PRECHT, 1960). The
position of the ET optimum of LDH activity in *Lepidosiren* is influenced by the
substrate, and its upper limit is determined by inhibitors (HOCHACHKA, 1968b).

Organ functions with Q_{10} values below 1 have been observed, which were often
descending slopes lying above inflection points at relatively low temperatures.
According to OHNESORGE and SCHMITZ (1968) the height of contraction of trans-
versely striated gut muscle of *Tinca tinca* goes up with increasing ET after the
addition of acetylcholine, but that of smooth muscle goes down. A declining slope
of a curve that extends over many ETs can also arise from inhibitory effects of a
control system, if these effects have greater Q_{10} values than those of the inhibited
organ functions.

The maximum point of an ET curve can vary with the experimental conditions,
e.g., the duration of the experiment. The inflection point of the digestive proteinase
of *Tethyum* (tunicata) is displaced to lower ETs in this way by longer incubation
periods due to the greater extent of inactivation (cf. FLOREY, 1970, Figs. 11—6).

The effect of high and low pretest temperatures can lead to increased enzyme activity
before an additional temperature increase begins to inactivate it. According to RUD-
NICK (1967) this occurs in the case of *Tubifex* catalase because the particle-bound
enzymes are released or made more accessible (cf. JANKOWSKY, 1967).

The interpretation of other authors is that enzyme activity is increased by the inactivation of a specific enzyme inhibitor or by loosening of labile bonds, by which more active groups are set free to become accessible to substrates (lit. in RUDNICK).

The inflection points of the curves can (even for activities of the same enzyme) be related to the habitat of the species. This is to be expected, particularly if the maximum can be seen as a measure of heat resistance (cf. e.g., LICHT in PROSSER, 1967), but in a completely displaced temperature range it may also be related to the lower limiting temperature (MUTCHMOR and RICHARDS, 1961). Even the microclimate of the species may be reflected, though the AT must be considered in such investigations (IVANOVIČ et al., 1968). In lizards the preferred body temperature of the species may also play a role. The heat resistance of different muscles of one animal may be equal; nevertheless the maximal isometric twitch tension may be different (LICHT et al., 1969; compare SCHMIDT-NIELSEN in DILL et al., 1964). In some cases there is no relation to the resistance of the animal: According to KIRBERGER (1953) the inflection point of the respiratory curve of *Eisenia foetida* exhibits no relation to AT but the heat resistance of the animal does. Acclimation of *Uca notata* to different ATs altered the thermostability of the skeletal muscle slightly but had little effect on the optimal temperature of contractibility (LICHT, 1964).

The position of the inflection point of oxygen consumption in *Formica polyctena* differs for ♂♂, ♀♀, ☿☿, and for different stages of development (SCHMIDT, 1968).

Organ functions as well as enzyme activities can have broad inflection points (e.g., MAKAREWICZ, 1968). The cirral beat of *Balanus improvisus* is maintained over a broad temperature range up to an upper limit, and then decreases, sooner or later, depending on the AT (PRECHT, 1949).

II 2. Changing Reaction Systems

H. LAUDIEN *

A. Effect of Temperature on Processes of Growth and Development

1. Constant-Temperature Conditions

Like all physiological processes, animal development takes place only within a definite temperature range. In order to describe the process one normally measures the time required by an animal to change from one stage to another at a given temperature. This development need not necessarily involve growth, i.e., increase in weight or volume (e.g., egg development), but it is related in all cases to a change in reaction systems.

If the temperature remains unchanged during the experiment, we are dealing with fully adapted organisms. Any adaptational phenomena which may be present will then be included in the measurements.

a) Lower Limiting Temperature

A temperature limit exists, below which an organism can survive, but cannot grow or develop. In the latter case, this limit is designated as the developmental null point. The location of this point on the temperature scale is difficult for several reasons. It is not easy to make exact determinations of processes which operate very slowly at low temperatures. Therefore SHELFORD (1927) suggested that the temperature at which development is first measurable should be taken as the null point. But this procedure is not advantageous, since through its use the value becomes dependent on the precision of the method of observation. Therefore the developmental null point is usually determined by graphical or arithmetic means. If one regards the speed of development in the normal temperature range as linearly dependent on temperature and one extends this slope to that co-ordinate value at which the speed is zero, then one obtains a temperature that can be designated as the developmental null point. But it must be remembered that the speed of development in general is not linearly dependent on temperature, especially in the region of the developmental null point. Rather, the curve approaches the developmental null point asymptotically (e.g., in the case of the caterpillar of the noctuid moth, *Heliothis zea*, MANGAT and APPLE, 1966). So it is possible for one author to determine a developmental null point of an animal species graphically, while another rears the same species at a still lower temperature.

---

* I am indebted to Mrs. M. E. VOGT for her assistance in proofreading.

The developmental null point of different stages of the same species can have different values. ZWÖLFER (1935) found a developmental null point of 3.2° C for the first larval stage of *Lymantria monacha*, 5.7° C for the second larval stage, 7.2° C for the third, 7.6° C for the fourth, 7.8° C for the fifth, and 8.4° C for the pupal stage. These results are consistent with the fact that larvae of *Lymantria monacha* in central Europe develop from April to June, that is at increasing environmental temperature. If development takes place in autumn the developmental null points of successive stages can form a decreasing series. This is true for *Bupalus piniarius*, whose third to fifth stages appear from August to October. Their null points fall at 24.5° C, 22° C and 18° C, respectively. (OLDIGES, 1959; cf. BLUNCK, 1923; HURPIN, 1955; for literature on various amphibia, see GERBERS, 1957; GANAROS, 1958; HAMDORF, 1961; KITTLAUS, 1961; LØVTRUP, 1961; KUSNETZOVA, 1969; HASEGAWA and CHIBA, 1969.)

In these cases increases in temperature above the developmental null point allow development to proceed. Below this point development ceases or is unmeasurably small. But it is also possible for higher temperatures to act as a stimulus as well as having an accelerating effect, so that subsequent development can proceed even at temperatures which would previously have been unfavorably low. According to WIESER (1965a), females of *Porcellio* do not begin to form broods below 8.6° C. But if they are briefly brought up to 20° C and then returned to 8° C the brood cycle has begun and continues, even though slowly, till the larvae emerge (cf. p. 454).

The environmental null point can be related to the geographical distribution of the species. For example the minimal temperatures for development of some species of anurans are precisely related to the water temperature during the reproductive season in the natural habitat of their brood (BALLINGER and McKINNEY, 1966). The species with the most northerly distribution had the lowest minimal temperatures. The minimal temperature of the antarctic fish *Trematomus bernacchi* is especially low. This species lives almost constantly under ice at $-1.9°$ C (WOHLSCHLAG, 1961).

In practice it does not matter greatly if different stages of e.g. harmful insects have different developmental null points. It is more important to know whether the animals can complete their development in a large proportion of cases at a given temperature.

Environmental changes caused by man, e.g. the warming of waters by the discharge of industrial cooling water, play a role in such processes. For example the development of the wood-boring isopod *Limnoria* is made possible in winter and is speeded in summer by the warming of water in harbors (ELTRINGHAM, 1967, cf. GIBBONS, 1970[1]).

The designation of a given temperature as the developmental null point is also complicated by the fact that such low temperatures may have effects on later stages. In practice therefore it is important for the development of eggs of harmful insects not only that embryonal development can proceed, but also that the animals can emerge from the eggshells and possess a sufficiently high vitality (HODSON and AL RAWY, 1958). Also the animal's fertility must not be injured too much by low temperatures in the course of development.

[1] GIBBONS, J. W.: Can. J. Zool. 48, 881—885 (1970).

The developmental null point can be affected by various other factors (relative humidity: e.g., HOWE, 1956, CANCELA DA FONSECA, 1958, SCHWERTFEGER, 1963; AMMAN,
1968; Salt content and oxygen supply: KINNE and KINNE, 1962, and many others.)

b) Central Temperature Range

Temperatures above the developmental null point increase the rate of development
up to an optimum (cf. for an unusual case of temporary temperature independence
the development in *Brachydanio rerio*, SCHIRONE and GROSS, 1968).

Usually an almost linear relation is found, over a wide temperature range, in which
the slope of the line reflects the closeness of the correlation between temperature
and developmental speed. This slope can have great biological significance. The
larvae of the lice *Eurygaster maura* and *Palomena prasina* hatch at the same time
of year (TISCHLER, 1939). But the first imagos of *E. maura* appear at least one
month earlier than those of *P. prasina*, because increasing temperatures speed the
development of *E. maura* more than that of *P. prasina* (cf. HOWE, 1962). So, each
species and even each race (HEUTS, 1956) and each developmental stage as well
can be distinguished by a curve of specific shape (cf. ed. 1, p. 86, HURPIN, 1955;
SINGH, 1962). For example, the first developmental stages of *Balanus balanoides*
(cleavage to formation of limb anlage) are strongly temperature dependent up to
ca. 14° C, while later development is independent of temperature between 3° C
and 12° C (CRISP, 1959). In the cottoid fish *Chinocottus analis*, gastrulation, circulatory system formation, hatching, and melanophore formation are especially
temperature dependent (HUBBS, 1966). After planaria are decapitated low temperatures have far less effect on the migration of neoblasts towards the incision site
than on the biochemical processes needed for the differentiation of the eyes
(BRØNDSTED and BRØNDSTED, 1961). *Bombyx mori* eggs were damaged less by
radiation if they were subsequently treated with cold (PAULOV, 1961).

Few observation have been made till now on whether, at given temperatures, successive
stages require different proportions of the total development time, or whether these
are always the same. In figures in which the durations or rates of development from
the time of egg-laying to the various stages are superimposed as a family of curves,
such values can be read off directly (e.g., QUEDNAU, 1957; HILTERHAUS, 1965). In the
development of *Ctenicera destructor* the time of water absorption depends on the stage.
But in relation to the total duration of embryonic development the length of the water
absorption period is relatively constant at all temperatures, and comprises ca. 4.4 to
4.7% of the total duration (DOANE, 1969).

c) Optimum Temperature

The highest rates of development are found at temperatures that are still relatively
normal. If the duration of a developmental stage or of the total time of development to the imago is measured, a curve is obtained with a minimum at the "optimal" temperature. The optimal temperatures may differ for each species or race
(PRESCOTT, 1957), for each developmental stage (KHAN, 1965; SWIFT, 1965;
KARANDINOS and AXTELL, 1967a) or for different individual processes as well (e.g.,
VAUGHN, 1953). In the case of harmful insects, the various physiological optima (optimum for most rapid development, for greatest hatching number, for lowest mortality,
etc.) are less significant for massive outbreaks than an ecological optimum is, in

which the various criteria are included with different weights. It would be useful to specify such an ecological temperature optimum into which the criteria to be observed in animal subjects, such as the rate of development, mortality, fertility, resistance to diseases, temperature dependence of behavioral patterns, etc. would enter: For *Gadus macrocephalus* a salinity of 15%, temperatures of 3–5° C and a wide range of oxygen concentrations were optimal (ALDERDICE and FORRESTER, 1971[2]). Until now, only the rate of development at the temperature of lowest mortality has often, somewhat arbitrarily, been designated as the optimum (cf. THOMPSON, 1959; COSTLOW et al., 1962; BERRYMAN and STARK, 1962; BURGES and CAMMELL, 1964; O'BANNON et al., 1966; STURROCK, 1966; LAKE, 1969).

This ecological temperature optimum becomes specifically graphic if the measured values are demonstrated through single points of a surface (ALDERDICE, in KINNE, 1972[3]).

In this sort of graphic presentation the reaction of the animal in dependence to two environmental factors (e.g. AT and ET) is shown very clearly. The dependence to many environmental factors can be demonstrated mathematically. In this connection it is important that various reactions of the animal can be projected one upon another; thus an ecological temperature optimum becomes visible.

d) Supraoptimal Temperatures

There is a zone between the optimum and the upper limiting temperature in which a relatively gradual decrease in the rate of development can be observed. It is difficult to decide whether this slowing is due to inhibition or to injury. Injury may take place over a broader temperature range, but its effect may be reduced by processes of repair.

If increasing the duration of an experiment does not lead to a constant slowing of the rate of development, several causes may be responsible for this effect. An inhibition may be present at a constant level. The optimum may have been displaced during the course of development or by acclimation. The establishment of a new equilibrium between injury and repair is also possible.

The upper limiting temperature is determined by the heat sensitivity of the species. Theoretically a temperature range must exist in which development remains unchanged. In contrast to the lower temperature zone however, this range in the warm end of the scale can either not be determined at all because all the animals die, or else it can be determined only with difficulty because it is very narrow and can differ for different processes in development. For example the eggs of *Leucania separata* can still develop to various extents at 35–41° C, but the larvae cannot emerge (AN-KUO et al., 1965). As for the lower limiting temperatures, so for the upper ones, there are adaptations to the natural temperature regimes. The upper limit for embryonal development of *Scaphiopus hammondii*, which lives in dry habitats, is 32.5° C. In the field the sensitive stages of development are passed before the seasonal increase in temperature takes place. In later stages the upper limit, which is also shifted through adaptation, lies at 39–40° C (BROWN, 1967).

---

[2] ALDERDICE, D. F., FORRESTER, C. R.: J. Fish. Res. Bd. Can. 28, 883—902 (1971).
[3] See p. 336.

e) Mathematical Formulation of the Relation between Rates of Development and Temperature

The observation that the relation of rates of development to temperatures can be presented as a regular curve, has led to attempts to formulate this relation mathematically. When one considers that a large number of different physical and chemical processes go into developmental phenomena, it is surprising that the relations are regular and that simple mathematical formulae can reproduce the course of the curves fairly exactly. These formulae are of interest in practice because they make it possible to determine the course of the curve from a few measurements. Therefore they need not necessarily serve to aid in understanding the processes (empirical formulae, DAVIDSON, 1944). On the other hand if relations are to be explained, and the processes are to be traced back to physical or chemical regularities, then such formulae must be distinguished from the empirical ones usually used with biological processes.

Arrhenius' equation. This equation states that $\ln k$ changes linearly with $1/T$ (k = rate constant, T = absolute temperature). The energy of the reacting material must exceed the energy amount of the initial material by a specific minimal amount; this is the activation energy (μcal/mole) which is temperature independent. Therefore simple chemical reaction steps, whose chemical conversion equations also indicate their mechanism, can be characterized by their activation energies. Even well-known reaction chains can have overall constant μ values. It is remarkable that complex biological reaction chains that have not been analyzed in detail also yield linear relations if the natural logarithms of the data points are plotted against $1/T$; though different temperature ranges may be characterized by different μ values. The probability must be still smaller that growth development processes, i.e. reaction systems that change in time, should yield straight lines when $\ln k$ is plotted against $1/T$. BLISS (1926) reports three μ values, which decrease with increasing temperature, for the prepupal development of *Drosophila melanogaster* over a temperature range of 12—30° C. The reproductive rate of single *Tetrahymena* cells between 17.5° C and 27.5° C is based on an activation energy μ = 14700 cal/mol. Below 17.5° C the value of μ = 32000 cal/mol (THORMAR, 1962). But it should be noted that μ values of developmental processes certainly do not characterize specific rate-determining reactions.

The RRT Rule. The reaction-rate-temperature rule was formulated by VAN'T HOFF (1884). The Q_{10} values which this formula yields are not suited for the characterization of chemical reactions because the rate increase goes down with increasing temperature even if the μ value remains constant. If there are no special requirements for precision, the RRT rule can be used in the approximately linear region below the optimum. In this case one usually finds that the rate of development increases 2—4 fold for a temperature increase of 10° C. It should be noted that this holds true for animals that are fully acclimated to a given temperature, because in these experiments the corresponding rates of development were usually determined by rearing at constant temperatures (AT = ET curve). If an acclimation were more prevalent or had a stronger effect, then such experimental procedures should lead to lower Q_{10} values (in case of a type 2 effect even a Q_{10} = 1). In investigations of constant systems ET curves were established principally, and examination of the effect of acclimation temperatures was neglected. The influence of ET-changes in changing reaction systems, i.e. during growth and development, has been investigated seldom (cf. p. 375). This omission is not too critical, since acclimation processes — recognizable by Q_{10} values corresponding to the van't Hoff rules — play only a subordinate role. It would be interesting to examine the growth or the development of animals that exhibit an extensive capacity adaptation with respect to energy metabolism.

Temperature Summation Rule. This rule is still widely used today in applied zoology, since it is very easy to apply and yields approximately correct values. The prediction of the spring emergence of adult coddling moth (*Laspeyresia pomonella*, TOUZEAU, 1966) may be given as an example. As soon as the daily average temperature in spring exceeds 10° C, the "effective temperature", i.e., the temperature obtained by subtracting 10° C from the daily temperature maximum, is noted. In this case 10° C is the developmental null point. If the sum of the effective temperatures of successive days reaches 400, the beginning of the adult hatching season may be expected. Some authors (BLUNCK, 1914; PEAIRS, 1914 i. a.) have based the dependence curve on a hyperbolic function. At constant temperatures in the laboratory the temperature summation rule:

$$t(T - K) = C$$

is valid (T = experimental temperature, C = thermal constant, t = developmental time, K = temperature at the developmental null point). This curve represents a hyperbola (cf. QUEDNAU, 1957). But it is incorrect to designate this product as the quantity of heat needed for development or the given formula as that of the heat-summation rule (UVAROV, 1931). A heat quantity must be measured and presented in calories. Furthermore an organism cannot take up heat from an environment in which its own body temperature prevails. Therefore the designation of temperature summation rule is better. If one wants to calculate the duration of development, the transposed formula reads:

$$t = \frac{C}{T - K} \; .$$

This formula was presented by REIBISCH (1902) for the development of fish eggs.

This rule is only valid in the central temperature range, the so-called favorable range in which rate of development depends on the temperature in linear, and the duration in hyperbolic, fashion (e.g. MANGAT and APPLE, 1966, but cf. KEIZ, 1959). Deviations at lower temperatures and at the optimum are not included. The validity in this limited range has been determined frequently (e.g., BLUNCK, 1923; BONNEMAISON, 1946; GOODON, 1956; WYNIGER, 1956; LARCZENKO, 1958; KAUSKOLEKAS and DECKER, 1966; CHIANG and SISSON, 1968; ARIDOV and PODOLER, 1968; ANNILA, 1969; HASEGAWA and CHIBA, 1969; MAXWELL and PARSONS, 1969; SUSKI, 1969, and many others). In many cases however the conclusion had to be reached that successive stages displayed differing thermal constants. For example HARRIES (1944) found that the earlier and later stages of embryonal development in *Eutettix* were dependent on temperature to different degrees. According to ZWÖLFER (1934) the formula is applicable to the development of the nun-moth *(Lymantria monacha)* only within a definite range, which changes with the stage of development. The rule is therefore valid for the total course of development only over a quite narrow temperature range. Different values can also be found for the temperature sums of development of the two sexes (BONNIER, 1926; KOZHANTSHIKOV, 1947). Furthermore, care must be taken that the duration of equivalent segments of development be compared. LILLELUND (1967) found that in the development of larval pike hatching was increasingly delayed as water temperature decreased, because at low temperatures the larvae first left the egg capsule at an advanced stage (cf. LECYK, 1965).

BĚLEHRÁDEK's *Formula*. The curve of developmental rate is most truly linear only in a central range, and with the usual scatter of data points it is difficult to determine whether an exponential or other type of relation may not fit the experimental results better. A formula which is valid over a somewhat wider temperature range reads (BĚLEHRÁDEK, 1935, 1957):

$$V = \frac{t^k}{K} \qquad \text{or} \qquad T = \frac{K}{t^k}$$

(V = rate, t = temperature, T = duration, K and k are constants).

When logarithms are taken the curve becomes a straight line ($\log V = \log K - k \log t$). This formula gives results that agree well with observed data (cf. MCLAREN, 1963; LASKER, 1964). MCLAREN (1965) objected to the classification of Bělehrádek's formula as an empirical formula.

Logarithmic Curves. The logarithmic formula originated by VERHULST (PEARL and READ, 1920; DAVIDSON, 1942, 1944):

$$V = \frac{K}{1 + e^{a-bt}}$$

(V = rate, t = temperature in °C, K, a, and b are constants, e = base of natural logarithms). This formula yields an S-shaped curve in which K corresponds to the upper asymptote; b and a are measures of the slope of the curve, and of the relative position of the origin of the curve on the abscissa, respectively. The constants K, a, and b are easy to determine if the formula is rewritten as:

$$\log e \frac{K - V}{V} = a - bx.$$

A straight line results, from which K, a, and b can be derived easily (PEARL, 1930; DAVIDSON, 1944). The data points match the course of the curve quite well. Nevertheless in the temperature range above the optimum, increasing divergences between the experimental results and the curve were found. Since the delaying effect that sets in at very high temperatures was not taken into account, the maximum temperatures observed were well below K. For example the optimum for the development of *Drosophila* eggs lies at 29.5° C and the value for K is 42° C. DAVIDSON (loc. cit.) believed that this deviation in the upper temperature range has no significance, since the section of the curve involved is often short, and higher temperatures usually cause injury over longer rearing periods (cf. BROWNING, 1952 a).

Catenary Formula. All of the formulae presented so far deal only with the region of the curve below the optimum, i.e. with the accelerating effects of temperature. If one wants to consider the retarding effects of temperatures above the optimum, which need not result from damage in all cases, JANISCH'S (1925) suggestion can be followed. This author starts from two exponential equations that describe the acceleration and retardation of development, respectively:

$$Y = m \cdot a^t \quad \text{or} \quad Y = m \cdot a^{-t}.$$

The combination of the two exponential curves results in a catenary:

$$Y = m/2 \, (a^t + a^{-t})$$

(Y = duration of development, m = duration of development at the optimum temperature, t = difference between temperature and optimum, a = constant).

The rate of development can be calculated as the reciprocal value of Y. The catenary gives a very good agreement between the data points and the course of the curve (ANDERSEN, 1930; RATHJEN, 1939; LARSEN and THOMSEN, 1940; QUEDNAU, 1957). But it has the disadvantage of being somewhat difficult to apply, since the developmental optimum and the duration of development at the optimum must be determined. But all empirical formulae — and the catenary is one of them — are important essentially for practical uses (DRUZHELYUBOVA and MAKAROVA, 1968). That is why the simple temperature summation rule is used most often for determining the duration of development. For this purpose it is not too significant whether it is an exact method or whether it gives only approximate values (cf. BRITZ and HÖHNE, 1955).

2. Effects of Variable Temperature on Processes of Growth and Development

When constant experimental temperatures are used, one is always dealing with animals that are fully acclimated to the temperature. By contrast, in the overwhelming majority of cases in nature, we find fluctuations in temperature (exceptions are found in caves, in groundwater, and in the deep seas), and it may be asked whether rearing at constant temperatures is not unbiological, in the sense

that it does not allow any assertions to be made about the environmental requirements of a species. Therefore experiments with changing temperatures are very
important.

a) Adaptational Phenomena of Developmental Processes

The simplest experiment involves a single change in experimental temperature. If
initiation or continuation of development is possible in this case, then the temperature change must have occurred between the limiting temperatures. If the results
of such experiments are compared with those in which the animals were kept constantly at the second temperature, one can determine whether an adaptation to
the first temperature has taken place. RYAN (1941) carried out such experiments.

He maintained eggs of *Rana pipiens* for the entire duration of development at a medium
temperature (controls). Other eggs were placed in a high or a low temperature at first,
and were moved to the medium temperature only after they had developed to a certain
stage. A comparison with the control animals showed that the eggs that were reared
first at the higher temperature developed more slowly at the medium temperature,
while the eggs pre-treated with low temperatures later developed faster than the controls. In this case a temperature adaptation corresponding to types 1—3 (cf. Constant
Systems, p. 334 f.), is involved (cf. ed. 1, p. 99 ff.). PIATT (1971)[4] investigated the capacity
of regeneration of *Ambystoma* embryos. Two animal groups, maintained at 5 and 14° C,
respectively, were cultivated after extirpation of the right otic vesicle or the right
forelimb disc in 18° C. In the 5° C test group the reconstruction of the extirpated parts
was better. This indicates temperature adaptation.

There are still many gaps in our understanding of such adaptational phenomena
of developmental processes. Therefore nothing definitive can be said yet about the
occurrence of such adaptations (cf. Discussion, p. 375).

b) Temperatures which Change Regularly during the Day

Temperatures which change during the course of the day resemble natural conditions more than a single change during the entire course of development does.
Therefore the results of such experiments are very important for practical use. In
such experiments animals as a rule are presented with a change in a 24-h rhythm,
e.g. 12 h of light and 20° C, 12 h darkness and 10° C, and the results are compared
with those of experiments in which the animals were presented with a constant
temperature of 15° C, i.e. the average temperature. The combination of light with
warmth corresponds to the natural conditions more than the combination of
darkness and warmth. The results of experiments with changing temperatures vary.
BRITZ and HÖHNE (1955) found an acceleration of development for *Anopheles
atroparvus*, as did MAKSIMOVIČ (1958) for *Liparis dispar*, CLARKE (1967) for *Locusta
migratoria*, (4th larval stage and imago) using temperature changes from 20—40° C,
REMMERT and WÜNDERLING (1970) for *Drosophila*, and BRAUNE (1971[5]) for *Leptopterna dolobrata*. Collembols react positively to changing temperatures (JOSSE and
VELTCAMP, 1970)[6]. Delays in development were found in experiments with *Adalia
bipunctata* (ELLINGSEN, 1969) and *Aedes sticticus* (TRPIŠ and HORSFALL, 1969).

[4] PIATT, J.: J. Embryol. Exp. Morphol. **25**, 339—345 (1971).
[5] BRAUNE, H.-J.: Oecologia (Berlin) **8**, 223—266 (1971).
[6] JOSSE, E. N. G., VELTKAMP, E.: Neth. J. Zool. **20**, 315—328 (1970).

In many cases the speed of development in changing temperatures did not differ from that at a constant temperature (*Trichogramma:* QUEDNAU, 1957; *Cordylophora:* FULTON, 1962; *Lygus hesperus:* CHAMPLAIN and BUTLER, 1967; 5th larval stage of *Locusta migratoria:* CLARKE, 1967; boll weevils: FYE et al., 1969; for earlier authors, see ed. 1, p. 95 ff.). Partial development only was observed in *Harpagoxenus sublaevis* exposed to changing temperatures (BUSCHINGER, 1966). In connection with this subject KAUFMANN (1932) called attention to some theoretical aspects related to the influence of constant or changing temperatures on rates of development.

According to the results described on p. 357, there is an optimum for developmental rate. Every deviation from this optimum (towards lower or higher temperatures) must make itself felt through a retardation of development. The designation of an average rearing temperature is therefore not sufficient. The degree of fluctuation must also be designated, because development is slowed in proportion to the extent of the temperature fluctuations. These considerations hold true for temperatures in the optimum range.

If temperatures are chosen in the range in which the rate of development depends on the temperature in an approximately linear fashion, then a theoretical difference exists between the developmental times calculated on the basis of a hyperbola on the one hand or of a catenary on the other. Proceeding from the hyperbolic formula, a regular change of e.g., $\pm$ 5° C around an average temperature leads to almost the same duration of development as rearing at the constant average temperature. A slight divergence is smaller than the constantly present scatter of data points (cf. QUEDNAU, 1957).

By contrast, another theoretical result is obtained on the basis of the catenary formula. KAUFMANN points out that, when the rearing temperature varies between two values, the calculation of the duration of development according to the exponential curve must necessarily lead to an acceleration of development in comparison with a constant average rearing temperature because a temperature increase cannot be cancelled out by a temperature decrease of the same amount.

These considerations are valid for the so-called favorable zone of the species. If the average temperature chosen is so low that practically no development is possible, then a regular fluctuation around this average temperature causes development to proceed at the high and stand still at the low temperature. Overall, the processes run their course, provided that the low temperatures have no damaging effect.

These theoretical considerations must be taken into account, if one wants to judge whether a temperature fluctuation in an experiment had an effect on the duration of development (KAUFMANN, 1932, p. 355, gives an introduction to the calculation of theoretical development time). The varying results of different experiments are probably produced in part by the different ways in which temperatures are chosen. KAUFMANN'S considerations have often not been heeded in judging the results of changing temperatures. Taking these effects into account, GRAINGER (1959) and KHAN (1965) found that changing temperature had no effect on embryonal development of *Rana temporaria* and *Acanthocyclops*, respectively. BRITZ and HÖHNE (1955) found an acceleration in the case of *Anopheles atroparvus* and MESSENGER and FLITTERS (1959) found some acceleration and some retardation in three different species of flies.

c) The Effect of Short-Term Treatment with Extreme Temperatures

Relatively high or low temperatures are often part of the natural environmental conditions to which a species is exposed. For example, *Locusta migratoria*, which inhabits tropical or sub-tropical regions, is exposed to very low temperatures at night, which have no inhibitory after-effects on development during the warm hours of the day (HUNTER-JONES, 1970). But if the time of exposure to low temperatures is extended to days, an increasing inhibition of development occurs (AHMAD, 1936; cf. HASE, 1927, 1930; WEISSMANN-STRUM and KINDLER, 1962).

Acceleration of development through the effects of low temperatures has been observed seldom, apart from the phenomenon of diapause. One example was given by PARKER (1930), who maintained eggs of the grasshopper *Melanoplus mexicanus* at 0° C and then moved them to the favorable temperature of 27—37° C. Eggs treated in this way developed more rapidly than those that were kept at high temperatures through the entire period (cf. BODINE, 1925).

As a rule high temperatures have an inhibiting effect on development. Even a temperature of 33° C for one day resulted in a delay in development of pupae of *Drosophila melanogaster* during subsequent rearing at 25° C (LUDWIG and CABLE, 1933). The two sexes reacted somewhat differently. Male pupae were most sensitive at an age of 1—3 days, females at 1—2 days (cf. JANISCH, 1930; HACKBARTH, 1939; MISSONIER, 1960). The effectiveness of high temperatures in delaying molting of *Rhodnius prolixus* is related to its uptake of nourishment (OKASHA, 1968).

It is often quite difficult to explain the harmfulness of high temperatures, because many different factors may exert an influence. MELLANBY (1938) was able to make a relatively simple relation seem probable. He examined the development of the golden fly *Lucilia sericata* and found that at 25—37° C the animals developed only to the prepupal diapause stage, because at this high temperature the pupation hormone was destroyed. If the temperature is lowered, puparia are formed after some time (cf. LARSEN, 1943; SAXENA, 1966). If the extreme temperatures affect the animals for a longer time the co-ordination between development and food supply can be disturbed.

In developing eggs, yolk will be used to maintain metabolic processes and structure even if the embryos are no longer developing because of low temperatures (KITTLAUS, 1961; cf. RICHARDS, 1965). Such low temperatures will therefore be tolerated for only a short time since otherwise the yolk supplies will not suffice for later development at more favorable temperatures. On the other hand, too large a yolk supply is also unfavorable. Such a relation can arise if the embryo develops too fast because temperatures are too high. This disturbance of the relative position of yolk and cellular area slows the rate of development and can also lead to developmental malformations (KITTLAUS, loc. cit.). In the temperature range of normal development, e.g., between 20° C and 30° C for *Leptinotarsa decemlineata*, yolk consumption and embryonic development are so mutually balanced that differentiation can proceed without disturbances. Short-term deviations from these temperatures can be tolerated without damage, especially if the temperature change does not take place too rapidly (cf. p. 386).

Since metabolism goes on at low temperatures which can still be tolerated, but which no longer allow development, oxygen use can be measured. Some authors have found that oxygen use is greater at low temperatures than at medium ones. Furthermore RICHARDS and SUANRAKSA (1962) observed a reduction in oxygen use and in weight loss when the environmental temperature was alternated between a very low temperature and one in the comfortable range, as compared to a situation in which the animals were always kept at a constant medium temperature. A daily transfer of only four hours to the temperature in the comfortable range ameliorated the development-delaying and oxygen use-increasing effects of low temperature. Apparently development-inhibiting factors exist which can be destroyed in a short time at favorable temperatures, so that development can then proceed even at low temperatures (cf. ed. 1, p. 93, RICHARDS, 1959).

The effect of irregularly changing temperatures has been examined very rarely. GRAINGER (1959) did so, working with *Rana temporaria* eggs, and KHAN (1965) with *Acanthocyclops viridis*. GRAINGER first measured the duration of development of certain successive stages at constant temperatures, and then after a single temperature change. The development times were in agreement in these two cases. Then he determined the development times when temperatures were changed with complete irregularity and compared them with calculated values for development times which would have been obtained if only temperature itself, and not the change in temperature, had had an effect. The comparison with the experimental values revealed a *significant delay* in development during gastrulation. KHAN (1965) conducted similar experiments. These revealed a slight increase (12.2%) in the rate of development from egg laying to the end of the fifth naupliar stage. Apparently temperature did not affect the rate of development in this case, but functioned as a stressor, to which certain segments of development reacted very intensively.

d) On the Applicability of Laboratory Experiments to Field Conditions

The question of whether, or to what extent, all these laboratory investigations can be transferred to conditions in the field is very important for applied zoology. It may be asked which field temperature should be used as a basis for comparison. In addition there is the difficulty of obtaining adequate field data. Weather stations usually record only the air temperature, but terrestrial animals are frequently exposed to completely different temperatures. The difference is still greater if the animals live within the soil. These deficiencies must be corrected to a sufficient extent, if temperature curves for the particular case cannot be taken. Such a condition exists when the observations are to be extended over a long time span, e.g., more than fifty years, to increase the reliability of statements about them. ANDREWARTHA (1944) tried to relate data on air temperature to the development of the grasshopper *Austoicetes*, by several methods.

The first method was fairly useful. The author calculated the average temperature for each two hour time segment and determined the progress of development for these temperatures from a curve that had been prepared by work in the laboratory. This method can be considered to be very precise and to form the basis for evaluating two further methods. The first of these was based on that part of the day in which the temperature exceeded the developmental null point. The maximum-temperature of this time segment, plus 12, was divided by two and the corresponding progress of development was read off the temperature-development curve. The use of this method, in which developmental null point certainly was, but the temperature course and the course of developmental curve were not taken into consideration, resulted in a deviation of ca. 6—24%. In the second method of comparison, the effective field temperature was calculated as the average of the minimal and maximal temperatures for the day. The expected progress of development was again read off the temperature-development curve. When this method, which took neither the developmental null point nor the course of the temperature- and development curves into consideration, was used, a deviation of more than 68% in one month resulted.

Average temperatures can be used as indicators of developmental progress only with reservations, and certainly only in those cases in which constant temperatures have led to approximately the same values as the equivalent averages of changing temperatures (cf. ed. 1, p. 98).

If one wishes to consider only one temperature for each day, the maximum value is the most useful. Deviations from the course of the curve can be caused by other meteorological influences (wind, rain, etc.), which makes it appear fundamentally doubtful that conclusions on field conditions can be drawn from laboratory experiments. An additional complication arises from the fact that the body temperatures of animals, especially under the influence of solar radiation, can differ substantially from the air temperature. This effect is heightened by active migration towards strongly insolated areas.

3. Life Span

The total life span of poikilothermic animals, from the start of development to the death of the adult by old age, depends on the environmental temperature, just as the durations of the individual phases of development (egg development, larval development, etc.) do. For the determination of this life span the animals must be able to go through their entire development from the embryo to egg-laying and to death, at the given temperature. This temperature range is often narrower than that in which embryonal or larval development is possible (cf. JANISCH, 1931). QUEDNAU (1957) ascertained the life span of the egg parasite *Trichogramma cacoeciae* at each of several temperatures. Temperatures of $25-27°$ C are considered to be the optimal range for this animal. Though the life span is not the longest at this point, the temperature requirements of the species are apparently best met (cf. BIEVER and MULLA, 1966). Trichogrammids reared at these temperatures possess high vitality and fecundity. The animals are therefore in a physiologically ideal condition. Deviations from this temperature towards higher or lower values resulted in harm, e.g. sterility in females. If a certain temperature limit is surpassed the insects no longer go through their cycle completely, because no more imagos develop. At temperatures under $10°$ C an additional complication arises for *Trichogramma* through the onset of prepupal diapause, which can continue for more than 300 days without causing harm. Subsequently, high temperatures can make further development possible.

A more specialized problem than the total life span is presented by the life expectancy of the imago, which is dependent on temperature (cf. e.g. NOLL, 1963; MARKKULA and PULLIAINEN, 1965; PARISE, 1966; BATEMAN, 1967; FILIPPONI and PASSARIELLO, 1969), in addition to various other factors (e.g. egg laying activity: SMITH, 1958; CLARKE and SARDESAI, 1959; size: BURCOMBE and HOLLINGSWORTH, 1970; population density: SEN-SARMA, 1964; relative humidity: PLATT et al., 1957; READSHAW, 1966, and others). Thus it is possible to speak of a theory of life expectancy ("rate of living", PEARL, 1928; cf. LAMB, 1968). HOLLINGSWORTH (1968) determined the life span of *Drosophila subobscura*. While the animals could theoretically survive for 2.7 years at $0°$ C (extrapolated from the results at higher temperatures) they died almost at once at $37°$ C. Such high temperatures, but also excessively low ones, probably lead to death for many reasons (cf. resistance), but they do not influence aging processes; therefore they will not be considered here. The dependence of life expectancy on temperature is not always constant. HOLL-INGSWORTH found a critical point in life expectancy between 28° C and 29° C (cf. SMITH, 1958; SHAW and BERCAW, 1962). According to investigations by CLARKE and SMITH (1961) the problem of life expectancy is found to be more complicated. These authors distinguish between processes of aging and dying. The rate of aging is said to be temperature independent. Male imagos of *Drosophila subobscura*, (after being reared at 20° C) were kept at 20° C for 4—36 days, and then moved to 26° C, and the length of time till their deaths was determined. Flies that had lived at 20° C for 4—24 days died after a total time of ca. forty days. The time of survival at 26° C was very variable (35.4—16.2 days). Following a longer exposure (28 to 56 days) to 20° C, death during the subsequent exposure to 26° C always took place after 16 days. The "dying" process therefore lasts for 16 days at 26° C and begins after 24 days, regardless of whether the animals are kept at 20° C or at 26° C. The authors also found that the dying process was reversible. Male flies that had been kept 12 days at 20° C, died in 3—4 days at 30° C. By contrast, if the exposure to 30° C is interrupted by a return to 20° C (1 day 30° C, 3 days 20° C, and so forth), the flies die only after 53—85 days, by which time they have been exposed to 30° C for a total of 8—17 days (cf. JOLLY and LEGAY, 1959). The aging process, on the other hand, is irreversible. By this process the vitality, i.e. the ability of the animals to resist various causes of death, is diminished. If this vitality falls below a certain threshold value (cf. HOLLINGSWORTH, 1968) then the temperature dependent dying process is initiated. The threshold value is temperature dependent. It is lower at lower temperatures. According to HOLLINGSWORTH (1969) vitality can be increased by rearing at 15° C. Also the genetical constitution plays an important role: Inbreeding *Drosophila* lived longer at 25° C; hybrids, however, at 30° C (WOODHAMS and HOLLINGSWORTH, 1971[7]). HÖLLDOBLER (1966) describes an interesting relation between aging and animal behavior. The dissolution of the fat body of male ants takes place rapidly at high temperatures (26—28°C), and the males display swarming behavior shortly after their winter torpor. If, by contrast, the populations are kept at a lower temperature (18° C), the males develop more slowly. Their social attachments are maintained, and they remain in their mother-nest for a second winter.

For applied research the length of time the animals live under the variable temperature conditions of the field, and the specific temperature value to be used as a basis for calculation, are important (cf. PILON, 1966). NASH (1936) observed the life span of imagos of *Glossina submorsitans* and *G. tachinoides* in northern Nigeria. Maximum temperatures, varying between 30° C and 40° C, proved to be decisive. In correlation with them, the life expectancy of the imagos was between 40 and 3 days. Special conditions are found in the case of insects that take no nourishment as imagos, or with those that suck blood and must fast till their first meal. In both cases the animals live on their reserves. Fats are the only energy source of *Glossina*, and the logarithm of the rate of utilization of fats is in linear proportion to the temperature. Premature exhaustion of the food reserve could be important for geographical distribution (BURSELL, 1961 b).

---

[7] WOODHAMS, C. A., HOLLINGSWORTH, M. J.: Exp. Gerontol. 6, 43—48 (1971).

4. Mortality

All individuals do not attain the life span that the rearing temperature makes possible. Some die even earlier due to various reasons. This mortality is temperature dependent, because temperatures outside the optimal range lead to a greater loss of vitality, the greater the deviation from the optimum is (Mc LAUGHLIN, 1962a; GUNSTREAM and CHEW, 1967; GORKHALI and BASIR, 1968; USUA, 1968; WANG and GILL, 1970). The significance of this mortality has been demonstrated, for instance in experiments with the pine moth (ZWÖLFER, 1930). Even at optimal conditions (17—18° C, 80—90% rel. hum.) 30% of the larvae died. The mortality of *Ips typographus* was greatest at the temperature at which development was most rapid (SCHIMITSCHEK, 1931). In other cases, no narrow temperature range of lowest mortality was found. The development of larvae of *Tribolium madens* proceeds with about the same low mortality between 20° C and 35° C (HOWE, 1962). At 17.5°C and 37.5° C a mortality of 100% is found. The curves representing summed mortality may have an *S*-shape. This indicates that the percentage of animals dying per unit of time does not remain constant. Early in imaginal life mortality is low, after which it rises rapidly, and then declines again (KARANDINOS and AXTELL, 1967b). Female *Dysdercus mendesi* exhibit a similar phenomenon (MENDES, 1956). The initial life expectancy at 20° C amounts to ca. 25 days, and during this time it sinks rapidly. If the animals possess enough vitality to survive this time period they may reach the age of 80 days. At this point mortality rises steeply, and all the animals die in a short time.

If mortality is related to vitality, those individuals that live the longest are physiologically superior, while others that do not live as long are physiologically somewhat weaker. The physiologically superior animals are not affected by conditions to which the weaker ones succumb. Such differences can arise even within one population (HACKBARTH, 1939). The question now arises whether these weaker individuals suffer harm as a result of definable environmental conditions or whether a random process is involved. JANISCH and other authors (e.g., V. SCHÖNIG, 1953; TISCHLER, 1967a) consider data obtained from animals that develop rapidly or are especially long-lived to be correct. According to them, individuals which develop more slowly or live for shorter times have been damaged by their rearing conditions and therefore yield incorrect values. The embryonal development of *Sitona lineata* runs its course much faster at a relative humidity of 100% than of 65%. But the curves for temperature dependence are, apart from a slight shifting of the optimum towards lower temperature, very similar and run almost parallel (ANDERSEN, 1931). However, it is also possible, as it always is with biological measurements, that natural variations are present, so that an animal that lives for only a short time belongs in essentially the same category as one that lives longer.

An evaluation of results on the basis of the first assumption is only dangerous if an extreme value is taken as a norm. If few experiments are made and the animals are kept under unfavorable conditions it is possible that the results will depend on the number of trials, because the probability of observing a longer-lived animal becomes greater as the number of experiments and the improvement of rearing conditions increase. Comparisons of results based on the two assumptions do not show fundamental differences. Nevertheless, in order to present experimental results as comprehensively as possible, the course of the curve should be supplemented by the individual data points, the extreme values, and the average value with the variation (e.g. KAUFMANN, 1939; v. SCHÖNIG, 1953), or both procedures can be combined, as e.g. KEIRANS and FAY (1968) do, who report the time elapsing till the appearance of the first pupa and also that till 90% of the animals pupate (cf. MOUNTFORD, 1966).

5. Fecundity

Like many other biological characteristics of poikilothermic animals, the number of offspring is often affected by temperature to a considerable extent. It is important for the preservation and propagation of each species that as many offspring as possible reach a stage at which they are able to reproduce. Therefore it is not only important that as many eggs as possible be laid, but that these eggs also be provided with an optimum amount of yolk[8], and that they be laid in places where favorable development is possible. It is more difficult to ascertain these circumstances than to make the relatively simple determination of the number of eggs that a female lays or develops, and consequently the former program has not been carried out often. If one must be limited to the two last possibilities, then the number of eggs laid is more significant. This number is dependent on various circumstances. It is larger if the females have mated (CLARKE and SARDESAI, 1959). Females that do not find favorable laying sites on food plants or in concealed places retain their eggs. Unfavorable relative humidity, as well as excessive population density, disease, etc. reduces the number of offspring. The number of eggs formed and of eggs laid can vary greatly. In the case of the pine moth *(Panolis flammea)* each female produces ca. 200 eggs at constant temperatures between 8° C and 27° C. At 18° C many of these eggs are also laid, but the farther the temperature lies from 18° C, the fewer are laid (ZWÖLFER, 1931; cf. MAKSIMOVIC, 1958; NOLL, 1963). Thus a pronounced peak in the dependence of egg laying activity on temperature can be established. If the curve declines with approximately the same slope at higher and lower temperatures, the highest value can be designated as the optimum. The curve is then approximately bell-shaped (e.g., *Panolis flammea:* ZWÖLFER, 1931; *Trichogramma cacoeciae:* QUEDNAU, 1957; *Myzus persicae:* BARLOW, 1962, i.a.). In other cases, in which the curve rises slowly as temperature rises and then falls sharply after the highest point is reached, the peak could be designated as the maximum (e.g. *Toxoptera graminium*, winged form, WADLEY, 1931; PRITCHARD, 1970, *Psylla pyri*, THANH-XUAN, 1970[9]). ♀♀ from *Pratylenchus penetrans* produced approximately the same number of offspring between 15 and 24° C; at 30° C this number is only slightly reduced (MAMIYA, 1971[10]).

At extremely high or low temperatures egg production and/or egg laying activity ceases (cf. e.g. ELDRIDGE, 1966; SOHLENIUS, 1968). In *Balanus balanoides*, a barnacle distributed in the boreal-arctic region, cessation occurs at 10° C (BARNES, 1963), and in mallophag *Goniodes colchici* (from pheasant), below 30° C (WILLIAMS,

---

[8] cf. for *Euchlanis dilatata:* KING, C.E.: Physiol. Zoöl. **43**, 206—212 (1970).
[9] THANH-XUAN, N.: Compt. Rend. D **271**, 2336—2338 (1970).
[10] MAMIYA, Y.: Nematologica **17**, 82—92 (1971).

1970[11]). Reaction of the gonads is, however, reduced by high temperatures in *Gillichthys mirabilis* (VLAMING, 1972[12]). In general these temperatures agree with the limits beyond which complete development can no longer take place. On the other hand, in the immediate neighborhood of the temperature limits, so few eggs are laid that the existence of a population cannot be guaranteed, and therefore the temperature limits are more of physiological interest, while the practical species range is more narrowly drawn (cf. DAS and DAS, 1967). According to OKASHA (1970[13]) *Rhodnius prolixus* maintained at 34° C after feeding produced no eggs. Apparently the corpus allatum is not active at high temperatures.

The fecundity, i.e. the total number of offspring of one female, must be distinguish- ed from the daily egg count, i.e. the number of offspring per day, which is also temperature dependent. More offspring are produced per day at high temperatures than at low ones (ANDERSEN, 1934; PARK and FRANK, 1948). Extreme high temper- atures lead to a reduction in the daily offspring rate (WEBER, 1931; HILTERHAUS, 1965). The production of offspring of aphids has been observed to begin sooner and end faster as temperature rises (BARLOW, 1962; cf. RIVARD, 1961).

Of course this means that there are also fluctuations in egg-laying activity under field conditions (e.g. PARKS, 1914). Laboratory experiments on the egg parasite *Trichogramma cacoeciae* (QUEDNAU, 1957) have shown that daily 8-h warming to the optimal temperature of 27° C can almost compensate for maintenance in the cold (15° C) for the remainder of the time. When exposure to cold lasts longer, egg- laying activity is sharply reduced. To be sure, the parasites live longer under these conditions, but they can make up for the omitted egg laying only at higher temper- atures. Animals reared under fluctuating conditions are more or less injured which can make it much more difficult to build up new populations (cf. BARLOW, 1962; MEUDEC, 1967). In other cases varying temperatures favor the attainment of large offspring numbers, because they lengthen the life span. Grain lice (*Schizaphis granium*, HEADLEE, 1914) lived substantially longer in fluctuating temperatures, which more than compensated for lower daily production of eggs (the same is true for *Oncopeltus fasciatus;* DINGLE, 1968).

The waxmoth *Galleria mellonella* (DESTOUCHES, 1921) produces twice as many off- spring in changing temperatures as in constant temperature (but, cf. MARKKULA and ROIVAINEN, 1961, for *Therioaphis maculata*, see KINDLER and STAPLES, 1970[14]).

Field investigations that could explain which distinct temperature points affect fecundity have seldom been carried out. Perhaps the results of STIRRET (1938) on *Pyrausta nubilalis* are of general significance. He found that the largest number of eggs was laid in those years in which the average daily temperature during the flight period was about 23° C (cf. PARKS, 1914). The use of low temperatures over longer periods often has the effect of increasing the number of eggs. If the temper- atures are lowered into a range in which the uptake and utilization of nourishment is still possible, but egg laying is not, and the exposure to cold does not last too long, an increase in egg production during subsequent optimal temperatures may result (*Sitona lineata*, ANDERSEN, 1935). Low temperatures may also exert stimulating

---

[11] WILLIAMS, R. T.: Australian J. Zool. **18**, 379—389 (1970).
[12] VLAMING, V. L. DE: Comp. Biochem. Physiol. **41**, 679—713 (1972).
[13] OKASHA, A. Y. K.: J. Exp. Biol. **53**, 37—45 (1970).
[14] KINDLER, S. D., STAPLES, R.: J. Econ. Entomol. **63**, 1198—1201 (1970).

effects. In the life of an individual organism the daily egg production normally increases rapidly after the onset of sexual maturity and then decreases slowly. If female *Tribolium confusum*, after laying eggs for 100 days at 27° C, are placed in 18° C temperatures for eight days, and then are reared at 27° C again, their egg production rises sharply in the following days (DICK, 1937). If the effects of low temperatures last too long, or if the temperatures used are too low, or if especially sensitive stages are treated, sterility may ensue (KVELLAND, 1965). Males of the hymenopteran *Aphytis lingnanensis* exhibit normal courtship and copulatory behavior after treatment with cold (30° F for 8 h). But their spermatozoa are killed. Similarly, the spermatozoids in previously mated females are killed by the same treatment. The treatment with cold has a permanent sterilizing effect on the males (DEBACH and RAO, 1968). Naturally, this leads to a sharp reduction in the number of offspring [see p. 383, cf. for high temperatures: *Dahlbominus fuscipennis* (hymenoptera), RIORDAN, 1957, *Arionater rufus* (pulmonata, LUSIS, 1966), *Locusta migratoria* (ALBRECHT and CASSIER, 1965), *Musca domestica*, MICHELSEN, 1960]. Thermal shocks ($-25°$ C or 100° C) resulted in a lengthening of the generation period between the first and second division of *Tetrahymena pyriformis* (NÉMETH, 1966). Damage to DPNH-oxidase and to lactic dehydrogenase has been suggested as a cause. Treating two-to-three-day-old imagos of *Oncopeltus fasciatus* with $-35°$ C temperature for three or seven minutes led to increased fecundity, as compared to the controls (SAINI and CHIANG, 1966).

So far the discussion has been confined to the effect of temperature on fecundity when it acts on the sexually mature imago: But temperature affects fecundity to a marked extent when it works on the juvenile stages, because the animals can grow larger and heavier when reared at low temperatures. This can have the effect of enabling the imagos to produce many more eggs due to their larger supply of reserve material (ZWÖLFER, 1934; BEKIR, 1935). Cold treatment of the diapausing pupae of *Mamestra brassicae* had a favorable effect on fecundity. It was greatest if the pupae spent two months at 5° C and then emerged towards the end of the third month at 20° C. According to BONNEMAISON (1961) this type of response explains the greater population density in northern regions. Fluctuating temperatures can be markedly more favorable than constant ones, as for example for *Trichogramma cacoeciae* (STEIN and FRANZ, 1960) whose females produced many more offspring after rearing in temperatures that fluctuated daily (16° C and 20° C) (cf. QUEDNAU, 1957; MESSENGER, 1964; STRONG and SHELDAHL, 1970[15]).

Temperature has a completely different significance for fecundity when it acts as an initiator of reproductive processes. High temperatures and long photoperiods act to promote gonadal activity in many fresh-water fish. Cold on the other hand probably acts as an inhibitor [(*Gasterosteus aculeatus*), SCHNEIDER and IMMELMANN, 1969; cf. ALEYEV, 1956 (various fishes); KORRINGA, 1957 *(Ostrea edulis)*; SASTRY, 1963, 1968 *(Aequipecten irradians)*; FISCHER, 1968 *(Lacerta sicula)*; LOFTS et al., 1968 *(Fundulus heteroclitus)*; AIKEN, 1969 *(Orconectes virilis)*; HYDER, 1969 *(Tilapia leucosticta)*; SAMEOTO, 1969 *(Acanthohaustorius milsi* and two other Haustoriidae)].

Different stages in development can vary greatly with respect to the fecundity of the later imagos. Female *Spodoptera littoralis* laid only 700 eggs per day per female

---

[15] STRONG, F. E., SHELDAHL, J. A.: Ann. Entomol. Soc. Am. **63**, 1509—1515 (1970).

after the larvae were reared at 30° C, in contrast to a maximum of 1200 after rearing at 22° C. Increasing the temperature during the pupal stage had a much smaller effect (RIVNAY and MEISNER, 1966).

According to HACKBARTH (1939) the same phenomenon of greater sensitivity of the immature stages is also found after treatment with low temperatures. Temperatures of 10° C (27° C is optimal) have no harmful effects on pupae or imagos of the Brazilian bean beetle *(Zabrotes)* with respect to later fecundity. Treatment of the younger stages for 3−8 days with the same low temperatures lowers their fertility down to the point of sterility. The first larval stage is the most sensitive. On the other hand the young larvae are fairly resistant to high temperatures. If the animals are exposed to heat as pupae, their reproductive capabilities are damaged.

Adaptations of reproductive rates to various temperatures have been investigated only infrequently. Such processes occur for example in *Paramecium aurelia* and *P. caudatum*, where several varieties and pairing-types exist, whose rates of reproduction are differentially temperature dependent. The reproductive optimum of these animals can be shifted to those temperatures at which they have been held for some time (PROPPER, 1963; cf. KOVALJEVA and SELIVANOVA, 1963). In animals that reproduce bisexually and parthenogenetically, or asexually, the point of switching between the two modes of reproduction sometimes does not lie at a definite temperature, but rather there is a fluid transition. PARK et al. (1965), working with individuals of *Hydra littoralis* that had been held at various temperatures, found a reduction in the number of bisexually reproducing animals and a shortening of the sexual period as temperatures fell. The marine sponge *Microciona* does not display such differences (SIMPSON, 1968). Certain oligochaetes form eggs only at high temperatures but grow especially well at low ones, so that the production of body substances is directed only towards growth or egg production, according to environmental circumstances (ASTON, 1968; cf. CRIDLAND, 1962; BÜCKLE, 1963; McLAREN, 1963).

Phenomena of adaptation with respect to the number of offspring have not been investigated often. In this connection, the investigation of BRUN (1966a, b) on the hermaphroditic nematode species *Caenorhabditis elegans* may be mentioned. These animals multiply well at 18° C. If animals that have been kept for long periods at 18° C are moved to 24.5° C, they do not multiply. Thus this high temperature lies above the reproductive resistance limit of the species. After maintenance at 22°, 23°, 23.5°, or 24° C, the rate of reproduction falls at first, for five or six generations, and then slowly rises. After 1000 generations reproduction at 24.5° C is possible. The resistance limit has been shifted to a higher temperature. If these animals are placed in 18° C, they reproduce far more (230 offspring per animal) than those that live constantly at 18° C (140 offspring per animal). Thus an adaptation of the function, resembling type 5, has taken place.

6. Nutritional Requirements and Synthetic Processes

a) Food Consumption

Like all active functions of poikilothermic animals, food consumption per unit time goes up with increasing temperature and falls after a maximum has been reached.

Food consumption also depends on the quality of the food, and on the age and physiological condition of the animal.

A different situation appears when one considers the amount of food required by an animal for the attainment of a specific stage. This is connected with the question of the degree of utilization of nourishment, i.e. of the relation of food uptake to production of feces (on this point, cf. JOHNSON, 1960; KINNE, 1960; PALOHEIMO and DICKIE, 1966; HOUSE, 1966). Generally valid rules cannot be formulated, because the temperature requirements of different species vary so greatly. Results following rearing at constant temperatures indicate the existence of several possibilities (on rearing in varying temperatures cf. PARKER, 1930; KAUFMANN, 1943):

1. The same quantity of food is used to reach a given stage at all temperatures that permit growth and development. PARKER (1930) found this type of response in the development of the grasshopper *Melanoplus mexicanus*. He reared the nymphs at 27°, 32° and 37° C. At the higher temperatures the amount of food taken up per unit time was greater, but development was faster, so that the total amount of food taken up during the entire development period was the same. According to LØVTRUP (1953) the amount of energy needed for a given segment of development of *Ambystoma mexicanus* does not depend on temperature. But this fact does not imply that development proceeds with the same efficiency at all temperatures, especially as it has been found that at higher temperatures reserve materials were lost without a corresponding uptake of oxygen (cf. LØVTRUP, 1959).

2. More food is used to reach a given stage at high temperatures than at low ones. FANKHÄNEL (1959) observed this effect in the caterpillars of the brown-tail moth *Euproctis chrysorrhoea*. He kept the caterpillars at 17°, 22.5°, and 28° C, and measured the total surface of oak leaves used from the time of departure from the winter nest till pupation. This leaf surface was greatest at 28° C and ca. 75% relative humidity. At this point it was ca. 70 cm². The intensity of utilization of nutrition of the nun moth *(Lymantria monacha)* goes down with increasing temperature (OLDIGES, 1959). This means that at higher temperature more food must be taken in and also absorbed (SATTLER, 1939; cf. HUTNER et al., 1957).

3. Less food is used at higher rearing temperatures than at lower ones. This type of response can be understood if one assumes that the time span of development and also of nutritional uptake is shortened by the more rapid maturation of the gonads (on the hormonal basis of temperature dependence of insect growth, cf. CLARKE, 1966). TITSCHAK (1925) reared clothes moths *(Tineola biselliella)* at 20° C and 30° C. The total uptake of food and production of feces were less at 30° C than at 20° C. After rearing at the lower temperature the pupae were heavier than at 30° C. As a result the female imagos that emerged laid more eggs (OLDIGES, 1959). With rising temperature (18 → 30° C) food consumption and also utilization are reduced in *Agrotis segetum* (BONGERS and WEISMANN, 1971[16]; cf. HAGSTRUM and WORKMAN, 1971[17]). This is also true for *Scotia segetum* (WEISMANN and PODMANICKÁ, 1970[18]). If one considers a rapid turnover of generations and low uptake of nutrition to be optimal, then a high rearing temperature is favorable in this case, but it is not advantageous if one wants to obtain a large number of offspring

[16] BONGERS, J., WEISMANN, L.: J. Insect Physiol. **17**, 2051—2059 (1971).
[17] HAGSTRUM, D. W., WORKMAN, E. B.: Ann. Entomol. Soc. Am. **64**, 668—671 (1971).
[18] WEISMANN, L., PODMANICKÁ, D.: Biológia (Bratislava) **25**, 769—778 (1970).

(cf. JÖHNSSEN, 1930; KAUFMANN, 1943; RATTAN and HAQUE, 1956). The question of the utilization of nutrition can be carried over to developing embryos. According to RICHARDS' (1959) results the energy use of embryos of *Oncopeltus fasciatus* increases with decreasing rearing temperatures. If the temperature falls below a definite threshold, the entire fat store of the eggs is used before embryonal development is concluded, which means that the embryo must die due to lack of nourishment (cf. ROSENBAUM, 1960; KITTLAUS, 1961).

4. A dependence with an optimum exists. Thus, for example, OLDIGES (1959) found a maximum absorption of food during the development of *Bupalus piniarius* caterpillars at 18° C. Since animals reared at this temperature displayed the highest pupal weight, laid the greatest number of eggs, and had the shortest duration of development it is possible to speak of a vital temperature optimum, which embraces a variety of criteria, for this species. Investigations of embryonal development of *Pleuronectes platessa* yielded similar results (RYLAND and NICHOLS, 1967). Rearing at various temperatures between 2.6° C and 9.8° C indicated that the optimum utilization of yolk took place at 6.5—8.5° C (cf. KINNE, 1960; FOXX and MARTIN, 1970). Salmon show optimal growth at 12° C. Higher temperature causes a better absorption in the colon, but also a higher energy supply. Energy is economized at lower temperatures (8° C); utilization however is reduced. Thus 12° C proves to be a compromise (ATHERTON and AITKEN, 1970[19]).

The effect of acclimation temperature on food utilization has been investigated relatively seldom. Such investigations encounter the basic difficulty that the time needed to ascertain the data to be measured after the transfer of an animal to a new temperature is fairly long, and that animals often refuse to feed immediately after a temperature change. WIESER (1965b) conducted such transfer experiments on *Porcellio scaber*. Animals acclimated for long periods at 8.6° C were placed in 20° C and their food uptake was compared with that of 20° C animals. The cold adapted isopods did not equal the higher food consumption of the 20° C animals until between the third and fourth day. Increased synthesis of digestive enzymes could not account for this. The same level of activity of proteolytic enzymes is found in the hepatopancreas of 8.6° C and of 20° C animals.

The question of whether the possibilities mentioned here are based on the presence or absence of adaptations remains to be investigated.

b) Respiration

A *U*-shaped dependence curve is often found for gas metabolism, with a minimum oxygen uptake and/or carbon dioxide output at the temperature of fastest development. For example a minimum was present in the caterpillar stage of various lepidoptera (KOZHANTSHIKOV and MASLOWA, 1935), in the pupae of *Ephestia kühniella, Tribolium confusum, Drosophila melanogaster*, (GROMADSKA, 1949), in the oxygen use of all stages of *Musca domestica* (LAUDANI, 1965), in the metamorphosis of all forms of *Formica polyctena* (SCHMIDT, 1968). RICHARDS (1964) measured the oxygen use and development of eggs of various insects and always found that as the minimal temperature is approached relatively more oxygen is

---

[19] ATHERTON, W. D., AITKEN, A.: Comp. Biochem. Physiol. **36**, 719—747 (1970).

used than at moderate temperatures. The curves for development and oxygen use ran parallel only in the favorable temperature region.

c) Synthetic Processes

Relatively little is known about the relation of temperature to the speed of synthetic processes. Till now, most effort has been directed towards attempts to understand the release of energy through the utilization of oxygen (e.g. LØVTRUP, 1959). But in this connection it is more advantageous to offer a specific foodstuff and to determine how fast the end products are formed. Such trials have been carried out by van HANDEL (1966) on *Aedes sollicitans*. Unfortunately the temperature at which the mosquitoes were reared and held before the trials is not unequivocally clear in this work. Therefore it is not certain whether an ET-dependence or an AT = ET-curve is present. Differently adapted animals can respond in different ways. Thus, according to MEWS (1957) cold-adapted frogs *(Rana temporaria)* incorporate fed glycocoll at a faster rate than warm-adapted ones (cf. JANKOWSKY, 1960; DAS, 1967; ANDERSON, 1970). At an experimental temperature of 25° C the rate of incorporation of ^{14}C-leucine into liver and gills, one hour after its injection, is higher in cold-adapted goldfish than in warm-adapted ones (Compensation, DAS and PROSSER, 1967). The cold-adapted animals had a higher rate of incorporation into muscle at ET = 15° C and 25° C, but not at 5° C. An increase in the amount of labelled protein could be due to either increased synthesis or decreased catabolism (cf. HASCHEMEYER, 1969).

These arbitrarily selected examples indicate that temperature adaptations of synthetic processes exist. There are relations of a special type between processes of synthesis, especially of protein biosynthesis, and processes of growth and development. If one considers the temperature dependence, expressed as a Q_{10} value, of growth and development, one finds relations corresponding to the van't Hoff rule ($Q_{10} = 2 - 4$). Here it should be emphasized again that the data measured are always related to the temperature at which the animals were maintained (AT = ET-curves). This signifies, as has been said, that temperature acclimation plays little or no role in these cases. This is also true for growth of fish cells *in vitro* ($Q_{10} = 2$, WOLF and QUIMBY, 1962). By contrast, the Q_{10} values of the frequently investigated processes of energy metabolism (e.g. oxygen use) are often reduced through capacity adaptation and even fall to $Q_{10} = 1$ in some cases, if one considers AT = ET-curves. It is now interesting to note that adaptation has also been demonstrated for synthetic processes, i.e. in processes that are related to maintenance metabolism in the adult animal, but are also related to growth and development in the immature animal. There is thus a discrepancy here. A distinct adaptation is found for processes of synthesis, as mentioned before, but no, or slight adaptation, respectively, with regard to growth and development, if the time needed to reach a certain stage is taken as a criterion. The experiments of HEITLINDEMANN (unpubl.) have yielded new aspects. She investigated the content of lactate, pyruvate, DNA and RNA and their basic compositions in larvae of *Tenebrio molitor* reared at different temperatures. Stage of development and bodyweight of the animals under investigation were always the same. The lactate-pyruvate quotient is said to be correlated with a stress situation (see p. 440). In this case there was no correlation to heat or cold resistance, but to velocity of development (max. at 30° C). If all processes participating in development were characterized by the same Q_{10} value, then animals of different temperatures, but of the same stage of development, independent of its velocity, would have the same chemical composition. This is true for substances which are synthesized by the organism and then subjected only to alteration, as well as for such substances (e.g. RNA) in which temperature determines the velocity of ana- and catabolism. The body composition of animals reared at different temperatures must be different, if the Q_{10} values of the individual processes are different too, which is not called an adaptation. This is true for yolk- and food-consumption, egg number etc. According to the results of HEITLINDEMANN, a biochemical adaptation of changing reaction systems could exist, if in connection with different rearing temperatures, specific enzymes, RNA-content etc. are found. Further experiments must be carried out.

7. The Effect of Temperature on Associations between Different Species

Symbioses may be said to exist when individuals of different species are brought together, not by a preference for the same environmental conditions, but by mutual relationships. Herbivores are attracted by their food plants, predators by their prey, and parasites by their hosts. In other cases (commensalism, proto-commensalism, and phoresis) one partner benefits from the relation while the other suffers no, or no substantial, harm. In mutualistic symbiosis both partners gain an advantage. Temperature frequently acts on these reciprocal relationships because the partners may be affected in different ways. Parasitism offers many examples. Thus, the development of *Nosema*, a dangerous epidemic among bees, is inhibited at temperatures above 37° C as the spores do not germinate. Bees reach such temperatures under intense solar radiation or at high work rates (SCHULZ-LANGNER, 1958). Different conditions may be found among various parasites on one host species. HURPIN (1968) distinguishes four categories among the parasites of a cockchafer. The pathogens may have a fairly narrow temperature range around 20° C (mycoses and bacterial diseases). Low temperatures may be as favorable as moderate ones of ca. 25° C (rickettsial diseases and *Nosema*). Normal temperature dependence is found with coccidioses. Viral diseases are temperature independent in the ranges investigated. According to McLAUGHLIN (1962b) elevated temperatures act as stimuli, with the result that bacterial infections of insects occur under conditions not found under normal life circumstances. Similar problems arise when animals parasitize plants. The nematode *Heterodera rostochiensis* can be reared on various nightshade plants, but the infection rate declines with increasing temperatures (FERRIS, 1957). Females of the root gall nematode *Meloidogyne* required a specific heat sum for maturation for each different host-parasite combination (DROPKIN, 1963; cf. 1969).

The virulence of parasites with a series of hosts is often reduced by temperatures which do not injure the intermediate hosts. Sporogeny of *Plasmodium berghei* takes place in *Anopheles* between ca. 16° C and 24° C. If the mosquitoes are briefly exposed to 28° C, sporogeny is hindered (VANDERBERG and YOLEI, 1966; cf. CHAO and BALL, 1962). The optimum temperature for the development of cercariae of

Schistosoma mansoni in *Australorbis glabratus* is 26—28° C. If the snails are maintained at 23—25° C, the resulting cercariae display low infectivity for mice. Snails kept at cool temperatures for one month have been found to be free of parasites (STIREWALT, 1954; cf. OLSON, 1966; on killing of endosymbionts by extreme temperatures SCHNEIDER, 1956; MUSGRAVE et al., 1963).

Similar circumstances are found in the case of the cestode, *Hymenolepis nana*. At 40° C their eggs in *Tribolium* develop only into abnormal larvae which are not infectious for mice. The optimum temperature for larval development is 30—32° C (HEYNEMANN, 1958). Heat-injured larvae of *Hymenolepis diminuta* cannot retract the scolex to form cysticercoids (VOGE, 1959). These parasites are apparently adapted for the presence of low temperatures in the intermediate host and for high temperatures in the homeothermal final host. Certain phases of development are determined by the prevailing temperature. For example, *in vitro*, *Schistocephalus solidus* grow at low temperatures but form gametes at high ones (SINHA and HOPKINS, 1967).

When parasites change hosts they must insert themselves into the mutual inter-action of the two hosts at a suitable place and a specific time. Trypanosomes, worm larvae, etc., often appear daily in the blood of birds or mammals at the time of day when the appropriate diptera tend to feed (cf. REMMERT, 1969). Temper-ature may determine the time of movement of the parasites from the tissues to the blood (e.g., HAWKING et al., 1967; BARDSLEY and HARMSEN, 1969), as well as the time of sucking of blood. The tick *Hyalomma* feeds on cold- and warm-blooded animals. When feeding on various mammals, birds, or reptiles at temperatures above 30° C the time of feeding is independent of temperature. Between 20° C and 30° C the time of feeding on reptiles is progressively prolonged and below 20° C they feed only occasionally (SWEATMAN, 1968, 1970; SWEATMAN and GREGSON, 1970). Temperature often plays an important role in predator-prey relationships due to the different dependencies on temperature of their reproductive capacities. The predator *Phytoseiulus* controls the bean mite most effectively at 20° C and less well at 15° C and 25° C (FORCE, 1967; cf. further examples in TISCHLER, 1955, p. 82f.; SUNDBY, 1966; BURNETT, 1970). *Aphelinus asychis* is a parasite of *Schiza-phis graminum*, *Rhopalosiphum maidis*, and *Sipha flava*. The latter was affected to a certain extent only at the lowest breeding temperature (RANEY et al., 1971[20]).

B. Morphological Effects of Normal and Extreme Temperatures

Internal and external factors influence the development of characteristics such as size, coloration, etc. In this setting only the environmental factor of temperature is of interest.

1. Mutations

Temperature can influence mutation rates in several ways. It can act directly as a mutagen, i.e., a cause of mutations. Heat- and cold-shocks that animals can tolerate only for short times can cause mutations, though to a much smaller extent

[20] RANEY, H. G., COLES, L. W., EIKENBARY, R. D., MORRISON, R. D., STARKS, K. J.: Ann. Entomol. Soc. Am. **64**, 169—176 (1971).

than e.g., X-rays. BUCHMANN and TIMOFEEFF-RESSOVSKY (1935/36) and PLOUGH (1941) reported an increase in the mutation rate caused by heat treatment (but cf. SHELDON, 1958). MUTO (1957) produced triploid toads *(Bufo vulgaris)* by temperature effects, thereby supporting FRANKHAUSER (1938 et seq.), and NEGRI (1964) used temperature shocks to obtain diploid eggs from *Discoglossus pictus* (cf. BEET-SCHEN, 1957; SWARUP, 1959 a, d). A certain number of triploid animals will develop when eggs of *Ambystoma*, after laying, are treated for a period of 15–18 h at 1–3° C (BRUST, 1970). The mutation of a designated gene in a prescribed direction has not yet been observed. In this respect the mutagenic effect of temperatures does not differ from that of other mutagens (but cf.: mutagenic specificity, DEMEREC, 1954).

In other cases temperature affects the speed of reaction. So SHELDON (1958) found that the mutation rate fell as temperature increased, and believed that this was the result of a longer development time at lower temperatures. An indirect effect is present when the mutation rate following radiation is affected by temperature. LÖBBECKE and OLTMANNS (1961) investigated the effect of cold ($-7°$ C and $+3°$ C) and warmth (30°, 35°, 40° C) on the relative frequency of somatic scale mutations caused by X-irradiation in *Ephestia kühniella*. Work with somatic mutations offers an advantage over observations on germ cell mutations, in that they already appear in the irradiated generation, and that many mutations can be produced and observed in each individual. As a disadvantage of this method it must be mentioned that it cannot be stated with certainty whether a genuine mutation or a modfication is present (cf. POHLEY ,1955). LÖBBECKE and OLTMANNS (1961) kept their experimental subjects (pupae of *Ephestia kühniella*) at the determined temperatures for six hours before the irradiation and found that the relative frequency of somatic mutations could be changed. After a six hour pretreatment at 3° C or 35° C, one mutant is $6 \times$ and $11 \times$, respectively, as frequent as another one. According to MÜLLER (1963), a one hour temperature pretreatment is not sufficient, a finding that could be explained by the supposition that the high or low temperatures in the six hours before radiation had accelerating and retarding effects, respectively, so that pupae at different states of development underwent irradiation. When only one-hour temperature pretreatments were used, the temperature effect did not appear.

2. Meiosis, Mitosis, and Cellular Processes

a) Mitosis and Meiosis

Temperature affects the course of mitosis and meiosis in many ways. During one cell division (mitosis) a chain of processes occurs, which can, according to studies on embryos of *Psammechinus miliaris* and *Echinus esculentus* (AGRELL, 1958, 1959) be influenced by temperature in various ways. During the cleavage stage of the embryos the course of metaphase is affected the least. The observation that the number of mitochondria fluctuates and reaches its lowest value precisely during metaphase is in agreement with this (AGRELL, 1955). By contrast, the mitochondrial number and correspondingly the temperature dependence is especially high during prophase. In the later course of embryonic development interphase is greatly prolonged in the early blastula and the temperature coefficient for all phases of division goes up (but cf. LØNNING, 1959). If the temperature goes to very high levels mitosis is blocked (STUMPF, 1959; cf. YAMAMOTO, 1964) or retarded (GEILEN-KIRCHEN, 1966; VERDONK and DE GROOT, 1970), even after the treatment. In the sea urchin *Allocentrotus fragilis*, normal egg development takes place between 7° C and 15° C. At 20° C nuclear division proceeds but cytoplasmic divisions stop (MOORE, 1959).

Temperature also affects development and growth on the level of parts of cells. BERLIN and DEAN (1967), using electron microscopy, found that in trout *(Salmo gairdneri)* acclimated to 5° C the Golgi apparatus was greatly enlarged and con-tained large amounts of particulate material (perhaps lipoproteins). Lizards *(Lacerta viridis)*, acclimated to different temperatures, exhibited differences of the large-celled layer of the hippocampus and in the cranial nerve nuclei (BOYCOTT and GUILLERY, 1959). Treatment with cold of the giant chromosomes of the nutri-tional cells of *Calliphora* during the growth phase between 16—64 n increases the number of non-reticular centers significantly (BIER, 1959).

A very important role is played by temperature as a tool in the production of synchron-ously dividing populations, e.g. of the ciliate *Tetrahymena pyriformis*. TAMURA et al. (1966) produced synchronization by rhythmically changing temperatures between 26° C and 34° C. BERNSTEIN and ZEUTHEN (1966) did so by a series of heat shocks, increasing the temperature from 28—34° C for twenty minutes each time. By observing the in-corporation of tritiated uridine, the two latter authors were able to confirm that RNA synthesis was blocked for thirty minutes after each temperature elevation (cf. HJELM and ZEUTHEN, 1967, influence upon size of cells, center-apparatus and division in the ciliate *Blepharisma* see BAI et al., 1969).

Temperature can influence the duration of meiosis (WILSON, 1959), the frequency of chiasma formation, and thereby the exchange of genes. PLOUGH (1917) found a minimum for crossing over at normal rearing temperatures, and RASMUSON (1957) found an increase when the experimental animals *(Drosophila melanogaster)* were kept at 31° C for 48 h at the age of 1—7 days (cf. WHITE, 1934, but see also ELLIOTT, 1955). Temperature dependent processes are important for gene flow within generations as ZIMMERING (1963) describes them: If male *Drosophila melanogaster* mature at 26° C or 18° C, there is a distinctly higher rate of loss of the Y-chromo-some in the higher temperature than if the lower temperature is used. There is no connection with the synaptic processes of meiosis.

Protein synthesis and growth are stimulated at low temperatures (15° C) by neurosecretion in *Locusta migratoria*. For this reason, partially empty corpora cardiaca are found at this temperature. At high temperatures, stimulation is not necessary and neurosecretory substances are not given off (CLARKE, 1966; cf. PASSANO, 1960; JENKINS, 1970, p. 322). Perhaps there is a connection between this phenomenon and the dependence of molt frequency on temperature. When *Tene-brio molitor* is reared at 25° C, 11—15 molts take place, but at 30° C, there are 15—23 molts (cf. ed. 1, p. 88; LUDWIG, 1956; LEONARD, 1970). It should be mentioned that temperature can influence processes of determination and induction during embryogenesis (cf. WOELLWARTH, 1961; FULDNER and UBISCH, 1965).

b) Relation of Gene Penetrance to Temperature

According to TIMOFEEFF-RESSOVSKY (1927) "penetrance" designates the frequency with which a gene is made manifest. This frequency depends on temperature, among other environmental factors. According to GOLDSCHMIDT (1955) the penetrance of an allele is certainly the result of the genetic effect that controls the kinetics of the chain reaction. In addition the reaction products of the gene's activities play a role. So it is conceivable that temperature affects the functions of a gene directly or by way of an effect on the activity of a reaction product. Temperature can have an effect in two ways. In the normal temperature range a longer exposure (possibly a complete rearing) of animals is possible. Extreme temperatures can only be used for short-term shocks, that must not be allowed to be fatal (e.g., RITOSSA, 1962; THOMPSON, 1967). At normal termperatures many animal species show differences in gene penetrance. In the case of the so-called temperature-mutants (DAVIS, 1950) the changed phenotype is expressed only at specific temperatures. PREER (1957) found a strain of *Paramecium aurelia* in the field with severely retarded growth and multiplication at $27°$ C, in contrast to normal animals, that reproduced up to $34°$ C. The author was able to establish that this difference was caused by a gene T. T (i.e. growth above $29°$ C) is dominant over t. At low temperatures both TT and tt animals multiply at the same rate. Only at temperatures above $27°$ C do the tt animals fall behind. BATTAGLIA and LAZZARETTO (1967) found another example in the copepod *Tisbe reticulata*. They observed a polychromatism controlled by three alleles at the same locus. Hybridization experiments showed that temperature worked as a particularly effective selective factor. Populations kept at $23-24°$ C exhibited a 60% frequency of a specific characteristic, while those kept at $18°$ C exhibited only a 40% frequency.

Temperature dependencies of this type have been observed in several species of *Drosophila* (reared at temperatures between $13°$ C and $31°$ C). Once again, observations on various alleles are especially interesting. AKITA (1955) reared fruit flies at temperatures between $17°$ C and $31°$ C and found that "vestigial" mutants resembled the wild type more as temperatures increased, while "vestigial-nipped", by contrast, exhibited greater penetrance at higher temperatures. The allele "vestigial-notched" was expressed most strongly at mid-range temperatures. The allele "vestigial-nick of green" showed the smallest degree of dependence on rearing temperature (cf. PIATKOWSKA, 1961; HANSEN and GARDNER, 1962; SCHARLOO, 1962; BORZEDOWSKA, 1963; PENNYCUIK and FRASER, 1964; TOBARI, 1966; MANGE, 1968; NAKASHINA-TANAKA, 1968; BUCKHOLD and SLATER, 1969). Other authors reared their experimental animals (*Drosophila* species) for many generations. DRUGER (1962) selected his animals for long or short wings (rearing temperatures: $16°$ C and $25°$ C). The selected differences were generally maintained under other temperature conditions, but they were expressed most markedly at the temperature in effect during the selection period. The observation of BEBAK (1965) should be mentioned in this connection. She reared a "vestigial" strain, whose trait had a particularly high penetrance at $18°$ C, at $28°$ C for 25 generations. Over the course of time more and more normal-winged flies appeared. The fact that the animals were still carriers of the vestigial trait was revealed when they were reared at $18°$ C again: "vestigial" types appeared at once (cf. PIATKOWSKA, 1963).

It can be concluded from these experiments that penetrance of a gene need not have a fixed magnitude even at constant temperature. This idea is supported by the investigation of DRUGER (1967), who was able to reverse the effect of temperature on the number of thoracic bristles (decreasing temperature produces more

bristles) by selection over generations (cf. PARSONS, 1959; BEARDMORE, 1960).
HARTMANN-GOLDSTEIN and SPERLICH (1963) concluded from experiments with
Drosophila subobscura that changes in temperature during development favored
the homozygotes. Temperature effects on gene penetrance are also found in vertebrates. LINDSEY (1962; cf. LINDSEY, 1954) reared stickleback *(Pungitius pungitius)*
at different temperatures and counted the number of vertebrae laid down in young
fish, the basal bones of the unpaired fins, the spines, the rays, and the scutes. The
result was that the number of serial skeletal elements increased as rearing temperature increased (cf. ed. 1, p. 121; WEISEL, 1955; LINDSEY and ALI, 1965;
LINDSEY, 1966).

ORSKA (1964) found two temperature sensitive periods for the number of fin-rays
in experiments with salmonids.

In addition to these morphologically manifested genes, lethal factors also show
temperature dependent penetrance. BRIGGS and HUMPHREY (1962) investigated a
semilethal gene *v* in the axolotl. At a rearing temperature of 25° C, 88% of the
offspring of *vv* animals died before the blastula stage, but at 14−20° C fewer than
1% died. At the lower temperatures the embryos developed to the gastrula or
neurula stage; 30% developed even further (cf. RIZKI, 1955). TARASOFF and
SUZUKI (1970)[21] investigated four temperature sensitive lethal-genes of the X-
chromosome of *Drosophila melanogaster* in relation to the time of operation.

c) Temporal Limitation of Temperature Sensitivity

Gene penetrance can frequently be affected by temperature only during a specific
stage of development that is limited in time. This segment of time can also be
called the sensitive period. KÜHN (1927) was able to fix the limits of this sensitive
period for the temperature-induced changes in wing coloration of a butterfly fairly
precisely. The effect was especially noticeable when the temperature shock was
administered 24−48 h after pupation (cf. ed. 1, p. 128; HARTMANN-GOLDSTEIN,
1967; GIBSON, 1969). *Drosophila* pupae, held at 23° C, are most sensitive to changes
in their posterior transverse veins and other traits at the age of 25 h (MILKMAN,
1962). This effect depends on the temperature as well as on the duration of the
heat treatment. The effect of temperature shock is often astonishingly specific. A
duration of $4^1/_4-4^1/_2$ h at 37.5° C is lethal for males, but the same duration at
38° C is not. At 36.5° C the rate of morphological changes reaches a maximum
after 200 min and then goes down (MILKMAN and HILLE, 1966). Very interesting
results emerged from a treatment of female pupae with temperatures of 40.5° C.
Very short shocks of 5−7 seconds duration, followed by specific periods at 23° C,
increased the resistance to subsequent treatment. This effect was reversed after a
few hours. The build-up of resistance is independent of temperature between 23° C
and 28° C. If the first shock is extended to 20−120 sec, then resistance to later
shocks lasts longer. A treatment time of 20 min induces the described morphological defects, and a specified period at 23° C followed by a second treatment has
no further effect. HILLE and MILKMAN (1966) explained these and some other experimental results by a comprehensive scheme, based on the concept that heat

---

[21] TARASOFF, M., SUZUKI, D.T.: Develop. Biol. **23**, 492—509 (1970).

treatments change the tertiary structure of a protein, which affects the formation of the transverse veins. The findings indicate that the structure of a single substance, in a partly linear multi-step series with different temperature dependence, is changed.

3. Effects on Sex Ratio

a) Indirect and Direct Influences

Sex ratios can often be changed in one direction or the other by different environmental factors. Temperature, as one such environmental factor, takes part in the process of sex determination. An indirect effect is present when certain temperatures affect the survival rate of males or females (larvae or imagos) in such a way as to shift the ratio (cf. e.g., ZWÖLFER, 1934; MAKSIMOVIC, 1958; WILKES, 1959; MALOGOLOWKIN, 1959; VASSILEVA-DRYANOVSKA and GENCHEVA, 1964; DAVIDE and TRIANTAPHYLLON, 1967). The observations of GÖSSWALD and BIER (1955) must also be counted as an indirect effect: they found that temperature determined the beginning of egg laying and the sex ratio of *Formica rufa*. Ovulation and egg-laying begin in spring, even at low nest temperatures (below 19.5° C). At these low temperatures the spermatheca (sperm pump) of the female cannot function, so that below 19.5° C only unfertilized eggs, which develop into males, are laid. Field observations support this investigation: Male *Formica rufa* originate chiefly from populations with inadequately warmed nests while females come from nests that are well insolated or densely populated. Low temperatures kill spermatozoa in the thecae of *Aphytis lingnanensis*, so that unfertilized eggs that develop exclusively into males are laid later at higher temperatures (DE BACH et al., 1955; DE BACH and RAO, 1968; cf. McMULLEN, 1967).

Those cases in which the gonads are directly influenced to develop into testes or ovaries are far more interesting. Thus, ANDERSON and HORSFALL (1965 a, b) found that larvae of *Aedes stimulans* always developed testes when they were maintained at 18° C. At 27° C, on the other hand, they formed ovaries. One could suggest that the temperature did not work directly on the gonadal anlage, but, for example, on an endocrine gland whose product then affected the development. In order to clear up this point, the authors carried out transplants of *Aedes stimulans* gonadal-discs into *Aedes vexans*, which develop into males even at high temperatures. But, at such temperatures the *Aedes stimulans* gonadal-discs always turned into ovaries, so that a direct influence of temperature can be considered as certain.

The same authors (ANDERSON and HORSFALL, 1963) also reared their experimental animals at temperatures between 18° C and 27° C. The higher the temperature, the more the male organs were reduced and replaced by female structures. In this way intersexes of various grades were developed; for example at 28.4° C intersexes with ovaries, but without functional mouthparts: Since the eggs however ripen only after a blood meal, these intersexes cannot form mature eggs. But the ovaries of these animals are capable of doing so, since they form well-developed eggs if they are implanted into normal animals. The gonads of these intersexes have thus become completely re-determined into ovaries through high temperatures (cf. BRUST and HORSFALL, 1965; BRUST, 1966; LAUGÉ, 1966; MOSBACHER, 1967). The influence of temperature on phenotypical sex-determination of *Ophryotrocha* was investigated by BACCI and VORIA (1970[22]).

b) Sensitive Periods

It may be asked, which stage of development must be affected by temperature for it to play a role in sex determination. In the hymenopterans *Trichogramma semifumatum* (BOWEN and STERN, 1966) and *Ooencyrtus submetallicus* (WILSON and WOOLCOCK, 1960) the sensitive phase occurs very early, during the time the egg is formed in the mother. *T. semifumatum* eggs, from which only females emerge, are formed by females, after heat treatment (75° F) during the pupal stage at the time of formation of the primordial germ cells and the maturation of the oocytes.

But the time of sex determination can also come later. In *Macrocyclops albidus* (MONAKOV, 1965) and *Carausius morosus* (BERGERARD, 1961 a, b) the sensitive period falls during embryonic development. In the latter, masculinization (by a temperature of 30° C) is maximal when the heat treatment begins 7—14 days after spawning and lasts about 30 days. Treatments of shorter duration produce inter-sexes of various grades. In this case they could be accounted for by the fact that *Carausius* females also have anlagen of the male reproductive organs, which degenerate around the 37th day of the embryonal development period (lasting 75 days in all), if the temperature remains below 30° C. At high temperatures the female sex organs degenerate and the male organs develop (on this point, cf. for *Agama agama*, CHARNIER, 1966; for *Rivulus marmoratus*, HARRINGTON, 1967).

The latest influence on sex is found in insects, in which the gonads may still be affected by events occurring in the larval stage (ANDERSON and HORSFALL, 1963, 1965 a, b; LAUGÉ, 1967, 1969).

c) Mechanisms of Influence

Temperatures can affect the direction of movement of the sex chromosomes in the maturation division. SEILER (1920) observed such a case in the butterfly *Talaeporia tubulosa*. The X-chromosome of the female is oriented in such a way during meiosis, that at low temperatures (3—5° C) it tends to move into the polar body. At higher temperatures, by contrast, it often remains in the egg, so that at fertilization an XX combination, that is to say a male, results. Similar assumptions could probably also be made about sex determinations in *Cyclops viridis* (METZLER, 1955, 1957). According to BIER (1954) another mechanism is found in *Formica rufa*. The female can form different types of egg. At low temperatures enhanced nucleolar formation and enlarged nurse cell nuclei are seen. This favors the development of females. Males on the other hand arise from summer eggs. Variations in nutrition of the larvae can alter this predisposition through egg types.

FAULHABER (1967) found another type of influence in a strain of *Drosophila simulans* that contained a sex ratio factor in chromosome 3. The expressivity of this factor is temperature dependent: At 16.5° C and 18° C, 2% of the offspring were male, at 20° C 7.7% were, at 24° C about 36%, and at 26—27° C, 50%.

---

[22] BACCI, G., VORIA, P.: Experientia 26, 1273—1274 (1970).

d) Sex Ratio in the Annual Cycle

According to BOILLON (1956) rising temperature affects the snail *Cepea nemoralis* to produce eggs at first, and then spermatozoa. However in this hermaphroditic species, this happens in the same animal. At 0° C no gametes are formed, at 6° C eggs are formed, and at 23° C spermatogenesis begins. So it is possible that temperature acts as a signal of the approach of favorable environmental conditions (for *Ophryotrocha* see BACCI and VORIA, 1970). Unfavorable temperature conditions lead to an increase in the proportion of males in nematodes (DAVIDE and TRIANTA-PHYLLON, 1967) which then fertilize the winter eggs of their mothers' generation (cf. SANFORD and BANTA, 1941; FRIES, 1964).

BUCHNER and KIECHLER (1966) observed a peculiar phenomenon in *Asplanchna*. They found a periodic alternation in bisexuality in their cultures between 3% and 30%. The peak values were reached every 2—3 weeks and the time interval between minimum and maximum grew shorter as the temperature increased.

4. Type of Reproduction

In animals with alternation of generations (metagenesis or heterogenesis; asexual or sexual reproduction), temperature can determine the type of reproduction. Hydroid polyps (GOETSCH, 1921/22), cultures of daphnids (FRIES, 1964), and aphids (LEES, 1959), can be propagated for years by the same reproductive means, if the external conditions are held constant. Similar observations can be made under natural conditions, since many species (e.g., daphnids) reproduce sexually (mictically) in the warmer parts of their area of distribution and by parthenogenesis in the cooler parts (geographical parthenogenesis).

In hydroids which reproduce asexually at lower temperatures gonadal development is induced by elevation of the temperature. At 7° C, secondary medusae are formed on the manubrium of the anthomedusa *Rathkea octopunktata*. A substantial increase in numbers occurs in this way, asexually. If the temperature rises to 9—12° C, the animals go over to sexual reproduction. The gonads mature. Sexually mature medusae become asexual again if they are kept at temperatures below 7° C (WERNER, 1958, 1961, 1963). In this species increasing temperature leads to a change from asexual to sexual reproduction. By contrast *Pelmatohydra* forms sex organs under the influence of low temperatures (cf. ed. 1, p. 123; GÜNZL, 1959; SUGIURA, 1965; CUSTANCE, 1966; artificial parthenogenesis through warmth: SUZUKI, 1966).

In species with heterogeny a relation is frequently found between temperature and the appearance of parthenogenic or sexual forms. According to MORTIMER (1936) clones of parthenogenic females of *Daphnia* produce an increased percentage of males and mictic (automictic) females under the influence of high or low temperatures (cf. SHULL, 1911). According to FRIES (1964) ephippial females, that lay winter eggs, are formed predominantly at low temperatures, males at higher temperatures. The effect of temperature in this case is so great that at 8° C nothing but winter eggs are produced, while at 30° C only males are.

In this connection the effect of temperature on the appearance of males (cf. p. 382) is not so interesting as its effect on the development of mictic (automictic) or amictic (apomictic) females. This point of view has been brought forward by, e.g.,

LEES (1959) (cf. BRUSLÉ, 1966). His experimental subject *(Megoura viciae)* can develop into five different types of female: wingless virginoparae, winged virginoparae, wingless oviparae, fundatrix, and their direct offspring the fundatrigenia. The length of illumination per day has a substantial effect on the appearance of the different types of female, and so does temperature. Long photoperiods and high temperatures lead to the formation of virginoparae, while short days and low temperatures result in the appearance of oviparae (cf. p. 383). At moderate temperatures (ca. 15° C) virginoparae or oviparae are produced. The offspring of one brood are always of one type, i.e. either virginoparae or oviparae, so it can be assumed that influences on the mother such as temperature and photoperiod determine the type of offspring. At 23° C only virginoparae are formed, even in short photoperiods (LEES, 1963; cf. SCHAEFERS and JUDGE, 1971[23]). According to BONNEMAISON (1964), who carried out similar investigations on the gray apple louse *Dysaphis plantaginea*, one can assume that at temperatures below 22° C a neurosecretion is produced, in proportion to the photoperiod, that induces development of sexuparae.

In species with different forms of parthenogenic females temperature can influence the transition from one type to another. In the alternation of generations of the vine louse *Dactylosphaera vitifolii*, wingless smooth-coated virgins appear on the leaves (gallicoles), and wingless tubercle-coated virgins appear on the roots (radicicoles) of the grape vine. The eggs of these two types of females can also be distinguished easily. Gallicoles lay smooth-shelled eggs, radicicoles lay rough-shelled. The two female- and egg-types are connected by intermediate forms. In the course of a growing season gallicoles appear predominantly at first, and later radicicoles. When the aphids were reared in constant darkness, using grape vine tissue culture as food, and at various constant temperatures a critical temperature of 26° C was found for egg-formation. At higher temperatures mostly gallicolous eggs are formed, at lower temperatures radicicolous (RILLING, 1962).

Metagenesis and heterogenesis are thus forms of alternation of generation that are frequently not controlled endogenously. The change of generation is often not obligatory, as earlier authors believed, but facultative. External factors affect the appearance of one or the other generation types. The causative mechanisms involved are very diverse. Temperature often has a great influence on the processes of determination.

5. Influence of Temperature on Form, Color and Size

a) Size and Form

The ichneumon fly *Habrobracon* (SCHLOTTKE, 1926) is a well-known example of the observation that size can be influenced by rearing temperature. This effect is not only felt on the size of the whole animal, but can also be seen at different levels. *Drosophila* giant chromosomes are distinctly larger if the larvae are reared at low temperatures (FAHRIG et al., 1968). This enlargement is paralleled by an increase in nucleic acid content, while the number of cells remains the same at all rearing temperatures. Even the nuclear-cytoplasm relation can be shifted by temperature

---

[23] SCHAEFERS, G. A., JUDGE, F. D.: J. Insect Physiol. **17**, 365—379 (1971).

changes, as older authors established in e.g., *Paramecium caudatum* (further examples in BĚLEHRÁDEK, 1935; HESSE and DOFLEIN, 1943; v. BERTALANFFY, 1951). Several effects of temperature on a ciliate *(Tetrahymena pyriformis)* have been observed. According to JAMES and READ (1957) the cell volume depends on the temperature. After rearing for one month at 10°C, the average volume was 16.250 μ^3, at 20°C, 12.350 μ^3, and at 30°C, 9.375 μ^3. The surface-to-volume ratio remained approximately the same. Temperatures above 30°C inhibit cell division, even if they operate for only short times, periodically alternating with normal temperatures. Under these conditions the cells grow to two to four times their normal size (SCHERBAUM and ZEUTHEN, 1955). The omitted cell divisions are made up as soon as the high temperatures fail to appear. In addition to size, the shape (LEHMANN, 1962) and the various cell organelles (FRANKEL, 1962; GAVIN, 1965; DINGLE, 1970) are affected. The amount of mitochondria in cold-adapted orfs and eels and thus the oxidative capacity is increased (JANKOWSKY and KORN, 1965; WODTKE unpubl.). During earlier developmental stages of *Rana pipiens* the DNA content of 23–25° C embryos is greater than in 8–10° C animals. In older stages the situation is reversed (GREGG and LØVTRUP, 1960). In this connection one may mention the temperature dependence of the proportion of nuclear attachments in segmented leucocytes, that have been viewed as sex indicators. BURMEISTER and BECKMANN (1966) found that after treatment at 16° C, *Xenopus laevis* had 0.4 ± 0.24 such bodies per 200 leucocytes, but after treatment at 26° C the number was 1.5 ± 0.51. The authors could not establish a correlation with sex.

If the cell number is kept constant, and the individual cells of the metazoan body grow larger at lower rearing temperatures, the body will grow larger. So one finds that frogs, kept in the cold, which are heavier than those that stay warm (CHAMBERS, 1908) also have larger cells (cf. ALVAREZ and COWDEN, 1966). The same phenomenon appears in coelenterates (KINNE, 1958), such as *Cordylophora*, which have higher and wider cells in the growth zone when reared at 10° C than at 20° C. Similarly the penetrants become longer and wider at the lower temperature (cf. ed. 1, p. 125, BOYCOTT and GUILLERY, 1959). Liver of goldfish from lower AT has a higher weight based on high protein content but not increased cell number (DAS, 1967).

WODTKE (unpubl.) found in experiments with *Anguilla vulgaris* that the liver was double the size and the amount of mitochondria-protein per total protein was twice as much in cold-adapted (AT = 7° C) than in warm-adapted (AT = 27° C) eels. Readaptation from 7 → 14° C or 27 → 14° C respectively, takes about 14 days. By this transformation the influence of temperature is completely or partly compensated.

The various parts of the body can react differently to the influence of temperature. RAUH (1963) found that at low temperatures the toes of three pure lines of *Brachionus capsuliflorus* (rotifera) were elongated, while the size of the rest of the body remained approximately the same. A further example was reported by HOSOI (1954) who found enlargement of the wings of *Culex* after rearing at low temperatures. By contrast, bees (MICHAILOV, 1927), wasps and flies (DEWITZ, 1920) are said to develop shorter wings at low temperatures.

A distinction must be made here between normal and extreme temperature ranges. FRIEDLAND and HARNLY (1945) observed an enlargement of the wings of *Drosophila melanogaster* at normal temperatures with a maximum at 28° C. At higher rearing temperatures the wings become smaller again (cf. GUPPY, 1969; further examples in UVAROV, 1931). The temperature sensitive period falls in the pupal stage (PANTELOURIS, 1957). Extreme temperatures produce such extreme changes that they must be regarded as crippling effects (SWARUP, 1959b, c; FOX et al., 1961). If butterflies (*Trichoplusiani*, GRAU and TERRIERE, 1967) are reared at 30° C (in contrast to 20° C) then 84% are formed with crippled wings. High temperatures also have similar effects on pupae of the house fly *Musca domestica* (BODENSTEIN, 1940; cf. RAVEN et al., 1955; LEGAY, 1959). According to RAVEN and VAN ERKEL (1955) $CaCl_2$ prevents the malformation following heat treatment in *Limnaea* eggs. In *Myzus persicae* the results of high temperature extremes can be prevented if the temperature continually swings back into a more favorable range in a 12-h rhythm (DE REGGI and DELMAS, 1965).

Temperature effects of this type also extend to traits that have been used as species characteristics. Thus CAYROL and LEGAY (1967) came to the conclusion from their investigations on the nematode *Ditylenchus myceliophagus* that biometric criteria could serve only for a first approximation to species determination and must always be confirmed by anatomical or other traits (cf. EVANS and FISHER, 1970). This view was supported by the findings of CASSAGNAU (1955) who found that when *Hypogastrura purpurascens* (collembola) were reared at 10° C or higher temperatures a fairly high percentage of a variant that resembled individuals of another species appeared. This was a case of ecomorphosis in which it is still not clear whether temperature alone or other factors as well caused the transformation. Such transformations can be carried out very rapidly, as in *Isotoma olivacea* (collembola) where the spines on the fifth abdominal segment that are regarded as taxonomic traits of the genus *Spinosotoma* are formed, fully sized, within one molt after the temperature changes from 3–23° C (cf. CASSAGNAU, 1956a, b; GYSELS and BRACKE, 1964).

b) Cyclomorphosis

Cyclomorphosis must be considered in this connection. This term refers to the phenomenon in which a species in the course of generations goes through a cyclic change of form. Such a course is especially prevalent among fresh-water plankton organisms (daphnia: MARONI, 1961; EL-MAGHRABY, 1965; rotifers: HALBACH, 1969, 1970a, b; ceratids), that appear fairly similar in winter, while in summer they form a variety of bodily processes, often considered as superfluous. These appear in spring when the water temperature has reached about 12–16° C and disappear in autumn at about the same temperature.

Some investigators (JACOBS, 1935; WESENBERG-LUND, 1939) regard the increases in the body processes in the summer forms as adaptations to the lower viscosity of the warmer water. Another concept holds (WOLTERECK, 1934, ed. 1, p. 124) that these structures should be seen as mechanisms for controlling movement and stabilization. A completely different significance has been developed through JACOBS' (1967, 1970) work. This author starts from the observation that in spring certain species of daphnia not only have no head processes, but that they also have a large and functional filtration apparatus and a fairly large brood space. The spring form is thus able to build up a dense population in a short time. They are adapted for a high reproductive capacity. The streamlined summer forms on the other hand are in a much better position than the spring phenotypes to survive in the presence of hungry fish (cf. ZARET, 1969). Accordingly the most extreme development of the summer form comes at the time when fish are feeding most heavily. The investigated species, *Daphnia galeata* is thereby adapted at different seasons to different environmental factors.

c) Color

It can be said quite generally that in species whose coloration is related to the temperature, darker individuals appear at lower, lighter at higher temperatures (PARTECKE, 1959; GOGALA and MICHIELI, 1966; PAPILLON, 1968). In extremely high temperatures again the intensity of coloration increases, from which it can be concluded that in these cases the intensity is affected by specific metabolic influences, which are accelerated by increasing temperature up to a maximum, beyond which they are inhibited. Many examples are found among insects (on the genetic bases of temperature modification in ladybirds see KOMAI, 1956). According to ZWEIFEL (1968) the dark coloration of anuran eggs contributes to their faster development in cool water through the absorption of heat. If *Microbracon pupae* (NARAYANAN et al., 1954) are held for 10—75 days at 10—13° C, the imagos will be dark, but after 25° C or higher temperatures they will be brown. For the green coloration of *Acrida bicolor* a temperature of at least 16—21° C is necessary (OKAY, 1956), and according to PAPILLON (1965, 1968) the proportion of green to dark larvae of *Schistocerca gregaria* is determined by photoperiod as well as by temperature. The ichneumon wasp *Habrobracon* (SCHLOTTKE, 1926) forms dark pigments during the pupal stage at low temperatures. The processes can still be affected while the pigments are being formed: if the pupae are transferred to a different temperature, the pigmentation is intensified if the temperature is lowered (and vice-versa). The change in color from brown to red of the larval epidermis of the large puss moth *(Cerura vinula)* that takes place before pupation can be delayed by cold and induced at an earlier time by warmth (BÜCKMANN, 1964). This color change is caused by reduction of xanthommatins to dihydroxanthommatin (further, see MOELLER, 1964; BIEVER and MULLA, 1966; ICKERT, 1968/70[24]).

d) Seasonal Dimorphism

The changes in temperature during the course of the year can cause individuals to take on quite different colors in different seasons. This phenomenon, known as seasonal dimorphism, was discovered in the 18th century by entomologists who brought egg masses laid by single butterflies into the laboratory and raised adults from them that differed so much in appearance that they had been previously regarded as different species. There are normally two generations of the map butterfly *Araschnia levana* each year. The first generation originates in spring from over-wintering caterpillars, the second generation originates in summer. The spring generation — *levana* — has a reddish color, and the summer generation — prorsa —

---

[24] ICKERT, G.: Entomol. Abhandl. 36, 121—192 (1968/1970).

is dark. For a long time the view prevailed that temperature alone caused the dimorphism, since for example the dark spring form of *Pieris napi* could be obtained from summer pupae after maintenance at low temperatures. REINHARDT (1969; cf. MÜLLER and REINHARDT, 1969) carried out further analyses of seasonal dimorphism in *Araschnia levana*. From these a complicated picture of relationships emerged, because day length also plays a role in the course of events (cf. HIDAKA and TAKAHASHI, 1967). When the photoperiod exceeds 15.5–16 h per day, the larvae give rise to quick-hatching (subitans) pupae, from which imagos emerge after a short time (10–18 days). If these animals develop at normal temperatures (12–24° C) then dark *prorsa*- types arise. If by contrast the caterpillars grow in short days of less than 15 h light per day, development is halted (diapause, see p. 391). The pupae resume development only if they are kept for a definite minimum period at low and then at normal (12–24° C) temperatures. These pupae then give rise to reddish *levana*-butterflies. REINHARDT was able to establish that the duration of the pupal stage determined the type of adult. If the pupal stage is ended quickly at high temperatures [quick-hatching (subitans) pupae are formed at temperatures above 30° C; warmth-sensitive latent pupae at 24–30° C], then *porima*-butterflies which have still less red color than *prorsa*-forms are formed. The butterflies can be arranged in the following order in terms of their length of development: At low temperatures, and therefore with long pupal stage, *superlevana*-forms arise. *Levana*-, *porima*-, and *prorsa*-types come next. At far higher temperatures the development time is lengthened again and *porima*-butterflies are formed.

e) Final Size

The final size reached frequently depends on the environmental temperature, but individual animal species react in so many different ways, that general statements cannot be made. In many cases low temperatures result in larger individuals. The discomedusan *Cyanea capillata* grows to a diameter of 50 cm in the North Sea, but in the Arctic it reaches 200 cm (further examples for marine animals in HESSE and DOFLEIN, 1943; cf. KINNE, 1956, 1958; KINNE and PAFFENHÖFFER, 1965, 1966). Insects can also react in this way. Thus GALLIARD and GOLVAN (1957) obtained the largest individuals of *Aedes aegypti* after slow growth at 17–20° C. (Examples for insects: *Tineola biselliella*, TITSCHAK, 1925; *Habrobracon juglandis*, SCHLOTTKE, 1926; *Drosophila melanogaster*, ALPATOV, 1932; *Anthrenus fasciatus*, HERFS, 1936; *Glossina morsitans*, BURSELL, 1961 a; *Drosophila*, TANTAWY, 1961; *Culex sp.*, LANG, 1963; Mosquitoes, BRUST, 1967; *Sideridis pallens*, TISCHLER, 1967a. Examples for vertebrates: RAY, 1960; HEMPEL and BLAXTER, 1961; SAGER, 1963; SWEET and KINNE, 1964; LIN and WALFORD, 1966.) In many cases the small body size found at higher temperatures can be regarded as a result of rapid development.

On the other hand, a small body size in certain animal species is correlated with low rearing temperatures, as in insects (ZWÖLFER, 1934; OLDIGES, 1959) or in the case of various amphibia and reptiles, which remain relatively small at the northern edge of their territory of distribution in Europe, but grow larger towards the south (HESSE and DOFLEIN, 1943). This phenomenon may be related to the fact that the species of northern regions that are often warmth-loving enjoy optimum growing conditions only rarely. Thus they remain small, and may be regarded as stunted forms. However one must always be careful to note whether the effect of temperature is also seen in experiments on racially identical material, so that divergences between differently reacting geographical races are not involved (cf. MARGALEF, 1955). Regeneration of amputated tails of *Rana temporaria* and *Rana ridibunda* tadpoles produced distinctly shorter regenerates at $14-16°$ C than at $18-20°$ C. There were more mitoses at the higher temperatures (PUKHALSKAYA, 1959). The observation of VERMA and ATWAL (1968) that the maximum rate of silk production of the mulberry silk moth, *Bombyx mori*, was reached at $25°$ C may be mentioned here also. STRAUCH (1971[25]) determined from the size of fossil animals the temperature of the biotop. However this is dubious because high temperatures may cause large or small individuals, according to the species.

On the utility and attempts at explanation of the dependence of size on temperature of poikilotherms, see the ed. 1, p. 126 (cf. for temperature dependence of growth in *Hydra:* STIVEN, 1962, in Crustacea: MILLER and VERNBERG, 1968; LOCK and McLAREN, 1970; in fishes: LE CREN, 1958; LASKER, 1964; BRETT et al., 1969; ATHERTON and AITKEN, 1970; PANDIAN, 1970).

C. Resting Stages in Development and Their Induction or Termination by the Effect of Temperature

Resting stages in development, caused by internal or external factors, are grouped together under the concept of dormancy. Only those phenomena of dormancy caused by temperature are of interest here. Other conditions that affect dormancy, such as photoperiod, strength of illumination, humidity, nutrition, certain other environmental factors such as, e.g., osmotic pressure (BRENY, 1957), etc., will not be discussed.

Dormancy can appear at too high or at too low temperatures; in the first case, a summer period is involved (Estivation, cf., e.g.., TOMBES, 1964; MASAKI and SAKAI, 1965; RAUTAPÄÄ and MARKKULA, 1966; RIEDEL, 1967), in the second, which has been investigated far more frequently, a winter period.

1. Quiescence

Below the developmental null point and between the upper limiting temperature and the lethal temperature there are two regions in which development cannot progress. Though the region at high temperatures is usually indistinct and damage can occur readily, the ability to endure fairly long periods at low temperatures without damage is widely distributed. When they are affected by the low temperatures, the animals' development stands still. These low temperatures may prevent movement, but must not do so in all cases. In the former case a cold torpor is present together with dormancy (cf. p. 441). When favorable temperatures return development is resumed at any time .This type of response is known as quiescence. There are no special adaptations for enduring periods of low temperature. Only the trait of being able to survive the unfavorable temperature without damage must be present (cf. for linkage of diapause and cold-resistance, PRECHT, 1964b; SIMAKOVA, 1971[26]). Following H. J. MÜLLER (1970), this type of response can be

---

[25] STRAUCH, F.: Palaeogeogr. Palaeoecol. **9**, 59—64 (1971).
[26] SIMAKOVA, T. P.: Ekologija **1**, 68—79 (1971).

called consecutive dormancy, because there is no provision to assure that the most resistant stages go through the unfavorable seasons. No special over-wintering stages are present (e.g., staphylinids, HEYDEMANN, 1956; RENKEN, 1956).

Thermal quiescence can appear together with other phenomena of dormancy. Thus many diapausing butterflies emerge in the laboratory after treatment with cold and subsequent maintenance at moderate temperatures, even in winter, but in the field they do not emerge because the temperatures are too low to allow development to stages capable of emerging. In such cases a thermal quiescence follows a diapause. This period can often be prolonged greatly without harm. Larvae of *Pyrausta* can remain in quiescence for up to 540 days and still pupate later, in over 80% of the cases, at 27° C (BAKER and JONES, 1934). Prepupae of *Trichogramma* survive up to 300 days at 10° C in a developmental rest phase. Nevertheless, the subsequent development to the imago is prolonged in proportion to the length of the quiescent period (QUEDNAU, 1957).

2. Diapause

a) Obligatory and Facultative Diapause, "Parapause"

In contrast to quiescence, diapause is a pause in or inhibition of development, distinguished by special physiological mechanisms, which can become functional only in certain stages. In juveniles, growth or development may be inhibited; in adults the gonads may be affected (examples of diapause in juvenile animals: LEES, 1955; MÜLLER, 1970; in imagos: *Ostrinia nubilalis:* CLOUTIER and BECK, 1963; *Musca autumnalis:* STOFFOLANO and MATTHYSSE, 1967; *Pyrrhocoris apterus:* HODEK, 1968; *Carabidae:* KREHAN, 1970). Similar resting periods in development are found in many different groups of animals (amphibia: BEBAK, 1958; WERNER, 1969; fishes: PETERS, 1965; insects: LEES, 1955; entomostraca: STORSS, 1966; mites: TESCHNER, 1961; BABENKO, 1967; DONDALE and LEGENDRE, 1970; nematodes: FUSHTEY and JOHNSON, 1966).

The principal physiological bases of diapause are of hormonal nature (cf. WIGGLESWORTH in BEAMENT et al., 1968; WOOLHOUSE, 1969; YAMASHITA and HASEGAWA, 1970[27]).

The inhibition of growth brings about an adaptation of the course of development of an animal species to the fluctuating environmental factors in the course of the year (prospective dormancy, MÜLLER, 1970). Since the mechanisms developed for this purpose are very varied, it is difficult to classify them. A fundamental distinction can be made between obligatory and facultative diapause and "parapause". The eggs of the opilionid *Mitopus morio* provide an example for "parapause". Though development begins at 20° C, as the contraction of the yolk within the eggs indicates, it stops quite soon (TISCHLER, 1967b). If the eggs are now transferred to 5° C, the embryos develop up to emergence. Thus a shift of the developmental optimum has taken place, so that 20° C is now above the optimum and development stops at this temperature. By contrast 5° C is in the favorable temperature range. This type of response is called a "parapause" (MÜLLER, 1970; cf.

[27] YAMASHITA, O., HASEGAWA, K.: J. Insect Physiol. **16**, 2377—2383 (1970).

OHNESORGE, 1960). It prevents embryonal development from running its course in autumn and the sensitive larvae, that could not survive the winter, from emerging too early. Instead its effects lead to the conclusion of embryonal development in spring, while the species passes through the cold season, in a resistant stage. *Mitopus* eggs are very resistant to cold (about the linkage of diapause and temperature-resistance, cf. PRECHT, 1964b; MINDER and CHESNEK, 1970). The embryonal development of the plant louse *Leptopterna dolobrata* provides an example of obligatory diapause. The eggs are laid in July and ca. 20 days later, independently of external factors, an obligatory diapause sets in. It can be divided into two phases on the basis of temperature dependency. A temperature-independent phase, which corresponds to the endogenous portion of the total resting period, lasts ca. 180 days (mesodiapause). Metadiapause, which is temperature-dependent, follows and ends after 18 days, at 16° C. But only a few larvae emerge. At 5° C metadiapause lasts longer (ca. 150 days) and many more larvae emerge (after transfer to higher temperatures). Under natural temperature conditions the developmental time is very long (ca. 11 months), causing the young larvae to appear at the same time as their food supply (freshly sprouted grass spikes, BRAUNE, 1971[28], cf. p. 396; BEHRENDT, 1963).

In many other cases, diapause is induced by external factors (e.g. photoperiod). We are then dealing with examples of facultative diapause, because, in the absence of external initiating factors, development is not inhibited. In the case of *Ischnodemus sabuleti*, the imagos that live through the winter lay eggs in spring from which nymphs emerge, which develop during the summer. Weather conditions determine the stage that is reached by fall. When the day length falls below a certain period the animals go into diapause. This can occur during any nymphal stage, and under very favorable conditions the animals molt into imagos before the diapausing day length is reached. In this case larval-diapause is omitted. Therefore nymphal diapause in *Ischnodemus* is facultative. The imagos appear in summer, but are not able to reproduce in the same year. The females must live through a winter to be able to form eggs and the males to become able to copulate. This is an obligatory imaginal diapause. Since larval diapause is facultative, but imaginal diapause is obligatory, *Ischnodemus* is a potentially univoltine species. In southern France it is univoltine, in the northern part of its distribution range bivoltin (TISCHLER, 1960). The butterfly *Iphiclides podalirius* is potentially multivoltine, but north of the Alps it is facultatively bivoltine. Unusually small individuals of the spring generations, found in central Europe, probably originate from the small amount of summer generation. They are poorly nourished forms, because the larvae lived under unfavorable conditions (WOHLFAHRT, 1955).

It may be generally stated that animal species able to undergo facultative diapause encounter the external inducing factors while they are in a sensitive state. In many cases a certain time period follows in which development continues but diapause can no longer be prevented. During this time the animals develop to the diapausing stage. Then development stops and diapause has begun. Certain environmental factors then cause the inhibition of development to be inactivated and diapause is thereby ended (for an exact classification, cf. BEHRENDT, 1963, p. 335 ff.). Under

---

[28] See p. 362.

natural conditions thermal quiescence may follow, because temperatures are too low for further development. The course of events need not always follow this outline strictly. For example in *Austroicetes cruciata* development proceeds very slowly during diapause (ANDREWARTHA, 1943). Diapausing larvae of *Diatraea saccharalis* do not become inactive, and even feed (KATIYAR and LONG, 1961).

In general, adaptation of annual rhythm to hard winter climate arises polyphyletically. Thus in carabids *Pterostichus*, very different forms of dormancy are found (THIELE, 1971[29]).

b) Induction of Diapause by Temperature

In most cases, diapause is induced by the lengthening or shortening of the photoperiod beyond certain limits (cf. MÜLLER, 1963; BECK, 1968). This is biologically meaningful since the photoperiod can be considered as a very precise indicator of time. But temperature can also act as a forerunner of future unfavorable conditions for development by itself (e.g. in animals living in the ground, cf. THIELE and KREHAN, 1969) or in conjunction with the photoperiod. Other causative factors have been recognized, as: substances in food; GAMBARO, 1954, 1958; rearing in isolation; IWAO, 1962; desiccation; BOULOT and GALLISSIAN, 1960, etc.

High temperatures quite commonly tend to promote continued development at unchanged rates and low temperatures, by contrast, to initiation of winter diapause (examples are found in CHURCH and SALT, 1952; LEES, 1955, p. 31; BONDARENKO and KUAY-JUAN, 1958; BECK, 1962; BABENKO, 1967; BRIAN and KELLY, 1967; KISHINO, 1969; SAUNDERS, 1971[30]). Pupae of *Sphinx pinastri* go into diapause after maintenance at $10-20°C$, at $23-35°C$ developmental rest is prevented in an increasing proportion of cases (GÖSSWALD, 1936). Diapause is induced in 95% of the individuals of *Hyphantria* in constant darkness at $22°C$, but at $28°C$ the proportion of diapausing animals goes down to 4% (JASIC, 1960; cf. HODEC, 1962; MÜLLER, 1963, p. 510). The diapause-inducing effect of low temperatures ($70°$ F) on *Pectinophora gossypiella* is so strong that other influences such as long day length, which tend to prevent diapause, remain ineffective (BULL and ADKISSON, 1960). Summer diapause is specifically favored by high temperatures (MASAKI and SAKAI, 1965).

In many other cases temperature and photoperiod work together in varying proportions. Thus with photoperiods of $9-12$ h *Mamestra brassica* always develops into diapause pupae at all temperatures, even at $28°$ C (BONNEMAISON, 1960). Beyond these limits high temperatures cause further rapid development, but low ones induce diapause (cf. BECK and HANEC, 1960; MÜLLER, 1963). This response is oriented towards field conditions in a biologically meaningful way since the danger that extremely low temperatures will follow is present when temperatures are low or moderate, not when they are high. Photoperiods and temperature factors can work against one another. According to LEES (1953) photoperiods of LD 12 : 12 and temperatures of $15°$ C induce diapause in 100% of the animals *(Metatetranychus ulmi)* while the same photoperiod at $25°$ C does so in 21%. Changing the temperature together with the transition from light to dark shows that high temperatures during the light phase have only a very slight effect in retarding diapause. In an

[29] THIELE, H. U.: Zool. Jahrb. System. Ökol. Geogr. **98**, 341—371 (1971).
[30] SAUNDERS, D.S.: J. Insect Physiol. **17**, 801—812 (1971).

experiment in which 12 h L and 25° C was followed by 12 h D and 15° C, diapause
was induced 96% of the time. Under the reversed conditions of 12 h L, 15° C and
12 h D and 25° C it was induced in 53%. In the case of *Nasonia vitripennis*, a given
photoperiod may act as a long day or as a short day, depending on the temperature
(SCHNEIDERMANN and HORWITZ, 1958; SAUNDERS, 1967).

c) Intensity of Diapause

In order to be biologically effective, diapause must not only be initiated, but must
also last long enough. It is customary to judge the intensity of diapause by its total
duration (ANDREWARTHA, 1952; but cf. JACOBSON, 1962; BEHRENDT, 1963, p. 339;
TAUBER et al., 1970). As a result of the variable intensity of diapause, if a group of
animals are examined together after a certain period at low temperatures followed
by transferral to temperatures which favor morphogenesis, they do not all emerge.
This effect has also been observed under natural temperature conditions. Eighty-
two percent of the larvae of the tortricid *Melissopus* end their diapause after one
winter, 13% after two winters, and 5% after three (DOHANION, 1942; cf. OHNE-
SORGE, 1960; BILIOTTI et al., 1964; DUMORTIER, 1967; SULLIVAN and WALLACE,
1967; BASEDOW and SCHÜTTE, 1971).

The intensity of diapause can be influenced by temperature, as for example in the
eggs of *Acheta commodus* (HOGAN, 1960) and *Gryllulus mitratus* (MASAKI, 1962). In
the latter species egg diapause is obligatory. High temperatures result in a high
intensity of diapause, low temperature in a low intensity. As a result of this effect
the eggs which are laid in fall, at relatively low temperatures, emerge from diapause
in spring, while the eggs laid in summer remain in diapause till the next spring.

d) Termination of Diapause

Most organisms with obligatory and facultative diapause require a change in environmental circumstances once their developmental rest phase has begun (unless
they are governed by an internal clock, p. 396). If they continue under conditions
which would normally be considered favorable for development, they die after some
time. If however they are transferred to different conditions, their diapause ends
and — brought once more into "normal" conditions — development proceeds. In a
large number of cases, low temperature acts as a diapause-ending factor, and does
so in a range in which biological processes would ordinarily be inhibited. Certain
processes must go on during the exposure to low temperatures which overcome the
inhibition of morphogenesis (diapause development, ANDREWARTHA, 1952). The
temperature dependence of these processes can differ fundamentally. On the one
hand the inhibition of development may be reduced at a rate inversely proportional
to the temperature; on the other hand a definite temperature optimum may be
present. The lower the temperature the faster egg diapause of *Acheta commodus* is
ended. At 12.8° C 20–60 days were needed, according to the intensity of diapause;
at − 7.2° C only about 5 h, and at − 16.5° C, 20 min (HOGAN, 1960; cf. BOHLE,
1969, p. 550 ff.).

According to SLIFER (1949), the egg diapause of *Melanoplus* is similarly ended by
the short-term effect of low temperatures, probably due to the detachment of a
wax layer from the hydropyle, enabling the egg to take up water (SLIFER and

KING, 1961; cf. ROEMHILD, 1965). The same effect — detachment of the wax layer, water uptake, and continuation of development — could be obtained by placing the eggs in xylol or other wax dissolving agents. Temperature can exert similar effects, since certain diapause states can be ended by short baths in hot water. In these cases the temperature must be close to the upper lethal limit (EMME, 1953; cf. BURGES, 1962). According to RAKSPHAL (1962) the yolk of diapausing eggs of *Gryllulus pennsylvanicus* becomes less viscous as a result of cold and therefore can be assimilated more rapidly by the embryo.

If the temperature dependence of the termination of diapause follows a curve with an optimum, specific biological processes must take part, at least in the neutralization of developmental inhibition. ANDREWARTHA (1943, 1952) found a graphic example in the diapause of *Austroicetes*. If the eggs of this grasshopper are held for 60 days at 10° C and then are moved to 25° C all the nymphs emerge after a relatively short time. Maintenance for 60 days at 6° C or 13.5° C results in emergence of a small percentage of nymphs; i.e. at 10° C all animals end their diapause in 60 days. This temperature can be regarded as the optimum for ending diapause (diapause development). At other temperatures inhibition of development ends only in those animals with a weak diapause intensity. Below 5° C and above 25° C diapause cannot be ended at all. This optimal temperature lies at ca. $+5°$ C for *Tetrix undulata* (SICKER, 1964). Low temperatures (5—15° C) can affect *Gryllulus campestris* over separate periods which summate to the required duration (FUZEAU-BRAESCH, 1965; cf. ROEMHILD, 1965).

The range of the optimum temperature for diapause development can vary in width and lie at different temperatures, according to the geographic distribution of the animal (BONNEMAISON, 1960; DANILEVSKII, 1965). Diapause of *Gilpinia polytoma* can be ended by temperatures of $-10°$ C, while $+10°$ C is too high (PREBBLE, 1941). Many European butterflies end diapause most rapidly at temperatures near 0° C (examples are found in LEES, 1955). Insects from warmer climates end their diapause at correspondingly higher temperatures. For example *Gryllulus commodus* ends its egg diapause most rapidly at 13° C (BROWNING, 1952b).

In many cases the optimal or other effective temperature must encounter a sensitive state or a sensitive period in the development of the animal in order to end diapause. At earlier or later times its effectiveness is reduced or totally abolished. Such processes have been investigated most often in diapausing eggs because their successive stages of development have been especially well characterized, morphologically. The diapausing stage of the eggs of *Locusta migratoria* occurs at the end of anatrepsis. This stage is reached after 10 days of development at 25° C. If a low temperature of 8° C is applied immediately afterwards, the treatment is completely ineffective in ending diapause. Thus while the embryo has reached the morphological stage of responsiveness after 10 days at 25° C, it has not reached the physiological stage. The latter stage is reached most quickly if physiological development is allowed to proceed for another 50 days at 25° C. The embryo continues to become increasingly more sensitive to low temperature until after only 60 days it ends diapause at 8° C and develops further at 33° C (LE BERRE, 1953; but cf. for other interpretations p. 396). The sensitive stage can occur very early in embryonal development. E.g. in *Gryllulus commodus*, it occurs immediately after the eggs are laid and long before the developmental rest period (BROWNING, 1952b;

cf. BLAKE, 1963) and in *Melanoplus bivittatus* (CHURCH and SALT, 1952), which can be completely prevented from diapausing by treatment with low temperatures, even though an obligatory diapause is probably present.

e) Relation between Developmental Arrest and the Endogenous Annual Clock

Induction and termination of developmental arrest by external factors lead to an adaptation of the life cycle of animal species to the changing environmental conditions during the course of the year. The appearance of sensitive developmental stages at a time when there is a high probability that extreme weather conditions would lead to the death of these stages, is prevented. It is a question whether other mechanisms have been developed which, alone or in combination with diapauses induced by external factors, lead to the introduction of periods of absolute or relative developmental arrest into the life cycle of a species. The obligatory diapause already mentioned, in which the resting phase must begin at a certain stage as a result of internal causes, and development subsequently proceeds — generally at higher temperatures — is a step in a fundamentally different direction. But frequently obligatory diapause is also dependent on the external world inasmuch as the developmental arrest is ended most rapidly at certain, i.e. optimal temperatures. In order to avoid the appearance of sensitive stages at unfavorable seasons it is safer to introduce a temperature independent phase before the ending of diapause. BEHRENDT (1963) found this situation realized in the egg development of *Aphis fabae*. The so-called mesodiapause lasted ca. 50 days at all temperatures tested between $-0.2°$ C and $16°$ C. This time seems to be sufficient — especially in co-operation with the subsequently initiated development of diapause — to prevent the appearance of young larvae in fall more effectively than the development of diapause at low temperatures alone. In this example we are dealing with a process of time measurement analogous to the principle of an hourglass. Once the process has been set in motion, i.e. once the animal has reached a certain stage, it runs its complete course independently of the external factor of temperature (cf. p. 392, BRAUNE, 1971[31]).

To solve the problem of overwintering, it would be still more useful if the animals were able to recognize a specific date. TISCHLER (1967a) describes one step in this direction. He transferred larvae of the noctuid species *Sideridis pallens* from their over-wintering temperature of $5°$ C at the beginning of January, the end of January and beginning of February, and in April to $20°$ C. The animals' speed of development at the high temperature was proportional to the length of time they had passed at $5°$ C (cf. CORBET, 1957; LOVE and GOODWIN, JR., 1959; TEULADE, 1963; RAAB, 1966; GUENNELON, 1966; WILDBOLZ and RIGGENBACH, 1969; ANNILA, 1970). The tendency to develop was strengthened to such an extent that those animals which were transferred earlier to $20°$ C required 6 weeks to pupate, the second group required 5 weeks, and those transferred in April needed only 3—4 weeks. The pupal stage that followed consistently lasted ca. 14 days for all three groups of animals. Through this type of response of the larvae the time of pupation is compressed into a shorter time span than that corresponding to given environmental conditions. As a result the adults emerge at approximately the same time

---

[31] See p. 362.

in different years, in spite of variations in temperature (cf. BROWNING, 1952c; CANARD, 1958; THANH-XUAN, 1967).

Similar phenomena are found in mollusks. PRECHT (1936) observed that egglaying in the posthorn snail *Planorbis corneus* was interrupted in fall (Oct.-Nov.) although the animals were kept at room temperature and long photoperiods, so that diapause was not being induced. SEGAL (1960), who maintained the slug *Limax flavus* for more than three years at 10°C and 20°C, also came to the same conclusion. At certain times of the year no eggs were laid (cf. BLAKE, 1958; I. PRECHT, 1967).

An annual period is found in the case of *Limax* since the adults live more than one year. Indications of comparable phenomena, somewhat in the principle of an unperiodic clock (for the difference between periodic and unperiodic clocks see PRECHT, 1964a), are also found in species which live only one year as adults. For example, according to PRECHT (1964a) low temperatures (5° C) can end diapause of the bug, *Ischnodemus sabuleti*, with great difficulty in November and December, but can do so easily in January or February. Stemming from several possible interpretations of these results, the question of how an endogenous component can play a role in ending diapause has been discussed most intensively. This endogenous component would have to inform the animal about the passage of time. Changing environmental temperatures should have as small an effect as possible, in order to enable the *Ischnodemus* imagos to terminate diapause at a specific time of year or after the passage of a certain time span (ca. several weeks).

These problems of time measurements extending over several years have been investigated very rarely, since it is very time consuming to maintain experimental animals for such long periods. To identify an endogenous component it must be determined that the process can run freely in forming the basis for a rhythm, i.e., that the phase length changes in the absence of external indicators of time. It must also be determined that, in accordance with the hourglass principle, the rhythm or the process of time measurement is not influenced by the surrounding temperature. SEGAL (1960) could determine that both were true for egg laying activity of the slug *Limax*. A temperature dependent process is not suitable for measuring time in poikilothermic animals unless a very rapid, complete adaptation of type 2 is present. Therefore metabolism (measurable by oxygen use) is not involved in the case of *Ischnodemus*, since there is no temperature adaptation in the sense of a compensation in the diapausing animals (PRECHT, 1964b).

The same problems are found in connection with circadian rhythms. Much more is known about them, including their dependence on temperature (cf. PRECHT, 1964a; HOFFMANN, 1968, 1970).

f) Distinction between the Concept of Diapause and Other Inhibitions of Development

A temporary resting phase in the development of an individual is designated as diapause. This means that development begins at first, and then ceases due to internal or external causes. To speak of imaginal diapause in connection with ovaries that do not develop immediately after the imaginal molt, or with male insects that are not able to copulate immediately, represents an extension of the concept. In these cases there is no pause in development of the individual, but rather a temporary delay at the point of transition between generations. It is interesting that such a delay does not appear in the newly fertilized egg, but that development of the individual must begin. Only then can diapause set in. It is not very meaningful to refer to certain phenomena as diapause, as some authors do. This refers to surviving stages, such as the cysts of protozoa, the gemmules of sponges, and the statoblasts of bryozoa. In these stages survival of unfavorable conditions is emphasized. An inhibition of development is naturally correlated with this (this is also true for anabiosis stages). In diapause the emphasis is more on the inhibition of development. At the same time one generally finds an increase in resistance, e.g. to extreme temperatures, in diapause. But the animal is not completely modified, as in the specialized resistant stages mentioned above. These stages are not eggs that are formed by sexual processes, nor forms that serve reproductive purposes. One does find differentiation in them, when they germinate, but not development, as in the eggs. In these resistant stages, body cells or fractions of the protoplasm are enclosed in protective shells. In contrast to the eggs, which begin development immediately, these forms begin differentiating only at or after germination. RASMONDT (1954, 1965) investigated the gemmules of fresh water sponges. RASMONDT assumes that the gemmules are prevented from germinating by a substance originating from the parent sponge. Investigations of the author[32] on *Ephydatia fluviatilis* showed that the inhibition of germination could easily be prevented by the application of extreme high or low temperatures for short periods. RASMONDT (1954) also overcame the inhibition of development by the use of various chemicals. Thus there are phenomenological correspondences with termination of diapause in insects, but the inhibition of germination in gemmules, etc. can nevertheless not be called diapause for this reason, because individuals of these animals do not survive, but rather parts of individuals differentiated for this specific purpose. This holds true for gemmules, statoblasts, and also the cysts of protozoa, in which the total protoplasmic body is often not enclosed. Thus the forms of dormancy include quiescence, diapause, and the formation of specialized resistant stages, which are classified in the following diagram:

---

[32] LAUDIEN, H.: Zool. Anz. **189**, 259—265 (1972).

References to this section, see p. 470.

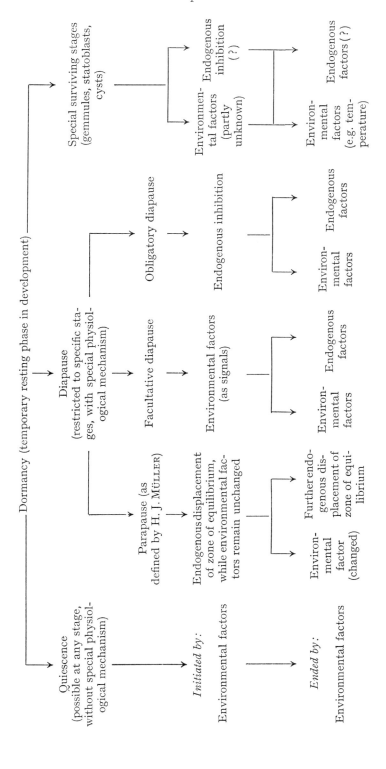

III. Limiting Temperatures of Life Functions[1]

H. PRECHT

This chapter will deal with problems of the resistance of animals or of selected life processes to extreme temperatures (excluding resistance of growth, development and reproduction), and not with the means by which poikilothermic animals frequently avoid extreme body temperatures. The extensive literature will be considered only as it bears on specific problems (see p. 302f., also SMITH, 1958, 1961; SCHOLANDER, 1959; PARKES and SMITH, 1960; PRECHT, 1963, 1964d; USHAKOV, 1963, 1966; POLYANSKY and USHAKOV, 1965; MERYMAN, 1966; ASAHINA, 1962a, 1965b, 1967, 1969; GLUSHANKOVA et al., 1967; E.I.F.A.C., 1969; MAZUR, 1970).

A. Limiting Temperatures

1. Factors which Influence Limiting Temperatures

Limiting temperatures are dependent on many kinds of factors, which must be taken into account when these temperatures are measured. These factors will be dealt with here only briefly (with arbitrarily selected examples), since these problems have only indirect relevance to the theme. Those reports which contribute to the understanding of resistance to temperature extremes will be considered more exhaustively later.

Beginning with single-celled organisms, it must be recognized that mass cultures may show low heat resistance; in *Paramecium* this appears to be related to the change in pH of the cultures (GARNER, 1934; cf. CHALKLEY, 1930).

For marine animals the salinity must be taken into account. Often higher salinity increases the resistance of poikilosmotic animals to both high and low temperatures, as for example in *Enchytraeus albidus* (KÄHLER, 1970); an optimum of salinity for high temperature resistance may exist (ALDERDICE). In *Enchytraeus* at least heat resistance changes also in the homeo-osmotic range at low salinities (KINNE, 1954; TODD and DEHNEL, 1960; MATUTANI, 1962; THEEDE and LASSIG, 1967; IVLEVA and also SCHLIEPER in TROSHIN, 1967; NAGABHUSHANAM and SAROJINI, 1969; WILLIAMS, 1970). In that sort of experiment the adaptation temperature (AT), as for *Enchytraeus*, can determine the outcome, likewise the high temperature at which the test is made (for example with guppies, ARAI et al., 1963; cf. DICKIE, 1958). With *Enchytraeus* and *Homarus americanus* increased salinity has less effect at higher than at lower ATs (KÄHLER, 1970; McLEESE, 1956). Also the ionic

---

[1] The following abbreviation will be used: CTM = critical thermal maximum (or minimum); see p. 406f. (see also p. 302).

composition of the water for the preliminary exposure can influence heat and cold resistance, for *Enchytraeus* in a manner difficult to interpret; both the concentration of ions and the AT were significant (ref. in KÄHLER, 1970; cf. THEEDE, 1970).

Also the adjustment to changed oxygen content can influence resistance to temperature extremes. The upper lethal temperature of *Homarus americanus* for example becomes lower as the oxygen content of the water decreases (McLEESE, 1956). Copepods *(Tigriopus japonicus)* acclimated at low oxygen saturation are significantly more heat resistant than those acclimated to high levels (MATUTANI, 1962; this is not true for *Carassius gibelio*, KREBS, unpubl.). We are here concerned with problems of change in general resistance (p. 432).

Changes of resistance to temperature extremes often depend on season (usually there is increased heat resistance in summer and enhanced cold resistance in winter); the importance of the AT for these changes will be considered later, but it is not the only critical factor (HOAR, 1955; cf. HART, 1952; SUOMALAINEN, 1958; TODD and DEHNEL, 1960; and others). According to PAYNE (1926), in several insects (pests of stored provisions, and water insects) the supercooling point and freezing point showed only slight modification with season (but see p. 413). For *Gadus morhua* the low limiting temperature was 2° C from October to June; it decreased in the course of the summer, perhaps because of better feeding conditions, to 0° C (WOODHEAD and WOODHEAD, 1959/60). A better *Zeitgeber* signal for season than temperature is day length. Annual cycles of resistance which are independent of AT should be tested to determine whether day length is significant, as it has frequently proved to be (cf. HUTCHISON, 1961; VOGEL, 1966), or whether an endogenous annual rhythm is present, which is relatively independent of *Zeitgeber* signals (compare TYLER, 1966). Goldfish maintained under controlled photoperiods (and constant ATs) for 6 weeks or longer were relatively more resistant to sudden elevation in temperature when the daily photoperiod had been long (16 h), and relatively more resistant to sudden chilling, when it had been short (8 h), but the magnitude of the effect varied with season (HOAR and ROBERTSON, 1959). The effectiveness of long versus short daylengths on heat resistance of *Chrysemys picta* was more pronounced at low ATs (HUTCHISON and KOSH, 1965; cf. KOSH and HUTCHISON, 1968; LICHT, 1968; BRATTSTROM, 1968; MAHONEY and HUTCHISON, 1969). In this sort of experiment on the effect of season it must be ascertained whether the change in photoperiod has simply brought the animal into another phase such as reproduction or dormancy, which is only indirectly related to a change of resistance to temperature extremes. Furthermore, in the changes in resistance by adaptation to varied temperatures which will be described later, it must be noted precisely whether the direct effect of an AT is realized or whether here too the temperature change has merely elicited a change of state. Russian authors in particular have investigated the relationship between resistance to extreme temperatures and these changes of animal state, which are often hormonally influenced. Heat resistance of the ciliary beat of gills of *Mytilus galloprovincialis*, investigated immediately after collection, fluctuates with the season (DREGOLSKAYA in USHAKOV, 1963), but not in accordance with temperature fluctuations in the sea water; in all cases the decrease in heat resistance coincides with the period of reproduction. Spawning of frogs begins in spring even though the animals have been kept in the dark at low temperatures (KASBOHM, 1967). At this time the heat resistance of the muscles changes in adults but not in young frogs (PASHKOVA in TROSHIN, 1967), even when the animals were kept at a constant ambient temperature. Resistance changes with life cycle also occur in unicellular organisms (POLYANSKY and SUKHANOVA in TROSHIN, 1967).

Diurnal changes in resistance may be significant, for example the critical thermal maximum (p. 400) of *Chrysemys picta* differs maximally by 1.4° C (this appeared in other experiments after an adaptation change of ~ 10° C). The highest value is at approximately the highest daytime temperature in nature (AT 20° C, 16 h daylight; KOSH and HUTCHISON, 1968; cf. MAHONEY and HUTCHISON, 1969; DUN-LAP, 1969.[2]) A tidal rhythm of resistance (such as occurs in algae, BIEBL, 1969) has not been observed for animals.

Changes of resistance to temperature extremes are often associated with entrance of the animal into dormancy. Insect winter diapause is usually characterized by high resistance to cold, although there need not be a causal relationship (PRECHT, 1964 e, see p. 417). ASTAUROV (in TROSHIN, 1967) made thorough investigations of the changes in heat resistance during embryogenesis of silkworm eggs [pre-diapause with initial maximum, estivation (diapause), hibernation and spring development]; these changes cannot be regarded as direct responses to environmental tempera-ture, but they can be considered indirectly as adaptations, since the temperature during estivation can be high (cf. ILYINSKAYA, 1968).

Resting stages which serve to bridge over unfavorable periods are often especially resistant to both extremes of temperature, for example the anabiosis of animals. Bdelloid rotifers build tubes, before actual drying out begins, into which the organs are stowed relatively free of tension (LINDAU, 1958; BURNS, 1964).

Further influences upon the heat and cold resistance of animals may depend on stage in development, age and size (ed. 1, p. 46, 54). In *Drosophila subobscura*, heat resistance decreases with age (HOLLINGSWORTH and BOWLER, 1966; cf. BOWLER, 1967; PLATT et al., 1957; DAVISON, 1969). Chill coma temperature for bees in summer decreases with age; worker bees have a lower chill coma tempera-ture than drones and queens (FREE and SPENCER-BOOTH, 1960; cf. SCHWALBACH and AGOSTINI, 1964; ATWAL, 1960). In *Dahlbominus fuscipennis* the aging effect on heat resistance rises with increasing AT (BALDWIN, 1954). Larvae of tropical fiddler crabs can withstand low temperature better than the adults (F. J. VERN-BERG, 1969). Resistance to high temperatures decreases with size in *Asellus inter-medius* and *Gammarus fasciatus* (SPRAGUE, 1963) and increases in *Hyalella azteca* [BOVEE, 1949; cf. TODD and DEHNEL, 1960; WAUTIER and TROIANI, 1960; BASE-DOW, 1969 *(Dinophilus)*; HUTCHISON, 1961; BRATTSTROM and LAWRENCE, 1962; HEATWOLE et al., 1968; KRAKAUER, 1970 (Amphibia); LOWE and VANCE, 1955 (Reptilia)]. In fishes the upper resistance limit is often independent of the size of the animals (TIMET, 1963; THIEDE, 1965), although not always (HART, 1952; IRVINE et al., 1957; SPAAS, 1959, 1960; KUTHALINGHAM, 1959). According to ALABASTER (1967) smolts of *Salmo salar* are more sensitive to high temperatures than parr when tested in fresh water (cf. HUNTSMAN, 1942; BISHAI, 1965; for changes in thyroid activity with age see ROBERTSON, 1948). Cold resistance of fish may

---

[2] cf. JOHNSON, C. R.: Comp. Biochem. Physiol. **41 A**, 727—730 (1972).

increase with age (HOAR, 1955; PITKOW, 1960; McCAULEY, 1963) or decrease
(GUNTER, 1947). Juvenile fish, *Aequidens portalegrensis*, are better able to tolerate
cold than older ones, which tolerate heat better (MORRIS, 1962). Body size is correlated significantly neither to heat tolerance nor to cold tolerance of guppies,
though males larger than 16 mm are relatively tolerant of heat and females larger
than 18 mm are relatively tolerant of cold. There was no significant correlation
between individual heat and cold coma temperatures (TSUKUDA, 1960c). Fish
eggs often have relatively narrow tolerance limits (FRY in ROSE, 1967, p. 390,
cf. E.I.F.A.C., 1969). Heat resistance of organs can depend on the size of the
animal (for example in *Rana temporaria*, CHERNOKOZHEVA in GLUSHANKOVA et al.,
1967).
Difference in resistance according to sex has often been demonstrated. In *Gammarus duebeni*, for example, the females are more sensitive to heat than the males,
in contrast to other species of *Gammarus* (SPRAGUE, 1963). In both sexes smaller
(younger) specimens of *G. duebeni* seem to be more resistant to high temperatures
than larger (older) ones (KINNE, 1959). Female *Idotea balthica* are more heat
resistant than males (FURCH, see p. 439; for insects, see PIELOU and GLASSER, 1954).
According to PITKOW (1960) male guppies are less cold tolerant than females in the
temperature range of "primary chill coma" (p. 404). There does not seem to be a
difference with sex in cold tolerance in the temperature range of "secondary chill
coma" (cf. GIBSON, 1954; HOAR, 1955; TSUKUDA, 1960c).
The kind of nutriment can also influence temperature resistance. Goldfish fed on
high melting point natural fat (lard) were most resistant to heat. Resistance to
cold showed no relationship to the melting points of natural fats (pilchard oil,
herring oil, and lard). Hydrogenation of pilchard oil used in the diets reduced
resistance of goldfish to both low and high temperatures. Fish fed on hydrogenated
cottonseed oil showed a similar reaction to high temperatures (HOAR and COTTLE,
1952). The temperature resistance of goldfish to heat and cold is increased by
addition of cholesterol or phospholipid to basic diets; the relative effectiveness of
these treatments varies with the season (IRVINE et al., 1957). Larvae of *Pseudo-
sarcophaga affinis* fed diets containing higher saturated fatty acids had a greater
heat resistance than those fed unsaturated fatty acids (HOUSE et al., 1958). In
Phormia terranovae a change in iodine number of the depot fats brought about by
feeding had no effect on heat resistance; heat adaptation was present (CHERRY,
1959). Moist foods may reduce cold resistance of insects by directly seeding freezing
if they have a higher freezing point than the insect tissues, but their effect as
foreign material probably overrides such action (SALT, 1953, 1958a; cf. HARRIS,
1962[3]). A fasting period also can affect resistance (FRY, 1957, p. 209; McLEESE,
1965; SUKHANOVA, 1965; POLYANSKY et al. in GLUSHANKOVA et al., 1967).
According to DOLLY and WHITE (1951) exposure to illumination lowers the upper
limiting temperature in *Eristalis*.

2. Limiting Temperatures and Their Measurement

For measuring upper and lower limiting temperatures, various methods have been
employed which, however, may give different values. In cases where the body

---

[3] cf. KRUNIC, M. D.: Can. J. Zool. 49, 863—865 (1971).

temperature comes slowly into conformity with that of the milieu, or where there is the possibility of change of the body temperature by water evaporation (compare EDNEY, 1957; HUTCHISON, 1961; PRECHT, 1963; CLOUDSLEY-THOMPSON, 1967), the body temperature should be measured and not only the temperature of the animal's milieu. Methods used for animals can be classified in two kinds:

1. The experimental animal is transferred from a milieu at normal temperature into an extreme experimental temperature (ET), either abruptly or by changing the temperature at a certain rate. The criterion employed is either the time of survival of the animals (or of a given per cent of them), or the temperature at which a given per cent of the animals survive for an arbitrary length of time (short or long) or indefinitely.

2. The heating or cooling is done at a constant rate and measurement is made of the temperature at which symptoms typical of resistance (which must be proved), or death of the animal, occur. — Definitions of temperature limits are dependent on the methods employed (FRY, 1957; HUTCHISON, 1961).

Method 1 was used in experiments with fish by FRY and co-workers and a special nomenclature developed (FRY in DILL et al., 1964; in ROSE, 1967; TYLER, 1966). In experiments with *Girella*, DOUDOROFF (1942, 1945) made a distinction between primary and secondary chill coma; the first resulted in death at extreme, low lethal temperatures (approaching 0° C), while at less severe low lethal temperatures the fish recovered from primary chill coma and were killed after prolonged exposure by a secondary chill coma. Sudden temperature changes can act as a stress which should be taken into account (p. 419). According to GREEN (1964), who placed nematodes *(Ditylenchus dipsaci)* for two hours in 46° C, there were no appreciable differences in kill caused by the rate of heating, and slow cooling caused only a slight additional mortality.

Method 2, which usually gives higher limiting temperature values (READ in PROSSER, 1967, p. 64), has been used by this author and his co-workers (cf. THIEDE, 1965). Even with this method, while the ET is changed very slowly, adaptation to higher temperatures may begin (for example in fish, COCKING, 1959); the second method then bears an increasing resemblance to the first (cf. SPAAS, 1959).

The AT must be taken into account in determining which symptoms are to be measured. In experiments with crucian carp, when temperature is raised continuously, jaw paralysis occurs in the warm-adapted fish, but not in the cold-adapted; this reaction in contrast to others is hardly dependent on the AT and appears at ETs which are not attained by the cold-adapted animals (CHRISTOPHERSEN and PRECHT, 1952a). In experiments using Method 2 the rate of temperature change often has an influence on the measured limiting values for intact animals and for organ functions (cf. WATANABE, 1950). For *Lebistes reticulatus* TSUKUDA and OSHAWA (1958) found that the heat coma temperature falls and the cold coma temperature rises hyperbolically as the rate of change of temperature decreases[4] (cf. PRECHT et al., 1966, p. 379f.).

In experiments with reptiles and amphibians the so-called critical thermal maximum or minimum (CTM) is often measured, that is the arithmetic mean of the

---

[4] The participation of a rapid reasonable resistance adaptation during the tests would have an opposite effect.

collective thermal points at which locomotory activity becomes disorganized and the animal loses the ability to escape (HUTCHISON, 1961).

With the aid of Method 2, limits of resistance of single organ functions can be obtained. As a suitable measure of heat resistance of *Xiphophorus helleri*, THIEDE (cf. PRECHT et al., 1966) used the lethal index, that is the mean derived from the limiting temperatures measured for single functions: irregularity of movements of gill covers, cessation of motion of pectoral fins, stoppage of gill covers, cessation of gill movements, stopping of heart contractions; these functions show, in the above order, increasing heat resistance. Some functions are more sensitive to heat than the gill-cover movements which are important for vital resistance, and are a measure of activity of the respiratory center (cf. PRECHT, 1959; CERF et al., 1958); other functions are more resistant. At lowered temperatures also the cessation of various organ functions can be tested. In *Xiphophorus*, for example, movement of the gill covers ceases sooner than contractions of the ventricle of the heart (PRECHT, 1959). Since the exact recording of the cessation of such organ functions is often difficult, because the frequency diminishes gradually until only small portions of an organ continue to show activity, THIEDE also measured the temperature which must be achieved to prevent the fish — returned to 19° C — from recovering after one day. THIEDE was able to delimit a region of cold narcosis in which fish no longer responded to mechanical stimulation.

In the investigation of organ functions, intact animals can be used, or if necessary, exposed or isolated organs or organ complexes; however the measured values must not be the same (p. 407). We investigated heat resistance of nerve-muscle preparations (M. gastrocnemius and N. ischiadicus) of *Rana temporaria* (PRECHT, 1960a). With single stimuli, end-plate conduction failed first, then muscle contraction on direct stimulation, and finally nerve conduction. Especially for nerve conduction the kind of stimulation was critical for the limiting value (cf. BATTLE, 1926; ORR, 1955).[5]

Values obtained for heat resistance of tissues of an animal can be quite different; they can differ even in various types of frog striated muscle fibers (RUMYANTSEV in TROSHIN, 1967).[6]

If the lower limiting temperature lies below zero, certain methods are suitable. Attention should be paid to the rate of cooling and warming and to the length of time the final low temperature is in effect — this also is true for resistance measurements on enzymes (cf. CHILSON et al., 1965). Very resistant species must not only be supercooled, but frozen, sometimes several times, to arrive at any visible effect (MERKER, 1957; VOGEL, 1966). Such a shifting of temperature generally does more damage than a constant low temperature (VOELKEL, 1925; and others).

Some life functions can continue at very low temperatures, for example enzyme reactions (KÜHNAU, 1964; BĚLEHRÁDEK in TROSHIN, 1967). Arctic chironomid larvae continued to take up oxygen down to − 26° C (SCHOLANDER et al., 1953, see p. 411).

When employing either method for determining limiting temperatures, it should be noted that experimental material which has been previously treated in the same

---

[5] cf. JENSEN, D. W.: Comp. Biochem. Physiol. **41 A**, 685—695 (1972).
[6] About difficulties in measuring the heat resistance of enzymes cf. KÜNNEMANN, H.: Mar. Biol. **18**, 37—45 (1973).

way usually shows more or less variation in the values for individuals; these individual differences were greater at the lower limits for *Musca domestica* than for *Aëdes aegypti* (KNIPLING-SULLIVAN, 1957; cf. I. PRECHT, 1967; USHAKOV et al., 1968).

3. Bases of Resistance and Possibilities of Changing Them

It will hardly be possible to give general explanations for heat and cold resistance of intact animals, because species are too different. Possibilities of change in resistance to temperature extremes whether by a change in life cycle, in AT, in season, or in other factors, will be discussed here only in so far as it will contribute to insight into the question of the causes of resistance. Dependence of resistance on AT will be discussed in more detail later.

a) Upper Temperature Limits

The theory that heat resistance of an animal depends upon the temperature at which membrane lipids are liquefied is too simple to serve as a general explanation. The tissue cholesterol/phospholipid ratio, which has been related by several workers to cell permeability and temperature resistance, shows very poor correlation with differences in temperature resistance of goldfish; resistance to both extremes of temperature was correlated more with actual content of cholesterol and phospholipid; on the contrary, in feeding experiments the tissue lipid content increased with time, while the resistance declined (IRVINE et al., 1957). VAN HANDEL (1967) pointed out that the body fat of most marine animals is highly unsaturated and has melting points much below environmental temperatures (cf. CHAPMAN in ROSE, 1967). The dependence of the amount, composition and saturation of neutral fats and lipids on the temperature of deposition and the effects on resistance of feeding fats of different melting points were discussed (p. 330). Fats in diet usually went to depots where they remained stored for some time, but under certain conditions ingested fats are metabolized promptly by the liver and may be found in organs such as liver, intestinal mucosa and blood; the lipids of nerve and muscle are modified much more slowly (HOAR and COTTLE, 1952, p. 47). In feeding experiments secondary effects must also be taken into account; these effects may bring about conditions in which the experiments do not always support the theory that the presence of double bonds in the lipids is important for maintaining the integrity of the plasma membrane, and thus favoring heat resistance. In comparing species, the melting point of the phospholipids can be the same, but the heat resistance different (FRAENKEL and HOPF, 1940), or contrariwise the heat resistance the same, but the iodine number different (USHAKOV and GLUSHANKOVA in TROSHIN, 1967). It should be noted that in the same cell, membranes of differential stability may be present; generally only the total lipid content of tissues is measured (BĚLEHRÁDEK in TROSHIN, 1967, p. 437). According to SLIFER (1932) in eggs of grasshoppers, iodine number shows little relationship to diapause, although during this phase resistance to temperature extremes is in general high; relationships to the egg-laying period of the species apparently do exist.

Water content has often been considered significant for heat resistance of animals. When the content of free cell water and thus the solvent volume *(Lösungsraum)* are decreased, and the bound water increased, there is usually lowered cell metabolism and heightened resistance to both temperature extremes. Withdrawal of water may have similar effects on heat and cold resistance of intact animals (e.g. for insects, p. 412; cf. LUDWIG, 1945; BALDWIN, 1954; HINTON, 1954; FRY, 1958; HUTCHISON, 1961; WEILL, 1962; CLOUDSLEY-THOMPSON in DILL et al., 1964, p. 459). However, mealworm larvae with high and low water contents (derived from different diets) behaved very similarly to high and low temperatures (MEL-LANBY, 1958). There are instances when the upper temperature limit was lower at low air humidity than at higher (PIELOU and GLASSER, 1954). This could be expected for insects which are not tolerant of dry conditions. Thus a statement can be made of a favorable combination of humidity and temperature; any deviation of either factor is unfavorable (UVAROV, 1931; PLATT et al., 1957; FLEMISTER in DILL et al., 1964). Cells and tissues can show reduced resistance to temperature extremes when partially dehydrated; e.g. the frog heart shows a reversible drop of temperature optimum on perfusion with a hypertonic solution (BĚLEHRÁDEK in TROSHIN, 1967, p. 438).

Water content of animal tissues can also be dependent on the AT (p. 327) but it appears not to be meaningful for the explanation of resistance adaptation in general (p. 432); for resistance of single processes in cells, water content can be significant (for example in the yeast *Torulopsis kefyr* investigated by us; see ed. 1, also FRY, 1958; HUTCHISON, 1961). When water content was above 17%, heat resistance of the apyrase system of carp muscle was independent of water content; below that heat resistance increased linearly with decreasing water content (PART-MANN and NEMITZ, 1959). Heat resistance of cell functions in the crucian carp can be dependent on the AT, although the content of free and bound water of the respective tissues is independent of it (SUHRMANN, 1955).

Cell phenomena have been associated with resistance of intact animals. The objection could be made that organ functions are more likely responsible, but it remains to be investigated which specific partial functions of organs break down first at temperature extremes; here too the resistance of certain cells (cell structures or molecules) may be critical. Many authors assume that the limit of viability of intact animals is determined by denaturation of proteins, specifically enzymes at high temperatures[7]. Many enzymes are inactivated at temperatures well above those for life resistance (e.g. see BASLOW and NIGRELLI, 1964), but this is not true in all cases. According to BOWLER (1963) oxygen consumption of hepatopancreas and nerve tissue of *Astacus pallipes* killed by heat shows a decline, that of other tissues does not (cf. READ in PROSSER, 1967, p. 95). In similar experiments with *Arianta arbustorum*, muscle and hepatopancreas showed a decline in oxygen consumption, but kidney did not (GRAINGER, 1969). The membrane-bound ATPase extracted from crayfish, which is inactivated at $33-36°$ C, shows a linear μ value, closely corresponding to that obtained for heat death of the whole animal[8]. The

---

[7] EVANS, P. R., BOWLER, K.: Sub-Cell Biochem. 2, 91—95 (1973); ROSENBERG, B., KEMENY, G., SWITZER, R. C., HAMILTON, T. C.: Nature (London) 232, 471—472 (1971).
[8] BOWLER, K., GLADWELL, R. T., DUNCAN, C. J.: In: ABRAHAMSSON, S. (Ed.): Freshwater crayfish. Lund 1973. These authors believe that heat death to the crab *Austropotamobius pallipes* is caused by denaturational changes in membrane Mg²⁺-ATPase, an increase of membrane permeability, a change of hemolymph Na⁺ and K⁺ affecting nerve and muscle resting potentials, and an increase in spontaneous activity in CNS as a consequence of high hemolymph K⁺.

temperature at which the calcium-activated adenosine triphosphatase from four
species of lizards was 50% irreversibly inactivated after a 15-min exposure *in
vitro* corresponded closely to the body temperatures at which these species were
incapacitated by heat (SCHMIDT-NIELSEN and DAWSON in DILL et al., 1964; LICHT
in PROSSER, 1967). FREEMAN (1950) found that the metabolic activity of goldfish
brain tissue approached zero at a temperature approximating the ultimate upper
lethal temperature reported for this species (BRACHET and BREMER, 1941; BRETT,
1956; KIRBERGER, 1953; WOLF and QUIMBY, 1962).
That the temperature sensitivity of muscle sarcosomes (mitochondria) is correlated
with the heat death point of intact *Calliphora erythrocephala* and the impairment
of oxidative phosphorylation with α-glycerophosphate suggests that a breakdown in
ATP synthesis may be one of the causes of heat death (DAVISON and BOWLER[9]).
In the fish *Trematomus bernachii* (high lethal temperature: 6° C, AT: − 1.9° C)
oxygen consumption of brain slices shows no further increase above 5° C (SOMERO
and DE VRIES, 1967; SOMERO et al., 1968; concerning oxygen consumption of the
intact fish compare WOHLSCHLAG in LEE, 1964/1965). The rate of reaction of an
isolated enzyme system actually increased linearly with ET up to approximately
30° C (cf. GREENE and FEENEY, 1970).
DINGLEY and SMITH (1968) found quick recovery of *Drosophila* after a brief exposure to 35° C (the partially denatured enzymes can be reactivated); after a
longer exposure to 32° C the denatured enzymes have been hydrolyzed; consequently, recovery takes more than 2 days instead of 2 h.
Protein of cells is often bound with other substances into complexes, and this
changes their heat resistance. According to USHAKOV and co-workers, who
attribute great significance to these denaturing processes for heat resistance at
several levels, changes in cell resistance take place thus: "Changes at the level of
the primary (molecular) structure of the protein play a decisive role in alterations
of the hereditarily fixed component in the level of heat resistance of cells and its
changes in the process of differentiation during embryogenesis. Changes at the
level of protein protoplasmic complexes and in the surrounding intracellular
media (i.e. changes at the supra-molecular levels of cell organization) are the main
mechanisms providing for alterations in the cellular heat resistance upon interaction of the organism with the media at different stages of the biological cycle
(reproduction, hibernation, etc.)" (in POLYANSKY and USHAKOV, 1965, p. 54,
USHAKOV, 1968). Those substances which bring about temporary change of resistance of proteins by combining with them (sucrose, glycerol, nucleic acids, ATP,
etc.) also prevent protein denaturation *in vitro* (cf. BRAUN et al. in TROSHIN, 1967).
The special problems of functions of thermophiles have been less well investigated
with heat-resistant animals than with microorganisms (cf. ZUBER, 1969).
Some substrates and cofactors increase heat resistance of enzymes, others decrease
it (GRISOLIA and JOYCE, 1960). The significance of protective substances for cold
resistance has often been investigated; such substances in general increase resistance to both temperature extremes. In this connection it should be mentioned
that in the embryonic development of eggs of *Bombyx mori* the heat resistance was
greater the smaller the content of glycogen (this was less distinct for glucose); this

---

[9] DAVISON, T. F., BOWLER, K.: J. Cell Physiol. **78**, 37—47 (1971), cf. pp. 49—57; Comp.
Biochem. Physiol. **45** A, 441—450 (1973).

was true for the period of pre-diapause, estivation and hibernation, but not of spring development. As heat resistance increases, glycogen becomes converted to polyhydric alcohols and other soluble carbohydrates (mainly glycerol), which play the role of antidenaturants in high cold and heat tolerance of estivating and hibernating eggs. The drop of heat resistance during the active phase of spring development is reciprocal to an increase in the rate of respiratory metabolism and is probably related to the hydration of the protoplasm at the expense of water produced during the oxidation of glycogen reserves consumed in the course of spring development (ASTAUROV in TROSHIN, 1967, p. 392; CHINO, 1958).

Impairment of organ functions has often been considered responsible for the lethal temperature of the intact animal, without ascertaining the components. THIEDE (1965) suggests the following two factors as instrumental in the heat death of fish: 1. Damage to the respiratory center; 2. inability of the fish to maintain its basic metabolism at high temperatures when ventilation has reached a critical value. Death by overloading the ventilatory volume ensues relatively late [for example, using Method (2), p. 404, according to HUGHES and ROBERTS, 1970; also with Method (1)], while for sudden death in THIEDE's experiments damage to the respiratory center has been held responsible. According to the position of the maximum of respiratory frequency three categories can be differentiated: 1. The maximum lies very near the lethal point (*Cyprinus carpio*, *Myxocephalus octodecimspinus*, etc.); 2. the maximum is reached several degrees Celsius before the lethal point; from then on, constant decline in frequency occurs (embryos of *Trutta fario*, *Platichthys flesus*); or 3. the upper limiting value is reached relatively early and then gradually decreases (goldfish, *Xiphophorus helleri* etc.) (references in THIEDE, 1965, p. 314f.).

WINTERSTEIN has explained the comparatively low upper lethal temperatures of numerous poikilothermic vertebrate animals by the inadequacy of oxygen supply to the tissues; his explanation has been used as an explanation for invertebrates as well (e.g. see MATUTANI, 1961). Experiments on fish *(Ameiurus)* and insects have shown that the upper limiting temperature is largely independent of the partial pressure of oxygen (BRETT, 1944; measurements according to Method (1), cf. COCKING, 1959; FRAENKEL and HERFORD, 1940). But WEATHERLY (1970) found that heat resistance of goldfish can be increased by superabundant oxygen. VON SKRAMLIK (1947) attributes great significance to vagus shock in the heat death of fish (cf. GRODZINSKI, 1948). Osmoregulation of aquatic animals may be impaired at both high and low temperatures; geographic races of the isopod *Cyatura* behaved differently in this respect (SEGAL and BURBANCK, 1963). BOWLER (1963) suggests that in *Astacus pallipes* at high temperatures a breakdown occurs in the cation pumping system in the cells. He perfused claw nerve-muscle preparations with blood of normal and of heat-killed animals and found that in the former the claw preparations functioned longer than in the latter. GRAINGER (1969) found a decrease in the Na concentration in the blood of heat-killed *Arianta arbustorum* and a very marked change in the Na/K ratio. It is suggested that this causes a failure in neuromuscular transmission, which leads to the death of the animal at high temperatures (cf. JOHANSEN, 1967).

It was previously mentioned that a change of temperature in the normal range can bring about disturbances in distribution of ions and in enzyme activity (p. 327).

The effect of sublethal temperatures is a still greater stress, which must be taken into account, especially with Method (1) (compare HOUSTON, 1962). According to TYLER (1966) when *Chrosomus* was transferred quickly to high temperatures, shock was contributory to the cause of death. When *Rhodeus amarus* was transferred from $20 \rightarrow 30°$ C, in which the fish survived only a limited time, one enzyme (G 6 PDH) showed constantly increasing activity and did not reach an end point during the experimental period (BRAUN et al., 1970). When in experiments with carps the AT was raised to the sublethal range, the measured ions (Na, K, Ca, Mg, except Cl) were more or less increased; in the course of days they gradually reached their former values (GRIGO, unpubl.). Heat influence on rainbow trout fingerlings is believed to cause the secretion of substances which increase the resistance of other trouts when tested in the same water (MCCAULEY, 1968). In flies high and low temperatures bring about changes in phosphate metabolism (HOPF, 1940). *Sarcophaga* larvae after brief exposure to heat died after a delay of several days. In *Rhodnius prolixus* several protein syntheses are inhibited by high temperatures, the formation of activation hormone of the brain is also stopped (OKASHA, 1964).

SCHLIEPER (in TROSHIN, 1967) on the basis of his experiments on isolated mussel gills concludes: "Every factor which reduces the activity and metabolic rate of the cells (e.g. high salinity, relatively increased Ca and Mg concentration of the external medium) causes an increase of the cellular resistance against extreme nonoptimal temperatures, but the addition of a second stress factor (e.g. prolonged starvation, unusual low salinity, low oxygen concentration, etc.) reduces the normal cellular thermal resistance range".

It must be borne in mind that heat death, particularly of vertebrate animals, is a complex phenomenon. The heterogeneity of the time-mortality curves of guppies suggests at least three loci of response to an upper lethal temperature; one site is probably related to osmoregulation (GIBSON, 1954; cf. FRY, 1957, in DILL et al., 1964, and in ROSE, 1967, p. 387 ff.; COCKING, 1959; TYLER, 1966).

One can postulate many other points at which the injurious action of high temperature might begin to attack the organism, which might be meaningful for heat death, but our knowledge of these is slight. Suggestions have been made: the physical gills in water insects (POPHAM, 1964), mistakes in the coding of proteins in temperatures outside the normal range (compare PROSSER, 1967, p. 362), damage to the structure of the mitochondria (JEFFERSON, 1945; POLYANSKY and SUKHANOVA in TROSHIN, 1967), accumulation of harmful metabolic products (cf. CLOUDSLEY-THOMPSON in DILL et al., 1964, p. 460), impairment of coordination of physiological processes (cf. SCHMIDT-NIELSEN in DILL et al., 1964), impairment of endocrine glands which affect heat resistance (p. 419), of neurosecretion in *Locusta* (CLARKE, 1966), and so on. PROSSER (1967, p. 355) warns against giving explanations "which cover our ignorance of the intimate mechanism".

b) Low Limiting Temperatures

Resistance to cold as a general concept can be divided into resistance to chilling (at lethal temperatures above the freezing point of body fluids), to supercooling and to freezing.

The mechanisms of *resistance to chilling* differ for different species. In cold resistance the method of determination also influences the values obtained. For example for guppies the length of exposure to cold was more relevant to lethal temperature than the test temperature itself (PITKOW, 1960). In experiments by THIEDE (1965, for method see p. 404) with *Xiphophorus helleri*, injury to the respiratory center by low temperatures was crucial. In *Gadus morhua*, in winter, osmoregulation ceases to function at $2°$ C; in summer the limit is lower, as has been mentioned (WOODHEAD and WOODHEAD, 1959/1960; cf. BRETT, 1952, 1956; WIKGREN, 1953; WILLIAMS, 1960; RASCHACK, 1967). Crucian carp, after transfer from AT $20°$ C, or $11°$ C to $3°$ C, showed approximately the same oxygen consumption, although the animals from the higher AT, in contrast to the others died after 5 days (personal report by MARINESCU). In guppies (AT $23-30°$ C) exposure to temperatures below $10°$ C produced primary chill coma while exposure to lethal temperatures above $10°$ C produced secondary chill coma. Death due to the former is probably caused by anoxic damage to a cold-depressed respiratory center; oxygen lack and osmoregulative failure are not lethal determinants in secondary chill coma (PITKOW, 1960). According to HOUSTON (1962) cold death of goldfish is not due entirely to breakdown of osmoregulation. The course of the curves (times to death at different low lethal temperatures in fish) often indicates that there are multiple causes of death (cf. FRY, 1957). In goldfish the extinction of reflexes at low temperatures may be meaningful (ROOTS in PROSSER, 1962; cf. BASLOW and NIGRELLI, 1964; but also HAZEL, 1969). The cold blocking temperatures of various responses of goldfish CNS (AT $= 25°$ C) were found to lie between $5°$ and $9°$ C (these values depended on AT). Some cerebellar responses were somewhat more resistant to cold than others (PETERSON and PROSSER[10]). Irreversible disturbances seem to occur in the nervous system of cold-immobilized cockroaches (COLHOUN, 1954).

In wasps the caste differences in temperature tolerance at $4°$ C appear due to resistance of the mitochondria (SCHWALBACH and AGOSTINI, 1964). The lower limit of enzyme reactions may lie above zero, since molecular movement corresponding to a certain thermal energy, combined with relatively inactive catalysts at these temperatures, is too slight to bring about an adequate rate of chemical reaction in the cell (ZUBER, 1969, p. 17).

Many animals have the capacity to undergo *supercooling* of the body temperature without damage. Insects for example survive down to $-20°$ or $-30°$ and more (SALT, 1950; SULLIVAN, 1965; ASAHINA in MERYMAN, 1966, 1969). Highly purified water completely freed from small particles can be supercooled to $-34°$ C (LING in ROSE, 1967). In hibernating larvae of *Bracon cephi* the glycerol content permits supercooling down to $-47°$ C (SALT, 1958a, in DILL et al., 1964; MAZUR in MERYMAN, 1966, p. 215, 1970).

The average supercooling point in eggs of *Zeiraphera diniana* (Lepid.) was $-51.3°$C (in February, BAKKE, 1969). The amount of supercooling is equal to the difference between the supercooling point and the freezing point (which in insects is about $-1°$ C). OHYAMA and ASAHINA (1969) observed in *Camponotus obscuripes* two supercooling points; a partial freezing within the alimentary canal occurred at $-8.5°$ C, and the whole body froze upon cooling to $-20.2°$ C. Many animals endure supercooling for long periods or permanently (e.g. insects; SØMME and ÖSTBYE, 1969; fishes of polar regions: compare, SCHOLANDER et al., 1957; SMITH, 1958; GOR-

---

[10] PETERSON, R. H., PROSSER, C. L.: Comp. Biochem. Physiol. 42 A, 1019—1037 (1972).

DON et al., 1962; MARR, 1962); sea urchin eggs endure it for only a limited period; if the exposure is longer, later development is disturbed (ASAHINA, 1967, p. 213). When caterpillars of *Monema flavescens* were kept at $-10°$ C, after 100 days the number of those surviving, and even more of those pupating, and those emerging as adults — declined when they were supercooled, and in approximately the same length of time, when they were kept frozen at $-10°$ C and at $-20°$ C (ASAHINA, 1955).

With lowering of supercooling temperature, with increase of duration of super-cooling, and with increase in amount of water present the probability increases of a molecular arrangement corresponding to ice crystals. SALT (1966) was able to predict the lethal freezing point in supercooled larvae of *Cephus cinctus* kept at defined low temperatures. Mean time to freeze doubled with each $0.53°$ C rise in temperature; it increased from 0.1 sec at $-30°$ C to more than a year at $-17°$ C (cf. SALT, 1969). With relatively quick chilling the lower limit of possible super-cooling is scarcely dependent on the rate of cooling; with very gradual lowering of temperature higher cooling may result (SALT, 1961a; CLARKE in ROSE, 1967, p. 337f.).

There is always the danger that seed crystals may enter the supercooled animal or come in contact with its surface and elicit freezing. Thus the supercooling point of moist insects is often very high (HODSON, 1937, p. 297ff.). Chitin exoskeleton of insects, wax coatings, egg cases, etc., furnish some protection. The body surface of many supercooled insects can be frozen, without ice formation spreading inside the body, but small lesions destroy the barrier. Soaking of the cuticle was of doubtful effectiveness, whereas boiling and use of detergents hastened inoculation (SALT, 1963; ASAHINA in MERYMAN, 1966, p. 453). Some supercooled arctic fish can come in contact with ice crystals without damage (GORDON et al., 1962).

Although osmotic and ionic regulation of supercooled *Fundulus* may not be so effective in the cold, they were by no means so poor as to cause death by osmotic imbalance (UMMINGER, 1969b, 1970c[11]).

If in experiments with decreasing temperature the supercooled solution freezes, temperature will increase until the freezing point is reached and then decrease again; in plants during this drop some small peaks can exist caused by biologically significant exotherm processes (WEISER, 1970).

Amazingly, some invertebrate animals survive *freezing*, or even repeated freezing and thawing — for example pupae of *Croesus septemtrionalis*, according to KO-ZHANTSHIKOV (1938), *Pterostichus brevicornis* in winter (MILLER, 1969). Arctic in-sects can be cold resistant even in the feeding stage (but see REMMERT and WIS-NIEWSKI, 1970); chironomid larvae of the Arctic can endure freezing of more than 90% of the body water without injury (SCHOLANDER et al., 1953; compare SMITH, 1961, p. 277ff.; LOSINA-LOSINSKY in TROSHIN, 1967). Mollusks in the tidal region can freeze regularly in winter at low tide; in *Littorina littorea* about 70% of the body water can be frozen; according to KANWISHER (in MERYMAN, 1966) extracellular ice is formed (cf. SØMME, 1966). In prepupae of the slug moth (*Monema flavescens*)

---

[11] However cf. UMMINGER, B.L.: Comp. Biochem. Physiol. 39, 625—632 (1971), also ALLANSON, B.R., BOK, A., VAN WYK, N.T.: J. Fish. Biol. 3, 181—185 (1971). About supercooling of reptiles cf. LOWE, C.H., LARDNER, P.J., HALPERN, E.A.: Comp. Bio-chem. Physiol. 39, 125—135 (1971).

ice formation occurred only in the blood within the body cavity (SHINOZAKI, 1962). Vertebrates do not survive total ice formation, or at least die later of the consequences (ed. 1, p. 59 ff., KALABUKHOV in PARKES and SMITH, 1960; SMITH, 1961, p. 270 f.). Many insects in certain stages survive freezing, but development suffers; half imagoes have been known to develop (cf. ASAHINA, 1969). Frost resistant species often have a good capacity to supercool. However the supercooling point is not always indicative of the resistance to freezing (e.g., diapause larvae of the corn borer, HANEC and BECK, 1960).

Upon formation of ice, oxygen consumption declines sharply (e.g. in diapausing cecropia pupae, SALT, 1958 b; cf. SCHOLANDER et al., 1953; ASAHINA in MERYMAN, 1966, Fig. 9); according to KANWISHER (1959) high salinity is responsible for decreases in oxygen uptake and, in addition, for the high degree of dehydration of body tissues and the hindrance to diffusion by the ice (ASAHINA, 1969, p. 19).

Damage by freezing has been much discussed. MERYMAN lists: 1. mechanical rupture of structural elements through ice crystal growth; 2. denaturation from electrolyte concentration; 3. pH changes; 4. dehydration sufficient to precipitate proteins from solution, or 5. to bring them into contact and permit abnormal cross-linking, or 6. the direct effect of the removal of structurally important water (cf. TAPPEL in MERYMAN, 1966; HANAFUSA in ASAHINA, 1967; MERYMAN in TROSHIN, 1967; BURNS, 1964, p. 30 ff.; UMMINGER, 1968; ASAHINA, 1969; MAZUR, 1970, p. 943). The causes of damage are probably different for different species.

Freezing can take place in cells or in the blood or intercellularly, which is in general better tolerated than ice formation within the cells. Air-filled intercellular spaces such as occur in plants are lacking in animals, but the cells are usually in contact with a fine layer of fluid (also in closed blood systems, e.g., the lymph in vertebrates) which can freeze. The magnitude of the extracellular space is not necessarily constant in fish (HOUSTON et al., 1968).

SALT (1969) mentions nucleative freezing, which raises the problem of the birth of an ice nucleus and the ice growth. This is to be distinguished from inoculative freezing which arises from the growth or propagation of external ice into the body, resulting in the freezing of body fluids (cf. ASAHINA, 1969).

The general view is advanced that, especially upon slow cooling, ice forms first in the blood and then between the cells in the more highly organized animals. The molecular configuration for formation of ice is less likely present in the cell plasma stroma than in the extracellular fluid (cf. BLOCH et al., 1963). The extracellular ice concentrates the fluid, and water is drawn out of the cells. It should be noted that supercooled water has a higher vapor pressure than ice. The plasma membrane acts as a barrier against seed crystals, but becomes less effective as the temperature goes down. Release of water from the cells depends on the permeability of the membrane and also on the speed of freezing; at a high rate of freezing, intracellular ice is readily formed.

The effects of the rates of freezing and thawing are difficult to evaluate. MERYMAN (1966, p. 65 f.) emphasizes that there is no simple relationship to the rate of survival. "Where the biochemical effects of freezing are a source of injury, the reaction rate will depend on the degree of dehydration and on temperature so that faster cooling rates provide less time for biochemical injury to develop; where intracellular crystallization is a lethal factor, it can be minimized by decreasing the rate of freezing ... Following slow freezing, thawing rates are generally inconsequential. Wherever rapid freezing is necessary for survival, rapid thawing also is almost universally necessary." Thus temperature ranges which might cause biochemical damage are passed through quickly, and growth of crystals is lessened. MAZUR (in MERYMAN, 1966, p. 285) gives verified principles.

Rate of cooling can be important only for certain temperature ranges, and for protozoans the effect also is dependent on the medium (HWANG et al., 1964; POLGE and SOLTYS in PARKES and SMITH, 1960). In the freeze-preservation of protozoans gradual freezing is usually employed, but for some methods quick freezing (DIAMOND, 1964). Slow cooling generally brings about formation of large, relatively stable crystals; sudden cooling gives rise to smaller unstable crystals, which tend to grow with prolonged exposure to low temperatures or on thawing. Therefore the duration of exposure to low temperatures can also be significant (see example, p. 412; also REY and SIMATOS in TROSHIN, 1967; and others). There are insects which tolerate exposure to extreme cold, but not if it is prolonged (KOZHANTSHIKOV, 1955).

Fertilized eggs of sea urchins are found to be more resistant to extracellular freezing than unfertilized eggs, which can withstand only slow freezing, but can also withstand rapid freezing after treatment with a solution of urea (increase of permeability); at fertilization a remarkable change takes place in the distribution of protein-bound SH groups (ASAHINA, 1962b, 1965a; ASAHINA and TANNO, 1963).

When either single cells or metazoans are transferred to extremely low temperatures (liquid gases), vitrification, as was formerly thought, usually does not occur, but very minute crystals form in the cells; quick thawing, which prevents formation of larger crystals, may under some conditions permit survival (compare ASAHINA et al., 1968). It has been possible to give optimal cooling rates for survival quotas of various cells, but this differs greatly with the specific cell types (compare MAZUR, 1970, p. 942). Some specimens, such as caterpillars of *Monema flavescens* in the overwintering prepupal stage (ASAHINA and AOKI, 1958), can tolerate transfer to liquefied gas better if they have been frozen previously at less extreme temperatures (approximately $-30°$ C); water has then been drawn out of the cells by extracellular formation of ice (cf. ASAHINA, 1959; in MERYMAN, 1966, Table III, 1969, p. 38f.; ASAHINA and TANNO, 1967; HWANG et al., 1964; LOSINA-LOSINSKY in TROSHIN, 1967).

For each case it must be ascertained which factor leads to damage after formation of ice. The damage might be attributed to mechanical factors; however when intracellular or extracellular ice is formed, the protoplasm becomes concentrated by water withdrawal. LOSINA-LOSINSKY (in TROSHIN, 1967) questions the general harmfulness of formation of ice in cells. He observed it in the salivary gland cells of the extremely cold-resistant diapause caterpillars of *Pyrausta nubialis* (cf. SALT, 1961b). Ice crystals are readily formed in the fat-body cells of the visceral layer of pupae of *Trichocampus populi;* this affects metamorphosis to imago unfavorably (TANNO in ASAHINA, 1967, p. 425). JANKOWSKY et al. (1969) were able to elicit at will extra- or extra- and intracellular ice formation in the freezing of polyps of two species of *Laomedea*. Ice formation in the cells is visible here as in other organisms (compare ASAHINA, 1967) as a sudden darkening of the cells (flashing, blackout), which is an indication of minute ice crystals, not visible in light microscopy. (In other cases large rosette or fern-like crystals are formed in the cells; cf. ASAHINA, 1962b, 1965b.) Formation of ice in the medium only, or in the cells, too was tolerated by the polyps only for limited periods, in the first case for a longer time;

apparently there was a quantitative and not a qualitative difference, although it might be assumed that ice crystals within the cells could damage the delicate structure mechanically. Further observations indicate that the extent of mechanical damage in cell structure is not necessarily a criterion for the survival of the cells (REY and SIMATOS in TROSHIN, 1967, p. 81; MERYMAN, 1966, p. 49). Lysosomes are believed to be especially sensitive to such damage (cf. THEEDE, 1970, and others). Noteworthy in our experiments on *Laomedea* polyps is the fact that a lowering of the experimental temperature one degree (from $-4°$ C to $-5°$ C) in each case definitely decreased the survival rate. Also, a blackout of the cells may appear upon slow warming of cells which have been quickly cooled to low temperatures (ASAHINA et al., 1968).

Careful observations have been made on the occurrence of ice crystals in meat (cf. LOVE in MERYMAN, 1966). Most animal muscles frozen commercially are in rigor mortis or in the relaxed state following rigor mortis. For these, quick freezing leads to intracellular ice formation, slow freezing to extracellular. LOVE and HARALDSON found that freezing of pre-rigor cod muscle at any speed always resulted in intracellular freezing, which is probably also true in other cases.

Also cold resistance of animals is subject to change. Most difficult to explain is a change in resistance to chilling; there is more information concerning resistance to supercooling and to freezing. Formation of ice in supercooled animals is enhanced by seed crystals, as was mentioned. This is not a hazard for bottom fish. They show no osmotic change in winter (cf. WOODHEAD and WOODHEAD, 1959/1960), in contrast to other fish from North Alaska and Labrador inlets, which by adding an antifreeze substance of unknown composition to their blood lower the freezing point from $-0.8°$ C to $-1.47°$ C to $-1.5°$ C, so that they become almost isotonic with sea water (SCHOLANDER et al., 1957; GORDON et al., 1962; compare THEEDE, 1967).

In *Trematomus*, concentrations of NaCl, urea and amino acids in the serum account for half of the freezing point depression of the serum. A protein-containing carbohydrate was isolated by DE VRIES and WOHLSCHLAG (1969) which accounted for 30% of the freezing point depression of the serum[12]. According to PROSSER et al. (1970) such substances, which favor supercooling, occur in fish in polar regions and in winter in fish of temperate salt water. Many subarctic and a few temperate-zone marine fish living at low temperatures have elevated concentrations of NaCl or Na in their muscle fibers. According to RASCHACK (1967, 1969) the freezing point in *Cottus scorpio* is lowered in winter more by the action of non-dissociated organic compounds than by changes in ionic content; in *Cottus bubalis* however the less pronounced concentration rise results from increased content of salts. According to PROSSER et al., fresh-water fish, in contrast to the marine forms, often have either reduced or unaltered plasma concentrations of Na and Cl, when acclimated to low temperatures (some fresh-water fish on adaptation to relatively low temperatures can also show other conditions) (cf. HICKMAN et al., 1964; HOUSTON et al., 1968; PRECHT, 1968, p. 505; MEINCKE, 1970). According to the authors the differences between fresh-water and salt-water fish are in connection with the maintenance of osmoregulation at low temperatures in hypo- and hypertonic media.

---

[12] cf. SHIER, W. T., LIN, Y., DE VRIES, A. L.: Biochim. Biophys. Acta **263**, 406—413 (1972); HARGENS, A. R.: Science **176**, 184—186 (1972); FEENEY, R. E., VANDENHEEDE, J., OSUGA, D. T.: Naturwiss. **59**, 22—29 (1971).

UMMINGER (1968, 1969 a, b, 1970 a, 1971) investigated the capacity of *Fundulus hetero-clitus* to be supercooled (cf. SMITH, 1968). In the cold, serum sodium, calcium, magnesium chloride and osmolarity increase in salt-water-adapted fish, but decrease in fresh-water-adapted ones. In salt water the level of serum potassium and inorganic phosphate remained relatively constant when the AT was lowered (from 20° C to −1.5° C), and the same was true for the levels of serum total non-protein nitrogen, total amino acids, urea, and non-glucose free carbohydrates. Serum total cholesterol levels increased by 69%, presumably due to increased lipid metabolism. The twentyfold increase in serum glucose in fresh-water-adapted killifish at 0.1° C is far greater than the three- to sixfold increase observed in salt-water-adapted fish at subzero temperatures (at ATs 20° C and 10° C the level of serum glucose is approximately the same, at 4° C slightly elevated)[13]. At −1.5° C, the glycogen content of the liver decreases appreciably, that of the muscles only slightly. Winter fish, which had higher levels of serum glucose in the subzero cold than summer fish, survived the subzero cold almost twice as long as did summer fish; glucagon appears significant (for changes of serum protein components see UMMINGER, 1970 b). Acclimation of goldfish to temperatures near freezing was characterized by a decrease in serum electrolytes, an increase in serum osmolarity, and an increase in serum glucose (which was osmotically insignificant). The increase in osmolarity was due to a buildup of yet unidentified, probably organic, constituents of the serum (UMMINGER).[14] For poikilosmotic sea animals there is no danger of freezing as long as the sea water milieu does not freeze.

In insects, fluctuations in osmotic pressure are most closely related to cold resistance. LOSINA-LOSINSKY (in TROSHIN, 1967, p. 95), who ascertained this numerous times, is of the opinion that it is difficult to say what causes the osmotic pressure in the insects to increase − formation of soluble sugars, amino acids, etc., or changes in the cell water binding capacity. In *Trichocampus populi* the overwintering pupae synthesized more trehalose (TANNO and ASAHINA, 1964; cf. DUTRIEU, 1961). In overwintering adults of the solitary bee *Ceratina flavipes* the level of fructose, glucose and trehalose showed an increase; this was in relation to super-cooling ability (TANNO, 1964; compare SØMME, 1967). However, many insects with relatively high sugar content exhibit little cold resistance (cf. ASAHINA, 1969, p. 34).

Glycerol was previously mentioned as an important protection for insects (and cysts of *Artemia salina*, CLEGG, 1962); less frequently sorbitol and mannitol have been found in insects (references in SMITH, 1961, p. 282f.; SALT in DILL et al., 1964, 1969; SØMME, 1965 b, 1969; ASAHINA in MERYMAN, 1966, 1969). These substances must lower the freezing point merely by raising osmotic pressure. With increased concentration of solutes in the hemolymph, significantly greater depressions of supercooling points than of freezing points were found in several species of insects with and without glycerol. SØMME (1967) believes that the effect on supercooling exhibited by glycerol is not unique, and that a number of solutes in the hemolymph may have a similar effect. Glycerol, according to SALT, protects also during freezing. It reduces the speed of ice crystal growth, changes the form of ice crystals, making them more rounded in aspect, and also has the effect of lessening the increase of concentration in electrolytes. In overwintering larvae of *Bracon cephi* the glycerol content amounted to 25% of the fresh body weight; there was a direct

---

[13] This cold-induced hyperglycemia of the killifish is not influenced by pituitary gland [UMMINGER, B.L.: Experientia (Basel) **27**, 701—702 (1971), cf. Physiol. Zoöl. **44**, 20—27 (1971)]. An increase in circulating levels of glucagon and a failure of the beta cells to release insulin presumably were important. Amer. Zool. **12** (3) (1972).

[14] Copeia **1971**, 689—691.

relationship between supercooling temperature and glycerol content (SALT, 1959).

In many diapausing insects the glycerol content increases in autumn[15]. It often remains high for the duration of diapause and declines at its end, at a rate corresponding to rise in temperature. Larvae of *Laspeyresia strobilella* lost glycerol during the spring, in spite of the fact that some of them remained in diapause (SØMME, 1965a). In prepupae of *Monema flavescens* glycerol content decreased upon transfer from 10° C to 20° C; from 10° C to 0° C there was no increase (after remaining at 0° C for a considerable period there was a decrease). The termination of the diapause had also in this case no relation to disappearance of glycerol, which may be significant, since cold weather may persist after early ending of diapause (TAKEHARA, 1964). The glycerol content of *Pterostichus brevicornis* increased with decreasing AT (BAUST and MILLER, 1970). In species which do not go through diapause but which contain glycerol, the level at any time of year may increase with declining AT and decrease with rising temperature (DUBACH et al., 1959; SØMME, 1964).

The significance of these substances as protection against temperature extremes (in particular, low temperatures) is often clear, although there are cold-resistant species which have them only in low concentrations or lack them, and cold-sensitive species which contain appreciable amounts of glycerol (SØMME, 1964, 1965a; ASAHINA in MERYMAN, 1966, Table 2, 1969, Table 3); this can also be true for individuals of the same species (KIRCHNER and KESTNER, 1969, e.g. for the spider *Araneus cornutus*). Protective substances do not always increase and decrease at the same time as resistance; this is also true for other factors to which seasonal changes in resistance have been attributed (HANEC, 1966; LOSINA-LOSINSKY in TROSHIN, 1967, p. 95[16]). Many forms which are able to undergo severe supercooling are not likely to be subjected to this problem in nature. Hibernating larvae of *Eurosta solidaginis* supercool only 8–10° C (4% glycerol, 4% sorbitol); yet in the spring they metamorphose to puparia that supercool more than 20° C (SALT).

After exposure to 4° C (3 weeks), 8 hibernating species of Ichneumonidae produced glycerol, 5 non-hibernating species did not (DUFFIELD and NORDIN, 1970).

Glycerol and other substances can be employed as protection against freezing (WILBUR and MCMAHAN, 1958; THEEDE, 1965; MERYMAN, 1966, p. 61 ff.); this has led to discussion as to whether such substances must penetrate into the cells in order to be effective (SHERMAN, 1963). Added macromolecular substances may also have a protective effect (MAZUR, 1970, p. 943). Protozoans can be kept alive for long periods by the addition of protective substances; *Trichomonas foetus* for example survived better at —95° C than at —28° C; at —95° C almost as many trichomonads were alive after 265 days as after 1 day; thereafter, however, they decreased in number; some were alive after 5.6 years (LEVINE and ANDERSEN, 1966). Some parasitic protozoans withstand low temperatures better without protective agents (DIAMOND, 1964; compare LAPIERRE and HIEN, 1968; BURNS, 1964).

In some overwintering pupae of *Papilio xuthus* an injection of glycerol did not increase the capacity to resist freezing (TANNO, 1963), but it did have this effect on prepupae of *Monema flavescens*, in winter only (ASAHINA and TAKEHARA, 1964; compare SHINOZAKI,

[15] cf. MANSINGH, A., SMALLMAN, B. N.: J. Insect Physiol. 18, 1565—1571 (1972).
[16] Glycerol content of adult insects may change during the cold season, however supercooling point and cold resistance remain constant (OHYAMA, Y., ASAHINA, E.: J. Insect Physiol. 18, 267—282, 1972).

1962), and in *Ephestia larvae*, although this species does not normally contain glycerol (SØMME, 1968b).

Protective substances have many hydroxyl groups, according to UMMINGER (1969b). It may be that these groups somehow interfere with water-to-water hydrogen bonding and thus retard the rate of ice crystal growth at the surface of ice nuclei (p. 420). Most additives appear to protect against osmotic effects rather than against intracellular freezing (compare MAZUR, 1970, p. 944, for detailed discussion).

Changes in water content can have effects on resistance to cold as well as to heat. Species of insects resistant to cold are usually lower in water content than those which are more sensitive (UVAROV, 1931; LUDWIG, 1945). The beetle *Pterostichus brevicornis* is much more resistant to cold in winter than in summer, to the extent of almost 30° C. Its water content decreases in winter by about 11% (the glycerol content rises from zero to 25%, MILLER, 1969).[17] Overwintering resting stages are usually low in water content, also due to the storing of reserve substances, and they are cold resistant (compare ed. 1, p. 57, SALT, 1956; SHINOZAKI, 1962; DOWNESS, 1965; but also MELLANBY, 1959). End of insect diapause may be marked by increase in water content (USCHATINSKAJA, 1949), or not, as in potato beetles (PRECHT, 1953). High resistance of anabiosis stages was mentioned (compare KLINGLER and LENGWEILER-REY, 1969). – It is noteworthy that many insects seek out moist winter quarters (HODSON, 1937; and ed. 1, p. 57ff.).

Progressive drying increased the cold resistance of some adult insects or their developmental stages; according to SALT this occurs often only after considerable water has been removed. It was caused by decreasing freezing points, corresponding directly to the concentrating of solutes; the amount of supercooling remained approximately constant (SALT, 1956). The tolerance of insects to desiccation is variable, but most insects will have died from desiccation before such treatment has increased their supercooling ability by more than 1–2° C (SALT, 1958a, p. 76; SALT, 1961a). HINTON (1960a, b, 1968) was able to dry larvae of a tropical chironomid *Polypedilum vanderplanki* almost completely. The larvae then survived temperatures over 100° C for some time, and a few of these animals subsequently metamorphosed. At a water content of 8% the larvae could be transferred from room temperature to −270° C without harm (see BURNS, 1964, p. 27).

It is not clear whether many insects undergo a change in cold tolerance by a change in proportion of free water to bound (cf. ASAHINA in MERYMAN, 1966, p. 460f., 1969; for mollusks see WILLIAMS, 1970). This proportion also may be genetically different, comparing species or races from different habitats; however this problem should be clarified with a wider range of experimental data (cf. PANTYUKHOV, 1956). Below the freezing point the body water becomes almost entirely removed in the form of ice crystals, as for example in *Monema* (SHINOZAKI, 1962).

Conditions for the formation and spreading of ice in the body can also be altered, for example in caterpillars of *Monema flavescens*, this occurs quickly in summer, slowly in winter. Also likelihood of intracellular ice formation can change with season of the year (cf. ASAHINA in MERYMAN, 1966).

It is pertinent to investigate the effects of hormones on changes in resistance to both extremes of temperature. HOAR proposed the working hypothesis that the hypophysis-thyroid system in fish participates in changes related to photoperiod or to AT, and specifically that thyroid hormone decreases heat resistance but increases cold resistance. Some investigations appeared to support this (HOAR and

---

[17] cf. BAUST, J.G.: Nat. New Biol. 236, 219—221 (1972).

EALES, 1963, cold resistance of goldfish; THIEDE, 1965, heat and cold resistance of *Xiphophorus helleri*). Experiments with thyroid inhibitors, which of course may have side effects, were inconclusive. However, TSH was found to increase the heat resistance of *Gambusia affinis*. Saline injections had the same effect (THEOBALD, 1959). In *Gadus morhua* sensitivity to cold increases in autumn when the thyroid becomes active (WOODHEAD and WOODHEAD, 1959). CHEVERIE and LYNN (1963) found no evidence for an influence of the thyroid gland on heat resistance of *Tanichthys albonubes*. Recent experiments by JOHANSEN (1967) demonstrated how unresolved the whole problem is, aside from the fact that the site of action of the hormones is not known. Treatment of hypophysectomized goldfish with luteinizing, thyrotropic, adrenocorticotropic, lactotropic (prolactin) hormones and with crude salmon pituitary extract appeared not to affect resistance to high temperature, but hypophysectomized fish were less resistant to a lethal high temperature than intact goldfish. The hypothesis is put forward that a prolactin-like hormone plays a role. Increase of heat resistance of goldfish by repeated brief exposures to heat (hardening?) disappeared after hypophysectomy (JOHANSEN, 1968). WEATHERLEY (1963) in several cases observed increase of heat resistance of perch 8 h after injection of ACTH (see STIER and TAYLOR, 1939).

According to USHAKOV and co-workers, hormones participate in changes in heat resistance when life phase is changed (such as breeding period). PASHKOVA is of the opinion that an increase in the function of the thyroid gland leads to an increase in cell thermostability of frogs, whereas an increase in the function of hypophysis and sexual glands reduces heat resistance (see USHAKOV, 1968, p. 155; v. OORDT and LOFTS, 1963; concerning changes of the activity of the thyroid gland of *Poecilia* during the gestation cycle see BROMAGE and SAGE, 1968). In winter, light and increased temperature did not affect activity of the gland or muscle heat resistance of frogs.

Simultaneous appearance of changes of hormone production with life cycle and of resistance is of course only indirect evidence for a hormone-resistance relationship, since hormones have many other effects. However, for example by injections with TSH and thyroid hormone, the heat resistance of frog muscle could be increased experimentally (PASHKOVA in USHAKOV, 1963; in TROSHIN, 1967; DZHAMUSOVA and PASHKOVA in GLUSHANKOVA, 1967).

As was mentioned, any temperature change in the normal range can be stressful. This is especially true for determinations of resistance by Method 1 (p. 404). In such experiments heed should be taken of the normal effects of hormonal action in response to any stress (p. 322), and it is conceivable that the phenomenon of hardening is of this kind. JOHANSEN, however, stated that ACTH injections are without effect. The adrenal cortex of *Bufo melanostictus* reacted to cold stress during winter hibernation less than at other times (MUKHERJI, 1968).

B. On the Problem of Hardening

Tests have been made with animals as to whether the distinction, proposed by ALEXANDROV for plants, between hardening and resistance adaptation (adjustment) is necessary (see PRECHT et al., 1966, several authors in TROSHIN, 1967). Hardening is a quick, usually transitory adaptation to high and low temperatures, while resistance adaptation refers to a more slowly progressing, longer-lasting adaptation of resistance to temperatures in the normal range. Heat-hardening, which has been more extensively investigated than cold-hardening, can be considered first. It can be observed in some invertebrate and vertebrate animals, and also in their tissues, that a brief exposure to sublethal high temperature increases heat resistance; this is apparent on a successive exposure (in a few cases lowered resistance has been observed; examples of each, BASEDOW, 1969). The process of resistance adaptation is often dependent on temperature, and thus may take place rapidly at high temperatures. The question arises of how hardening can be distinguished from rapidly occurring resistance adaptation. Such a distinction would readily be made in animals which show hardening but no resistance adaptation; however there are no examples so far (see I. PRECHT, 1967).

It was mentioned that temperature changes have manifold effects which can be considered as consequences of stress. Is it possible that a stress might have the favorable effect of producing a rapid increase in resistance? Hardening would then be adaptation to shock[18].

If the high temperatures were so chosen that the animals could just be kept alive, and if the temperature change acted as a stress, an initial peak of the resistance-time curve would indicate hardening; later another rise would indicate lasting resistance adaptation. BASEDOW attempted in vain to demonstrate such a peak at the beginning of adaptation; this does not rule out hardening, since it could require more time, and it would not then be clearly distinguishable as a peak from later resistance adaptation. BASEDOW was unable to delimit the two phenomena by using measurements exclusively of temperature resistance of several invertebrates and vertebrates. Other kinds of shocks could be tried in support, but this has seldom been done in resistance measurements made up to this time. Allowance must be made for the fact that under lasting stress the adaptation to it may extend over considerable time. High or low ATs may cause a permanent stress (see the table of GRONOW, p. 330).

In this connection it should be pointed out that the increase of heat resistance in early larvae of the fish *Tilapia niloticus* (BISHAI, 1965) and of larvae of *Tenebrio molitor* (MELLANBY, 1954) is limited to high ATs, but this observed limitation (also known for capacity adaptation) to certain ranges of AT, is not a direct indication of a delimited shock adaptation.

It would be more convincing if brief cold shocks increased resistance, since this effect could not be explained as a temperature-dependent resistance adaptation with $Q_{10} > 1$ (but see p. 437).

Sublethal cold shocks were without effect upon the high and low limits of temperature resistance of *Idus idus*, *Xiphophorus helleri* and young *Anguilla vulgaris* (BASEDOW). SAMOKHVALOVA (1938) found a rise in cold resistance in various Cyprinodonts with sublethal cold shocks, which is in contrast to the findings of TSUKUDA (1959) on *Lebistes*. Sudden transfer to low temperatures and continued exposure for a time raises cold resistance of *Anguilla* and *Xiphophorus* rather quickly; here there may be an effect of stress. The mechanism of cold adaptation in both fish species is not strongly temperature-dependent. In high temperatures, the speed of adaptation of heat resistance diminishes, and cold resistance goes down sharply. This, according to BASEDOW, is evidence against a shock adaptation. In the beetle *Tribolium confusum*, SØMME (1968a) found a marked increase in cold resistance within a few hours, when AT was decreased;

---

[18] A stressor may cause damages which the organism tries to repair, or, after SELYE, it may increase the ability of the organism to resist further damage. Therefore (also in the case of heat hardening) we can speak of a shock adaptation. A differentiation between shock effect and stress, as is usual in medicine (shock is called the first phase of a stress with special symptoms), is not made here; our knowledge of these problems in lower animals is too limited.

this may be related to the idiosyncracies of insect resistance adaptation (p. 438; for long-lasting effects see p. 440). Indications of cold-hardening *per se* were not present. Heat resistance of end-plate conduction in nerve-muscle preparations of *Rana temporaria* increased after they had been kept several hours at low temperatures; heat resistance of the muscle was unchanged (PRECHT, 1960a). When snails *(Planorbis corneus)* or their isolated antennae were transferred to high or low temperatures (22° → 5° C, 5° → 22° C) the cold resistance of the antennal ciliated epithelium changed within one to two hours (PRECHT et al., 1966; see PRECHT and CHRISTOPHERSEN, 1965). ATWAL (1960) found that short adaptation periods of a few hours at low temperatures increased cold resistance more than longer periods, since the latter were damaging (pupae of *Ephestia kühniella*); this can simulate hardening.

BASEDOW's experiments did not entirely rule out hardening *per se* (see the experiments with varied preliminary exposures, p. 438). Our experiments with *Rhodeus amarus* are interesting: the fishes were heated once a day until they exhibited a symptom which was indicative of the presence of heat resistance (e.g. cessation of gill movement) and then for recovery were quickly replaced into their former container (AT). In spite of the very short duration of the daily test, heat resistance of fishes from AT = 10° C slowly increased, while that of fishes from AT = 20° or higher ATs did not; the resistance of the latter was already high at the beginning of the test, due to heat adaptation (PRECHT and KÜNNEMANN, unpubl.). No increase in heat resistance of AT = 10° C-fishes occurred, if an inhibitor of protein synthesis (e.g. cycloheximide) was injected (KÜNNEMANN, unpubl.).

The results of FRIEDRICH (1967) on isolated *Mytilus* gills supported it; the resistance of the ciliary beat was investigated. Heat shocks increased the resistance to freezing as well as the heat resistance of the epithelium, as did cold shocks; exposure to low oxygen partial pressure had similar effects. Resistance adaptation of the ciliary epithelium is reasonable at both temperature extremes (according to experiments of SCHLIEPER). When the gills are kept at high temperatures, a prolonged "undershoot" of cold resistance, because of hardening, should be observed before the reasonable cold adaptation occurs. (In *Mytilus* gills investigated by THEEDE, 1970, pre-treatment by heat shock increased only heat resistance, not resistance to freezing.)

Fish *(Girella)* were investigated by DOUDOROFF (1942): when he lowered the AT (26° → 14° C) they showed increased heat resistance for the first five days. Heat resistance of *Amoeba proteus* fell steeply after the culture temperature was lowered (20° → 4° C), but only transitorily (POLYANSKY et al. in GLUSHANKOVA et al., 1967).

Just as for resistance adaptation the terms reasonable and paradoxical may also be used for hardening (p. 429); however, this could cause difficulties as follows: if AT is altered, examination can be made whether reasonable or paradoxical heat or (and) cold adaptation are present. After providing a short heat shock, heat-hardening may be reasonable or paradoxical; when cold resistance is also changed, this cannot be termed cold hardening, because this would only be possible in experiments where animals were exposed to low sublethal temperatures. Another difficulty is to distinguish a paradoxical hardening from damage caused by sublethal temperatures.

C. Regulations by Control Systems

Regulation is present when resistance to temperature extremes is changed by higher centers; in these cases the problem of adaptation to different ATs need not arise (p. 323). An example is the excitability of the isolated foot of *Lymnea stagnalis;* after removal of the pedal ganglion the heat resistance as measured by excitability goes down (BENTHE, 1954). In several instances isolated organs have demonstrated temperature limits other than those exhibited in situ. Frog muscles retaining nervous connection with the telencephalon and diencephalon are less thermotolerant than those wholly isolated from the organism. The nervous associations with the spinal cord do not affect the level of thermostability (PASHKOVA in POLYANSKY and USHAKOV, 1965). The isolated heart is more tolerant to heat than the heart in situ in *Lebistes reticulatus* and vice versa in *Oryzias latipes* (TSUKUDA, 1961; see also ORR, 1955; GRODZINSKI, 1955; RICHARDS, 1963). The isolated heart of *Mytilus californianus* is more resistant to heat and cold than when in situ (PICKENS, 1965). Here too the process should be termed regulation, only if the higher center changes the resistance, thus imposing true regulation (p. 421).

D. Resistance Adaptations

1. Genetic Adaptations

a) Distribution Limited by Temperature

The aspects of temperature-influenced limits of distribution worked out by ecologists will be mentioned only briefly; for ecologists the classification into constant and changing reaction systems (p. 302) is arbitrary. They are obliged to consider temperature as only one factor in climate and to take other biotic and abiotic factors into consideration. Correlations can be presented, but there is little information on causal relationships.

There are many examples of the significance of temperature for the dispersion limits of animals. Coral reefs for example are limited in occurrence to warm seas where the temperature does not go below 20° C; abyssal animals do not tolerate high temperatures (see VERNBERG and VERNBERG, 1970). The distribution of many human parasites is temperature-limited (see TISCHLER, 1969). The distribution of many species is in accordance with the lines indicating annual minimal, annual average temperatures, or certain annual fluctuations, but some of these relationships may be accidental (see UVAROV, 1931; TISCHLER, 1952; MAELZER, 1965; ed. 1, p. 164f.).

Additionally, life niches on land and in water, which were briefly described in ed. 1 (p. 162f.) can be very different in their temperature conditions (see DILL et al., 1964; RUTTNER, 1962; BERÉNYI, 1967; TISCHLER, 1955, 1965). Furthermore macroclimatic conditions need not be critical for distribution, and it is usually not possible to associate a uniform temperature with a macroclimate. Differing from this in several respects is ecoclimate (habitat climate), and still more limited is microclimate (see PLATT, 1961; GEIGER, 1962; HERFS, 1962, and others). Temperature differences may be considerable in a very small space. For example, in a xeric locality MAZEK-FIALLA (1941) measured differences of nearly 7° C in ground temperature in extremely small areas and differences in solar reflection of 23%.

There are also cases in which limits of distribution of a species run parallel to climatic changes in the present or in historic time. This has been suggested as an explanation for the broken distribution pattern of butterflies and other animals in North Europe. Mackerel have only recently penetrated into the East Finland gulfs, and herring and cod into the Caribbean Sea, which is probably a result of temperature changes (RENSCH, 1950). Climatic changes may also leave some species in a favorable ecoclimate as relicts, for example the thermophilic *Mantis religiosa* on Kaiserstuhl mountain in Southern Germany.

Some species make use of microclimates to extend their range — for example, the occupants of houses and pests of stored provisions. Thus bedbug *(Cimex lectularius)* by making use of human dwellings has become widely distributed (HERTER, 1953, p. 317). The same is true of crickets *(Acheta domesticus)* in rubbish heaps in cities. In many habitats, conditions of the milieu are not constant, as in the deposits of litter at the edge of a beach. Some species are present especially during certain months (REMMERT, 1960), others are present during the whole year, but are subject to definite maxima. This is not attributable to endogenous annual rhythm or to day length, but is conditioned by other factors such as available foodstuffs, competition, etc. Thus apparently the northern, cold-tolerant fly *Heterocheila buccata* in winter displaces the imagoes of the southern *Fucellia intermedia* from the beach; in summer the opposite occurs. — Animals of the tidal region are often exposed to low temperatures at low tide; farther and farther north, this fauna diminishes visibly (see GERLACH, 1965; THEEDE, 1967).

The presence of hosts does not always coincide with the presence of parasites; each can be subject to different temperature conditions. Temperature limits differing for host and parasite may be related to the fact that in vertebrate animals the synthesis of antibodies by the host is dependent on temperature (BISSET, 1948). In *Drosophila* the incidence of tumor formation decreased at high temperatures (GHELELOVITCH, 1953). For the dispersion of an animal species the temperature limits of its predators can also be significant (see for example BOURNE and HAYS, 1968).

Lethal temperature is not necessarily the determinant of the limits of dispersal of a species. Resistance to high temperatures is often greater than the habitat temperatures make necessary; but this does not rule out some relationship between lethal temperature and that of the habitat (see BRETT, 1956; SPRAGUE, 1963). Embryonic and reproductive processes, especially spermatogenesis, usually show narrower temperature limits than those required for survival itself; these processes are often limited to favorable seasons of the year (see LARSEN, 1943).

HUTCHINS (1947) differentiates several possibilities for marine organisms: Either lethal temperatures are crucial for expansion of range to the north and south, in winter toward the poles, in summer toward the equator, or temperature limits of reproduction are important, in summer toward the poles and in winter toward the equator; or there are animals subject to both influences. Distribution of *Balanus amphitrite* is cited as a complicated case; the northern limit is set by two factors: Inability to survive average monthly temperatures below 7.2° C, and absence of reproduction when the average monthly summer temperature does not rise above 18.3° C (ORTON, 1920; CRISP and SOUTHWARD, 1958; WERNER, 1958; and others). For distribution it is also crucial whether a species can escape extreme temperatures, or can survive unfavorable periods in a resistant state such as summer or winter diapause. Often each stage of development of a species has a different resistance to temperature extremes, so that a distinct lethal temperature cannot be given for the species. It is then important that the most resistant stage occurs in the most unfavorable season (JOHNSON, 1967; SCHWERTFEGER, 1963, p. 105). When animal stages in nature are exposed to similar temperature conditions, the most sensitive stage determines the distribution of the species insofar as resistance to temperature extremes is crucial. Eggs in air, eggs in water, young pupae, established pupae, larvae and imagoes of *Aëdes aegypti*, in that order, show decreasing resistance to both heat and cold (BAR-ZEEV, 1957; see also p. 403).

There are examples in nature where the lethal temperature is unquestionably crucial. Eggs, larvae and pupae of *Buprestidae*, *Cerambycidae* and *Scolytidae*, which develop under the bark of fallen trees, are often exposed on the sunny side to lethal temperatures (GRAHAM, 1924; SCHWERTFEGER, 1963, p. 164). According to BAILEY (1955) fish undergo the hazard of heat death since they are less able than reptiles, for example, to leave a region of high temperature in extreme biotopes. More apparent are the hazards of low temperature limits, even when these are

above zero. After a comparatively cold winter (lowest temperature 7° C) the beach
on the Bermuda Islands was strewn with dead fish and other animals (VERRILL,
1901; see GALLOWAY, 1941; GUNTER, 1941; SIMPSON, 1953; SCHOLANDER et al.,
1953; VERNBERG and VERNBERG, 1970). CRISP (1964) reports a high percentage of
mortality in mussels of tidal regions of South England after the winter of 1962/1963;
especially affected were celtic and southerly forms, less so the northerly (see
KÜHLMORGEN-HILLE, 1963; ZIEGELMEIER, 1970). It is well known how dangerous
to life for many animals a heavy frost can be, especially if it comes suddenly in
spring to animals that are already warm-adapted, or which have already come
to the end of the annual resistant phase. Especially readily reduced in numbers by
low temperatures are those soil-dwelling animals which are most active in late
autumn and early spring, such as earthworms, *Tipulid* larvae, *Bibionidae* and young
Elaterid larvae (LARSEN, 1949; TISCHLER, 1955, p. 93). Also insects overwintering
above ground may suffer from severe cold; examples are pupae of *Pieris brassicae*
(NORDMANN, 1954) or *Psylliodes chrysocephals*, an insect which after reduction in
numbers by cold requires several generations to regain its former population level
(GODAN, 1947). Abnormal winters can bring about total annihilation of some
species (NEWCOMBER, 1920). Often late frosts are only indirectly damaging to insects, in depriving them of feeding sources, such as blossoms of fruit trees; fruit
pests may be decimated, but not those insects which live on buds, leaves and young
shoots (examples in TISCHLER, 1965, p. 60).

Based on limits imposed by temperature, the rather arbitrary distinction has been
made, as has been mentioned, into eury- (labile) thermal and steno- (constant)
thermal species, and subdivision has been made: Cold and warm stenotherms (see
for example HERTER, 1953; ZUBER, 1969). EKMAN (1953) distinguishes for marine
organisms "reproductive" and "vegetative" thermolability and stenothermy. Stenothermy is often present only during the reproductive phase; thus distinction can
be made between habitats in which the animals do and do not reproduce. SCHLIE-
PER and BLÄSING (1952) warn against conclusions on distribution based only on
experimental investigation of limits to life set by extreme temperatures.

According to GERE (1964) cave animals do not readily lose the capacity to survive
temperatures fluctuating widely above or below those which prevail in their usual
habitats.

b) Genetically Conditioned Temperature Limits

In order to exist, each species is more or less adapted to its milieu. Some adaptations
for survival at temperature extremes have been mentioned. The fact that a species
is especially cold-resistant in winter, and heat-resistant in summer can be looked
upon as an adaptation, as can the fact that for insects with one generation in a
year, often only a certain stage can overwinter (SALT in DILL et al., 1964).

Naturally, in general, high and low temperature limits suited to the habitat are
hereditary. Limits too narrow would risk the continuance of the species; but it is
not to be expected that resistance as a rule would be much greater than is required
for the habitat of a species, unless it might be a species that had penetrated or
been carried from warmer regions into colder, and retained the original limits
(compare SCHLIEPER, 1966). The Antarctic fish investigated by WOHLSCHLAG do not

survive a few degrees above zero (see SOMERO and DEVRIES, 1967; MARR, 1962; ARMITAGE, 1962); this is also true for *Euphausia superba* (MCWHINNIE in LEE, 1964 [19]). Desert grasshoppers were found at temperatures above 60°C (SPEYER, 1937, p. 90; see HERTER, 1953; FRAENKEL, 1961; CLOUDSLEY-THOMPSON in DILL et al., 1964, p. 458; and others). Warm water fishes have low temperature limits well above zero, as every aquarium hobbyist knows. The malacostracan *Thermosbaena mirabilis* from warm brackish water (habitat temperature 37° C to 47° C) showed a low lethal temperature of 30° C (BARKES, 1959; for limiting temperatures for vertebrate animals, see FRY in ROSE, 1967). In experiments, tests must be made regularly (heeding the influences on resistance already mentioned, to determine the presence of non-genetic resistance adaptation, the limits to which are of course set by heredity, or whether a genetic adaptation which has evolved in the history of the species or race is present, or both. Unfortunately much of the work in this field has been done without making these distinctions, or not making them clearly, so that a logical summary of the results in this chapter is of dubious value; this uncertainty we wish to emphasize. On the other hand cross-breeding experiments with fish and other animals have been undertaken in order to set exactly the genetic basis for heat resistance (MINAMORI, 1957; VERNBERG and MERINEY, 1957). According to SCOTT (1964) heat resistance of *Esox lucius* and *E. masguinongy* is the same, but in F_1 hybrids it is higher. BROWN (1969) hybridized geographic races of *Scaphiopus hammondii*, the tadpoles of which show hereditary difference in heat resistance (CTM). Strains of *Balantidium coli* which were resistant to varying degrees showed change in resistance after conjugation (SVENSSON, 1955).

Selection has naturally played a fundamental role in the evolution of genetic adaptation. Such selection phenomena can be imitated experimentally (see AMOSOVA in GLUSHANKOVA et al., 1967), and with rapidly multiplying animals like protozoans care must be taken that selection of resistant animals during treatment is not mistaken for non-genetic adaptation. Competition experiments with overpopulated cultures have also been carried on, for example with *Drosophila* (ed. 1, p. 171), but in such experiments lethal temperature is not necessarily limiting (see MILKMAN, 1965). In some animals *(Drosophila)* mutations for specific temperature sensitivity are known (SUZUKI et al., 1967).

We cannot go further here into consideration of the less extreme cases (than those mentioned) where the lethal temperatures of the species under comparison show a relationship to temperature of the habitat (geographic distribution, occurrence at different depths in water, etc.) (BRETT, 1956; SOUTHWARD, 1964; FRY and BULLARD in DILL et al., 1964; SAMEOTO, 1969 [20]). According to SCHLIEPER (1960) genetic resistance adaptation is present rather in marine invertebrates in the more cold stenothermic species of deep water, and non-genetic adaptation in the eurythermal forms of surface water where temperature fluctuations are far greater. In fish (HART, 1952) and also in other animals, high lethal temperature correlates better with distribution of species than of races; therefore, when races are widely distributed, the correlation is partly negated (for racial differences see MCCAULEY, 1958; GLUSHANKOVA et al., 1967, p. 99).

---

[19] Further examples in VERNBERG, W. B., VERNBERG, F. J.: 1972, see p. 303.
[20] cf. GRAHAM, J. B.: Science 172, 861—863 (1971).

HUTCHISON (1961) found, in general, a fairly good correlation of the CTM with the habitat and distribution of many species of *salamanders* (see DUNLAP, 1968; BROWN, 1969). The CTM for *Diemictylus viridescens* after being kept at the same adaptation temperature for some time, showed differences for animals from different regions[21].

Relationships between the habitat and the preferred temperature (LICHT et al., 1966; LICHT in PROSSER, 1967) or the limits of activity (SPAAS, 1959; MUTCHMOR in PROSSER, 1967) of species have been found. In lizards the extent of heat resistance appears to correlate with the level of body temperature employed for activity (SCHMIDT-NIELSEN in DILL et al., 1964; for factors limiting the northern distribution of lizards see HOCK, ibid.). LICHT et al. (1969) calls attention to the fact that heat resistance relationships to macroclimate, which is usually considered without qualification, are misleading if the lizards seek out microclimates which differ from the macroclimate.

With regard to genetically conditioned heat and cold resistance, populations of the much investigated crab genus *Uca* from the tropics and temperate zone show distinct correlations with the temperature of the habitat, and this holds for comparisons of species and races (F. J. VERNBERG and W. B. VERNBERG, 1967; MILLER and VERNBERG, 1968; F. J. VERNBERG, 1969; JOHNSON, 1952; KINNE, 1953; SOUTHWARD, 1964; for insects: TAKSDAL, 1967; SULLIVAN, 1965; BATEMAN, 1967). *Nodilittorina granularis* from the coast of the Japan Sea, compared with animals from the Pacific coast (localities at the same longitude, but different climate especially in winter), were more tolerant to both heat and cold in winter (OHSAWA and TSUKUDA, 1956). *Niphargus longicaudatus* from a surface biotope are more resistant then individuals from a subterranean biotope; also non-genetic heat adaptation occurred (MATHIEU, 1968; GINET and MATHIEU, 1968). Parasites of warm-blooded animals are reported often to be more heat resistant than those from fish (W. B. VERNBERG, 1969).

In single organ functions a relationship can often be demonstrated between resistance to temperature extremes and the habitat of the animals as well as to their lethal temperatures, although the range of tolerance of many organ functions is greater than that of the viability of the animal, so that a genetic adaptation can be present where it is not of biological significance, and selection cannot be of direct influence; an example is the upper limit of the loss of muscle excitability of cottoid fish (KUSAKINA, 1963; see LICHT, 1964). Gastrocnemius muscles of *Rana sylvatica* from Alaska continued to be excitable by the nerve down to $-3°$ C or $-5°$ C; those from Mexico to about $0°$ C (MILLER and DEHLINGER, 1969). SCHLIEPER (1966; in TROSHIN, 1967; in PROSSER, 1967) found a correlation between the genetically determined high and low temperature limits of the gill epithelium, measured by ciliary beat, and the place of occurrence of the mussels (geographic distribution, water depth). Exceptions could be accounted for by the history of the species. Thus *Congeria cochleata* from the Kiel Canal was extremely heat resistant; this is an eurythermic species brought in from West Africa (SCHLIEPER et al., 1967). Of the mussels from North Sea shallows investigated in winter, the gill epithelium of *Mytilus edulis* showed the highest resistance to freezing; that of *Cardium edule* showed distinctly less. Even lower in resistance is the gill epithelium of *Mya arenaria*, a species that lives deeper in the sand than *Cardium*. The slight resistance of *Modiolus modiolus* is related to its occurrence in cold deep water, as

---

[21] BRATTSTROM, B. H. (Copeia 1971, 554—557) investigated the CTM of several skinks.

is that of *Ostrea edulis*, which also occurs below the tidal zone (THEEDE, 1967, 1969, 1970; but see ZHIRMUNSKY and PISAREVA, 1961). According to THEEDE (1965) tolerance to varying salinity can be a primary determinant of distribution; euryhaline species have undergone non-specific increased resistance, and therefore are often resistant also to temperature extremes.

Finally, corresponding observations can be made not only for heat resistance of cells, but also for proteins and enzyme activities, which in general have far greater tolerance than do organ functions or whole animals. In the gamete stage of development the resistance of separate enzymes may be greater than that of the cells (discussion in PROSSER, 1967, p. 423ff.). The same enzymes in different species can show markedly different heat resistance; for H_4 LDH the limit in fish is about 60° C, the same for lower reptiles and mammals, but for higher reptiles it is in the region of 80° C (WILSON et al., 1964). The fructose-diphosphate aldolase of the trout is more heat resistant than that of antarctic fishes but it is more heat sensitive than that of rabbits (KOMATSU and FEENEY, 1970[22]). It is necessary in this kind of experiment to rule out the effects of AT or of seasonal differences, before concluding that there is genetic adaptation; it is also an incorrect deduction that non-genetic adaptation must be lacking, since in this area, outside the vital limits, selection which would favor such capacities cannot have its effect. Genetic or nongenetic adaptation of proteins can in some instances per se be important, in others it is perhaps merely a side effect. ALEXANDROV (1969) believes that the adaptation of even extremely resistant proteins to the temperature of the milieu can be explained as follows: For selection, heat resistance of the proteins is not the direct point of attack, but rather their conformational flexibility, which is in harmony with the mean environmental temperature and which is important for catalytic activity, allosteric regulation, mechanochemical properties, resistance to denaturing agents and proteinases, etc.

USHAKOV and co-workers look upon heat resistance of proteins as characteristic of species; racial differences are rarer. The thermostability of various proteins is closely associated with reproductive and environmental temperatures (latitudinal distribution of the species, depth of their distribution in the sea and peculiarities of their ecology), but is, to a great extent, independent of position of animals in phylogenetic series (USHAKOV, 1964; see KUSAKINA, 1963; several authors in TROSHIN, 1967; USHAKOV and PASHKOVA in GLUSHANKOVA, 1967; RIGBY and MASON, 1967). In such investigations it needs to be borne in mind that "the level of cellular heat resistance of adult poikilotherms does not remain constant, and undergoes ontogenetic changes in connection with reproduction, season and metamorphosis. These changes occur due to changes in the function of endocrine glands and, in most cases, are not associated with the changes in temperature of the habitat. Hence, as a rule, they are not connected directly with adaptations of the adult organisms to temperature" (USHAKOV, 1963, p. 42). Heat resistance of collagen also shows a relationship to the habitat of species (LOVE, 1970, p. 214ff). Corresponding relationships can exist with the preferred temperature of the species (LICHT in PROSSER, 1967) — for example in the heat resistance of myosin adenosine triphosphatase of lizards, but not in alkaline phosphatase, which after the temper-

---

[22] cf. KOHONEN, J., TIRRI, R., LAGERSPETZ, K. Y. H.: Comp. Biochem. Physiol. **44**B, 819—821 (1963).

ature optimum is reached, does not show a steep decline in activity caused by denaturation (but see Table 3 of LICHT).

Naturally there are exceptions in organ and cell functions and in enzyme activities. The heat resistance of glyceraldehyde-3-phosphate dehydrogenase of an antarctic fish *(Dissostichus mawsoni)* was very similar to that of the rabbit, but the inactivation by ATP or AMP at low temperatures differed (GREENE and FEENEY, 1970).

In mussels for example, there is a correlation between the capacity to survive anaerobic conditions and the heat resistance of aspartic-glutamic transaminase, but no correlation between their resistance and the temperature of the species niche (READ in PROSSER, 1967). In two geographic races of the lizard *Phrynocephalus helioscopus* the heat resistance of hemoglobin showed the opposite reaction to that of other proteins (USHAKOV in PROSSER, 1967). The heat resistance of low molecular weight proteins of eye lens nuclei of closely related species of *Thunnus*, in particular, was different (SMITH, 1970).

Since many proteins are quite resistant to low temperatures, correlations between cold resistance and the lower limiting temperature for the species are found less readily. Activity of brain cholinesterase of two "southern" fishes showed a steep rise of Q_{10} values below 10° C, those of "northern" fishes (goldfish, blackfish, killifish) below 2° C (BASLOW and NIGRELLI, 1964). Since usually the whole range of tolerance is shifted (see for example PETERSEN, 1948, ed. 1, p. 175), one must also consider a correlation of cold resistance of a species with the heat sensitivity of enzyme activity, or with the position of the temperature optimum for activity. μ values of apyrase activity, either in the range from 5—15° C or from the respective chill-coma temperatures to temperatures 15° C higher, were greater in insect species having lower chill-coma temperatures; apyrase rate maxima occurred at lower temperatures in species having lower chill-coma temperatures (MUTCHMORE and RICHARDS, 1961). Comparisons of different organs of an animal can readily be made. *Lethocerus americanus* ceases to fly below 10° C, to run below 2° C (KENNEY and RICHARDS, 1955); the apyrase activity of the leg muscles shows higher μ values and lower temperature optimum than the enzyme activity of the flight muscles.

2. Non-Genetic Adaptations

a) Occurrence and Meaning

Investigations have often been made of the change in resistance after a change in acclimation temperature; this may influence the heat and cold resistance of animals

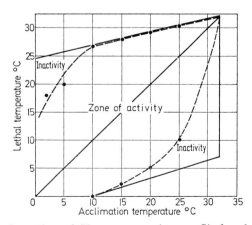

Fig. 1. Resistance adaptation of *Homarus americanus*. Circles show temperatures of zero activity (from MCLEESE and WILDER, 1958)

(the resistance to chilling, supercooling and freezing); also the position of the super-cooling and freezing point may be changed as well as the capacity to undergo supercooling and freezing. Not only lethal temperatures are changed, but also other temperature limits, for example the temperatures at which normal activity ceases (Fig. 1), the chill-coma temperature, and others.

In several animals there is adaptation only to one temperature limit, in some insects to low temperatures, in the snail *Deroceras reticulatus* to high temperatures (I. PRECHT, 1967).

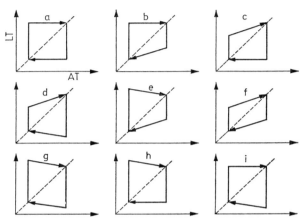

Fig. 2. Possibilities of the dependence of heat and cold resistance on the adaptation temperature (scheme). a No heat adapt., no cold adapt. b No heat adapt., reasonable cold adapt. c Reasonable heat adapt., no cold adapt. d Reasonable heat adapt., para-doxical cold adapt. e Paradoxical heat adapt., reasonable cold adapt. f Reasonable heat adapt., reasonable cold adapt. h Paradoxical heat adapt., no cold adapt. i No heat adapt., paradoxical cold adapt. *A T* adaptation temperatures, *L T* lethal temperatures or another index for the measured heat and cold resistance (from THIEDE, 1965). Examples: a *Dixippus morosus* (I. PRECHT); b *Ischnodemus sabuleti* (PRECHT); c Sum-mer animals of *Nodilittorina granularis* (OHSAWA), tropical *Uca* (VERNBERG and TASHIAN), some plants in summer experiments of LANGE; d *Torulopsis kefyr* (CHRISTO-PHERSEN and PRECHT); e Several winter plants (LEVITT and LANGE); f Fishes (BRETT), *Homarus americanus* (McLEESE and WILDER); g Succinodehydrogenase activity of *Xiphophorus helleri* (PRECHT) (from PRECHT in TROSHIN, 1967)

Resistance adaptation is termed a *reasonable* acclimation whenever heat resistance and sensitivity to cold increase with rising AT; the opposite case is termed *para-doxical* (Figs. 2, 3). Difficulties in the use of the terms reasonable and paradoxical for hardening (insofar as this is a separable characteristic, making a differentiation from resistance adaptation necessary) are mentioned (p. 421). A reasonable adap-tation to both temperature extremes occurs in animals as simple as ciliates (VOGEL, 1966; POLYANSKY and SUKHANOVA in TROSHIN, 1967), and in crabs (Fig. 1; see also KINNE in DILL et al., 1964), fishes (FRY in DILL et al., 1964; in ROSE, 1967) and other animals (see PRECHT, 1963, 1964d; in TROSHIN, 1967; PRECHT et al., 1966; JACOBSON and WHITFORD, 1970). Adaptation of the chill-coma temperature also occurs, as already mentioned, in strongly stenothermic antarctic fish and later stages of arctic insects, though not in those stages which in nature are exposed for long periods to low temperatures (MELLANBY, 1940). The effect of AT on heat resistance can change in the course of development (for example in *Tilapia nilotica* according to BISHAI, 1965; see also MORRIS, 1962). *Drosophila subobscura* shows the capacity for heat adaptation only up to the age of 56 days (BOWLER and HOL-LINGSWORTH, 1966). In *Littorina littorea* the extent of heat adaptation is said to be greatest in animals which are least resistant (NEWELL et al.[23]).

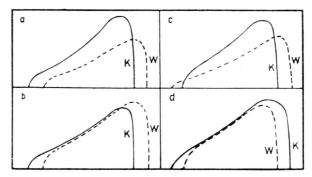

Fig. 3. Different types of temperature adaptation (scheme). *W* warm-adapted organisms *K* cold-adapted organisms. Ordinates: the measured capacity. The left part of the curves has been less investigated and may run differently. Usually the resistance of the measured capacity is not given, rather the life resistance, which often shows a corresponding resistance adaptation (for instance as the oxygen consumption); abscissae: experimental temperatures. a Compensation in the normal range of temperature, reasonable heat and reasonable cold adaptation; b No compensation, reasonable heat and reasonable cold adaptation; c Compensation, reasonable heat and paradoxical cold adaptation; d No compensation, paradoxical heat and reasonable cold adaptation (from PRECHT in TROSHIN, 1967)

In some fishes no further increase in heat resistance can be observed when a high AT is exceeded. Also in the low AT-range, heat resistance may remain unaltered (for example in lobsters, McLEESE, 1965; see DUNLAP, 1968). This is true also for many insects, whose cold resistance is influenced by ATs over the entire range; this supports the concept of the independence of heat and cold adaptation (MELLANBY, 1954). Also when adaptation of the same kind is present over the entire range, the effect of the adaptation can change with the AT, for example in heat resistance of woodlice (EDNEY, 1964; see MELLANBY, 1958).

According to MELLANBY (1939, 1960) cold adaptation in *Cimex lectularius, Blatta orientalis* and larvae of *Aëdes aegypti* does not take place at low temperatures; in mosquitoes it is absent at 10° C, at which temperature the larvae were in chill coma. For the chill-coma temperature of *Blatella germanica* according to COLHOUN (1960), ATs below 15° C were not more effective than 15° C. According to ATWAL (1960) no further effects could be expected after the ATs reached chilling temperatures; cold injury may then outweigh acclimation benefit. According to PAYNE (1926) in larvae of *Synchroa punctata* cold resistance could be further enhanced by a stay at

[23] NEWELL, R.C., PYE, V.I., AHSANULLAH, M.: J. exp. Biol. **54**, 525—533 (1971).

$-3°$ C. As in heat adaptation, cold adaptation may be lacking in the lower AT-range. Some insect stages underwent no cold adaptation at ATs of $0°$ C and $5°$ C, but did show adaptation (hardening?) when kept at the fluctuating temperatures of the natural environment (SALT, 1958a, see p. 439).

SEGAL (1961) was able to give an optimal AT for *Limax flavus* for survival at low temperatures. Cold adaptation for diapausing larvae of the corn borer was optimal at $0°$ C; at $-9°$ C it was better than at $-30°$ C (LOSINA-LOSINSKY in TROSHIN, 1967).

Paradoxical adaptation has less often been noted for survival limits of animals; EDWARDS (1958) found for imagoes of *Tribolium confusum* reasonable adaptation to cold and paradoxical to heat (animals adapted to $38°$ C appeared to be damaged). In the copepod *Tigriopus japonicus* there are adaptation regions with reasonable and with paradoxical heat adaptation; in addition to the AT before the tests, the temperature of culture was also important. Paradoxical phenomena are said to correlate with in-activation of aerobic metabolism by cold (MATUTANI, 1961; see KIRO in GLUSHANKOVA et al., 1967). According to BALDWIN (1954) in the parasitic hymenopteran *Dahlbominus fuscipennis* the course of heat adaptation in the upper temperature range is reasonable, at lower temperatures paradoxical. Such a switch (also for cold resistance) could be correlated with a maximum formation of protective substances at certain temperatures (p. 416f.). Decrease of heat resistance at extremely high ATs that cause damage is hardly attributable to paradoxical adaptation.

In order to improve the understanding of paradoxical adaptations, a brief examination of possible mechanisms of resistance adaptation will be made; these are not well known. We may consider as causes the factors already mentioned: water content, protective substances, stable proteins, perhaps thyroid hormones, etc. In general they tend to change resistance to both temperature extremes in the same direction. For example let us (theoretically) consider the content of free water. Change in this with AT would be able to bring about only a reasonable adaptation to *one* temperature extreme; in animals exposed to cold in nature the quota of free cell water should decrease with lower AT, in order to cause increase of cold resistance (side effect: increased heat resistance); in animals often exposed to heat, the cell water content should decrease with rising AT, in order to increase the heat resistance (side effect: increased cold resistance). Changes of this kind can explain the increased resistance of certain life stages (e.g. anabiosis) to both temperature limits, but never a reasonable adaptation in both directions, such as many fish and other animals show. If reasonable resistance adaptation is to be explained as conditioned by the factors named above, this can be so for only one temperature extreme. In the other direction some other more sensitive mechanisms must be dealt with, in order for adaptation to run reasonably in this direction too.

Survival at extreme temperatures is best guaranteed by reasonable heat and cold adaptation, whether considerable temperature fluctuations occur at the same season of the year or at different seasons, but there is often an additional change of resistance caused by other factors such as day-length (see e.g. HUTCHISON, 1961; FRY in ROSE, 1967, p. 382; TYLER, 1966; LASHBROOK and LIVEZEY, 1970). CTM of *Hyla labialis* appears to show heat adaptation only during long-day periods, or when the light-dark hours are $12-12$, not during short days, which are not normal for the tropical animal (MAHONEY and HUTCHISON, 1969). In *Asellus inter-medius* AT alone was responsible for the seasonal variation of heat resistance (SPRAGUE, 1963; see TODD and DEHNEL, 1960; FREE and SPENCER-BOOTH, 1960; OHSAWA, 1956[24]).

---

[24] cf. HILL, B. J., ALLANSON, B. R.: Mar. Biol. **11**, 337—343 (1971).

It was difficult to detect regularity in the occurrence of heat adaptation in insects (references in I. PRECHT, 1967; see CLARKE in ROSE, 1967). Reasonable cold adaptation was seen in water-living larvae of *Aëdes, Culex, Theobaldia* and *Corethra*. The temperature of small bodies of water can fluctuate greatly, and there is no opportunity for the animals to avoid temperature extremes by seeking out a favorable microclimate. Reasonable cold adaptation occurs also in many terrestrial animals, particularly in those which may be exposed to low temperatures. This, e.g., is lacking in larvae of *Tenebrio molitor* and imagoes of *Carausius morosus;* these animals originated in warmer climates.

Comparing CTM values of species, BRATTSTROM (1968) found that tropical anurans do not have a narrow range of acclimation. Instead the entire thermal regime of the more southern species is higher than for northern forms. Species with restricted geographic range, such as *Bufo exsul* and *B. nelsoni,* have little or poor ability to acclimate (for exceptions see BRATTSTROM, 1970). High altitude forms from the United States and Mexico perform similarly, in regard to their ability to acclimate, as temperate forms of equivalent thermal latitudes (see BRATTSTROM, 1960; BROWN, 1969). Northern species of *Uca* have a greater capacity for cold adaptation than tropical forms (see VERNBERG and TASHIAN, 1959; MILLER and VERNBERG, 1968; F. J. VERNBERG, 1969). Even in geographic races of species of *Uca* there is a relationship between the temperature fluctuations of the habitat and extent of resistance adaptation (F. J. VERNBERG and W. B. VERNBERG, 1967). Heat adaptation of gill epithelium of intertidal mussel species was greater than in subtidal species (SCHLIEPER in TROSHIN, 1967).

Distinction has been made between non-specific (general) and specific adaptations (see PRECHT et al., 1966, p. 382f.). Warm-adapted animals are often resistant to several other factors (lack of oxygen, unusual salinity, poisons, narcotics and so on), but entirely non-specific resistance probably occurs rarely; this is found also by ecologists (cf. KROGERUS, 1939; TESCH, 1956). When reasonable adaptation to both extremes of temperature has taken place, the general rise of resistance of warm-adapted animals to heat does not include enhanced resistance to the other temperature extreme, that is to cold. The general resistance can also be combined with cold resistance. PITKOW (1960) found that guppies adapted to 23° C survived not only cold but also lack of oxygen better than fishes adapted to 30° C. If this is to be termed a non-specific increase of resistance with falling AT, heat resistance must be excluded.

In the experiments of SUMNER and DOUDOROFF (1938) on *Gillichthys,* the time course of the change in resistance to KCN did not run in parallel to that of the change in heat resistance when the AT was raised from 20→30° C. Change in sensitivity to urethane, after change of AT, took place much more rapidly than change in tolerance to KCN and to lack of oxygen, in experiments with fish by SUMNER and WELLS (1935). The concept of a general resistance thus appears to be an over-simplification. Insects often show cold adaptation and no heat adaptation; the resistance in general then is not changed. According to VERNBERG et al. (1963), whether the cold- or warm-adapted animals in a species are especially resistant to other factors is dependent on temperature relationships in the biotope.

In *Paramecium caudatum* resistance to ethanol increases with higher AT, but the resistance to threshold concentrations of $CaCl_2$, KCl and KCN decreases; concentrations above lethal levels have the opposite effect. A threshold concentration of KCN inhibits the cytochrome respiratory system; in high concentration it acts as a protoplasmic poison (POLYANSKY and SUKHANOVA, and IRLINA in TROSHIN, 1967).

Resistance to heat and to cold are apparently independent of each other, which is evident from several considerations, among them different ionic effects and specifically of salinity (see VOGEL, 1966; KÄHLER, 1970), from feeding experiments (p. 403) and from direct observation (POLYANSKY, 1959) etc. In general, the existence of a single mechanism for heat and cold adaptation which would change both heat and cold resistance, seems most improbable. *Umadaptation* experiments of reasonable adaptation to both temperature extremes have shown that the increase in resistance to one extreme and the decrease in resistance to the other do not always correspond in time course (see p. 438). This is also evident in experiments in which the previous treatment was inconstant (p. 439). In the development of an animal the capacities for heat and for cold adaptation do not necessarily appear at the same time (for example in *Aquidens portalegrensis*, MORRIS, 1962). As stated previously, many animals are adapted to only one temperature limit (see p. 429).

b) Adaptation on Different Levels

The dependence on AT of functions of separate organs can be different and can also differ from resistance of the whole animal. In Fig. 1, activity limits of the lobster show a dependence on AT which differs from survival limits. In bees cold lethargy shows reasonable resistance adaptation, and the low lethal temperature is independent of the AT (FREE and SPENCER-BOOTH, 1960).

It is to be expected that paradoxical phenomena would be found more often in those cell and organ functions which are not directly involved in survival resistance than in those so involved. Thus many examples of "side effects" not important for survival can be shown for the theoretical possibilities given in Fig. 2.

Organ functions: Gill cover movements of *Xiphophorus helleri*, which signal the activity of the respiratory center, show a reasonable resistance adaptation to both cold and heat; a similar adaptation to both temperature extremes is found for other organ functions with a wider range of tolerance (PRECHT, 1959, 1962; THIEDE, 1965). Preparations of N. ischiadicus—M. gastrognemius of *Rana temporaria* were examined with regard to the conductivity and irritability of the nerves, transmission at the end-plate, and contractile response of the muscle to direct stimulation. The end-plate had the narrowest temperature range and showed a reasonable cold- and a paradoxical heat-adaptation; the contractions of the muscle had a small reasonable heat- and a clearer reasonable cold-adaptation, as was also found for transmission at the end-plates below 0° C. The conduction of the nerves adapts paradoxically to heat; cold resistance of the nerve could not be measured (PRECHT, 1960a; see BISHOP and GORDON in PROSSER, 1967, p. 277). A decrease in muscle thermostability (gastrocnemius and rectus abdominalis of *Bombina bombina*) with cold adaptation was not accompanied by alterations in the heat resistance of muscle proteins under investigation (USHAKOV and GLUSHANKOVA, 1970). PERTTUNEN and LAGERSPETZ (1956) observed reasonable resistance adaptation for heart beat of *Corethra plumicornis* to both temperature extremes.

Functions of cells: Several authors observed reasonable adaptations of isolated ciliary epithelium of mussels to both temperature extremes (SCHLIEPER et al., 1960; VERNBERG et al., 1963; LAGERSPETZ and DUBITSCHER, 1966; THEEDE, 1967; for snails see PRECHT and CHRISTOPHERSEN, 1965).

Employing the method previously mentioned (p. 353) we determined the heat resistance of oxygen consumption of muscle of 3 races of *Xiphophorus helleri;* reversible and irreversible denaturing effects could be distinguished. In a red race (1) the reversible and irreversible denaturations did not show heat adaptation; in a green iridescent race (2) the irreversible denaturation adapted paradoxically to heat and the reversible denaturation showed no heat adaptation. In a grey race (3) the heat resistance of their irreversible denaturation did not depend on AT but their reversible denaturation adapted to heat in a reasonable way. In this race the cold resistance of the oxygen consumption, the measurement of which corresponded to that of the irreversible denaturation in the determination of heat resistance, showed a paradoxical adaptation. The succinic dehydrogenase activity of muscle tissue in races (1) and (3), determined by the Thunberg method, showed a paradoxical resistance adaptation to heat and cold (PRECHT, 1962, in TROSHIN, 1967; see also SUHRMANN, 1955). Also the oft-cited work of BENTHE (1954) is an example of resistance adaptation of isolated organs (see TSUKUDA, 1961).

BRETT (1946) found that a continuous low oxygen saturation of the water in the acclimation tank inhibited resistance adaptation of *Ameiurus nebulosus* up to 23 h at least; during this time fishes from normal aerated water were fully acclimated. With rising AT an increase in heat resistance was found for the oxygen consumption of the white muscle of *Carassius gibelio* (with succinate) adapted in aerated water as well as in water with low oxygen concentration (KREBS, unpubl.).

USHAKOV and co-workers found greater changes in heat resistance of tissues of higher metazoans (and also of their proteins, see for example VINOGRADOVA in POLYANSKY and USHAKOV, 1965), even more with change of stage in life cycle in the course of a year, and less with changes in AT, although the latter can have appreciable effect on vital limits of the animals (see AMASOVA in USHAKOV, 1963). In "lower" metazoans dependence of cellular heat resistance on AT can occur as it can in higher animals, possibly in early stages of ontogenesis, which are often relatively sensitive to heat. Later this dependence on AT will be present in general only in elements controlling the integrative functions of the nervous system (USHAKOV, 1964, 1968; several authors in TROSHIN, 1967). However, AT had an influence on cell thermostability of *Nereis diversicolor* (IVLEVA in TROSHIN, 1967).

Heat adaptation can be shown in the action of digestive enzymes (MEWS, 1957, *Helix pomatia*), which is significant, since the influence of the cell milieu is not present during measurement. Heat resistance of serum proteins in goldfish also is dependent on AT (OHSAWA and TSUKUDA, 1964)[25]. It should be clarified whether, in resistance changes of cellular enzymes, it is the molecules of the enzymes themselves which participate, or the pattern of differentially resistant isoenzymes (see p. 345), or whether effects of the cell milieu impinge. It might be assumed that with changes in proportions of isoenzymes present according to AT, capacity and resistance adaptation of the intact animal could be attributed to a common origin.

Interpretations in terms of enzymes are suggested by SOMERO (1969b): "the sharp decreases in enzyme-substrate affinity which frequently occur at the extremes of an organism's habitat temperature may be important in establishing thermal tolerance limits for organisms"; as mentioned, the K_m values are often dependent on AT and therefore may be of significance for resistance and capacity adaptation. However, the temporal course of the two forms of adaptation may not correspond. There are animals which show no capacity adaptation of oxygen consumption, but

[25] cf. TSUKUDA, H., OHSAWA, W.: Annot. zool. jap. **44**, 90—98 (1971), also LAGERSPETZ, K. Y. H., KOHONEN, J., TIRRI, R.: Comp. Biochem. Physiol. **44**B, 823—827 (1973).

do show resistance adaptation, at least to one temperature extreme (see I. PRECHT, 1967).

It should be mentioned that the results may depend on the method used; e.g. the concentration of pyruvate had an influence on the thermal optimum of lungfish muscle LDH (*Lepidosiren paradoxa*, HOCHACHKA and SOMERO, 1968).
Resistance adaptation in pure proteins has been investigated, either by isolating the proteins before or after adaptation of the animal; in the first case the isolated proteins were exposed to varying ATs (p. 434, PRECHT in TROSHIN, 1967). Many of these "adaptations" of the very resistant proteins have no biological meaning, since they do not influence resistance limits of life.

The question now arises, in connection with resistance adaptation, whether the influence of higher systems (CNS, endocrine glands), as in capacity adaptations, can have direct or after-effects (secondary effects, p. 348). When AT is changed the higher systems are affected first, organ or cell functions are affected secondarily; in organs or tissues which have been isolated, after-effects are demonstrable, but not direct effects. According to LAGERSPETZ and TIRRI (1968), the influence of AT on resistance of organs under nervous control may act via transmitter substances, for example in the heart beat of *Anodonta* (p. 349). Though one might think of participation of endocrine glands in resistance adaptation, this will not be discussed here, since the effect of hormones on resistance is not yet completely clarified (see p. 352, PRECHT, 1964 d). Brook trout were acclimated heterogeneously for three weeks by exposing the head and tail simultaneously to two different temperatures (cf. p. 436). The heat resistance of the whole organism depended on the mean thermal history of the total tissue regardless of the anatomical site (FAHMY[26]).

An example of a direct effect is given by the experiments of BENTHE (1954). Removal of the pedal ganglion from the foot of *Lymnaea stagnalis* diminishes the adaptation effect on the excitability of the foot of animals adapted at 3° C and at 12° C, and removal of the ganglion increases the adaptation effect between 12° C- and 21° C-adapted animals (see p. 324). This shows perhaps a combination of direct and after-effect. In the eurythermic fish *Oryzias latipes* the effects of heat acclimation on the heart beat are markedly greater when the nervous system participates than in the isolated heart (TSUKUDA, 1961).

Direct effects may also occur as in capacity adaptation of fish blood (p. 350, see also SAVITZKY, 1964). In the measurement of heat resistance of oxygen consumption of muscle by the method described on p. 353, reversible and irreversible denaturations can be distinguished. In particular, reversible denaturations in tissues of *Idus idus* show reasonable heat adaptation. Serum of cold-adapted carp lowers heat resistance of oxygen consumption of muscle of *Idus*, but serum of warm-adapted carp has far less of a decreasing effect, or even an increasing effect (especially in reversible denaturations). The effect of serum from cold-adapted carp in lowering oxygen consumption is greater the more heat resistant the muscle is through its own heat adaptation. Carp serum has a similar effect on carp muscle (PRECHT, 1964 c). Destruction of the cell membranes increases the effect of the serum from cold-adapted carps. Cold resistance of oxygen consumption of carp muscle is not differentially influenced by serum, regardless of whether it is from cold-adapted or from warm-adapted carp (PRECHT, 1965).

---

[26] FAHMY, F. K.: Can. J. Zool. **50**, 1035—1037 (1972).

After resistance adaptation has been demonstrated in an isolated organ or tissue, the question still remains whether the AT has had an effect upon the organ or tissue itself or through higher centers, so that only after-effects are present. In the eel, resistance adaptation apparently takes place in muscles of parts of the body only; however, one can make the objection (p. 349) that sections of the nerve cord could function as higher centers and thus control the resistance of the muscle (PRECHT in TROSHIN, 1967).

c) The Time Course of Resistance Adaptation

Resistance adaptation, like capacity adaptation, is normally an adjustment to a new lasting AT. The change of the heat lethal index mentioned (p. 405) for *Xiphophorus* required 13 days and 7 days respectively when the AT was raised (19° → 31°C and 25° → 31° C), but 18 days when the temperature was lowered (31° → 19° C) (for further examples see PRECHT et al., 1966). In some cases the temperature dependence of the adaptation mechanism can be considerable (see SPOOR, 1955; DICKIE, 1958); in other cases it may not be present, for example in the ciliate *Zoothamnium hiketes* (VOGEL, 1966) and *Gammarus salinus* (FURCH, see p. 439). In different species of anurans, the change of heat resistance after a change in AT can be more rapid when the temperature is increased than when it is decreased, or the reverse can be true, or no differences may be measured. Changes of water content in an animal's body may have considerable influence (BRATTSTROM, 1968). Resistance adaptation can take place quickly. Rate of adjustment of heat adaptation is not correlated with latitude, though mid-latitude frogs seem to be able to adjust to new ATs the fastest (BRATTSTROM, 1970).

With high ATs (p. 430), the rate of adaptation may decrease. In the fish *Girella* when the AT was changed from 20 → 26° C, the change in cold resistance took place more slowly than upon transfer from 26° C back to 20° C (DOUDOROFF, 1942). In *Lebistes reticulatus* (ATs: 18°, 23°, 28° C) after all possible acclimations *(Umadaptationen)*, the least time was taken for an adaptation to 23° C, for both cold and heat adaptation (TSUKUDA, 1960 b).

In some cases, a latent period of longer or shorter duration precedes the change of resistance to temperature extremes after alteration of AT; the curve is then sigmoid, as for goldfish or the American lobster (BRETT, 1944, 1946; MCLEESE, 1956). In other instances the latent period is very brief, as for *Xiphophorus*, or it does not occur. Then change in resistance is at first manifested quickly; it then declines (logarithmic function of time). This type of curve was presented for example by DOUDOROFF (1942) for *Girella*, by TSUKUDA (1960 b) for *Lebistes* and by HUTCHISON (1961) for urodeles. TSUKUDA asserts that the method of investigation determines the course of the curve obtained.

There are known instances where the adaptation to a new temperature occurs within one day or even sooner (examples in PRECHT et al., 1966, p. 392; LEVINS, 1969)[27]. The time required depends naturally on the difference between beginning and final AT (cf. BALLINGER and SCHRANK, 1970).

---

[27] After transferring frogs *(Rana temporaria)* from AT 5° to 25° C, the thermal maximum of oxygen consumption increased at a constant rate and reached a steady level in 5 or 6 days; in frogs placed in a colder temperature, it decreased very rapidly during the first day, and thereafter rose slightly (the level on the first day was reached again after several days). Also the partial compensation of the oxygen consumption of winter frogs was essentially complete during the first day following the transfer of the animals to a new AT (ET = 22° C; HARRI, M. N. E.: Physiol. Zoöl. **46**, 148—156, 1973).

When adaptation times are very brief, it should be ascertained in each case whether other phenomena (such as hardening) are contributory. Short times, sometimes only a few hours (after rise in AT, but also after lowering), were found especially for insects; the findings of COLHOUN (1960) on the very quick adaptation of *Blatta germanica* do not point to hardening as adaptation to shock (p. 420). The time of a re-adaptation *(Umdaptation)* may be shorter after a decrease ($25° \rightarrow 15°$ C) than after an increase of AT ($15° \rightarrow 25°$ C); the Q_{10} value for this process is < 1 (as in frogs, see also FURCH, p. 439).

In the resistance adaptation taking place more gradually in most animals (in spite of previously mentioned direct effects via the blood), protein syntheses probably participate; yet, in quick heat adaptation of *Drosophila* this was demonstrably not the case (DINGLEY and SMITH, 1968). In both of these resistance adaptations, perhaps so different in mechanism, lasting changes occur in resistance after a change in AT, not merely adaptation to a shock, as in hardening.[28]

It is difficult to demonstrate the participation of protein synthesis in resistance adaptation. LAUDIEN (1973)[29] investigated the ciliated epithelium of tentacles of *Heliosoma nigricans*. Long-term adapted animals showed cold adaptation of the epithelium but no heat adaptation. During *Umadaptation* of the whole animal ($30° \rightarrow 20°$ C) as well as of isolated tentacles (after adding a protein synthesis suppressor, actinomycin D, to the external medium), cold resistance of the epithelium was increased above the level of cold resistance of those animals which had not previously been treated (there was no influence on heat resistance). There appears to have been hardening by actinomycin D with secondary influence on cold resistance. In untreated animals no further increase in cold resistance after the 3rd or 4th day could be measured; however it was present in those animals which had been treated with actinomycin. Similar complications occurred in fishes; however clear indication was given that protein synthesis participates in heat adaptation (*Rhodeus amarus*, change in AT: $10° \rightarrow 20°$ C). These results also show the importance of proteins for the heat resistance of intact fishes (KÜNNEMANN).[30]

Relevant here are experiments of MILKMAN (in PROSSER, 1967). He exposed *Drosophila* pupae to high temperatures. Brief exposures, followed by a return to physiological temperatures, produced increase of resistance by changes in the tertiary structure of proteins; longer exposures produced later defects in wing veins and still longer exposures caused death. In acclimation experiments on pure proteins (p. 435, PRECHT in TROSHIN, 1967) changes in molecules were the only relevant possibility. Further investigations must be awaited before all these phenomena can be coordinated. It is difficult to attribute such phenomena to regulations or to adaptations.

KÜNNEMANN (1973) found that heat resistance of enzymes can be changed (besides by a new synthesis, by changing the tertiary or quaternary structure or by com-

---

[28] In the experiments of PASSIA (p. 345), the configuration of a protein is probably changed by ET (so that a W-figure results); a comparable effect, which influences the heat resistance of insects to extreme temperatures, may be caused by the change of AT after a relatively short time. It may be possible that the distinction between direct responses and adaptations becomes uncertain.
[29] Zool. Anz., **189**, 244—256 (1973).
[30] KÜNNEMANN, H.: Mar. Biol. 18, 260—271 (1973).

plex bounds in the sense of USHAKOV) also by accompanying substances of the cell milieu. For the interpretation of heat adaptation, this must be taken into account.

Difficult to interpret is the observation of ATWAL (1960) on pupae of *Ephestia kuehniella*, that exposure to low ATs (down to 10° C) for 4 h increases cold resistance more than exposure for either shorter or longer times. In races of *Drosophila subobscura* the extent of heat acclimation did not differ in inbreds and hybrids, but the time of acclimation was usually shorter in hybrid males (BOWLER and HOLLINGSWORTH, 1966).

Although the adaptation to the upper temperature limit in the experiments of SUMNER and DOUDOROFF (1938) was completed in one day when the fish were transferred from 20 → 30° C, re-adaptation took more than 23 days if the animals had been kept at 30° C for 46 days, and only 10 days if they had been kept at 30° C for only one day. Thus "anchoring" of the adaptation effect has occurred, without becoming perceptible in the temperature limit. According to MELLANBY (1939), cold adaptation of bedbugs does not differ if they have lived at 30° C for several generations; they adapt to 15° C within the same amount of time as those animals kept at 30° C for only one day.

As has been mentioned, the time courses of capacity and of resistance adaptation can differ considerably (p. 436); the question has been discussed (p. 432) whether the mechanisms of heat- and of cold-adaptation can be distinguished, as well as the mechanisms of heat- and cold-resistance. Pertinent to this problem are experiments on the time course of *Umadaptation*. VOGEL (1966) demonstrated that in *Zoothamnium hiketes*, with reasonable acclimation to both temperature extremes, increase of magnesium in the sea water medium raises heat tolerance, but not cold resistance. Thus in this one-celled animal heat and cold resistance are independent of each other (see KÄHLER, 1970). That there is no single regulatory mechanism for simultaneous heat and cold adaptation is evident in the fact that in experiments with a change of AT the gain or loss in heat resistance does not run parallel in time with decrease or increase in cold hardiness. In similar experiments of THIEDE on *Xiphophorus*, bearing on this problem, closer correspondence of time course was demonstrated, but not so exact that the result could be a basis for generalizations (see also Figs. 1 and 2 in TSUKUDA, 1960b; also BRETT, 1946; COLHOUN, 1960; KÄHLER, 1970). According to TSUKUDA (1960c) the fluctuations in heat and cold resistance in individual guppies were not correlated. DOUDOROFF (1942) found that the difference of the "lower median tolerance limits" of *Girella* acclimated to 20° C or to 28° C, was 4.3° C; the "upper median tolerance limits" were however not different. BASEDOW's results have already been mentioned (p. 420).

d) Varied Preliminary Exposures

In general when the preliminary exposure has been variable, the deviation of resistance to extreme temperatures from animals kept in the middle temperature is greater than of oxygen consumption etc. (p. 331). Fish (HEATH, 1963; BERKHOLZ, 1966) and crustaceans (EDNEY, 1964; FURCH, see p. 439) displayed, after variable preliminary exposure, distinct increase in heat resistance. HEATH observed that, at certain times of the day, trout searched out different temperatures, sometimes very high. By this exposure their resistance could be increased more than was possible in animals adapted to the previous constant high temperature. This indicates that resistance adaptation is not the only determinant. Cold resistance of fish from fluctuating temperatures may be less than that of animals from a median constant AT (SUMNER and DOUDOROFF, 1938). EDNEY's observations that woodlice which had been exposed to periodic changes in temperature (10° ↔ 30° C) were as heat resistant as animals maintained at 30° C, and as cold resistant as 20° C animals, also favor the assumption of a separate mechanism of reasonable adaptation to each temperature extreme (p. 432f.); however the objection can be made that when other factors participate (for example shock adaptation) the effect of adaptation to heat and cold resistance may be superimposed all the more so, as those factors can be of different influence upon each. Reference is made to the interpretations of increased heat resistance on p. 347; however the entire period of *Umadaptation* need not be taken into consideration, but only the first 12 h if equal periods of cold and heat are used. EDNEY's figures and FURCH's experiments on *Idotea* and *Gammarus* show that the present interpretations are not conclusive and that additional phenomena must be taken into consideration.

In experiments in which *Idotea balthica* (♂♂) were transferred from adaptation temperature 20° → 8° C, the total *Umadaptation* period of heat resistance actually was longer than that in the experiment 8° → 20° C; however, during the first 12 h after temperature change, change of resistance occurred at a faster rate when the temperature was lowered than when it was raised. Yet a periodic change of temperature (12 h 8° C, 12 h 20° C) resulted in a marked increase of heat resistance well above the value obtained for the animals kept at 14° C for several days. For these results the interpretation of BERKHOLZ (p. 347) is not appropriate. Since adaptation occurs rapidly, already the 12-h cycle change is effective and leads to cyclical changes of heat resistance (see POUGH and WILSON, 1970). Additionally, the shock of temperature change results in an increase in heat resistance (heat hardening?) which is more distinct in quick temperature changes than in slow ones. In similar experiments (12 : 12, 8° ↔ 20° C, slow transition) *Gammarus salinus* showed no increase of heat resistance in comparison to animals kept at 14° C, but in faster transition increase of resistance was shown (probably because of shock-effect by quick temperature change). This animal lives in a biotope which is normally exposed to considerable fluctuations of environmental temperature (FURCH).[31]

GRONOW (Mar. Biol., in print) showed that not only in mammals but also in fishes a stress situation is characterized by an increase of the lactate-pyruvate quotient and other biochemical changes. However, the data of p. 320 demonstrate that not every rise of this quotient increases heat resistance (*Idus idus* adapts reasonably to high temperatures). The mentioned investigations of BASEDOW (p. 420) also show that not every stress has an influence upon the resistance to extreme temperatures; nevertheless heat resistance may be increased (see also p. 421). Further experiments are necessary.

Between the ATs 22° and 14° C larvae of *Tenebrio molitor* show a reasonable cold but no heat adaptation, which exists in the AT range between 30° and 22° C. A preliminary exposure of the larvae to varied ATs (22° ↔ 14° C) effects a higher cold resistance than in animals adapted to the constant low AT (the cold re-

---

[31] FURCH, K.: Mar. Biol. **15**, 12—34 (1972); cf. CORNELIUS, P. F. S.: J. exp. mar. Biol. Ecol. 9, 43—53 (1972).

sistance of the AT 30° larvae shows peculiarities, not discussed here). Heat resistance after varied preliminary exposures (ATs 30°↔ 14° C, 30°↔ 22° C) is less than that of the AT 30° C animals. — If a high lactate-pyruvate ratio is typical for a stress situation (as in vertebrates), a constant AT of 30° C is most favorable (this is the optimal temperature for development of the larvae); however lower values are measured when the larvae are kept in changing temperatures. The following values for the lactate-pyruvate ratio were found: AT 5° C: 19.4; 14° C: 17.8; 22° C: 11.8; 30° C: 4.5; 38° C: 17.2; 22°↔ 14° C: 10.6; 30°↔ 22° C: 4.4; 30°↔ 14° C: 4.1 (HEITLINDEMANN, unpubl). Heat resistance is not correlated with a high ratio; the high cold resistance of the AT 22°↔ 14° C larvae must have other causes, too. The mentioned values were measured at the same stage of development (it may be that these investigations show a biochemical adaptation in changing reaction systems; see p. 375). These values (and others) were not only measured in animals adapted to different ATs, but also during the days after a change of AT; they became constant gradually, so that one can speak of an adaptation (see p. 329). The shortest time of development seems to be correlated with the lowest ratio of lactate-pyruvate and a high n-RNA/DNA quotient.

Bugs *(Ischnodemus sabuleti)* show a reasonable cold but no heat adaptation; after a varied preliminary exposure to different temperatures the cold resistance of the bugs was like that of animals adapted to a middle AT (HEITLINDEMANN, unpubl.). *Carausius morosus* cannot adapt to heat or cold (I. PRECHT, 1967); a pretreatment with changing ATs does not alter heat resistance (BRAUNE, unpubl.). In *Blaberus craniifer*, transition from constant ATs to changing ATs was marked by considerable increase in weight.

Decisive are the position of the temperature range, the width of the fluctuations, the speed of daily temperature changes and the dimension of temperature fluctuations to which the species under investigation are exposed in nature.

e) Long-Lasting Influences

SMITH (1956, 1957) makes the distinction (when heat resistance in *Drosophila subobscura* with increased AT is increased) between "long lasting developmental acclimatization during pre-adult life, and a transitory physiological acclimatization in adults". ATs present during development can be influential for the entire life cycle. In guppies the effect of the temperature at which the parents had been held was apparent in the young (TSUKUDA, 1960a; see also GIBSON, 1954; MATUTANI, 1961; FRY in ROSE, 1967, p. 387).

Intermediate cases are those in which unusually prolonged effects occur in adult animals. After the AT had been raised for two or three days, SUMNER and DOUDOROFF (1938) observed increased heat resistance of fish lasting a month (cf. LOEB and WASTENEYS, 1912; DICKIE, 1958). Prolonged exposure to high ATs affects heat resistance of *Paramecium caudatum* for several generations (POLYANSKY and ORLOVA, 1957). When warm-adapted *Tribolium confusum* was exposed to 12° C for 3 h, its cold resistance still showed an increase after 4 days (SØMME, 1968a).

References to this section, see p. 470.

IV. Activity, Behavior etc.

H. LAUDIEN

A. Temperature and Behavior

Well-coordinated behavior is possible only between temperature limits which usually are closer together than the limits for life. This is especially clear if one observes complicated instinctual actions rather than simple movements. Various temperature zones can be distinguished on this basis. The life zone for the egg moth *Lymantria monacha* (v. ARNIM, 1936) extends from $-10.5°$ C to $+45.5°$ C. Survival is possible within these borders, but between $-10.5°$ C and $-0.5°$ C the animals are in a state of cold torpor and between $43.5°$ C and $45.5°$ C, in a state of heat torpor. When the temperature is in these ranges, they are incapable of locomotion, food uptake, etc. Temperatures between $-0.5°$ C and $43.5°$ C allow active movements (active zone).

1. Lower Temperature Limits

The lower temperature limits lie at different values for different types of movement. Individuals of *Dendroctonus pseudotsugae* lie completely still at temperatures below $-1°$ C (cold torpor or winter torpor), which does not mean that they are dead. Movements are first seen at $0°$ C, but they can be called well coordinated only at temperatures above $10°$ C. At this temperature the animals can fly, but to take off actively they require at least $22°$ C (RUDINSKY and VITE, 1956; cf. MERKER and WILD, 1954; ATKINS, 1959). Therefore, for the fraction of the insect population that is in flight and can thus be captured in certain cases, it is not the average daily temperature that is important, but the time span over which the flight threshold temperature is exceeded (TAYLOR, 1963; PORTER and GOYMERAC, 1970). COCKBAIN (1961) examined the threshold temperature values of various flying species of alienicoles of *Aphis fabae* in a wind tunnel and found a minimum temperature of $6.5°$ C for the wing beat, $13°$ C for horizontal flight and $15°$ C for upward flight. Closely related species have different threshold values. *Drosophila subobscura* still shows movement at $+13°$ C while *Drosophila obscura* is already in cold torpor at this temperature (LEUTHOLD, 1962). The threshold temperature may differ for flights undertaken for different purposes. COUTURIER et al. (1955), working with the cockchafer *Melolontha melolontha*, distinguished between flights for nourishment, flights for egg laying and flights after the eggs were laid in the forest. The threshold temperature for the feeding flights lies at $+4°$ C, but for the egg flights it is $10°$ C.

The lower limiting temperature is often influenced markedly by the temperatures prevailing during the periods of activity. For the digging wasp *Microbembix*, which is active on dune surfaces by day, when they are hot, the lower limiting temperature is 23° C. For *Geopinus*, which lives in the same habitat but is active by night, it is + 4° C (CHAPMAN et al., 1926). Animals from very cold habitats have low limiting temperatures (McWHINNIE in ROSE, 1967).

Somewhat more complicated activities such as, e.g., feeding movements, require higher levels of warmth. At 4—5° C the pulmonate snail, *Rumina decollata*, burrows into the ground; food uptake is only possible above 9° C (FRÖMMING, 1956). The lower limit for the oyster drill, *Urosalpinx cinerea*, is at ca. 7.5° C (HANKS, 1957). At this temperature its movements are still somewhat disturbed, as interruptions in food uptake demonstrate. These pauses cease only above 10° C. At critical temperatures the function of food uptake may fail because of differences between the limits of various subunits of the activity. For example, although hydra kept in the cold can seize their prey, they cannot kill or eat it, because the discharge of their penetrants is only possible at higher temperatures (BURNETT et al., 1960). Individuals of *Triturus helveticus* can still snap at prey at +1° C, but even at +2°C they cannot swallow it (JOLY, 1958), so that food intake into the gut is only possible at +3° C and higher. By contrast, trout still feed in supercooled water when expanding ground ice tears animals on which they feed loose from the bottom (NEEDHAM and JONES, 1959; cf. OSTDIEK and NARDONE, 1959; TYNEN, 1970).

As a rule the temperature thresholds for complicated instinctive actions (stalking of prey, sexual behavior, fighting behavior) are fairly high. Mating activities of the water bug, *Aphelocheirus aestivalis*, take place only at temperatures above 15° C (OHM, 1956), but those of small spiders which mature sexually during winter probably appear at 0° C (BUCHE, 1966; cf. HEYDEMANN, 1960; GRESSITT in GRESSITT, 1967). Although ants hunt for insects only at temperatures above 9° C, all activity outside their nest ceases at or below 5° C (WELLENSTEIN, 1957; cf. AYRE, 1958). Spawning in the brook lamprey, *Petromyzon fluviatilis*, — which goes through a courtship display at this time — takes place only at temperatures above 10° C (HAGELIN and STEFFNER, 1958). MACKENZIE (1961) found the same temperature threshold for copulation in the oyster drill *Euplura caudata*. Several flies *(Orygma, Coelopa)* copulate at room temperature only when the size of their cage allows flying swarms to form. At 23—26° C they copulate in small vessels as well (REMMERT, 1958; cf. SHOREY, 1966).

The swarming behavior of *Chironomus salinarius* depends on temperature as well as on the light intensity. It does not occur below 9° C (KOSKINEN, 1969). In this way swarming is synchronized (cf. HEATH, 1968). Bees begin their scouting flights and wagging dances only if the temperature at the food source exceeds 20° C (PFLUMM, 1969[1]). In dependency to water temperature *Lepomis-♂* show different fighting activity. Cold water (11—13° C) acts as inhibitor (SMITH, 1970[1]). The fighting fish *Betta splendens* build their foam nests at temperatures of 28° C, but not at 20° C, and anurans begin to call only above a certain limit. This limit lies at 8° C (SCHNEIDER, 1967) for the tree frog *Hyla arborea*, and at 10—12° C for the aquatic frog *Rana esculenta* (WAHL, 1969; cf. LÖRCHER, 1966).

---

[1] cf. SMITH, R. J. F.: Anim. Behav. 18, 575—587 (1970).

Precautions are often found in connection with especially important biological activities, such as those which serve to protect the animal, to assure that they can be executed quickly enough when the temperature is low. The arboreal *Anolis lineatopus*, whose habitat lacks hiding places, cannot conceal itself as terrestrial lizards do. As temperature falls its reaction speed decreases, and the danger of being captured by predators grows. As compensation for the slower reaction, the flight distance increases as temperature goes down (RAND, 1964). Bees cannot defend their nest when their body temperature is low, because they are not able to sting. But they demonstrate a form of behavior that at first frightens off intruders. If their nest is opened in winter when the temperature is at or below 0° C, all the outer bees of the winter cluster stretch out their stings. The greater the disturbance, the faster this response appears (on the average, after 13 seconds). The warmer inner bees become alarmed and then move to the surface of the cluster and attack the intruder (MORSE, 1967).

The threshold temperature is often correlated with geographical distribution. For example the lower limiting temperature for activity of the Australian gecko *Diplodactylus vittatus* lies at 17° C for a northern population and at 13° C for a southern one (BUSTARD, 1968). In correlation with its occurrence in warm brackish water, even temperatures of 35° C are unfavorably low for *Thermosbaena mirabilis* (BARKER, 1959).

2. Constant Temperatures in the Moderate Range

At levels of warmth between the upper and lower limits the frequency or speed of activities increases with increasing temperatures in the majority of cases. The designation of average or moderate temperatures is intended to apply to that range in which the animal movements of interest here are possible (cf. discussion see FISHER in PROSSER, 1958). The pretreatment temperature as well as the experimental temperature is significant. The value of many important and interesting investigations has been reduced because the acclimation temperature has either not been observed or has not been reported. For laboratory investigations, the room temperature may perhaps be assumed to be the AT. Due to the variability of natural temperatures, an exact analysis is not possible for field investigations. These must be supplemented by laboratory experiments under defined conditions. In the following section the form of dependence on the experimental temperature (so far as possible at constant AT) will first be discussed.

In a series of cases, a simple increase in activity was observed with an increase in temperature. GLASER (1925) found a dependence which agreed with Arrhenius' formula. Between 6.6° C and 20° C the μ value for the swimming speed of *Paramecium* was 16000, and between 15° C and 40°C, 8000. AT was not reported, but presumably was room temperature (in addition, cf., e.g., for running speed of ants at field temperatures, BODENHEIMER and KLEIN, 1930; for running activity of *Tenebrio molitor* AT 19–23° C, PERTTUNEN and PALOHEIMO, 1964; for crawling speed of *Musca domestica* larvae, AT has no effect, HAFEZ. 1950).

BLOCK (1966) found a similar simple dependence for the chirping rate of the cricket *Oecanthus fultoni*. The observations were undertaken in the natural habitat over the space of one month, which led to a relatively high variability. Nevertheless, the dependence could be clearly shown to follow a sigmoid course. FRINGS and FRINGS (1957) had already found the same curve for the chirping rate of the grasshopper *Neoconocephalus ensiger* (in field experiments on three successive nights). The observation that the flight pitch of certain female dipterans (e.g. *Chironomus plumosus*, field experiments, RÖMER and ROSIN, 1969[2]) increases with temperature belongs in the same category. The frequencies which attract males increase to the same extent. The flight pitches of the females retain their specific attractive effect for males of the same species between ca. 8—25° C. The attraction is favored by the regular flight pattern of the females which produces only slight variations in the flight pitch. Sounds produced by their own flight or by other males are not disturbing because they cannot be heard. Both sounds, the dipteran flight pitch and the grasshopper's chirping, are acoustic signals and show the same dependence on temperature (cf. ZWEIFEL, 1959; BRÄUNINGER, 1964; ESCH, 1964; FINKE, 1968; BAR-ILAN et al., 1969; KUTSCH, 1969; HEINZMANN, 1970).

If the speed or frequency of activity at first increases with increasing temperature and then falls again when a specific degree of warmth is exceeded, a temperature range of optimal or maximal response can be said to exist. Fundamentally different types of relations may exist. The frequency or intensity of many activities falls at the same rate when temperature levels rise or fall (bell-shaped curve). In other cases, such as the frequency of opercular movements in fish (THIEDE, 1965) or the beat of the cirri in *Balanus* (PRECHT, 1949), the maximum value is displaced towards higher temperatures. Since it lies close to the upper limiting temperature it must be considered as part of the problem of resistance. The maximum values for modes of behavior whose temperature dependence resembles a bell-shaped curve are found at moderate temperatures. This temperature range of optimal responses, whose position may vary according to the distribution of the species (e.g., ELWI ABDEL HAMID, 1959), can be quite narrow or fairly broad. The flying activity of the bark beetle (*Blastophagus piniperda*, AT = 4—5° C, PERTTUNEN and HÄYRINEN, 1969a) is greatest at 25° C, during all swarming periods, and declines rapidly at higher or lower temperatures. This type of temperature dependence is shown even more clearly by *Acanthoscelides obtectus* which develops in bean plants (AT = 25° C, PERTTUNEN and HÄYRINEN, 1969b; cf. for more material on this type of reaction: gathering activity of *Crematogaster scutellaris*, field experiments, SOULIE, 1955; speed of response of toads twelve hours after capture, MARTOF, 1962; locomotion in *Sitophilus granarius*, BARLOW and KERR, 1969). The optimum temperature for spontaneous take-off is 30° C (cf. ARNOUX et al., 1958), and this activity is reduced quickly on both sides of the temperature scale. This type of response affects the species' mode of life. In southern Europe the outdoor temperatures reach a level suited for spontaneous flight. Therefore females of *Acanthoscelides obtectus* fly from the granaries to the bean fields and lay their eggs in the bean pods. The young caterpillars burrow into the beans and are carried into the granaries with the harvested crops. In the northern area of their European distribution (France, Germany) the temperatures needed for spontaneous flight do not occur or occur far too seldom. Therefore the flight to the fields is omitted in this region and the entire life cycle takes place in the beans stored in the granary.

---

[2] cf. RÖMER, F.: Rev. Suisse Zool. 77, 603—616 (1970).

When the temperature range of maximal response is narrow, the animals are dependent on a very specific temperature. This is certainly not very advantageous since under natural conditions there is danger that these temperatures will not be present at the right time. It is better to widen the zone of optimal temperature or even be completely independent of temperature over as wide a range as possible. Where an approximate or complete temperature independence is observed, several causes may be responsible. For one, the process may be temperature independent in the biological range, or an adaptation may move the values for the functions measured at different temperatures closer together. Tests must always be carried out to learn whether the peak of the dependence curve is broader when AT is held constant and ET is changed, or when measurements are made at ET = AT. Here, only ET dependence is considered. Several species of diptera of the genus *Hippelates* (AT = 80° F, KARANDINOS and AXTELL, 1967b) showed a broad peak of activity. The maximal swimming speed of fish (cruising speed) can be included in this category (cf. FRY in ROSE 1967, p. 390 ff.).

The most favorable situation, i.e. a broad range of temperature independence, is illustrated by the flying speed of female *Aedes aegypti* (AT = 27° C, ROWLEY and GRAHAM, 1968). At 90% relative humidity and temperatures of 18°, 21°, 27°, and 32° C the rates in m/min ($\pm$SE) were 27.4 $\pm$ 0.86, 26.8 $\pm$ 1.04, 28.4 $\pm$ 0.80, 28.0 $\pm$ 1.61. Temperatures higher than 32° C were progressively less favorable because very high temperatures developed within the thorax, which is heated by the flight motion (cf. YURKIEWICZ and SMYTH, 1966). LEE and BADHAM (1963) investigated the effect of field temperature on the movement and behavior of the lizard *Amphibolurus barbatus*. Between 30° C and 40° C activity was fairly independent of temperature. The highest speed of movement appeared above 40° C, because the animals were exerting themselves to find shade. The range between 30° C and 40° C can be called the range of comfort for this species. Modes of particular biological importance such as feeding or threat displays are fairly independent of this favored range, so that they can be carried out even at lower temperatures. But the range of 30—40° C is especially advantageous, and the animals try to reach a temperature within this span as soon as possible in the morning by intensive basking. The pumping rate of the mussel *Branchidontes recurvus* (NAGABHUSHANAM and SAROJINI, 1965) also demonstrated temperature independence. The rate changed only slightly between 20° C and 32° C. In bees dancing-activity increases only about 10% with a temperature rise of 10° C (v. FRISCH, 1965). The observation of HECKROTTE (1967) on the snake *Thamnophis sirtalis* is interesting in this connection. The crawling speed during hunting and displays was nearly independent of temperature. The maximal crawling speed, e.g. in escape, was temperature dependent. This example shows that the motivation as well as the type of motion is significant (for other examples on temperature independence, cf. MCLEESE and WILDER, 1958; SERFATY and WAITZENEGGER, 1964; GOLIK and PIENKOWSKI, 1969).

FISHER and SULLIVAN (1958) found a completely different curve for the temperature dependence of a motion. They observed the spontaneous swimming movements of trout at various temperatures and found two maxima in the frequency of movements. The first maximum lay between 9° C and 15° C. This is the temperature range which fish, subjected to the same pretreatment, prefer in a temperature gradient. Lesions in the brain (especially in the dorsal part of the cerebellum) abolished the activity peak at 9–15° C. The spontaneous movement of the operated fish rose to a low maximum, which lay at the same temperature at which the intact animals reach their second maximum. In this case then, two processes

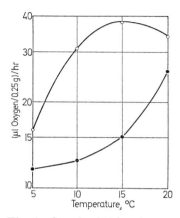

overlap. The first maximum is the result of control mechanisms in the brain and corresponds to the preferred temperature optimum. The second maximum results from an acceleration of movement due to the reaction-rate-temperature law and to inhibition at excessively high temperatures.

It is difficult to assign a cause for the temperature dependence of speeds of motion or frequencies of activities. One must start from the fact that resting metabolism increases with temperature and that the resting oxygen consumption shows the same dependence. Possibly oxygen uptake or transport to the tissues is the limiting factor. In this case there will be an optimum amount of oxygen available for movement at a certain temperature. As temperature increases the resting oxygen use increases more rapidly than the capacity for oxygen uptake. Less and less oxygen will be available for active movement. The activity curve is governed by the difference between the possible uptake of oxygen

Fig. 1. Standard (closed symbols) and routine (open symbols) rates of oxygen consumption of *Gammarus oceanicus* at four temperatures. For further explanation see text (from HALCROW and BOYD, 1967)

and its use for resting metabolism. However, only activity curves with a maximum at high temperatures can be explained in this way. This interpretation is not adequate for cases of approximate or complete temperature independence, or for dependence curves with two maxima (FISHER in PROSSER, 1958; p. 33 ff.; HALCROW and BOYD, 1967; FRY in ROSE, 1967).

3. Responses to Temperature Change

So far the discussion has dealt with experiments in which the animals were exposed to various constant temperatures. Additional complications arise if the temperature changes during an experiment, because the change can have a stimulating effect. NICHOLSON (1934) observed the activity of *Lucilia* at constant and rising temperatures. When constant temperatures, to which the flies had been exposed for twelve hours before the experiment, were used, an approximately bell-shaped curve with a maximum at 30° C resulted (AT = ET curve). Rising temperatures (from +1° C up to 45° C in six hours) caused the curve to rise more rapidly so that similar levels of activity were reached earlier than in the experiments with constant temperature. In this case the temperature increase acted on the reaction rate, and as a stimulus. Towards the end of the experiment at ca. 42° C an especially intense increase in activity took place, probably as a result of efforts to escape. A similar increase in flying activity with rising temperature was observed around

40° C. According to observations by GUNN and HOPF (1942), rising or falling temperatures strongly activated the running activity of *Ptinus tectus*. They changed the temperature at a rate of 14° C/h or 3—7.5° C/h. In each case a marked increase in running activity resulted. A sudden fall in temperature also resulted in increases in activity of diplopods (CLOUDSLEY-THOMPSON, 1951), and of the larch leafcutter, *Zeiraphera diniana* (MEYER, 1969). Even very slow temperature changes can produce responses. Snails react to a rate of 0.1° C/h by increasing their activity (DAINTON, 1954). In these cases temperature acts not only on the reaction rate, but has also a stimulating effect.

The finding that activity increases when temperature either rises or falls partially contradicts the statements of other authors, who found increases in activity with rising temperature and decreases with falling ones. These reactions are responses to the changes in temperature and therefore are transitory (cf. p. 454). An activity level is subsequently attained which corresponds to the prevailing temperature (effect of temperature on reaction rate). However, this statement is not valid for all temperatures at which activity is possible, but only for a central range.

The running and jumping activity of the grasshopper *Chorthippus albomarginatus* is affected in various ways by rapid temperature discontinuities, depending on the temperature level and on the direction of change. Increases or decreases in lower temperature range affect only the level of activity. In the central temperature range, an increase leads to a transitory fall in activity, while a temperature reduction leads to a transitory activity increase. As a result of this pattern the animals in their natural habitat remain close to a certain temperature (in the case of this species, 37° C). Decreases in temperature allow them to be active, increases to remain *in situ*. In this way the grasshoppers can stay still and sun themselves at relatively high temperatures. Changes from extremely high temperatures towards the chosen temperature of 37° C result in a reduction in activity, so that the animals remain where they are (GÄRDEFORS, 1964; cf. DIGBY, 1958).

Neurophysiological investigations of this problem go back to KERKUT and TAYLOR (1958). They found an increase in the electrical activity of the central nervous system of *Astacus fluviatilis* and *Periplaneta americana* when temperatures fall, and a decrease with increasing warmth.

Changes in nervous system activity brought about by temperature increases or decreases not only affect the frequency of locomotor behavior such as running, jumping or flying, but may also induce more specialized types of behavior, such as cleaning movements. If the leg of a cockroach, e.g. *Periplaneta americana*, is touched with a hot needle, the leg is taken to the mouth and cleaned (HERTER, 1953). This may be due to a direct effect on receptors on the body surface. But cleaning movements may also appear when the entire animal is warmed slowly. BEECKEN (1934) warmed bees in a vessel and found that the duration of cleaning depended on the rate of change of temperature (for spiders: see ENGELHARD, 1964). Cleaning activity took place only at rates of change between 0.25° C and 6.75° C/min. This so-called warmth-cleaning (examples in LANG, 1932; HERTER, 1953, p. 24) was investigated in greater detail by HOPPENHEIT and LAUDIEN (1969; LAUDIEN, 1969, 1970), who worked with the cockroach *Blaberus craniifer*. The results indicate that the animals carry out allochthonous cleaning behavior, i.e. behavior that is not elicited by soiling, when they are disturbed or when the tem-

perature rises. They are disturbed by the process of catching them in their rearing
cage and placing them in the observation vessel. At first they are so agitated by
their capture and introduction into the brightly lit, unfamiliar surroundings that
they only run about, attempting to escape. Within a few minutes the strong agitation is reduced and the animals begin to clean themselves. As excitement is reduced
further the antennae and forelegs are cleaned first, then the middle legs, and
finally the hind legs and abdomen. If the temperature is raised (1° C/4 min) after
they have grown quiet, they soon begin to clean themselves again, and this time
they begin with the abdomen and hind legs. At higher temperatures, i.e. with
greater excitation, cleaning of the middle legs, forelegs and antennae becomes more
frequent. At extremely high temperatures, i.e. over 37° C, the animals do nothing
except to run around in an agitated manner. The frequency sequence of the actions
of cleaning due to warmth is the reverse of that due to the disturbance of being
caught. So one can say that the excitation-cleaning of *Blaberus* results from a
disturbance or from increasing temperature.

In the case of *Blaberus* the temperature increase induces cleaning activity. But it
is also possible for a given temperature to facilitate the appearance of cleaning
activity due to other causes. Bees clean themselves at their feeding stations even if
there is no visible sign of soiling. PFLUMM (1969) interprets this behavior in terms
of a disinhibition hypothesis (v. IERSEL and BOL, 1958), as a result of reciprocal
inhibition of a tendency to gather food from a specific source and a tendency to
move to new food sources, whereby a continually present cleaning-tendency is
released from inhibition. This cleaning at the feeding station takes place most
frequently at 20° C. The question of whether cleaning is more frequent at higher
temperatures remains open, since under these circumstances the bees fly off
immediately after feeding. They may clean themselves during their homeward
flight, in which case the frequency of their activity cannot be observed.

4. Upper Limiting Temperatures

The upper limiting temperature also differs for different activities. Escape movements, intended to remove animals from extremely high temperatures, are carried
out with special intensity near the upper lethal limit. The beetle *Dendroctonus
pseudotsugae* runs very actively at temperatures above 34° C and dies at 43—49° C
(RUDINSKY and VITE, 1956). Larvae and imagos of the beetle genus *Ochthebius* are
still able to move and to leave unsuitable habitats at 41° C (BEIER, 1956). Some
species of aphids only fly at very high temperatures, *Schizaphis graminum* for
example up to 41° C (DRY and TAYLOR, 1970[3]; cf. HALGREN, 1970[4]).

The temperature limits of other activities, which do not have biologically protective functions of this type, are usually narrower. In winter balanids are no longer
fertilized at 15° C (RITZ and CRISP, 1970). The water frogs *Rana esculenta* do not
call if the temperature rises above 36.5° C (WAHL, 1969; cf. SCHNEIDER, 1967). At
higher temperatures female *Polistes* stop laying eggs and leave their nest
(STRAMBI, 1963). Ant lion larvae (larvae of *Euroleon nostras*) construct traps and

---

[3] DRY, W. W., TAYLOR, L. R.: J. Anim. Ecol. **39**, 493—504 (1970).
[4] HALGREN, L. A.: Ann. Entomol. Am. **63**, 712—715 (1970).

catch prey between 15° C and 40° C. At higher temperatures they withdraw into the deeper, cooler layers of sand (GEILER, 1966).

5. Adaptation Phenomena of Movements

The effect of pretreatment temperatures on the rates of active movement or on behavioral patterns can be tested, in the same way as their effect on oxygen consumption by the whole animal or by single organs or tissues, or on many other physiological processes. The problem has been treated in earlier chapters, and at this point only a few examples will be given as evidence that investigations of this type may also play an important role with respect to active movement and instinctive behavior of animals.

As stated earlier, rates of movement may increase with temperature up to a maximum, and then fall. Several authors have determined the maximum swimming speed (cruising speed) of fish subjected to different pretreatment temperatures, and in many cases they found that the maximum lay close to the temperature of acclimation (cf. FRY in DILL et al., 1964, p. 722; FRY in ROSE, 1967, p. 392).

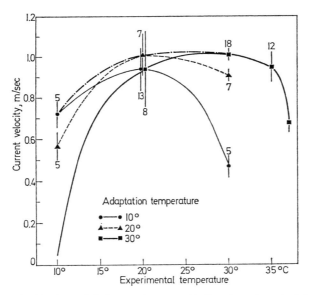

Fig. 2. Current velocity to maintain maximum activity at various test temperatures in *Palaemonetes vulgaris*. The AT = ET curve is subsequently added. The small numbers refer to the size of each sample (from McFARLAND and PICKENS, 1967; in PRECHT, 1968)

As the illustration for *Palaemonetes vulgaris* shows, the ET = AT curve lies at relatively high values (cf. NICHOLSON, 1934; GUNN and HOPF, 1942). In this way a rate of motion is attained that is both high and fairly similar under different maintenance temperature regimes. Such a plateau is often found in a moderate temperature range, whose position may vary from species to species. At temperatures between 2° C and 25° C, acclimated lobsters moved away from a bright lamp that was moved after them at a constant distance. From 0° C to near 10° C and from slightly over 20—25° C the running speed increased, but in the range between 10° C and 20° C it remained at approximately the same value (McLEESE and WILDER, 1958). The optimum for speed of motion of the nematode *Ditylenchus dipsaci* is clearly dependent on the pretreatment temperature. When AT = 10° C the optimum lay at 15° C, when AT = 20° C it lay at 20° C, and when AT = 30° C it lay at 25° C (CROLL, 1967; cf. ROOTS, 1961). The optimum for swimming speed of trout, mentioned earlier, is shifted when AT changes (FISHER in PROSSER, 1958; FISHER and SULLIVAN, 1958). The running speed and orientation reactions of *Sitophilus granarius* exhibited adaptational phenomena (BARLOW and KERR, 1969; cf. on swimming movements of medusae, MAYER, 1914; cold adaptation of alarm reaction of mosquito larvae, MELLANBY, 1958; on running activity in *Hemigrapsus*, SYMONS, 1964, optimum of feeding activity of *Daphnia*, KIBBY, 1971[5]).

The effect of pretreatment temperature is also seen in more complicated forms of behavior, as for example in feeding by the grasshopper *Nomadacris septemfaciata*, which depends on the temperature as well as on the time of day and year. The optimum for speed of feeding lies at a lower ET after exposure to low AT than after a higher AT (CHAPMAN, 1957).

Experiments by v. DALWIGK (1973[6]) deal with the effect of AT on the processes of recovery following exhaustion of an instinctive activity. The author tested the ability of *Aeschna cyanea* larvae to snap at prey and found that animals acclimated at 10° C recovered faster at 20° C than at 15° C, and recovered faster at 15° C than at 10° C. Animals acclimated at 15° C and 20° C recovered most rapidly at 15° C, and recovered faster than AT = 10° C animals at all ETs. Animals kept at varying AT (16 h light at 20° C and 8 h darkness at 10° C per day) recovered to a lesser extent than those kept at constant AT (10°, 15°, or 20° C).

v. DALWIGK also tested the jumping spider *Epiblemum scenicum*. When the temperature during the experiment and during the recovery period was held constant at 23° C, spiders acclimated to 18°C recovered more rapidly than spiders acclimated to 23° C, and these in turn recovered more rapidly than spiders acclimated to 28° C. This indicates that compensation for the temperature effect takes place, and shows that the acclimation or pretreatment temperature must also be taken into consideration when dealing with behavioral phenomena. NIELSEN and DREISIG (1970[7]) found that the duration of stridulation of *Orthoptera ensifera* depends on an "action-specific energy". This amount of energy is correlated with temperature during the rest phase.

6. Effect of Temperature on Daily Activity Rhythms of Behavior

The diurnal periodicity of many biological processes is often based on an endogenous circadian periodicity which is synchronized by external indicators of time (Zeitgebers). Temperature is an important external factor in this connection for poikilotherms, far more so than for homeotherms (cf. SWEENEY and HASTINGS,

---

[5] KIBBY, H. V.: Oceanogr. **16**, 580—581 (1971).
[6] See p. 319.
[7] NIELSEN, E. T., DREISIG, H.: Behaviour **37**, 205—252 (1970).

1960; BÜNNING, 1964; REMMERT, 1965). Temperature periodicities frequently do not synchronize endogenous rhythms, but exert a direct influence on the activity. Species of the genus *Cicindeia* hunt at different times of day, depending on temperature conditions (RENSCH, 1957). The initiation of activity of *Sitona* and *Chymomyza* is determined by temperature (CALIUS and MANGLITZ, 1968; NUORTEVA and HACKMAN, 1970). For this reason *Sitona* is nocturnal in summer because of the higher temperatures, but is diurnal in spring and fall (cf. FISCHER and ROSIN, 1968). Temperature decrease causes mating behavior in *Ostrinia nubilalis* (LOUGHNER and BRINDLEY, 1971[8]). The activity of lizards is governed to a great extent by the optimal temperature (e.g. HIRTH, 1963). The fact that some forms of behavior cannot be carried out at high or low temperatures is naturally also important. The diurnal rhythm may also be lost. For example the emergence rhythm of *Culex nigripalpus* is abolished at low temperatures (NAYAR, 1968).

If the periodicity follows a temperature cycle, a Zeitgeber or a masking effect may be present. In order to ascertain that an external rhythm is acting as a Zeitgeber, it is often necessary to measure the endogenous periodicity under constant conditions, before and after the external cycle is allowed to affect the process. If temperature is acting as a synchronizer the endogenous periodicity under constant conditions begins at the point to which it was taken by the external cycle. If temperature only has a masking effect, the cycle begins at a point corresponding to a continuation of the original circadian rhythm (HOFFMANN, 1969a). HOFFMANN (1969a) has gathered examples of experiments on the roles of temperature cycles as Zeitgebers. According to HOFFMANN (1969b) five individuals of *Lacerta sicula* were definitely completely synchronized by cyclic temperature fluctuations of as little as 0.9° C, and another five were probably completely synchronized (total number of experimental animals: 15).

Different constant temperature levels can exert different effects on rhythms. Constant temperatures have no effect on the frequency of spontaneous running activity of *Tenebrio molitor*, but the total amount of activity is strongly temperature dependent (LOHMANN, 1964; cf. BÜNNING, 1958). Between 20° C and 35° C the pattern of activity of the snake *Thamnopis radix* changes markedly with its environmental temperature. At low temperatures the activity is at a maximum around noon; in a central range (around 28° C) the pattern is bimodal with activity in the morning and evening, and at 35° C there is a major peak in the morning, followed by a minor peak towards evening (HECKROTTE, 1960, cited by SWEENEY and HASTINGS, 1960).

Rapid temperature changes from one level to another may have direct effects on the process of measurement of time. When temperature is lowered 5° C, a maximum of emergence is found after 10—11 h in a population of *Clunio* from Tromsö (Norway) (PFLÜGER and NEUMANN, 1971[9]). The "internal clock" can be stopped by the temporary action of extremely low temperatures. Maintenance of *Ephestia kühniella* at 0.5° C for 12 h set the cycle back by this period (MORIARTY, 1959). However the beginning of activity was delayed in *Blatella germanica* (DREISIG and NIELSEN, 1971[10]). The temperature level which acts thus, varies from species to

[8] LOUGHNER, G. E., BRINDLEY, T. A.: Ann. Entomol. Soc. Am. **64**, 1091—1094 (1971).
[9] PFLÜGER, W., NEUMANN, D.: Oecologia (Berlin) **7**, 262—266 (1971).
[10] DREISIG, H., NIELSEN, E. T.: J. Exp. Biol. **54**, 187—198 (1971).

species. Temperatures as high as 10°C have such an effect on thermophilic cock-roaches (ROBERTS, 1962), while the "clock" of a water strider *Velia currens* continues to run though the animals are in a state of cold torpor (EMEIS, 1959). According to BÜNNING (1958) the internal clock does not merely stop at low temperatures, but is set back to zero. The oscillator suggested by the author takes up a rest position as a result of cold. When temperatures are changed in the normal range, it must be remembered that the rhythm often operates the more reliably, the smaller the deviations are from a species specific temperature optimum. The possibility also exists that the process which the internal clock regulates is directly influenced by temperature. Stepwise temperature changes in an experiment must be distinguished from rectangular pulses. Temperature increases speed the rhythm of *Drosophila pseudoobscura* and reductions retard it. Rectangular pulses could be observed either to speed or to retard the rhythm, depending on the time of day. If a rectangular pulse lasted 12 h, the shift in periodicity could be calculated as a rough approximation, on the basis of the two stepwise temperature changes. This indicates that the effect of the stepwise changes ends after a few hours (ZIMMERMAN et al., 1968). An increase in temperature causes *Orchestia platensis* to change its preferred escape direction in a way which would be expected from an acceleration of its internal clock (JANKOWSKY, 1969; cf. PARDI, 1957; NAYLOR, 1963). On the other hand, in the case of *Tenebrio* a reduction of 5° C shortens the period immediately following exposure, while the second period is lengthened. From then on the initial value is re-established (LOHMANN, 1964).

From these experiments we can conclude that measurement of time is influenced by temperature but that regulations or adaptations always take effect quickly to maintain a constant period length (cf. p. 449). Results of research on *Rana temporaria* indicate that its circadian rhythm is temperature independent due to a rapid ideal compensation (KASBOHM, 1967), which need not be the case for other metabolic functions tested at the same time (as e.g. in the case of *Orchestia*, JANKOWSKY, l.c.). This is also true for the time sense of the goldfish (ROZIN, 1965), and the duration of swimming of the amphipod *Synchelidium* after elevation of the hydrostatic pressure (ENRIGHT, 1967; cf. BLEST, 1958; ROTH and SZABO, 1970; v. LOH and GOLENHOFEN, 1970).

7. Effect of Temperature on Learning and Memory

An animal's capacity for learning can be tested by its ability to form conditioned reflexes and to learn paths in mazes. In poikilotherms this ability is often temperature dependent in such a fashion that learning is slower or poorer at low temperatures than at high ones. APPLEWHITE (1968a) conditioned the protozoan *Spirostomum* to mechanical shocks and found that the speed of conditioning was the same at 25° C and 37° C, but that it was lower at 15° C. HERZ et al. (1964) tested the development of a conditioned reflex — retraction of the anterior segments on exposure to the conditioning stimulus, vibration — in the earthworm. In the course of fifty trials, the positive responses of animals maintained in cold as well as in warm temperatures increased, but this increase was significant only in the latter group. Individuals of *Dipsosaurus dorsalis* (KREKORIAN et al., 1968) learned paths through two different mazes with less success at lower temperatures. The animals learned much faster at 32° C than at 27° C. It must be remembered that the temperature of active individuals in the field is ca. 42° C, and therefore the more favorable temperature was in the direction of the preferred temperature. Unfortunately there have been few attempts to discover whether the speed of learning may be affected by the acclimation temperature as well as by the experimental temperature. One finding of ALLOWAY (1969) suggests that this may be true; larvae and adults of *Tenebrio molitor* learned their way in a T-maze better when they were held at low temperatures before the training session and between trials. This may be explained by the fact that the metabolic rate of the animals kept in the cold is more intense at the higher experimental temperatures, as a result of compensation, than the rates of animals kept in warm temperatures, and therefore the former can learn more effectively. According to FRENCH (1942) goldfish kept at +2° C for 20 h just before the experiment learned very poorly (cf., on the duration of adaptation, PROSSER and FAHRI, 1965).

Low temperatures are often favorable for the retention of learned material, e.g. memory. The protozoan *Spirostomum* mentioned earlier (APPLEWHITE, 1968a, b) forgot its conditioning to mechanical shocks faster when it was transferred to high temperatures immediately after the trial. Experiments with *Tenebrio molitor* produced the same results (ALLOWAY and ROUTTENBERG, 1967; ALLOWAY, 1969). The experimental animals learned a path through a maze faster on the second trial when they were kept at 2° C in the meantime. Similarly, goldfish made fewer errors on the 2nd, 3rd, 4th and 5th experimental days when they were kept at 4° C rather than 16° or 28° C between learning sessions, which always took place at 22° C (FRENCH, 1942).

To explain these results FRENCH hypothesized a forgetting process, whose speed is a function of the temperature (but, cf. JONES, 1945). ALLOWAY (1969) suggested that low temperatures protect the memory traces from disturbances (cf. APPLEWHITE, 1968b).

In considering temperature limits one must distinguish between the limits for learning and those for the execution of the learned response. The minimum temperature required by fish for formation of a conditioned reflex (cessation of respiratory movements after stimulation by light) changes with the acclimation temperature (PROSSER and FAHRI, 1965). Goldfish acclimated to 15° C could not be trained below 6–7° C, and those acclimated to 25° C could not be trained below 12–13° C. Therefore the minimum conditioning temperature is AT-dependent (resistance adaptation). The temperature limits change quite rapidly (6–72 h) after a change in AT. The ability to carry out learned reflex is lost at a certain temperature. If goldfish are maintained at 30°, 25°, 15° or 5° C and trained at their AT, the conditioned reflex (cessation of opercular movement after light stimulation) is blocked at 20°, 15°, 10° or 1° C, respectively (ROOTS and PROSSER, 1962; cf. AREFYEVA, 1965). The temperature used during conditioning also has significant effects on the blocking of conditioned reflexes.

B. Temperature Senses

In considering the effect of temperature on the responses of poikilothermic animals, one can distinguish several categories. Temperature acts:

1. Directly, on the rates of reactions (basal metabolism, growth and development, locomotion, etc.): As a rule increasing temperatures at first lead to an increase, but a peak rate for behavior is reached at moderate temperatures and a relatively slow decline follows.

2. Directly, as a limiting factor for metabolic processes, action, locomotion etc. It is usually true for the dimension of temperature at which a function is stopped or reduced, that complicated instinctive behavior has narrow limiting temperatures, simple locomotion wider ones, and certain organ functions still more extensive limiting temperatures.

3. Directly, as an inducer, by initiating a given reaction on a certain state. For example the brood cycle of *Porcellio* is initiated by the action of temperature at 20° C (WIESER, 1965a). Diapause can be initiated by temperature. Low temperatures frequently act to end diapause, which leads some authors to refer to a "breaking off" rather than an "ending" of diapause.

4. Directly, as a stressor. This is especially true of sublethal temperatures, but also of quick temperature changes in the central range. Thus accompanying phenomena that are also associated with other forms of stressors must be considered too.

5. Directly, in terms of a stimulating effect that begins immediately after temperature change takes place. Cockroaches showed increased frequency of action after transfer from AT 26° C to a lower temperature of 20° C (HOPPENHEIT and LAUDIEN, 1969). Aeschna larvae snapped more frequently at a dummy after they had been transferred from AT 20° C to 15° C and 10° C, respectively, than when the experiment had been carried out at 20° C (v. DALWIGK, 1973[11]).

6. Indirectly, through the sense organs: Problems of temperature sense and animal behavior are involved here. The fact that very small temperature differences, which have no direct effect, can be perceived and cause a response, can be confirmed by neurophysiological and behavioral experiments.

1. Temperature Receptors and Their Functions

One can speak of a sense of temperature in a strict fashion only if sense organs or at least sensory cells, for which temperature is the adequate stimulus, can be identified. This identification is not easy since many sense organs also react to inadequate stimuli. As a rule sense organs have extremely low thresholds with respect to an adequate stimulus and therefore one can often assume that a sense of temperature is present if the sense organ, or in behavioral studies the whole animal, responds to very slight changes in temperature.

Reaction to temperature differences are found even among protozoa. Of course, a genuine temperature sense is not present. Rather the reaction occurs because protoplasm is capable of being stimulated by heat (KOEHLER, 1934).

Many attempts have been made to elicit responses from metazoa by approaching them with warmed or cooled probes (examples are found in HERTER, 1962). Another method consists of allowing the animals to move within a stimulus field and to observe certain reactions (turning responses, fright or escape responses). A temperature difference of 0.05° C was sufficient to elicit a response from the

---

[11] See p. 319.

nematode, *Nippostrongylus* (PARKER and HALEY, 1960). For hirudinea, a difference of 1.5° C was sufficient (KAISER, 1954; cf. HERTER, 1929).

Still another method consists of conditioning animals to respond to temperature differences. Bees (HERAN, 1952), goldfish (BARDACH, 1956) and roach (*Leuciscus rutilus*, SPÄTH, 1967) can distinguish differences of 2° C. Marine teleosts responded to increases of the temperature of the surrounding water of 0.03—0.07° C (BULL, 1936), and several American fresh-water fish responded to such changes of 0.05° C (BARDACH and BJORKLUND, 1957).

Attempts have been made to localize the thermoreceptors of arthropods by removing specific temperature-sensitive body appendages, and of poikilothermic vertebrates by sectioning nerves. Various isopods (*Oniscus murarius* and *Porcellio scaber*) respond to a warmed glass rod at a distance of 1—2 cm. The response disappears after the antennae are amputated (HERTER, 1925). According to BAUER (1955) *Lithobius forficatus* are distributed within temperature gradients in a bimodal fashion with equal peaks. After the terminal antennnal segments are amputated, the higher temperatures in the gradient are preferred. The tarsi were also important for positive responses to the lower temperatures. If the basal antennal sections are also removed the temperature which causes escape reactions is shifted from 39—46° C (cf. on the millipede *Blaniulus guttulatus*, KLINKEL, 1955; on shrimp, COSTA, 1966).

A series of investigations carried out on insects (cf. HERTER, 1953; MURRAY in LOWENSTEIN 1962) showed that antennae, maxillary palps, tarsi, and abdominal appendages (cerci, ovipositors) were especially sensitive to temperature. According to GEBHARDT (1953), the maxillary palps of *Dorcus parallepipipedus* were the most sensitive, but amputation of the antennae of *Lygaeus equestris* produced only a slight change in their escape-inducing temperature. From this fact it was concluded that in this species, the receptors are probably located on the abdomen (cf. for a comparable situation in *Dysdercus fasciatus*, MADGE, 1965). The number of receptors present is important in determining the escape-reaction temperature of the fire brat *Pyrrhocoris apterus*. This temperature was higher after amputation of both antennae. The effect was only half as great after unilateral amputation. The escape temperature for bees, on the other hand, rose only after bilateral removal of the antennae (HERAN, 1952; cf. on *Stomoxis calcitans*, ŽDÁREK and POSPIŠIL, 1965).

Various insect receptors have been identified as temperature sense organs, but there is often no proof that temperature is their adequate stimulus. CAPPE DE BAILLON (1932) considered the wartlike elevation on the twelfth antennal segment of the walking stick, *Menexenus*, to be a thermal receptor because the phasmids attempted to escape when a warm substance was brought near this organ. SLIFER (1951, 1954) found sites with thin cuticle on the head, thorax and abdomen, which he believed to be temperature sense organs. MAKINGS (1964) tested the effect of operative removal of these structures on orientation towards a source of warmth and found no change in the response (cf. DUNHAM, 1962, see p. 457). The antennae of *Rhodnius* bear a large number of very fine, relatively thick-walled trichoid sensilla, each of which is connected to a group of ca. 6 sensory cells. These are probably thermal receptors (WIGGLESWORTH and GILLETT, 1934). BERNARD et al. 1970[12]) found sensory cells in the maxillal bristle of *Triatoma infestans*, which

---

[12] BERNARD, J., PINET, J.M., BOISTEL, J.: J. Insect Physiol. 16, 2157—2180 (1970).

reacted to temperature and/or humidity changes. It is probably the CNS where the present sorts of stimuli are determined. Among the vertebrates, no unimodal temperature receptors have been found so far in fish, rather only thermally sensitive ones. The temperature stimuli in the investigations carried out until now have always been combined with mechanical stimulation of the fish skin, e.g. by currents from pipettes, by active swimming movements or by touching with thermal probes (SPÄTH, 1967). The temperature stimuli were probably perceived by sensitive free nerve endings in the uppermost layer of epidermis (on the effect of pharmacological agents, cf. RUFF and ZIPPEL, 1966). It has been shown by sectioning of nerves, that the spinal cord, and not the lateral line system, is significant for thermoperception (DIJKGRAAF, 1940, 1943; cf. SULLIVAN, 1954). From the results of behavioral experiments it can be concluded that warm and cold receptors exist. DIJKGRAAF conditioned minnows, *Phoxinus laevis*, to respond to warmth or to cold.

By means of electrophysiological experiments, the ampullae of Lorenzini of selachians have been found to be very sensitive to temperature changes. Behavioral experiments have shown that they act as receptors for detecting the muscle action currents of their prey (HENSEL, 1955; DIJKGRAAF and KALMIJN, 1966; AGALIDES, 1969).

In amphibia, as in fish, thermal stimuli probably act on free nerve endings (cf. BABAK, 1914; MORGAN, 1922; DODT, 1955; MURRAY, 1956).

Among the reptiles, behavioral and electrophysiological studies have shown that the crotalidae possess receptors for radient heat. These receptors are located in the so-called pit organs and function in locating prey. Blinded vipers with intact pit organs track and strike prey animals as well as intact snakes do, but individuals with blocked pit organs make no catches (BULLOCK and DIECKE, 1956; cf. TERASHIMA et al., 1970[13]; OTTO, 1972[14]; STORCH, 1972[15]).

Specialized temperature receptors have been found in only a few cases. Only behavioral experiments are capable of determining whether a receptor actually perceives temperature stimuli. According to SCHOONHOVEN (1967) three specialized nerve cells, which can be stimulated by reduction in temperature, are found in the third antennal segment of three different butterflies. Temperature increases inhibit their spontaneous activity. Still another cell of this type is probably located in the maxillary palp (cf. DETHIER and SCHOONHOVEN, 1968).

Cold receptors are found on the antennae of *Apis mellifica* (according to LACHER, 1964), of *Periplaneta americana* (according to LOFTUS, 1966, 1968, 1969), and of locusts (according to WALDOW, 1970; for *Triatoma infestans* see BERNARD et al., 1970). The temperature receptors of *Periplaneta* are fairly large sensilla, which occur in small numbers (usually one per antennal segment, and about twenty per antenna) and do not respond to vibrations, touch or light, but which do respond to sudden cooling by rapid increase of spontaneous firing frequency. The extent of the increase depends on the difference between the initial and final temperatures. Subsequently the frequency falls to a constant value. If the temperature is quickly

---

[13] TERASHIMA, S., GORIS, R. C., KATSUKI, Y.: J. Ultrastruct. Res. **31**, 494—506 (1970).
[14] OTTO, J.: Kybernetik **10**, 103—106 (1972).
[15] STORCH, V.: Z. Mikrosk.-Anat. Forsch. **85**, 55—84 (1972).

raised back to its initial level the impulses stop briefly. These findings led to the conclusion that cooling is the effective stimulus.

Among the poikilothermic vertebrates, only the pit organs of the Crotalidae mentioned earlier have been recognized as specialized temperature sense organs. Their response indicates that they are warmth receptors. Temperature elevation raises their spontaneous firing frequency and cooling reduces it. Their threshold value in both behavioral and electrophysiological experiments lies at about 0.002° C.

It must be pointed out that the functions of many sense organs may be temperture dependent to the extent that the picture of the world which they present to the animal is affected (cf. SPÄTH, 1967; CLARKE in ROSE, 1967). Lucerne plant lice can distinguish two plant species from each other at 27° C, but not at 10° C (SCHALK et al., 1969). According to CAMPBELL (1969) the maximum sensitivity to hearing in various lizards lies at the species-specific preferred temperature. The threshold values for sucrose stimulation of flies are lowest at the temperature at which they were reared (DETHIER and ARAB, 1958). Nerves show a comparable temperature dependence (cf. CABANAC et al., 1967; SHEA et al., 1969).

2. Behavior Leading to Avoidance of Unfavorable Temperatures

a) Temperature Choice as a Problem in Orientation

Before an animal's behavior is considered, its field of stimulation must be analyzed to learn which physical entities are actually present.

An animal body can take up heat energy from its surroundings by absorption of radiation or by conduction (heat transfer by convection plays only a subordinate role in this process). Either radiation or conduction may be more significant, depending on the circumstances. It is possible for an animal to be exposed only to radiation. In such a case the stimulus field has definite directions, so that warmed or shaded sites can be distinguished, but its gradient is quite narrow (example: stimulation by solar radiation). Under other conditions, heat energy is furnished by conduction. In such cases the stimulus fields often have fairly steep gradients. As the distance to the heat source increases, temperature decreases, and consequently a cold and a warm direction can be distinguished (example: temperature gradient apparatus). Similar distinctions can be made for light stimulus fields (PRECHT, 1942).

Types of response: In observing the behavior of animals in thermal stimulus fields, the types of response must be described before analysis can be attempted. PRECHT (1942) distinguishes between simple (orientation) and composite (with orientation and locomotion components) responses. Composite responses include taxes and menotaxes. Phobo-, tropo-, and telotaxis can be distinguished as primary categories (cf. KÜHN, 1919; JANDER, 1963; LINDAUER, 1964). According to FRAENKEL and GUNN (1961) the kineses can be added to this group. The latter belong to category 1 (p. 454), whether they involve running speed (orthokinesis) or the number of directional changes (clinokinesis). Taxes belong to category 6. These two categories can overlap in actual behavior. The sign of the taxis (+ or −) may be determined by observing the animals while they are searching with the aid of their temperature sense for the preferred stimulus range. In this range animals are at rest or pursue other activities which are not governed by thermal taxes. The question of whether the speeds of movements of these activities, like the thermo-taxes on both sides of the preferred temperature range, are dependent on ET in the sense of point 1 has not been investigated (cf. RODE, 1969).

Phobic reactions occur when the animal turns away from the correct course towards the preferred temperature region, i.e. moves in the direction of unfavorable temperatures again. Thus the sensitivity to temperature changes over time can be used to initiate countermeasures when temperatures become less favorable, even in the normal temperature range. The phobic reaction therefore is not restricted to specific temperatures, but rather is a response to temperature changes outside the preferred range. Phobotaxis can be induced easily in stimulus fields with gradients. In a stimulus field without gradients the sense organs are shielded from environ-mental influences changing with time. FRAENKEL and GUNN (1961) refer to this mechanism as clinotaxis, which includes phobotaxis (KOEHLER in TINBERGEN, 1956). Phobic reactions can certainly be distinguished meaningfully from specific escape reactions at high stimulus intensities (e.g., escape reactions of *Paramecium* at the border between cold and warm water, BRAMSTEDT, 1935). But it is improbable that such escape reactions occur only at high temperatures. Perhaps they would be difficult to recognize as such at lower temperatures because of their slow speed. A bimodal distribution within the temperature gradient, such as BAUER (1955) found for *Lithobius*, could thus be accounted for on the basis of an accumulation in the vicinity of an upper or lower escape temperature (cf. MADGE, 1965). At low humidity (20−60%) WATERHOUSE et al. (1971[16]) found a distribution curve of dwelling and egg-laying with two peaks in *Triboleum castaneum* and *T. destructor*, but not at high humidity (60−95%).

In investigating types of responses among animals it must be remembered that behavior may be based on different reaction principles, acting in series. The reac-tion principles depend on stimulus, sense organ, and mode of behavior in ques-tion.

b) Orientation in Temperature-Stimulus Fields with Gradients

Concept of "Preferred Temperature": If insects or other freely moving animals are presented with an environment with a temperature gradient, they accumulate in a distinct zone. The temperature of this zone has been called the preferred temper-ature, the temperature preferendum, or the temperature optimum. A zone of thermal indifference can also be recognized, because the animals within this zone behave indifferently with respect to temperature stimulation (HERTER, 1962).

When the results of such work are reported the average should not be presented alone. A measure of the distribution of the individual data points around the average should also be given. For example, the two temperature values between which 50% of the data points lie (e.g. NIELSEN and NIELSEN, 1959), may be used, or the standard deviation (SD) which is customary in statistics. The latter includes a range (average ± SD) in which circa 2/3 of the observed values fall (cf. GÄRDE-

---

[16] WATERHOUSE, F. L., ONYEARU, A. K., AMOS, T. G.: Oikos (Kbh.) 22, 131—135 (1971).

FORS, 1966). If this value is to be used, the individual values must be normally distributed. One often finds, however, a skewed distribution in which the number of data points within a temperature range of a given width decreases much faster at higher temperatures than at low ones (e.g. ZWICKY, 1949), since a degree Celsius in cold conditions is not the equivalent of a degree in warm conditions.

It is possible, that an alarm reaction which might occur at high temperatures only leads to the steep decline in the warmth (JANETSCHEK in GRESSITT, 1967). As a result, the average lies near the highest temperature on the curve. Therefore, in order to present the information obtained from the experimental results completely, one must report several values. These include the average value (mean preferred temperature), the standard deviation (preferred temperature range) and the peak value of the distribution curve (peak of preferred temperature range, e.g. YOUDEOWEI, 1968).

BAUER (1955) found *Lithobius* to be distributed bimodally within the temperature gradient, an unusual situation. The animals accumulated at about 8° and 30° C. The former value is probably the preferred temperature because it agrees with the habitat temperature (for another possible interpretation, see p. 458).

Temperature Selection Apparatus: The determination of the preferred temperature (PT) under natural conditions is undoubtedly the best method and would probably yield the most correct values, but it is very difficult to contrive a temperature gradient in a natural habitat. Therefore apparatus have been built to permit PT to be determined in the laboratory. These can be classified as temperature selection apparatus for animals in water, under the ground, on the ground and in the air. HERTER (1962) called these devices temperature organs, to emphasize that they were characterized by regular gradations in units of heat energy. This term is based on the nature of the apparatus. But it is preferable to focus on the action of the animal and to call the device a temperature selection apparatus. ROZIN and MAYER (1961) employed a fundamentally different method for determining PT, in which goldfish could adjust their own environmental temperature by regulating the flow of warm and cold water (cf. KEMP, 1969). Apparatus with horizontal or vertical gradients have been built for animals living in water. Vertical stratification is aided by the fact that water rises when warmed (GRAHAM, 1949; BRETT, 1952; FERGUSON, 1958). However, one must run tests to see whether the animal moves towards a specific temperature or a specific depth (BRETT, 1952; McCAULEY and POND, 1971[17]). It is difficult to establish a horizontal gradient because water masses at different temperatures always tend to mix due to the upwelling of the warmer water. This handicap is less evident in very thin water layers (up to several millimeters). Therefore experiments can be done easily with small organisms (protozoa, small invertebrates, cf. on investigations of nematodes: RODE, 1965). Investigations of larger animals, e.g. fish, require additional equipment.

FRY (1958), ZAHN (1960) and others have tried to attain stable temperature conditions with the aid of water perfusion. In such cases the water flow and temperature gradient are coupled, and one must always make tests to see if the animals are orientating thermo- or rheotactically. Similar difficulties arise when one tries

---

[17] McCAULEY, R. W., POND, W. L.: J. Fish. Res. Bd. Can. **28**, 1801—1804 (1971).

to present an animal with air at definite temperatures because warm air rises and so the air masses of different temperatures are continually mixed. Frequently, to avoid this difficulty, only the substrate temperature is measured (e.g., HERTER, 1953, 1962; KRÜGER, 1952; COULIANOS, 1955; GRAHAM et al., 1965; ÖSTBYE, 1970 and many others). It is easy to establish a temperature gradient in a metal bar, since one is working with a good heat conductor. Heating at one end and cooling at the other produces a very stable temperature gradient. Thigmotactic selection of the ends of the bar can be prevented by using KRÜGER's (1952) round apparatus.

Analysis of the Search for Preferred Temperatures: The temperature gradient in the temperature selection apparatus is established and maintained by convection, while radiation very probably plays a minor role at most. The taxic orientation in this gradient may be directed or random. An accumulation or cessation of movement at a given temperature that results only from chance cannot be called a thermotaxis. Phobic orientations and random escape reactions are seen in a range above the PT and are especially common. These lead to a change in direction or to a reversal if they are repeated often enough. A negative thermotaxis seems to be missing in *Sarcoptes scabiei*, which exposes it to the danger of heat death in the gradient (MELLANBY et al., 1942; the same is true for larvae of *Ascia monuste*, NIELSEN and NIELSEN, 1959). An animal can explore the gradient on all sides in this way. During phobic orientation in very steep gradients such exploration may be pursued with movable appendages (e.g. antennae) alone, in which case the animal's trail may run fairly straight. The trial alone therefore cannot be used to classify the type of orientation present. The distance separating thermoreceptors, especially in small animals and less steep gradients, is generally not great enough to permit tropotactic orientation. This depends, however, on the ability to distinguish thermal differences. Newly emerged females of *Culex fatigans* respond to gradients of 0.05° C/cm in the supraoptimal range of 29—30° C and *Formica rufa* workers distinguish temperature differences of 0.25° C (HERTER, 1953). For *Rana esculenta* CABANAC and JEDDI (1971[18]) proved that the internal temperature apparently influences the behavior in the temperature selection apparatus. Internal heating with an isolated electric resistance was followed by a displacement of the animal towards the cold.

Occurrence of PT: The occurrence of a PT has been investigated in a large number of animal species. Tabular compilations of ascertained values are found in UVAROV (1931) and HERTER (1953) for insects, in BRATTSTROM (1963) for body temperatures in active amphibia, in BRATTSTROM (1965) for reptiles, in LANKIEWIEZ (1964) for various fresh-water fish, and in THIELE and LEHMANN (1967) for carabidae.

Alterations in PT by External and Internal Factors

PT as a Species Characteristic: Even if a number of individuals of an animal species under identical conditions search out the same temperature with a certain degree of variability, one can speak of a species characteristic only with certain reservations, because many influences may modify the PT. Females and males of the

---

[18] CABANAC, M., JEDDI, E.: Physiol. Behav. **7**, 375—380 (1971).

same species may have different PT values. For example female *Porcellio scaber*, *Oniscus asellus*, and *Armadillidium* prefer temperatures ca. 4° C lower than the males (BARLOW et al., 1957). The same situation is found (with a temperature difference of 0.8° C) in the case of the butterfly *Ascia monuste* (NIELSEN and NIELSEN, 1959; cf. ELGER, 1969).

PT often changes during development, especially if the various stages live in different habitats. Larvae of *Boreus hiemalis* live under moss plants in summer. Their PT is 34° C. The adults, which are found on snow in winter, accumulate at ca. 10° C (HERTER, 1953). Among larvae of *Musca domestica*, which live in dung, young larvae in feeding stages preferred high temperatures between 30° C and 37° C, while older larvae preparing to pupate preferred temperatures below 15° C. They actively sought out the appropriate zones of dungheaps (THOMSEN and THOMSEN, 1937; HAFEZ, 1950, 1953; OMARDEEN, 1958). Caterpillars of *Ascia* seem to have no temperature preferences, while the imagoes show distinct thermotactic behavior (NIELSEN and NIELSEN, l.c.). Before they begin to take nourishment nymphs of *Oncopeltus fasciatus* sun themselves all through the day, but later this behavior is confined to the morning hours (BARRETT and CHIANG, 1967). According to JAKOVLEV (1961) the PT of larvae of the field cricket *Liogryllus campestris* is higher than that of adults. Young of the perch *Girella nigricans* prefer high temperatures of 28—31° C. Older fish choose regions with cooler water as their habitat (NORRIS, 1963; cf. LOWE and HEATH, 1969). The opposite situation is found for *Rana arvalis*, *R. pipiens* and *Alytes*, whose larvae prefer lower temperatures than the adults (RÜHMEKORF, 1958).

Therefore PT can be used as a species trait only with caution. But these doubts carry less weight if one is satisfied with comparisons and regards the values as being only relative.

Changes During the Course of the Day and Year: Daily day-night rhythms, which are often correlated with changes in temperature, may be the occasion for changes in PT. REMMERT (1960) investigated tiger beetles, *Cicindella campestris*, and found a much higher PT during the day, i.e., the activity phase, than during the night, when the animals remain in hiding. The daily rhythms in PT- and activity changes are maintained in constant light and are independent of the environmental temperature (cf. REGAL, 1967). NIELSEN and NIELSEN (1959) surmised that the PT of *Ascia monuste* increases towards evening, when its activity normally ends.

Changes in PT throughout the year may be due to endogenous factors or to changing environmental conditions. For example the PT of the lizard *Sceloporus* is altered by photoperiod (BALLINGER et al., 1969). If changes in PT values appear even when the animals are kept in various constant temperatures, the changes must be due to endogenous factors. SULLIVAN and FISHER (1954) found that the PT of *Salvelius fontinalis* fell rapidly in December-January from ca. 12 to 8° C and then rose to the old value in February-March. Changes of PT of this type probably govern the seasonal migration of flounder in the North Sea (ZAHN, 1963; cf. on the effect of temperature on fish migration: BIRMAN, 1958; NORTHCOTE, 1958; WOODHEAD and WOODHEAD, 1958; POSTUMA, 1960; MCCRACKEN, 1963; JEAN, 1964; MARTIN and JEAN, 1964; MACLEAN and GEE, 1971[19]; on the migration of

---

[19] MACLEAN, J.A., GEE, J.H.: J. Fish. Res. Bd. Can. 28, 919—923 (1971).

Loligo: SUMMERS, 1969). Significant relations between PT changes and an animal's
biology are found, e.g., in *Ips typographus.* Such changes induce these beetles to
look for winter quarters. In summer they prefer temperatures of 27–33° C, but in
autumn the PT falls to a point near the null point for movement, +7° C. At that
time they search out hiding places that are cool, but protected from cold. In late
winter PT rises again and induces the animals to leave their winter quarters to
search for trees for feeding and egg-laying (MERKER and ADLUNG, 1957, 1958).
ZAHN (1958) also observed a low PT in autumn for the red forest ant, *Formica
rufa.* Individual groups responsible for various tasks display various temperature
requirements. Thus, all the animals do not gather in the deep nest chambers
during the winter. Rather a few ants remain at moderate depths in the nest,
so that they can note the reappearance of light earlier, communicate the information to deeper parts of the nest, and so terminate the winter rest (cf.
KNEITZ, 1966).

Adaptation of PT: PT is often influenced by previous temperature. Therefore
seasonal changes in PT may be based on changes in the environmental temperatures. In the majority of cases PT increases as AT goes up, and examples of this
type of reaction are found in various groups of animals. GRANT (1955) tested
various groups of oligochaetes and confirmed the presence of this response, as did
SCHULZ (1954) (for nematodes: RODE, 1969; for snails: cf. RISING and ARMITAGE,
1969). MESKE (1961) examined diplopods and chilopods. PT values of *Lithobius*
fell between 16.3° C and 22.6° C, depending on pretreatment temperature (and on
the soil moisture, cf. p. 463). For *Geophilus longicornis* the values were between
12.8° C and 17.3° C (cf. GROMYSZ-KALKOWSKA, 1967; PT of *Orthomorpha gracilis*
(Diplopoda) was 20.6 C; AT had no effect). Pretreatment temperatures below
39° C had no effect on the preference of the scorpion *Leiurus quinquestriatus* for
any temperature. After two days of exposure to temperatures between 39° and
44° C an increase in PT appeared (ABUSHAMA, 1964). Observations on insects have
been reported by, e.g., BODENHEIMER and SCHENKIN (1928), UVAROV (1931),
AGRELL (1947), GRAHAM (1958), COULIANOS (1958). The results of BODENHEIMER
and SCHENKIN (cf. DORNER and MÜLLER, 1962) indicate that among land animals
– especially insects – PT and air moisture can scarcely be considered in isolation
from one another. The PT of *Tribolium confusum* rose initially with increasing AT,
up to 26°C. But animals that had lived at still higher temperatures for more than
one month turned towards lower temperatures in the selection apparatus; presumably because they then preferred a range with higher relative humidity. Experiments on beetles *(Tribolium, Tenebrio, Sitophilus,* etc.) showed rapid changes
in PT. Their response was markedly influenced by the end of the gradient (warm
or cold) at which they were introduced (YINON and SHULOV, 1970).

Fish have been investigated most intensively with regard to this problem. Compilations of a part of the experimental results are found in FERGUSON (1958) and
ZAHN (1962). FRY (in ROSE, 1967) treats the subject in a chapter of a book. Therefore only a very brief presentation will be given here. DOUDOROFF (1938) found
that PT increased as AT rose, up to a point at which PT equalled AT. This specially
marked temperature value is called the "final preferendum" (FRY, 1947; FRY and
HART, 1948). At still higher AT the PT of *Girella nigricans* did not continue to rise

(DOUDOROFF, l.c.). In other cases PT is not affected by AT (cf. McCAULEY and
TAIT, 1970), and examples have been recognized in which PT fell as AT rose (e.g.
in *Salmo gairdneri*, GARSIDE and TAIT, 1958). ZAHN (1962) tried to organize these
observations into a system based on the angle at which the AT-PT curve inter-
sects the slope that connects points of equal AT and PT. The reasons for the differences in types of response are not known. There is also no explanation for the fact
that warm water fish acclimated to a specific temperature select their PT with
much less precision than animals tested immediately after their capture; or for
the fact that the temperatures they select in the apparatus have only a slight relation to their behavior in the field (NORRIS, 1963). Experimental results on amphibia
are found in STRÜBING (1954). LUCAS and REYNOLDS (1967), working with *Rana
pipiens* larvae, found an effect of AT on PT (cf. SCOTT, 1943 cited by FRY in ROSE,
1967; LILLYWHITE, 1971[20]), but in general, while amphibia (e.g. LICHT and BROWN,
1967) as well as reptiles do display a species-specific preferred temperature range,
AT only rarely induces changes in it (cf. LICHT, 1968).

Effect of Moisture: The water content of the air is an especially important environmental factor influencing PT. Since animals at high temperatures transpire more
strongly, thereby using more water, and are exposed to greater danger of desiccation, PT is closely connected to their water budget. It has often been observed
that immediately after their introduction into the experimental chamber animals
prefer high temperatures but later, if no provision has been made to permit them
to make up for water losses by drinking, they prefer low ones (PALMÈN, 1954). On
the other hand high temperatures may be preferred if the body water content is
high. Thus, the distribution maximum of the tropical earthworms, *Hyperiodrilus
africanus* and *Endrilus eugeniae,* is at 28.5° C in uniformly moist soil, but after the
animals are kept one hour in water the PTs go to 37—38° C (MADGE, 1969). So it is
conceivable that preference for a specific relative humidity and temperature
regulates the adjustment of a specific water content and transpiration rate.
JAKOVLEV and KRÜGER (1954), who observed various acridid grasshoppers at
100% relative humidity in the temperature selection apparatus, noted a decrease
of 2—3° C in PT below the value at lower humidities. They believed that the high
rates of transpiration promoted by high temperatures are not possible in air
saturated with water vapor, which disturbs the animals' water budget more than
any other factor. To avoid such disturbances, they seek out a temperature that
lies close enough to the normal PT but that does not require as high a rate of
transpiration. Thus the unavoidable losses of water are smaller and more tolerable.
In the vicinity of the PT the transpiration curve is horizontal, i.e. transpiration is
optimal (JAKOVLEV, 1961; cf. JAKOVLEV and KRÜGER, 1954). Similar results were
found for carabidae (SCHMIDT, 1956, 1957, 1970; cf. BONGERS, 1969). The connection between temperature and humidity is at times so close, especially in the case
of soft-skinned animals, that one has to speak of a thermohydro preferendum. In
some cases humidity alone is the deciding factor, as long as the temperature does
not reach extreme values. Nematodes, *Ditylenchus dipsaci*, placed in gradients of
2—30° C, gathered at the area of highest humidity (WALLACE, 1961), and at temperatures between 2° C and 29° C the salamander *Aneides lugubris* preferred areas

---

[20] LILLYWHITE, H. B.: Comp. Biochem. Physiol. **40 A**, 213—227 (1971).

with a soil moisture content of 25% (ROSENTHAL, 1957). Relatively few examples
are known of cases where air humidity or water content have no effect (e.g. MADGE,
1961; CHAPMAN, 1963). The PT of individual social insects is highly dependent
on the water economy of the whole colony. When the rate of use of brood food
in their nest goes up, ants that are in the sun outside the nest move towards
lower temperatures, whereby water loss through evaporation is limited (ZAHN,
1958).

Effect of Lack of Food: The effect of lack of food on PT shows clearly that temperature is only one of the environmental factors to which animals respond. Hungry
ticks, *Ornithodorus erraticus*, run around without reacting to temperature differences (EL-ZIADY, 1958). Snails, *Biomphalaria glabrata*, distribute themselves over
the entire available temperature scale (12–38° C) when they are hungry and
gather at an otherwise unsuitable temperature if food is present there (CHERNIN,
1967). NIELSEN and NIELSEN (1959) observed an increase in PT of about 4° C in a
butterfly after feeding. Temporary ectoparasites on birds and mammals, depending on whether they are hungry or fed, exhibit different PTs, which are close to
their hosts' surface temperatures or have lower values. In this way hungry animals
are directed towards their host (HERTER, 1942, 1952).

Effect of Other Individuals and of Light: PT can be modified by the search for other
members of the species or for a specific degree of illumination as well as by the
search for food. SULLIVAN and FISHER (1954) found an effect of light on the temperature selection behavior of *Salvelinus fontinalis*. The PT of larvae of the beetle
Epilachna varivestis was shown to be dependent on illumination. In dim light
(1 foot-candle) PT was 21–24° C, and in bright light (40 foot-candle), 32–38° C.
The PT at the lower illumination is close to the temperature in which the reproductive rate is high and mortality is low. Therefore the lower PT is biologically
correct (KING and RILEY, 1960).

In many animal species, especially in those that live in social aggregations, single
individuals tend to join groups of the same species. This behavior may influence
the PT (e.g. LAVIE and ROTH, 1953; LIKVENTOV, 1960; BRATTSTROM, 1963; ELLIS,
1963). Thus the PT of *Conophthorus coniperda* increased in proportion to the number of animals examined at one time (HENSON, 1960). Bugs exhibited definite
group behavior in the temperature selection apparatus. The group stayed together
as it moved through the temperature gradient, but if a specific temperature was
exceeded, the group dispersed. This temperature was 30° C for *Dysdercus inter-
medius* (YOUDEOWEI, 1968) and *Oncopeltus fasciatus* (BONGERS, 1969). At low temperatures collemboles formed aggregations (JOOSSE and VELTCAMP, 1970[21]).

The Biological Significance of PT

A temperature range which is sought out as actively as PT often must have a
special biological meaning. Therefore many attempts have been made to establish
correlations between this delineated temperature range and various other functions. Experiments on lizards indicate that body temperatures are held at a level
at which a broad range of activity is possible (LICHT, 1965; cf. SCHOENER, 1970)

---

[21] See p. 362.

and the sense organs are especially sensitive (CAMPBELL, 1969). This temperature level is the preferred range in which the rate of activity is often fairly uniform and markedly different from the rate outside this range (LEE and BADHAM, 1963). It is not truly accurate to say that animals in the preferred range are at rest. Though there certainly is more running activity outside this range because tendencies to escape are activated, many other modes of behavior are not exhibited at all except within the range. Accordingly animals often do not rest in the PT range. Rather they display their broadest gamut of activity in precisely this region, though they can do this in the temperature selection apparatus only if their natural surroundings are reproduced here. CHRISTIANSEN and BAKKE (1968) placed beetles, *Hylobius abietis*, in a selection apparatus with a habitat including sand and pine twigs and found an optimum for feeding movements at 20—25° C. Egg-laying was also favored in this range.

Neodiprion cocoons in the field are found only in areas of the ground with favorable temperatures (STARK and DAHLSTEN, 1966). Spiders build their nests only in certain temperature ranges (ENGELHARDT, 1964; cf. BUCHE, 1966). After feeding, individuals of *Lygosoma laterale* gather side by side over warmed subterranean sites. Constrictors hold the body segments that contain food over the source of warmth, thereby speeding digestion (REGAL, 1966; on the question of "activity", cf. HEYDEMANN, 1957; SCHÄFER, 1971). If specific activities such as, e.g., the collecting activity of *Crematogaster scutellaris*, take place outside the PT range as well, then, according to SOULIÉ (1955, 1956), exposure to the PT acts as a stimulus. The animals need the exposure to the PT to enable them to be active at other temperatures.

PT is related to metabolism. The results of HERFS (1963) show effects on anabolism in the leaf-roller, *Tortrix pronubana*, whose larvae develop most rapidly at 27° C (and 70—90% relative humidity), the preferred temperature (cf. CHAPMAN, 1965). The PT of rainbow trout is ca. 10.4° C (SCHMEING-ENGBERDING, 1953) and they utilize food most efficiently at ca. 9.2° C (CORNELIUS, 1933). YOUNG *Girella nigricans* display the greatest appetite at 26° C, their PT (NORRIS, 1963; cf. for *Oncorhynchus nerka*: BRETT, 1971[22]). Special relationships between oxygen use and the PT have been demonstrated many times. The oxygen uptake of fish has been shown to be temperature independent in a temperature range beginning with the PT. This plateau was absent in narcotized fish (SCHMEING-ENGBERDING, 1953). HERTER (1940) found a similar situation in lizards, whose rate of respiratory motion went down in the PT range (cf. REICHLIG, 1957; SCHULZE, 1959; HUDSON and BERTRAM, 1966). Narcotized animals did not show this behavior. LICHT and BASU (1967) observed maxima in certain metabolic functions at the PT in two different species of lizards. The ability of testis explants to take up nucleotides from culture medium was greatest at the PT of the intact animal. These results have been confirmed by the work of MOBERLY (1968), who found a metabolic maximum at the PT of the green iguana.

Among social insects special conditions prevail because the metabolism of the entire colony and not only of the individual is significant (e.g., HERAN, 1952; LUKOSCHUS, 1956; ZAHN, 1958).

---

[22] BRETT, J. R.: Am. Zool. 11, 99—113 (1971).

c) Orientation in Temperature-Stimulus Fields without Gradients

α) Animals as Sources of Heat

The search for natural sources of warmth is important for those animals that hunt for birds or mammals as hosts or prey. HOWLETT (1910) and KRIJGSMAN (1930) observed orientation towards sources of warmth in bloodsucking midges and in *Stomoxys calcitrans*, respectively (cf. MILNE and MILNE, 1963). The sensitivity of the sense organs to small differences determines the distance from the source at which orientation is still possible (cf. p. 454, HOMP, 1939). Orientation may be phobic or tropic. The bug *Rhodnius prolixus* carries out searching movements with its motile antennae. The animal's course of movement is fairly straight and only one sense organ is needed for orientation (WIGGLESWORTH and GILLETT, 1934). If unilateral removal of a sense organ causes an animal to move in circles, its orientation is probably tropotactic. HOMP (1939) observed *Pediculus vestimenti* circling around a source of warmth and interpreted this behavior as the result of persistence of a stimulus disequilibrium (thermomenotaxis on the principle of tropotaxis). The pit vipers *(Crotalus)* probably orientate themselves towards their prey on telotactic principles. The instinctive action of striking at prey is initiated when heat radiation coming from a specific point in front of the snake reaches both pit organs at the same time (BULLOCK and COWLES, 1952). *Oncopeltus* nymphs sitting in a circle around a source of heat all maintained the same distance from the source (BARRETT and CHIANG, 1967). They might have been orientating towards the source of heat or to the substrate temperature.

Infrared radiation from the thorax of night-flying butterflies is said to aid the males in finding the females (CALLAHAN, 1965).

β) The Sun as a Heat Source

It is often difficult to determine whether taxes or instincts are responsible for the behavior initiated by the warmth of solar radiation. Grasshoppers, *Schistocerca gregaria*, regulate their body temperatures by holding their long body axis at a given angle to the rays of the sun (FRAENKEL, 1930, cf. CASSIER, 1965). A similar type of orientation is found among lizards that turn their backs or sides to the sun in the morning and also try to take up as much warmth as possible by spreading out their ribs (LEE and BADHAM, 1963; HEATH, 1965; BRATTSTROM, 1965; DEWITT, 1967; McGINNIS and DICKSON, 1967; BRADSHAW and MAIN, 1968, and many others). An active change in coloration may aid the absorption of heat. If towards noon it becomes too hot, the grasshoppers turn their narrow sides towards the sun, whereby the amount of solar radiation striking the body surface is reduced to low levels. In these cases the entire animal assumes an oriented position. Other animals orient only parts of their body. Butterflies orient their wings according to the sun. If heat absorption is desired the wings are spread out so that the rays strike as perpendicularly as possible. If it is too warm, the wings are folded together and the long axis is also turned in the direction of the sun's rays, so that the rays strike as small a surface as possible (e.g., CLENCH, 1966). In such cases one must always investigate the possible participation of the sense of vision (or of a dermal light sense) in orientation. In the butterfly, *Argynnis paphia*, orientation for sunning and for protection against overheating is regulated separately. Sunning is controlled optically. This is particularly clear in temperatures below the optimum range. Protection against overheating is controlled tropotactically. In a heat radiation field formed by two intersecting pencils, the animals try to maintain a laterally symmetrical distribution of stimulation, even though the absorption of heat is thereby maximized (VIELMETTER, 1957, 1958; cf. LOZINA-LOZINSKY, 1953).

d) Temperature as an Inducer of Tactic or Instinctive Behavior

Temperature often has effects on tactic orientation towards other types of stimuli. According to HERTER (1953) tsetse flies, *Glossina morsitans*, exhibit negative phototactic orientation at temperatures above 30° C. If they are given a choice between 42° C in the light and 49° C in the dark they prefer the dark and are killed by the high temperatures. Under natural conditions their orientation leads them towards shade where they find both lower temperatures and their host animals (cf. LAARMAN, 1958; SULLIVAN, 1959; FRANCIA and GRAHAM, 1967; WILSON, 1968).

By contrast, the thermoregulatory behavior of the cicada, *Magicicada cassini*, cannot be considered as a simple thermotaxis (HEATH, 1967). These animals seek out sunny spots in their habitat and move to the lower surface of leaves when it is too warm. In the morning, when the air temperature is still low, the wings are spread out at the sides (possibly instinctive behavior), so the sun's rays can strike the abdomen, which is appropriately oriented. Later in the day the wings are folded over the abdomen and shade it. The long body axis is then parallel to the sun's and the head is turned away from the sun. In this example simple thermotaxes or phototaxes are mixed almost inseparably with complicated forms of behavior. Such a phenomenon can often be demonstrated when the behavior induced by temperature leads to a change in location. The cydnid *Brachypelta aterrima* burrows into the earth when it is too warm or too cold (SHORR, 1957; cf. NAGY, 1959; FOULK, 1968), and in winter the burrowing isopod *Limnoria lignorum* and the amphipod *Chelura terebrans* leave burrows lying above the low water line in large numbers before the ebb tide. In this way the animals are protected from death due to extremely low air temperatures (NAIR and LEIVESTAD, 1958; cf. IANSSEN, 1960). In these cases a specific temperature level caused the behavior (cf. as further examples: burrowing of *Grylloblatta*, HENSON, 1957; of elaterid larvae, LA FRANCE, 1968; abandonment of their nests by marine turtles, MROSOVSKY, 1968). The larvae of the leaf-roller, *Tortrix pronubana*, prevent extreme increases in body temperature by eating their way deep into the ovary (HERFS, 1963). In these and in many other cases (see p. 457 f., cf. HERTER, 1953, p. 26; on regulation of body temperature in winter crabs, WILKENS and FINGERMAN, 1965; searching for sunny sites by marine turtles, BOYER, 1965; GIBBONS, 1967; regulation of body temperature in *Dipsosaurus*, McGINNIS and DICKSON, 1967; search for optimal temperatures by fish, BARLOW, 1958, and tadpoles, BRATTSTROM, 1963) the behavior is probably instincitive and temperature can be looked at as the releasing factor.

Arctic *Aedes nigripes* and *A. impiger* sun themselves without feeding in the focal point of parabolic flowers, where temperatures up to 6° C higher than the surrounding ones prevail (HOCKING and SHARPLIN, 1965). Certainly no taxis could guide them to the place. The same circumstances are found with regard to the so-called phobotaxes. A specific environmental situation, e.g., a pH value or a temperature, releases an innate behavior pattern that is not oriented with respect to the direction of the stimulus (cf. LORENZ, 1967).

e) Temperature Regulation by Instinctive Behavior

Instinctive behavior is certainly present when complicated activities, which are not directed by temperature though they are often initiated by it, prevent further changes in body temperature. Examples are found among insects and crabs. When there is danger of overheating or of desiccation, Brachyuran decapods of various species produce foam bubbles which they press out between their mouth parts with the respiratory water. The volume of foam is large enough to envelop the animal completely and probably acts as a protective shield against heating by solar radiation (ALTEVOGT, 1968; JANSEN, 1970).

Butterflies can reduce their body temperature by drinking water, which is rapidly conducted through the gut and given off. If it is cooler than the body it can contribute to temperature reduction (CLENCH, 1966, p. 1027). ADAMS and HEATH (1964) observed similar behavior in the sphingid moth, *Pholus achemon*, which regurgitates a drop of fluid, moves it between the palps and swallows it again. This behavior results in a distinct cooling effect. Other butterflies give off fluid from the gut and take it up again (CLENCH, 1966). The fluid is probably cooled outside the body by evaporation.

Among social insects a large number of instinctive actions serve to regulate temperature directly or indirectly. Tropical termites lay out their nests in such a way that the noon sun strikes the narrow side. In this way excessive heating of the structure is prevented (on temperature regulation in the termite colony, cf. LÜSCHER, 1961). Certain species of ants place their domed nest structures precisely so that maximal warming is possible. They elevate the nest over cool earth and use material with low heat capacity. Other species build nests over flat stones that warm quickly when the sun shines on them. In summer *Formica rufa* maintain temperatures of 23—29° C at a depth of 15—20 cm in their domes. When the sun shines the openings in the dome are widened and if there is a threat of cold weather they may be closed completely (STEINER, 1924). But according to ZAHN (1958) the activity of heat-transferring ants, which take up warmth by basking on top of the dome and give it up again in the nest, is much more important (see p. 296). Temperature is responsible for the initiation of this and other modes of behavior.

Additonal examples of social temperature regulation are found among the hymenopterans, *Polistes*, *Vespa*, and *Apis mellifica*. Field wasps, *Polistes gallica*, cool their nest to 35.5° C, the optimal temperature for the brood, by water transport and ventilation (STEINER, 1930). *Vespa* regulates its nest temperature through water transport and through heat production by means of rapid wing beats (WEYRAUCH, 1936; cf. ISHAY et al., 1967; see p. 296). At temperatures below 13° C, honey bees, *Apis mellifica*, form a winter cluster within which extreme reductions in temperature are prevented by thermogenesis (cf. ROTH, 1965; MORSE, 1967; see p. 297). In summer the temperature in the brood nest area is adjusted to 33—36° C (indifference range of regulation, WOHLGEMUTH, 1957; cf. SIMPSON, 1961; LENSKY, 1964). In cooler weather the worker bees crowd together on the honeycomb and cover the brood cells with their bodies (v. FRISCH, 1965). As a countermeasure against heat, water is brought in and spread on the honeycomb or is drawn out as a film between the proboscis and peristome (LINDAUER, 1954). The bees are informed of the population's water requirements at all times. This information results from the constant exchange of food among the population, by which the concentration of the honey bladder contents is equalized among all the bees (KIECHLE, 1961).

References to Section "Animals"

ABUSHAMA, F. T.: Animal Behav. **12**, 140—153 (1964).
ADAMS, P. A., HEATH, J. E.: J. Res. Lepid. **3**, 69—72 (1964).
AGALIDES, E.: Trans. N. Y. Acad. Sci. 2, **31**, 1083—1102 (1969).
AGRELL, I.: Arkiv Zool. **10**, 1—48 (1947).
AGRELL, I.: Exp. Cell Res. **8**, 232—234 (1955).
AGRELL, I.: Arkiv Zool. **11**, 383—392 (1958).
AGRELL, I.: Arkiv Zool. **12**, 291—300 (1959).
AHMAD, T.: J. Animal Ecol. **5**, 67—93 (1936).
AIKEN, D. E.: Can. J. Zool. **47**, 931—935 (1969).
AKITA, Y.: Jap. J. Zool. **11**, 407—423 (1955).
ALABASTER, J. S.: Water Res. **1**, 717—730 (1967).
ALBRECHT, F. O., CASSIER, P.: Compt. Rend. Acad. Sci. (Paris) **260**, 6449—6451 (1965).
ALEKSIUK, M.: Can. J. Zool. **48**, 1155—1161 (1970).
ALEXANDER, A. J., EWER, D. W.: J. Exp. Biol. **35**, 349—359 (1958).
ALEXANDROV, V. Y.: Current Mod. Biol. **3**, 9—19 (1969).
ALEYEV, J. C.: Dokl. Akad. Nauk SSSR **110**, 491—493 (1956).
ALLOWAY, T. M.: J. Comp. Physiol. Psychol. **69**, 1—8 (1969).
ALLOWAY, T. M., ROUTTENBERG, A.: Science **158**, 1066—1067 (1967).
ALPATOV, W. W.: J. Exp. Zool. **63**, 85—111 (1932).
ALTEVOGT, R.: Zool. Anz. **181**, 298—402 (1968).
ALTMAN, P. L., DITTMER, D. S. (Ed.): Environmental biology. Bethesda: Fed. Am. Soc. Exp. Biol. 1966.
ALVAREZ, M. R., COWDEN, R. R.: Z. Zellforsch. Mikroskop. Anat. **75**, 240—249 (1966).
AMMAN, G. D.: Ann. Entomol. Soc. Am. **61**, 1606—1611 (1968).
ANDERS, F., DRAWERT, F., ANDERS, A., REUTHER, K. H.: Z. Naturforsch. **19**b, 495—499 (1964).
ANDERSEN, K. T.: Z. Morphol. Ökol. Tiere **17**, 649—676 (1930).
ANDERSEN, K. T.: Monogr. Pfl. Sch. **6** (1931).
ANDERSEN, K. T.: Biol. Zbl. **54**, 478—486 (1934), **55**, 571—590 (1935).
ANDERSON, J. F., HORSFALL, W. R.: J. Exp. Zool. **154**, 67—107 (1963), **158**, 211—222 (1965a).
ANDERSON, J. F., HORSFALL, W. R.: Science **147**, 624—625 (1965b).
ANDERSON, T. R.: Comp. Biochem. Physiol. **33**, 663—687 (1970).
ANDREWARTHA, H. G.: Bull. Entomol. Res. **34**, 1—17 (1943), **35**, 31—41 (1944).
ANDREWARTHA, H. G.: Biol. Rev. **27**, 50—107 (1952).
AN-KUO, C., WEI-HSIUNG, F., CHIH-HUI, C., HSING-CHEN, C.: Acta Entomol. Sinic. **14**, 238 (1965).
ANNILA, E.: Ann. Zool. Soc. Zool. Botan. Fennicae Vanamo **6**, 161—208 (1969).
ANNILA, E.: Ann. Entomol. Fennicae **36**, 186—190 (1970).
ANTHONY, E. T.: J. Exp. Biol. **38**, 93—107 (1961).
APPLEWHITE, P. B.: Nature **219**, 91—92, 1265—1266 (1968a, b).
ARAI, M. N., COX, E. T., FRY, F. E. J.: Can. J. Zool. **41**, 1011—1015 (1963).
AREFYEVA, T. A.: Fiziol. Ž (Kiev) **11**, 45—51 (1965).
ARIDOV, Z., PODOLER, H.: Israel J. Entomol. **3**, 1—15 (1968).
ARMITAGE, K. B.: Biol. Bull. **123**, 225—232 (1962).
ARNIM, FRHR. V.: Z. Angew. Entomol. **22**, 533—557 (1936).

ARNOUX, J., LABEYRIE, V., MAISON, P.: Compt. Rend. Acad. Sci. (Paris) **247**, 2443—2445 (1958).
ASAHINA, E.: Zool. Mag. (Tokyo) **64**, 280—285 (1955).
ASAHINA, E.: Nature **184**, 1003—1004 (1959).
ASAHINA, E.: Bull. Marine. Biol. Stat. Asamushi, Tohoku Univ. **10**, 251—252 (1962).
ASAHINA, E.: Low Temp. Sci. Ser. B. (Sapporo) **20**, 45—56 (1962b), **23**, 65—70 (1965a).
ASAHINA, E.: Federation Proc. **24**, 183—187 (1965b).
ASAHINA, E. (Ed.): Cellular injury and resistance in freezing organisms. (Int. Conference Low Temp. Sci.) Proc. II, Hokkaido Univ., Japan (1967).
ASAHINA, E.: Advan. Insect Physiol. **6**, 1—49 (1969).
ASAHINA, E., AOKI, K.: Nature **182**, 327 (1958).
ASAHINA, E., HISADA, Y., EMURA, M.: Contr. Inst. Low Temp. Sci., Hokkaido Univ., Sapporo, B., **15**, 36—50 (1968).
ASAHINA, E., TAKEHARA, I.: Low Temp. Sci. Ser. B. (Sapporo) **22**, 79—90 (1964).
ASAHINA, E., TANNO, K.: Exp. Cell Res. **31**, 223—225 (1963).
ASAHINA, E., TANNO, K.: Low Temp. Sci. Ser. B. (Sapporo) **25**, 105—110 (1967).
ASTON, R. J.: Z. Zool. (London) **154**, 29—40 (1968).
ATHERTON, W. D., AITKEN, A.: Comp. Biochem. Physiol. **36**, 719—747 (1970).
ATKINS, M. D.: Can. Entomol. **91**, 283—291 (1959), **99**, 181—187 (1967).
ATWAL, A. S.: Can. J. Zool. **38**, 131—141 (1960).
AUERBACH, M.: Z. Fischerei (N.F.) **6**, 605—620 (1957).
AUTRUM, H., HAMDORF, K.: Z. Vergleich. Physiol. **48**, 266—269 (1964).
AYRE, G. I.: Insectes Sociaux **5**, 147—157 (1958).
BABÁK, E.: Bl. Aquar. Terr. Kd. **25**, 115—117 (1914).
BABENKO, L. V.: Wiadomosci Parazytol **13**, 517—524 (1967).
BACCI, G., VORIA, P.: Experientia **26**, 1273—1274 (1970).
BAI, A. R. K., SRIHARI, K., SHADAKSHARASWAMY, M., JYOTHY, P. S.: J. Protozool. (Lawrence, Can.) **16**, 738—743 (1969).
BAILEY, R. M.: Ecology **36**, 526—528 (1955).
BAKER, W. A., JONES, L. G.: Techn. Bull. U.S. Dept. Agric. **460** (1934).
BAKKE, A.: Norsk. Entomol. Tidsskr. **16**, 81—83 (1969).
BALDWIN, W. F.: Can. J. Zool. **32**, 157—171 (1954).
BALDWIN, J., HOCHACHKA, P. W.: Biochem. J. **116**, 883—887 (1970).
BALLINGER, R. E., HAWKER, J., SEXTON, O. J.: J. Exp. Zool. **171**, 43—48 (1969).
BALLINGER, R. E., McKINNEY, C. O.: J. Exp. Zool. **161**, 21—28 (1966).
BALLINGER, R. E., SCHRANK, G. D.: Physiol. Zoöl. **43**, 19—22 (1970).
BANERJEE, S. K., DE, N. C.: Indian J. Physiol. Allied Sci. **23**, 53—55 (1969).
BARÁNSKA, J., WLODAWER, P.: Comp. Biochem. Physiol. **28**, 553—570 (1969).
BARDACH, J. E.: Am. Naturalist **90**, 309—317 (1956).
BARDACH, J. E., BJORKLUND, R. G.: Am. Naturalist **91**, 233—251 (1957).
BARDSLEY, J. E., HARMSEN, R.: Can. J. Zool. **47**, 283—288 (1969).
BAR-ILAN, A. R., SHULOV, A., PENER, M. P.: Physiol. Zoöl. **42**, 411—428 (1969).
BARKER, D.: Hydrobiologia **13**, 209—235 (1959).
BARLOW, C. A.: Can. J. Zool. **40**, 145—156 (1962).
BARLOW, C. A., KERR, W. D.: Can. J. Zool. **47**, 217—224 (1969).
BARLOW, C. A., KUENEN, D. J., DUK, W.: Proc. Koninkl. Ned. Akad. Wetenschap., C **60**, 240—250, 251—254 (1957).
BARLOW, G.: Ecology **39**, 580—587 (1958).
BARNES, H.: J. Marine. Biol. Assoc. U.K. **43**, 717—727 (1963).
BARRETT, I., HESTER, F. J.: Nature **203**, 96—97 (1964).
BARRETT, R. W., CHIANG, H. C.: Ecology **48**, 590—598 (1967).
BARTHOLOMEW, G. A.: Copeia **1966**, 241—250.
BARTHOLOMEW, G. A., LASIEWSKI, R. C.: Comp. Biochem. Physiol. **16**, 573—582 (1965).
BARTHOLOMEW, G. A., TUCKER, V. A.: Physiol. Zoöl. **37**, 341—354 (1964).
BARTHOLOMEW, G. A., TUCKER, V. A., LEE, A. K.: Copeia **1965**, 169—173.
BAR-ZEEV, M.: Bull. Entomol. Res. **48**, 593—599 (1957).
BASEDOW, T.: Intern. Rev. Ges. Hydrobiol. **54**, 765—789 (1969).

BASEDOW, T., SCHÜTTE, F.: Nachrichtenbl. Deutsch. Pflanzenschutzd. (Braunschweig) 23, 4—8 (1971).
BASLOW, M. H., NIGRELLI, R. F.: Zoologica 49, 41—51 (1964).
BASTIAN, J., ESCH, H.: Z. Vergleich. Physiol. 67, 307—324 (1970).
BATEMAN, M. A.: Australian J. Zool. 15, 1141—1161 (1967).
BATTAGLIA, B., LAZZARETTO, I.: Nature 215, 999—1001 (1967).
BATTLE, H. J.: Trans. Roy. Soc. Can. 20, 127—143 (1926).
BAUER, K.: Zool. Jahrb. (Physiol.) 65, 267—301 (1955).
BAUST, J. B., MILLER, L. K.: J. Insect Physiol. 16, 979—990 (1970).
BEAMENT, J. W.: J. Exp. Biol. 35, 494—519 (1958).
BEAMENT, J. W. L., TREHERNE, E., WIGGLESWORTH, V. B. (Eds.): Advan. Insect Physiol. New York 1964, 1968.
BEARDMORE, J. A.: Heredity 14, 411—422 (1960).
BEBAK, B.: Bull. Acad. Polon. Sci. Ser. Sci. Biol. 6, 367—369 (1958).
BEBAK, B.: Acta Biol. Cracov. Zool. 7, 187—208 (1965).
BECK, S. D.: Biol. Bull. 122, 1—12 (1962).
BECK, S. D.: Insect Photoperiodism, New York 1968.
BECK, S. D., HANEC, W.: J. Insect Physiol. 4, 304—318 (1960).
BEECKEN, W.: Arch. Bienenkd. 15, 213—275 (1934).
BEETSCHEN, J.-C.: Compt. Rend. Acad. Sci. (Paris) 244, 2095—2097 (1957).
BEHRENDT, K.: Zool. Jahrb. (Physiol.) 70, 309—398 (1963).
BEHRISCH, H. W.: Biochem. J. 115, 687—696 (1969).
BEHRISCH, H. W., HOCHACHKA, P. W.: Biochem. J. 111, 287—295, 112, 601—607 (1969).
BĚLEHRÁDEK, J.: Temperature and living matter. Berlin 1935.
BĚLEHRÁDEK, J.: Ann. Rev. Physiol. 19, 59—82 (1957).
BEIER, M.: Thalass. Jugosl. 1, 193—242 (1956).
BEKIR, M.: Z. Angew. Entomol. 21, 501—522 (1935).
BENTHE, H.-F.: Z. Vergleich. Physiol. 36, 327—351 (1954).
BENTLEY, P. J.: Science 152, 619—623 (1966).
BERÉNYI, D.: Mikroklimatologie. Mikroklima der bodennahen Atmosphäre. Stuttgart 1967.
BERG, K., JÓNASSON, P. M., OCKELMANN, K. W.: Hydrobiologia 19, 1—39 (1962).
BERGERARD, J.: Bull. Biol. Fr. Belg. 95, 273—300 (1961a).
BERGERARD, J.: Compt. Rend. Acad. Sci. (Paris) 253, 2149—2151 (1961b).
BERKHOLZ, G.: Z. Wiss. Zool. 174, 377—399 (1966).
BERLIN, J. D., DEAN, J. M.: J. Exp. Zool. 164, 117—132 (1967).
BERNARD, J., PINET, J. M., BOISTEL, J.: J. Insect Physiol. 16, 2157—2180 (1970).
BERNSTEIN, E., ZEUTHEN, E.: Compt. Rend. Trav. Lab. Carlsberg 35, 501—517 (1966).
BERRYMAN, A. A., STARK, R. W.: Ecology 43, 722—726 (1962).
BERTALANFFY, L. v.: Theoretische Biologie, Vol. 2, Bern 1951.
BIEBL, R.: Protoplasma 67, 451—472 (1969).
BIER, K.: Biol. Zbl. 73, 170—190 (1954).
BIER, K.: Chromosoma 10, 619—653 (1959).
BIEVER, K. D., MULLA, M. S.: Mosquito News 26, 416—419 (1966).
BILIOTTI, E., DEMOLIN, G., HAM, R.: Compt. Rend. Acad. Sci. (Paris) 258, 706—707 (1964).
BIRCH, L. C.: Ecology 28, 17—25 (1947).
BIRK, Y., HARPAZ, I., ISHAAYA, I., BONDI, A.: J. Insect Physiol. 8, 417—429 (1962).
BIRMAN, I. B.: Dokl. Akad. Nauk SSSR 122, 146—148 (1958).
BISHAI, H. M.: Hydrobiologia 25, 473—488 (1965).
BISSET, K. A.: J. Pathol. Bacteriol. 60, 87—92 (1948).
BLACK, E. C., KIRKPATRICK, D., TUCKER, H. H.: J. Fish. Res. Board Can. 23, 1—13 (1966).
BLAKE, G. M.: Bull. Entomol. Res. 49, 751—775 (1958).
BLAKE, G. M.: Nature 198, 462—463 (1963).
BLAŽKA, P.: Zool. Jahrb. (Physiol.) 65, 357—504 (1955).
BLAŽKA, P.: Physiol. Zoöl. 31, 117—128 (1958).

BLEST, A. D.: Behaviour **13**, 297—318 (1958).
BLISS, C. I.: J. Gen. Physiol. **9**, 467—495 (1926).
BLOCH, R., WALTERS, D. H., KUHN, W.: J. Gen. Physiol. **46**, 605—615 (1963).
BLOCK, B. C.: Ann. Entomol. Soc. Am. **59**, 56—59 (1966).
BLUNCK, H.: Z. Wiss. Zool. **111**, 76—151 (1914).
BLUNCK, H.: Z. Wiss. Zool. **121**, 171—391 (1923).
BODENHEIMER, F. S., KLEIN, H. Z.: Z. Vergleich. Physiol. **11**, 345—385 (1930).
BODENHEIMER, F. S., SCHENKIN, D.: Z. Vergleich. Physiol. **8**, 1—15 (1928).
BODENSTEIN, G.: Arch. Entw. Mech. Org. **140**, 614—655 (1940).
BODINE, J. H.: J. Exp. Zool. **42**, 91—109 (1925).
BOGERT, C. M.: Zoologica **38**, 63—64 (1953).
BOHLE, H. W.: Zool. Jahrb. (Anat.) **86**, 493—575 (1969).
BOILLON, J.: Nature **177**, 142—143 (1956).
BOKDAWALA, F. D., GEORGE, J. C.: J. Animal Morph. Physiol. **14**, 60—68, 223—230, 231—241 (1967).
BONDARENKO, N. V., KUAY-IUAN, K.: Dokl. Akad. Nauk SSSR **119**, 1247—1250 (1958).
BONGERS, J.: Oecologia **2**, 223—231 (1969).
BONNEMAISON, L.: Ann. Epiphyt. (Entomol.) **12**, 115—143 (1946).
BONNEMAISON, L.: Bull. Soc. Entomol. Fr. **65**, 73—78 (1960).
BONNEMAISON, L.: Bull. Soc. Entomol. Fr. **66**, 128—133 (1961).
BONNEMAISON, L.: Compt. Rend. Acad. Sci. (Paris) **259**, 1768—1770 (1964).
BONNIER, G.: Brit. J. Exp. Biol. **4**, 186—191 (1926).
BORZEDOWSKA, B.: Folia Biol. (Warsaw) **11**, 231—252 (1963).
BOUDINOT, M.: Comp. Biochem. Physiol. **37**, 601—603 (1970).
BOULOT, M., GALLISSIAN, A.: Compt. Rend. Acad. Sci. (Paris) **250**, 3403—3404 (1960).
BOURNE, J. R., HAYS, K. L.: J. Econ. Entomol. **61**, 321—322 (1968).
BOVEE, E. C.: Biol. Bull. **96**, 123—128 (1949).
BOWEN, W. R., STERN, V. M.: Ann. Entomol. Soc. Am. **59**, 823—834 (1966).
BOWLER, K.: J. Cell Comp. Physiol. **62**, 119—132, 133—146 (1963).
BOWLER, K.: Entomol. Exp. Appl. **10**, 16—22 (1967).
BOWLER, K., HOLLINGSWORTH, M. J.: Genet. Res. **6**, 1—12 (1965).
BOWLER, K., HOLLINGSWORTH, M. J.: Exp. Geront. **2**, 1—8 (1966).
BOYCOTT, B. B., GUILLERY, R. W.: Nature **183**, 62—63 (1959).
BOYCOTT, B. B., GRAY, E. G., GUILLERY, R. W.: Proc. Roy. Soc. (B) **154**, 151—172 (1961).
BOYER, D. R.: Ecology **46**, 99—118 (1965).
BRACHET, J., BREMER, F.: Arch. Intern. Physiol. **51**, 195 (1941).
BRADSHAW, S. D., MAIN, A. R.: J. Zool. **154**, 154—221 (1968).
BRÄUNINGER, H. D.: Z. Vergleich. Physiol. **48**, 1—130 (1964).
BRAMSTEDT, F.: Zool. Anz. **112**, 257—262 (1935).
BRATTSTROM, B. H.: Anat. Rec. **137**, 343 (1960).
BRATTSTROM, B. H.: Herpetologica **18**, 38—46 (1962).
BRATTSTROM, B. H.: Ecology **44**, 238—255 (1963).
BRATTSTROM, B. H.: Am. Midland Naturalist **73**, 376—422 (1965).
BRATTSTROM, B. H.: Comp. Biochem. Physiol. **24**, 93—111 (1968).
BRATTSTROM, B. H.: Comp. Biochem. Physiol. **35**, 69—103 (1970).
BRATTSTROM, B. H., LAWRENCE, P.: Physiol. Zoöl. **35**, 148—156 (1962).
BRAUN, K., KÜNNEMANN, H., LAUDIEN, H.: Marine Biol. **5**, 59—70 (1970).
BRENY, R.: Mém. Acad. Roy. Belg. Sci. **30**, 3, 1—88 (1957).
BRETT, J. R.: Publ. Ont. Fish. Res. Lab. **63**, 1—49 (1944), **64**, 9—28 (1946).
BRETT, J. R.: J. Fish. Res. Board Can. **9**, 265—323 (1952).
BRETT, J. R.: Quart. Rev. Biol. **31**, 75—87 (1956).
BRETT, J. R., SHELBOURN, J. E., SHOOP, C. T.: J. Fish. Res. Board Can. **26**, 2363—2394 (1969).
BRIAN, M. V., KELLY, A. F.: Insectes Sociaux **14**, 13—24 (1967).
BRIGGS, R., HUMPHREY, R. R.: Develop. Biol. **5**, 127—146 (1926).
BRITZ, L., HÖHNE, W.: Z. Angew. Zool. **42**, 209—234 (1955).

BROMAGE, N. R., SAGE, M.: J. Endoc. **41**, 303—311 (1968).
BRØNDSTED, A., BRØNDSTED, H. V.: J. Embryol. Exp. Morph. **9**, 159—166 (1961).
BROWN, H. A.: Copeia **1967**, 365—370, **1969**, 138—147.
BROWN, W. D.: J. Cell Comp. Physiol. **55**, 81—84 (1960).
BROWNING, T. O.: Australian J. Sci. Res. **5**, 96—111, 112—127, 344—353 (1952 a, b, c).
BRUN, J.-L.: Ann. Biol. Anim. Biochem. Biophys. **6**, 127—158, 267—300 (1966 a, b).
BRUSLÉ, S.: Compt. Rend. Acad. Sci. (Paris), Sér. D **212**, 1550—1553 (1966).
BRUST, R. A.: Mosquito News **26**, 512—514 (1966).
BRUST, R. A.: Can. Entomol. **99**, 986—993 (1967).
BRUST, R. A., HORSFALL, W. R.: Can. J. Zool. **43**, 17—53 (1965).
BRUST, V. V.: J. Exp. Zool. **175**, 37—68 (1970).
BUCHE, W.: Z. Morph. Ökol. Tiere **57**, 329—448 (1966).
BUCHMANN, W., TIMOFEEFF-RESSOVSKY, N. W.: Z. Indukt. Abstamm. Vererb.-L. **70**, 130—137 (1935).
BUCHNER, H., KIECHLER, H.: Naturwissenschaften **53**, 708 (1966).
BUCKHOLD, B., SLATER, J. V.: Radiation Res. **37**, 567—576 (1969).
BUDDENBROCK, E. v.: Z. Wiss. Zool. **164**, 173—187 (1960).
BÜCKLE, W.: Zool. Jahrb. (Physiol.) **70**, 130—137 (1963).
BÜCKMANN, D.: Naturwissenschaften **51**, 344 (1964).
BÜNNING, E.: Biol. Zbl. **77**, 141—152 (1958).
BÜNNING, E.: The physiological clock. New York 1964.
BUFFINGTON, J. D.: Comp. Biochem. Physiol. **30**, 865—878 (1969).
BUIKEMA, A. L., JR., ARMITAGE, K. B.: Herpetologica **25**, 194—206 (1969).
BULL, D. L., ADKISSON, P. L.: J. Ecol. Entomol. **53**, 793—798 (1960).
BULL, H. O.: J. Marine Biol. Ass. **21**, 1—27 (1936).
BULLOCK, T. H.: Biol. Rev. **30**, 311—342 (1955).
BULLOCK, T. H., COWLES, R. B.: Science **115**, 541—543 (1952).
BULLOCK, T. H., DIECKE, F. P. J.: J. Physiol. **134**, 47—87 (1956).
BURCOMBE, J. V., HOLLINGSWORTH, M. J.: Gerontologia **16**, 172—181 (1970).
BURGES, H. D.: Bull. Entomol. Res. **53**, 193—213 (1962).
BURGES, H. D., CAMMELL, M. E.: Bull. Entomol. Res. **55**, 313—325 (1964).
BURKHARDT, D.: Biol. Zbl. **78**, 22—62 (1959).
BURMEISTER, J., BECKMANN, A.: Acta Haematol. **35**, 246—252 (1966).
BURNETT, A. L., LENTZ, T., WARREN, M.: Ann. Soc. Roy. Zool., Belg. **90**, 247—267 (1960).
BURNETT, T.: Physiol. Zoöl. **43**, 155—165 (1970).
BURNS, M. E.: Cryobiology **1**, 18—39 (1964).
BURR, M. J., HUNTER, A. S.: Comp. Biochem. Physiol. **29**, 647—652 (1969), **37**, 251—256 (1970).
BURSELL, E.: Bull. Entomol. Res. **51**, 39—46, 583—598 (1961 a, b).
BUSCHINGER, A.: Insectes Sociaux **13**, 311—322 (1966).
BUSTARD, H. R.: Copeia **1967**, 733—758.
BUSTARD, H. R.: Copeia **1968**, 606—612.
CABANAC, M., HAMMEL, T., HARDY, J. D.: Science **158**, 1050—1051 (1967).
CALDWELL, R. S.: Comp. Biochem. Physiol. **31**, 79—93 (1969).
CALDWELL, R. S., VERNBERG, F. J.: Comp. Biochem. Physiol. **34**, 179—191 (1970).
CALHOON, R. E., JAMESON, D. L.: Copeia **1970**, 124—134.
CALIUS, C. D., MANGLITZ, G. R.: J. Econom. Entomol. **61**, 391—394 (1968).
CALLAHAN, P. S.: Ann. Entomol. Soc. Am. **58**, 727—745 (1965).
CAMERON, J. N.: Comp. Biochem. Physiol. **32**, 175—192 (1970).
CAMPBELL, H. W.: Physiol. Zoöl. **42**, 183—210 (1969).
CANARD, M.: Ann. Ecole Natl. Supér. Agron. (Toulouse) **6**, 187—267 (1958).
CANCELA DA FONSECA, I. P.: Broteria (Lisboa) **27**, 145—152 (1958).
CAPPE DE BAILLON, P.: Compt. Rend. Acad. Sci. (Paris) **195**, 557—559 (1932).
CAREY, F. G., TEAL, J. M.: Proc. Natl. Acad. Sci. U. S. **56**, 1464—1469 (1966).
CAREY, F. G., TEAL, J. M.: Comp. Biochem. Physiol. **28**, 194—204, 205—213 (1969 a, b)
CARLSEN, H.: Z. Vergl. Physiol. **35**, 199—208 (1953).

CASSAGNAU, P.: Compt. Rend. Acad. Sci. (Paris) **240**, 1483—1485 (1955), **242**, 1531—1534 (1956a), **243**, 603—605 (1956b).
CASSIER, P.: Compt. Rend. Acad. Sci. (Paris) **261**, 3679—3682 (1965).
CAYROL, J.-C., LEGAY, J.-M.: Ann. Epiphytics **18**, 193—211 (1967).
CERF, J.A., OTIS, L.S., TAKAGI, S.F.: Am. J. Physiol. **192**, 453—456 (1958).
CHALKLEY, H.W.: Physiol. Zoöl. **3**, 425—440 (1930).
CHAMBERS, R.: Arch. Mikr. Anat. **72**, 607—661 (1908).
CHAMPLAIN, R.A., BUTLER, G.D., JR.: Ann. Entomol. Soc. Am. **60**, 519—521 (1967).
CHAO, J., BALL, G.H.: J. Parasitol. **48**, 252—254 (1962).
CHAPMAN, R.F.: Brit. J. Anim. Behav. **5**, 60—75 (1957).
CHAPMAN, R.F.: Behaviour **24**, 283—317 (1965).
CHAPMAN, R.N., MICKEL, C.E., PARKER, J.R., MILLER, G.E., KELLY, E.G.: Ecology **7**, 416—426 (1926).
CHARNIER, M.: Compt. Rend. Soc. Biol. (Paris) **160**, 620—622 (1966).
CHERNIN, E.: J. Parasitol. **53**, 1233—1240 (1967).
CHERRY, L.M.: Entomol. Exp. Appl. **2**, 68—70 (1959).
CHEVERIE, J. CH., LYNN, W.G.: Biol. Bull. **124**, 153—162 (1963).
CHIANG, H.C., SISSON, V.: J. Econom. Entomol. **61**, 1406—1410 (1968).
CHILSON, O.P., COSTELLO, L.A., KAPLAN, N.O.: Federation Proc. **24**, 55—65 (1965).
CHINO, H.: J. Insect Physiol. **2**, 1—12 (1958).
CHIU, K.W., PHILLIPS, J.G., MADERSON, P.F.A.: Biol. Bull. **136**, 347—354 (1969).
CHRISTIANSEN, E., BAKKE, A.: Z. Angew. Entomol. **62**, 83—89 (1968).
CHRISTOPHERSEN, J., PRECHT, H.: Biol. Zbl. **71**, 313—326, 585—601 (1952a, b).
CHURCH, N.S.: J. Exp. Biol. **37**, 171—185 (1960), **37**, 186—212 (1960).
CHURCH, N.S., SALT, R.W.: Can. J. Zool. **30**, 173—184 (1952).
CLARKE, J.M., SMITH, J.M.: Nature **190**, 1027—1028 (1961).
CLARKE, K.U.: J. Insect Physiol. **5**, 23—36 (1960).
CLARKE, K.U.: J. Insect Physiol. **12**, 163—170 (1966).
CLARKE, K.U.: Bull. Entomol. Res. **57**, 259—270 (1967).
CLARKE, K.U., SARDESAI, J.B.: Bull. Entomol. Res. **50**, 387—405 (1959).
CLEGG, J.S.: Biol. Bull. **123**, 295—301 (1962).
CLENCH, H.K.: Ecology **47**, 1021—1034 (1966).
CLOUDSLEY-THOMPSON, J.L.: Proc. Zool. Soc. Lond. **121**, 253—277 (1951).
CLOUDSLEY-THOMPSON, J.L.: Cold Spring Harbor Symp. Quant. Biol. **25**, 345—355 (1961).
CLOUDSLEY-THOMPSON, J.L.: J. Zool. (Lond.) **152**, 43—54 (1967).
CLOUTIER, E.J., BECK, S.D.: Ann. Entomol. Soc. Am. **56**, 253—255 (1963).
COCKBEIN, A.J.: Entomol. Exp. Appl. **4**, 211—219 (1961).
COCKING, A.W.: J. Exp. Biol. **36**, 203—216, 227—226 (1959).
COGGER, H.G., HOLMES, A.: Proc. Linn. Soc. Lond. **85**, 328—333 (1961).
COLBERT, E.H., COWLES, R.B., BOGERT, C.M.: Bull. Am. Mus. Nat. Hist. **87**, 327—374 (1946).
COLHOUN, E.H.: Nature **173**, 582 (1954).
COLHOUN, E.H.: Entomol. Exp. Appl. **3**, 27—37 (1960).
CORBET, P.S., Entomology **5**, 403—418 (1957).
CORDIER, D., WORBE, J.F.: Compt. Rend. Séance Soc. Biol. **148**, 1253—1256 (1954).
CORNELIUS, W.O.: Z. Fischerei **31**, 535—566 (1933).
COSTA, H.H.: Hydrobiologia **28**, 583—588 (1966).
COSTLOW, J.D., JR., BOOKOUT, C.G., MONROE, R.: Physiol. Zoöl. **35**, 79—93 (1962).
COULIANOS, C.-C.: Oikos **6**, 71—77 (1955).
COULIANOS, C.-C.: Entomol. Tidskr. **78**, 265—268 (1958).
COUTURIER, A., ROBERT, P., ANTOINE, F., BLAISINGER, P.: Ann. Inst. Nat. Rech. Agron. C, **6**, 19—60 (1955).
COWLES, R.B.: Science **135**, 670 (1962), **158**, 1340—1341 (1967).
CRIDLAND, C.C.: Hydrobiologia **20**, 155—166 (1962).
CRISP, D.J.: J. Animal. Ecol. **28**, 119—132 (1959).
CRISP, D.J.: Helgoländer Wiss. Meeresuntersuch. **10**, 313—327 (1964).

CRISP, D. J., RITZ, D. A.: J. Exp. Marine Biol. Ecol. **1**, 236—256 (1967).
CRISP, D. J., SOUTHWARD, A. D.: J. Marine Biol. Ass. U. K. **37**, 157—208 (1958).
CROLL, N. A.: Nematologica **13**, 385—389 (1967).
CUSTANCE, D. R. N.: Experientia **22**, 588—589 (1966).
DAINTON, B. H.: J. Exp. Biol. **31**, 165—187 (1954).
D'AJELLO, V., BETTINI, S., GRASSO, A.: Riv. Parasitologia **28**, 71—78 (1967).
DALTON, J. C., HENDRIX, D. E.: Am. J. Physiol. **202**, 491—494 (1962).
DALTON, T., SNART, R. S.: Comp. Biochem. Physiol. **27**, 591—595 (1968).
DANIELLI, J. F.: Symp. Soc. Exp. Biol. **8**, 502—516 (1954).
DANIELS, J. M., ARMITAGE, K. B.: Hydrobiologia **33**, 1—13 (1969).
DANILEVSKII, A. S.: Photoperiodism and seasonal development of insects. Edinburgh-London 1965.
DAS, A. B.: Comp. Biochem. Physiol. **21**, 469—485 (1967).
DAS, A. B., PROSSER, C. L.: Comp. Biochem. Physiol. **21**, 449—467 (1967).
DAS, G. M., DAS, S. C.: Bull. Entomol. Res. **57**, 433—436 (1967).
DAUSCHER, H., FLINDT, R.: Z. Vergleich. Physiol. **62**, 291—300 (1969).
DAVIDE, R. G., TRIANTAPHYLLON, A. C.: Nematologica **13**, 102—110 (1967).
DAVIDSON, J.: Australian J. Exp. Biol. Med. Sci. **20**, 233—239 (1942).
DAVIDSON, J.: J. Animal Ecol. **13**, 26—38 (1944).
DAVIES, P. S.: J. Marine Biol. Ass. U. K. **46**, 647—658, **47**, 61—74 (1966/67).
DAVIES, P. S., WALKEY, M.: Comp. Biochem. Physiol. **18**, 415—425 (1966).
DAVIS, B. D.: Experientia **6**, 41—50 (1950).
DAVISON, T. F.: J. Insect Physiol. **15**, 977—988 (1969).
DAWSON, W. R., TEMPLETON, J. R.: Physiol. Zoöl. **36**, 219—236 (1963).
DAWSON, W. R., TEMPLETON, J. R.: Ecology **47**, 759—765 (1966).
DEAN, J. M.: Comp. Biochem. Physiol. **29**, 185—196 (1969).
DEAN, J. M., BERLIN, J. D.: Comp. Biochem. Physiol. **29**, 307—312 (1969).
DEAN, J. M., VERNBERG, F. J.: Biol. Bull. **129**, 19—22 (1965).
DEAN, J. M., VERNBERG, F. J.: Comp. Biochem. Physiol. **17**, 19—22 (1966).
DE BACH, P., FISHER, W., LANDI, J.: Ecology **36**, 743—753 (1955).
DE BACH, P., RAO, S. V.: Ann. Entomol. Soc. Am. **61**, 332—337 (1968).
DEHNEL, P. A.: Biol. Bull. **118**, 215—249 (1960).
DEHNEL, P. A., STONE, D.: Biol. Bull. **126**, 354—372 (1964).
DELSOL, M., FLATIN, J.: Compt. Rend. Séanc. Soc. Biol. **150**, 938—940 (1956).
DEMEREC, M.: Ann. Rep. Dir. Dep. Gen. (Cold Spring Harbor N. Y.) Carnegie Inst. **53**, 225 (1954).
DÉMEUSY, N.: Biol. Bull. **113**, 245—253 (1957).
DE REGGI, L., DELMAS, J. C.: Compt. Rend. Acad. Sci. (Paris) **260**, 6427—6429 (1965).
DESTOUCHES, L.: Compt. Rend. Acad. Sci. (Paris) **172**, 998—999 (1921).
DETHIER, V. G., ARAB, Y. M.: J. Insect Physiol. **2**, 153—161 (1958).
DETHIER, V. G., SCHOONHOVEN, L. M.: J. Insect Physiol. **14**, 1049—1054 (1968).
DE VRIES, A. L., WOHLSCHLAG, D. E.: Science **163**, 1073—1075 (1969).
DE WILDE, M. A., HOUSTON, A. H.: J. Fish. Res. Board Can. **24**, 2267—2281 (1967).
DE WITT, C. D.: Physiol. Zoöl. **40**, 49—66 (1967).
DEWITZ, J.: Zool. Jahrb. (Allgem.) **37**, 305—314 (1920).
DIAMOND, L. S.: Cryobiology **1**, 95—102 (1964).
DICK, J.: Ann. Appl. Biol. **24**, 762—796 (1937).
DICKER, S. E., ELLIOT, A. B.: J. Physiol. (London) **190**, 359—370 (1967).
DICKIE, L. M.: J. Fish. Res. Board Can. **15**, 1189—1211 (1958).
DIGBY, P. S. B.: J. Exp. Biol. **32**, 279—298 (1955), **35**, 1—19 (1958).
DIJKGRAAF, S.: Z. Vergleich. Physiol. **27**, 587—605 (1940), **30**, 252 (1943).
DIJKGRAAF, S., KALMIJN, A. J.: Z. Vergleich. Physiol. **53**, 187—194 (1966).
DILL, D. B., ADOLPH, E. F., WILBER, C. G. (Eds.): Adaptation to the environment. In: Handbook of physiology. 4, Am. Physiol. Soc. Washington, D. C. 1964.
DINGLE, A. D.: J. Cell Sci. **7**, 463 (1970).
DINGLE, H.: Am. Naturalist **102**, 149—163 (1968).
DINGLEY, F., SMITH, J. M.: J. Insect Physiol. **14**, 1185—1194 (1968).

DOANE, J. F.: Ann. Entomol. Soc. Am. **62**, 567—572 (1969)
DODT, E.: Pflügers Arch. Ges. Physiol. **260**, 225—238 (1955).
DOHANION, S. M.: J. Econom. Entomol. **35**, 406 (1942).
DOLE, J. W.: Copeia **1967**, 141—149.
DOLLEY, W. L., JR., WHITE, J. D.: Biol. Bull. **100**, 90—94 (1951).
DONDALE, C. D., LEGENDRE, R.: Compt. Rend. Acad. Sci. (Paris), Sér. D, **270**, 2483—2485 (1970).
DORNER, R. W., MÜLLER, M. S.: Ann. Entomol. Soc. Am. **55**, 36—39 (1962).
DORSETT, D. A.: J. Exp. Biol. **39**, 579—588 (1962).
DOUDOROFF, P.: Biol. Bull. **75**, 494—509 (1938), **83**, 213—244 (1942), **88**, 194—206 (1945).
DOWNES, J. A.: Ann. Rev. Entomol. **10**, 257—274 (1965).
DROPKIN, V. H.: Phytopathology **53**, 663—666 (1963), **59**, 1632—1637 (1969).
DRUGER, M.: Genetics **47**, 209—222 (1962), **56**, 39—47 (1967).
DRURY, D. E., EALES, J. G.: Can. J. Zool. **46**, 1—9 (1968).
DRUZHELYUBOVA, T. S., MAKAROVA, L. A.: Zool. Ž. **47**, 73—78 (1968).
DUBACH, P., SMITH, F., PRATT, J., STEWART, C. M.: Nature **184**, 288—289 (1959).
DUCHÂTEAU, G., FLORKIN, M.: Arch. Intern. Physiol. Biochim. **63**, 213—221 (1955).
DUDEL, J., RÜDEL, R.: Pflügers Arch. Ges. Physiol. **301**, 16—30 (1968).
DUFFIELD, R. M., NORDIN, J. H.: Nature **228**, 381 (1970).
DUMORTIER, B.: Ann. Epiphyties **18**, 387—400 (1967).
DUNHAM, J.: Physiol. Zoöl. **35**, 297—303 (1962).
DUNLAP, D. G.: Physiol. Zoöl. **41**, 432—439 (1968).
DUNLAP, D. G.: Copeia **1969**, 852—854.
DUNLAP, D. G.: Comp. Biochem. Physiol. **31**, 555—570 (1969a), **38**, 1—16 (1971).
DU PAUL, W. D., WEBB, K. L.: Comp. Biochem. Physiol. **32**, 785—801 (1970).
DUTRIEU, J.: Compt. Rend. Séanc. Acad. Sci. **253**, 3071—3073 (1961).
DWARAKANATH, S. K.: Comp. Biochem. Physiol. **38**, 351—358 (1971).
EDGREN, R. A.: Proc. Soc. Exp. Biol. Med. **87**, 20—23 (1954).
EDNEY, E. B.: The water relations of terrestrial arthropods. Cambridge (1957).
EDNEY, E. B.: Physiol. Zoöl. **37**, 364—377, 378—394 (1964).
EDNEY, E. B., BARRASS, R.: J. Insect Physiol. **28**, 469—481 (1962).
EDWARDS, D. K.: Can. J. Zool. **36**, 363—382 (1958).
E. I. F. A. C. (European Inland Fisheries Advisory Comission): Water Res. **3**, 645—662 (1969).
EKBERG, D. R.: Biol. Bull. **114**, 308—316 (1958).
EKBERG, D. R.: Comp. Biochem. Physiol. **5**, 123—128 (1962).
EKMAN, S.: Zoogeography of the sea. London 1953.
ELDRIDGE, B. F.: Science **151**, 826—828 (1966).
ELGER, R.: Oecologia **2**, 162—197 (1969).
ELLENBY, C., SMITH, L.: Comp. Biochem. Physiol. **21**, 51—57 (1967).
ELLINGSEN, I.-J.: Norsk Entomol. Tidsskr. **16**, 121—125 (1969).
ELLIOTT, C. G.: Heredity **9**, 385—398 (1955).
ELLIS, P. E.: Anim. Behav. **11**, 142—151 (1963).
ELLORY, J. C., SMITH, M. W.: Biochim. Biophys. Acta **193**, 137—145 (1969).
EL-MAGHRABY, A. M.: Crustaceana **8**, 37—47 (1965).
ELTRINGHAM, S. K.: J. Appl. Ecol. **4**, 521—529 (1967).
ELWI ABDEL HAMID, M.: Anz. Öst. Akad. Wiss., Math.-Nat. Kl. **1959**, 46—58.
EL-ZIADY, S.: Ann. Entomol. Soc. Am. **51**, 317—336 (1958).
EMEIS, D.: Z. Tierpsychol. **16**, 129—154 (1959).
EMME, A. M.: Dokl. Akad. Nauk SSSR., N.S. **88**, 381—384 (1953).
ENGELHARDT, W.: Z. Morph. Ökol. Tiere **54**, 219—392 (1964).
ENRIGHT, J. T.: Science **156**, 1510—1512 (1967).
ESCH, H.: Z. Vergl. Physiol. **43**, 305—335 (1960).
ESCH, H.: Z. Vergl. Physiol. **48**, 534—546, 547—551 (1964).
EVANS, A. A. F., FISHER, J. M.: Nematologia **16**, 113—122 (1970).
EVANS, E. E., COWLES, R. B.: Proc. Soc. Exp. Biol. Med. **101**, 482—483 (1959).

EVANS, R. M., PURDIE, F. C., HICKMAN, C. P., JR.: Can. J. Zool. **40**, 107—118 (1962).

EYRING, H., EYRING, E. M.: Can. J. Zool. **40**, 107—118 (1962).

EYRING, H., EYRING, E. M.: Modern chemical kinetics, pp. 1—114. New York: Reinhold Publ. Corp. 1963.

FAGERLUND, U. H. M.: Gen. Comp. Endocr. **8**, 197—207 (1967).

FAHRIG, R., SIEGER, M., ANDERS, F.: Zool. Anz. (Suppl. Bd) **31**, 565—578 (1968).

FANKHÄNEL, H.: Beitr. Entomol. **9**, 303—322 (1959).

FARKAS, T., HERODEK, S.: Acta Biol. Hung. **10**, 85—90 (1959).

FAULHABER, S. H.: Genetics **56**, 189—213 (1967).

FERGUSON, R. G.: J. Fish. Res. Board Can. **15**, 607—624 (1958).

FERRIS, J. M.: Phytopath. **47**, 221—230 (1957).

FILIPPONI, A., PASSARIELLO, S.: Riv. Parassitol. **30**, 295—310 (1969).

FINKE, C.: Z. Vergleich. Physiol. **58**, 398—422 (1968).

FISCHER, J., ROSIN, S.: Rev. Suisse Zool. **75**, 538—549 (1968).

FISCHER, K.: Z. Vergleich. Physiol. **61**, 394—419 (1968), **66**, 273—293 (1970).

FISHER, K. C., SULLIVAN, C. M.: Can. J. Zool. **36**, 49—63 (1958).

FLITTERS, N. E.: Ann. Entomol. Soc. Am. **61**, 36—38 (1968).

FLOREY, E.: Lehrbuch der Tierphysiologie. Stuttgart 1970.

FORCE, D. C.: J. Econom. Entomol. **60**, 1308—1311 (1967).

FORTUNE, P. Y.: Nature **178**, 98 (1956).

FORTUNE, P. Y.: J. Exp. Biol. **35**, 824—831 (1958).

FOULK, J. D.: J. Med. Entomol. **5**, 223—229 (1968).

FOX, H. M.: Bull. Soc. Zool. Fr. **80**, 288—298 (1955).

FOX, W., GORDON, C., FOX, M. H.: Zoologica **46**, 57—71 (1961).

FOXX, R. M., MARTIN, L.: Commun. Behav. Biol. A **5**, 85—87 (1970).

FRAENKEL, G.: Z. Vergleich. Physiol. **13**, 300—313 (1930).

FRAENKEL, G.: Ecology **42**, 604—606 (1961).

FRAENKEL, G., HOPF, H. S.: Biochem. J. **34**, 1085—1092 (1940).

FRAENKEL, G. S., GUNN, D. L.: The orientation of animals. New York 1961.

FRAENKEL, G. S., HERFORD, G. V. B.: J. Exp. Biol. **17**, 386—395 (1940).

FRAENKEL, J.: Compt. Rend. Trav. Lab. Carlsberg **33**, 1—52 (1962).

FRANCIA, F. C., GRAHAM, K.: Can. J. Zool. **45**, 985—1002 (1967).

FRANKEL, H. M., STEINBERG, G., GORDON, J.: Comp. Biochem. Physiol. **19**, 279—283 (1966).

FRANKHAUSEN, G.: Quart. Rev. Biol. **20**, 20—78 (1945).

FREE, J. B., SPENCER-BOOTH, Y.: J. Exp. Biol. **35**, 930—937 (1958).

FREE, J. B., SPENCER-BOOTH, Y.: Entomol. Exp. Appl. **3**, 222—230 (1960).

FREED, J.: Comp. Biochem. Physiol. **14**, 656—659 (1965).

FREEMAN, J. A.: Biol. Bull. **99**, 416—424 (1950).

FRENCH, J. W.: J. Exp. Psychol. **31**, 79—87 (1942).

FREWIN, N.: Diss. Univ. Dublin, Trinity College (1969).

FRIED, G. H., SCHREIBMAN, M. P., KALLMAN, K. D.: Comp. Biochem. Physiol. **28**, 771—776 (1969).

FRIEDLAND, B. L., HARNLY, M. H.: Biol. Bull. **88**, 247—253 (1945).

FRIEDRICH, L.: Kieler Meeresforsch. **23**, 105—126 (1967).

FRIES, G.: Z. Morph. Ökol. Tiere **53**, 475—516 (1964).

FRINGS, H., FRINGS, M.: J. Exp. Zool. **134**, 411—425 (1957).

FRISCH, K. v.: Tanzsprache und Orientierung der Bienen. Heidelberg 1965.

FRÖMMING, E.: Zool. Jahrb. (Ökol.) **84**, 577—602 (1956).

FROMM, P. O.: Physiol. Zoöl. **29**, 234—240 (1956).

FRY, F. E. J.: Univ. Toronto Stud. Biol. Ser. **55**, 1—62 (1947).

FRY, F. E. J.: Ann. Biol. **33**, 205—219 (1957).

FRY, F. E. J.: Ann. Rev. Physiol. **20**, 207—224 (1958).

FRY, F. E. J.: Proc. Indo-Pacific Fish. Coun. III, 37—42 (1958).

FRY, F. E. J., HART, J. S.: J. Fish. Res. Board Can. **7**, 169—175 (1948).

FULDNER, P., UBISCH, L. v.: Wilh. Roux Arch. **155**, 693—700 (1965).

FULTON, C.: J. Exp. Zool. **151**, 61—78 (1962).

FUSHTEY, S. G., JOHNSON, P. W.: Nematologica **12**, 313—320 (1966).
FUZEAU-BRAESCH, S.: Compt. Rend. Soc. Biol. **159**, 1048—1052 (1965).
FYE, R. E., PATANA, R., McADA, W. C.: J. Econom. Entomol. **62**, 1402—1405 (1969).
GÄRDEFORS, D.: Entomol. Exp. Appl. **7**, 71—84 (1964), **9**, 395—401 (1966).
GAHLENBECK, H., BARTELS, H.: Z. Vergleich. Physiol. **59**, 323—340 (1968).
GALLIARD, H., GOLVAN, Y. J.: Ann. Parasit. Hum. **32**, 563—579 (1957).
GALLOWAY, J. C.: Copeia **1941**, 118—119.
GALTON, V. A., INGBAR, S. H.: Endocrinology **70**, 622—632 (1962).
GALVAO, P. E., TARASANTCHI, J., GUERTZENSTEIN, P.: Am. J. Physiol. **209**, 501—506 (1965).
GAMBARO, P.: Bull. Zool. **21**, 163—169 (1954).
GAMBARO IVANCICH, P.: Arch. Zool. Ital. **42**, 207—219 (1958).
GANAROS, A. E.: Biol. Bull. **114**, 188—195 (1958).
GARNER, M. R.: Physiol. Zoöl. **7**, 408—434 (1934).
GARSIDE, E. T., TAIT, J. S.: Can. J. Zool. **36**, 563—567 (1958).
GATES, D. M.: Energy exchange in the biosphere. New York: 1962.
GATES, D. M.: Environ. Res. **3**, 132—144 (1970).
GAVIN, R. H.: J. Protozool. **12**, 307—318 (1965).
GEBHARDT, H.: Zool. Jahrb. (Physiol.) **63**, 558—592 (1953).
GEIGER, R.: Das Klima der bodennahen Luftschicht, 4. Ed. Braunschweig 1962.
GEILENKIRCHEN, W. L. H.: J. Embryol. Exp. Morph. **16**, 321—337 (1966).
GEILER, H.: Z. Morph. Ökol. Tiere **56**, 260—274 (1966).
GELINEO, S.: Bull. Acad. Serb. Sci., Arts Cl. Sci. Math.-Nat., Sci. Nat. (No. 11), **39**, 3—18 (1967), (No. 12), **44**, 1—7 (1969).
GELINEO, S., KOLENDIC, M.: Glas srp. Akad. Nauka (Prirod.-mat. nauka) **6**, (208), 19—23 (1953).
GERBERS, L.: Zool. Jahrb. (Physiol.) **67**, 373—406 (1957).
GERE, G.: Ann. Univ. Sci. Budapestinensis Rol. Eötvös Nom., Sect. Biol. **7**, 95—103 (1964).
GERLACH, S. A.: Proc. 5th marine biol. Symp. Botanica Gothoburgensia **3**, 81—92 (1965).
GHELELOVITCH, S.: Compt. Rend. Acad. Sci. (Paris) **237**, 1445—1447 (1953).
GIBBONS, W.: Can. J. Zool. **45**, 585 (1967).
GIBSON, J.: Experientia **25**, 1198—1199 (1969).
GIBSON, M. B.: Can. J. Zool. **32**, 393—407 (1954).
GINET, R., MATHIEU, J.: Ann. Spéléologie **23**, 425—440 (1968).
GLASER, O.: J. Gen. Physiol. **7**, 177—188 (1925).
GLICK, J. L., BRONK, J. R.: Biochim. Biophys. Acta **82**, 165—167 (1964).
GLUSHANKOVA, M. A., ZHIRMUNSKY, A. V., POLYANSKY, G. I., USHAKOV, B. P. (Ed.): Variability in cellular heat resistance of animals in ontogenesis and phylogenesis. Acad. Sci. USSR, Sci. Counc. Probl. Cytology (1967).
GODAN, D.: Nachr. Bl. Deut. Pflanzenschutzdienst (Berlin) (N. F.) **1**, 101—104 (1947).
GÖSSWALD, K.: Z. Angew. Entomol. **22**, 521—532 (1936).
GÖSSWALD, K., BIER, K. H.: Naturwissenschaften **42**, 133—134 (1955).
GOETSCH, W.: Biol. Zbl. **41**, 374—381 (1921), **42**, 231—240 (1922).
GOGALA, M., MICHIELI, S.: Biol. Vestnik. **14**, 83—90 (1966).
GOLDSCHMIDT, R. B.: Theoretical Genetics. California 1955.
GOLIK, Z., PIENKOWSKI, R. L.: Entomol. Exp. Appl. **12**, 133—138 (1969).
GOODON, E. J.: Med. Parasitol. 234—238 (1956).
GORDON, M. S.: J. Exp. Biol. **36**, 227—252 (1959).
GORDON, M. S.: Science **159**, 87—90 (1968).
GORDON, M. S., AMDUR, B. H., SCHOLANDER, P. F.: Biol. Bull. **122**, 52—62 (1962).
GORKHALI, C. P., BASIR, M. A.: Indian J. Helminth. **20**, 25—29 (1968).
GRAHAM, S. A.: J. Econom. Entomol. **17**, 377—383 (1924).
GRAHAM, W. M.: Can. J. Res. D **27**, 270—288 (1949).
GRAHAM, W. M.: Anim. Behav. **6**, 131—237 (1958).

GRAHAM, W. M., ONYEARU, A. K., WATERHOUSE, T. L.: Can. Entomol. **97**, 880—886 (1965).

GRAINGER, J. N. R.: Zool. Anz. **163**, 267—277 (1959).

GRAINGER, J. N. R.: Z. Wiss. Zool. **163**, 317—341 (1960).

GRAINGER, J. N. R.: Zool. Anz. (Suppl. Bd) **24**, 60—72 (1961).

GRAINGER, J. N. R.: Proc. Roy. Irish. Acad. **64**B, 25—32 (1964).

GRAINGER, J. N. R.: Zool. Anz. (Suppl. Bd) **32**, 479—487 (1969).

GRAINGER, J. N. R.: Comp. Biochem. Physiol. Biol. **29**, 665—670 (1969).

GRANT, W. C., JR.: Ecology **36**, 412—417 (1955).

GRAU, P. A., TERRIERE, L. C.: Ann. Entomol. Soc. Am. **60**, 549—552 (1967).

GREEN, C. D.: Ann. Appl. Biol. **54**, 381—390 (1964).

GREENE, F. C., FEENEY, R. E.: Biochim. Biophys. Acta **220**, 430—442 (1970).

GREGG, J. R., LØVTRUP, S.: Exp. Cell Res. **19**, 619—620 (1960).

GRESSITT, H. J. L. (Ed.): Entomology of Antarctica. 10, Antarctic Res. Ser. Am. Geophys. Union (1967).

GRIGG, G. C.: Australian J. Zool. **13**, 407—411 (1965).

GRIGG, G. C.: Comp. Biochem. Physiol. **23**, 139—148 (1967); **28**, 1203—1223 (1969).

GRISOLIA, S., JOYCE, B. K.: Biochem. Pharmakol. **3**, 167—168 (1960), cf. Med. Exp. **2**, 291—294 (1960).

GRODZINSKI, Z.: Bull. Acad. Sci. Cracovie, B II, 255—288 (1948).

GRODZINSKI, Z.: Zool. Pol. **6**, 187—208 (1955).

GROMADSKA, M.: Stud. Soc. Sci. Torun. E. (Zool.) **2**, 27—36 (1949).

GROMYSZ-KALKOWSKA, K.: Folia Biol. (Warsaw) **15**, 101—115 (1967).

GROMYSZ-KALKOWSKA, K., STOJALOWSKA, W.: Folia Biol. (Warsaw) **14**, 379—389 (1966).

GUENNELON, G.: Ann. Epiphyt. **17** (Hors-Ser. 2) 1—135 (1966).

GÜNZL, H.: Naturwissenschaften **46**, 337 (1959).

GUNN, D. L., HOPF, H. S.: J. Exp. Biol. **18**, 278—289 (1942).

GUNSTREAM, S. E., CHEW, R. M.: Ann. Entomol. Soc. Am. **60**, 434—439 (1967).

GUNTER, G.: Ecology **22**, 203—208 (1941).

GUNTER, G.: Science **106**, 472 (1947).

GUPPY, J. C.: Can. Entomol. **101**, 1320—1327 (1969).

GYSELS, H., BRACKE, E.: Natuurw. Tijdschr. **46**, 17—33 (1964).

HAARMANN, W.: Arch. Exp. Path. Pharm. **180**, 167—182 (1936).

HACKBARTH, W.: Z. Morph. Ökol. Tiere **35**, 469—534 (1939).

HADLEY, N. F.: J. Exp. Biol. **53**, 547—558 (1970).

HAFEZ, M.: Parasitology **40**, 215—236 (1950).

HAFEZ, M.: J. Exp. Zool. **124**, 199—225 (1953).

HAGELIN, L. O., STEFFNER, N.: Oikos **9**, 221—238 (1958).

HALBACH, U.: Naturwissenschaften **56**, 142—143 (1969).

HALBACH, U.: Oecologia **4**, 262—318, 176—207 (1970 a, b).

HALCROW, K.: Limnol. Oceanogr. **8**, 1—8 (1963).

HALCROW, K., BOYD, C. M.: Comp. Biochem. Physiol. **23**, 233—242 (1967).

HAMDORF, K.: Z. Vergl. Physiol. **44**, 523—549 (1961).

HANDEL, E. v.: J. Exp. Biol. **44**, 523—528 (1966), **46**, 487—490 (1967).

HANE, S., ROBERTSON, O. H., WEXLER, B. C., KRUPP, M. A.: Endocrinology **78**, 791—800 (1966).

HANEC, W.: J. Insect Physiol. **12**, 1443—1449 (1966).

HANEC, W., BECK, S. D.: J. Insect Physiol. **5**, 169—180 (1960).

HANEGAN, J. L., HEATH, J. E.: J. Exp. Biol. **53**, 349—362, 611—627, 629—632 (1970 a, b, c).

HANKS, J. E.: Biol. Bull. **112**, 330—335 (1957).

HANNON, J. P., VIERECK, E. (Eds.): Comparative physiology of temperature regulation. Arctic Aeromed. Lab., Fort Wainwright, Alaska (1962).

HANSEN, A. M., GARDNER, E. J.: Genetics **47**, 587—598 (1962).

HANSON, R. C., STANLEY, J. G.: Comp. Biochem. Physiol. **33**, 871—879 (1970).

HARRIES, F. H.: J. Agr. Res. **69**, 127—136 (1944).

HARRINGTON, R. W., JR.: Biol. Bull. **132**, 174—199 (1967).
HARRIS, P.: Can. Entomol. **94**, 774—780 (1962).
HART, J. S.: Univ. Toronto, Biol. Ser. **60**, Publ. Ont. Fish. Res. Lab. 1—79 (1952).
HARTMANN-GOLDSTEIN, I. J.: Gen. Res. **10**, 143—159 (1967).
HARTMANN-GOLDSTEIN, I. J., SPERLICH, D.: Genetics **48**, 863—869 (1963).
HARWOOD, R. F., TAKATA, N.: J. Insect Physiol. **11**, 711—716 (1965).
HASAN, R., QASIM, S. Z.: Proc. Indian Acad. Sci. (B) **53**, 230—239 (1961).
HASCHEMEYER, A. E. V.: Biol. Bull. **135**, 130—140 (1968).
HASCHEMEYER, A. E. V.: Comp. Biochem. Physiol. **28**, 535—552 (1969).
HASE, A.: Arb. Biol. Reichsanstalt **15**, 109—133 (1927).
HASE, A.: Z. Parasitenk. **2**, 368—418 (1930).
HASEGAWA, T., CHIBA, T.: Jap. J. Appl. Entomol. Zool. **13**, 124—128 (1969).
HAVSTEEN, B.: J. Biol. Chem. **242**, 769—771 (1967).
HAWKING, F., MOORE, P., GAMMAGE, K., WORMS, M. J.: Trans. Soc. Trop. Med. Hyg. **61**, 674—683 (1967).
HAZEL, J. R.: Life Sci. **8** (2), 775—784 (1969).
HAZEL, J., PROSSER, C. L.: Z. Vergleich. Physiol. **67**, 217—228 (1970).
HEADLEE, T. J.: J. Econom. Entomol. **7**, 413—417 (1914).
HEATH, A. G., PRITCHARD, A. W.: Physiol. Zoöl. **35**, 323—329 (1962).
HEATH, J. E.: Univ. Calif. Publ. Zool. **64**, 97—135 (1965).
HEATH, J. E.: J. Exp. Biol. **47**, 21—33 (1967).
HEATH, J. E.: Am. Midland. Naturalist **77**, 64—76 (1967), **80**, 440—448 (1968).
HEATH, J. E., ADAMS, P. A.: Nature **205**, 309—310 (1965).
HEATH, J. E., ADAMS, P. A.: J. Exp. Biol. **47**, 21—33 (1967).
HEATH, J. E., GASDORT, E., NORTHCUTT, G.: Comp. Biochem. Physiol. **26**, 509—518 (1968).
HEATH, J. E., WILKIN, P. J.: Physiol. Zoöl. **43**, 145—154 (1970).
HEATH, W. G.: Science **142**, 486—488 (1963).
HEATWOLE, H., DE AUSTIN, S. B., HERRERO, R.: Comp. Biochem. Physiol. **27**, 807—815 (1968).
HEBB, C., MORRIS, D., SMITH, M. W.: Comp. Biochem. Physiol. **28**, 29—36 (1969).
HECKROTTE, C.: Copeia **1967**, 759—763.
HEINICKE, E. A., HOUSTON. A. H.: J. Fish. Res. Board Can. **22**, 1455—1476 (1965).
HEINRICH, B.: Science **168**, 580—582 (1970).
HEINRICH, B., BARTHOLOMEW, G. A.: J. Exp. Biol. **55**, 223—239 (1971).
HEINZMANN, U.: Oecologia **5**, 19—55 (1970).
HEMPEL, G., BLAXTER, J. H. S.: Z. Naturforsch. **16** b, 227—228 (1961).
HENSEL, H.: Z. Vergleich. Physiol. **37**, 509—526 (1955).
HENSON, W. R.: Nature **179**, 637 (1957).
HENSON, W. R.: Yale J. Biol. Med. **33**, 128—132 (1960).
HERAN, H.: Z. Vergleich. Physiol. **34**, 179—206 (1952).
HERFS, A.: Zoologica **34**, 90 (1936).
HERFS, W.: Z. Angew. Entomol. **51**, 42—54 (1962).
HERFS, W.: Z. Pflanzenk. **70**, 73—81 (1963).
HERTER, K.: Zool. Bausteine I, 1. Berlin (1925).
HERTER, K.: Z. Vergleich. Physiol. **10**, 248—271 (1929), **28**, 358—388 (1940).
HERTER, K.: Z. Parasitenk. **12**, 552—591 (1942).
HERTER, K.: Zool. Anz. **148**, 139—155 (1952).
HERTER, K.: Der Temperatursinn der Insekten. Berlin 1953.
HERTER, K.: Der Temperatursinn der Tiere. Wittenberg 1962.
HERZ, M. J., PEEKE, H. V. S., WYERS, E. J.: Anim. Behav. **12**, 500—507 (1964).
HESSE, R., DOFLEIN, F.: Tierbau und Tierleben. 2. Ed. Jena 1943.
HEUSNER, A., STUSSI, T.: Insectes Sociaux **11**, 239—266 (1964).
HEUTS, M. J.: Publ. Staz. Zool. Napoli **28**, 44—61 (1956).
HEYDEMANN, B.: Entomol. Blätter **52**, 138—150 (1956).
HEYDEMANN, B.: Zool. Anz. (Suppl. Bd) **20**, 332—347 (1957).
HEYDEMANN, B.: Akad. Wiss. Lit., Math.-Naturw. Kl. **11**, 745—913 (1960).

HEYNEMANN, D.: Exp. Parasitol. **7**, 374—382 (1958).

HICKMAN, C. P., JR., MCNABB, R. A., NELSON, J. S., BREEMEN, E. D., VAN, COMFORT, D.: Can. J. Zool. **42**, 577—597 (1964).

HIDAKA, F., TAKAHASHI, H.: Ann. Zool. Jap. **40**, 200—204 (1967).

HILL, C. W., FROMM, P. O.: Gen. Comp. Endocr. **11**, 69—77 (1968).

HILLE, B., MILKMANN, R.: Biol. Bull. **131**, 346—361 (1966).

HILTERHAUS, V.: Z. Angew. Zool. **52**, 257—295 (1965).

HINTON, H. E.: Ann. Mag. Natur. Hist. **12**, 158—160 (1954).

HINTON, H. E.: J. Insect Physiol. **5**, 286—300 (1960a).

HINTON, H. E.: Nature **188**, 336—337 (1960b).

HINTON, H. E.: Proc. Roy. Soc. B **171**, 43—57 (1968).

HIRTH, H. F.: Ecol. Monogr. **33**, 83—112 (1963).

HJELM, K. K., ZEUTHEN, E.: Compt. Rend. Trav. Lab. Carlsberg **36**, 127—160 (1967).

HOAR, W. S.: Trans. Roy. Soc. Can. **49**, 25—34 (1955).

HOAR, W. S., COTTLE, M. K.: Can. J. Zool. **30**, 41—48, 49—54 (1952).

HOAR, W. S., EALES, J. G.: Can. J. Zool. **41**, 653—669 (1963).

HOAR, W. S., RANDALL, D. J. (Ed.): Fish physiology, I—IV, New York, London 1969.

HOAR, W. S., ROBERTSON, G. B.: Can. J. Zool. **37**, 419—428 (1959).

HOCHACHKA, P. W.: Gen. Comp. Endocr. **2**, 499—505 (1962).

HOCHACHKA, P. W.: Comp. Biochem. Physiol. **25**, 107—118 (1968a), **27**, 609—611 (1968b).

HOCHACHKA, P. W., HAYES, F. R.: Can. J. Zool. **40**, 261—270 (1962).

HOCHACHKA, P. W., LEWIS, J. K.: J. Biol. Chem. **245**, 6567—6573 (1970).

HOCHACHKA, P. W., SOMERO, G. N.: Comp. Biochem. Physiol. **27**, 659—668 (1968).

HOCKING, B., SHARPLIN, C. D.: Nature **206**, 215 (1965).

HODEK, I.: Čas. Čsl. Spol. Entomol. **59**, 297—313 (1962).

HODEK, I.: Acta Entomol. Bohem. **65**, 422—435 (1968).

HODSON, A. C.: Ecol. Mon. **7**, 271—315 (1937).

HODSON, A. C., ALRAWY, M. A.: Trans. X. Int. Congr. Entomol. **2**, 61—65 (1958).

HÖLLDOBLER, B.: Z. Vergleich. Physiol. **52**, 430—455 (1966).

HOFF, J. H. VAN'T: Etudes de dynamique chimique. Amsterdam 1884.

HOFFMANN, K.: Oecologia **3**, 184—206 (1969a).

HOFFMANN, K.: Z. Vergleich. Physiol. **58**, 225—228 (1968), **62**, 93—110 (1969b).

HOFFMANN, K.: Verhand. Deut. Zool. Ges. **64**, 266—273 (1970).

HOGAN, T. W.: Australian J. Biol. Sci. **13**, 527—540 (1960).

HOLLINGSWORTH, M. J.: Nature **218**, 869—870 (1968).

HOLLINGSWORTH, M. J.: Exp. Geront. **4**, 49—55 (1969).

HOLLINGSWORTH, M. J., BOWLER, K.: Exp. Geront. **1**, 251—257 (1966).

HOMP, R.: Z. Vergleich. Physiol. **26**, 1—34 (1939).

HOPF, H. S.: Biochem. J. **34**, 1396—1403 (1940).

HOPPENHEIT, M., LAUDIEN, H.: Zool. Anz. **182**, 303—311 (1969).

HOSOI, T.: Jap. J. Med. Sci. Biol. **7**, 129—134 (1954).

HOUSE, H. L.: Ann. Entomol. Soc. Am. **59**, 1263—1267 (1966).

HOUSE, H. L., RIORDAN, D. F., BARLOW, J. S.: Can. J. Zool. **36**, 629—632 (1958).

HOUSTON, A. H.: Can. J. Zool. **40**, 1169—1174 (1962).

HOUSTON, A. H., DE WILDE, M. A.: J. Exp. Biol. **49**, 71—81 (1968).

HOUSTON, A. H., DE WILDE, M. A.: Comp. Biochem. Physiol. **28**, 877—885 (1969).

HOUSTON, A. H., MADDEN, J. A.: Nature **217**, 969—970 (1968).

HOUSTON, A. H., MADDEN, J. A., DE WILDE, M. A.: Comp. Biochem. Physiol. **34**, 805—818 (1970).

HOUSTON, A. H., REAVES, R. S., MADDEN, J. A., DE WILDE, M. A.: Comp. Biochem. Physiol. **25**, 563—581 (1968).

HOWE, R. W.: Ann. Appl. Biol. **44**, 356—368 (1956), **50**, 649—660 (1962).

HOWLETT, F. M.: Parasitology **3**, 479—484 (1910).

HUBBS, C.: Copeia **1966**, 29—42.

HUDSON, J. W., BERTRAM, F. W.: Physiol. Zoöl. **39**, 21—29 (1966).

HUGGEL, H.: Z. Vergleich. Physiol. **42**, 63—102 (1959).

HUGGEL, H.: Rev. Suisse Zool. **59**, 242—247 (1951), **68**, 111—119 (1961).
HUGGEL, H., KLEINHAUS, A., HAMZEHPOUR, M.: Rev. Suisse Zool. **70**, 286—290 (1963).
HUGGINS, S. E., PERCOCO, R. A.: Proc. Soc. Exp. Biol. Med. **119**, 678—682 (1965).
HUGHES, G. M., ROBERTS, J. L.: J. Exp. Biol. **52**, 177—192 (1970).
HÜLSEN, J., ZERBST, E.: Pflügers Arch. Ges. Physiol. **279**, R 40 (1964).
HUNTER, A. S.: Comp. Biochem. Physiol. **11**, 411—417; **16**, 7—12; **19**, 171—177; **24**, 327—333 (1964—1968).
HUNTER, A. S., CEDIEL, N.: Comp. Biochem. Physiol. **37**, 243—249 (1970).
HUNTER-JONES, P.: Bull. Entomol. Res. **59**, 707—718 (1970).
HUNTSMAN, A. G.: J. Fish. Res. Board Can. **5**, 485—501 (1942).
HURPIN, B.: Ann. Inst. Nat. Res. Agron., C **6**, 529—534 (1955).
HURPIN, B.: J. Invert. Path. **10**, 252—262 (1968).
HUTCHINS, L. W.: Contr. No. **374** from Woods Hole Oceangr. Inst., 325—335 (1947).
HUTCHISON, V. H.: Physiol. Zoöl. **34**, 92—125 (1961).
HUTCHISON, V. H., KOSH, R. J.: Herpetologica **20**, 233—238 (1965).
HUTNER, S. H., BAKER, H., AARONSON, S., NATHAN, H. A., RODRIGUEZ, E., LOCKWOOD, S., SANDERS, M., PETERSEN, R. A.: J. Protozool. **4**, 259—269 (1957).
HWANG, S. W.: Nematologica **16**, 305—308 (1970).
HWANG, S., DAVIS, E. E., ALEXANDER, M. T.: Science **144**, 64—65 (1964).
HYDER, M.: Nature **224**, 1112 (1969).
IANSSEN, C. R.: Arch. Neerl. Zool. **13**, 500—510 (1960).
IDLER, D. R., SZEPLAKI, B. J.: J. Fish. Res. Board Can. **25**, 2549—2560 (1968).
IERSEL, J. J. A. VAN, BOL, A. C. A.: Behaviour **13**, 1—88 (1958).
ILYINSKAYA, N. B.: Věstn. Čsl. Společ. Zool. **32**, 217—222 (1968).
IRVINE, D. G., NEWMAN, K., HOAR, W. S.: Can. J. Zool. **35**, 691—709 (1957).
ISHAY, J., BYTINSKI-SALZ, H., SHULOV, A.: Israel J. Entomol. **2**, 45—106 (1967).
ISHAY, J., RUTTNER, F.: Z. Vergleich. Physiol. **72**, 423—434 (1971).
ISTENIC, L.: Raspr. Slov. Akad. Znan. Umet. — Diss. Acad. Scient. Art. Slov. (Cl. 4: Hist. Nat.) **7**, 201—236 (1963).
IVANOVIČ, J., JANKOVIČ, M., MARINKOVIČ, D., BARDIČ, F.: Ž. Evol. Biokhim. Fiziol. **4**, 342—347 (1968).
IWAO, S.: Mem. Coll. Agric. Kyoto Univ. **84**, 1—80 (1962).
JACOBS, J.: Zool. Anz. (Suppl. Bd) **30**, 290—296 (1967).
JACOBS, J.: Oecologia **5**, 96—126 (1970).
JACOBS, W.: Ergebn. Biol. **11**, 131—218 (1935).
JACOBSON, E. R., WHITFORD, W. G.: Comp. Biochem. Physiol. **35**, 439—449 (1970).
JACOBSON, L. A.: Can. Entomol. **94**, 889—892 (1962).
JAKOVLEV, V.: Zool. Anz. (Suppl. Bd) **24**, 92—96 (1961).
JAKOVLEV, V., KRÜGER, G.: Biol. Zbl. **73**, 633—650 (1954).
JAMES, T. W., READ, C. P.: Exp. Cell Res. **13**, 510—516 (1957).
JANDER, R.: Ann. Rev. Entomol. **8**, 95—114 (1963).
JANISCH, E.: Arch. Ges. Physiol. **209**, 414—436 (1925).
JANISCH, E.: Z. Morph. Ökol. Tiere **17**, 339—416 (1930), **22**, 287—348 (1931).
JANKOWSKY, H.-D.: Z. Vergleich. Physiol. **43**, 392—410 (1960).
JANKOWSKY, H.-D.: Zool. Anz. **172**, 233—239 (1964a).
JANKOWSKY, H.-D.: Helgoländer Wiss. Meeresuntersuch. **9**, 412—419 (1964b), **13**, 402—407 (1966), **18**, 317—362 (1968).
JANKOWSKY, H.-D.: Zool. Jahrb. (Physiol.) **73**, 251—260 (1967).
JANKOWSKY, H.-D.: Kieler Meeresforsch. **25**, 205—214 (1969).
JANKOWSKY, H.-D., KORN, H.: Naturwissenschaften **52**, 642—643 (1965).
JANKOWSKY, H.-D., LAUDIEN, H., PRECHT, H.: Marine Biol. **3**, 73—77 (1969).
JANSEN, P.: Forma et Functio **2**, 58—100 (1970).
JASIČ, J.: Proc. Conf. Sci. Probl. Plant Protect. (Budapest) **2**, 367—370 (1960).
JEAN, Y.: Fish. Res. Board Can. **21**, 429—460 (1964).
JEFFERSON, G. T.: Nature **156**, 111—112 (1945).
JENKINS, P. M.: Control of growth and metamorphosis. Oxford 1970.
JOB, S. V.: Proc. Indian Acad. Sci. **45**, 302—303 (1957).

JOB, S.V.: Marine Biol. 2, 121—126 (1969).
JÖHNSSEN, A.: Z. Angew. Entomol. 16, 87—158 (1930).
JOHANSEN, P.H.: Can. J. Zool. 45, 329—345 (1967), 46, 805—806 (1968).
JOHN, K.O.: J. Anim. Morphol. Physiol. 14, 131—139 (1967).
JOHNSON, C.G.: Entomol. Exp. Appl. 3, 238—240 (1960).
JOHNSON, D.S.: J. Animal Ecol. 21, 118—119 (1952).
JOHNSON, N.E.: Ann. Entomol. Soc. Am. 60, 199—204 (1967).
JOHNSTON, P.V., ROOTS, B.I.: Comp. Biochem. Physiol. 11, 303—309 (1964).
JOLLY, M.S., LEGAY, J.M.: Compt. Rend. Acad. Sci. (Paris) 248, 858—860 (1959).
JOLY, J.: Bull. Soc. Zool. Fr. 83, 128—131 (1958).
JONES, F.N.: J. Exp. Psychol. 35, 76—79 (1945).
JUNGREIS, A.M.: Comp. Biochem. Physiol. 24, 1—16 (1968).
JUNGREIS, A.M., HOOPER, A.B.: Comp. Biochem. Physiol. 26, 91—100 (1968), 32, 417—432, 433—444 (1970).
KÄHLER, H.H.: Marine Biol. 5, 315—324 (1970).
KAISER, F.: Zool. Jahrb. (Physiol.) 65, 59—90 (1954).
KAMMER, A.E.: J. Exp. Biol. 48, 88—109 (1968).
KAMMER, A.E.: Z. Vergleich. Physiol. 72, 364—369 (1971).
KANEHISA, K.: Jap. J. Appl. Entomol. Zool. 9, 301—302 (1965).
KANUNGO, M.S., PROSSER, C.L.: J. Cell Comp. Physiol. 54, 259—263. 265—274 (1959).
KANWISHER, J.W.: Biol. Bull. 116, 258—264 (1959); 130, 96—105 (1966).
KARANDINOS, M.G., AXTELL, R.C.: Ann. Entomol. Soc. Am. 60, 1055—1062, 1252—1255 (1967 a, b).
KARGER, W.: Pflügers Arch. Ges. Physiol. 274, 331—339 (1962).
KARLSON, P.: Deut. med. Wschr. 86, 668—672 (1961).
KASBOHM, P.: Helgoländer Wiss. Meeresuntersuch. 16, 157—178 (1967).
KATIYAR, K.P., LONG, W.H.: J. Econom. Entomol. 54, 285—287 (1961).
KAUFMANN, O.: Z. Morph. Ökol. Tiere 25, 353—361 (1932).
KAUFMANN, O.: Z. Angew. Entomol. 24, 185—386 (1939).
KAUFMANN, O.: Arb. Physiol. Angew. Entomol. 10, 105—117 (1943).
KAUSKOLEKAS, C.A., DECKER, G.C.: Ann. Entomol. Soc. Am. 59, 292—298 (1966).
KEIRANS, J.E., FAY, R.W.: Mosquito News 28, 338—341 (1968).
KEISTER, M., BUCK, J.: J. Insect Physiol. 7, 51—72 (1961).
KEIZ, G.: Naturwissenschaften 46, 499 (1959).
KEMP, F.D.: Anim. Behav. 17, 446—451 (1969).
KEMP, P., SMITH, M.W.: Biochem. J. 117, 9—15 (1970).
KENNEDY, J.S.: Trans. Roy. Entomol. Soc. Lond. 89, 385—540 (1939).
KENNEY, J.W., RICHARDS, A.G.: Entomol. News 66, 29—36 (1955).
KERKUT, G.A., RIDGE, A.P.: Comp. Biochem. Physiol. 3, 64—70 (1961).
KERKUT, G.A., TAYLOR, B.J.R.: Behaviour 13, 259—279 (1958).
KHAN, M.F.: Proc. Irish Acad., B 64, 117—130 (1965).
KIECHLE, H.: Z. Vergleich. Physiol. 45, 154—192 (1961).
KING, E.W., RILEY, R.C.: Ann. Entomol. Soc. Am. 53, 591—595 (1960).
KINNE, O.: Z. Wiss. Zool. 157, 427—491 (1953).
KINNE, O.: Zool. Anz. 152, 10—16 (1954).
KINNE, O.: Z. Morph. Ökol. Tiere 45, 217—249 (1956).
KINNE, O.: Zool. Jahrb. (Physiol.) 67, 407—486 (1958).
KINNE, O.: Veröffl. Inst. Meeresforsch. Bremerhaven 6, 177—202 (1959).
KINNE, O.: Physiol. Zoöl. 33, 288—317 (1960).
KINNE, O.: Oceanogr. Marine Biol. A. Rev. 1, 301—340 (1963), 2, 281—339 (1964a).
KINNE, O.: Helgoländer Wiss. Meeresuntersuch. 9, 433—458 (1964b), 11, 131—156 (1965).
KINNE, O., KINNE, E.M.: Can. J. Zool. 40, 231—253 (1962).
KINNE, O., PAFFENHÖFER, G.-A.: Helgoländer Wiss. Meeresuntersuch. 12, 329—341 (1965), 13, 62—72 (1966).
KIRBERGER, CH.: Z. Vergleich. Physiol. 35, 175—198 (1953).
KIRCHNER, W., KESTNER, P.: J. Insect Physiol. 15, 41—53 (1969).

KISHINO, K.: Jap. J. Appl. Entomol. Zool. **13**, 52—60 (1969).
KITCHING, J. A.: J. Exp. Biol. **31**, 68—75 (1954).
KITTLAUS, E.: Deut. Entomol. Z. N. F. **8**, 41—62 (1961).
KLICKA, J.: Physiol. Zoöl. **38**, 177—189 (1965).
KLINGLER, J., LENGWEILER-REY, V.: Z. Pflanzenkr. **76**, 193—208 (1969).
KLINKEL, H.: Z. Angew. Entomol. **37**, 401—436 (1955).
KNEITZ, G.: Insectes Sociaux **13**, 285—296 (1966).
KNIPLING, E. B., SULLIVAN, W. N.: J. Econom. Entomol. **50**, 368—369 (1957).
KNIPPRATH, W. G., MEAD, J. F.: Lipids **1**, 113—117 (1966), **3**, 121—128 (1968).
KOEHLER, O.: Zool. Anz. (Suppl. Bd) **7**, 74—84 (1934).
KOMAI, T.: Adv. Gen. (Ed.: DEMEREC, M.) Vol. 8, New York 1956.
KOMATSU, S. K., FEENEY, R. E.: Biochim. Biophys. Acta **206**, 305—315 (1970).
KOMATSU, S. K., MILLER, H. T., DE VRIES, A. L., OSUGA, D. T., FEENEY, R. E.: Comp. Physiol. **32**, 519—527 (1970).
KONISHI, J., HICKMAN, C. P., JR.: Comp. Biochem. Physiol. **13**, 433—442 (1964).
KORRINGA, P.: Année Biol. 3, **33**, 1—17 (1957).
KOSH, R. J., HUTCHISON, V. H.: Copeia **1968**, 224—246.
KOSKINEN, R.: Ann. Zool. Fennicae **6**, 145—149 (1969).
KOVALJEVA, N. E., SELIVANOVA, G. V.: Citologija **5**, 273—278 (1963).
KOZHANTSHIKOV, I. W.: Bull. Entomol. Res. **29**, 253—262 (1938).
KOZHANTSHIKOV, I. W.: Rev. Appl. Entomol. **35**, 342 (1947).
KOZHANTSHIKOV, I. W.: Dokl. Akad. Nauk SSSR, N. S. **103**, 517—519 (1955).
KOZHANTSHIKOV, I. W., MASLOWA, E.: Zool. Jahrb. (Physiol.) **55**, 219—230 (1935).
KRAKAUER, T.: Comp. Biochem. Physiol. **33**, 15—26 (1970).
KRAMER, G.: Z. Morph. Ökol. Tiere **32**, 752—783 (1937).
KREHAN, J.: Oecologia **6**, 58—105 (1970).
KREKORIAN, C. O'NEIL, VANCE, V. J., RICHARDSON, A. M.: Anim. Behav. **16**, 429—436 (1968).
KRIJGSMAN, B. J.: Z. Vergleich. Physiol. **11**, 702—729 (1930).
KROGERUS, R.: Verh. 7. intern. Kongr. Ent. **2**, 1215—1231 (1939).
KRÜGER, F.: Zool. Anz. (Suppl. Bd) **16**, 263—267 (1952).
KRÜGER, F.: Helgoländer Wiss. Meeresuntersuch. **8**, 333—356 (1963), **14**, 302—325 (1966).
KRÜGER, F.: Marine Biol. **5**, 145—153 (1970).
KRÜGER, G.: Zool. Anz. (Suppl. Bd) **24**, 80—83 (1961).
KRÜGER, G.: Z. Wiss. Zool. **167**, 87—104 (1962).
KÜHLMORGEN-HILLE, G.: Ann. Biol. **20**, 98—99 (1963).
KÜHN, A.: Die Orientierung der Tiere im Raum, Jena 1919.
KÜHN, A.: Nachr. Ges. Wiss. Göttingen, Math.-Phys. Kl. 120—141 (1927).
KÜHNAU, J.: Naturwiss. Rdsch. **17**, 465—467 (1964).
KÜNNEMANN, H., LAUDIEN, H., PRECHT, H.: Marine Biol. **7**, 71—81 (1970).
KUSAKINA, A. A.: Federation Proc. Transl. **22**, T 123—126 (1963).
KUSNETZOVA, Y. I.: Zool. Ž. **48**, 1349—1357 (1969).
KUTHALINGAM, M. D. K.: Curr. Sci. **28**, 75—76 (1959).
KUTSCH, W.: Z. Vergleich. Physiol. **63**, 335—378 (1969).
KVELLAND, I.: Hereditas **54**, 88—100 (1965).
LAARMAN, I. I.: Trop. Geogr. Med. **10**, 293—305 (1958).
LÁBOS, E., LUKACSOVICS, F.: Ann. Inst. Biol. (Tihany), Hung. Acad. Sci. **35**, 13—24 (1968).
LACHER, V.: Z. Vergleich. Physiol. **48**, 587—623 (1964).
LA FRANCE, J.: Can. Entomol. **100**, 801—807 (1968).
LAGERSPETZ, K. Y. H., DUBITSCHER, I.: Comp. Biochem. Physiol. **17**, 665—671 (1966).
LAGERSPETZ, K. Y. H., IMPIVAARA, H., SENIUS, K.: Comp. Gen. Pharmacol. **1**, 236—240 (1970).
LAGERSPETZ, K. Y. H., TIRRI, R.: Ann. Zool. Fennicae **5**, 396—400 (1968).
LAGET, P., GUÉRIN, J., VANNIER, J.: Compt. Rend. Soc. Biol. **149**, 2160—2162 (1955).
LAKE, P. S.: Hydrobiologia **33**, 342—351 (1969).

LAMB, M. J.: Nature **220**, 808—809 (1968).
LANG, C. A.: J. Insect Physiol. **9**, 279—286 (1963).
LANG, J.: Biol. Zbl. **52**, 582—584 (1932).
LANKIEWICZ, Z.: Folia Biol. (Warsaw) **12**, 95—140 (1964).
LAPIERRE, J., HIEN, T. V.: Comp. Rend. Soc. Biol. (Paris) **162**, 622—625 (1968).
LARCZENKO, K. J.: Roczn. Nauk. Poln. A. **78**, 27—41 (1958).
LARIMER, J. L.: Comp. Biochem. Physiol. **22**, 683—700 (1967).
LARSEN, E. B.: Biol. Medd. Dansk. Vidensk. Selsk. **19**, 1—52 (1943).
LARSEN, E. B.: Det. Dansk. Vidensk. Sel. Biol. Medd. XIX, **3**, 1—49 (1943/46).
LARSEN, E. B.: Oikos **1**, 186—207 (1949).
LARSEN, E. B., THOMSEN, M.: Vid. Medd. Dansk. Nat. For. **104**, 1—25 (1940).
LASHBROOK, M. K., LIVEZEY, R. L.: Physiol. Zoöl. **43**, 38—46 (1970).
LASKER, R.: Copeia **1964**, 399—405 .
LAUDANI, U.: Boll. Zool. **32**, 751—758 (1965).
LAUDIEN, H.: Zool. Anz. **182**, 311—322 (1969).
LAUDIEN, H.: Z. Tierpsychol. **27**, 136—149 (1970).
LAUGÉ, G.: Bull. Soc. Zool. Fr. **91**, 661—686 (1966).
LAUGÉ, G.: Compt. Rend. Soc. Biol. (Paris) **161**, 16—21 (1967).
LAUGÉ, G.: Bull. Soc. Zool. Fr. **94**, 341—362 (1969).
LAVIE, P., ROTH, M.: Physiol. Comp. Oecol. **3**, 57—62 (1953).
LE BERRE, J.-R.: Bull. Biol. **87**, 227—241 (1953).
LECREN, E. D.: J. Animal. Ecol. **27**, 287—334 (1958).
LECYK, M.: Zool. Pol. **15**, 101—110 (1965).
LEE, A. K., BADHAM, J. A.: Copeia **1963**, 387—394.
LEE, M. O. (Ed.): Biology in the Antarctic Seas. Am. Geophys. Union, Wash. D.C. **1, 2** (1964/65).
LEES, A. D.: Ann. Appl. Biol. **40**, 449—486 (1953).
LEES, A. D.: The physiology of diapause in arthropods. London 1955.
LEES, A. D.: J. Insect Physiol. **3**, 92—117 (1959), **9**, 153—164 (1963).
LEGAY, J. M.: Ann. Inst. Natl. Rech. Agron. C, **10**, 409—421 (1959).
LEHMANN, D. L.: J. Protozool. **9**, 325—326 (1962).
LEHMANN, J.: Int. Rev. Ges. Hydrobiol. **3**, 413—429 (1970a), **5**, 763—781 (1970b).
LEICHT, R.: Marine Biol. **3**, 28—45 (1969).
LENT, C. M.: Biol. Bull. **134**, 60—73 (1968).
LENSKY, Y.: J. Insect Physiol. **10**, 1—12 (1964).
LEONARD, D. E.: Can. Entomol. **102**, 239—249 (1970).
LEUTHOLD, W.: Vierteljahresschr. Naturforsch. Ges. Zürich **107**, 147—154 (1962).
LEVINE, N. D., ANDERSEN, F. L.: J. Protozool. **13**, 199—202 (1966).
LEVINS, R.: Am. Naturalist **103**, 483—499 (1969).
LEWIS, H. W.: Genetics **45**, 1217—1231 (1960).
LICHT, P.: Comp. Biochem. Physiol. **13**, 27—34 (1964).
LICHT, P.: Physiol. Zoöl. **38**, 129—137, 252—257 (1965).
LICHT, P.: Am. Midland Naturalist **79**, 149—158 (1968).
LICHT, P., BASU, S. L.: Nature **213**, 672—674 (1967).
LICHT, P., BROWN, A. G.: Ecology **48**, 598—611 (1967).
LICHT, P., DAWSON, W. R., SHOEMAKER, V. H.: Copeia **1966**, 162—169.
LICHT, P., DAWSON, W. R., SHOEMAKER, V. H.: Z. Vergleich. Physiol. **65**, 1—14 (1969).
LICHT, P., DAWSON, W. R., SHOEMAKER, H., MAIN, A. R.: Copeia **1966**, 97—110.
LIKVENTOV, A. V.: Zool. Ž. **39**, 53—62 (1960).
LILLELUND, K.: Arch. Fisch. **17**, 95—113 (1967).
LILLYWHITE, H. B.: Copeia **1970**, 158—168.
LIN, R. K., WALFORD, R. L.: Nature **212**, 1277—1278 (1966).
LINDAU, G.: Z. Morph. Ökol. Tiere **47**, 489—528 (1958).
LINDAUER, M.: Z. Vergleich. Physiol. **36**, 391—432 (1954).
LINDAUER, M.: Fortschr. Zool. **16**, 58—140 (1964).
LINDSEY, C. C.: Can. J. Zool. **32**, 87—98 (1954), **40**, 1237—1247 (1962).
LINDSEY, C. C.: Nature **209**, 1152—1153 (1966).

LINDSEY, C. C., ALI, M. Y.: Can. J. Zool. **43**, 99—104 (1965).
LOCK, A. R., McLAREN, I. A.: Limnol. Oceanogr. **15**, 638—640 (1970).
LOCKER, A.: Z. Vergleich. Physiol. **41**, 249—266 (1958a).
LOCKER, A.: Experientia **14**, 407 (1958b).
LOCKER, A.: Biol. Zbl. **78**, 383—390 (1959).
LOCKER, A.: Pflügers Arch. Ges. Physiol. **275**, 238—255 (1962).
LOCKER, A., WEISH, P.: Zool. Anz. (Suppl. Bd) **28**, 365—378 (1965).
LOCKWOOD, A. P. M.: J. Exp. Biol. **37**, 614—630 (1960).
LOEB, H., WASTENEYS, H.: Z. Exp. Zool. **12**, 543—557 (1912).
LÖBBECKE, E.-A., OLTMANNS, O.: Z. Vererbungslehre **92**, 246—251 (1961).
LÖRCHER, K.: Naturwissenschaften **53**, 559—560 (1966), cf. Oecologia **3**, 84—124 (1969).
LOFTS, B., PICKFORD, G. E., ATZ, J. W.: Biol. Bull. **134**, 74—86 (1968).
LOFTUS, R.: Z. Vergleich. Physiol. **52**, 380—385 (1966), **59**, 413—455 (1968), **63**, 415—433 (1969).
LOH, D. V., GOLENHOFEN, K.: Pflügers Arch. Ges. Physiol. **318**, 35—50 (1970).
LOHMANN, M.: Z. Vergleich. Physiol. **49**, 341—389 (1964).
LOHMANN, M.: Z. Vergleich. Physiol. **55**, 307—332 (1967).
LØNNING, S.: Arkiv Zool. **2**, **12**, 359—381 (1959).
LOOSANOFF, V. L.: Biol. Bull. **114**, 57—70 (1958).
LORENZ, K.: Naturwissenschaften **54**, 377—388 (1967).
LOVE, G. J., GOODWIN, M. H.: Ecology **40**, 198—205 (1959).
LOVE, R. M.: The chemical biology of fishes. London-New York 1970.
LØVTRUP, S.: Compt. Rend. Trav. Lab. Carlsberg Ser., Chim. **28**, 400—425 (1953).
LØVTRUP, S.: J. Exp. Zool. **140**, 383—394 (1959), **147**, 227—232 (1961).
LOWE, C. H., HEATH, W. G.: Physiol. Zoöl. **42**, 53—59 (1969).
LOWE, C. H., JR., VANCE, V. J.: Science **122**, 73—74 (1955).
LOWER, W. R., WILLETT, J. D., HANSEN, E. L.: Comp. Biochem. Physiol. **34**, 473—479 (1970).
LOZINA-LOZINSKIJ, L. K.: Dokl. Akad. Nauk SSSR, N. S. **93**, 369—372 (1953).
LUCAS, E. A., REYNOLDS, W. A.: Physiol. Zoöl. **40**, 159—171 (1967).
LUDWIG, D.: Physiol. Zoöl. **18**, 103—135 (1945).
LUDWIG, D.: Ann. Entomol. Soc. Am. **49**, 12—15 (1956).
LUDWIG, D., CABLE, R. M.: Physiol. Zoöl. **6**, 493—508 (1933).
LÜSCHER, M.: Sci. Am. **205**, 138—145 (1961).
LUKOSCHUS, F.: Insectes Sociaux **3**, 185—193 (1956).
LUMPER, L.: HOPPE-SEYLER/THIERFELDER, Analyse, 10. Aufl., Bd. VI/A, 15—55 (1964).
LUSIS, O.: Proc. Malac. Soc. Lond. **37**, 19—26 (1966).
LYNN, W. G., McCORMICK, J. J., GREGOREK, J. C.: Gen. Comp. Endocr. **5**, 587—595 (1965).
MACKAY, W. C., BEATTY, D. D.: Comp. Biochem. Physiol. **26**, 235—245 (1968).
MACKENZIE, C. L., JR.: Ecology **42**, 317—338 (1961).
MADARIAGA, M. A., MUNICO, A. M., RIBERA, A.: Comp. Biochem. Physiol. **35**, 63—68 (1970).
MADDWELL, S. H. P.: J. Exp. Biol. **41**, 163—176 (1964).
MADGE, D. S.: Nature **190**, 106—107 (1961).
MADGE, D. S.: Entomol. Exp. Appl. **8**, 135—152 (1965).
MADGE, D. S.: Pedobiologia **9**, 188—214 (1969).
MAELZER, D. A.: J. Theoret. Biol. **8**, 141—162 (1965).
MAHER, M. J.: Endocrinology **74**, 994—995 (1964).
MAHER, M. J.: Gen. Comp. Endocr. **5**, 320—325 (1965).
MAHER, M. J.: Copeia **1967**, 361—365.
MAHON, E. F., HOAR, W. S., TABATA, S.: Can. J. Zool. **4**, 449—464 (1962).
MAHONEY, J. J., HUTCHISON, V. H.: Oecologia **2**, 143—161 (1969).
MAKAREWICZ, W.: J. Marine Biol. Ass. U. K. **48**, 535—542 (1968).
MAKINGS, P.: J. Exp. Biol. **41**, 473—497 (1964), **48**, 247—263 (1968).

MAKSIMOVIĆ, M.: Inst. Biol. Monogr. **3**, 1—115, Belgrad 1958.

MALESSA, P.: Mar. Biol. **2**, 143—158 (1969).

MALOGOLOWKIN, C.: Am. Naturalist **93**, 365—368 (1959).

MANGAT, B. S., APPLE, J. W.: J. Econom. Entomol. **59**, 1005—1006 (1966).

MANGE, E. J.: Genetics **58**, 399—413 (1968).

MAREK, M.: Comp. Biochem. Physiol. **35**, 615—622 (1970).

MARGALEF, R.: P. Inst. Biol. Appl. **19**, 13—94 (1955).

MARKKULA, M., PULLIAINEN, E.: Ann. Entomol. Fennicae. **31**, 39—45 (1965).

MARKKULA, M., ROIVAINEN, S.: Ann. Entomol. Fennicae. **27**, 30—45 (1961).

MARONI, A.: Boll. Zool. **28**, 441—448 (1961).

MARR, J.: Discovery Repts. **32**, 433—464 (1962).

MARTIN, W. R., JEAN, Y.: J. Fish. Res. Board Can. **21**, 215—238 (1964).

MARTOF, B. S.: Physiol. Zoöl. **35**, 38—46 (1962).

MARUSIC, E., MARTINEZ, R., TORETTI, J.: Proc. Soc. Exp. Biol. Med. **122**, 164—167 (1966).

MARZUSCH, R.: Z. Vergl. Physiol. **34**, 75—92 (1952).

MASAKI, S.: Kontyû (Tokyo) **30**, 9—16 (1962).

MASAKI, S., SAKAI, T.: Jap. J. Appl. Entomol. Zool. **9**, 191—205 (1965).

MASSEY, B. D., SMITH, C. L.: Comp. Biochem. Physiol. **25**, 241—255 (1968).

MATHIEU, J.: Bull. Soc. Zool. Fr. **93**, 595—603 (1968).

MATUTANI, K.: Publ. Sato Marine Biol. Lab. **9**, 379—411 (1961).

MATUTANI, K.: Physiol. Ecol. (Japan) **10**, 59—62, 63—67 (1962).

MAXWELL, C. W., PARSONS, E. C.: J. Econom. Entomol. **62**, 1310—1313 (1969).

MAYER, A. G.: Pap. Tortugas Lab. **6**, 3—24 (1914).

MAZEK-FIALLA, K.: Z. Wiss. Zool. **154**, 170—246 (1941).

MAZUR, P.: Science **168**, 939—949 (1970).

McCAULEY, R. W.: Can. J. Zool. **36**, 655—662 (1958).

McCAULEY, R. W.: J. Fish. Res. Board Can. **20**, 483—490 (1963), **25**, 1983—1986 (1968).

McCAULEY, R. W., TAIT, J. S.: J. Fish. Res. Board Can. **27**, 1729—1733 (1970).

McCRACKEN, F. D.: J. Fish. Res. Board Can. **20**, 551—586 (1963).

McERLEAN, A. J., BRINKLEY, H. J.: J. Fish. Biol. **3**, 97—114 (1971).

McFARLAND, W. N., PICKENS, P. E.: Can. J. Zool. **43**, 571—585 (1965).

McGINNIS, S. M., DICKSON, L. L.: Science **156**, 1757—1759 (1967).

McLAREN, J. A.: J. Fish. Res. Board Can. **20**, 685—727 (1963).

McLAREN, J. A.: J. Gen. Physiol. **48**, 1071—1079 (1965).

McLAUGHLIN, R. E.: J. Insect Path. **4**, 279—284, 344—352 (1962a, b).

McLEESE, D. W.: J. Fish. Res. Board Can. **13**, 247—272 (1956); **22**, 385—394 (1965).

McLEESE, D. W., WILDER, D. G.: J. Fish. Res. Board Can. **15**, 1345—1354 (1958).

McMULLEN, R. D.: Can. Entomol. **99**, 578—586 (1967).

McNAPP, R. A., PICKFORD, G. E.: Comp. Biochem. Physiol. **33**, 783—792 (1970).

McWHINNIE, D. J., CORTELYOU, J. R.: Am. Zool. **7**, 857—868, cf. 843—855 (1967).

MEINCKE, K. F.: Marine Biol. **6**, 281—290 (1970).

MEISNER, H. M., HICKMAN, C. P., JR.: Can. J. Zool. **40**, 127—130 (1962).

MELLANBY, K.: Parasitology **30**, 392—402 (1938).

MELLANBY, K.: Proc. Roy. Soc., B. **127**, 473—489 (1939).

MELLANBY, K.: J. Animal Ecol. **9**, 296—301 (1940).

MELLANBY, K.: Nature **173**, 582—583 (1954), **181**, 1403 (1958).

MELLANBY, K.: Entomol. Exp. Appl. **1**, 153—160 (1958).

MELLANBY, K.: Adv. Sci. **1959**, 409—417.

MELLANBY, K.: Bull Entomol. Res. **50**, 821—823 (1960).

MELLANBY, K., JOHNSON, C. G., BARTLEY, W. C., BROWN, P.: Bull. Entomol. Res. **33**, 267—271 (1942).

MENDES, L. O. T.: Bragantia (Sao Paulo) **15**, 43—53 (1956).

MERKER, E.: Die ökologischen Ursachen der Massenvermehrung des großen Fichtenborkenkäfers in Südwestdeutschland. Freiburg 1957.

MERKER, E., ADLUNG, K. G.: Naturwissenschaften **44**, 122—123 (1957).

MERKER, E., ADLUNG, K. G.: Zool. Jahrb. (Physiol.) **68**, 325—334 (1958).

MERKER, E., WILD, M.: Beitr. Entomol. **4**, 451—468 (1954).
MERYMAN, M. T. (Ed.): Cryobiology. London-New York 1966.
MESKE, C.: Z. Vergleich. Physiol. **45**, 61—77 (1961).
MESSENGER, P. S.: J. Econom. Entomol. **57**, 71—76 (1964).
MESSENGER, P. S., FLITTERS, N. E.: Ann. Entomol. Soc. Am. **52**, 191—204 (1959).
METZLER, S.: Naturwissenschaften **42**, 518 (1955).
METZLER, S.: Zool. Jahrb. (Physiol.) **67**, 81—110 (1957).
MEUDEC, M.: Compt. Rend. Soc. Biol. (Paris) **160**, 1693—1696 (1967).
MEUWIS, A. L., HEUTS, M. J.: Biol. Bull. **112**, 97—107 (1957).
MEWS, H.-H.: Z. Vergleich. Physiol. **40**, 345—355, 356—362 (1957).
MEYER, D.: Rev. Suisse Zool. **76**, 93—141 (1969).
MEYER, H., PREISS, B., BAUER, S.: Biochem. J. **76**, 27—35 (1960).
MICHAILOW, A. S.: Arch. Bienenk. **8**, 1—15 (1927).
MICHELSEN, A.: Oikos **11**, 250—264 (1960).
MILKMAN, R.: J. Gen. Physiol. **45**, 777—799 (1962).
MILKMAN, R. D.: Genetics **51**, 87—96 (1965).
MILKMAN, R., HILLE, B.: Biol. Bull. **131**, 331—345 (1966).
MILLER, D. A., STANDISH, M. L., THURMAN, A. E., JR.: Physiol. Zoöl. **41**, 500—506 (1968).
MILLER, D. C., VERNBERG, F. J.: Am. Zool. **8**, 459—463 (1968).
MILLER, L. K.: Science **166**, 105—106 (1969).
MILLER, L. K., DEHLINGER, P. J.: Comp. Biochem. Physiol. **28**, 915—921 (1969).
MILNE, L., MILNE, M.: Die Sinneswelt der Tiere und des Menschen. Hamburg-Berlin 1963.
MILSTEAD, W. W. (Ed.): Lizard ecology, a symposium. Kansas City 1967.
MINAMORI, S.: J. Sci., Hiroshima Univ., B 1, **17**, 65—119 (1957).
MINDE, J. F., CHESNEK, S. I.: Zool. Ž. **49**, 855—861 (1970).
MISSONNIER, J.: Compt. Rend. Acad. Sci. (Paris) **251**, 1424—1426 (1960).
MOBERLY, W. R.: Comp. Biochem. Physiol. **27**, 1—20 (1968).
MOELLER, J.: Arch. Hydrobiol. **60**, 358—365 (1964).
MONAKOV, A. V.: Zool. Ž. **44**, 606—608 (1965).
MOORE, A. M.: Biol. Bull. **117**, 150—153 (1959).
MORAN, V. C., EWER, D. C.: J. Insect Physiol. **12**, 457—463 (1966).
MORAWSKA, B., ZALESKA, A.: Acta Biol. Exp., Vars. **18**, 39—53 (1958).
MOREIRA, G. S., VERNBERG, W. B.: Marine Biol. **1**, 282—284 (1968).
MORGAN, A. H.: J. Exp. Zool. **35**, 83—114 (1922).
MORGAREIDGE, K. R., WHITE, F. N.: Nature **223**, 587—591 (1969).
MORIARTY, F.: J. Insect Physiol. **3**, 357—366 (1959).
MORRIS, R. W.: Am. Naturalist **96**, 35—50 (1962).
MORRIS, R. W.: Physiol. Zoöl. **34**, 217—227 (1961), **38**, 219—227 (1965), **40**, 409—423 (1967).
MORSE, R. A.: J. Econom. Entomol. **59**, 1091—1093 (1967).
MORTIMER, C. H.: Zool. Jahrb. (Allgem.) **56**, 323—388 (1936).
MOSBACHER, G. C.: Zool. Anz. (Suppl. Bd) **30**, 509—521 (1967).
MOUNTFORD, M. D.: Nature **211**, 993—994 (1966).
MROSOVSKY, N.: Nature **220**, 1338—1339 (1968).
MÜLLER, H. J.: Fortschr. Zool. **16**, 500—523 (1963).
MÜLLER, H. J.: Nova Acta Leopoldina, N. F., **191**, 1—27 (1970).
MÜLLER, H. J., REINHARDT, R.: Entomol. Ber. **1969**, 93—100.
MÜLLER, J.: Z. Vererbungsl. **94**, 101—111 (1963).
MUKHERJI, M.: Acta Histochem. **29**, 297—303 (1968).
MURRAY, R. W.: J. Exp. Biol. **33**, 798—805 (1956).
MURRAY, R. W. In: LOWENSTEIN, O. (Ed.): Adv. Comp. Physiol. Biochem. **1**, 117—175 (1962).
MURRAY, R. W.: Comp. Biochem. Physiol. **18**, 291—303 (1966).
MURRISH, D. E., VANCE, V. J.: Comp. Biochem. Physiol. **27**, 329—337 (1968).
MUSGRAVE, A. J., ASHTON, G. C., HOMAN, R.: Can. J. Zool. **41**, 1245—1261 (1963).

MUTCHMOR, J. A., RICHARDS, A. G.: J. Insect Physiol. **7**, 141—158 (1961).
MUTO, Y.: J. Sci., Hiroshima Univ., B **1**, 17, 143—199 (1957).
NAGABHUSHANAM, R., SAROJINI, R.: Indian J. Phys. All. Sci. **19**, 1—4 (1965).
NAGABHUSHANAM, R., SAROJINI, R.: Hydrobiologia **34**, 126—134 (1969).
NAGY, B.: Acta Zool. Hung. **5**, 369—391 (1959).
NAIR, N. B., LEIVESTAD, H.: Nature **182**, 814—815 (1958).
NAKASHINA-TANAKA, E.: Genetica ('s-Grav.) **38**, 447—458 (1968).
NARAYANAN, N. S., ANGALET, G. W., RAO, B. R. S., D'SOUZA, G. I.: Nature **173**, 503—504 (1954).
NASH, T. A. M.: Bull. Entomol. Res. **27**, 273—279 (1936).
NAYAR, J. K.: J. Med. Entomol. **5**, 39—46 (1968).
NAYLOR, E.: J. Exp. Biol. **40**, 669—679 (1963).
NEEDHAM, P. R., JONES, A. C.: Ecology **40**, 465—474 (1959).
NEGRI, M.: Ric. Sci. Ser. 2, Parte 2: Rend., Sez. B **4**, 485—494 (1964).
NÉMETH, G. P.: Acta Biol. Acad. Sci. Hung. **16**, 311—317 (1966).
NÉMETHY, G., SCHERAGA, H. A.: J. Chem. Phys. **36**, 3382—3400, 3401—3417(1962a,b)
NÉMETHY, G., SCHERAGA, H. A.: J. Phys. Chem. **66**, 1773—1789 (1962c).
NEWCOMBER, E. J.: J. Econom. Entomol. **13**, 441—442 (1920).
NEWELL, R. C., NORTHCROFT, H. R.: J. Zool. **151**, 277—298 (1967).
NEWELL, R. C., PYE, V. I.: Comp. Biochem. Physiol. **34**, 367—384, 385—398 (1970).
NICHOLSON, A. J.: Bull. Entomol. Res. **25**, 85—99 (1934).
NIELSEN, E. T., NIELSEN, H. F.: Ecology **40**, 181—185 (1959).
NOLL, J.: Nachrbl. Deut. Pflanzenschutzdienst (Berlin) N. F. **17**, 9—24 (1963).
NOPP, H., FARAHAT, A. Z.: Z. Vergleich. Physiol. **55**, 103—118 (1967).
NORDMAN, A.: Notulae Entomol. **34**, 99—106 (1954).
NORRIS, K. S.: Ecology **34**, 265—287 (1953).
NORRIS, K. S.: Ecol. Monogr. **33**, 23—62 (1963).
NORTHCOTE, T. G.: Nature **181**, 1283—1284 (1958).
NUMANOI, H.: Jap. J. Zool. **7**, 613—641 (1938).
NUORTEVA, P., HACKMAN, W.: Ann. Zool. Fennicae. **7**, 267—269 (1970).
O'BANNON, J. H., REYNOLDS, H. W., LEATHERS, C. R.: Nematologica **12**, 483—487 (1966).
O'CONNOR, J. M.: Proc. Irish Acad. **61**B, 187—200 (1960), **63**B, 201—205 (1964).
ÖSTBYE, E.: Nytt. Mag. Zool. **18**, 75—79 (1970).
OHLENBUSCH, H.-D., PRECHT, H.: Z. Wiss. Zool. **164**, 364—373 (1960).
OHM, D.: Zool. Beitr. (N.F.) **2**, 254—386 (1956).
OHNESORGE, B.: Beitr. Entomol. **10**, 854—871 (1960).
OHNESORGE, F.-K., SCHMITZ, G.: Z. Vergleich. Physiol. **58**, 171—184 (1968).
OHSAWA, W.: J. Inst. Polytechn., Osaka City Univ., D, **7**, 197—217 (1956).
OHSAWA, W., TSUKUDA, H.: J. Inst. Polytechn., Osaka City Univ., D, **7**, 189—196 (1956).
OHSAWA, W., TSUKUDA, H.: J. Biol. Osaka Univ., D, **15**, 31—38 (1964).
OHYAMA, Y., ASAHINA, E.: Low Temp. Sci. B, **27**, 153—160 (1969).
OKASHA, A. Y. K.: Nature **204**, 1221—1222 (1964).
OKASHA, A. Y. K.: J. Expl. Biol. **48**, 455—463 (1968).
OKAY, S.: Arch. Intern. Physiol. **64**, 80—91 (1956).
OLDIGES, H.: Z. Angew. Entomol. **44**, 115—166 (1959).
OLSON, R. E.: J. Parasitol. **52**, 327—334 (1966).
OMARDEEN, T. A.: Bull. Entomol. Res. **48**, 349—357 (1958).
OORDT, P. G. W. J. VAN, LOFTS, B.: J. Endocr. **27**, 137—146 (1963).
ORR, P. R.: Physiol. Zoöl. **28**, 290—294, 294—302 (1955).
ORSKA, J.: Zool. Pol. **13**, 49—76 (1964).
ORTON, J. H.: J. Marine Biol. Ass. U. K. (N. S.) **12**, 339—366 (1920).
OSTDIEK, J. L., NARDONE, R. M.: Am. Midland Naturalist **61**, 218—229 (1959).
PALMÉN, E.: Ann. Entomol. Fennicae. **20**, 1—13 (1954).
PALOHEIMO, J. E., DICKIE, L. M.: J. Fish. Res. Board Can. **23**, 869—908 (1966).
PALUMBO, S. A., WITTER, L. B.: Can. J. Microbiol. **15**, 995—1000 (1969).

PANDIAN, T. J.: Marine Biol. **5**, 1—17 (1970).
PANTELOURIS, E. M.: J. Genet. **55**, 507—510 (1957).
PANTYUKHOV, G. A.: Zool. Ž. **35**, 1312—1324 (1956).
PAPILLON, M.: Compt. Rend. Acad. Sci. (Paris) **260**, 6446—6448 (1965).
PAPILLON, M.: Bull. Biol. Fr. Belg. **102**, 85—139 (1968).
PARDI, L.: Z. Tierpsychol. **14**, 261—275 (1957).
PARISE, A.: Atti Acc. Naz. Lincei. Mem. **40**, 502—508 (1966).
PARK, T., FRANK, M. B.: Ecology **29**, 368—374 (1948).
PARK, H., SHARPLESS, N., ORTMEYER, A. B.: J. Exp. Zool. **160**, 247—254 (1965).
PARKER, G. E.: Copeia **1967**, 610—616.
PARKER, J. C., HALEY, A. J.: Exp. Parasitol. **9**, 92—97 (1960).
PARKER, J. R.: Bull. Univ. Montana Agr. Exp. Sta. **223** (1930).
PARKES, A. S., SMITH, A. U. (Ed.): Recent research of freezing and drying. Oxford 1960.
PARKS, T. H.: J. Econom. Entomol. **7**, 417—421 (1914).
PARRY, G.: Biol. Rev. **41**, 392—444 (1966).
PARSONS, P. A.: Genetics **44**, 1325—1333 (1959).
PARSONS, R. H., ALVARADO, R. H.: Comp. Biochem. Physiol. **24**, 61—72 (1968).
PARTECKE, H. J.: Zool. Beitr. (N.F.) **5**, 37—116 (1959).
PARTMANN, W., NEMITZ, G.: J. Lebensmittelunters.-Forsch. **110**, 109—114 (1959).
PARVATHESWARARAO, V.: Comp. Biochem. Physiol. **21**, 619—626 (1967).
PARVATHESWARARAO, V.: Endocr. Exp. **2**, 173—178 (1968a).
PARVATHESWARARAO, V.: Proc. Indian Acad. Sci., B, **68**, 225—231 (1968b).
PASSANO, L. M.: Biol. Bull. **118**, 129—139 (1960).
PAULOV, S.: Folia Biol. **7**, 281—284 (1961).
PAVLIDIS, T., ZIMMERMANN, W. F., OSBORN, J.: J. Theoret. Biol. **18**, 210—211 (1968).
PAYNE, N. M.: Quart. Rev. Biol. **1**, 270—282 (1926), cf. J. Ecol. **7**, 99—106 (1926),
  8, 194—196 (1927).
PEAIRS, L. M.: J. Econom. Entomol. **7**, 174—179 (1914).
PEARL, R.: The rate of living. London 1928.
PEARL, R.: Introduction to medical biometry and statistics. Philadelphia 1930.
PEARL, R., READ, L. J.: Proc. Natl. Acad. Sci. (Wash.) **6**, 275—288 (1920).
PEARSON, O. P.: Copeia **1954**, 111—116.
PENNYCUIK, P. R., FRASER, A.: Australian J. Biol. Sci. **17**, 764—770 (1964).
PERTTUNEN, V., HÄYRINEN, T.: Ann. Entomol. Fennicae. **35**, 105—122, 190—204
  (1969a, b).
PERTTUNEN, V., LAGERSPETZ, K.: Arch. Soc. Zool.-Bot. Fennicae Vanamo **11**, 65—70
  (1956).
PERTTUNEN, V., PALOHEIMO, L.: Ann. Entomol. Fennicae. **30**, 156—172 (1964).
PETERS, N., JR.: Wilh. Roux Arch. **156**, 75—87 (1965).
PETERSEN, B.: Entomol. Tidsskr. **69**, 135—141 (1948).
PFLUMM, W.: Z. Vergleich. Physiol. **64**, 1—36 (1969).
PIATKOWSKA, B.: Bull. Acad. Pol. Sci., Sci. Biol. **9**, 299—301 (1961), **11**, 237—239
  (1963).
PICKENS, P. E.: Physiol. Zoöl. **38**, 390—405 (1965).
PIELOU, D. P., GLASSER, R. F.: Can. J. Zool. **32**, 30—38 (1954).
PILON, J. G.: Can. Entomol. **98**, 789—794 (1966).
PITKOW, R. B.: Biol. Bull. **119**, 231—245 (1960).
PLATNER, W. S.: Am. J. Physiol. **161**, 399—405 (1950).
PLATT, R. B.: Microclimate. In: Encyclop. Biol. Sci. Reinhold Publ. Corp. 1961.
PLATT, R. B., COLLINS, C. L., WITHERSPOON, I. P.: Ecol. Monogr. **27**, 303—324 (1957).
PLOUGH, H. H.: J. Exp. Zool. **24**, 147—209 (1917).
PLOUGH, H. H.: Cold Spring Harb. Symp. Quant. Biol. **9**, 127—137 (1941).
POCRNJIĆ, Z.: Arh. Biol. Nauka **17**, 139—148 (1965).
POHLEY, H.-J.: Biol. Zbl. **74**, 474—480 (1955).
POLYANSKY, G. I.: XVth Intern. Congr. Zool. Sect. IX, 26—27 (1959).
POLYANSKY, G. I., ORLOVA, A. F.: Zool. Ž. **36**, 1630—1646 (1957).

POLYANSKY, G. I., USHAKOV, B. P. (Ed.): Heat resistance of cells of animals. Inst. Cytol., Akad. Sci. USSR, Coll. art **8**, Moscow 1965.

POPE, C. H.: The giant snakes. New York 1961.

POPHAM, E. J.: Proc. XII. Int. Congr. Entomol 324 (1964).

PORA, E. A., ABRAHAM, A. D., SILDAU-RUSU, N.: Marine Biol. **1**, 33—35 (1967).

PORTER, C. H., GOYMERAC, W. L.: Mosquito News **30**, 54—56 (1970).

POSTUMA, K. H.: Arch. Neerl. Zool. **13**, 592—595 (1960).

POTTS, D. C., MORRIS, R. W.: Marine Biol. **1**, 269—276 (1968).

POUGH, F. H., WILSON, R. E.: Physiol. Zoöl. **43**, 194—205 (1970).

PREBBLE, M. L.: Can. J. Res., D. Zool. Sci. **19**, 295—454 (1941).

PRECHT, H.: Zool. Anz. **115**, 80—89 (1936).

PRECHT, H.: Z. Wiss. Zool. **156**, 1—128 (1942).

PRECHT, H.: Z. Vergleich. Physiol. **26**, 696—739 (1939), **35**, 326—343 (1953), **42**, 365—382 (1959), **44**, 451—462 (1961 b).

PRECHT, H.: Z. Naturforsch. **4**b, 26—35 (1949).

PRECHT, H.: Biol. Zbl. **70**, 71—85 (1951).

PRECHT, H.: Z. Wiss. Zool. **164**, 336—353, 354—363 (1960 a, b), **167**, 73—86 (1962).

PRECHT, H.: Zool. Anz. (Suppl. Bd) **24**, 38—60 (1961 a).

PRECHT, H.: Naturw. Rdsch. **16**, 9—16 (1963), **17**, 438—442 (1964 b).

PRECHT, H.: Zool. Anz. **172**, 87—95, 306—318 (1964 a, d), **175**, 301—310 (1965).

PRECHT, H.: Zool. Jahrb. (Physiol.) **71**, 313—327 (1964 c).

PRECHT, H.: Helgoländer Wiss. Meeresuntersuch. **9**, 392—411 (1964 e), **18**, 487—548 (1968).

PRECHT, H., BASEDOW, T., BERECK, R., LANGE, F., THIEDE, W., WILKE, L.: Helgoländer Wiss. Meeresuntersuch. **13**, 369—401 (1966).

PRECHT, H., CHRISTOPHERSEN, J.: Z. Wiss. Zool. **171**, 197—209 (1965).

PRECHT, H., LINDNER, E.: Helgoländer Wiss. Meeresuntersuch. **13**, 354—368 (1966).

PRECHT, I.: Z. Wiss. Zool. **176**, 122—172 (1967).

PREER, J. R., JR.: J. Genet. **55**, 375—378 (1957).

PRESCOTT, D. M.: J. Protozool. **4**, 252—256 (1957).

PRITCHARD, G.: Australian J. Zool. **18**, 77—89 (1970).

PROPPER, A.: Arch. Zool. Exp. **105**, 259—271 (1963).

PROSSER, C. L.: Physiological adaptation. Am. Physiol. Soc. Washington 1958.

PROSSER, C. L. (Ed.): Molecular mechanisms of temperature adaptation. Publ. Am. Ass. Adv. Sci. **84** (1967).

PROSSER, C. L., BROWN, F. A., JR.: Comparative animal physiology. 2nd ed., Philadelphia 1961.

PROSSER, C. L., FARHI, E.: Z. Vergleich. Physiol. **50**, 91—101 (1965).

PROSSER, C. L., MACKAY, W., KATO, K.: Physiol. Zoöl. **43**, 81—89 (1970).

PROSSER, C. L., PRECHT, H., JANKOWSKY, H.-D.: Naturwissenschaften **52**, 168—169 (1965).

PUKHALSKAYA, E. C.: Bull. Exp. Biol. Med. **47**, 85—88 (1959).

PUNT, A.: Arch. Neerl. Zool. **7**, 205—212 (1945).

QUEDNAU, W.: Mitt. Biol. Bundesanstalt Land-Forstwirtsch. Berlin-Dahlem, Berlin, Hamburg **90** (1957).

RAAB, R. L.: Ann. Entomol. Soc. Am. **59**, 160—165 (1966).

RAFFY, A.: Compt. Rend. Soc. Biol. (Paris) **148**, 1796—1798 (1954).

RAKSPHAL, R.: Can. J. Zool. **40**, 179—194 (1962).

RAND, A. S.: Ecology **45**, 863—864 (1964).

RAO, G. M. M.: Can. J. Zool. **47**, 131—134 (1969).

RAO, K. P.: Helgoländer Wiss. Meeresuntersuch. **14**, 439—450 (1966).

RAO, K. P., BULLOCK, T. H.: Am. Naturalist **88**, 33—34 (1954).

RASCHACK, M.: Naturwissenschaften **54**, 97 (1967).

RASCHACK, M.: Int. Rev. Hydrobiol. **54**, 423—462 (1969).

RASMONDT, R.: Ann. Soc. Zool. Belg. **85**, 173—181 (1955).

RASMONDT, R.: Compt. Rend. Acad. Sci. (Paris) **261**, 845—847 (1965).

RASMUSON, M.: Hereditas **43**, 589—594 (1957).

RASMUSSEN, R. A., RASMUSSEN, L. E.: Trans. N. Y. Acad. Sci. 2, **29**, 397—413 (1967).
RATHJEN, W.: Z. Morph. Ökol. Tiere **35**, 14—83 (1939).
RATTAN, L., HAQUE, E.: Indian Entomol. **17**, 317—325 (1956).
RAUH, F.: Z. Morph. Ökol. Tiere **53**, 61—106 (1963).
RAUTAPÄÄ, J., MARKKULA, M.: Ann. Entomol. Fennicae **32**, 146—152 (1966).
RAVEN, C. P., DE ROON, A. C., STADHOUDERS, A. M.: J. Embryol. Exp. morph. **3**, 142—159 (1955).
RAVEN, C. P., VAN ERKEL, G. A.: Exp. Cell Res. Suppl. **3**, 294—303 (1955).
RAY, C.: J. Morph. **106**, 85—108 (1960).
READ, K. R.: Biol. Bull. **127**, 489—498 (1964).
READSHAW, J. L.: Bull. Entomol. Res. **56**, 685—700 (1966).
REAVES, R. S., HOUSTON, A. H., MADDEN, J. A.: Comp. Biochem. Physiol. **25**, 849—860 (1968).
REEVES, R. B.: Fed. Proc. **28**, 1204—1208 (1969).
REGAL, P. J.: Copeia **1966**, 588—590.
REGAL, P. J.: Science **155**, 1551—1553 (1967).
REGGI, DE L., DELMAS, J. C.: Compt. Rend. Acad. Sci. (Paris) **260**, 6427—6429 (1965).
REIBISCH, J.: Wiss. Meeresunters. (N. F.) **6**, 213—231 (1902).
REICHLIG, H.: Zool. Jahrb. (Physiol.) **67**, 1—64 (1957).
REINHARDT, R.: Zool. Jahrb. (Physiol.) **75**, 41—75 (1969).
REMANE, A., SCHLIEPER, C.: Die Biologie des Brackwassers. Stuttgart (Binnengewässer **22**) (1958).
REMMERT, H.: Naturwissenschaften **45**, 498 (1958).
REMMERT, H.: Z. Morph. Ökol. Tiere **49**, 504—520 (1960).
REMMERT, H.: Biol. Zbl. **79**, 577—584 (1960).
REMMERT, H.: Hdb. Biol. (Ed. GESSNER, F.) Frankfurt a. M., **5**, 335—411 (1965).
REMMERT, H.: Oecologia **3**, 214—226 (1969).
REMMERT, H., WISNIEWSKI, W.: Oecologia **4**, 111—112 (1970).
REMMERT, H., WÜNDERLING, K.: Oecologia **4**, 208—210 (1970).
RENKEN, W.: Z. Morph. Ökol. Tiere **45**, 34—106 (1956).
RENSCH, B.: Zool. Anz. **158**, 33—38 (1957).
RENSCH, B.: Hdb. Biol. (Ed. GESSNER, F.) Frankfurt a. M., **5**, 125—172 (1950).
RICHARDS, A. G.: Biol. Zbl. **78**, 308—314 (1959).
RICHARDS, A. G.: J. Insect Physiol. **9**, 597—606 (1963).
RICHARDS, A. G.: Physiol. Zoöl. **37**, 199—211 (1964).
RICHARDS, A. G.: Z. Naturforsch. **20**b, 347—349 (1965).
RICHARDS, A. G., SUANRAKSA, S.: Entomol. Exp. Appl. **5**, 167—178 (1962).
RICHARDS, S. A.: Biol. Rev. **45**, 223—264 (1970).
RIEDEL, M.: Bayer. Landwirtsch. Jahrb. **44**, 387—429 (1967).
RIGBY, B. J., MASON, P.: Australian J. Biol. Sci. **20**, 265—271 (1967).
RILLING, G.: Naturwissenschaften **49**, 90—91 (1962).
RING, K., CHRISTOPHERSEN, J.: Arch. Mikrobiol. **48**, 50—65 (1964).
RIORDAN, D. F.: Can. J. Zool. **35**, 603—608 (1957).
RISING, T. L., ARMITAGE, K. B.: Comp. Biochem. Physiol. **30**, 1091—1114 (1969).
RITOSSA, F.: Experientia **18**, 571—573 (1962).
RITZ, D. A., CRISP, D. J.: J. Marine Biol. Ass. U. K. **50**, 223—240 (1970).
RIVARD, I.: Can. J. Zool. **39**, 869—876 (1961).
RIVNAY, E., MEISNER, J.: Bull. Entomol. Res. **56**, 623—634 (1966).
RIZKI, M. T. M.: Genetics **40**, 130—136 (1955).
ROBERTS, J. L.: Physiol. Zoöl. **30**, 232—242, 242—255 (1957).
ROBERTS, J. L.: Helgoländer Wiss. Meeresuntersuch. **9**, 459—473 (1964), **14**, 451—465 (1966).
ROBERTS, S. K.: J. Cell Comp. Physiol. **59**, 175—186 (1962).
ROBERTSON, O. H.: Physiol. Zoöl. **21**, 282—295 (1948).
RODE, H.: Pedobiologica **5**, 1—6 (1965), **9**, 405—425 (1969).
RÖMER, F., ROSIN, S.: Rev. Suisse Zool. **76**, 734—740 (1969).
ROEMHILD, G.: J. Insect Physiol. **11**, 1633—1639 (1965).

ROLAND, C.: Compt. Rend. Acad. Sci. (Paris) **290**, 914—916 (1969).

ROOTS, B. I.: Fed. Proc. **20**, 209 (1961).

ROOTS, B. I.: Comp. Biochem. Physiol. **25**, 457—466 (1968).

ROOTS, B. I., JOHNSTON, P. V.: Comp. Biochem. Physiol. **26**, 553—560 (1968).

ROOTS, B. I., PROSSER, C. L.: J. Exp. Biol. **39**, 617—629 (1962).

ROSE, A. H. (Ed.): Thermobiology. London 1967.

ROSENBAUM, R. M.: Devel. Biol. **2**, 427—445 (1960).

ROSENTHAL, C. M.: Univ. Calif. Publ. Zool. **54**, 371—420 (1957).

ROTH, A., SZABO, T.: J. Exp. Biol. **52**, 707—719 (1970).

ROTH, M.: Ann. Abeille **8**, 5—77 (1965).

ROWLEY, W. A., GRAHAM, C. L.: J. Insect Physiol. **14**, 1251—1257 (1968).

ROZHAJA, D.: Rec. Trav. Fac. Lett. Sci., Pristina **1**, 257—263 (1963).

ROZIN, G.: Science **149**, 561—563 (1965).

ROZIN, P. N., MAYER, I.: Science **134**, 942—943 (1961).

RUDINSKY, J. A., VITE, J. P.: Forest Sci. **2**, 258—267 (1956).

RUDNICK, M. H.: Zool. Jahrb. (Physiol.) **73**, 227—250 (1967).

RÜHMEKORFF, E.: Z. Morph. Ökol. Tiere **47**, 20—36 (1958).

RUFF, P. W., ZIPPEL, U.: Acta Biol. Med. Ger. **16**, 395—403 (1966).

RUTTNER, F.: Grundriß der Limnologie. 3. Ed., Berlin 1962.

RYAN, F. J.: J. Exp. Zool. **88**, 25—54 (1941).

RYLAND, J. S., NICHOLS, J. H.: Nature **214**, 529—530 (1967).

SAGE, M.: Gen. Comp. Endocr. **10**, 304—309 (1968).

SAGER, H.: Z. Wiss. Zool. **168**, 321—375 (1963).

SAINI, R. S., CHIANG, H. C.: Ecology **47**, 473—477 (1966).

SALT, R. W.: Can. J. Res., D. **28**, 285—291 (1950).

SALT, R. W.: Can. Entomol. **85**, 261—269 (1953), **95**, 1190—1202 (1963).

SALT, R. W.: Can. J. Zool. **34**, 283—294 (1956), **36**, 265—268 (1958b), **37**, 59—69 (1959), **39**, 349—357 (1961b), **44**, 947—952 (1966).

SALT, R. W.: Proc. 10th intern. Congr. Ent. **2**, 1956, 73—77 (1958a).

SALT, R. W.: Ann. Rev. Entomol. **6**, 55—74 (1961a).

SALT, R. W.: Symp. Soc. Exp. Biol. **23**, 331—350 (1969).

SAMEOTO, D. D.: J. Fish. Res. Board Can. **26**, 1321—1345, 2283—2298 (1969).

SAMOKHVALOVA, G. V.: Compt. Rend. Acad. Sci. U. S. S. R. (N. S.) **20**, 475—478 (1938).

SANDISON, E. E.: J. Exp. Marine Biol. Ecol. **1**, 271—281 (1967).

SANFORD, K. K., BANTA, A. M.: Genetics **26**, 166 (1941).

SAROJA, K., RAO, K. P.: Z. Vergleich. Physiol. **50**, 35—54 (1965).

SASTRY, A. N.: Biol. Bull. **125**, 146—153 (1963).

SASTRY, A. N.: Physiol. Zoöl. **41**, 44—53 (1968).

SATTLER, H.: Z. Angew. Entomol. **25**, 543—587 (1939).

SAUNDERS, D. S.: Science **156**, 1126—1127 (1967).

SAUNDERS, D. S.: J. Insect Physiol. **14**, 433—450 (1968).

SAUNDERS, D. S.: Dormancy and survival. Symp. Soc. Exp. Biol. **23**, 300—329 (1969).

SAUNDERS, R. L.: J. Fish. Res. Board Can. **20**, 373—386 (1963).

SAVITZKY, J. P.: Biochim. Biophys. Acta **80**, 183—192 (1964).

SAXENA, P. N.: Proc. Zool. Soc. **19**, 145—151 (1966).

SCHAEFER, M.: Biol. Zbl. **90**, 579—609 (1971).

SCHALK, J. M., KINDLER, S. D., MANGLITZ, G. R.: J. Econom. Entomol. **62**, 1000—1003 (1969).

SCHALLER, F., ZINKLER, D.: Naturwissenschaften **50**, 385 (1963).

SCHARLOO, W.: Arch. Neerl. Zool. **14**, 431—512 (1962).

SCHEIL, H.-G.: Marine Biol. **6**, 158—166 (1970).

SCHERBAUM, O., ZEUTHEN, E.: Exp. Cell Res. (Suppl.) **3**, 312—325 (1955).

SCHIMITSCHEK, E.: Z. Angew. Entomol. **18**, 460—491 (1931).

SCHIRONE, R. C., GROSS, L.: J. Exp. Zool. **169**, 43—52 (1968).

SCHLIEPER, C.: Kieler Meeresforsch. **16**, 180—185 (1960).

SCHLIEPER, C.: Zool. Anz. (Suppl. Bd) **29**, 239—242 (1966).

SCHLIEPER, C.: Helgoländer Wiss. Meeresuntersuch. **14**, 482—502 (1966).

SCHLIEPER, C., BLÄSING, J.: Arch. Hydrobiol. **47**, 288—294 (1952), cf. Zool. Jahrb. (Physiol.) **64**, 112—152 (1953).

SCHLIEPER, C., FLÜGEL, H., RUDOLF, J.: Experientia **16**, 470—472 (1960).

SCHLIEPER, C., FLÜGEL, H., THEEDE, H.: Physiol. Zoöl. **40**, 345—370 (1967).

SCHLOTTKE, E.: Z. Vergleich. Physiol. **3**, 692—736 (1926).

SCHMEING-ENGBERDING, F.: Z. Fischerei (N. F.) **2**, 125—155 (1953).

SCHMIDT, D., SANTA, N.: Stud. Cercet. Biol., Zool. **21**, 229—237 (1969).

SCHMIDT, G.: Biol. Zbl. **75**, 178—205 (1956).

SCHMIDT, G.: Z. Angew. Entomol. **40**, 390—399 (1957).

SCHMIDT, G. H.: Zool. Jahrb. (Physiol.) **66**, 273—294 (1956).

SCHMIDT, G. H.: Z. Angew. Entomol. **61**, 61—109 (1968).

SCHMIDT, G. H.: Naturwiss. Med. **35**, 41—50 (1970).

SCHNEIDER, G.: Z. Vergleich. Physiol. **49**, 195—269 (1964).

SCHNEIDER, H.: Z. Morph. Ökol. Tiere **44**, 555—625 (1956).

SCHNEIDER, H.: Z. Vergleich. Physiol. **57**, 174—189 (1967).

SCHNEIDER, L., IMMELMANN, K.: Naturwissenschaften **56**, 93 (1969).

SCHNEIDERMANN, H. A., HORWITZ, J.: J. Exp. Biol. **35**, 520—551 (1958).

SCHOENER, T. W.: Ecology **51**, 408—418 (1970).

SCHÖNIG, R. v.: Beitr. Entomol. **3**, 627—652 (1953).

SCHOLANDER, P. F.: 21. Intern. Congr. Physiol. Sci., Buenos Aires, Symp. and spec. lectures, 77—80 (1959).

SCHOLANDER, P. F., DAM, L. VAN, KANWISHER, J. W., HAMMEL, H. T., GORDON, M. S.: J. Cell Comp. Physiol. **49**, 5—24 (1957).

SCHOLANDER, P. F., FLAGG, W., WALTERS, V., IRVING, L.: Physiol. Zoöl. **26**, 67—92 (1953).

SCHOONHOVEN, L. M.: J. Insect Physiol. **13**, 821—826 (1967).

SCHULTZE, D.: Z. Wiss. Zool. **172**, 104—133 (1965).

SCHULZ, W.: Z. Wiss. Zool. **158**, 31—78 (1954).

SCHULZE, W. G.: Z. Angew. Entomol. **44**, 64—101 (1959).

SCHULZ-LANGNER, E.: Z. Bienenforsch. **4**, 67—86 (1958).

SCHWALBACH, G., AGOSTINI, B.: Z. Zellforsch. **62**, 113—120 (1964).

SCHWARTZKOPFF, J.: Biol. Zbl. **74**, 480—497 (1955).

SCHWERDTFEGER, F.: Autökologie. Hamburg, Berlin 1963.

SCOTT, D. P.: J. Fish. Res. Board Can. **21**, 1043—1049 (1964).

SEGAL, E.: Cold Spring Harb. Symp. Quant. Biol. **25**, 505 (1960).

SEGAL, E.: Am. Zool. **1**, 235—244 (1961).

SEGAL, E.: Nature **195**, 674—675 (1962).

SEGAL, E., BURBANCK, W. D.: Physiol. Zoöl. **36**, 250—263 (1963).

SEGAL, E., RAO, K. P., JAMES, T. W.: Nature **172**, 1108—1109 (1953).

SEIBT, U.: Z. Vergleich. Physiol. **57**, 77—102 (1967).

SEILER, J.: Arch. Zellforsch. **15**, 249—268 (1920).

SELVARAJAN, V. R.: Proc. Indian Acad. Sci. **55**, 91—98 (1962).

SEMENOVA, G.: Citologia (Moscow) **9**, 949—957 (1967).

SEN-SARMA, P. K.: Proc. Nat. Inst. Sci. India, B. **30**, 300—314 (1964).

SERFATY, A., LAFFONT, J.: Hydrobiologia **26**, 409—420 (1965).

SERFATY, A., WAITZENEGGER, M.: Hydrobiologia **23**, 281—286 (1964).

SHAW, R. F., BERCAW, B. L.: Nature **196**, 454—457 (1962).

SHEA, S., SIGAFOOS, D., SCOTT, D., JR.: Comp. Biochem. Physiol. **28**, 701—708 (1969).

SHELDON, B. L.: Australian J. Biol. Sci. **11**, 85—94 (1958).

SHELFORD, V. E.: Bull. Illinois Nat. Hist. Surv. **16**, 307—440 (1927).

SHERMAN, J. K.: J. Cell Comp. Physiol. **61**, 67—83 (1963).

SHIELDS, J. L., PLATNER, W. S., NEUBESIER, R. E.: Am. J. Physiol. **199**, 942—944 (1960).

SHINOZAKI, J.: Contrib. Inst. Lond. Temp. Sci., B., **12**, 1—52 (1962).

SHOREY, H. H.: Ann. Entomol. Soc. Am. **59**, 502—506 (1966).

SHORR, H.: Z. Morph. Ökol. Tiere **45**, 561—602 (1957).

SHULL, A. F.: J. Exp. Zool. **10**, 117—166 (1911).

SICKER, W.: Z. Morph. Ökol. Tiere **54**, 107—140 (1964).
SIMPSON, A. C.: Extr. J. Conseil intern. expl. de la mer **19**, 150—177 (1953).
SIMPSON, J.: Science **133**, 1327—1333 (1961).
SIMPSON, T. L.: J. Exp. Marine Biol. Ecol. **2**, 252—277 (1968).
SINGH, M. P.: Proc. Zool. Soc. (Calcutta) **15**, 27—37 (1962).
SINGH, T. P., SATHYANESAN, A. G.: Acta Zool. **49**, 47—56 (1968).
SINHA, D. D., HOPKINS, C. A.: Parasitology **57**, 555—566 (1967).
SITBON, G.: Compt. Rend. Acad. Sci. (Paris) **266**, 1305—1307 (1968).
SKRAMLIK, E. v.: Jenaische Z. Med. Naturwiss. **78**, 208—218 (1947).
SLICHER, A. M., PICKFORD, G. E.: Physiol. Zoöl. **41**, 293—297 (1968).
SLIFER, E. H.: Physiol. Zoöl. **5**, 448—456 (1932).
SLIFER, E. H.: J. Exp. Zool. **110**, 183—195 (1949).
SLIFER, E. H.: Proc. Roy. Soc. Lond., B 138, **892**, 414—437 (1951).
SLIFER, E. H.: Ann. Entomol. Soc. Am. **47**, 255—264 (1954).
SLIFER, E. H., KING, R. L.: J. Heredity **52**, 39—44 (1961).
SMIT, H.: Comp. Biochem. Physiol. **21**, 125—132 (1967).
SMITH, A. C.: Comp. Biochem. Physiol. **33**, 1—14 (1970).
SMITH, A. U.: Biol. Rev. **33**, 197—253 (1958).
SMITH, A. U.: Biological effects of freezing and supercooling. Monogr. Physiol. Soc. **9**, Baltimore 1961.
SMITH, J. M.: J. Genet. **54**, 497—507 (1956).
SMITH, J. M.: J. Exp. Biol. **34**, 85—96 (1957).
SMITH, J. M.: Nature **181**, 496—497 (1958).
SMITH, M. W.: J. Physiol. (London) **182**, 559—573 (1966a), **183**, 649—657 (1966b).
SMITH, M. W.: Experientia **22**, 252 (1966c), **23**, 548—549 (1967a).
SMITH, M. W.: Biochem. J. **105**, 65—71 (1967b).
SMITH, M. W.: Comp. Biochem. Physiol. **35**, 387—401 (1970).
SMITH, M. W., COLOMBO, V. E., MUNN, E. A.: Biochem. J. **107**, 691—698 (1968).
SMITH, M. W., KEMP, P.: Biochem. J. **114**, 659—661 (1969).
SMITH, M. W., MORRIS, D.: Experientia **22**, 678—679 (1966).
SMITH, R. N.: The biochemistry of freezing resistance of some Antarctic fish. Abstr. from SCAR Symp. on Antarctic Ecol., Cambridge 1968.
SOHLENIUS, B.: Pedobiologia **8**, 137—145 (1968).
SOLOMON, K., ALLANSON, B. R.: Comp. Biochem. Physiol. **25**, 485—492 (1968).
SOMERO, G. N.: Biochem. J. **114**, 232—241 (1969a), Am. Naturalist **103**, 517—530 (1969b).
SOMERO, G. N., GIESE, A. C., WOHLSCHLAG, D. E.: Comp. Biochem. Physiol. **26**, 223—233 (1968).
SOMERO, G. N., HOCHACHKA, P. W.: Biochem. J. **110**, 395—400 (1968).
SOMERO, G. N., HOCHACHKA, P. W.: Nature **223**, 194—195 (1969).
SOMERO, G. N., DE VRIES, A. L.: Science **156**, 257—258 (1967).
SØMME, L.: Can. J. Zool. **42**, 87—101 (1964), **43**, 765—770, 881—884 (1965 a, b).
SØMME, L.: Nytt. Mag. Zool. **13**, 52—55 (1966).
SØMME, L.: J. Insect Physiol. **13**, 805—814 (1967).
SØMME, L.: Norsk. Entomol. Tidsskr. **15**, 134—136 (1968a).
SØMME, L.: Entomol. Exp. Appl. **11**, 143—148 (1968b).
SØMME, L.: Norsk. Entomol. Tidsskr. **16**, 107—111 (1969).
SØMME, L., ÖSTBYE, E.: Norsk Entomol. Tidsskr. **16**, 45—48 (1969).
SOTAVALTA, O.: Ann. Soc. Zool. Vanamo **16**, 1—22 (1954).
SOULIÉ, J.: Compt. Rend. Soc. Biol. (Paris) **149**, 806—808 (1955).
SOULIÉ, J.: Insectes Sociaux **3**, 431—438 (1956).
SOUTHWARD, A. J.: J. Marine Biol. Ass. U. K. **37**, 49—66 (1958).
SOUTHWARD, A. J.: Helgoländer Wiss. Meeresuntersuch. **10**, 391—403 (1964).
SPAAS, J. T.: Hydrobiologica **14**, 155—176 (1959), **15**, 78—88 (1960).
SPÄTH, M.: Z. Vergleich. Physiol. **56**, 431—462 (1967).
SPEYER, W.: Entomologie. Wiss. Forsch.-Ber. Naturwiss. R. **43**, 194 (1937).

SPITZER, K. W., MARVIN, D. E., JR., HEATH, A. G.: Comp. Biochem. Physiol. **30**, 83—90 (1969).
SPOOR, W. A.: Biol. Bull. **108**, 77—87 (1955).
SPRAGUE, J. B.: J. Fish. Res. Board Can. **20**, 387—415 (1963).
STANGENBERG, G.: Pflügers Arch. Ges. Physiol. **260**, 320—332 (1955).
STARK, R. W., DAHLSTEN, D. L.: Ecology **47**, 488—489 (1966).
STEBBINS, R. C., BARWICK, R. E.: Copeia **1968**, 541—547.
STEIN, W., FRANZ, J.: Naturwissenschaften **47**, 262—263 (1960).
STEINER, A.: Z. Vergleich. Physiol. **2**, 23—56 (1924).
STEINER, A.: Z. Vergleich. Physiol. **11**, 461—502 (1930).
STEPHENS, G. C.: Physiol. Zoöl. **30**, 55—69 (1957).
STEVENS, E. D.: Comp. Biochem. Physiol. **25**, 615—625 (1968).
STEVENS, E. D., FRY, F. E.: Can. J. Zool. **48**, 221—226 (1970).
STEVENS, E. D., FRY, F. E.: Comp. Biochem. Physiol. **38**, 203—211 (1971).
STIER, T. J. B., TAYLOR, H. E.: J. Cell Comp. Physiol. **14**, 309—312 (1939).
STIEVE, H.: Z. Vergleich. Physiol. **46**, 249—275 (1963).
STIREWALT, M. A.: Exp. Parasitol. **3**, 504—516 (1954).
STIRRETT, G. M.: Sci. Agr. **18**, 462—484 (1938).
STIVEN, A. E.: Ecology **43**, 325—328 (1962).
STOFFOLANO, J. G., JR., MATTHYSSE, J. G.: Ann. Entomol. Soc. Am. **60**, 1242—1246 (1967).
STORSS, R. G.: Ecology **47**, 368—374 (1966).
STOWER, W. J., GRIFFITHS, J. F.: Entomol. Exp. Appl. **9**, 127—178 (1966).
STRAMBI, A.: Compt. Rend. Acad. Sci. (Paris) **256**, 5642—5643 (1963).
STRAUB, M.: Z. Vergleich. Physiol. **39**, 507—523 (1957).
STRÜBING, H.: Z. Morph. Ökol. Tiere **43**, 357—386 (1954).
STUMPF, H.: Biol. Zbl. **78**, 116—142 (1959).
STURROCK, R. F.: Ann. Trop. Med. Parasitol. **60**, 100—105 (1966).
SUGIURA, Y.: Biol. Bull. **128**, 493—496 (1965).
SUHRMANN, R.: Biol. Zbl. **74**, 432—448 (1955).
SUKHANOVA, K. M.: Acta Protozool. (Warzawa) **3**, 153—163 (1965).
SULLIVAN, C. M.: J. Fish. Res. Board Can. **11**, 153—170 (1954).
SULLIVAN, C. M., FISHER, K. C.: Biol. Bull. **107**, 278—288 (1954).
SULLIVAN, C. R.: Can. Entomol. **91**, 213—232 (1959).
SULLIVAN, C. R.: Can. Entomol. **97**, 978—993 (1965).
SULLIVAN, C. R., WALLACE, D. R.: Can. Entomol. **99**, 835—850 (1967).
SUMMERS, W. C.: Biol. Bull. **137**, 202 (1969).
SUMNER, F. B., DOUDOROFF, P.: Biol. Bull. **74**, 403—429 (1938).
SUMNER, F. B., WELLS, N. A.: Biol. Bull. **69**, 368—378 (1935).
SUNDBY, R. A.: Entomophaga **11**, 395—404 (1966).
SUOMALAINEN, P.: Verh. Intern. Ver. Limnol. **13**, 873—878 (1958).
SUSKI, Z. W.: Pol. Pismo Entomol. **33**, 579—602 (1969).
SUZUKI, D. T., PITERNICK, L. K., HAYASHI, S., TARASOFF, M., BAILLIE, D., ERASMUS, U.: Proc. Nat. Acad. Sci. (Wash.) **57**, 907—912 (1967).
SUZUKI, S.: Sci. Rep. Tôhoku Univ. **4**, **52**, 153—161 (1966).
SVENSSON, R.: Exp. Parasitol. **4**, 502—525 (1955).
SWAN, L. W.: Ecology **33**, 109—111 (1952).
SWARUP, H.: J. Zool. Soc. India **10**, 108—113 (1959a).
SWARUP, H.: J. Zool. Soc. India **11**, 1—6, 7—10 (1959b, c).
SWARUP, H.: J. Genet. (Calcutta) **56**, 129—142 (1959d).
SWEATMAN, G. K.: J. Med. Entomol. **5**, 429—439 (1968).
SWEATMAN, G. K.: J. Med. Entomol. **7**, 71—78 (1970).
SWEATMAN, G. K., GREGSON, J. D.: J. Med. Entomol. **7**, 575—584 (1970).
SWEENEY, B. M., HASTINGS, J. W.: Cold Spring Harb. Symp. Quant. Biol. **25**, 87—104 (1960).
SWEET, J. G., KINNE, O.: Helgoländer Wiss. Meeresuntersuch. **11**, 49—69 (1964).
SWIFT, D. R.: J. Fish. Res. Board Can. **22**, 913—917 (1965).

SYMONS,P.E.K.: Ecology **45**, 580—591 (1964).
TAKEHARA,I.: Low Temp. Sci., B, **22**, 71—78 (1964).
TAKSDAL,G.: Entomol. Exp. Appl. **10**, 377—387 (1967).
TAMURA,S., TOYOSHIMA,Y., WATANABE,Y.: Jap. J. Med. Sci. Biol. **19**, 85—96 (1966).
TANNO,K.: Low Temp. Sci., B, **21**, 41—53 (1963), **22**, 51—57 (1964).
TANNO,K., ASAHINA,E.: Low Temp. Sci., B, **22**, 59—70 (1964).
TANTAWY,A.O.: Genetics **46**, 227—238 (1961).
TASHIAN,R.E.: Zoologica **41**, 39—47 (1956).
TASHIAN,R.E., VERNBERG,F.J.: Zoologica **43**, 89—93 (1958).
TAUBER,M., TAUBER,C.A., DENYS,C.J.: J. Insect Physiol. **16**, 949—955 (1970).
TAYLOR,L.R.: J. Animal. Ecol. **32**, 99—117 (1963).
TAYLOR,R.E.,JR., BARKER,S.B.: Gen. Comp. Endocr. **9**, 129—134 (1967).
TEAL,J.M.: Physiol. Zoöl. **32**, 1—14 (1959).
TEMPLETON,J.R.: Physiol. Zoöl. **33**, 136—145 (1960).
TESCH,F.-W.: Biol. Zbl. **75**, 397—407 (1956).
TESCHNER,D.: Naturwissenschaften **48**, 724 (1961).
TEULADE,P.: Compt. Rend. Acad. Sci. (Paris) **256**, 4507—4509 (1963).
THAN-XUAN,N.: Ann. Soc. Entomol. Fr. (N.S.) **3**, 151—164 (1967).
THEEDE,H.: Kieler Meeresforsch. **21**, 153—166 (1965).
THEEDE,H.: Naturw. Rdsch. **20**, 468—475 (1967).
THEEDE,H.: Limnologica **7**, 119—128 (1969).
THEEDE,H.: Habil. Schr. Math.-Naturw. Fak., Univ. Kiel (1970).
THEEDE,H., LASSIG,J.: Helgoländer Wiss. Meeresuntersuch. **16**, 119—129 (1967).
THEOBALD,P.U.K.: Catholic Univ. Am. Biol Stud. **50**, 1—37 (1959).
THIEDE,W.: Z. Wiss. Zool. **172**, 305—346 (1965).
THIELE,H.-U., KREHAN,J.: Entomol. Exp. Appl. **12**, 67—73 (1969).
THIELE,H.-U., LEHMANN,H.: Z. Morph. Ökol. Tiere **58**, 373—380 (1967).
THIESSEN,C.I., MUTCHMOR,J.A.: J. Insect Physiol. **13**, 1837—1842 (1967).
THOMPSON,S.R.: Genetics **56**, 13—22 (1967).
THOMPSON,T.E.: Oikos **9**, 246—252 (1959).
THOMSEN,E., THOMSEN,M.: Z. Vergleich. Physiol. **24**, 343—380 (1937).
THORMAR,H.: Exp. Cell Res. **28**, 269—279 (1962).
THORNBURN,C.C., MATTY,A.J.: Comp. Biochem. Physiol. **8**, 1—12 (1963).
THORNBURN,C.C., MATTY,A.J.: J. Endocr. **36**, 221—229 (1966).
THORSON,G.: Zool. Anz. (Suppl. Bd) **15**, 276—327 (1952).
THORSON,T.B.: Ecology **38**, 100—116 (1955).
TIMET,D.: Thalass. Jugosl. **2**, 5—21 (1963).
TIMOFEEFF-RESOWSKY,N.W.: Genetics **12**, 128—165 (1927).
TINBERGEN,N.: Instinktlehre. Berlin, Hamburg 1956.
TISCHLER,W.: Z. Morph. Ökol. Tiere **35**, 251—287 (1939).
TISCHLER,W.: Klima, Witterung und Tierwelt. In: Klima, Wetter, Mensch. Heidelberg 1952.
TISCHLER,W.: Synökologie der Landtiere. Stuttgart 1955.
TISCHLER,W.: Z. Wiss. Zool. **163**, 168—209 (1960).
TISCHLER,W.: Agrarökologie. Jena 1965.
TISCHLER,W.: Z. Morph. Ökol. Tiere **59**, 54—74 (1967a).
TISCHLER,W.: Biol. Zbl. **86**, 473—484 (1967b).
TISCHLER,W.: Grundriß der Humanparasitologie. Jena 1969.
TISHLER,P.V.: Endocrinology **72**, 673—676 (1963).
TITSCHAK,E.: Z. Wiss. Zool. **124**, 213—251 (1925).
TOBARI,J.: Genetics **53**, 249—259 (1966).
TODD,M.-E., DEHNEL,P.A.: Biol. Bull. **118**, 150—172 (1960).
TOEWS,D.P., HICKMAN,C.P.,JR.: Comp. Biochem. Physiol. **29**, 905—918 (1969).
TOMBES,A.S.: J. Insect Physiol. **10**, 997—1003 (1964).
TOUTAIN,J.: Ann. Endocr. (Paris) **22**, 886—897 (1961).
TOUZEAU,J.: Rev. Zool. Agric. Appl. **65**, 41—49 (1966).
TRIBE,M.A., BOWLER,K.: Comp. Biochem. Physiol. **25**, 427—436 (1968).

TROSHIN, A. S. (Ed.): The cell and environmental temperature. Proc. Int. Symp. Cytoecology, Leningrad 1963; Engl. Ed. by PROSSER, C. L., Oxford 1967.

TRPIŠ, M., HORSFALL, W. R.: Ann. Zool. Fennicae. **6**, 156—160 (1969).

TSUKUDA, H.: J. Inst. Polytechn., Osaka Univ. (D) **10**, 95—104 (1959), **11**, 43—54, 55—62 (1960b, c).

TSUKUDA, H.: Biol. J., Nara Women's Univ. **10**, 11—14 (1960a).

TSUKUDA, H.: J. Biol., Osaka City Univ. **12**, 15—45 (1961).

TSUKUDA, H., OHSAWA, W.: J. Inst. Polytechn., Osaka Univ. (D) **9**, 69—76 (1958).

TUCKER, V. A.: J. Exp. Biol. **44**, 77—92 (1966).

TYLER, A. V.: Can. J. Zool. **44**, 349—364 (1966).

TYNEN, M.: Nature **225**, 587 (1970).

UMMINGER, B. L.: Yale Sci. Magaz. **42**, 6—10 (1968).

UMMINGER, B. L.: Am. Zool. **9**, 1092 (1969a), cf. 588—589, **7**, 731 (1967), **8**, 764—765 (1968).

UMMINGER, B. L.: J. Exp. Zool. **172**, 283—302, 409—423 (1969b), **173**, 159—174 (1970a).

UMMINGER, B. L.: J. Fish. Res. Board Can. **27**, 404—409 (1970b).

UMMINGER, B. L.: Nature **225**, 294—295 (1970c).

UMMINGER, B. L.: Comp. Biochem. Physiol. **38**, A, 141—145 (1971).

USCHATINSKAJA, R. S.: Dokl. Akad. Nauk SSSR **68**, 1101—1104 (1949).

USHAKOV, B. P.: Zool. Jahrb. (Syst.) **87**, 507—524 (1960).

USHAKOV, B. P. (Ed.): Problems of cytoecology of animals. Inst. Cytol., Akad. Sci. USSR, coll. art. **6**, Moscow 1963.

USHAKOV, B. P.: Physiol. Rev. **44**, 518—560 (1964).

USHAKOV, B. P.: Helgoländer Wiss. Meeresuntersuch. **14**, 466—481 (1966).

USHAKOV, B. P.: Marine Biol. **1**, 153—160 (1968).

USHAKOV, B. P., AMOSOVA, L., PASHKOVA, L., CHERNOKOZHEVA, I.: J. Exp. Zool. **167**, 381—390 (1968).

USHAKOV, B. P., GLUSHANKOVA, M. A.: Citologija (Moscow) **12**, 510—515 (1970).

USHAKOV, B. P., KUSAKINA, A. A.: V[th] intern. Congress Biochem., Moscow, 1961.

USUA, E. J.: J. Econom. Entomol. **61**, 1091—1093 (1968).

UVAROV, B. P.: Transact. Entomol. Soc. Lond. **79**, 1—232 (1931).

VALEN, E.: Acta Physiol. Scand. **42**, 358—362 (1958).

VANDERBERG, J. P., YOELI, M.: J. Parasitol. **52**, 559—564 (1966).

VASSILEVA-DRYANOVSKA, O. A., GENCHEVA, E.: Dokl. Bolgar. Akad. Nauk **17**, 1055—1057 (1964).

VAUGHN, C. M.: Am. Midland Naturalist **49**, 214—228 (1953).

VERDONK, N. H., DEGROOT, S. J.: Proc. Koninkl. Ned. Akad. Wetenschap. C. **73**, 171—185 (1970).

VERMA, A. N., ATWAL, A. S.: Beitr. Entomol. **18**, 249—258 (1968).

VERNBERG, F. J.: Biol. Bull. **117**, 163—184, 582—593 (1959).

VERNBERG, F. J.: Ann. Rev. Physiol. **24**, 517—546 (1962).

VERNBERG, F. J.: Am. Zool. **9**, 333—341 (1969).

VERNBERG, F. J., COSTLOW, J. D., JR.: Physiol. Zoöl. **39**, 36—52 (1966).

VERNBERG, F. J., MERINEY, D. K.: J. Elisha Mitchell Sci. Soc. **73**, 351—362 (1957).

VERNBERG, F. J., SCHLIEPER, C., SCHNEIDER, D. E.: Comp. Biochem. Physiol. **8**, 271—285 (1963).

VERNBERG, F. J., TASHIAN, R. E.: Ecology **40**, 589—593 (1959).

VERNBERG, F. J., VERNBERG, W. B.: Helgoländer Wiss. Meeresuntersuch. **9**, 476—487 (1964).

VERNBERG, F. J., VERNBERG, W. B.: J. Elisha Mitchell Sci. Soc. **82**, 30—34 (1966a).

VERNBERG, F. J., VERNBERG, W. B.: Comp. Biochem. Physiol. **19**, 489—524 (1966b).

VERNBERG, F. J., VERNBERG, W. B.: Oikos **18**, 118—123 (1967).

VERNBERG, F. J., VERNBERG, W. B.: The animal and the environment. New York 1970.

VERNBERG, W. B.: Exp. Parasitol. **11**, 270—275 (1961).

VERNBERG, W. B.: Am. Zool. **9**, 357—365 (1969).

VERNBERG, W. B., HUNTER, W. S.: Exp. Parasitol. **11**, 34—38 (1961).

VERNBERG, W. B., VERNBERG, F. J.: Comp. Biochem. Physiol. **17**, 363—374 (1966).
VERNBERG, W. B., VERNBERG, F. J.: J. Exp. Marine Biol. Ecol. **2**, 113—123 (1968a).
VERNBERG, W. B., VERNBERG, F. J.: Comp. Biochem. Physiol. **26**, 499—508 (1968b).
VERRILL, A. H.: Am. J. Sci. (4) **12**, 88 (1901).
VIELMETTER, W.: Zool. Anz. (Suppl. Bd) **20**, 102—106 (1957)
VIELMETTER, W.: J. Insect Physiol. **2**, 13—37 (1958).
VOELKEL, H.: Arb. Biol. Anst. Land. Forstw. **13**, 129—172 (1925).
VOGE, M.: J. Parasitol. **45**, 175—181 (1959).
VOGEL, G., JOHN, H., KRAUSE, H.: Pflügers Arch. Ges. Physiol. **267**, 414—416 (1958),
    cf. 2. intern. Symp. (Mech. Erregung), Berlin 1959, 122—128.
VOGEL, W.: Z. Wiss. Zool. **173**, 344—378 (1966).
VOGT, T., JAMESON, D. L.: Copeia **1970**, 135—144.
WADLEY, F. M.: Ann. Entomol. Soc. Am. **24**, 325—395 (1931).
WAHL, M.: Oecologia **3**, 14—55 (1969).
WALDOW, U.: Z. Vergleich. Physiol. **69**, 249—283 (1970).
WALLACE, H. R.: Nematologica **6**, 222—236 (1961).
WALTHER, J. B.: Zool. Anz. (Suppl. Bd) **28**, 353—358 (1965).
WANG, Y-H. E., GILL, G. D.: J. Econom. Entomol. **63**, 1666—1668 (1970).
WARBURG, M. R.: Physiol. Zoöl. **38**, 99—109 (1965).
WATANABE, J.: Physiol. Zoöl. **23**, 258—264 (1950).
WAUTIER, J., TROIANI, D.: Ann. Stat. Centr. Hydrobiol. Appl. **8**, 7—50 (1960).
WEATHERLEY, A. H.: Proc. Zool. Soc. **141**, 527—555 (1963).
WEATHERLEY, A. H.: Biol. Bull. **139**, 229—238 (1970).
WEBER, H.: Z. Morph. Ökol. Tiere **23**, 575—753 (1931).
WEDEMEYER, G.: Comp. Biochem. Physiol. **29**, 1247—1251 (1969).
WEILL, R.: Compt. Rend. Acad. Sci. (Paris) **254**, 4345—4346 (1962).
WEISEL, G. F.: Ecology **36**, 1—6 (1955).
WEISER, C. J.: Science **169**, 1269—1278 (1970).
WEISSMAN-STRUM, A., KINDLER, S. H.: Nature **196**, 1231—1232 (1962).
WELLENSTEIN, G.: Z. Angew. Entomol. **41**, 368—385 (1957).
WERNER, B.: Helgoländer Wiss. Meeresuntersuch. **6**, 137—170 (1958), **7**, 206—237
    (1961).
WERNER, B.: Veröff. Inst. Meeresforsch. Bremerhaven, Sonderbd., 153—177 (1963).
WERNER, J. K.: Copeia **1969**, 592—602.
WESENBERG-LUND, C.: Biologie der Süßwassertiere. Wirbellose Tiere. Wien 1939.
WEYRAUCH, W.: Z. Vergleich. Physiol. **23**, 51—63 (1936).
WHITE, M. J. D.: J. Genet. **29**, 203—215 (1934).
WIESER, W.: Z. Vergleich. Physiol. **47**, 1—16 (1963).
WIESER, W.: Helgoländer Wiss. Meeresuntersuch. **9**, 356—370 (1964).
WIESER, W.: Pedobiologia **5**, 304—331 (1965a).
WIESER, W.: Zool. Anz. (Suppl. Bd) **28**, 359—364 (1965b).
WIESER, W., FRITZ, H., REICHEL, K.: Z. Vergleich. Physiol. **70**, 62—79 (1970).
WIESER, W., NOPP-PAMMER, E.: Zool. Anz. (Suppl. Bd) **31**, 131—139 (1968), cf. Comp.
    Biochem. Physiol. **24**, 1015—1025 (1968).
WIGGLESWORTH, V. B., GILLETT, J. D.: J. Exp. Biol. **11**, 120—139 (1934).
WIKGREN, B. J.: Acta Zool. Fennicae **71**, 3—102 (1953).
WILBUR, K. M., MCMAHAN, E. A.: Ann. Entomol. Soc. Am. **51**, 27—32 (1958).
WILDBOLZ, T., RIGGENBACH, W.: Mitt. Schweiz. Entomol. Ges. **42**, 58—78 (1969).
WILHOFT, D. C.: Gen. Comp. Endocr. **7**, 445—451 (1966).
WILKENS, J. L., FINGERMAN, M.: Biol. Bull. **128**, 133—141 (1965).
WILKES, A.: Can. J. Gen. Cyt. **1**, 102—109 (1959).
WILLIAMS, A. B.: Bull. Biol. **119**, 560—571 (1960).
WILLIAMS, R. J.: Comp. Biochem. Physiol. **35**, 145—161 (1970).
WILSON, A. C., KAPLAN, N. O., LEVINE, L., PESCE, A., REICHLIN, M., ALLISON, W. S.:
    Federation Proc. **23**, 1258—1266 (1964).
WILSON, D. M.: Adv. Insect Physiol. **5**, 289—338 (1968).
WILSON, F., WOOLCOCK, L. T.: Australian J. Zool. **8**, 153—169 (1960).

WILSON, J. Y.: Heredity **13**, 263—267 (1959).
WILSON, K. J., LEE, A. K.: Comp. Biochem. Physiol. **33**, 311—322 (1970).
WILSON, L. F.: Ann. Entomol. Soc. Am. **61**, 1490—1495 (1968).
WINTER, CH.: Zool. Anz. (Suppl. Bd) **30**, 234—239 (1967).
WOELLWARTH, C. V.: Embryologia **6**, 219—242 (1961).
WOHLFAHRT, T. A.: Zool. Anz. (Suppl. Bd) **18**, 133—137 (1955).
WOHLGEMUTH, R.: Z. Vergleich. Physiol. **40**, 119—161 (1957).
WOHLSCHLAG, D. E.: Copeia **1961**, 11—18.
WOHLSCHLAG, D. E.: Ecology **41**, 287—292 (1960), **44**, 557—564 (1963).
WOLF, K., QUIMBY, M. C.: Science **135**, 1065—1066 (1962).
WOLTERECK, R.: Z. Abstammungsl. **67**, 173—196 (1934).
WOODHAED, P. M. J., WOODHAED, A. D.: Proc. Linnae Soc. Lond. **169**, 63—64 (1958).
WOODHAED, P. M. J., WOODHAED, A. D.: Proc. Zool. Soc. **133**, 181—199 (1959/60).
WOOLHOUSE, H. W. (Ed.): Dormancy and survival. Symp. Soc. Exp. Biol. Cambridge 1969.
WOYNÁROVICH, E.: Acta Biol. Debrećina **2**, 155—168 (1963).
WYNIGER, R.: Mitt. Schweiz. Entomol. Ges. **29**, 41—57 (1956).

YAMAMOTO, M.: Sci. Rep. Tôhoku Univ. 4, **30**, 179—186 (1964).
YARON, Z.: Gen. Comp. Endocr. **12**, 604—608 (1969).
YINON, U., SHULOV, A.: Entomol. Exp. Appl. **13**, 107—121 (1970).
YOUDEOWEI, A.: Entomol. Exp. Appl. **11**, 68—80 (1968).
YURKIEWICZ, W. J., SMYTH, T., JR.: J. Insect Physiol. **12**, 189—194 (1966).

ZAHN, M.: Zool. Beitr. (N. F.) **3**, 127—194 (1958).
ZAHN, M.: Int. Rev. Ges. Hydrobiol. Hydrog. **45**, 455—460 (1960).
ZAHN, M.: Zool. Beitr. (N. F.) **7**, 15—25 (1962).
ZAHN, M.: Umschau **63**, 711 (1963).
ZAIN-UL-ABEDIN, M., KABORSKI, B.: Can. J. Physiol. **44**, 505—507 (1966).
ZARET, T. M.: Limnol. Oceanogr. **14**, 301—303 (1969).
ŽDÁREK, I., POSPIŠIL, I.: Acta Entomol. Bohem. **62**, 421—427 (1965).
ZEBE, E.: Ergebn. Biol. **24**, 247—286 (1961).
ZERBST, E.: Habil. Schr. Med. Fak., FU Berlin 1966.
ZERBST, E., SCHMIDT, U., MÜLLER, E.: Pflügers Arch. Ges. Physiol. **289**, R. 26 (1966).
ZHIRMUNSKY, A. V., PISAREVA, L. N.: Problems of cytology and protistology. Inst. Cytocol. Leningrad (1961).
ZIEGELMEIER, E.: Helgoländer Wiss. Meeresuntersuch. **21**, 9 (1970).
ZIMMERING, S.: Genetics **48**, 133—138 (1963).
ZIMMERMAN, W. F., PITTENDRIGH, C. S., PAVLIDIS, T.: J. Insect Physiol. **14**, 669—684 (1968).
ZUBER, H.: Naturw. Rdsch. **22**, 16—22 (1969).
ZWEIFEL, R. G.: Copeia **1959**, 322—327.
ZWEIFEL, R. G.: Bull. Am. Mus. Hist. **140**, 1—64 (1968).
ZWICKY, K.: Rev. Suisse Zool. **56**, 316—321 (1949).
ZWÖLFER, W.: Biol. Zbl. **50**, 724—759 (1930).
ZWÖLFER, W.: Z. Angew. Entomol. **17**, 475—562 (1931), 20, 1—50 (1934), **21**, 333—384 (1935).

Homeothermic Organisms

H. Hensel, K. Brück, and P. Raths

I. Homeothermy and Poikilothermy

The achievement of "thermodynamic freedom" (BURTON and EDHOLM, 1955) by *homeothermic* (constant-temperature or warm-blooded) animals, who can maintain a practically constant internal temperature regardless of changes in external temperature, is undoubtedly the most significant advance in regard to the manifold relations between temperature and living beings. In contrast to the poikilotherms, whose vital processes depend entirely on external temperature conditions, homeotherms — birds and mammals — live at practically the same level of intensity during the entire year, and in all climatic regions; thus, they have a high level of reactivity, and can make use of their strengths and abilities whenever they desire. Their high internal temperature prevents them from freezing, which is the fate of poikilotherms under cold conditions unless they can retire to deep waters, to frost-free layers of the soil, or to warmer climatic zones, or unless they can survive in desiccated, encapsulated form. The geographic distribution of homeotherms is thus not so limited as that of poikilothermic forms of life, especially plants, which cannot change locality in order to avoid unfavorable temperatures. In regard to the direct influence of temperature, warm-blooded life is possible practically everywhere on earth; in regard to other factors, however, there are naturally a number of practical limitations, among them the problem of nutrition, which is also dependent upon temperature.

There is no clear-cut distinction between homeothermy and poikilothermy. On the one hand, many poikilothermic forms of life show rudimentary signs of temperature regulation. This is particularly true for certain species of reptiles which have a striking ability to regulate their body temperature within narrow limits by behavioral means and, to a certain extent, even by metabolic changes (TEMPLETON 1970). For example, the "heliotherms" among the reptiles regulate shuttling between sun and shade and orienting positively or negatively to the sun's rays. However, their body temperature may vary by 20° C and more according to internal and external conditions. Furthermore, reptiles may abandon normal behavioral temperature regulation while defending a territory or avoiding a predator (DE WITT, 1967). On the other hand, some warm-blooded animals have extremely imperfect control over their body temperature. In the course of ontogeny, homeothermy was developed only gradually, and a certain group of warm-blooded animals is poikilothermic for a certain time after birth. A special position is occupied by those warm-blooded animals which go into *hibernation* or other forms of immobility under certain conditions, and thus spend a period of "vita minima" at body temperatures of a few degrees above the freezing point. Warm-blooded animals with relatively great fluctuations in body temperature are called heterothermic; however, this term cannot be clearly defined, since it includes all the stages between homeothermy and poikilothermy. The basic differences in body temperature control in the cold existing between poikilotherms, homeotherms and hibernating homeotherms are shown in the schematic diagram (Fig. 1). With falling environmental temperature the body temperature of poikilotherms declines continuously. In some insects it may fall below the freezing point, with death occurring after a sudden increase in body temperature. The body temperature of homeotherms is maintained at normal over a wide range of cold stress; but with cold stress progressing beyond a critical point, homeothermy breaks down and death ensues. Large and well-insulated homeotherms can survive much lower environmental temperatures than hibernators. The body temperature of hibernating homeotherms approaches freezing point but is prevented from falling below it.

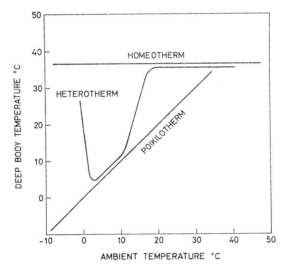

Fig. 1. Schematic representation of changes in deep body temperature of large homeotherms, poikilotherms (insect), and hibernators, in response to environmental temperature. Adapted from EISENTRAUT (1960)

In designating homeothermic animals with the old term "warm-blooded", we must remember that many "cold-blooded" animals occasionally attain and even exceed the temperatures of warm-blooded animals (p. 299). The decisive point here is not the level of the temperature, but rather its *regulation*, which is one of the most remarkable physiological accomplishments of the organism. This is made possible mainly by specific peripheral and central nervous structures which constantly detect the organism's temperature fluctuations and attempt to keep them in balance by means of appropriate countermeasures. One of these means is the regulation of heat production from exothermic metabolic processes: whereas in poikilotherms metabolism rises and falls with the external temperature, in homeo-therms it moves, within certain limits, counter to the external temperature. In addition, homeotherms have means for regulating their thermal resistance, and for regulated absorption of heat via evaporation of water. It is of fundamental importance that with decreasing body size the surface/volume quotient increases, and thus likewise the relative heat-releasing area. Since neither heat production nor thermal insulation can be increased indefinitely, this imposes a lower limit on the size of warm-blooded animals, the smallest of which can live only in warmer zones.

Of the physical properties of the substances in the homeotherm organism, those of *water* play a special part in regard to temperature. The greatest part of the body substance is formed of water, which has a high specific heat of $1 \text{ cal g}^{-1} \text{ °C}^{-1}$; thus it buffers fluctuations in the external temperature and permits a considerable transport of heat through the flowing blood. Another important property of water is its high evaporation heat (580 cal g^{-1}), which provides the body with effective protection against overheating.

Homeotherms differ from poikilotherms not only in temperature regulation, but also in the level of their *energy metabolism*. According to BENEDICT (1932), the energy turnover of a resting rabbit weighing 2.5 kg at a body temperature of 37° C is $1.9 \text{ kcal kg}^{-1} \text{ h}^{-1}$, whereas that of a rattlesnake with the same weight and body temperature is only $0.3 \text{ kcal kg}^{-1} \text{ h}^{-1}$. The resting metabolism of small mammals at 37° C body temperature is $2.3 - 6.7$ times higher than that of lizards of compar-able body size and body temperatures of 37° C (TEMPLETON, 1970). Thus homeo-therms are *tachymetabolic* (fast metabolism) and poikilotherms *bradymetabolic* (slow metabolism). ADOLPH (1951) has shown that not only the regulatory centers of homeotherms are responsible for such differences; according to him, the differ-ences remain even after the thermoregulatory mechanism has been blocked and body temperature artificially lowered. In homeothermic vertebrates (mammals, birds), O_2 consumption (Fig. 2) and cardiac frequency are much higher in the

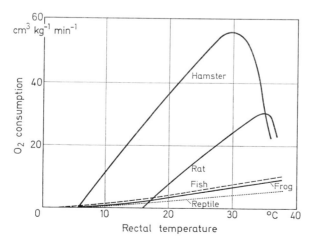

Fig. 2. Oxygen consumption of homeothermic and poikilothermic vertebrates as function of body temperature. From ADOLPH (1951)

body-temperature range between 10 and 20° C than in poikilothermic vertebrates (fish, amphibians, reptiles). The same is true of the frequency of the isolated heart, in which there is naturally no possibility of any interference on the part of central regulatory processes (Fig. 3). Other processes, however, such as respiratory frequency, which varies widely in individual species, show no striking differences. Thus, a homeotherm is not simply "a poikilotherm with a particular hypothalamic function superimposed" (ADOLPH, 1951), but rather has different structures in regard to temperature behavior, down to the level of tissue metabolism and enzymatic processes.

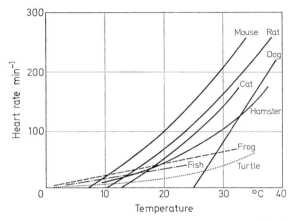

Fig. 3. Frequency of isolated hearts of homeothermic and poikilothermic vertebrates as function of temperature. From ADOLPH (1951)

II. Body Temperatures

A. The Temperature Field of the Body

1. Radial and Axial Temperature Gradients

Warm-blooded animals have an extremely complicated spatial temperature field; furthermore, it is not time-constant. In accordance with the basic laws of heat flow, the external parts of the body have a lower mean temperature than do the internal parts; likewise, the temperature decreases in the longitudinal axis of the extremities, thus producing a *radial* and an *axial* temperature gradient (BURTON, 1934; BRÜCK and HENSEL, 1953; WISSLER, 1961, 1970; BROWN, 1965). The situation is further complicated by the differing heat production of individual organs, by irregularities in geometric forms, by changes in insulation and evaporation, and by convective heat transport via the blood. Attempts have been made to distinguish between the *core* of the body and its *shell*; the former is defined as the homeothermic part of the body, and the latter as the remaining part, which participates more or less in the fluctuations of the external temperature. However, no clear-cut dividing line can be drawn between the two parts. In homeotherms, the core generally consists of the interior of the thorax and abdomen, the brain, and part of the skeletal muscles. Under moderate changes in ambient temperature, the shell has been calculated as comprising 20% (DU BOIS, 1951) to 35% (BURTON, 1935) of the human body. However, during marked chilling, the shell amounts to as much as 50% of the total body; this is equivalent to a mean layer thickness of 2.5 cm.

2. Skin and Subcutaneous Temperature

In *man*, the *skin temperatures* vary most with external factors (Fig. 4), and also show the greatest topographic differences (MURLIN, 1939; SHEARD, WILLIAMS, and HORTON, 1941; DU BOIS and HARDY, 1941). In unclothed humans after 3 h of cooling at a room temperature of 5° C (50% relative humidity, wind 10−20 cm sec^{-1}), WEZLER and NEUROTH (1949) found local differences in skin temperature of up to 15° C (13−28° C); temperatures were lowest in the fingers and toes, and highest in the trunk and forehead. After 3 h of warming up to 50° C, the differences amounted to only 2.5° C (35−37.5° C). Here the *mean* or *integral* skin temperature of the body was 22 and 37° C respectively, and the temperature gradient between core and surface averaged 15 and 1° C respectively. With normal clothing and room temperatures between 15 and 20°C, the mean skin temperatures were between 32 and 35° C. Mean skin temperatures of human subjects as a function of ambient temperatures between 10 and 50° C and wind velocities of 1−5 m/sec have been measured by MITCHELL et al. (1968). Their results are in good agreement with the theoretical equation given by IAMPIETRO (1961) for the estimation of

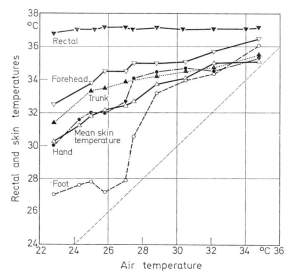

Fig. 4. Skin and rectal temperatures of man at various ambient temperatures. From HARDY and DU BOIS (1938)

mean skin temperature from air temperature and velocity. The topography of skin temperatures of the human head, especially in the cold, was measured by EDWARDS and BURTON (1960).

The skin temperatures of *mammals* and *birds* are strongly dependent upon the ability of their hair or feathers to insulate heat. Numerous measurements of skin temperatures made by veterinary physiologists in domestic agricultural animals have revealed conditions similar to those in human beings under their clothing. According to LANDSIEGEL (1937) cattle have a mean skin temperature of 33.7° C (33.4−34.8° C) on the trunk and head; at 15° C room temperature in calves it is considerably higher, i.e. 36.2° C. The mean gradient between rectal and skin temperature is 4.9° C. At the distal extremities of the body, e.g. the tip of the tail or ears, the temperatures are as much as 10° C lower. The better the thermal insulation of the body covering, the more uniform and more independent of ambient temperature the skin temperatures will be. In sheep with a hair length of 8 cm, HOFFMANN (1938) found no measurable differences in skin temperature at ambient temperatures of 17° C and 4° C. According to BLAXTER (1965), the temperature of the trunk in sheep kept at an environment of −9° C, is 29° C, while the fluctuating ear temperature falls periodically to 0° C and the foot temperature reaches 3° C. At an air temperature of 14° C, the mean skin temperature of the trunk is 29.8° C

in pigs, 33.9° C in goats, and 36.3° C in sheep; at an external temperature of
−12° C it sinks by 14.4° C in pigs, but by only 4.5° C in goats and sheep (LEE et al.,
1941b). Pigs kept at environmental temperatures of −30° C reach skin temperatures lower than 5° C (IRVING and KROG, 1955). For further references see
WHITTOW (1971).

The feathered parts of birds' skin have very high temperatures, owing to the excellent insulation provided by the feathers. In chickens, according to WILSON and
PLAISTER (1951), skin temperatures on the neck, breast, humerus and femur are
between 38.5 and 41° C at room temperatures of 23−32° C, only 0.7−1.8° C lower
than the rectal temperature. The femur has the warmest skin area. In cold environments, skin temperatures beneath the plumage generally exceed 30° C and, in a
significant proportion of cases, 35° C (KALLIR, 1930; WILSON, HILLERMAN, and
EDWARDS, 1952; BARTHOLOMEW and DAWSON, 1954; IRVING and KROG, 1955;
STEEN and ENGER, 1957; DAWSON and TORDOFF, 1959). Skin temperatures in areas
not covered with feathers, such as the comb and wattles, vary to a much greater
extent, and are from 3−6° C lower than the cloacal temperature, depending upon
the ambient temperature. The greatest fluctuations are found in the extremities,
just as in mammals: at an ambient temperature of 23° C, the skin temperature in
the lower leg is 10° C lower than the cloacal temperature. For temperature distribu-
tions in mammals and birds under arctic conditions see p. 727.

The *subcutaneous temperature* increases continuously with increasing depth (BRÜCK
and HENSEL, 1953). The temperature of the *muscles* of the extremities at rest is
lower than the rectal temperature, and can fluctuate considerably with changes in
external temperature (READER and WHYTE, 1951; DANIELSON and KINARD, 1951).
In unclothed humans after 2 h at 20° C room temperature, BRÜCK and HENSEL
(1953) found values which were lower than the rectal temperature by 10° C in the
hand and foot, by 5° C in the lower leg, and by 2.5° C in the thigh. The temperature in the thigh muscles of dogs was around 4° C below the rectal temperature
(HORVATH, RUBIN, and FOLTZ, 1950). In water at 16° C, the muscle temperature
in the human hand sinks to 23° C within 40 min (BRÜCK and HENSEL, 1953).

3. Core Temperature

Temperatures in the interior of the body are also by no means uniform (Table 1).
In mice, rats, guinea-pigs, cats, dogs and calves, the temperatures of the inner
organs, including the brain, vary by 0.2−1.2° C under normal room conditions and
under anesthesia (WALTHER, BISHOP, and WARREN, 1941). Even within the individual organs, temperature differences of up to 0.9° C are found. KALLIR (1930)
found differences of 1−2° C in the inner organs of crows, pigeons and sparrows. The
brain has considerable local temperature gradients amounting to 1.4° C. In general,
the cortex is cooler than the basal regions (SEROTA and GERARD, 1938; HAYWARD,
1967), and the incoming blood is cooler than the central brain tissue. Under normal
living conditions the brain temperature of unrestrained cats fluctuates within
0.5° C, rising initially with external cooling and falling with external warming
(KUNDT, BRÜCK, and HENSEL, 1957; OGATA, 1959). Brain temperature decreases
during sleep and rises during arousal and states of emotion (DELGADO and HANAI,
1966). In some species, such as cat, dog and sheep (TAYLOR, 1966; BAKER and

HAYWARD, 1968; HAYWARD and BAKER, 1969; MAGILTON and SWIFT, 1969), the blood supplying the hypothalamus passes an extracranial carotid rete which is bathed by venous blood returning from various parts of the head. Variations in the temperature of this venous blood are transferred by countercurrent heat exchange to the arterial blood passing through the rete to the circle of Willis. Thus hypothalamic temperature might deviate from the central blood temperature under the influence of external thermal stimuli.

In striking contrast to previous views is the finding that liver temperatures in unanesthetized man are 0.2–0.6° C lower than rectal temperatures (GRAF, PORJÉ, and ALLGOTH, 1955; GRAYSON and KINNEAR, 1963). It is suggested that the liver loses heat via the respiratory tract through the diaphragm. The temperature of the testes in the scrotum is several degrees lower than that of the abdominal cavity (p. 670).

Blood temperature also shows considerable local differences. In internal organs with a high metabolic rate, the outflowing blood is warmer than the inflowing, so that the blood here has a cooling effect. The situation is just the opposite in the outer parts of the body, especially in the skin. The temperature of the venous blood in the extremities may be several degrees lower than that of the arterial blood, and even the arterial blood shows marked short-term differences in temperature (dogs: HORVATH, RUBIN, and FOLTZ, 1950), and can sink to 21.5° C in human extremities at cold room temperatures (BAZETT, 1949; BAZETT et al., 1948a, b). The cooling effect of air on the pulmonary blood is very slight. MATHER, NAHAS, and HEMINGWAY (1953) using dogs found that even at a room temperature of −18° C there was a temperature difference of no more than 0.03° C between blood in the pulmonary artery and in the pulmonary vein. These findings have been confirmed in dog and man by AFONSO et al. (1962a, b). Blood temperature in the human common carotid artery remains within 0.2° C of oral temperature; it is lowered by 0.2 to 0.5° C by cooling the face, possibly due to cooling from the internal jugular vein (RUBENSTEIN, MEUB, and ELDRIDGE, 1960).

Table 1. Internal temperatures of the body

Organ	Pigeon[a] °C	Calf[b] °C	Organ[c] (man)	Temp. diff. (rectum) °C	Organ[c] (man)	Temp. diff. (A. femoralis) °C
Liver	41.0	37.6	Sternum	−2.5	V. subclavia	−0.35
Stomach		37.5	Oral cavity	−0.4	V. cava cran.	−0.1
Pancreas	40.5		Esophagus	−0.25	V. cava caud.	±0
Kidney	41.0	37.3	Aorta	−0.25	Right ventricle	±0
Heart	40.0	37.2	Stomach	±0	A. pulmonalis	±0
Lung	40.4	36.7	Liver	±0	V. jugularis	+0.25
Cerebrum	39.0		Hypothalamus	+0.3	V. hepatica	+0.25
Spinal cord		37.4	Uterus	+0.3	Rectum, deep	+0.25
V. poplitea		35.0				

[a] From KALLIR (1930).
[b] From WALTHER, BISHOP, and WARREN (1941).
[c] From ASCHOFF, GÜNTHER, and KRAMER (1971).

4. Temperature of Body Orifices

The temperature of the body orifices (rectum, vagina, mouth, auditory meatus) is of particular interest, since it is usually considered to be representative of the body's interior temperature. In *man*, the rectal temperature averages 0.26° C more than the urine temperature, and 0.5° C more than the mouth temperature (HEP-BURN et al., 1933). The oral cavity has relatively large local variations in temperature (LUX and LUX, 1933), as do the rectum and vagina (cows: KRISS, 1921; man: BENEDICT and SLACK, 1911).

In man, a rectal temperature of 37° C is considered to be "normal"; however, the physiological range covers no less than 2° C (p. 518). Children usually have somewhat higher temperatures during the day than adults, with a maximum at the age of 1 year (DU BOIS, 1951). Numerous investigators have strongly criticized the usefulness of the rectal temperature as a criterion of the internal temperature of the human body, since rectal temperature depends in part upon the surrounding venous plexus and other local influences (MEAD and BONMARITO, 1949; GLASER, 1949; GRANT, 1951). Likewise, the temperature of the mouth can be considerably affected by external disturbances; according to HORVATH, MENDUKE, and PIERSOL (1950), it has no significant correlation with rectal temperature. Measurement of urine temperature provides more reliable data, but is not suitable for general use. From the point of view of temperature regulation it would perhaps be advisable to take the temperature of the hypothalamus as a reference value; however, for methodological reasons this can be done only in exceptional cases. Thus, it is scarcely possible to draw conclusions about the temperature of any given part of the body from the temperature of any other part. Likewise, calculating the *mean temperature* of the body (e.g. 0.67 rectal temperature $+0.33$ mean skin temperature, BURTON, 1935) can hardly be expected to produce better results. The temperature of the tympanic membrane (BENZINGER and TAYLOR, 1963) or the external auditory meatus (COOPER, CRANSTON, and SNELL, 1964) in man has been suggested as being closely related to central or hypothalamic temperature. However, cooling and warming the skin of the head causes temperature changes of the tympanic membrane of up to 0.4° C by the returning venous blood (WURSTER, 1968).

The rectal temperatures of *mammals* are usually between 36 and 39° C at moderate ambient temperatures. A selection is shown in Table 2. Rectal temperatures in *birds* are between 40 and 43° C. (Additional literature is given by KANITZ, 1925; BALDWIN and KENDEIGH, 1932; WISLOCKI, 1933; MORRISON and RYSER, 1952; PROSSER and BROWN, 1961; UDVARDY, 1963; WHITTOW, 1970, 1971; and others.) However, certain orders of birds virtually show body temperatures in the mammalian range (Table 3).

Various attempts have been made, e.g. by RODBARD (1950), to establish a relation between body size and rectal temperature. However, using the most reliable measurements available for rectal temperature and body weight in 56 species of mammal (from mice weighing 9 g to whales weighing 10^7 g) MORRISON and RYSER (1952) showed that the mean rectal temperature was 37.8° C $\pm$ 0.4° C in all weight classes, with a scattering of 1.23° C within the groups. According to these examinations there is thus no correlation between body size and core temperature.

Table 2. Rectal temperatures of various homeothermic mammals

Species	Mean °C	Variation °C
Shrew[a]	35.7	
Rat[b]	38.1	37.5—38.6
Rabbit[b]	39.4	38.6—40.1
Cat[b]	38.6	38.1—39.2
Dog[b]	38.6	38.1—39.2
Sheep[b]	39.1	38.3—39.9
Horse[b]	37.7	37.2—38.2
Cattle[b]	38.6	38.0—39.2
Camel[c]	36.4	34.9—37.8
Chimpanzee[c]	37.0	36.3—37.8
Man[d]	37.0	36.2—37.8
Elephant[c]	36.2	35.7—36.7
Seal[c]	38.3	
Whale[c]	36.5	36.0—37.0

[a] From KENDEIGH (1945a).
[b] From DUKES (1952).
[c] From WISLOCKI (1933).
[d] From DU BOIS (1951).

Table 3. Body temperatures for birds of various orders

Order	No. of species sampled	Range °C
Sphenisciformes (penguins)	6	37.0—38.9
Struthioniformes (ostrich)	1	39.2
Casuariiformes (casuaries and emu)	4	38.8—39.2
Apterygiformes (kiwis)	3	37.8—39.0
Tinamiformes (tinamous)	1	40.5
Gaviiformes (loons)	1	39.0
Podicipediformes (grebes)	4	38.5—40.2
Procellariiformes (albatrosses, shearwaters, petrels, and allies)	13	37.5—41.0
Pelicaniformers (tropic birds, pelicans, frigate-birds, and allies)	9	39.0—41.3
Ciconiiformes (herons, storks, ibises, flamingos, and allies)	12	39.5—42.3
Anseriformes (screamers, swans, geese, and ducks)	28	40.1—43.0
Falconiformes (vultures, hawks, and falcons)	12	39.7—42.8
Galliformes (megapodes, curassows, pheasants, and hoatzins)	22	40.0—42.4
Gruiformes (cranes, rails, and allies)	7	40.1—41.5
Charadriiformes (shorebirds, gulls, auks, and allies)	39	38.3—42.4
Columbiformes (sand-grouse, pigeons, and doves)	5	40.0—42.3
Cuculiformes (cuckoos and plantain eaters)	2	41.9—42.3
Strigiformes (owls)	9	39.2—41.2
Caprimulgiformes (goatsuckers, oilbirds, and allies)	5	37.6—42.4
Apodiformes (swifts and hummingbirds)	25	35.6—44.6
Coraciiformes (kingfishers, motmots, rollers, bee-eaters, and hornbills)	1	40.0
Piciformes (woodpeckers, jacamars, toucans, and barbets)	10	39.0—43.0
Passeriformes (perching birds)	101	39.2—43.8

From DAWSON and HUDSON (1970). The values presented are the extremes of the means for the active phase of the daily cycles of the species sampled.

The same conclusion was reached by IRVING and KROG (1954), who found no dependence of rectal temperature on body size in arctic mammals ($10^2 - 10^6$ g) or birds ($10 - 2 \cdot 10^3$ g). Likewise, there are no significant differences in body temperature between tropical and arctic animals.

B. Body Temperature and External Temperature

Fig. 5 shows the body temperature of various species as a function of air temperature. In general, there is a relation between the evolutionary level and the extent of homeothermy (DAWSON, 1972). Relatively imperfect is the temperature regulation in monotremes (MARTIN, 1902; ROBINSON and MORRISON, 1957; PARER and METCALFE, 1967), their rectal temperature and metabolic rate being considerably lower than those of higher homeotherms. More efficient is the regulation in

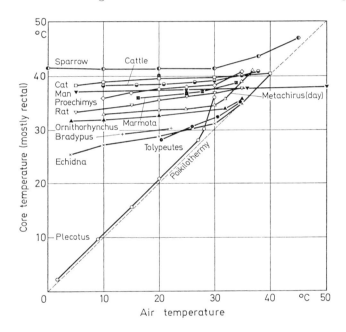

Fig. 5. Mean rectal temperatures as function of ambient temperature: *Echidna* (spiny anteater), *Ornithorhynchus* (duck bill platypus) and cat from MARTIN (1902); *Metachirus* and *Proechimys* (opossum) from MORRISON (1946); *Bradypus* (sloth) from SAWAYA (1941); *Tolypeutes* (armadillo) from EISENTRAUT (1932); *Plecotus* (bat) from EISENTRAUT (1940); *Marmota* (woodchuck) from BENEDICT (1938); rat from BIERENS DE HAAN (1922); cattle from REGAN and RICHARDSON (1935); sparrow from KENDEIGH (1944); man from WEZLER and NEUROTH (1949)

marsupials or Metatheria (MORRISON, 1946, 1965; BARTHOLOMEW, 1956; MORRISON and PETAJAN, 1962; DAWSON, 1969, 1972; DAWSON, DENNY, and HULBERT, 1969; DAWSON and HULBERT, 1970; MACMILLEN and NELSON, 1969; DAWSON and BENNETT, 1971). The tropic marsupials have a higher temperature and a better regulation during their nocturnal activity than during the day (Fig. 6). Various primitive placentals or Eutheria, such as insectivores (EISENTRAUT, 1956a; MORRISON, 1957; HILDWEIN, 1970) and edentates (WISLOCKI, 1933; WISLOCKI and ENDERS, 1935; ENDERS and DAVIS, 1936; BRITTON and ATKINSON. 1938; SAWAYA, 1941; EISENTRAUT, 1956a; ENGER, 1957; JOHANSEN, 1961a) also show low body temperatures, whereas in the more advanced mammals as well as in birds the highest degree of thermoregulatory efficiency and a high body temperature is seen.

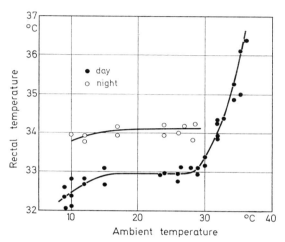

Fig. 6. Body temperature at various ambient temperatures in the opossum *(Metachirus caudatus)*. From MORRISON (1946)

In the higher homeotherms, changes in activity generally cause greater fluctuations in rectal temperature (up to several degrees) than do normal changes in ambient temperature. However, in some larger mammals living in hot climates, fluctuations of more than 3° C occur with changing environmental temperature (Fig. 7). This is by no means to be thought of as a temperature regulation defect but rather as a special phylogenetic adaptation (p. 722). Homeothermy is best developed in the carnivores, the equids, and man. (Ref. on quantitative responses over the full range of ambient temperatures in birds and mammals see HAMMEL, 1968; WHITTOW, 1970, 1971).

In hibernating or heterothermic animals, temperature regulation during the period of wakefulness is generally not so good as it is in the other higher mammals (p. 698). Temperature regulation in some species of bats during sleep is very poor and can be considered as nearly poikilothermic (p. 694.)

C. Periodic Fluctuations in Body Temperature

1. Diurnal Fluctuations

There is a more or less marked *diurnal fluctuation* of temperature in the core of homeotherms (Fig. 8). In the species that are active mainly during the day, the

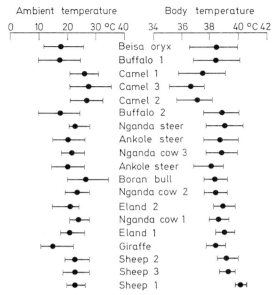

Fig. 7. Deep body temperatures of various African mammals and corresponding ambient temperatures (means, maxima and minima recorded during each period of measurement). From Bligh (1965)

maximum temperature is found in the afternoon and the minimum in the morning. Birds usually have their maximum temperature between noon and 3 p.m. and their minimum between midnight and 3 a.m. The maximum in mammals is usually be-

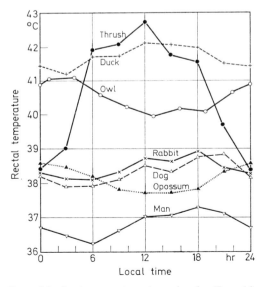

Fig. 8. 24-hour rhythm of body temperature in animals. *Proechimys* (opossum) from Morrison (1946); others from Simpson and Galbraith (1905)

tween 6 p.m. and 9 p.m. and the minimum between 3 a.m. and 6 a.m. (SIMPSON and
GALBRAITH, 1905; BALDWIN and KENDEIGH, 1932; BARTHOLOMEW and DAWSON,
1958). In *nocturnal animals*, the diurnal course of temperature is usually the opposite; this is true both of birds (SIMPSON and GALBRAITH, 1905; BALDWIN and
KENDEIGH, 1932; FARNER, CHIVERS, and RINEY, 1956) and mammals (HERTER,
1934; MORRISON, 1946). There are wide differences in the extent of fluctuation. It
appears to be greater in smaller birds than in larger species, whereas the mean
temperatures obviously do not depend upon size. In mammals, no connection can
be seen between body size and the amplitude of diurnal fluctuation. According to
tests made by MELLETTE et al. (1951) on humans under precisely standardized
conditions, the fluctuation of rectal temperature averages 1.20° C (37.16−38.36° C)
in young women and 1.49° C in young men (36.83−38.32° C), with amplitudes
ranging from 0.7−2.1° C. The diurnal temperature fluctuation is absent in human
neonates and develops during the first weeks of life (JUNDELL, 1904; HELL-
BRÜGGE, 1960). In children, VAN DER BOGERT and MORAVEC (1937) found a
fluctuation of 1.7° C, which is greater than in adults. For further data on diurnal
fluctuation of body temperature in man, see ASCHOFF (1966, 1967, 1970),
ASCHOFF, GERECKE, and WEVER (1967), and CONROY and MILLS (1970).

Temperature is only *one* of many physiological parameters which are subject to
diurnal rhythms (for ref. see ASCHOFF, 1960; FOLK, 1966; MENAKER and ESKIN,
1968; HALBERG, 1969). If external conditions (light, temperature etc.) are kept
constant, the rhythm usually remains but in many cases changes its phase and
period in regard to the earth's rotation (free running period). As a rule, constant
light causes a lengthening of the free running period for a nocturnal (night active)
organism and a shortening for diurnal (day active) organisms (HOFFMANN, 1965).
The deviation in the length of periods from a 24-h cycle might amount to about
± 2 h. In bats hibernating at 10° C in complete darkness there is still a circadian
rhythm of body temperature (MENAKER, 1961).

The occurrence of free running oscillations of body temperature and other physiological parameters shows that the organism does not simply react in a passive way
to external periodic events but possesses an *endogenous* component to be periodic.
Since these free running periods differ around 24 h, they are referred to as *circadian*
(about a day). There is still much discussion about the nature and origin of endogenous circadian rhythms. These periodic events can be synchronized with the
earth's rotation (local time) by means of *synchronizers* (time-givers, entraining
agents, clues or cues). Usually the most effective synchronizer is the lighting
regimen but any other periodic meteorological or ecological event (temperature,
humidity, noise, feeding etc.) might act in this way (ASCHOFF, 1965; SOLLBERGER,
1965; FOLK, 1966; HALBERG, 1969). In mammals and birds it is easy to cause a
phase shift in diurnal rhythm in regard to local time by shifting the phase of the
synchronizer; in contrast, the frequency of the period can be influenced only to a
limited extent.

A phase shift in all synchronizers is known to cause a corresponding phase shift
in the temperature curve in *man* as well, most completely when there is a change
in local time during transmeridian travel. A matter of high practical importance is
the time course of adaptation to the new rhythm. HAUS et al. (1968) found that
rhythm adaptation following a flight from east to west (Minnesota, U.S.A. −

Japan, phase shift $\Delta \Phi = 135° = 9$ h), involving social synchronizer delay, seems to be faster (7 days) than that following a flight from west to east (12 days) involving synchronizer advance — despite the circumstance that rhythm advance is associated with return to familiar home setting.

Human subjects isolated from their environment for a span of one week up to 6 months continue to exhibit a circadian rhythm of internal temperature desynchronized from the 24-h local time; the free running periods vary between 24.4 to 26 h but in some cases the variability is much greater. During the free running period, a desynchronization between temperature rhythm and other biological rhythms, such as sleep and wakefulness (Fig. 9), can occur (ASCHOFF and WEVER,

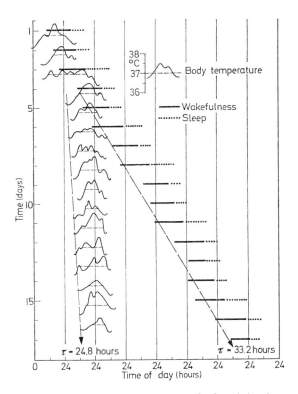

Fig. 9. Circadian rhythms of body temperature and of activity in a subject, enclosed in isolation in an underground bunker without time cues. In order to give a clear impression of the drift of the activity cycle against local time, the scale on the abscissa (time of day) has been drawn 8 times. From ASCHOFF, GERECKE, and WEVER (1967)

1962; ASCHOFF, GERECKE, and WEVER, 1967; COLIN et al., 1968). The changes in temperature rhythm of subjects living in unusual environments, such as the Arctic, or of people whose working regime is shifted against local time vary over a wide range, depending on the interference of natural and artificial synchronizers (for ref. see HALBERG, 1969).

2. Longer Temperature Periods

The *menstrual cycle* evokes a period fluctuation in temperature which has been closely examined, particularly in humans (HARDY, SCHORR, and DuBOIS, 1947; MARSHALL, 1963). Immediately after menstruation, the rectal temperature decreases and reaches a low point shortly before ovulation. The latter is then accompanied by a sudden rise of around 0.5° C, and remains at this level until the next menstruation. The most common time for the temperature rise is 13 days before the onset of menstruation but there are wide individual variations. WRENN, BITMAN, and SYKES (1958) have measured long-term temperature changes in cows during the reproductive cycle. The values were lowest just before the estrus, high on the day of heat, low again at the time of ovulation and high during the luteal phase of the cycle. If pregnancy occurs, the temperature remains at this high level. The causes of these fluctuations are not yet known.

III. Principles of Thermoregulation

A. Concept of Regulation

Homeothermy is part of the ability of organisms to maintain homeostasis or stability (ASHBY, 1960, 1966). Internal body temperature is regulated in higher vertebrates with a certain degree of accuracy. This means that internal temperature is compared in some way with a reference, and when a difference occurs an appropriate response is made tending to reduce that difference.

It has become fashionable to describe the thermoregulatory system in terms of cybernetics and control engineering. On the one hand, these models are certainly of heuristic value and may advance the formation of clear concepts. On the other hand, such analogies can also be misleading since the principles actually used by control engineers are only part of the theoretical possibilities of control systems (BROWN and BRENGELMANN, 1970; BLIGH and HENSEL, 1973). Of course, since the models in question are purely formal, they cannot replace the task of physiological research to investigate empirically the system of temperature regulation. A marked difference between the biologist and the control systems engineer is that the former desires to describe an already existing regulatory process whereas the latter designs a system to regulate or control some quantity.

In the terminology of control theory (HARDY, 1961; GRODINS, 1963; YAMAMOTO and BROBECK, 1965; MILHORN, 1966; HAMMEL, 1968; BROWN and BRENGELMANN, 1970), stability of a controlled variable, such as temperature, is achieved by means of a *negative feedback control system*.

Figure 10A shows a generalized diagram of a system as it is commonly used as a model for thermoregulation (MITCHELL, SNELLEN, and ATKINS, 1970). It consists of a *controlled system*, the body, which may be subjected to a *disturbance*, environmental heat or cold or metabolic heat. This disturbance causes a change in the *controlled variable*, some body temperature. The value of the controlled variable is detected by a *sensor*, and a corresponding *feedback* signal is generated. Any deviation of the controlled variable from the reference value constitutes a *load error* which activates a *controller*. The controller institutes a *control action* in such a way as to counteract the deviation of the controlled variable. This model includes a reference in the form of a signal, generated physiologically, against which the feedback signal is compared. The value of this reference signal is then the *set-point* of the system.

Figure 10B shows a feedback system similar to that of Fig. 10A, but which does not incorporate a reference signal. Here there are two types of sensors of the con-

trolled variable, one which responds with increased activity to rises in the controlled variable, and one which responds with increased activity to falls in the
controlled variable. There is good evidence that such sensors (warm-sensors and
cold-sensors) exist in the core and in the periphery of homeotherms (p. 567). The
load error is generated by comparing the feedback from both sets of sensors. An
elevation of body temperature causes the warm-sensor activity to dominate and a
depression causes cold-sensor activity to dominate. When the information from
the cold sensors exactly balances that from the warm sensors, there is a zero
output from the comparator, or zero load error.

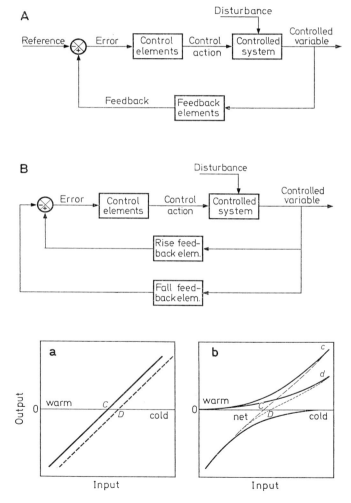

Fig. 10. *A*, general negative feedback control system with a reference; *B*, negative
feedback control system without a reference. *a*, input/output relationship for the
system *(A)* with reference. *b*, input/output relationship for the system *(B)* without
reference. Further explanation see text. From MITCHELL, SNELLEN, and ATKINS (1970)

In the first model (Fig. 10 A) a shift in the set-point is brought about by changing the reference signal, while in the second model (Fig. 10 B) the set-point can be shifted by changing the gain to one feedback. For instance, if the sensitivity of warmth feedback decreases (Fig. 10 *b*, line *d*), the "set-point" is shifted to a higher value (*D*). This "set-point" of the system is, however, not a reference point in the sense of a reference signal generated by a definable neural structure.

Automatic control systems are characterized by some fundamental properties: 1. the decisive variable being processed in the system is *information*, not matter or energy, 2. the flow of information is unidirectional, 3. the system includes feedback loops, 4. input and output signals are reversed (negative feedback).

B. Biological Temperature Regulation

Figure 11 shows a schematic control system diagram of temperature regulation in homeotherms. Two populations of thermosensors are known, one in the core (central nervous system) and the other in the skin, each consisting of a dual set of

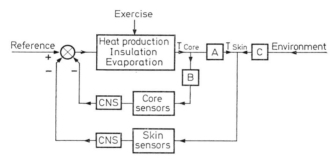

Fig. 11. Control system diagram of temperature regulation in homeotherms. Boxes *A* and *B* represent heat transfer characteristics within the body, box *C* represents characteristics of heat transfer to the environment

elements sensitive to warming and to cooling. There is also some evidence for other thermosensors elsewhere in the body. The signals from the thermosensors are integrated within the central nervous system which corresponds to the controller. The controlled temperature is compared with a reference input in the central nervous system; any deviation leads to various control actions (changes in heat production, thermal insulation, and evaporative heat loss) which tend to reduce the deviation of temperature. Thermal disturbances have two main sources, namely, internal heat generation by excercise and environmental heat or cold. Signals from external temperature disturbances can rapidly be transferred to the central nervous system by means of cutaneous thermoreceptors via afferent nervous pathways, before the disturbance has reached the core of the body. Important in this respect is the ability of cutaneous thermoreceptors to respond not only to the temperature (*T*) but also to the rate of a temperature change (dT/dt). The *proportional* and *rate* control is so effective that rapid external cooling or warming of homeotherms may result in a transient opposite change of internal temperature.

Many investigators have assumed that the *controlled temperature* is that of the hypothalamus. This assertion may be difficult to sustain now that there is evidence of other thermosensors and of the convergence of their afferent pathways. Doubt as to whether one specific temperature such as that of the hypothalamus is the controlled variable stems from various lines of evidence: 1. sweating in man can be best correlated not with mean skin temperature, nor with deep body temperature, but with mean body temperature calculated from the weighted mean of representative skin temperatures and deep body (rectal) temperature, 2. various control actions (metabolism, vasomotor responses) in homeothermic animals are correlated neither with hypothalamic nor with skin temperature alone but with a function of both (p. 574), 3. sheep possessing a carotid rete (p. 512), and thus subject to disturbances of hypothalamic temperature nevertheless have an exceptionally fine control of deep body temperature (BLIGH et al., 1965). It would thus seem very likely that the controlled variable is not a single temperature but a function of various temperatures (HENSEL, 1952a; BRÜCK and WÜNNENBERG, 1967; BROWN and BRENGELMANN, 1970; Chapter VI, p. 574). This function, which constitutes the integrated load error signal, varies with species and circumstances, in particular with the ratio of the heat transfer characteristics A and C (Fig. 11).

In fact, the biological system of thermoregulation is more complicated than any engineer's blueprint; it is grossly nonlinear in the mathematical sense, has no single regulated variable and a high redundancy. Further, it contains multiple sensors, multiple feedback loops and multiple outputs (p. 574). Therefore a quantitative model of the thermoregulatory system requires considerable simplification but may still be useful. Such models have been presented by WYNDHAM and ATKINS, 1960; WISSLER, 1963; SMITH and JAMES, 1964; STOLWIJK and HARDY, 1966; STOLWIJK, 1970, and others. Usually one discriminates between *autonomic* (or *physiological*) and *behavioral* regulations, ascribing the former to the hypothalamus and the latter to the cerebral cortex. However, a sharp distinction on this basis has proved impossible. Behavioral responses to heat and cold exposure modify the relations between organism and environment and thereby modify the need for autonomic thermoregulatory responses, but this sparing action does not necessarily imply central nervous coordination between behavioral and autonomic thermoregulations.

In man, behavioral thermoregulation is associated with cerebral cortex activities involved in the conscious perception of thermal comfort and discomfort which can be related, subjectively, to changes in skin temperature; but recent experiments have also clearly demonstrated the involvement of deep body temperature in the conscious sensation of thermal comfort. Likewise, behavioral thermoregulation in rats, dogs and monkeys is not only dependent on skin temperature but also on the temperature of the hypothalamus (p. 616). A distinction might properly be made between those *autonomic* functions which can be activated by direct efferent influences from the hypothalamic or other centers more directly concerned with particular effector functions, and the more complex *behavioral* responses, the primary activation of which is effected by the hypothalamic thermoregulatory centers, but which involve complex muscular activity patterns requiring cortical integration (BLIGH and HENSEL, 1973).

Table 4 summarizes various autonomic and behavioral components of thermo-regulation. Their physical importance can be derived from the general laws of *stationary heat flux* in the organism, the details of which will be dealt with on p. 544. Since the body continuously produces heat, a steady state is only possible if there is a continuous heat flux along a temperature gradient from the body core to the environment. It is useful to treat the heat flux from the core to the body surface and from the surface to the environment separately. The heat flux density (H) in an element of the body surface, i.e. the thermal energy passing through a unit area in unit time in the steady state (BURTON, 1934) can be written as

$$H = h_i \left(T_i - T_s\right). \tag{1}$$

h_i is an internal heat transfer coefficient, T_i the temperature of a point inside the body and T_s the temperature of the surface element. The heat flux density in the same surface element is also

$$H = h_a \left(T_s - T_a\right) + H_e \tag{2}$$

Table 4. Control actions of temperature regulation in homeotherms (in brackets: behavioral responses in man)

Physical factor	Autonomic regulation	Behavioral regulation
Ambient temperature T_a		Migration Seeking sun or shade, etc. (Artificial heating and cooling)
Heat production H	Shivering Nonshivering thermo-genesis	Active movements Food intake: Specific dynamic action (Warm and cold food)
Internal thermal resistance including integument R_i	Cutaneous blood flow Erection of hair and feathers	(Clothing)
External thermal resistance R_a	Respiration: dry heat loss	Nest building Seeking shelter Seeking ground surfaces of various thermal conductivity, wind, water, etc. Air movement by fanning (Ventilation)
Water evaporation H_e	Sweat secretion Respiration: evaporative heat loss Secretion of nasal and oral glands	Moistening of body surface with water, saliva or nasal fluid (Moistening of skin and clothing)
Geometric factor		Body posture Huddling of several individuals

where h_a is an external heat transfer coefficient, T_a the ambient temperature and H_e the heat absorption by evaporation, expressed as heat flux density. For simplicity it is assumed that the heat flux by conduction, convection and radiation is proportional to the temperature difference. This is not quite correct for radiation but has proved empirically as a good approximation in the biological temperature range (BURTON and EDHOLM, 1955). Since the heat flux density from the core to the surface and from the surface to the environment is identical in the steady state, we obtain from Eqs. (1) and (2) for T_i

$$T_i = T_a + H R_i + R_a (H - H_e) . \tag{3}$$

The reciprocal values $1/h_i$ and $1/h_a$ are replaced by the symbols R_i and R_a. The signs R_i and R_a have the meaning of an internal and external thermal resistance or insulation. Among others, thermal resistance is also dependent on the geometry of the body.

The organism is able to change actively the values of H, R_i and H_e, to some lesser extent also R_a. Thermal disturbances arise mostly from T_a, R_a and H. It is important to realize that the heat production of the body acts not only as a controller output but can also cause thermal disturbances. In man, heat production during extreme work can be 20 times higher than the basic metabolism. This excess heat must be compensated by other physiological factors, such as sweating.

If the aim of regulation is to keep T_i constant, then it follows from Eq. (3) that homeotherms can activate their heat production as well as their thermal resistance for temperature regulation. Equal ambient and core temperatures ($T_a = T_i$) are only possible if the sum of the second and third term on the right side in Eq. (3) becomes 0. Since H, R_i and R_a are always positive, only the term $H - H_e$ remains which becomes negative if $H_e > H$. For the condition $T_a = T_i$ the evaporative heat loss is $H_e = H (1 + R_i/R_a)$. The amount of H_e can be further increased, so that some homeotherms can even exist under the condition $T_a > T_i$. Heat absorption by evaporation of water is thus the only means by which temperature can be maintained at higher ambient temperatures.

C. Disturbances of Thermoregulation

From the diagram (Fig. 11) we can derive various possibilities of changing the core temperature of the body: 1. external heat or cold stress overrides the thermoregulatory system; in this case the properties of the system are unchanged but the quantitative capacity of the control actions is insufficient, 2. primary impairment of the controller output, such as paralysis of cutaneous blood vessels, artificial metabolic increases etc., 3. changes or disturbances of the controller itself.

An increase in core temperature above a standard range is called *hyperthermia*, whereas a corresponding decrease is called *hypothermia*. Both deviations can be brought about either by external heating or cooling of the organism (1) or by changes in the central nervous controller (3). *Fever* is a rise of internal temperature involving such changes in the controller (shift of set-point or change of gain, p. 522). A corresponding hypothermia has no special term; it may be called "*central hypothermia*".

D. Interference of Regulations

There is considerable interference between thermoregulation and other biological control systems. In many cases we find loops of two or more regulatory systems having a common output, e.g. energy production, water metabolism, or cutaneous circulation. This involves the possibility of competition between various systems, in which case temperature regulation turns out to be very powerful. Many homeotherms living in the cold keep their body temperature at high levels even during starvation, instead of saving energy by hypothermia. Pigeons without food in cold environments maintain a constant body temperature and a constant metabolic rate 4 times higher than the basal metabolism until they die from starvation (STREICHER, HACKEL, and FLEISCHMANN, 1950). Exceptions from this rule are seen in hibernators and some birds which become hypothermic during periods in which food is scarce (p. 688). Water metabolism may interfere with temperature regulation as well, as is seen in the heat where sweat secretion is maintained at high rates even if the body becomes dehydrated. Exceptions have been found in some mammals living in hot environments; they save water by periodically becoming hyperthermic (p. 722). Other examples are the competition between thermoregulation and regulation of blood pressure during heat exhaustion (p. 656) and the regulation of core temperature in the cold in spite of cold injury of the peripheral parts of the body.

IV. Production of Body Heat

A. Basal Metabolism

Production of heat from exothermic chemical reactions, especially in the skeletal muscles and certain internal organs, is one of the variable quantities by which homeothermic organisms can regulate their core temperatures. At high external temperatures and during heavy physical labor the formation of heat becomes more and more of a handicap, since performance of mechanical work by the muscles is always connected with an unavoidably high production of heat, amounting to two to ten times their mechanical energy, depending on the efficiency (LEHMANN, 1953). Even in the absence of external work or cold exposure a relatively large amount of heat is produced, which, when measured under certain standard conditions (rest, calm, neutral external temperature, psychological relaxation), is designated as basal metabolism. Its determination in animals is often difficult due to uncertainty that the condition of "rest" is satisfied. Basal metabolism is appreciably higher than the level needed to maintain body temperature in a warm environment. Consequently it is not determined by a requirement for heat (cf. "surface area law", p. 531).

Heat production in homeotherms depends on a large number of factors, such as species, body size, age, sex, nutrition, external temperature, activity, the activity of endocrine glands, acclimatization, etc., which we will consider here only to the extent that they have a direct significance for temperature regulation.

B. Heat Production and Body Size

Heat production in homeotherms is clearly related to body size and in fact can be presented as a power function of body weight or volume

$$M = k \cdot W^n \tag{1}$$

in which M represents the total heat formation per unit time, k a constant which depends on the geometric body form, W the body weight and n an exponent. In double logarithmic form this function yields a straight line

$$\log M = k' + n \log W . \tag{2}$$

According to the investigations of various authors n lies between 0.75 and 0.82 (KLEIBER, 1947, 1961; MORRISON, 1948; PEARSON, 1948; among others). In intraspecific investigations n is said to lie closer to 0.82, but in interspecific ones closer to 0.75. KLEIBER (1947) gives the formula

$$\log M = 1.83 + 0.756 \log W \pm 0.05 \tag{3}$$

for the metabolism of twenty-six different species of animals from the mouse to the
elephant. It agrees quite well with the experimental values (Fig. 12). According
to this formula, the average resting heat production of homeotherms amounts to
approximately $3 \, kcal \cdot kg^{0.75} \cdot h^{-1}$. The quantity $W^{3/4}$ is often designated as the
"metabolic body size". Even when the relation is extended to very small mammals,
down to shrews weighing 3.4 g (MORRISON, 1948; PEARSON, 1948), the rule has been
confirmed. MORRISON's measurements yield the value of $n = 0.73$.

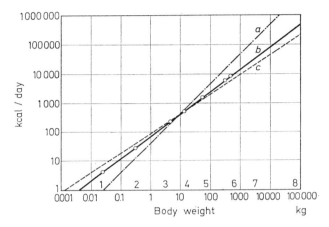

Fig. 12. Heat production as a function of body weight, presented in log-log form; *a* in
proportion to weight, *b* actual values *1* mouse, *2* rat, *3* rabbit, *4* dog, *5* man, *6* cow,
supplemented, *7* elephant, *8* whale. *c* in proportion to surface area. From KLEIBER
(1947)

Equation (3) signifies that metabolism increases at a considerably slower rate than
body weight, so that larger warm-blooded organisms produce less heat per unit
weight than small ones. In very small homeotherms the intensity of metabolism is
extraordinarily high. At external temperatures of $24-28°$ C the metabolic rate of
the shrew *Sorex c. cinereus* corresponds to 65 times the basal metabolic rate of man
or 200 times that of the elephant. The asymptotic rise in the curve of Fig. 13 leads
to a lower limit of ca. 2.5 g for the body size of mammals at tropical temperatures
(PEARSON, 1948), since at lower sizes the uptake of nutrition could not cover the
extremely high energy demands. Shrews weighing 3-4 g or similarly small birds
utilize approximately their own body weight in food in 24 h. The lower limit of
body size of birds is somewhat lower due to their better heat insulation. The relation given by Eq. (3) is strictly valid only for mature representatives of the
various species. If young, especially newborn, forms are included the relation is
far more complicated. The age dependency of the metabolic rate makes it impossible to delineate the metabolic rate for individual species so that it includes the
neonates as well as adults by a single exponential function of weight, even if a
higher exponent than KLEIBER's is used (BRÜCK, 1961, 1970).
As demonstrated in Fig. 14, in newborn infants up to one week of age the BMR is
lower than predicted by KLEIBER's equation, i.e., BMR is lower in a 3 kg human

newborn than in a 3 kg rabbit. In the weight range of 5—20 kg, on the other hand, BMR is higher in human infants than in adult animals of the same weight group. In this period BMR is nearly linear in relationship to body weight; [n = 1 in the Eq. (3)]. During the period which corresponds to the weight range of 20—70 kg

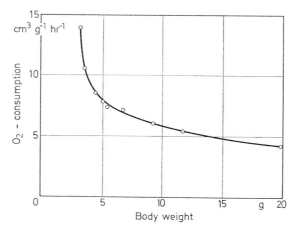

Fig. 13. Oxygen use by the smallest mammals (shrews) at external temperatures of 24—28° C, as a function of body weight. From PEARSON (1948)

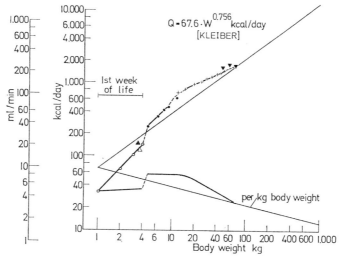

Fig. 14. Relation of BMR to weight and age. Upper curves: Standard metabolic rate expressed in kcal/day and in ml oxygen uptake/minute in relation to body weight. Lower curves: The same data but expressed in relation to unit of body weight. The straight lines represent basal metabolic rates according to KLEIBER's equation. o Newborn and premature infants first week of life (BRÜCK, 1961), △ 0—6 hours, full-term infants (HILL and RAHIMTULLA, 1965), ▲ 18—30 hours, ● Boys and girls, 1—36 months (LEE and ILIFF, 1956), + Boys, 13—15 years (LEWIS et al., 1943), ▼ Men, 14—30 years. From BRÜCK (1970)

the metabolic rate approaches the Kleiber curve; here, n would be smaller than 0.75 and come close to 0.46, an exponent calculated on the basis of heat flow; see Fig. 86 on p. 673.

An exponent close to 1, or somewhat larger than 1 would also have to be used to calculate BMR for infants weighing 1–4 kg, i.e., the range including premature infants. This means that in the neonatal period the BMR per unit body weight is almost independent of weight, whereas a *decrease* should be expected with increasing weight according to KLEIBER's equation (Fig. 14).

A similar relationship between postnatal BMR and the Kleiber curve may be derived from data obtained from the guinea pig (BRÜCK and WÜNNENBERG, 1965), the rhesus monkey (DAWES et al., 1960), the rat (TAYLOR, 1960), and the pig (MOUNT, 1955, 1968).

The deviation of BMR from the Kleiber curve in newborn individuals (cf. Fig. 14) has been ascribed to the elevated content of extracellular fluid (ECF), which makes up as much as 44% of body weight in a 4000 g newborn infant, and even 58% in a 1000 g premature infant; the corresponding value for the adult ECF is 16.5%. With a correction made on this basis the neonatal BMR would come very close to the Kleiber curve (BRÜCK, 1970).

To interpret the elevated metabolic rate found at an older age (cf. Fig. 14) two factors have been suggested so far: 1. The high specific metabolic rate of the brown adipose tissue (see below), and 2. certain peculiarities in the endocrine system. Here, particular attention has been focussed on the thyroid function. After thyroidectomy the BMR has been shown to be reduced much more in juvenile rats than in adults (GRAD, 1953); this would mean that at least part of the excess in BMR over the figure predicted by the Kleiber curve (Fig. 14) may be related to an increased thyroid function. In three-to-nine-year-old children thyroid activity has, indeed, been shown to be increased, when thyroid turnover is compared with the adult (HADDAD, 1960).

Many points of view on the origin of the empirically observed power function relation between body weight and basal metabolism have been brought forward which cannot be discussed here. The suggestion of RUBNER that "impulses proceeding from the skin as a result of cooling stimulate the cells to activity" has attracted special notice and has been widely disseminated in connection with the concept of "surface area law". However objections were soon raised, since heat production does not correspond exactly to the body surface area ($n = 0.67$), but to a higher power of body weight. Furthermore heat losses as well as the metabolism-regulating impulses which proceed from the skin definitely do not depend on surface area alone but on many other factors as well, including blood flow through the skin, local variations in its sensitivity to cold, fur thickness, subcutaneous fat layers, etc. (cf. KESTNER, 1934; GALVÃO, 1947, 1948; ASCHOFF, 1948; int. al.). On the other hand, the fact remains that the surface/volume ratio and with it the relative size of the heat emitting area increases with decreasing body size, so that under the same conditions a small homeotherm must maintain metabolism at a higher intensity in order to maintain the same core temperature at equivalent external temperatures. A quantitative theoretical example may illustrate the significance of body size for the heat budgets of animals. Assume that spherical animals are insulated by an air layer of 1/4 their body radius. The difference between the core temperature and the external temperature may be taken as 10° C. According to the laws of stationary heat flow an animal weighing 4 g would have to maintain a heat production of 33 kcal · kg^{-1} · h^{-1} in order to maintain this temperature gradient. For an animal weighing 2 g a heat production of 50 kcal · kg^{-1} · h^{-1} would be needed, and one weighing 1 g would need 80 kcal · kg^{-1} · h^{-1}, i.e. 80 times the rate of human basal metabolism. These values, which are derived entirely from theory, agree quite well with the order of magnitude of figures obtained from measurements of shrews (cf. also Fig. 86).

Even though smaller animals must metabolize more rapidly in order to regulate their temperature under equivalent conditions, it does not follow that thermo-regulation must be the direct cause of this relation. From an evolutionary point of view however, the thermoregulatory aspect could provide an explanation of the exponential relation between body weight and basal metabolism. If the chance for survival of animals of various sizes becomes greater as their metabolism is more precisely adjusted to their body surface area, then it must be granted that there is a selection value in the relation between basal metabolism and body surface, or an even better fitting exponential function of body weight (cf. Fig. 86). The reduction in metabolism with increasing body size could then be further inter-preted as a genetic adaptation.

A high rate of basal metabolism per unit body weight can be compensated up to a point by an increase in surface area that is independent of body weight, perhaps in that of the ears, whose surface in elephants living under tropical conditions amounts to more than two square meters.

It follows from this that still other explanations for the "surface area rule" of basal metabolism must be brought into consideration. A critical discussion of the pros and cons of various explanations is found in KLEIBER (1961). A thorough theo-retical discussion of the problem from the point of view of a "theory of biological similarity" is found in GÜNTHER (1971).

In every discussion of the question of causes it must be kept in mind that the metabolism of many poikilotherms is also proportional to a similar power function of body weight. WEIMOUTH et al., (1944) found an exponent of $n = 0.8$ in marine crustacea.

C. Heat Production and External Temperature

Countless experiments on the most diverse homothermic animals (GIAJA, 1938; BRODY, 1945; int. al.), have agreed in demonstrating that heat production changes to a marked degree as external temperature changes. Two factors can be dis-tinguished here: 1. *Indirect* effects of temperature on heat production, which act on temperature regulation and are the most prominent effects in homeotherms; they lead to an inverse relation between temperature and heat production. 2. *Direct* temperature effects following the van't Hoff-Arrhenius rule, which work primarily in extreme temperature ranges in which internal temperature can no longer be regulated. Both factors combine to produce a course of heat production which is presented schematically in Fig. 15. At a certain external temperature, which varies according to species and acclimatization (p. 624), heat production is at a minimum ("zone of thermal neutrality" or "metabolically indifferent zone").

The minimum t_3-t_4 can embrace a narrrower or wider temperature range (Fig. 16). If the external temperature falls below the value t_3, metabolism increases approximately linearly as the external temperature decreases (HERRINGTON, 1941; KENDEIGH, 1944; SCHOLANDER et al., 1950; GELINEO, 1954; int. al.). At t_2 the extent of regulation becomes insufficient (i.e. the "zone of regulation" is exceeded),

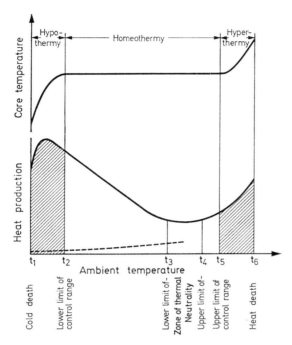

Fig. 15. Common course of heat production and body temperature of homeotherms at different external temperatures. Dashed line: heat production of poikilotherms. From curves according to GIAJA (1938)

and the core temperature begins to fall. Heat production climbs up to a maximum value at first ("métabolisme du sommet" according to GIAJA) and then falls as the body temperature decreases; further, paralysis of the regulatory centers may occur (cf. p. 662). GIAJA (1938) designates the ratio maximum metabolism/ basal metabolism as the "metabolic quotient". Its size indicates the organism's "thermogenic reserve". Most metabolic quotients vary between three and seven, depending on the species. At temperatures above t_4, heat production climbs. At first this is probably due predominantly to activation of regulatory processes (increase of respiratory frequency, etc.). At t_5 regulation becomes ineffective and the core temperature rises, which, according to the temperature-reaction rate law, leads to a further elevation of heat production. It must however be remembered that the optimal temperature for many enzyme reactions may already have been exceeded (p. 406); this would cause a deviation from the simple logarithmic relationship between temperature and metabolism as described

by the van't Hoff-Arrhenius equation. Nevertheless GELINEO (1934) found a 60%
increase in metabolism in the rat when rectal temperature was raised from 37 to
40° C, and DuBois (1921) reports an increase in heat production in man of 13% per
degree temperature. Following experimental lesions of the diencephalon the temperature-dependent increase of heat production may be absent. From this finding,
DONHOFFER (1966) concludes that the latter phenomenon may not be fully explained by the temperature-reaction rate law, but rather contains a regulatory component.

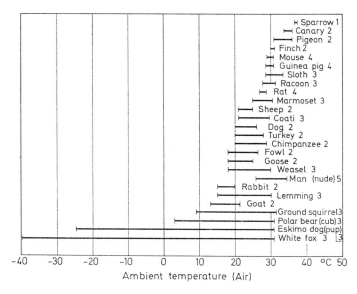

Fig. 16. Temperature regions of minimal heat production (zones of thermal neutrality):
1 from KENDEIGH (1944); *2* from BRODY (1945); *3* from SCHOLANDER et al. (1950);
4 from HERRINGTON (1941); *5* from WINSLOW (1941) and from WEZLER and NEUROTH
(see Fig. 17)

Figure 17 shows heat production in various warm-blooded organisms as a function
of external temperature. Numerous experiments of this type have yielded the
following regularities: Increased heat production follows exposure to cold sooner
and at greater rates: 1. The smaller the organism, 2. the poorer the heat insulation
of the body covering, 3. the greater the heat transfer coefficient of the surrounding
medium (water, wind). Since birds of the temperate and arctic zones have very
good insulation over almost the entire body, due to their plumage, the larger species
can endure very low temperatures without taxing their metabolism excessively
(cf. p. 659). The heat insulation of geese and chickens is substantially better than
that of guinea fowl. As a result their heat production in the cold increases much less
than that of the latter (GIAJA, 1931). The significance of insulation for heat production was demonstrated by experiments of GIAJA (1929) on plucked geese, hens, and
guinea fowl. Their metabolism was substantially increased over the normal level.
In many small birds (blackbirds, chaffinches, goldfinches, green finches, robins,

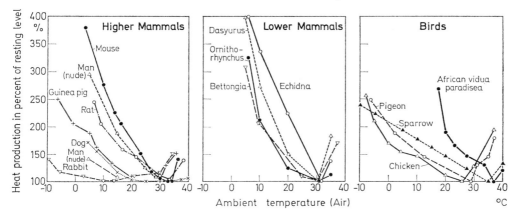

Fig. 17. Heat production as a function of external temperature. Man: Individuals with marked and slight increases of metabolism in the cold; from WEZLER and NEUROTH (1949); lower mammals from MARTIN (1902); sparrow from KENDEIGH (1944); others from BRODY (1941)

etc.) heat production at rest seems to be insufficient in the cold. Such birds show a greater tendency to move about and an increased food intake under these conditions. Many small mammals show similar behavior (GROEBBELS, 1928).

In mammals whose heat insulation is, on the average, poorer, the increase in metabolism in the cold is usually more strongly marked at equivalent body size. Polar animals by contrast increase their metabolism much less strongly than tropical animals (Fig. 18). The metabolism of the arctic fox does not begin to rise till temperatures of ca. −40° C are reached; for larger polar mammals the corresponding

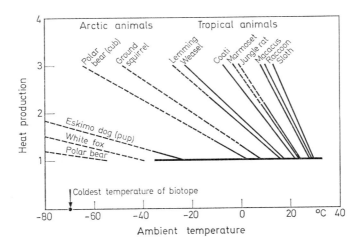

Fig. 18. Heat production of arctic and tropical animals at rest, as a function of external temperature. Minimal metabolism is always set equal to 1. Solid lines: average measured values. Dashed lines: extrapolated values. The response of human metabolism corresponds to that of the tropical animals. From SCHOLANDER et al. (1950)

value is $-50°$ C; and the tropical sloth begins to elevate its metabolism at an external temperature of $28°$ C (cf. p. 727). So, paradoxical as it may seem, the heat production of many tropical animals in their native surroundings is in part taxed more heavily than that of some polar animals during the arctic winter (IRVING, 1951).

In unclothed humans there are great differences in metabolic rates under variable external temperatures (HARDY and DuBois, 1938; WEZLER and THAUER, 1942; WEZLER and NEUROTH, 1949). In still air and at moderate humidity metabolism begins to increase at external temperatures below $26-28°$ C. The slope of the curves is quite variable, depending on the individual and on adaptation (see p. 621). In small, poorly insulated organisms the upper limit of the zone of neutrality (t_4 in Fig. 15) is close to the lower limit, t_3. One can establish the rule that the zone of thermal neutrality is shifted towards higher temperatures and also becomes narrower with decreasing body size and decreasing heat insulation.

D. The Modes and Sites of Extra Heat Production

The ability to produce additional heat in a cool environment is one of the characteristic features of homeothermy. Three principal modes of heat production exist which are responsible for its increase with decreasing environmental temperature: 1. Voluntary muscle activity; 2. Involuntary tonic or rhythmic muscle activity, which may be invisible and only detectable by electromyography (GÖPFERT et al., 1952, 1953; GLICKMAN et al., 1967). With higher intensity, however, the rhythmic activity is accompanied by a characteristic visible tremor known as "shivering"; 3. Nonshivering thermogenesis (NST). The existence of this mode of heat production was originally shown by the demonstration of a cold-induced increase in oxygen uptake which persists after administration of curare; this drug is known to block the neuromuscular transmission in muscle and hence prevents shivering.

Normally, in adult man and in larger adult mammals shivering is quantitatively the most important involuntary mechanism of thermoregulatory heat production. As shown by Fig. 19 and Fig. 95 (p. 686) there is a clear relationship between muscle activity, as judged from electromyography, and heat production in man during exposure to a cold environment for several hours. The extra heat production attained by shivering is about $100-200\%$ of BMR, i.e. the metabolic quotient amounts to $2-3$ (see Fig. 19); according to some observations in man (BEHNKE and YAGLOU, 1951; ADOLPH and MOLNAR, 1946), the metabolic quotient may be increased to 5 under extreme cold exposure. In other species, e.g. the dog, metabolic quotients in the order of 5 are quite common (BEHMANN and BONTKE, 1958; int. alt.).

Topographic studies on shivering revealed that shivering does not involve the whole musculature at once (for literature see HEMINGWAY, 1963). According to studies by GÄRTNER and GÖPFERT (1966) shivering begins in the musculature of the shoulder girdle and of the thoracic wall and thereafter extends to the legs. This agrees with the observations by DENNY-BROWN et al. (1935) and our own observations (BRÜCK and WÜNNENBERG, unpublished) in which shivering, beginning in the masseter muscle and in the musculature of the shoulder girdle, extended to the thigh musculature and eventually to the musculature of the lower leg. Heat production does not increase considerably before the large muscle mass of the thigh is activated. On the other hand, GOLENHOFEN (1965), including the lower arm in his topographic studies, found that shivering commenced in the lower arm, and that this activation was accompanied by a negligible increase in heat production; with prolonged cold exposure shivering extended to the larger proximal muscle groups while vanishing in the lower arm. As a similar activation was seen in the myogram of the lower arm muscles during a mental arithmetic task, GOLENHOFEN termed the initial cold activation of the arm musculature "the affective component" in the cold defense reaction.

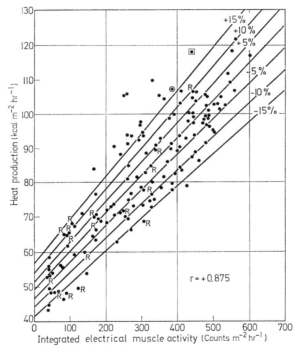

Fig. 19. Relationship between 142 individual values for heat production and integrated electrical activity of muscle. One subject stamped his feet for 2.6 min (⊙), and for 5.25 min (⊡). R indicates the values obtained during the last 15 min of the 4-hr exposure when the subjects were requested to relax. From GLICKMAN et al. (1967)

Nonshivering thermogenesis (NST) develops to a greater extent only after long-term cold exposure (i.e. cold adaptation), as was first shown by HART et al. (1956) in the rat. Evidence was later provided for the development of NST through cold adaptation in various mammalian species and it has also been shown that NST is a most important and effective mechanism of heat production in the neonates of a number of mammalian species (MOORE and UNDERWOOD, 1960, 1963; DAWKINS and HULL, 1964; BRÜCK and B. WÜNNENBERG, 1966; including the human infant: BRÜCK, BRÜCK, and LEMTIS, 1958; BRÜCK, 1961; DAWKINS and SCOPES, 1965; Review: BRÜCK, 1970).

The elicitation of NST is mediated by the sympathetic nervous system (Review: HIMMS-HAGEN, 1967). Evidence for this has been obtained by the demonstration of an inhibition of NST through sympathectomy (FREUND and JANSEN, 1923),

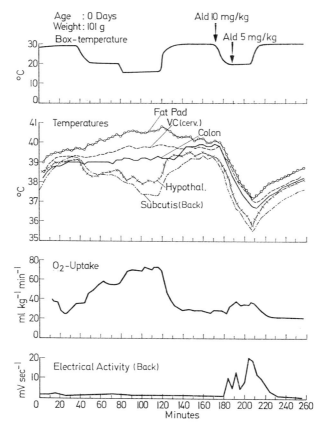

Fig. 20. Course of study in a newborn guinea pig. Age: 0 days. Weight: 101 g. In the first part of the experiment only nonshivering thermogenesis *(NST)* occurs during cold exposure. After blockade of *NST* by administration of pronethalol (alderlin = ald) shivering takes place and can be recognized by the increased electric activity of the back muscles. Note: *Increasing* temperature in the area of the interscapular adipose tissue (fat pad) and in the cervical part of the vertebral canal ("VC [cerv.]") *before* but parallel decrease of all temperatures *after* blockade of *NST*. For further explanation see text. From BRÜCK and WÜNNENBERG (1966)

through ganglionic blockade (HSIEH, CARLSON, and GRAY, 1957; MOORE and UNDERWOOD, 1960, 1962, 1963), and through sympathetic stimulation (HULL and SEGALL, 1965b). In this respect classic studies by HSIEH, CARLSON, and GRAY (1957) are of particular importance. In the cold-adapted, curarized rat, cold-induced oxygen consumption, after suppression by ganglionic blocking agents, can be restored by noradrenaline. NST can also be blocked by adrenergic beta-receptor blocking agents such as Pronethalol (Alderlin) and Propranolol (BRÜCK and WÜN-NENBERG, B., 1965 b; HEIM and HULL, 1966). An interesting phenomenon has been observed in studies of the blockade of NST in the newborn guinea pig: When NST was blocked, shivering occurred in place of it, even at environmental temperatures at which normally no shivering was seen (Fig. 20). This observation led to the supposition that shivering is suppressed by NST (BRÜCK and WÜNNENBERG, B., 1965; BRÜCK and WÜNNENBERG, W., 1966; see also p. 569).

Further, only recently, brown adipose tissue has been shown to be an important site of NST (SMITH and HOYER, 1962; DAWKINS and HULL, 1964; BRÜCK and WÜNNENBERG, B., 1965c; BRÜCK and WÜNNENBERG, W., 1966; Review: SMITH and HORWITZ, 1969) whereas no confirmation has been obtained for a previous concept, according to which the liver was thought to play an important role as the site of NST. In comparison with NST, shivering is a less economical form of heat production, since it inevitably increases convective heat loss due to the body oscillations; moreover, it interferes with body movement.

This economic aspect becomes more important the smaller the organism is (high surface/volume ratio) and the poorer its thermal insulation is. Thus it is satisfying, from a teleological point of view, to find that the maximum extent of NST available in one species is inversely related to its order of body size (Fig. 21). Extrapolating the regression line in Fig. 21 one would arrive at the conclusion that subjects with body weights above 10 kg lack the capacity for NST (see also p. 624).

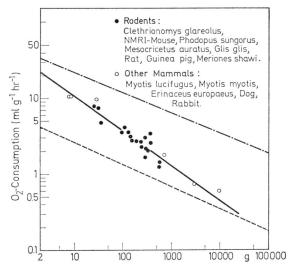

Fig. 21. Maximum amount of nonshivering thermogenesis, measured as increase in oxygen uptake following noradrenaline injection, in relation to body weight (various species in the adult age). The lower dashed line represents minimum oxygen uptake at neutral temperature. The upper dashed line represents the calculated maximum oxygen uptake for exercise. From HELDMAIER, G. (1971)

E. Brown Adipose Tissue and Its Significance for Nonshivering Thermogenesis

From its first description in 1551 by GESNER (quoted from SMITH and ROBERTS, 1964) until almost recently the function of brown adipose tissue was obscure (review by JOHANSSON, 1959). For some time it has been preferably studied in hibernators and was denoted "hibernating gland" by many authors, although no evidence had been obtained that it possesses an endocrine function.

Only a few years ago it was shown that the *in vitro* metabolic activity of brown adipose tissue was much higher than that of white fat (JOEL, 1965; SMITH and HOIJER, 1962) and that the arousal of hibernators is preceded by a local temperature rise in brown adipose tissue (SMITH and HOCK, 1962; SMALLEY and DRYER, 1963; JOEL et al., 1964). Thus it was presumed that the main function of this tissue is to be seen in the thermoregulatory heat production. At the same time it has been redetected that the occurrence of brown adipose tissue is by no means restricted to the hibernators but that it can be found in many nonhibernating mammalian species, mainly, however, in the neonatal stage; in adult animals it is present only after cold acclimatization.

The cells of the brown adipose tissue (Fig. 22) are characterized by a number of small fat droplets; the nucleus is round and centrally located. The fat droplets are surrounded by a large number of mitochondria (NAPOLITANO, 1965; HULL, 1966; RAFAEL and HOHORST, 1968). In contrast, the white fat cell has only one large fat drop with a small plasma skirt which contains the nucleus (signet ring form).

Location and Mass of Brown Adipose Tissue. In rodents most of the brown adipose tissue is located around the neck and on the back between the scapulas; some is further located in the axillas, in the inguinal region, on the surface of the heart and kidneys, and adjacent to the large vessels within the thoracic cavity. In the newborn kitten as well as in the human neonate there is no compact interscapular fat pad, but the adipose tissue in this region is more diffusely distributed between the sheets of the musculature. Otherwise it is distributed as in rodents.

The mass of brown adipose tissue in the newborn rabbit and guinea pig makes up as much as 5–7% of the body weight (DAWKINS and HULL, 1964; BRÜCK and WÜNNENBERG, B., 1966). In adults, brown adipose tissue amounts to much less than in the neonate. In cold-adapted rats the total amount of brown adipose tissue has been found to be about 1% (SMITH and ROBERTS, 1964) and in adult cold-adapted guinea pigs to be less than 1% (For Lit. see BRÜCK, 1970).

No brown adipose tissue has been found in the newborn piglet (DAWKINS and HULL, 1964) or in the newborn miniature pig (BRÜCK, WÜNNENBERG, W. and ZEISBERGER, 1969).

The brown adipose tissue is richly vascularized and supplied by sympathetic nerves. Of particular importance for its function (see p. 543) are the vascular connections of the interscapular adipose tissue to other organs. According to studies in the rat by SMITH and ROBERTS (1964) the venous blood of the cervical and thoracic interscapular adipose tissue is partly drained into the inner vertebral sinus (plexus venosus vertebralis internus) which surrounds the spinal cord like a heat exchanger (Fig. 23). A similar vascular connection between the interscapular adipose tissue and the vertebral canal has been found in the human neonate (AHERNE and HULL, 1964).

The thermogenetic function of brown adipose tissue is shown by a continuous recording of the temperature within the adipose tissue as environmental temperature decreases (Fig. 20). Here, the temperature in the interscapular adipose tissue of a newborn guinea pig rises while the other temperatures fall during the decrease of environmental temperature. Since a simultaneous blood flow increase has been demonstrated in this tissue (BRÜCK and WÜNNENBERG, B., 1965c; HEIM and HULL, 1966a), it can be concluded that extra heat is produced there under cold exposure.

As for the mechanism of NST, several possibilities have been considered during the last few years (for detailed description see SMITH and HORWITZ, 1969; PRUSINER, CANNON, and LINDBERG (Ed.), 1970). At present a concept prevails which may

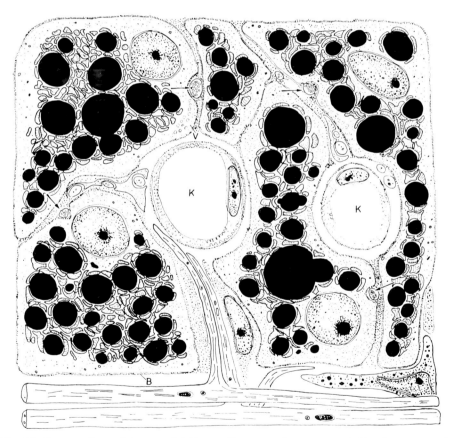

Fig. 22. Innervation of brown fat tissue. Paravascular nerves at the lower margin. *B* basal membrane; *K* capillaries; *V* vesiculation. In the intercellular space, sections of axons are shown, some with Schwann sheath cells or their processes (stippled). Arrows: epicellular nerve endings with synaptic vesicles. For a clearer view, the intercellular collagen fibrils were not drawn in. From BARGMANN, HEHN, and LINDNER (1968)

be briefly outlined: noradrenaline or adrenaline stimulates an enzyme, adenyl cyclase, which stimulates the synthesis of cyclic AMP (adenosine monophosphate). Cyclic AMP activates triglyceride lipase, which in turn splits off free fatty acids from the lipid reserves. This process also takes place, though to a lesser extent, in the white fat tissue. The significant property of the brown fat is that it can oxidize these free fatty acids in very large amounts, as required, and thereby form heat. This process is controlled so precisely that heat production can literally be increased by more than 100% in seconds, and can be lowered again nearly as rapidly. The excess oxidation is made possible by uncoupling, i.e. the coupling which normally exists between the formation of energy-rich phosphate compounds and the oxidative catabolism of substrate is abolished or loosened. In other words the P/O ratio, normally 3.0, is more or less strongly reduced. This uncoupling is based on an effect of free fatty acids on the mitochondria. The peculiarity of brown fat consists of this — its mitochondria are uncoupled two to four times more easily by free fatty acids than are the mitochondria of other tissue (RAFAEL and HOHORST, 1968; LINDBERG,O. [Ed.], 1970). Part of the free fatty acids does leave the brown fat and is oxidized at other sites, possibly in the musculature. Recently it has been suggested that the ability of other organs to oxidize free fatty acids is enhanced by a substance released from the brown fat, which may be a hormone or a metabolite (HIMMS-HAGEN, 1969).

Special attention has been directed to the biologic importance of the typical localization of brown adipose tissue. According to this localization certain organs

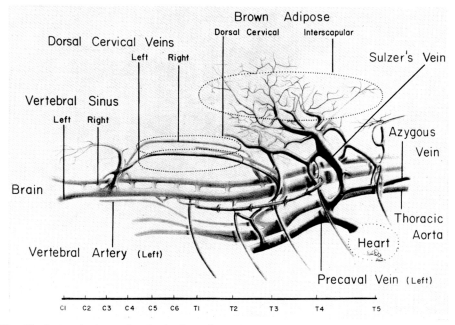

Fig. 23. Lateral view of polyvinyl replica of vascular drainage from brown fat pad of intrascapular and dorsal cervical regions. Note the venous connections between the fat pads and the vertebral sinus (plexus venosus vertebralis internus) which surrounds the spinal cord like a heat exchanger. From SMITH and ROBERTS (1964)

are preferentially supplied with metabolic heat, such as the thoracic and cervical spinal cord, the heart, and the intrathoracic organs in general. Similarly the kidneys are surrounded by brown adipose tissue. The heat flow to the cervical spinal cord (Fig. 23) became particularly interesting when it was shown that thermoreceptive structures exist in the cervical spinal cord which control shivering (BRÜCK and WÜNNENBERG,W., 1966; WÜNNENBERG, W., and BRÜCK, 1968). This thermoreceptive area receives heat from the interscapular and cervical brown adipose tissue, as can be inferred from the fact that the cervical spinal cord temperature runs parallel to that of the adipose tissue (Fig. 20), in contrast to the other temperatures which drop on cold exposure. It has been further demonstrated that shivering remains suppressed as long as the brown adipose tissue produces heat sufficient to maintain the cervical spinal cord temperature above the shivering threshold (Fig. 20, BRÜCK and WÜNNENBERG,W., 1966, Review: BRÜCK, 1970). This means that the more economical NST (cf. p. 539) can be utilized to its full amount before shivering occurs.

This interlocked control of shivering and nonshivering thermogenesis is demonstrated by a block diagram (Fig. 24). Here, a controller is thought to be localized

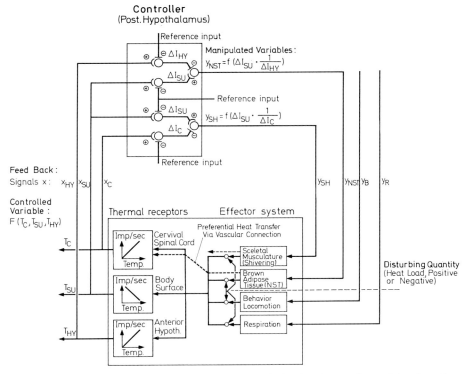

Fig. 24. Block diagram of the interlocked control of shivering and nonshivering thermogenesis in the guinea pig. The plus and minus symbols indicate facilitation or inhibition, respectively, in a neurophysiologic sense. T represents temperature, I impulse frequencies in nerves. Indices, C cervical spinal cord; Hy hypothalamus; SU body surface. From BRÜCK and WÜNNENBERG (1970)

in the posterior hypothalamus. It receives information (x_{su}, x_c, x_{Hy}) on the temperature of at least three sites of the organism — the body surface (T_{su}), the cervical spinal cord (T_c), and the anterior hypothalamus (T_{Hy}). It is assumed that the impulse frequency is reciprocally related to the temperature in the cutaneous receptors, as one would expect from cutaneous *cold* receptors. For the internal receptors, a direct proportionality to the temperature could be demonstrated (WÜNNENBERG,W. and BRÜCK, 1968, 1970). These input signals are compared with three corresponding reference signals. Each pair of the three resulting signals, namely, ΔI_{su} and ΔI_c on one hand, and ΔI_{su} and ΔI_{Hy} on the other, would then determine the output signals (manipulated variables), y_{sh} and y_{NST}, which in turn control the extent of shivering and nonshivering thermogenesis, according to Eqs. (4) and (5).

$$y_{SH} = f\left(\Delta I_{SU} \cdot \frac{1}{\Delta I_c}\right), \tag{4}$$

$$y_{NST} = f\left(\Delta I_{SU} \cdot \frac{1}{\Delta I_{Hy}}\right). \tag{5}$$

Thus, a *multiplication* of the input signals or their reciprocal values must occur in the controller (cf. p. 572).

V. Heat Exchange with the Environment

Under equilibrium heat flow conditions the amount of heat formed in the body per unit time is equal to the amount given off from the body surface, the temperature field within the body remaining constant over that time. For homeothermal organisms however this is true only for integrated values over longer periods, while the instantaneous values are in equilibrium only in most unusual circumstances. Usually the heat content of the body fluctuates widely as a result of *accumulation* and *discharge* of heat and of changes in the temperature field. The heat content of individual humans (specific heat of the body $= 0.83$ cal $g^{-1} \, °C^{-1}$) may change by as much as several hundred kcal.

In addition to the regulation of heat production discussed earlier, homeotherms may act on heat flow by modifying *thermal resistance* and *heat absorption* through the evaporation of water from the external and internal surfaces. At any given rate of heat production the thermal resistance is actively set at such a level that the heat formed is in equilibrium with that given off in a prescribed temperature gradient, $T_i - T_a$. T_i represents the "set point" of the core temperatu reand T_a the environmental temperature.

A. Pathways of Heat Transport

1. Heat Flow from the Body Interior to the Skin Surface

Heat formed within the body reaches the skin surface partly by *conduction* through the tissues and partly by *convection* via the blood. In the epidermis, which lacks blood vessels, heat is transported exclusively by conduction (Fig. 25). Due to the relatively poor heat conductivity of body tissues (Tab.5), heat transport through conduction is small in comparison with the maximum convective transport possible through the blood. The heat flux density (H) (cal cm^{-2} sec^{-1}) through conduction and convection perpendicularly through a layer of tissue with a temperature difference $T_1 - T_2$ obeys the equation

$$H = c \, (T_1 - T_2) \, . \tag{1}$$

c is the *thermal conductance* (cal cm^{-2} $sec^{-1} \, °C$).

H can also be expressed by the equation:

$$H = \lambda' \, \frac{T_1 - T_2}{x} \tag{2}$$

in which x is the thickness of a planar layer of constant width perpendicular to the flow of heat, and λ' (cal cm^{-1} sec^{-1} °C^{-1}) is the *apparent thermal conductivity* of the material composing the tissue. (For details see GRÖBER, ERK, and GRIGULL, 1955; CARSLAW and JAEGER, 1959; ARPACI, 1966; for measurement of thermal conductivity of tissues, see GOLENHOFEN, HENSEL, and HILDEBRANDT, 1963).

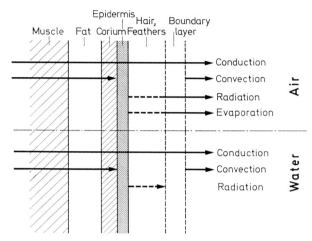

Fig. 25. Heat transport from the body to the environment

Table 5. Thermal conductivity of various tissues and materials

Tissue		Thermal conductivity 10^{-3} cal · cm^{-1} · sec^{-1} · °C^{-1}
Tissues:		
Skeletal muscle (man)[a]	$n = 45$	1.21 ± 0.04
Liver (man)[a]	$n = 28$	1.13 ± 0.04
Brain (man)[a]	$n = 12$	1.17 ± 0.02
Brain (dog)[a]	$n = 5$	1.15 ± 0.06
Adipose tissue (man)[a]	$n = 4$	0.51 ± 0.06
Skin (finger pad, man)[a]	$n = 37$	1.10 ± 0.08
Skin (premature infant)[a]	$n = 7$	1.33 ± 0.03
Blood (mean)[a]		1.14
Hair and feathers:		
Wool[b]		0.057
Feathers[b]		0.057
Fur (rabbit)[b]		0.060
Other materials:		
Air[c]		0.056
Water (20° C)[c]		1.4
Silver[c]		1000

[a] From GOLENHOFEN, HENSEL, and HILDEBRANDT (1963).
[b] From PFLEIDERER and BÜTTNER (1940).
[c] From D'ANS and LAX (1943).

The *circulation* of the blood acts as a universal heat exchange system (Fig. 26). The blood ejected from the left heart gives off a large part of its heat to the exterior of the body and flows back cooled. Blood is warmed primarily in the active skeletal musculature and in the internal organs. In addition there is an extensive exchange of heat between arteries and veins, especially in the appendages (venae comitantes), on a countercurrent principle, by which the arteries lose heat and the veins acquire it (p. 555).

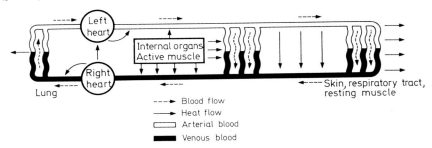

Fig. 26. Heat exchange by circulation

2. Heat Flow between Body and Environment

The heat flux density (H) from the body surface to the environment is the sum of partial fluxes by *conduction* (H_k), *convection* (H_c), *radiation* (H_r) and *evaporation* (H_e)

$$H = H_k + H_c + H_r + H_e. \tag{3}$$

The total heat flux from the body to the environment also depends on the area of the "effective surface" of the body. This area is smaller than the geometric surface and can also vary markedly depending on body posture (BÜTTNER, 1938). Substantial amounts of heat are also given off from the inner surface of the *respiratory pathways* through conduction, convection, and evaporation. On the other hand, the heat loss from the surface of the lungs is very slight (MATHER, NAHAS, and HEMINGWAY, 1953; AFONSO et al., 1962a, b). Since the laws governing the flow of heat between the body and the environment have long been well known, we will limit ourselves here to a few essential points and for the rest refer the reader to the numerous comprehensive presentations (BÜTTNER, 1938; THAUER, 1939; PFLEIDERER and BÜTTNER, 1940; HARDY, 1949; BURTON and EDHOLM, 1955; HARDY, 1963; HARDY, GAGGE, and STOLWIJK, 1970; WHITTOW, 1970, 1971; int. al.).

Heat Conduction and Convection. The body surface is surrounded by a layer of *still air* (boundary layer) in which no heat convection takes place. The boundary layer becomes thinner when the curvature of the body surface or air pressure increases and above all when wind speed rises. In the absence of wind the still layer is 4–8 mm thick, while at wind speeds of only 2 m sec^{-1} it is reduced to 1 mm, whereby its resistance to heat conduction is reduced to 0.2–0.12 (BÜTTNER, 1938; BURTON and EDHOLM, 1955; CARLSON and BÜTTNER, 1957; COLIN and HOUDAS, 1967; COLIN et al., 1970).

Hair, feathers, and protective clothing are means for the retention of a thick
layer of resting air on the skin that cannot be blown away by the wind. After the
same principle heat insulating coverings are made. They consist of a loose inner
layer which holds air closely (woolly hair, down, in man-made clothes loose
textures, padding, etc.) and an outer smooth layer (straight hair, tectrices, in
man-made clothes wind-proof outer wear). Due to effective heat insulation the
temperature gradient is very steep, so that the surface is at almost the same
temperature as the exterior air. Thermal resistance is sometimes measured, for
purposes concerned with the physiology of clothing, in "clo-units", where 1 clo
$= 0.18°$ C m² hr kcal⁻¹. By way of comparison, a business suit has ca. 1 clo, heavy
arctic clothing up to 5 clo, and a 1 cm thick layer of fat 0.8 clo (BURTON and
EDHOLM, 1955). Beyond the boundary layer heat is also transported by *convection*.
Heat flux density, H_c, from the surface to the surrounding layer is described by
the equation

$$H_c = h_c\,(T_s - T_a) \tag{4}$$

where T_s is the surface temperature, T_a the air temperature, and h_c a heat
transfer coefficient for conduction and convection (cal cm⁻² sec⁻¹ °C⁻¹) .
h_c increases with several factors, including increasing curvature of the surface and
increased air pressure, and increases still more with wind speed. Heat is also given
off from the mucous membranes of the respiratory tract, by conduction and convection, to the respired air as it passes by.

Radiation. For the heat flux density (H_r) by radiation from the body to the
environment we can write according to Stefan-Boltzmann's law:

$$H_r = \sigma\varepsilon_s\,(T_s{}^4 - T_r{}^4) \tag{5}$$

where σ is the Stefan-Boltzmann constant ($1.38 \cdot 10^{-12}$ cal cm⁻² sec⁻¹ °K⁻⁴), T_s(°K)
the temperature of the body surface and T_r(°K) the mean radiant temperature
of the surroundings. ε_s is the radiant emittance of the body surface, i.e., the
emission of radiation as a proportion of a perfect emitter (black body) at the same
temperature. For human skin at body temperature (infrared, $\lambda = 3 - 60\,\mu m$,
maximum at $10\,\mu m$), ε is practically that of a black body ($\varepsilon = 1$) and independent
of color (BÜTTNER, 1938; HARDY, 1939; BUCHMÜLLER, 1961; GÄRTNER and
GÖPFERT, 1964; GÄRTNER, LING, and GÖPFERT, 1964; MITCHELL, 1970). It varies
between 0.941 and 0.976 according to the site (GÄRTNER and GÖPFERT, 1964) and
between 0.93 and 0.97 with changes in cutaneous blood flow (GÄRTNER, LING, and
GÖPFERT, 1964). Fur and feathers have values of $\varepsilon = 0.9$ (GATES, 1968), while the
figure for polished metal surfaces is $\varepsilon < 0.1$. When the skin is covered with fur,
feathers, or clothing, most of the heat radiation from the body will be absorbed
within the integument. Heat radiation from the surface of the integument is low
because of its low radiant temperature (GATES and PORTER, 1970).

For small temperature differences ($T_s - T_r$), the above equation can be linearized

$$H_r = h_r(T_s - T_r) \tag{6}$$

where h_r is a heat transfer coefficient for radiation and T the temperature in °C
(COLIN et al., 1970).

In contrast to the findings in the infrared range, the values for the absorptance α under the condition of solar irradiation ($\lambda = 0.3 - 3\,\mu m$) are dependent on the color of the body surface. Absorptance is the ratio of radiant flux absorbed to the incident flux. For white and black human skin, α varies between 0.55 and 0.84 (BLUM, 1945). Short-wave infrared ($0.8 - 1.5\,\mu m$) penetrates deeper into human skin than does infrared in the longwave range (PFLEIDERER and BÜTTNER, 1940; BUCHMÜLLER, 1961). Under clear sky conditions the energy uptake of the human body from solar radiation can be 5 times higher than the heat production (ADOLPH, 1947a). According to the color and the material of clothing or integument, part of the solar radiation will be reflected from the surface and the rest absorbed in the more superficial layers. The dorsal side of darker mammals has a solar absorptance between 0.75 and 0.9; usually the ventral side is lighter and therefore has an absorptance around 0.6 (GATES, 1968). Considerable variations of radiation penetration with color of coat have been found in cattle (HUTCHINSON and BROWN, 1969).

Evaporation. Due to the high heat of vaporization of water ($580\ \mathrm{cal\ g^{-1}}$) its evaporation from the surface of the skin and from the mucous membranes of the respiratory tract is a very effective means of heat loss. A *boundary layer* on the skin exists for evaporation as well, through which water vapor moves exclusively by *diffusion*. The heat flux density (H_e) through evaporation from completely wetted skin follows, by way of approximation, the equation

$$H_e = h_e (P_s - P_a)$$

where P_s is the vapor pressure of the skin surface, P_a the vapor pressure of the air and h_e the evaporative heat loss coefficient, which depends on the curvature of the surface, on the air pressure and on the wind speed, among other factors. A result which is significant for homeothermy can be seen, i.e. that water can evaporate from the skin even if the air is saturated with water vapor, provided that $P_s > P_a$. This condition can exist if the skin temperature is high. Water vapor is given off from the surface of the respiratory tract according to the same laws. Extensive theoretical studies of evaporative heat loss have been made by RAPP (1970) and SIBBONS (1970).

Heat Loss in Water. Heat loss in water follows a completely different course from heat loss in air. The hair- and feather-coats are saturated and lose their heat insulating properties, except in the case of aquatic birds, whose lubricated tectrices prevent the entry of water. Following surgical removal of the coccygeal gland *(Glandula uropygialis)* ducks are unable to maintain their rectal temperature after remaining only ten minutes in water at 10°C, due to the saturation of their plumage (HSIANG-CH'UANHOU, 1928). The transfer of heat from the body surface takes place almost entirely by conduction and convection. As the heat transfer coefficients in water may be 200 or more times greater than the coefficients in air, the body loses large amounts of heat even in moderately cool water. In this situation the only effective means of heat insulation for mammals consists of the development of a thick layer of blubber. For birds the lubrication of the plumage, a blubber layer (in penguins), and enlargement of the air sacs (diving birds, PRENGLOWITZ, 1933) serve this purpose. When heat is lost by contact with solid bodies, for example by lying on the ground, it is transferred mainly by conduction, and partly by radiation.

Heat Transfer in Natural Environments. Animals under outdoor conditions are subject to a variety of factors that determine positive or negative heat fluxes between the environment and the body surface. The most complicated of these factors is radiation which can be direct, reflected, or scattered sunlight as well as infrared radiation from ground, vegetation, and atmosphere (Fig. 27). An account

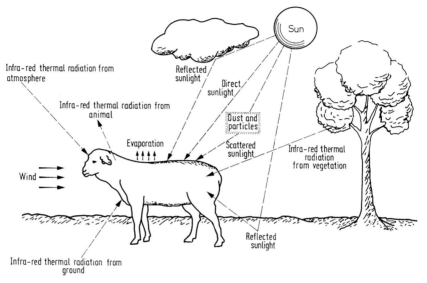

Fig. 27. The streams of energy flow to and from an animal organism in its natural environment. From GATES (1968)

of the energy budget of animals in natural environments is given by GATES (1968) and by GATES and PORTER (1970); the latter have also calculated heat fluxes for several birds and mammals under various climatic conditions and designed climate diagrams showing relations between air temperature, radiation absorbed, and wind speed for constant-body and radiant-surface temperatures at actual values of metabolic and water-loss rates. The diagrams are based on body dimension (D), metabolism (M), expiratory moisture loss (E_{ex}), thickness of feathers or fur (d_f), thickness of fat (d_b), deep body temperature (T_b) and radiant surface temperature (T_r). An example for the Kentucky cardinal *(Richmondena cardinalis)* is shown in Fig. 28. The sets of lines (1—4) indicate the combinations of temperature (ordinate) and radiation absorbed (abscissa) at which the animal maintains constant body and radiant surface temperatures under conditions varying from full sunlight (sun) to clear sky at night (sky plus ground) and wind speeds varying from $10-1000$ cm sec^{-1}. Set 1 of lines stands for the physiological state of cold defense (high metabolism, low evaporation, thick plumage, low body and surface radiation temperatures), sets 2 and 3 for neutral conditions and 4 for a state of heat defense (high evaporation, thin plumage, high body and surface temperature). The diagram shows, for example, that at the high temperature limit, the cardinal would not withstand full sunlight in still air for extended periods of time at an air temperature exceeding 20° C. However, the animal can withstand very high temperatures up to 50° C in still air at night when exposed to clear sky. The lower temperature limit at night in still air is −35° C.

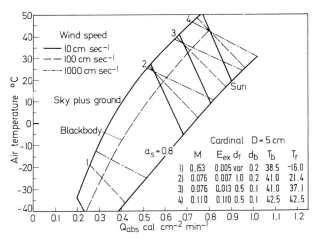

Fig. 28. Climate diagram for a cardinal *(Richmondena cardinalis)* showing relations between air temperature, radiation absorbed, and wind speed for constant body and radiant surface temperatures at actual values of metabolic and water loss rates. Further explanation see text. From GATES and PORTER (1970)

Comfort Equations. It is clear that the climatic conditions for thermal comfort and discomfort (p. 613) cannot be expressed by a single factor, such as air temperature. Numerous attempts have been made to describe the combined effects of environmental variables, such as air temperature, air humidity, mean radiant temperature, air velocity and clothing by certain indices or equations which are significant for both physiologists and engineers in judging thermal comfort and in predicting heat exchange of the human body. The criteria for the determination of such comfort indices are either the votes of subjects assessing their thermal comfort, or the state of the thermoregulatory system as assessed by measurement or calculation. HOUGHTEN and YAGLOU (1923), KOCH, JENNINGS, and HUMPHREY (1960), NEVINS et al. (1966), MCNALL et al. 1967, and MCNALL, RYAN, and JAAX, 1968, have exposed subjects to different combinations of two environmental factors (e.g. air temperature and relative humidity) while keeping constant all other factors which influence thermal comfort. Large numbers of subjects have been asked to vote upon their thermal sensation or their thermal comfort according to special scales. A comfort line has been determined, i.e. a line representing all combinations of the two variables which give optimal thermal comfort. For example, the *"effective temperature"* (HOUGHTEN and YAGLOU, 1923) combines the effect of air temperature and humidity in a room without radiation and air movement for a resting subject with standard clothing. 20° C and 100% relative humidity will lead to the same comfort as 25° C and 35% relative humidity, both combinations being defined as an effective temperature of 20° C.

These methods have given valuable results, the practical application of which are limited, however, as they are valid only when the other parameters are kept constant. Therefore, comfort equations have been developed for a larger number of variables. The *"humid operative temperature"* (NISHI and GAGGE, 1971), for instance, is a biophysical index of comfort which includes air temperature, radiant heat, humidity, air movement and clothing. It is determined on the basis of the heat balance equation of the body and corresponds well with comfort votes by subjects. A *general comfort equation* has been derived by FANGER (1970a, b) which allows for any activity level and any clothing to calculate all combinations of the environmental variables (air temperature, air humidity, mean radiant temperature, and air velocity) which will create optimal thermal comfort. The equation is based on heat balance calculations as well as on comfort votes by subjects. It should be emphasized, however, that under certain conditions, such as thermal transients or unusual topography of skin temperature, the actual experience of thermal comfort might deviate from the values predicted by the above comfort equations (p. 614).

B. Physiological Regulation of Heat Flux

1. Blood Flow through the Skin and Mucous Membranes

By alterations in the flow of blood through the skin and mucous membranes of the respiratory tract the *apparent thermal conductivity* of the tissues can be raised to levels several times higher than the true thermal conductivity. The outflow of heat from the skin surface is also affected, since skin temperature changes with changes in blood flow. With increasing skin temperature, heat outflow via conduction, convection, and radiation rises, as does heat loss via evaporation which goes up due to the higher vapor pressure, P_s, of the skin.

The significance of skin blood flow is particularly great for *man*, in whom it can be modified over the entire body surface, and above all in the ends of the extremities. Under indifferent environmental conditions blood flow in the fingers may fluctuate in a proportion of 1 : 10 (JOHANSEN and TØNNESEN, 1969). This seems largely a function of arteriovenous anastomoses (VANGGAARD, 1969) which allow heat transport to increase above the corresponding values of the capillary flow. With changing temperature conditions the blood flow through the human hand changes by a factor of 30 (FORSTER, FERRIS, and DAY, 1946; FERRIS et al., 1947). In extreme cases it changes in the fingers from $0.2-120 \text{ cm}^3$ 100 cm^{-3} tissue min^{-1}, that is by 600-fold (WILKINS, DOUPE, and NEWMAN, 1938). In this case the thermal conductance of the skin changes by a factor of 4 (BAZETT, 1952; BRÜCK and HENSEL, 1953). The reciprocal of the thermal conductance, the *thermal resistance* of the body shell, changes from $0.02-0.14°$ C m^2 h kcal^{-1} or from $0.1-0.7$ clo-units with changes from very warm to very cold surroundings (Fig. 29).

Other mammals, with the exception of a few thinly hirsute species, show less thermoregulatory variability of blood flow through the skin than man does

(EDERSTROM, 1954). In birds it seems to be of little significance in the feather-covered areas of the skin, but very important in the bare regions (DAWSON and HUDSON, 1970). Thermal conductance in birds changes very markedly with the

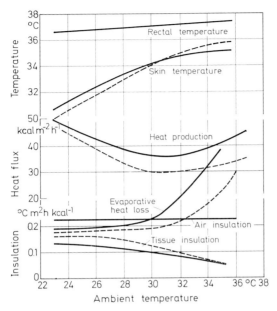

Fig. 29. Rectal temperature, mean skin temperature, heat production, evaporative heat loss, thermal insulation of boundary layer of air around the body, and thermal insulation of body shell in man. Solid lines: mean values for men; dashed lines: mean values for women. From HARDY and DU BOIS (1940)

adjustment of the plumage (Fig. 30). The flow of blood through the *ears* of many animals (elephant, BUDDENBROCK, 1937; rabbit and hare, v. DOBBEN-BROEKEMA and DIRKEN, 1950; HONDA, CARLSON, and JUDY, 1963; SCHMIDT-NIELSEN et al., 1965; dog, HEMINGWAY, 1938; STRÖM, 1950; cow, BEAKLEY and FINDLAY, 1955) can be changed drastically for thermoregulatory purposes. According to HONDA, CARLSON, and JUDY (1963) the blood flow in the rabbit's ear under neutral conditions may fluctuate between almost 0 and 200 cm³ 100 cm⁻³ min⁻¹. Vessels in the horn of the goat dilate and constrict with changing ambient temperature (TAYLOR, 1966). The *swimming flippers* of the fur seal *(Callorhinus ursinus)* in subarctic water may be as cold as the water, but may increase nearly to body temperature during exertion on land. Acquisition of unwettable adult body fur and development of effective regulation of circulation of heat to the flippers are essential before the young seal can enter the aquatic environment (BARTHOLOMEW and WILKE, 1956; IRVING et al., 1962; IRVING, 1964).

The flow of blood through the *flight membrane* of some bats also seems to be related to temperature regulation. COWLES (1947b) and REEDER and COWLES (1951), working with *Myotis yumanensis*, *Myotis velifer*, and *Macrotus californicus*, observed a distinct vasodilatation, due to opening of precapillary sphincters, as soon as the rectal temperature rose above 39° C. For many animals blood flow through the tongue is important for heat transport. In the dog, which pants with its tongue hanging out, blood flow through it can increase sixfold in warm surroundings (EDERSTROM, 1954). All appendages to the body integument, whose blood flow can be changed within particularly wide limits, have a high surface volume ratio (Table 6), and are thereby particularly suited for the emission of heat.

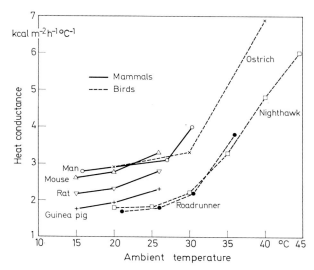

Fig. 30. Thermal conductance in various birds and mammals. Man, mouse, rat, and guinea pig from HERRINGTON (1941); Roadrunner *(Geococcyx californianus)* from CALDER and SCHMIDT-NIELSEN (1967); Ostrich *(Struthio camelus)* from CRAWFORD and NIELSEN (1967); Common nighthawk *(Chordeiles minor)* from LASIEWSKI and DAWSON (1964)

Table 6. Surface/Volume

Organism or organ	Surface/Volume cm^{-1}
Man	
Adult (70 kg)	0.2
Child, 1 year (9.1 kg)	0.5
Newborn infant (3 kg)	0.6
Premature infant (1.5 kg)	0.8
Finger	2.1
Dog	
Whole body (10 kg)	0.5
Tongue	3.6
Rabbit	
Whole body (2 kg)	0.7
Ear	5.6

Various homeotherms have *counter-current heat exchange* systems in the extremities and appendages. A more or less intricate web of parallel arteries and veins allows the exchange of heat between the arterial blood flowing to the periphery and the returning venous blood (Fig. 31). Thus the arterial blood is pre-cooled and peripheral heat loss is reduced to a minimum, while the nutritive circulation to the tissues is sustained. In contrast to this mechanism, vasoconstriction in the cold would reduce heat loss as well as blood flow.

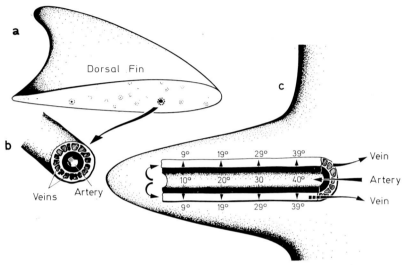

Fig. 31. Countercurrent heat exchange system in the fin of the porpoise. *a* section of the dorsal fin where the veins surround each artery. *b* close-up of the artery surrounded by the multiple venous channel. *c* hypothetical temperature gradient in this concentric countercurrent system. From SCHOLANDER and SCHEVILLE (1955)

Countercurrent heat exchange systems have been found in the extremities of man, where the returning blood passes through the venae comitantes along the arteries, thereby producing axial temperature gradients up to $0.35°$ C cm^{-1} (BAZETT et al., 1948a, b), in the extremities of Panamanian sloths (*Bradypus griseus* and *Choloepus hoffmanni*; SCHOLANDER and KROG, 1957), and in the fins of porpoises (SCHOLANDER and SCHEVILLE, 1955). Body heat is conserved at the cost of keeping the appendage cold. There are also separate superficial veins associated with the countercurrent system; their function might be as follows: if the animal needs to conserve heat, blood circulation through the fins would be slow and the venous return would preferentially pass through the countercurrent veins; on the other hand, if maximal cooling is needed, as during exercise in warm water, this could be accomplished by a high rate of blood flow through the fins with a venous return through the superficial veins.

2. Hair, Feathers, and Air Sacs

The pelage of mammals and to an even greater extent the plumage of birds are excellent heat insulators, which man also uses in a variety of ways. COWLES

(1947a) hypothesized that hair and feathers originally developed in phylogeny as protection against solar radiation and only secondarily took over a role in protection against cold. In fact, body coats of homeotherms can play an important role for protection against radiation. The short, glossy coat of light color that is found in mammals living in hot and arid regions not only reflects solar radiation but also allows ready convection from the skin surface and heat loss by infrared radiation (MacFarlane, 1964, 1968; Hafez, 1968). The other extreme is a type of coat with long hairs intermixed with finer fibers, giving optimal thermal insulation in the cold.

The envelope of air that surrounds a bird's body varies considerably with the thickness and tightness of the plumage. In the tawny owl, which has a particularly thick and loose plumage, each cm² of body surface is overlain by an air volume of 4.7 cm³, while in diving birds with a closely adhering feather coat there is only 0.77 cm³. Plumage makes up 88% of the body volume of the tawny owl, but only 46% of that of the diver (v. Buddenbrock, 1937), while the volume of the air sacs is greater in the diver (15.8% of the feather-less body, in contrast to 12.5% in the dove). According to Prenglowitz (1933) this is related to heat insulation. Penguins which, being highly specialized swimming and diving birds, have a closely adherent plumage can achieve adequate thermal protection in the polar climate with the help of an additional subcutaneous layer of blubber. Fig. 32 shows the thicknesses and heat resistances of the coats of a series of tropical and arctic

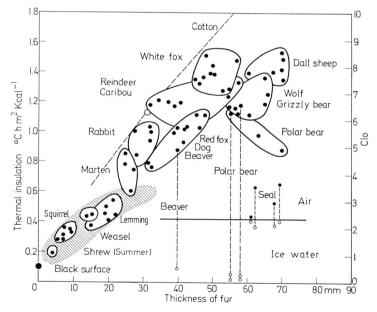

Fig. 32. Insulation in relation to winter fur thickness in a series of arctic mammals. The insulation in tropical mammals is indicated by the shaded area. In the aquatic mammals (seal, beaver, polar bear) the measurements in 0° C air are connected by vertical broken lines with the same measurements taken in ice water. From Scholander et al. (1950a)

mammals. The heat insulation of the larger polar animals is superior to that of man's arctic clothing, which goes up to 5 clo or $0.9°$ C m² h kcal⁻¹.

Thermal resistance can be raised within certain limits by "bristling" of hair and "fluffing" of feathers. These reactions are due to involuntary tensing of smooth muscles (arrectores pilorum) connected to the hair follicles. The muscles are attached to individual hairs or feathers and set the shafts, which are planted obliquely into the skin while at rest, into an upright position. Quantitative values for the insulating properties of various species may be found in DAWSON and HUDSON (1970). Wetting with water, sweat, saliva, or nasal gland secretion reduces the thermal resistance sharply and at the same time increases the absorption of heat by evaporation. The limits of these changes are set by the heat conductivity of water at the minimal and maximal thickness of the body covering under saturation.

3. Ventilation Volume

The ventilation volume of respiration, which changes the "dry" as well as the "moist" heat loss, has a very different significance for *sweating* than for *non-* or *weakly*-sweating homeotherms. For man, whose ability to secrete sweat is very effective, and for the strongly sweating animals (Equidae) heat loss via the respiratory tract plays a minor role. In hot surroundings the water vapor lost from the human respiratory tract amounts to ca. 30% of the total water loss, while the respiratory volume, according to ROBINSON (1949), increases by only 0.25—0.5 fold per degree increase in rectal temperature. The situation is completely different in the case of birds, which lack sweat glands, and for most mammals with no or weak sweat secretion (mouse, rat, rabbit, cat, dog, sheep, pig, cow, int. al.). Panting in the furred mammal and the feathered bird is the most efficient and direct route for dissipating body heat by evaporating water. Evaporative heat loss from the airways represents nearly the same amount of heat removed from the body core, while heat entering the core from the external environment is limited by the insulated exterior surface. Even neglecting the wastage of sweat dripping off the body surface, sweating in thermally insulated animals is less efficient in removing core heat from the body.

Ventilation volume is typically increased by an increase in respiratory frequency (polypnoe, panting), accompanied by a reduction in volume of the individual breaths. In the dog the volume of each breath is reduced to 0.2 times the normal value (ROBINSON and LEE, 1941b), while the minute volume of respiration may be raised 27-fold. At a respiratory frequency of 275 min⁻¹ the volume of one breath is about the same as that of the dead air space, so that no exchange of blood gas takes place (ALBERS, 1961). At higher respiratory frequencies, elevated CO_2 output can lead to an intense respiratory alkalosis (ANREP and HAMMOUDA, 1933; BIANCA and FINDLAY, 1962; HALES and WEBSTER, 1967). Birds can oppose such alkalosis by increasing the ventilation of the air sacs, thus forming a functional shunt system allowing a regulated bypass of the lungs (CALDER and SCHMIDT-NIELSEN, 1968; SCHMIDT-NIELSEN et al., 1969).

The development and effectiveness of polypnoe vary within wide limits in individual species. In many animals, respiratory frequency is increased only as their core temperatures rise (sheep, pig, birds), while others pant as soon as the external temperature goes up (rabbit, dog, cat, cow; LEE, 1948). BLIGH (1963) observed that the sheared sheep, when transferred from a moderately cold (20° C or less) to a warm (42° C) environment, does not increase its respiration for up to 1−2 h even though skin temperature rises rapidly. Respiratory frequency remains at about 20 min^{-1} throughout this period, and then rises abruptly to the panting condition normally associated with the higher ambient temperature. Dogs and cats pant with their mouths open, their tongues hanging out, and with copious secretion of saliva (ANTAL and KIRILČUK, 1969; SHARP et al., 1969; HAMMEL and SHARP, 1971; ADAMS, 1971); polypnoe is more effective for them than for animals which pant with closed mouths (rabbit, rat, sheep), but many of these (e.g., rabbit and sheep) go over to open-mouthed panting when their rectal temperatures exceed 40° C (LEE, ROBINSON, and HINES, 1941; LEE and ROBINSON, 1941). For many mammals, especially the dog, the panting frequency corresponds to the resonance frequency of the respiratory system, which reduces the muscular effort required to low levels (CRAWFORD, 1962). Cattle are able to increase their respiratory ventilation with little increase in energy expenditure (WHITTOW and FINDLAY, 1968).

Evaporative water loss of birds received considerable attention in recent years (DAWSON, 1958; LASIEWSKI, ACOSTA, and BERNSTEIN, 1966a, b; CALDER and SCHMIDT-NIELSEN, 1967, 1968; DAWSON and HUDSON, 1970). Birds augment evaporative water loss in hot environments by panting, a process involving vigorous movements of the thoracic cage at frequencies falling between 40 and 700 min^{-1}. The manner in which panting develops varies with species. In birds such as the domestic fowl (HUTCHINSON, 1955), the breathing rate increases steadily with increasing heat load and, as body temperature exceeds some characteristic value between 41 and 44° C, open-mouthed panting commences. Other species, such as the ostrich (SCHMIDT-NIELSEN et al., 1969) and the roadrunner, *Geococcyx californianus* (CALDER and SCHMIDT-NIELSEN, 1967), respond to heat stress by abruptly changing from a relatively low breathing rate to a high one characteristic for panting. The panting frequency matches the resonance frequency of the respiratory system. Depending on species and heat load, the evaporative heat loss can amount to 150% of heat production (for comparative data see DAWSON and HUDSON, 1970).

Some birds, e.g., pelicans, ducks, geese, domestic fowl, owls, pigeons, roadrunners, and caprimulgids (Ref. see DAWSON and HUDSON, 1970), augment panting by rapidly fluttering the gular area. Blood vessels in this area become conspicuously engorged during heat stress. Gular fluttering is driven by flexing of the hyoid apparatus. In many birds the fluttering frequency which may amount to 1000 min^{-1} is independent of heat load and breathing rate; it is suggested that the frequency matches the resonance frequencies of the hyoid-gular structures (BARTHOLOMEW, LASIEWSKI, and CRAWFORD, 1958; DAWSON and HUDSON, 1970).

4. Salivary Cooling

A number of mammals neither sweat nor pant, but nevertheless regulate their body temperatures through evaporation. They use saliva for cooling, but in contrast to panting the saliva is spread onto the body surface from where it evaporates. Relatively little is known about saliva spreading as a mechanism for temperature regulation during heat stress; yet it is the only known mechanism for increasing evaporative cooling that has been observed in rodents (HERRINGTON, 1940; HUDSON, 1962; LEE, 1963; CARPENTER, 1966), in several species of marsupials (ROBINSON and MORRISON, 1957), in bats, particularly the Megachiroptera (BARTHOLOMEW, LEITNER, and NELSON, 1964), and in the elephant (KUNO, 1956). In addition, saliva spreading occurs in a few mammals that also sweat or pant, such as the cat (ROBINSON and LEE, 1941a) and opossum (HIGGINBOTHAM and KOON, 1955). Recent studies have revealed the crucial contribution of the salivary glands to thermal tolerance (HAINSWORTH, 1967; HAINSWORTH, STRICKER, and EPSTEIN, 1968; HART, 1971). More detailed investigations of the activation of salivary glands during heat stress have been performed by HAINSWORTH and STRICKER (1970, 1971) and by HAMMEL and SHARP (1971).

5. Water Loss from the Skin

Extraglandular Water Loss. A distinction is made between a glandular "sensible" water loss from the skin by way of the *sweat glands* and an extraglandular "insensible" water loss by way of diffusion through the *skin*. The proportion of extraglandular water loss and its physiology have not been adequately explained at this time. We refer the reader to the works of BUETTNER (1953, 1959, 1969a, b, 1971), ZÖLLNER, THAUER, and KAUFMANN (1955), HEERD and OHARA (1960), HEERD and OPPERMANN (1966) and OPPERMANN and HEERD (1970). It may take place essentially by a process of diffusion of water vapor from the skin capillaries through the uppermost skin layers, following the laws of diffusion (vapor pressure difference). According to BUETTNER (1971), in some areas of skin which are devoid of active sweat glands the rate of water vapor diffusion is much greater than it is through the general skin surface. Under thermally indifferent conditions in human subjects, evaporative water loss was found to be a linear function of the vapor pressure of the ambient air, the mean diffusion resistance being 2.9 Torr min $cm^2 \mu g^{-1}$ (HEERD and OPPERMANN, 1966; OPPERMANN and HEERD, 1970). In birds and mammals extraglandular water loss follows the same rules. REEDER and COWLES (1951) suggest that increased blood flow through the flight membranes of bats also leads to increased diffusion of water vapor.

Sweat Glands. The significance of *atrichial* (i.e. those which open directly onto the skin) or *eccrine* sweat glands for human temperature regulation is impressively demonstrated in individuals who are born without sweat glands and depend only on the extraglandular fraction (ca. $20-30$ g h^{-1}) which can absorb approximately 25% of the resting heat production. These individuals cannot regulate their core temperature in summer or while doing the slightest physical work (SUNDERMAN, 1941; SHELLEY, HORVATH, and PIUSBURY, 1950). Water loss from the skin may be very small. Its lower limit is determined by the physical laws of water vapor diffusion, and its upper limit by the working capacity of the sweat glands. The latter may be extraordinarily high in man, up to 4 l h^{-1} in extreme cases. In general the secretion from the sweat glands is hypotonic to the blood (salt concentration $0.2-0.3\%$), and therefore requires active work. Sweat contains all the salts of the blood, principally NaCl. But the salt concentration of sweat can vary between 0.84 and 0.03% depending on the region of the body, individual constitution, heat load, and acclimatization (ROBINSON, 1949; KUNO, 1956; cf. p. 639 [Chapter VIII]). Most investigators found that the quantity of sweat secretion was fairly independent of water intake. It did not rise significantly when large quantities were drunk (LADELL, 1945); similarly it was not reduced when the body was moderately dehydrated (PITTS, JOHNSON, and CONSOLAZIO, 1944; ADOLPH, 1947a). However, recent experiments at ambient temperatures of 44° C and 50% rel. humidity have shown that ingestion of 1—3 l of water (before entering the heat chamber) will induce a significant increase in sweat rate-up to 50% of the value without drinking (CAGE et al., 1970).

Birds have no sweat glands; in *mammals* we find great variations in their number and functions. They are absent in monotremata, edentata, proboscidea, and many rodents and lagomorphs (mice, rats, rabbits). Some marsupials, the dog, goat, sheep, cow, and all equids possess *epitrichial* (i.e. those which occur in association with the hair follicles) or *apocrine* sweat glands over the entire body, but their functional significance is very variable (BLIGH and ALLEN, 1970). Eccrine glands are found principally on the hairless skin of the soles and palms, and in primates over the rest of the body as well (BIEDERMANN, 1930). The sweat glands of the dog do not become active if the rectal temperature increases but do so on local warming of the skin above 38.5° C (AOKI and WADA, 1951). Their function seems to be to guard against local overheating. According to BLIGH (1961) and BLIGH and ALLEN (1970) the apocrine sweat glands in sheep seem to have other than thermoregulatory functions. In the cow, which has functionally active sweat glands over the entire body surface (RAGSDALE et al., 1950), cutaneous and respiratory evaporation is almost adequate to cool the body at environmental temperatures of up to 40° C (MCLEAN, 1963). If evaporation is reduced by spraying oil on the skin or by an impermeable polyethylene coat, rectal temperature climbs markedly (RIEK and LEE, 1948; DOWLING, 1958). Sweat secretion from the skin is just as effective in equids as in man (ADOLPH and DILL, 1938), as is indicated by the fact that panting, which is very pronounced in all the above-mentioned animals, is absent in them.

Sweating and panting are complementary, in the sense that animals with a low capacity for sweating normally have a high capacity for panting (Fig. 33). BIANCA (1968) assumes that the link for this inverse relationship between sweating and panting is provided by skin temperature, which at the same time responds to

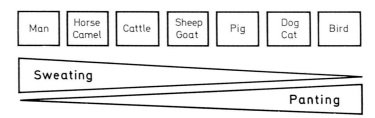

Fig. 33. Relative importance of sweating and panting in various species. The diagram shows the general trend, not the exact quantitative relationship of thermal sweating and panting. From BIANCA (1968)

sweating and acts as a stimulus for panting. Thus, a poorly sweating, and, therefore, warm skin tends to be associated with a high respiratory activity and vice
versa. A comparative study of sweating and panting in various mammals, with
some considerations of the evolutionary aspect has been made by BLIGH and
ALLEN (1970). For further references on mechanisms of water loss in birds and
mammals see WHITTOW (1970, 1971).

6. Total Evaporation and Air Temperature

After the presentation given on p. 549 no special explanation should be needed for
the fact that it is not the quantity of fluid produced that is decisive for temperature regulation but rather the quantity of water that actually *evaporates* on the
body surface. Wet skin from which sweat runs in streams is not a sign of good
regulation. Instead it indicates that there is a misproportion between the amounts
of water produced and evaporated. A theoretical treatment of heat loss by sweat
secretion is found in TIMBAL et al. (1969). The behavior of total water evaporation
at different external temperatures and moderate air humidity shows profound differences between sweating and non-sweating homeotherms (Fig. 34). In man,
who may be taken as a typical example of a sweating species, the quantity of water
evaporated remains approximately the same up to a "critical temperature range"
which, according to KUNO (1956), lies between 27° and 32° C. It makes up 25–35%
of the total heat loss. The outflow of heat through radiation, conduction, convec-

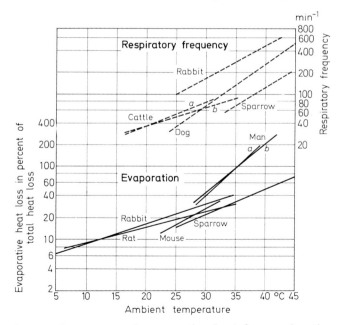

Fig. 34. Respiratory frequency and evaporative heat loss as function of ambient
temperature. Respiratory frequency: Rabbit from LEE, ROBINSON, and HINES (1941);
dog from RICHET (1898); cattle *a* from REGAN and RICHARDSON (1935); cattle *b* from
GAALAAS (1945); sparrow from KENDEIGH (1944). Evaporation: Sparrow from KEN-
DEIGH (1944); others from BRODY (1945)

562 Homeothermic Organisms

tion, and insensible water loss becomes insufficient for resting nude men at an external temperature of 31° C and for normally clothed men at 29° C. At this point sweat secretion sets in abruptly at an average critical skin temperature of ca. 35° C (ZÖLLNER, THAUER, and KAUFMANN, 1955). Above the critical temperature the curve of evaporation rises approximately exponentially with external temperature. At an external temperature of 36° C, where heat flow via conduction, convection, and radiation is close to zero, heat absorption via water evaporation reaches 100% of the total heat production. At higher temperatures evaporation again increases sharply, so that all of the heat formed, which now includes heat taken up from the environment, is absorbed. In weakly sweating animals the evaporation curve does not show this abrupt shift. The slope of the curve is only 5% per degree elevation of temperature in the rat, 6% in the rabbit, and 10% in the mouse (BRODY, 1945). At an external temperature of 36° C water evaporation accounts for only 35—60% of heat production in the rabbit; therefore its core temperature must rise.

Circulation also shows typical variations related to sweat secretion. In strongly sweating homeotherms (man, horse, donkey) the pulse frequency rises when it is warm, since blood flow through the skin is augmented, while in the cow for example (Fig. 35) pulse frequency goes down in a hot environment.

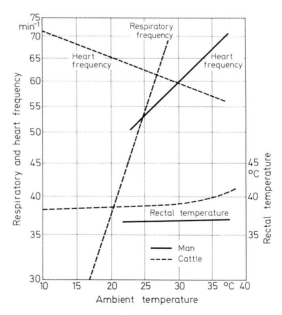

Fig. 35. Rectal temperature, heart and respiratory frequency in man and cattle as function of ambient temperature. From BRODY (1945)

7. Body Posture

Body posture can bring about substantial changes in the effective surface. If a man were to roll up completely into a sphere his body surface area would be reduced to one half its original size. In hot weather, raccoons lie on their backs and stretch all four legs, exposing the belly which has very short hair. Dogs in an outstretched position lose approximately 30% more heat than they would lose in a normal sitting posture. Pigs in hot environments lie down on their side with their snouts facing the wind. Under heat stress, chickens hold their wings slightly separated from their body so that air can circulate past the less insulated under-surface. They may also lie down with their head and neck stretched forward or sit on the soil. The thick insulation of the dorsal feathers and the still air above them impede the flow of heat into the bird, while the squashing of the breast feathers facilitates heat loss to the cool soil (HAFEZ, 1968).

In birds the most conspicuous adjustments in the cold involve protection of the head, legs, and feet. Penguins may sleep with the neck fully retracted and the head buried beneath the axilla (STONEHOUSE, 1967). In chickadees *(Parus atricapillus)* and gray jays, these sites comprise a relatively high percentage of the total body surface, and they are protected in individuals resting in cold environments by tucking the head beneath the feathers on the back. Ptarmigans *(Lagopus leucurus)* and ravens *(Corvus corax)*, animals weighing several hundred grams, do not appear to be protected by any similar postural adjustment (VEGHTE and HER-REID, 1965). Birds evidently can reduce heat loss through the unfeathered legs and feet by bringing the ventral surface of the body down on them and enclosing them with the ventral portion of the plumage. Domestic fowl were able to reduce heat production by a third through adaptation of this posture (DEIGHTON and HUT-CHINSON, 1940). It allowed the mourning dove *(Zenaidura macroura)* to maintain its tarso-metatarsi about 5° C warmer than when these segments of the legs were unprotected in an environment of 4° C (BARTHOLOMEW and DAWSON, 1954). Ob-viously, postural adjustments are possible only for birds at rest.

The relative surface area can also be reduced by the *crowding together of several animals.* In a warm environment rats spread their newborn young singly about the ground (STIGLER, 1930). An isolated mouse has a higher metabolic rate than one within a huddled group (PEARSON, 1947). Similarly, chicks use 15% less oxygen when several are crowded together (KLEIBER and WINCHESTER, 1933). During cold weather pigs huddle together while on their bellies to conserve body heat (HAFEZ, 1968; MOUNT, 1968). The metabolic increase of a single newborn pig at 20° C ambient temperature is about 5 times higher than that of a pig within a group (MOUNT, 1968, see p. 680). Huddling is characteristic of altricial nestlings and of young precocial birds even outside the nest (see, for example, LEHMANN, 1941; BARTHOLOMEW and DAWSON, 1952). It does not appear so widespread among adult birds. However, gentoo penguins *(Pygoscalis papua)*, tree creepers *(Certhia brachydactyla)* and other species are reported to congregate at night, huddling in tight groups (MURPHY, 1936; NIETHAMMER, 1952; LÖHRL, 1955).

8. Parental Behavior

Nest Building. Thermoregulation by nest building in mammals is an example of regulation of parental behavior to produce homeothermy as long as the own thermoregulatory capacity of the young is too small (cf. p. 682). Maternal nest-building occurs in the pig, rabbit and rodents (LEHRMAN, 1961; HAFEZ, 1968). The sow selects a definite area for nest building, 1–3 days before parturition, carrying grass and/or straw in her mouth for considerable distances. The rabbit builds a straw nest and lines it with hair plucked from her own body. Nest building is initiated by peripheral body cooling and is inhibited by peripheral heating. At ambient temperatures above 27° C, nest building in rats is inhibited, except in nursing mothers. Nests built at low temperatures are compact and close-knit, while those constructed at high temperatures are loose and scattered.

Incubation and Broodiness. Careful regulation of the temperature and humidity of the immediate environment of avian eggs is necessary (cf. p. 672). The female selects the nest site, usually in the shade or buried in the soil. The Megapodes bury eggs in soil which is warmed by solar radiation, volcanic activity, or decay of vegetative material. Temperature is tested frequently by the bird which rams its head into the mound (p. 615). The time spent sitting on the eggs is greater at lower ambient temperatures than at higher ones. The average temperature of eggs and mounds of birds in the Arctic (p. 729), measured in the nest, was about the same as that of those found in milder climates. It is thus evident that thermoregulation plays a role in the modification of the bird's behavior towards the egg.

The transference of heat from the birds to the eggs takes place through *"brood patches"*, areas on the ventral side of the bird which become bare during the reproductive season and contain elaborate capillary circulatory systems (MARLER and HAMILTON, 1966). Some birds possess one large brood patch, while others such as the herring gull *(Larus argentatus)* have three separate patches. Stages in egg setting are stopping, ruffling the belly feathers, wagging, quivering, shifting the eggs, etc., all serving to make the contact with the patches and the isolation from the environment as perfect as possible.

At times the danger to the eggs from overheating may be more important than the more common problem of keeping them warm. On very hot days, some birds cool their eggs by standing over them with no actual contact. Desert birds which place their nests on the ground shade their eggs to avoid fatal overheating of the embryo. Likewise, the nestlings must be kept cool and provided with water after birth, often a major problem where food with a high water content is not easily available.

VI. Nervous and Hormonal Factors in Temperature Regulation

A. Central Nervous Structures of Temperature Regulation

1. Extirpation Experiments

Early concepts on the thermoregulatory significance of the various parts of the central nervous system were mainly based on extirpation and truncation experiments. These led to the conclusion that the hypothalamic region plays a leading role in temperature regulation of homeotherms. Justified objections were raised against the early acute extirpation experiments (Lit. see THAUER, 1939) on the ground that the deficits observed in each case could be considered as results of the well-known neural shock. Highly developed operational techniques, however, have made it possible to undertake more conclusive chronic transection experiments. In THAUER's and PETERS' experiments rabbits survived transection of the brain at the border between the hypothalamus and midbrain for $6^{1}/_{2}$ weeks. Cats survived similar procedures for up to six months (Lit.: BARD and MACHT, 1958). In the experiments on cats described by BARD and MACHT, transections of the telencephalon at its border with the diencephalon had no recognizable effect on thermoregulation. But if the diencephalon was isolated by incisions in the region of the midbrain or still more caudally, the cats were no longer able to regulate their body temperatures. They were "essentially poikilothermic" (BARD and MACHT, 1958).

In the animals with truncations at these low, or even lower levels (bulbospinal cats), it is true that some "clonic jerking movements, chiefly of the legs, began when the rectal temperature had fallen to ca. 33° C, and became more violent as the body temperature dropped further" (BARD and MACHT, 1958). In certain cases, with a further fall of body temperature to ca. 29° C, fine rapid tremors quite like normal shivering appeared, and those movements ceased when the cats were warmed. These latter observations appear to be — in essence — in accordance with the earlier findings by THAUER and PETERS (1938), in rabbits: these authors emphasized that some ability of regulatory power against cooling was left in their chronic rabbits with truncation of the brain stem at levels below the hypothalamus; these animals were able to maintain constant body temperature within an external temperature range of between 14—15 and 28—30° C, a small range, it is true, when compared with the normal control range (= tolerated ambient temperature range; see p. 680) which extends to −45° C in the intact rabbit (see p. 659). In more recent studies no fine rapid tremors quite like normal shivering were seen in mesencephalic cats (BARD, WOODS, and BLEIER, 1970); these authors also state that the motor activity evoked by cold in previous studies in mesencephalic cats had no thermoregulatory effectiveness.

Typical panting was evoked in all the cats with truncations at the midbrain level, but only when the rectal temperature reached levels of $41-44°$ C (normal rectal temperature $38.5-39.5°$ C). All animals exhibited cutaneous vasodilatation, but all these heat dissipation mechanisms were obviously ineffective for maintaining the body temperature within the normal range (BARD and MACHT, 1958). Thus there is sufficient evidence for the concept that the hypothalamus plays an important role in the control of both defense reactions against cold and heat dissipation mechanisms.

There are however other accounts, according to which processes *protecting against warming* are still effective in the dog and cat after excision of the hypothalamus; although effective reactions against cold were abolished in these cases as well as in the truncation experiments described above (KELLER, 1938).

Some of the remaining observations are not in complete agreement with this view of the unique significance of the hypothalamic region, especially with regard to processes protecting against cold. Thus, no gross disturbances in thermoregulation have been found in humans whose hypothalamus was destroyed by tumors (Lit. in THAUER, 1939). A nearly normal regulatory increase in heat formation was also observed in an anencephalic human neonate, which survived for ten days, even though no hypothalamic tissue could be detected morphologically. In this case the thermoregulation of a "mid-brain organism" was almost intact (BETZ and HENSEL, 1964). On the other hand, CROSS et al. (1966) described an anencephalic case in which, though the morphological findings were similar, every thermoregulatory reaction was absent. This child however lived for only a few hours. Failure of thermoregulation in neonates has, however, been observed even when the CNS was much less severely damaged (see BRÜCK, K., and BRÜCK, M., 1960; MESTYAN et al., 1962), this is to say, the loss of thermoregulation may not have been a specific effect of the brain anomaly. More recently, however, a case of a large cyst centered on the anterior hypothalamic region has been demonstrated in a young woman whose body temperature had been observed to be around $33.5°$ C when she was exposed to room temperatures of $22°$ C (COOPER, 1970).

A marked limitation of thermoregulation is also to be expected following transection of the spinal cord, since on the one hand central nervous regulation of effectors (skeletal muscles, which form heat through shivering; sweat glands; vasomotor muscles) and on the other hand the afferents from the peripheral receptors (especially cutaneous thermoreceptors) in the caudally located dermatomes are thereby interrupted. Corresponding disturbances of thermoregulation have been detected in humans and described (GUTTMANN, SILVER, and WYNDHAM, 1958).

The stated aim of the described extirpation experiments was to identify "centers of thermoregulation". MEYER (1913), HASAMA (1929), and several other authors, who tried to separate a "cold center" from a "warm center", went the furthest in this direction. Today however it is widely agreed that the central nervous elements of thermoregulation are by no means crowded together into clearly recognizable anatomical centers in the region of the hypothalamus. We are faced to a far greater extent with a larger functional unit,in which both lower and higher levels of the CNS play a role (POPOFF, 1934; THAUER, 1935, 1939, 1964; THAUER and PETERS, 1938). It is hard to approach such a system through crude transection and extirpation experiments. The "plasticity" and "hierarchic construction" of the CNS make it difficult to draw conclusions from such experiments. Furthermore, in interpreting such experiments it must be remembered that substantial differences in results from various species are to be expected due to the different grades of "cephalization". Above all, inferences on the relation of animal experiments to conditions in man are rendered very doubtful by this consideration. The lasting value of the classical extirpation experiments consists of the information they can provide the neurologist and neurosurgeon concerning the sites of lesions and in their pointing out the areas in which the experimenter can carry out his work with more subtle methods.

Today the investigation of the central nervous structures of temperature regulation is based on searches for the answers to the following questions:

1. Where are thermoreceptive structures located in the CNS?
2. Where and how is "thermal information" from the various thermoreceptor fields integrated?
3. How and over which pathways are the "processed" output signals switched over to the final control elements, i.e. the effector systems?
4. How and where is the so-called set point of body temperature determined or altered?

2. Thermoreceptive Structures (Other than Cutaneous Thermoreceptors)

In addition to the cutaneous thermoreceptors, whose function is described in Chapter VII, the existence of internal thermoreceptors has been long suspected and has been postulated on the basis of regulation theory (Chapter III). The search for internal receptors was at first centered on the hypothalamic region. In numerous investigations on greatly varying species (cats, dogs, monkeys, rabbits, goats, guinea pigs, and many others) mechanisms of heat loss (vasodilatation, sweat secretion, panting) could be stimulated by circumscribed warming of the hypothalamic region (Lit. in RANSON and MAGOUN, 1939; THAUER, 1939; BLIGH, 1966; ANDERSSON, 1970). As the methods of thermal stimulation became more refined, the concept developed that the thermosensitive region is concentrated in the anterior hypothalamus. Thus it was possible, by warming the *regio hypothalamica anterior* of the dog, to suppress shivering that had been elicited by external cooling; by contrast, warming the posterior hypothalamus had no such effect (HEMINGWAY et al., 1940). On the other hand, cooling the anterior hypothalamus of *unanesthetized* animals could — provided the ambient temperatures were suitable (cf. p. 572) — elicit defense mechanisms against cold including increased heat production (DON-HOFFER et al., 1959; BETZ et al., 1960, 1962; HAMMEL, HARDY, and FUSCO, 1960; ANDERSSON et al., 1964) and peripheral vasoconstriction (KUNDT, BRÜCK, and HENSEL, 1957a).

Nonshivering thermogenesis (cf. p. 536) could be elicited by cooling the anterior
hypothalamus of the guinea pig (BRÜCK and SCHWENNICKE, 1971); on the other
hand, warming completely suppressed NST induced by external cooling (BRÜCK
and WÜNNENBERG,W., 1970). By means of all the described experiments the existence of thermoreceptive structures could be demonstrated, and through refinements of technique the regio praeoptica and the anterior hypothalamic region were
indicated as the sites of these structures; but is was not possible to decide whether
the elements of these structures were activated by warmth or by cold. For example
the elicitation of shivering could, in a formal sense, be due as easily to stimulation
of internal cold receptors as to reduced stimulation of internal warmth receptors
which act to inhibit shivering induced by cutaneous cold receptors. Further
clarification was provided by the electro-physiological investigations of NAKAYAMA
et al. (1963). By microelectrode recordings from individual neurons in the anterior
hypothalamus they succeeded in demonstrating that the frequency of impulses
increased with local warming (Fig. 36). The Q_{10} of this reaction was as high as 10,

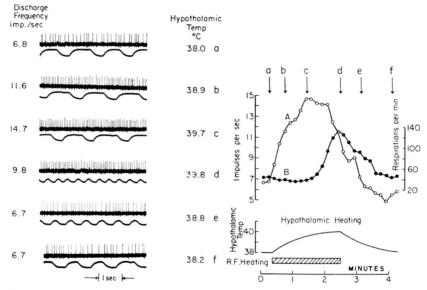

Fig. 36. Discharge of a neuron in the preoptic region, (Curve A) and change of respiratory rate (Curve B) in relation to hypothalamic temperature. From NAKAYAMA et al.
(1963)

which justifies one in speaking of a specific thermal sensitivity. In later investigations cold-activated units were also found, but they remained in the minority
(Lit. EISENMAN and JACKSON, 1967; HELLON, 1970; GUIEU and HARDY, 1971).
Through systematic testing of the brain stem, units which can be thermally activated have recently been demonstrated in the posterior hypothalamus (WÜNNENBERG
and HARDY, 1972) and in the midbrain (NAKAYAMA and HARDY, 1969; CABANAC,
1970), but the physiological significance of these elements has not been clarified
yet.

However, through extensions of thermal stimulations to other areas in recent years it has been learned that the spinal cord as well as the brain stem has thermo-receptive functions (Lit. THAUER, 1970). Thus, violent shivering was induced in dogs by local cooling of the spinal cord, while the brain and body surface temper-ature were kept constant (SIMON et al., 1964). Conversely, panting (JESSEN, 1967) and vasodilatation (JESSEN, MEURER, and SIMON, 1967) could be induced by warming. Heat loss processes are activated when the temperature of the spinal cord is raised to levels which occur under physiological conditions. Shivering, induced in guinea pigs by external cooling, could be suppressed rapidly by circum-scribed warming of a section ($C5-T2$) of the spinal cord (Fig. 37), even in muscles

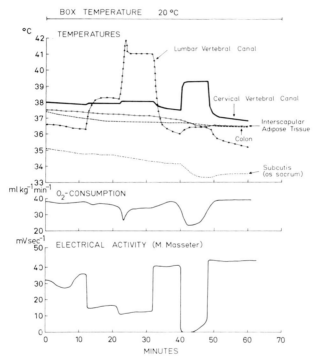

Fig. 37. Change in electrical muscle activity (as a measure of shivering), and oxygen uptake during local heating of the cervicothoracic (C_5-T_1), and lumbar spinal cord, respectively, in an unanesthetized guinea pig (3 weeks old, warm-adapted); ambient temperature was kept sufficiently cool to stimulate shivering. Note that shivering was entirely suppressed by increasing the cervical spinal cord temperature by ca. 1.5° C. Lumbar heating was much less effective. From BRÜCK and WÜNNENBERG (1966)

whose motor neurons lay outside the thermally stimulated regions (cf. p. 584). By contrast, warming the lumbar regions of the cord had a much smaller effect (BRÜCK and WÜNNENBERG, W., 1966).

Through extirpation experiments and electrical measurements (Fig. 38) it could be shown that ascending pathways from the thermosensitive cervicothoracic regions run in the ventromedial area of the spinal cord (in the region of the tractus spinothalamicus) to the posterior hypothalamus (WÜNNENBERG and BRÜCK, 1967, 1968 a,b, 1970; WÜNNENBERG, 1969). Finally, single fibers could be detected in the posterior hypothalamus of the rabbit whose impulse frequency increased with warming of the cervicothoracic section of the spinal cord (WÜNNENBERG and HARDY, 1972). Recently increased frequency of impulses has also been detected in

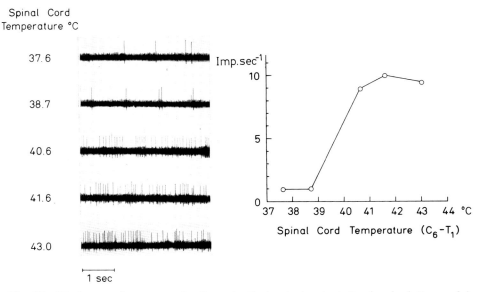

Fig. 38. Discharge of a neuron in the spinothalamic tract at the level of the nuclei olivares in response to temperature changes in the cervicothoracic spinal cord (C_5—T_1). Guinea pig, 3 weeks old; Nembutal anesthesia. From BRÜCK and WÜNNENBERG (1970)

ascending pathways in the dog, as reactions to thermal stimulation of the spinal cord, including stimulation by cold as well as by warmth. However, the units which could be activated by warmth were also in the majority here, and were far more sensitive to stimulation than the cold-activated units (SIMON and IRIKI, 1971).

In addition to cutaneous thermoreceptors and the central nervous structures discussed here there may be still other "thermosensors". It has been postulated that such hypothetical elements may exist in muscles (ROBINSON et al., 1965; STOLWIJK and HARDY, 1966). Recently evidence has been obtained for the existence of intra-abdominal thermal receptors (RAWSON, QUICK, and GOUGHLIN, 1969; RIEDEL, SIAPLAURAS, and SIMON, 1973).

The microanatomy of the thermosensitive central nervous structures has not been characterized at this time. Therefore, as might be expected, no concepts of the mechanisms of thermal perception have been proposed. The only attempts made till now at microanatomical characterizations of thermosensitive structures have been confined to the cutaneous thermoreceptors (see Chapter VII, p. 600).

According to investigations of RAUTENBERG (1969) on the pigeon, cooling the spinal cord of *birds* also leads to an increase in heat formation, while warming leads to an increase in evaporative heat loss and vasodilatation in the feet.

The dispersion of thermosensitive structures through different regions of the organism is in harmony with the concept advanced in Chapter III, p. 524, which states that the thermoregulatory system represents a "multiple input system".

3. Thermointegrative Structures and Processing of Temperature Information in the Central Nervous System

On the basis of numerous experimental findings (Lit. in BENZINGER, 1964, 1969) a central role in the processing of afferent thermal signals is ascribed to the area hypothalamica posterior, which was also designated as the Krehl-Isenschmidt center by BENZINGER. Thus, connective pathways between the anterior and posterior hypothalamus were detected some time ago in the dog, which mediate suppression of the shivering induced by warming of the anterior hypothalamus (Lit. in HEMINGWAY, 1963). In experiments carried out quite recently (WÜNNEN-BERG and HARDY, 1972) by means of recordings from single fibers in the posterior hypothalamus, neurons were detected which responded to thermal stimulation in the regio praeoptica as well as in the cervical region of the spinal cord. Neurons that are excited through stimulation of cutaneous thermoreceptors have so far been found only in the ventro-basal thalamus (LANDGREN, 1960) and in the anterior hypothalamus (WIT and WANG, 1968a; HELLON, 1970). However these findings do not exclude the possibility that afferents from the cutaneous thermo-receptors terminate in the posterior hypothalamus. This region clearly has not been thoroughly examined yet in this respect.

As is described in greater detail on p. 580, microinjection of noradrenaline into hypothalamic regions lying caudal to the thermosensitive regio praeoptica in the guinea pig causes a displacement of the threshold at which processes protecting against cold are initiated. Injection of noradrenaline into the thermosensitive regio praeoptica does not have this effect. These findings led to the hypothesis that the former region contains nervous elements ("reference units", see p. 582) which determine the adjustment of the "set point" of body temperature (Chapter VIE, p. 586). This could be considered as a further integrative function of the hypo-thalamus. With regard to the regulation of control processes by the various thermoreceptive structures, the following concepts have been brought forward (see HEMINGWAY, 1963; BENZINGER, 1970). The cold-activated cutaneous thermo-receptors, which are more numerous and more widely distributed than the warmth receptors, initiate and sustain processes protecting against cold. These reactions are opposed by warmth-activated central thermoreceptive structures. This concept is supported by the following experiments: local warming of the anterior hypo-thalamus leads to suppression of metabolic reactions elicited by external cooling (see p. 567), but cooling this region induces more vigorous metabolic reactions only if the skin temperatures are below the zone of thermal neutrality (BETZ et al., 1960, 1962; ANDERSEN, ANDERSSON, and GALE, 1962; BRÜCK and SCHWENNICKE, 1972). This bridling effect of internal warmth-activated thermoreceptors can be demonstrated most obviously by inactivating them through electrocoagulation.

A dramatic increase in shivering (ANDERSSON et al., 1965) or nonshivering thermogenesis (WÜNNENBERG and BRÜCK, 1968c) appears at once (Fig. 39). The warmth-activated spinal thermoreceptor structures have a corresponding in- hibitory function (BRÜCK and WÜNNENBERG,W., 1966, 1968a) (Fig. 37). Conversely, heat loss processes are stimulated by internal warmth-activated thermoreceptors and inhibited by cold-activated cutaneous thermoreceptors according to the concept described above (BENZINGER, 1964, 1969). There is some evidence, however, that some drive comes, in addition, from central cold receptors and from cutaneous warmth-receptors; these drives would then be counteracted by cutaneous warmth receptors and central and/or peripheral cold receptors, respectively. This scheme certainly represents a simplification; consideration will have to be given to functional principles which vary from species to species.

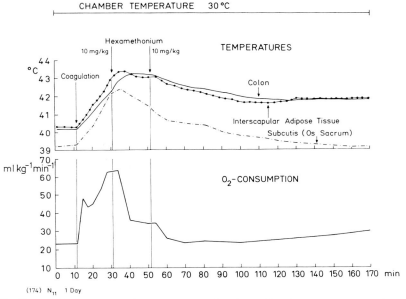

Fig. 39. Dramatic increase of oxygen uptake (nonshivering thermogenesis, NST) and of all body temperatures measured after electrocoagulation of the preoptic area (i.e. the site of thermosensitive structures) in a guinea pig, 1 day old. The ganglionic blocking agent hexamethonium bromide was given, in order to stop NST and prevent further increases of body temperature. From BRÜCK and WÜNNENBERG (1970)

The processing of signals from the cutaneous and central thermoreceptors is partly multiplicative. The hyperbolic form of the threshold curves for shivering (BRÜCK and WÜNNENBERG, 1967b) in guinea pigs (Fig. 68) is an expression of such a multiplicative processing of information. A hyperbolic curve was also obtained for the threshold curves for *nonshivering* thermogenesis in the guinea pig (BRÜCK and SCHWENNICKE, 1972) as Fig. 40 shows. According to these and other findings NST can be approximately described by Eq. (1) (BRÜCK and SCHWENNICKE, 1972):

$$NST = K \cdot (T_{Hy0} - T_{Hy}) (T_{Su0} - T_{Su}) - a \qquad (1)$$

where T_{Hy} and T_{Su} are the temperature in the hypothalamus and the subcutaneous temperature in the back, respectively; T_{Hy0} and T_{Su0} are the respective reference temperatures (cf. p. 586) which are represented by the asymptotes in Fig. 40. The

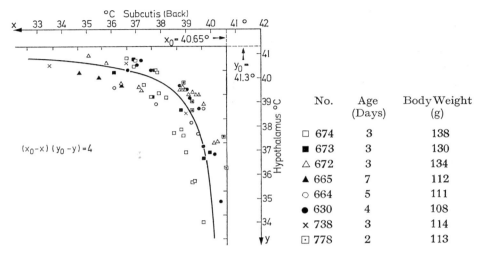

No.	Age (Days)	Body Weight (g)
□ 674	3	138
■ 673	3	130
△ 672	3	134
▲ 665	7	112
○ 664	5	111
● 630	4	108
× 738	3	114
⊡ 778	2	113

Fig. 40. Plot of the pairs of threshold temperatures for nonshivering thermogenesis as obtained from 8 unanesthetized guinea pigs by changing hypothalamic and body surface temperature independently. From a group of hyperbolas of the general equation, $(x_0-x) \cdot (y_0-y) = C^2$, one was chosen which best fitted the group of threshold temperatures. From BRÜCK and SCHWENNICKE (1971)

expression is limited to the following conditions: $0 \le NST \le 100$; $(T_0 - T) > 0$. A formally identical relation was also ascertained for shivering in the guinea pig (BRÜCK and WÜNNENBERG, 1968, 1970).

In man (BENZINGER, 1969, l.c. Fig. 13) and the dog (CHATONNET, CABANAC, and MOTTAZ, 1964, Lit. in CABANAC, 1970) hyperbolic curves were also obtained when the metabolic threshold responses were plotted as a function of the average skin temperature and the tympanic or hypothalamic temperature. Accordingly, in a theoretical work by STOLWIJK and HARDY (1966), thermoregulatory heat production in man (ΔM) was also described as the product of two temperature deviations from two reference values.

$$\Delta M = 60 \cdot (T_{HC} - 36.6) \, (\overline{T}_S - 34.1) \text{ kcal/hr.,} \tag{2}$$

where T_{HC} = head core temperature, and $\overline{T}_S$ = mean skin temperature.

By contrast, in the dog a straight line for shivering threshold is obtained if the data of HAMMEL (1965) are used (Fig. 41, lower left diagram), thus indicating additive rather than multiplicative processing of the input signals. Here, however, it seems conceivable that a flat (large c^2) hyperbola would be obtained if one extended the ranges from body surface (air) temperature and hypothalamus temperature to the extremes. The straight line in Fig. 41 might then appear as a linear approximation to the center section of a flat hyperbola.

The characteristic feature of the multiplicative processing of the input signals is to be seen in the fact that beyond a certain temperature (represented by one of the two asymptotes cf. Fig. 41) it is not possible to evoke a control reaction regardless of how far the other temperature is displaced from its normal level. By contrast, in the additive system, theoretically, any displacement of the one temperature can be compensated for by an opposite change of the other temperature; such a system may also be described as a proportional control system with a variable set point, the latter being a function of the skin temperature (HAMMEL, 1965; see also Chapter III, p. 521 and BROWN and BRENGELMANN, 1970).

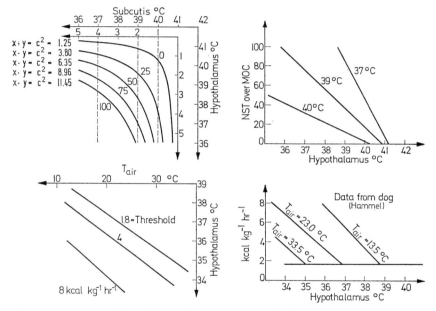

Fig. 41. Comparison of "multiplicative" with an "additional" model. Upper left: The family of hyperbolas is derived from equation 1 on page 572 and this in turn is based on studies in the guinea pig in which hypothalamic and body surface temperature were changed independently of each other and related to nonshivering thermogenesis *(NST)*. The figures beside each hyperbola give *NST* in percent. Upper right: The three curves are obtained by plotting the data from the upper left diagram following the vertical dashed lines. Lower right: Shivering heat production related to air temperature and hypothalamus temperature in the dog, as published by HAMMEL (In: Physiological controls and regulations. YAMAMOTO, W.S., and J.R. BROBECK, eds.: W.B. Saunders Comp. Philadelphia and London 1965; reproduction by courtesy of the author). Note absence of change in the slope of the curves. Lower left: diagram derived from the lower right diagram. From BRÜCK and SCHWENNICKE (1971)

A multiplicative processing of input signals is also significant for heat loss processes, as is shown for example by the hyperbolic form of the threshold curve of panting in dogs (Fig. 42). The curvature of the curve represented in Fig. 42 is parallel to the threshold curve for shivering (Fig. 83) and *NST* (Fig. 40) (on this point, cf. CABANAC, 1970). At hypothalamic temperatures below 40.3° C (Fig. 42) panting can be prevented by reduction of the skin temperature. At temperatures above 40.3° C (this temperature would correspond to the asymptote) even a great reduction in skin temperature is clearly ineffective in inhibiting panting. Further, it can be seen from Fig. 42, that at high skin temperatures (bath temperatures) panting can no longer be inhibited by decreasing the hypothalamic temperature.

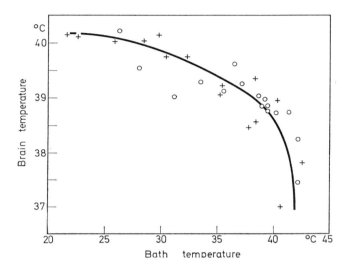

Fig. 42. Threshold of heat tachypnea in two dogs submerged (except for the head) in a water bath. For further explanation see text. From CHATONNET, CABANAC, and MOTTAZ, (1964)

Sweat secretion in man has also been shown to be controlled by both inner and cutaneous thermal receptors (BENZINGER, 1964, 1969; STOLWIJK and HARDY, 1966; WYNDHAM, 1966; WYNDHAM and ATKINS, 1968; BULLARD et al., 1970; NADEL, BULLARD, and STOLWIJK, 1971). At skin temperatures above 33° C, according to BENZINGER (1964, 1969), the rate of sweat secretion can be described, to a good approximation, as a function of the internal temperature (tympanic temperature). Reduction of the skin temperature to below 33° C (excitation of cutaneous cold receptors) must be counteracted by increasing elevation of the internal temperature (excitation of central warmth receptors) if equal rates of sweat secretion are to be maintained (Fig. 43). In contrast to BENZINGER's view BULLARD et al. (for the original literature see BULLARD et al., 1970) found that sweat secretion still depended on skin temperature in the range above 33° C. Also in WYNDHAM's studies (1966), and those by WYNDHAM and ATKINS (1968) skin temperature was found to have much more influence on the slope of the relationship between core temperature and sweat rate than in BENZINGER's studies. In studies by WURSTER, McCOOK, and RANDALL (1966) skin temperature changes in the range above 33° C were shown to increase sweat secretion before any increase occurred in tympanic temperature. The latter results would be more in harmony than BENZINGER'S with recent results on cutaneous warm receptors which are activated at temperatures far above 33° C (see Chapter VII, p. 606). As for the modifications of the sweat secretion by direct temperature effects on the effector system see p. 584.

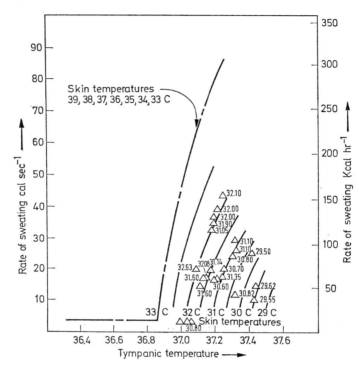

Fig. 43. Human sweat rate in relation to tympanic temperature and mean skin temperature. Note change of slope and threshold with decreasing skin temperature; compare with Fig. 41 upper and lower right diagram. The slope of the present curves would be about intermediate between the two types of models. At skin temperatures above 33° C sweating follows tympanic temperature. From BENZINGER (1969)

SNELLEN (1966) described sweat secretion in man at rest and at work as a function of the increase in average body temperature. Formally, this finding also means only that the internal temperature and the surface temperature work together in controlling the rate of sweat secretion, and conversely that the effect of an increasing internal temperature can be opposed within a certain temperature range by reduction in skin temperature. A comprehensive quantitative description of sweat secretion in man — including the effect of rate of change of mean skin temperature — has recently been given by NADEL, BULLARD, and STOLWIJK (1971).

Internal and external receptors also work together to regulate changes in thermoregulatory cutaneous blood flow. Thus it has been shown that increases in cutaneous blood flow of human neonates can be brought about by increases in skin temperatures, but only if the rectal temperature is above 36.5° C (Brück, Brück, and Lemtis, 1957a; Brück, 1961). Similar findings have been reported by Cooper, Johnson, and Spalding (1964) for adults. Similarly, the findings of Kundt, Brück and Hensel (1957) and Jacobson and Squires (1970) on the cat revealed that the surface temperature and hypothalamic temperature worked together to control peripheral blood flow.

With regard to the processing of the signals from the various *inner* receptors, recent results by Jessen and Mayer (1971) and Jessen and Simon (1971) are pertinent. Studying unanesthetized dogs at constant ambient temperature of 18° C these authors found a nearly linear increase in metabolic heat, of approximately the same slope when they cooled either the hypothalamus or the spinal cord (Fig. 44); the threshold temperature for the elicitation of this cold-induced heat production was about 1° C lower in the case of spinal cord cooling. Slight warming of one of the areas to 39.3° C inhibited the effect of cooling the other thermoreceptive area. This interdependence of the effects caused by stimulation of these two thermosensitive areas is similar to the multiplicative interaction between cutaneous and inner receptors as described above (cf. Fig. 40). Selective warming of the spinal cord increased evaporative heat loss as did selective warming of the hypothalamus (Fig. 44); in both cases the heat sensitivity considerably exceeds the cold sensitivity. In contrast to the behavior of extra heat production, "slopes and thresholds for the increase of respiratory evaporative heat loss during selective heating of either the spinal cord or the hypothalamus remained nearly unaffected even with intensive simultaneous cooling of the corresponding area" (Jessen and Simon, 1971). The principle of the information processing has yet to be revealed in this case.

In the guinea pig the effects of warming either the spinal cord (area between C5 and T2) or the preoptic region on respiratory rate were approximately equal; however, warming both areas together produced an effect on respiratory rate which was about twice that of heating only one of the areas (Brück and Herrmann, 1972). Here, the thermal information from the two sites seems to be added.

All the control actions described in the foregoing paragraphs must be ultimately executed by certain nervous structures. As repeatedly said, the hypothalamus seems to play a leading role in this respect, but by no means do we believe that it is the only place where all the multiplications and additions are carried out. There are certainly "higher structures" involved. Attempts to describe the thermoregulatory integrative actions on the basis of single unit activities have been made by Hammel (1965, 1968), by Hardy and Guieu (1971), and by others (cf. Fig. 24 in this book).

Grüsser, Klinke, and Kossow (1968) have published some stimulating ideas on the performance of additions and multiplications in the nervous system.

The fact that information from signals originating in quite diverse areas of the body is processed, be it multiplicatively or additively, has the following consequences: The value controlled by the thermoregulatory system is not a circumscribed interior body temperature, such as that of the hypothalamus, but rather is a function of various local temperatures (Brück and Wünnenberg, 1967b, 1970). Therefore the regulatory system tends to keep an approximately average body temperature, and not the body core temperature, constant (on this point, cf. SNELLEN, 1966 and Section VI E).

The recognition of multiplicative or additive information processing, in a system with multiple inputs as described above, permits the interpretation of several well-known phenomena, as for example the following: Swimming in the North Sea

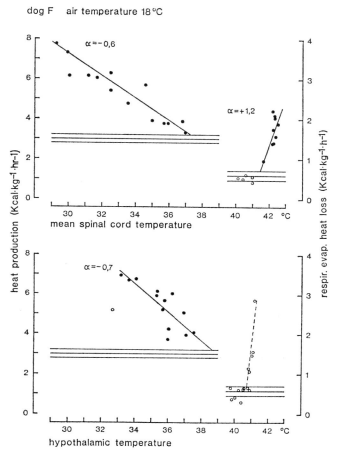

Fig. 44. Heat production (shivering) and respiratory evaporative heat loss in response to changing hypothalamic temperature (below) and spinal cord temperature (above) in one dog. The horizontal lines indicate mean initial values ± S.D. From JESSEN and MEYER (1971)

causes the internal body temperature to fall to 35° C or less within 15 min, especially in thin individuals. But the shivering induced by this fall disappears very quickly on exposure to the solar radiation which is reflected from all directions on the sandy beach. This radiation causes skin temperature to rise rapidly, while the internal temperature requires hours to reach 37° C (PIRLET, 1962). Since ΔT_s

[Eq. (2)] goes to 0 in this process the shivering threshold is exceeded even though the internal temperature is sharply reduced (cf. Figs. 40, 41, and 68).

B. Control of Effector Systems

After description of the processing of the input signals and the formal relations between the input- and effector-magnitudes in the previous section a question remains concerning the path by which the *output signals* leave the "central controller" in order to bring about the final "adjustments" of the peripheral control-elements. Contrary to a view which has been often expressed this control is exerted exclusively via nervous pathways; hormonal factors play a role only in long-term alterations of the system (Chapter VIII, p. 644). Two nervous systems take part in this transmission:

1. the spinal and supraspinal motor system, and
2. the sympathetic nervous system.

Shivering is controlled via the motor system, whose peripheral portions (α and γ motor neurons) and supraspinal pathways (tr. cerebrospinalis and reticulospinales) are very well known. The central portion of the "shivering pathway" was described in greater detail by HEMINGWAY (1963). Descending tracts leave the posterior hypothalamus and run caudally through the midbrain tegmentum and the pons close to the rubrospinal tracts. In the medulla oblongata the pathway comes close to the ventrolateral surface.

Nonshivering thermogenesis is controlled by the sympathetic system, as discussed in Chapter IV. The pathway in the brainstem and spinal cord has not been determined more precisely at this time.

Vasomotor and sudomotor control. It has long been known that the sweat glands are controlled via cholinergic sympathetic pathways. The thermoregulatory control of blood flow varies regionally (Fox, GOLDSMITH, and KIDD, 1962; GOLENHOFEN, 1971a, b: HILLE, 1965). At least three functionally different regions can be distinguished (Fig. 45).

a) Extremities (finger, hand, ears, lips, nose)
b) Trunk and proximal limbs
c) Head and brow.

Blood flow through the extremities is controlled exclusively via noradrenergic sympathetic nerves. An increase in sympathetic tone causes vasoconstriction, and a decrease in tone vasodilatation. Inactivation of the sympathetic system leads to an almost maximal dilatation (on local temperature effects, see below). The resting level of blood flow is low, i.e. the sympathetic tone is usually fairly high. In the trunk and proximal parts of the limbs the maximal increase in blood flow appearing as a result of heat stress is far higher than that following inactivation of the nerves. At first this led to the hypothesis that specific vasodilatatory nerves existed. Then, Fox and HILTON (1958) showed that a tissue hormone, bradykinin, which is strongly vasodilatatory, is secreted together with the sweat. Vasomotor nerves have only a slight effect in the forehead, i.e. there is no vasoconstriction in response to cold stress. However vasodilatation does occur together with sweat secretion in

response to heat exposure (cf. Lit. in BLAIR, GLOVER, and RODDIE, 1961; GOLEN-
HOFEN, 1971).

With the discovery of bradykinin, our previous assertion that thermoregulatory
effectors are controlled exclusively by nervous pathways becomes somewhat
limited. In this connection it may be asked whether hormones released from the
adrenal glands, especially adrenaline, have a thermoregulatory function. This does
not seem to be the case. Only extreme thermal stress, exceeding the capacity for resistance, leads to the secretion of adrenaline (see Chapter VIII, p. 647). Of course
circulating adrenaline will then be able to take part in stimulating vasoconstriction
in cutaneous blood vessels. Adrenaline secretion has been induced experimentally
by cooling the regio praeoptica of the goat; is was accompanied by an increase in
shivering (Lit. in ANDERSSON, 1970).

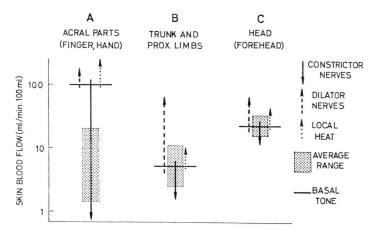

Fig. 45. Schematic representation to show some regional differences in human skin
blood flow. Horizontal lines indicate maximum blood flow after acute denervation.
Downward arrows indicate influence of noradrenergic constrictor nerves. Upward
dashed arrows indicate the influence of dilatator nerves; the dotted arrows indicate
direct temperature effect on blood flow. Dotted fields represent average range of
blood flow. From GOLENHOFEN (1972)

C. Central Transmitting Substances and Thermoregulation

In 1943 VON EULER, LINDER, and MYRIN showed that intraventricular injection
of adrenaline in the rabbit led to increases in body temperature of 1—2° C. FELD-
BERG and MYERS took up these findings again in 1963 and carried out similar
investigations in the cat. In this case, however, intraventricular injection of
noradrenaline always led to decreases in body temperature, which were maintained
for hours. Serotonine had an antagonistic effect. On this basis FELDBERG and
MYERS (1964) introduced what they termed a "new theory of thermoregulation".
Mechanisms for warming the body and dissipating heat were said to be somehow
determined by the relative concentrations of the two chemicals in the hypothalamic
area. Some support for this concept was provided by the detection of a high concentration of catecholamine and 5-hydroxytryptamine in this area (VOGT, 1954).

FELDBERG and MYERS stimulated numerous works in this field, which however yielded very contradictory results at first. In most of the subsequent works intraventricular or even intrahypothalamic injection of noradrenaline led to increases in body temperature, as in the experiments of von EULER. Attempts were made to explain the different results on the basis of species differences (FELDBERG et al., 1967), and of uncomparable environmental conditions, but unanimity on this question has not been reached to this day (Lit. in LOMAX, 1970; BLIGH et al., 1971). On the other hand it has been recognized that in order to ascertain the significance of chemical substances for thermoregulation one must select the appropriate parameters to measure and must give the injections with precise methods at an exact location in a site relevant to thermoregulation. In this way it could recently be shown that noradrenaline had no effect in the guinea pig (Fig. 46) when it was

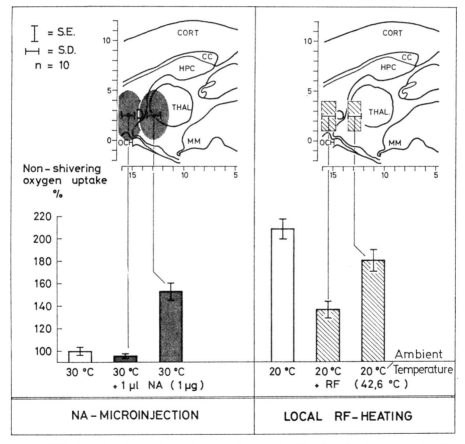

Fig. 46. Effects of local noradrenaline injection, and of local radio-frequency heating, in two frontal planes of the diencephalon on nonshivering thermogenesis (positions of heating electrodes or injecting cannulas are indicated in inset sagittal sections). Note that noradrenaline was only effective if injected in the caudal area, whereas local heat was more effective in suppressing nonshivering thermogenesis (evoked by external cooling). From ZEISBERGER and BRÜCK (1971b)

injected into the thermosensitive regio praeoptica; but injection into a more caudal region of the hypothalamus led to increases in nonshivering thermogenesis, and in the temperature of the brown fat tissue (cf. p. 540), as well as to a substantial increase in the temperature of the whole body (ZEISBERGER and BRÜCK, 1971). Injections of acetylcholine however were more effective if they were given in the thermosensitive regio praeoptica. The effect of acetylcholine consisted of a sudden suppression of nonshivering thermogenesis that had been induced by external cooling; i.e., ACh has the same effect as local warming (= excitation of thermosensitive units) of the anterior hypothalamus (ZEISBERGER and BRÜCK, 1973). When noradrenaline was injected into the more caudal regions of the hypothalamus, NST and shivering appeared at higher value pairs of surface- and central temperatures, i.e., the threshold temperatures lay above and to the right of the threshold hyperbolas (Fig. 47). On the basis of these findings it was suggested that noradenaline acted on the postulated reference units thereby changing the set point (cf. p. 586) of the thermoregulatory system.

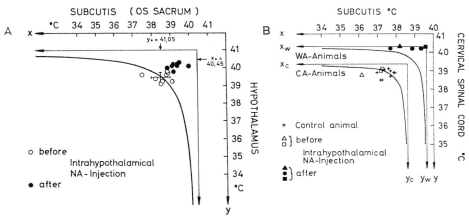

Fig. 47. Shift of threshold temperatures for nonshivering thermogenesis *(A)* and shivering *(B)* following intrahypothalamic injection of noradrenaline (as for the site of injection see Fig. 46) in cold-adapted *(A)* and cold- and warm-adapted *(B)* guinea pigs. The values represented by cross symbols in *B* were obtained from a control animal, in which the threshold temperatures were determined several times during an experiment lasting 10 hours; here it is obvious that the threshold temperatures are fairly stable and easily reproducible. The hyperbola in *A* corresponds to a threshold curve previously determined in a series of cold-adapted guinea pigs. The hyperbolas in *B* correspond to those depicted in Fig. 68. Note that following noradrenaline injection the cold-adapted animals' shivering threshold shifts to a level close to that found normally in warm-adapted animals. From ZEISBERGER and BRÜCK (1971a)

D. Peripherally Induced Processes

1. Direct Effects of Temperature on Cutaneous Blood Flow

Isolated and denervated limb areas continue to respond to thermal stimuli by vasomotor adjustments (PERKINS et al., 1948; PAPPENHEIMER et al., 1948). Direct effects of temperature on the blood vessels (ASCHOFF, 1943), the formation of vasoactive substances in the skin, and blockage of cutaneous reflexes have been considered as explanations for this phenomenon (Review in THAUER, 1965).

A direct effect of temperatures on the vessels regulating blood flow could be ascertained in later investigations on fiber preparations from arteries. Thus KEATINGE (1958), working with isolated segments of the a. ulnaris of the ox, showed that the constrictive effect of catecholamines is abolished at temperatures below 10° C. In recent investigations of fiber preparations from arteries with diameters of 0.3 to 0.5 mm this temperature dependence of the catecholamine-induced constriction was characterized in greater detail (SAMS and WINKELMANN, 1969). In a temperature range between 41° and 30° C the tension developed after the administration of noradrenaline increased up to fivefold with falling temperature. As the temperature fell still further the development of tension diminished and at 15° C it reached the initial level again. On the level of the whole organism this indicates that when the temperature of the extremities is high, central constrictive impulses have no, or only a slight, effect. Maximal sensitivity to sympathetic constrictive stimuli can be expected at a local temperature of 30° C. If the local temperature sinks to lower levels the central constrictive stimuli must gradually become ineffective again. These results can serve to explain the observations of CROCKFORD, HELLON, and PARKHOUSE (1962). Using strain gauge plethysmography they showed that cutaneous blood flow in the lower arm increased appreciably under the influence of locally limited heat radiation. This response was also found in patients with cervical sympathectomy or complete brachial plexus tears. The direct effect of temperature on the regulating blood vessels has also been advanced as an explanation of the so-called cold vasodilatation (Lewis' Hunting Reaction). This phenomenon appears when, for example, a hand is immersed in ice water. At first a maximal vasoconstriction sets in and the temperature of the finger tips falls to a level close to that of the water. After some time (order of magnitude: minutes; time depends on the general thermal situation) a sudden intense increase in the finger temperature takes place due to vasodilatation of the cutaneous vessels. The warmed vessels then respond to the nervous or hormonal constrictive stimuli again, and a new phase of vasoconstriction sets in (KEATINGE, 1970). If the hand remains in cold water this series of events is repeated periodically. The reaction described here takes place in the same manner after complete denervation (Lit. in KEATINGE, 1970; THAUER, 1965), but it must be said that the direct effect of temperature on the vessels cannot explain all the details. Therefore other supplementary factors (e.g. thermally induced local fluctuations in metabolism) have been suggested to account for them; but no proof whatever has yet been forthcoming (KEATINGE, 1970).

The Lewis reaction has been considered as a form of protection against local cold injury. However KEATINGE (1970) came to the conclusion , that "Cold vasodilatation may be of some value in preventing frostbite, but its delay in onset and subsequent intermittency make it inefficient in doing so. The main practical consequence of cold vasodilatation is therefore the adverse one of increasing heat loss. This is particularly serious during whole body immersions in water below 10 to 12° C, and it would be useful to have some means of preventing this dilatation".

As stated in Chapter VIII, the Lewis reaction takes place at higher skin temperatures after a period of local cold adaptation. Only after such an adaptive modification does the reflex actually attain a positive functional significance.

2. Direct Effects of Temperature on Sweat Secretion

Modifications in sweat secretion, like cutaneous blood flow, are influenced by local
temperature conditions. This long-suspected effect was demonstrated by BULLARD
et al. (1970) who showed that sweat secretion in a skin surface area of ca. 10 cm²
was appreciably changed with local temperature variations in this area. This
effect is strictly limited to the area affected by the temperature change. It has been
proposed that the site affected by temperature is the neuro-glandular junction. This would provide an immediate explanation for the finding that a temperature increase can cause a local increase in the rate of sweat secretion only
in the presence of an adequate central drive. This local temperature effect is of
considerable quantitative significance. If the centrally controlled rate of sweat
secretion is set equal to E_{SW} (sweat rate/cm²), the local effect can be expressed by
a factor L, and a local sweating rate, $E_L = E_{SW} \cdot L$ is obtained. As BULLARD et al.
(1970) showed, this factor takes on values between 0.03 and 2.24, when the local
skin temperature varies between 30° and 40° C. As an immediate result, sweat
secretion practically ceases at skin temperatures below 30° C, due to the local
factor, even if the central drive is maximal. At this point it must be recalled once
more (cf. p. 575) that the quantity E_{SW} is not determined by the temperature of
the hypothalamus alone, but also by that of the cutaneous thermoreceptors. Thus
the skin temperature has a twofold influence on the control of the rate of sweat
secretion; firstly via the cutaneous thermoreceptors — afferent nerves — central nervous integration-efferent nerves, and secondly via the direct effect of temperature on
the neuro-glandular junction. The local effect can be interpreted as a peripheral
amplifier or attenuator mechanism in the effector system of a regulatory system.

3. Direct Effects of Temperature on Anterior Horn Cells and Shivering

Cooling the spinal cord leads to stimulation of the motor neurons and thereby to
shivering (KLUSSMANN, 1969; PIERAU and KLUSSMANN, 1971). Thus the spinal cord
is not only a site of thermosensitive structures which transmit impulses to the
posterior hypothalamus via ascending pathways (cf. p. 569) but it also contains
thermosensitive motor neurons which can reinforce the effector mechanisms of the
system regulating against cooling. This reinforcement can be compared to the
local reinforcement of the central drive of the sweat glands, though not in a
quantitative sense, since the temperature of the spinal cord is much more stable
than that of the skin.

The thermosensitivity of the anterior horn cells may account for the tremor
observed in hypothermic cats after truncation of the brain stem caudal to the
hypothalamus (see p. 565).

4. Initiation of Thermoregulatory Responses by Thermal Stimulation of Limited Skin Areas ("Reflex Control")

In Section VI A all discussions of the skin temperature referred to the average temperature. At that time the fact that cutaneous thermoreceptors are definitely not distributed equally over the skin was never mentioned. However, variations in the importance of different areas of the skin, resulting from such uneven distribution, were, very early, repeatedly emphasized. Thus for example the particular thermoreceptive properties of the trigeminal area have always been acknowledged (REIN, 1930; BADER and MACHT, 1948; HENSEL, 1952), but a systematic quantitative investigation of the thermoreceptive sensitivity of different skin areas has only begun in recent times. A sensitivity coefficient has been calculated (NADEL, MITCHELL, and STOLWIJK, 1973) based on the thigh sweating rate in man; the relative sensitivity of the facial skin to heating was 3–4 times that of the thigh; the chest sensitivity was about 1.5 times that of the thigh, while those for the abdomen and thigh were the same. Such a differentiation could provide a basis for the calculation of an average skin temperature, in which the individually measured skin temperatures were weighted according to their thermoregulatory significance as well as their surface areas.

In the ram, warming the skin of the scrotum (which, according to investigations of the rat, is richly provided with warmth receptors; see Chapter VII, p. 559) to temperatures above 36° C induces panting, which continues even after hypothermia of the body core sets in. Denervation of the scrotal skin prevents this panting. Warming a shorn area of the trunk, equal in area to the scrotal skin, had no effect on respiration. In completely shorn animals, whose skin temperatures were 3–7° C lower than before shearing, panting could not be induced by warming the scrotal skin (WAITES, 1962). It may be that the stimulus from the scrotal warmth receptors is neutralized by activation of cold receptors in the rest of the skin or that supporting impulses from warmth receptors in other areas of the body are lacking. These results are in accord with the multiple input hypothesis (cf. p. 571).

The induction of thermoregulatory responses from a limited skin site has often been contrasted, e.g. reflex sweating, reflex shivering, etc. with "central" sweating, shivering, etc. On the basis of a multiple input theory of thermoregulation these phenomena require no special treatment. Such reflex responses (which have till now been classified with the "locally induced phenomena") would be of a special nature only if it could be shown that they were transmitted from axon- or spinal-reflex arcs. In fact, GÄRTNER and GÖPFERT (1966), cooling one side of a human subject in a radiation climatic chamber, found reflex tone of the skeletal musculature more pronounced on the cold exposed side. Otherwise there is only scant evidence of such reflex phenomena at this time, and one should not expect to find that significant thermoregulatory reactions are mediated entirely on a spinal level in homeothermic forms of life whose CNS is marked by a high degree of cephalization. One might however expect such spinal mechanisms to act in the presence of lesions of higher segments, just as a medullary type of respiration can appear after lesions of higher control centers. In fact, McCOOK et al. (1970) were able to

induce sweat secretion by local warming of the lower half of the body of a para-
plegic with complete transverse paralysis at the T3 level. This was seen as proof
of a spinal sweat secretion reflex.

E. Body Temperature as the Controlled Variable. Displacements of the "Set Point"

A question often discussed in the past concerns the temperature which should be
regarded as the actual body temperature, i.e. in terms of regulatory technology,
which intra-organismal temperature is the variable that is controlled. The question
is no longer put in this form, since it has been recognized that the thermoregulatory
system represents a multiple input system (Section VI A), in which the temperatures
of numerous sites are sensed and integrated for the control of effector processes.
From this point of view the temperature of the hypothalamus represents the
actual controlled value just as much or just as little as does the rectal temperature. To a much greater extent, the regulated quantity ("controlled variable", T) must be considered formally as a function of various local temperatures,
$T = f(T_S, T_{Hy}, T_C \ldots)$ (where index S refers to "skin", Hy to hypothalamus,
C to spinal cord). In this way the regulatory system tends to maintain an approximate constancy of an average body temperature more than that of a limited area
of the body core. In such a situation it makes no sense to ask about a single
"set point temperature" of the regulatory system. The "set point" of body temperature, T_{set}, must be viewed far more as a function of various reference temperatures, T_{ref}: $T_{set} = f(T_{ref, Sh}, T_{ref, Ev}, \ldots)$; where Sh signifies shivering and Ev,
evaporative heat loss. T_{set} is to be thought of as a complex value; in the case of
nonshivering thermogenesis (NST), e.g. $T_{ref, NST}$ would be determined by at least
two temperatures, x_0 and y_0 (Fig. 40), which are represented by the two asymptotes
of the threshold hyperbola (BRÜCK and WÜNNENBERG, 1967, 1970).

The formal consideration would be simplified if one could assume uniform reference
temperatures for all regulatory processes (shivering, nonshivering thermogenesis,
vasoconstriction, evaporative heat loss), but this is not the case. In the investigations on guinea pigs discussed in detail in Chapter VIII, p. 625, the shivering
threshold could be displaced to lower temperatures and the panting threshold displaced simultaneously to higher temperatures by long-term thermal stress. This
finding could only be accounted for if different reference temperatures (= asymptotes) were specified for each of the thermoregulatory effector systems. This in turn
cannot be reconciled with the assumption of a uniform system for determining reference values. We must instead assume that several elements exist which determine reference values, act independently of one another, and can be independently
influenced. How should one conceive of these postulated reference determining
elements? HAMMEL (1965, 1968), HARDY (1965) and others considered the thermo-
insensitive units found in the exploration of the hypothalamus to be such elements.
As shown in greater detail on p. 582 the thresholds for shivering and for nonshivering thermogenesis can be displaced to higher temperatures by local injections of noradrenaline into the hypothalamus (Fig. 47). Since this effect
appears only if the NA is injected into an area which has no effect on NST and

shivering under thermal stimulation, it has been suggested that it is the site of the reference determining cells, postulated by HAMMEL (1965) and HARDY (1965), whose function is influenced by noradrenaline (BRÜCK et al., 1970; ZEISBERGER and BRÜCK, 1971 a).

In spite of the apparently complicated state of affairs depicted just now, in practice easily accessible sites can be satisfactorily found, especially in larger organisms, whose temperature can be viewed as representative of that of the whole organism, provided that comparable conditions can be adhered to (cf. Chapter II, p. 513). The rectal temperature or the sublingual temperature is often used for this purpose in medical practice. BENZINGER introduced the tympanic temperature, which responds to changes in ambient temperature with less inertia. Fluctuations and deviations of these representative temperatures from an empirically arrived at "normal value" may be due to the following causes:

1. The control system is overtaxed; this refers to hypo- and hyperthermia.
2. The standard conditions of measurement no longer apply: the temperature field has changed (see under: Hyperthermia of work, p. 591).
3. The "set point" (set point = function of various reference temperatures; see above) of the regulatory system has been displaced.

The most striking example of a set point displacement is the temperature discontinuity exhibited by hibernators during the onset of hibernation and in arousal (see Chapter XI). The body temperature of hibernators is usually controlled at 5° C during the sleep period, while in summer, values between 35° and 37° C are maintained very precisely. Hibernators may represent a suitable model for the elucidation of set point displacements, which are much narrower in extent in non-hibernators.

1. Periodic Fluctuations of Body Temperature

As is known, the body temperature of man and other species fluctuates in a daily rhythm with an amplitude of ca. 1° C in man (Chapter II, p. 518). Long-term fluctuations of body temperature are connected with estrus, and with the human menstrual cycle. In the second half of this cycle the temperature is several tenths of a degree higher. It has long been suspected (ASCHOFF, 1955) that the daily rhythmical fluctuations must involve changes in the form of set point displacements and recently this view has continued to receive more support (ASCHOFF, 1970). The temperature increase in the second half of the menstrual cycle represents an effect of progesterone. Recent experiments on the rat have shown that the temperature effect of progesterone is mediated by the anterior pituitary (FREEMAN et al., 1970). Future investigations will be aimed at discovering whether the effect of progesterone can be traced to an influence on the postulated reference units.

2. Fever

Fever as a pathological condition lies outside the scope of our presentation. A few short references may suffice here, insofar as they concern the regulatory process. A displacement of the set point is present in fever. Therefore neither a passive

hyperthermia resulting from increased metabolism nor a failure of temperature
regulation is involved. Fox and McPHERSON (1954) using a climatic chamber,
compared the temperature regulation of a normal subject with that of a subject
with fever, and found exactly the same qualitative and quantitative responses
to cooling and warming, except that in the subject with fever all the processes
revolved around a higher body temperature. PARK and PALMES (1948) investigated
the degrees of fever produced in humans by the injection of standardized pyrogenic
substances and determined that they remained unchanged at ambient temperatures between 25° and 43° C. If the pyrogens were injected under cold conditions
the rectal temperature, vasoconstriction, and the metabolic rate increased. In a
hot environment injections led to vasoconstriction and inhibition of sweat secretion. These results indicate that *fever can occur only if temperature regulation is
functioning*. In fact, appropriate experiments have shown that only animals with
intact temperature regulation can have fever (see THAUER, 1942; BARD et al.,
1970).

In a typical acute attack of fever, such as is found in malaria, the regulatory
centers are adjusted to a higher "set point". Normal temperatures then have the
effect of cold, and all the physiological responses to cold appear: increased metabolism, vasoconstriction and a decreased skin temperature, shivering (shaking with
fever), and subjective feelings of cold (Fig. 48). The core temperature rises to the
new "set point", which is then maintained through regulation, like a normal
temperature. Cooling and warming induce the appropriate counteracting responses. An increased heat production at the height of the fever is partly explained
as a direct effect of the increased temperature on metabolism which, according to

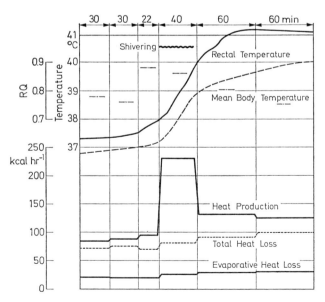

Fig. 48. Heat budget of a man during feverish shivering. Heat production is elevated;
heat loss is not increased. The core temperature and the average body temperature
rise. From DU BOIS (1936)

the reaction-rate-temperature law, must increase by 10—15% per degree of body temperature (Du Bois, 1921, cf. Chapter IV, p. 533).

During voluntary work of the same intensity as at the onset of fever (Fig. 48), the rectal temperature scarcely changed. An elevated metabolism is not a necessary precondition for fever. Wells and Rall (1948) demonstrated this by injecting pyrogenic substances into curarized dogs, which developed fevers of the same intensity as normal animals, in spite of their severely limited ability to increase their metabolism. In this case vasoconstriction was more intense. In cold surroundings,

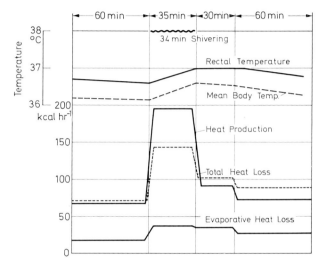

Fig. 49. Increase in heat formation through voluntary shivering. Heat production was adjusted to the same level as that of the shivering shown in Fig. 48. The body temperature rose only slightly in this case since heat loss increased. From du Bois (1936)

curarized animals did not develop such high fevers, because the fever reactions were not quantitatively adequate to raise the body temperature to the corresponding level. The impact on metabolism produced by removal of the adrenal medulla and thyroid from rabbits also makes no difference to the intensity of fever (Grant and Hirsch, 1950). The impulses causing subjective feelings of cold and shivering arise predominantly from the *cold receptors* of the skin. These phenomena can be largely suppressed by hot baths at temperatures above 40° C. Fever is distinguished from hyperthermia principally by the fact that the regulatory processes are hardly burdened at all, while in hyperthermia they are working at the limit of their capacity. For this reason high fever temperatures can often be astonishingly well tolerated, while an exogenous hyperthermia of the same intensity severely exhausts the organism. The investigation by Du Bois (1951) of hundreds of patients suffering from fever revealed that most of their temperatures fell between 40 and 40.5° C. Above 41° C the frequency curve declines sharply, which also indicates that regulation in fever is directed towards a quite definite value.

As fever abates, the processes observed during its increase are reversed. The regulatory centers are set at lower values, so that the body is no longer "feverish" but "hyperthermic", with all the regulatory processes opposed to excess warming: decreased metabolism, vasodilatation with increased skin temperature, sweating, and feelings of warmth. As a result the core temperature falls to the normal level again.

The theoretical concept of fever presented here has been corroborated in recent years by two experiments: In one it was shown that fever elicited in the goat by intravenous injections of pyrogens could be suppressed by warming the anterior hypothalamus (ANDERSEN, HAMMEL, and HARDY, 1961). By this method the body is, so to speak, persuaded that the higher set point has already been reached. It was thus shown that fever is not due to toxic injury of the regulatory system. In addition, fever was induced in rabbits by injecting a pyrogen (leucocyte pyrogen) into the anterior hypothalamus; injections of other substances and at other sites had no effect (COOPER, CRANSTON, and HONOUR, 1967). Single unit studies surprisingly showed that pyrogen significantly reduces the sensitivity of preoptic warmth receptors (CABANAC, STOLWIJK, and HARDY, 1968; EISENMANN, 1969, 1970; WIT and WANG, 1968b); there is, however, no evidence so far that the postulated thermo-insensitive neurons ("Q_{10}-1-cells", EISENMANN, 1970) are affected by pyrogen. At first glance the change in sensitivity of the warmth receptors seems to explain perfectly the fever response, but as shown by EISENMANN (1970) the preoptic thermal receptors did not change their firing rates at normal core temperature following pyrogen administration. Thus, some additional factor is needed for a sound neuronal concept of fever and for this reason the search for an influence of pyrogen on thermo-insensitive cells has not been given up (EISENMANN, 1970).

3. Long-Term Adaptation

Displacements of set point in the course of long-term thermal adaptations are described in greater detail in Chapter VIII.

4. Central Nervous Diseases

Some rare cases of central nervous diseases have been described, in which the core temperature is held at a low level in spite of external warming (THAUER, 1939; SUNDERMAN and HAYMAKER, 1947). More recently, studies were made on a woman who had a large cyst centered on the anterior hypothalamic region (cf. p. 566; for literature see COOPER, 1970). The resting internal temperature was maintained at 33.5° C when the patient was exposed to normal room temperature (22° C). Immersing the left arm in hot water caused considerable reflex dilatation in the opposite hand (cf. p. 585); internal body temperature did not rise by more than 0.16° C during the immersion experiment. Similar observations were made in some elderly people who had suffered one or more episodes of hypothermia. In all these cases a heat dissipation mechanism was activated at a body temperature considerably below the normal level. According to our present concepts on temperature regulation this must be thought of as a clear-cut case of a "set-point deviation".

5. Psychogenic Effects (Emotional Reactions). Vigilance

Temperature regulation can also be modified in *man* by psychological effects (KLEITMAN, 1939; GODELL et al., 1949). Emotionally induced temperature increases are known which are very similar to episodes of fever, especially in "stage fright" among actors, public speakers, musicians, etc. All the typical signs of a genuine attack of fever are present: beginning with feelings of cold, paleness of the skin, elevated muscle tone, a slight tremor that can build up to intense shivering and chattering of the teeth, and an increased rectal temperature; towards the end of the episode there are feelings of warmth, reddening of the face, an outbreak of sweating, and a decrease in rectal temperature (EBBECKE, 1948).

If *rabbits* were lightly restrained in their normal position without being squeezed their rectal temperatures fell by up to 2.7° C at ambient temperatures of 25° C. According to the investigations of GRANT (1950) hypothermia is maintained for a few hours at most and then gradually abates. It may be induced merely by bringing the animals into unfamiliar surroundings. It is brought about by intense vasodilatation in the ears and panting, while shivering is suppressed and basal metabolism remains unchanged. As the degree of hypothermia lies far outside the animal's normal range of temperature fluctuations, it must result from a genuine modification of temperature regulation. GRANT (1950) interprets this *"emotional hypothermia"* as a cortical effect on the thermoregulatory centers. The hypothermia produced by stretching a rabbit out (p. 562) is also said to be based on this phenomenon and not on an increase in surface area.

During sleep the body temperature tends to fall slightly. This has been taken as an indication that the set point is displaced to lower temperatures (HAMMEL, 1965, 1968). The fall in body temperature accompanying light anesthesia can also be seen as an expression of a shift in the threshold for the induction of thermoregulatory processes. As is well known, shivering appears only at lower body temperatures during anesthesia. One can thus generalize that the set point for body temperature increases with increasing vigilance.

F. Body Temperature during Work

As is known, the body temperature of homeotherms increases during physical work and activity. This increase is especially intense in very active small birds. BALDWIN and KENDEIGH (1932) observed increases of 4° C in the wren *(Troglodytes aedon)* during short periods of vigorous muscular motion. As soon as the animal rests quietly or is prevented from moving by being restrained the body temperature falls markedly. The human rectal temperature also rises by up to 4° C after very heavy physical exertion. Thus ROBINSON (1949) found a rectal temperature of 41° C in two American racers after a three mile run.

In 1938 NIELSEN attributed this phenomenon to displacement of the set point of the regulatory system. His chief argument was that after equilibrium was attained the temperature level during physical work depended only on the intensity of the work, and not on the external temperature, as would have to be the case if the increased body temperatures were due to passive heat retention. In recent years serious objections have been raised both against this finding and against the interpretation (BLEICHERT et al., 1966, 1968; BLEICHERT, KITZING, and BEGEMANN, 1966; KITZING et al., 1971). BLEICHERT and his co-workers (KITZING, KUTTA, and BLEICHERT, 1970) showed that the independence of external conditions was not given for heavy work and for somewhat longer experimental durations, or at best was given only for a very narrow temperature range. On the other hand, HAMMEL (1968) indicated that as work began sweat secretion started even though the skin and internal temperatures had not changed. If this were due to a set point displacement it would have to be considered a displacement towards *lower* temperature levels.

The undisputed fact, that the body temperature rises, was a serious problem for a theory which regarded the deep body temperature as the value being regulated. One would have had to postulate either a failure of thermoregulation or a surprisingly large load error or a displacement of the set point. This is not necessary with a multiple input theory, according to which internal temperature would be expected to rise under equilibrium conditions since during work, even in a warm environment, evaporative heat loss would cause the skin temperature to be lower than it is in the zone of thermal neutrality under resting conditions, as many investigations have shown (B. NIELSEN, 1969; ROBINSON et al., 1965; and others). Since the quantity being regulated in a multiple input system is $T = f(T_{Hy}, T_S \ldots)$ (see p. 586) the load error, $T - T_{set}$ may be small in spite of a greatly increased internal temperature (e.g., T_{Hy}).

Although, as explained earlier, the rectal temperature can be taken as an approximate measure of T (= controlled variable) under resting conditions, it is no longer valid during work. As a result of altered conditions of heat formation and of evaporative heat loss the temperature field must necessarily be different, and an increasing rectal temperature cannot be taken as proof for either an accumulation of heat or of "set point displacement".

With respect to thermoregulatory relations during physical labor, the center of interest today is occupied by questions concerning the stimulation of heat loss mechanisms, especially sweat secretion. Sweat secretion can appear as work begins (SALTIN, GAGGE, and STOLWIJK, 1970), even before an increase can be recognized in customarily measured skin- or internal temperatures. The presence of thermoreceptors in the skeletal musculature has been postulated as an explanation for this phenomenon (BEAUMONT and BULLARD, 1963; ROBINSON et al., 1965; STOLWIJK and HARDY, 1966; B. NIELSEN, 1969b). As ROBINSON et al. (1965) demonstrated the rate of sweat secretion in the initial phase of work is correlated quite well with the increase in muscle temperature. However, no direct proof of the existence of the postulated muscle receptors has been obtained at this time. In addition, it has recently been suggested that when work is begun sweat secretion is stimulated by sympathetic excitation based on non-thermal factors (NADEL, MITCHELL, and STOLWIJK, 1971).

VII. Cutaneous Thermoreception

Cutaneous thermoreceptors in homeotherms are involved in conscious temperature sensations but they may be even more important in connection with behavioral and thermoregulatory responses. Thus the physiology of thermoreceptors includes various aspects, namely 1. temperature sensations in human subjects, 2. afferent impulses from units responding to thermal stimulation in animals and, to some extent, in man, 3. behavioral responses to thermal stimulation in animals, and 4. thermoregulatory reflexes in animals and human subjects. We will restrict ourselves to a short presentation here and refer the reader to comprehensive works on thermoreceptors (HENSEL, 1952a, 1966, 1973; ZOTTERMAN, 1959) and patterns of behavior of homeotherms under thermal stimulation (HERTER, 1952; HARDY, GAGGE, and STOLWIJK, 1970; WHITTOW, 1970, 1971; CABANAC, 1971). Thermosensitive structures in the central nervous system are dealt with on p. 567.

A. Thermal Sensations in Man

1. Structure of Sensation

It is relatively easy to discriminate the phenomenal qualities of *warm* and *cold* from the spectrum of cutaneous sensations. Both qualities form a sensory continuum of various intensities: – "indifferent" – "lukewarm" – "warm" – "hot" – "heat pain" on the warm side and "indifferent" – "cool" – "cold" – "cold pain" on the cold side. Whether the sensation of *"heat"* is only a more intense sensation of warmth or a mixture of various qualities is not quite clear. (For references see HENSEL, 1952a.) Perhaps "heat" might be a quality of its own, its neurophysiological correlate being the activity of particular "heat" fibers excited by high temperatures (IGGO, 1959).

Heat pain, which begins at temperatures of 45° C, is a burning "sharp" pain that is localized in the outermost skin surface, while cold pain, which can be induced by persistent temperatures below 17° C, has a "dull" character, is poorly localizable, and radiates intensely into surrounding areas. Both components of pain arise from pain nerves, not from thermoreceptors; blood vessel spasms probably contribute to the dull cold pain (WOLFF and HARDY, 1941; ABRAMSON et al., 1966).

Thermal *comfort* and *discomfort* are "pleasant" and "unpleasant" emotional feelings (HARDY, 1970; STEVENS, ADAIR, and MARKS, 1970) which are not necessarily and unambiguously correlated with temperature sensations. For example, cooling of the hand can be comfortable when the body is hyperthermic, while the same amount of local cooling leads to cold discomfort under hypothermic conditions. In physiological terms, thermal comfort reflects an integrated state of the thermoregulatory system (for details see p. 613).

2. Thermal Sensations and Temperature

Because of their intracutaneous site, thermoreceptors have neither the temperature of the skin surface nor that of the blood. Any reliable metrics of thermal stimuli has thus to take account of the temperatures in different layers of the skin. By means of fine thermocouples the intracutaneous temperature field has been measured directly under stationary and unstationary conditions (HENSEL, 1952a).

When a cutaneous area, such as the hand or foot, is adapted to a constant temperature of 25° C, linear temperature increases will cause a sequence of sensations from "cool" to "warm". After a constant temperature level is reached the intensity of sensation decreases considerably. On linear cooling with a similar slope, a cold sensation starts at the same temperature at which a warm sensation occurs when the temperature is rising. Starting with a neutral temperature of 33.5° C, the threshold (ΔT) deviates the more from this point, the more slowly the temperature is changed. By plotting the rate of change (dT/dt) versus the thermal threshold (ΔT), a hyperbolic function is obtained (Fig. 50). Similar results have been found by KENSHALO, HOLMES, and WOOD (1968). At high and low temperatures the threshold rate of change finally reaches zero, which means that steady temperature sensations occur at constant skin temperatures. With controlled

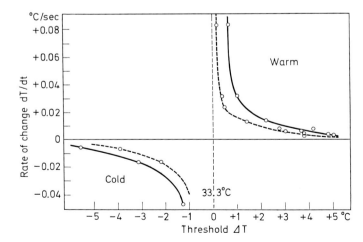

Fig. 50. Average thresholds (ΔT) of warm and cold sensations on the forearm (20 cm²) as a function of rate of temperature change (dT/dt). Initial temperature 33.3° C. Dashed lines: thresholds; solid lines: distinct sensations. From HENSEL (1952a)

course, no static discharge is possible without energy but this energy may be derived from chemical processes within the receptor, the temperature being only a controlling factor for metabolic rates. A general survey on the effects of temperature on membrane potentials of excitable cells was given by SPERELAKIS (1970).

The fundamental mechanisms of thermoreceptor excitation are still unknown. Several hypothetical models have been proposed but so far these are only formal in nature. In order to describe formally the behavior of thermoreceptors such a model should at least account for 1. the difference between warm and cold receptors, 2. the static response at constant temperatures and 3. the dynamic response to temperature changes.

The course of the receptor discharge at constant temperatures and particularly the effect of temperature changes suggest that at least two interacting processes are involved. SAND (1938) has assumed an exciting (E) and an inhibiting (I) process,

Table 7. Specific thermoreceptors in mammals

Species	Nerve	Receptive field	Type of receptor
Man	Radial	Hairy skin, hand dorsum	Cold[a]
Monkey	Median, ulnar saphenous	Hairy and glabrous skin, arm and leg	Cold[b,c,d,e,f]
	Saphenous, radial	Hairy skin, hand and foot dorsum	Warm[b,g]
Dog	Infraorbital	Hairy and marginal skin, face	Cold[e,h,i] Warm[i]
Cat	Lingual	Mucous membrane, tongue	Cold[j,k] Warm[j,l]
	Infraorbital	Hairy and marginal skin,	Cold[i,m,n] Warm[i,o]
	Saphenous	Hairy skin, leg	Cold, warm[i,p]
Rat	Infraorbital	Hairy and marginal skin, face	Cold[n]
	Saphenous	Hairy skin, leg	Cold, warm[i]
	Scrotal	Hairy skin, scrotum	Cold, warm[e]
Hamster	Infraorbital Saphenous	Hairy skin, face and leg	Cold, warm[q]

[a] From HENSEL and BOMAN (1960).
[b] From IGGO (1963).
[c] From IGGO (1969).
[d] From KENSHALO and GALLEGOS (1967).
[e] From PERL (1968).
[f] From HENSEL and IGGO (1971).
[g] From HENSEL (1969).
[h] From HENSEL (1952a).
[i] From IRIUCHIJIMA and ZOTTERMAN (1960).
[j] From ZOTTERMAN (1936).
[k] From HENSEL and ZOTTERMAN (1951a).
[l] From DODT and ZOTTERMAN (1952a).
[m] From HENSEL (1952b).
[n] From BOMAN (1958).
[o] From HENSEL and KENSHALO (1969).
[p] From HENSEL, IGGO and WITT (1960).
[q] From RATHS and HENSEL (1967).

the condition for the initiation of impulses being $E > I$. This model would account principally for the static behavior of cold as well as of warm receptors since the static frequency curves of both are of similar shape with static maxima at certain temperatures. The time dependence of the discharge could be introduced by the assumption that E and I change exponentially with different time constants k_E and k_I when the temperature is changed between two levels. Thus the temporal course of the impulse frequency would result from the difference of two exponential functions. For lingual cold receptors k_E was found to be 0.3–2.2 sec (HENSEL, 1953), whereas k_I was smaller than 0.1 sec. By varying k_E and k_I, the dynamic response of both cold and warm receptors could be described. For further details see HENSEL (1952a) and ZOTTERMAN (1953).

A more recent model (ZERBST and DITTBERNER, 1970; ZERBST, 1972) is based on the assumption that a chemical reaction of transmitter release is coupled with the diffusion transport of the transmitter to the excitable membrane. In the membrane compartment the transmitter is inactivated by a chemical process and also removed by diffusion. Temperature alters the velocity coefficients of the chemical reactions whilst the diffusion coefficients are unaltered by temperature. This model was found to agree well with the stimulus response characteristics of cold receptors.

2. Morphological Structures

Cold and Warm Spots. In human skin, cold spots seem to be distributed more densely than warm spots. The highest density of thermoreceptors is found in the face (trigeminal area). Investigations on the topography of warm spots in the external skin of human subjects have been made by REIN (1925b), of cold spots by STRUGHOLD and PORZ (1931). For further references see v. SKRAMLIK (1937) and HENSEL (1952a).

Little is known about the topography of the thermoreceptors of other *mammals* or of *birds*. From investigations of temperature regulation, of behavior patterns (p. 615) and from electrophysiological findings (p. 605), we can conclude that they are arranged most densely in the areas of the mouth and nose served by the trigeminal nerve and are distributed singly over the rest of the skin.

Electrophysiological investigations have shown that in most cases single cold and warm fibers, respectively, innervate one peripheral spot in the skin (PERL, 1968; IGGO, 1969; HENSEL and KENSHALO, 1969; HENSEL, 1969; HENSEL and IGGO, 1971). Only for primates KENSHALO and GALLEGOS (1967) have reported single cold fibers which innervate up to 8 multiple spots, the whole field amounting to about 1.7 cm². Fig. 54 shows a map of the tip of the cat's nose with specific cold and warm spots, each of them innervated by a single fiber.

Morphology of Thermoreceptors. v. FREY's (1895) hypothesis that specific corpuscular nerve terminals were the anatomical substrate of thermal receptors has started numerous attempts to identify histologically the underlying neural structures of cold and warm spots in human subjects. The results of these endeavors were negative (for references see HENSEL, 1952a). Cold and warm sensitive cutaneous areas in man were later found without any encapsulated or corpuscular nerve endings (HAGEN et al., 1953; WEDDELL, PALMER, and PALLIE, 1955; WED-

DELL and MILLER, 1962). This rules out v. FREY's original concept but does not disprove, of course, the functional specialization of cold and warm receptors.

By means of direct combination of electrophysiological and electron microscopical methods it has become possible to identify the morphological substrate of cold

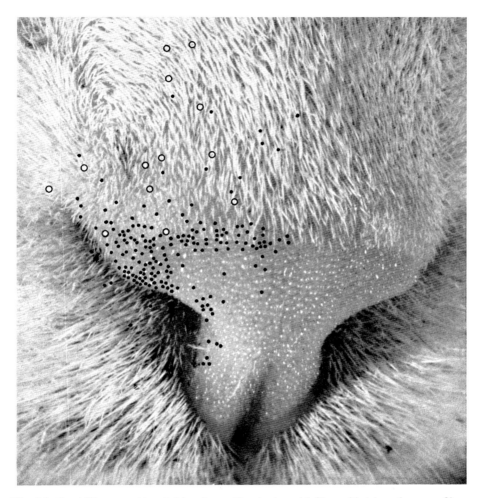

Fig. 54. Spot-like receptive fields of specific single cold fibers (dots) and warm fibers (circles) in the nasal region of the cat. Results from several preparations of the lateral nasal branches of the infraorbital nerve. From HENSEL (1973)

receptors in the cat's nasal region (KENSHALO et al., 1971; ANDRES, v. DÜRING, and HENSEL, published in HENSEL, 1973). In the hairy skin the receptive structures at the site of specific cold spots (cf. Fig. 54) are served by thin myelinated axons dividing into several non-myelinated terminals within the stratum papillare (Fig. 55). The axon terminals are accompanied by unmyelinated Schwann cells as

far as the epidermal basement membrane. A continuous connection between the
basement membrane of the epidermis and that of the nerve terminals is seen. The
receptive endings which penetrate a few microns deep into the basal epidermal cells
contain numerous mitochondria as well as an axoplasmatic matrix with fine
filaments and microvesicles.

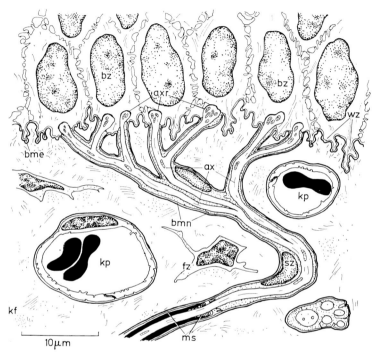

Fig. 55. Semi-schematic diagram of nerve endings at the site of an electrophysio-
logically identified cold receptor in the hairy skin of the cat's nose: *(ms)* myelin sheath
of afferent nerve fiber; *(sz)* non-myelinated Schwann cell accompanying axon terminals
(ax) to the basement membrane *(bme)* of epidermis; *(bmn)* basement membrane of
nerve terminals; *(axr)* receptive endings with mitochondria; *(bz)* basal epidermal
cells; *(wz)* root feet of basal cells; *(kp)* capillaries; *(fz)* fibrocyte; *(kf)* collagen
fibrils of stratum papillare. From ANDRES, v. DÜRING, and HENSEL, published in HEN-
SEL (1973)

Specific warm receptors have not been identified yet but from present evidence it
can be concluded that they are situated in a deeper layer. This corresponds well
with the results of functional measurements of the depth of cold and warm receptors (BAZETT and McGLONE, 1930; BAZETT, McGLONE, and BROCKLEHURST,
1930; HENSEL, STRÖM, and ZOTTERMAN, 1951).

Afferent Innervation. The specific cold fibers in the tongue of the cat are fairly thin
myelinated fibers belonging to the A, δ group according to ERLANGER and GAS-
SER's (1937) classification (ZOTTERMAN, 1936). In the external skin of dogs, cats,
and rats, except in the facial region, most specific cold and warm receptors are

served by unmyelinated C fibers, with mean conduction velocities around
0.9 m sec^{-1} (HENSEL, IGGO, and WITT, 1960; IRIUCHIJIMA and ZOTTERMAN, 1960;
BESSOU and PERL, 1969). In the infraorbital region of dogs and cats, most cold
fibers are myelinated, their mean conduction velocity being 14 m sec^{-1} (IGGO,
1969). The specific cold fibers in the extremities of monkeys are partially myelinated (average conduction velocities in different regions ranging from 6–11 m sec^{-1};
PERL, 1968; IGGO, 1969; HENSEL and IGGO, 1971) and partially unmyelinated
(mean conduction velocity 0.7 m sec^{-1}), whereas the warm fibers are entirely
unmyelinated with mean conduction velocities of 0.7 m sec^{-1} (HENSEL and IGGO,
1971). The distribution of thermal fibers in man is not known; on the basis of
available evidence (HENSEL and BOMAN, 1960) it seems probable that part of the
cold fibers are myelinated and that the warm fibers are non-myelinated. We can
conclude that the cutaneous cold fibers with the highest conduction velocities of
20 m sec^{-1} (IGGO, 1969) have diameters of 3–4 μm (MARUHASHI, MIZUGUCHI, and
TASAKI, 1952), whereas the diameters of the slowly-conducting C fibers are in the
range of 1 μm. (Central pathways see p. 608.)

3. Electrophysiology of Thermoreceptors in Homeotherms

Cold Receptors. At constant skin temperatures in the normal range all cutaneous
cold receptors exhibit a static discharge with a constant impulse frequency. The
temporal sequence of impulses can be more or less regular, or it can consist of
periodic bursts of 2–10 impulses separated by silent intervals (Fig. 56). Such
bursts have been described for cold fibers in the lingual nerve of cats (HENSEL and
ZOTTERMAN, 1951a; DODT, 1953), the infraorbital nerve of cats (HENSEL and
KENSHALO, 1969; HENSEL and WURSTER, 1970) and dogs (IGGO, 1969), and the

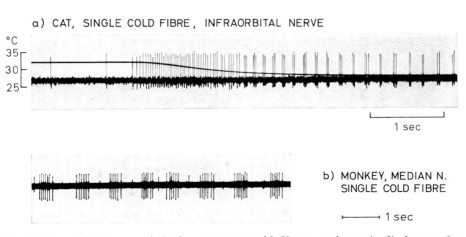

Fig. 56. Burst discharges of single cutaneous cold fibers: *a*, dynamic discharge of a
fiber from the cat's infraorbital nerve and temperature of the thermode when cooling
the nose from 32—27° C. At constant temperatures of 32 and 27° C the discharge was
regular. From HENSEL and WURSTER (1970). *b*, static discharge of a fiber from the
median nerve in the monkey at constant temperature of the hairy skin of the forearm.
From IGGO (1969)

saphenous and trigeminal nerves of monkeys (IGGO, 1963, 1969, 1970; IGGO and
IGGO, 1971; KENSHALO and GALLEGOS, 1967; POULOS, 1971).
The temperature range of static activity varies for different cold fibers, the extremes being about 5° C and 43° C. Low thermal limits of activity were found in
cold receptors served by C fibers (HENSEL, IGGO, and WITT, 1960), whereas for
A, δ fibers the lowest temperature was about 10° C. The static impulse frequency of
individual cold fibers rises with temperature, reaches a maximum, and falls again
at high temperatures (Fig. 58). Sometimes a second maximum is seen at low
temperatures in connection with the transition from regular to burst discharges.
Myelinated cutaneous cold fibers have static temperature ranges of more than 20° C
and maximum frequencies of 6–20 imp/sec, whereas in non-myelinated cold fibers
the static ranges and maximum frequencies may be somewhat smaller. The temperatures of the maximum discharge are different for each unit and vary in most
cases between 18 and 34° C. The average temperatures for the maximum activity
of various cutaneous cold receptor populations are rather similar, ranging from
26–30° C in monkeys, cats and rats. Only in the lip of dogs was the average value
near 35° C (IGGO, 1969).
It is not possible to discriminate between static temperatures below and above the
maximum, say, 18 and 34° C (Fig. 58) by observing the average frequency of cold
impulses. However, the burst discharge might carry additional temperature information independent of the average impulse frequency and thus allow differentiation
between lower and higher temperatures. For example, in monkeys the average
impulse frequency of cutaneous cold fibers has a positive temperature coefficient
in the low temperature range, while the number of impulses in a burst as well as
the ratio of impulses within a burst to the number of bursts per second has a
negative temperature coefficient (IGGO, 1969).
On sudden cooling to a lower temperature level, the cold receptors respond with a
transient overshoot in frequency, followed by adaptation to the new static dis-

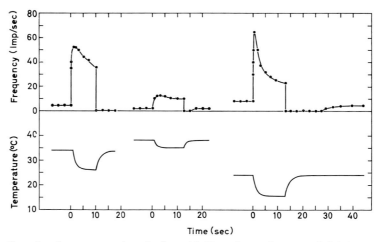

Fig. 57. Impulse frequency of a single cold fiber from the superficial branch of the
radial nerve in man and skin temperature as a function of time when applying thermal
stimuli to the dorsum of the hand. From HENSEL and BOMAN (1960)

charge (Fig. 53 and 57). When the skin is warmed up again to the initial level, a transient decrease in frequency or silent period is seen, after which the frequency rises again and finally reaches the initial static value. The highest dynamic over-shoot observed in a single cutaneous cold fiber was 300 imp/sec (IGGO, 1969), the highest ratio of static to dynamic frequencies 1 : 30. Cutaneous cold receptors in the trigeminal area of cats often respond with a burst discharge on cooling even when the static discharges are fairly regular (Fig. 56a).

The higher the rate of cooling at a given temperature, the higher is the dynamic overshoot. When equal temperature changes are applied at various adapting temperatures, the dynamic overshoot is a function of temperature and follows approximately the shape of the static activity curve (HENSEL and ZOTTERMAN, 1951a; IGGO, 1969; KENSHALO et al., 1971). From the slope of the static and dynamic frequency curves we can derive the static and dynamic differential sensitivity $(\Delta v/\Delta T)$, i.e., the change in frequency (Δv) for a small change in temperature (ΔT). Of course, the dynamic differential sensitivity increases with the rate of temperature change. Some extreme values of dynamic differential sensitivities obtained with high rates of change are given in Table 8.

Table 8. Properties of single cold and warm fibers from the nasal area of cats[a]

Property	Cold fibers	Warm fibers
Number of units	26	22
Static temperature limits	5—43° C	30—48° C
Maximum static frequency (average)	9 impulses/sec	36 impulses/sec
Temperature of static maximum (average)	27° C	46° C
Maximum static differential sensitivity (average)	— 1 impulses/sec° C	+14 impulses/sec° C
Highest dynamic differential sensitivity	—50 impulses/sec° C	+70 impulses/sec° C
Highest dynamic frequency	240 impulses/sec	200 impulses/sec

[a] From HENSEL and KENSHALO (1969).

Long-term adaptation to various ambient temperatures has little effect on the properties of cold receptors. When cats are exposed for several months to ambient temperatures of 5° and 35° C, respectively, their nasal temperatures being about 14° and 35° C, no statistically significant difference is seen in the static and dynamic characteristics of cold receptor populations in the nasal area. The only possible change might be a slight increase in the occurrence of bursts on dynamic cooling in the warm-adapted cats (HENSEL and SCHÖNER, unpublished).

Relatively little is known about cold receptor activity in *birds*. Recording from lingual and laryngo-lingual nerve fibers serving the tongue of the chicken, KIT-CHELL, STRÖM, and ZOTTERMAN (1959) observed a high static activity of cold fibers at 20° C and a much lower one at 44° C. Dynamic cooling led to a transient increase, dynamic warming to a transient inhibition of the discharge. NECKER (1972) found cold and warm sensitive receptors in the beak region of pigeons. Cold receptors increased their static activity from 0 to about 3 sec^{-1} when the constant temperature level was lowered from 36 to 20° C. The static discharge of warm receptors rose from 0 to about 60 sec^{-1} in the temperature range between 20 and 44° C. Both groups of receptors showed very little or no dynamic responses on rapid temperature changes.

Warm Receptors. The quantitative relationship between temperature stimulus and activity of cutaneous warm receptors has been studied in the infraorbital nerve of cats (HENSEL, 1968; HENSEL and KENSHALO, 1969; HENSEL and HUOPANIEMI, 1969) and, to some extent, in the radial and saphenous nerves of monkeys (HENSEL, 1969; HENSEL and IGGO, 1971). In addition, a few quantitative data are available from unmyelinated warm fibers in the cat saphenous nerve (HENSEL, IGGO, and WITT, 1960; STOLWIJK and WEXLER, 1971) and the scrotal nerve of rats (IGGO, 1969).

In contrast to the findings in other cutaneous areas, except the rat's scrotum, warming the nasal region of cats elicits a mass discharge of warm impulses in the lateral nasal branch of the infraorbital nerve serving the apical hairy skin of the nose. Fig. 58 shows the average static frequencies of warm and cold receptor populations in the infraorbital nerve of the cat as a function of skin temperature.

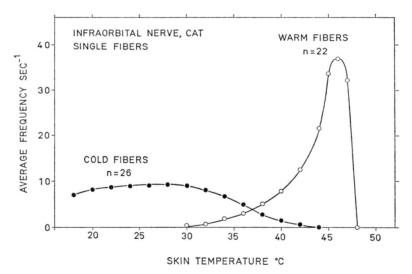

Fig. 58. Average frequency of the static discharge as a function of temperature for populations of single warm and cold fibers from the cat's infraorbital nerve supplying the nose. From HENSEL and KENSHALO (1969)

There is some overlap in the static temperature ranges of both groups, the curves crossing near 37° C. The population of infraorbital warm receptors is surprisingly homogenous, the maximum discharges occurring at temperatures between 45 and 47° C. In the whole temperature range the sequence of impulses is fairly regular, in contrast to the burst discharge of cold receptors. When the temperature exceeds that at which activity is maximal the impulse discharge stops rather abruptly, this

inhibition being completely reversible. A summary of quantitative data for populations of cold and warm receptors in the nasal area of cats is given in Table 8.

In primates, specific warm receptors have been found in the dorsum of the hand
and foot. Their spot-like fields are served by non-myelinated fibers (HENSEL and
IGGO, 1971) which show a rather regular discharge pattern at temperatures below
the static maximum. The primate warm receptors differ in some respect from those
in the cat's nose. One group seems to have a temperature dependence of static
activity similar to that found for infraorbital warm receptors, the maximum
occurring near 45° C, but another group has its maximum discharge at temperatures between 40 and 42° C. At higher temperatures the discharge slows down and
changes from a regular to a more irregular or a burst pattern.

On sudden temperature changes the warm receptors behave in the opposite way
than do the cold receptors, in that they respond to warming with an overshoot and
to cooling with transient inhibition (Fig. 53). The highest dynamic response of a
single warm fiber from the cat's nose was 200 imp/sec, that is, 5.5. times higher
than the average static maximum of 36 imp/sec.

The diagram in Fig. 59 shows the static frequency as well as the dynamic maximum
on sudden warming as a function of temperature for 3 single warm fibers. At
temperatures below the static threshold, dynamic responses can still be elicited.
When the adapting temperature increases between 28 and 44° C, the overshoot
caused by equal amounts of warming (ΔT) becomes higher, the slope of the dynam-

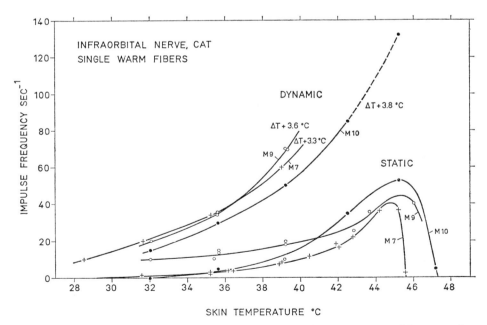

Fig. 59. Static impulse frequency and dynamic peak frequency on sudden warming
as a function of skin temperature for 3 single warm fibers from the cat's nose. The
amount of warming (ΔT) is indicated on the dynamic curves. The dashed part of the
curve is extrapolated from warming by 2.8° C. From HENSEL and HUOPANIEMI (1969)

ic curves being steeper than that of the static ones. These results are analogous to those found for cold receptors.

Adequate and Inadequate Stimulation. From the neurophysiological data we can conclude that the excitation of cutaneous thermoreceptors is dependent 1. on the absolute temperature (T) and 2. on the rate of temperature change (dT/dt) or the temporal gradient.

The hypothesis that *spatial* temperature gradients (dT/dx) are the adequate stimulus of thermosensitive nerve endings could not be verified by cooling the tongue of the cat with reversed gradients (HENSEL and ZOTTERMAN, 1951b; HENSEL and WITT, 1959). When a cold receptor on the surface of the tongue is cooled from above, its discharge frequency will increase. On cooling the tongue from the lower surface, however, the resting discharge of the cold receptor will increase as well, in spite of a reversed spatial temperature gradient. Thus we come to the conclusion that in accordance with sensory experiments (p. 596) the adequate stimulus of thermoreceptors is temperature and its temporal change *per se* and not the direction or slope of a spatial gradient.

A *"paradoxical"* excitation at extreme temperatures has been found in cold fibers from the lingual nerve of cats (DODT and ZOTTERMAN, 1952b). While at temperatures between 40 and 45° C no static discharge of these fibers is seen, they start discharging again at lingual temperatures above 45° C. On further warming, the frequency of this "paradoxical" excitation rises and reaches a maximum at about 50° C. Above this value the receptors will be damaged.

No "paradoxical" discharge of cutaneous warm receptors has been observed as yet; however, as DODT and ZOTTERMAN (1952a) and DODT (1953) have reported, fibers in the chorda tympani sensitive to warming respond with a phasic burst of impulses when the tongue is cooled rapidly by more than 8° C.

For references on the action of *chemical substances* and changes in blood supply on thermoreceptor activity see HENSEL (1973).

Central Pathways of Thermal Afferents. In mammals afferent pathways run from the thermoreceptors via the dorsal roots and the spinal ganglion to ganglion cells in the corresponding segment of the spinal cord, over the contralateral spino-thalamic tract to the thalamus, and from there to the gyrus postcentralis of the cerebral cortex. Temperature fibers from the trigeminal region terminate predominantly in the nucleus tractus spinalis and also go from there to the thalamus. Many investigators, such as KERSLAKE and COOPER (1950) and COOPER and KERSLAKE (1953), suggest that there are also extraspinal temperature pathways in the sympathetic system.

Afferent impulses from cold receptors in the trigeminal region have been recorded in the spinal trigeminal nucleus of the cat (FRUHSTORFER and HENSEL, 1973), and in the thalamus of cats (LANDGREN, 1960, 1970) and monkeys (POULOS and BENJAMIN, 1968; POULOS, 1971) as well as in the somatosensory cortex of cats (LANDGREN, 1957a, b, 1970). In addition to a substantial central convergence of different modalities, especially mechanical and thermal, single neurons may be found in all the nuclear regions up to the cerebral cortex which respond specifically to peripheral stimulation by cold. The activity of peripheral cold receptors in response to static and dynamic temperature stimuli can still be clearly recognized in the quantitative behavior of single neurons in the thalamus (POULOS and BENJAMIN, 1968).

4. Comparative Electrophysiology of Thermoreceptors

Hibernators. Peculiar properties of thermoreceptors are found in hibernators. European hamsters *(Cricetus cricetus)* have numerous myelinated and unmyelinated cold fibers in various cutaneous nerves but only a few warm fibers (RATHS and HENSEL, 1967). In the infraorbital nerve serving the face, small myelinated (A, δ) cold fibers prevail. They show a static activity in the range of $3-22.5°$ C (extreme values -5 to $35°$ C) and a long latency on dynamic temperature changes; the unmyelinated (C) cold fibers have a static discharge between -2.3 and $27.5°$ C (extremes -5 to $32°$ C). In both groups the maximum static activity is found at skin temperatures of $4-5°$ C which is much lower than the corresponding temperatures in homeotherms. The activity of cold receptors at very low temperatures corresponds well with the fact that hibernating animals show arousal reactions when cold stimuli of $0-5°$ C are applied to the skin. The characteristic properties of cold receptors in hamsters seem to be phylogenetic in origin since they do not differ in the hibernating and non-hibernating state of the animal.

Reptiles. Rattlesnakes *(Crotalus)* and related species of pit vipers *(Crotalidae)* possess a pair of sense organs below and in front of the eyes, the so-called *facial pits*, which are among the most sensitive thermoreceptors of any animal. They act as directional distance-receptors and make it possible for the snake to strike at warm prey without using the eye, nose or tongue for orientation (NOBLE and SCHMIDT, 1937). Each pit consists of a cavity about 5 mm deep, equally wide but narrowing at the opening. Above the bottom and separated from it by a narrow air space a densely innervated membrane of $10\,\mu$m thickness is stretched out between the walls of the pit. An open connection between the air space beneath the membrane and the open air allows pressure equilibration on both sides of the membrane. The warm-sensitive receptors are distributed over the membrane and consist of tree-like structures of unmyelinated nerve terminals (BULLOCK and FOX, 1957).

Under resting conditions there is an irregular steady discharge of nerve impulses from the pit membrane (BULLOCK and COWLES, 1952; BULLOCK, 1953; BULLOCK and DIECKE, 1956). Rapid warming by infrared radiation by only $0.002°$ C at the nerve endings causes a significant increase in impulse frequency, while cooling causes an inhibition of the resting discharge (Fig. 60). In contrast to the warmth receptors in mammals, the pit receptors are practically insensitive to constant temperatures (T) but extremely sensitive to the rate of change (dT/dt). This has the biologically important consequence that slow variations in air temperature are compensated, and only the more rapid changes in heat radiation are perceived. The sensitivity to rapid temperature changes is enhanced by the thinness of the receptive membrane resulting in a very small heat capacity. When an animal $10°$ C warmer than the background appears for 0.5 sec at a distance of 40 cm in front of the snake, the heat radiation is sufficient to cause a significant speeding of the receptor discharge in the pit organ. Behavioral experiments show that under these conditions the snake is in fact able to discover warm prey by infrared reception.

Fishes. Specific thermoreceptors have not yet been identified by electrophysiological methods in fishes. SPÄTH (1967) has investigated the response of cutaneous mechanoreceptors in the fresh-water fish *Leuciscus rutilus* to combined mechanical and thermal stimulation. These nerve endings had no spontaneous discharge and did not respond to thermal stimuli alone. When stimulated mechanically at various temperatures, the response passed through a maximum between 18 and 22° C when the fish were chronically preadapted to 5 and 15° C, while the response rose monotonically from 6–30° C when they were preadapted to 25° C. It is not known whether these receptors are involved in temperature discrimination.

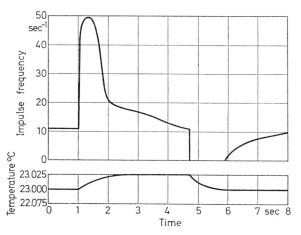

Fig. 60. Discharge frequency of 2 or 3 fibers from the supramaxillar nerve of *Crotalus* during small temperature changes at the facial pit. From BULLOCK (1953)

Elasmobranchs, such as rays and sharks, possess the so-called *ampullae of Lorenzini*. These sense organs consist of little capsules beneath the skin of the head which continue into long canals ending at the skin surface. The capsules and canals are filled with a jelly-like substance, and the sensory receptors are situated within the capsule. Single nerve fibers supplying the ampullae in rays *(Raja)* and dogfishes *(Scyliorhinus)* show a steady activity (Fig. 61) at constant temperatures between 0 and 30° C with an average frequency maximum near 19° C. Rapid cooling caused a transient overshoot, rapid warming a transient inhibition of the discharge (SAND, 1938; HENSEL, 1955, 1956; MURRAY, 1959, 1962). In some single fibers cooling by 3° C led to a dynamic overshoot of about 100 impulses sec^{-1}. The responses to cooling are similar to those found in specific cold receptors in mammals. Long-term adaptation to 4° C and 18° C for 2 months did not result in a significant change of the static properties of the receptors (HENSEL and NIER, 1971). It remains an open question whether the ampullae of Lorenzini have a thermoreceptive function, since they also respond to mechanical stimuli (HENSEL, 1956;

MURRAY, 1957, 1960a) and to weak electrical currents (MURRAY, 1959, 1960b, 1965). Elasmobranchs are, in fact, able to detect the bioelectric fields of their prey, and it seems justified to ascribe this function to the ampullae of Lorenzini (KALMIJN, 1971).

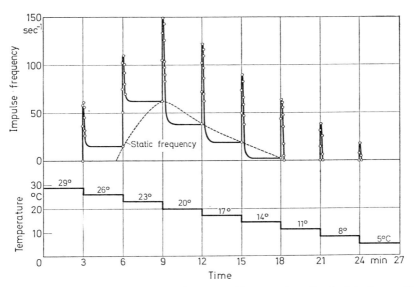

Fig. 61. Impulse frequency of a single nerve fiber serving the isolated mandibular ampullae of Lorenzini in *Scyliorhinus* during stepwise cooling by 3° C in the temperature range from 29—5° C. From HENSEL (1955)

Invertebrates. Cold receptors have been located on the antennae of cockroaches *(Periplaneta americana)*, about 20 per antenna and rarely more than one per segment (Fig. 62A, B). They consist of a delicate hair-like structure (sensillum) emerging from a ring-shaped wall. Recordings with microelectrodes have shown that at constant temperatures the cold receptor is continuously active, the average maximum frequency of its discharge being 16 impulses sec^{-1} at temperatures near 28° C (LOFTUS, 1966, 1968, 1969). At higher and lower temperatures the steady frequency becomes lower. When the receptor is rapidly cooled, its discharge frequency rises steeply up to 300 impulses per second and then declines gradually to a much lower constant level (Fig. 62, C). On rapid warming, the opposite response is seen, namely, a transient inhibition of the receptor discharge, followed by a gradual restoration of the steady activity. The cold receptor is thus sensitive to constant temperatures (T) as well as to the rate of temperature changes (dT/dt). Electrophysiological evidence has also been presented for the presence of thermosensitive structures in the antennae of migratory locusts *(Locusta migratoria migratorioides)*, (WALDOW, 1970) and of honey bees *(Apis mellifica)* (LACHER, 1964).
Caterpillars of various moths *(Lasiocampidae, Saturniidae, Sphingidae)* have cold receptors in the antenna and in the maxillary palp. Electrophysiological in-

vestigations by means of microelectrodes have shown that 3 cells located in the third antennal segment and probably not more than one cell in the maxillary palp were sensitive to cooling (SCHOONHOVEN, 1967). At constant room temperatures a static activity was seen which increased in frequency when the temperature was lowered. During rapid cooling the frequency rose steeply to a dynamic maximum up to 300 impulses/sec, while rapid warming caused a transient inhibition of the discharge. The fact that only a few cells out of 20 or 30 show the typical response to cooling strongly suggests a specific thermoreceptive function.

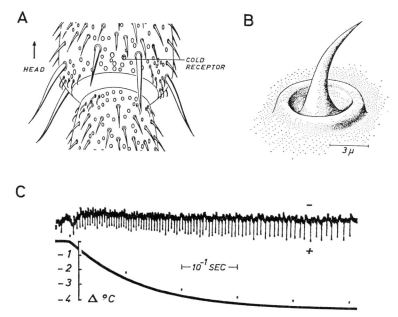

Fig. 62. Cold receptor of the cockroach *(Periplaneta americana)*. *A*, segment of antenna with cold receptor. *B*, cold sensillum at higher magnification. *C*, impulse discharge from single cold sensillum when cool air is blown over the antenna. From LOFTUS (1968)

C. Comparison of Various Approaches to Thermoreceptor Function

1. Temperature Sensation and Neural Activity

Very few measurements of afferent impulses from cutaneous thermoreceptors in human subjects have been made as yet (HENSEL and BOMAN, 1960). The results prove that specific cold receptors exist but do not allow the properties of thermoreceptor populations to be correlated quantitatively with the data on temperature sensation. If thermoreceptors in man are comparable with those in homeothermic animals, there would be a good agreement between thermoreceptor activity and the facts of human temperature sensation. At skin temperatures near 33° C cold and warm receptors are continuously active (Fig. 57) but no conscious temperature sensation is felt. The latter begins only when a relatively high number of thermal impulses per unit of time reaches the central nervous system. Thus the threshold of thermal sensations is correlated with an integrative central process rather than with the activity of single peripheral receptors. This *"central threshold"* may possibly be expressed as the magnitude of a slow cortical potential (P_T), which, in turn, is dependent on the average impulse frequency (v) and the number (n) of simultaneously active receptors. Since the frequency of single temperature fibers is a function of absolute temperature (T) and rate of change (dT/dt), while the number of active units depends on the surface area (F), the condition for conscious temperature sensations (E_T) can be expressed as follows

$$E_T \leftrightarrow P_T \leftrightarrow f(n, v) \leftrightarrow \varphi(T, dT/dt, F) \, .$$

This concept is consistent with the observations that in human subjects the warm thresholds fluctuate statistically and the relative number of positive responses is a probability function of the temperature stimulus (EIJKMAN and VENDRIK, 1963; VENDRIK and EIJKMAN, 1968; VENDRIK, 1970). This suggests that, according to the "detection theory", the sensory threshold for warmth is dependent on the signal-to-noise ratio of nervous activity. The sub-threshold static discharge of cutaneous thermoreceptors can be considered as "internal noise" of the system. JÄRVILEHTO (1973) has found that the central threshold for stimulation of a single cold spot in human subjects corresponded to a frequency of 80 sec^{-1} in the single fiber serving the cold receptor.

The presence of separate cold and warm spots in human skin and the electrophysiological findings of specific cutaneous cold and warm receptors support the theory that the sensory qualities of "cold" and "warmth" can be ascribed to a *dual* set of receptors. The assumption that only one cold-sensitive receptor system accounts for all facts of human temperature sensation would be incompatible with several experimental findings (DARIAN-SMITH, JOHNSON, and DYKES, 1973; JOHNSON, DARIAN-SMITH, and LAMOTTE, 1973; ZENZ et al., 1973).

2. Thermal Comfort and Temperature Regulation

Afferent impulses from cutaneous thermoreceptors are not only a source of conscious information but also a very important input for the system of *temperature regulation* (THAUER, 1958; HARDY, 1961; HENSEL, 1966; HAMMEL, 1968). They can signal thermal disturbances from the skin before the central temperature of the body has been influenced. This function of cutaneous receptors is favored by the fact that they have a high dynamic sensitivity and thus respond in particular to rapid thermal disturbances.

The emotional feeling of *thermal comfort* and *discomfort* (p. 594) does not reflect the activity of peripheral thermoreceptors alone but rather an *integrated* state of the thermoregulatory system. The interaction of peripheral and central factors as the basis for thermal comfort has been investigated by CHATONNET and CABANAC (1965), GAGGE, STOLWIJK, and HARDY (1967), HARDY (1970), GAGGE (1971) and HARDY, STOLWIJK, and GAGGE (1971). According to HARDY (1970), the physiological correlate of thermal comfort is a composite of (1) the signals evoking temperature sensations, (2) those for thermoregulation, and (3) those sensations arising from the thermoregulatory activities of vasomotor control of skin blood flow, sweating and shivering (Fig. 63). In the steady state the physiological conditions of thermal comfort or thermal neutrality can be described as follows: 1. Internal body temperature 36.6—37.1° C, i.e. generally on the warm side of the set point (36.6° C) for the internal body temperature. 2. Mean skin temperature 33—34.5° C for men and 32.5—35° C for women. 3. Local skin temperature variable over the body but generally between 32 and 35.5° C. 4. Temperature regulation active and completely accomplished by vasomotor control of blood flow to the skin — no sweating or shivering present. Blood flow to hands and feet high, producing reversed thermal gradients along the extremities. (Climatic conditions of thermal comfort see p. 551.)

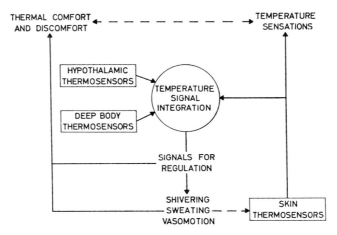

Fig. 63. Physiological correlates of thermal comfort. Adapted from HARDY, STOL-WIJK, and GAGGE (1971)

However, under certain conditions, deviations from this general scheme are possible. For example, the distal parts of the extremities play a dominant role for thermal comfort, irrespective of their weight for mean skin temperature. Strong peripheral vasoconstriction can be associated with cold discomfort, even if temperature regulation is balanced. On the other hand, the body can lose heat without cold discomfort when the temperature topography of the extremities is artificially kept at a certain pattern (THAUER, 1953). Another example is the effect of local drafts on thermal comfort (NEVINS, 1971). Furthermore, during exposure to thermal transients, cold discomfort may disappear in spite of hypothermia of the body and peripheral cold sensation (GAGGE, STOLWIJK, and HARDY, 1967).

3. Behavioral Responses in Animals

Thermal Preferenda, Selected Temperatures. If small mammals, such as insectivores, rodents, and bats, are placed in a temperature gradient they usually select a specific thermal preferendum where they come to rest with every sign of "thermal comfort" (BACCINO, 1935; BODENHEIMER, 1941; HERTER, 1952; STINSON and FISHER, 1953; HART, 1971). If the gradient is generated in the floor of the experimental cage, the animals usually orientate themselves according to the ground temperature and not by the air temperature. The preferred temperatures seem to be essentially species- or race- specific characteristics. Sometimes they also depend to a certain extent on the ecoclimate of the animal's site of origin. The selected temperatures of *newborn* rodents are many degrees higher than those of adult animals; accordingly the female selects a site for giving birth which corresponds to the preferred temperatre of the young and brings the young from other temperature ranges to this spot (HERTER, 1952).

A peculiar behavior is seen in *birds* of the family *Megapodidae*, such as the Australian malleebird *(Lipoa)*. They use the thermal sensitivity of their face and mouth to control the surrounding temperature of eggs during hatching. The eggs are laid into a mound where heat is generated by fermentation of rotting vegetation and by irradiation from the sun. For long periods of time the male is busy covering and uncovering the eggs, thereby keeping the temperature constant at $34 \pm 1°$ C for two months (FRITH, 1957).

Only very tentative conclusions can be drawn from preferred temperatures about the presence of and function of a *temperature sense*. On this point PROSSER (1954) remarks "Temperature selections in an ecological gradient need not necessarily involve specific thermoreceptors". In some cases the participation of pain receptors cannot be excluded, as in the experiments of STINSON and FISHER (1953) in a gradient with ground temperatures ranging from 6—50° C. Furthermore we know from human physiology that thermal comfort does not depend on the sense of temperature alone, but also on the integrated state of the total thermoregulatory system (p. 613). Finally we must remember that the stimulating effect of temperature on the thermal receptors varies, depending on the condition of the skin and pelt. Consequently behavior patterns allow only uncertain conclusions to be drawn about the functions of the temperature sense.

Local Thermal Conditioning. Behavioral measurements of the thresholds for thermal stimulation of the face have revealed a temperature sensitivity comparable with that of the human forearm (KENSHALO, DUNCAN, and WEYMARK, 1967; BREARLEY and KENSHALO, 1970). When the thresholds (ΔT) were plotted as a function of adapting temperature (T), and the larger surface area in human experiments was taken into account, the curves for cat and man were similar (Fig. 64). In contrast to the face, other cutaneous regions, such as the back and thigh, were less sensitive to cooling and highly insensitive to warming, the cats responding only to noxious heat (KENSHALO, 1964; KENSHALO, DUNCAN, and WEYMARK, 1967).

It should be emphasized that *thermoregulatory reflexes* from cutaneous cold and warm receptors may be elicited even when no conscious sensations or behavioral responses are present, e.g., during sleep. Warming the leg of unanesthetized cats in water of 40° C causes a marked reflex vasodilatation in the ear. The peripheral origin of this reflex is proved by the fact that the central body temperature, as measured in the hypothalamus, drops at the same time (KUNDT, BRÜCK, and HENSEL, 1956). However, as shown by KENSHALO (1964), warming the cat's leg to 40° C does not lead to any behavioral response. Thus the behavioral threshold for thermal stimuli seems to be considerably higher than the threshold for thermoregulatory reflexes.

Thermoregulatory Motivation. The interaction of peripheral and central thermo-sensitive structures involved in thermal comfort or motivation can be demon-strated by operant conditioning methods in animals. (Integration of peripheral

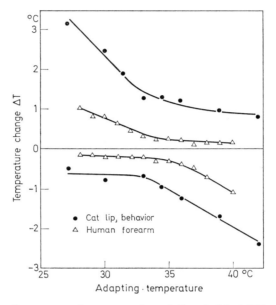

Fig. 64. Behaviorally measured warm and cool thresholds (ΔT) of the cat's upper lip (1.7 cm²) and thermal thresholds on human forearm (14.4 cm²) as a function of adapting temperature (T). (Cat data from BREARLEY and KENSHALO, 1970; human data from KENSHALO, 1969)

and central processes in thermoregulation, see also p. 571.) Mice (REVUSKI, 1966), rats (CARLTON and MARKS, 1958; WEISS and LATIES, 1961; LIPTON, 1968; CORBIT, 1970; MURGATROYD and HARDY, 1970), cats (WEISS, LATIES, and WEISS, 1967), pigs (BALDWIN and INGRAM, 1967), and monkeys (CARLISLE, 1966a; ADAIR and STITT, 1971; MYERS, 1971) can be trained to regulate their body temperature behaviorally in cold and warm environments by pressing a lever for warm and cold air, respectively, or by switching a heat lamp on and off. When the local temperature of the hypothalamus is artificially kept at low levels, the animal will increase the external heating rates, while an increased hypothalamic temperature will lead to a higher cooling rate (SATINOFF, 1964; CARLISLE, 1966b; BALDWIN and INGRAM, 1967; MURGATROYD and HARDY, 1970; CORBIT, 1970). Similar results have been found in squirrel monkeys when the midbrain temperature was dis-placed (ADAIR and STITT, 1971). Some of these experiments suggest a multi-plicative integration of peripheral and central factors (cf. p. 572).

VIII. Long-Term Thermal Adaptation

In addition to the fast regulatory processes taking place within seconds, minutes or hours, which were dealt with earlier, homeothermic organisms are able to carry out long-term adjustments to changing climatic conditions. These processes taking place over periods of days, months, perhaps years, are called long-term adaptations.

Adaptation has been defined as the modification of organisms, including their regulatory processes, taking place during *prolonged* or *repeated* action of an environmental stimulus ("stressor", SELYE, 1950; ADOLPH, 1956, 1964). Physiological adaptation is part of the ability of the organism to maintain homeostasis or stability (ASHBY, 1960) under more extreme environmental conditions. Adaptation may thus enhance the ability to survive and it appears to be favorable for the well-being of the organism. It must be admitted here, however, that it may be difficult in many cases to decide whether an adaptive change is favorable or not. A general outline of the possibilities of adaptative modifications is given in Table 9.

Table 9. Possible adaptive modifications

Morphologic modifications	Functional modifications	
e.g. growth of fur	Change in capacity of effector systems (e.g. increased cold-induced heat production)	Alteration in the regulatory system (e.g. shift in "set point")

From BRÜCK (1969).

Under natural conditions man and animals are confronted by complex environmental situations which must be described in terms of a number of parameters such as temperature, barometric pressure, humidity, solar radiation etc.

To distinguish the adjustment to such complex climatic adaptation from adaptation to a single factor the term acclimatization has been frequently used; there is, however, a large degree of confusion about terminology, in spite of considerable efforts to clarify the situation (see FOLK, who gives an historical account of this terminology; see also Fed. Proc. **22**, 1963, where the problems of terminology are discussed by a number of scientists). We will therefore speak of adaption or acclimatization but avoid the term acclimation which may be easily replaced by "single factor adaptation" (cf. Glossary of the I.U.P.S. Thermal Physiology Commission).

To approach *climatic adaptation* from an experimental point of view, separation of the several factors composing climate is required. In the context of this book, we are mostly interested in the modifications of organisms occurring in response to changes in ambient temperature, i.e. in *thermal adaptation*.

ADOLPH (1956) has shown that the adaptive modifications occurring during prolonged cold exposure are typical for cold and quite different from those modifications taking place when the organism is exposed to a different stressor such as lack of oxygen, water deprivation and so on. ADOLPH (1956) has thus emphasized the *stressor-specificity* of physiological adaptation, whereas SELYE (1950) paid particular attention to certain *similarities* and *common factors* in adaptation; thus, his investigations led to the concept of "unspecific adaptation". Studying thermal adaptation will lead us to both of these phenomena, *stressor-specific* and *stressor-unspecific* adaptation.

Moreover, there are modifications which may be beneficial under extreme temperature conditions, although they are not the result of stimulation by cold but due to another "stressor". To give an example, cold-resistance is increased in animals and people whose physical fitness was increased through exercise (HEBERLING and ADAMS, 1961; STRØMME and HAMMEI, 1967). This phenomenon is called "*positive cross adaptation*". There are, on the other hand, examples of "*negative cross adaptation*". Thus adaptation to low levels of oxygen in the inspired air, as encountered at high altitudes, has been shown to decrease cold resistance (ADOLPH, 1956; FREGLY, 1954; HARRIS, MEFFERD, and RESTIVO, 1960).

Adaptive modifications in the organism and in its functions may occur in the span of one individual life, or only in the course of several generations; this latter phenomenon is called *genetic adaptation*. Here, however, we will be concerned mainly with *non-genetic adaptation* to changes in environmental temperature.

If a stressor acts intermittently the adaptive modifications may be expressed as a function of intensity, duration, and frequency of exposure, $A = f(I, t, n)$ (ADOLPH, 1956). This equation is valid only within certain limits. If a certain intensity of stressor action is exceeded the formation of *specific adaptation* is impeded. Rather the organism is pushed into a dangerous situation which has been called an alarm reaction by SELYE (1950), and which is identical with what was previously called an "emergency reaction" by CANNON (1928). Under these conditions, at best, some unspecific adaptation may occur (SELYE, 1950).

The time course of the development of the various adaptive modifications may be different. Further, the stressor may exert its influence on the organism in different ways, i.e. there may be:

1. a single-step change in environmental temperature which is eventually kept constant
2. a steadily increasing stimulation
3. an intermittent stimulation.

It follows from this that a large number of different combinations of adaptive changes must exist. A few examples of these are shown in Fig. 65.

Further, adaptive modifications will be different depending on whether the whole body or only parts of it are exposed to the adverse environment.

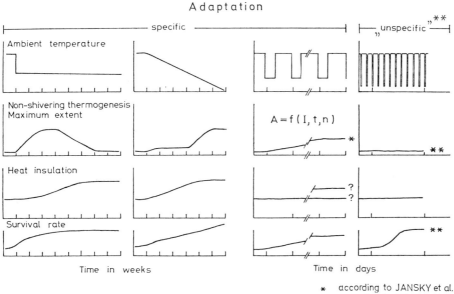

Fig. 65. Effect of mode of adaptation (intensity, duration, and frequency of stress) on the development of cold adaptation in small animals which are capable of nonshivering thermogenesis. Only a few examples have been selected from a large number of possible cases. From Brück (1969)

Besides the adaptive modifications which are the result of a response to one or several stressors there are structural and functional alterations to be found which are based on an endogenous cycling factor. These "circannian" periodic events (molt, shedding of hair, hair growth) may be synchronized to the seasons by an external stimulus such as duration of daylight which would have the function of a timing agent (synchronizer, *Zeitgeber*). Such phenomena characterized by endogenous cycling processes which are *triggered* by an external factor have been referred to as "seasonal modifications" (Hart, 1964) and these play a very important role for the survival of organisms in a changing environment. They are substantially different from adaptive processes in that they are not *reactions* of the organism to a stressor; rather, the seasonal modifications have been shown to occur even in artificial environments (*zeitgeberfreie* environments) without any stimulus being present. Readers interested in this field are referred to the excellent review by Hart (1964). Here, we will give only a brief account of this phenomenon in the appropriate paragraphs.

One phenomenon involved in many adaptive processes has been called *habituation*; this may be described as follows: A sensory stimulus evokes — besides the stressor-specific reaction (sweat secretion, cold-induced heat production) — a response which is not typical for the stimulus, such as an increase in blood pressure, heart rate, etc. If the sensory stimulus is repeatedly applied at intervals of minutes, hours or days, then the response to this stimulus gradually diminishes or disappears (GLASER, WHITTOW, 1957; GLASER and GRIFFIN, 1962; GLASER, 1966).

Thus, habituation may be spoken of if a stressor-unspecific response vanishes during repeated stimulation (cf. BRÜCK, 1972b). The mechanism underlying habituation has been intensively studied in the rat (GLASER and GRIFFIN, 1962; GLASER, 1966). The tails of rats were submerged in water of 4°C and the increase of heart rate was observed. This response was shown to diminish considerably when the procedure was frequently repeated within a few days. At first, it was assumed that this change was due to a modification in the function of the peripheral receptors, but later it could be demonstrated that habituation was impeded when certain lesions were made in the frontal area of the brain. Thus is was concluded that habituation is a process mediated by changes within the central nervous system.

The reduced heart rate in the example of habituation described in this paragraph is a result of a diminished response of the sympathetic system; with respect to this reduced sympathetic reaction, habituation appears to be closely related to what has been described as "unspecific adaptation" a few paragraphs below (p. 624). The significance of habituation for cold adaptation in man, particularly local cold adaptation, will be discussed later.

Behavioral Adaptation. Behavior extends the resources of the organism beyond the range of its internal physiological reactions. Animals may achieve behavioral adaptation by avoiding stressors, seeking preferenda and shelters, building artificial protection, by changes in posture and motor activity, changes in food and water selection, and by a wide variety of other behavioral patterns.

The adaptation of man to various environments is largely a matter of "culture" and behavior, including artificial and technical means, by which he can exist under environmental conditions that no other organism would endure. On the other hand, his capacity for physiological adaptation is rather limited. In many cases, where physiological adaptations were assumed to occur, they turned out to be merely behavioral. For instance, the cold adjustment of Eskimos consists mainly in avoiding the cold. The possibilities of adaptive behavioral response are innumerable and they cannot be discussed here extensively. There is no sharp limit between physiological and behavioral adaptation. Both mechanisms act together reciprocally.

A. Adaptation to Cold

1. In Animals

Changes in Heat Insulation. The most striking and at the same time the most effective protection against cold is the formation of a thick cover of hair or feathers or of a thick layer of subcutaneous fat. According to SCHOEPFER (1937) the relative capacity for heat flow reduction of the summer- and winter-coats of squirrels is expressed by the ratio 1 : 1.2. For the dog, the ratio is 1 : 1.5, and for the wild hare 1 : 1.7. Far greater differences are probably found among the large polar animals, whose resistance to cold is based primarily on the heat insulation of their body covering, and not on increased heat production. Winter fur growth is partly an effect of a circannian periodicity (see p. 619) and is most probably controlled by the duration of daylight (HART, 1964). Thus, HÉROUX, DEPOCAS, and HART (1959) showed that in white rats kept outdoors during the winter, pelage insulation increased, but it remained unchanged in controls kept in a constant temperature room maintained at 6° C. However, other functions such as metabolism did change under cold-room conditions.

Under natural conditions a relation of fur growth to day length is very useful, since it allows the animal to initiate fur growth, which requires a certain amount of time, some time before the onset of winter cold ("preadaptation").

Now, the question arises whether cold per se is able to stimulate hair growth. This question can be answered positively on the basis of several experimental studies. Thus, increased pelage insulation has been evoked by cold exposure in the mouse (BARNETT, 1959; HARRISON, 1959, 1963), in the young pig (WEAVER and INGRAM, 1969), and in cats; the latter were exposed to the cold (+5°C) for several months; after this period the mean weight of the hairs of a shaved area of 100 cm² was found to be 48% [$P < 0.001$] more than that of a control group which was kept warm (HENSEL and SCHÖNER, personal communication). Similar results have been obtained in mice (HARRISON, 1963). Further, in cocks exposed to cold for 15 months the feather weight increased significantly (FISCHER et al., quoted from CHAFFEE and ROBERTS, 1971). Obviously severe cold exposure, lasting for at least several months, is needed for a substantial stimulation of hair or feather growth. It is perhaps for this reason that some authors failed to demonstrate this effect of cold exposure.

Peripheral Blood Flow. Exposure to cold produces a less intense restriction of blood flow through the extremities of cold-adapted animals than through those of warm-adapted ones. HÉROUX and ST. PIERRE (1957) were able to demonstrate an increase in the number of capillaries in the skin of rats that had been exposed to cold for lengthy periods. This adaptive change lessens the danger of local cold damage to the tissues. However, as HÉROUX (1961) showed, a more intense exposure to cold was required to produce such changes in the blood vessels than to produce the changes in metabolism which will be discussed later. He found a metabolic cold adaptation in rats held at +6° C for only a few hours per day, but found no increase in vascularization and no thickening of the epidermis on the ear.

Heat Production. Fig. 66 gives a general view of the possibilities for cold-adaptive changes of energy metabolism, in the form of metabolic curves of several mammals. On closer examination of these curves, three factors of metabolic modification can be distinguished in varying degrees, namely

1. increases in basal metabolism
2. an increase in the slope of the rise in metabolism with falling environmental temperature, and
3. an elevated peak metabolism.

In older works the elevated basal metabolism was considered first and almost exclusively. Increases in basal metabolism as adaptations to cold were demonstrated in mammals and birds, e.g. in the rat (GELINEO, 1934; SCHWABE et al., 1938; RING, 1939; HART, 1950; SELLERS and YOU, 1950; SELLERS et al., 1951), the rabbit (RING, 1939; LEE, 1942), the golden hamster (ADOLPH and LAWROW, 1951), the dog (GELINEO, 1954), the pigeon (RAUTENBERG, 1969) and others. The increases in basal metabolism could go as high as 50% (GELINEO, 1934; and others).

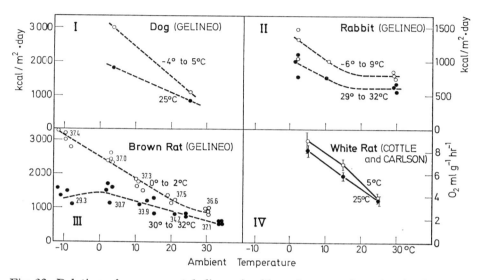

Fig. 66. Relation of energy metabolism of cold- and warm-adapted animals to environmental temperature. The numbers written on the curves give the adaptation temperature; the values given in the circles (for the brown rat) indicate rectal temperatures. Cold-adapted animals reach higher energy metabolism levels under cold stress and their rectal temperatures fall less steeply. From a compilation by HART (1957)

The lower limit of the zone of thermal neutrality (t_2 in Fig. 15, p. 633) can be displaced downwards by 2—4° C through increases in basal metabolism (GELINEO, 1953). Of course this increase, as well as increased hair or feather covers, means that a greater heat load will exist in a warm environment. Therefore we are dealing here with a type of adaptation which may be preferentially advantageous under conditions of constant cold (apart from the effect of the coat in protecting against direct solar irradiation; cf. FOLK, 1966; see also p. 556).

Increases in the *increment of heat production* do not carry this disadvantage with them. A steeper slope of heat production enables the internal temperature to remain stable in spite of reduced vasoconstrictor action. By this means the reduction in peripheral vasoconstriction described above can be compensated for metabolically. Finally, increases in peak metabolism broaden the range of regulation, i.e. lead to a displacement of t_2 (see Fig. 15) to lower temperatures. Such a displacement is the most obvious expression of "adaptation to new environmental conditions".

Mechanism of Cold-induced Increase in Heat Production. A decisive step in the investigation of metabolic cold adaptation was made when COTTLE and CARLSON

(1956), HÉROUX, HART and DEPOCAS (1956), and HSIEH, CARLSON, and GRAY (1957) succeeded in demonstrating the development of a new mechanism of cold-induced heat production in the course of cold adaptation. In their now classical studies in the rat they showed that during prolonged cold exposure the mechanical activity of the skeletal musculature, which was monitored by electromyography, is gradually reduced, while, at the same time, the increment of heat production and the summit metabolism increase considerably (Fig. 67). Thus they had demonstrated, for the first time, the development of nonshivering thermogenesis (NST). At first it was thought that NST replaces shivering, but later it could be shown that shivering can still be evoked when the cold stimulus is strong enough. Thus, these animals have two mechanisms of heat production at their disposal.

The ability of the organism to actuate the more economical NST first before shivering occurs has already been described in Section IV.E, p. 543. As described in Section IV.E, p. 540, brown adipose tissue is an important site of NST. As first

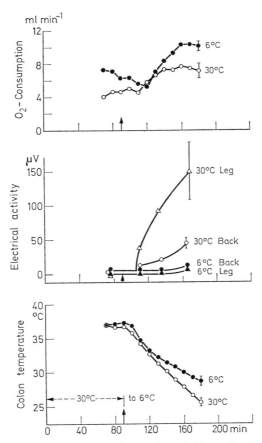

Fig. 67. Mean oxygen uptake, muscle electrical activity and colonic temperature of anesthetized warm- ("30° C") and cold-adapted ("6°C") rats. Note that large increases in oxygen uptake in cold-adapted rats are not accompanied by electrical activity (non-shivering thermogenesis). From HÉROUX, HART, and DEPOCAS (1956)

shown by SMITH (1961), SMITH and HOYER (1962), SMITH and ROBERTS (1964)
(for further literature see: SMITH and HORWITZ, 1969) the development of NST
is accompanied by an increase in the mass of brown adipose tissue, and there is a
striking quantitative relationship between the amount of NST and the percentage
mass of brown adipose tissue (Reviews: BRÜCK, 1970; CHAFFEE and ROBERTS,
1971). NST does not take more than a few weeks to develop to its maximum level
when the animal is exposed to fairly sharp cold; this is at about 5° C for the rat.
With intermittent cold exposure (cf. Fig. 68) the amount of NST which is evocable within three weeks is 90% and 60% of the maximum, when the daily exposure
amounts to 12 and 6 h, respectively (JANSKÝ, BARTUNKOWA, and ZEISBERGER,
1967).

In animals protected by a heavy coat of fur it may be difficult or impossible to
evoke NST under laboratory conditions. Thus, JANSKÝ et al. (1969) demonstrated
a slight amount of NST (about 10% of BMR) in rabbits which had been exposed
to 1° C for a prolonged period of time. In shaved animals exposed to +5° C,
however, they demonstrated NST amounting to 50—60% of BMR. Assuming
that we were to expose an unshaved rabbit to a temperature decreasing steadily to
say —60° C, we may anticipate that adaptation would develop as sketched in
Fig. 65, example 2: Thermal insulation would increase while NST remained
unchanged during the first part of adaptation; only when temperature had come
down to extremely low values would NST develop in addition.

The problem of metabolic cold adaptation appeared to be essentially solved with
the demonstration of NST. As shown in Section IV. D and Fig. 21, NST is confined, however, to organisms of small size, i.e. to smaller species and to the neonates
of somewhat larger species. Moreover, animals have been found which do not possess NST, even though small and young; thus, in the newborn and the cold-
adapted young miniature pig cold-induced heat production has been shown to be
strictly related to shivering (BRÜCK, WÜNNENBERG, and ZEISBERGER, 1969).
Furthermore, the whole class of birds does not appear capable of nonshivering
thermogenesis (CHAFFEE and ROBERTS, 1971). Here, the question arises, whether
such groups are able to enhance the capacity of cold-induced heat production by
increased shivering activity. In the young miniature pig such increased shivering
activity has been observed in the course of a 4—6 week cold exposure (12 h daily
at 0° C) (BRÜCK, WÜNNENBERG, and ZEISBERGER, 1969); but no other evidence
for such a mechanism is to be found in the literature (cf. CHAFFEE and ROBERTS,
1971).

"Unspecific" Cold Adaptation. LE BLANC et al. (1967) exposed rats repeatedly to
an extremely cold environment (—15° C), but only for very short periods — 10 min
— just bearable for the unadapted animals. After 18 exposures applied within two
days (i.e. 18 times 10 min = 3 h total exposure time) the rats showed significantly
increased cold-tolerance as measured in terms of survival time at environmental
temperatures of —5° C to —20° C. There was no indication of increased peak
metabolism nor could any development of NST be demonstrated, and there was,
of course, no indication of any growth of fur. It is for this lack of any modification
known to be specific for cold stress that this type of adaptation has been considered by LeBLANC (1969) "unspecific" adaptation (Fig. 65, Type 4). The only
modification to be demonstrated was the ability to maintain increased heat pro-

duction (shivering) for a prolonged period of time. Moreover, it could be shown that these animals had at the same time acquired increased tolerance to hypoxia i.e. there was no negative cross-adaptation between cold and hypoxia as described for specific cold adaptation. The characteristic feature of unspecific adaptation may be seen in a diminished adrenomedullary and sympathetic responsiveness (LeBlanc, 1969), that is to say in a "diminished alarm reaction" in Selye's (1950) terminology. This type of adaptation has very great practical implications, since it may provide a model for the effects of physical therapy which is thought to increase resistance to many environmental factors including noxious stimuli (cf. Brück, 1969, 1972b).

Seasonal Modifications in Metabolic Rate. As described above under natural conditions fur grows some time before the winter arrives. It is thus to be assumed that, as the increased thermal insulation protects the animal from cold stress, metabolic adaptations may be found to be scarce or even missing.

In fact, in many species the increment in heat production when related to ambient temperature has been found to be smaller in winter than in summer; at the same time, the lower limit of the neutral temperature range (t_3 in Fig. 15, p. 533) is found to be shifted to lower values, as one would expect. This behavior would correspond to the end phase of example 1 in Fig. 65: Due to the increased thermal insulation metabolic rate has come back to the initial value.

Divergent results have been obtained for the behavior of basal metabolic rate. In previous papers *BMR* has been reported to be increased in winter while Chaffee and Roberts (1971) state, on the basis of more recent studies, that little or no difference is to be found in *BMR* of animals and birds studied in summer and winter.

Shift of Threshold Temperature for Thermal Defense Reactions. Only recently another type of thermal adaptation, namely a shift of threshold temperatures for the elicitation of thermal defense reactions, has been emphasized (Brück and Wünnenberg, 1967; Brück et al., 1970). A number of results are to be found scattered in the literature which allow the existence of such a mechanism to be anticipated. Figure 68 demonstrates the shift of shivering threshold as found in young guinea pigs which had been exposed to $+5°$ C for $4-6$ weeks.

The biological significance of this shift may be seen in the fact that it facilitates the suppression of shivering, which is brought about through the transport of heat from the brown interscapular adipose tissue to the cervical spinal cord area which controls shivering — as described in detail on p. 543. Moreover, it may be considered advantageous, from an energetic point of view, to maintain body temperature at a somewhat lower level, as this would decrease the metabolic cost of temperature regulation; assuming as an example that such a cold-adapted guinea pig were exposed to an environmental temperature of $10°$ C it would save an amount of energy corresponding to 11% of *BMR* as compared to an animal without such a shift in shivering threshold (Brück and Wünnenberg, 1967). Shivering threshold shift has been shown to be accompanied by a corresponding shift of threshold for heat polypnea (Brück et al., 1970). In a warm environment this modification would be disadvantageous, since the adapted animals would have to actuate and maintain respiratory heat dissipation at a lower level of body temperature than the non-cold-adapted animals.

Under certain conditions, i.e. if the animals are exposed only intermittently to a
cool environment (12 h at +3° C, 12 h at a warm environment) heat dissipation
threshold does not shift down along with the shivering threshold. In other words,
there exists a mode of adaptation which is characterized by a "widening" of that
body temperature range at which neither thermoregulatory heat production nor
respiratory heat dissipation can be elicited.

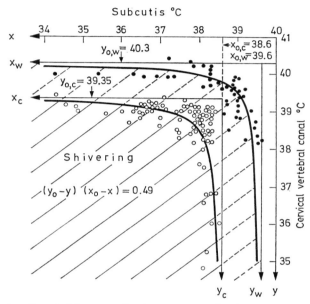

Fig. 68. Shivering threshold curves for two groups of guinea pigs (aged 4—8 weeks)
reared at different environmental temperatures. The values were obtained by independent changes of the body-surface temperature and the temperature in the cervical
vertebral canal. The asymptotes correspond to the respective reference temperature.
This diagram shows, for instance, that at a certain body-surface temperature, which
corresponds to a subcutaneous temperature of 37° C, shivering begins in warm-
adapted animals (solid dots) if the hypothalamic temperature drops below 40° C.
In the cold-adapted animals (circles), however, shivering does not occur until hypothalamic temperature has reached a value slightly below 39° C. Consequently, the set
point of the temperature regulatory system is shifted to a lower level in the cold-
adapted group. From BRÜCK and WÜNNENBERG (1970)

Hence, when these animals are exposed to an environment with fluctuating temperatures (such as day-night alterations) they can make use of the heat storing
capacity of their bodies to a much greater extent than non-adapted animals, i.e.
use it as a heat buffer. Thus the "widening of the interthreshold zone" provides
the basis for a more economical temperature regulation (BRÜCK et al., 1970).
The behavior in the intermittently cold-adapted guinea pigs resembles that of the
camel as previously described by SCHMIDT-NIELSEN et al. (1957). They have shown
that the camel's body temperature fluctuates considerably in summer, when the
animals are deprived of water. Under these conditions body temperature may drop

to values as low as 35° C during the night, and increase to more than 40° C during the day before sweating occurs.

As for the underlying mechanism of the set point shift described above, the following possibilities may be discussed:

1. alteration in the function of peripheral cold receptors
2. functional alteration of internal thermosensors
3. change in some central nervous structures serving as "reference system" which determines the "set point" of the control system (see HAMMEL, 1965, 1968; ZEISBERGER, and BRÜCK, 1971a; this volume p. 586).

In studies in cats adapted to +5° C for several months, no indication of a functional change in facial cutaneous thermal receptors was obtained, although the facial skin temperature was about 15° C under cold room conditions compared with 35° C in the controls (HENSEL and SCHÖNER, unpublished, this book p. 605).

One may thus assume that it is a change according to 2. and/or 3. which is responsible for the shift in threshold temperature. Recently, it has been shown in the guinea pig that the set point can be shifted by microinjection of noradrenaline into a hypothalamic area which does not contain the thermosensors but possibly contains the reference units (ZEISBERGER and BRÜCK, 1971a; this book p. 582); these results would favor assumption 3.

Restraint Hypothermia. HÉROUX and HART (1945) described an additional cold adaptive modification. If a rat is prevented from moving about freely in the cold its rectal temperature sinks markedly. This effect, which is designated as restraint hypothermia, is abolished by cold adaptation.

Physiological Values which Remain Constant During Cold Adaptation. According to the experiments of ADOLPH and LAWROW (1950, 1951) on the rat and golden hamster, *the lethal temperature*, i.e. the body temperature which leads to death by hypothermia, remains unchanged over the course of cold adaptation. In addition the responses of respiratory frequency, of heart frequency, and of the rate of use of oxygen to falling body temperature were investigated in experimental hypothermia. The temperature characteristics of all these functions did not change measurably in the course of cold-adaptation.

On the other hand, cold-adapted dogs were reported to die at a lower rectal temperature (14.9° C) than that at which controls died (18.6° C) (COVINO and BEAVERS, 1957).

From this it may be concluded that cold adaptation involves only those processes and systems that are directly related to thermoregulatory processes. As soon as thermoregulation ceases in deep hypothermia, the differences between warm- and cold-adapted animals disappear. This is only true, however, for the entire animal.

By using appropriate methods, many differences can be found between the *isolated* tissues of warm- and cold-adapted animals. In recent decades the experimental results in this area have reached gigantic proportions. So far however they have not led to a decisive deepening of our understanding of cold adaptation on this basis.

2. Cold-Adaptation in Man

Since man is unable to develop a hair coat providing substantial heat insulation he is devoid of the most powerful physiological cold-adaptive modification (cf. p. 620). On the other hand, owing to his ability, as an intelligent being, to provide himself with appropriate housing and cooling, he is able to cope with so murderous a climate as that encountered in the winter of the polar zones. Heat insulation of his houses and his clothing is so excellent that the inhabitants of these polar zones, e.g. the Eskimos, are rarely subjected to severe cold stress. According to STEFANSON the Eskimo lives in a microclimate corresponding to that of Sicily. In other words, man seems to rely a great deal on what has been called "behavioral adaptation". It is thus not surprising that no gross physiological cold adaptive modifications were found in men who were living in an Antarctic camp for several months (EDHOLM and LEWIS, 1964), and that so experienced an investigator as DuBois stated in 1953: "The great adaptation to heat is well known but the adaptation of men to cold seems to be so small that its existence has been doubted". Nevertheless the layman would insist that man is able to "get accustomed" to a rough climate, and the physiologists in their turn did not cease to search for alterations which might put this general belief on a solid scientific basis. The question of where and how primitive man could survive before he had invented clothing, housing and fire, was another challenge to the investigation of the physiological properties of man which enabled him to withstand whole body exposure to cold environment.

Certainly, man could only have developed in a tropical climate; yet even the tropical rain forest and the tropical savanna climates do not provide environmental temperatures above the critical temperature (cf. Fig. 15, p. 533) of modern man at all times of the day or year (HAMMEL, 1964). Among the physiologists interested in human ecology HICKS and his colleagues (HICKS, MATTERS, and MITCHELL, 1931; HICKS and MATTERS, 1933; HICKS, 1964) were the first who succeeded in finding and investigating a group of primitive men, the Australian aborigines, shortly before they adopted civilization. These people, who did not wear any clothing, used to sleep in the open air, although the air temperature dropped to as low as $+4°$ C and the radiation temperature of the sky was about $20°$ C below the ambient temperature. The cooling power of this environment was moderately reduced in that they slept between small fires. This degree of cold stress does not appear to be acceptable to the European accustomed to sleeping in a warm microenvironment (quoted from HAMMEL, 1964; see also SCHOLANDER et al., 1958). The responses of the central Australian aborigines to cold exposure are compared with those found in white controls in Fig. 69. The natives do not increase heat production over the initial value whereas there is a cold-induced metabolic reaction of about $40—50\%$ in the white controls. In the natives, rectal temperature, mean body temperature, mean skin temperature, and particularly foot temperature drop much more than in the white controls. In plotting metabolic rate against mean body temperature (Fig. 70) it becomes obvious that, in the temperature range under consideration, the metabolic rate is positively correlated with temperature in the natives.

It must be inferred from this that the shivering threshold is shifted to a lower temperature level in the natives. In this respect, their behavior resembles that of the cold-adapted guinea pigs described on p. 625.

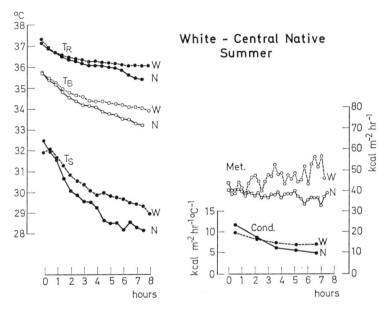

Fig. 69. Comparison of temperature changes and metabolic responses between control whites (W) and central Australian natives (N). T_R rectal temperature, T_B mean body temperature, T_S mean skin temperature. *Cond.* conductance. From HAMMEL et al. (1959)

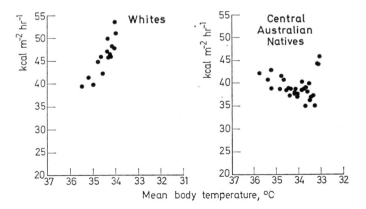

Fig. 70. Relationship between metabolic rate and mean body temperature in control whites and central Australian natives. Note increasing metabolic rate in the controls, falling metabolic rate in the natives with decreasing mean body temperature. From HAMMEL et al. (1959)

Conductance (Fig. 69) is smaller in the natives, or, to use the reciprocal value, heat insulation of the body shell is greater in the natives than in the white controls. This must be due to a reduced blood flow to the skin and appendages since the sub-cutaneous fat layer is smaller in the natives than in the white controls (HAMMEL, 1964).

In summary, the central Australian natives overcome the cold stress to which they are exposed during the night

1. by an increased functional heat insulation through extremely reduced peripheral blood flow;
2. by tolerating some drop in body temperature without reacting with a costly increased heat production and without shivering which would interfere with quiet sleep.

The most striking feature of this *"insulative-hypothermic"* adaptation (HAMMEL, 1964) is the ability to tolerate the low foot temperatures all night long. In the experience of the civilized inhabitant of the earth such low foot temperatures are accompanied by severe pain. Thus, "habituation" to painful stimuli (see p. 636) seems to play a major part in the cold tolerance of the central Australian aborigine.

Subsequently, a number of other ethnic groups experiencing intermittent or continuous cold stress have been studied (see review by HAMMEL, 1964).

The Bushmen of the Kalahari and nomadic Lapps were shown to behave similarly to the central Australian aborigines. The Alacaluf Indians (living in the West Patagonian channels), another lightly clothed ethnic group, did not behave like these three groups. They are accustomed to exposure to a very unpleasant climate with the mean maximum temperature in January being 13° C, and in July 7° C. The climate is further characterized by frequent rain falls mixed with sleet or snow. According to HAMMEL et al. (1960), (for literature see HAMMEL, 1964) their resting metabolic rate was about $50-60$ kcal/m^2 h, i.e. $25-50\%$ higher than the ordinary basal metabolic rate. This high resting metabolic level is sufficient to prevent rectal temperature from falling below 36° C and at the same time to keep body surface temperature within the range commonly found in civilized controls under basal conditions. An elevated metabolic rate was also found in Eskimos, although they are usually well clothed. An enlarged capacity for cold-induced heat production, which has been shown to be an important mechanism of cold-adaptation in animals (see p. 621), has not been demonstrated in man.

In sum, there are two principal types of adjustment to living lightly clothed in a moderately cold environment:

1. The hypothermic type in which some hypothermia, possibly accompanied by painful cooling of the extremities, is tolerated while metabolic expenditures, which would require increased *food* supply, are kept small.

2. A type in which the price of costly metabolic reactions is paid in return for a more comfortably warm body surface.

In an environment with fluctuating temperatures, as encountered in subtropical regions (hot days — warm nights), adjustments of type 1 appear to be very economical, since the heat lost during the night is easily made up for by solar radiation next morning. This type thus makes use, to the largest possible extent, of the heat storing capacity of the body. In a constantly cool environment this type of adaptation might be detrimental to the organism as it might easily remain in a hypothermic state even during the day; moreover, maintaining a reduced blood flow to the extremities would result in nutritional disturbances and finally lead to gangrene. Here, the second type seems to be much more suitable. On the other hand, the second type of adjustment would be disadvantageous in an environment with high day temperatures where the problem of dissipating heat would be rendered more difficult by the increased basal metabolic rate. Nor would this type be very advantageous in an environment in which scarcities in food supply are encountered.

By no means does a substantial increase in the capacity for cold-induced increase in heat production — so convincingly demonstrated in many animal species (see p. 623) — appear to be a characteristic feature of cold adaptation in man. In any case, shivering appears to be avoided as far as possible, even at the expense of some hypothermia and pain in the extremities. (Another way of avoiding shivering is the development of nonshivering thermogenesis, which is restricted, however, to small individuals, as shown above; see p. 539). The questions as to whether the two types of cold adjustment as described are to be thought of as the result of a process of individual adjustment to the stressor "cold", or if the physiological modifications found in these ethnic groups are characteristics of the aborigines as a race have not yet been fully answered; finally they might be induced by some factors other than cold. The *BMR* of Eskimos has been shown to decrease when they lived for a year on white man's diet. The higher resting metabolism of the Eskimos has thus been linked with the higher percentage of protein in their natural diet (RODAHL, 1952). It does not seem possible, however, to go the other way around, i.e. to increase *BMR* substantially in white men by giving them high protein food. Thus, it must be inferred that some inherited factor determines the *BMR* in the Eskimo.

Cold-adaptation in Civilized Man. Under conditions comparable to those described above for the studies in the ethnic groups living in their native biotopes, a group of Norwegian youths was studied during a 6-weeks stay, under primitive conditions, in the mountain plateau of south central Norway situated above the tree line (SCHOLANDER et al., 1958). They were requested to expose themselves as much as possible to cold wearing only light summer clothing. They were allowed to hike, hunt and fish, so they could relieve cold stress by exercise. The test studies were performed with the subject inside a light sleeping bag at an air temperature of about $+3°$ C (quoted from HAMMEL, 1964). The course of metabolic rate and body temperature in the men throughout the night is compared with that of a control group in Fig. 71.

The metabolic response of the cold-adapted group was a little higher than that in the control group. A striking difference existed in the behavior of foot temperature: it dropped considerably more in the controls than in the adapted group, whereas no significant change was to be seen in the rectal temperature. Subjectively, the higher skin temperature corresponded to a somewhat increased feeling of thermal comfort (cf. p. 613) in the adapted group, which was thus able to sleep for a while, whereas the controls did not. The type of adaptation found in this group of students may be referred to as metabolic adaptation according to the definition of HAMMEL (1964), although it is not exactly identical with any type of cold adjustment found in the other ethnic groups.

HEBERLING and ADAMS (1961) showed that the cold-induced vasoconstriction may be reduced, and metabolic response may be enhanced at the same time, by physical training. Thus, the functional modifications obtained in the Norwegian group during their stay in the mountains were ascribed to increased fitness rather than to cold. If this interpretation is correct one would have an example of positive cross adaptation between the stressors "cold" and "physical exercise" (ADAMS and HEBERLING, 1958; STRØMME and HAMMEL, 1967).

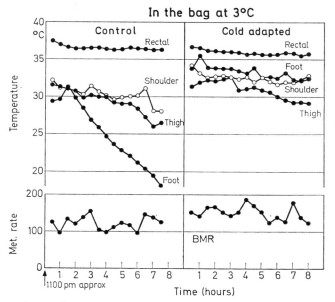

Fig. 71. Comparison of temperature changes and metabolic responses of a cold-adapted and a control student during a nightly cold exposure. From SCHOLANDER et al. (1958)

Another group of students was observed over a period of 4 weeks of which they spent 8 h daily in a climatic chamber adjusted at 8° C, wearing light clothing (DAVIS, 1961). As shown by Fig. 72, rectal temperature, heat production and electrical muscle activity (shivering) were all reduced at the end of the period of study, i.e. their behavior moved in the direction of that found in the central Australian aborigines: they had developed hypothermic adaptation to some degree.

These studies were originally designed to demonstrate the development of nonshivering thermogenesis in man. The results have been interpreted in this way by the authors, but this does not seem to be justified. Nevertheless, in a subsequent study (JOY, 1963), in which the noradrenaline test (cf. p. 539) was used to determine the amount of nonshivering thermogenesis in a group of students similarly treated as that of DAVIS (1961), oxygen uptake rose by 15—20% over BMR following the noradrenaline infusion; this is a small amount of NST when compared with that of newborn animals and infants, and with cold-adapted animals, in which *NST* amounts to 100—300% of *BMR* (see p. 538); moreover, to evoke this response in man doses of noradrenaline had to be applied which caused cardiac arrhythmias and extra systoles. Recently, a similar metabolic response (+17.5%) has been demonstrated in members of a Japanese ethnic group, the Ainus, living in a cold area. In this study noradrenaline had been administered subcutaneously, and no cardiac disturbances had been reported (ITOH, DOI, and KUROSHIMA, 1970).

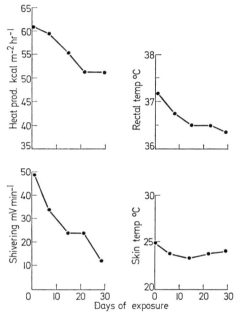

Fig. 72. Alterations in mean shivering heat production, mean skin temperature, and rectal temperature in subjects exposed 8 hours daily in a cold room. From DAVIS (1961)

Similar results were obtained by GÄRTNER (1970) who studied a group of male subjects under laboratory conditions. They were exposed 1 h daily to an air temperature of 10° C for 3 months. At the end of this period the subjects were exposed to a standard cold test consisting of exposure to an environmental temperature dropping from 32 to 15° C within 1 h. The cold-adapted group showed considerably less reduction in conductance and slightly less increase in muscle activity although metabolic rate was slightly higher. Mean body temperature tended to decrease more steeply and rectal temperature was slightly lower in the cold-adapted group at the end of the test study. On the whole, these results do not seem to be essentially different from those obtained by DAVIS (1961).

Recently it has been possible to demonstrate a downward shift of the shivering threshold in male Europeans who were placed, wearing only a bathing suit, in a climatic chamber, the temperature of which decreased from 28 to −5° C within one hour. In the first test, peak values for shivering and oxygen uptake were within the range of 100 −300% of BMR. With increasing numbers of cold tests the onset of shivering was retarded and took place at a decreased esophageal and mean skin temperature (BRÜCK, 1972a).

It appears from this description that artificial cold-adaptation results in modifications which resemble most closely the type of hypothermic adaptation found in the ethnic groups. These studies with artifical cold-adaptation can only be seen as a beginning. It may turn out that various types of adaptation can be elicited by choosing special adaptation conditions.

Furthermore, an adaptive modification in the direction towards a hypothermic type of adaptation has been demonstrated in a group of soldiers living for several months in an Arctic station (LEBLANC, 1956). Finally the Amas, the Korean and Japanese female pearl divers, must be mentioned here (HONG, 1963). They may dive several hours a day even though the water temperature may be as low as 10° C. Their shivering threshold has been shown to be markedly shifted to a lower temperature level as shown by Fig. 73.

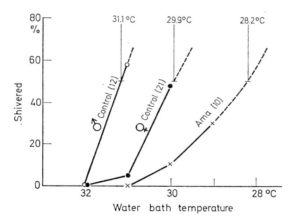

Fig. 73. Shivering related to water temperature in "Amas" (female pearl divers) and controls. Note shivering threshold considerably shifted to lower water temperatures in the Amas. From HONG (1963)

Adaptation to Local Cold Exposure. In everyday life, local cold exposure, particularly of the hands, which are difficult to protect appropriately from cooling without interfering with manual dexterity, is certainly more frequently encountered by civilized man than whole body cold exposure. Not only under arctic conditions, but even in the temperate climate zone, outdoor activities in winter may be severely hampered by numbness of the fingers, and finally by severe pain. Professionally cold-exposed men, such as fishermen, fish filleters, or farmers, who are used to working with bare hands in the cold are duly admired by indoor people for their ability to stand the cold exposure of the hands which is presumed to be very painful. Do they actually feel pain as much as presumed, or is pain diminished, be it by an increased blood flow to the hands, or by an alteration in peripheral perception? Or do they just not mind the pain? Answers to these and similar questions have been sought by a number of investigators studying various ethnic groups which were naturally cold exposed, special professional groups, and indoor people before and after local cold exposure applied under laboratory conditions. Less pain or no pain was experienced during severe exposure of hands or fingers to cold by all groups accustomed to exposing their hands to cold. The control subjects in these studies all reported severe pain, and some fainted (for literature see HAMMEL, 1964). There is also general agreement about the results of experiments in which hands or fingers were repeatedly exposed to the cold, under experimental conditions, usually by immersion in water of 0−4° C: pain being initially reported as severe or even "unbearable" diminished in the course of the daily exposures (GLASER and WHITTOW, 1957; GLASER, HALL, and WHITTOW, 1959; EAGAN, 1962; GLASER, 1966). This phenomenon ("habituation", cf. p. 620) has been found to be strictly stressor-specific, i.e. if the hand had been exposed repeatedly to the cold, exposure to hot water (47° C) was still painful, although it was possible to reduce heat pain by repeated heat exposure (GLASER, HALL, and WHITTOW, 1959); in these studies it has also been shown that the change in pain sensation was strictly confined to the exposed hand.

The influence of repeated cold exposure on manual dexterity was studied using a sensory discrimination test for the fingers (V-test by MACKWORTH, 1953, 1955). Striking improvement in dexterity of the repeatedly cold exposed hand was found. A diminished numbness was also found when men in their second year in the Antarctic were compared with new arrivals. This difference disappeared within the first 6 weeks of residence in the Antarctic (MASSEY, 1959, quoted from HAMPTON, 1969). The cold exposure of this group of men was that imposed by their work and life at the Antarctic base.

The observed differences in cold sensitivity and manual dexterity may be at least partly (cf. EAGAN, 1963) ascribed to differences in the blood flow to the extremities, which may be inferred from temperature measurements. When the hand is immersed in ice water, after a period of dropping, temperature begins to increase, indicating the occurrence of vasodilatation. This cold-induced vasodilatation (LEWIS' reaction; cf. p. 583) has been shown to occur sooner and at higher finger temperatures in naturally cold-adapted groups (Arctic Indians, ELSNER et al.; British fish filleters, NELMS and SOPER, 1962; Gaspé fishermen, LEBLANC, HILDES, and HÉROUX, 1960) as well as in people artificially adapted by daily immersions of the hands or fingers in ice water for periods of a few weeks (GLASER, HALL, and WHITTOW, 1959; GLASER and WHITTOW, 1957; EAGAN, 1963), Fig. 74. Remarkably, in a group of men experimentally starved for five days, hardly any cold-induced vasodilatation was to be seen (EAGAN, 1963); BMR was reduced in this group (Fig. 74); these and similar results suggest that finger temperature, and consequently local cold tolerance, depends on physical fitness. This may partly explain why one suffers more from cold extremities during the onset of a disease.

From the increased finger temperature observed during immersion in cold water it was concluded that sympathetic vascular tone is reduced during cold adaptation; this phenomenon is not restricted to the exposed extremity but can be seen in the contralateral extremity as well (Fig. 74b). Moreover, as repeatedly shown, immersion of one hand in cold water is followed by an increased heart rate and by an increase in systolic and diastolic blood pressure. This response is gradually diminished with repeated immersions (GLASER, 1966). Thus, diminished vasocon-

striction in the adapted cold-exposed hand is only part of an overall diminution
of the responsiveness of the sympathetic system (cf. "Habituation", p. 620).
Moreover, this adaptive modification could be prevented by frontal lesions of the
brain in rats and by certain drugs in man (GLASER and GRIFFIN, 1962). This
strongly suggests that the central nervous system is involved in the modification
of the cold-induced vascular responses which was previously thought of as a
vascular modification localized in, and confined to, the exposed extremity.

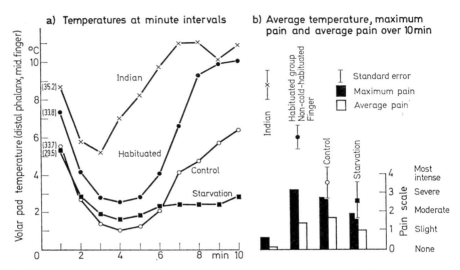

Fig. 74. a Average course of finger temperatures in four groups during immersion of
finger in ice water (for details see text). b Average finger temperature of 4 groups
after a 10-minute-immersion in ice water and record of pain sensation. In the "habituated" group the finger *contralateral to the recurrently immersed finger* was tested.
Note, even though 10-min-temperatures are higher in the habituated group there is
no difference in pain record, but much less pain, in the Indian group. From EAGAN
(1963)

In a series of studies EAGAN (1963) showed that, although temperature in the
finger recurrently immersed in ice water, did not differ from that measured in the
non-adapted contralateral finger when both were subjected simultaneously to a
test immersion, pain was markedly reduced only in the finger which had been
recurrently cold exposed. Thus, pain reduction might be thought of − in contrast
to the circulatory alterations − as a strictly local phenomenon, conceivably due to
a functional change in nociceptors.
This is an unproven proposition, however. It might well be that central modifications are linked with the reduction of pain.

B. Adaptation to Heat

Not only the coldest but also the warmest, and even the most warm and humid,
places on earth are inhabited by living organisms. With the aid of extremely vari-

able behavior patterns and physiological adaptations a large number of homeotherms have succeeded in settling and maintaining themselves in such extreme
circumstances. Examples may be found in the Chapters V, IX, and XII of this
part of the book.

In this chapter the question will be discussed of whether and how homeothermal
forms of life including man, normally living in moderate climates, react to intermittent heat loads by modifications in morphology or functions, thereby making
life under such increased heat loads easier or possible.

Adaptation of an individual man or animal to previously unaccustomed heat
exposure might take the following forms.

1. The fur- or feather-coat might be reduced.
2. The basal metabolic rate could be reduced.
3. The capacity and efficiency of evaporative mechanisms for heat loss (panting,
   sweating) could be increased.
4. A "hyperthermic" adaptation to intermittent heat loads, similar to the "hypothermic" cold-adaptation (see p. 630), i.e. a simple toleration of elevated body
   temperature, would be conceivable.

   Which of these possible mechanisms are actually found in animals and in man?

1. Heat Adaptations in Animals

Changes in Fur- and Feather Coats. Seasonal changes in the thickness of fur and
feather coats are generally known and have been thoroughly investigated in many
species. Shedding of the winter fur or feather covering is primarily a photoperiodic process, i.e. it is determined by the ratio of daylight to darkness. Here one
might ask if heat — acting as "stressor" — exerts a modifying influence. This is
generally assumed to be true, but only a few experimental investigations of the
question have been carried out. THWAITES (1967) for example reports that the
growth of fur in sheep was definitely reduced in the course of only two weeks of
heat stress. Distinct differences in the density of hair growth were also found in
pigs which had been reared from birth at either $+5°$ C or $35°$ C (WEAVER and
INGRAM, 1969).

In the authors' observations on guinea pigs which had been kept at $30°$ C for
several weeks, a considerable thinning of the pelt could be ascertained by simple
observation. However, some of this thinning bordered on a pathological loss of
hair, since it appeared in an irregular, partly patchy, pattern. It must be remembered that, depending on the habitat, an excessive reduction of the pelt may not
provide an advantage but may even be disadvantageous with respect to heat
tolerance. The pelt may offer a substantial degree of protection against the sun,
especially to large animals (e.g., ungulates; cf. MACFARLANE, 1964) which cannot
withdraw from solar irradiation by seeking shelter or shady places. Part of the
solar energy is reflected from the pelt surface, while another portion is absorbed in
the pelt and, even at very low wind speeds, is given off to the environment again
by convection. Furthermore, the animal's skin is protected against UV-irradiation,
which could lead to inflammatory reactions, and finally to malignant changes.
The best protection against UV- and heat radiation is provided by a short reflect-
ing coat which does not create too great a barrier to heat outflow from the animal

to its environment. Therefore a qualitative change in the pelt must also be considered as a possible form of heat adaptation (see MacFarlane, 1964). Nevertheless it should be mentioned here as an oddity that domesticated merino sheep tolerate desert heat well in spite of their retention of a thick wool coat. In winter the fleece protects against heat loss, since a gradient of 40 or 50° C is easily maintained between body temperature and the environment. The same gradient, reversed during summer, holds a 40° C differential between the hot surface and cool skin across the coat (MacFarlane, 1964). In this case, heat dissipation takes place predominantly via the bare extremities and the respiratory pathway.

Basal Metabolism. A marked reduction in basal metabolism during the course of heat adaptation has been demonstrated recently in the hamster (Cassuto and Chaffee, 1966; Cassuto and Amit, 1968). The basal metabolic rates of adult hamsters that had been heat adapted at 35° C for two to four weeks amounted to 0.77 ml $g^{-1} h^{-1}$ while in the controls kept at 23°C it amounted to 1.09 ml $g^{-1} h^{-1}$. The value for the controls is close to the predicted value given by Kleiber's formula (1961); the *BMR* of the heat-adapted animals is ca. 29% less than this predicted value.

By contrast, Gelineo (1935) found no unequivocal reduction of the *BMR* in rats maintained at 36° C. At first their *BMR* rose by about 35%; after about one month the initial value was reached again, but the rate did not go below this level even after seven months of exposure to 36° C. No clear support for a reduction in *BMR* could be found in birds (King and Farner, 1964). The question of whether basal metabolism is reduced in heat-adapted humans is still in dispute. At least, considerable changes in *BMR* during heat adaptation are not to be expected. Nevertheless, Collins and Weiner (1968) point out that even a slight reduction may have great biological significance. Even a 10% fall, which is not easy to measure, could enable an organism to tolerate an increase in environmental temperature of ca. 2° C without changing sweat secretion or vasodilatation. This would be a significant gain in "thermal comfort" in hot regions.

However, a lowered basal metabolism is a characteristic of a group of animals included among the estivators. The periodic reductions of basal metabolism in these cases are connected with dormancy or torpidity and therefore belong in a special category (see also Chapter XI). Since the hamster is a hibernator it appears that heat-adaptation based on reduction of basal metabolism is found only among the heterotherms, i.e. hibernators and estivators, and that such a reduction does not represent a general principle of heat adaptation.

Efficiency and Capacity of Evaporative Heat Loss. Outstanding examples of efficient evaporative heat losses are represented by the camel and the donkey, both of which, having the ability to sweat, can lose considerable amounts of heat by evaporation. The high capacity of this heat loss mechanism is made possible by the low NaCl-content of the sweat and by an ability to tolerate a high level of dehydration and increase in the osmolarity of the body fluids (cf. p. 723).

In these cases however we are clearly dealing with innate functional mechanisms. In man this mechanism is less efficient but can be most decidedly improved by heat adaptation. In non-sweating species, which give off evaporative heat by an increase in respiration, an elevated maximal frequency of breathing can be observed. Furthermore, the curve of respiratory frequency is displaced towards lower body temperatures, whereby hyperthermia can be opposed more effectively (BIANCA, 1961; FINDLAY, 1963).

Hyperthermia and Hypothermia. Toleration of higher body temperature is especially evident in the camel: In the dehydrated state, according to the investigations of SCHMIDT-NIELSEN et al. (1957), it does not activate evaporative heat loss until the body temperature reaches 40—41° C, while in well-hydrated conditions the body temperature is maintained at 39° C. The camel therefore tolerates a certain degree of hyperthermia added to intense dehydration. But the camel's high resistance to heat can be understood completely only if one takes into account the hypothermia which it tolerates during nocturnal cooling. According to SCHMIDT-NIELSEN et al. (1957), the body temperature at that time falls to 35° C, but no shivering occurs. This hypothermia represents — so to speak — a preparation for the heat load to be expected the following day. This form of temperature regulation, which might at first glance seem to be defective, is actually a significant survival factor in an environment marked by widely fluctuating temperatures. The donkey shows similar temperature relations, but to a lesser extent (see also p. 722: Antelopes). The question of whether these relations, which are probably genetically based in the animals mentioned in this paragraph, can be acquired in other species through repeated exposure to heat has not been adequately investigated.

Rearing newborn animals in warm surroundings may lead to distinctly recognizable changes in body form, which one would not really expect to encounter except over the course of generations. For example, the tails of mice which had been held since birth at 32° C were significantly longer than those of controls reared at 20° C (HARRISON, 1959). WEAVER and INGRAM (1969) found corresponding changes in the tail, lengthened extremities, and enlarged ears, in pigs that had been reared at 35° C. These changes, which correspond to the "rules" of ALLEN and BERGMANN (see p. 717) facilitate the output of heat in warm environments.

2. Heat Adaptation in Man

Heat acclimatization in man is very important and plays a significant role during seasonal climatic fluctuations, for life in tropical and desert climates, and for the adaptation of men who work under hot conditions. The most fundamental modification is to be seen in changes in the secretion of sweat and in the water and salt metabolism. Thus, if men are exposed to a constant heat load — be it by external heat exposure, or by exercise — for a prolonged period of time or undergo repeated exposures within a certain segment of time the sweat secretion rate, as measured under heat stress, rises from day to day and may finally amount to more than 1 kg h^{-1} °C^{-1} which is about three times the initial value (Fig. 75) (LADELL, 1951; WYNDHAM, 1951; Fox et al., 1963a). The sweat secretion begins at lower body core and surface temperatures (Fox et al., 1963b; ROBINSON et al., 1943) and the volume at a given body temperature increases (LADELL, 1945; HORVATH and SHELLY, 1946, 1951; EICHNA et al., 1950). Thus, in cybernetic terms (cf. Chapter VI) sweating threshold is shifted to lower temperature values in the course of heat adaptation. On the basis of the described modifications in sweat secretion the classic observations by ADOLPH (1947a), WYNDHAM (1951), and WYNDHAM et al. (1954) (Fig. 76) can easily be interpreted: Due to the increasing capacity of evaporative heat loss, body temperature can stabilize at a lower level and, consequently, heart rate and peripheral blood flow are prevented from rising to a level at which hyperthermic circulatory collapse is likely to occur. Circulatory stability may be supported, in addition, by an increase in blood volume (SUNDERMAN, 1938; SPEALMAN, NEWTON, and POST, 1947; FOX, 1965).

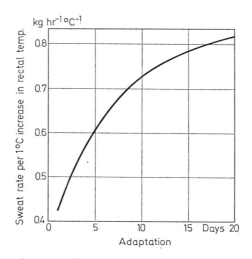

Fig. 75. Time course of heat acclimatization in man, measured by sweat formation per degree increase in rectal temperature. Average value from several individuals, work = 87 kcal m^{-2} h^{-1}, external temperature = 37.8° C, relative humidity = 78%. From LADELL (1951)

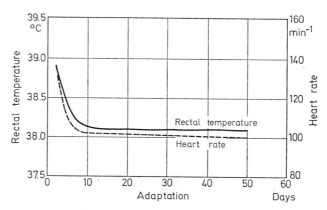

Fig. 76. Time course of acclimatization of men in desert climates. Rectal temperatures and pulse frequency were measured after a standard work load. From ADOLPH (1947)

The NaCl content of the sweat falls drastically in the course of heat adaptation and in extreme cases reaches 0.03%, i.e. ca. 1/25 of the NaCl-concentration of the plasma (CONN, JOHNSTON, and LOUIS, 1946). This will decrease the susceptibility to heat exhaustion due to sodium chloride depletion. It may also be mentioned that the number of active sweat glands increases; more sweat glands are found in inhabitants of tropical regions than those of colder climates (KUNO, 1956).

As a further modification the spatial distribution of sweat secretion changes in the course of heat adaptation in such a way that the rate of secretion, as shown in Fig. 77, increases to a greater extent in the region of the extremities (HÖFLER et al., 1966; HÖFLER, 1968); due to the smaller radii of curvature the evaporative and convective emission of heat is favored here (by a reduced thickness of the boundary layer, especially at low wind speeds; cf. p. 547). This modification, however, has been observed only during adaptation to *moist*-hot climates. During continuous heat exposure a constant decrease in the rate of sweat secretion is found in the

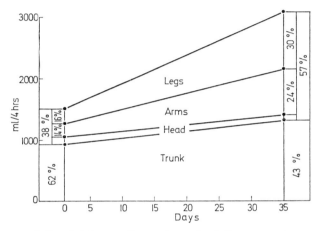

Fig. 77. Changes of the total sweating rate and of the percentage contribution of trunk, head, arms, legs, in the course of a 35-day heat adaptation period. After HÖFLER (1958)

course of one to three hours. This phenomenon, which was at first designated as "sweat gland fatigue", is directly related to a moistening of the skin (RANDALL, PEISS, 1957; HERTIG et al., 1961). However the mechanism has not yet been explained in detail. Fox et al. (1967) investigated the local rate of secretion of sweat on the arm and found that in individuals *adapted to a moist-hot climate* "sweat gland fatigue" could only be observed to a slight extent. According to these and the above-mentioned findings of HÖFLER (1968), which may be closely connected, one can state that the specially demanding conditions of life in moist-hot climates present a stimulus for the development of thermoregulatory modifications which is not found in dry heat.

Peripheral blood flow (hand, forearm) increases in relation to body temperature by 10—30% in the course of heat adaptation (Fox et al., 1963b). This facilitates heat transport to the body surface.

Though it has often been assumed that energy metabolism may fall below the level of basal metabolism no definite support for this assumption is to be found even in the more recent literature (COLLINS and WEINER, 1968). Some changes in metabolism reported occasionally (Literature: ROBINSON, 1949; RADSMA, 1950; MUNRO, 1949) may be explained on the basis of release of an increased muscle tone during heat adaptation. It is known that relaxation exercises (HINDMARSH, 1927; PICKWORTH, 1927) or hypnosis (v. EIFF, 1951) can lead to a marked reduction in muscle tone.

BURTON et al. (1940) and MUNRO (1949) found that after lengthy acclimatization in the Indian tropics the rectal temperatures of white men did not differ from those in Europe. ADAM and FERRES (1954) reported a significant, though very small, elevation of about 0.2° C in the average rectal temperatures of whites after eighteen months of heat acclimatization in Singapore, as compared to levels measured in Oxford. In contrast, FOX et al. (1963a) found pre-heat body temperatures, on the average, 0.19° C lower after artificial heat acclimatization. Thus, pre-heat body temperature does not seem to be a significant parameter in heat adaptation in man (in contrast to the camel, see p. 639).

During the first days of heat stress the uptake of water, which is controlled by the sensation of thirst, falls far short of water loss, and the body becomes dehydrated. This is probably related to the fact that the sweat is nearly isotonic with the blood at first, so that the osmotic pressure of the blood remains constant. As a result the feeling of thirst is reduced (ADOLPH and DILL, 1938).

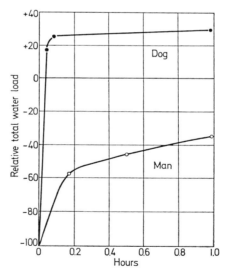

Fig. 78. Water load, as % of the initial water deficit, created by previous water deprivation. Each curve represents the cumulative water ingested. From ADOLPH (1957)

This problem does not appear in panting animals, since no electrolytes are lost as long as saliva does not drop from the tongue. (The electrolytes which might accumulate on the tongue through evaporation are swallowed.) Indeed, even the non-heat-adapted dog makes up for the water lost due to panting quantitatively and completely at the first opportunity (Fig. 78).

According to our present understanding the adequate stimulus for the sensation of thirst is the elevated osmotic pressure or elevated NaCl content of the blood. ANDERSSON (1952a, b) could demonstrate that certain regions in the hypothalamus of the goat are connected with the stimulation of thirst. When 0.1 ml of weakly hypertonic (1.5%) NaCl solution was injected into the anterior medial region of the hypothalamus of an unanesthetized animal, it drank several additional liters of water, even though it had not taken up any water spontaneously. Injections in the lateral or posterior region of the hypothalamus or of isotonic or hypotonic solutions had no effect.

In the course of acclimatization, man becomes much thirstier after a given loss of sweat, so that henceforth he can usually make up for the loss of water (ADOLPH and DILL, 1938). According to extensive American desert experiments (ADOLPH et al., 1947), thirst is not always adequate for this purpose, so that even after heat acclimatization there is a danger of increasing dehydration. While the chief danger in the tropics consists of restricted evaporation of water, in the desert it is the loss of water from the body, which reduces the power of resistance against heat, and in the course of time leads, through progressive reduction in blood volume and increase of pulse rate and rectal temperature, to heat collapse (*Dehydration exhaustion*, ADOLPH et al., 1947). In contrast to the effective acclimatization to heat, there is *no acclimatization to dehydration* or any kind of water-sparing alterations in body functions in the heat. The only remedy is to provide as much water as is lost. Today, according to ADOLPH et al. (1947), the view that less water should be drunk to produce heat adaptation must finally be buried; on the contrary, harmful dehydration can only be avoided if one deliberately drinks more than thirst requires. This is illustrated by a report on observations and investigations by MACFARLANE (1969) on aboriginal nomads "who are accustomed to hunt and walk up to 100 km in summer desert Conditions. Men, women and children drank rapidly. Normally hydrated men drank 1 l in 10—38 sec and 2 l in 36—45 sec, but Europeans found it hard to ingest 1 l in 150 sec; when 5% dehydrated they took 1 l in 35 sec, however. After a water load of 3% of body weight the aboriginals excreted twice as much in 3 h as the Europeans and at any time their urine flow rates were 3—10 times greater than in Europeans in the same environment." A large urine flow is very important, particularly in the case of diseases of the descending urinary pathways and the bladder, where the highly concentrated urine may worsen inflammatory processes.

According to the equation $A = F(I, t, n)$ (cf. p. 618), complete adaptation, as measured by the increase in sweat secretion following a stimulus of sufficient intensity, can be attained even after a short exposure. Thus an increase of the sweating rate to maximal values was attained after only twelve days of daily one-hour intensive heat stresses (oral temperature $= 38.5°$ C under resting conditions). Body temperature reached during heat stress is the deciding factor for the increase in sweating rate (FOX et al., 1963). However if the exposure to heat is suspended for only one or a few days, the degree of adaptation subsides (WENZEL, 1961; LADELL, 1964). After an interruption of exposure of ca. three weeks, a complete disappearance of heat adaptation is to be expected.

Now and then some doubt is expressed as to whether the modifications as seen in experimental heat adaptation, namely increased sweat rate and increased water intake, would persist when a subject is exposed for a prolonged period of time, say for several months or years, to hot climatic conditions. It has been claimed that such long-term adaptation leads to a reduction in sweat rate and water intake and that the resistance to heat is brought about by some unknown mechanism. To clear up this situation, EDHOLM (1969, 1972) has studied two groups, one of which consisted of men who were stationed in England and had not experienced hot climatic conditions for at least one year; they were thus considered to be un-acclimatized to heat. The second group was drawn from men who had been station-ed for nine months in a hot and frequently humid climate. The two groups were then examined under various environmental conditions. When exposed to hot con-ditions the acclimatized men on the average drank more than the unacclimatized on each day and their calculated sweat rate was also higher. It can thus be con-cluded that the physiological modifications as seen in relatively short-lasting adaptation experiments are not overcome by a different type of adaptation in the long run.

Life in Dry- and Moist-hot Climates. The greatly elevated increase in capacity for evaporative heat loss may be fully utilized in dry-hot desert climates, but not in humid-hot climates, since this loss depends on the difference between the vapor pressure of the skin surface and that of the environment (see p. 549), while the vapor pressure in the humid-hot climate is so high, on the average, that the volume of sweat which an acclimatized man can produce cannot evaporate com-pletely. Therefore the maintenance of the highest possible rate of sweat secretion cannot be viewed as an adaptation to a moist-hot climate. The optimal response seems to be to secrete exactly as much sweat as can be evaporated. In fact a person newly arrived in moist-hot tropical regions tends to sweat excessively, i.e. sweat can be seen to drop from many parts of the body. A habituated person, on the other hand, adjusts the secretion rate to the volume of evaporation in such a way that no sweat is lost by rolling off the body (LADELL, 1964; COLLINS and WEINER, 1968). This is aided by the observed preferential increase in sweat secretion rate on the extremities (HÖFLER, 1968 , s. also p. 641). However, the avoidance of any intense physical effort is quite decisive (LADELL, 1964).

C. Endocrine System and Thermal Adaptation

The physiological processes which form the basis for adaptation are still obscure in spite of intensive research during recent years. The participation of glands of internal secretion, namely the adrenal cortex and the thyroid, has been long sus-pected. But the extent to which these hormonal adjustments are the causes of the phenomena of adaptation or are simply one expression of far more comprehensive processes is far from clear.

1. Thyroid

The thyroid of many mammals and birds undergoes a typical yearly fluctuation in activity. In summer the colloid is filled with stored material, the epithelium is low, and the total thyroid weight is reduced, as an expression of its lowered activity. In winter the thyroid is enlarged, the colloid holds little stored material, and the epithelium is in an active, hyperplastic condition (SPÖTTEL, 1929; RIDDLE, SMITH and BENEDICT, 1932; WATZKA, 1934; MILLER, 1939; HOEHN, 1949, among others). The "summer sterility" of many animals accompanies the summer hypoplasia of the thyroid (p. 671).

During long-term cold exposure, thyroid activity changes very distinctly. If hens' eggs are incubated at 36° C, the weight of the chicks' thyroids is increased three-fold in comparison with control animals incubated at 39° C. Similarly chicks which are maintained for several weeks at 4° C show an enlargement of the thyroid gland. Similar results have been obtained in many experiments on guinea pigs and rats. If the animals are kept in the cold for longer periods, thyroid activity rises sharply in the course of the first days and weeks, as the colloid shrinks and the epithelium becomes hyperplastic (STARR and ROSKELLEY, 1940; LESSER, WINZLER, MICHAEL-SON, 1949; PICHOTKA, 1952a, b). The hyperplasia reaches its maximum after ca. six weeks, and after ten weeks of cold exposure it goes back towards the original condition (STARR and ROSEKELLEY, 1940; PICHOTKA, 1952a, b). Alterations in thyroid activity paralleling these morphological changes can be demonstrated by means of radioactive iodine tests (LEBLOND et al., 1943; SCHACHNER, GIERLACH, KREBS, 1949). Furthermore, increased utilization of thyroxine in the cold, e.g. in shorn sheep which were exposed to winter weather, could be shown (FREINKEL and LEWIS, 1957) (Fig. 79).

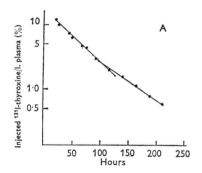
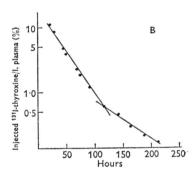

Fig. 79. Increased turnover of thyroxine in a shorn sheep subjected to severe cold *(B)*, when compared with an unshorn animal *(A)*. The curves represent the disappearance of circulating radioactivity following the administration of ^{131}I-thyroxine. From FREINKEL and LEWIS (1957)

The investigations of LEBLANC and VILLEMAIRE (1970) have produced substantial-ly new points of view on the significance of the thyroid for the development of cold-adaptive modifications. They were able to stimulate the formation of brown fat in the rat by the daily administration (subcutaneous) of 50 µg/kg of thyroxine for five weeks (Table 10). The growth of the (interscapular) fat tissue was stimulated still more strongly by the administration of thyroxine combined with noradrenaline. The weight of these tissues attained by use of the latter treatment was quite close to the weight found in the cold-adapted rat. The thyroxine- and thyroxine-norad-renaline treated animals should subsequently be able to react to stimulation of the sympathetic system or to the injection of noradrenaline with an acute increase in heat formation. As Table 10 shows, this actually occurred. Surprisingly the reaction is even stronger than in cold-adapted animals; this finding has not been explained more fully at this time.

Table 10. Interscapular brown fat weight, basal oxygen consumption, metabolic response to noradrenaline, and cold resistance test at —25° C in various groups studied

Treatment	Inter-scapular Brown Fat	Oxygen Consumption			Colonic Temp. Drop After 2.5 hr at —25° C
		Before NA	After NA	Increase	
	mg	ml O$_2$/min per m^2			° C
Control (C)	193 $\pm$ 14 (14)	144 $\pm$ 6 (11)	188 $\pm$ 6	44 $\pm$ 6	9.4 $\pm$ 0.7 (6)
Noradrenaline (NA)	276 $\pm$ 15 (16)	148 $\pm$ 8 (11)	241 $\pm$ 18	93 $\pm$ 16	7.5 $\pm$ 1.3 (5)
L-Thyroxine (T$_4$)	369 $\pm$ 23 (18)	182 $\pm$ 10 (11)	304 $\pm$ 20	122 $\pm$ 15	4.6 $\pm$ 0.8 (6)
T$_4$ + NA	518 $\pm$ 44 (13)	184 $\pm$ 7 (11)	393 $\pm$ 24	209 $\pm$ 21	1.9 $\pm$ 0.3 (7)
Cold-acclimated (6 C)	663 $\pm$ 30 (8)	162 $\pm$ 7 (8)	241 $\pm$ 12	79 $\pm$ 9	0.5 $\pm$ 0.1 (7)

Values are means $\pm$ sem. Numbers in parentheses indicate numbers of animals. Noradrenaline dose = 30 μg/100 g body wt.
From LEBLANC and VILLEMAIRE (1970).

By contrast, no answer whatever has been found to the question of what the effects of the cold-induced increase in thyroxine secretion rate are in larger species, such as man, which do not form brown fat and are incapable of nonshivering thermogenesis. The possible influence of an increase in thyroxine secretion rate on other thermoregulatory mechanisms, such as for example peripheral vasomotor activity or the effectiveness of shivering, is still to be tested.

The heat tolerance of thyroidectomized animals is increased (for literature, see COLLINS and WEINER, 1968); a reduction in thyroid activity has therefore been regarded as a possible mechanisms for heat adaptation. On the other hand, no definite signs of lowered levels of functioning could be found in heat-adapted rats by means of radioactive iodine tests (LEBLOND et al., 1943). This is in agreement with the finding mentioned above (see p. 638), that the basal metabolism of heat-adapted rats does not fall. However, CASSUTO and AMIT (1968) found an indirect indication of a *reduction in thyroid activity* in hamsters, whose basal metabolism was shown to fall in the course of heat-adaptation (see p. 638). The difference between basal metabolisms of heat-adapted animals and controls was reduced from 27% to 12% by two administrations of thyroxine, 48 and 24 h before the metabolism test. By giving thyroxin plus noradrenaline the difference was almost completely eliminated. From this it was concluded "that the thyroid gland is hypoactive in heat-acclimated animals". At this point the warning of HÉROUX, which is still valid today, must be mentioned concerning the equating of the functional conditions of heat- and cold-adapted animals with the pathological conditions of hypo- and hyperthyroidism. In these disturbances of function the thyroxine level (PBI) is lowered or raised, which is certainly not the case in thermal adaptation. An elevation of the metabolic rate in cold-adaptation has been demonstrated. A participation of thyroid activity in the processes of heat adaptation may yet be demonstrated through subtler methods of investigation.

2. Adrenals

The Adrenal Medulla. Catecholamines, which occur in the organism in the form of adrenaline and noradrenaline, have recently been shown to be links in the chain of events controlling non-shivering thermogenesis (see Chapter IV, p. 542). Accordingly, giving exogenous adrenaline and noradrenaline produces nonshivering thermogenesis. In the intact organism the regulation of *NST* takes place exclusively through the noradrenaline released by the sympathetic nerve endings, as the excretion of noradrenaline- and adrenaline-derivatives in the urine shows. An increased secretion of adrenaline from the gland is seen only when animals are exposed to very severe cold stress and no further increase of noradrenaline is possible (LEDUC, 1961; JOHNSON, 1966). For these reasons the secretion of adrenaline from the adrenal medulla is seen today as a last defensive reaction against cold, set in motion only in emergencies; a view which corresponds completely with CANNON'S (1928) now historic concepts. It can therefore be understood why, under a cold stress that can be tolerated with ease by the cold-adapted rat, no appreciable deterioration in thermoregulation can be seen after removal of the adrenal medulla (POULIOT, 1966). Accordingly the release of adrenaline from the adrenal medulla is more likely to occur in non-cold-adapted animals than in cold-adapted ones.

Therefore an obvious question to ask is, whether a marked elevation in catecholamine level, whether resulting from greater liberation by the sympathetic nerve endings or from secretion by the adrenal medulla, contributes to the development of cold-adaptive modifications, especially to the formation of brown fat and to nonshivering thermogenesis. According to the investigations of LEBLANC and POUILOT (1964) this possibility must be taken into account. They showed that daily repeated injections of noradrenaline could, in the course of a few weeks, elicit changes corresponding to those of cold-adaptation, i.e. in rats so treated nonshivering thermogenesis can be induced by exposure to cold or by noradrenaline injection. It was later shown that the effect of repeated noradrenaline injections could be increased still more by combining it with thyroxine (cf. p. 645).

Changes opposite to those brought about by cold-adaptation, namely shifts in the thresholds for initiating shivering and non-shivering thermogenesis (see Fig. 68), have recently been elicited by local injections of noradrenaline into the hypothalamus of guinea pigs (ZEIBERGER and BRÜCK, see Fig. 47). It is therefore possible that some of the adaptive changes are brought about through thermal activation or inhibition of some noradrenergic neuronal systems in the brain.

The sympathetico-adrenal system of non-adapted organisms tends to over-react to stresses (SELYE, 1950). As has already been noted, this overshoot reaction is abolished by the effect of repeated stimulation ("Habituation" see p. 620). For this purpose *very short* but *intensive* stressor effects apparently suffice, e.g. ten minutes of exposure of rats to $-15°$ C nine times a day (LEBLANC et al., 1967; LEBLANC, 1969; also see p. 624). Nevertheless, the mechanism by which the stressor reaction grows weaker as cold resistance increases is still completely unclear.

Adrenal Cortex. Cold- and heat-stress lead — as do all other possible stressors — to an activation of the adrenal cortex. As we now know, glucocorticoids are preferentially released in this process and aldosterone only to a slight extent. SELYE (1950) has designated these reactions as alarm reactions. The stressor-induced increase in glucocorticoid secretion goes down with continuous application of the stress and appears only in a weakened form if the stress is applied repeatedly. The behavior is therefore similar to that of the adrenal medulla (see above). Today it is assumed that the increased glucocorticoid secretion is needed for the formation of specific adaptive traits and that with the establishment of these traits the adrenal cortex function is normalized again (HÉROUX and HART, 1954; GANONG and FORSHAM, 1960).

In spite of many investigations focussing on the problem, no support has been found for the existence of a specific role of glucocorticoids in thermal adaptation (HÉROUX, 1955). Therefore one would not expect to achieve thermal adaptation by the administration of exogenous glucocorticoids.

The case is different for aldosterone. Here a direct and specific relation to heat adaptation can be recognized. Aldosterone stimulates the reabsorption of NaCl, thereby lowering its level not only in the urine but also in the sweat (Lit. COLLINS and WEINER, 1968), thus contributing to one of the most characteristic modifications within the compass of heat adaptation. Indeed an elevated level of aldosterone has been demonstrated in the plasma of various sweating species and of man in the course of development of heat adaptation (Lit. COLLINS and WEINER, 1968). By contrast the NaCl content of sweat is greatly elevated in cases of adrenal insufficiency (M. ADDISON), which explains the known high degree of sensitivity to heat in the patients. The NaCl content of sweat may even be used for diagnosis of this condition.

3. Hypothalamo-Hypophyseal System

The rates of secretion of hormones from the thyroid and adrenal cortex are governed by so-called trophic hormones (ACTH-adrenocorticotropic hormone, and TSH-thyroid stimulating hormone) from the adenohypophysis (anterior lobe of the pituitary); these in turn are under the influence of so-called releasing factors which are formed in the neurons of the hypothalamus. The activation of releasing factors results from the stimulation of thermoreceptors in the skin and from stimulation of thermosensitive structures of the anterior hypothalamus (ANDERSSON, 1970).

It can therefore be understood at once that after the hypophyseal stalk is sectioned, cold activation of the thyroid ceases, as UOTILA (1939) showed long ago. In addition we know that the liberation of releasing factors in the hypothalamus is influenced in turn by the higher central nervous system, especially by the limbic system. In the last analysis then, thermal adaptation can be traced back to modifications in the central nervous system. The details here are still largely unknown. The center of gravity of future research will probably lie in this area.

D. Artificial Thermal Adaptation

Being adapted to hostile hot or cold environment is fine, but acquiring cold or heat adaptation is certainly a stressful process. Thus, not only for the sake of a better understanding of the adaptive mechanisms but also for the sake of circumventing the stressful prolonged or repeated exposure to the stressor, attempts have been made to find out how cold or heat adaptation might be produced by means other than exposure to cold or heat. One such possibility for acquiring cold adaptation to some degree has already been mentioned: Physical exercise increases both cold resistance (cf. p. 632) and heat resistance (GISOLFI and ROBINSON, 1969). More interest has been attracted, however, by the belief that adaptation might be brought about by application of drugs and hormones, or by special diets. We have already mentioned (p. 645) that application of a combination of thyroxine and noradrenaline brings about modifications which resemble those of cold adaptation in small animals, such as development of brown adipose tissue and nonshivering thermogenesis (LEBLANC and VILLEMAIRE, 1970). No information is available which would show that heat and cold resistance in man are improved by application of hormones.

As for food supply, it is clear that severe undernutrition will certainly decrease the degree of resistance to a stressor. On the other hand, in the Eskimo (see p. 631) protein-fat-rich nutrition has been shown to favor a special type of cold-adaptation characterized by an increased *BMR*. In Caucasians the same regimen failed, however, to produce cold-adaptation. An interesting interdependence between thermal adaptation and nutrition has been shown in rats: A group of these animals previously adapted to intermittent feeding and fasting was subjected to 5° C for 3 weeks and fed *ad libidum*. During the time of exposure to cold these rats consumed an average of 587 g of food and gained a mean of 33 g of body weight. Control rats of the same age consumed an average of only 540 g of food and *lost* 10 g of body weight. Thus animals adapted previously to intermittent feeding and fasting reacted to cold in a more favorable way (HOLÉCKOWÁ 1964).

In a series of studies carried out by HÉROUX (1969) the influence of two diets [Master Laboratory Cubes (MLC) and a "purified diet", called thyroxine-free diet (T_4F)] on the development of cold resistance was studied. The composition of the two diets is given in Table 11. Under severe cold stress experienced outdoors during the winter, the group fed on T_4F showed considerably increased cold resistance. They had developed about twice as much brown adipose tissue as the group fed on MLC, and this may explain the increased cold resistance. The author finally stated "that the superiority of the purified thyroxine-free diet cannot be attributed to the absence of thyroxine or to any other single factor, or combination of factors".

Table 11. Composition of the two diets

	MLC	T$_4$F
Protein (%)	20	27
Fat (%)	3	14
Carbohydrate (%)	60	56
Fiber (%)	5	
Thyroxine (µg/g)	0.58	
Energy value (kcal/g)	4.26	4.95

From HÉROUX (1969).

In man, KREIDER and BUSKIRK (1957) showed that when a food supplement of normal composition was eaten before retiring at night in the cold, body temperatures (skin, rectal) were maintained at a higher level than when no supplement was eaten. Alteration in the food composition did not seem to have a significant influence on this effect (KREIDER, 1961).

Elucidation of the influence of diet on long-term modifications in the thermoregulatory system seems to require further studies.

IX. Temperature Limits of Life

Homeotherms are substantially more capable of resisting cold than heat, if the normal rectal temperature is taken as the point of departure. With the possible exception of man, the number and effectiveness of the mechanisms protecting against loss of heat are also far greater. In contrast to the poikilotherms, whose temperature resistance can be essentially characterized by their lethal temperatures, *a number of critical temperature ranges* can be distinguished for homeotherms: 1. the external temperature at which the capacity for temperature regulation is exceeded and the core temperature changes; 2. the core temperature at which the regulatory centers are paralyzed and 3. the lethal temperature. Special conditions are also found, in which *local* temperature damage takes place while overall temperature regulation is maintained. Accordingly, the tolerance shown by homeotherms for extreme temperatures is determined by the interaction of a large number of factors.

A. Limits of Regulation in Heat

As is explained on p. 526, *heat absorption* through water evaporation is the only effective protection against overheating at high external temperatures — except for behavioral regulation by moving to cooler surroundings. Therefore heat tolerance depends on the extent to which the quantities of heat formed in the body (H_i) and taken up from the environment (H_a) can be absorbed by water evaporation (H_e). If

$$H_e < H_i + H_a \tag{1}$$

hyperthermia sets in. Since water evaporation is determined by the difference between the vapor pressures of the skin and air, the tolerance limits depend not only on capacity for water loss (sweat glands, panting), but also on the humidity of the air (Fig. 80).

1. Weakly Sweating Animals

The heat tolerance of many weakly sweating mammals is limited. The point at which water loss becomes insufficient and the core temperature rises is often reached at 30° C, even in dry air. The condition of many animals is already critical

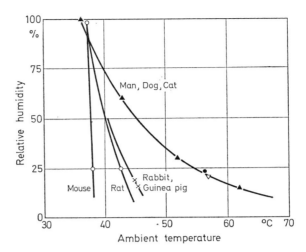

Fig. 80. Mean tolerance limits of various mammals for a 3-hour exposure to various air temperatures and relative humidities. From ADOLPH (1947b)

at 35° C, and in others the rectal temperature is at least sharply elevated. Weakly sweating animals exhibit quite variable behavior in intense heat (Tab. 12). According to ADOLPH (1947b) the dog and the cat, which pant with open mouths and copious secretion of saliva, achieve nearly the same heat tolerance as strongly sweating men (Fig. 81). Water loss through respiration is far less effective for rabbits and guinea pigs, while rats and mice increase their respiratory volume only slightly, and are therefore most likely to reach their limits of tolerance. Although sweat glands are anatomically present in all mamals, ADOLPH (1947b) found no

Table 12. Maximum respiratory frequencies in hot environments

Species	Respiratory frequency min^{-1}
Cattle[a]	100
Sheep[b]	240
Cat[c]	250
Roadrunner[d]	356
Dog[e]	500
Rock dove[f]	650
Rabbit[g]	720

[a] From McLEAN (1963).
[b] From LEE and ROBINSON (1941).
[c] From HESS and STOLL (1944).
[d] From CALDER and SCHMIDT-NIELSEN (1967).
[e] From RICHET (1898).
[f] From CALDER and SCHMIDT-NIELSEN (1966).
[g] From LEE, ROBINSON and HINES (1941).

discernible sweat secretion in any of the species mentioned above, even in intense heat. In long desert marches the heat resistance of dogs resembles that of man, though the dog's rectal temperature rises to higher levels. Since the dog scarcely sweats at all, its skin temperature rises to as much as 46° C in the sun, at external

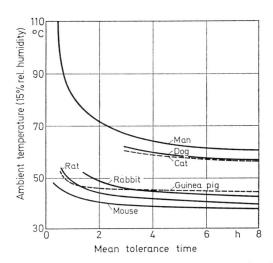

Fig. 81. Mean tolerance time of various mammals as a function of ambient temperature. Man from BLOCKLEY and TAYLOR (1949); animals from ADOLPH (1947b)

temperatures of 42° C, while that of man only rises to 34° C (DILL, BOCK, and EDWARDS, 1933). Nevertheless the dog's panting can give off enough heat, even in the face of this negative heat flow. The dog enjoys one advantage with respect to man in that it replaces all the water it has lost at once, through drinking (cf. p. 643). The heat resistance of the sheep is still greater. At an external temperature of 45° C with 40% relative humidity, the rectal temperature of the merino sheep remains below 40° C (RIEK et al., 1950), and even a seven-hour exposure to an air temperature of 43.3° C with 65% rel. hum. is tolerated (LEE and ROBINSON, 1941; MACFARLANE, 1964).

Body size plays an important role in determining tolerance to the short-term effects of heat. The larger the animal is, the more slowly it is warmed, due to its greater heat capacity. When heat flux is in equilibrium this factor plays no role; only the effectiveness of regulation matters. At low relative humidities of ca. 15% the following average tolerance limits are found: Man 59.4° C, dogs and cats 56° C, guinea pigs 43.9° C, rabbits 41.7° C, rats 38.6° C, mice 37.2° C. If enough water is provided these temperatures can be borne for long periods (Fig. 81).

Because of their lack of sweat glands, *birds* have a rather limited capacity for keeping their body temperature constant in hot environments. However, some birds, particularly the large ones or those with a well-developed gular flutter (p. 558), can prevent overheating during exposure to ambient temperatures which exceed the upper lethal body temperatures by 4° C or more (DAWSON and HUDSON,

1970). The ostrich possesses an impressive capacity for temperature regulation in hot environments, maintaining body temperature essentially unchanged at 39.3° C in ambient temperatures between 25° C and 51° C (CRAWFORD and SCHMIDT-NIELSEN, 1967). This performance is in marked contrast to the significant rise in body temperature that most other birds undergo when subjected to heat stress, even during inactivity at low humidities (ref. see DAWSON and HUDSON, 1970). Since birds readily tolerate elevations of body temperature on the order of 4° C (BARTHOLOMEW and DAWSON, 1958), moderate hyperthermia conveys distinct advantages in terms of temperature regulation and water economy. Hyperthermia increases non-evaporative heat loss to and reduces heat gain from the environment, thus lowering requirements for evaporative cooling and, consequently, for rapid water loss in heat (KING and FARNER, 1961; DAWSON and HUDSON, 1970).

The degree of hyperthermia differs among the species tested. Body temperatures of small passerines (< 50 g) often closely approached ambient temperatures of $44-45°$ C, their mean body temperatures being $44-44.7°$ C (LASIEWSKI, ACOSTA, and BERNSTEIN, 1966a), in line with their capacity for dissipating, through evaporation, all the heat they produce, but little more. Larger birds (> 100 g), such as the white-necked raven *(Corvus cryptoleucus)*, generally remained $1-3°$ C cooler than the environment during exposures of a few hours to $44-46°$ C. Rock doves (315 g) had body temperatures averaging 43.1° C after resting for several hours at an ambient temperature approximating 51° C (CALDER and SCHMIDT-NIELSEN, 1966).

2. Man and Strongly Sweating Mammals

Man and the strongly sweating mammals (Equids) possess highly effective regulation under hot conditions. The fact may serve for comparison that, at an effective temperature of 27° C (corresponding, for example, to 35° C with 35% rel. hum.), which is a critical limit for some animals even at rest, a man can maintain a constant rectal temperature even while working at a rate of 180 kcal h⁻¹ (LIND, 1963). The heat tolerance of the donkey and probably that of the other equids is in no way inferior to that of man (ADOLPH and DILL, 1938). In dry air man can endure astonishingly high temperatures for short periods. As early as 1775 BLAGDEN reported an experiment in which a man remained a quarter of an hour at an air temperature of 120° C without suffering harm. More recent experiments (BLOCKLEY and TAYLOR, 1949; LIND and HELLON, 1957; IAMPIETRO, MAGER, and GREEN, 1961; PROVINS et al., 1962; LEITHEAD and LIND, 1964; KUZNETS et al., 1968) have confirmed man's great heat tolerance. According to BLOCKLEY and TAYLOR (1949) the tolerance times presented in Fig. 81 were recorded in dry heat up to 115° C. At an air temperature of 115° C the average skin temperature rose to 42° C. In this case the body surface acts as a buffer against overheating of the internal organs. Respiratory frequency, oxygen use, and pulse frequency rose to three times their resting values. More recently 7 men have tolerated 205° C in a hot room for periods of up to 20 min (MURRAY and ROSS, 1965). The mean skin temperature rose to 43°C and rectal temperature increased by 0.9°C. Heat acclimatized mine workers can maintain approximately 50% of their normal work capacity in water-saturated air at 34°C (effective temperature 34°C) (WYNDHAM et al., 1959), and CAPLAN and

LINDSAY (1946) report that Indian miners, who often work at their tolerance limits, could still operate at 43% of their normal capacity after two hours at temperatures of 50° C with 38% rel. hum. (effective temperature 36° C). BELL and WALTERS (1969) indicate that 6 min is the limit of safety (level at which 5% collapsed) for non-acclimatized young men working at a rate of 280 kcal h^{-1} in a room at 53° C with 70% rel. hum.

Human heat tolerance is difficult to assess, since it depends on a large variety of individual parameters (LEITHEAD and LIND, 1964; WYNDHAM, 1970), such as clothing, race, age, sex, physique, nutrition and acclimatization (p. 639). Many attempts have been made to integrate, into a single index, the effects of two or more of the several factors that influence heat exchange between man and his environment. Among various *heat stress indices*, the following are mentioned: 1. effective temperature (HOUGHTEN and YAGLOU, 1923), combining air temperature and humidity, 2. corrected effective temperature, with allowance for radiant

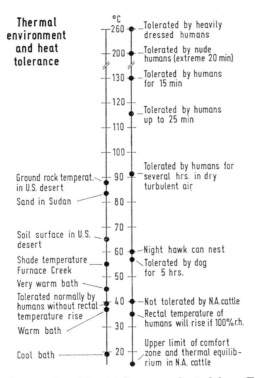

Fig. 82. Thermal environment and heat tolerance, adapted from FOLK (1966)

heat (BEDFORD, 1948), 3. predicted four-hour sweat rate (P_4SR), where the sweat rate is the criterion of thermal stress (MCARDLE et al., 1947), and 4. the Belding-Hatch index (BELDING and HATCH, 1955) which is based on the amount of sweat that must be evaporated to maintain thermal equilibrium. Data on various hot environments and heat tolerance are shown in Fig. 82.

The *volume of sweat* that a man can secrete over short periods amounts to more than 4 l h^{-1} and the daily volume to 18 l (Eichna et al., 1945a, b; Ladell, 1945). If the volume of 4 l h^{-1} were completely evaporated it would correspond to a heat absorption of ca. 580 kcal h^{-1}, i.e. 30 times the basal metabolic rate. Such volumes of sweat however cannot be secreted for extended periods even if sufficient fluid is provided. Near the limits of tolerance the sweat glands become exhausted to a very considerable extent in the course of a few hours, with the result that hyperthermia must set in. The regular daily secretion of 10−12 l of sweat is not unusual among workers in hot conditions (Leithead and Lind, 1964). In addition to acute failure of sweat secretion there is a subacute form which may appear after repeated heat stress, and also a chronic failure which gradually, in the course of weeks and months, leads to heatstroke (Malamud, Haymaker, and Custer, 1946; Leithead and Lind, 1964). Man's *circulation* is strongly stressed by heat, especially if physical work is being done at the time. The cardiac output increases by 1−2 l as a result of heat alone, while increases of 15 l and more have been observed during work. Pulse frequency rises in the heat up to as much as 180 min^{-1} and still more when a threat of heat collapse exists. Furthermore the blood volume is increased by displacement of plasma from the tissues (Glickmann et al., 1941; Bass and Henschel, 1956).

B. Hyperthermia

1. General Heat Injuries

Heat syncope (heat collapse; exercise-induced heat exhaustion) is primarily a failure of the circulatory and not of the thermoregulatory system. Vasodilatation in the heat leads to a tendency to peripheral venous pooling and to hypotension. Any stress imposed thereafter, such as sudden postural change, prolonged standing, or strenuous exercise, may be enough to provoke syncope through a fall in blood pressure. Failure to maintain adequate cerebral blood flow leads to a sensation of giddiness or to unconsciousness. Heat syncope may occur at rectal temperatures well below 40° C. During fever with an elevated set-point of the core temperature (p. 588), the tendency towards heat syncope is considerably reduced in comparison with passive hyperthermia. The state of heat syncope is an example of a meshed-loop feedback system (p. 527) where the regulation of blood pressure (tendency for peripheral vasoconstriction) is overrun by thermoregulation (tendency for vasodilatation).

Water depletion exhaustion (dehydration) is due to inadequate replacement of water losses in prolonged sweating, and is characterized by thirst, fatigue, giddiness, oliguria, pyrexia, and in advanced stages by delirium and death (Leithead and Lind, 1964). The critical limit for man is passed when the volume of water lost amounts to 12% of the body weight, while dogs tolerate a water loss of 15% and cats one of 20%. Still higher values are found in camels, sheep, and donkeys, which survive acute losses of 27−32% (MacFarlane, 1964). At lower ambient temperatures the lethal limit of dehydration lies at 20−30% of the body weight for other mammals as well (Adolph, 1947a, b).

Salt depletion heat exhaustion occurs if salt intake is inadequate to replace losses of sodium and chloride in thermal sweat. It is commonly associated with circumstances which provoke high rates of sweating, and is characterized by fatigue, nausea, vomiting, giddiness, muscle cramps, and in the late stages by circulatory failure (LEITHEAD and LIND, 1964). The loss of NaCl by sweating may amount to 3 g h^{-1} or 20 g per day. Plasma volume is diminished as the salt depletion becomes more severe. The resulting hemoconcentration may be the basis of giddiness and oligemic shock.

Heat cramps may be due to salt depletion, or to a combination of salt depletion and water intoxication (LADELL, 1949; BLACK, 1957). They occur in workers who are sweating heavily and at the same time drinking large amounts of unsalted fluids. Heat cramps are extremely painful attacks of muscular spasms, most attacks lasting for less than 1 min. The immediate treatment of choice in severe cases is intravenous administration of normal saline.

Heatstroke (heat apoplexy, thermoplegia, sunstroke) is a state of thermoregulatory failure usually of sudden onset, following exposure to a hot environment, and characterized by disturbance of the central nervous system (disorientation, mania, delirium, generalized convulsions and coma), by generalized anhidrosis, and a rectal temperature above 40.6° C. It is frequently fatal (LEITHEAD and LIND, 1964). The treatment of choice is cooling the patient as rapidly as possible, for example in a bath with ice water. In animals without sweat glands, heatstroke is characterized by a decrease in polypnea (RANDALL and HIESTAND, 1939).

The pathogenesis of heatstroke is unknown. Since heatstroke is one of the most urgent of all medical emergencies, immediate treatment has priority over investigation. The major question which is unanswered is whether the cessation of sweating precedes or follows hyperthermia. MALAMUD, HAYMAKER, and CUSTER (1946), in 157 fatal cases of heatstroke, found that pathologic changes in the central nervous system were most conspicuous. They consisted in degeneration of neurones and replacement by glia in various parts of the brain, edema and petechial hemorrhages. Hemorrhages and necrosis were also seen in the heart, kidneys, liver and adrenal glands.

2. General Heat Death

The *upper lethal temperatures* of all homeotherms fall within a fairly narrow temperature range which lies only a few degrees above the normal rectal temperature (Tab. 13). In reporting lethal temperatures special attention must be paid to the *time factor*. The immediate cause of death in general heat death of homeotherms is not known with certainty. Damage to the central nervous system has been considered as well as circulatory and cardiac failure (ADOLPH, 1947b; NEWMAN and WOLSTENCROFT, 1960; LEITHEAD and LIND, 1964). If an animal is warmed to a point above its critical temperature and is then immediately recooled, death may nevertheless ensue as much as 26 h later (ADOLPH, 1947a, b).

3. Local Effects of Heat

Local effects of heating may be tolerated even at temperatures above the general lethal limit. The tissue changes are fundamentally the same as those found in

Table 13. Upper lethal body temperatures

Species	Lethal temperature °C	Remarks
Wren[a]	46.8	long-term exposure
Fowl[b]	45.5—47	
Bat (Myotis sodalis)[c]	42.0	
Bat (Myotis sodalis)[c]	48.0	tolerated at least 15 min
Mouse[d]	43.3	
Rat[d]	42.5	50% lethal
Guinea pig[d]	42.8	50% lethal
Rabbit[d]	43.4	50% lethal
Cat[d]	43.4	50% lethal
Dog[d]	41.7	50% lethal
Man[e]	45.0	short-term survival
Man[f]	43.0	usually fatal

[a] From BALDWIN and KENDEIGH (1932).
[b] From FULLER and HIESTAND (1947).
[c] From HENSHAW (1970).
[d] From ADOLPH (1947b).
[e] From LEITHEAD and LIND (1964).
[f] From HERRINGTON (1949).

higher poikilothermic vertebrates. Local heating of the skin first causes dilatation of the capillaries (erythema, 1st degree burns).Damage to the vessel walls leads to leakage of plasma into the tissues, development of edema, and separation of the epidermis (2nd degree). Finally protein coagulation and cellular necrosis occurs (third degree). According to measurements made by GUILLEMIN et al. (1952) human skin begins to redden after application of heat for 30—120 sec at 45° C, and to blister at 53—57° C. More extensive local burns result in severe *injury to the entire organism*. Toxic products of protein degradation are absorbed from the burned tissues, and the extensive loss of plasma from the burned surfaces leads to a reduction in blood volume and to collapse.

It is remarkable that many organs and parts of the body can be warmed to temperatures far higher than the general body lethal temperature without being harmed. The cat tolerates local heating of the cerebral cortex to 46° C for 30 min with no disturbance in its functions (TESCHAN and GELLHORN, 1949). Lung tissue of dogs tolerates temperatures up to 43° C for 1 h without significant permanent disturbance of structure or function (REED, 1965). Dogs survived hepatic hyperthermia of 46° C (REED, MANNING, and HOPKINS, 1964), and in the dog's limb tissue temperatures of 45° C were tolerated for 1 h with complete recovery of function (REED and HOPKINS, 1962). Tissue temperatures above 46° C resulted in marked edema.

C. Limits of Regulation in Cold

The limits of regulation in cold are extraordinarily diverse among the homeotherms. They depend primarily on the *heat insulation* provided by the body covering. The temperature regulation of rabbits, which is ordinarily effective at temperatures as low as −45° C, fails at much higher temperatures if their hair is removed (GRANT, 1950). The same limited regulation is found in the hedgehog, which has no insulating coat of hair (GROEBBELS, 1928). Doves resist temperatures of −40° C for days, but if deprived of feathers they freeze after 20−30 min (STREICHER, HACKEL, and FLEISCHMANN, 1950). Body size also plays an important role due to the associated changes in surface-volume ratio. On the other hand the extent of *increased metabolism* in the cold, expressed by the metabolic quotient of GIAJA (p. 533), is not a measure of the effectiveness of regulation. Under some circumstances animals with high metabolic quotients may have very poor regulation in the cold (e.g., the hedgehog), while the quotients of some very cold-resistant homeotherms (e.g., the polar bear) are not especially high.

1. Limits of Regulation in Animals

The limits of regulation of some arctic mammals and of many birds lie at levels that are below the lowest temperatures occurring on earth (ca. −88° C) (Table 14). The cold-resisting regulation of tropical animals is much less perfect at all temperatures than that of arctic animals, primarily due to the poorer heat insulation of the hair or plumage of the former. The extent of change in their metabolic rate is not an important factor (see Fig. 18).

Table 14. Lowest ambient temperatures tolerated for at least 1 hr without hypothermia

Species	Ambient temperature ° C	Difference between ambient and rectal temperature ° C
Man, naked[a]	− 1	38
Guinea pig[b]	− 15	55
Rat[b]	− 25	65
Sparrow[b]	− 30	70
Rabbit[b]	− 45	85
Fowl[b]	− 50	90
White fox[c]	− 80	120
Goose[b]	− 90	130
Duck[b]	−100	140

[a] From ADOLPH and MOLNAR (1946).
[b] From GIAJA and GELINEO (1933).
[c] From IRVING and KROG (1954).

Due to the better heat insulation provided by their plumage, *of birds* cold resistance is substantially greater than that of mammals of the same body size, except for some arctic mammals. Even small birds perform astonishingly well in this respect. Sparrows weighing ca. 30 g can endure temperatures as low as −40° C for 8 h (KENDEIGH, 1944), with no reduction in their rectal temperatures. VEGHTE (1964) found that in Alaska the daytime levels of body temperature in the gray jay remained essentially unchanged between January, when the ambient temperatures lay between −30° C and −40° C, and July, when they approximated 10−15° C. The smaller the animal and the higher the metabolic rate the more the *supply of nourishment* becomes the limiting factor for regulation in the cold (ROWAN, 1925; McGOWAN, 1969). Ducks survive temperatures of −40° C without feeding for 7−16 days, while doves survive under similar conditions for 2−6 days. Their glycogen reserves are exhausted after the first 8 h of this period. The doves' cloacal temperature does not change over the entire period and their metabolic rate is maintained constantly at four times the basal level. Therefore death occurs primarily as a result of starvation and not of freezing (STREICHER, HACKEL, and FLEISCHMANN, 1950). This holds still more strongly for small birds, which lose considerable amounts of weight after only a few hours in the cold if they are deprived of food (KENDEIGH, 1944, 1945b, 1949; SCHILDMACHER, 1952). Their glycogen reserves last for only about 1 h (KENDEIGH, 1944). In this connection it is important that the period of daylight in winter be long enough to permit the bird to store enough nourishment to last through the resting period in the dark. According to KENDEIGH (1945b) the survival time of fasting sparrows at −14° C in winter is only 19 h, compared to 61 h at 34° C. It should be noted that even so the animals are substantially acclimatized, since the survival time at −14° C during the summer months is only 11 h. Very small birds, such as the wren (*Troglodytes aedon*, 11 g), survive less than 5 h without food at −14° C. Such a bird species cannot exist at this temperature, but must migrate to warmer surroundings.

In *water* the cold resistance of warm-blooded animals is very limited, despite their elevated metabolic levels, since they lose the largest part of their protection against cold. Swimming birds with water repelling plumage are exceptions to this rule, as are aquatic mammals (whales, seals) and the polar bear whose thick blubber layers enable it to resist water temperatures of 0° C indefinitely. In water of 20−30° C the normal regulation of rats, guinea pigs, and rabbits fails, but a new thermal equilibrium at lower rectal temperatures can be maintained for long periods. Large dogs remain at normal temperatures in water of 20° C, and some can still keep their rectal temperatures constant for at least 5 h at 0° C (SPEALMAN, 1946), which indicates that a thick, long coat of hair provides a distinct protection against cold, even in water (WOLFF and PENROD, 1950).

2. Limits of Regulation in Man

The limits of regulation are reached quickly by *unclothed* humans. Reduction of the dermal blood flow cannot increase the heat insulation of the tissues greatly, since even without blood flow, they are good conductors of heat in comparison with hair or feathers (p. 546). Experiments of ADOLPH and MOLNAR (1946) have shown that at air temperatures of −1° C a resting nude man can maintain a constant rectal temperature for at least 1−2 h. During this time heat production rises, due to shivering and increased metabolism, up to 370 kcal h^{-1}, i.e. 5 times the basal metabolic rate, while the body loses 700 kcal in the first hour. Skin temperatures fall very sharply, so that after 1−2 h a gradient of 22° C exists between average skin temperature and rectal temperature. The pulse frequency rises by about 30%

of the resting value. A combination of cold ($-1°$ C) with a slight wind can be tolerated voluntarily for about 1 h, on the average. Endurance of such conditions is limited by acute discomfort, cold pain, fatigue of cold shivering, and cold stupor.

The limiting temperatures for regulation by man in _water_ are very significant for the practical problem of survival of shipwrecked men or downed aviators. MOLNAR (1946) compiled the experiences of shipwrecked men. His data indicate that survival is possible for up to 1 h at water temperatures of $-1°$ C, for an average period of about 10 h at $15-21°$ C, and for far longer periods above $21°$ C. $18°$ C is a critical temperature, below which survival times are sharply reduced. The question of swimming and of wearing clothes is very important in this connection. Most people without specific advice are likely to regard clothing as simply an encumbrance to swimming, and many of them will remove clothing before or just after entering the water. Furthermore, there is a natural tendency to keep warm by swimming. The rectal temperature of unclothed subjects in water of $5-15°$ C falls at a significantly greater rate than those of subjects wearing clothes (KEATINGE, 1969). Further, in contrast to previous opinions, more recent experiments by KEATINGE (1969) have shown that in water of $5-20°$ C working generally causes a greater fall in rectal temperature than keeping still; working had an opposite effect only at water temperatures above $25°$ C. Figure 83 shows the physiological

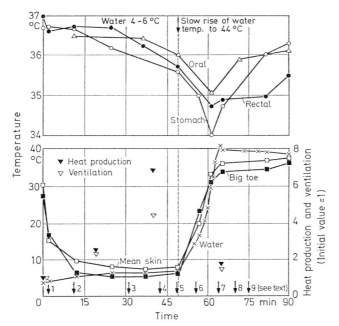

Fig. 83. Changes in deep and skin temperatures of subject chilled and rewarmed in water. 1, shaking chills, painful especially around neck, goose flesh; 2, less pain, face and ears blue, hands cyanotic, body pinkish blue; 3, breathing deeply, toes painful; 4, toes painfully numb; 5, toes numb, still shaking; 6, shaking violently, feels much colder than before, toes painful; 7, feeling warmer, still shivering; 8, comfortable for first time; 9, very comfortable, shivering gone. From BEHNKE and YAGLOU (1951)

events following immersion of a volunteer in water of 6° C for 52 min. Rectal temperature fell to 35.7° C, and a further drop to 34.7° C occurred after the beginning of rewarming before the temperature started to rise.

D. Hypothermia

In contrast to the narrow temperature range of hyperthermia, homeothermal life can continue to exist uninjured in the face of intense cooling. Recently the use of artificial hypothermia in medicine gave a strong impetus to the investigation of the life processes of homeotherms at low body temperatures. For references to numerous works, especially in the clinical area, the reader may turn to several comprehensive presentations (DRIPPS, 1956; HEGNAUER, 1959; STARKOV, 1960; DARBINIAN and PORTNOI, 1961; THAUER and BRENDEL, 1962; SWAN, 1963; BLAIR, 1964; MENGES, 1968).

1. The Course of Hypothermia in Animal Experiments

The most thorough investigations have been carried out on dogs, to which we chiefly refer. When cooling begins the regulatory processes are first stimulated to higher levels of activity. Shivering and heat production reach levels seven times as high as that of basal metabolism, while the blood sugar level rises and the liver glycogen stores are depleted (KRAMER and REICHEL, 1944). At a rectal temperature of 30−34° C shivering and heat production are at a maximum (Fig. 84); at lower temperatures they gradually diminish due to paralysis of the regulatory centers. Some dogs still display an elevated muscle tone and slight shivering at a brain temperature of 20° C (PENROD, 1949). Once the centers are paralyzed by cooling the life processes are reduced with further cooling, in parallel with the temperature.

If the regulatory processes are suppressed by narcosis and muscle relaxers, then oxygen use, heart rate, respiratory volume, etc. diminish with the falling rectal temperature from the time cooling starts (Fig. 85). Most investigators find a linear relation between rectal temperature and oxygen use, heart rate, respiratory volume, and other processes, others a more exponential curve. A mathematical interpretation of the curves based on the reaction-rate-temperature law or similar equations could scarcely be meaningful since complicated interrelations between regulatory processes and direct temperature effects

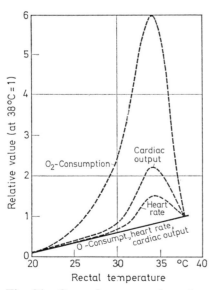

Fig. 84. General course of various physiological parameters in the dog during hypothermia. Dashed lines: Thermoregulation active. Solid lines: without thermoregulation. Adapted from KRAMER and REICHEL (1944) and HEGNAUER and D'AMATO (1954)

are involved. At least three different temperature coefficients would have to be considered: 1. local tissue processes, 2. nervous centers and 3. peripheral cold receptors and afferent nerves. The uncontrollable effects of the anesthesia also play a role.

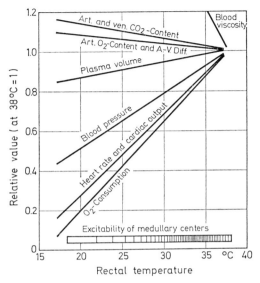

Fig. 85. Physiological parameters of the hypothermic dog without thermoregulation as function of rectal temperature. The schematized curves are based on numerical data from various authors

Cardiovascular Functions. As body temperature falls, the *heart rate* slows due to direct effects of cold on the sinus node. The reduction in frequency follows a course approximately parallel to the reduction in the organism's use of oxygen. At 18° C systole lasts 5—6 times as long as normal (HEGNAUER, FLYNN, and D'AMATO, 1951; D'AMATO, 1956), while the duration of the relaxation phase increases even more than that of the contraction phase (PENEFSKY, 1968; KAUFMANN, HOM-BURGER, and TRITTHART, 1969). Coronary blood flow also falls, but the myocardium receives an adequate oxygen supply (PENROD, 1951; BERNE, 1954; HANSEN et al., 1956; GEROLA, FREINBERG, and KATZ, 1959). Since the stroke volume of the heart remains approximately constant in hypothermia, the reduced minute volume must be due entirely to bradycardia. One may say that the cardiac activity is *"normal"* for the hypothermic condition of the organism.

At temperatures between 19° and 26° C ventricular fibrillation may appear (HEGNAUER and D'AMATO, 1954). Various causative factors have been discussed, such as catecholamine release or plasma potassium changes. According to BEAVERS and COVINO (1959) the ventricular fibrillation may be eliminated if the reduced calcium level of the plasma is brought back to normal by the intravenous adminis-tration of calcium chloride. KEARNS and MURNAGHAN (1969) suggest that hypo-thermia sensitizes the ventricles to fibrillate directly, possibly by inhibiting the electrogenic pump via its effect on metabolism. This suggestion agrees with the observation that in deep hypothermia the potassium content of the heart muscle falls, the sodium level rises, and the membrane potential is reduced (GLITSCH, 1969). Sensitivity towards catecholamines and sympathetic nervous stimulation is increased (SCHNEIDER and GILLIS, 1966), and conversely sympathectomy prevents ventricular fibrillation in the hypothermic heart (NIELSEN and OWMAN, 1969). At temperatures between 19° and 14° C *asystole* sets in (HEGNAUER and COVINO, 1956). Asystole may occur, through impairment of sinoatrial conduction, while the pacemaker is still active (TORRES and ANGELAKOS, 1964).

In deeper hypothermia the blood pressure falls appreciably. At 20° C the blood viscosity is 2—3 times its normal level. Furthermore the plasma volume is reduced and a hemoconcentration appears (HEGNAUER, SHRIBER, and HATERIUS, 1950; D'AMATO and HEGNAUER, 1953; SIRCAR, 1954; D'AMATO, 1954; KLUSSMANN, LÜTCKE, and KOENIG, 1959; FEDOR and FISHER, 1959; POPOVIC and KENT, 1965; KANTER, 1968).

Respiration. During hypothermia the *respiratory rate* falls and respiration finally ceases due to paralysis of the respiratory centers. The temperature range in which apnoe occurs varies greatly and is affected by anesthesia, among other factors. PENROD observed spontaneous respiration in unanesthetized dogs at a rectal temperature of 11.8° C and at a brain temperature of 18° C. If ventilation is adequate, the O_2 and CO_2 contents of the *blood* remain approximately normal down to the lowest temperatures of 15—18° C (GROSSE-BROCKHOFF and SCHOEDEL, 1943; KRAMER and REICHEL, 1944; PENROD, 1949). The cold-induced leftward displacement of the oxygen dissociation curve and the consequent reduction in oxygen pressure (BROWN and HILL, 1923) are substantially compensated by a rightward displacement caused by acidosis of the blood (falling pH) (SEVERINGHAUS, 1959). In fact all the more recent experiments lead to the important conclusion that the oxygen supply to the tissues is normal over the entire range of hypothermia down to the lowest levels of cooling, since the oxygen provided and the oxygen demand go down in parallel (KRAMER and REICHEL, 1944; PENROD, 1949; BIGELOW et al., 1950; ROSENHAIN and PENROD, 1951; GÄNSHIRT et al., 1954; HEGNAUER and D'AMATO, 1954; PEIRCE et al., 1958 int. al.). Of course this requires an adequate oxygen saturation of the blood in the lungs.

Nervous System. As temperatures fall the higher *brain functions* are slowed first. At temperatures below 28° C the electroencephalogram shows reduced frequencies and potentials, while the thalamic and medullary centers are inhibited only at lower temperatures (PENROD, 1949; BRENDEL, 1957; DZAHNELIDSE, 1966). The EEG of the dog is extinguished at 18—20° C (GÖKHAN and ANGELAKOS, 1963), but in monkeys electrical activity of the cortex can still be detected at 19° C (BRYCE-SMITH, EPSTEIN, and GLEES, 1960). The hypothermal changes include reduced accommodation on electrical stimulation (KOIZUMI, USHIYAMA, and BROOKS, 1960) and a marked lengthening of the latency of the cortical evoked potential (BOAKES, KERKUT, and MUNDAY, 1967). Occasionally tonic potentials also appear (HIRSCH et al., 1963). It can be assumed that the changes in brain activity are not based on hypoxia, since the oxygen supply usually remains normal (ROSOMOFF and HOLADAY, 1954; ROSOMOFF, 1956; BERING et al., 1956) even though in deep hypothermia below 15° C local reductions in oxygen pressure may appear in the brain tissues (BYON and ADOLPH, 1961). Changes in brain metabolism, brought about by cold, may be the essential cause (GÖKHAN and ANGELAKOS, 1963). The rate of conduction in the *peripheral nerves* is sharply reduced (DE JONG, HERSHEY, and WAGMAN, 1966).

Liver and Kidney. Liver metabolism, bile secretion, and glucose utilization are reduced (BRAUER, LEONG, and HOLLOWAY, 1954; RINK, RUECKERT, and SLOCUM, 1956; WYNN, 1956), but the relative changes in bile composition are very slight (VANLERENBERGHE et al., 1967). In the kidney glandular filtration as well as tubular reabsorption is reduced; as a result of these changes urine volume and glucose excretion rise (KANTER, 1959).

Other Homeotherms. The remaining homeotherms display a similar pattern in hypothermia. In *birds* shivering sets in at first, but is gradually paralyzed as the temperature sinks further. In the chick, which is fairly resistant to cold, shivering stops at a cloacal temperature of 20° C, respiration stops at 15° C, and cardiac activity at 10° C (RANDALL, 1943). Newborn and young animals in general can withstand more intense hypothermia than adults (p. 687). The lowest body temperatures are tolerated by hibernators (p. 667).

Rewarming. The method of choice for relieving acute general hypothermia is by warming as rapidly as possible in a bath at 40–44° C (ALEXANDER, 1946; HATERIUS and MAISON, 1948; PENROD, 1949; BEHNKE and YAGLOU, 1951; D'AMATO, KRONHEIM, and COVINO, 1960; KEATINGE, 1969 int. al.). Slow warming is not desirable due to the sharp subsequent fall in rectal temperature (p. 661). Spontaneous rewarming does not take place at low temperatures since the regulatory centers are paralyzed. They become active again only when the rectal temperature reaches 24–28° C (PENROD, 1949). In general the course of recovery is approximately a mirror image of the course of cooling, but there are various functions, e.g. the brain potentials, whose recovery takes place at substantially higher temperatures than those of paralysis during cooling (GÄNSHIRT et al., 1954). Apparently the ability of such processes to recover is damaged by the preceding cooling. KEATINGE (1969) contains references to the revival of sub-cooled humans on the basis of recent experiences.

2. Hypothermia in Man

As experiences with naturally or artificially cooled individuals show, man's internal temperature can be reduced by ca. 12° C (to 25° C), without incurring irreversible damage to the life processes. Below this temperature the continuation of life is in danger unless special precautions are taken. In the absence of anesthesia, metabolism can be increased to eleven times its basal level. The maximum is reached below 34° C (FAY and SMITH, 1941), and below 30° C shivering and metabolism are substantially reduced.

The first *experimental* cooling of humans for medical purposes, down to rectal temperatures of 23.5° C, was performed by FAY (1940). His work and that of others show that in man amnesia usually occurs at a rectal temperature of about 34° C (FAY, 1940, 1945; FAY and SMITH, 1941). Dysarthria ensues at about 30 to 34° C, and close to 27° C spontaneous movements and response to verbal commands are lost. Artificial hypothermia finds wide applications in surgery (Lit. see THAUER and BRENDEL, 1962; SWAN, 1962; BLAIR, 1964; MENGES, 1968). It has also proved its value in brain injuries and brain embolisms (CHEPKII, TRESHCHINSKII, and SVIRYAKIN, 1967). More recently hypothermia has been combined with extra-corporeal circulation to reduce body temperature to ca. 10° C (NIAZI and LEWIS, 1958; BJÖRK, 1960; DREW, 1961; DUBOST, BLONDEAU, and PIWNICA, 1962; LANGDON and KINGSLEY, 1964 int. al.). The physiological changes during such cooling fundamentally correspond to the series of changes described in the previous section. Down to ca. 25° C spontaneous respiration remains intact. Acidosis in the range between 28 and 20° C amounted to a pH reduction of -0.0153 per °C, and at 20° C the partial pressure of CO_2 rose to nearly twice the value at 38° C, while the CO_2 content remained the same (AUSTIN, LACOMBE, and RAND, 1964).

3. General Cold Death

The *lethal temperatures* for homeotherms (except for hibernators) are usually between 15 and 20° C for long exposures, i.e. far above the freezing point of tissues. The lethal limit appears to be lower for mammals than for birds. The *time factor* is also of great significance for the lower temperature limit. Table 15 presents some lethal temperatures for adult birds and mammals (young animals, p. 687).

By using a special technique for cooling and rewarming it is possible to lower the temperatures of homeotherms far below the usually lethal levels of $15-20°$ C and to revive them unharmed. ANDJUS (1955) cooled rats to a colonic temperature of $-3°$ C for ca. 40 min and then rewarmed them by microwave diathermy. Such animals survived in large numbers and showed no damage over a time span of up to two years. Hamsters were cooled to still lower temperatures (LOVELOCK and SMITH, 1956, 1959; SMITH, 1961). The immediate survival rate of animals cooled to colon temperatures of $-5°$ C for $30-70$ min was almost 100%. If the period of cooling did not exceed 40 min, about one half of the animals survived for more than one year. In the cooled state $30-40\%$ of the body water was frozen; however if a greater proportion turned to ice the survival rate fell sharply. Pregnant females, cooled in this manner, later gave birth to normal young, which matured without difficulty and reproduced. Recent experiments on cooled bats (e.g. *Myotis sodalis*) led to the astonishing result that cardiac activity did not stop even at body temperatures of $-5°$ C (KALABUKHOW, 1958; HENSHAW, 1965).

The *cause* of acute *cold death* is not adequately known at this time. Doubtless much depends on the species and on the time course of cooling. In accidental hypothermia, when the regulatory processes which resist cooling are not switched off, the transitory increase in metabolism may lead to a relative anoxia (McNICOL and SMITH, 1964). If this stage is survived, death usually results from ventricular fibrillation and asystole (p. 663). But heart failure is not the only cause of cold death. Thus ADOLPH (1948) found that sometimes hypothermal rats could not be revived even though the heart was still beating. The hypothermal state of homeotherms is not a steady state; even when the life processes seem to be fully extinguished at low temperatures, time-dependent changes in tissue metabolism are taking place which lead to irreversible damage after some time. This is demonstrated by experimental results on hypothermic rats and hamsters, which could be revived

Table 15. Lower lethal body temperatures

Species	Lethal temperature ° C	Remarks
Wren[a]	32	
Fowl[b]	23	
Bat (Myotis sodalis)[c]	—5	Still heartbeat
Rat[d]	13	Cardiac arrest
Rat[e]	—3	Supercooled, artificial hypothermia, revival
Hamster[f]	—5	Supercooled, artificial hypothermia, revival
Guinea pig[g]	17.5—21	
Cat[h]	14—16	
Dog[i]	15—18	
Dog[j]	0—1.5	Extracorporeal circulation, revival
Rhesus[k]	14	
Man[h]	24—26	Average, accidental hypothermia
Man[l]	18	Lowest, accidental hypothermia, revival
Man[m]	9	Artificial hypothermia, 45 min cardiac arrest, revival

[a] From BALDWIN and KENDEIGH (1932).
[b] From MORENG and SHAFFNER (1951).
[c] From HENSHAW (1970).
[d] From ADOLPH (1948).
[e] From ANDJUS (1955).
[f] From SMITH (1961).
[g] From GOSSELIN (1949).
[h] From CRISMON and ELIOT (1947).
[i] From PENROD (1949).
[j] From GOLLAN et al. (1955).
[k] From SIMPSON (1902).
[l] From LAUFMAN (1951).
[m] From NIAZI and LEWIS (1958).

for long periods only if the duration of cooling was not substantially greater than one nour (SMITH, 1961). The average survival time of hypothermic mice falls off exponentially as the body temperature falls (USINGER, 1962). The oxygen consumption of artificially perfused homeotherm hearts at 4° C does not remain constant, but diminishes in the course of time (ARNOLD and LOCHNER, 1965). In deep hypothermia between 0° and 5° C electrolyte changes take place in the brain in which intracellular sodium- and water content rises due to an inhibition of cation transport (MESSMER et al., 1966). As a result, death may ensue through edema of the brain (DONALD and KERR, 1964; MESSMER et al., 1966; REULEN et al., 1966).

4. Local Effects of Cold

The direct effects of cold on the tissues of homeotherms are similar in many respects to those found in poikilothermic vertebrates. The temperature dependence of local tissue processes is also similar, since central temperature regulation does not affect them. The delayed injuries following local action of cold consist of reddening of the skin (cold erythema) due to injury of the capillaries (first degree freezing), development of edema and separation of the epidermis (second degree), and finally necrosis (third degree).

Some tissues are damaged even at temperatures *above the freezing point*. Cooling tissues *in situ* induces a strong local and reflex vasoconstriction, which sometimes extends far into the normal tissue (MEINERS, 1952). Intense cold pain is felt down to tissue temperatures of 0° C (ABRAMSON et al., 1966; KEATINGE, 1969). Vaso-constriction may be interrupted in rhythmic patterns by dilatations (Lewis-reaction, p. 583), which rewarm the tissues. This reaction ceases at very low temperatures and in general hypothermia. In this stage long-term cold exposure may lead to injury, even at temperatures between 1 and 15° C (SAYEN et al., 1960). In man these injuries are prone to appear on the feet ("immersion foot"). In some cases a chronic loss of sensation is found, which possibly results from direct injury of the nerves by cold rather than from ischemia (KEATINGE, 1969).

The *freezing point* of tissues of homeotherms usually lies at −1 to −2° C. Experiments on the fingers of volunteers showed that the skin could freeze when the fingers were immersed in brine at −1.9° C, the true freezing temperature of tissue being −0.53° C (KEATINGE and CANNON, 1960). Cooling periods of up to 7 min caused no lasting damage, while cooling of 20 min was followed by blistering of the skin. The cause of local cold injury is not sufficiently understood at this time, but it seems to be based less on the mechanical effect of ice crystals than on an increased electrolyte concentration (MERYMAN, 1957). In most tissues, ice crystals form outside cells, and most of the intracellular water is drawn out osmotically during freezing, leaving a high intracellular concentration of electrolytes. (With respect to the effect of local cold and ice formation, see p. 410.) Unlike cold injury due to prolonged exposure to temperatures above the freezing-point, freezing injury results largely from damage to small blood vessels with consequent failure of the tissue's blood supply. When frozen tissue is thawed a high blood flow returns to it immediately, but the walls of the vessels are damaged and allow plasma to escape rapidly. Later thrombosis and subsequent necrosis of tissue can develop (LANGE, BODY, and LOEWE, 1945; CRISMON and FUHRMAN, 1947; KREYBERG, 1949). The strongest evidence that necrosis following frostbite is due to this arrest of blood flow rather than to direct effects of freezing on tissues other than blood vessels is provided by the fact that frostbitten tissue, which will necrose *in situ*, will survive if grafted to a normal area (KREYBERG, 1950; SJOSTROM, WEATHERLEY-WHITE, and PATON, 1964; WEATHERLEY-WHITE, SJOSTROM, and PATON, 1964).

Slow rewarming used to be regarded as the best way to treat frostbite, but animal experiments have shown that there is less necrosis of tissue if the frozen part is *rewarmed suddenly* in hot water of 40° C (LEWIS, 1951; CRISMON, 1951; SHUMAKER and LEMBKE, 1951; FERRER, 1955; FUHRMAN and FUHRMAN, 1957). Evidence that this reduces the damage to blood vessels is provided by ATURSON (1966). Clinical evidence (MILLS, WHALEY, and FISH, 1960; McDADE, 1962; KEATINGE, 1969) is consistent with the view that rapid rewarming is the only proven means of reducing tissue loss. Even tissue that appears to be irreparably damaged a few days after frostbite may largely recover if it is left alone and premature surgery is avoided.

5. Deep Cooling of Organs and Tissues

Numerous investigations on the reactions of homeotherm organs and tissues to low temperatures have been carried out in connection with their preservation for purposes of transplantation and the preservation of sperm for artificial insemination. At this point we will discuss the special relations of homeotherms only briefly and for the remainder refer the reader to the general exposition on p. 410.

The preservation of seminal fluid of agriculturally useful animals at very low temperatures has now become a routine procedure. Frozen bull semen can be preserved for 7 years at −80 to −90° C, without loss of high motility and fertilizing capacity by the spermatozoa (POLGE, 1957). Embryonal chicken hearts, pretreated with glycerine, could be brought back to normal activity after several weeks at −79° C. Even extreme cooling to −196° C was tolerated for short periods (REY, 1958). HAUSCHKA, MITCHELL, and NIEDERRUEM (1959) were able to keep various normal and neoplastic mammalian tissues alive for one to two years at −79° C in media containing 10−15% glycerine. Ovarian tissue of rats, which was treated with 15% glycerine and held at −79° C for 70 weeks, resumed its functioning after autotransplantation, though with noticeable defects (PARKES, 1958). Deep-cooled ovaries of mice can form fertilizable ova from which normal offspring develop (PARROTT, 1960). Human skin tissue (TAYLOR, 1957; REY, 1959 int. al.) and corneal tissue (RYCROFT, 1955; BILLINGHAM, 1957; HENAFF, 1960 int. al.) can be maintained for several months at −79° C after treatment with glycerine or ethylene glycol, and be successfully transplanted. Nerve fibers and ganglion cells can also recover and develop action potentials on electrical stimulation after several days at −76° C (PASCOE, 1957). Additional literature on deep-cooling of organs and tissues is found in SMITH (1961).

X. Temperature and Development

A. Spermatogenesis

Spermatogenesis, fertilizing capacity, and motility of the sperm of homeothermic species are strongly temperature dependent. Chicken sperm survive temperatures of $-15°$ C and can still move at $6°$ C, but they are irreversibly damaged by heating to $50°$ C for more than two minutes. They use oxygen most rapidly at $40.8°$ C, and are damaged least by temperatures of $20°$ C (WINBERG, 1941). Fertilized rabbit ova remain capable of development for up to 150 h at $10°$ (CHANG, 1947). POLGE et al. (1949) demonstrated that spermatozoa can be cooled to $-79°$ C and thawed again without losing their capacity for fertilization. Similarly, ovarian tissue can be re-implanted successfully after several days of cooling to $-79°$ C (p. 669).

The *spermatogenesis* of those *mammals* whose testes lie in a scrotum is strongly inhibited by slight temperature increases, even by elevation to the level of the rectal temperature. MOORE (1926) could obtain complete degeneration of the seminal tubules of the ram by thermally insulating the scrotum for 90 days. The same effect could be produced in goats, guinea pigs, and rabbits by artificial warming of the scrotal sack (MOORE, 1926; FUKUI, 1923) and in rats, guinea pigs, rabbits, and sheep by repositioning the descended testes into the abdominal cavity (MOORE, 1926; KNAUS, 1897). Temporary sterility can also be produced by artificial hyperthermia in *man* (MACLEOD and HOTCHKISS, 1941).

Furthermore it is known that *chryptorchism* in man and in mammals with permanently descended testes invariably leads to sterility and degeneration of the seminal tubules while the endocrine functions of the gonads are evidently undamaged. On the basis of these findings, MOORE (1926) developed the view that the scrotal sack serves as an *organ for cooling the testes*. In fact the temperatures in the scrotum of mammals are $2-10°$ C lower, depending on the external temperature, than the rectal or abdominal temperature (Lit. in KNAUS, 1932).

As has been shown more recently [for literature see HUNDEIKER and KELLER (1962), HUNDEIKER (1971), WAITES (1970)] this cooling system is formed in mammals and man by the following elements: 1. a countercurrent vascular system consisting of the venous plexus pampiniformis surrounding the internal spermatic artery, 2. the existence of warm receptors in the scrotal skin (cf. Chapter VII, p. 585 and p. 599), 3. the existence of sweat glands in the scrotal skin, 4. the wide variability of blood flow through the scrotal skin, 5. the ability of the scrotal skin to contract and relax.

KNAUS (1932) showed that not only spermatogenesis but also the fertilizing capacity and motility of sperm stored in the epididymis is injured by the temperatures of the abdominal cavity. If isolated rabbit epididymis is transplanted into the abdomen, the sperm lose their motility in five days, while normally their capacity for fertilization lasts 40 days and their motility substantially longer. Evidently the chemical potential energy stored in the sperm is exhausted very rapidly at the higher temperature. Thus, according to KNAUS (1932), the mammalian scrotum also serves to maintain the sperm at low temperatures after they are formed.

The fact that *birds*, whose body temperatures are especially high, have *intra-abdominal* testes seems to stand in a certain contradiction to this: However in their case spermatogenesis is said to proceed only at night, when their body temperature is at a minimum, and during the reproductive period the testes of some species descend to a position between the abdominal air sacks, which, according to COWLES and NORDSTRÖM (1946), may have a cooling effect.

Among the *mammals* a permanent *intra-abdominal* position of the testes is found, according to WISLOCKI (1933), in the monotremata, in some insectivores, e.g., the Centetidae (tenrecs), Macroscelididae (elephant shrews), Chrysochloridae (golden moles), and in sloths, elephants, hyraxes (Hyracoidae), sirenia, and whales. In fact the body temperatures of some of these animals, especially the monotremes and the edentates, are far lower than the rectal temperatures of higher mammals, but that of the elephant which also has intra-abdominal testes is 36° C, which is by no means low. Conversely, the temperatures of the American marsupials, which all have descended testes, are very low. In addition a *periodic descent* during the reproductive season is known in many mammals, such as some insectivores, e.g. Talpidae (moles), Soricidae (shrews), Solenodontidae, Erinaceidae (hedgehogs), and in all rodents and chiroptera. All hibernators belong to this category. Therefore, even if there is no strict relation between body temperature, spermatogenesis, and descent of the testis, the facts speak clearly for the idea that the scrotal sack has an important cooling function in most mammals.

The fertility of most animals fluctuates *seasonally*. In cold climates it is usually highest in summer (STEVENSON, 1946). In moderate to warm climatic zones it often declines in summer (BRODY, 1941; MERCIER, 1946), though not in heat acclimatized animals, such as the Indian zebu *(Bos indicus)* (WILSON, 1946). Sterility during the hot season is *not* due to a direct effect of temperature on spermatogenesis, but rather, according to BOGART and MAYER (1946), to a temporary hypofunction of the thyroid. The summer sterility of the ram is relieved by giving thyroxin, while inhibition of the thyroid by thiouracil during the autumn fertility maximum leads to sterility.

B. Effects of Temperature on Growth

Due to its intrauterine site embryonal development in *mammals* is completed under very constant temperature conditions, while the egg temperature of *birds* fluctuates considerably during the brood period, thereby exerting varying effects on embryonal development. The average egg temperature during the brood period of wild birds is said to be 34° C, but this figure includes great variations. For example the average temperature of eggs of Passerinae is 33.6° C, those of Galliformes 36.4° C. Temperature differences as great as 12° C have been measured between eggs at the center and edge of wild ducks' nests (HUGGINS, cited by BRODY, 1945).

Various segments of the *growth curve* of chicken embryos are influenced differently by temperature (HENDERSON and BRODY, 1927; HENDERSON, 1930; ROMANOFF, 1935, 1936; ROMANOFF and SOCHEN, 1936; ROMANOFF, SMITH, and SULLIVAN, 1938; BRODY, 1945). With increasing age and the development of homeothermy the growth rate becomes less dependent on temperature.

The *brood temperatures* of birds show distinct optima with respect to normal embryonal development, hatching rates, and survival rates, int. al. The optimal external temperatures for the duck, chicken, and turkey lie between 36.5 and 37.5° C (ROMANOFF, 1943; BRODY, 1945). In these cases as the size and heat production of the embryo increase the temperature within the egg rises, up to 40° C on the 18th day (BAROTT, 1937; ROMANOFF, 1941). Colder and warmer temperatures have unfavorable effects on developmental processes. Temperatures above the optima lead to inhibition of development and delay the resorption of albumen by the embryo (ROMANOFF, 1943). The growth of chicks after hatching is also inhibited by heat (KLEIBER and DOUGHERTY, 1934). Cold brooding temperatures lead primarily to a reduction of the post-hatching survival rate and to a marked increase in the number of malformations. These effects can be elicited by only a few days of exposure to low temperatures (BRODY, 1945).

As for mammals, no marked weight differences were found in groups of mice reared at 21° and 33° C, respectively (HARRISON, 1963). Studies in pigs have also shown that growth is not substantially influenced by the ambient temperature; thus body weight did not significantly differ in two groups of pigs which were kept for two months from weaning on at ambient temperatures of +5° and +35° C, respectively, provided that the feeding regimen was adjusted to the increased food demand in the cold. Considerable differences were obtained, however, in the body shape, length of extremities and of the tail, and in the magnitude of the ears of these pigs (WEAVER and INGRAM, 1969). These changes in shape are interesting in connection with the problem of thermal adaptation (cf. Chapter VIII) since overall heat insulation is greatly dependent on such changes.

C. Ontogenetic Development of Homeothermy

Considering the complexity of the thermoregulatory system, one might expect its development to require a fairly long time span within ontogenesis. In mammals this system is needed for the first time immediately after birth. This raises the question of whether the organism is already equipped with a fully organized thermoregulatory system at birth, or whether the development of this system is set in motion or completed only as a result of extra-uterine thermal stress (cf. ADOLPH, 1968). An additional problem exists in the case of birds, i.e. whether thermoregulatory reactions are already present in the intra-oval stages, enabling the egg to counteract cooling to a certain extent in the absence of the brooding bird. It should be said at once that, to our knowledge, there have been no investigations of the latter problem.

In evaluating temperature regulation systems over the course of ontogenesis one must remember the trivial fact that neonates of homeothermic species are always smaller than their parents. One must therefore bear in mind that the range of ambient temperature within which they can regulate ("tolerated ambient temperature range") is smaller than that of the adults. This phenomenon should not be confused with a poorer "quality" of regulation (see below). In order to understand the ontogenetic processes of development of temperature regulation, changes of size must therefore be eliminated or otherwise included in the calculations. Only then can changes with age *per se*, and in certain cases maturation processes, be recognized; as will be seen, when such factors are taken into account one can even find definite adjustments of the temperature regulatory system to the smaller body size (BRÜCK, 1964, 1970).

1. The Significance of Body Size and Shape

At a given ambient and body temperature, heat loss per unit of body weight is reciprocally related to body size; this is due to the fact that 1. the surface-volume-ratio (Fig. 86), 2. the absolute thickness of the body shell, and 3. the radius of curvature of the body are functions of body size; thus to maintain the same tem-

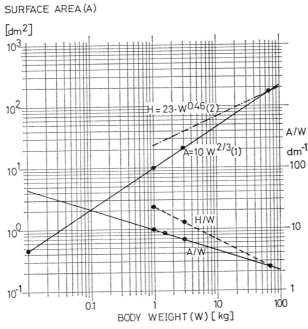

Fig. 86. Body surface area *(A)* as a function of body weight *(W)* according to Meeh's equation (1); assuming a specific gravity of 1, *W* [kg] is put in place of volume, *V* [dm³]. For the "shape factor" [10 in Eq. (1)] cf. MEEH, GÜNTHER. Eq. (2) predicts the heat loss, *H* (in arbitrary units) required to maintain a constant overall thermal gradient for body weights from 1—70 kg. Eq. (2) is based on the size-dependent total maximum heat insulation according to Fig. 87. For further details see text

perature gradient between core and air the small individual would have to produce more heat per unit of body weight, and the increment of metabolic rate with decreasing ambient temperature would have to be larger. On the other hand, in a warm (radiating) environment the small individual is much more subjected to hyperthermia than is a larger organism; again due to the difference in surface-volume ratio.

The change in the surface-volume index (A/W) with increasing body size is shown by Fig. 86. The values given for a 1500 g premature infant, a 3000 g full-term infant, and an adult man show a proportion of $3.7 : 2.9 : 1$. This means that the neonate's heat production per unit body weight has to be increased by these factors, at least, in order to compensate for the greater heat loss. However, considering the different values of maximum thermal insulation, heat loss per unit body weight must be expected to be approximately 5 times as high in the 3000 g infant, than in the adult.

The Eq. (2) in Fig. 86 approximately predicts, in arbitrary units, the heat production which would be required per unit body weight to maintain a constant thermal gradient between the body core and the environment in the body weight range of $1-70$ kg. This equation is based on the data on total maximum insulation as given in Fig. 87, i.e. it makes due allowance for the fact that heat loss per unit body

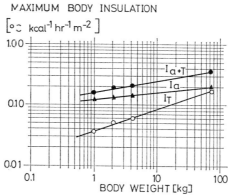

Fig. 87. Maximal possible thermal insulation of the body shell, I_T, of the "still air layer" on the body surface, I_a, and the sum of both, I_{a+T} in man from birth to adulthood. The values for the newborn infants are taken from HEY, KATZ, and O'CONNELL (1970). The adult values are the values generally agreed upon (for Literature see HARDY 1961, and this book, Chapter V)

surface area increases with diminishing body size. In view of the so-called surface rule discussed in Chapter IV, p. 531, it is interesting that the exponent of Eq. (2) is much smaller than 0.67. Its size will vary from species to species, and the larger the differences in heat insulation (thickness of fur, and layer of subcutaneous fat) between the newborn and the adult the smaller it will be. It should be mentioned here that the values predicted from Eq. (2) in Fig. 87 fit the heat production and temperature gradients measured in the newborn infant satisfactorily (BRÜCK, 1961, 1970).

Change in body shape may require, in addition, an age-dependent change of the "shape factor" (cf. GÜNTHER, 1971) in the exponential allometric equations predicting body surface area from body weight, since, with advancing age the growing animal not only changes its size but also its shape. In many species the young animal tends to have a slender body with proportionally longer legs, and this increases the relative surface/mass ratio (BIANCA, 1970). By contrast the extremities of human neonates are shorter and all their contours more rounded; their surface/volume ratio is thereby slightly reduced, i.e. the shape factor in Eq. (1) tends to increase with increasing age.

The ratio between body weights of newborns and adults varies greatly among species. In the guinea pig this ratio may be as low as $1:7$ to $1:10$, while it is $1:20$ in the full-size human neonate, $1:70$ to $1:90$ in premature infants as well as in the rat and $1:100$ in the pig. Accordingly, the surface-volume ratio and maximum heat insulation (Fig. 86) show quite different alterations, during postnatal development, from species to species. Thus, the surface-volume ratio of a fertile female rat weighing 300 g would be 15 dm^{-1}, in comparison with 60 dm^{-1} in the 5 g newborn rat; i.e. it is increased by a factor of 4, while this factor is only 2.9 when the full-size human neonate is compared with its mother, and the factor would be even smaller in the case of the guinea pig. One must keep in mind that the neonates of small species such as the rat, and to a greater extent the mouse, which weigh 5 and $1-2$ g at birth, respectively, are at or below the critical limiting weight for homeothermic species (cf. Fig. 13).

Minimal Conductance; Maximal Insulation. As was shown in Chapter V, the heat efflux from the body core to the environment is determined by the heat insulation of the body shell (I_T), or the resistance (R_T), the hair- or feather coat (I_{cl}), and finally by the marginal layer of air in contact with the body surface (I_a). The thickness of the air layer depends on the radius of curvature of the heat-emitting body surface (cf. Chapter V, p. 548). I_T is reduced in the newborn, not only because of the lesser thickness of the body shell, but also due to the absence of a significant layer of fat and the higher heat conductivity of the tissues, which usually contain more water in the newborn. In many neonates which are born without hair or feathers, I_{cl} approaches zero. The quantity I_a amounts to $0.18°\,C \cdot m^2 \cdot hr \cdot kcal^{-1}$ in adult humans; for neonates weighing 4.2 and 1 kg, respectively, the corresponding values are 0.146 and 0.120 (HEY, KATZ, and O'CONNELL, 1970; see Fig. 87).

The characteristics determined by body and shape must always be kept in view if one wants to describe the ontogenetic development of thermoregulation. Precisely because of this changing body size and shape, measurement of the relation between body- and ambient temperature is by no means sufficient for evaluating the temperature regulatory system. A comparative consideration of the final effector systems of thermoregulation is required.

2. Comparison of Effector Systems

Heat Production. As is shown in Fig. 88, at least two different groups of neonates can be distinguished on the basis of their metabolic responses to cold; in the rat (GELINEO and GELINEO, 1951) one finds a thermoregulatory increase in heat formation when ambient temperature falls, even on the first day of life; the magnitude of this reaction increases in the following days. In the ground squirrel (*Citellus citellus;* GELINEO and SOKIC, 1953) on the other hand a decrease in ambient temperature in the period immediately after birth leads to a decrease in heat formation, just as in the poikilotherms (cf. Arrhenius' law, p. 305). Only in the course of 3—4 weeks does cooling lead to a more distinct increase in heat formation. Recently another species pair with corresponding differences in metabolic reactions, the lemming and the hamster, have been investigated. The lemming's reaction is similar to that of the rat, while the hamster exhibits thermoregulatory metabolic reactions only in the course of 2—3 weeks of life (HISSA, 1968).

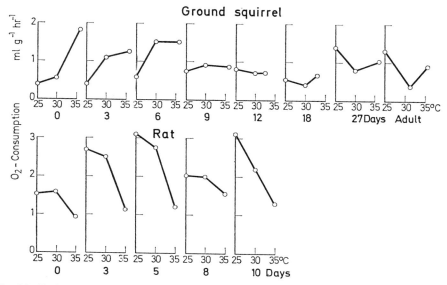

Fig. 88. Relation of metabolism to external temperature at different stages of development. Ground squirrel *(Citellus citellus)* from GELINEO and SOKIC (1953), rat from GELINEO and GELINEO (1951)

In human neonates thermoregulatory metabolic reactions can be identified immediately after birth, even in premature infants weighing 1 kg at birth (BRÜCK, BRÜCK, and LEMTIS, 1958; BRÜCK, 1961, 1970), as Fig. 89 shows.

Various observations suggest that birds also include species which show no cold-induced metabolic reactions after hatching. The wren (KENDEIGH, 1939) and other insessorial birds (PEMBREY, 1895) may be of this type. The decisive factor for thermoregulatory effectiveness and accordingly the thermostability of the newborn is the magnitude of the *BMR* and peak metabolism in relation to the heat efflux (*H*) or to the quantity H/W in Fig. 86. Metabolic rates as high as those required for conformation with Fig. 86, Eq. (2), have not been found in any species examined to this time. At best, e.g., in the guinea pig, the *BMR* immediately after birth already exhibits the value predicted by the Kleiber function ($H = 0.67 \ W^{0.75}$ kcal/ day) for an adult animal of the same size (Fig. 90); in many species, however, the

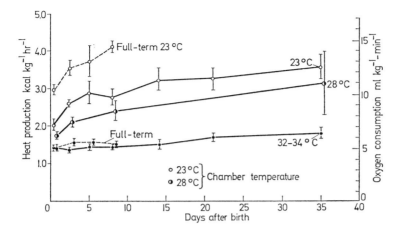

Fig. 89. Heat production in full-size (dashed lines) and premature (solid lines) infants in relation to environmental temperature and age. Mean values $\pm 2\ S.E.$ From BRÜCK (1961)

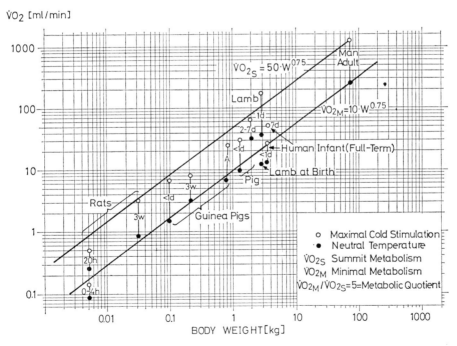

Fig. 90. Minimum and maximum (summit) metabolism related to age and body weight in various homeothermic species. The lower line corresponds to KLEIBER's mouse-to-elephant-curve. Data for rats: TAYLOR (1960); guinea pigs: BRÜCK and WÜNNENBERG (1965a); lamb: ALEXANDER (1962); pig: MOUNT (1968); human infant: BRÜCK (1961)

BMR exceeds the Kleiber curve in the course of the following weeks and months. The value of the metabolic quotient at birth is at best (for example, again, in the guinea pig) the same as that observed in adults, i.e., 4−5 (Fig. 90). The peak metabolism also remains within the range specified by the upper curve in Fig. 90. At the time of birth, the *BMR* and peak metabolism of the rat are substantially lower than either of the limiting curves drawn in Fig. 90, but in the course of the first few days they reach levels within the limiting region of Fig. 90. At the age of three weeks the rat's reactions are such as would be expected of an adult animal of equal size.

In the first hours of life the *BMR* and peak metabolism of the *human newborn* also fall below the corresponding limiting curve (Fig. 90); in the course of the first week (or even of the first two days; HILL and RAHIMTULLA, 1965) they increase, without however reaching the upper limiting curve within that time. In the pig these values also increase in magnitude during the first 2−3 days of life, the peak metabolism climbing almost to the upper limiting curve of Fig. 90 at the age of 2−5 days (MOUNT, 1968). Taking into consideration the heat production *per unit weight*, one finds that the peak metabolism of the human newborn, 4−5 kcal kg^{-1}, already approaches the maximal value of the adult, 5−6 kcal kg^{-1} h^{-1} (ADOLPH and MOLNAR, 1946; BEHNKE and YAGLOU, 1951).

In the guinea pig, peak metabolism per unit body weight, ca. 20 kcal kg^{-1}h^{-1} (BRÜCK and B. WÜNNENBERG, 1965a), is even higher than in the adult where it amounts to only 10−12 kcal kg^{-1} h^{-1}; i.e. the "metabolic reduction" (see Fig. 14) in *BMR* is paralleled by a corresponding reduction in peak metabolism. Immediately after birth the pig's maximum oxygen uptake is 27 ml kg^{-1} min^{-1} (MOUNT, 1968). This is greater than the value of 17 ml kg $^{-1}$min^{-1} which is to be expected for a 100 kg pig with a metabolic quotient of 5, according to Fig. 90.

The four species represented in Fig. 90 indicate the major possible modes of development. Dogs, cats, and rabbits appear to behave similarly to the rat. We would expect the larger species, such as the cow and horse, to behave similarly to the pig. Among the smaller species the guinea pig represents a kind of Goliath, with respect to the capacity of its metabolic system; this has already been noted in the earlier work of GINGLINGER and KAISER (1929).

Nonshivering Thermogenesis (NST). Neonates of many species increase their heat production, on exposure to cold, without shivering, i.e. in these cases nonshivering thermogenesis is involved (see p. 636). Shivering begins only in the event of extreme cold stress. The absence of shivering led to the incorrect conception that neonates of these species had no thermoregulatory metabolic reactions at all at the time of birth (cf. BRÜCK, 1964, 1970).

It could be demonstrated that nonshivering thermogenesis in the guinea pig is eliminated in the course of a few weeks when the animal is held in a warm environment. Rearing in a cold environment partly prevents this deterioration (Fig. 94; ZEISBERGER et al., 1967).

In contrast to the newborn of man (KARLBERG, MOORE, and OLIVER, 1965; BRÜCK, 1961), guinea pigs (BRÜCK and B. WÜNNENBERG, 1965a, b; 1966), rabbits (DAWKINS and HULL, 1964; HULL and HARDMAN, 1970), lambs (ALEXANDER, 1968), cats, rats, and dogs (MOORE and UNDERWOOD, 1960, 1962, 1963), non-shivering thermogenesis apparently does not occur in neonates of larger species.

Thus, for example, no *NST* could be demonstrated in the calf (JENKINSON, NOBLE, and THOMPSON, 1968). The pig, even though quite small at birth, likewise does not make use of *NST*; on the other hand its shivering mechanism is very well developed (MOUNT, 1959, 1968a; BRÜCK, WÜNNENBERG, and ZEISBERGER, 1969; cf. also Fig. 95). In agreement with the absence of *NST* in the pig, no brown fat could be detected.

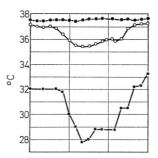

Vasomotor Activity. Thermal conductance of the body shell or its reciprocal value, thermal tissue insulation, varies with blood flow to the body surface and to the extremities. It has been asserted for a long time that the neonate is not able to constrict skin blood vessels properly, and thus, to reduce peripheral blood flow. During the last decade, however, it has been shown that the newborn infant responds to changes in the environmental temperature, even to local skin temperature changes, with vasomotor reactions (BRÜCK, BRÜCK, and LEMTIS, 1957, 1958b; BRÜCK, 1961; CELANDER, 1960; YOUNG, 1962), as shown by Fig. 91.

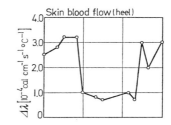

The efficiency of vasoconstriction with regard to increasing tissue insulation is enhanced with increasing size, since insulation is a function of the absolute thickness of the body shell and the layer of subcutaneous adipose tissue, as mentioned earlier (cf. Fig. 87). Vasomotor changes in heat insulation have also been demonstrated in the newborn pig (MOUNT, 1963, 1964, 1968).

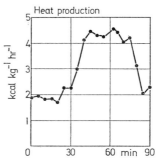

Details on vasomotor control will not be presented here, since this area has recently been treated in monographs; for the pig, by MOUNT, 1968; for the human neonate, BRÜCK, 1961, 1970.

Fig. 91. Simultaneous response of skin blood flow (heel) and heat production to a drop in environmental temperature. Study in a 7-day-old infant, birth-weight 3290 g. Top: rectal temperature •; average skin temperature o; chamber temperature ••. From BRÜCK (1961)

Sweat Secretion and Heat Tachypnea. In newborn full-term infants, sweat secretion has been observed during the first few hours of life when they were exposed to environmental temperatures of 35—37° C; sweat secretion begins at rectal temperatures of 37.5—38° C (BRÜCK, 1961). Quantitative determinations of sweat secretion in the neonate have been performed recently (HEY and KATZ, 1969; FOSTER, HEY, and KATZ, 1969). They found that the neonate born within 3 weeks of term is capable of more than doubling its total evaporative water loss when the environmental temperature and the rectal temperature exceed 36 and 37.5° C, respectively. The maximum amount of evaporative water loss increases during the first 10 days of life, and the threshold body temperature for sweating decreases simultaneously. Thus the ability to defend the body against overheating is enhanced in the postnatal period.

In the newborn guinea pig, an animal in which increasing respiratory rate is the prevailing mechanism of heat dissipation (see p. 560), heat tachypnea can be evoked both by external heat exposure and by local heating of the hypothalamus and the spinal cord (cf. Section VI A). With extreme heat exposure the respiratory rate increased from 60 (resting level) to 200–300 min; in addition, newborn guinea pigs were observed to lick their paws when they were exposed to intense heat, thus increasing evaporative heat loss.

BIANCA and HALES (1970) demonstrated that the newborn calf could lose evaporative heat effectively by sweating. The rate of moisture loss from the skin per unit of metabolic body size ($W^{0.75}$) at high environmental temperatures was even greater in the newborn than in the older animals. In the sheep, which is one of the species that loses heat predominantly from the respiratory pathways, effective heat-induced panting could be recognized — as in the guinea pig — immediately after birth (ALEXANDER and WILLIAMS, 1962). When newborn pigs were exposed to heat stress for one to three hours, the respiratory rate increased from ca. 30 to 200–300 (MOUNT, 1968).

It would be interesting to inquire whether vasomotor, respiratory, or sweating thermoregulatory reactions can be demonstrated at birth in species whose young do not display metabolic thermoregulatory reactions at that time (hamster, ground squirrel, see above). To our knowledge no investigations of this question have been carried out.

Behavioral Regulation. Behavioral regulation can be demonstrated at birth in many species. Newborn rats, mice, and pigs lie close to one another in the nest, thereby reducing the surface/volume ratio for the litter as a whole (cf. Chapter V, p. 563). MOUNT (1968) described still another behavior pattern of pigs with relevance for thermoregulation. On cold surfaces they lie down in such a way that as little thermal contact as possible exists between their ventral surface and the substrate, thereby reducing conductive heat loss. By contrast, they lie flat on their bellies on warm surfaces or at high ambient temperatures.

Since human neonates are born singly and with poorly developed motor abilities, they cannot make use of these behavioral patterns. All they can do is to cry, thereby inducing indirect behavioral regulation, and also to supplement non-shivering thermogenesis by bodily activity, which forms heat by muscular contraction.

3. Neutral Zone, Range of Regulation ("Tolerated Ambient Temperature Range")

According to the exposition in Chapter IV, p. 536, the lower limit of the zone of thermal neutrality is shifted to higher temperature values for small species. Since the minimal heat production immediately after birth lies *below*, or at best on, the Kleiber curve for almost all species (cf. Fig. 90), the lower limit of the neutral zone for newborns is usually at a higher level than it is for the parents or for mature adults of the same size as the newborn. For example this limit for human neonates is 32–34° C, compared to 28° C for adults (Fig. 92). In correspondence with the increase in basal metabolism per unit weight that takes place over the course of hours, days, or weeks, the neutral zone is shifted towards the adult value. As a rule however this value is not attained completely until a much later stage of development, namely when the pelt and the corresponding heat insulation have grown to conditions resembling those of the adult. The similarity of the zone of neutrality

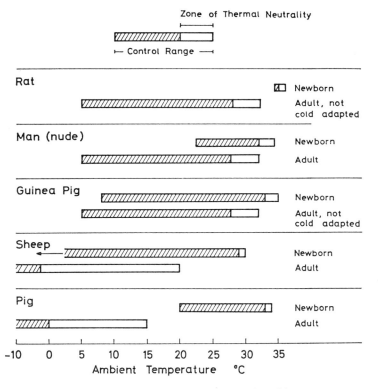

Fig. 92. Zone of thermal neutrality (□) and "tolerated ambient temperature range" (□ + ▨) of newborn and adult animals of various species. Rat: TAYLOR (1960); COTTLE and CARLSON (1956); man: ADOLPH and MOLNAR (1946); BRÜCK (1961); guinea pig: BRÜCK and WÜNNENBERG (1965); sheep: ALEXANDER (1962); pig: MOUNT (1968)

of neonates, that are very small in comparison with their parents and that have only moderate heat insulation, to the adult range signifies a considerable accomplishment by their metabolic systems, i.e., the neonates must produce substantially more heat per unit of body mass than the adults (cf. p. 675). Such a metabolic compensation is not found in animals, such as the rat, that are born in a very immature condition and at a small body size [cf. Fig. 86, Eq. (2)]. In these cases the approach of the neutral zone to the adult range is brought about to a large extent by an increase in size, which however takes place quite rapidly.

As a rule the range of regulation, i.e. the range of ambient temperatures within which the deep body temperature can be maintained at a constant level for at least a few hours, is also narrower in the newborn (see "tolerated ambient temperature range" in Fig. 92). For example an unclothed adult human can maintain his body temperature for at least 1–2 h at an ambient temperature of about 5° C (ADOLPH and MOLNAR, 1946), while this lower limit for the human newborn is in the region of 20–23° C (BRÜCK, BRÜCK, and LEMTIS, 1958a; BRÜCK, 1961). If the neonate were to have the same range of regulation as its parent, its maximal heat formation would have to conform to Eq. (2) in Fig. 86, i.e. it would have to lie above the upper limiting line for adults in Fig. 90. Such a situation has not been observed in any species examined thus far. Evidently the metabolic capacity per unit body mass can be increased only to a very limited degree within one species, (i.e. the ratio between maximal heat production by the neonate and by the adult of a species apparently cannot exceed a definite limiting value). Accordingly the lower limit of the range of regulation will be closer to the adult values for neonates that are large, in proportion to their parents, than for those with smaller birth sizes. In species whose young approach the limiting lower weight for homeotherms (rats, mice) the range of regulation immediately after birth is only slightly wider than the zone of thermal neutrality (see Fig. 92). Thus, if they are exposed individually the extra heat production induced by cold might be considered, at first glance, as a useless (since ineffective) reaction.

Due to their behavioral regulation, i.e. huddling together with their litter mates in the nest, isolation of an individual from the nest is quite an unnatural situation. When they are huddled together, the surface-volume ratio of the whole group is markedly reduced; under these conditions the seemingly small amount of extra heat production in the single individual becomes an effective tool in establishing a reasonably extended control range for the whole group. According to GELINEO's (1951) studies the rats commonly build their nests at 16° C. For the whole group, one may thus consider a temperature of around 16° C or even lower to be the lower limit of the range of regulation.

4. Deep-Body Temperatures; Stability of Body Temperature; Quality of the Neonatal Control System; "Set Point"

The average deep-body temperature of neonates of many species, measured at a neutral ambient temperature, does not differ significantly from that of their parents; this is true for (full-term) human neonates (BRÜCK, BRÜCK, and LEMTIS, 1958; BRÜCK, 1961, 1964, 1970, 1971), for the guinea pig (GINGLINGER and KAYSER, 1929; BRÜCK and WÜNNENBERG, 1965a), for the pig (MOUNT, 1968), and certainly for the newborn of all larger mammals as well. Reductions of body temperatures, of one to two degrees C, are observed only in the first hours; this so-called initial temperature drop has been attributed to the substantial amounts of heat which the wet neonates may lose immediately after birth (cf. BRÜCK, 1961).

However, when the neonates of *small species*, such as the cat (author's observation, unpublished), dog, and rat are exposed to the neutral ambient temperature very early in life, their body temperatures are several degrees lower than the adults'. Very small human premature infants (birth weight less than 1500 g) also maintain a rectal temperature of 35–36°C, i.e. 1–2 degrees lower than that of more mature neonates or adults, when they are exposed to conditions which do not evoke increases in metabolism (BRÜCK, PARMELEE, and BRÜCK, 1962; BRÜCK, 1970). In such cases one could say that the thermoregulatory system is aimed at a somewhat lower "set point". It remains to be investigated whether the lowering of the "set point" is due to a cold-adaptive shift or is determined by the stage of development. Since it is possible to displace the set point for body temperature to lower values through cold adaptation (Chapter VIII, p. 625), it is of course impossible to regard the reduced body temperature of the newborn as simply an expression of underdeveloped thermoregulation, as is often done.

From the point of view of control engineering the quality of a system is determined by the degree of deviation from the set point under stress. Naturally the stress must not exceed the capacity of the range of regulation. It has been shown that the full-term neonate within its zone of regulation, which is indeed narrow in comparison with the adult's, (cf. Fig. 92), can maintain its body temperature with the same precision and at the same level as the adult (Fig. 91; BRÜCK, 1961, 1964, 1970).

At a time when precise information about the zones of thermal neutrality of many neonates was not available, the "quality" or "degree of maturity" of thermoregulation of many animals with very narrow ranges of regulation was tested by choosing temperature stresses that lay far outside these ranges, e.g. 20° C for the rat. In such a case the body temperature must of course fall rapidly (induced hypothermia). Merely as a result of increasing size and the widening of the range of regulation that goes with it, this fall in body temperature gradually becomes smaller until one can finally observe temperature constancy under the chosen conditions. This phenomenon has often led to the concept of a maturation process involving the central nervous or endocrine system and has been taken as proof that homeothermy in the species under consideration is not fully developed before a certain number of days after birth have elapsed. In fact the only change to occur may have been that the discrepancy between the efficiency of the effector system and the body size and heat insulation was reduced (cf. BRÜCK, 1961, 1970). Therefore considering only body temperature can lead to significant misjudgements of the degree of maturity at birth and the processes of maturation after birth. In this connection it must also be remembered that thermal adaptation can lead to substantially increased fluctuations in the body temperature of adult animals (Chapter VIII, p. 626). Accordingly, relatively wide fluctuations of body temperature in cold or heat exposed newborn animals cannot be interpreted simply as "immaturity of regulatory centers". An evaluation of thermoregulatory maturity requires a thorough analysis of the thermoregulatory system. In this way it could be shown that thermoregulatory reactions, i.e. increases in heat production and vasoconstriction, followed changes in the temperature of the body surface, while the core temperature remained constant, as is shown in Fig. 91. (With regard to the initiation of thermoregulatory processes by thermoreceptors on the body surface, see p. 571 and Fig. 24.) As a result of the poorer heat insulation of their body shell the average skin temperature of neonates is higher than that of adults at the same core and ambient temperature (BRÜCK, 1961). For example in the zone of thermal neutrality the skin temperature of neonates is over 36° C and that of adults is 32–33° C. As Figs. 91 and 93 show, at this increased temperature level metabolic

thermoregulatory reactions and vasoconstriction are induced in the neonate to
their full extent. The same temperature constellation in an adult would induce
sweating. This shift in relation between skin temperature and effector activity
(Fig. 93) can be seen as an adjustment of the thermoregulatory system to the
smaller body size (BRÜCK, 1964). It is the prerequisite for the ability of small
neonates to maintain their body temperatures, within their given range of regulation, with the same precision and at the same level as the adult.

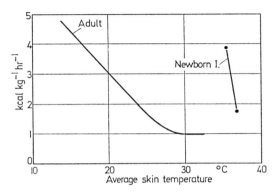

Fig. 93. Relationship between average skin temperature and heat production in adults
and newborn infants. Adult curve based on data from ADOLPH and MOLNAR (1946);
infant curve based on mean values from 4 full-term infants aged 7 to 8 days. From
BRÜCK (1961)

We may speculate that this adjustment, i.e. the shift to the right of the newborn
curve in Fig. 93, is due to an alteration of the "reference signal input" (cf. Fig. 24).
Cyclic Changes in Body Temperature. Circadian rhythmicity in body temperature
is not present in the human newborn infant; it is only developing during the first
few months of life (HELLBRÜGGE, 1960; see also p. 516).

5. Thermal and Other External Effects on the Development of
   Temperature Regulation

For a long time neonatal physiology has been looked upon as the physiology of
deficient or incompletely developed functional systems. Hence, any such system
would have to become a perfectly working system during postnatal development.
However, temperature regulation does not fit such a concept, as we have demonstrated. On the contrary, the temperature regulatory system of many mammals,
including man, not only possesses a high degree of perfection at the time of birth,
but is well adjusted to the smaller body size.
The existence of *NST* and brown adipose tissue at the time of birth may be
thought of as such a compensatory mechanism for the smaller body size, since the
thermoregulatory efficiency of *NST* is greater than that of shivering (cf. p. 539).
Loss of Nonshivering Thermogenesis. In relatively mature newborns, such as the
guinea pig, the amount of brown adipose tissue and the extent of *NST* are greatest

at the time of birth; both vanish within a few weeks (Fig. 94). This involution of brown adipose tissue and *NST* can be retarded and partly inhibited by rearing the animals in a cold environment. After it has disappeared through exposure to a warm environment, *NST* can again be evoked by exposing the older or even adult guinea pig to a cool environment (cf. Chapter VIII, p. 622). As for the maximum extent of *NST*, it does not make any difference whether the animal was exposed to a cool or warm environment during the newborn period. With increasing age,

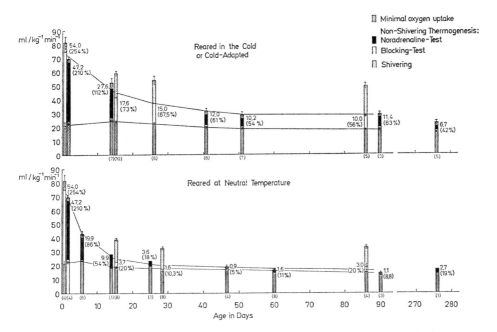

Fig. 94. Reduction of the maximum extent of nonshivering thermogenesis (as determined by two methods) with increasing age and the dependence of this process on the environmental temperature at which the guinea pigs were reared. The figures beside the columns give the *NST* in ml kg^{-1}min^{-1} and in percent of minimal oxygen uptake. Figures below the columns indicate the number of animals. The bars indicate S.E. From Zeisberger et al. (1967)

however, the extent of evocable *NST* becomes smaller and smaller. Since the extent of *NST* is already at its maximum at birth, there is no possibility of improving cold resistance by adaptive modifications in the metabolic system. On the contrary, we come to the following conclusion. Regarding the preponderance of *NST*, the newborn resembles a cold-adapted more than a warm-adapted adult subject, although the intrauterine environment has protected the fetus from experiencing cold stimuli. On the other hand, we may conclude that cold adaptation in the adult organism (as far as it produces brown adipose tissue and *NST*) is partly based on the re-establishment of a neonatal functional mechanism (Brück, 1964, 1970).

GLASS, SILVERMAN, and SINCLAIR (1968) obtained some results which permit us to assume that in newborn infants a process of degradation of *NST* occurs similar to the temperature-dependent process as described for the guinea pig. By contrast, in the rat the maximum extent of *NST* increases during the first 2—3 weeks of age (HSIEH and CARLSON, 1969) and this is presumably linked with a corresponding growth of brown adipose tissue. Such behavior may be expected to occur in other neonates which, like the rat, are born in a relatively immature stage.

As in the guinea pig atrophy of the brown fat and the decline of nonshivering thermogenesis is retarded by cold stress, so the postnatal development of brown fat in the rat is evidently enhanced by cold environmental conditions. This can be deduced from the classical experiments of GELINEO (1950), GELINEO and GELINEO (1951). They maintained rats' nests at ambient temperatures of 10 to 16° C or 28—30° C. At the end of the first two weeks of life the metabolic reactions of the young reared at the colder temperatures were higher.

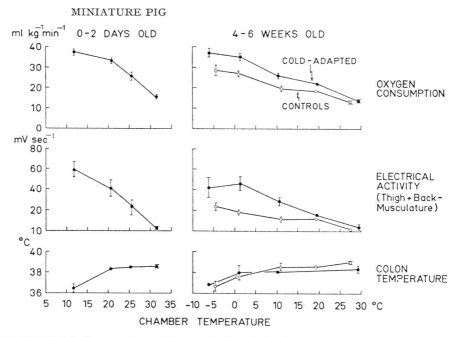

Fig. 95. Metabolism and muscle electrical activity in miniature pig in relation to environmental temperature, age and temperature of adaptation. From BRÜCK, WÜNNENBERG, and ZEISBERGER (1969)

It can be assumed that, as in the rat, a postnatal increase in brown fat and *NST* takes place in small premature human infants. This would explain the gradual increase in the maximal cold-induced heat production (shown in Fig. 89) that can be observed during the first weeks of life of premature infants.

In the miniature pig, which is not able to make use of NST, the maximum extent of shivering is reduced in the course of postnatal development, and this reduction like the reduction in NST of the guinea pig is affected by the rearing temperature, as Fig. 95 shows. So, even though the mechanism of heat production in the two species differs, there is a similarity with respect to the postnatal change in magnitude of cold-induced heat production. Whatever the mechanism of heat production may be, in both cases the extra heat production is maximal at the time of birth; its decline can be retarded to a greater or lesser degree by rearing the animals in the cold, but it does not seem to be possible to raise it to a higher level.

D. Tolerance of Hypothermia by Homeothermic Species in the Course of Development

For reasons set out before, the neonate is in far greater danger than the adult of falling into a hypothermic condition. On the other hand it has long been known that newborn birds and mammals are very resistant to cold. Respiratory and cardiac activities cease at lower temperatures in the newborn. Thus, investigations by ADOLPH (1951) showed that the hearts of newborn cats stopped at body temperatures between 5 and 10° C, while those of adult cats stopped at 15—20° C.

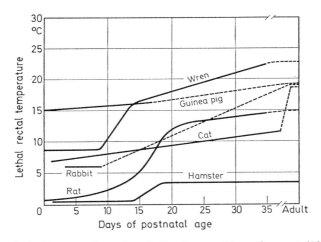

Fig. 96. Lower lethal temperatures in relation to age. From ADOLPH (1951)

The lowest colonic temperature which the animals can survive (lethal temperature) is 3—14° C lower for the newborn than for the adults, as Fig. 96 shows (ADOLPH, 1951); i.e. "biological zero" (ADOLPH, 1968) is at a lower point on the temperature scale for the newborn. At this time nothing is known with certainty about the causes of age-dependent cold tolerance. There is a correlation between developmental stage and tolerance of hypothermia, i.e. the less mature the animal is at birth, the greater is its tolerance for hypothermia, as a rule. Tolerance to high body temperature scarcely changes with age (ADOLPH, 1968).

XI. Hibernation and Related Phenomena

A. Distribution of Hibernation and Other Lethargic States

Some warm-blooded animals do not maintain homeothermy to the furthest limits, but instead undergo *hibernation* periodically. In this condition their body temperature can fall to 0—4° C, permitting them to survive for months without feeding. This condition differs from hypothermia in that, at the lower limit of tolerance, regulation begins again, and even at ambient temperatures below 0° C hibernation homeothermy ("second homeothermy" — Wyss, 1932) is maintained. Waking and active rewarming is possible at any time.

Hibernators are found in 6—8 orders of mammals (HERTER, 1956; EISENTRAUT, 1956; KAYSER, 1961). Among the insectivores are the hedgehogs: European hedgehog *(Erinaceus europaeus)*, Afghan large-eared hedgehog *(Hemiechinus megalotis)*, Asian eared hedgehog *(Hemiechinus auritus)*, North African *(Para- echinus aethiopicus)* and Algerian *(Aethechinus algirus)* hedgehogs; in the family of bristly hedgehogs *(Centetidae)* from Madagascar are the tenrec *(Centetes ecaudatus)* and several related genera (HERTER, 1962, 1963 a, b, 1964 a, b). Among the rodents the family *Sciuridae* (squirrels) is distinguished by numerous represen- tatives: the European, Asian, and North American marmots (Alpine marmot, *Marmota marmota;* steppe marmot, *M. bobac;* woodchuck, *M. monax);* and ground squirrels *(Citellus citellus, C. tridecemlineatus, C. lateralis, C. suslicus* and many others). Several species of ground squirrels are said not to hibernate, e.g. *C. leucurus* and *C. leptodactylus.* Nevertheless the isolated heart of *C. leucurus* displays hiber- nator characteristics: It does not ceases to beat until between 2° and 0°C (LYMAN, 1964). The same family also includes the hibernating chipmunks (American Eastern chipmunk, *Tamias striatus;* burunduk; *Eutamias asiaticus).* Among the American pocket mice *(Heteromyidae), Perognathus hispidus* and *P. longimembris* may be named, and among the hamsters *(Cricetidae)* the European hamster *(Cricetus cricetus),* the Syrian golden hamster *(Mesocricetus auratus)* and the Asian *Cricetulus triton.* In the family of jumping mice *(Zapodidae)* we find, among others, the birch mouse *(Sicista betulina),* and the jumping mouse *(Zapus hud- sonius),* among the jerboas *(Dipodidae) Dipus sagitta,* among the dormice *(Gliridae)* the common dormouse *(Glis glis),* the garden dormouse *(Eliomys quer- cinus),* the tree dormouse *(Dryomys nitedula),* and the hazel mouse *(Muscardinus avellanarius),* as well as several other related genera.

Among the *bats (Chiroptera)* most microchiroptera are probably able to hibernate after undergoing cold adaptation. Among the *prosimians (Prosimii)* of Madagascar lethargic conditions resembling hibernation are found in the genera *Cheirogaleus*, *Microcebus* and *Altililemur*. The same conditions are found among the *monotremata*, the echidna or spiny anteater *(Tachyglossus)* and the duck-billed platypus *(Ornithorhynchus)* during the Australian winter, as well as in small Australian and New World *marsupials (Marsupialia):* dormouse phalangers *(Cercaertus nanus)*, and opossums *(Didelphis)*, among others.

According to EISENTRAUT (1933) the semi-lethargic condition, related to hibernation, found among some *Carnivora* is known as *winter sleep*. It occurs in the bear, badger, raccoon *(Procyon lotor)*, and skunk *(Mephitis)* and is associated with restricted food intake and reductions in physiological functions. Body temperatures decrease only slightly: The minimal value for the American black bear *(Ursus americanus)* and the grizzly is 31.5° C (compared to 37.5° C in summer). The heart rate sinks from $40-70$ min^{-1} in resting sleep to $8-10$ min^{-1} in lethargy (FOLK et al., 1966). In the polar bear the heart rate goes down from 60 min^{-1} to 27 min^{-1} (FOLK, BREWER, and SANDERS, 1970). According to KAYSER (1961) deep hibernation could not develop in such large animals due to the laws of reduction of metabolism (Brody-Procter equation). The rate of cooling is too slow and thermogenesis would not be sufficient for rapid rewarming.

In some mammals temporary torpor also appears in summer, after the reproductive period — especially if food and water are scarce and the weather is cool (below 22° C). This so-called *summer sleep* or *estivation* has been observed in dormice, ground squirrels, in species of *Marmota* and *Perognathus*, in *Cercaertus nanus* (dormouse phalanger), *Microdipodops pallidus (Heteromyidae)* and perhaps even in the African Aarsvark *(Orycteropus)* (cf. EISENTRAUT, 1956; KAYSER, 1961; BARTHOLOMEW and CADE, 1957; BARTHOLOMEW and HUDSON, 1962; BARTHOLOMEW and MACMILLEN, 1961; AMBID, 1971). It is homologous to hibernation (EISENTRAUT, 1956; CADE, 1964), except for the absence of cold adaptation and the smaller degree of thermogenesis during arousal. Finally the ambient temperature is not of major significance: Estivation of ground squirrels and dormice sets in at ambient temperatures between 26° and 33° C. The body temperature necessarily falls far less than it does in hibernation. According to KALABUKHOV (1956) the ground squirrel's estivation is an adaptation to the lack of food and water in the arid summers of the steppe regions; nevertheless its initiation presupposes an accumulated store of Vitamin C in the liver, gonads, adrenals, and thyroid. The periods of sleep last for several hours, days, or weeks. In extremely dry regions the ground squirrel's estivation extends directly into hibernation.

CADE (1964) reviewed the occurrence of lethargic conditions among the rodents from a systematic-phylogenetic point of view. From this it appears that in the families of hibernators estivation or *circadian torpor* always appears as well. For example in the super family *Dipodoidea* there are some 20 species which hibernate (species of *Alactaga*, *Zapus*, and *Alactagulus*) or which hibernate and undergo diurnal torpor in summer *(Sicista betulina, S. subtilis)*. Among the *Heteromyidae* (12 species of *Perognathus*, species of *Microdipodops* and *Dipodomys*, 30 species of *Sciuridae*, and 7 of *Gliridae*) hibernation and estivation, or one of the two appears. Within the *Cricetidae* there are some 10 species which hibernate and 2–3 genera *(Gerbillus, Meriones)* with other conditions of torpor. According to HERTER (1961) the European hedgehog, established in New Zealand since 1885, undergoes hibernation as well as estivation, while the bristly hedgehogs of Madagascar *(Setifer setosus, Echinops telfairi, Centetes ecaudatus)* fall into diurnal torpor or hibernate (HERTER,1962, 1963a, b, 1964a, b). The daytime torpor of bats is also distinguished from hibernation only by degrees, as is the *nocturnal torpor* of some birds. From the point of view of thermoregulation, all forms of lethargy involve a *shift in the set point* for body temperature (cf. Chapter VI, p. 586).

B. Torpidity among Birds

Lethargy, accompanied by reduction in core temperature, has been observed among birds belonging to various systematic groups, partly under natural and partly under laboratory conditions. DAWSON and HUDSON (1970) present the minimal body temperatures of various species in a tabular summary. Among the *hummingbirds (Trochilidae)* several representatives *(Calypte anna, C. costae, Eulampis jugularis)* survive cooling to 8° C; the minima given for 7 other species *(Selasphorus rufus, Archilochus alexandri)* are 15−22.5° C, and for 2 more *(Hylocharis cyanus, Chlorestes notatus)* 31° C. The *swifts (Apodidae)*, which are related to the hummingbirds, tolerate less severe cooling: the common European swift, *(Apus apus)* can be cooled for several days to 27° C (young birds to 20° C), *Aëronautes saxatilis* for two to three days to 20° C. In the equally closely related family of the *nightjars* (goatsuckers, *Caprimulgidae*), the lowest value for *Caprimulgus europaeus* is 7.0° C, that of the Californian *Phalaenoptilus nuttallii* (Poorwill) is 4.8−6.0° C; by contrast only 18−19.2° C was measured in *Chordeiles acutipennis* and *C. minor*, and 29.6° C in *Eurostopodus guttatus*.

In other avian families only slight reductions in body temperatures could be measured: 22.8° C in the South African *Colius striatus* (colies or mousebirds, *Coliidae*), 22° C in the Inca dove *(Scardafella inca, Columbidae)*, 32.6° C in *Crotophaga ani (Cuculidae)*, and 34° C in one species of vulture *(Cathartes aura, Cathartidae)*. In addition reports on torpidity in three species of swallows are given in the compilations of EISENTRAUT (1956) and KAYSER (1961). Lack of food plays an important role in the initiation of torpor in all birds. Darkness is also significant for hummingbirds (LASIEWSKI, 1964) and for *Cathartes*. However hunger probably — as in mammalian hibernation — acts only as an inducing factor, (cue), without causing a breakdown in thermoregulation.

The average body temperature of *Colius striatus* at an ambient temperature of 5° C is 35.7° C at night and 39° C in the day (BARTHOLOMEW and TROST, 1970). Lack of food intensifies this rhythm to the point that *Colius* — even at a room temperature of 19.2° C — reaches a minimum body temperature of 22.8° C. KOSKIMIES (1948) observed a similar phenomenon in the common European swift: At an air temperature of 19°, rectal temperatures of 20−29° C were measured; young birds were cooled more drastically than adults. As periods of starvation were extended the daily poikilothermic phase became continually longer and more intense, and rewarming became weaker. The core temperature of various species of hummingbirds during resting sleep without torpor is 26.6−40.5° C, compared to 39.5−44.6° C during the day. In cool weather they fall into a daily rhythm of lethargy, thereby evading the thermoregulatory disadvantages of their small body size (nocturnal torpor). This is especially significant for the survival of species inhabiting the South and Central American high plateaus and mountains with their cold nights.

Torpor of longer duration, a type of hibernation, probably occurs in the American poorwill (*Phalaenoptilus nuttallii*). JAEGER (1948, 1949) found an animal in a rock niche during three winters in a state of lethargy. Its body temperature was 18° C, and active rewarming was possible. It is not known how long the phases of lethargy last, but on the basis of the small monthly weight loss a duration of 85 days is believed to be probable, by other authors as well (e.g. KAYSER, 1961). According to LIGON (1967), after cold adaptation the poorwill also can tolerate deeper body temperatures than 18° C and rewarm itself actively. This recalls similar conditions in bats.

Observations on hummingbirds (*Calypte anna*), in which nocturnal torpor is absent during the breeding season, provide a strong argument for the facultative nature of torpor (HOWELL and DAWSON, 1954). In addition, birds including the common European swift during hunger coma, can arouse spontaneously or under stimulation without application of external warmth and return to their normal body temperature. Low ambient temperatures do not hinder this process, but the bird can wake only if its body temperature has not fallen below a minimal value. This value, which is far above the lethal limit, lies at 13° C for *Calypte anna* and *C. costae*, 15—20° C for *Apus apus*, and 28.5° C for the Inca dove (cf. DAWSON and HUDSON, 1970).

According to the work of HUXLEY, WEBB, and BEST (1939), when Ecuadorian hummingbirds are exposed to cold they first fall into a partial torpor in which respiration is intensified, the feather coat is fluffed out, and the head is bent back. Subsequently the body temperature declines with an *S*-shaped course (Fig. 97). Shivering is not seen. Plotting the temperature logarithmically against time results in a certain distortion, but it can thus be seen that the speed of cooling is inversely

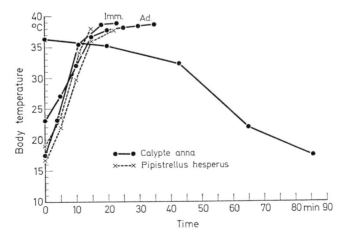

Fig. 97. Rates of arousal from two hummingbirds (*Calypte anna*) weighing 4.6 and 2.65 g, and two bats (*Pipistrellus hesperus*) weighing 4.5 and 3.7 g. The descending curve shows entry into torpor of an immature *Calypte anna* at an ambient temperature of 2° C. From BARTHOLOMEW, HOWELL, and CADE (1957)

proportional to the body weight (Fig. 98). The total set of reactions thus indicates the presence of true poikilothermy in torpor. Nevertheless, it could be demonstrated that the hummingbird *Eulampis jugularis* is poikilothermic only at ambient temperatures between 30° C and 18° C, and that at lower temperatures it becomes homeothermic again and is able to stabilize its body temperature at 18—20° C (HAINSWORTH and WOLF, 1969). Although low temperatures are said to be ineffective as stimuli for waking it is known that the poorwill reacts to sudden cooling of the air from 6—4° C by shivering.

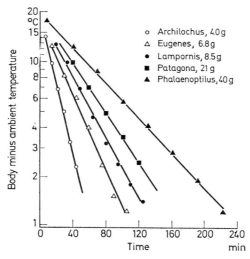

Fig. 98. Body temperature of birds during entry into torpor plotted as the log of the gradient between body and ambient temperatures vs time. From LASIEWSKI, WEATHERS, and BERNSTEIN (1967)

During *arousal* the increase in body temperature follows the same course as it does in small bats (Fig. 97). In comparison to cooling, rewarming takes place four times as fast. It is linear up to a body temperature of 35° C, is faster for small than for large birds (Fig. 99), and in various species it is accompanied by shivering of varying intensity. According to BARTHOLOMEW, HOWELL, and CADE (1957), however, the first phase of arousal of swifts takes place without shivering.

All physiological functions are strongly reduced during torpor. The heart rate of the poorwill declines from 500 min^{-1} to 10 min^{-1} (at a body temperature of 6° C) and in the hummingbird *Patagona gigas* (cooled to 21.4° C) it goes from 300 min^{-1} to 60 min^{-1} ($Q_{10} = 2.5$). In deep torpor (body temperature about 7° C) functional disturbances of the heart, in the form of atrio-ventricular dissociations, appear in hummingbirds (LASIEWSKI, 1964). During *arousal* the heart is under strong sympathetic stimulation: In *Patagona gigas* the rate increases to more than 1000 min^{-1}; the Q_{10} is approximately 8 (LASIEWSKI, WEATHERS, and BERNSTEIN, 1967).

Respiration during torpor is slow and irregular. The poorwill's oxygen use falls from 2 ml/g · h to 0.06 ml/g · h at 4.8° C and that of *Patagona gigas* (Fig. 100) falls from $2.7\ \text{ml/g} \cdot \text{h}$ (at an ambient temperature of $30°\ C$) to $0.25\ \text{ml/g} \cdot \text{h}$ (at an ambient temperature of $22°\ C$). The metabolism of the hummingbirds in torpor is twice as high as that of bats. Nevertheless lethargy represents a considerable saving of energy. Fat reserves of 10 g may suffice for 100 days of uninterrupted torpor for the poorwill (BARTHOLOMEW, HOWELL, and CADE, 1957; LASIEWSKI, WEATHERS, and BERNSTEIN, 1967).

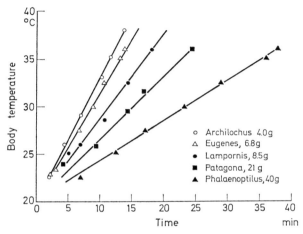

Fig. 99. The increase of body temperature during arousal from torpor for hummingbirds and the poorwill at an ambient temperature of $21°$—$23°\ C$. From LASIEWSKI, WEATHERS, and BERNSTEIN (1967)

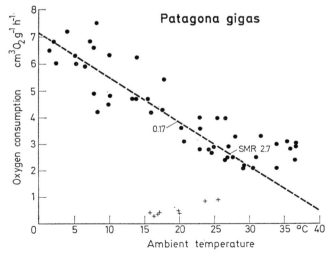

Fig. 100. Relationship between oxygen consumption and ambient temperature for awake (•) and torpid (+) hummingbird *(Patagona gigas)*. The regression line, fitted by the least-squares method to homeothermic values below thermal neutral temperatures, is taken as an estimate of minimum conductance. *SMR* standard metabolic rate. From LASIEWSKI, WEATHERS, and BERNSTEIN (1967)

C. The Lethargy of Bats

The bats, comprising approximately 1000 species, are found in tropical, sub-tropical, and temperate regions (KULZER et al., 1970). As nocturnal animals, they spend the day sleeping, suspended by their hind legs. The large flying foxes *(Megachiroptera)* exhibit no unusual daily rhythmical fluctuations in body temperature, while the small flying foxes and the microchiroptera exhibit normal homeothermic characteristics only when awake and are lethargic during sleep *(diurnal torpor* – EISENTRAUT, 1934).

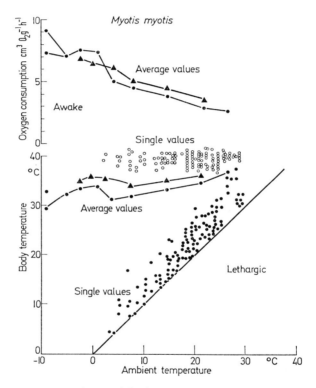

Fig. 101. Oxygen consumption and body temperature in the bat *Myotis myotis* at different ambient temperatures: o awake and • torpid, single temperature values. • Average values in winter 45-min-experiments, and ▲ average values in summer 45-min-experiments. Adapted from KULZER (1965) and MEJSNAR and JANSKY (1967)

The metabolism of *microchiropterans* of all climatic zones increases when the ambient temperature falls while they are awake, summer and winter, so that their body temperature falls only slightly (Fig. 101). Accordingly, diurnal torpor does not reflect any loss of thermoregulation, but results to a much greater extent from a displacement of the core temperature set point. The animals are also able to rewarm themselves actively at any time.

As the season progresses and the weather grows colder, the diurnal body temperature becomes progressively lower till diurnal torpor gradually merges into *hibernation*. But even then the *circadian rhythm* of nocturnal temperature increases if maintained for days or weeks (MENAKER, 1961; POHL, 1961). In principle hibernation can be distinguished from daytime torpor only by the intervening cold adaptation. After habituation to room temperatures the winter-hardy little brown bat *Myotis lucifugus* is unable, even in winter, to wake actively from hibernation at low ambient temperatures. On the other hand it can do so throughout the summer after three weeks of cold adaptation (MENAKER, 1962; HENSHAW, 1970). *Eptesicus fuscus* and *Myotis myotis* react in a similar fashion (POHL, 1961). After 14 days of cold acclimation even the tropical Australian bat, *Myotis adversus*, which ordinarily does not hibernate, was able to pass from diurnal torpor into a lethargic state lasting three weeks and then to arouse spontaneously (KULZER et al., 1970).

The BMR of bats, like that of other mammals, increases after *cold adaptation*. Therefore the metabolism of waking animals in winter — in spite of the higher level of readiness for hibernation — is as high as in summer (MEJSNAR and JANSKY, 1967), or even higher (HOLYOAK and STONES, 1971). Probably the ability of the microchiroptera to "reprogram" their thermoregulation and to become cold adapted first made it possible for them to colonize the temperate regions (KULZER, 1965).

As a result of climatic and ecological adaptation, each species of bat goes into torpor only above a certain temperature limit: Tropical and sub-tropical forms do so only when the ambient temperature is not lower than $17°$ C (KULZER, 1965). The minimal body temperature which can be attained under these conditions, $17°$ C, is also the limit below which spontaneous arousal is no longer possible. The core temperature of the small *flying foxes* of New Guinea, *Nyctimene albiventer* and *Paranyctimene raptor*, whose body weight scarcely reaches 30 g is only reduced to $25°$ C during diurnal torpor (BARTHOLOMEW, DAWSON, and LASIEWSKI, 1970). It is probably for the same reason that *Plecotus subflavus* prefers temperatures between $14°$ C and $18°$ C for hibernation in Florida and *Rhinolophus blasii* in Yugoslavia prefers $14.2—16.2°$ C (cf. DAVIS, 1970). If cold adaptation is inadequate a sharp reduction in ambient temperature during diurnal torpor or hibernation leads to hypothermia. The sub-tropical *Tadarida pumila* and *T. condylura* suffer cold death at $5°$ C and *Taphozous melanopogon* dies at body temperatures as high as $12—15°$C (KULZER, 1965).

During the *onset of torpor* the body temperature falls at a rate of $0.1—0.2°$ C min^{-1} (KULZER et al., 1970). The frequency of the *heart* beat is reduced abruptly and thereafter exhibits phases of tachy- and bradycardia. According to KULZER (1967) the rate in *Myotis* is $250—450$ min^{-1} during awake state (880 min^{-1} in excitation), $120—180$ min^{-1} in diurnal torpor and 18 min^{-1} in deep hibernation. After temperature regulation is "reprogrammed" the body temperature in the "steady state" is about $1°$ C higher than the ambient temperature. The reduction in metabolism follows the reaction-rate-temperature law (KAYSER, 1950; HOCK, 1951). KAYSER (1950) calculated an activation energy of 19000 cal mol^{-1} for *Plecotus* and *Vesperugo* and HENSHAW (1968), working with *Myotis lucifugus*, found a Q_{10} of 5.2 between $35°$ C and $25°$ C and one of 2.1 at lower temperatures.

After the hibernating bat reaches a body temperature of 1–2° C, further cooling is prevented by *regulatory processes* such as shivering and increases in heart and respiratory rates and metabolism (HOCK, 1951; KULZER, 1969), and perhaps by vasomotor reactions as well (HENSHAW, 1968). If the "second homeothermy" is overtaxed, the animal arouses. The minimal temperature is characterized by minima in oxygen use and heart rate. The rate for *Myotis myotis* at 5.5° C (ambient temperature = 4.6° C) is 18 beats min^{-1}, and for *Nyctalus noctula* at 5.5° C (ambient temperature = 3.0° C) it is 48 beats min^{-1} (KULZER, 1967). Various species *(Plecotus auritus, Nyctalus noctula, Lasiurus borealis, Myotis lucifugus, M. sodalis)* maintain a constant body temperature even at air temperatures of −5° C and below (cf. DAVIS, 1970). If thermoregulation fails after several hours in extreme cold, and arousal is not induced hypothermia with "supercooling" can set in *(Plecotus auritus, Nyctalus noctula, Myotis daubentoni)*. Active arousal is no longer possible, but the animals survive after passive warming (DAVIS and REITE, 1967). Pricking with a needle leads to a sudden hard freezing, and due to the release of latent heat of fusion the body temperature rises from −5° C up to −1.0° C to −0.5° C.

KULZER et al. (1970) investigated the speed of *arousal* in 27 European, African, and Australian species. At an air temperature of 20° C their body temperature rose at rates of 0.1–1.6° C min^{-1}, depending on the species. The most vigorous thermogenesis is displayed by the well-known hibernating families of *Vespertilionidae*, *Rhinolophidae*, and *Molossidae*. The investigations of KALLEN (1960), KULZER (1967, 1969), and MEJSNAR and JANSKY (1970) and the reviews of HENSHAW (1970), DAVIS (1970), and LYMAN (1970) reveal that during arousal — as in other hibernators — all the indications of generalized sympathetic discharge are present: sudden doubling of the heart rate, abdominal vasoconstriction, and activation of the brown fat tissue.

D. True Hibernation

1. Readiness for Hibernation

Hibernation, like instinctive behavior, is elicited by *internal stimuli* (HERTER, 1956). Readiness to respond is subject to an endogenous *annual* and *daily rhythm*. It is limited in each case to the cold season of the northern or southern hemisphere, and is initiated during resting sleep, i.e. by day for nocturnal animals (e.g. hamsters) and at night for diurnal animals (e.g. ground squirrels). Conversely, periodic spontaneous arousal usually takes place during the animal's normal time of activity, insofar as a sizable difference between circadian rhythm and the 24 h terrestrial cycle has not abolished their synchronization. During torpor the "internal clock" continues to run in various species of ground squirrels, the hedgehog, common dormouse, and the bat (cf. SUOMALAINEN and SAARIKOSKI, 1970). The *average* duration of one period in the bat is 25 h (Fig. 102).

In the annual cycle, hibernation competes with the *reproductive drive* (cf. SAURE, 1969); animals that are in heat or nursing do not become lethargic. The bat is an exception to this rule; it copulates before the onset of hibernation or during the short waking phases (cf. WISMATT, 1969); ovulation however does not take place till spring. Fertilization and embryonal development take place during the hibernation period only in species of *Miniopterus*. Lengthening the duration of cold exposure or castration in spring increases rodents' readiness to hibernate by suppressing their sexual drive.

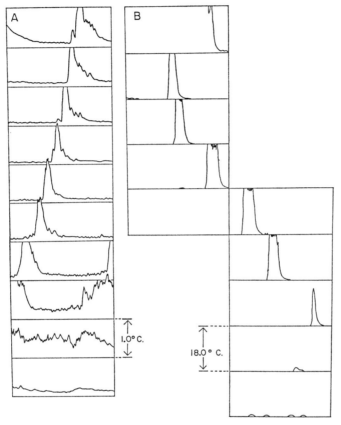

Fig. 102. Body temperature rhythms of *(A)* a short period summer bat — *Myotis lucifugus* (period about 22 hours) and *(B)* a long period winter bat (period about 30 hours from day 3 on). Both recordings were made in constant darkness and at a constant ambient temperature of 8.5° C. Each block is 24 hours with day 1 at the top. In *B* day 6 and subsequent days have been placed to the right of day 5 (in which no peak occurred) in order to make the period length more readily apparent. From MENAKER (1961)

As in the case of instinctive processes, hibernation is preceded by a set of *internally elicited actions*, which result in an increased level of *readiness for hibernation* (HERTER, 1956). These include: preparing winter quarters, bringing in food, and migration to wintering caves (by bats). At the level of readiness present at that time the actual *inducers* (cues) might include lack of food and water as well as falling ambient temperature. The role of declining length of daylight is in dispute, and

the importance of cold was previously overestimated. Basically hibernation in
most of the species examined has been observed at ambient temperatures between
−5° C and +25° C, and estivation appears at 28−33° C (POHL, 1961; AMBID,
1971). However it appears that prolonged lethargy at high body temperature and
without food uptake is uneconomical from the point of view of energetics and is even
threatening to survival. Hibernation is energy-sparing and purposeful only at low
body temperatures. Many authors have noted that the most effective ambient
temperatures are in the range between 10° C and 4° C. As their readiness for hibernation progresses hamsters, golden hamsters, and bats clearly prefer continually
cooler ambient temperatures (HERTER, 1956; GUMMA, SOUTH, and ALLEN, 1967;
DAVIS, 1970).

When readiness for hibernation appears the autumnal increase in food intake is
reduced by a decreased desire for food — even if more than optimal quantities are
available. Consequently, weight is lost continuously throughout the winter. Therefore hibernation is preceded by two reversals in feeding behavior, which may be
connected to changes in hypothalamic functions (MROSOVSKY, 1970).

2. Onset of Hibernation

If readiness for hibernation is present most species do not require cold adaptation
to alter their temperature set point (cf. FOLK, FOLK, and KREUZER, 1970). Nevertheless a season-dependent reduction in energy metabolism and body temperature precedes the onset of torpor (KAYSER, 1953). Hibernation begins with
slight or marked *fluctuations of body temperature* (BENEDICT and LEE, 1938;
LYMAN and CHATFIELD, 1955), which contain circadian components in the case of

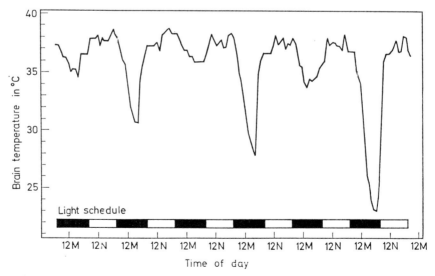

Fig. 103. A 6-day segment of the brain-temperature history of a summer ground
squirrel *(Citellus beecheyi)* during which time it took its first 3 test drops. Brain
temperature plotted every hour. On the time axis midnight and noon are indicated by
M and N, respectively. From STRUMWASSER (1959a)

the Californian ground squirrel (STRUMWASSER, 1960), hedgehog (SOIVIO, TÄHTI, and KRISTOFFERSSON, 1968), or common dormouse (WYSS, 1932). The temperatures fall to lower and lower levels from day to day ("test drops", Fig. 103), till finally the daily period of waking is completely omitted. In some species the fall in body temperature takes place smoothly (Fig. 106), in others it undulates slightly and lags behind the falling energy metabolism. Fluctuations in metabolism and temperature are brought about by brady- and tachycardia, phases of shivering, and changes in posture during sleep (STRUMWASSER, 1959b; LYMAN, 1965). The final value to which this "controlled cooling" leads is reached in 8 h by the golden hamster (LYMAN, 1965) and in 11−12 h by the ground squirrel and hedgehog (TWENTE and TWENTE, 1965; SUOMALAINEN and SAARIKOSKI, 1970).

Challenges to hibernation homeothermy, which is directed towards temperatures of 0.5−5° C, are met by increases in heart rate, muscle tone, oxygen consumption, and respiratory rate as well as by vasomotor reactions. If its regulation is overtaxed, the animal arouses (cf. KAYSER, 1961; STRUMWASSER, 1959a; LUECKE and SOUTH, 1971).

3. Arousal

According to EISENTRAUT (1956) hibernators can be divided into long-term and short-term hibernators. The first group includes the dormice, marmots, and bats, since they can sleep for two to four months without interruption. The sleep phases of ground squirrels and hedgehogs last several weeks, and those of hamsters about five days. Short-term sleepers can be wakened more easily than long-term sleepers by a great variety of sensory stimuli. The length of the sleep phases and the waking threshold increase from autumn to midwinter and then decline. The hibernation bouts of some ground squirrels and bats are shortened by low ambient temperatures (KAYSER, 1950; DAVIS, 1970) but those of *Citellus lateralis* are lengthened (TWENTE and TWENTE, 1965).

The cause of *spring arousal* is undoubtedly associated with the development of the reproductive state and with a general increase in excitement. The cause of *spontaneous periodic arousal* is not known with certainty. Probably only small amounts of urine are formed during deep hibernation due to structural changes in the glomeruli of the kidneys and the reduced blood pressure (ZIMNY, 1968; MOY, 1971; ZATZMAN and SOUTH, 1971; ZIMNY and LEVY, 1971). As a result, disturbances (accumulation of ketone bodies, Mg, and Ca in the blood, take place in the "milieu intérieur". According to TWENTE and TWENTE (1968) the sensitivity of a ground squirrel *(C. lateralis)* to arousal stimuli, e.g. NaCl injections, increases continually with the duration of sleep.

LYMAN and his co-workers (cf. LYMAN and O'BRIEN, 1969) have reported in various works that the first result of an *arousal stimulus* is a volley of muscle action potentials and intensified respiration, followed by an increase in heart rate. The 13-lined ground squirrel responds in the same way to minimal doses of various acids, bases, or acetylcholine applied through a chronically implanted aortic tube. The responsiveness to "internal stimuli" or changes in the "milieu intérieur" is even higher in hibernation. However this is only partly related to an increase in sensi-

tivity of the muscles towards acetylcholine (Table 16), since an infusion of acetylcholine and other drugs remains ineffective in hibernating spinal animals (C_1) not
only after poisoning with curare but also after the administration of hexamethonium
or after destruction of the spinal cord. Using the same preparation, LYMAN and
O'BRIEN (1969) could demonstrate that the muscle responses to painful thermal
stimuli to the back (below $-12°$ C and above $44-64°$ C) grew continally stronger
as body temperature fell. The amplitude and duration of the action potential
volleys increased.

Table 16. Threshold response of isolated muscle (Extensor digiti quarti) of *Citellus
tridecemlineatus* (μg ACh/100 ml)

	n	$\bar{x}$	S.D.
Normal active	31	585	± 591
Eserinized active	19	155	± 136
Denervated active	6	36.4	± 20.9
Normal hibernator	16	127	± 155
Eserinized hibernator	9	35.2	± 15.8

From LYMAN and O'BRIEN (1969).

Periodic arousal thus occurs (firstly) during the activity phase of circadian rhythm
but (secondly) only then when the stimulation of (unknown) receptors by disturbances of the "milieu intérieur" becomes so great that it leads to the activation of
protective reflexes and to waking. RATHS and HENSEL (1967) have demonstrated
the ability of the cold- and mechano-receptors of hibernating hamsters to discharge
at skin temperatures below 0° C.

The *process of arousal* is an extremely dramatic physiological event, which brings
about a complete rewarming up to $36-39°$ C in the course of one to four hours. It
is characterized by an explosive glycogen-, fat-, and protein-catabolism as well as a
rapid increase in heart and respiratory rates and oxygen consumption. LYMAN and
CHATFIELD (1955) present as an example the increase in respiratory frequency of
the golden hamster to 30 min^{-1} in the first 90 min and to 100 min^{-1} in 150 mins.
Energy metabolism climbs to 500 times the hibernating value, and the heart rate
reaches 550 min^{-1}. According to ADOLPH and RICHMOND (1955) the amount of
energy used is equivalent to that used in about ten days of hibernation.

With the exception of species of *Tamias* and *Peromyscus* (WANG and HUDSON,
1971), the anterior part of the body is warmed more rapidly than the posterior part
(Fig. 104). Thorotrast representation of the circulation (LYMAN and CHATFIELD,
1950) as well as investigations carried out with rubidium-86 (JOHANSEN, 1961;
BULLARD and FUNKHOUSER, 1962; RAUCH and HAYWARD, 1970) have confirmed
that arousal stimuli lead to a vasomotor constriction in the abdomen and skin, so
that the thorax and head are warmed first. The restriction is lifted almost at once
when a head temperature of approximately 30° C and maximum oxygen use has
been attained. In the ground squirrel and hedgehog the vasoconstriction can be
prevented by poisoning the adrenergic β-receptors (LYMAN, 1965; BRÅNEMARK
and JOHANSSON, 1967).

In the *first phase* of arousal (up to a head temperature of about 15° C) there is no shivering, though muscle tone is present. In the bat, hedgehog, ground squirrel, and marmot, among others, thermogenesis is due largely to the brown fat (cf. JOEL, 1965; SMITH and HORWITZ, 1969) and the heart, both of which are under strong sympathetic stimulation. Probably heat production in both organs is intensified still more by uncoupling of the respiratory chain (ZIMNY and MORELAND, 1968).

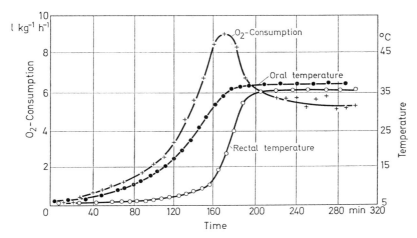

Fig. 104. Body temperature and oxygen consumption of the golden hamster *(Meso-cricetus auratus)* during arousal from hibernation. From LYMAN (1948)

A three-week period of cold adaptation brings about a further increase in thermo-genetic capacity in the bat (HENSHAW, 1970). In the *second phase* "non-shivering" thermogenesis is supported by massive muscular shivering. Each phase can be inhibited separately: the first by sympathicolytic drugs and ganglionic blockade, the second by curare (LYMAN, 1965; MEJSNAR and JANSKY, 1970). The contribu-tion of nonshivering thermogenesis has been estimated at 20—80%. Due to the vasomotor constriction the liver plays only a minor role in arousal (LYMAN and CHATFIELD, 1950).

4. Temperature Regulation

During hibernation animals assume a *sleeping posture*. Bats hang down, others roll into a ball. The ears and tail are laid against the body or concealed, the nose is laid on the belly, and the pelt is fluffed up; the effective surface for heat loss in intense cold is therefore minimal.
The *thermal sensitivity* of the skin is maintained during hibernation. According to RATHS and HENSEL (1967) the cold fibers of the hamster are still excitable at −5° C, partly tonically and partly phasically. They fail to function only when the skin is frozen solid (−6° C to −13° C). Multi-fiber preparations of the infraorbital nerve exhibit maximal activity at 4—5° C, which is close to the lowest hibernation temperature for this species.

Waking hamsters, 13-lined ground squirrels, and marmots respond to cooling or warming of the hypothalamus with the same thermoregulatory reactions that other mammals show (MALAN, 1969; WILLIAMS and HEATH, 1970; HARTNER et al., 1971). In the case of *Citellus lateralis* cooling during hibernation leads to arousal, and warming leads to resumption of torpor (HELLER and HAMMEL, 1972). The lower the ambient temperature, the higher is the threshold temperature for the central effect of cooling. The hibernating marmot *(Marmota flaviventris)* responds to stimulation of the diencephalon by cold (2–3° C) by increasing its heart rate, and to stimulation by warmth by decreasing it (HARTNER et al., 1971). HELLER and HAMMEL (1972) suggest that the thermosensitive cells lie in the midbrain instead of the diencephalon and that the hypothalamus plays no role in temperature regulation during hibernation. On the other hand, the model brought forward earlier by HAMMEL (1967) as well as by LUECKE and SOUTH (1971) was based on hypothalamic processes, in which cell populations with high and with low Q_{10}s were thought to act alternately.

Lesions in the posterior hypothalamus interfere with readiness for hibernation and with temperature regulation in the hamster (MALAN, 1966, 1969) and cause apathy and sleepiness in the ground squirrel (SATINOFF, 1970). Destruction of sites in the area praeoptica or in the nucleus ventromedialis makes arousal impossible, but none of these operations induces hibernation. *Injections of noradrenaline* into the third ventricle or the area praeoptica of *Marmota flaviventris* also lead only to a reduction of the core temperature by 2–5.4° C, but not to hibernation (JACOBS et al., 1971).

Various authors (cf. HERTER, 1956; RATHS, 1958) have determined that the *reflex processes* in deep hibernation correspond to the abilities shown by oblongata- or midbrain animals. The functions of higher centers are restored only when the brain is warmed during arousal.

The *EEG investigations* also tend to indicate that processes of hibernation are regulated more by the midbrain than by the diencephalon. The reviews of SOUTH et al. (1969) as well as SATINOFF (1970) indicate that hibernation probably sets in during the spindle stages of normal "slow wave" sleep. Down to brain temperatures of 20° C it is still possible to distinguish paradoxical from slow cortical sleep. In the hamster, golden hamster, marmot, and hedgehog the continuous activity of the cortex is extinguished between 20° C and 15° C, but in the birch mouse and ground squirrel it is maintained in deep hibernation (cf. PUTKONEN, SARAJAS, and SUOMALAINEN, 1964). Evidence of hypothalamic bioelectric activity at head temperatures between 4° C and 7° C has been brought forward for *Citellus beecheyi* (STRUMWASSER, 1959c) and *Marmota flaviventris* (SOUTH et al., 1969). At a head temperature of 6° C bioelectric rhythms can be detected only from the pons and parts of the midbrain of the hamster and golden hamster. During arousal, as the temperature rises, the rhinencephalon and diencephalon are activated first, and the neocortex last (CHATFIELD and LYMAN, 1954; RATHS, 1958).

5. Respiration

Pulmonary ventilation during hibernation is slow, irregular, and partly periodic. However, according to KAYSER (1961) the frequently observed CHEYNE-STOKES respiration is not typical for torpor. In the reviews of HERTER (1956), EISENTRAUT (1956) and KAYSER (1961) the approximate minimum frequency is given as 4 min^{-1} for the hamster, 1 min^{-1} for the golden hamster, $0.5-4$ min^{-1} for the ground squirrel, 7 min^{-1} for the hedgehog, and $0.07-0.25$ min^{-1} for the marmot. If hibernation homeothermy is challenged, respiration speeds up. The irregular and periodic ventilation indicates a reduction in sensitivity of the respiratory center, as does the fact that the respiration of hibernating hamsters, golden hamsters, hedgehogs, marmots, and dormice increases only when the CO_2 concentration of the inspired air exceeds $2.5-4\%$ (cf. KAYSER, 1961).

As a result of the increased solubility of gases at low temperatures and the reduced sensitivity of the respiratory center the blood contains more CO_2 during hibernation than during waking (cf. KAYSER, 1961; CLAUSEN, 1966; CLAUSEN and ERSLAND, 1968; AMBID, 1971). As the body temperature falls, the arterial CO_2 tension decreases. Since the lactic acid content of the blood and organs falls drastically at the same time (cf. AMBID, 1971) the proportion of bound carbonic acid (HCO_3^-) increases (BARTELS et al., 1969). According to PERSON (1952), however, the CO_2 binding capacity of the blood of *Citellus suslicus* rises appreciably only at the beginning and begins to sink again at body temperatures below 15° C. In parallel with these changes the pH value of the blood rises early in hibernation and then declines to the normal value. This normalization is related to a metabolic or respiratory acidosis. However, the increased respiration supporting the "second homeothermy" compensates an extreme drop in blood pH. According to BAUMBER et al. (1971) the level of ketone bodies in the plasma of *Marmota flaviventris* rises by 40%. Various authors have therefore observed increases as well as decreases in pH during hibernation (cf. CLAUSEN, 1966; AMBID, 1971).

In spite of the disappearance of the ketone bodies from the blood during *arousal*, the CO_2 binding capacity decreases. AMBID (1971) removed blood from *Eliomys quercinus* through an aortic tube and determined levels of 65.2 Vol.-% at 7° C, 49.2 Vol.-% at 15° C, and 56.75 Vol.-% at 37° C (57 Vol.-% CO_2 during waking). Hyperventilation, reduced solubility at higher temperatures, and the increased lactic acid level all contribute to the decrease in CO_2 content (FERDMANN and FEINSCHMIDT, 1932). The lactic acid content in the plasma of *Eliomys* increases from 5 mg-% during hibernation to 95 mg-% during arousal. Acidosis in the hedgehog during arousal was described by CLAUSEN (1966): between 15° C and 34° C the pH value lay between 7.21 and 7.31.

The *oxygen dissociation curve* of the hedgehog's blood is displaced to the left during hibernation, as compared to the summer condition. This displacement is in addition to the well-known effects of cold (BARTELS et al., 1969). The investigations of BURLINGTON and WHITTEN (1971) on various species of ground squirrels indicate that it is related to a 50% decline in the 2,3-diphosphoglyceric acid content of the erythrocytes. Together with the alkalosis present at the same time, in the midrange of hibernation body temperatures ($15-10°$ C) it signifies a further leftward shift in the dissociation curve, making oxygen uptake in the lungs much easier and the discharge of oxygen in the tissues more difficult. According to MUSACCHIA and VOLKERT (1971) the arterial pO_2 value in *Citellus tridecemlineatus* rises from 64.9 to 87.7 mm Hg (Torr), while the venous pO_2 (in the vena cava) falls from 27.9 to 6.3 mm Hg. Thus the tissues remove extraordinarily large amounts of oxygen from the blood in spite of the leftward shift in the dissociation curve. The low level of lactic acid in the blood and organs also indicates that there is no lack of oxygen in the tissues during torpor.

During *arousal* the arterial oxygen content of *Eliomys quercinus* (AMBID, 1971) falls from 17 Vol.-% (7° C) to 15 Vol.-% (15° C), 12 Vol.-% (33° C), and finally to 10 Vol.-% (37° C), while at the same time the lactate level in the plasma rises rapidly. Hypoxemia also develops during arousal in the hedgehog (CLAUSEN, 1966). One could therefore surmise that the displacement of the oxygen dissociation curve is more likely to be an adaptation to arousal than to torpor.

The *resistance* of various organs, e.g. the brain, of hibernators to hypoxia has been demonstrated (KAYSER and MALAN, 1963). During arousal anaerobic liberation of energy partially compensates for the oxygen deficit (MOKRASCH, GRADY, and GRISOLIA, 1960). According to STEWART, ZIMNY, and WEST (1970), the brain does not actually suffer from hypoxia, since the succinic dehydrogenase activity (*in vitro*) rises at low temperatures or in hypothermia (3° C), and the efficiency of oxygen utilization is thereby increased.

6. Metabolism

The *basal metabolic* rate (BMR) of numerous hibernators is lower than that of other mammals even in summer (cf. KAYSER, 1961; HILDWEIN and MALAN, 1970), but there are also a number of exceptions (HENSHAW, 1968, 1970; HUDSON, DEAVERS, and BRADLEY, 1972). According to HUDSON (1969) the magnitude of metabolism is not a criterion for hibernators.

Following *cold adaptation*, the basal metabolism of hibernators, like that of other homeotherms, increases (cf. KAYSER, 1939, 1961; HOLYOAK and STONES, 1971). Nevertheless by autumn BMR has sunk to 65—68% of its June/July level (*Marmota, Cricetus, Citellus* — KAYSER, 1939), probably for reasons of a complex nature. An increase in the proportion of "inactive" tissue (HAYWARD, 1965), a decrease in thermal conductance (HUDSON, DEAVERS, and BRADLEY, 1972) as a result of increased fat deposition, and reduction of the core temperature by 1—2° C may be among them. The core temperature reduction may be due to a greater displacement of the set point during the circadian rhythm, or to fluctuations in thermogenesis (cf. HERTER, 1956; KAYSER, 1961).

The *capacity for thermogenesis* of hibernators is excellent at all seasons, even better than that of other homeotherms (KAYSER, 1957, 1961). Nevertheless, according to KAYSER they are characterized by inadequate physical regulation. This concept may be considered doubtful, since the fluctuations in total thermal conductance indicate the presence of control. However, hibernators do respond to cold more by a reduction in body temperature than by an increase in insulation. For this reason they, especially ground squirrels, are also well suited to inhabit hot, arid regions (HUDSON, DEAVERS, and BRADLEY, 1972).

During *hibernation* the metabolism of larger species is reduced to 1/30 of the summer BMR, that of smaller species to 1/100 (Table 17). In the poikilothermic state the reaction-rate-temperature law governs thermogenesis, and Newton's law of cooling governs heat loss. Consequently, since their heat capacity and production are lower and their surface area is relatively large, small animals grow colder more rapidly than large ones. The smaller the animal is, the smaller is the difference between the body temperature and the ambient temperature. The larger the animal is, the higher is the body temperature at which heat production is in equilibrium with heat loss (KAYSER, 1964a; SOUTH and HOUSE, 1967).

Table 17. Energy metabolism of hibernators

Species	Summer		Hibernation		Ratio
	Weight	Metabolism	Weight	Metabolism	
	g	kcal/kg.day	g	kcal/kg.day	
Marmota marmota	1868	50	2146	2.0	25:1
Erinaceus europaeus	684	84	600	1.8	47:1
Citellus citellus	227	108	275	2.0	54:1
Glis glis	127	120	130	1.7	70:1

From KAYSER (1950).

According to numerous studies by KAYSER (cf. 1964a), in *summer* the law relating metabolic reduction to a surface proportionality is approximately valid: Q [kcal/animal · 24 h] $= 63.6\ W^{0.62}$ [kg]. When the (modified) BRODY equation is applied to *torpor* the minimal metabolic level must be considered in each case, which corresponds to a body temperature of 4.5° C in the bat, 7.8° C in the ground squirrel, 9.5° C in the common dormouse, and 31° C in the black bear. Data taken from 16 species with weights between 5.2 and 60000 g have yielded the formula $Q = 3.2\ W^{1.03}$, i.e. a definite weight proportionality. If the temperature dependence of the constants is taken into account and the metabolic rates are recalculated for a temperature of 10° C, assuming a Q_{10} of 2.5, a "corrected, temperature-independent hibernation metabolic rate" is obtained, amounting to $Q = 2.5\ W^{0.84}$, with the weight exponent varying between 0.90 and 0.78 (KAYSER, 1964a). However, in the same year KAYSER (1964b) proposed the formulas $Q = 2.16\ W^{1.02}$ and $Q = 2.09\ W^{0.69}$, respectively, if referred to a temperature of 10°C. Hibernators are in a state of *poikilothermy* in which the reaction-rate-temperature law is valid only when in a thermal steady state. While the body temperature is falling variations in vasomotor tone, in oxygen consumption, and in heart rate, as well as phases of shivering, show that the cooling is "controlled". According to HENSHAW (1968, 1971) changes in thermal conductance in bats have the same significance.

The *respiratory quotient* during hibernation is approximately 0.7, since energy is obtained from fat, and only to a slight extent from carbohydrates. During arousal the RQ rises to 0.767 in the ground squirrel, to 0.846 in the hamster, to 0.831 in the golden hamster, and at times can approach a value of 1 (cf. KAYSER, 1961, 1964).

The fall increase in food intake produces a large *fat reserve* before the onset of hibernation, which composes 30—50% of the body weight and is distributed through the brown fat, blood, and organs (liver, heart) in addition to the white fat depots. These reserves are used up during the course of the winter. The hardening point of the white fat of hibernators lies between 5° C and −25° C, while in other mammals it lies between 15° C and 40° C. In the golden hamster and ground squirrel it falls from about 0° C to −20° C as a result of long-term cold exposure. In this process the proportion of unsaturated fatty acids and the iodine number (about 5–10%) both rise (FAWCETT and LYMAN, 1954). The proportion of unsaturated fatty acids in the brown fat of the garden dormouse also rises (AMBID, 1971). Each *arousal process* mobilizes a small amount of white fat and a very large proportion of brown fat. Due to sympathetic activation of lipolysis the level of free fatty acids and glycerol in the blood rises precipitously (Fig. 105) (SUOMALAINEN and SAARIKOSKI, 1967; AMBID, 1971; AMBID, BERLAN, and AGID, 1971). Towards the end of arousal lipolysis is reduced, and replenishment of the brown fat body begins. Since this organ's capacity for lipogenesis is high (BAUMBER and DENYES, 1964), and energy is provided by carbohydrates (TASHIMA, ADELSTEIN, and LYMAN, 1970), the process is completed in a few hours (AMBID, 1971).

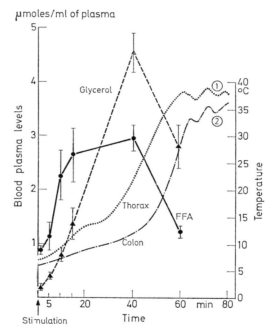

Fig. 105. Body temperature, and content of free fatty acids *(FFA)* and glycerol in the plasma of the garden dormouse *(Eliomys quercinus)* during arousal from hibernation $(\bar{x} \pm S.E., n\ 5\text{—}12)$. From AMBID (1971)

The second energy reserve for hibernation is the elevated level of *glycogen* in the liver, musculature, and heart (cf. KAYSER, 1961; SARAJAS, 1967). During each arousal this store is partly depleted. In spite of the subsequent replenishment during the waking period, the total glycogen reserve slowly declines during the winter. The *blood sugar* level begins to fall immediately upon the onset of torpor, probably as a result of stimulation of glycogen- and fat synthesis under the influence of insulin (SUOMALAINEN, 1950; SCHEUFLER and RATHS, 1967) and a decreased sympathetic effect or reduced phosphorylase activity in the liver (HANNON and VAUGHAN, 1961; SARAJAS, 1967). Hypoglycemia increases with the length of a period of torpor (GALSTER and MORRISON, 1970; AMBID, 1971) and is also dependent on the body temperature. In the hamster (RATHS, 1961), hedgehog (SARAJAS, 1967) and garden dormouse (AMBID, 1971) the blood sugar level at a body temperature of 6° C is higher than at 15° C or 20° C. This effect may be related to an increase in sympatheticoadrenal activity (RATHS, 1961), energy savings (AMBID, 1971) or to an inefficient utilization of glucose (BAUMBER and DENYES, 1965; TASHIMA, ADELSTEIN, and LYMAN, 1970) in deep hibernation. During arousal the blood sugar level of the hedgehog (SAARIKOSKI and SUOMA-LAINEN, 1970) and the garden dormouse (AMBID, 1971) falls briefly at first and then rises to the normal level or higher (110—185 mg-%). Arousal hyperglycemia can be suppressed by a ligature which prevents the liver from providing glycogen (LYMAN and LEDUC, 1963).

Protein metabolism is limited and qualitatively changed during hibernation. This affects cell division and maturation, growth, enzyme activity, blood coagulation, and immunological processes in very complex fashions. During the late phase of arousal the level of free amino acids in the blood is increased 2—4 fold (KRISTOF-FERSSON and BROBERG, 1968; KLAIN and WHITTEN, 1968).

ATP-formation and *utilization* during hibernation are in such an equilibrium that the reserves are maintained and remain available for arousal (ZIMNY, 1960). Although various enzymes are known to be inhibited (cf. ROBERTS and CHAFFEE, 1972), ATP synthesis continues to take place by oxidative phosphorylation. The P/O ratio is unchanged or even increased, i.e. coupling becomes tighter or more efficient (cf. SOUTH and HOUSE, 1967; ROBERTS and CHAFFEE, 1972). This effect is consistent with a decrease in the activation energy of oxidative phosphorylation in the heart mitochondria of the golden hamster (SOUTH, 1960). According to STEWART, ZIMNY, and WEST (1970) the temperature dependence of ATP-cleavage is also slight. Low temperatures have only a small effect on the Mg-ATPase in the brain of the ground squirrel.

7. Heart and Circulation

Information on the condition of the heart and circulation during the various stages of hibernation has been obtained largely through the use of chronically implanted aortic tubes, ECG electrodes and telemetric methods. From these it has been learn-ed that at the onset of torpor (Fig. 106) the heart is subject to a distinct tonic stimu-lation by the vagus (cf. LYMAN, 1965; WANG and HUDSON, 1971). Atropine retards the reduction in heart rate in *Marmota monax* and eliminates the vagally induced dysrhythmias and asystoles in *Citellus tridecemlineatus*. However, atropine has no effect at temperatures below 10° C. TURPAJEW (1948) found the same limit for the effect of acetylcholine on isolated ground squirrel hearts, as did BIEWALD and RATHS (1959) for vagal stimulation *in situ* in the hibernating hamster. It is believed that changes in ionic equilibrium may be responsible (BIEWALD and RATHS, 1967; BIEWALD, 1967). Subsequently, in deep hibernation, the heart is controlled only via the sympathetic system. The heart rate is slowest when thermoregulation is under the least stress. For example the average rate is 5−6 min⁻¹ in the hedgehog and woodchuck and 2−8 min⁻¹ in the ground squirrel (cf. KAYSER, 1961). As the ambient temperature falls, the heart rate increases, e.g. to 20−30 min⁻¹ in the marmot (FOLK, FOLK, and KREUZER, 1970) (Fig. 107). Stimulation by cold at

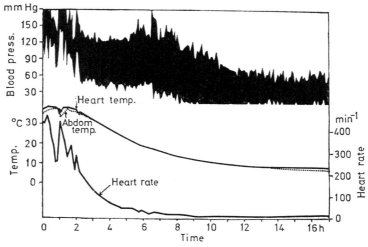

Fig. 106. Blood pressure, heart and abdominal temperature, and heart rate of ground squirrel *(Citellus tridecemlineatus)* entering hibernation. Blood pressure in dark area is highest, systole and lowest diastole recorded every four minutes for a one-minute period. Note declines in heart rate and blood pressure, followed by body temperature. From LYMAN and O'BRIEN (1960)

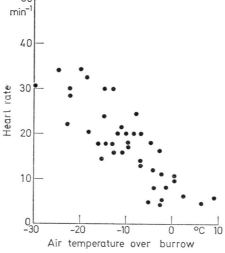

Fig. 107. Air temperature over burrow, and heart rates of hibernating woodchuck *(Marmota monax)*, recorded by implanted radio-capsule. Adapted from FOLK, FOLK, and KREUZER (1970)

levels above the minimal temperature, which places no stress on the "second homeothermy", induces only brief periods of tachycardia, which become briefer as time passes as a result of "adaptation" (LUECKE and SOUTH, 1971).

At the onset of hibernation blood pressure falls sharply, as a result of bradycardia and dilatation of the *blood vessels of the skin* (STRUMWASSER, 1959b; TUCKER, 1965) (Fig. 106), but as the body temperature falls the early condition of parasympathetic excitation yields to sympathetic control (RATHS, 1953; SCHEUFLER and RATHS, 1967). LYMAN (1965) could still measure a systolic arterial pressure of 40—90 and a diastolic pressure of 7—40 mm Hg at the level of the renal artery of the 13-lined ground squirrel during deep hibernation when the heart rate had fallen to 3—4 per min. Infusing with β-TM-10 or hexamethonium has led to circulatory collapse and frequently to death during torpor. Regulation of blood flow during hibernation can be recognized through fluctuations in pressure (LYMAN, 1965), as well as periodic changes in temperature of the brain and skin of the back (STRUMWASSER, 1959c).

Several authors have observed a notched R-wave in the ECG and an atrioventricular dissociation at the beginning of *arousal* (cf. LYMAN, 1965). Immediately after arousal the blood pressure of the 13-lined ground squirrel falls for 10—30 min, and the subsequent rise is somewhat slower than that of the heart rate. Both maxima coincide with the peak levels of respiration and metabolism. The restriction of blood flow through the skin, abdominal organs and muscles of the posterior part of the body is eliminated only for one to two minutes at a time by β-TM-10 or acetylcholine. After its physiological termination it can be reinstated only momentarily by infusions of noradrenaline (LYMAN, 1965). During arousal there is a simultaneous increase in blood flow through the brown adipose tissue, myocardium, diaphragm, and the muscles of the anterior part of the body. The Rb-86 method has shown that blood flow increased 6-fold in the brown fat and 16.3-fold in the anterior musculature of the hedgehog (JOHANSEN, 1961), while in the bat *Myotis lucifugus* it increased 11.7-fold in the interscapular brown fat, 6.32-fold in the remaining brown fat, and 3.3-fold in the anterior musculature (RAUCH and HAYWARD, 1970).

8. Hormonal Changes

The endocrine activity of hibernators — like that of other wild mammals — exhibits a marked annual *rhythm* (cf. MOGLER, 1958; POPOVIC, 1960; KAYSER, 1961; GABE et al., 1964; CANGUILHEM and BLOCH, 1967; HOFFMAN, 1968; AMBID, 1971). From late summer to January the anterior lobes of the pituitary, the gonads, thyroid, and adrenal cortex undergo a progressive involution, and from then till spring, in spite of hibernation, the activity of these glands slowly increases. The conditions of the parathyroids, pancreas, and zona glomerulosa of the adrenals are very variable. The adrenal medulla is progressively depleted of catecholamines during winter (cf. CANGUILHEM and BLOCH, 1967).

A brief increase in activity of the adrenal medulla and cortex in September/October has been described in various hibernators, and has been interpreted as a symptom of *cold adaptation* (KAYSER, 1961; PETROVIC and DAVIDOVIC, 1965; BOULOUARD, 1971). Autumnal cold also activates the thyroid of the garden dormouse, but not that of the ground squirrel (HOFFMAN and ZARROW, 1958; LACHIVER and PETRO-

VIC, 1960; LACHIVER, 1964). In addition, secretion of corticosterone is converted to secretion of cortisol in the garden dormouse (BOULOUARD, 1969) and noradrenaline to adrenaline in the hedgehog and ground squirrel (SUOMALAINEN and UUSPÄÄ, 1958; SMIT-VIS, 1962; PETROVIC and DAVIDOVIC, 1968). Hibernation can be prevented by removal of the thyroid (CANGUILHEM and MALAN, 1969) or adrenals (POPOVIC, 1960; CANGUILHEM and PETROVIC, 1968), and replacement of the missing hormones allows it to occur again. Only extirpation of the gonads promotes readiness for hibernation.

In torpor itself biogenesis of hormones is at a low level due to the low body temperature. However histological, histochemical, and karyometric investigations as well as serum and urine analyses indicate that aldosterone, parathormone, thyroxine, and renin are secreted (MOGLER, 1958; KAYSER, 1961; RATHS, 1964; CANGUILHEM and BLOCH, 1967; SUOMALAINEN and NYHOLM, 1965; ZIMNY and LEVY, 1971). Insulin and noradrenaline are stored during hibernation but it seems possible that there is an adaptation of their secretion to thermoregulatory requirements (SCHEUFLER and RATHS, 1967). The numerous works of SUOMALAINEN (1944—1956) and other authors (cf. SCHEUFLER and RATHS, 1967) on the increased β-cell activity during hibernation may be interpreted in the same way. The concept put forward by AMBID (1971), who proposed that there were no hormonal regulations in torpor, stands in contrast. According to HUIBREGTSE et al. (1971) the adrenal cortex of the hibernating ground squirrel cannot be activated by ACTH even *in vitro*.

The enormous metabolic activity during *arousal* would be inconceivable without a general hormonal activation. The data of the authors mentioned as well as the results of AGID, GABE, and MARTOJA (1967) show that the first phase of arousal is controlled only by sympathetic nervous stimulation. A series of hormones are mobilized one after the other, during the subsequent increase in body temperature beginning with catecholamines and glucagon, followed by corticosteroids, and finally by insulin as well as thyroxine.

9. Attempts at Artificial Induction of Hibernation

It has already been pointed out that hibernation is bound to forms of instinctive behavior that are restricted to specific times of the day and year. For this reason measures that abolish homeothermy, such as adrenalectomy and hypophysectomy or transection of the brain stem cannot be compared to hibernation; furthermore they lead to death within a few days if the animal is exposed to cold. This is also true of hypothermias that are forcibly induced with the aid of anesthetics, phenothiazine derivatives (MOGLER, 1958), reduced atmospheric pressure (GIAJA, 1940) or simply by violent cooling. Injections of insulin or $MgSO_4$ and insulin lead to hypothermia in some hibernators, which may last as long as ten days in the hedgehog (SUOMALAINEN, 1939). Exposure to helium (80% He + 20% O_2), the so-called "Helox"-method of FISCHER and MUSACCHIA (1968), produces an enormous increase in heat loss in the golden hamster, resulting in a very deep hypothermia. Remarkably, the acid-base balance of the blood is maintained in this case (MUSACCHIA, VOLKERT, and BARR, 1971). This and all the other methods mentioned share a common characteristic: The animals treated in these ways are not able to rewarm themselves actively at low ambient temperatures. Far more often they die after one to three days. The differences between these conditions and hibernation are clear. However, according to DAWE and SPURRIER (1969) as well as DAWE, SPURRIER, and ARMOUR (1970), tranfusion of preserved blood, taken from hibernating woodchucks or 13-lined ground squirrels, into waking ground squirrels in summer can induce true lethargy.

XII. Temperature and Geographic Distribution of Homeotherms

A. General Remarks

In this book (p. 422) it was emphasized that temperature is only one of many factors upon which the geographic distribution of organisms depends. This is especially true of warm-blooded animals, whose homeothermy has made them independent, to a certain degree, of external temperature influences. The very fact that with negligible exceptions homeotherms live in the open air means that they must be able to cope with greater temperature fluctuations than must aqueous animals. A direct limitation in the distribution of warm-blooded animals as a result of temperature is usually found only in extreme ranges. Thus, most authors in this field have emphasized that the temperature factor in the distribution of homeotherms should not be overestimated (e.g. MARCUS, 1933; NIETHAMMER, 1952). In addition, warm-blooded animals, especially birds, are able to evade seasonal fluctuations in temperature because of their mobility.

In colder climates, homeothermy necessitates the constant consumption of large amounts of food in order to maintain temperature. In contrast, poikilothermic animals can survive without food in the cold for several months, owing to their low metabolism. Thus, with the exception of certain hibernators, warm-blooded animals are highly dependent upon their *nutritional area*, which in turn also depends upon the temperature. These indirect temperature limits for the distribution of homeotherms, which in most cases are probably of decisive importance, are thus ultimately determined by the distribution of poikilothermic organisms (cf. p. 422).

Man is able to live permanently everywhere on the surface of the earth owing to the technical and civilizational means at his disposal, which can be considered as an extreme form of "behavioral regulation". To this extent, man is not subjected to any limitation of distribution based on temperature. However, such limitations appear when it is a question of permanent living areas for whole populations, if only the natural aids of the particular biotope are available. In this case, the distribution of man is not so widespread as that of certain homeothermic animals (deserts, polar regions, mountains). More recent investigations have shown that human physiological adaptability to warm and cold climates is relatively uniform among various ethnic groups (HAMMEL, 1964; LADELL, 1964; BAKER and WEINER, 1966), although definite differences exist (examples see p. 628). In many cases, apparent genetic adaptations of indigenous populations may, in fact, depend on their better knowledge of living in a particular climate.

B. Temperature Conditions in Various Biotopes

Certain general considerations have already been discussed (p. 422). In the case of homeotherms, it is likewise insufficient to list the *mean annual* temperature as a thermic characterization of a biotope. The *seasonal* fluctuations in temperature are of great importance as well, since they provide a certain measure as to whether a warm-blooded animal with a given temperature regulation will be able to tolerate a certain climate. The extent of annual fluctuation increases more and more from the equator to the poles; if the geographic latitude remains the same, the extent of fluctuation depends especially on whether the area has a maritime or a continental location. At a northern latitude of 52°, Valentia (Ireland) has a mean annual fluctuation of 7.8° C (7.3° C in January, 15.1° C in July), whereas Nerchinsk (eastern Siberia) has a mean annual fluctuation of 51.8° C (−33.6° C in January, 18.2° C in July) (KÖPPEN, 1931). Tibet, with a mean annual temperature of 20° C, has annual fluctuations of up to 77° C (40° C in the summer, −37° C in the winter) (CLARKE, 1954). *Diurnal fluctuations* are usually of only slight significance as a limiting factor, with the exception of the tremendous fluctuations and extremely high daily temperatures in subtropical desert areas. The temperatures of the *macroclimate*, measured 2 meters over the ground, are decisive for man, the larger mammals, and birds. Here, microclimatic influences are naturally less important, compared with the importance of the ecoclimate and microclimate for many poikilotherms (for examples, cf. p. 422). Occasionally, however, the ecoclimate and microclimate are also of considerable importance for homeotherms, e.g. for small rodents in deserts.

In view of the extraordinary adaptability of homeotherms, general observations on various climatic zones can be made only if the temperature clearly appears as an influencing factor; this, however, is the case only under extreme conditions. Otherwise, estimating the direct effect of temperature on the distribution of warm-blooded animals is a difficult question which can be answered only in individual cases, since it requires extensive knowledge (usually not available) of the thermophysiology of the particular species.

If we take the stress caused by *temperature regulation* in homeotherms as a basis, we can discern the following typical large-scale climatic zones (cf. KÖPPEN, 1931):

1. *Tropical climate* with mean annual temperatures of 25 to 30° C, continuous heat, strong sunshine, high humidity, slight daily fluctuations of up to 6° C, even slighter annual fluctuations in temperature (only 0.5−1° C at the equator), and absence of nonperiodic temperature changes.

2. *Desert climate* in hot subtropical desert zones with mean annual temperatures of over 20° C, medium annual fluctuations (10−20° C), vast diurnal fluctuations of up to 50° C with extremely high daily temperatures and ground temperatures of more than 70° C, extremely dry air, and practically complete absence of water (extreme values for Death Valley, California: air temperature 56.6° C, relative humidity 5%).

3. *Moderate climate* with mean annual temperatures from 0−20° C, moderate diurnal fluctuations, large annual fluctuations of 20−50° C, depending upon

continentality, especially marked by cold winters, and nonperiodic (advective) temperature changes.

4. *Polar climate* with annual means between 0 and $-20°$ C, cool summers, cold to extremely cold winters, depending upon continentality (extreme values: Verkhoyansk, in northeast Siberia, with $-70°$ C; the Northwest Territories of Canada, with $-65°$ C; the Antarctic, with $-88°$ C), and areas in which the soil is constantly frozen. The same is true of high mountain regions, where the mean temperature decreases by around $1°$ C for each 150 meters of increase in altitude. The mean temperature of the Arctic and Antarctic Ocean is around -1.5 °C.

C. Temperature and Migration

Temperature is an important factor in the migrations of homeotherms living in moderate and cold zones. However, this factor often manifests itself indirectly in the form of *food deficiency*, forcing animals to migrate although they would be able to resist the temperature *per se* (e.g. wild ducks). Small birds, such as titmice (Paridae), which feed mainly on hibernating insects during the cold season of the year, can also be found in moderate climate zones in the winter, whereas much larger birds must migrate to warmer zones in order to find food. Within certain limits, birds do not react to temperature fluctuations; however, if the fluctuations exceed these limits, the bird will respond by migrating. Tolerance towards cold varies with seasonal acclimatization (KENDEIGH, 1945b, BOWEN, 1946). It is greatest in the winter, and declines up to April. For this reason, birds in May will try to evade a drop in temperature to which they would not have reacted in March or April (BOWEN, 1946). Another decisive factor in regard to migration is the *length of the day*, since it determines the amount of food, and thus the tolerance to cold (KENDEIGH, 1945b, 1949; SCHILDMACHER, 1952; HESSE, ALLEE, and SCHMIDT, 1951). Migration is intimately associated with the reproductive cycle, and both are strongly influenced by the *photoperiod* (WOLFSON, 1959; HART, 1964). In *Junco hyemalis*, ROWAN (1925, 1929) was first to show that increased artificial day length caused gonad recrudescence and birds were stimulated to migrate in late fall or winter, months ahead of time. On the other hand, the actual migration of juncos and white-throated sparrows (WOLFSON, 1958) will occur even when they are held at winter day lengths of 9 hours. Day length merely governs the rate at which the migratory state develops and hence when it occurs. In spite of migration, small birds are often decimated in exceptionally cold winters; however, the species especially threatened by cold are especially fast in reproducing again. According to PEITZMEIER (1947), 3 abnormally severe winters can be compensated for by 2 favorable years.

In warm-blooded animals, migration ranges from a few centimeters to thousands of kilometers. The shortest distances are covered by small mammals which burrow into the earth under the influence of heat or cold. More extensive moves are made within the range of the ecoclimate by many animals living in high mountain regions, and also by high-mountain hibernators, such as marmots, which move down to lower altitudes in the winter. The part played by temperature is not always clear in regard to more extensive migrations of mammals and birds, including the biannual passage of the latter. In many cases, temperature has an indirect influence via lack of food, but there are often other factors involved (cf. HART, 1964).

D. Special Characteristics of Chiroptera

According to EISENTRAUT (1947), the temperature factor is of particular importance for the distribution of bats. Of around 150 species, ca. 90% inhabit tropic and subtropic zones, 7 species are found in tropic, subtropic, and moderate zones (e.g. *Rhinolophus, Myotis, Pipistrellus*, and *Eptesicus*), whereas only 1 species, *Eptesicus nilssonii*, inhabits the Arctic zone in northern Russia and Norway. According to HERTER (1952), the preferred temperatures of bats are very high: 39—43° C (ground temperature). This reflects their need for warmth and explains their preference for very warm microclimates, such as hot attics, etc. In moderate zones, Chiroptera go into seasonal *hibernation* (p. 694), which means that they must find protected winter domiciles (cellars, mines, caves, etc.). Ringing tests have shown that these migrations cover distances of from 50—1150 kilometers (EISENTRAUT, 1947); the temperature factor here is not always clear.

According to EISENTRAUT (1947), the unique means of reproduction found in bats is also a result of adaptation to climate. Mating takes place in the autumn, but the spermatozoa remain preserved during the entire winter, and ovulations and fertilization do not take place until spring (WISMATT, 1969). If hibernating bats are brought into a warm place, ovulation and embryonic development occur earlier (EISENTRAUT, 1937). The gestation period in bats is also dependent upon temperature.

Bats can survive in the cold with no food for several weeks, i.e. under conditions which would cause small homeotherms to perish within 1—2 days. On the other hand, it is impossible for bats to be active at lower ambient temperatures; temperatures slightly below the freezing point are the lowest which they can tolerate, whereas small birds of similar size can survive at temperatures considerably below the freezing point.

E. Physiological Causes of Geographic Temperature Limits

Very *small* warm-blooded animals, such as the small species of the hummingbird, cannot survive in cold climates because of their large surface/volume ratio, even if they have enough to eat (cf. p. 528). The smallest species of hummingbird, and the smallest of all birds, *Chaeterocercus bombus*, is found only in the vicinity of the equator in Ecuador and northern Peru (HESSE, ALLEE and SCHMIDT, 1951). Other small birds, if they have enough food, can stand considerably cold temperatures, but are unable to obtain enough food in the winter. An important limiting factor here is the short *length of the day*, which does not give the birds enough time to search for food. Wrens *(Troglodytes aedon)* can go without food for only a few hours in severe cold; thus they are unable to survive a long period of darkness (KENDEICH, 1945b). Small animals, e.g. titmice *(Parus)*, nuthatches *(Sitta)*, jays *(Garrulus)*, or hamsters *(Cricetus)*, ground squirrels *(Citellus)*, and voles *(Microtus terrestris)*, can compensate to a certain extent for periods of deficient food supply by collecting and storing food.

Because of their poor *heat insulation*, small mammals cannot stand such low temperatures as birds (cf. Fig. 28, p. 551). According to HESSE, ALLEE, and SCHMIDT (1951), the hare is the smallest European animal which remains completely exposed to cold during the winter. All smaller mammals, such as mice, shrews, martens and squirrels, require some domicile to protect them. This is especially true of small arctic mammals, which otherwise could not survive the polar winter (SCHOLANDER et al., 1950a). In contrast, the warm temperatures of the tropics present no difficulties for small homeotherms, except that they cannot tolerate intensive sunlight because of their sensitivity to heat (cf. p. 653).

Animals with slender *body forms* and large surfaces, long necks and legs, or animals with large ears, tails, or other appendages, e.g. giraffes, gazelles, kangaroo rats, desert foxes, are not suited for cold climates, since they release too much heat, and thus would run the risk of local freezing in their extremities. In this regard, birds, which have no protruding body parts with blood flow, such as ears, nose or tail, are more advantageously shaped than mammals. In addition, the legs of birds, which contain no muscles, are extremely insensitive to cold. Cold-adapted polar birds can stand temperatures of -40 to $-50°$ C without their legs freezing. In contrast, a seagull which was kept at room temperature suffered severe freezing of the feet at $-20°$ C (SCHOLANDER et al., 1950b). Warm-blooded animals with poorly developed integument, or partial or complete lack of integument, such as man, apes, certain moles *(Heterocephalus philippsi)*, vultures, ostriches, and marabous, also have only slight tolerance towards cold.

The part played by *preferred temperatures* (thermal preferenda) in the distribution of homeotherms has not yet been adequately explained: if they have any importance at all, it is probably only in regard to small mammals, especially rodents (HART, 1971) and bats. HERTER (1952), who studied this question extensively, came to the following conclusion: "Usually there is a certain parallel to be found between the preferred temperature level and the borders of the area in which a given type of animal occurs, in the sense that types with higher preferred temperatures do not occur so frequently in the direction of the poles, and do not live at such high altitudes as related types with lower preferred temperatures. Unfortunately, however, we do not have complete data on the geographic habitats of many small mammals." The world-wide distribution of many mammals and birds permits us to conclude that preferred temperatures obviously do not play a dominant part in the distribution of homeotherms. Indeed, homeothermy can be defined as the ability to exist at temperatures other than the preferred one. Thus, mice live in cold-storage depots in Hamburg at $-6°$ C if they have an abundance of food (MOHR, 1931); this is far below their preferred temperature, which, according to HERTER (1952), is around $37°$ C ground temperature.

For birds, the stage of development which is most sensitive to temperature is the *egg*, which thus may become the limiting factor in their geographic distribution (HESSE, ALLEE, and SCHMIDT, 1951; GRAHAM and HESTERBERG, 1948; CLARKE, 1954; and others). However, this influence is strongly decreased by placing the brooding period in the most favorable season of the year, or by special protection of eggs, as is seen in polar birds (p. 729).

F. Ecogeographical Rules

The tropical zones with their optimal temperatures produce a tremendous number of species, but the number of individual animals in each species is usually small. In the direction of the least habitable regions (desert, polar zones, high mountains) the number of species decreases progressively, but in the direction of the polar regions there is often a marked increase in the number of individual animals. The arctic lemmings are among the most numerous of all mammals, and other species living in cold zones, such as seals and seabirds, have large numbers of individual members (MARCUS, 1931).

Within particular homeothermic species we find *hereditary ecogeographical varia-tions*. The forms inhabiting cold zones are usually larger than in warmer regions *(Bergmann's rule)*; many examples of this have been demonstrated in regard to mammals and birds (MARCUS, 1931; TISCHLER, 1949, 1952; MAYR, 1949, 1956; HESSE, ALLEE, and SCHMIDT, 1951; DANSEREAU, 1957; DARLINGTON, 1957; GEORGE, 1962; LATTIN, 1967). Likewise, various exceptions are known (cf. MARCUS, 1931; MAYR, 1949; TISCHLER, 1952); according to RENSCH (1939), this is true of around 16% of paleoarctic birds. Apparent exceptions to Bergmann's rule always require a precise analysis of the local climatic conditions. The smaller forms of the cockatoo *(Cacatua galerita triton)* live partly in more northerly areas, where it has been shown that the ecoclimate is actually warmer (MAYR, 1949). Sometimes there is competition between other factors influencing body size, e.g. lack of food, and the effects of climate, so that in high mountain regions small races may originate, contrary to Bergmann's rule (DAVIS, 1938). Additional effects of cold climates are to be seen in stocky body shapes, shortened extremities, tails, and ears *(Allen's rule*, Fig. 108), and increasing lack of pigmentation in the integument, leading to the white fur and feathers of many polar animals *(Gloger's rule)*.

Fig. 108. Left: head of arctic fox *(Alopex lagopus)*; middle: red fox *(Vulpes vulpes)*; right: desert fox *(Megalotis zerda)*. From HESSE, ALLEE, and SCHMIDT (1951)

Warmer climates and high humidity often produce darker colors as the result of blackish eumelanin formation, whereas in dry climates the reddish or yellowish-brown pheomelanin is predominant, producing the characteristic colors of many desert animals. In addition, the relative heart weight of small mammals and birds in cold climates is greater *(Hesse's rule)*. Birds in cold zones lay larger numbers of eggs (RENSCH, 1936; KIPP, 1948), have a larger gastrointestinal tract and longer wings (RENSCH, 1936), and migrate further and more frequently.

Mammals in colder climates possess thicker fur and overhair *(Rensch's rule)*, and have a larger number of young (RENSCH, 1936).

The causes of these changes and their biological significance are not clear (cf. RENSCH, 1954; MAYR, 1956; RÖHRS, 1962). In addition to the hereditary characteristics, there are *modifications* resulting from the direct effect of temperature, as can be shown experimentally. Mice (SUMNER, 1911) and chicks (ALLEE and LUTHERMAN, 1940) which were raised at a room temperature of 6° C have greater weight, stockier bodies, shorter extremities, tails, and ears, and a higher relative heart weight than do animals raised at 21–25° C. As MAYR (1956) rightly says, the ecogeographical rules are purely empirical generalizations describing parallelisms between morphological variations and physiogeographic features. Their physiological interpretation, however, seems difficult. This is especially true for Bergmann's rule. The importance of small variations in body size for temperature regulation and, therefore, its selective value has been doubted (SCHOLANDER, 1955; RÖHRS, 1962), the more since there are other and more powerful mechanisms of thermal adaptation. Another objection was made by KIPP (1948) who pointed out that the increase in procreation rate (number of eggs and young) in colder regions counteracts the selective effect of climate. (Modifications s. also p. 639.)

G. Eurythermic and Stenothermic Species

If we apply the concepts *eurythermic* (wide temperature limits) and *stenothermic* (narrow temperature limits), which are completely valid only for poikilotherms (cf. p. 424), to homeotherms, we must remember that these terms have a completely different meaning in this new context. In regard to cold-blooded animals they designate the range of *body temperatures* within which life is possible, whereas in regard to warm-blooded animals they designate the *external temperatures* at which homeothermy can be maintained. If body temperature is considered, then practically all homeotherms are stenothermic in the warm range, but if the external temperature is taken as a criterion, many warm-blooded animals are extremely eurythermic, precisely because of their homeothermy.

Among the *eurythermic* animals are ravens *(Corvus corax)*, whose habitat extends "from the icy wastes of Greenland to the heart of the Sahara" (MAYR, 1949). HESSE, ALLEE, and SCHMIDT (1951) also include the stonechat *(Saxicola)*, which is found from Greenland to Spain, and the tiger, which ranges from the tropical jungles of India to the snow-covered mountains of Central Asia at altitudes of 4000 meters; its habitat covers 60 degrees of latitude. Further examples of eurythermic homeotherms are the Bactrian camel, which can tolerate temperatures between −37° C and +40° C in the deserts of Tibet, and the elephant, which is found at high altitudes in Africa, and which is not affected in the least by snowstorms (KRUMBIEGEL, 1943).

Among the *heat-stenothermic* animals, HESSE, ALLEE, and SCHMIDT (1951) list buffalos, giraffes, chevrotains *(Tragulus)*, hippopotami, the anthropoid apes, and vultures. Unclothed, man is also heat-stenothermic; it is only because of his artificial aids that man is the most eurythermic of all creatures.

Cold-stenothermic animals, according to HESSE, ALLEE, and SCHMIDT (1951), include the snow-leopard *(Felis uncia)*, the ibex, and the llama. We should not like to decide whether their preference for cold climates is really due to the temperature factor, since from the physiological point of view it is difficult to imagine that a warm-blooded animal can thrive only in a cold climate, and we know that many polar animals easily become acclimatized to hot summers. Polar bears in zoos lie, "paradoxically enough, in the blazing sun at noon even in the middle of the summer, and in the winter it is often almost impossible to get them to enter the water. They apparently become accustomed to the warm climate very quickly, and prefer it to the cold of their native habitat. This is true of many other species as well" (HEDIGER, 1949).

As this example shows, the distribution area of most homeotherms is not so large as it could be on the basis of their ability to adapt themselves: their *"ecological valency"* is greater than their actual distribution. This is also true of mammal fossil findings (cf. MARCUS, 1933).

H. Adaptation to Various Climates

1. Tropical Climate

Man. The hot-wet climate of the tropics is at once a difficult and an easy climate to which a man can adapt (LADELL, 1964). It is difficult because convective cooling is almost impossible for much of the day and evaporative cooling is limited. It is easy because it is probably the primordial climate in which man evolved. The tropical environment places the slightest thermal demand on a quiescent person, so that human life is possible with a minimum of clothing, food, and shelter. With sufficient protection against intense solar radiation, and if properly acclimatized, humans will have no particular difficulty living in the tropics. Not only are permanent settlements possible but advanced civilizations may floursih, as is shown by the Mayas, the Hindus, and the Javanese (in Djakarta, Java, the mean temperature in January is 26.5° C, in October 27.5° C, with average daily fluctuations from 23 to 29° C).

The adaptation of *temperate-climate men* to the tropics is less a question of direct climate influence than one of the related accompanying conditions (tropical diseases, social and psychological factors, etc.). As LADELL (1964) says, "the most important adaptations to a tropical climate are mental and cultural. Temperate-climate men are frightened of the tropics". There is no physiological reason why people from cooler zones, after proper acclimatization, should not thrive in a tropical environment; experiences in the Panama Canal Zone and in northern Australia have shown that this is possible. Where permanent settlements in the tropics have failed, the reason was not usually the climate itself, but rather the continuation of habits brought from the cooler zones in which the settlers originally lived (LEE, 1950; LEE and PENDLETON, 1951; LADELL, 1964).

If direct sunlight is avoided, the heat stress is not unduly great. This is shown by the slight number of cases of heatstroke compared, for instance, with the notorious hot areas of Mesopotamia. In tropical climates, a person resting in the shade can manage to release his heat production via radiation, conduction, and convection without sweating (Table 18). Direct sunlight or physical exertion will cause an additional supply of heat which must be absorbed by sweat evaporation. Because of the high humidity, it is important for the sweat to be able to evaporate without hindrance; this is best achieved by wearing no clothing, or at most thin, loose-fitting, well-ventilated garments (YAGLOU and RAO, 1947; LADELL, 1964).

It is not clear whether genetically fixed differences in heat tolerance exist between various *ethnic groups;* at least they must be relatively small. Extensive studies of Nigerians (LADELL, 1955, 1957) revealed that although they were slightly more tolerant of a standard hot-humid climate than were unacclimatized Caucasians, they sweated less than either residentially or artificially acclimatized Europeans. The rectal temperatures of the Nigerians were slightly lower, under given conditions, but their skin temperatures were higher than in Europeans (THOMSON, 1954). When trained to work in a standard climate, however, the Nigerians sweated progressively more, the time required for maximum heat adaptation being shorter than that for Europeans. At this point the Africans sweated exactly the same amount as Europeans had done after artificial acclimatization (LADELL, 1964). Acclimatized Bantu and Europeans showed practically the same heat tolerance and performance (WYNDHAM et al., 1953; WYNDHAM, 1966), and the course of deacclimatization was similar in Caucasians (EICHNA et al., 1945b) and Bantu (WYNDHAM and JACOBS, 1957). THOMSON (1954) found the number and distribution of functioning sweat glands to be identical in Europeans and Africans. The difference between the Africans and the residentially acclimatized Europeans represents "climatic know-how"; there is no advantage in developing high sweat rate in a climate where evaporative cooling is difficult, and the Africans instinctively avoid this by taking frequent rests after bouts of hard work which allow the body to cool (LADELL, 1964).

Although the heavily *pigmented skin* of tropical populations is thermally disadvantageous (cf. p. 551), melanin by its absorptive properties protects the underlying cells from the deleterious effects of ultraviolet radiations. However, since direct exposure to solar radiation is avoided as much as possible in tropical areas, the adaptive significance of dark skin is doubtful (BLUM, 1961). In spite of first appearances its geographical distribution does not parallel actual exposure of indigenes to the sun.

Animals. Optimal temperature conditions, i.e. those putting no stress on the organism's temperature regulation, are reflected by the extensive disappearance of such mechanisms in tropical climates, especially the weakly developed or entirely lacking integument. The loss of heat insulation is carried so far that many tropical animals are subjected to greater cold-stresses by the temperature fluctuations of their ecoclimate than are arctic animals during the polar winter (IRVING 1951). Most animals in hot climates are lacking in subcutaneous fat; if fat is present, it is accumulated in limited body areas, so that it cannot interfere with heat release, e.g. the humps of the camel and the Zebu *(Bos indicus),* the fat-tailed sheep.

The heat tolerance of tropical animals is not particularly great. If homeotherms of tropical and warm zones are exposed to intensive heat via sunlight, they will soon suffer a heatstroke. Animals living in wooded areas are especially sensitive to heat, e.g. apes and elephants (HESSE, ALLEE, and SCHMIDT, 1951; KRUMBIEGEL, 1943). For additional data on the adaptation of domestic animals to tropical climates, cf. p. 733.

Although the available information on thermoregulatory adaptations by *birds* in hot-humid environments is extremely meager, it appears that specific genetically fixed adaptations are relatively uncommon and unimportant. The thermoregulatory devices which have evolved with flight, such as panting, gular fluttering and temporary hyperthermia (p. 654), seem to have preadapted birds adequately for successful existence in hot-humid climates. It must be emphasized that because of elevated lower critical temperatures, higher shell conductivity, and, in some hummingbirds, small body size, nights and cool seasons may impose genuine problems of heat conservation (KING and FARNER, 1964).

2. Desert Climate

Man. As Table 18 shows, the main characteristic of desert climate is the organism's high gain of heat as a result of *radiation* from the sun, the sky, and the ground (cf. Fig. 27, p. 550). Light-colored garments will reduce this heat gain by approximately one half (ADOLPH, 1947a). In contrast to the tropics, the dry air of the desert permits considerable heat loss via evaporation; quantities of sweat amounting to as much as $1.7 \, l \, h^{-1}$ have been measured in the desert (ADOLPH and DILL, 1938). The main limiting factor in the desert is *lack of water*. Without water, a human can survive in the desert for only 1—2 days, but with sufficient water, man

Table 18. Heat exchange of man in tropical and desert climate by conduction, convection and radiation

Environment	clothed kcal h^{-1}	naked kcal h^{-1}
Tropics (night)	— 80	— 90
Tropics (jungle)	— 60	— 75
Tropics (sun)	+ 80	+120
Desert (sun)	+210	+380

Negative values: heat loss to environment; positive values: heat gain from environment. From ADOLPH (1947).

is a physiologically well-equipped desert inhabitant, whose heat tolerance is better than that of most desert animals. The hot dry areas of the earth have been settled by man since time immemorial, and have produced very advanced civilizations lasting for thousand of years, especially those of the Egyptians and the Sumerians. It is often overlooked that because of the excessive daily temperature fluctuation, the nights and early morning hours can be very cold, so that there is considerable stress imposed by cold as well as by heat (Fig. 109). Pictures of nomadic Arabs in flowing robes have led to the widespread but erroneous assumption that the Bedouins wear woollen clothing as protection against the heat, whereas in truth these regions have a rather cold climate during most of the year. In warmer regions, e.g. along the Persian Gulf, less clothing is worn.

Animals. Two factors limit the life of homeothermic animals in the desert: *heat* and *lack of water*. Thus the number of species which can stand these extreme conditions is quite small. The larger animals, e.g. gazelles *(Gazella loderi)*, chigetai *(Equus hemionus)* (ANDREWS, 1924), and waterbuck *(Kobus defassa ugandae)*, live only on the borders of the actual desert, but may occasionally be found deep in the interior of the desert. These animals require water in order to regulate their

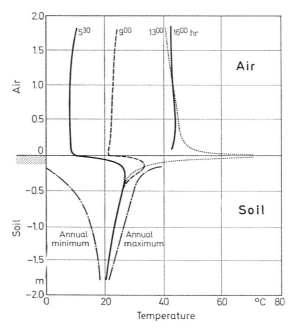

Fig. 109. Air and soil temperatures in various levels on the hottest day in Tucson, Arizona. From ADOLPH (1947a)

temperature in the heat, and thus are dependent upon sources of water (TAYLOR, SPINAGE, and LYMAN, 1969). However, the East African eland antelope *(Tauro-tragus oryx)* and the oryx antelope *(Oryx beisa)* can survive indefinitely in hot semideserts without drinking (TAYLOR, 1969a,b). The oryx is a true desert animal remaining exposed to the hot sun throughout the day. It can survive extremely high body temperatures up to 45° C, the brain being cooled by a countercurrent heat exchange system between the cerebral arteries and the veins supplying the nasal cavity (cf. p. 512). The heat storage of the body during the day reduces the need for evaporative cooling and thus saves water. Additional water-saving mechanisms are low metabolic rates and, during the nocturnal fall of body temperature, a higher extraction rate of oxygen from the inspired air. Therefore, the respiratory volume and the amount of water lost through respiratory eva-poration decrease.

Camels have developed a high degree of adaptation to desert conditions (SCHMIDT-NIELSEN, 1959; MACFARLANE, 1964; FOLK, 1966). The mechanisms involved are 1. heat tolerance by physical rejection of environmental heat, by heat storage up to 6° C during the day (SCHMIDT-NIELSEN et al., 1957a) cf. p. 517 (Ch. II), and by depression of metabolism when dehydrated; 2. water economy in the production of small volumes of concentrated urine, low rates of fecal water loss, efficient sweating, and reduction of sweat rate when lacking water (SCHMIDT-NIELSEN et al., 1956, 1957b); 3. relative preservation of plasma and interstitial fluid volumes during water deprivation, and mobilization of water from the cells and alimentary tract (MACFARLANE, MORRIS, and HOWARD, 1963); and 4. eclectic tastes in food, fat storage, and urea recycling for protein synthesis. Table 19 shows the main features of desert ungulates that help survival in arid regions, Table 20 gives a comparison of camel, sheep, and cattle with respect to water metabolism in the desert.

Table 19. Main features of desert ungulates that help survival in hot arid regions

Protection from heat radiation	Short reflecting coat Light color	Gazelle Eland
Evaporative cooling	Apocrine sweating Sweating and respiratory evaporation	Camel, donkey Zebu, Merino
Metabolism	Lowered during dehydration Relatively low and low when heated Higher oxygen utilization, thus low respiration	Camel, Eland, Oryx Merino, Zebu Oryx
Heat storage	Large fluctuation in body temperature	Camel, Eland, Oryx
Water reserves	Alimentary reserve, mainly rumen Extracellular volume	Camel Sheep
Water economy	Low renal flow normally Low renal flow when dehydrated Colon water reabsorption	Camel Sheep Zebu
Urea economy	Renal reabsorption	Camel, sheep, Zebu
Behavior	Feed in the sun Eat xerophytes, thorns, and salt plants Feed without water Active walkers between waterings	Antelope, camel, Zebu, Merino Antelope, goat, camel, donkey, sheep Eland, Oryx, camel, donkey, Zebu, Merino Antelope, camel, Zebu, Blackhead Persian and Somali sheep, donkey, Merino

Adapted from MACFARLANE (1964) and TAYLOR (1969a, b).

According to HESSE, ALLEE, and SCHMIDT (1951), the desert fox *(Megalotis zerda)* is the largest mammal whose permanent habitat is the desert; its water requirements are met by the liquid in the animals it preys upon. Its subterranean and nocturnal mode of life prevents excessive warming. Jack rabbits *(Lepus cali-fornicus)* live on the water they obtain from green plants and succulents; however, they cannot subsist on dry food (ARNOLD, 1942). On hot days jack rabbits tend to rest in the shade. In this situation, according to SCHMIDT-NIELSEN (1964), the large ears of the animal are in radiation exchange with the sky where the temperature may be assumed to be about 25° C lower than that of the ears.

Table 20. Relation of body fluids to survival of ruminants without water in desert environments with daily maximum temperatures of 40° C

Phenomena	Camel	Merino sheep	Shorthorn cattle
Rate of weight loss % per day	2.0	4.5	7.0
Percentage of fluid lost from plasma	4.5	8.0	10.0
Days survival (weight loss 28—32%)	12—15	6—8	3—4
Max. urine concentration osmoles/liter	3.8	3.1	2.6
Max. fecal dehydration % water	38	45	60
Max. plasma sodium mEq/liter	202	185	170
Water loss as % of weight lost	85	74	66
Initial water replacement as % of weight lost	60	72	84

From MACFARLANE (1968).

Since both metabolic rate and relative body surface increase with decreasing body size (p. 529), the heat load in hot environments becomes more severe for smaller animals (Fig. 110). Therefore, they cannot afford to use water for heat regulation in a desert climate (SCHMIDT-NIELSEN, 1954). Accordingly, most desert rodents are nocturnal, burrowing animals that rarely if ever are exposed to the full heat of the desert day.

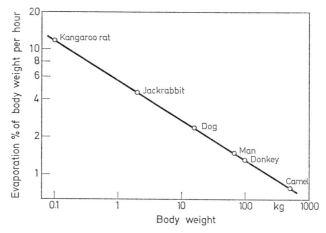

Fig. 110. Double-logarithmic plot of relation between body size and the evaporation estimated to be necessary for the maintenance of a constant body temperature in a hot desert climate. Adapted from SCHMIDT-NIELSEN (1954)

The failure of most *birds* to utilize shelters afforded by underground burrows limits the abilities of these animals to evade the extreme conditions of desert environments. With relatively few exceptions, they appear to lack special mechanisms for contending with heat and aridity. Their survival on hot days in deserts depends primarily upon the ability to tolerate elevations of body temperatures of as much as 4° C above normal levels (p. 654), and upon behavioral patterns which tend to reduce heat stress. All desert species of birds must rely on preformed water from the surface or from succulent plants, fruits, insects, or vertebrate prey for maintenance of water balance (DAWSON and SCHMIDT-NIELSEN, 1964).

Maximal adaptation to desert climate is found in the kangaroo rat *(Dipodomys)* whose physiological behavior has been extensively studied (SCHMIDT-NIELSEN, 1964 a, b). Actually, these animals are just as sensitive to heat as other small mammals; their water loss via evaporation corresponds to that of other rodents (SCHMIDT-NIELSEN and SCHMIDT-NIELSEN, 1950b). They protect themselves against the extremely high ground temperature (Fig. 109) by living in underground burrows the entrances of which they close, thus saving evaporation water. Kangaroo rats can survive on a completely *water-free* diet, since the oxidation water of the carbohydrates they consume is sufficient for their water requirements (Fig. 111), even when the absolute humidity is only 2 grams of H_2O per m^3;

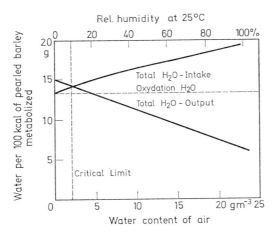

Fig. 111. Water balance of kangaroo rats *(Dipodomys)* at various atmospheric humidities at 25° C. The calculations are for 25 g pearl barley (dry weight), corresponding to a metabolic value of 100 kcal. Adapted from SCHMIDT-NIELSEN and SCHMIDT-NIELSEN (1952)

this is the minimum humidity during the driest period in the desert (SCHMIDT-NIELSEN and SCHMIDT-NIELSEN, 1950c). There is no possibility that they consume dew, since it is absent in many desert regions. They survive on oxidation water only by dint of extreme economy measures. The water content of their feces is 45%, that of rats 68% (SCHMIDT-NIELSEN and SCHMIDT NIELSEN, 1951). The concentrative ability of their kidneys is four times greater than that of the human organ. According to SCHMIDT-NIELSEN and SCHMIDT-NIELSEN (1952), the urea concentration in *Dipodomys* can reach 22.8% (6% in man), and the NaCl concentration 8.7% (2.1% in man). Glomerular filtration in the kidney is completely normal, but tubular reabsorption is very high (SCHMIDT-NIELSEN, 1951). If kangaroo rats are forced to drink water by being given a protein-rich diet (soy beans), which produces very little oxidation water, they can, thanks to their kidneys, thrive on sea water with a salt content of 3.4% just as well as on fresh water (SCHMIDT-NIELSEN and SCHMIDT-NIELSEN, 1950a). AMES and VAN DYKE (1950) found that *Dipodomys* secretes more antidiuretic hormone from the hypophysis than do other animals.

3. Polar Climate

Man. The cold adaptation of the circumpolar tribes of Greenland, North America, and Siberia is based on *behavioral patterns* and not on physiological mechanisms (cf. p. 628). The climate is so severe that it permits scarcely more than mere survival; thus, there have never been advanced civilizations in the Arctic region as long as the populations were dependent only on the natural resources of this area.

Eskimo houses and clothing display an almost perfect adaptation to what is probably the world's most severe environment. Eskimo weapons, boats and sledges are equally remarkable. According to WULSIN (1949), the houses of snow or earth are so built that the sleeping quarters are located in the highest, i. e. warmest part of the dome-shaped edifice. The entrance is at the bottom, through an open tunnel. Ventilation is regulated by an opening in the dome, so that cold air from below can flow in only to the extent that warm air flows out at the top. Thus the natural temperature stratification of the air is maintained. When the hut is heated by burning seal blubber, the inside temperature becomes so high that the occupants sweat while sitting, even if practically naked. But even without heating, a well-built igloo is comfortably warm on the inside at external temperatures of -40 to $-60°$ C.

The Eskimos' genuine clothing consists of two layers, usually of caribou skin; the inner layer is worn with the fur inside, the outer layer with the fur outside. In addition, arctic-fox and polar-bear hides are used, as well as sealskins for boots and overcoats. Eskimo clothing is tailored to fit the body, but fits very loosely. The management of clothing in the presence of moisture demands special precautions. Since wet clothes lose part of their thermal insulation, it is important to avoid sweating into the garments, and the Eskimos habitually loosen their clothing when at work out of doors, or take part of it off, to avoid the appearance of sensible perspiration. They strip when indoors, for the same reason, and often sit almost naked when the temperature of the house is high.

Animals. Various studies have provided extensive information on the physiological adaptation of homeotherms to the polar climate (for references cf. IRVING, 1964). The *rectal temperatures* of polar animals are the same as those of tropical animals, and do not change even in extreme cold (IRVING, 1951; IRVING and KROG, 1954). Small polar animals, such as lemmings *(Myodes lemmus)* and weasels *(Mustela rixosa)*, must seek protection in holes, caves, etc. in order to survive the polar winter. There are no hibernators in the actual polar zone, because of the excessive cold; such animals are found only in the somewhat warmer subboreal areas. Thus, the ground squirrel *(Citellus)* is found in Fairbanks, Alaska, where it hibernates under the snow in the unfrozen layers of the earth (CLARKE, 1954).

Other terrestrial mammals, such as the arctic hare *(Lepus timidus)*, the arctic fox *(Alopex lagopus)*, the wolf *(Canis lupus)*, the polar bear *(Thalarctos mariti- mus)*, the Dall sheep *(Oxis dalli)*, and the reindeer *(Rangifer tarandus* and *R. caribou)*, survive the polar winter without external protection. Their adaptation is based exclusively on the excellent *thermal insulation* provided by their ex- tremely thick fur (cf. Fig. 32, p. 556) and by their layer of winter fat, which also serves as a food-store (SCHOLANDER et al., 1950a,b; IRVING, 1951; GRIFFIN, HAMMEL, and RAWSON, 1956; HART, 1956; LENTZ and HART, 1960). In contrast, *heat production* scarcely increases in any of the larger polar animals, no matter how cold it is. In these animals, no signs of cold adaptation have been found in regard to their basal metabolism. The "critical temperature", at which metabolism begins to rise, is below $-50°$ C external temperature for larger arctic animals (cf. Fig. 18, p. 535). Accordingly, they show less metabolic increase during the polar winter than do certain tropical animals of the same size at $+20°$ C. At an external temperature of $-50°$ C, Eskimo dogs have a constant rectal temperature of 38° C, whether awake or asleep. Not even confinement in a cold-chamber at $-80°$ C will cause a change of rectal temperature in arctic foxes or Eskimo dogs. In the Alaska ptarmigan *(Lagopus lagopus alascensis)*, the threshold is around $-50°$ C (IRVING and KROG, 1954).

Polar animals are stocky in build, with relatively short extremities and short noses, ears, and tails (cf. Fig. 108). Arctic hares, arctic foxes, and polar bears have hair on the pads of their feet and toes; the snowy owl *(Nyctaea scandiaca)* and the Alaska ptarmigan *(Lagopus lagopus alascensis)*, the only nonmigratory arctic birds, have feathers on their legs (HESSE, ALLEE, and SCHMIDT, 1951). Ac- cording to IRVING and KROG (1955), Eskimo dogs have skin temperatures up to 37°C under the fur at ambient temperatures of $-30°$C (Fig. 112). The temperature of the nose and paws may sink to below 5° C; this is also true of the extremities of reindeer and polar birds. In man such temperatures would cause intolerable cold pain. The cold appendages emit only a small amount of heat and thus serve as effective means of heat conservation. These tissues are particularly adapted to low temperatures; for instance, marrow fats from the phalanges of the legs of several species of arctic wild mammals had low melting points and were soft when cold, white fat from the femur melted at about body temperature and was brittle when cold (IRVING, SCHMIDT-NIELSEN, and ABRAMSON, 1956). Further, con- duction of impulses in the nerve from the cold bare part of the leg in herring gulls *(Larus argentatus)* was blocked by cold around 2 to 5° C, whereas the central part of the nerve from the warm part of the leg under the feathers failed at 8 to 13° C. The nerve conduction at low temperatures is a result of local cold adaptation, but in nerves from hen *(Gallus domesticus)*, this adaptability was absent (CHATFIELD, LYMAN, and IRVING, 1953).

The *white color* that is seen in many polar animals bears no relation to their heat radiation from the body surface. Heat radiation from the body is in the infrared range and thus independent of visible color. In fact, infrared emissivities have been found identical for black and white fur (HAMMEL, 1956). The black raven found throughout the Arctic is undisturbed by cold, and so are Antarctic birds and seals which are not white (IRVING, 1964).

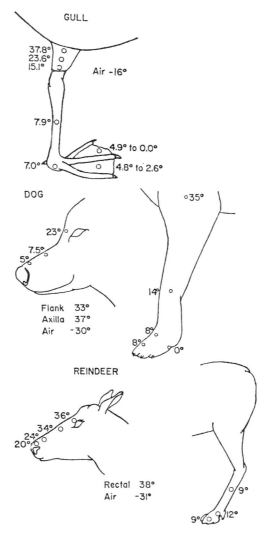

GULL

37.8°
23.6°
15.1°

Air −16°

7.9°

4.9° to 0.0°

7.0°

4.8° to 2.6°

DOG

°35°

23°

7.5°
5°

14°

Flank 33°
Axilla 37°
Air −30°

8°
8°
0°

REINDEER

36°
34°
24°
20°

Rectal 38°
Air −31°

9°

9° 12°

Fig. 112. Topographic distribution of superficial temperatures in a dog, a reindeer, and the leg of a gull in arctic winter. From IRVING and KROG (1955)

Thermal insulation in *aquatic* mammals and swimming birds is provided mainly by their subcutaneous fat, which, together with the skin, can amount to as much as 50% of the total body weight in seals. Harbor seals *(Phoca vitulina)* live in North Atlantic and Pacific waters which represent the coldest inhabited environment. Their skin cools within 1 to 2 degrees of the water. The critical temperature of harp seals *(Phoca groenlandica)* in winter was not reached in ice water, while it was higher in summer (IRVING and HART, 1957; HART and IRVING, 1959). Since records of blubber thickness did not show variations corresponding to season, other factors, such as vascular insulation, may account for the seasonal changes. Special behavioral and physiological adaptations have been developed by polar birds for thermal protection of the *eggs*. In arctic nests, the temperature among the eggs was maintained at 35° C (IRVING and KROG, 1956). The highest degree of adaptation is found in antarctic penguins, such as the emperor penguin *(Aptenodytes forsteri)* and the Adelee penguin *(Pygoscelis adeliae)*. They can brood even in the middle of the polar winter, carrying their eggs in a special fold under their feathers. The temperature of the egg was found to be approximately 35° C (EKLUND and CHARLTON, 1959), and the incubated eggs of the south polar skua *(Cotharacta maccormicki)* are even warmer (EDLUND, 1961).

I. Temperature and Productivity of Domestic Animals

Cold. The critical ambient temperatures at which the metabolic rate rises in laboratory experiments on animals which are at rest lie between 15 and 21° C for cattle, sheep, pigs, and chickens (BRODY, 1945; BLAXTER, MACGRAHAM, and WAINMANN, 1959). Under normal conditions and with sufficient food, however, this temperature limit has no practical significance. Since their metabolic rate is already rather high, these animals can stand much lower temperatures with no increase in physiological stress. Temperatures as low as − 40° C were tolerated without ill effect by horses, cows, and sheep outdoors during the winter (BRODY, 1945). Obviously the formation of a properly insulating winter coat and an increase in subcutaneous fat are sufficient to provide adequate protection against the cold. Sheep are the most populous and widely distributed domestic ruminants, extending from the arctic circle to the most southerly tip of South America, and from tropical lowlands to hot, arid regions (MOULE, 1968). They are successfully kept in climates with temperatures reaching extreme lows of − 50° C. In general, the reactions of sheep in fleece to cold temperatures are quite mild in relation to those resulting from high temperatures. Clipped sheep and lambs, of course, or animals exposed to wind and rain are more susceptible to cold stress (JOYCE and BLAXTER, 1964; TERRILL, 1968). Swine can live outdoors at environmental temperatures of − 30° C without signs of discomfort (IRVING, PEYTON, and MONSON, 1956). These bare-skinned animals are protected by the thermal insulation of their subcutaneous fat. The blood flow of the skin is reduced by vasoconstriction and the skin thus allowed to cool down to temperatures below 5° C. Evidently the cold skin prevents heat loss and serves metabolic economy. As a consequence, the cost of feeding pigs is not noticeably elevated in the cold Alaskan winter. However, young pigs are more sensitive to cold, and in the newborn animal the critical ambient temperature for metabolic rise is as high as 35° C (MOUNT, 1968; cf. p. 681). Dairy goats are sensitive to cold and begin to shiver at environmental temperatures of − 5° C (TERRILL, 1968).

In cattle, the *milk yield* is relatively unaffected within the temperature range between 0° and 21° C (Fig. 113). At temperatures lower than 5° C, the yield decreases slowly, together with an increase in feed consumption. The minimal critical temperature of milk yield for Jersey cattle is around 2° C, whereas Holsteins are not greatly affected at even − 13° C (HAFEZ, 1968). Maximum *wool growth* in sheep occurs in the summer, minimum growth in the winter. This seasonal cycle is primarily dependent on *photoperiodism* and not on temperature.

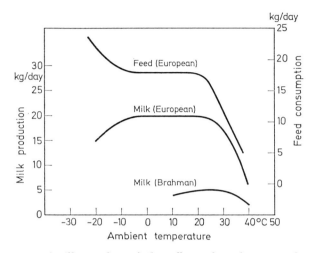

Fig. 113. Diagrammatic illustration of the effect of environmental temperature on milk production and feed consumption of European and Brahman cattle. Adapted from HAFEZ (1968)

Sheep subjected to a reversed, seasonal daylength rhythm developed a reversed, seasonal wool growth cycle with maximal growth in the winter and minimum in the summer, while the natural wool-growth cycle was not obviously altered by heating the sheep in winter and cooling them in summer (MORRIS, 1961). Severe cold, however, may lead to a decline in wool growth (HUTCHINSON and WODZICKA-TOMASZEWSKA, 1961).

In general, animals lose more weight during the winter when kept out of doors than similar animals kept indoors. Food supplied *ad lib* may be consumed in larger quantities, but yet still be insufficient to prevent a retardation of growth. The growth of well-fed cattle, however, is not reduced at temperatures well below freezing. Pigs given the same amount of food at different temperatures retain appreciably less of their dietary nitrogen under colder environmental conditions (FULLER, 1965). In conclusion one can say that in ordinary stock-farming the problems resulting from cold are less severe than those resulting from hot environments.

Heat. Since a number of domestic animals, particularly cattle, have a rather limited heat tolerance, serious problems may result from high environmental temperatures (LEE, 1959, 1961; FINDLAY, 1961; MACFARLANE, 1964; HAFEZ, 1968). Respiratory rate in European cattle starts increasing at temperatures as low as 15° C, and at 30° C it has reached more than three times the initial value (BIANCA,

1968). At this temperature level a marked rise in rectal temperature begins, which reaches about 40° C at 35° C ambient temperature (cf. Fig. 34, p. 561).

European cattle *(Bos taurus)* are less tolerant towards heat than the indigenous cattle breeds in tropical zones, such as Banteng *(Bibos banteng)* and Brahman cattle *(Bos indicus)*. A combination of physiological adaptations has been achieved by crossing functionally different types of animals. Santa Gertrudis, for example, is a crossbreed between European Shorthorn and Brahman cattle which has been successful in producing good beef animals in both wet and dry regions within the tropics. A comparison of features of Shorthorn and Brahman cattle in hot environments is shown in Table 21. With respect to coat, sweating mechanisms, food and water economy, Indian cattle are better adapted to heat than European breeds (MACFARLANE, 1968). At higher ambient temperatures, water turnover, heat production, and rectal temperatures of Indian cattle are considerably lower than those of European cattle at the same temperature (WORSTELL and BRODY, 1953; MACFARLANE and HOWARD, 1966; YOUSEF, HAHN, and JOHNSON, 1968).

Table 21. Comparison of features of Merino sheep, Brahman cattle and Shorthorn cattle in hot environments

Function in hot environments	Merino (Ovis aries)	Brahman (Bos indicus)	Shorthorn (Bos taurus)
Coat	Wool	Short, shiny	Longer
Behavior in sun	(longwave radiation)	(reflecting)	(heat absorbing)
Sweat glands	Small +	Large +++	Smaller ++
Respiratory cooling	Important ++	Slight +	Moderate ++
Foraging in the sun	Common	Common	Rare
Range of foodstuff eaten	Restricted	Wide	Restricted
Rumen yield of metabolites	+++	+++	++
Intestinal absorption of metabolites	+++	+++	++
of water	+++	++	+
Efficiency of conversion	+++	+++	++
Metabolic rate and water turnover	Low	Higher	Highest
Response to vasopressin	+++	++	+

+++ Major activity   ++ Moderate   + Minor
From MACFARLANE (1968).

Of great practical importance is the correlation between hot environments and *milk yield* (McDOWELL, 1972). As seen in Fig. 113, yield decreases markedly at temperatures above 25° C. There are also breed differences in the maximal critical temperatures. For example, the milk yield declines in Holstein cattle at 21° C, in Brown Swiss and Jerseys at 24° to 27° C and in the Brahman at 32° C, the relative humidities being 50—60%. At 10° C, the milk yield in Brown Swiss is about 6 times higher than in Brahmans, whereas the ratio decreases to about 3:1 at 40° C (Fig. 113).

Merino sheep are well adapted to dry and hot environments. In still air above 35° C the surface of the tip wool heats in the sun to 90° C. The surface reflects 30—50% of the total radiation. Another fraction of heat is given off by free convection and infrared radiation (cf. Fig. 27, p. 550), the rest passes through the fleece which has a high thermal resistance. On the side of the animal exposed to the sun, the skin temperature is about 42° C in these circumstances, while on the shaded side it is near the rectal temperature of 39—40° C (MACFARLANE, 1964). Other physiological features of sheep (Tables 20 and 21) that are favorable in hot arid conditions are: low water turnover, high water reabsorption and concentrating power of the kidney as well as high sensitivity to both the antidiuretic and potassium-excreting actions of vasopressin (MACFARLANE, MORRIS, and HOWARD, 1956; MACFARLANE, HOWARD, and SIEBERT, 1967; MACFARLANE, 1964, 1965, 1968). With respect to water metabolism, merino sheep range between cattle and camel (Table 20).

The *growth* of domestic animals after weaning can be stunted by high environmental temperatures (HAFEZ, 1968). European breeds of cattle show a marked depression of growth at temperatures constantly maintained above 24° C. The effects become progressively more severe until weight gain ceases entirely at temperatures of 29—32° C. Puberty is delayed in Shorthorn cattle and Brahmans reared at 27° C compared to those reared at 10° C. The most rapid growth of pigs occurs at temperatures between 16° and 21° C (FULLER, 1965). Extremely high environmental temperatures may also cause disturbances of gestation and fetal development, such as smaller litter size, abortion, and fetal dwarfism (YEATES, 1958; HAFEZ, 1967). For effects of heat on male fertility see p. 670.

Within the range of 9—29° C, *egg production* in domestic fowl is not reduced (HAFEZ, 1968). Heat stress, especially sharp changes in temperature, affect both egg weight and shell quality; above 32° C shell thickness decreases. The optimal storage temperature for the maintenance of blastoderm viability is 10—12° C, the maximal storage time being 2 weeks; hatchability is lowered if eggs are stored at 20—35° C, and embryogenesis is distorted.

References to Section "Homeothermic Organisms"

ABRAMSON, D. I., TUCK, S., JR., LEE, S. W., RICHARDSON, G., CHU, L. S.: Arch. Phys. Med. **47**, 300 (1966).

ADAIR, E. R., STITT, J. T.: J. Physiol. (Paris) **63**, 191 (1971).

ADAM, J. M., FERRES, H. M.: J. Physiol. (London) **125**, 21 P (1954).

ADAMS, T.: In: WHITTOW, G. C. (Ed.): Comparative physiology of thermoregulation, Vol. II, p. 151. New York-London: Academic Press 1971.

ADAMS, T., HEBERLING, E. J.: J. Appl. Physiol. **13**, 226 (1958).

ADOLPH, E. F.: Physiology of man in the desert. New York: Interscience 1947 a.

ADOLPH, E. F.: Am. J. Physiol. **151**, 564 (1947 b).

ADOLPH, E. F.: Am. J. Physiol. **155**, 378 (1948).

ADOLPH, E. F.: Am. J. Physiol. **166**, 75 (1951).

ADOLPH, E. F.: Am. J. Physiol. **166**, 92 (1951).

ADOLPH, E. F.: Am. J. Physiol. **184**, 18 (1956).

ADOLPH, E. F.: Regulation of water metabolism in stress. In: Homeostatic Mechanisms, Brookhaven Symposia and in Biology: No 1, 1957.

ADOLPH, E. F.: In: Handbook of physiology, Sect. IV, p. 27. Washington, D. C.: Am. Physiol. Soc. 1964.

ADOLPH, E. F.: Origins of physiological regulations. New York-London: Academic Press 1968.

ADOLPH, E. F., DILL, D. B.: Am. J. Physiol. **123**, 369 (1938).

ADOLPH, E. F., LAWROW, J. W.: Am. J. Physiol. **161**, 359 (1950).

ADOLPH, E. F., LAWROW, J. W.: Am. J. Physiol. **166**, 62 (1951).

ADOLPH, E. F., MOLNAR, G. W.: Am. J. Physiol. **146**, 507 (1946).

ADOLPH, E. F., RICHMOND, J.: J. Appl. Physiol. **8**, 48 (1955).

AFONSO, S., HERRICK, J. F., YOUMANS, W. B., ROWE, G. G., CRUMPTON, C. W.: Am. J. Physiol. **203**, 278 (1962 a).

AFONSO, S., ROWE, G. G., CASTILLO, C. A., CRUMPTON, C. W.: J. Appl. Physiol. **17**, 706 (1962 b).

AGID, R., GABE, M., MARTOJA, M.: Compt. Rend. **161**, 459 (1967).

AHERNE, W., HULL, D.: Proc. Roy. Soc. Med. **57**, 1172 (1964).

ALBERS, C.: Pflügers Arch. Ges. Physiol. **274**, 125 (1961).

ALEXANDER, G.: Australian J. Agric. Res. **13**, 100 (1962).

ALEXANDER, G., WILLIAMS, D.: Australian J. Agric. Res. **13**, 122 (1962).

ALEXANDER, G., WILLIAMS, D.: J. Physiol. (Lond.) **198**, 251 (1968).

ALEXANDER, L.: Combined Intelligence Objectives Sub-committee, Item No. 24, File No. 26—37 (1946).

ALLEE, W. C., LUTHERMAN, C. Z.: Ecology **21**, 29 (1940).

ALRUTZ, S.: Scand. Arch. Physiol. **7**, 321 (1897).

AMBID, L.: Modifications métaboliques en relation avec l'hibernation et les réveils périodiques du lerot (Eliomys quercinus L.). Thèse p. obtenir le grade de docteur des-sciences naturelles. Toulouse, No. d'ordre **447**, 1 (1971).

AMBID, L., BERLAN, M., AGID, R.: J. Physiol. (Paris) **63**, 505 (1971).

AMES, R. G., VAN DYKE, H. B.: Proc. Soc. Exp. Biol. (N. Y.) **75**, 417 (1950).

ANDERSEN, H. T., ANDERSSON, B., GALE, C.: Acta Physiol. Scand. **54**, 159 (1962).

ANDERSEN, H. T., HAMMEL, H. T., HARDY, J. D.: Acta Physiol. Scand. **53**, 247 (1961).

ANDERSSON, B.: Nord. Med. **47**, 663 (1952 a).

ANDERSSON, B.: Experientia **8**, 157 (1952 b).

ANDERSSON, B.: In: HARDY, J. D., GAGGE, A. P., STOLWIJK, J. A. J. (Eds.): Physiological and behavioral temperature regulation, p. 634. Springfield, Ill.: Thomas 1970.

ANDERSSON, B., BROOK, A. H., GALE, C. C., HÖKFELT, B.: Acta Physiol. Scand. **61**, 393 (1964).

ANDERSSON, B., GALE, C. C., HÖKFELT, B., LARSSON, B.: Acta Physiol. Scand. **65**, 45 (1965).

ANDJUS, R. K.: J. Physiol. (Lond.) **128**, 547 (1955).

ANDREWS, R. C.: Nat. History **24**, 150 (1924).

ANREP, V., HAMMOUDA, M.: J. Physiol. (Lond.) **77**, 16 (1933).

ANTAL, J., KIRILČUK, V.: Pflügers Arch. Ges. Physiol. **308**, 25 (1969).

AOKI, T., WADA, M.: Science **114**, 123 (1951).

ARNOLD, G., LOCHNER, W.: Pflügers Arch. Ges. Physiol. **284**, 169 (1965).

ARNOLD, J. F.: Univ. Arizona, Agric. Exp. Sta., Tech. Bull. **98**, 51 (1942).

ARPACI, V. S.: Conduction heat transfer. Massachusetts-Palo Alto-London-Don Mills, Ontario: Addison-Wesley 1966.

ASCHOFF, J.: Pflügers Arch. Ges. Physiol. **247**, 132 (1943).

ASCHOFF, J.: Naturwissenschaften **35**, 235 (1948).

ASCHOFF, J.: Klin. Wschr. **33**, 545 (1955).

ASCHOFF, J.: Cold Spring Harb. Symp. Quant. Biol. **25**, 11 (1960).

ASCHOFF, J.: Circadian clocks. Amsterdam: North-Holland 1965.

ASCHOFF, J.: Ärztl. Prax. **18**, 1569 (1966).

ASCHOFF, J.: In: BROWN, A. H., FAVORITE, F. G. (Eds.): Life Science Space Research, 5, p. 159. Amsterdam: North-Holland 1967.

ASCHOFF, J.: In: HARDY, J. D., GAGGE, A. P., STOLWIJK, J. A. J. (Eds.): Physiological and behavioral temperature regulation, p. 905. Springfield, Ill.: Thomas 1970.

ASCHOFF, J., GERECKE, U., WEVER, R.: Japan J. Physiol. **17**, 450 (1967).

ASCHOFF, J., GÜNTHER, B., KRAMER, K.: Energiehaushalt und Temperaturregulation. München-Berlin-Wien: Urban & Schwarzenberg 1971.

ASCHOFF, J., WEVER, R.: Naturwissenschaften **49**, 337 (1962).

ASHBY, W. R.: Design of a brain. 2nd Ed. New York: Wiley 1960.

ASHBY, W. R.: Ann. Rev. Physiol. **28**, 89 (1966).

ATURSON, G.: Acta Chir. Scand. **131**, 402 (1966).

AUSTIN, W. H., LACOMBE, E. H., RAND, P. W.: J. Appl. Physiol. **19**, 893 (1964).

BACCINO, M.: Compt. Rend. **119**, 1246 (1935).

BADER, M. E., MACHT, M. B.: J. Appl. Physiol. **1**, 215 (1948).

BAKER, M. A., HAYWARD, J. N.: J. Physiol. (Lond.) **198**, 561 (1968).

BAKER, P. T., WEINER, J. S.: The biology of human adaptability. Oxford: Clarendon Press 1966.

BALDWIN, B. A., INGRAM, D. L.: J. Physiol. (Lond.) **191**, 375 (1967).

BALDWIN, S. P., KENDEIGH, S. C.: Sci. Publ. Cleveland Mus. Nat. Hist. **3**, 1 (1932).

BARD, P., MACHT, M. B.: The behavior of chronically decerebrate. In: Ciba Foundation Symposium on the neurological basis of behavior. Boston: Little, Brown & Co. 1958.

BARD, P., WOODS, J. W., BLEIER, R.: In: HARDY, J. D., GAGGE, A. P., STOLWIJK, J. A. J. (Eds.): Physiological and behavioral temperature regulation, p. 519. Springfield, Ill.: Thomas 1970.

BARGMANN, W., HEHN, G. v., LINDNER, E.: Z. Zellforsch. **85**, 601 (1968).

BARNETT, S. A.: Quart. J. Exp. Physiol. **44**, 35 (1959).

BAROTT, H. G.: U. S. Dept. Agr. Tech. Bull. 553 (1937).

BARTELS, H., SCHMELZLE, R., ULRICH, S.: Resp. Physiol. **7**, 278 (1969).

BARTHOLOMEW, G. A.: Physiol. Zool. **29**, 26 (1956).

BARTHOLOMEW, G. A., CADE, T. J.: J. Mammal. **38**, 60 (1957).

BARTHOLOMEW, G. A., DAWSON, W. R.: Condor **54**, 58 (1952).

BARTHOLOMEW, G. A., DAWSON, W. R.: Ecology **35**, 181 (1954).

BARTHOLOMEW, G. A., DAWSON, W. R.: Auk **75**, 150 (1958).

BARTHOLOMEW, G. A., DAWSON, W. R., LASIEWSKI, R. C.: Z. Vergl. Physiol. **70**, 196 (1970).

BARTHOLOMEW, G. A., HOWELL, T. R., CADE, T. J.: Condor **59**, 145 (1957).

BARTHOLOMEW, G. A., HUDSON, J. W.: Physiol. Zool. 35, 94 (1962).
BARTHOLOMEW, G. A., LASIEWSKI, R. C., CRAWFORD, E. C., JR.: Condor 70, 31 (1968).
BARTHOLOMEW, G. A., LEITNER, P., NELSON, J. E.: Physiol. Zool. 37, 179 (1964).
BARTHOLOMEW, G. A., MCMILLEN, R. E.: Physiol. Zool. 34, 177 (1961).
BARTHOLOMEW, G. A., TROST, CH. H.: Condor 72, 141 (1970).
BARTHOLOMEW, G. H., WILKE, F.: J. Mammal. 37, 327 (1956).
BASS, D. E., HENSCHEL, A.: Physiol. Res. 36, 128 (1956).
BAUMBER, J., DENYES, A.: Canad. J. Biochem. 42, 1397 (1964).
BAUMBER, J., DENYES, A.: Can. J. Biochem. 43, 747 (1965).
BAUMBER, J., SOUTH, F. E., FERREN, L., ZATZMAN, M. L.: Life Sci. 10, 463 (1971).
BAZETT, H. C.: Am. J. med. Sci. 218, 483 (1949).
BAZETT, H. C.: J. Appl. Physiol. 4, 245 (1951/52).
BAZETT, H. C., LOVE, L., NEWTON, M., EISENBERG, E., DAY, R., FORSTER, R.: J. Appl. Physiol. 1, 3 (1948a).
BAZETT, H. C., MCGLONE, B.: Am. J. Physiol. 93, 632 (1930).
BAZETT, H. C., MCGLONE, B., BROCKLEHURST, R. J.: J. Physiol. (Lond.) 69, 88 (1930).
BAZETT, H. C., MENDELSON, E. S., LOVE, L., LIBET, B.: J. Appl. Physiol. 1, 169 (1948b).
BEAKLEY, W. R., FINDLAY, J. D.: J. Agric. Sci. (Cambridge) 45, 373 (1955).
BEAUMONT, W., VAN, BULLARD, R. W.: Science 141, 643 (1963).
BEAVERS, W. R., COVINO, B. G.: J. Appl. Physiol. 14, 60 (1959).
BEDFORD, T.: Basic principles of ventilation and heating. London: Lewis 1948.
BEHMANN, F. W., BONTKE, E.: Pflügers Arch. Ges. Physiol. 266, 408 (1958).
BEHNKE, A. R., YAGLOU, C. P.: J. Appl. Physiol. 3, 591 (1951).
BELDING, H. S., HATCH, T. F.: Heat. Pip. Air Condit. 27, 129 (1955).
BELL, C. R., WALTERS, J. D.: J. Appl. Physiol. 27, 684 (1969).
BENEDICT, F. G.: Carnegie Inst., Wash. Publ. 425, 1932.
BENEDICT, F. G.: Carnegie Inst., Wash. Publ. 503, 1938.
BENEDICT, F. G., LEE, R. C.: Carnegie Inst., Wash. Publ. 497, 1938.
BENEDICT, F. G., SLACK, E. P.: Carnegie Inst. Wash. Publ. 155 1911.
BENZINGER, T. H.: Symp. Soc. Exp. Biol. 18, 49 (1964).
BENZINGER, T. H.: Physiol. Rev. 49, 671 (1969).
BENZINGER, T. H., TAYLOR, G. W.: In: Temperature, its measurement and control in science and industry, Vol. III, part 3, p. 111—120. New York: Reinhold 1963.
BERING, E. A., JR., TAREN, J. A., MCMURREY, J. D., BERNHARD, W. F.: Surg. Gynec. Obstet. 102, 134 (1956).
BERNE, R. M.: Circulation 2, 236 (1954).
BESSOU, P., PERL, E. R.: J. Neurophysiol. 32, 1025 (1969).
BETZ, E., BRÜCK, K., HENSEL, H., JARAI, I., MALAN, A.: Pflügers Arch. Ges. Physiol. 272, 76 (1960).
BETZ, E., BRÜCK, K., HENSEL, H., JARAI, I., MALAN, A.: In: Biometeorology, p. 675. Oxford-London-New York-Paris: Pergamon Press 1962.
BETZ, E., HENSEL, H.: Pflügers Arch. Ges. Physiol. 281, 17 (1964).
BIANCA, W.: Int. J. Biometeor. 5, 5 (1961).
BIANCA, W.: In: HAFEZ, E. S. E. (Ed.): Adaptation of domestic animals, p. 97. Philadelphia: Lea & Febiger 1968.
BIANCA, W.: Intern. J. Biometeor. Suppl. 14, 119 (1970).
BIANCA, W., FINDLAY, J. D.: Res. Vet. Sci. 3, 38 (1962).
BIANCA, W., HALES, J. R. S.: Brit. Vet. J. 126, 45 (1970).
BIEDERMANN, W.: Ergebn. Biol. 6, 427 (1930).
BIERENS DE HAAN, J. A.: Roux Arch. 50, 1 (1922).
BIEWALD, G.-A.: Wiss. Z. Univ. Halle, Math.-Nat. Reihe 16, 239 (1967).
BIEWALD, G.-A., RATHS, P.: Pflügers Arch. Ges. Physiol. 268, 530 (1959).
BIEWALD, G.-A., RATHS, P.: Wiss. Z. Univ. Halle, Math.-Nat. Reihe 16, 227 (1967).
BIGELOW, W. G., LINDSAY, W. K., HARRISON, R. C., GORDON, R. A., GREENWOOD, W. F.: Am. J. Physiol. 160, 125 (1950).
BILLINGHAM, R. E.: Proc. Roy. Soc. B 147, 530 (1957).
BJÖRK, V. O.: Thoraxchirurgie 8, 271 (1960).

BLACK, D. A. K.: Essentials in fluid balance. Oxford: Blackwell 1957.

BLAGDEN, C.: Phil. Trans. B 13, 604 (1775).

BLAIR, D. A., GLOVER, W. E., RODDIE, I. C.: J. Appl. Physiol. 16, 119 (1961).

BLAIR, E.: Clinical hypothermia. New York: McGraw-Hill Book Company 1964.

BLAXTER, K. L.: J. Univ. Newcastle - upon - Tyne Agric. Soc. 19, 3 (1965).

BLAXTER, K. L., MCGRAHAM, N., WAINMANN, F. W.: J. Agric. Sci. 52, 41 (1959).

BLEICHERT, A., BEHLING, K., GEBBERS, J. O., KITZING, J., SCARPERI, M., SCARPERI, S.: Pflügers Arch. Ges. Physiol. 319, R 100 (1970).

BLEICHERT, A., KITZING, I., BEGEMANN, F.: Naturwissenschaften 53, 88 (1966).

BLIGH, J.: Nature 189, 582 (1961).

BLIGH, J.: J. Physiol. (Lond.) 168, 764 (1963).

BLIGH, J.: J. Physiol. (Lond.) 176, 145 (1965).

BLIGH, J.: Biol. Rev. 41, 317 (1966).

BLIGH, J., ALLEN, T. E.: In: HARDY, J. D., GAGGE, A. P., STOLWIJK, J. A. J. (Eds.): Physiological and behavioral temperature regulation, p. 97. Springfield, Ill.: Thomas 1970.

BLIGH, J., COTTLE, W. H., MASKREY, M.: J. Physiol. (Lond.) 212, 377 (1971).

BLIGH, J., HARTHOORN, A. M.: J. Physiol. (Lond.) 176, 145 (1965).

BLIGH, J., HENSEL, H.: In: TROMP, S. W. (Ed.): Progress in human biometeorology. Amsterdam: Swets & Zeitlinger (in press).

BLIGH, J., INGRAM, D. L., KEYNES, R. D., ROBINSON, S. G.: J. Physiol. (Lond.) 176, 136 (1965).

BLIX, M.: Upsala Läk.-Fören. Förh. 18, 87 (1882—1883).

BLOCKLEY, W. V., TAYLOR, C. L.: Heat. Pip. Air Condit. 21, 111 (1949).

BLUM, H. F.: Physiol. Rev. 25, 483 (1945).

BLUM, H. F.: Am. J. Med. Sci. 242, 812 (1961).

BOAKES, R. J., KERKUT, G. A., MUNDAY, K. A.: Life Sci. 6, 457 (1967).

BODENHEIMER, F. S.: Physiol. Zool. 14, 186 (1941).

BOGART, R., MAYER, D. T.: Am. J. Physiol. 147, 320 (1946).

BOGERT, F., VAN DER, MORAVEC, C. L.: J. Pediat. 10, 466 (1937).

BOMAN, K. K. A.: Acta Physiol. Scand. 44, Suppl. 149 (1958).

BOULOUARD, R.: Mem. Mus. Nation. Hist. Nat. Paris, Serie A, 60, 77 (1969).

BOULOUARD, R.: J. Physiol. (Paris) 63, 77 (1971).

BOWEN, W.: Proc. Acad. Sci. New Hampshire 1, 11 (1946).

BRÅNEMARK, P.-J., JOHANSSON, B. W.: Acta Physiol. Scand. 73, 300 (1968).

BRAUER, R. W., LEONG, G. F., HOLLOWAY, R. J.: Am. J. Physiol. 177, 103 (1954).

BREARLEY, E. A., KENSHALO, D. R.: J. Comp. Physiol. Psychol. 70, 1 (1970).

BRENDEL, W.: Verhandl. Deut. Ges. Kreislaufforsch. 23, 33 (1957).

BRITTON, S. W., ATKINSON, W. E.: J. Mammal. 19, 94 (1938).

BRODY, S.: In: Temperature, its measurement and control, p. 462. New York: Reinhold 1941.

BRODY, S.: Bioenergetics and growth. New York: Reinhold 1945.

BROWN, A. C.: Bull. Math. Biophys. 27, 67 (1965).

BROWN, A. C., BRENGELMANN, G. L.: In: HARDY, J. D., GAGGE, A. P., STOLWIJK, J. A. J. (Eds.): Physiological and behavioral temperature regulation, p. 684. Springfield, Ill.: Thomas 1970.

BROWN, W. E. L., HILL, A. V.: Proc. Roy. Soc. B 94, 297 (1923).

BRÜCK, K.: Biol. Neonat. (Basel) 3, 65 (1961).

BRÜCK, K.: In: JONXIS, J. H. P., VISSER, H. K. A., TROELSTRA, J. A. (Eds.): The adaptation of the newborn infant to extra-uterine life, p. 229. Leiden: H. E. Stenfert Kroese N. V. 1964.

BRÜCK, K.: Arch. Phys. Ther. 21, 218 (1969).

BRÜCK, K.: In: LINDBERG, O. (Ed.): Brown adipose tissue, p. 117. New York: Am. Elsevier Publ. Comp. 1970 a.

BRÜCK, K.: In: STAVE, U. (Ed.): Physiology of the perinatal period, Vol. I, p. 493. New York: Appleton-Century-Crofts 1970 b.

Brück, K.: In: OPITZ, H., SCHMID, F. (Hrs.): Handbuch für Kinderheilkunde, Teil 1, S. 23. Berlin-Heidelberg-New York: Springer 1971.

BRÜCK, K.: Pflügers Arch. Ges. Physiol. 335, R 51 (1972a).

BRÜCK, K.: Med. Monatsschrift 26, 350 (1972b).

BRÜCK, K., BRÜCK, M.: Klin. Wschr. 38, 1125 (1960).

BRÜCK, K., BRÜCK, M., LEMTIS, H.: Pflügers Arch. Ges. Physiol. 265, 55 (1957).

BRÜCK, K., BRÜCK, M., LEMTIS, H.: Pflügers Arch. Ges. Physiol. 267, 382 (1958a).

BRÜCK, K., BRÜCK, M., LEMTIS, H.: Pflügers Arch. Ges. Physiol. 266, 518 (1958b).

BRÜCK, K., HENSEL, H.: Pflügers Arch. ges. Physiol. 257, 70 (1953).

BRÜCK, K., HERRMANN, A.: Einfluß lokaler Temperaturänderungen in der area prä-optica und im Bereich C5—T1 des Rückenmarks auf die Wärmetachypnoe des Meerschweinchens. Inaug.-Diss. Gießen 1972.

BRÜCK, K., PARMELEE, A. H., BRÜCK, M.: Biol. Neonat. (Basel) 4, 32 (1962).

BRÜCK, K., SCHWENNICKE, H. P.: Int. J. Biometeor. 15, 156 (1971).

BRÜCK, K., WÜNNENBERG, B.: Pflügers Arch. Ges. Physiol. 282, 362 (1965a).

BRÜCK, K., WÜNNENBERG, B.: Pflügers Arch. Ges. Physiol. 282, 376 (1965b).

BRÜCK, K., WÜNNENBERG, B.: Pflügers Arch. Ges. Physiol. 283, 1 (1965c).

BRÜCK, K., WÜNNENBERG, B.: Federation Proc. 25, 1332 (1966).

BRÜCK, K., WÜNNENBERG, W.: Pflügers Arch. Ges. Physiol. 290, 167 (1966).

BRÜCK, K., WÜNNENBERG, W.: Pflügers Arch. Ges. Physiol. 293, 215 (1967a).

BRÜCK, K., WÜNNENBERG, W.: Pflügers Arch. Ges. Physiol. 293, 226 (1967b).

BRÜCK, K., WÜNNENBERG, W.: In: Biokybernetik, Part I, p. 154. Leipzig: Karl-Marx-Univ. 1968.

BRÜCK, K., WÜNNENBERG, W.: In: HARDY, J. D., GAGGE, A. P., STOLWIJK, J. A. J. (Eds.): Physiological and behavioral temperature regulation, p. 562. Springfield, Ill.: Thomas 1970.

BRÜCK, K., WÜNNENBERG, W., GALLMEIER, H., ZIEHM, B.: Pflügers Arch. Ges. Physiol. 321, 159 (1970).

BRÜCK, K., WÜNNENBERG, W., ZEISBERGER, E.: Federation Proc. 28, 1035 (1969).

BRYCE-SMITH, R., EPSTEIN, H. G., GLEES, P.: J. Appl. Physiol. 15, 440 (1960).

BUCHMÜLLER, K.: Pflügers Arch. Ges. Physiol. 272, 360 (1961).

BUDDENBROCK, W. v.: Grundriß der vergleichenden Physiologie, Bd. 2, 2. Aufl., Berlin: Verlag Gebrüder Bornträger 1937.

BÜTTNER, K.: Physikalische Bioklimatologie. Leipzig: Akad. Verlagsgesellschaft 1938.

BUETTNER, K.: J. Appl. Physiol. 6, 229 (1953).

BUETTNER, K. J. K.: J. Appl. Physiol. 14, 269 (1959).

BUETTNER, K. J. K.: Federation Proc. 28, 528 (1969a).

BUETTNER, K. J. K.: Physiologist 12, 187 (1969b).

BUETTNER, K. J. K.: J. Physiol. (Paris) 63, 216 (1971).

BULLARD, R. W., BANERJEE, M. R., CHEN, F., ELIZONDO, R., McINTYRE, B. A.: In: HARDY, J. D., GAGGE, A. P., STOLWIJK, J. A. J. (Eds.): Physiological and behavioral temperature regulation, p. 597. Springfield, Ill.: Thomas 1970.

BULLARD, R. W., FUNKHOUSER, G. E.: Am. J. Physiol. 203, 266 (1962).

BULLOCK, T. H.: Federation Proc. 12, 666 (1953).

BULLOCK, T. H., COWLES, R. B.: Science 115, 541 (1952).

BULLOCK, T. H., DIECKE, F. P. J.: J. Physiol. (Lond.) 134, 47 (1956).

BULLOCK, T. H., FOX, W.: Quart. J. Micr. Sci. 98, 219 (1957).

BURLINGTON, R. F., WHITTEN, B. K.: Comp. Biochem. Physiol. 38A, 469 (1971).

BURTON, A. C.: J. Nutr. 7, 497 (1934).

BURTON, A. C.: J. Nutr. 9, 261 (1935).

BURTON, A. C., EDHOLM, O. G.: Man in cold environment. London: Edward Arnold 1955.

BURTON, A. C., SCOTT, J. C., McGLONE, B., BAZETT, H. C.: Am. J. Physiol. 129, 84 (1940).

BYON, Y. K., ADOLPH, E. F.: J. Appl. Physiol. 16, 827 (1961).

CABANAC, M.: In: HARDY, J. D., GAGGE, A. P., STOLWIJK, J. A. J. (Eds.): Physiological and behavioral temperature regulation, p. 549. Springfield, Ill.: Thomas 1970.

CABANAC, M.: J. Physiol. (Paris) 63, 189 (1971).

CABANAC, M., STOLWIJK, J. A. J., HARDY, J. D.: J. Appl. Physiol. 24, 645 (1968).

CADE, T. J.: Ann. Acad. Sci. Fennicae A 4 Biol. **71**, 79 (1964).
CAGE, G. W., WOLFE, S. M., THOMPSON, R. H., GORDON, R. S., JR.: J. Appl. Physiol. **29**, 687 (1970).
CALDER, W. A., SCHMIDT-NIELSEN, K.: Proc. Nat. Acad. Sci. (Wash.) **55**, 750 (1966).
CALDER, W. A., SCHMIDT-NIELSEN, K.: Am. J. Physiol. **213**, 883 (1967).
CALDER, W. A., SCHMIDT-NIELSEN, K.: Am. J. Physiol. **215**, 477 (1968).
CANGUILHEM, B., BLOCH, R.: Arch. Sci. Physiol. **21**, 27 (1967).
CANGUILHEM, B., MALAN, A.: J. Physiol. (Paris) **61**, Suppl. 2, 232 (1969).
CANGUILHEM, B., PETROWIC, A.: J. Physiol. (Paris) **60**, Suppl. 2, 35 (1968).
CANNON, W. B.: Ergebn. Physiol. **27**, 380 (1928).
CAPLAN, A., LINDSAY, J. K.: Bull. Inst. Mining Met. No. 480 (1946).
CARLISLE, H. J.: Physiol. Psychol. **61**, 388 (1966a).
CARLISLE, H. J.: Nature **209**, 1324 (1966b).
CARLSLAW, H. S., JAEGER, J. C.: Conduction of heat in solids. Oxford: Clarendon Press 1959.
CARLSON, L. D., BÜTTNER, K. J. K.: Federation Proc. **16**, 609 (1957).
CARLTON, P. L., MARKS, R. A.: Science **128**, 1344 (1958).
CARPENTER, R. E.: Univ. Calif., Berkeley, Publ. Zool. **78**, 1 (1966).
CASSUTO, Y., AMIT, Y.: Endocrinology **82**, 17 (1968).
CASSUTO, Y., CHAFFEE, R. R. J.: Am. J. Physiol. **210**, 423 (1966).
CELANDER, O.: Acta Paediat. Scand. **49**, 488 (1960).
CHAFFEE, R. R. J., ROBERTS, J. C.: Ann. Rev. Physiol. **33**, 155 (1971).
CHANG, M. C.: J. Gen. Physiol. **31**, 385 (1947).
CHATFIELD, P. O., LYMAN, CH. P.: Electroenceph. clin. Neurophysiol. **6**, 403 (1954).
CHATFIELD, P. O., LYMAN, C. P., IRVING, L.: Am. J. Physiol. **172**, 639 (1953).
CHATONNET, J., CABANAC, M.: Int. J. Biometeor. **9**, 183 (1965).
CHATONNET, J., CABANAC, M., MOTTAZ, M.: Compt. Rend. **158**, 1354 (1964).
CHEPKII, L. P., TRESHCHINSKII, A. I., SVIRYAKIN, V. T.: Eksp. Khir. Anesteziol. **12**, 65 (1967).
CHRISTENSEN, W. R.: Am. J. Physiol. **148**, 86 (1947).
CLARKE, G. L.: Elements of ecology. New York-London: Wiley 1954.
CLAUSEN, G.: Arb. Univ. Bergen, Mat.-Nat. 1966, No. 6, 1 (1966).
CLAUSEN, G., ERSLAND, A.: Resp. Physiol. **5**, 221 (1968).
COLIN, J., HOUDAS, Y.: J. Appl. Physiol. **22**, 31 (1967).
COLIN, J., TIMBAL, J., BOUTELIER, C., HOUDAS, Y., SIFFRE, M.: J. Appl. Physiol. **25**, 170 (1968).
COLIN, J., TIMBAL, J., GUIEU, J., BOUTELIER, C., HOUDAS, Y.: In: HARDY, J. D., GAGGE, A. P., STOLWIJK, J. A. J. (Eds.): Physiological and behavioral temperature regulation, p. 81. Springfield, Ill.: Thomas 1970.
COLLINS, K. J., WEINER, J. S.: Physiol. Rev. **48**, 785 (1968).
CONN, J. W., JOHNSTON, M. W., LOUIS, L. H.: Federation Proc. **5**, 230 (1946).
CONROY, R. T. W. L., MILLS, J. N.: Human circadian rhythms. London: J. & A. Churchill 1970.
COOPER, K. E.: In: HARDY, J. D., GAGGE, A. P., STOLWIJK, J. A. A. (Eds.): Physiological and behavioral temperature regulation, p. 224. Springfield, Ill.: Thomas 1970.
COOPER, K. E., CRANSTON, W. I., HONOUR, A. J.: J. Physiol. (Lond,) **191**, 325 (1967).
COOPER, K. E., CRANSTON, W. I., SNELL, E. S.: J. Appl. Physiol. **19**, 1032 (1964).
COOPER, K. E., JOHNSON, R. H., SPALDING, J. M. K.: J. Physiol. (Lond.) **174**, 46 (1964).
COOPER, K. E., KERSLAKE, McK., D.: J. Physiol. (Lond.) **119**, 18 (1953).
CORBIT, J. D.: In: HARDY, J. D., GAGGE, A. P., STOLWIJK, J. A. J. (Eds.): Physiological and behavioral temperature regulation, p. 777. Springfield, Ill.: Thomas 1970.
COTTLE, W. H., CARLSON, L. D.: Proc. Soc. Exp. Biol. (N. Y.) **92**, 845 (1956).
COVINO, B. G., BEAVERS, W. R.: Am. J. Physiol. **191**, 153 (1957).
COWLES, R. B.: Science **103**, 74 (1947a).
COWLES, R. B.: Science **105**, 362 (1947b).
COWLES, R. B., NORDSTRÖM, A.: Science **104**, 586 (1946).
CRAWFORD, E. C., JR.: J. Appl. Physiol. **17**, 249 (1962).

CRAWFORD, E. C., JR., SCHMIDT-NIELSEN, K.: Am. J. Physiol. **212**, 347 (1967).

CRISMON, J. M.: Bull. Vasc. Surg. **1951**, 110.

CRISMON, J. M., ELIOT, W. H.: Stanford Med. Bull. **5**, 115 (1947).

CRISMON, J. M., FUHRMAN, F. A.: J. Clin. Invest. **26**, 268 (1947).

CROCKFORD, G. W., HELLON, R. F., PARKHOUSE, J.: J. Physiol. (Lond.) **161**, 10 (1962).

CROSS, K. W., GUSTAVSON, J., HILL, J. R., ROBINSON, D. C.: Clin. Sci. **31**, 449 (1966).

D'AMATO, H.: Am. J. Physiol. **178**, 143 (1954).

D'AMATO, H. E.: In: DRIPPS, R. D. (Ed.): The physiology of induced hypothermia, p. 146. Washington, D. C.: Nat. Acad. Sci.-Nat. Res. Council Pub. 451. 1956.

D'AMATO, H. E., HEGNAUER, A. H.: Am. J. Physiol. **173**, 100 (1953).

D'AMATO, H. E., KRONHEIM, S., COVINO, B. G.: Am. J. Physiol. **198**, 333 (1960).

DANIELSON, R. N., KINARD, F. W.: J. Appl. Physiol. **4**, 373 (1951).

D'ANS, J., LAX, E.: Taschenbuch für Chemiker und Physiker. Berlin: Springer 1943.

DANSEREAU, P.: Biogeography, an ecological perspective. New York: Ronald 1957.

DARBINIAN, T. M., PORTNOI, V. F.: Exsp. Khir. Anest. **6**, 52 (1961).

DARIAN-SMITH, I., DYKES, R. W.: In: DUBNER, R., KAWAMURA, Y.: (Eds.) Oral-facial sensory and motor mechanisms, p. 7. New York: Appleton-Century-Crofts, Meredith. Corp. 1971.

DARIAN-SMITH, I., JOHNSON, K. O., DYKES, R. W.: J. Neurophysiol. **36**, 325 (1973).

DARLINGTON, P. J.: Zoogeography. New York-London: Wiley 1957.

DAVIS, T. R. A.: J. Appl. Physiol. **16**, 1011 (1961).

DAVIS, W. B.: J. Mammal. **19**, 338 (1938).

DAVIS, W. H.: In: Biology of bats, p. 265. New York: Academic Press Inc. 1970.

DAVIS, W. H., REITE, O. B.: Biol. Bull. **132**, 320 (1967).

DAWE, A. R., SPURRIER, W. A.: Science **163**, 298 (1969).

DAWE, A. R., SPURRIER, W. A., ARMOUR, J. A.: Science **168**, 497 (1970).

DAWES, G. S., JACOBSON, H. N., MOTT, J. C., SHELLEY, H. J.: J. Physiol. (Lond.) **152**, 271 (1960).

DAWKINS, M. J. R., HULL, D.: J. Physiol. (Lond.) **172**, 216 (1964).

DAWKINS, M. J. R., SCOPES, J. W.: Nature **206**, 201 (1965).

DAWSON, T. J.: Comp. Biochem. Physiol. **28**, 401 (1969).

DAWSON, T. J.: In: BLIGH, J., MOORE, R. E. (Eds.): Essays on temperature regulation, p. 1. Amsterdam: North-Holland Publishing Comp. 1972.

DAWSON, T. J., DENNY, M. J. S., HULBERT, A. J.: Comp. Biochem. Physiol. **31**, 645 (1969).

DAWSON, T. J., HULBERT, A. J.: Am. J. Physiol. **218**, 1233 (1970).

DAWSON, W. R.: Physiol. Zool. **31**, 37 (1958).

DAWSON, W. R., BENNETT, A. F.: J. Physiol. (Paris) **63**, 239 (1971).

DAWSON, W. R., HUDSON, J. W.: In: WHITTOW, G. C. (Ed.): Comparative physiology of thermoregulation, p. 224. New York-London: Academic Press 1970.

DAWSON, W. R., HUDSON, J. W.: In: WHITTOW, G. C. (Ed.): Comparative physiology of thermoregulation. I. Invertebrates and non-mammalian vertebrates, p. 287. New York-London: Academic Press 1970.

DAWSON, W. R., SCHMIDT-NIELSEN, K.: In: Handbook of physiology, Sect. 4, p. 481. Washington, D. C.: American Physiological Society 1964.

DAWSON, W. R., TORDOFF, H. B.: Condor **61**, 388 (1959).

DEIGHTON, T., HUTCHINSON, J. C. D.: J. Agr. Sci. **30**, 141 (1940).

DE JONG, R. H., HERSHEY, W. N., WAGMAN, I. H.: Anesthesiology **27**, 805 (1966).

DELGADO, J. M. R., HANAI, T.: Am. J. Physiol. **211**, 755 (1966).

DENNY-BROWN, D., GAYLOR, J. B., UPRUS, V.: Brain **58**, 233 (1935).

DE WITT, C. T.: Physiol. Zool. **40**, 49 (1967).

DILL, D. B., BOCK, A. V., EDWARDS, H. T.: Am. J. Physiol. **104**, 36 (1933).

DOBBEN-BROEKEMA, M. V., DIRKEN, M. N.: Acta Physiol. Pharmacol. Neerl. **1**, 562 (1950).

DODT, E.: Acta Physiol. Scand. **27**, 295 (1953).

DODT, E., ZOTTERMAN, Y.: Acta Physiol. Scand. **26**, 345 (1952a).

DODT, E., ZOTTERMAN, Y.: Acta Physiol. Scand. **26**, 358 (1952b).

DONALD, D. E., KERR, F. W. L.: J. Surg. Res. **4**, 243 (1964).

DONHOFFER, Sz.: Helgoländer Wiss. Meeresunters. **14**, 541 (1966).

DONHOFFER, Sz., FARKAS, M., HAUG-LÁSZLÓ, A., JARAI, I., SZEGVÁRY, GY.: Pflügers Arch. Ges. Physiol. **268**, 273 (1959).

DOUGLAS, W. W., RITCHIE, J. M., STRAUB, R. W.: J. Physiol. (Lond.) **150**, 266 (1960).

DOWLING, D. F.: Australian J. Agr. Sci. **9**, 579 (1958).

DREW, C. E.: Brit. Med. Bull. **17**, 37 (1961).

DRIPPS, R. D.: The physiology of induced hypothermia. Washington, D.C.: Nat. Acad. Sci.-Nat. Res. Council Pub. No. 451, 1956.

DU BOIS, E. F.: J. Am. Med. Ass. **77**, 352 (1921).

DU BOIS, E. F.: Basal metabolism in health and disease. Philadelphia: Lea & Febiger 1936.

DU BOIS, E. F.: West. J. Surg. **59**, 476 (1951).

DU BOIS, E. F.: Abstr. 19. Internat. Physiol. Congr. Montreal, S. 120, 1953.

DU BOIS, E. F., HARDY, J. D.: In: Temperature, its measurement and control in science and industry, p. 537. New York: Reinhold 1941.

DUBOST, C., BLONDEAU, P., PIWNICA, A.: J. Cardiovasc. Surg. (Torino) **3**, 4 (1962).

DUKES, H.: The physiology of domestic animals. New York: Comstock 1952.

DZAHNELIDZE, Ts. SH.: Soobshch. Akad. Nauk Gruz. SSR **42**, 749 (1966).

EAGAN, C. J.: Federation Proc. **22**, 947 (1963).

EBAUGH, F. G., JR., THAUER, R.: J. Appl. Physiol. **3**, 173 (1950).

EBBECKE, U.: Klin. Wschr. **26**, 609 (1948).

EDERSTROM, H. E.: Am. J. Physiol. **176**, 347 (1954).

EDHOLM, O. G.: Proc. Roy. Soc. Med. **62** (2), 1175 (1969).

EDHOLM, O. G.: In: ITOH, S., OGATA, K., YOSHIMURA, H. (Eds.): Advances in Climatic Physiology. Tokyo: Igaku Shoin Ltd. Berlin-Heidelberg-New York: Springer 1972.

EDHOLM, O. G., LEWIS, H. E.: In: Handbook of Physiology, Sect. IV. Adaptation to the environment, p. 435. Washington, D.C.: American Physiological Society 1964.

EDWARDS, M., BURTON, A. C.: J. Appl. Physiol. **15**, 209 (1960).

EICHNA, L. W., ASHE, W. F., BEAN, W. B., SHELLEY, W. B.: J. Industr. Hyg. **27**, 59 (1945 a).

EICHNA, L. W., BEAN, W. B., ASHE, W. F., NELSON, N.: Bull. Johns Hopkins Hosp. **76**, 25 (1945 b).

EICHNA, L. W., PARK, C. R., NELSON, N., HORVATH, S. M., PALMES, E. D.: Am. J. Physiol. **163**, 585 (1950).

EIFF, A. W., v.: Z. Exp. Med. **117**, 261 (1951).

EIJKMAN, E., VENDRIK, A. J. H.: Biophys. J. **3**, 65 (1963).

EISENMAN, J. S.: Am. J. Physiol. **216**, 330 (1969).

EISENMAN, J. S.: In: HARDY, J. D., GAGGE, A. P., STOLWIJK, J. A. J. (Eds.): Physiological and behavioral temperature regulation. p. 507. Springfield, Ill.: Thomas 1970.

EISENMAN, J. S., JACKSON, D. C.: Exp. Neurol. **19**, 33 (1967).

EISENTRAUT, M.: Z. Vergl. Physiol. **18**, 174 (1932).

EISENTRAUT, M.: Mitt. Zool. Mus. Berlin **19**, 48 (1933).

EISENTRAUT, M.: Z. Morphol. Ökol. Tiere **29**, 231 (1934).

EISENTRAUT, M.: Biol. Zbl. **57**, 59 (1937).

EISENTRAUT, M.: Biol. Zbl. **60**, 199 (1940).

EISENTRAUT, M.: Biol. Zbl. **66**, 236 (1947).

EISENTRAUT, M.: Z. Säugetierkunde **21**, 49 (1956 a).

EISENTRAUT, M.: Der Winterschlaf mit seinen ökologischen und physiologischen Begleiterscheinungen. Jena: VEB Gustav-Fischer-Verlag 1956 b.

EISENTRAUT, M.: Bull. Mus. Compar. Zool. Harvard Coll. **124**, 31 (1960).

EKLUND, C. R.: Bird-Banding **32**, 187 (1961).

EKLUND, C. R., CHARLTON, F. E.: Am. Scie. **47**, 80 (1959).

ENDERS, R. K., DAVIS, D. E.: J. Mammalogy **17**, 165 (1936).

ENGER, P. S.: Acta Physiol. Scand. **40**, 161 (1957).

ERLANGER, J., GASSER, H.S.: Electrical signs of nervous activity. Philadelphia (Pa.): Philadelphia University Press 1937.

EULER, U.S., v., LINDER, E., MYRIN, S.O.: Acta Physiol. Scand. 5, 85 (1943).

FANGER, P.O.: In: HARDY, J.D., GAGGE, A.P., STOLWIJK, J.A.J. (Eds.): Physiological and behavioral temperature regulation, p. 152. Springfield, Ill.: Thomas 1970a.

FANGER, P.O.: Thermal comfort. Copenhagen: Danish Technical Press 1970b.

FARNER, D.S., CHIVERS, N., RINEY, T.: Emu 56, 199 (1956).

FAWCETT, D., LYMAN, C.P.: J. Physiol. (Lond.) 126, 235 (1954).

FAY, T.: N. Y. St. J. Med. 40, 1351 (1940).

FAY, T.: Ass. Res. Nerv. Dis. Proc. 24, 611 (1945).

FAY, T., SMITH, G.W.: Arch. Neurol. Psychiat. (Chicago) 45, 215 (1941).

FEDOR, E.J., FISHER, B.: Am. J. Physiol. 196, 703 (1959).

FELDBERG, W., HELLON, R.F., LOTTI, V.J.: J. Physiol. (Lond.) 191, 501 (1967).

FELDBERG, W., MYERS, R.D.: J. Physiol. (Lond.) 173, 226 (1964).

FERDMANN, D., FEINSCHMIDT, O.: Ergebn. Biol. 8, 1 (1932).

FERRER, M.I.: Cold injury. New York: Josiah Macy Foundation 1955.

FERRIS, B.G., FORSTER, R.E., PILLION, E.L., CHRISTENSEN, W.R.: Am. J. Physiol. 150, 304 (1947).

FINDLAY, J.D.: Agric. Progr. 36, 7 (1961).

FINDLEY, J.D.: Federation Proc. 22, 688 (1963).

FISCHER, B.A., MUSACCHIA, X.J.: Am. J. Physiol. 215, 1130 (1968).

FOLK, G.E.: Environmental Physiology. Philadelphia: Lea & Febiger 1966.

FOLK, G.E., JR., BREWER, M.C., SANDERS, D.: Arctic 23, 130 (1970).

FOLK, G.E., JR., FOLK, M.A., KREUZER, F.: Acta Theoriol. Bialowieza 15, 373 (1970).

FOLK, G.E., JR., FOLK, M.A., SIMMONDS, R.C., BREWER, M.C.: Bull. Ecol. Soc. Am. 47, 203 (1966).

FORBES, W.H., DILL, D.B., HALL, F.G.: Am. J. Physiol. 130, 739 (1940).

FORSTER, R.E., FERRIS, B.G., DAY, R.: Am. J. Physiol. 146, 600 (1946).

FOSTER, K.G., HEY, E.N., KATZ, G.: J. Physiol. (Lond.) 203, 13 (1969).

FOX, R.H.: In: DEHOLM, O.G., BACHARACH, A.L. (Eds.): The Physiology of Human Survival. Chapter 3. London-New York: Academic Press 1965.

FOX, R.H., GOLDSMITH, R., HAMPTON, I.F.G., HUNT, T.J.: J. Appl. Physiol. 22, 39 (1967).

FOX, R.H., GOLDSMITH, R., KIDD, J.D.: J. Physiol. (Lond.) 161, 298 (1962).

FOX, R.H., GOLDSMITH, R., KIDD, D.J., LEWIS, H.E.: J. Physiol. (Lond.) 166, 530 (1963a).

FOX, R.H., GOLDSMITH, R., KIDD, D.J., LEWIS, H.E.: J. Physiol. (Lond.) 166, 548 (1963b).

FOX, R.H., HILTON, S.M.: J. Physiol. (Lond.) 142, 219 (1958).

FOX, R.H., McPHERSON, R.K.: J. Physiol. (Lond,) 125, 21P (1954).

FREEMAN, M.E., CRISSMAN, J.K., JR., LOUW, G.N., BUTCKER, R.L., INSKEEP, F.K.: Endocrinology 86, 717 (1970).

FREGLY, M.J.: Am. J. Physiol. 176, 267 (1954).

FREINKEL, N., LEWIS, D.: J. Physiol. (Lond.) 135, 288 (1957).

FREUND, H., JANSEN, S.: Pflügers Arch. Ges. Physiol. 200, 96 (1923).

FREY, M., v.: Ber. Sächs. Ges. Akad. Wiss. 47, 166 (1895).

FRITH, H.J.: C.S.I.R.O. Wildl. Res. 2, 101 (1957).

FRUHSTORFER, H., HENSEL, H.: Naturwissenschaften 60, 209 (1973).

FUHRMAN, F.A., FUHRMAN, G.J.: Medicine (Baltimore) 36, 465 (1957).

FUKUI, N.: Japan. Med. World 3 (1923).

FULLER, F.D., HIESTAND, W.A.: Turtox News 25, 148 (1947).

FULLER, M.F.: Brit. J. Nutr. 19, 531 (1965).

GAALAAS, R.F.: J. Dairy Sci. 28, 555 (1945).

GABE, M., AGID, R., MARTOJA, M., SAINT GIRONS, M.C., SAINT GIRONS, H.: Arch. Biol. (Liège) 75, 1 (1964).

GÄNSHIRT, H., HIRSCH, H., KRENKEL, W., SCHNEIDER, M., ZYLKA, W.: Arch. Exp. Path. Pharmakol. 222, 431 (1954).

GÄRTNER, W.: Z. Physik. Med. 1, 149 (1970).
GÄRTNER, W., GÖPFERT, H.: Pflügers Arch. Ges. Physiol. 280, 224 (1964).
GÄRTNER, W., GÖPFERT, H.: Pflügers Arch. Ges. Physiol. 290, 335 (1966).
GÄRTNER, W., LING, K., GÖPFERT, H.: Pflügers Arch. Ges. Physiol. 280, 236 (1964).
GAGGE, A. P.: J. Physiol. (Paris) 63, 373 (1971).
GAGGE, A. P., STOLWIJK, J. A. J., HARDY, J. D.: Environ. Res. 1, 1 (1967).
GALSTER, W. A., MORRISON, P.: Am. J. Physiol. 218, 1228 (1970).
GALVÄO, P. E.: Am. J. Physiol. 148, 478 (1947).
GALVÄO, P. E.: J. Appl. Physiol. 1, 395 (1948).
GANONG, W. F., FORSHAM, P. H.: Ann. Rev. Physiol. 22, 579 (1960).
GATES, D. M.: In: HAFEZ, E. S. E. (Ed.): Adaptation of domestic animals, p. 46.
    Philadelphia: Lea & Febiger 1968.
GATES, D. M., PORTER, W. P.: In: HARDY, J. D., GAGGE, A. P., STOLWIJK, J. A. J. (Eds.):
    Physiological and behavioral temperature regulation, p. 177. Springfield, Ill.:
    Thomas 1970.
GELINEO, S.: Compt. Rend. 115, 865 (1934).
GELINEO, S.: Ann. Physiol. Physicochim. Biol. 10, 1083 (1934).
GELINEO, S.: Compt. Rend. 147, 134 (1953).
GELINEO, S.: Arch. Biol. Nauk. 6, 235 (1954).
GELINEO, S.: Compt. Rend. 148, 1114 (1954).
GELINEO, S.: Bull. Acad. Serbe Sci. 18, 97 (1957), Cl. Sci. Math. Nat. n° 5.
GELINEO, S., GELINEO, A.: Bull. Acad. Serbe Sci. 4, 197 (1950).
GELINEO, S., GELINEO, A.: Bull. Acad. Serbe Sci. Cl. Sci. Méd. 3, 119 (1951).
GELINEO, S., GELINEO, A.: Bull. Acad. Serbe Sci. Cl. Sci. Méd. 3, 149 (1951).
GELINEO, S., SOKIC, P.: Compt. Rend. 147, 138 (1953).
GELINEO, S., SOKIC, P.: Bull. Acad. Serbe Sci. 12, 1 (1953).
GEORGE, W.: Animal geography. London: Heinemann 1962.
GEROLA, A., FREINBERG, H., KATZ, L. N.: Am. J. Physiol. 196, 719 (1959).
GIAJA, J.: Compt. Rend. 100, 1225 (1929).
GIAJA, J.: Ann. Physiol. 7, 13 (1931).
GIAJA, J.: Nutrition (Paris) 9/10, 576 (1938).
GIAJA, J.: Bull. Acad. Serbe Sci. (B) Cl. Sci. nat. 6, 185 (1940).
GIAJA, J., GELINEO, S.: Arch. Intern. Physiol. 37, 20 (1933).
GINGLINGER, A., KAYSER, C.: Ann. Physiol. Physicochim. Biol. 5, 710 (1929).
GISOLFI, C., ROBINSON, S.: J. Appl. Physiol. 26, 530 (1969).
GLASER, E. M.: J. Physiol. (Lond.) 109, 366 (1949).
GLASER, E. M.: The physiological basis of habituation. London: Oxford University
    Press 1966.
GLASER, E. M., GRIFFIN, J. P.: J. Physiol. (Lond.) 160, 429 (1962).
GLASER, E. M., HALL, M. S., WHITTOW, G. C.: J. Physiol. (Lond.) 146, 152 (1959).
GLASER, E. M., WHITTOW, G. C.: J. Physiol. (Lond.) 136, 98 (1957).
GLASS, L., SILVERMAN, W. A., SINCLAIR, J. C.: Pediatrics 41, 1033 (1968).
GLICKMAN, N., HICK, F. K., KEETON, R. W., MONTGOMERY, M. M.: Am. J. Physiol. 134,
    165 (1941).
GLICKMAN, N., MITCHELL, H. H., KEETON, R. W., LAMBERT, E. H.: J. Appl. Physiol.
    22, 1 (1967).
GLITSCH, H. G.: Pflügers Arch. Ges. Physiol. 307, 29 (1969).
GODELL, H., GRAHAM, D. T., WOLFF, H. S.: Proc. Ass. Res. Nerv. Ment. Dis. 29, 418
    (1949).
GÖKHAN, N., ANGELAKOS, E. T.: J. Appl. Physiol. 18, 69 (1963).
GÖPFERT, H., EIFF, A. W. v., HOWIND, C.: Z. Exp. Med. 120, 308 (1953).
GÖPFERT, H., STUFLER, R.: Pflügers Arch. Ges. Physiol. 256, 161 (1952).
GOLENHOFEN, K.: Pflügers Arch. Ges. Physiol. 285, 124 (1965).
GOLENHOFEN, K.: In: COMÈL, M., LASZT, L. (Eds.): Physiology of blood and lymph
    vessels, Symposia Angiologica Santoriana, p. 97 (253). Basel: Karger-Verlag 1971 a.

GOLENHOFEN, K.: Haut. In: BAUEREISEN, E., (Hrsg.): SCHÜTZ-TRENDELENBURG, Lehrbuch der Physiologie. Physiologie des Kreislaufs I, Berlin-Heidelberg-New York: Springer 1971 b.

GOLENHOFEN, K., HENSEL, H., HILDEBRANDT, G.: Durchblutungsmessung mit Wärmeleitelementen. Stuttgart: Thieme 1963.

GOLLAN, F., TYSINGER, D. S., JR., GRACE, J. T., KORY, R. C., MENEELY, G. R.: Am. J. Physiol. 181, 297 (1955).

GOSSELIN, R. E.: Am. J. Physiol. 157, 103 (1949).

GRAD, B.: Am. J. Physiol. 174, 481 (1953).

GRAF, W., PORJÉ, L. G., ALLGOTH, A. M.: Gastroenterologia 83, 233 (1955).

GRAHAM, S. A., HESTERBERG, G.: J. Wildlife Management 12, 9 (1948).

GRANT, R.: Am. J. Physiol. 160, 285 (1950).

GRANT, R.: Ann. Rev. Physiol. 13, 75 (1951).

GRANT, R., HIRSCH, J. D.: Am. J. Physiol. 161, 528 (1950).

GRAYSON, J., KINNEAR, T.: Federation Proc. 22, 775 (1963).

GRIFFIN, D. R., HAMMEL, H. T., RAWSON, K. S.: Comparative physiology of thermal insulation. Arctic Aeromedical Laboratory Project 8-7951, Report No. 6, 1956.

GRODINS, F. S.: Control theory and biological systems. New York: Columbia Univ. Press 1963.

GROEBBELS, F.: Pflügers Arch. Ges. Physiol. 218, 98 (1928).

GRÖBER, H., ERK, S., GRIGULL, U.: Wärmeübertragung. Berlin-Göttingen-Heidelberg: Springer 1955.

GROSSE-BROCKHOFF, F., SCHOEDEL, W.: Arch. Exp. Pathol. Pharmakol. 201, 417 (1943).

GRÜSSER, O. J., KLINKE, R., KOSSOW, K. D.: Studium Generale 21, 1052 (1968).

GRUNDIG, J.: Z. Biol. 89, 547 (1930).

GÜNTHER, B.: In: GAUER-KRAMER-JUNG (Hrsg.): Physiologie des Menschen, Bd. 2, S. 117. Berlin-Wien: Urban & Schwarzenberg 1971.

GUIEU, J. D., HARDY, J. D.: J. Physiol. (Paris) 63, 253 (1971).

GUILLEMIN, V., BENJAMIN, F., CORNBLEET, T., GROSSMAN, M. I.: J. Appl. Physiol. 4, 920 (1952).

GUMMA, M. R., SOUTH, F. E., ALLEN, J. N.: Animal. Behav. 15, 534 (1967).

GUTTMANN, L., SILVER, J., WYNDHAM, C. H.: J. Physiol. (Lond.) 142, 406 (1958).

HADDAD, H. M.: Pediatrics 57, 391 (1960).

HAFEZ, E. S. E.: Adaptation of domestic animals. Philadelphia: Lea & Febiger 1968.

HAGEN, E., KNOCHE, H., SINCLAIR, D., WEDDELL, G.: Proc. Roy. Soc. B.141, 279 (1953).

HAHN, H.: Beiträge zur Reizphysiologie. Heidelberg: Scherer 1949.

HAINSWORTH, F. R.: Am. J. Physiol. 212, 1288 (1967).

HAINSWORTH, F. R., STRICKER, E. M.: In: HARDY, J. D., GAGGE, A. P., STOLWIJK, J. A. J. (Eds.): Physiological and behavioral temperature regulation, p. 611. Springfield, Ill.: Charles C. Thomas Publ. 1970.

HAINSWORTH, F. R., STRICKER, E. M.: J. Physiol. (Paris) 63, 257 (1971).

HAINSWORTH, F. R., STRICKER, E. M., EPSTEIN, A. N.: Am. J. Physiol. 214, 983 (1968).

HAINSWORTH, F. R., WOLF, L. L.: Am. Zoologist 9, 1100 (1969).

HALBERG, F.: Ann. Rev. Physiol. 31, 675 (1969).

HALES, J. R. S., WEBSTER, M. E. D.: J. Physiol. (Lond.) 190, 241 (1967).

HAMMEL, H. T.: J. Mammal. 37, 375 (1956).

HAMMEL, H. T.: In: Handbook of physiology, Sect. IV., p. 413. Washington, D.C. American Physiological Society 1964.

HAMMEL, H. T.: In: YAMAMOTO, W. S., BROBECK, J. R. (Eds.): Physiological controls and regulations, Chapt. 5, p. 71. Philadelphia-London: W. B. Saunders 1965.

HAMMEL, H. T.: Temperature regulation and hibernation. In: Mammalian hibernation III. Edinburgh: Oliver and Boyd 1967.

HAMMEL, H. T.: Ann. Rev. Physiol. 30, 641 (1968).

HAMMEL, H. T., ELSNER, R. W., leMESSURIER, D. H., ANDERSEN, T. H., MILAN, F. A.: J. Appl. Physiol. 14, 605 (1959).

HAMMEL, H. T., HARDY, J. D., FUSCO, M. M.: Am. J. Physiol. 198, 481 (1960).

HAMMEL, H. T., SHARP, F.: J. Physiol. (Paris) 63, 260 (1971).

HAMPTON, I. F. G.: Brit. Antarctic Surv. Bull. **19**, 9 (1969).
HANNON, J. P., VAUGHAN, D. A.: Am. J. Physiol. **201**, 217 (1961).
HANSEN, A. T., HAXHOLDT, B. F., HUSEFELDT, E., LASSEN, N. A., MUNCK, O., SØRENSEN,
    H. R., WINKLER, K.: Scand. J. Clin. Lab. Invest. **8**, 182 (1956).
HARDY, J. D.: Am. J. Physiol. **127**, 454 (1939).
HARDY, J. D.: In: NEWBURGH, L. H. (Ed.): Physiology of heat regulation and the
    science of clothing, p. 79. Philadelphia-London: Saunders 1949.
HARDY, J. D.: Physiol. Rev. **41**, 521 (1961).
HARDY, J. D.: Temperature, its measurement and control in science and industry.
    New York: Reinhold 1963.
HARDY, J. D.: In: YAMAMOTO, W. S., BROBECK, J. R. (Eds.): Physiological controls and
    regulations, Chapt. 6, p. 98. Philadelphia-London: W. B. Saunders 1965.
HARDY, J. D.: In: HARDY, J. D., GAGGE, A. P., STOLWIJK, J. A. J. (Eds.): Physiological
    and behavioral temperature regulation, p. 856. Springfield, Ill..: Thomas 1970
HARDY, J. D., DU BOIS, E. F.: J. Nutr. **15**, 482 (1938).
HARDY, J. D., DU BOIS, E. F.: Proc. Nat. Acad. Sci. (Wash.) **26**, 389 (1940).
HARDY, J. D., GAGGE, A. P., STOLWIJK, J. A. J.: Physiological and behavioral tempera-
    ture regulation. Springfield, Ill.: Thomas 1970.
HARDY, J. D., GUIEU, J. D.: J. Physiol. (Paris) **63**, 264 (1971).
HARDY, J. D., OPPEL, TH. W.: J. Clin. Invest. **16**, 535 (1937).
HARDY, J. D., OPPEL, TH. W.: J. Clin. Invest. **17**, 771 (1938).
HARDY, J. D., SHORR, E., DU BOIS, E. F.: Federation Proc. **6**, 122 (1947).
HARDY, J. D., STOLWIJK, J. A. J., GAGGE, P.: In: WHITTOW, G. C. (Ed.): Comparative
    physiology of thermoregulation, Vol. II, p. 327. New York-London: Academic
    Press 1971.
HARRIS, N. O., MEFFERD, R. B., JR., RESTIVO, S. R.: Am. J. Physiol. **198**, 476 (1960).
HARRISON, G. A.: Am. Naturalist **93**, 392 (1959).
HARRISON, G. A.: Federation Proc. **22**, 691 (1963).
HART, J. S.: Canad. J. Res. (D) **28**, 280 (1950).
HART, J. S.: Can. J. Zool. **34**, 53 (1956).
HART, J. S.: Rev. Can. Biol. **16**, 133 (1957).
HART, J. S.: In: Handbook of physiology, Sect. IV, p. 295. Washington, D.C.: Am.
    Physiol. Soc. 1964.
HART, J. S.: In: WHITTOW, G. C. (Ed.): Comparative physiology of thermoregulation,
    Vol. II, p. 1. New York-London: Academic Press 1971.
HART, J. S., HÉROUX, O., DEPOCAS, F.: J. appl. Physiol. **9**, 404 (1956).
HART, J. S., IRVING, L.: Can. J. Zool. **37**, 447 (1959).
HARTNER, W. C., SOUTH, F. E., JACOBS, H. K., LUECKE, R. H.: Cryobiology **8**, 312 (1971).
HASAMA, B.: Arch. Exp. Pathol. Pharmakol. **146**, 129 (1929).
HATERIUS, H. O., MAISCN, G. L.: Am. J. Physiol. **152**, 225 (1948).
HAUS, E., HALBERG, F., NELSON, W., HILLMAN, D.: Federation Proc. **27**, 224 (1968).
HAUSCHKA, T. S., MITCHELL, J. T., NIEDERRUEM, D. J.: Cancer Res. **19**, 643 (1959).
HAYWARD, J. N.: Proc. Soc. Exp. Biol. (N. Y.) **124**, 155 (1967).
HAYWARD, J. N., BAKER, A.: Brain Res. **16**, 417 (1969).
HAYWARD, J. S.: Can. J. Zool. **43**, 297 (1965).
HEBERLING, E. J., ADAMS, TH.: J. Appl. Physiol. **16**, 226 (1961).
HEDIGER, H.: Exotische Freunde im Zoo. Basel: Reinhardt 1949.
HEERD, E., OHARA, K.: Pflügers Arch. Ges. Physiol. **272**, 25 (1960).
HEERD, E., OPPERMANN, CH.: Pflügers Arch. Ges. Physiol. **291**, 174 (1966).
HEGNAUER, H. H.: Ann. N. Y. Acad. Sci. **80**, 315 (1959).
HEGNAUER, A. H., COVINO, B. G.: In: DRIPPS, R. D. (Ed.): The physiology of induced
    hypothermia, p. 327. Washington, D.C.: Nat. Acad. Sci.-Nat. Res. Council Publ.
    451 (1956).
HEGNAUER, A. H., D'AMATO, H. E.: Am. J. Physiol. **178**, 138 (1954).
HEGNAUER, A. H., FLYNN, J., D'AMATO, H.: Am. J. Physiol. **167**, 69 (1951).
HEGNAUER, A. H., SHRIBER, W. J., HATERIUS, H.O.: Am. J. Physiol. **161**, 455 (1950).
HEIM, T., HULL, D.: J. Physiol. (Lond.) **186**, 42 (1966a).

HEIM, T., HULL, D.: J. Physiol. (Lond.) **187**, 271 (1966b).

HELDMAIER, G.: Z. Vergl. Physiol. **73**, 222 (1971).

HELLBRÜGGE, T.: Cold Spring. Harb. Symp. Quant. Biol. **25**, 311 (1960).

HELLER, C. H., HAMMEL, H. T.: Comp. Biochem. Physiol. **41** A, 349 (1972).

HELLON, R. F.: In: HARDY, J. D., GAGGE, A. P., STOLWIJK, J. A. J. (Eds.): Physiological and behavioral temperature regulation, p. 463. Springfield, Ill.: Thomas 1970.

HEMINGWAY, A.: Am. J. Physiol. **122**, 511 (1938).

HEMINGWAY, A.: Physiol. Rev. **43**, 397 (1963).

HEMINGWAY, A., RASMUSSEN, TH., WIKOFF, H., RASMUSSEN, A. T.: J. Neurophysiol. **3**, 329 (1940).

HENAFF, F.: In: PARKES, A. S., SMITH, A. U. (Eds.): Recent research in freezing and drying, p. 295. Oxford: Blackwell 1960.

HENDERSON, E. W.: Univ. Missouri Agr. Exp. Sta. Res. Bull. **149** (1930).

HENDERSON, E. W., BRODY, S.: Univ. Missouri Agr. Exp. Sta. Res. Bull. **99** (1927).

HENSEL, H.: Ergebn. Physiol. **47**, 166 (1952a).

HENSEL, H.: Pflügers Arch. Ges. Physiol. **256**, 195 (1952b).

HENSEL, H.: Acta Physiol. Scand. **29**, 109 (1953).

HENSEL, H.: Z. Vergl. Physiol. **37**, 509 (1955).

HENSEL, H.: Pflügers Arch. Ges. Physiol. **263**, 48 (1956).

HENSEL, H.: Allgemeine Sinnesphysiologie, Hautsinne, Geschmack, Geruch: Berlin: Heidelberg-New York: Springer 1966.

HENSEL, H.: Pflügers Arch. Ges. Physiol. **302**, 374 (1968).

HENSEL, H.: Pflügers Arch. Ges. Physiol. **313**, 150 (1969).

HENSEL, H.: In: Handbook of Sensory Physiology, Vol. II, p. 75. Berlin-Heidelberg-New York: Springer 1973.

HENSEL, H., BOMAN, K. K. A.: J. Neurophysiol. **23**, 564 (1960).

HENSEL, H., HUOPANIEMI, T.: Pflügers Arch. Ges. Physiol. **309**, 1 (1969).

HENSEL, H., IGGO, A.: Pflügers Arch. Ges. Physiol. **329**, 1 (1971).

HENSEL, H., IGGO, A., WITT, I.: J. Physiol. (Lond.) **153**, 113 (1960).

HENSEL, H., KENSHALO, D. R.: J. Physiol. (Lond.) **204**, 99 (1969).

HENSEL, H., NIER, K.: Pflügers Arch. Ges. Physiol. **323**, 279 (1971).

HENSEL, H., STRÖM, L., ZOTTERMAN, Y.: J. Neurophysiol. **14**, 423 (1951).

HENSEL, H., WITT, I.: J. Physiol. (Lond.) **148**, 180 (1959).

HENSEL, H., WURSTER, R. D.: J. Neurophysiol. **33**, 271 (1970).

HENSEL, H., ZOTTERMAN, Y.: Acta Physiol. Scand. **23**, 291 (1951a).

HENSEL, H., ZOTTERMAN, Y.: J. Neurophysiol. **14**, 377 (1951b).

HENSEL, H., ZOTTERMAN, Y.: J. Physiol. (Lond.) **115**, 16 (1951c).

HENSHAW, R. E.: Physiology of hibernation and acclimatization in two species of bats (Myotis lucifugus and Myotis sodalis). Ph. D. Thesis. Univ. of Iowa, Iowa City, Iowa 1965.

HENSHAW, R. E.: J. Theoret. Biol. **20**, 79 (1968).

HENSHAW, R. E.: Thermoregulation in bats. In: About bats. p. 188. Dallas, Texas: South. Methodist Univ. Press. 1970.

HEPBURN, J. S., EBERHARD, H. M., RICKETTS, R., RIEGER, C. L.: Arch. Intern. Med. **52**, 603 (1933).

HÉROUX, O.: Am. J. Physiol. **181**, 75 (1955).

HÉROUX, O.: Can. J. Biochem. **38**, 517 (1960).

HÉROUX, O.: Federation Proc. **28**, 955 (1969).

HÉROUX, O., DEPOCAS, F., HART, G. S.: Can. J. Biochem. **37**, 473 (1959).

HÉROUX, O., HART, J. S.: Am. J. Physiol. **177**, 219 (1945).

HÉROUX, O., HART, J. S.: Am. J. Physiol. **178**, 453 (1954).

HÉROUX, O., HART, J. S., DEPOCAS, F.: J. Appl. Physiol. **9**, 399 (1956).

HÉROUX, O., PIERRE, J. ST.: Am. J. Physiol. **188**, 163 (1957).

HERRINGTON, L. P.: Am. J. Physiol. **129**, 123 (1940).

HERRINGTON, L. P.: In: Temperature, its measurement and control, p. 446. New York 1941.

HERRINGTON, L. P.: In: NEWBURGH, L. (Ed.): Physiology of heat regulation and the science of clothing, p. 262. Philadelphia-London: Saunders 1949.
HERTER, K.: Z. Vergl. Physiol. 20, 511 (1934).
HERTER, K.: Der Temperatursinn der Säugetiere. Leipzig: Geest & Portig 1952.
HERTER, K.: Winterschlaf. In: Handb. Zool., Bd. 8/4. Berlin: De Gruyter 1956.
HERTER, K.: Zool. Beitr. Berlin N. F. 6, 347 (1961).
HERTER, K.: Zool. Beitr. Berlin N. F. 7, 239 (1962).
HERTER, K.: Zool. Beitr. Berlin N. F. 8, 125 (1963a).
HERTER, K.: Zool. Beitr. Berlin N. F. 9, 237 (1963b).
HERTER, K.: Zool. Beitr. Berlin N. F. 10, 161 (1964a).
HERTER, K.: Zool. Beitr. Berlin N. F. 10, 189 (1964b).
HERTIG, B. A., RIEDESEL, L. M., BELDING, H. S.: J. Appl. Physiol. 16, 647 (1961).
HESS, W. R., STOLL, W. A.: Helv. Physiol. Acta 2, 461 (1944).
HESSE, R., ALLEE, W. C., SCHMIDT, K. P.: Ecological animal geography. New York: Wiley 1951.
HEY, E. N., KATZ, G.: J. Physiol. (Lond.) 200, 605 (1969).
HEY, E. N., KATZ, G., O'CONNELL, B.: J. Physiol. (Lond.) 207, 683 (1970).
HICKS, C. S.: In: Handbook of physiology, Sect. 4, p. 405. Washington, D.C.: Am. Physiol. Soc. 1964.
HICKS, C. S., MATTERS, R. F.: Australian J. Exp. Biol. Med. Sci. 12, Part 3 (1933).
HICKS, C. S., MATTERS, R. F., MITCHELL, M. L.: Australian. J. Exp. Biol. Med. Sci. 8, Part 1 (1931).
HIGGINBOTHAM, A. C., KOON, W. E.: Am. J. Physiol. 181, 69 (1955).
HILDWEIN, G.: Arch. Sci. Physiol. 24, 55 (1970).
HILDWEIN, G., MALAN, A.: Arch. Sci. Physiol. 24, 133 (1970).
HILL, J. R., RAMINTULLA, K. A.: J. Physiol. (Lond.) 180, 239 (1965).
HILLE, H.: Zur Durchblutung der Terminalstrombahn. Heidelberg: Dr. Alfred Hüthig 1965.
HIMMS-HAGEN, J.: Pharmacol. Rev. 19, 367 (1967).
HIMMS-HAGEN, J.: J. Physiol. (Lond.) 205, 393 (1969).
HINDMARSH, E. M.: Australian. J. Exp. Biol. Med. Sci. 4, 225 (1927).
HIRSCH, H., BREUER, M., BUCH, K. G. v., DOHMEN, M., KÖRNER, K., RÜMMELE, H.: Pflügers Arch. Ges. Physiol. 277, 251 (1963).
HISSA, R.: Ann. Zool. Fennicae 5, 345 (1968).
HOCK, R. J.: Biol. Bull. 101, 289 (1951).
HÖFLER, W.: J. Appl. Physiol. 25, 503 (1968).
HÖFLER, W., SCHMOLL, D., VOIGT, R., ACQUAT, D.: Pflügers Arch. Ges. Physiol. 292, R 79 (1966).
HOEHN, E. O.: Am. J. Physiol. 158, 337 (1949).
HOFFMANN, K.: In: ASCHOFF, J. (Ed.): Circadian clocks, p. 87. Amsterdam: North-Holland (1965).
HOFFMANN, R.: Inaug.-Diss. Tierärztl. Hochschule Hannover 1938.
HOFFMAN, R. A.: Federation Proc. 27, 999 (1968).
HOFFMAN, R. A., ZARROW, M. X.: Acta Endocr. 27, 77 (1958).
HOLEČKOVÁ, E.: Physiol. Bohemoslov. 13, 78 (1964).
HOLYOAK, G. W., STONES, R. C.: Comp. Biochem. Physiol. 39A, 413 (1971).
HONDA, N., CARLSON, L. D., JUDY, W. V.: Am. J. Physiol. 204, 615 (1963).
HONG, S. K.: Federation Proc. 22, 831 (1963).
HORVATH, S. M., MENDUKE, H., PIERSOL, G. M.: J. Am. med. Ass. 144, 1562 (1950).
HORVATH, S. M., RUBIN, A., FOLTZ, E. L.: Am. J. Physiol. 161, 316 (1950).
HORVATH, S. M., SHELLEY, W. B.: Am. J. Physiol. 146, 336 (1946).
HOUGHTEN, F. C., YAGLOU, C. P.: ASHVE Trans. 29, 361 (1923).
HOWELL, T. R., DAWSON, W. R.: Condor 56, 93 (1954).
HSIANG-CH'UAN HOU: Chinese J. Physiol. 11, Nr. 4 (1928).
HSIEH, A. C. L., CARLSON, L. D., GRAY, G.: Am. J. Physiol. 190, 247 (1957).
HUDSON, J. W.: Univ. Calif., Berkeley, Publ. Zool. 64, 1 (1962).
HUDSON, J. W.: Depressed metabolism, p. 231. New York: Amer. Elsevier 1969.

HUDSON, J. W., DEAVERS, D. R., BRADLEY, R. S.: Symp. zool. Soc. Lond. **31**, 191 (1972).
HUIBREGTSE, W. H., GUNVILLE, R., UNGAR, F.: Comp. Biochem. Physiol. **38** A, 763 (1971).
HULL, D.: Brit. Med. Bull. **22**, 92 (1966).
HULL, D., HARDMAN, M. J.: In: LINDBERG, O. (Ed.): Brown adipose tissue, p. 97. New York: Am. Elsevier Publ. Comp. 1970.
HULL, D., SEGALL, M. M.: J. Physiol. (Lond.) **181**, 458 (1965 b).
HUNDEIKER, M.: Fortschr. Med. **89**, 1403 (1971).
HUNDEIKER, M., KELLER, L.: Morph. Jahrb. **105**, 26 (1963).
HUTCHINSON, J. C. D.: J. Agr. Sci. **45**, 48 (1955).
HUTCHINSON, J. C. D., BROWN, G. D.: J. Appl. Physiol. **26**, 454 (1969).
HUTCHINSON, J. C. D., WODZICKA-TOMASZEWSKA, M.: Animal. Breed. Abstr. **29**, 1 (1961).
HUXLEY, J. S., WEBB, C. S., BEST, A. F.: Nature **143**, 683 (1939).
IAMPIETRO, P. F.: J. Appl. Physiol. **16**, 405 (1961).
IAMPIETRO, P. F., MAGER, M., GREEN, E. B.: J. Appl. Physiol. **16**, 409 (1961).
IGGO, A.: Quart. J. Exp. Physiol. **44**, 362 (1959).
IGGO, A.: Acta Neuroveget. (Wien) **24**, 225 (1963).
IGGO, A.: In: KENSHALO, D. R. (Ed.): The skin senses, p. 84. Springfield, Ill: Thomas 1968.
IGGO, A.: J. Physiol. (Lond.) **200**, 403 (1969).
IGGO, A.: In: HARDY, J. D., GAGGE, A. P., STOLWIJK, J. A. J. (Eds.): Physiological and behavioral temperature regulation, p. 391. Springfield, Ill.: Thomas 1970.
IGGO, A., IGGO, B. J.: J. Physiol. (Paris) **63**, 287 (1971).
IRIUCHIJIMA, J., ZOTTERMAN, Y.: Acta Physiol. Scand. **49**, 267 (1960).
IRVING, L.: Federation Proc. **10**, 543 (1951).
IRVING, L.: In: Handbook of physiology, Sect. 4, p. 361. Washington, D. C.: Am. Phys. Soc. 1964.
IRVING, L., HART, J. S.: Can. J. Zool. **35**, 497 (1957).
IRVING, L., KROG, J.: J. Appl. Physiol. **6**, 667 (1954).
IRVING, L., KROG, J.: J. Appl. Physiol. **7**, 355 (1955).
IRVING, L., KROG, J.: Physiol. Zool. **29**, 195 (1956).
IRVING, L., PEYTON, L. J., BAHN, C. H., PETERSON, R. S.: Physiol. Zool. **35**, 275 (1962).
IRVING, L., PEYTON, L., MONSON, M.: J. Appl. Physiol. **9**, 421 (1956).
IRVING, L., SCHMIDT-NIELSEN, K., ABRAHAMSEN, N.: Physiol. Zool. **30**, 93 (1956).
ITOH, S., DOI, K., KUROSHIMA, A.: Int. J. Biometeor. **14**, 195 (1970).
JACOBS, H. K., SOUTH, F. E., HARTNER, W. C., ZATZMAN, M. L.: Cryobiology **8**, 313 (1971).
JACOBSON, F. H., SQUIRES, R. D.: In: HARDY, J. D., GAGGE, A. P., STOLWIJK, J. A. J. (Eds.): Physiological and behavioral temperature regulation, p. 581. Springfield, Ill.: Thomas 1970.
JAEGER, E. C.: Condor **50**, 45 (1948).
JAEGER, E. C.: Condor **51**, 105 (1949).
JÄRVILEHTO, T.: Ann. Acad. Sci. Fenn., Ser. B, **184**, 9 (1973).
JANSKY, L., BARTUNKOWA, R., ZEISBERGER, E.: Physiol. Bohemoslov. **16**, 366 (1967).
JANSKY, L., BARTUNKOVA, R., KOCKOWA, J., MEJSNAR, J., ZEISBERGER, E.: Federation Proc. **28**, 1053 (1969).
JESSEN, C.: Pflügers Arch. Ges. Physiol. **297**, 53 (1967).
JESSEN, C., MAYER, E. TH.: Pflügers Arch. Ges. Physiol. **324**, 189 (1971).
JESSEN, C., MEURER, K.-A., SIMON, E.: Pflügers Arch. Ges. Physiol. **297**, 35 (1967).
JESSEN, C., SIMON, E.: Pflügers Arch. Ges. Physiol. **324**, 217 (1971).
JOEL, C. D.: In: RENOLD, A. E., CAHILL, G. F., JR. (Eds.): Handbook of physiology, Sect. 5, p. 59. Washington, D. C.: Am. Physiol. Soc. 1965.
JOEL, C. D., TREBLE, D. H., BALL, E. G.: Federation Proc. **23**, 271 (1964).
JOHANNSEN, B.: Metabolism **8**, 221 (1959).
JOHANSEN, K.: Physiol. Zool. **34**, 126 (1961 a).
JOHANSEN, K.: Acta Physiol. Scand. **52**, 379 (1961b).

JOHANSEN, K., TØNNESEN, K. H.: Acta Physiol. Scand. **76**, 21 A (1969).
JOHNSON, D. G.: Acta Physiol. Scand. **68**, 129 (1966).
JOHNSON, K. O., DARIAN-SMITH, I., LaMOTTE, C.: J. Neurophysiol. **36**, 347 (1973).
JOY, R. J. T.: J. Appl. Physiol. **18**, 1209 (1963).
JOYCE, J. P., BLAXTER, L. K.: Res. Vet. Sci. **5**, 506 (1964).
JUNDELL, J.: Jahrb. Kinderheilk. **59**, 521 (1904).
KALABUKHOV, N. I.: The hibernation of animals. Charkow: Gorki State Univ. Press. 1956.
KALABUKHOV, N. I.: Usp. Sovrem. Biol. **46**, 217 (1958).
KALLEN, F. C.: Bull. Mus. Comp. Zool. Harvard Coll. **124**, 373 (1960).
KALLIR, E.: Z. Vergl. Physiol. **13**, 231 (1930).
KALMIJN, A. J.: J. Exp. Biol. **55**, 371 (1971).
KANITZ, A.: Tabul. Biol. **2**, 9 (1925).
KANTER, G. S.: Am. J. Physiol. **196**, 866 (1959).
KANTER, G. S.: Am. J. Physiol. **214**, 856 (1968).
KARLBERG, P., MOORE, R. E., OLIVER, T. K., JR.: Acta Paediat. Scand. **54**, 225 (1965).
KAUFMANN, R., HOMBURGER, H., TRITTHART, H.: Pflügers Arch. Ges. Physiol. **305**, 1 (1969).
KAYSER, CH.: Ann. Physiol. **15**, 1087 (1939).
KAYSER, CH.: Mammalia **14**, 105 (1950).
KAYSER, CH.: Ann. Biol. **29**, 109 (1953).
KAYSER, CH.: Rev. Can. Biol. **16**, 303 (1957).
KAYSER, CH.: The physiology of natural hibernation. Oxford-London-New York-Paris: Pergamon Press 1961.
KAYSER, CH.: Helgoländer Wiss. Meeresunters. **9**, 158 (1964a).
KAYSER, CH.: Arch. Sci. Physiol. **18**, 137 (1964b).
KAYSER, CH., MALAN, A.: Experientia **19**, 441 (1963).
KEARNS, J. B., MURNAGHAN, M. F.: J. Physiol. (Lond.) **203**, 51 P (1969).
KEATINGE, W. R.: J. Physiol. (Lond.) **142**, 395 (1958).
KEATINGE, W. R.: Survival in cold water. Oxford-Edinburgh: Blackwell 1969.
KEATINGE, W. R.: In: HARDY, J. D., GAGGE, A. P., STOLWIJK, J. A. J. (Eds.): Physiological and behavioral temperature regulation, p. 231. Springfield, Ill.: Thomas 1970.
KEATINGE, W. R., CANNON, P.: Lancet **1960**, 11.
KELLER, A. D.: J. Neurophysiol. **1**, 543 (1938).
KENDEIGH, S. C.: J. Exp. Zool. **82**, 419 (1939).
KENDEIGH, S. C.: J. Exp. Zool. **96**, 1 (1944).
KENDEIGH, S. C.: J. Mammal. **26**, 86 (1945a).
KENDEIGH, S. C.: J. Wildlife Management **9**, 217 (1945b).
KENDEIGH, C. S.: Auk. **66**, 113 (1949).
KENSHALO, D. R.: J. Physiol. (Lond.) **172**, 439 (1964).
KENSHALO, D. R.: In: NEFF, W. D. (Ed.): Contributions to sensory physiology, Vol. 4, p. 19. New York: Academic Press 1970.
KENSHALO, D. R., BREARLEY, E. A.: J. Comp. Physiol. Pyschol. **70**, 5 (1970).
KENSHALO, D. R., DUNCAN, D. G., WEYMARK, C.: J. Comp. Physiol. Psychol. **63**, 133 (1967).
KENSHALO, D. R., GALLEGOS, E. S.: Science **158**, 1064 (1967).
KENSHALO, D. R., HENSEL, H., GRAZIADEI, P., FRUHSTORFER, H.: In: DUBNER, R., KAWAMURA, Y. (Eds.): Oral-facial sensory and motor mechanisms, p. 23. New York: Appleton-Century-Crofts, Meredith. Corp. 1971.
KENSHALO, D. R., HOLMES, CH. E., WOOD, P. B.: Perception Psychophysics **3**, 81 (1968).
KESTNER, O.: Pflügers Arch. Ges. Physiol. **234**, 290 (1934).
KING, J. R., FARNER, D. S.: In: MARSHALL, A. J. (Ed.): Biology and comparative physiology of birds, Vol. 2, p. 215. New York: Academic Press 1961.
KING, J. R., FARNER, D. S.: In: Handbook of physiology, Sect. IV, p. 603. Washington, D. C.: Am. Physiol. Soc. 1964.
KIPP, F. A.: Biol. Zbl. **67**, 250 (1948).

KITCHELL, R. L., STRÖM, L., ZOTTERMAN, Y.: Acta Physiol. Scand. **46**, 133 (1959).
KITZING, J., KUTTA, D., BLEICHERT, A.: Intern. Z. Angew. Physiol. **23**, 159 (1966).
KLAIN, G. J., WHITTEN, B. K.: Comp. Biochem. Physiol. **27**, 617 (1968).
KLEIBER, M.: Physiol. Rev. **27**, 511 (1947).
KLEIBER, M.: The fire of life. New York-London: Wiley 1961.
KLEIBER, M., DOUGHERTY, J. E.: J. Gen. Physiol. **17**, 701 (1934).
KLEIBER, M., WINCHESTER, C. F.: Proc. Soc. Exp. Biol. (N.Y.) **31**, 158 (1933).
KLEITMAN, N.: Sleep and wakefulness. Chicago: Univ. of Chicago Press 1939.
KLUSSMANN, F. W.: Pflügers Arch. Ges. Physiol. **305**, 295 (1969).
KLUSSMANN, F. W., LÜTCKE, A., KOENIG, W.: Pflügers Arch. Ges. Physiol. **268**, 515 (1959).
KNAUS, H.: Klin. Wschr. **2**, 1897 (1932).
KOCH, W., JENNINGS, B. H., HUMPHREY, C. M.: ASHRAE Trans. **66**, 264 (1960).
KÖPPEN, W.: Grundriß der Klimakunde. Berlin-Leipzig: De Gruyter 1931.
KOIZUMI, K., USHIYAMA, J., McC. BROOKS, CH.: J. Neurophysiol. **23**, 421 (1960).
KOSKIMIES, J.: Experientia **4**, 274 (1948).
KRAMER, K., REICHEL, H.: Klin. Wschr. **23**, 192 (1944).
KREIDER, M. B.: J. Appl. Physiol. **16**, 239 (1961).
KREIDER, M. B., BUSKIRK, R. E.: J. Appl. Physiol. **11**, 339 (1957).
KREYBERG, L.: Physiol. Rev. **29**, 156 (1949).
KREYBERG, L.: Acta Path. Microbiol. Scand. Suppl. **91**, 40 (1950).
KRISS, M.: J. Agr. Res. **21**, 1 (1921).
KRISTOFFERSSON, R., BROBERG, S.: Ann. Acad. Sci. Fennicae. A4 Biol. **130**, 1 (1968).
KRUMBIEGEL, I.: Der afrikanische Elefant. Monogr. Wildsäugetiere, Bd. 9. Leipzig: Schöps 1943.
KULZER, E.: Z. Vergl. Physiol. **50**, 1 (1965).
KULZER, E.: Z. Vergl. Physiol. **56**, 63 (1967).
KULZER, E.: Umschau Wiss. Techn. Heft 7, 195 (1969).
KULZER, E., NELSON, J. E., McKEAN, J. L., MÖHRES, F. P.: Z. Vergl. Physiol. **69**, 426 (1970).
KUNDT, H. W., BRÜCK, K., HENSEL, H.: Naturwissenschaften **44**, 496 (1957a).
KUNDT, H. W., BRÜCK, K., HENSEL, H.: Pflügers Arch. Ges. Physiol. **264**, 97 (1957b).
KUNO, Y.: Human perspiration. Springfield, Ill.: Thomas 1956.
KUZNETS, E. T., CHADOV, V. I., ZHARIKOVA, G. S., SADOVNIKOVA, L. V., PEREPLETCHI-KOVA, B. S., INSHAKOVA, V. M., MORDOVSKAYA, L. G.: Kosm. Biol. Med. **2**, 11 (1968).
LACHER, V.: Z. Vergl. Physiol. **48**, 587 (1964).
LACHIVER, F.: Ann. Acad. Sci. Fennicae. A4 Biol. **71**, 285 (1964).
LACHIVER, F., PETROVIC, V.: J. Physiol. (Paris) **52**, 140 (1960).
LADELL, W. S. S.: Brit. Med. Bull. **1945** III, 175.
LADELL, W. S. S.: Lancet **1949** II, 836.
LADELL, W. S. S.: J. Physiol. (Lond.) **115**, 296 (1951).
LADELL, W. S. S.: Brit. J. Ind. Med. **12**, 111 (1955).
LADELL, W. S. S.: Trans. Roy. Soc. Trop. Med. Hyg. **51**, 189 (1957).
LADELL, W. S. S.: In: Handbook of physiology, Sect. IV, p. 625. Washington, D. C.: Am. Physiol. Soc. 1964.
LANDGREN, S.: Acta Physiol. Scand. **40**, 202 (1957a).
LANDGREN, S.: Acta Physiol. Scand. **40**, 210 (1957b).
LANDGREN, S.: Acta Physiol. Scand. **48**, 255 (1960).
LANDGREN, S.: In: HARDY, J. D., GAGGE, A. P., STOLWIJK, J. A. J. (Eds.): Physiological and behavioral temperature regulation, p. 454. Springfield, Ill.: Thomas 1970.
LANDSIEGEL, K.: Inaug.-Diss. Tierärztl. Hochschule Hannover 1937.
LANGDON, L., KINGSLEY, D. P. E.: J. Clin. Path. **17**, 257 (1964).
LANGE, K., BOYD, L. J., LOEWE, L.: Science **102**, 151 (1945).
LASIEWSKI, R. C.: Physiol. Zool. **37**, 212 (1964).
LASIEWSKI, R. C., ACOSTA, A. L., BERNSTEIN, M. H.: Comp. Biochem. Physiol. **19**, 445 (1966a).

LASIEWSKI, R. C., ACOSTA, A. L., BERNSTEIN, M. H.: Comp. Biochem. Physiol. **19**, 459 (1966 b).
LASIEWSKI, R. C., DAWSON, W. R.: Condor **66**, 477 (1964).
LASIEWSKI, R. C., WEATHERS, W. W., BERNSTEIN, M. H.: Comp. Biochem. Physiol. **23**, 797 (1967).
LATTIN, G., DE: Grundriß der Zoogeographie. Stuttgart: Fischer 1967.
LAUFMAN, H.: J. Am. Med. Ass. **147**, 1201 (1951).
LEBLANC, J.: J. Appl. Physiol. **9**, 395 (1956).
LEBLANC, J.: J. Appl. Physiol. **17**, 950 (1962 a).
LEBLANC, J.: Federation Proc. **28**, 996 (1969).
LEBLANC, J., HILDES, J. A., HÉROUX, O.: J. Appl. Physiol. **15**, 1031 (1960).
LEBLANC, J., POULIOT, M.: Am. J. Physiol. **207**, 853 (1964).
LEBLANC, J., ROBINSON, D., SHARMAN, D. F., TOUSIGNANT, P.: Am. J. Physiol. **213**, 1419 (1967).
LEBLANC, J., VILLEMAIRE, A.: Am. J. Physiol. **218**, 1742 (1970).
LEBLOND, C. P., GROSS, J., PEACOCK, W., EVANS, R. D.: Am. J. Physiol. **140**, 671 (1943).
LEDUC, J.: Acta Physiol. Scand. **53**, Suppl. 183, 1 (1961).
LEE, A. K.: Univ. Calif., Berkeley, Publ. Zool. **64**, 57 (1963).
LEE, D. H. K.: Ann. Rev. Physiol. **10**, 365 (1948).
LEE, D. H. K.: New Engl. J. Med. **243**, 723 (1950).
LEE, D. H. K.: Animal. Breed. Abstr. **27**, 1 (1959).
LEE, D. H. K.: Ann. N. Y. Acad. Sci. **91**, 608 (1961).
LEE, D. H. K., PENDLETON, R. L.: Geographical Rev. **41**, 124 (1951).
LEE, D. H. K., ROBINSON, K. W.: Proc. Roy. Soc. Queensland **53**, 189 (1941).
LEE, D. H. K., ROBINSON, K. W., HINES, H. J. G.: Proc. Roy. Soc. Queensland **53**, 129 (1941).
LEE, R. C.: J. Nutr. **23**, 83 (1942).
LEE, R. C., NICHOLAS, F. C., RITZMAN, E. G.: J. Nutr. **21**, 321 (1941).
LEHMANN, G.: Praktische Arbeitsphysiologie. Stuttgart: Georg Thieme 1953.
LEHMANN, V. W.: North Am. Fauna **57**, 1 (1941).
LEHRMAN, D. S.: In: YOUNG, W. C. (Ed.): Sex and internal secretion, Vol. II, Chapt. 21. Baltimore, Maryl.: Williams & Wilkins 1961.
LEITHEAD, C. S., LIND, A. R.: Heat stress and heat disorders. London: Cassell 1964.
LELE, P. P.: J. Physiol. (Lond.) **126**, 191 (1954).
LENTZ, C. P., HART, J. S.: Can. J. Zool. **38**, 679 (1960).
LESSER, A. J., WINZLER, R. J., MICHAELSON, J. B.: Proc. Soc. Exp. Biol. (N. Y.) **70**, 571 (1949).
LEWIS, R. B.: Am. J. Med. Ass. **222**, 300 (1951).
LIGON, J. D. (1967): cf. DAWSON, W. R., HUDSON, J. W. (1970).
LIND, A. R.: J. Appl. Physiol. **18**, 51 (1963).
LIND, A. R., HELLON, R. F.: J. Appl. Physiol. **11**, 35 (1957).
LINDBERG, O.: Brown adipose tissue. New York: Am. Elsevier Publ. Comp. 1970.
LIPTON, J. M.: Behaviour **3**, 165 (1968).
LÖHRL, H.: Vogelwarte **18**, 71 (1955).
LOFTUS, R.: Z. Vergl. Physiol. **52**, 380 (1966).
LOFTUS, B.: Z. Vergl. Physiol. **59**, 413 (1968).
LOFTUS, B.: Z. Vergl. Physiol. **63**, 415 (1969).
LOMAX, P.: Intern. Rev. Neurobiol. **12**, 1 (1970).
LOVELOCK, J. E., SMITH, A. U.: Proc. Roy. Soc. B **145**, 427 (1956).
LOVELOCK, J. E., SMITH, A. U.: Ann. N. Y. Acad. Sci. **80**, 487 (1959).
LUECKE, R. H., SOUTH, F. E.: In: SOUTH, F. E., HANNON, J. P., WILLIS, J. R., PENGELLEY, E. T., ALPERT, N. R. (Eds.): Hibernation and hypothermia, p. 577. Amsterdam: Elsevier 1972.
LUX, F., LUX, W.: Deut. Mschr. Zahnheilk. **51**, 535 (1933).
LYMAN, C. P.: J. Exp. Zool. **109**, 55 (1948).
LYMAN, C. P.: J. Mammal. **45**, 122 (1964).

LYMAN, C. P.: In: Handbook of physiology, Sect. 2, Chapt. 56, p. 1967. Washington, D.C.: Am. Physiol. Soc. 1965.

LYMAN, C. P.: In: Biology of bats, p. 301. New York: Academic Press Inc. 1970.

LYMAN, C. P., CHATFIELD, P. O.: J. Exp. Zool. **114**, 491 (1950).

LYMAN, C. P., CHATFIELD, P. O.: Physiol. Rev. **35**, 403 (1955).

LYMAN, C. P., LEDUC, E. H.: J. Cell. Comp. Physiol. **41**, 471 (1953).

LYMAN, C. P., O'BRIEN, R. C.: Bull. Mus. Comp. Zool. Harvard Coll. **124**, 353 (1960).

LYMAN, C. P., O'BRIEN, R. C.: Symp. Soc. Exp. Biol. **23**, 489 (1969).

MacFARLANE, W. V.: In: Handbook of physiology, Sect. IV, p. 509. Washington, D.C.: Am. Physiol. Soc. 1964.

MacFARLANE, W. V.: In: CURTIS, D. R., McINTYRE, A. K. (Eds.): Studies in physiology, p. 191. Berlin-Heidelberg-New York: Springer 1965.

MacFARLANE, W. V.: In: HAFEZ, E. S. E. (Ed.): Adaptation of domestic animals, p. 264. Philadelphia: Lea & Febiger 1968.

MacFARLANE, W. V.: J. Physiol. (Lond.) **205**, 13 P (1969).

MacFARLANE, W. V., HOWARD, B.: J. Agr. Sci. **66**, 297 (1966).

MacFARLANE, W. V., HOWARD, B., SIEBERT, B. D.: Australian J. Agr. Res. **18**, 947 (1967).

MacFARLANE, W. V., MORRIS, R. J. H., HOWARD, B.: Nature **178**, 304 (1956).

MacFARLANE, W. V., MORRIS, R. J. H., HOWARD, B.: Nature **197**, 270 (1963).

MACKWORTH, N. H.: J. Appl. Physiol. **5**, 533 (1953).

MACKWORTH, N. H.: Proc. Roy. Soc. B **143**, 392 (1955).

MacLEOD, J., HOTCHKISS, R. S.: Endocrinology **28**, 780 (1941).

MacMILLEN, R. E., NELSON, J. E.: Am. J. Physiol. **217**, 1246 (1969).

MAGILTON, J. H., SWIFT, C. S.: J. Appl. Physiol. **27**, 18 (1969).

MALAMUD, N., HAYMAKER, W., CUSTER, R. P.: Military Surg. **99**, 397 (1946).

MALAN, A.: J. Physiol. (Paris) **58**, 565 (1966).

MALAN, A.: Arch. Sci. Physiol. **23**, 47 (1969).

MARCUS, E.: In: Handbuch der Geographischen Wissenschaften, Allgemeine Geographie, Teil II, S. 81. Potsdam: Athenaion 1933.

MARÉCHAUX, E. W., SCHÄFER, K. E.: Pflügers Arch. Ges. Physiol. **251**, 765 (1949).

MARKS, L. E., STEVENS, J. C.: Perception and Psychophysics 4, 220 (1968).

MARLER, P. R., HAMILTON, W. J.: Mechanisms of animal behavior. New York: Wiley 1966.

MARSHALL, J.: Brit. Med. J. **5323**, 102 (1963).

MARTIN, C. J.: Phil. Trans. B **195**, 1 (1902).

MARUHASHI, I., MIZUGUCHI, K., TASAKI, I.: J. Physiol. (Lond.) **117**, 129 (1952).

MASSEY, P. M. O.: J. App. Physiol. **14**, 616 (1959).

MATHER, G. W., NAHAS, G. G., HEMINGWAY, A.: Am. J. Physiol. **173**, 390 (1953).

MAYR, E.: Systematics and the origin of species, p. 88. New York: Columbia Univ. Press 1942. (Fourth printing 1949).

MAYR, E.: Evolution 10, 105 (1956).

McARDLE, B., DUNHAM, W., HOLLING, H. E., LADELL, W. S. S., SCOTT, J. W., THOMSON, M. L., WEINER, J. S.: Spec. Rep. Ser. Med. Res. Coun. (Lond.) **47**, 391 (1947).

McCOOK, R. D., RANDALL, W. C., HASSLER, C. R., MIHALDZIC, N., WURSTER, R. D.: In: HARDY, J. D., GAGGE, A. P., STOLWIJK, J. A. J. (Eds.): Physiological and behavioral temperature regulation, p. 627. Springfield, Ill.: Thomas 1970.

McDADE, H.: Appalachia **24**, 233 (1962).

McDOWELL, R. E.: Improvement of livestock production in warm climates. San Francisco, California: W. H. Freemann and Company Publ. 1972.

McEWAN JENKINSON, D., NOBLE, R. C., THOMPSON, G. E.: J. Physiol. (Lond.) **195**, 639 (1968).

McGOWAN, J. D.: Auk **86**, 142 (1969).

McKKERSLAKE, D., COOPER, K. E.: Clin. Sci. **9**, 31 (1950).

McLEAN, J. A.: J. Physiol. (Lond.) **167**, 427 (1963).

McNALL, P. E., JR., JAAX, J., ROHLES, F. H., NEVINS, R. G., SPRINGER, W. S.: ASHRAE Trans. **73**, 1 (1967).

McNall,P.E.,Jr., Ryan,P., Jaax,J.: ASHRAE Trans. 74, 1 (1968).
McNicol,M.W., Smith,R.: Brit. Med. J. 5374, 19 (1964).
Mead,J., Bonmarito,C.L.: J. Appl. Physiol. 2, 97 (1949).
Meeh,K.: Z. Biol. 15, 428 (1879).
Meiners,S.: Pflügers Arch. Ges. Physiol. 254, 557 (1952).
Mejsnar,J., Jánský,L.: Physiol. Bohemoslov. 16, 147 (1967).
Mejsnar,J., Jánský,L.: Can. J. Physiol. Pharmacol. 48, 102 (1970).
Mellette,H.C., Hutt,B.K., Askovitz,S.I., Horvath,S.M.: J. Appl. Physiol. 3, 665 (1951).
Menaker,M.: J. Cell. Comp. Physiol. 57, 81 (1961).
Menaker,M.: J. Cell. Comp. Physiol 59, 163 (1962).
Menaker,M., Eskin,A.: In: Hafez,E.S.E. (Ed.): Adaptation of domestic animals, p. 141. Philadelphia: Lea & Febiger 1968.
Menges,G.: Die gelenkte Hypothermie in der operativen Medizin. Heidelberg: Hüthig 1968.
Mercier,E.: J. Dairy Sci. 29, 556 (1946).
Meryman,H.T.: Physiol. Rev. 37, 233 (1957).
Messmer,K., Brendel,W., Reulen,H.J., Nordmann,K.J.: Pflügers Arch. Ges. Physiol. 288, 240 (1963).
Mestyán,G., Varga,F., Fohl,E., Heim,T.: Arch. Dis. Childh. 37, 466 (1962).
Meyer,H.H.: Verhandl. Kongr. Inn. Med. 30, 15 (1913).
Milhorn,H.T.,Jr.: The application of control theory to physiological systems. Philadelphia, Pa.: Saunders 1966.
Miller,D.S.: J. Exp. Zool. 80, 259 (1939).
Mills,W.J., Whaley,R., Fish,W.: Alaska Med. 2, 114 (1960).
Mitchell,D.: In: Hardy,J.D., Gagge,A.P., Stolwijk,J.A.J. (Eds.): Physiological and behavioral temperature regulation, p. 25. Springfield, Ill.: Thomas 1970.
Mitchell,D., Snellen,J.W., Atkins,A.R.: Pflügers Arch. Ges. Physiol. 321, 293 (1970).
Mitchell,D., Wyndham,C.H., Atkins,A.R., Vermeulen,A.J., Hofmeyr,H.S., Strydom,N.B., Hodgson,T.: Pflügers Arch. Ges. Physiol. 303, 324 (1968).
Mogler,R.K.-H.: Z. Morphol. Ökol. Tiere 47, 267 (1958).
Mohr,E.: Die Säugetiere Schleswig-Holsteins. In: Der Faunist, S. 93. Altona: Naturwissenschaftlicher Verein 1931.
Mokrasch,L.C., Grady,H.J., Grisolia,S.: Am. J. Physiol. 199, 945 (1960).
Molnar,G.W.: J. Am. med. Ass. 131, 1046 (1946).
Moore,C.R.: Quart. Rev. Biol. 1, 4 (1926).
Moore,R.E., Underwood,M.C.: Lancet 1960, 1277.
Moore,R.E., Underwood,M.C.: J. Physiol. (Lond.) 161, 30 (1962).
Moore,R.E., Underwood,M.C.: J. Physiol. (Lond.) 168, 290 (1963).
Moreng,R.E., Shaffner,C.S.: Poultry Sci. 30, 255 (1951).
Morris,L.: Nature 190, 102 (1961).
Morrison,P.R.: J. Cell. Comp. Physiol. 27, 125 (1946).
Morrison,P.R.: J. Cell. Comp. Physiol. 31, 281 (1948).
Morrison,P.R.: Austalian J. Zool. 13, 173 (1965).
Morrison,P.R., Petajan,J.H.: Physiol. Zool. 35, 52 (1962).
Morrison,P.R., Ryser,F.A.: Science 116, 231 (1952).
Moule,G.R.: In: Hafez,E.S.E. (Ed.): Adaptation of domestic animals, p. 18. Philadelphia: Lea & Febiger 1968.
Mount,L.E.: J. Physiol. (Lond.) 147, 333 (1959).
Mount,L.E.: J. Physiol. (Lond.) 168, 698 (1963).
Mount,L.E.: J. Physiol. (Lond.) 173, 96 (1964).
Mount,L.E.: The climatic physiology of the pig. London: Edward Arnold Ltd. 1968.
Moy,R.M.: Am. J. Physiol. 220, 747 (1971).
Mrosovsky,N.: Psychosomat. Res. 14, 239 (1970).
Munro,A.F.: J. Physiol. (Lond.) 110, 356 (1949).

MURGATROYD, D., HARDY, J. D.: In: HARDY, J. D., GAGGE, A. P., STOLWIJK, J. A. J. (Eds.): Physiological and behavioral temperature regulation, p. 874. Springfield, Ill.: Thomas 1970.

MURLIN, J. R.: Ergebn. Physiol. **42**, 153 (1939).

MURPHY, R. C.: Oceanic birds of South America, Vol. I. New York: MacMillan 1936.

MURRAY, R. W.: Nature **179**, 106 (1957).

MURRAY, R. W.: J. Physiol. (Lond.) **145**, 1 (1959).

MURRAY, R. W.: J. Exp. Biol. **37**, 417 (1960a).

MURRAY, R. W.: Nature **187**, 957 (1960b).

MURRAY, R. W.: Advan. Comp. Physiol. Biochem. **1**, 117 (1962).

MURRAY, R. W.: J. Physiol. (Lond.) **180**, 592 (1965).

MURRAY, R. H., ROSS, J. C.: Federation Proc. **24**, 280 (1965).

MUSACCHIA, X. J., VOLKERT, W. A.: Am. J. Physiol. **221**, 128 (1971).

MUSACCHIA, X. J., VOLKERT, W. A., BARR, R. E.: Radiation Res. **46**, 353 (1971).

MYERS, R. D.: In: WHITTOW, G. C. (Ed.): Comparative physiology of thermoregulation, Vol. II, p. 283. New York-London: Academic Press 1971.

NADEL, E. R., BULLARD, R. W., STOLWIJK, J. A. J.: J. Appl. Physiol. **31**, 80 (1971a).

NADEL, E. R., MITCHELL, J. W., STOLWIJK, J. A. J.: Proc. Intern. Union Physiol. Sc. Nr. 1224 Munich 1971b.

NADEL, E. R., MITCHELL, J. W., STOLWIJK, J. A. J.: Pflügers Arch. Ges. Physiol. **340**, 71 (1973).

NAKAYAMA, T., HAMMEL, H. T., HARDY, J. D., EISENMAN, J. S.: Am. J. Physiol. **204**, 1122 (1963).

NAKAYAMA, T., HARDY, J. D.: J. Appl. Physiol. **27**, 848 (1969).

NAPOLITANO, L.: In: Handbook of physiology, Sect. V, p. 109. Washington, D.C.: Am. Physiol. Soc. 1965.

NECKER, R.: J. Comp. Physiol. **78**, 307 (1972).

NELMS, J. D., SOPER, D. J. G.: J. Appl. Physiol. **17**, 444 (1962).

NEVINS, R. G.: J. Physiol. (Paris) **63**, 356 (1971).

NEVINS, R. G., ROHLES, F. H., SPRINGER, W., FEYERHERM, A. M.: ASHRAE Trans. **72**, 283 (1966).

NEWMAN, P. P., WOLSTENCROFT, J. H.: J. Physiol. (Lond.) **152**, 87 (1960).

NIAZI, S. A., LEWIS, J.: Ann. Surg. **147**, 264 (1958).

NIELSEN, B.: Acta Physiol. Scand. Suppl. **323**, 1 (1969b).

NIELSEN, K. C., OWMAN, CH.: Acta Physiol. Scand. **76**, 73 (1969).

NIELSEN, M.: Skand. Arch. Physiol. **79**, 193 (1938).

NIETHAMMER, G.: Fortschr. Zool. **9**, 368 (1952).

NISHI, Y., GAGGE, A. P.: J. Physiol. (Paris) **63**, 365 (1971).

NOBLE, G. K., SCHMIDT, A.: Proc. Am. Phil. Soc. **77**, 263 (1937).

OGATA, M.: Kyushu J. Med. Sci. **10**, 61 (1959).

OPPERMANN, C., HEERD, E.: Pflügers Arch. Ges. Physiol. **318**, 51 (1970).

PAPPENHEIMER, J. R., EVERSOLE, S. L., JR., SOTO-RIVERA, A.: Am. J. Physiol. **155**, 458 (1948).

PARER, J. T., METCALFE, J.: Resp. Physiol. **3**, 136 (1967).

PARK, C. R., PALMES, E. D.: Med. Dept., Field Res. Lab. Proj. 6-4-12-06, Ft. Knox, Ky. 1948.

PARKES, A. S.: Advan. Sci. (Lond.) **15**, 46 (1958).

PARROTT, D. M. V.: J. Reprod. Fertility **1**, 230 (1960).

PASCOE, J. E.: Proc. Roy. Soc. B **147**, 510 (1957).

PEARSON, O. P.: Ecology **28**, 127 (1947).

PEARSON, O. P.: Science **108**, 44 (1948).

PEIRCE, E. C., II., DABBS, C. H., ROGERS, W. K., RAWSON, F. I., TOMPKINS, R.: Surg. Gynec. Obstet. **107**, 339 (1958).

PEITZMEIER, J.: Ornithol. Forsch. **1**, 22 (1947).

PEMBREY, M. S.: J. Physiol. (Lond.) **18**, 363 (1895).

PENEFSKY, Z. J.: Am. J. Physiol. **214**, 730 (1968).

PENROD, K. E.: Am. J. Physiol. **157**, 436 (1949).

PENROD, K. E.: Am. J. Physiol. **164**, 79 (1951).
PERKINS, J. F., LI, M. C., HOFFMANN, F., HOFFMANN, E.: Am. J. Physiol. **155**, 165 (1948).
PERL, E. R.: J. Physiol. (Lond.) **197**, 593 (1968).
PERSON, R. S.: Tr. Inst. Morphol. Zhivotn. Akad. Nauk SSSR **6**, 173 (1952).
PETROVIC, V. M., DAVIDOVIC, V.: J. Physiol. (Paris) **57**, 678 (1965).
PETROVIC, V. M., DAVIDOVIC, V.: J. Physiol. (Paris) **60**, 514 (1968).
PFLEIDERER, H., BÜTTNER, K.: Bioklimatologie. Lehrbuch der Bäder- und Klimaheilkunde, Vol. II, p. 676. Berlin: Springer 1940.
PICHOTKA, J.: Arch. Exp. Pathol. Pharmakol. **215**, 299 (1952a).
PICHOTKA, J.: Arch. Exp. Pathol. Pharmakol. **215**, 317 (1952b).
PICKWORTH, F. A.: Proc. Roy. Soc. B **101**, 163 (1927).
PIERAU, F. K., KLUSSMANN, F. W.: J. Physiol. (Paris) **63**, 380 (1971).
PIRLET, K.: Pflügers Arch. Ges. Physiol. **275**, 71 (1962).
PITTS, G. C., JOHNSON, R. E , CONSOLAZIO, F. C.: Am. J. Physiol. **142**, 253 (1944).
POHL, H.: Z. Vergl. Physiol. **45**, 109 (1961).
POLGE, C.: Proc. Roy. Soc. B **147**, 498 (1957).
POLGE, C., SMITH, A. U., PARKES, A. S.: Nature **164**, 666 (1949).
POPOFF, N. F.: Pflügers Arch. Ges. Physiol. **234**, 137 (1934).
POPOVIC, V.: Bull. Mus. Comp. Zool. Harvard Coll. **124**, 105 (1960).
POPOVIC, V. P., KENT, K. M.: Am. J. Physiol. **209**, 1069 (1965).
POULIOT, M.: Acta Physiol. Scand. **68**, 164 (1966).
POULOS, D. A.: In: DUBNER, R., KAWAMURA, Y. (Eds.): Oral-facial sensory and motor mechanisms, p. 47. New York: Appleton-Century-Crofts, Meredith Corp. 1971.
POULOS, D. A., BENJAMIN, R. M.: J. Neurophysiol. **31**, 28 (1968).
PRENGLOWITZ, R.: Zool. Jahrb. Abt. System. Ökol. **64**, 129 (1933).
PROSSER, C. L.: Ann. Rev. Physiol. **16**, 103 (1954).
PROSSER, C. L., BROWN, F. A., JR.: Comparative animal physiology. Philadelphia-London: W. B. Saunders 1961.
PROVINS, K. A., HELLON, R. F., BELL, C. R., HIORNS, R. W.: Ergonomics **5**, 93 (1962).
PRUSINER, S., CANNON, B., LINDBERG, O.: In: LINDBERG, O. (Ed.): Brown adipose tissue, p. 283. New York-London-Amsterdam: Am. Elsevier Publ. Comp. 1970.
PUTKONEN, P., SARAJAS, H. S., SUOMALAINEN, P.: Ann. Acad. Sci. Fennicae A 5 Medica **106**, 1 (1964).
QUIMBY, E. H., WERNER, S. C., SCHMIDT, C.: Proc. Soc. Exp. Biol. (N. Y.) **75**, 537 (1950).
RADSMA, W.: Acta Physiol. Pharmacol. Neerl. **1**, 112 (1950).
RAFAEL, J., KLAAS, D., HOHORST, H. J.: Hoppe-Seyler's Z. Physiol. Chem. **349**, 1711 (1968).
RAGSDALE, A. C., THOMPSON, H. J., WORSTELL, D. M., BRODY, S.: Univ. Missouri, Agr. Exp. Sta., Res. Bull. **460** (1950).
RANDALL, W. C.: Am. J. Physiol. **139**, 56 (1943).
RANDALL, W. C., HIESTAND, W. A.: Am. J. Physiol. **127**, 761 (1939).
RANDALL, W. C., PEISS, C. N.: J. Invest. Derm. **28**, 435 (1957).
RANSON, S. W., MAGOUN, H. W.: Ergebn. Physiol. **41**, 56 (1939).
RAPP, G. M.: In: HARDY, J. D., GAGGE, A. P., STOLWIJK, J. A. J. (Eds.): Physiological and behavioral temperature regulation, p. 55, Springfield, Ill.: Thomas 1970.
RATHS, P.: Z. Biol. **106**, 109 (1953).
RATHS, P.: Z. Biol. **110**, 62 (1958).
RATHS, P.: Z. Biol. **112**, 282 (1961).
RATHS, P.: Experientia **20**, 178 (1964).
RATHS, P., HENSEL, H.: Pflügers Arch. Ges. Physiol. **293**, 281 (1967).
RAUCH, J., HAYWARD, J. S.: Can. J. Physiol. Pharmacol. **48**, 269 (1970).
RAUTENBERG, W.: Z. Vergl. Physiol. **62**, 221 (1969).
RAUTENBERG, W.: Z. Vergl. Physiol. **62**, 235 (1969).
RAWSON, R. O., QUICK, K. P., GOUGHLIN, R. F.: Science **165**, 919 (1969).
READER, S. R., WHYTE, H. M.: J. Appl. Physiol. **4**, 396 (1951).
REED, W. A.: Am. J. Physiol. **208**, 451 (1965).
REED, W. A., HOPKINS, L.: Am. J. Physiol. **203**, 1062 (1962).

REED, W. A., MANNING, R. T., HOPKINS, L. T.: Am. J. Physiol. **206**, 1304 (1964).
REEDER, W. G., COWLES, R. B.: J. Mammal. **32**, 389 (1951).
REGAN, W. M., RICHARDSON, G. A.: J. Dairy Sci. **19**, 11 (1935).
REIN, H.: Z. Biol. **82**, 189 (1925 a).
REIN, H.: Z. Biol. **82**, 513 (1925 b).
REIN, H.: Z. Biol. **89**, 319 (1930).
RENSCH, B.: Arch. Naturg. N. F. **5**, 317 (1936).
RENSCH, B.: Arch. Naturg. N. F. **8**, 89 (1939).
RENSCH, B.: Neuere Probleme der Abstammungslehre. Stuttgart: Enke 1954.
REULEN, H. J., AIGNER, P., BRENDEL, W., MESSMER, K.: Pflügers Arch. Ges. Physiol. **288**, 197 (1966).
REVUSKY, S. H.: Psychosomat. Sci. **6**, 209 (1966).
REY, L. B.: J. Embryol. Exp. Morph. **6**, 171 (1958).
REY, L. R.: Conservation de la vie par le froid. Actualités sci. industr. No. 1279. Paris: Hermann 1959.
RICHET, CH.: Dict. Physiol. par Ch. Richet **3**, 81 (1898).
RIDDLE, O., SMITH, G. C., BENEDICT, F. G.: Am. J. Physiol. **101**, 88 (1932).
RIEDEL, W., SIAPLAURAS, G., SIMON, E.: Pflügers Arch. **340**, 59 (1973).
RIEK, R. F., HARDY, M. H., LEE, D. H. K., CARTER, H. B : Australian J. Agr. Res. **1**, 217 (1950).
RIEK, R. F., LEE, D. H. K.: J. Dairy Res. **15**, 219 (1948).
RING, R. C.: Am. J. Physiol. **125**, 244 (1939).
RINK, R. A., GRAY, I., RUECKERT, R. R., SLOCUM, H. C.: Anesthesiology **17**, 377 (1956).
ROBERTS, J. C., CHAFFEE, R. R. J.: In: SMITH, R. E., SHIELDS, J. L., HANNON, J. P., HORWITZ, B. A. (Eds.): Proc. Internat. Sympos. Environ. Physiol: Bioenergetics and temperature regulation. Bethesda, Maryland: Faseb 1972.
ROBINSON, K. W., LEE, D. H. K.: Proc. Roy. Soc. Queensland **53**, 159 (1941 a).
ROBINSON, K. W., LEE, D. H. K.: Proc. Roy. Soc. Queensland **53**, 171 (1941 b).
ROBINSON, K. W., MORRISON, P. R.: J. Cell. Comp. Physiol. **49**, 455 (1957).
ROBINSON, S.: In: NEWBURGH, L. H. (Ed.): Physiology of heat regulation and the science of clothing, p. 193. Philadelphia-London: Saunders 1949.
ROBINSON, S., MEYER, F. R., NEWTON, J. L., TS'AO, C. H., HOLGERSEN, L. O.: J. Appl. Physiol. **20**, 575 (1965).
ROBINSON, S., TURRELL, E. S., BELDING, H. S., HORVATH, S. M.: Am. J. Physiol. **140**, 168 (1943).
RODAHL, K.: J. Nutr. **48**, 359 (1952).
RODBARD, S.: Science **111**, 465 (1950).
RÖHRS, M.: In: KURTH, G. (Hrsg.): Evolution und Hominisation, S. 49. Stuttgart: Fischer 1962.
ROMANOFF, A. L.: J. Agr. Sci. **25**, 318 (1935).
ROMANOFF, A. L.: Poultry Sci. **15**, 311 (1936).
ROMANOFF, A. L.: Science **94**, 218 (1941).
ROMANOFF, A. L.: Anat. Rec. **86**, 143 (1943).
ROMANOFF, A. L.: Poultry Sci. **22**, 148 (1943).
ROMANOFF, A. L., SMITH, L. L., SULLIVAN, R. A.: Cornell Univ. Agr. Exper. Sta. Memoir. **216** (1938).
ROMANOFF, A. L., SOCHEN, M.: Anat. Rec. **65**, 59 (1936).
ROSENHAIN, F. R., PENROD, K. E.: Am. J. Physiol. **166**, 55 (1951).
ROSOMOFF, H. L.: Surgery **40**, 328 (1956).
ROSOMOFF, H. L., HOLADAY, D. A.: Am. J. Physiol. **179**, 85 (1954).
ROWAN, W.: Nature **115**, 494 (1925).
ROWAN, W.: Proc. Boston Soc. Nat. Hist. **39**, 151 (1929).
RUBENSTEIN, E., MEUB, D. W., ELDRIDGE, F.: J. Appl. Physiol. **15**, 603 (1960).
RUBNER, M.: Z. Biol. **19**, 535 (1883).
RUBNER, M.: Biochem. Z. **148**, 222 (1924).
RYCROFT, B. W.: Corneal grafts. London: Butterworth 1955.
SAARIKOSKI, P. L., SUOMALAINEN, P.: Ann. Acad. Sci. Fennicae A 4 Biol. **171**, 1 (1970).

SAMS, W. M., JR., WINKELMANN, R. K.: Am. J. Physiol. **216**, 112 (1969).
SAND, A.: Proc. Roy. Soc. B **125**, 524 (1938).
SALTIN, B., GAGGE, A. P., STOLWIJK, J. A. J.: J. Appl. Physiol. **28**, 318 (1970).
SARAJAS, H. S. S.: Ann. Acad. Sci. Fennicae. 4 A Biologica **120**, 1 (1967).
SATINOFF, E.: Am. J. Physiol. **206**, 1389 (1964).
SATINOFF, E.: Progr. Physiol. Psychol. **3**, 201 (1970).
SAURE, L.: Aquilo, Ser. Zool. **9**, 1 (1969).
SAWAYA, P.: Arch. Cirurg. Clin. Exp. **5**, 235 (1941).
SAYEN, A., MELOCHE, B. R., TEDESCHI, G. C., MONTGOMERY, H.: Clin. Sci. **19**, 243 (1960).
SCHACHNER, H. G., GIERLACH, Z. G., KREBS, A. T.: Proj. Rep. No. 6-64-12-02, Med. Dept. Field, Res. Lab., Fort Knox, Ky. 1949.
SCHEUFLER, K., RATHS, P.: Wiss. Z. Univ. Halle, Math.-Nat. Reihe **16**, 253 (1967).
SCHILDMACHER, H.: Biol. Zbl. **71**, 238 (1952).
SCHMIDT-NIELSEN, B.: Am. J. Physiol. **167**, 824 (1951).
SCHMIDT-NIELSEN, B., SCHMIDT-NIELSEN, K.: Am. J. Physiol **160**, 291 (1950a).
SCHMIDT-NIELSEN, B., SCHMIDT-NIELSEN, K.: Am. J. Physiol. **162**, 31 (1950b).
SCHMIDT-NIELSEN, B., SCHMIDT-NIELSEN, K.: Ecology **31**, 75 (1950c).
SCHMIDT-NIELSEN, B., SCHMIDT-NIELSEN, K.: J. Cell. Comp. Physiol. **38**, 165 (1951).
SCHMIDT-NIELSEN, B., SCHMIDT-NIELSEN, K., HOUPT, T. R., JARNUM, S. A.: Am. J. Physiol. **185**, 185 (1956).
SCHMIDT-NIELSEN, B., SCHMIDT-NIELSEN, K., HOUPT, T. R., JARNUM, S. A.: Am. J. Physiol. **188**, 477 (1957b).
SCHMIDT-NIELSEN, K.: In: CLOUDSLEY-THOMPSON, J. L. (Ed.): Biology of deserts, p. 182. London: Inst. Biol. 1954.
SCHMIDT-NIELSEN, K.: Sci. Am. **201**, 140 (1959).
SCHMIDT-NIELSEN, K.: Desert animals: Physiological problems of heat and water. London: Oxford University Press 1964.
SCHMIDT-NIELSEN, K., DAWSON, T. J., HAMMEL, H. T., HIND, D.: Hvalrådets Skrifter **48**, 125 (1965).
SCHMIDT-NIELSEN, K., KANWISHER, J., LASIEWSKI, R. C., COHN, J. E., BRETZ, W. L.: Condor **71**, 341 (1969).
SCHMIDT-NIELSEN, K., SCHMIDT-NIELSEN, B.: Physiol. Rev. **32**, 135 (1952).
SCHMIDT-NIELSEN, K., SCHMIDT-NIELSEN, B., JARNUM, S. A., HOUPT, T. R.: Am. J. Physiol. **188**, 103 (1957).
SCHNEIDER, F. H., GILLIS, C. N.: Am. J. Physiol. **211**, 890 (1966).
SCHOEPFER, E.: Zit. b. BUDDENBROCK, W. v.: Grundriß der vergleichenden Physiologie, Bd. 2, S. 886. Berlin: Gebr. Bornträger 1937.
SCHOLANDER, P. F.: Evolution **9**, 15 (1955).
SCHOLANDER, P. F., HAMMEL, H. T., HART, J. S., LEMESSURIER, D. H., STEEN, J.: J. Appl. Physiol. **13**, 211 (1958).
SCHOLANDER, P. F., HAMMEL, H. T., LANGE ANDERSEN, K., LØYNING, Y.: J. Appl. Physiol. **12**, 1 (1958).
SCHOLANDER, P. F., HOCK, R., WALTERS, V., IRVING, L.: Biol. Bull. **99**, 259 (1950a).
SCHOLANDER, P. F., HOCK, R., WALTERS, V., JOHNSON, F., IRVING, L.: Biol. Bull. **99**, 237 (1950b).
SCHOLANDER, P. F., KROG, J.: J. Appl. Physiol. **10**, 405 (1957).
SCHOLANDER, P. F., SCHEVILLE, W. F.: J. Appl. Physiol. **8**, 279 (1955).
SCHOLANDER, P. F., WALTERS, P., HOCK, R., IRVING, L.: Biol. Bull. **99**, 225 (1950a).
SCHOONHOVEN, L. M.: J. Insect Physiol. **13**, 821 (1967).
SCHWABE, E. L., EMERY, E. E., GRIFFITH, F. R.: J. Nutr. **15**, 199 (1938).
SELLERS, E. A., YOU, S. S.: Am. J. Physiol. **163**, 81 (1950).
SELLERS, E. A., YOU, S. S., THOMAS, N.: Am. J. Physiol. **165**, 481 (1951).
SELYE, H.: The physiology and pathology of exposure to stress. Montreal: Acta Inc. Medical Publ. 1950.
SEROTA, H. M., GERARD, R. W.: J. Neurophysiol. **1**, 115 (1938).
SEVERINGHAUS, J. W.: Ann. Acad. Sci. N. Y. **80**, 384 (1959).
SHARP, F., SMITH, D., THOMPSON, M., HAMMEL, H. T.: Life Sci. **8**, 1069 (1969).

SHEARD, CH., WILLIAMS, M.M.D., HORTON, B.T.: In: Temperature, its measurement and control in science and industry, p. 557. New York: Reinhold 1941.

SHELLEY, W.B., HORVATH, P.N., PIUSBURY, D.M.: Medicine (Baltimore) **29**, 195 (1950).

SHUMAKER, H.B., LEMBKE, R.: Bull. Vasc. Surg. **1951**, 77.

SIBBONS, J.L.H.: In: HARDY, J.D., GAGGE, A.P., STOLWIJK, J.A.J. (Eds.): Physiological and behavioral temperature regulation, p. 108. Springfield, Ill.: Thomas 1970.

SIMON, E., IRIKI, M.: Pflügers Arch. Ges. Physiol. **328**, 103 (1971).

SIMON, E., RAUTENBERG, W., THAUER, R., IRIKI, M.: Pflügers Arch. Ges. Physiol. **281**, 309 (1964).

SIMPSON, S.: J. Physiol. (Lond.) **28**, 37 P (1902).

SIMPSON, S., GALBRAITH, J.J.: J. Physiol. (Lond.) **33**, 225 (1905).

SIRCAR, P.: Proc. Soc. Exp. Biol. (N.Y.) **87**, 194 (1954).

SJOSTROM, B., WEATHERLEY-WHITE, R.C.A., PATON, B.C.: J. Surg. Res. **4**, 12 (1964).

SKRAMLIK, E.v.: Arch. Psychol. Erg.-Bd. **4**, Teil 1 und 2 (1937).

SMALLEY, R.L., DRYER, R.L.: Science **140**, 1333 (1963).

SMIT-VIS, J.H.: Arch. Neerl. Zool. **14**, 513 (1962).

SMITH, A.U.: Biological effects of freezing and supercooling. London: Edward Arnold 1961.

SMITH, P.E., JAMES, E.W.: Arch. Environm. Hlth. **9**, 323 (1964).

SMITH, R.E.: Physiologist **4**, 113 (1961).

SMITH, R.E., HOCK, R.J.: Science **140**, 199 (1963).

SMITH, R.E., HOIJER, D.J.: Physiol. Rev. **42**, 60 (1962).

SMITH, R.E., HORWITZ, B.A.: Physiol. Rev. **49**, 330 (1969).

SMITH, R.E., ROBERTS, J.C.: Am. J. Physiol. **206**, 143 (1964).

SNELLEN, J.V.: Acta Physiol. Pharmacol. Neerl. **14**, 99 (1966).

SOIVIO, A., TÄHTI, H., KRISTOFFERSSON, R.: Ann.. Zool. Fennicae **5**, 224 (1968).

SOLLBERGER, A.: Biological rhythm research. Amsterdam: Elsevier 1965.

SOUTH, F.E.: Am. J. Physiol. **198**, 463 (1960).

SOUTH, F.E., BREAZILE, J.E., DELLMANN, H.D., EPPERLY, A.D.: In: Depressed metabolism, p. 277. New York: Am. Elsevier 1969.

SOUTH, F.E., HOUSE, W.A.: In: Mammalian hibernation III, p. 305. New York: Am. Elsevier 1967.

SPÄTH, H.: Z. Vergl. Physiol. **56**, 431 (1967).

SPEALMAN, C.R.: Am. J. Physiol. **146**, 262 (1946).

SPEALMAN, C.R., NEWTON, M., POST, R.L.: Am. J. Physiol. **150**, 628 (1947).

SPERELAKIS, N.: In: HARDY, J.D., GAGGE, A.P., STOLWIJK, J.A.J. (Eds.): Physiological and behavioral temperature regulation, p. 408. Springfield, Ill.: Thomas 1970.

SPÖTTEL, W.: Z. Anat. **89**, 606 (1929).

STARKOV, P.M. (ed.): The problem of acute hypothermia. London-Oxford-New York-Paris: Pergamon Press 1960.

STARR, P., ROSKELLEY, R.: Am. J. Physiol. **130**, 549 (1940).

STEEN, J., ENGER, P.S.: Am. J. Physiol. **190**, 157 (1957).

STEFANSSON, V.: Arctic manual. New York: Macmillan Comp. 1944.

STEVENS, J.C., ADAIR, E.R., MARKS, L.E.: In: HARDY, J.D., GAGGE, A.P., STOLWIJK, J.A.J. (Eds.): Physiological and behavioral temperature regulation, p. 892. Springfield, Ill.: Thomas 1970.

STEVENSON, W.G.: Can. J. Comp. Med. **10**, 137 (1946).

STEWART, L.C., ZIMNY, M.L., WEST, H.: J. Neurochem. **17**, 285 (1970).

STIGLER, R.: Arch. Exp. Pathol. Pharmakol. **152**, 68 (1930).

STINSON, R.H., FISHER, K.C.: Can. J. Zool. **31**, 404 (1953).

STOLWIJK, J.A.J.: In: HARDY, J.D., GAGGE, A.P., STOLWIJK, J.A.J. (Eds.): Physiological and behavioral temperature regulation, p. 703. Springfield, Ill.: Thomas 1970.

STOLWIJK, J.A.J., HARDY, J.D.: Pflügers Arch. Ges. Physiol. **291**, 129 (1966a).

STOLWIJK, J.A.J., WEXLER, I.: J. Physiol. (Lond.) **214**, 377 (1971).

STONEHOUSE, B.: Advanc. Ecol. Res. **4**, 131 (1967).
STREICHER, E., HACKEL, D. B., FLEISCHMANN, W.: Am. J. Physiol. **161**, 300 (1950).
STRÖM, G.: Acta Physiol. Scand. **21**, 271 (1950).
STRØMME, S. B., HAMMEL, H. T.: J. Appl. Physiol. **23**, 815 (1967).
STRUGHOLD, H., PORZ, R.: Z. Biol. **91**, 563 (1931).
STRUMWASSER, F.: Am. J. Physiol. **196**, 7 (1959 a).
STRUMWASSER, F.: Am. J. Physiol. **196**, 15 (1959 b).
STRUMWASSER, F.: Am. J. Physiol. **196**, 23 (1959 c).
STRUMWASSER, F.: Bull. Mus. Comp. Zool. Harvard Coll. **124**, 285 (1960).
SUMNER, F. B.: Am. Naturalist **45**, 90 (1911).
SUNDERMAN, F. W.: Arch. Intern. Med. **67**, 709 (1941).
SUNDERMAN, F. W., HAYMAKER, W.: J. Am. Med. Sci. **213**, 562 (1947).
SUNDERMAN, F. W., SCOTT, J. C., BAZETT, H. C.: Am. J. Physiol. **123**, 199 (1938).
SUOMALAINEN, P.: Ann. Acad. Sci. Fennicae A **45**, 1 (1939).
SUOMALAINEN, P.: Arch. Soc. Zool.-Bot. Fennicae **5**, 35 (1950).
SUOMALAINEN, P., NYHOLM, P.: B. Hanström Zool. Papers, 269 (1956).
SUOMALAINEN, P., SAARIKOSKI, P.-L.: Experientia **23**, 457 (1967).
SUOMALAINEN, P., SAARIKOSKI, P.-L.: Comment. Biol. Soc. Sci. Fennicae **30**, 1 (1970).
SUOMALAINEN, P., UUSPÄÄ, V. J.: Nature **182**, 1500 (1958).
SWAN, H.: Hypothermia. Philadelphia, Pa.: Saunders 1962.
TASHIMA, L. S., ADELSTEIN, S. J., LYMAN, C. P.: Am. J. Physiol. **218**, 303 (1970).
TAYLOR, A. C.: Proc. Roy. Soc. B **147**, 466 (1957).
TAYLOR, C. R.: Physiol. Zool. **39**, 127 (1966).
TAYLOR, C. R.: The eland and the oryx. Sci. Am. **220**, 89 (1969 a).
TAYLOR, C. R.: Am. J. Physiol. **217**, 317 (1969 b).
TAYLOR, C. R., SPINAGE, C. A., LYMAN, C. P.: Am. J. Physiol. **217**, 630 (1969).
TAYLOR, P. M.: J. Physiol. (Lond.) **154**, 153 (1960).
TEMPLETON, J. R.: In: WHITTOW, G. C. (Ed.): Comparative physiology of thermo-
    regulation, Vol. I, p. 167. New York-London: Academic Press 1970.
TERRILL, C. E.: In: HAFEZ, E. S. E. (Ed.): Adaptation of domestic animals, p. 246.
    Philadelphia: Lea & Febiger 1968.
TESCHAN, P., GELLHORN, E.: Am. J. Physiol. **159**, 1 (1949).
THAUER, R.: Pflügers Arch. Ges. Physiol. **236**, 102 (1935).
THAUER, R.: Ergebn. Physiol. **41**, 607 (1939).
THAUER, R.: Pflügers Arch. Ges. Physiol. **246**, 372 (1942).
THAUER, R.: Arbeitsphysiologie **15**, 175 (1953).
THAUER, R.: Klin. Wschr. **36**, 989 (1958).
THAUER, R.: Naturwissenschaften **51**, 73 (1964).
THAUER, R.: In: Handbook of physiology, Sect. II, Vol. III, p. 1921. Washington,
    D. C.: Am. Physiol. Soc. 1965.
THAUER, R.: In: HARDY, J. D., GAGGE, A. P., STOLWIJK, J. A. J. (Eds.): Physiological
    and behavioral temperature regulation, p. 412. Springfield, Ill.: Thomas 1970.
THAUER, R., BRENDEL, W.: Progr. Surg. (Basel) **2**, 73 (1962).
THAUER, R., EBAUGH, F. G.: Pflügers Arch. Ges. Physiol. **225**, 27 (1952).
THAUER, R., PETERS, G.: Pflügers Arch. Ges. Physiol. **239**, 483 (1938).
THOMSON, M. L.: J. Physiol. (Lond.) **123**, 225 (1954).
THUNBERG, T.: Skand. Arch. Physiol. **11**, 382 (1901).
THWAITES, C. J.: Int. J. Biometeor. **11**, 297 (1967).
TIMBAL, J., COLIN, J., GUIEU, J.-D., BOUTELIER, C.: J. Appl. Physiol. **27**, 726 (1969).
TISCHLER, W.: Grundzüge der terrestrischen Tierökologie. Braunschweig: Vieweg 1949.
TISCHLER, W.: In: Klima, Wetter, Mensch, p. 259. Heidelberg: Quelle & Meyer 1952.
TORRES, J. C., ANGELAKOS, E. T.: Am. J. Physiol. **207**, 199 (1964).
TUCKER, V. S.: J. Cell. Comp. Physiol. **65**, 405 (1965).
TURPAJEW, T. M.: Dokl. Akad. Nauk SSSR **60**, 8 (1948).
TWENTE, J. W., TWENTE, J. A.: J. Appl. Physiol. **20**, 411 (1965).
TWENTE, J. W., TWENTE, J. A.: Comp. Biochem. Physiol. **25**, 467 (1968).

UDVARDY, M. D. F.: Auk **80**, 191 (1963).
UOTILA, U. U.: Endocrinology **15**, 605 (1939).
USINGER, W.: Pflügers Arch. Ges. Physiol. **275**, 646 (1962).
VANGGAARD, L.: Acta Physiol. Scand. **76**, 13 A (1969).
VANLERENBERGHE, J., TRUPIN-BAR, N., GUISLAIN, R., BEL, C.: J. Physiol. (Paris) **59**, 111 (1967).
VEGHTE, J. H.: Physiol. Zool. **37**, 316 (1964).
VEGHTE, J. H., HERREID, C. F.: Physiol. Zool. **38**, 267 (1965).
VENDRIK, A. J. H.: In: HARDY, J. D., GAGGE, A. P., STOLWIJK, J. A. J. (Eds.): Physiological and behavioral temperature regulation, p. 819. Springfield, Ill.: Thomas 1970.
VENDRIK, A. J. H., VOS, J. J.: J. Appl. Physiol. **13**, 435 (1958).
VENDRIK, H., EIJKMAN, E. G.: In: KENSHALO, D. R. (Ed.): The skin senses, p. 178. Springfield, Ill.: Thomas 1968.
VOGT, M.: J. Physiol. (Lond.) **123**, 451 (1954).
WAITES, G. M. H.: Quart. J. Exp. Physiol. **47**, 314 (1962).
WAITES, G. M. H.: In: JOHNSON, A. D., GOMES, W. R., VANDERMARK, N. L. (Eds.): The testis. Development, Anatomy and Physiology, Vol. I, p. 241. New York-London: Academic Press 1970.
WALDOW, U.: Z. Vergl. Physiol. **69**, 249 (1970).
WALTHER, J., BISHOP, F. W., WARREN, S. L.: In: Temperature, its measurement and control in science and industry, p. 474. New York: Reinhold 1941.
WANG, L., HUDSON, J. W.: Comp. Biochem. Physiol. **38A**, 59 (1971).
WATZKA, M.: Z. Mikr.-Anat. Forsch. **36**, 67 (1934).
WEATHERLEY-WHITE, R. C. A., SJOSTROM, B., PATON, B. C.: J. Surg. Res. **4**, 17 (1964).
WEAVER, M. E., INGRAM, D. L.: Ecology **50**, 710 (1969).
WEBER, E. H.: In: Wagners Handwörterbuch der Physiologie, Bd. III, Teil 2, S. 481. Braunschweig: Vieweg 1846.
WEDDELL, G., MILLER, S.: Ann. Rev. Physiol. **24**, 199 (1962).
WEDDELL, G., PALMER, E., PALLIE, W.: Biol. Rev. **30**, 159 (1955).
WEISS, B., LATIES, V. G.: Science **133**, 1338 (1961).
WEISS, B., LATIES, V. G., WEISS, A. B.: Arch. Int. Pharmacodyn. **165**, 467 (1967).
WELLS, J. S., RALL, D. P.: Proc. Soc. Exp. Biol. (N.Y.) **68**, 421 (1948).
WENZEL, H. G.: In: Handbuch der gesamten Arbeitsmedizin, Bd. **1**, S. 554. München-Berlin-Wien: Urban & Schwarzenberg 1961.
WEYMOUTH, F. W., CRISMON, V. E., HALL, H., BELDING, S., FIELD, J.: J. Physiol. Zool. **17**, 50 (1944).
WEZLER, K., NEUROTH, G.: Z. Exp. Med. **115**, 127 (1949).
WEZLER, K., THAUER, R.: Z. Luftfahrtmed. **7**, 237 (1942).
WHITTOW, G. C. (ed.): Comparative physiology of thermoregulation, Vol. I. New York-London: Academic Press 1970.
WHITTOW, G. C. (ed.): Comparative physiology of thermoregulation, Vol. II. New York-London: Academic Press 1971.
WHITTOW, G. C., FINDLEY, J. D.: Am. J. Physiol. **214**, 94 (1968).
WILKINS, R. W., DOUPE, J., NEWMAN, H. W.: Clin. Sci. **3**, 403 (1938).
WILLIAMS, B. A., HEATH, J. E.: Am. J. Physiol. **218**, 1654 (1970).
WILSON, S. G.: J. Agr. Sci. **36**, 246 (1946).
WILSON, W. O., HILLERMAN, J. P., EDWARDS, W. H.: Poultry Sci. **31**, 843 (1952).
WILSON, W. O., PLAISTER, T. H.: Am. J. Physiol. **166**, 572 (1951).
WIMSATT, W. A.: Symp. Soc. Exp. Biol. **23**, 511 (1969).
WINBERG, H.: Arkiv. Zool. (Stockh.) **33 A**, (1941).
WINSLOW, C. E. A.: In: Temperature, it's measurement and control in science and industry, p. 509. New York: Reinhold 1941.
WISLOCKI, G. B.: Quart. Rev. Biol. **8**, 385 (1933).
WISLOCKI, G. B., ENDERS, R. K.: J. Mammal. **16**, 328 (1935).
WISSLER, E. H.: J. Appl. Physiol. **16**, 734 (1961).

WISSLER, E. H.: In: HARDY, J. D. (Ed.): Temperature, its measurement and control in science and industry, p. 53. New York: Reinhold 1963.
WISSLER, E. H.: In: HARDY, J. D., GAGGE, A. P., STOLWIJK, J. A. J. (Eds.): Physiological and behavioral temperature regulation, p. 367. Springfield, Ill.: Thomas 1970.
WIT, A., WANG, S. C.: Am. J. Physiol. **215**, 1151 (1968a).
WIT, A., WANG, S. C.: Am. J. Physiol. **215**, 1160 (1968b).
WITT, I.: Acta Neuroveg. (Wien) **25**, 208 (1963).
WITT, I., HENSEL, H.: Pflügers Arch. Ges. Physiol. **268**, 582 (1959).
WOLFF, R. C., PENROD, K. E.: Am. J. Physiol. **163**, 580 (1950).
WOLFF, S., HARDY, J. D.: J. Clin. Invest. **20**, 521 (1941).
WOLFSON, A.: In: GORBMAN, A. (Ed.): Comparative endocrinology, p. 38. New York: Wiley 1958.
WOLFSON, A.: Publ. Am. Ass. Advan. Sci. No. 55 (1959).
WORSTELL, D. M., BRODY, S.: Mo. Agr. Exp. Sta. Res. Bull. No. **515**, 1 (1953).
WRENN, T. R., BITMAN, J., SYKES, J. F.: J. Dairy Sci. **41**, 1071 (1958).
WÜNNENBERG, W.: Pflügers Arch. Ges. Physiol. **312**, R 118 (1969).
WÜNNENBERG, W., BRÜCK, K.: Pflügers Arch. Ges. Physiol. **294**, R 84 (1967).
WÜNNENBERG, W., BRÜCK, K.: Pflügers Arch. Ges. Physiol. **299**, 1 (1968a).
WÜNNENBERG, W., BRÜCK, K.: Nature **218**, 1268 (1968b).
WÜNNENBERG, W., BRÜCK, K.: Pflügers Arch. Ges. Physiol. **300**, R 45 (1968).
WÜNNENBERG, W., BRÜCK, K.: Pflügers Arch. Ges. Physiol. **321**, 233 (1970).
WÜNNENBERG, W., HARDY, J. D.: J. Appl. Physiol. **33**, 547 (1972).
WULSIN, F. R.: In: NEWBURGH, L. H. (Ed.): Physiology of heat regulation and the science of clothing, Chapt. I, Part 1, p. 3. Philadelphia, Pa.-London: Saunders 1949.
WURSTER, R.: Pflügers Arch. Ges. Physiol. **300**, R 47 (1968).
WURSTER, R. D., McCOOK, R. D., RANDALL, W. C.: J. Appl. Physiol. **21**, 617 (1966).
WYNDHAM, C. H.: J. Appl. Physiol. **4**, 383 (1951).
WYNDHAM, C. H.: In: BAKER, P. T. (Ed.): The biology of human adaptability. Oxford: Clarendon Press 1966.
WYNDHAM, C. H.: J. Appl. Physiol. **20**, 31 (1966).
WYNDHAM, C. H.: In: HARDY, J. D., GAGGE, A. P., STOLWIJK, J. A. J. (Eds.): Physiological and behavioral temperature regulation, p. 324. Springfield, Ill.: Thomas 1970.
WYNDHAM, C. H., ATKINS, A. R.: 3rd. Int. Conference on Medical Electronics, Paper No. 27 (1960).
WYNDHAM, C. H., ATKINS, A. R.: Pflügers Arch. Ges. Physiol. **303**, 14 (1968).
WYNDHAM, C. H., BOUWER, W. v. D. M., PATERSON, H. E., DEVINE, M. G.: Arch. Ind. Hyg. Occupational Med. **7**, 234 (1953).
WYNDHAM, C. H., JACOBS, G. E.: J. Appl. Physiol. **11**, 197 (1957).
WYNDHAM, C. H., STRYDOM, N. B., COOKE, H. M., MARITZ, J. S.: Appl. Physiol. Lab. Reps. 1 (1959).
WYNDHAM, C. H., STRYDOM, N. B., MORRISON, J. F., DU TOIT, F. D., KRAAN, J. G.: J. Appl. Physiol. **6**, 681 (1954).
WYNN, V.: Clin. Sci. **15**, 297 (1956).
WYSS, O.: Pflügers Arch. Ges. Physiol. **229**, 599 (1932).
YAGLOU, C. P., RAO, M. N.: J. Industr. Hyg. **29**, 140 (1947).
YAMAMOTO, W. S., BROBECK, J. R.: Physiological controls and regulations. Philadelphia, Pa.: Saunders 1965.
YEATES, N. T. M.: J. Agr. Sci. **51**, 84 (1958).
YOUNG, I. M.: Clin. Sci. **22**, 325 (1962).
YOUSEF, M. K., HAHN, L., JOHNSON, H. D.: In: HAFEZ, E. S. E. (Ed.): Adaptation of domestic animals, p. 233. Philadelphia: Lea & Febiger 1968.
ZATZMANN, M. L., SOUTH, F. E.: Cryobiology **8**, 310 (1971).
ZEISBERGER, E., BRÜCK, K.: Pflügers Arch. Ges. Physiol. **322**, 152 (1971a).
ZEISBERGER, E., BRÜCK, K.: J. Physiol. (Paris) **63**, 464 (1971b).

ZEISBERGER, E., BRÜCK, K.: In: SCHÖNBAUM, E., LOMAX, P. (Eds.): The pharmacology of thermoregulation, p. 232. Basel: Karger 1973.

ZEISBERGER, E., BRÜCK, K., WÜNNENBERG, W., WIETASCH, C.: Pflügers Arch. Ges. Physiol. **296**, 276 (1967).

ZENZ, M., FRUHSTORFER, H., NOLTE, H., HENSEL, H.: Pflügers Arch. Ges. Physiol. Suppl. **339**, 171 (1973).

ZERBST, D.: Analyse der Nachrichtenverarbeitung durch biologische Rezeptoren. Leipzig: VEB Thieme 1972.

ZERBST, E., DITTBERNER, K.-H.: Pflügers Arch. Ges. Physiol. **319**, R 126 (1970).

ZIMNY, M. L.: Bull. Mus. Comp. Zool. Harvard Coll. **124**, 457 (1960).

ZIMNY, M. L.: Comp. Biochem. Physiol. **27**, 859 (1968).

ZIMNY, M. L., LEVY, E. D., JR.: Z. Zellforsch. **118**, 326 (1971).

ZIMNY, M. L., MORELAND, J. E.: Can. J. Physiol. Pharmacol. **46**, 911 (1968).

ZÖLLNER, G., THAUER, R., KAUFMANN, W.: Pflügers Arch. Ges. Physiol. **260**, 261 (1955).

ZOTTERMAN, Y.: Skand. Arch. Physiol. **75**, 105 (1936).

ZOTTERMAN, Y.: Ann. Rev. Physiol. **15**, 357 (1953).

ZOTTERMAN, Y.: In: Handbook of physiology, Vol. I, Sect. 1, p. 431. Washington, D. C.: Am. Physiol. Soc. 1959.

Subject Index